OXFORD MATHEMATICAL MONOGRAPHS

Series Editors

E. M. FRIEDLANDER   I. G. MACDONALD   H. MCKEAN
R. PENROSE   J. T. STUART

OXFORD MATHEMATICAL MONOGRAPHS

Lie Groups, Convex Cones, and Semigroups

JOACHIM HILGERT
KARL HEINRICH HOFMANN
Technische Hochschule Darmstadt

AND

JIMMIE D. LAWSON
Louisiana State University

CLARENDON PRESS · OXFORD
1989

Oxford University Press, Walton Street, Oxford OX2 6DP
Oxford New York Toronto
Delhi Bombay Calcutta Madras Karachi
Petaling Jaya Singapore Hong Kong Tokyo
Nairobi Dar es Salaam Cape Town
Melbourne Auckland
and associated companies in
Berlin Ibadan

Oxford is a trade mark of Oxford University Press

Published in the United States
by Oxford University Press, New York

British Library Cataloguing in Publication Data
Hilgert, Joachim
Lie groups, convex cones and semigroups.
1. Lie groups & Lie algebra
I. Title   II. Hofmann, Karl Heinrich   III. Lawson,
Jimmie D.
512'.55
ISBN 0-19-853569-4

Library of Congress Cataloging in Publication Data
Hilgert, Joachim.
Lie groups, convex cones, and semigroups/Joachim Hilgert, Karl
Heinrich Hofmann, and Jimmie D. Lawson.
(Oxford mathematical monographs)
Bibliography. Includes index.
1. Lie groups.   2. Convex bodies.   3. Semigroups.   I. Hofmann,
Karl Heinrich.   II. Lawson, Jimmie D.   III. Title.   IV. Series.
QA387.H535   1989   512'.55–dc20   89-9289
ISBN 0-19-853569-4

Typeset by the authors using TeX, with
original line drawings by W. A. Ruppert,
Universität für Bodenkultur, Vienna

Printed in Great Britain

Preface

The research project leading to this book and its production was aided by many persons and agencies.

The *Stiftung Volkswagenwerk* lent its financial support to our Workshops in 1984, 1985, and 1987; in 1987, Karl Heinrich Hofmann was a VW-Foundation Fellow in the program *Akademiestipendien der Stiftung Volkswagenwerk*. The *National Science Foundation* granted summer research support and travel money to Jimmie D. Lawson in the years 1984 through 1988. *Der Hessische Minister für Wissenschaft und Kunst* supported the 1987 Workshop by a grant for supplies and appointed Lawson visiting research professor during the month of June 1988. For all Workshops on its premises, the *Technische Hochschule Darmstadt* gave money, supplied the environment, and provided the logistics. The *Mathematische Forschungsinstitut Oberwolfach* hosted the Workshop in 1986. The *Vereinigung der Freunde der Technischen Hochschule Darmstadt* funded electronic equipment without which the production of this book would not have been possible. *Tulane University* in New Orleans was the host institution to Karl Heinrich Hofmann during his sabbatical in the fall of 1986. To all of these agencies and institutions the authors express their heartfelt thanks.

WOLFGANG ALEXANDER FRIEDRICH RUPPERT from Vienna spent the winter term 1987-8 at Darmstadt as a fellow of the Alexander von Humboldt-Foundation. His recent work on congruences found its way into Chapter V. He took an intense interest in the development of the book and contributed numerous improvements to the text. He drew all the Figures in this book. Thank you, WAF! The *Darmstadt Seminar "Sophus Lie"*, notably NORBERT DÖRR, ANSELM EGGERT, KARL-HERMANN NEEB, KARLHEINZ SPINDLER, CHRISTIAN TERP, and WOLFGANG WEISS contributed much through proofreading and by providing a "caisse de resonance".

The book was typeset by the authors in TEX, primarily at the *Technische Hochschule Darmstadt*, but also at *Louisiana State University* and at *Tulane University*. We have learned TEX from JOHN HILDEBRANT at Louisiana State University in Baton Rouge and from SIDNEY A. MORRIS of La Trobe University in Melbourne. He visited the Technische Hochschule Darmstadt in the Spring of 1986 and contributed forcefully to the introduction of TEX to the mathematicians at THD. The macros used for this book are built around a nucleus of macros which he wrote and kindly permitted us to use. We have used plain TeX on VAXes at the Technische Hochschule Darmstadt, Louisiana State University, Tulane University, and the University of New Orleans, and the version ST-TeX written by KLAUS GUNTERMANN of THD for the Atari 1040ST. The help we received from the System Manager for Computing at the Fachbereich Mathematik der Technischen Hochschule Darmstadt, KLAUS-THOMAS SCHLEICHER, through these years has been invaluable. We also thank GUDRUN SCHUMM of THD for her assistance in managing the laser printer and WOLFGANG WEIKEL for sharing with us the programs he wrote for editing, file management, transmission, and PC-operation. MICHAEL MISLOVE of Tulane University indefatigably assisted Hofmann during his sabbatical with all computer related problems. He also introduced him to the Chemical Engineering Department of Tulane University. The first TeX program at Tulane was mounted and operated with the assistance of ANIL MENAWAT and MICHAEL HERMANN on the computer of this department. First printouts were done at the University of New Orleans through the generosity of its Computer Science Department and the patient help of WILLIAM A. GREENE. Also, NEAL STOLTZFUS at Louisiana State University was very helpful in the management of file transfer between Tulane University and LSU.

Dr. MARTIN GILCHRIST of Oxford University Press has organized the publication of this book and the preparation of our final files for printing at the facilities of Oxford University Press. The Copy Editors and the Assistant Editor have carefully scrutinized a hard copy. The elimination of numerous typographical errors is due to their effort. American spelling came most naturally to all of us. We are grateful that our publisher allowed us to leave this orthography where it deviates from the British one and that, in addition, he permitted certain aberrations from the format standards of the series whose modifications would have upset our pagination.

We thank these numerous people for their contributions to the production of this book.

J.H.
K.H.H.
J.D.L.

Darmstadt and Baton Rouge,
January 1989

Contents

Chapter IV. The local Lie theory of semigroups

Chapter V. Subsemigroups of Lie groups

Chapter VI. Positivity

Chapter VII. Embedding semigroups into Lie groups

Appendix

Reference material

Introduction

This book focuses on a new aspect of the theory of Lie groups and Lie algebras, namely, the consideration of semigroups in Lie groups. The systematic development of a Lie theory of semigroups is motivated by their recent emergence in different contexts. Notably, they appeared at certain points in geometric control theory and in the theory of causal structures in mathematical physics. Beyond that, it is becoming increasingly clear that the broader perspective of considering not just the analytic subgroups of a Lie group, but the appropriate subsemigroups as well, leads to a fuller and richer theory of the original Lie group itself. Hence it is appropriate to consider this work as a new branch of Lie group theory, too.

Historically, the rudiments of a Lie theory of semigroups can be detected in Sophus Lie's own work. If the language had been available at the time, he could have expressed one of his basic results in this sentence: *The infinitesimal generators of a local semigroup of local differential transformations of some euclidean domain is a convex cone in a vector space.* However, in Lie's own diction, any family of transformations of a set which is closed under composition is called a group, irrespective of the presence of an identity or the existence of inverses. In fact, Lie attempted for a while to deduce the existence of an identity and the inverse from his other assumptions until the first concrete examples credited to Friedrich Engel showed the futility of such efforts. The word semigroup belongs to the vocabulary of the 20th century. There were some initiatives to deal with Lie semigroups such as the attempts by Einar Hille in the early nineteen-fifties which also made their way into the the book by Hille and Phillips, and the studies of Charles Loewner on certain types of subsemigroups of Lie groups extending into the nineteen-sixties. By and large, these efforts remained somewhat isolated and they were either aborted or ignored, or both.

It may appear surprising that further systematic investigations of semigroups in Lie groups were not undertaken. However, the technical obstacles are considerable, and incisive results of both generality and mathematical depth did not quickly appear on the horizon. Indeed the traditional tools of Lie theory were inadequate for dealing with the new theory. One needed to introduce the geometry of convex sets; certain techniques and ideas from geometric control theory also turned out to be quite useful. Additionally, specialized methods appropriate to the circumstances had to be introduced and developed. Only in very recent years has a significant body of results begun to emerge. A notable example is the investigation of invariant cones in Lie algebras due to KOSTANT, VINBERG, PANEITZ, and OL'SHANSKIĬ.

Besides having to cope with these technical obstacles and a historical scarcity of external stimuli, the Lie theory of semigroups often found itself in a no-man's land. Semigroup theorists have tended to regard subsemigroups of groups as a branch of group theory, while group theorists have concentrated on the subgroups of a group and paid scant attention to the subsemigroups. This has been a serious oversight. The Lie theory of semigroups is an interesting, rich, and applicable branch of study. This book is a first attempt to present a systematic Lie theory of semigroups. Numerous examples are also included. Apart from some background theory which we felt we should provide, its contents are of recent origin.

Although a strong motivation for this book is the development of a useful and applicable Lie theory of semigroups, major lines of applications will be deferred to later volumes. Nevertheless, let us briefly illustrate the emergence of semigroups in the context of geometric control. Let Ω denote a set of smooth vector fields on a manifold M; each vector field $X \in \Omega$ determines a local flow, say, $t \mapsto F_t(X)$ which associates with any $m \in M$ the unique largest solution $x: I \to M$, $I \subseteq \mathbb{R}$, $x(t) = F_t(X) \cdot m$ of the initial value problem

$$(1) \qquad\qquad \dot{x}(t) = X\big(x(t)\big), \qquad x(0) = m.$$

In order to keep this illustration short let us assume that each $X \in \Omega$ is complete in the sense that (1) has a solution for all $t \geq 0$ and all $m \in M$. Then $F_t: M \to M$ is a smooth self-map of M for each $t \geq 0$. Now let us consider a function $c: [0, T] \to \Omega$, called a steering function, which is piecewise constant. Typically, we are thinking that such a function selects for each interval of constancy $[t_{k-1}, t_k[$ a vector field $X_k = c(t)$ with $t \in [t_{k-1}, t_k[$, and that each jump at time t_k, $k = 0, 1, \ldots, n$, $t_n = T$ represents a sudden switch which redirects the trajectory from one vector field X_k to the next vector field X_{k+1}. A solution of the initial value problem

$$(2) \qquad\qquad \dot{x}(t) = c(t)\big(x(t)\big), \qquad x(0) = m$$

is then a concatenation of solutions

$$\dot{x}_k(t) = X_k\big(x_k(t)\big), \quad x_k(t_{k-1}) = x_{k-1}(t_{k-1}) \qquad \text{for} \qquad t_{k-1} \leq t \leq t_k$$

with $t_0 = 0$, $t_n = T$ and $x_1(0) = m$. In the terms of the flows we have

$$(3) \quad x(t) = F_{t-t_{k-1}}(X_k) F_{t_{k-2}}(X_{k-1}) \cdots F_{t_1}(X_1)(m) \qquad \text{for} \qquad t_{k-1} \leq t < t_k,$$

and for $k = 1, \ldots, n-1$. A typical problem in systems theory is to determine the points of M which are attainable from a point $m \in M$ by traversing one of these trajectories obtained from the system Ω and all piecewise constant steering functions. This problem is then clearly tantamount to the question of which elements of M are in the orbit $S \cdot m$ of m where S is the semigroup generated (under composition) by all $F_t(X)$, $t \geq 0$, $X \in \Omega$.

If, in particular, M is the underlying manifold of a Lie group G and all vector fields $X \in \Omega$ are left-invariant, that is, if Ω is a subset of the Lie algebra $\mathfrak{g}$ of all left-invariant vector fields, then $F_t(X) = \exp t \cdot X$, and $S = S(\Omega)$ is the semigroup of all elements $\exp t_1 \cdot X_1 \exp t_2 \cdot X_2 \cdots \exp t_n \cdot X_n$ in G. Indeed, in several decisive parts of the general theory of semigroups in Lie groups, the framework of geometric control theory will organize our procedures.

The simplest special case, of course, is that of the the group $G = \mathbb{R}^n$, in which case we may also write $\Omega \subseteq \mathfrak{g} = \mathbb{R}^n$. Then S is the additive semigroup $\sum \{\mathbb{R}^+ \cdot X \colon X \in \Omega\}$ which is stable under multiplication by non-negative scalars. The example of the group $\mathbb{R}^n$ and its subsemigroups demonstrates right away that a treatment of *all* subsemigroups is unreasonable. It is clearly those semigroups which are generated by the rays $\mathbb{R}^+ \cdot x \subseteq S$ that are amenable to a general theory. Therefore, the idea of an infinitesimally generated semigroup in a Lie group will be crucial, and the whole theory will eventually have to concentrate on them.

The title of the book features the word *semigroup*. This word means different things to different people. For many a *functional analyst*, a semigroup is a strongly continuous one-parameter family of bounded operators on a Banach or a Hilbert space. Then semigroup theory is the description of the infinitesimal generation of these semigroups by unbounded closed operators and, as a branch of ergodic theory, the study of their behavior for large parameter values. For the *algebraist*, semigroup theory is a vast body of structure theory involving ideals, equivalence and order relations, idempotents, and generalized inverses, in short, a theory blending algebra with order. In *topological semigroup theory* a prevalent image is that of a compact semigroup, whose one outstanding feature is a minimal ideal full of idempotents.

None of these images is pertinent in the context of this book. While one-parameter semigroups do indeed play a crucial role here, they are only the raw material from which a distinctly multiparameter theory is built. We deal primarily, albeit not exclusively, with subsemigroups of Lie groups. In algebraic semigroup theory one has a whole subtheory characterizing semigroups which are embeddable in groups, but there is an inclination to consider those semigroups of little semigroup theoretical interest thereafter. And as far as compact semigroups are concerned, as soon as they are contained in a group, they are themselves compact groups; thus they instantaneously become the topic of classical group theory.

A helpful preliminary idea of the type of semigroup which shall occupy us in this book is that of a closed convex cone in $\mathbb{R}^n$. In fact, the theory of such cones is basic and thus needs much initial attention. Thus in the kind of Lie theory we have to deal with, *geometry of convex sets* is added to *linear algebra, calculus, global analysis, and topology*.

To highlight by comparison and contrast the main concerns of this book, let us recall the basic components of the theory of Lie groups. Traditional Lie group theory deals, firstly, with the *infinitesimal* structure theory of Lie groups and their subgroups. The basic tool is linear algebra applied to Lie algebras. Secondly, it deals with the *local* structure theory by means of the exponential function in which an amazing wealth of information is encoded. This approach to Lie theory uses analysis on open sets in $\mathbb{R}^n$, that is, calculus of several real variables. Finally, one has to deal with the *global* structure of Lie groups by means of global differential geometry, analysis on manifolds, and algebraic topology. The structure of a Lie group is uniquely determined by two data, one infinitesimal, one global: its Lie algebra and its universal covering homomorphism whose kernel is the fundamental group.

These features of classical Lie group theory roughly correspond to the lines of its historic development: Sophus Lie, its creator, developed the idea of infinitesimal transformations of a local group of transformations and thus the concept of infinitesimal generators of a (local) Lie group, and he invented for their analysis the type of algebra which now bears his name. The tools of analysis available to him in the later decades of the nineteenth century allowed him to develop an infinitesimal and local theory constrained to open domains of euclidean space. He was able to inspect examples of global groups. Indeed the groups of geometry provided an ample supply even at that time. A systematic treatment of the global theory, however, required the tools of topology and global differential geometry that soon became available through the work of Henri Poincaré, Georg Frobenius, Elie Cartan, Hermann Weyl, Heinz Hopf, and numerous other mathematicians.

Even more distinctly than in the case of groups, the Lie theory of semigroups falls into at least three parts:

1) *the infinitesimal theory,*

2) *the local theory,* and

3) *the global theory.*

The *infinitesimal theory* deals with those subsets of Lie algebras which are the exact infinitesimal generating sets of (local) subsemigroups of Lie groups. The tools belong to classical Lie algebra theory *and* to the theory of convex bodies and cones. The *local theory* has the task of characterizing local infinitesimally generated semigroups in a Lie group and must lead to the Fundamental Theorem in the sense of Lie. Historically, the direction of constructing, for a given Lie algebra, a (local) group with the given algebra as tangent set at the origin was hard. The corresponding task is much harder in the case of semigroups. Finally, the *global theory*, perhaps the least developed portion of the Lie theory of semigroups at this time, is concerned with the structure of infinitesimally generated subsemigroups of Lie groups and, in particular, with the global variant of the Fundamental Theorem: If a set of infinitesimal generators is given which is already known to be the tangent set of a local semigroup, is it always the tangent set of a (global) subsemigroup of a Lie group? Since one discovers very quickly that the answer to this question is negative, it converts immediately to the hard question: Which local infinitesimal generating sets are global? More accurately: Given a subset W in the Lie algebra $\mathfrak{g}$ of a Lie group G such that W is the precise set of tangent vectors at the origin of some local subsemigroup in G, what are necessary and sufficient conditions that there is a (global!) subsemigroup S in G whose set $\mathbf{L}(S)$ of infinitesimal generators is exactly W?

These outlines were drawn following the contours of classical Lie group theory. Yet in developing a Lie theory of semigroups, one recognizes very quickly that the analogy with Lie group theory does not carry very far at all. This may account for the apparent fact that most previous attempts at a Lie theory of semigroups were abandoned sooner or later.

However, there are more pieces to this puzzle. Up to this point we have considered subsemigroups of Lie groups as the proper territory of a Lie semigroup theory. But in looking back at classical vistas of Lie *group* theory, we find other views on a possible Lie *semigroup* theory just as natural: Given a topological semigroup, say, on a manifold with or without boundary, introduce a suitable differentiable structure and study the objects so obtained in the abstract! Clarify to which extent the semigroups arising in this fashion can be embedded into Lie groups—at least locally in the vicinity of an identity! Even on the historical plane, this viewpoint is natural because it is close to Sophus Lie's original vantage point. As a consequence we have to face a fourth aspect which we might call

4) *the abstract Lie semigroup and embedding theory.*

We shall address this issue, too, and find that our original attitude is justified. Any reasonably defined Lie semigroup can be embedded into a Lie group at least locally on a neighborhood of the identity. This is reassuring. Yet many interesting problems remain open in the entire theory.

Let us now look at the lay-out of the book and highlight some of its results. We begin with a fundamental fact which was, in a way, known to Sophus Lie, which is explicitly and clearly stated in Loewner's work, and appears in some form in a variety of contexts where semigroups in Lie groups have been considered. Let us consider a subsemigroup S of a Lie group G and its exponential function $\exp: \mathbf{L}(G) \to G$. In order to skip technicalities—which eventually we shall have to face squarely— we shall assume for now that S is closed. We define

$$\mathbf{L}(S) = \{X \in \mathbf{L}(G): \exp \mathbb{R}^+ \cdot X \subseteq S\}.$$

Then the set $W = \mathbf{L}(S)$ is topologically closed; it is stable under addition and is closed under multiplication by non-negative scalars in the finite dimensional real vector space $\mathbf{L}(S)$. We shall call such sets *cones* or, more frequently, *wedges*. Indeed, W will contain a largest vector subspace $W \cap -W$ called the *edge*, which in general is not zero and plays a crucial role in the overall theory; this is one reason why we prefer the terminology of "wedge" (another is that the word "cone" sometimes refers to not necessarily convex objects). But in the literature the terminology "cone" is so prevalent that we have decided to use the two terms synonymously. Those wedges, whose edge is zero, will be called *pointed cones*.

We have to prepare adequate background information on wedges. Chapter I serves this purpose. We deal with the structure theory of wedges in two ways: Firstly in terms of duality, secondly in terms of geometry. If W is a wedge in a finite dimensional real vector space L then *the dual wedge W^** is the set of all functionals ω in the dual $\widehat{L}$ of L satisfying $\langle \omega, x \rangle \geq 0$ for all $x \in W$. Frequently we can realize the dual wedge in L itself; this happens as soon as we are given, through natural circumstances, a nondegenerate bilinear symmetric form B on L (for instance, a scalar product, or a Cartan-Killing form) in which case we have $W^* = \{y \in L: \quad B(x, y) \geq 0 \text{ for all } x \in W\}$. We are particularly interested in wedges with interior points; this means that the dual is a pointed cone. The geometry of such wedges is determined by the structure of their boundary. A helpful concept is that of a face. A special type of face is particularly suited for duality, namely, the concept of an *exposed face*. We shall analyse this concept in terms of duality in great detail; at this point it suffices to understand its geometrical meaning. A support hyperplane of a wedge with inner points is a hyperplane meeting the wedge non-trivially and bounding a closed half-space containing the wedge. An exposed face is the intersection of a support hyperplane with the wedge (or the whole wedge). A non-zero point on the wedge is an *exposed point* if it lies on a one-dimensional exposed face. Unfortunately, if W has a non-zero edge then W has no exposed points. We need focus on the next best object, namely, those points $x \in W$ for which $x + (W \cap -W)$ is an exposed face. These points are called E^1-points. If W is pointed, then the E^1-points are exactly the exposed points. Of even greater importance are the so-called C^1-points. A point x is a C^1-*point* of a wedge with inner points if there is one and only one support hyperplane of the wedge containing x. In an *arbitrary* wedge, a point is called a C^1-point if it is a C^1-point of W in the vector space $W - W$ in which W does have inner points. There is a close correspondence between the C^1-points of W and the E^1-points of W^* which is encapsuled in the so-called Transgression Theorem (I.2.35).

Two types of wedges are particularly familiar: polyhedral and Lorentzian ones. A wedge is *polyhedral* if it is the intersection of finitely many closed half-spaces; it is *Lorentzian* if it is one half of the solid double cone defined by a Lorentzian form. A boundary point of a polyhedral cone is either a C^1-point or a E^1-point or neither of the two; each non-zero boundary point of a Lorentzian cone is both a C^1- and an E^1-point. This is the starting point of a small theory of *round* cones which we shall develop because we need it later in the infinitesimal Lie theory of semigroups.

There are several results in the first Chapter which are applied later. Some of them are of independent interest. The first is a classical theorem of MAZUR'S saying that *the set of C^1-points $C^1(W)$ of a convex closed set W with inner points in a* separable *Banach space is a dense G_δ in the boundary ∂W*. This result is non-trivial even in the case of finite dimensional vector spaces. Since the C^1-points play a central role, we give a complete proof of the Density Theorem. In the finite dimensional situation this implies a dual result due to STRASZEWICZ which says that *a finite dimensional cone W is the closed additive span of $E^1(W)$*, the set of its $E^1(W)$-points.

A further tool of crucial importance is a theorem on ordinary differential equations due to BONY and BREZIS. It deals with the invariance of closed sets under flows. For a brief discussion let A denote a closed subset of a finite dimensional vector space and let U be an open subset containing A. Let X be a vector field on U satisfying a local Lipschitz condition. Then X defines a local flow $(t, u) \mapsto F_t(u)$ on U via $F_t(u) = x(t)$, where x is a solution of the initial value problem $\dot{x}(t) = X\big(x(t)\big)$, $x(0) = u$. We say that A is invariant under the flow F if $F_t(a) \in A$ for all $a \in A$ and all $t \geq 0$ such that $F_t(a)$ is defined. The point is that the invariance of A under F can be expressed in terms of X and the geometry of A. For this purpose we need a definition.

1. **Definition.** A *subtangent vector* of a subset W of a topological vector space L at a point $w \in L$ is a vector x such that there are elements w_n and numbers r_n such that

    (i) $\lim w_n = w$, $w_n \in W$,

    (ii) $0 \leq r_n \in \mathbb{R}$,

    (iii) $x = \lim r_n(w_n - w)$.

The set of all subtangent vectors of W at w will be denoted $L_w(W)$. If $w = 0$, we shall write $L(W)$ instead of $L_0(W)$. We shall call x a *tangent vector of W at w* if both x and $-x$ are subtangent vectors. The set of tangent vectors of W at w, denoted $T_w(W)$, therefore is $L_w(W) \cap -L_w(W)$.

    (It is no problem to define a subtangent vector x of a subset W of a differentiable manifold M at a point $w \in M$. Under such circumstanceas, x is an element of the tangent space $T(M)_w$ of M at w.)

2. **Theorem.** *If A is a closed subset of a finite dimensional vector space L and U an open subset containing A, and if X is a vector field on U satisfying a local Lipschitz condition, then A is invariant under the local flow defined by X if and only if*

$$X(a) \in L_a(A) \qquad \text{for all} \qquad a \in A.$$

                                                             ■

    (The theorem, by the way, remains intact for closed subsets and vector fields on differentiable manifolds.)

    Our primary applications of this theorem concerns wedges in finite dimensional vector spaces and their invariance under linear flows. In fact we shall prove the following Invariance Theorem for Wedges:

3. **Theorem.** *Let W be a generating wedge in a finite dimensional vector space L and $X: L \to L$ a linear map. Then the following conditions are equivalent:*

    (1) $e^{t \cdot X} W \subseteq W$ *for all* $t \in \mathbb{R}^+$ *(respectively, for all* $T \in \mathbb{R}$*).*

    (2) $X(w) \in L_w(W)$ *(respectively,* $X(w) \in T_w(W)$*) for all* $w \in W$*.*

    (3) $X(c) \in L_c(W)$ *(respectively,* $X(c) \in T_c(W)$*) for all* $c \in C^1(W)$*.*

    (4) $X(e) \in L_e(W)$ *(respectively,* $X(e) \in T_e(W)$*) for all* $e \in E^1(W)$*.*   ■

    The equivalence of (1) and (2) is a rather immediate consequence of the Bony–Brezis Theorem; the equivalence of (3) with these conditions requires the Mazur Density Theorem—but that is not enough; the proof further requires a result which we shall call the Confinement Theorem which says that a flow confined in a wedge by the tangent hyperplanes in all C^1-points cannot seep out through the corners. A duality argument will establish the equivalence of (4) with the other conditions.

———

The reading of Chapter I will not demand many prerequisites. We have some cause to formulate the theory of wedges without restriction of the dimension as far as this generality can be sustained painlessly. *However*, for a first reading little is lost to the reader who prefers to restrict attention to the finite dimensional case. In this situation, most of the material is elementary, yet not trivial. For the Mazur Density Theorem, some background in functional analysis is required such as familiarity with Baire category arguments. The Bony–Brezis Theorem demands some knowledge on ordinary differential equations. Given all of this, however, the first chapter is self-contained.

———

The second and third chapter are devoted to *the infinitesimal Lie theory of semigroups*; they deal with those wedges in finite dimensional Lie algebras which arise as tangent sets of semigroups and local semigroups in Lie groups.

Let S denote again a closed subsemigroup of a finite dimensional real Lie group G and set $W = \mathbf{L}(S)$. It is not hard to verify that $\mathbf{L}(S) = L_0\big(\exp^{-1}(S)\big)$. We have observed above that W is a wedge in the Lie algebra $L = \mathbf{L}(G)$. More generally, if U is an open neighborhood of the identity in G and $S \subseteq U$ a subset satisfying $SS \cap U \subseteq S$, then the set $W = L_0\big(\exp^{-1}(S)\big)$ of subtangent vectors of the pull-back of S under the exponential function is always a wedge. But how does such a wedge relate to the *Lie algebra* structure of L? The answer is not part of the classical repertory. It was discovered independently by OL'SHANSKIĬ and by HOFMANN and LAWSON, that *every subtangent wedge W of a local semigroup in a Lie algebra satisfies*

$$(*) \qquad\qquad e^{\operatorname{ad} x}W = W \qquad \textit{for all} \qquad x \in W \cap -W,$$

where $\operatorname{ad} x\colon L \to L$ as usual is the inner derivation of L given by $(\operatorname{ad} x)(y) = [x, y]$. Recall, in this context, that every derivation D of L gives rise to an automorphsm e^D of the Lie algebra L. It is clear that every pointed cone trivially satisfies condition $(*)$, and that this condition implies that the edge $W \cap -W$ of the wedge is a Lie subalgebra.

All of this, once understood, is comparatively easy to establish. It is much harder to accomplish Sophus Lie's Fundamental Theorem for a local theory of semigroups by showing that, conversely, *if a wedge W in the Lie algebra $L(G)$ of a Lie group satisfies* $(*)$, *then there is an open neighborhood U of the identity in G and a subset $S \subseteq U$ with $SS \cap U \subseteq S$ such that $W = L_0(\exp^{-1} S)$*. This result is the core of the entire *local Lie theory of semigroups* and was established by the authors. A whole chapter is devoted to a proof of this fact, namely, Chapter IV.

However, this carries us beyond the infinitesimal theory, but it amply justifies the terminology of calling *Lie wedge* any wedge in a Lie algebra satisfying condition $(*)$ above.

Condition $(*)$ has a drawback. An infinitesimal equivalent of the semigroup property should be expressed in terms of the Lie bracket alone and not with the aid of a convergent power series. An immediate corollary of Theorem 3 is the key to such a reformulation.

4. Corollary. *If W is a wedge in a finite dimensional Lie algebra L, then for any element $y \in L$, the following conditions are equivalent:*

(1) $e^{\operatorname{ad} y} W = W$.

(2) $[w, y] \in T_w(W)$ *for all* $w \in W$.

(3) $[c, y] \in T_c(W)$ *for all* $c \in C^1(W)$.

(4) $[e, y] \in T_e(W)$ *for all* $e \in E^1(W)$. ∎

This allows us to conclude

5. Theorem. (The Characterization Theorem for Lie wedges) *For a wedge W in a finite dimensional Lie algebra L, the following conditions are equivalent:*

(1) W *is a Lie wedge.*

(2) $[w, W \cap -W] \subseteq T_w(W)$ *for all* $w \in W$.

(3) $[c, W \cap -W] \subseteq T_c(W)$ *for all* $c \in C^1(W)$.

(4) $[e, W \cap -W] \subseteq T_e(W)$ *for all* $e \in E^1(W)$. ∎

Let us pause to inspect condition (2) for the elements $w \in W \cap -W$ of the edge. For each such element, $T_w(W) = W \cap -W$. Thus we note once more that the edge of a Lie wedge is a Lie subalgebra. In particular if W happens to be a vector space—which is the case precisely when $W = -W = W \cap -W$—then Theorem 5 expresses nothing else but the fact that a vector space is a Lie wedge if and only if it is a Lie subalgebra.

The Lie wedge condition $(*)$ is of the type of an invariance condition which suggest the concept of an invariant wedge which will engage much of our energy in this book.

6. Definition. A wedge W in a Lie algebra L is *invariant* if

$(**)$ $$e^{\operatorname{ad} x} W = W \qquad \text{for all} \qquad x \in L.$$

From Corollary 4 we obtain

7. Theorem. (The Characterization Theorem for Invariant Wedges—Elementary Version) *For a wedge W in a Lie algebra L, the following conditions are equivalent:*

(1) W *is invariant.*

(2) $[w, L] \subseteq T_w(W)$ *for all* $w \in W$.

(3) $[c, L] \subseteq T_c(W)$ *for all* $c \in C^1(W)$.

(4) $[e, L] \subseteq T_e(W)$ *for all* $e \in E^1(W)$. ∎

If, in condition (2) we consider once again only elements $w \in W \cap -W$ in the edge, we find that

$$[W \cap -W, L] \subseteq W \cap -W,$$

that is, that the edge of an invariant wedge is always an ideal. More trivially, $(**)$ directly implies that $W - W$ is an ideal, too. In particular, a vector space is an invariant wedge if and only if it is an ideal. Thus in a very immediate sense, Lie wedges generalize subalgebras, invariant wedges generalize ideals.

———————————

One of the familiar properties of Lie group theory is that local Lie subgroups of a Lie group are ruled smoothly by local one parameter subsemigroups. Sometimes this is expressed as the theorem of the "existence of canonical coordinates of the first kind". It is one of the unpleasant surprises that even in the simplest examples, nice infinitesimally generated local subsemigroups in Lie groups fail to be ruled by local one-parameter semigroups. Such examples exist in 3-dimensional Lie groups such as the Heisenberg group, the group $\mathbb{R}^2 \rtimes SO(2)$ of euclidean motions of the plane, in $Sl(2, \mathbb{R})$ and in $SO(3)$. We shall discuss such examples explicitly and in detail in various parts of the book. The one-parameter subsemigroups of a closed subsemigroup S of G are $t \mapsto \exp t \cdot X$ with $X \in \mathbf{L}(S)$. Even if S is algebraically generated by $\exp \mathbf{L}(S)$ and $\mathbf{L}(S)$ is the exact set of subtangent vectors of $\exp^{-1}(S)$ at 0, in general there are arbitrarily small elements of the form $\exp X_1 \cdots \exp X_n$ with $X_1, \ldots, X_n \in \mathbf{L}(S)$ which *cannot* be written in the form $\exp Y$ with some $Y \in \mathbf{L}(S)$. If S happens to be a group, this is always the case. This deficiency in the Lie theory of semigroups is a fact of life we have to live with whether we like it or not.

However, if W is an invariant wedge in the Lie algebra $\mathbf{L}(G)$, then there is always an open neighborhood B of 0 in $\mathbf{L}(G)$ which is mapped homeomorphically onto an identity neighborhood U in G under the exponential function such that $\big(\exp(B \cap W)\big)^2 \cap U \subseteq \exp(B \cap W)$. In other words, *invariant wedges always define local subsemigroups which are ruled smoothly by one-parameter subsemigroups.* However, not every subsemigroup which is locally ruled by one parameter subsemigroups is invariant. Hence the hunt is on for those Lie wedges which belong to local semigroups behaving more like local groups. The Campbell–Hausdorff multiplication which is give near 0 by $X * Y = X + Y + \frac{1}{2}[X, Y] + \frac{1}{12}[X[X, Y]] + \frac{1}{12}[Y, [Y, X]] + \cdots$ in terms of a universally defined infinite series in Lie monomials allows us a definition in terms the Lie algebra L:

8. **Definition.** A wedge W in a finite dimensional real Lie algebra L is a *Lie semialgebra* if there is an open neighborhood B of 0 such that the series for $X * Y$ converges for all $(X, Y) \in B \times B$ and such that

(†) $$(B \cap W) * (B \cap W) \subseteq W.$$

There is minute variance in terminology among the authors here. Because the idea was first introduced by Hofmann, in his papers Lawson has called a Lie semialgebra also a *Hofmann wedge*. In this book we shall use the term Lie semialgebra.

We shall see that all closed half-spaces in a Lie algebra bounded by a subalgebra are Lie semialgebras. It follows that all finite intersections of such half-spaces are Lie semialgebras. After our characterization of Lie semialgebras below we shall be able to drop the word "finite".

In $\mathrm{sl}(2,\mathbb{R})$ for instance the set of all matrices

$$\begin{pmatrix} x & y \\ z & -x \end{pmatrix} \qquad x \in \mathbb{R}, \quad y,z \in \mathbb{R}^+$$

is a Lie semialgebra. We shall show in a careful discussion of $\mathrm{sl}(2,\mathbb{R})$ that the exponential function of $\mathrm{Sl}(2,\mathbb{R})$ maps this semialgebra diffeomorphically onto the semigroup of all matrices with determinant 1 and non-negative entries. This is not obvious.

According to our general guideline division of Lie theory into an *infinitesimal*, a *local*, and a *global* one, the concept of a Lie semialgebra is an infinitesimal one. But its definition is local. Indeed the Campbell–Hausdorff multiplication on suitable neighborhoods of 0 in a Lie algebra is *the* prime tool for any local theory. This will become abundantly clear in our discussion of the local theory in Chapter IV. Typically, if B is a Campbell–Hausdorff neighborhood in a Lie algebra L with a multiplication $*: B \times B \to L$ well defined, then we can define local left translations $\lambda_x: B \to L$ by $\lambda_x(y) = x * y$. The derivative $d\lambda_x(0): L \to L$ at 0 is another universal analytic function which is indispensible for the infinitesimal theory. It is given by $d\lambda_x(0) = g(\mathrm{ad}\,x)$ where the power series $g(X)$ is uniquely defined by the equation $g(X)(1 - e^{-X})/X = 1$ yielding

$$g(X) = 1 + \frac{1}{2}X + \sum_{n=1}^{\infty} \frac{b_{2n}}{(2n)!} X^{2n}$$

with the Bernoulli numbers b_n.

But there is, fortunately, also a purely infinitesimal characterization of Lie semialgebras:

9.   Theorem. (The Characterization Theorem of Lie Semialgebras) *For a wedge W in a Lie algebra L, the following conditions are equivalent:*

(1) *W is a Lie semialgebra.*

(2) *Suppose that B is any open neighborhood of 0 on which an analytic function $*: B \times B \to L$ is defined which extends the Campbell–Hausdorff multiplication. Then $(B \cap W) * (B \cap W) \in W$.*

(3) *$[w, T_w(W)] \subseteq T_w(W)$ for all $w \in W$.*

(4) *$g(\mathrm{ad}\,x)(W) \subseteq L_w(W)$ for all $w \in B \cap W$ where B is any Campbell–Hausdorff neighborhood of 0 in L.*

If W is generating, these conditions are also equivalent to

(5) *$[c, T_c(W)] \subseteq T_c(W)$ for all $c \in C^1(W)$.*   ∎

A vector space is a Lie semialgebra if and only if it is a subalgebra. Thus the generalizations of Lie subalgebras into the semigroup domain fan out in different directions. A characterization of Lie semialgebras in terms of E^1-points is conspicuously absent for good reason.

Condition (5) however, allows us to introduce an effective tool for working with generating Lie semialgebras. Indeed for each C^1-point c of a generating Lie semialgebra W there is a unique real number $\lambda(c)$ such that $[c, y] - \lambda(c) \cdot y \in T_c(W)$ for all $y \in L$. The function $\lambda : C^1(W) \to \mathbb{R}$ is called *the characteristic function of the Lie semialgebra W*. Since W is invariant if and only if the characteristic function vanishes identically, it is a measure for the deviation from invariance.

The relationship between the different kind of wedges W in a Lie algebra L which we have now seen is tabulated as follows:

$$\text{Lie wedge} \iff [x, W \cap -W] \subseteq T_x(W),$$
$$\text{Lie semialgebra} \iff [x, T_x(W)] \subseteq T_x(W),$$
$$\text{invariant wedge} \iff [x, L] \subseteq T_x(W),$$

for all $x \in W$, respectively, for all $x \in C^1(W)$ if W is generating.

For Lie semialgebras, the hypothesis that W be generating is much less stringent than in the case of Lie wedges in general. In fact, one might say that it is no restriction of generality whatsoever because we shall see that *for any Lie semialgebra W in a Lie algebra L, the linear span $W - W$ is,* in striking contrast with the case of mere Lie wedges, *a Lie subalgebra*.

We notice in passing that condition (2) of Theorem 9 says, among other things, that in an *exponential* Lie algebra, that is, a Lie algebra in which the Campbell–Hausdorff multiplication has an extension to a function $* : L \times L \to L$, making $(L, *)$ into a simply connected Lie group, a wedge W is a Lie semialgebra if and only if $(W, *)$ is a subsemigroup of $(L, *)$. The class of exponential Lie algebras is a class of solvable algebras which contains the class of all nilpotent ones properly.

———————

Just as in the case of the Lie theory of groups, many different aspirations motivate us. Firstly we want a *general theory* such as it is exemplified in Sophus Lie's Fundamental Theorems which regulate the connection between Lie algebras and local groups, that is, the connection between the infinitesimal and the local theory. This line of thought is also illustrated by all results which connect the local and the global theory such as the theorems on the existence of analytic subgroups for every Lie subalgebra in the Lie algebra of a Lie group, the existence of simply connected Lie groups for each Lie algebra, to name a few. Secondly, we desire to *classify* relevant objects and to develop an explicit *structure theory*. An example is the classification of semisimple Lie algebras and Lie groups. In the solvable and nilpotent case, where a complete classification is not only impossible but unreasonable, still a powerful structure theory and methodology is available. Thirdly, on the side of these two aspects of the theory, overlapping and closely linked, there is the entire complex of ideas having to do with representation theory.

In this book, no effort is made to attempt a systematic representation theory for Lie semigroups even though some rudiments occur. A *general Lie theory* of semigroups, however, is available now; its final form may not be attained yet. In this frame work, Lie wedges do indeed play a crucial role as will emerge in the second half of the book. As far as *the structure and classification theory* is concerned, next to nothing is known on Lie wedges in general. The situation, as we shall see, is much better for Lie semialgebras and invariant wedges. After the work of

KOSTANT, VINBERG, OL'SHANSKIĬ, PANEITZ, HILGERT, and HOFMANN, a rather complete structure theory is available for invariant cones. Chapter 3 is entirely devoted to it and we shall summarize below the salient features of this theory. The structure theory of Lie semialgebras builds up to one major result which shows that Lie semialgebras have a strong tendency to be invariant.

10.   Theorem.   (The Tangent Hyperplane Subalgebra Theorem) *Let W be a generating Lie semialgebra in a Lie algebra L and let λ denote its characteristic function. Then at any boundary point $x \in C^1(W)$ at which $\lambda(x)$ is non-zero the tangent hyperplane T_x is a Lie subalgebra.*   ∎

The status of our knowledge may be summarized in still rather vague terms as follows: i) We know all Lie semialgebras up to dimension 3 and most of them up to dimension 4, and there are far-reaching general principles which make this knowledge necessary and fruitful in the general theory; ii) we know and can construct Lie semialgebras of special types notably in the domain of solvable metabelian Lie algebras; iii) the Tangent Hyperplane Subalgebra Theorem indicates that generating Lie semialgebras in a Lie algebra either belong to special types of non-invariant semialgebras or else are invariant. In the direction of a general classification, much remains to be done.

Frequently one class of Lie algebras emerges in this context, namely, that of almost abelian Lie algebras. Here a Lie algebra L is called *almost abelian* if there is an abelian ideal I of codimension ≤ 1 such that L/I acts on I by scalar multiplication. An almost abelian Lie algebra L is characterized by the fact that every vector subspace is a subalgebra and every wedge $W \subseteq L$ is a Lie semialgebra (which is invariant only if $W - W$ is an ideal). Almost abelian Lie algebras give the prototype of a class of metabelian Lie algebras in which non-invariant Lie semialgebras abound.

———————————

One methodological key for all deeper structure results on Lie semialgebras and invariant wedges is surprisingly classical: The idea of Cartan algebras and root decompositions. We emphasize that we do not restrict our attention to a semisimple theory. Thus we have to use these tools in the most general frame work. Let us make the following observation first. We have already observed that it is no loss of generality to assume that a Lie semialgebra is generating. The edge of a Lie semialgebra trivially contains a unique largest Lie algebra ideal; it is only feasible to consider the factor algebra modulo this ideal and thus to assume that the Lie semialgebra is *reduced*, that is, does not contain any non-zero ideal. As a first result on Cartan subalgebras we shall firstly show that *any Cartan algebra in a Lie algebra supporting a reduced generating Lie semialgebra is necessarily abelian*. A second structure invariant which emerges as extremely useful for the investigation of Lie semialgebras in a Lie algebra L is the so-called *base ideal*, that is the sum of all 1-dimensional ideals. For a linear map $\alpha: L \to \mathbb{R}$ we define $M_\alpha(L) = \{y \in L: [x,y] = \alpha(x) \cdot y \text{ for all } x \in L\}$ and say that α is a *base root* if $M_\alpha(L) \neq \{0\}$. In particular, $M_0(L)$ is the center $Z(L)$ of L and $M(L)$ is the direct sum of all base root spaces $M_\alpha(L)$.

For each base root α we define a subset $C_\alpha^1(W) = \{c \in C^1(W): M_\alpha(L) \not\subseteq T_c(W)\}$. Then α and the characteristic function λ of W agree on $C_\alpha^1(W)$. Thus if

W is reduced and $C_\alpha^1(W)$ is dense in $C^1(W)$, then $M(L) = M_\alpha(L)$. In particular, if W is invariant and reduced (hence pointed) then $M(L) = Z(L)$ and every abelian ideal is central. We have remarked that in Chapter I we make precise the intuitive idea of a *round cone*. We shall show that *for every round Lie semialgebra there is a base root α such that its characteristic function is $\alpha|C^1(W)$ and $M(L) = M_\alpha(L)$.* This result exemplifies the interplay between base roots and characteristic function. We summarize:

11.   Theorem. (The First and Second Cartan Algebra Theorems) *If W is a reduced generating Lie semialgebra in L then all Cartan algebras of L are abelian. If, in addition, the characteristic function of W is $\alpha|C^1(W)$ for some base root α then either W is invariant or L has rank 1 (that is, all Cartan subalgebras have dimension 1).*

12.   Theorem. (The Rank 1 Structure Theorem) *Let W be a reduced generating Lie semialgebra in a Lie algebra of rank 1. Then one and only one of the following situations is possible:*

(i) *$L \cong \mathbb{R}$ and W is one of the two half-lines.*

(ii) *$L \cong \mathrm{sl}(2, \mathbb{R})$ and W is one of a set of fully described Lie semialgebras in $\mathrm{sl}(2)$.*

(iii) *L is metabelian (that is $[L, L]$ is abelian) and the structure of L as well as the possible W in L are precisely described.*   ∎

As a corollary, we note

13.   Corollary. *Suppose that W is a round generating Lie semialgebra in a Lie algebra which is not semisimple. Then W is invariant or L is almost abelian (and W is an arbitrary round wedge).*   ∎

This will allow us to conclude that Lorentzian Lie semialgebras are either invariant or else span an almost abelian Lie algebra. We shall classify Lorentzian Lie semialgebras completely.

The next Cartan algebra theorem deals exclusively with algebras, but it is basic for the theory of invariant wedges. In order to understand its formulation we need to know the concept of a compactly embedded subalgebra of a Lie algebra. Indeed we shall say that a subalgebra K is *compactly embedded* into a Lie algebra if the analytic group generated by $e^{\mathrm{ad}\, K}$ in $\mathrm{gl}(L)$ has a compact closure. An element $x \in L$ is called *compact* if $\mathbb{R} \cdot x$ is compactly embedded. Let us denote with $\mathrm{comp}\, L$ the set of all compact elements of the Lie algebra L. The interior of a set M in a topological space is written $\mathrm{int}\, M$.

14.   Proposition. (The Third Cartan Algebra Theorem) *Let H be a Cartan algebra of L and let $x \in L$. Denote $\mathcal{H}$ the set of all compactly embedded Cartan algebras of L. Then the following conditions are equivalent:*

(1) $x \in \mathrm{int}(\mathrm{comp}\, L)$.

(2) $\ker(\mathrm{ad}\, x) = \bigcup \{K \colon x \in K \in \mathcal{H}\}$.

(3) $\ker(\mathrm{ad}\, x) \subseteq \mathrm{comp}\, L$.

Moreover, all $K \in \mathcal{H}$ are conjugate under inner automorphisms of L. All compactly embedded Cartan algebras are abelian. Hence all Cartan algebras of L are abelian if $\mathcal{H} \neq \emptyset$.   ∎

In the following theorem, $\mathcal{H}$ continues to denote the conjugacy class of the compactly embedded Cartan algebras of L.

15.   Theorem.   (The Fourth Cartan Algebra Theorem) *Let W be an invariant pointed generating cone in a Lie algebra L. Then the following conclusions hold:*

  (i) $\operatorname{int} W \subseteq \operatorname{comp} L$.

  (ii) $\mathcal{H} \neq \emptyset$, *and thus all Cartan algebras of L are abelian.*

  (iii) *If $H \in \mathcal{H}$, then $H \cap \operatorname{int} W \neq \emptyset$.*   ∎

This theorem clearly points the way for a general theory of invariant cones: We fix a compactly embedded Cartan algebra H. Such Cartan algebras must exist if there are invariant generating pointed cones. Fixing an arbitrary compactly embedded Cartan algebra is no restriction, because all of them are conjugate and each one intersects the interior of any invariant pointed cone with inner points. Then we have

16.   Theorem.   *If W_1 and W_2 are invariant generating pointed cones with $W_1 \cap H = W_2 \cap H$, then $W_1 = W_2$.*   ∎

In other words, invariant pointed generating cones are uniquely determined by their intersection with any fixed compactly embedded Cartan algebra. From here on we have the task of determining which pointed cones in a given compactly embedded Cartan algebra are the traces of invariant cones. This labor is arduous and requires the machinery of root space decomposition. More about this in a moment.

━━━━━━━

First let us record a series of theorems which give sufficient conditions for a Lie semialgebra to be invariant.

17.   Theorem.   (The First Invariance Theorem) *Let W be a reduced generating Lie semialgebra in the Lie algebra L and assume that the center of L and the interior of W intersect. Then W is invariant.*   ∎

18.   Theorem.   (The Second Invariance Theorem) *Let W be a generating Lie semialgebra in the Lie algebra L which is not semisimple and has rank at least 2. If W is reduced and its characteristic function is the restriction of a base root then W is invariant. In particular, if W is round then W is invariant.*   ∎

19.   Theorem.   (The Third Invariance Theorem) *Suppose that L is a Lie algebra without any hyperplane subalgebras. In particular, this applies to any semisimple Lie algebra without any $\mathrm{sl}(2, \mathbb{R})$-factor. Then any generating Lie semialgebra is invariant.*   ∎

We shall see that in special Lie algebras more can be said. For instance, in any compact Lie algebra, a generating Lie semialgebra is necessarily invariant, and in a semisimple compact Lie algebra there are no proper generating Lie semialgebras.

Aside from the Cartan algebra and the invariance theorems, there is another string of significant results. If a wedge W in a Lie algebra L contains the commutator algebra $[L, L]$, then it is, as a moments reflection shows, an invariant wedge. We call those wedges *trivial*.

20.   Theorem. (The First Triviality Theorem) *Every generating Lie semialgebra in a nilpotent Lie algebra is trivial.*   ∎

21.   Theorem. (The Second Triviality Theorem) *Every generating Lie semialgebra in the underlying real Lie algebra of a complex Lie algebra is necessarily trivial.*   ∎

22.   Theorem. *An invariant wedge W in a solvable Lie algebra L is trivial if one of the following conditions is satisfied:*

(i) int W *meets the nilradical.*

(ii) $W \cap [L, L] \subset -W$.   ∎

Note that (ii) is satisfied if $W \cap [L, L] = \{0\}$.

━━━━━━━━━━━━━━━━━━━

We remarked that there is a strong tendency for a generating semialgebra to be invariant. Invariant pointed cones W in a Lie algebra L, we saw, are determined completely by their intersection $C = W \cap H$ with a fixed compactly embedded Cartan algebra H. This justifies the position that invariant pointed cones are classified as soon as we have the means to say exactly which cones C in H are the traces of invariant cones. Internally, however, as an abelian Lie algebra, H is nothing but a vector space. Thus we have to import more structure into H, using the fact that it is a compactly embedded Cartan algebra in L. We sketch how we do this.

Firstly we record a purely Lie algebraic fact. If L is a Lie algebra we denote with $\mathrm{Inn}(L)$ the analytic subgroup of $\mathrm{Gl}(L)$ generated by $e^{\mathrm{ad}\, L}$ and with $\mathrm{INN}(L)$ its closure.

23.   Proposition. *If H is a compactly embedded Lie algebra of a Lie algebra L, then there is one and only one maximal compactly embedded subalgebra $K(H)$ containing H. Moreover, L decomposes into a direct sum $L = K(H) \oplus P(H)$ of $K(H)$-modules under the adjoint action. The closure of $e^{\mathrm{ad}\, H}$ in $\mathrm{INN}(L)$ is a maximal torus T, and the closure Γ of the analytic subgroup of $\mathrm{INN}(L)$ generated by $e^{\mathrm{ad}\, K(H)}$ is the unique maximal compact subgroup of $\mathrm{INN}(L)$ containing T. The normalizer $N = \{\gamma \in \mathrm{INN}(L) : \gamma(H) = H\}$ is contained in Γ and the centralizer $Z = \{\gamma \in \mathrm{INN}(L) : \gamma(h) = h \text{ for all } h \in H\}$ has finite index in N.*   ∎

In particular, this allows us to define the *Weyl group* $\mathcal{W}(L, H)$ of L with respect to H which is, after all, nothing but the classical Weyl group of the compact group Γ with respect to the maximal torus T.

It is not hard to see that a cone C in H is of the form $W \cap H$ with an invariant pointed cone W only if the following condition is satisfied:

(I) $\qquad\qquad\qquad$ C is mapped into itself by $\mathcal{W}(L, H)$.

We shall see that this necessary condition is also sufficient if L is a compact Lie algebra. However, in general, (I) is not a sufficient condition. In order to formulate the next condition we have to resort to the real root decomposition of L with respect to the Cartan algebra H. Indeed, L is a representation space for the torus T and as such decomposes into isotypic components

$$L = H \oplus H^+, \qquad H^+ = \sum_{\omega \in \Omega^+} L^\omega.$$

This means that we find linear forms $\omega \colon H \to \mathbb{R}$ and vector subspaces L^ω equipped with a complex structure $I \colon L^\omega \to L^\omega$, $I^2 = -1$ such that $h \in H$ and $x \in L^\omega$ we have

$$[h, x] = \omega(h) \cdot Ix.$$

The ω's appear in pairs. The selection of one representative yields what we call the set Ω^+ of positive real roots, and their selection also determines the complex structure I on H^+. For every *root vector* $x \in L^\omega$ the element $Q(x) = [Ix, x]$ is in H. Thus the vectors $Q(x)$, $x \in L^\omega$, $\omega \in \Omega$ endow the vector space H with additional structure. In particular, if $x \in L^\omega$ is a root vector, and $h \in H$ then $(\operatorname{ad} x)^2 h = \omega(h) \cdot Q(x) \in H$. Thus if x is any root vector, then $(\operatorname{ad})^2$ maps H into itself and, in fact, induces a rank 1 operator on H.

Each root space L^ω is either contained in $K(H)$ or in $P(H)$. In the first case we say that ω is a *compact root*, in the second a *non-compact root*.

If L contains a pointed invariant cone W, then it turns out that $K(H)$ has a non-zero center Z and that $K(H)$ is the centralizer of Z in L. We fix an arbitrary non-zero element $z \in Z$. It will be shown that the roots may be selected in such a fashion that a root ω is compact if $\omega(z) = 0$ and non-compact if $\omega(z) > 0$. We shall see that the following condition is also satisfied for $C = W \cap H$:

(⋆) $\qquad\qquad$ $(\operatorname{ad} x)^2 C \subseteq C$ for all $x \in L^\omega$ for all non-compact roots ω.

We can now define $\mathcal{C}(L, H) \subseteq \operatorname{Hom}(H, H)$ to be the closed convex cone generated by all rank 1 operators $(\operatorname{ad} x)^2 | H$ with $x \in L^\omega$, for all *non-compact* roots ω. Then condition (⋆) is equivalent to the condition

(II) $\qquad\qquad\qquad$ C is mapped into itself by $\mathcal{C}(L, H)$.

Condition (I) has to do with the compact part $K(H)$ and Condition (II) with the non-compact part $P(H)$ of L. Technical as they may seem, both conditions are very handy in the concrete situations. The principal result states that the two conditions (I) and (II) are both necessary and sufficient:

24.   Theorem. *Let L be a Lie algebra with a compactly embedded Cartan algebra H such that $K(H)$ has a non-trivial center whose centralizer in L is $K(H)$. Then a pointed cone C in H is the trace of a (unique) invariant pointed cone W in L if and only if both conditions (I) and (II) are satisfied.*   ∎

The proof of this result requires patience. One needs to deal with a Levi decomposition of L. The first part of the proof treats the solvable case. In this case condition (II) alone is relevant. It turns out that *a solvable Lie algebra supports an invariant generating pointed cone only if the solvable length of L is at most* 3. Many structural details emerge. After we have accomplished this, we are left with reductive Lie algebras. The compact case alone requires the Convexity Theorem of KOSTANT which says that *in any compact Lie algebra, the orthogonal projection of an orbit of an element in a Cartan subalgebra into this algebra is the closed convex span of the orbit of this element under the Weyl group.* Subsequently, the root decomposition has to be inspected very carefully. We follow ideas initiated by PANEITZ and by KUMARESAN and RANJAN in the case of a simple algebra. The proof itself, of course, reveals much of the internal apparatus of invariant pointed cones.

―――――――

The reader will find the prerequisites in Chapters II and III increasingly demanding. The first sections of Chapter II leading us through the characterization of Lie wedges and Lie semialgebras are comparatively elementary and are self contained on the level of the linear algebra of Lie algebra theory and some basic calculus resting, of course, on the material which was prepared in Chapter I. This leads us through the discussion of the finite dimensional Lie algebras; an occasional reference is made here to convex analysis such as it is found in ROCKAFELLAR'S text on this subject. The later sections begin to call increasingly on familiarity with Lie algebra theory and in Chapter III a certain knowledge of the basic techniques and facts of the theory of semisimple Lie algebras becomes indispensible. We hope that enough accessible references are provided to keep the book sufficiently self-contained. Most of the substance of Chapters II and III is new, although the theory of invariant cones owes the decisive impulses to the work of VINBERG, OL'SHANSKIĬ, and PANEITZ. Much of the material is published here for the first time.

―――――――

The first half of the book, Chapters I, II, and III, is concerned with the *infinitesimal theory* of semigroups in Lie groups and the necessary preparations. The second half is devoted to the local and global theory. The local theory appears in Chapter IV.

Let L be the Lie algebra of a Lie group. We consider open balls B around 0 on which the Campbel–Hausdorff series converges and defines a local group structure (the so-called C-H neighborhoods). A subset S is a *local semigroup* with respect to B if $S * S \cap B \subseteq S$. Since a local semigroup in the Lie group can be pulled back isomorphically (at least locally) by means of the local inverse of the exponential mapping into the Lie algebra equipped with the Campbell–Hausdorff multiplication, it is really no loss in generality to consider local semigroups in Lie algebras. Then one has readily at hand the structure of the Lie algebra to work with.

To each local semigroup S we assign its set of subtangent vectors $\mathbf{L}(S) = L_0(S)$. As has been previously remarked, $\mathbf{L}(S)$ turns out to be a Lie wedge (that is, a wedge invariant under the induced action of its edge). There is a useful alternate characterization of the subtangent vectors for the case that S is a local semigroup: $X \in \mathbf{L}(S)$ if and only if $\mathbb{R}^+ \cdot X \cap B \subseteq \overline{S}$. Thus, if S is locally closed, the subtangent vectors correspond to the local one parameter semigroups of S. The Surrounding Wedge Theorem guarantees that our notion of a tangent wedge is a reasonable one. This theorem states that *given any slightly larger wedge W that "surrounds" $\mathbf{L}(S)$, there exists a C-H neighborhood B such that $S \cap B \subseteq W$.*

The assignment to each local semigroup S of its tangent Lie wedge $\mathbf{L}(S)$ is highly non-injective (unlike the situation for local Lie groups). To remedy this we must restrict our attention to special classes of local semigroups. We say a local semigroup is *strictly infinitesimally generated* if it is the smallest local subsemigroup locally containing its tangent wedge. This semigroup is then uniquely determined by its tangent wedge and the given CH-neighborhood B. Analogous results apply to the closures of these local semigroups.

The most difficult part of Lie's Fundamental Theorem(s) in the case of local Lie groups is establishing the existence of a local Lie group that corresponds to a given local Lie algebra. This one can do by showing that the Campbell–Hausdorff multiplication on a C-H neighborhood gives the desired solution. Even in the presence of this result, the solution of the corresponding problem in the setting of local semigroups requires a great deal of effort and special machinery and is the major focus of Chapter IV. The objective is to reverse the correspondence of the previous paragraph and associate with each Lie wedge a local semigroup with that wedge as its tangent wedge. Indeed we establish Lie's Fundamental Theorem in the following form:

25. **Theorem.** (Lie's Fundamental Theorem for Semigroups) *Let W be a subset of a finite dimensional real Lie algebra L. The following are equivalent:*

(1) *W is a Lie wedge.*

(2) *There is an open neighborhood B of 0 in L and a local semigroup S with respect to B such that $\mathbf{L}(S) = W$.* ∎

There are some special cases of the theorem which allow more elementary proofs. The simplest case is that of a pointed cone W, which is trivially a Lie wedge. Here a very direct proof is possible. But even then the rather technical nature of the problem shows up if one insists on a description of the *smallest* local semigroup S with respect to B satisfying $\mathbf{L}(S) = W$. Already this special situation reveals some of the distinctive aspects of the Lie theory of semigroups as opposed to that of groups. If we equip L with a norm such that $\|[x, y]\| \leq \|x\| \cdot \|y\|$ (which is always possible) and if we let B denote the open ball of radius $\frac{1}{2} \log 2$, then for every Lie subalgebra W of L the set $S = W \cap B$ is a local group satisfying $\mathbf{L}(S) = W$ and, additionally, $-S = S$. The emphasis here is on the fact that the neighborhood B of reference can be chosen once and for all, housing local groups for all Lie subalgebras. Not so in the case of Lie wedges! The construction is such that in each individual case, the wedge W determines its own neighborhood B

of reference. An example is furnished by any Lie algebra containing a hyperplane which is not a subalgebra (e.g. so(3), which we may conveniently visualize as the euclidean space $\mathbb{R}^3$ with the vector product $\times$ as Lie algebra product). Let W_n be an ascending sequence of pointed cones whose union is one of the open half-spaces determined by the hyperplane plus $\{0\}$. Let W denote the closure of this union. If a simultaneous neighborhood of reference B would exist in which we could find local subsemigroups S_n with respect to B with $W_n = \mathbf{L}(S_n)$, then we could assume each S_n to be strictly infinitesimally generated, and thus the sequence S_n would be ascending and $S = \bigcup S_n$ would be a local semigroup with respect to B with $W = \mathbf{L}(S)$. In particular, W would be a Lie wedge. But we know from the infinitesimal theory that a closed half-space is a Lie wedge if and only if the bounding hyperplane is a Lie algebra. Hence our assumption that a simultaneous neighborhood of reference B exists for all W_n is refuted. Intuitively, as the cones W_n "open up", the respective neighborhoods of reference B_n shrink. Thus even very simple examples show that considerable complications arise with the Fundamental Theorem in the semigroup case.

Chapter IV also serves to introduce some useful machinery from the theory of geometric control and to develop some of the significant connections between that theory and the Lie theory of semigroups. The basic results from geometric control concerning the notion of the "accessibility" property translate in the context of local semigroups to the Dense Interior Theorem, which asserts that *the local semigroup generated by a family of local one parameter semigroups has dense interior in itself, where the interior is taken in the smallest Lie algebra containing all the local one parameter semigroups.* The control theoretic notion of a reachable set, suitably adapted to the context under consideration, is most useful in the study of the fine structure of a local semigroup, in particular in considering how it is generated. It is eventually shown in the Edge of the Wedge Theorem that *if one starts with a Lie wedge whose members are interpreted as left invariant vector fields on a Campbell-Hausdorff neighborhood, then the local reachable sets are local semigroups which have the original wedge as tangent wedge.* Thus, in the end, results from the Lie theory of local semigroups yield information of a control theoretic nature.

―――――――――

The discourse of Chapter IV which eventually secures a proof of the Fundamental Theorem 25 builds up a considerable degree of technicality. We do not promise light reading, particularly if one is unfamiliar with control theoretic notions and language. But the required techniques do not transcend standard calculus on open sets of $\mathbb{R}^n$, the theory of ordinary differential equations, and basic topology. The problems arising in the context of local reachability and "local rerouting", as we shall call it, are conceptually intricate and require a certain degree of patience. The saving grace is that the Fundamental Theorem itself is easily formulated and understood and that simple examples show the complications one has to expect. In a first perusal of the material, the reader may prefer to absorb the more straightforward special cases discussed in an independent fashion in Section 4.

We found it not much more difficult to carry out the program of this chapter for completely normable topological Lie algebras, the so-called *Dynkin algebras.* Thus we have in fact a local Lie theory of semigroups in Lie groups without dimensional restriction. A mild price has to be paid in the form of some modest additional geometric hypotheses on the wedges which are automatically satisfied in the finite dimensional case.

———

Finally, the book turns to the *global theory*. In Chapter V we squarely face the demand for a discourse on subsemigroups of Lie groups amenable to a Lie theoretic approach. We preface this chapter with a section on generalities which link subsemigroups and preorders of groups. These matters are elementary but have to be discussed because we need them later. The first serious topic is the struggle with the question which objects we should call Lie semigroups and which analytic subsemigroups. After considerable vacillation we have come to the conclusion not to speak of either. We face the fact that we must subsume Lie group theory under our theory. Even there the question of analytic subgroups is delicate due to the fact that they frequently fail to be closed. We take as a starting point the theorem of YAMABE which says that *a subgroup of a finite dimensional real Lie group is analytic if and only if it is arcwise connected.* For a subset X of a group G we write

$$\langle X \rangle = X \cup X^2 \cup X^3 \cup \cdots$$

for the subsemigroup algebraically generated by X. A subsemigroup S of a Lie group G will be called, by want of a better idea, *preanalytic* if the subgroup $\langle S \cup S^{-1} \rangle$ generated by S in G is path connected. This group is analytic by Yamabe's Theorem and thus has an inherent Lie group structure whose underlying topology in general will be finer than the topology induced by that of G on $\langle S \cup S^{-1} \rangle$. This group, endowed with its Lie group structure will be denoted $G(S)$. Semigroups other than preanalytic ones are rarely of interest to us. With any preanalytic semigroup S we can associate an infinitesimal object contained in $\mathbf{L}\big(G(S)\big) \subseteq \mathbf{L}(G)$ as follows:

$$\mathbf{L}(S) = L_0\big((\exp_{G(S)})^{-1}(S)\big),$$

where we carefully choose to use the exponential function $\exp_{G(S)}\colon \mathbf{L}\big(G(S)\big) \to G(S)$. After the identification of $\mathbf{L}\big(G(S)\big)$ with a subalgebra of $\mathbf{L}(G)$, we obtain it by restricting the exponential function $\exp_G\colon \mathbf{L}(G) \to G$. This definition shows some of the subtleties involved. A simple but illuminating example is the 2-torus $G = \mathbb{R}^2/\mathbb{Z}^2$ whose exponential function $\exp_G\colon \mathbb{R}^2 \to G$ is simply the quotient map. Let us consider the preanalytic subsemigroup $S = \big(\mathbb{R}^+\cdot(1, \sqrt{2}) + \mathbb{Z}^2\big)/\mathbb{Z}^2$. Then $\mathbf{L}\big(G(S)\big) = \mathbb{R}\cdot(1, \sqrt{2}) \cong \mathbb{R}$ and $\mathbf{L}(S) = \mathbb{R}^+\cdot(1, \sqrt{2}) \cong \mathbb{R}^+$. But let us observe that $L_0\big((\exp_G)^{-1}(S)\big)$ is $\mathbb{R}^2$ in which subtangent set the local fine structure of S has become obliterated. The definition of $\mathbf{L}(S)$, therefore, is judiciously chosen, and if S happens to be a subgroup, then by Yamabe's Theorem it is a preanalytic subsemigroup if and only if it is an analytic subgroup. In this case $\mathbf{L}(S)$ is the traditional Lie algebra associated with this analytic group. The set $\mathbf{L}(S)$ has a very useful characterization: Indeed, *a vector $X \in \mathbf{L}(G)$ is a member of $\mathbf{L}(S)$ if and only if* $\exp \mathbb{R}^+\cdot X \subseteq \mathrm{cl}_{G(S)}\, S$, *where the operation* $\mathrm{cl}_{G(S)}$ *indicates closure in the Lie group $G(S)$ (and not in the topology of G!).*

The crucial concept which replaces, in the case of semigroups, the concept of an analytic subgroup is that of an infinitesimally generated semigroup:

26.   Definition. A subsemigroup S of a Lie group G is called *infinitesimally generated* if all of the following conditions are satisfied:

(i) S is preanalytic.

(ii) $\exp \mathbf{L}(S) \subseteq S \subseteq \mathrm{cl}_{G(S)} \langle \exp \mathbf{L}(S) \rangle$.

(iii) $G(S) = \langle \exp \mathbf{L}(S) \cup \exp - \mathbf{L}(S) \rangle$.

The semigroup S is called *strictly infinitesimally generated* if

$$S = \langle \exp \mathbf{L}(S) \rangle.$$

Every strictly infinitesimally generated subsemigroup is infinitesimally generated. It is an open problem as far as we are concerned whether or not condition (iii) is perhaps a consequence of (i) and (ii).

The following theorem shows for the first time the efficiency of these concepts.

27.   Theorem. (The Infinitesimal Generation Theorem) *Let S be an infinitesimally generated subsemigroup of a finite dimensional Lie group G. Then $S_0 \overset{\text{def}}{=} \langle \exp \mathbf{L}(S) \rangle$ is the unique largest strictly infinitesimally generated subsemigroup of S, and the following conclusions hold:*

(i) $G(S_0) = G(S)$ *and* $S_0 \subseteq S \subseteq \mathrm{cl}_{G(S)}(S_0)$.

(ii) $\mathbf{L}(S_0) = \mathbf{L}(S)$.

(iii) $\mathrm{int}_{G(S)} S_0 = \mathrm{int}_{G(S)} S$, *and this set is a dense ideal of S.*   ∎

A few comments are in order. The statement that S generates G as a group is tantamount to $G(S) = G$. If this situation prevails, then condition (iii) says not only that S has non-empty interior, but that this interior is in fact dense. This density aspect is a consequence of a theorem which we established already in Chapter IV. Results of this type are generally established in geometric control theory. They are based on the Theorem of Frobenius giving the integrability of a distribution on a manifold. Other aspects of Theorem 27 elucidate the fact that every infinitesimally generated semigroup S has always a canonical *strictly infinitesimally generated* semigroup S_0 attached to it which has the same closure and the same interior in G(S) and which has the same infinitesimal object as S.

If S is strictly infinitesimally generated or closed in $G(S)$, then S is invariant in G, that is, satisfies $gSg^{-1} = S$ for all $g \in G$ if and only if $\mathbf{L}(S)$ is an invariant wedge in $\mathbf{L}(G)$. A group G is said to be *preordered*, respectively, *partially ordered* if it is endowed with a reflexive and transitive relation, respectively, with such a relation which is, in addition, antisymmetric, such that both left and right translations are order preserving. The set S of non-negative elements for a group preorder is an invariant semigroup, and every invariant semigroup defines, in a simple fashion familiar from the additive group of reals, a group preorder. The semigroup S has no invertible elements other than **1** precisely when the preorder is a partial order. In this fashion the theory of invariant wedges and cones in Lie algebras is directly linked with the question of preordered and partially ordered connected Lie groups.

——————————

Any semigroup S with identity in a group G has attached to it two natural groups: One is the group $\langle S \cup S^{-1} \rangle$ generated by S in G and the other is the group $H(S) \overset{\text{def}}{=} S \cap S^{-1}$ of units. If S is an infinitesimally generated subsemigroup of a Lie group G, then the subgroup generated by S is analytic and its Lie algebra is the Lie algebra $\langle\!\langle \mathbf{L}(S) \rangle\!\rangle$ generated in $\mathbf{L}(G)$ by $\mathbf{L}(S)$. This is comparatively easy because we incorporated condition 26(iii) in the definition of an infinitesimally generated semigroup. Much harder is the result that $H(S)$ *is analytic and in fact closed in* $G(S)$. Moreover, $L\big(H(S)\big) = \mathbf{L}(S) \cap -\mathbf{L}(S)$. This follows from a more general theorem which describes useful neighborhoods of $H(S)$ in S:

28.   Theorem.   *Let S be an infinitesimally generated subsemigroup of a Lie group G. Then for any open neighborhood U of $\mathbf{1}$ in G there is a proper right ideal I of S which is closed in S such that $S \subseteq UH(S) \cup I$.*   ∎

Examples illustrate that there is little if any room for generalizations. In partiular, one cannot expect a two-sided ideal I to work in place of a right ideal (unless $H(S)$ happens to be compact, which is a different story). The significance of the ideal is, intuitively, that, a one-parameter semigroup or even a concatenation of one-parameter semigroups, once inside I, cannot leave I ever to return into the vicinity of a unit. This property is used in the sequel at certain strategic points.

——————————

In the Infinitesimal Generation Theorem we have observed that for many purposes it is reasonable to assume that a subsemigroup of a Lie group has a non-void interior. A maximal subsemigroup of a group G is a subsemigroup S such that $S \neq G$ and $S \subset T \subseteq G$ for a semigroup T implies $T = G$. If $G = \mathbb{R}^n$, then the maximal subsemigroups with non-void interior are exactly the closed half-spaces. The maximal subsemigroups as such pertain to the structure of $\mathbb{R}^n$ as a vector space over $\mathbb{Q}$ of continuum dimension and are, in general, not an object of a Lie theory of semigroups. The maximal semigroups with non-void interior definitively are. We do not know very much about them in general. In $\widetilde{\mathrm{Sl}}(2,\mathbb{R})$, the universal covering group of $\mathrm{Sl}(2,\mathbb{R})$, we can identify all maximal subsemigroups, and they are all half-space semigroups, bounded by a closed subgroup. This remains true for all connected Lie groups up to dimension 3. In higher dimensional Lie algebras this is no longer correct. One example is as follows: The Cartan–Killing form in $\mathrm{sl}(2,\mathbb{R})$ is Lorentzian and thus defines two invariant pointed generating cones; fix one of them, say W. Form the semidirect product $\mathrm{sl}(2,\mathbb{R}) \rtimes_{\mathrm{Ad}} \mathrm{Sl}(2,\mathbb{R})$ with respect to the adjoint action. The subset $W \times \mathrm{Sl}(2,\mathbb{R})$ is a subsemigroup of G which turns out to be maximal. However, in many Lie groups, maximal subsemigroups with non-void interior *are* half-space semigroups. In this direction we have the following result:

29.   Theorem.   *Let G be a connected Lie group such that $G/\mathrm{Rad}G$ is compact and let M be a maximal subsemigroup with non-void interior. Then the boundary of M is a subgroup, and if N is the unique largest normal subgroup of G contained in*

M, then N is a connected closed Lie subgroup of G and $(G/N, M/N)$ is isomorphic either to $(\mathbb{R}, \mathbb{R}^+)$ or to the pair consisting of the two dimensional non-abelian Lie group and a half-space subsemigroup bounded by a non-normal one-parameter subgroup. ∎

In particular, this theorem applies to all solvable connected Lie groups and shows that the maximal subsemigroups with non-empty interior are half-space semigroups whose Lie wedges accordingly are half-space semialgebras. It is not at all the case that every closed infinitesimally generated subsemigroup is an intersection of maximal ones but our knowledge of the maximal subsemigroups is nevertheless quite useful.

Chapter V contains a large section in which a long catalogue of examples is discussed, mostly in great detail. In this section it is illustrated how semigroups arise naturally in Lie groups (notably in the case of contraction semigroups in a very general sense of the word) and what pathologies are to be expected. In a section on divisible subsemigroups of Lie groups it is shown that *a closed semigroup is divisible if and only if its exponential function maps its Lie wedge surjectively onto the semigroup.* We do not know in general whether the Lie wedge of such a semigroup is always a Lie semialgebra; but we show this to be the case if the semigroup contains no units other than the identity. Finally we discuss congruence relations on open subsemigroups of a Lie group under the assumption that the identity is in the closure of the semigroup. The interior of an infinitesimally generated semigroup generating the Lie group satisfies this condition. We show that *any congruence with closed congruence classes on such a semigroup produces local foliations of the semigroup induced by ideals in the Lie algebra—except at the elements of a thin set of singular points.*

The prerequisites for this chapter remain in the range of general Lie group theory and topological group theory. An occasional reference to outside sources is needed, but one is justified in calling the chapter largely self-contained. The principal results are presented in bite-size chunks.

With Chapter VI we remain still in the global Lie theory of subsemigroups of Lie groups. We address the question of Lie's Fundamental Theorem in the following form: *Let G denote a connected Lie group and W a Lie wedge in $\mathbf{L}(G)$. What are the conditions which guarantee the existence of an infinitesimally generated subsemigroup S of G such that $\mathbf{L}(S) = W$?* Such Lie wedges W shall be called *global in G.* If W is a subalgebra then it is global in G, but Lie wedges can fail to be global for any number of reasons. A first obstruction may be that G is too far from being simply connected. It turns out that the chances for a Lie wedge to be global in G are optimal if the fundamental group of G is finite. The most familiar case is the simply connected one. But there are perfectly good Lie wedges in the Heisenberg algebra which fail to be global in the simply connected Heisenberg group which is diffeomorphic to $\mathbb{R}^3$. Thus the obstructions cannot be tied alone to the global topological structure of G.

Inevitably, global concepts enter. We consider the tangent bundle $T(G)$ of the given Lie group, and we identify the Lie algebra $\mathfrak{g} = \mathbf{L}(G)$ with $T(G)_1$ in the usual fashion. Every left translation λ_g induces a linear isomorphism from $d\lambda_g(\mathbf{1}): \mathfrak{g} \to T(G)_g$. In particular, for a given wedge $W \subseteq \mathfrak{g}$ we obtain a wedge $\Theta(g) = d\lambda_g(\mathbf{1})(W)$ in $T(G)_g$. We call the assignment $g \mapsto \Theta(g)$ a *left-invariant wedge field*.

We recall that, in the elementary theory of wedges and their duality in Chapter I, the dual object of a wedge W in, say, the Lie algebra $\mathfrak{g}$ consisted of all functionals $\omega: \mathfrak{g} \to \mathbb{R}$ with $\langle \omega, X \rangle \geq 0$ for all $X \in W$. It is now not unreasonable that we should look for a sort of dual object of a wedge field. Such a dual object should consist of a section $\omega: G \to \widehat{T}(G)$ of the cotangent bundle which assigns to each $g \in G$ an element of $\widehat{T}(G)_g$, the dual of $T(G)_g$, that is a linear function $\omega(g): T(G)_g \to \mathbb{R}$ which relates to the given wedge field Θ in such a fashion that

$$(\text{P}) \qquad \langle \omega(g), X \rangle \geq 0 \text{ for all } X \in \Theta(g).$$

The sections of the cotangent bundle $\widehat{T}(G)$ are called 1-*forms* or briefly *forms*, and we shall always assume here that all forms are smooth. We shall say that a form ω is W-positive if (P) is satisfied for the left invariant wedge field Θ obtained from W by left translations as explained above. Of course, the zero form which associates with each point the zero functional is W-positive. In order to express positivity in a non-trivial fashion we formulate strict positivity at a point as follows: A form ω is called *strictly W-positive at g* if

$$\langle \omega(g), X \rangle > 0 \text{ for all } X \in \Theta(g) \setminus -\Theta(g),$$

that is, for all elements X of the wedge $\Theta(g)$ not in the edge.

The question arises in the first place whether forms exist which are, say, strictly W-positive at $\mathbf{1}$. This we shall answer in the affirmative. But we have to do more. We recall that a 1-form ω is called *exact* if there is a smooth function $f: G \to \mathbb{R}$ such that $\omega = df$ or, in other words, $\langle \omega(g), X \rangle = X(f)$ where the elements X of $T(G)_g$ are considered as derivations operating on smooth functions. A 1-form ω is called *closed* if its exterior derivative $d\omega$ vanishes identically, that is, if for any pair X, Y of smooth vector fields (that is, smooth cross sections of the tangent bundle $T(G)$) we have the relation $\omega([X, Y]) = X(\omega(Y)) - Y(\omega(X))$, where the smooth function $\omega(X)$, say, is defined by $\omega(X)(g) = \langle \omega(g), X(g) \rangle$. It is a consequence of DE RHAM'S Theorem and some basic algebraic topology that *every closed 1-form is exact on G if and only if the fundamental group of G is finite. Now we can formulate the following fundamental theorem:*

30. **Theorem.** (The Globality Theorem) *Let G be a connected Lie group and W a Lie wedge generating $\mathfrak{g}$ as a Lie algebra. Then the following conditions are equivalent:*

(1) *W is global in G.*

(2) *There exists a W-positive exact form on G which is strictly W-positive at $\mathbf{1}$.* ∎

The hypothesis that W should generate the Lie algebra $\mathfrak{g}$ is adopted at this point in order to allowing an easy formulation at the expense of optimal generality which we shall attain in Chapter VI. The present formulation still captures the spirit of the final result. In principle, this theorem is one answer to Lie's Fundamental Theorems in a global version: *The issue of the existence of global semigroups in a given Lie group G for a given Lie wedge W in the Lie algebra $\mathfrak{g}$ is the issue of the existence of exact 1-forms on G which are compatible with W in a precisely specified form.* If the fundamental group of G is finite, then exactness of the form may be replaced by closedness.

The tangent bundle $T(G)$ is trivial; therefore the cotangent bundle is trivial, too. Vector fields and 1-forms as well may thus be identified with smooth functions on G taking values in $\mathfrak{g}$ and $\widehat{\mathfrak{g}}$, respectively. It is not entirely simple to calculate, however, what our conditions mean if such identifications are made. We shall do these calculations. They force us to keep careful records of all identifications made in the process.

The Globality Theorem allows some immediate applications. Let us consider some samples.

31.   Corollary.   *Let G be a Lie group and $W_1 \subseteq W_2$ two Lie wedges in $\mathfrak{g}$. Suppose that the following conditions satisfied:*

$$W_1 \setminus -W_1 \subseteq W_2 \setminus -W_2.$$

Then W_1 is global in G if W_2 is global in G and $W_1 \cap -W_1$ is the Lie algebra of a closed subgroup of G.   ∎

The hypothesis is the precise expression of the geometric assumption that the edge of W_1 is the intersection of the edge of W_2 with W_1.

Information of this sort is more useful than it may appear at first, since it allows us to conclude new globality information from given one.

32.   Corollary.   *If G is a connected Lie group with finite fundamental group and if the Lie wedge W in $\mathfrak{g}$ satisfies*

$$[\mathfrak{g},\mathfrak{g}] \cap W \subseteq -W,$$

then W is global in G.   ∎

The hypothesis says that the commutator algebra meets the wedge in its edge only. In particular, this implies that in a Lie group with finite fundamental group every trivial Lie wedge is global.

The problem of finding an exact 1-form ω compatible with a given Lie wedge is tantamount to finding smooth functions $f: G \to \mathbb{R}$ which are "locally monotone" with respect to the local partial order given by the left invariant wedge field Θ. The insight that the existence of such functions is crucial for the existence of semigroups with prescribed Lie wedge is due to VINBERG and OL'SHANSKIĬ.

The preparations for the Globality Theorem, not unexpectedly, require some circumspection. We have to begin with a careful analysis of piecewise smooth curves compatible with a given wedge field. This eventually gives us the following theorem on the semigroup generated by a given Lie wedge:

33.   **Theorem.**   *Let G be a Lie group, H a closed connected subgroup, and W a Lie wedge in $\mathfrak{g}$ with $\mathfrak{h}$ as edge. Consider the semigroup $S \overset{\text{def}}{=} \langle \exp W \rangle$ generated by W and the semigroup $S(W)$ of all elements in G which are endpoints of piecewise smooth trajectories $t \mapsto x(t)$ starting from $\mathbf{1}$ and having their forward derivatives $\dot{x}(t)$ in $\Theta\big(x(t)\big)$. Then these semigroups are related by*

$$S \subseteq S(W) \subseteq \mathrm{cl}_{G(S)}\, S.$$

In particular, $\mathrm{cl}_{G(S)}\, S(W)$ is the smallest $G(S)$-closed subsemigroup containing $\exp W$.   ∎

The semigroup S is generated, so to speak, algebraically, while the semigroup $S(W)$ is defined analytically. In terms of geometric control, the points of S are those which are attainable with piecewise constant steering functions. It is with the semigroup $S(W)$ that we perform the necessary constructions. As we have already seen in the local theory, the presence of the edge of the wedge is the source of considerable complications. We organize our dealing with such complications by resorting to the quotient manifold G/H. The wedge field transported down to this level actually becomes a field of pointed cones, and it is easier to work with that where possible. In the establishing globality of a Lie wedge W we have to verify that the trajectories ending in $S(W)$ have no chance of returning, in some devious way, to the group H after they have left the vicinitiy of this group. This requires exact "timekeeping" so that we know "at which time" a trajectory leaves certain suitable chosen neighborhoods of the group H ("the Escape Theorem").

Conversely, the proof that the stated conditions on the existence of certain forms are necessary requires an excursion into the domain of monotone functions and measures. It is relatively easy to obtain monotone Borel functions and then to smooth them in a standard fashion. However, it is the issue of strict W-positivity of a form at the origin which requires scrutiny and is settled with the application of some measure theory.

The reading of this chapter requires some knowledge of global differential geometry, measure theory, an occasional reference to de Rham's Theorem and basic algebraic topology.

———————

A Lie theory for semigroups was asked for. We have proposed a theory of local and global infinitesimally generated subsemigroups in finite dimensional real Lie groups. The reader was taken on an occasional foray into infinite dimensions. Chapter VII finally presents an *abstract Lie and embedding theory* of semigroups. This program has two parts. Firstly, we need a theory for the forming of quotient groups of semigroups, locally and globally, in the topological setting. This is the appropriate generalisation of the construction of the group of integers from the semigroup of natural numbers. Secondly, we need a suitable calculus on manifolds with very irregular boundaries in order to initiate an abstract theory of differentiable semigroups. Even the simplest examples of differentiable semigroups, namely the additive semigroups of wedges in finite dimensional vector spaces, show that the boundaries must be allowed to have singularities for the differentiable structure. We proceed with the following definitions (for which an infinite dimensional version is also provided in the text):

34.   Definition. A subset A of a topological space X is called *admissible* if the interior of A in X is dense in A. A T_3-space X is called a *manifold with generalized boundary* if there exists a natural number n such that each point of X possesses a neighborhood which is homeomorphic to some admissible subset of $\mathbb{R}^n$.

The theory of local quotients of semigroups is fraught with too many technicalities to be summarized in this introduction. Suffice it to say that under rather mild hypotheses, a local semigroup possesses a local group of quotients in which it locally embeds as an admissible subset, and furthermore this local group of quotients is unique up to local isomorphism. The following proposition, a corollary of these results, identifies one important setting in which this machinery can be applied.

35.   Proposition. *Let S be a locally compact locally cancellative local semigroup with identity and suppose that S is homeomorphic to an admissible subset of $\mathbb{R}^n$. Then S is locally embeddable into a finite dimensional Lie group G such that the image under this embedding is an admissible subset of G. Accordingly, S is locally topologically isomorphic to a local subsemigroup on an admissible subset of a finite dimensional real Lie algebra with a multiplication induced from the Campbell–Hausdorff multiplication.* ∎

It is noteworthy that no assumption on differentiability enters into this proposition. Not surprisingly, this result rests on a local version of the solution to Hilbert's Fifth Problem due to R. JACOBY, to which we must refer.

The results to which we have just alluded show that the consideration of only local semigroups in Campbell–Hausdorff neighborhoods of Lie algebras in Chapter IV was not so restrictive as it might have at first appeared. They also allow the theory developed there to be carried over to more general settings.

These methods retain their significance in the global setting, even in the absence of an identity. For instance *for any cancellative topological semigroup S on a finite dimensional connected topological manifold there is a functorially constructed simply connected Lie group G^S, a covering semigroup $\sigma_S \colon C(S) \to S$, and a local homeomorphism $\gamma_S \colon C(S) \to G^S$ which is a morphism of semigroups such that both S and $C(S)$ carry unique analytic structures making the respective multiplication functions and the functions γ_S and σ_S analytic.*

Another basic result on global embeddings is the following: *If S is a cancellative topological semigroup on a finite dimensional connected topological manifold which happens to be algebraically embeddable in a group, then the free group $G(S)$ on S admits uniquely the structure of a Lie group such that the embedding of S into $G(S)$ is an analytic embedding.* Thus the embedding problem in this case reduces to an algebraic problem. What we describe in Chapter VII on forcing analyticity from algebraic and topological assumptions is a systematic pursuit of a theory initiated by BROWN and HOUSTON.

The second approach is the introduction of differentiable semigroups on a manifold with generalized boundary due to GRAHAM. This requires the development of differential calculus on such manifolds based on the classical idea of the strong derivative of functions of several variables. Given such a calculus, it is no problem to define strongly k-differentiable (local) semigroups.

36. **Theorem.** *Let S be a C_s^k local semigroup for $k \geq 1$. Then there exists a Lie algebra L and a C_s^k diffeomorphism of some neighborhood of the identity of S into a Campbell–Hausdorff neighborhood of L transforming the given semigroup multiplication into the Campbell–Hausdorff multiplication.* ∎

This theorem is not restricted to finite dimensions. It secures the connection between a general theory of differentiable semigroups on manifolds with generalized boundary and the Lie theory of semigroups displayed in the first six chapters of this book. In the vicinity of the identity, a differentiable semigroup (or local semigroup) is locally isomorphic to a local subsemigroup of a Lie group. Thus our theory applies.

However more is true. Differentiable semigroups on manifolds with generalized boundary can be developed quite systematically. An overview is given in Chapter VII. This theory is due to GRAHAM. In the spirit of Sophus Lie's original ideas one can associate with a differentiable monoid a Lie algebra, and this construction is functorial as expected. Thus a global theory of differentiable monoids carries along an infinitesimal theory of Lie algebras and thereby a local Lie group theory. This allows us to consider differentiable semigroups as being embedded in Lie groups locally and gives rise to the notion of a local log function appropriate for this setting.

The global theory presented in Chapters V, VI, and VII is far from perfection. We have seen necessary and sufficient conditions for a Lie wedge W in the Lie algebra $\mathbf{L}(G)$ of a Lie group G to be the tangent wedge of a subsemigroup S of G. However, for all we know, it may be possible that there always exists some topological semigroup S_W with identity generated by its one parameter subsemigroups and possessing an identity neighborhood isomorphic to a local subsemigroup of G which has W as tangent wedge. In his dissertation, W. WEISS will show that this is the case whenever W is pointed. Examples show that there might be semigroups S_W which are differentiable in the sense of Chapter VII even though they are not embeddable into any Lie group. Further research is needed to clarify the questions which arise in this context. The volumes following this book will have more to say on the examples.

The logical interdependence

The panorama of the Lie theory of semigroups which we show in this book can be viewed from different vantage points. Our general division into the infinitesimal, the local, the global theory, and the abstract Lie semigroup and embedding theory hints at several possible approaches a reader may take. For instance, Chapter VII may be read independently by anyone seeking an orientation on the possibilities of an abstract theory of differentiable semigroups and its links with classical Lie group theory. No prerequisites from other parts of the book are required.

A reader interested in a local Lie semigroup theory and Lie's Fundamental Theorems can go directly to Chapter IV without having to pass through the preceding parts of the book. The global theory described in Chapters V and VI may also be approached directly, but at least a perusal of the local theory of Chapter IV is advisable; special results introduced and proved in Chapter IV are definitely used in these later parts, but the full power of the Fundamental Theorem of the local theory is not used. It is important to realize this fact since the Fundamental Theorem is an end in itself and its proof is long and technical. Also, the long discussion of examples in Section 4 of Chapter V does make frequent references to material presented in Chapters II and III. Lastly, one can obviously occupy one's attention with the study of the first three chapters without ever having to resort to material in the rest of the book. The infinitesimal Lie theory of semigroups really is contained in Chapters II and III, but Chapter II seriously requires the background material collected in Chapter I. Much of Chapter II pertains to Lie semialgebra theory. Even though any invariant wedge is, in particular, a Lie semialgebra, Chapter II is not a prerequisite for Chapter III. Thus, once our general terminology is accepted, Chapter III on invariant wedges in Lie algebras can be read independently from the remainder of the book.

Chapter I

The geometry of cones

Before we can develop a Lie theory of cones and wedges in a Lie algebra, we have to develop the necessary geometric and analytical background theory of cones in vector spaces. What we need is the theory of cones in finite dimensional real vector spaces. A portion of this theory we shall present in a more general frame work so that later on we can occasionally speak of Lie theory in infinite dimensions. However, the more casual reader will not lose much information in specializing immediately to the finite dimensional case. This Chapter will have to deal with some non-trivial aspects of cones, and they remain non-trivial in the finite dimensional case.

Just as in the case of topological vector spaces, duality is an invaluable tool in handling cones and wedges. Therefore, a discussion of duality will be our first order of business. The geometry of a cone is largely determined by its "facial structure". In the presence of an effective duality theory, it is a certain subclass of faces which takes the lead. These are the so called exposed faces which we treat in the second section. The examples given in this section should be helpful for our intuitive understanding of the cones and wedges with which we have to deal in most parts of this book. In the third section we shall deal with cones in Banach spaces and in finite dimensional spaces in particular. We shall present Mazur's theorem on the density of the set of those points in the boundary in which the support hyperplane is unique. Certain refinements of the density theorem will be crucial in later applications. In a subsequent section we test the ideas presented so far for the classes of polyhedral and Lorentzian cones, and develop a theory of "round" cones. In our later investigations, the concepts of a subtangent and a tangent vector of a set in a point will be used time and again. They will be introduced in the fifth section, and we shall show how they link up with the previous discussions. More importantly, we shall present a version of the theory due to Bony and Brezis on the invariance of closed sets on a manifold under flows. We are mainly interested in the invariance of wedges in a finite dimensional vector space expressed in terms of the vector field generating the flow. A good deal of the infinitesimal Lie theory of semigroups will later call upon these results.

1. Cones and their duality

We will deal with locally convex topological vector spaces L over $\mathbb{R}$. Our primary objective is finite dimensional real vector spaces; occasionally we do want to consider Banach spaces. However, some of the structure is more systematically and more lucidly presented in a framework of appropriate generality. For this purpose we recall that two topological vector spaces L and $\widehat{L}$ are said to be *in duality* if there is a continuous bilinear function

$$(x, \omega) \mapsto \langle \omega,\, x \rangle \colon L \times \widehat{L} \to \mathbb{R}$$

such that $\langle \omega,\, x \rangle = 0$ for all $x \in L$ implies $\omega = 0$ and, dually, that $\langle \omega,\, x \rangle = 0$ for all $\omega \in \widehat{L}$ implies $x = 0$. Thus $\widehat{L}$ may be identified with a point-separating vector subspace of the topological dual of L, and L may be identified with a point separating vector subspace of the topological dual of $\widehat{L}$. The duality endows both L and $\widehat{L}$ with possibly new locally convex topologies which are coarser than the given ones, namely the topology of pointwise convergence of all functionals $x \mapsto \langle \omega,\, x \rangle \colon L \to \mathbb{R}$ with $\omega \in \widehat{L}$ on L, and the topology of pointwise convergence of all functionals $\omega \mapsto \langle \omega,\, x \rangle \colon \widehat{L} \to \mathbb{R}$ with $x \in L$ on $\widehat{L}$. We will call these topologies the *weak topologies on L and $\widehat{L}$*, respectively.

The most frequent example will be a Banach space L and its topological dual $\widehat{L}$, where $\langle \omega,\, x \rangle$ is simply the evaluation $\omega(x)$. The associated weak topologies of this duality are the weak topology on L and the weak $*$-topology on $\widehat{L}$. But here again the case we shall encounter most frequently in our own applications is that of a *finite dimensional real vector space L*. In this case, there is one and only one vector space topology on L as well as on the dual $\widehat{L} \stackrel{\text{def}}{=} \text{Hom}(L, \mathbb{R})$. The weak topologies agree in this case with the given topologies.

We recall at this point, that the *weakly continuous* linear functionals of the locally convex vector space L in duality with $\widehat{L}$ are *precisely* the functionals $x \mapsto \langle \omega \mid x \rangle$ with $\omega \in \widehat{L}$. (See for instance [Ru73], Theorem 3.10.)

After these general remarks we formulate our first definition. We shall denote the set of non-negative real numbers with $\mathbb{R}^+$.

I.1.1. **Definition.** A subset W of a real topological vector space is called a *cone* or, synonymously, *wedge*, if it satisfies the following conditions:

   (i) $W + W \subseteq W$,

   (ii) $\mathbb{R}^+ \cdot W \subseteq W$,

   (iii) $\overline{W} = W$, that is, W is closed in L.

   The subset $H(W) \overset{\text{def}}{=} W \cap -W$, the largest vector subspace contained in W is called the *edge of the wedge*, and a wedge will be called *pointed* if its edge is singleton. A wedge will be called *generating*, if $L = W - W$.

   A few remarks are in order because the terminology diverges widely in the literature. While the name "cone" is prevalent in the literature, we prefer the designation "wedge", because it suggests the presence of vector subspaces in W which play a significant role in the Lie theory of semigroups. We shall therefore use the terms "cone" and "wedge" synonymously. In the case that W is pointed, we shall usually speak of a "pointed cone", but recall that the word "cone" in this book does not mean the absence of a non-trivial edge unless it is accompanied by the adjective "pointed". We also remark, that the word "cone" in the literature often refers to subsets of a vector space which are closed under addition and positive scalar multiplication; but these play such a little role here that we include the property of being topologically closed in our definition.

EI.1.1.   Exercise.   For a wedge L in a topological vector space L consider the following conditions:

   (1) W is generating.

   (2) The interior $\operatorname{int} W$ is non-empty.

   Then $(2) \Rightarrow (1)$, but not conversely. If $\dim L$ is finite, then $(1) \Leftrightarrow (2)$.   ■

   One notices that *a subset W of a topological vector space is a wedge if and only if it is closed, convex, contains 0, and is additively closed.* In $L = \mathbb{R}$, the possible wedges are $\{0\}$, $\mathbb{R}^+$, $-\mathbb{R}^+$, and $\mathbb{R}$. The half-line $\mathbb{R}^+$ is the prototype of a cone. This will become even more apparent in the development of the duality theory of wedges which we are about to begin.

I.1.2.   Definition.   If L and $\widehat{L}$ are topological vector spaces in duality, then for a subset $W \subseteq L$ we shall set

$$W^* = \{\omega \in \widehat{L} : \langle \omega, x \rangle \geq 0 \quad \text{for all} \quad x \in W\},$$

and

$$W^{\perp} = \{\omega \in \widehat{L} : \langle \omega, x \rangle = 0 \quad \text{for all} \quad x \in W\}.$$

   The following observation should be obvious from the definitions:

   For any subset W of L whatsoever, W^* is a wedge in $\widehat{L}$ and $W^{\perp}$ is a closed vector subspace of $\widehat{L}$. If $W = -W$, then $W^* = W^{\perp}$. If $W_1 \subseteq W_2$, then $W_1^* \supseteq W_2^*$ and $W_1^{\perp} \supseteq W_2^{\perp}$.

I.1.3.   Definition. The wedge W^* will be called the *dual wedge of* W *with respect to the duality between* L *and* $\widehat{L}$, and $W^\perp$ will be called *the annihilator of* W *with respect to the duality between* L *and* $\widehat{L}$. If no confusion is possible, we will allow ourselves to simply talk of *duals* and *annihilators*.

Again we pause for some remarks. The concepts which we just introduced will almost always be applied to the case that W is a wedge in L. Let us notice a certain divergence between the duality theory of vector spaces and cones: The prototype of a vector space is $\mathbb{R}$, the prototype of a cone is $\mathbb{R}^+$. Thus by all rights the dual of a cone should be defined by "homomorphisms" of a cone into $\mathbb{R}^+$. But we recall that the class of wedges does not exclude vector subspaces, and if we were to define the dual of a wedge in this fashion, the dual of any vector space would be zero. This might very well be a desirable outcome for some purposes; it certainly is not for ours. On the other hand, if W has inner points, then an additive and positively homogeneous function $W \to \mathbb{R}^+$ extends uniquely to an element of W^*; in this regard it appears quite fair that we should call the wedge W^* the dual of the wedge W. But we do remember that, in general, the dual of a wedge according to our definition depends on a given duality.

If we start with a subset W of $\widehat{L}$, then the corresponding definitions will give us a wedge W^* and a closed vector subspace $W^\perp$ in L. In particular, starting with $W \subseteq L$ we obtain a wedge W^{**} in L which contains W. This is the point for recalling the theorem of Hahn and Banach in locally convex vector spaces. *Every closed convex subset of a locally convex topological vector space is the intersection of all closed half-spaces containing it.* The vector space topology of a vector space L with duality is clearly locally convex. We therefore have

I.1.4.   Proposition. *Let* W *be a weakly closed wedge in a topological vector space* L *in duality with* $\widehat{L}$. *Then* $W^{**} = W$. *If* $\widehat{L}$ *is the topological dual of* L, *then every wedge* W *is automatically weakly closed, hence satisfies* $W^{**} = W$.   ∎

Since W^* is always weakly closed as the intersection of half-spaces defined by functionals coming from L, we have $W^{***} = W^*$ under any circumstances. The duality assignment $W \mapsto W^*$ establishes an involutory containment reversing bijection between the set of all weakly closed wedges in L and the weakly closed wedges in $\widehat{L}$.

I.1.5.   Corollary. *Let* L *be a finite dimensional real vector space and* $\widehat{L}$ *its dual. Then* $W \mapsto W^*$ *establishes an involutory containment reversing bijection between the set of wedges in* L *and the set of wedges in* $\widehat{L}$.   ∎

In this chapter, we shall denote with $\operatorname{cl} W$ the closure of a subset $W \subseteq L$ with respect to the weak topology. We observe that for any subset W of L, the set $W^{\perp\perp}$ is the weakly closed linear span of W and the set W^{**} is the weakly closed convex hull of W. If $\widehat{L}$ is the topological dual of L, then $W^{\perp\perp}$ is the closed linear span and W^{**} is the smallest weakly closed wedge containing W. Furthermore, $W^\perp = (\operatorname{cl} W)^\perp$ and $W^* = (\operatorname{cl} W)^*$.

We are now on our way to a duality theory of wedges. It will be understood that we consider two topological vector spaces L and $\widehat{L}$ in duality.

I.1.6. **Proposition.** *Let $\{W_j : j \in J\}$ be a family of weakly closed wedges. Then*

(i) $\left(\bigcap_{j \in J} W_j\right)^* = \mathrm{cl}\left(\sum_{j \in J} W_j^*\right)$

(ii) $\left(\sum_{j \in J} W_j^*\right)^* = \bigcap_{j \in J} W_j$.

Proof. In view of $W^* = (\mathrm{cl}\, W)^*$ for all subsets W of L , conclusion (ii) follows from statement (i) by duality. We therefore prove (i).

If we set $D = \bigcap_{j \in J} W_j$, then $D \subseteq W_i$ for all $i \in J$, whence $W_i^* \subseteq D^*$ for all $i \in J$, and thus $\mathrm{cl}\left(\sum_{j \in J} W_j^*\right) \subseteq D^*$. Conversely, for each $i \in J$, we have $W_i^* \subseteq \sum_{j \in J} W_j^*$ and thus $\left(\sum_{j \in J} W_j^*\right)^* \subseteq W_i^{**} = W_i$ for all $i \in J$ by Proposition I.1.4., whence $\left(\sum_{j \in J} W_j^*\right)^* \subseteq D$, and thus $D^* \subseteq \mathrm{cl}\left(\sum_{j \in J} W_j^*\right)$ by duality. $\blacksquare$

The edge $H(W)$ and the weakly closed span $\mathrm{cl}(W - W)$ of a wedge are dual concepts as we shall point out in the next proposition. We shall frequently refer to this simple fact.

I.1.7. **Proposition.** *For a wedge W the following conditions hold:*

(i) $W^\perp = (W - W)^\perp = H(W^*)$. *If W^* is pointed, then $\mathrm{cl}(W - W) = L$.*

(ii) *If W is weakly closed, then $H(W)^\perp = cl(W^* - W^*)$ (which equals $W^* - W^*$ if L is finite dimensional!). If W is generating, then W^* is pointed.*

(iii) $W^{*\perp} = H(W)$, *if W is weakly closed.*

In particular, if L is finite dimensional, a wedge W is pointed if and only if its dual W^* is generating and vice versa.

Proof. (i) We observed earlier that $W^\perp = (W - W)^\perp = \left(\mathrm{cl}(W - W)\right)^\perp$. Furthermore, $\omega \in W^\perp$ if and only if $\langle \omega, x \rangle = 0$ holds for all $x \in W$, and this is equivalent to both $\langle \omega, x \rangle \geq 0$ and $\langle -\omega, x \rangle \geq 0$ for all $x \in W$; but this holds precisely when $\omega \in W^* \cap -W^* = H(W^*)$. The remainder follows.

(ii) We have $H(W)^\perp = H(W)^* = (W \cap -W)^* = \mathrm{cl}(W^* - W^*)$ if W is weakly closed by Proposition I.1.6.i. Since $W^* - W^*$ is a vector space, it is automatically closed if $\hat{L}$ is finite dimensional. The remainder follows.

(iii) If W is weakly closed, then so is $H(W) = W \cap -W$, whence $H(W)^{\perp\perp} = H(W)$, and since $W^{*\perp} = \left(\mathrm{cl}(W^* - W^*)\right)^\perp$, (iii) follows from (ii) by passing to annihilators. $\blacksquare$

We note that $H(W) \subseteq W^{*\perp}$ holds for any wedge W .

In the following sections we will heavily use an operation which transforms a subset of L into a weakly closed subwedge of W^*, plus several other operations which are derived from it. It is very important that these operations and their geometrical meaning is understood. We introduce them in the following manner:

I.1.8. **Definition.** (i) For any subset M of L we define

$$\mathrm{op}_W(M) = \mathrm{op}(M) = M^\perp \cap W^*$$

and call this wedge the *opposite wedge of M (with respect to the wedge W)*. If $M = \{x\}$ we write $\mathrm{op}\, x$ as an abbreviation for $\mathrm{op}\{x\}$.

(ii) We set

$$L_M(W) = \mathrm{op}_W(M)^*,$$

and call this set the *subtangent space of W at the set M*. Again we may replace $L_M(W)$ by L_M if the context is clear, and if M is the singleton set $\{x\}$, we write

$$L_x(W) = L_x.$$

(iii) For any subset M of a wedge W, we set

$$T_M(W) = \mathrm{op}(M)^{\perp},$$

and call this set the *tangent space of W at the set M*. If no confusion is possible, we shall simply write T_M instead of $T_M(W)$, and if M is a singleton set $\{x\}$, then we write T_x in lieu of T_M.

The terminology for L_M and T_M is motivated by the concrete examples which we will depict presently; later developments (see, in particular, Section 5 below) will justify this terminology amply. For the developments in the subsequent section we need technical information about these concepts. In the remainder of this section we make the necessary preparations.

I.1.9.   Proposition.   *For a subset M of a wedge W the following conclusions hold:*

(i) $\mathrm{op}(M) = W^* \cap -M^*$.

(ii) *If W is weakly closed, then*

(a) $L_M(W) = \mathrm{cl}(W - M^{**}) = \mathrm{cl}(W + M^{\perp\perp})$, *and*

(b) $T_M(W) = H\big(L_M(W)\big)$.

Proof.   (i) From the definitions we know $M^{\perp} = M^* \cap -M^*$, and thus $\mathrm{op}(M) = M^{\perp} \cap W^* = W^* \cap M^* \cap -M^*$, which equals $W^* \cap -M^*$, since $W^* \subseteq M^*$.

(ii) From (i) we have $(W^* \cap M^{\perp})^* = (W^* \cap -M^*)^*$, which by Proposition I.1.6.i and by duality equals $\mathrm{cl}\big(W^{**} + (-M^*)^*\big) = \mathrm{cl}(W - M^{**})$, and this proves the first equality of (a). The proof of Proposition I.1.7.i shows that $M^{\perp} = M^* \cap -M^*$; hence $M^{\perp\perp} = (M^* \cap -M^*)^{\perp} = \mathrm{cl}(M^{**} - M^{**})$. Thus $\mathrm{cl}(W - M^{**}) = \mathrm{cl}\big(W + (M^{**} - M^{**})\big)$, since M and therefore M^{**} is contained in W. The last expression equals $\mathrm{cl}\big(W + \mathrm{cl}(M^{**} - M^{**})\big) = \mathrm{cl}(W - M^{\perp\perp})$, which establishes the second equality of (a).

But now (b) follows from (a) via Proposition I.1.7.i.   ■

We observe that in Proposition I.1.9(ii)(a) we have $\mathrm{op}(M)^* = \mathrm{cl}(W - M)$ as soon as M, too, is a weakly closed wedge.

Let us note the following observation right away:

I.1.10.   Remark.   If $x \in W$,

$$L_x = \mathrm{cl}(W - \mathbb{R}^+ \cdot x) \qquad \text{and} \qquad L_x^* = x^{\perp} \cap W^*.$$

The following information is useful:

I.1.11. **Proposition.** *If W_J, $j \in J$ is a family of weakly closed wedges, we set*

$$W = \bigcap_{j \in J} W_j.$$

Then the following conclusions hold:

(i)
$$H(W) = \bigcap_{j \in J} H(W_j).$$

(ii)
$$L_x(W) \subseteq \bigcap_{j \in J} L_x(W_j),$$

(iii)
$$T_x(W) \subseteq \bigcap_{j \in J} T_x(W_j),$$

for each $x \in W$.
Conclusion (i) *does not require that the W_j are weakly closed. Equality holds in* (ii) *and* (iii) *if*

(†)
$$L_x(W) \supseteq (x^\perp \cap \bigcup_{j \in J} W_j^*)^*.$$

Proof. (i) $H(\bigcap W_j) = \bigcap W_j \cap -\bigcap W_j = \bigcap W_j \cap \bigcap -W_j = \bigcap(W_j \cap -W_j) = \bigcap H(W_j)$.

On account of $T_x = H(L_x)$ and in view of (i) above, (iii) is a consequence of (ii) so that we have to prove (ii). We calculate $\bigcap L_x W_j = \bigcap(x^\perp \cap W_j^*)* = (\sum(x^\perp \cap W_j^*))* = (\bigcup(x^\perp \cap W_j^*)^*$, since for any family of wedges V_j we have $(\sum V_j)* = (\bigcup V_j)^*$; indeed, more generally, for any subset X of L we note $(\sum\{\mathbb{R}\cdot x : x \in X\})^* = X^*$. But $\bigcup(x^\perp \cap W_j^*) = x^\perp \cap \bigcup W_j^*$, and

(††)
$$x^\perp \cap \bigcup W_j^* \subseteq x^\perp \cap \sum W_j^*.$$

Thus Proposition I.1.6.ii implies $\bigcap L_x W_j \supseteq \mathrm{cl}(\mathbb{R}\cdot x + (\sum W_j^*)^*) = \mathrm{cl}(\mathbb{R}\cdot x + \bigcap W_j^{**}) = \mathrm{cl}(\mathbb{R}\cdot x + \bigcap W_j)$ by duality, since the W_j are weakly closed. On the other hand, $L_x(\bigcap W_j) = \mathrm{cl}(\mathbb{R}\cdot x + \bigcap W_j)$ by Definition I.1.8. Thus the inclusion in (ii) follows. If (†) is available, equality follows, because the chain of containments can be reversed. ∎

I.1.12. **Corollary.** *If W is a weakly closed wedge, E a closed vector subspace of L, and $x \in W \cap E$, then $T_x(W \cap E) \subseteq T_x(W) \cap E$. Equality holds if*

(∗)
$$L_x(W \cap E) \supseteq (\mathrm{op}_W(x) \cup E^\perp)^*.$$

Proof. We apply Proposition I.1.11 to the family $\{W, E\}$ of weakly closed wedges and notice that $(x^\perp \cap (W^* \cup E^*))^* = (x^\perp \cap W^*) \cup E^\perp)^*$. Condition (†) of Proposition I.1.11, under the present circumstances, reads $\mathbb{R}.x + (W^* + E^*)^* \subseteq (x^\perp \cap (W^* \cup E^*))^*$. Hence (∗) implies (†), and the corollary follows. ∎

The example of a circular cone W in 3-space and a tangent plane E shows that in Corollary I.1.12 (and thus, a fortiori, in Proposition I.1.11(ii) and (iii)) equality does not hold automatically. A partial converse will be given in Proposition II.2.18 in the next section.

It seems important to form a good geometric intuition for the concepts we dealt with. In the case that L is finite dimensional, the following procedure can help: In a finite dimensional real vector space L we can consider a scalar product $\langle \bullet \mid \bullet \rangle$ and thus make it in a Hilbert space. Example: $L = \mathbb{R}^n$, and $\langle x \mid y \rangle = \sum_{j=1}^{n} x_j y_j$. Once we fix such a scalar product, we can identify the dual $\widehat{L}$ with L itself by associating with $x \in L$ the functional $y \mapsto \langle x \mid y \rangle$. In such a situation, the dual W^* of a wedge is located in L itself, in fact

$$W^* = \{y \in L : \langle x \mid y \rangle \geq 0 \quad \text{for all} \quad x \in W\}.$$

This allows a geometric interpretation of W^*: Indeed, for $x \neq 0$, the set $H_x \overset{\text{def}}{=} \{y \in L : \langle x \mid y \rangle \geq 0\}$ is a closed half-space containing x and being bounded by the hyperplane orthogonal to the vector x. Thus $W^* = \bigcap \{H_x : x \in W\}$. For any subset M of L, the annihilator $M^\perp$ is simply the set of all vectors perpendicular to all vectors in M. Let us indicate this situation in Figure 1, in which we illustrate some of the concepts which we have introduced.

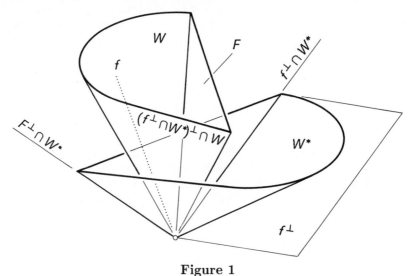

Figure 1

It is a very good exercise to work out some concrete examples of the operations $W \mapsto L_x(W) = cl(W - \mathbb{R}^+ \cdot x)$ in $\mathbb{R}^3$. If x is an interior point of W, then quite generally, 0 will be an interior point of $W - \mathbb{R}^+ \cdot x$ and thus this set is L. If x is a boundary point—which is the most interesting situation—then one might visualize L_x as the closure of an ascending union of wedges $W - n \cdot x$, $n = 1, \ldots$. Figure 2 may serve as an illustration.

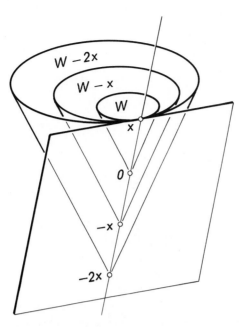

Figure 2

We conclude the section by remarking that the choice of a positive definite quadratic form for the purpose of identifying the dual of a finite dimensional vector space with itself is distinguished only for geometrical reasons. In fact every non-degenerate quadratic form serves the same theoretical purpose, and in later chapters we shall use indefinite non-degenerate quadratic forms exactly in this capacity. (See Section 4 below for the first time we use this device.)

2. Exposed faces

If W is a subset of a vector space L which is closed under addition and non-negative scalar multiplication, then a subset F of W is called a *face* of W if it has the same two properties and if, in addition the relations $x+y \in F$ and $x, y \in W$ imply $x \in F$ (and then, by symmetry, also $y \in F$). Notice that every face contains 0, since it is stable under the multiplication with 0. Thus if both x and $-x$ are in W, then $x + (-x) = 0 \in F$, and so $x \in F$, that is, every face contains the edge $H(W)$ of W. In Figure 1 of Section 1, the subsets f and F are faces of W. Faces play an important role in the geometry of cones in general; in the context of duality, however, and in view of our applications, a particular type of face is much more important. Indeed we have to deal with the so-called exposed faces. They will play a crucial role, and in this section we shall provide the necessary background. We continue to assume that we have a pair of topological vector spaces L and $\widehat{L}$ in duality.

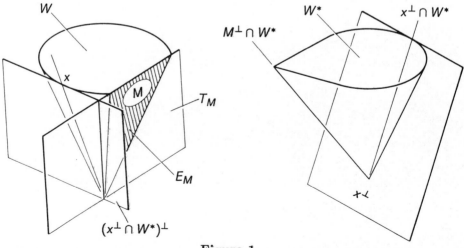

Figure 1

I.2.1.   **Definition.** A subset F of a wedge W is called an *exposed face*, if $F = \mathrm{op}\,\mathrm{op}\,F$, i.e. if
$$F = (F^\perp \cap W^*)^\perp \cap W.$$
A point $x \in W$ will be called *exposed*, if $x \neq 0$ and $\mathbb{R}^+ \cdot x$ is an exposed face, that

is, if

$$\mathbb{R}^+ \cdot x = (x^\perp \cap W^*)^\perp \cap W.$$

A point $x \in W$ is called an E^1-*point*, if $\mathbb{R}^+ \cdot x + H(W)$ is an exposed face, and the set of all E^1-points will be denoted $E^1(W)$.

If W is pointed, then a point $x \in W$ is exposed if and only if it is an E^1-point.

EI.2.1. **Exercise.** Every exposed face is a face. ∎

The definition of an exposed face is now given in terms of duality. Our first observation focuses on an geometric aspect:

I.2.2. **Proposition.** *For weakly closed wedges* $F \subseteq W$

$$\text{op op } F = \text{cl}(F - W) \cap W.$$

Further, F *is an exposed face if and only if*

$$F = \text{cl}(F - W) \cap W.$$

Proof. First we note that $(\text{op } F)^\perp = H((\text{op } F)^*)$ by Proposition I.7.i., and that $(\text{op } F)^* = \text{cl}(W - F)$ by Proposition I.1.9.ii.a. Hence $\text{op op } F = \text{cl}(F - W) \cap \text{cl}(W - F) \cap W = \text{cl}(F - W) \cap W$, since $\text{cl}(W - F) \supseteq W$. This proves the first assertion and the second is an immediate consequence. ∎

For our later applications it is important that we characterize exposed faces in a variety of ways.

I.2.3. **Proposition.** *For a wedge* $F \subseteq W$ *the following statements are equivalent:*

(1) F *is an exposed face of* W.

(2) *There is an exposed face* Φ *of* W^* *such that* $F = \text{op } \Phi$.

(3) *There is a subset* $\Phi \subseteq W^*$ *such that* $F = \text{op } \Phi$.

Proof. $(1) \Rightarrow (2)$: We assume that F is an exposed face and set $\Phi = opF$. Then $\text{op op } \Phi = \text{op op op } F = \text{op } F = \Phi$, since F is exposed. Thus Φ is an exposed face of W^*.

$(2) \Rightarrow (3)$: This is trivial.

$(3) \Rightarrow (1)$: We assume $F = \text{op } \Phi$ with a subset $\Phi \subseteq W^*$. We want to show that $\text{op op } F = F$, that is $\text{op op op } \Phi = \text{op } \Phi$. However, this identity is true quite generally: Firstly, the passage to the opposite wedge is containment reversing:

$$(*) \qquad \Psi_1 \subseteq \Psi_2 \implies \text{op } \Psi_1 \supseteq \text{op } \Psi_2.$$

Secondly,

$$(**) \qquad \Psi \subseteq \text{op op } \Psi,$$

as is readily verified. If, in $(**)$, we substitute $\Psi = \text{op } \Phi$, we obtain $\text{op } \Phi \subseteq \text{op op op } \Phi$; on the other hand if, in $(**)$, we substitute Φ for Ψ and then apply $(*)$, we find the reverse containment $\text{op } \Phi \supseteq \text{op op op } \Phi$. This proves the claim. ∎

According to this proposition, the exposed faces are exactly the opposite wedges of some subset of the dual wedge. Certain aspects of the previous results are summarized in the following corollary:

I.2.4.   Corollary.   *If L and $\widehat{L}$ are topological vector spaces in duality, then the functions* op *between the sets of exposed faces of L and $\widehat{L}$ are inverses of each other; in particular, they establish order reversing bijections between these sets.* ∎

I.2.5.   Proposition.   *Let M be any subset of a wedge W. Then* op op M *is the smallest exposed face of W containing M.*

Proof.   Let F be an exposed face of W containing M. Then op $M \supseteq$ op F by $(*)$ above. In the same fashion we have op op $M \subseteq$ op op F. But op op $F = F$ since F is exposed. Hence op op $M \subseteq F$, which proves the assertion. ∎

I.2.6.   Definition.   For any subset M of a wedge W we set

$$E_M(W) = \text{op op } M = (M^\perp \cap W^*)^\perp \cap W$$

and call this set the *exposed face generated by M*. If no confusion is possible, we shall simply write E_M, and for a singleton set $M = \{x\}$ we replace E_M by E_x.

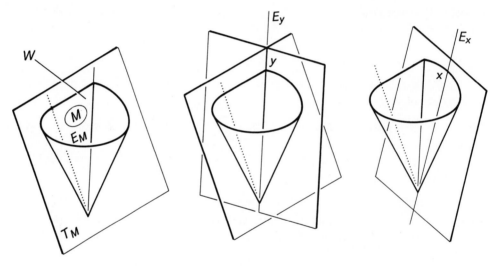

Figure 2

The three sets $L_M = (\text{op } M)^*$, $T_M = (\text{op } M)^\perp$ and $E_M = \text{op op } M$ are closely related. (See Definition I.1.8.)

I.2.7.   Proposition.   *For a subset M of a wedge W the following relations hold:*

(i) $M \subseteq E_M = T_M \cap W \subseteq T_M = H(L_M) \subseteq L_M$.

(ii) $\text{op } T_x = \text{op } E_x = \text{op } x$.

(iii) $L_M = L_{E_M} = L_{T_M}$.

(iv) $T_M = T_{E_M} = T_{T_M}$.

Proof.   (i) The first inclusion is clear. Next $E_M = \text{op op } M = (\text{op } M)^\perp \cap W = T_M \cap W$. This shows the second relation, and the next inclusion is trivial. Further $T_M = (\text{op } W)^\perp$ is the edge of $L_M = (\text{op } M)^*$ by Proposition I.1.7. The rest of (i) is clear. (ii) Since $x \in E_x \subseteq T_x$, the containments $\text{op } T_x \subseteq \text{op } E_x \subseteq \text{op } x$ are clear.

We must show $\operatorname{op} x \subseteq \operatorname{op} T_x$. But $\operatorname{op} x = L_x^*$ from $L_x = (\operatorname{op} x)^*$ and duality. From Proposition I.1.7.iii we know $L_x^* \subseteq H(L_x)^\perp$, but $H(L_x) = T_x$, whence $\operatorname{op} x \subseteq T_x$, which we had to show. Statements (iii) and (iv) are direct consequences from (ii) via the definitions. ∎

The containment relations are illustrated in the following diagram:

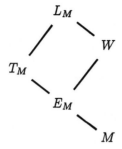

We notice in passing that, by duality, $(L_M)^* = \operatorname{op} M$.

I.2.8.   Proposition.   *Each exposed face F of a wedge W contains $H(W)$, and for each $x \in W$, the following statements are equivalent, if W is weakly closed:*

(1) $x \in H(W)$.

(2) $T_x = H(W)$.

(3) $E_x = H(W)$.

(4) $T_x = E_x$.

(5) E_x *is a vector space.*

Proof.   The vector space $H(W)$ is annihilated by all functionals in W^*, whence $H(W) \subseteq W^{*\perp} \cap W = \operatorname{op} W^* \subseteq \operatorname{op} \operatorname{op} F$, since $\operatorname{op} F \subseteq W^*$. But since F is an exposed face, we have $\operatorname{op} \operatorname{op} F = F$. This proves the first assertion.

Now let $x \in H(W)$. Then $W^* \subseteq H(W)^\perp \subseteq x^\perp$, and thus $W^* = x^\perp \cap W^* = \operatorname{op} x$. Then $T_x = (\operatorname{op} x)^\perp = W^{*\perp} = H(W)$ by Proposition I.1.7.iii. Thus (1) implies (2). If (2) is satisfied, then $E_x = T_x \cap W = H(W) \cap W = H(W)$ in view of Proposition I.2.7.i. Thus (3) follows. If (3) is satisfied, then $x \in T_x = H(W)$, whence $T_x = H(W) = E_x$, since we already know that (1) implies (2), and (2) implies (3). Trivially, (4) implies (5) and if (5) is satisfied, then $E_x \subseteq H(W)$ since $H(W)$ is the largest vector subspace of W. Thus $x \in E_x \subseteq H(W)$ implies (1).   ∎

The associated pointed cone

Proposition I.2.8 reflects the more general fact, that the facial structure of W is the same as that of a pointed cone, namely, the image of W in the factor space $L/H(W)$.

I.2.9.   Proposition.   *Let W be a wedge in a topological vector space L. Then the image $W/H(W)$ of W in the topological vector space $L/H(W)$ is a pointed cone, and the following conclusions hold:*

(i) *W is generating L if and only if $W/H(W)$ is generating in $L/H(W)$.*

(ii) *A wedge F is an exposed face of W if and only if F contains $H(W)$ and $F/H(W)$ is exposed in $W/H(W)$.*

(iii) *A vector subspace T of L is a tangent space of W at x if and only if it contains $H(W)$ and $T/H(W)$ is a tangent space of $W/H(W)$ at $x+H(W)$.*

Proof.   As the image of a wedge under a linear map, $W/H(W)$ is closed under addition and non-negative scalar multiplication. Since $w \in W$ implies $w + H(W) \subseteq W$, the set W is saturated for the coset map $p: L \to L/H(W)$, that is, $W = p^{-1}p(W)$. Hence $W/H(W) = p(W)$ is closed in the quotient topology, since W is closed. Thus $W/H(W)$ is a wedge. Since $p^{-1}H\big(p(W)\big)$ is a vector space contained in W, we have $p^{-1}H\big(p(W)\big) \subseteq H(W)$ and thus $H(p(W)) \subseteq p(H(W)) = \{0\}$. Hence $W/H(W)$ is a pointed cone.

Condition (i) is immediate. Let us prove condition (ii). First we have to specify a dual pairing for $L/H(W)$. We proceed in the standard fashion and consider the dual pairing of $L/H(W)$ and $H(W)^{\perp}$ given by $(x + H(W), \omega) \mapsto \langle \omega \mid x \rangle$. Every functional from W^* annihilates $H(W)$, whence $W^* \subseteq H(W)^{\perp}$. The dual of $W/H(W)$ in $\big(L/H(W)\big)^{\char`\^} = H(W)^{\perp}$ is simply W^* itself. If $F \subseteq L$, then the opposite wedge of $\big(F + H(W)\big)/H(W)$ in this dual is op F. The opposite wedge of any set $G \subseteq W^*$ in $L/H(W)$ is $(\mathrm{op}\,G)/H(W)$, where op G is the opposite wedge in W which, as we know from Proposition I.2.8 must contain $H(W)$ as it is an exposed face by Proposition I.2.3.

Now if F is an exposed face of W, then $F = \mathrm{op}\,\mathrm{op}\,F$. But then $F/H(W)$ is also the opposite of op F in $L/H(W)$, and vice versa. This proves (ii).

Next let T be a vector subspace. If $T = T_x$, then $H(W) \subseteq W^{*\perp} \subseteq (x^{\perp} \cap W^*)^{\perp} \subseteq T$ by Proposition I.2.8, and $T/H(W)$ is the annihilator of op $x \subseteq H(W)^{\perp}$ in $L/H(W)$. But as we have noted, op x is also the opposite wedge of $x + H(W) \in L/H(W)$ in the dual W^* of $W/H(W)$ inside the vector space $H(W)^{\perp}$ which is in duality with $L/H(W)$. Thus $T/H(W) = T_{x+H(W)}\big(W/H(W)\big)$. Conversely, if T contains $H(W)$ and $T/H(W) = T_{x+H(W)}\big(W/H(W)\big)$, then, tracing our steps back we argue that $T = T_x(W)$.   ∎

I.2.10.   Definition. *If W is a wedge in a topological vector space L, we say that $W/H(W)$ is the associated pointed cone in $L/H(W)$.*

The passage from a wedge to its associated pointed cone is a systematic way of reducing many problems to pointed cones.

I.2.11.   Corollary. *The function $F \mapsto F/H(W)$ which associates with an exposed face F of a wedge W its image $F/H(W)$ is an isomorphism of the lattice of exposed faces of W onto the lattice of exposed faces of the associated pointed cone $W/H(W)$.*   ∎

We recall that these lattices are antiisomorphic to the lattice of exposed faces of W^* by Corollary I.2.4.

The pointed cone associated with a wedge W is obtained in a canonical fashion. Under suitable circumstances it is possible to realize it inside W as a subcone, although no longer in a canonical fashion. We recall that a vector subspace H of a topological vector space L is said to *split in L* if there is a closed vector subspace V such that the function $(h, v) \mapsto h + v: H \times V \to L$ is an isomorphism of topological vector spaces. Any vector subspace V with this property is called a *complement* for H.

We remark that a closed vector subspace H in a locally convex topological vector space L will certainly split if its dimension or its codimension is finite. In particular, in a finite dimensional vector space, all vector subspaces split.

I.2.12. **Proposition.** *Let W be a wedge in a topological vector space L, and suppose that $H(W)$ splits in L. Let V be any complement for $H(W)$ in L. Then*

$$W = H(W) \oplus (V \cap W),$$

where $V \cap W$ is a pointed cone and where the direct sum refers to the fact that

$$(h, v) \mapsto h + v: H(W) \times (V \cap W) \to W$$

is an isomorphism of wedges.

Proof. By hypothesis, the function $(h, v) \mapsto h + v: H(W) \times V \to L$ is a linear and topological isomorphism. Hence it will map $H(W) \times (V \cap W)$ algebraically and topologically isomorphically onto $H(W) + (V \cap W)$. We have to show that this map is surjective and that $V \cap W$ is a pointed cone. Let $w \in W$. Then $w = h + v$ with some $h \in H(W)$ and $v \in V$. Now $v = w - h \in W + H(W) = W$, and thus $W \subseteq H(W) + (V \cap W)$ which is the first thing we had to show. Clearly, $V \cap W$ is a wedge, since V is closed in L. Since $H(V \cap W) \subseteq H(W) \cap V = \{0\}$, this wedge is pointed. ∎

I.2.13. **Corollary.** *Under the hypotheses of the preceding proposition, the coset map $x \mapsto x + H(W)$ from L to $L/H(W)$ maps $V \cap W$ isomorphically onto the associated pointed cone $W/H(W)$.*

Proof. The function $v \mapsto v + H(W): V \cap W \to W/H(W)$ is bijective in view of Proposition I.2.12 above. Since it is the restriction and corestriction of the continuous and continuously invertible map $V \to L/H(W)$, the assertion follows. ∎

I.2.14. **Corollary.** *In a finite dimensional vector space L, every wedge W decomposes into a direct sum $H(W) \oplus (V \cap W)$ of its edge and a pointed cone $V \cap W$ isomorphic to the associated pointed cone, and every vector space complement V for $H(W)$ gives rise to such a complementary cone $V \cap W$.* ∎

Support hyperplanes

We now recall an important geometric concept which we shall have occasion to invoke quite often in the sequel, and that is the concept of a support hyperplane of a wedge at a point x. Geometrically, a support hyperplane of a wedge W at a point $x \in W$ is a closed hyperplane containing x and bounding a closed half-space containing W.

I.2.15.   Definition. Let W be a wedge in a topological vector space L and $x \in W$. Then a *support hyperplane of W at x* is the zero set $\omega^{-1}(0) = \omega^{\perp}$ of a functional $\omega \in W^*$ which is non-zero on $W - W$ and satisfies $\langle \omega \mid x \rangle = 0$.

According to this definition, a hyperplane T of L is a support hyperplane of W at x if and only if there is an $\omega \in x^{\perp} \cap W^* = \operatorname{op} x$ which is non-zero on $W - W$. The elements of W^* vanishing on $W - W$ are precisely the elements of $H(W^*)$ according to Proposition I.1.7.i. Thus we have the following remark:

I.2.16.   Proposition. *For a wedge W and an $x \in W$, the set of support hyperplanes of W at x is exactly*

$$\{\omega^{\perp} : \omega \in (x^{\perp} \cap W^*) \setminus H(W^*) = \operatorname{op} x \setminus H(W^*)\}.$$

If W is generating, then this set is simply

$$\{\omega^{\perp} : \omega \in \operatorname{op} x \setminus \{0\}\}.$$

∎

We shall apply the idea of support hyperplanes mainly to generating wedges. According to the definition, a hyperplane containing $W - W$ is not a support hyperplane. If W is generating then no non-zero functional vanishes on W.

I.2.17.   Proposition. *Let W be a generating wedge and x a point of W with $T_x \neq L$. Then the following conclusions hold:*

(i) *T_x is the intersection of the set of all support hyperplanes of W at x.*

(ii) *L_x is the intersection of the set of all half-spaces containing W and being bounded by support hyperplanes of W.*

Proof.   By definition we have $T_x = (\operatorname{op} x)^{\perp}$ and $L_x = (\operatorname{op} x)^*$. Further,

$$(\operatorname{op} x)^{\perp} = \bigcap_{\omega \in \operatorname{op} x} \omega^{\perp} = \bigcap_{\omega \in (\operatorname{op} x) \setminus \{0\}} \omega^{\perp},$$

and

$$(\operatorname{op} x)^* = \bigcap_{\omega \in \operatorname{op} x} \omega^* = \bigcap_{\omega \in (\operatorname{op} x) \setminus \{0\}} \omega^*.$$

However, in view of Proposition I.2.16, this is in fact the assertion.   ∎

With this proposition we can prove a partial converse of Corollary I.1.12.

I.2.18.   Proposition. *Let W be a generating wedge in a finite dimensional vector space L and E a vector subspace meeting the interior of W. Then*

(1) $T_x(W \cap E) = T_x(W) \cap E$ *and* $E_x(W \cap E) = E_x(W) \cap E$ *for all* $x \in W \cap E$,

and

(2) $\qquad\qquad L_x(W \cap E) = L_x(W) \cap E \qquad$ *for all* $\qquad x \in W \cap E$.

Proof. The proof of (2) is completely analogous to the proof of the first equation in (1), and the second equation in (1) follows from the first in view of Proposition I.2.7(i). Therefore, we prove the first equation of (1). For $x \in E \cap \operatorname{int} W$ there is nothing to prove. If $x \in E \cap \partial W$, then we find a support hyperplane T of W at x. Since E meets $\operatorname{int} W$, then $T \cap E$ is a support hyperplane of $W \cap E$ in E at x. If V is a vector space complement for E in T, then $L = E \oplus V$ since $E \not\subseteq T$. Now let S be a completely arbitrary support hyperplane of $W \cap E$ in E at x. Then $S \oplus V$ is a support hyperplane of W in L at x. If $\mathcal{T}$ denotes the set of all support hyperplanes of W at x, and if $\mathcal{S}$ denotes the set of all support hyperplanes of $W \cap E$ in E at x, then $T_x(W) = \bigcap_{H \in \mathcal{T}} H \subseteq \bigcap_{S \in \mathcal{S}} S \oplus V = (\bigcap_{S \in \mathcal{S}} S) \oplus V = T_x(W \cap E) \oplus V$ in view of Proposition I.2.17. Thus we have $T_x(W) \cap E \subseteq T_x(W \cap E)$. The reverse containment was shown in Corollary I.1.12. ∎

The following proposition is the analog of Proposition I.2.9 for support hyperplanes.

I.2.19. **Proposition.** *(i) A hyperplane T of L is a support hyperplane of the wedge W if and only if it contains $H(W)$ and $T/H(W)$ is a support hyperplane of $L/H(W)$ at $x + H(W)$.*

(ii) The assignment $T \mapsto T/H(W)$ from the set of closed vector subspaces T containing $H(W)$ onto the set of closed vector subspaces of $L/H(W)$ induces a bijection from the set of support hyperplanes of W onto the set of support hyperplanes of the associated pointed cone $W/H(W)$.

Proof. (i)Let T be a hyperplane in L. If T is a support hyperplane of W at $x \in W$, then $T = \omega^\perp$ for some $\omega \in \operatorname{op} x \setminus H(W^*) \subseteq H(W)^\perp$. Hence $H(W) \subseteq T$. If we interpret ω as a functional on $L/H(W)$ in the obvious way, we have interpreted $T/H(W)$ as the zero-set of this functional. In this fashion we have proved that $T/H(W)$ is a support hyperplane of $W/H(W)$ at $x + H(W)$. The converse we show by retracing our steps. Part (ii) is a consequence of part (i). ∎

The algebraic interior

It is important for us to note that in the case of finite dimensions, additional characterizations of exposed faces are possible. This is due to the fact that finite dimensional wedges are "generated" by inner points in the lattice of faces. We make this precise. First it is useful to speak of the algebraic interior of a wedge in a manner which we explain in the following definition:

I.2.20. **Definition.** If L is a vector space and W a convex subset, then we write

$$\operatorname{algint} W = \{x \in W : (\forall y \in W - W)(\exists t > 0)x + t \cdot y \in W\},$$

and we call this set the *algebraic interior of W*.

EI.2.2.   Exercise.   If x is in the algebraic interior of W, then for each $y \in W$ there is a $t > 0$ such that $x + t \cdot y$ is in the algebraic interior of W. If W is closed under addition and non-negative scalar multiplication, then $W + \text{algint}\, W \subseteq \text{algint}\, W$. If $(\text{algint}\, W) \cap H(W) \neq \emptyset$, then W is a vector space.   ∎

I.2.21.   Proposition.   *Let L be a topological vector space and W a wedge in L. Then the following conclusions hold:*

(i)   *If the vector space $W - W$ is finite dimensional, then*

$$\text{algint}\, W = \text{int}_{W-W}\, W,$$

the topological interior of W in the vector space $W - W$.

(ii)   *If $\omega \in \text{algint}\, W^*$, then $\langle \omega \mid x \rangle > 0$ for all $x \in W \setminus H(W)$, and if $W^* - W^*$ is finite dimensional, then the converse implication is true, too.*

Proof.   (i) For any wedge in any topological vector space, we readily observe $\text{int}\, W \subseteq \text{algint}\, W$. Conversely, let $w \in \text{algint}\, W$, and let $e_1, \ldots, e_n$ be a basis of $W - W$. Then by Definition I.2.20 there exist positive numbers $r_1, \ldots, r_n$ such that $w \pm r_k \cdot e_k \in W$ for $k = 1, \ldots, n$. Then the convex hull of these $2n$ points is a neighborhood of w in the vector space $W - W$, and it is entirely contained in W. Thus $w \in \text{int}_{W-W}\, W$.

(ii) We assume that there is an $x \in W \setminus H(W)$ such that $\langle \omega \mid x \rangle = 0$ with $\omega \in W^*$ and show that $\omega \notin \text{algint}\, W^*$. By Proposition I.2.17, our assumption is tantamount to saying that $x^\perp$ is a support hyperplane of W^* in ω. But then we find a $\nu \in W^* \setminus x^\perp$. Now $\langle \nu \mid x \rangle > 0$. Hence $\langle \omega + t \cdot \nu \mid x \rangle = t \langle \nu \mid x \rangle < 0$ for all $t < 0$, whence $\omega + t \cdot \nu \notin W^*$ for these t, and this shows $\omega \notin \text{algint}\, W^*$.

If $W^* - W^*$ is finite dimensional, then $\text{algint}\, W^* = \text{int}_{W^* - W^*}\, W^*$ by (i) above, and then the relation $\omega \in W^* \setminus \text{algint}\, W^*$ implies the existence of a support hyperplane $x^\perp$ of W^* at ω, which means that we have an $x \in W \setminus H(W)$ with $\langle \omega \mid x \rangle = 0$. This completes the proof.   ∎

The algebraic interior of a wedge may be empty. Let L be the vector space $\mathbb{R}^{(\mathbb{N})}$ of all sequences with finite support. Let $(\mathbb{R}^+)^{(\mathbb{N})}$ be the wedge of all sequences of finite support with non-negative entries. Then $W - W = L$ but $\text{algint}\, W = \emptyset$. By Exercise 1 in Section 1 and Proposition I.2.21 above we know that *every finite dimensional wedge has a non-empty algebraic interior.*

I.2.22.   Lemma.   (i) *Let L be a vector space and W a subset closed under addition and non-negative scalar multiplication. If F and G are faces of W and G contains a point of the algebraic interior of F, then $F \subseteq G$.*

(ii) *Let F be an exposed face of a wedge W in a topological vector space L (in duality with $\widehat{L}$), then $F = E_x$ for any $x \in \text{algint}\, F$.*

Proof.   (i) Fix an $x \in (\text{algint}\, F) \cap G$. Consider an arbitrary $f \in F$ and set $y = f - x$. Then $y \in F - F$ and thus there is an $r \in\,]0, 1[$ such that $w = x - r \cdot y \in F$ by Definition I.2.20, that is, $w = x - r \cdot (f - x) = (1 + r) \cdot x - r \cdot f \in F \subseteq W$. Now $x = \frac{1}{1+r} \cdot w + (1 - \frac{1}{1+r}) \cdot f$ is a point of G, and $w, f \in W$. Hence, by the definition of a face, f is in G.

(ii) Let $x \in \text{algint}\, F$. Then any face G of W containing x contains F by (i) above. Now E_x is such a face, hence $F \subseteq E_x$. But F is exposed, and E_x is the smallest exposed face containing x, whence $E_x \subseteq F$. The assertion follows.   ∎

Exposed faces of finite dimensional wedges

Now we are ready for additional characterisations of exposed faces in the case of finite dimensional wedges. We recall that we defined $\operatorname{op} x = x^\perp \cap W^*$ for $x \in L$.

I.2.23.   Proposition. *Let W be a finite dimensional wedge in a topological vector space L. Let F be a proper subset of W which is closed under addition and non-negative scalar multiplication. Then the following statements are equivalent:*

(1) *F is an exposed face of W.*

(2) *$F = \operatorname{op} \omega$ for each $\omega \in \operatorname{algint} \operatorname{op} F$.*

(3) *There is an $\omega \in W^* \setminus H(W^*)$ such that $F = \operatorname{op} \omega$.*

(4) *There is a support hyperplane T of W at some x with $F = T \cap W$.*

(5) *$F = E_x$ for all $x \in \operatorname{algint} F$.*

(6) *There is an $x \in W$ such that $F = E_x$.*

Proof.   Since $\dim W$ is finite and F is a proper subset of W, then $\operatorname{op} F \neq \{0\}$ and so $\operatorname{algint}(\operatorname{op} F) \neq \emptyset$. Now we assume (2) and claim that $\operatorname{op} F$ is not a vector space. Indeed, assume that $\operatorname{op} F$ is a vector space. Then, being a face $\operatorname{op} F$ equals $H(W^*)$. Then $W = (W - W) \cap W = H(W^*) \cap W^{**} = \operatorname{op} H(W^*) = \operatorname{op} \operatorname{op} F$. By (2) and Proposition I.2.3, $\operatorname{op} \operatorname{op} F = F$. Hence $W = F$ and thus F is not a proper subset of W. The claim is then established. Consequently, by Lemma I.2.21 (or by Exercise EI.2.2), $H(W^*) \cap \operatorname{op} F = \emptyset$. Hence any $\omega \in \operatorname{algint}(\operatorname{op} F)$ is contained in $W^* \setminus H(W^*)$. This proves (2)$\Rightarrow$(3).

Likewise, $\operatorname{algint} F \neq \emptyset$, and thus (5)$\Rightarrow$(6).

The remaining implications do not require the hypothesis of W being finite dimensional. (6)$\Rightarrow$(1) is trivial. (1)$\Rightarrow$(2): $\operatorname{op} F$ is an exposed face by Proposition I.2.3. If we have an $\omega \in \operatorname{algint}(\operatorname{op} F)$, then $\operatorname{op} F = E_\omega(W^*) = \operatorname{op} \operatorname{op} \omega$ by Lemma I.2.22. From this we derive $\operatorname{op} \operatorname{op} F = \operatorname{op} \operatorname{op} \operatorname{op} \omega = \operatorname{op} \omega$; but by condition (1) we know $F = \operatorname{op} \operatorname{op} F$, and thus $F = \operatorname{op} \omega$, as asserted. (3)$\Rightarrow$(4): Take $T = \omega^\perp$ and recall Proposition I.2.16. (4)$\Rightarrow$(3): Let ω be in $\operatorname{op} x \setminus H(W^*)$ such that $T = \omega^\perp$ in view of Proposition I.2.16. (3)$\Rightarrow$(1): Clear from Proposition I.2.3. (1)$\Rightarrow$(5): Let $x \in \operatorname{algint} F$. Then $F = E_x$ by Lemma I.2.22, as asserted.   ∎

Geometrically, the previous proposition says, among other things, that in finite dimensional wedges an exposed proper face of a wedge is obtained exactly by intersecting the wedge with some support hyperplane.

C^1-points and their duality theory

We saw in Definition I.2.1, that the idea of an exposed face immediately gave us a special type of point in a wedge, the so-called *exposed* points or E^1-points. With the aid of the concept of a support hyperplane we single out a new type of special points on a wedge.

I.2.24.   Definition. If W is a generating wedge then a C^1-*point* of W is a point x such that there is one *and only one* support hyperplane of W at x. If W is an arbitrary wedge then a C^1-*point* is a C^1-point of the generating wedge W in $W - W$. The set of all C^1-points of W is denoted $C^1(W)$.

I.2.25.   Proposition. *The following statements are equivalent for a point x in a generating wedge W:*

(1) *x is a C^1-point of W.*

(2) $\operatorname{op} x$ *is a half line.*

(3) $\operatorname{op} x \setminus \{0\} \subseteq E^1(W^*)$.

(4) $\operatorname{algint} \operatorname{op} x \subseteq E^1(W^*)$.

(5) *T_x is a hyperplane.*

Proof.   Since W is generating, $H(W^*) = \{0\}$. From Proposition I.2.16 we know that x is a C^1-point if and only if the set

$$\{\omega^\perp : \omega \in \operatorname{op} x \setminus \{0\}\}$$

is singleton. This implies that the wedge $\operatorname{op} x$ is 1-dimensional and thus must be a line or a half line. However, if $\operatorname{op} x$ is a vector space, then $\operatorname{op} x \subseteq H(W^*) = \{0\}$. Thus $\operatorname{op} x$ cannot be a line, and (2) follows. Now $\operatorname{op} x$ is an exposed face by Proposition I.2.3. Hence, if $\operatorname{op} x$ is a half-line $\mathbb{R}^+ \cdot \omega$, say, then $\omega \in E^1(W^*)$ by Definition I.2.1. Hence (2) implies (3). Finally, suppose that (3) holds. Then $\operatorname{op} x$ is an exposed face contained in $E^1(W^*) \cup \{0\}$. Let μ and ν be non-zero points in $\operatorname{op} x$, then let ω be their midpoint $\frac{\mu+\nu}{2}$. Then $\omega \in \operatorname{op} x \setminus \{0\} \subseteq E^1(W^*)$. Thus $\mathbb{R}^+ \cdot \omega$ is an exposed face, and thus $\mu, \nu \in \mathbb{R}^+ \cdot \omega$ follows by the definition of a face. Thus μ and ν are scalar multiples of each other. Since all non-zero elements of $\operatorname{op} x$ are scalar multiples of each other, x is necessarily a C^1-point by our initial remarks. Thus (1) follows. Conditions (3) and (4) are clearly equivalent, since the algebraic interior of a half-line is its relative interior in its span. The equivalence of (1) and (5) is immediate from I.2.17.   ∎

I.2.26.   Proposition. *For a functional ω in the dual W^* of a generating cone W, consider the following conditions:*

(1) *ω is an E^1-point of W^*.*

(2) $\operatorname{algint} \operatorname{op} \omega \subseteq C^1(W)$.

   Then (1) *implies* (2), *and if* $\operatorname{algint} \operatorname{op} \omega \neq \emptyset$, *then both conditions are equivalent.*

Proof.   We begin by the following remarks: Fix an $\omega \in W^*$. For any $x \in \operatorname{algint} \operatorname{op} \omega$ we have $E_x = \operatorname{op} \omega$ in view of Lemma I.2.22. But $\operatorname{op} x = \operatorname{op} \operatorname{op} \operatorname{op} x = \operatorname{op} E_x = \operatorname{op} \operatorname{op} \omega = E_\omega$. Thus

(*) $E_x = \operatorname{op} \omega$ and $E_\omega = \operatorname{op} x$.

We recall that condition (1) is equivalent to

$(**)$ $$E_\omega = \mathbb{R}^+ \cdot \omega.$$

Now we prove the asserted equivalences:

$(1) \Rightarrow (2)$: Let $x \in \text{algint op}\,\omega$. The support hyperplanes of W at x are the zero sets of the functionals $\mu \in \text{op}\,x \setminus \{0\}$. From $(*)$ and $(**)$ above it follows that there is only one support hyperplane of W at x, that is, $x \in C^1(W)$.

$(2) \Rightarrow (1)$: Since we now assume that $\text{algint op}\,\omega \neq \emptyset$, we can pick an $x \in \text{algint op}\,\omega$. By hypothesis (2), $x \in C^1(W)$, which by Proposition I.2.16 means that $\text{op}\,x$ is a half line. By $(*)$ we have $\text{op}\,x = E_\omega$. But then the half-line $\mathbb{R}^+ \cdot \omega$ must be all of E_ω, and thus $(**)$ holds. This proves (1). ∎

I.2.27.   Corollary.   *If W is a finite dimensional generating wedge, then the following conditions are equivalent:*

(1) $\omega \in W^*$ *is an E^1-point.*

(2) $\text{algint op}\,\omega \subseteq C^1(W)$. *Here, $\text{algint op}\,\omega$ is simply the interior of $F \overset{\text{def}}{=} \text{op}\,\omega$ in $F - F$.*

Also, the following conditions are equivalent:

$(1')$ $x \in W$ *is a C^1-point.*

$(2')$ $\text{algint op}\,x \subseteq E^1(W^*)$. *Here $\text{algint op}\,x$ is simply the open half line $]0, \infty[\cdot \omega$ with $\text{op}\,x = \mathbb{R}^+ \cdot \omega$.*

Proof.   This is clear from the preceding two propositions in view of the fact that for finite dimensional wedges, the algebraic interior is the topological interior relative to the linear span of the wedge. ∎

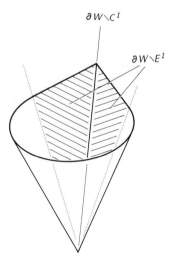

$\partial W \setminus C^1$

$\partial W \setminus E^1$

The preceding results tell us that every $C^1(W)$-point x of W will necessarily give us a ray $\text{op}\,x$ of $E^1(W^*)$ "orthogonal" to x. Dually, a E^1-point ω of W^* does not, in general, guarantee a C^1-point of W "orthogonal" to ω, but it will, if $\text{op}\,\omega$ is finite dimensional. Thus, as long as we stay within the domain of finite dimensional vector spaces, the duality is fully guaranteed. But even then we observe a certain asymmetry. We will later investigate the special class of wedges for which this asymmetry disappears.

We observe that neither the C^1-points nor the E^1-points need form closed sets.

The concept of a C^1-point was really meaningful for generating wedges only. Of course, one could generalize the idea by defining a point x of an arbitrary wedge W to be a C^1-point if and only if it is a C^1-point of W with respect to $W - W$ (dually paired with $\widehat{L}/(W - W)^\perp$ in the usual and obvious fashion). Certainly in the case of finite dimensional vector spaces, this is a completely satisfactory generalisation. However, we shall see that in our applications we shall very frequently be allowed to consider generating wedges without serious loss of generality. In view of the duality theory between C^1-points and E^1-points this means, that E^1 points will be considered primarily in the case of pointed cones W, because after Proposition I.1.7.i, if W is generating, then W^* is pointed.

The semiprojective space of a wedge, bases of cones

For the specifics, recall that an *action* $(g, x) \mapsto g{\cdot}x : G \times X \longrightarrow X$ of any group G on a set X is a function satisfying $1{\cdot}x = x$ and $(gh){\cdot}x = g{\cdot}(h{\cdot}x)$, and that any action decomposes the set X into *orbits* $G{\cdot}x$, $x \in X$. The coset map $x \mapsto G{\cdot}x : X \to X/G$ is called the *orbit map*. If X is a topological space and G a group of homeomorphisms, then the set X/G equipped with the quotient topology is called the *orbit space*. In this case *the orbit map is both open and continuous*. Now we apply this for the action of the multiplicative group $\mathbb{P}$ of positive reals acting on a punctured topological vector space L. In passing we remember that $\mathbb{P}$ is isomorphic to the additive group of reals $\mathbb{R}$ under $r \mapsto e^r : \mathbb{R} \to \mathbb{P}$.

Let W be a non-singleton wedge in a topological vector space L. The orbit space $(W \setminus \{0\})/\mathbb{P}$ is written $\Pi(W)$. We let $\pi : W \setminus \{0\} \to \Pi(W)$ denote the orbit map. If Y is a $\mathbb{P}$-stable subset of $W \setminus \{0\}$, we shall set $\Pi(Y)$ for the subspace of all $\mathbb{P}{\cdot}y \in \Pi(W)$ with $y \in Y$.

The semiprojective space of a wedge is reminiscent of the projective space associated with a vector space. However, we note that the preceding definition applies, in particular, to a finite dimensional vector space W. In this case, $\Pi(W)$ does not give the projective space associated with W, but rather a sphere. In fact, the semiprojective space reveals a significant distinction between vector spaces and wedges which are not vector spaces, as we shall see presently. First we observe, however, that in the case of pointed cones W, the semiprojective space allows a representation as a convex subset of W which is compact if $\dim W$ is finite.

I.2.28.   Proposition. *Let W be a pointed cone in a topological vector space L which is in duality with $\widehat{L}$. Suppose that $\omega \in \operatorname{algint} W^*$. Let A denote the affine hyperplane $\omega^{-1}(1)$. Then $B = W \cap A$ is a closed convex subset of W such that $W = \mathbb{R}^+{\cdot}B$. The function $b \mapsto \mathbb{P}{\cdot}b : B \to \Pi(W)$ is a homeomorphism. If $\dim W$ is finite, then $\omega \in \operatorname{algint} W^*$ always exists and B is compact. In particular, $\Pi(W)$ is homeomorphic to a compact finite dimensional cell.*

Proof.   As the intersection of two closed convex sets, B is closed and convex. Since $\omega \in \operatorname{algint} W^*$, we know $E_\omega = W^*$ by Lemma I.2.22. Since by definition, $E_\omega = \operatorname{op} \operatorname{op} \omega$ we conclude $\operatorname{op} \omega = \{0\}$, that is, $\omega^\perp \cap W = \{0\}$. Now let $0 \neq$

$x \in W$. Then $\langle \omega \mid x \rangle > 0$. Hence $y \stackrel{\text{def}}{=} \langle \omega \mid x \rangle^{-1} \cdot x$ is the unique point in $\mathbb{P} \cdot x$ meeting $B = \omega^{-1}(1) \cap W$. This shows at the same time that $W = \mathbb{R} \cdot B$ and that $p = (b \mapsto \mathbb{P} \cdot b) : B \to \Pi(W)$ is bijective. It is clearly continuous, and since the function $x \mapsto \langle \omega \mid x \rangle^{-1} \cdot x : W \setminus \{0\} \to B$ is clearly continuous, also p^{-1} is continuous. This proves the first part of the proposition.

Now suppose that W is finite dimensional. We must show that B is compact. For this purpose it is no loss of generality to assume $L = W - W$. Then W is generating and thus has inner points by Exercise EI.1.1. Since the action by $\mathbb{P}$ preserves inner points, by the preceding we find some $w \in B \cap \text{int}\, W$. Now $B - w$ is a closed convex neighborhood of 0 in $A - w$ which we have to prove to be bounded. If not, then it contains a half-ray $\mathbb{R}^+ \cdot z$ for some $z \neq 0$, so $w + \mathbb{R}^+ \cdot z \subseteq B \subseteq W$. If $r \in \mathbb{R}^+$, then $\frac{1}{n}(w + nr \cdot z) \in W$ for $n = 1, \ldots$ Then $r \cdot z = \lim_{n \to \infty} \frac{1}{n}(w + nr \cdot z) \in W$ since W is closed. We see $\mathbb{R}^+ \cdot z \subseteq (A - w) \cap W = \{0\}$, and this is a contradiction. The proposition is proved. ∎

It is easy to verify that we have a converse of the following form: *If A is an affine closed hyperplane not containing 0 and B a compact convex subset of A, then $W = \mathbb{R}^+ \cdot B$ is a pointed wedge.* In the finite dimensional situation this establishes a correspondence between convex pointed cones and compact convex sets. We need not elaborate further at the moment.

I.2.29. **Definition.** (i) The space $\Pi(W)$ will be called *the semiprojective space associated with W*.

(ii) Any set B obtained in the way explained in Proposition I.2.28 is called a *base* of the cone W.

Now we complete easily the topological description of the possible semi-projective spaces of wedges in a finite dimensional vector space.

I.2.30. **Proposition.** *Let W be a generating wedge in an n-dimensional vector space . Then the semiprojective space $\Pi(W)$ associated with W is homeomorphic to an $(n-1)$-sphere if W is a vector space and to a compact $(n-1)$-cell otherwise. The topological space $W \setminus \{0\}$ is homeomorphic to $\mathbb{P} \times \Pi(W)$.*

Proof. If W is a vector space, then $W = L$. We may assume that L is a finite dimensional Hilbert space equipped with a scalar product. If S is the unit sphere, then S is a compact subspace of $W \setminus \{0\}$ which is mapped bijectively and continuously onto $\Pi(W)$, and this proves that the target space is homeomorphic to an $(n-1)$-sphere. Now suppose that $W \neq L$. We claim that $\Pi(W)$ is an $(n-1)$-cell. There are many ways to prove this claim. We know for instance that $W = H(W) \oplus W_1$ with a pointed cone W_1 by Proposition I.2.12. Let B be a compact base of W_1 according to Definition I.2.29. Then B is the boundary of a 0-neighborhood B_1 in W_1. If we take a unit ball B_0 around 0 in $H(W)$ with boundary S_0, then the 0-neighborhood $B_0 \oplus B_1$ in W has the boundary $(B_0 \oplus B) \cup (S_0 \oplus B_1)$. It is an exercise to verify that this boundary is homeomorphic to an $(n-1)$-cell as a "cylinder with one lid". The orbit map once more maps this boundary homeomorphically onto $\Pi(W)$ which shows that the semiprojective space is an $(n-1)$-cell.

The function $(r, w) \mapsto r \cdot w : \mathbb{P} \times (W \cap S) \to W \setminus \{0\}$ is a continuous bijection. Its inverse is continuous as a restriction of the continuous map $x \mapsto$

$(\|x\|, \|x\|^{-1}{\cdot}x): L \setminus \{0\} \to \mathbb{P} \times S$. Hence it is an homeomorphism, and since $W \cap S$ and $\Pi(W)$ are homeomorphic, the final assertion follows. ∎

Sums of two wedges

We shall need conditions which guarantee that the sum of two wedges is closed and therefore again a wedge.

The first remark is a variation of the theme that in Frechet spaces every absorbing closed convex set containing 0 is a neighborhood of 0. It establishes a similar result for finite dimensional wedges.

I.2.31.   Proposition. *Let W be a wedge in a finite dimensional vector space and $U \subseteq W$ a compact convex set containing 0 such that $W \subseteq \mathbb{R}^+{\cdot}U$. Then U is a neighborhood of zero in W.*

Conversely, if a subset W of a finite dimensional vector space is convex and stable under non-negative scalar multiplication, and if it has a compact neighborhood of 0, then it is a wedge.

Proof.   We begin with the hypothesis that U is a compact set containing 0 such that $W = \mathbb{R}{\cdot}U$. We fix a norm for L and set $B = \{b \in W : \|x\| = 1\}$. By hypothesis, for each $x \in B$ there is a unique *largest positive* real number $f(x)$ such that $f(x){\cdot}x \in U$. The compactness of U implies the existence of a non-negative number c such that $f(x) \in [0, c]$ for all $x \in B$. The homeomorphism $(x, r) \mapsto r{\cdot}x : B \times \mathbb{P} \to W \setminus \{0\}$ (considered in the proof of Proposition I.2.30) maps the graph of $f: S \to [0, c]$ homeomorphically onto the boundary ∂U of U in W. Since ∂U is compact, the graph of f is compact, hence closed. Thus f is continuous by the Topological Lemma recorded after this proposition. Hence f is bounded away from 0, which establishes the claim that U is a neighborhood of 0.

Now we prove the converse: We assume that U is a compact neighborhood of 0 in W. Let V be an open ball around 0 such that $W \cap V \subseteq U$. Now let $x \in \overline{W}$. Then $x = \lim w_n$ with $w_n \in W$. Next we take a positive number r such that $x \in rV$. Then, eventually, all w_n are in $W \cap rV = r(W \cap V)$, hence in the compact set rU. Then $x = \lim w_n \in rU \subseteq W$. ∎

Topological Lemma. *Let X' be a topological space, Y' a compact Hausdorff space, and let F be a closed subspace of $X' \times Y'$. If $X \subseteq X'$ and $Y \subseteq Y'$ are subsets such that $F \cap (X \times Y')$ is the graph of a function $f: X \to Y$, then f is continuous.*

Proof.   Let $(x_j)_{j \in J}$ be a net in X converging to $x \in X$, and let y be any cluster point in Y' of the net $(f(x_j))_{j \in J}$ in Y'. If we can show that $y = f(x)$, then by the compactness and Hausdorff property of Y' we have shown that $f(x) = \lim_{j \in J} f(x_j)$, and this will prove the continuity of f. Now the net of the elements $\big(x_j, f(x_j)\big)$ is contained in $f \subseteq F$ and has the cluster point (x, y). Since F is closed in $X' \times Y'$ we have $(x, y) \in F$. But also $(x, y) \in X \times Y'$. Consequently $(x, y) \in F \cap (X \times Y')$, and this means $y = f(x)$ by hypothesis. ∎

It will be important in the applications to have certain information on the sum of two wedges. While the sum of two vector subspaces is a simple matter, delicate questions arise in the context of a sum of two wedges—even if one is a vector space.

I.2.32.   Proposition. *Let V and W be two wedges in a Banach space L and suppose that $V \cap -W$ is a vector space. Then $V + W$ is closed, hence a wedge.*

In particular, if I is a vector space and $W \cap I$ is a vector space, then $W + I$ is a wedge.

Proof.   We shall factor the closed vector subspace $V \cap -W$. If the claim is true in $L/(V \cap -W)$ for the quotients $V/(V \cap -W)$ and $W/(V \cap -W)$, then $(V + W)/(V \cap -W)$ is closed, and thus $V + W$ is closed. It is therefore no loss of generality if we assume that $V \cap -W = \{0\}$. Now let $x = \lim(v_n + w_n)$ with $v_n \in V$ and $w_n \in W$. We must show that $x \in V + W$. Obviously, we may assume that all v_n and w_n are non-zero. Then we define unit vectors e_n and f_n in terms of the equations $v_n = \|v_n\| \cdot e_n$ and $w_n = \|w_n\| \cdot f_n$. The existence of the limit guarantees that there is a real bound c such that $\|v_n + w_n\| \leq c$. Thus

$$\left\| e_n + \frac{\|w_n\|}{\|v_n\|} \cdot f_n \right\| \leq \frac{c}{\|v_n\|}.$$

We distinguish two cases. Case 1: The sequence v_n is bounded. Since $v_n + w_n$ converges, the sequence w_n is bounded, too. After picking convergent subsequences and renaming them we may assume that $v = \lim v_n$ and $w = \lim w_n$ exist. But then $x = v + w \in V + W$ and the claim is proved in this case. Case 2: $\|v_n\|$ is unbounded. Then

$$\lim\left(e_n + \frac{\|w_n\|}{\|v_n\|} \cdot f_n \right) = 0.$$

Again we may assume without loss of generality that $e = \lim e_n$ and then $g = \lim(\|w_n\|/\|v_n\|) \cdot f_n$ exist so that $e + g = 0$. Then $g = -e \in W \cap -V = \{0\}$. The conclusion $e = 0$ however contradicts the assumption that the e_n and thus also e are unit vectors. Case 2 is therefore ruled out, and the first assertion of the proposition is proved. The second one, however, is an immediate consequence of the first.   ∎

In the following proposition we are dealing with the case that, for a wedge W and a vector space I in a finite dimensional vector space L, the sum $W + I$ is assumed to be closed. While every element $x \in W + I$ is a sum $w + i$ of elements in W and I, respectively, it is by no means obvious that all small elements $x \in W + I$ are sums of *small* elements $w \in W$ and $i \in I$. The following proposition, however, asserts exactly that.

I.2.33.   Proposition. *Let W be a wedge and I a vector subspace in a finite dimensional vector space L. Then there are compact convex subsets $S_W \subseteq W$ and $S_I \subseteq I$ containing 0 such that the following conditions are satisfied for $U = S_W + S_I$:*

(i) *$U \cap W = S_W$.*

(ii) *There is an $\omega \in \widehat{L}$ which is strictly positive on $W \setminus \{0\}$ and for which $\omega(S_W) = [0, 1]$ and $\omega(S_I) = [-1, 0]$.*

(iii) *For each non-zero $x \in W$ there is a positive r such that $r \cdot x \in U$.*

(iv) *$W + I$ is closed if and only if U is a neighborhood of 0 in $W + I$.*

Proof.   We let F denote the smallest face of W containing $W \cap I$. We claim that $F = \{x \in W : (\exists w \in W)\ \ x + w \in I \cap W\} = W \cap ((I \cap W) - W)$; indeed one readily verifies that this set is a face and is contained in any face containing $I \cap W$. We set $H = (F - F) + I$. We claim that

(1) $H = F + I$,

(2) $F = H \cap W$, and

(3) $H = (W + I) \cap -(W + I)$.

In order to prove (1) we first note that $F + I \subseteq H$, trivially. Thus we have to show that $-F \subseteq F + I$. Thus let $x \in F$. There we find elements $i \in I \cap W \subseteq F$ and $w \in W$ such that $x + w = i$, and since F is a face, we know $w \in F$. Hence $-x = w - i \in F + I$ which proves the claim.

Next we show (2): Trivially, $F \subseteq H \cap W$. Conversely, let $h \in H \cap W$. Then by (1), we have $H = -H = -F - I = -F + I$; hence there are elements $f \in F$ and $i \in I$ such that $h = -f + i \in W$. Then $h + f = i \in I \cap W \subseteq F$. Since F is a face of W, $f \in F$ follows.

For a proof of (3) we note first that clearly $H \subseteq (W + I) \cap -(W + I)$. Conversely, let $h \in (W + I) \cap -(W + I)$. Then $h = w + i = -w' - i'$ with suitable elements $w, w' \in W$ and $i, i' \in I$. Now $w + w' = -i' - i \in W \cap I \subseteq F$. Since F is a face, $w \in F$, whence $h = w + i \in F + I \subseteq H$.

There is a vector subspace J in I such that $H = (F - F) \oplus J$ and $I = ((F - F) \cap I)) \oplus J$. The proof is simple: Since $H = (F - F) + I$, there a vector subspace $J \subseteq I$ such that $H = (F - F) \oplus J$. Then any element $i \in I$ is of the form $f - f' + j$ with suitable elements $f, f' \in F$, $j \in J$. Hence $f - f' = i - j \in I \cap (F - F)$. This shows that $I = ((F - F) \cap I) \oplus J$. In particular, $J \cap W \subseteq I \cap W \subseteq F$, whence $J \cap W = \{0\}$. Thus $J^{\perp} \cap \text{algint } W^* \neq \varnothing$. Hence we can pick a functional $\omega \in \widehat{L}$ in $(\text{algint } W^*) \cap J^{\perp}$.

At this point we assume that W is pointed and finish the proposition first in this case. Then $B = \omega^{-1}(1) \cap W$ is a compact base of W and $\omega(J) = \{0\}$. Now we select an element $i \in B \cap \text{algint}(W \cap I)$. Then every face containing i must contain $W \cap I$ and therefore F, thus F is the smallest face containing i. Hence $i \in \text{algint } F$, because otherwise by the Theorem of Hahn and Banach, i would be contained in a face of F contained in the boundary of F in $F - F$. Then we let $\mathbb{I} = [0, 1]$ and notice that 0 is, within $F - F$, an inner point of the set $U_F = F_1 - \mathbb{I} \cdot i$, where $F_1 = F \cap \omega^{-1}(\mathbb{I})$. Next we take an arbitrary compact convex neighborhood U_J of 0 in J and set $U_H = U_F \oplus U_J$. Then U_H is a neighborhood of 0 in H and $U_H = F_1 + (-\mathbb{I} \cdot i \oplus U_J)$.

Now we define $S_W = W \cap \omega^{-1}(\mathbb{I})$ and $U = U_H + S_W$. Then U is a compact convex set in $W + I$ such that with $S_I = -\mathbb{I} \cdot i \oplus U_J$ we have

(4) $$U = S_W + S_I,$$

and

(5) $$\omega(S_W) = \mathbb{I} \qquad \text{and} \qquad \omega(S_I) = -\mathbb{I}.$$

Now let $w \in W \cap U$. Then $\omega(w) \in \mathbb{I}$ and thus $w \in W \cap \omega^{-1}(\mathbb{I}) = S_W$. Trivially, $S_W \subseteq W \cap U$. Thus

(6) $$S_W = U \cap W.$$

For the case of pointed cones W, claims (i) and (ii) are now established. Next we prove (iii) in this case. For this purpose, we take $x \in W + I$. Then there are elements $w \in W$ and $i \in I$ such that $x = w + i$. We now pick a positive r in such a fashion that $r\omega(w) \in [0, 1/2]$ and $2r \cdot i \in U_H$, which is possible, since U_H is a neighborhood of 0 in H. Then, by the definition of U_H, there exist elements $f \in F_1$, $s \in \mathbb{I}$, and $j \in U_J$ such that $2i = f + (s \cdot i + j)$. Now $r.x = (r.w + \frac{1}{2} \cdot f) + \frac{1}{2} \cdot (s \cdot i + j) \in S_W + S_I$. This finishes the proof of (iii) for pointed cones W.

Now we turn to the case of a general wedge W. We find a vector space complement of $H(W)$ in L which is adjusted to I in such a fashion that $I = (I \cap H(W)) \oplus I'$ with $I' = I \cap V$. Then Proposition I.2.12 shows $W = H(W) \oplus W'$ with a pointed cone $W' = W \cap V$. The proposition now applies to the vector space V with the pointed cone W' and the vector subspace I' and yields compact convex subsets $S'_W \subseteq W'$ and $S'_I \subseteq I'$ such that $U' = S'_W + S'_I$ satisfies the conclusions of the proposition. Now we pick a compact convex neighborhood C of 0 in $H(W)$ and define $S_W = S'_W \oplus C$, $S_I = S'_I$ and, accordingly, $U = C \oplus U' = S_W + S_I$. It follows that $U \cap W' = U'$ and indeed $U \cap W = U \cap (H(W) \oplus W') = C \oplus (U \cap W') = C \oplus S'_W = S_W$. This proves (i). Conclusion (ii) is immediate from the definitions, since $\omega(C) = \{0\}$. We quickly check condition (iii) in the general case: Let $x \in W + I$, that is, $x = w + i$ with $w \in W$ and $i \in I$. We decompose w in the form $h + w'$ with unique elements $h \in H(W)$ and $w' \in W'$. By the result which we proved for V we find a positive real number r so that $r \cdot (w' + i) \in U'$; but since U' is convex we can makes $r > 0$ smaller, if necessary to achieve $r \cdot h \in C$, too. But then $r \cdot x = r \cdot h + r \cdot (w' + i) \in C \oplus U' = U$, and this proves (iii) in the general case.

Finally, (iv) follows as a consequence of the preceding Propositions I.2.31 and 32. ∎

Obviously, if we have one compact convex neighborhood U of 0 in $W + I$ with the properties described in the proposition, then we have arbitrarily small ones, since the $r \cdot U$, $r > 0$ form a basis for the neighborhoods of 0 in $W + I$ with the same properties.

It is instructive to visualize the situation of Proposition I.2.33 in $\mathbb{R}^3$ with a Lorentzian cone W and a tangent line I a half-line of which is contained in W. In this case $W + I$ is not closed, and the subset U we constructed in this proposition is not a neighborhood of 0 in $W + I$.

The canonical function from $C^1(W)$ to $\Pi(E^1(W^*))$

We now describe the relation between the C^1-points of a wedge and the E^1-points of its dual in terms of an actual function. The main result, the so-called Transgression Theorem will be very useful in our later developments.

I.2.34.   Definition. (i) Let W be a generating wedge in a topological vector space L (in duality with $\widehat{L}$). Let $\perp$ denote the binary relation $\{(x, \omega) : \langle \omega \mid x \rangle = 0\}$ on $L \times \widehat{L}$. We shall write R for the restriction of this "orthogonality" relation $\perp \mid ((W \backslash \{0\}) \times (W^* \backslash \{0\}))$. As is usual we shall write $R(x) = \{\omega \in W^* \backslash \{0\} : xR\omega\}$ and $R^{-1}(\omega) = \{x \in W \backslash \{0\} : xR\omega\}$.

(ii) The function $\sigma : C^1(W) \to \Pi\big(E^1(W^*)\big)$ given by $\sigma(x) = R(x)$ will be called *transgression function from C^1 to E^1*.

If B is any base of the pointed cone W^*, then the function $\sigma^B : C^1(W) \to B \cap E^1(W^*)$ given by $\{\sigma^B(x)\} = \sigma(x) \cap B$ or, equivalently, by $\sigma(x) = \mathbb{P} \cdot \sigma^B(x)$, $\sigma^B(x) \in B$ is called *the transgression function for the base B*.

If we interpret R as multivalued function from $W \backslash \{0\}$ to $W^* \backslash \{0\}$, then we can say that R maps $C^1(W)$ into $E^1(W)$, and, in the finite dimensional case, that $R^{-1}(\omega)$ meets $C^1(W)$ for each E^1-point ω. There is a very convenient way to convert the restriction of R to $C^1(W)$ into a function, namely, by considering the quotient space of a punctured vector space $L \backslash \{0\}$ modulo the relation which identifies two vectors if one results from the other through a positive scalar multiplication.

We make the following observation: *Let W be a generating wedge in L. If $R \subseteq C^1(W) \times E^1(W^*)$ is the binary relation of Definition I.2.34 above, then for each $x \in C^1(W)$, the set $R(x) = \operatorname{op} x \backslash \{0\}$ is an element of $\Pi(W^*)$.*

I.2.35.   Theorem. (The Transgression Theorem) *Let W be a generating wedge in a finite dimensional vector space L. Let B denote an arbitrary compact base of the pointed cone W^*. Then the following conclusions hold:*

(i) *The transgression function*

$$\sigma : C^1(W) \to \Pi\big(E^1(W^*)\big), \qquad \sigma(x) = x^\perp \cap W^* \backslash \{0\},$$

is continuous and surjective.

(ii) *The transgression function*

$$\sigma^B : C^1(W) \to B \cap E^1(W^*), \qquad \{\sigma^B(x)\} = B \cap x^\perp,$$

for the base B is surjective and continuous.

(iii) *In all points of $C^1(W) \cap E^1(W)$, the transgression functions σ and σ^B are open.*

Proof. (i) That σ is surjective is simply a reformulation of Corollary I.2.27. The space $\Pi(W^*)$ is compact by Proposition I.2.30. The binary relation R on $W \backslash \{0\} \times W^* \backslash \{0\}$ of Definition I.2.34 has a closed graph, and it induces on $W \backslash \{0\} \times \Pi(W^*)$ a binary relation $\overline{\sigma} = \big\{(x, R(x)) : x \in W \backslash \{0\}\big\}$ which still has a closed graph. The graph of the function σ is simply $\overline{\sigma} \cap \big(C^1(W) \times \Pi\big(E^1(W^*)\big)\big)$. The continuity of the transgression function is now a consequence of the Topological Lemma following Proposition I.2.31.

This completes the proof of part (i) of Theorem I.2.35. Part (ii) is equivalent to (i) by Proposition I.2.28. By this same proposition it suffices to prove part (iii) for the transgression function σ^B. Thus we must show that σ is open in all points x of the set $C^1(W) \cap E^1(W)$. For this purpose let x be simultaneously a C^1- and an E^1-point of W and U an open neighborhood of x in L. We have to show that $\sigma^B(U \cap C^1(W))$ is a neighborhood of $\sigma^B(x)$ in $B \cap E^1(W^*)$. We prove this claim by contradiction. Indeed, if the claim is false, then there is a sequence of elements ω_n, $n = 1, \dots$ in $B \cap E^1(W^*)$ converging to $\sigma^B(x)$ but satisfying $\omega_n \notin \sigma^B(U)$. Since σ^B is surjective, we find points $x_n \in C^1(W)$ such that, firstly, $\sigma^B(x_n) = \omega_n$ and, secondly, $\mathbb{R}^+.x_n \cap U = \emptyset$. We may equip the finite dimensional vector space L with a norm and assume that all x_n are in fact unit vectors. Then, after having passed to a converging subsequence in view of the compactness of the unit sphere, and after having renamed the sequence we now assume that the x_n converge to the unit vector x'. In particular, $x' \neq 0$. Since U and hence $\mathbb{P}.U$ are open and $x_n \notin \mathbb{P}.U$ we conclude also $x' \notin \mathbb{P}.U$. Furthermore, $\langle \sigma^B(x) \mid x' \rangle = \lim \langle \omega_n \mid x_n \rangle = 0$ as $\sigma^B(x_n) = \omega_n$ and thus $\omega_n \in x_n^\perp$. Thus $x' \in \sigma^B(x)^\perp \cap W = \operatorname{op} \sigma^B(x) = \operatorname{op} \mathbb{P}.\sigma^B(x) = \operatorname{op} \operatorname{op} x = E_x$. Now we use the hypothesis that x is an E^1-point, which implies that $E_x = \mathbb{R}.x$. Thus $x' \in \mathbb{P}.x \subseteq \mathbb{P}.U$, and this is a contradiction. ∎

Problem. Prove or disprove the Transgression Theorem for infinite dimensional wedges.

The full power of the Transgression Theorem will become more evident in the next section, where we shall discuss Mazur's Density Theorem which will show that under the circumstances of the Transgression Theorem, $C^1(W)$ is a dense subset of the boundary of W. While part (i) of the Transgression Theorem has a canonical form inasmuch as it does not depend of the choice of a base B, but our applications as a rule fall back on the non-canonical version in part (ii).

The following corollary should be clear:

I.2.36.   Corollary.   *If W is a generating wedge in a vector space L with duality, then the transgression function $\sigma: C^1(W) \to \Pi(E^1(W^*))$ induces a continuous surjective function*

$$\Sigma: \Pi(C^1(W)) \to \Pi(E^1(W^*)), \qquad \Sigma(\mathbb{P}.x) = x^\perp \cap W^* \setminus \{0\}.$$

In all points of $\Pi(C^1(W) \cap E^1(W))$ the function Σ is open. ∎

By a slight abuse of notation, we shall also call Σ the *transgression function*. No confusion will be possible.

I.2.37.   Corollary.   *If W is a generating wedge in a finite dimensional vector space L, then the function Σ induces a homeomorphism from the semiprojetive space $\Pi(C^1(W) \cap E^1(W))$ of all C^1-E^1-points of W onto the space $\Pi(C^1(W^*) \cap E^1(W^*))$ of all C^1-E^1-points of W^*.*

Proof.   If $x \in C^1(W) \cap E^1(W)$ then $\operatorname{op} x$ is a half-ray, as x is a C^1-point, and $\operatorname{algint} \operatorname{op} x = \operatorname{op} x \setminus \{0\}$ consists of C^1-points as x is an E^1-point (see Corollary I.2.27). Thus $\sigma(x) \in \Pi(C^1(W^*) \cap E^*(W^*))$ and thus Σ maps the semiprojective space of C^1-E^1-points of W into the semiprojective space of C^1-E^1-points of W^*, and thus induces a continuous and open function

$$\Pi(C^1(W) \cap E^1(W)) \to \Pi(C^1(W^*) \cap E^1(W^*)).$$

The transgression function from $\Pi\big(C^1(W^*)\big)$, however, induces the inverse of this function. This proves the corollary. $\blacksquare$

3. Mazur's Density Theorem

In our development of the duality theory of wedges and their facial structure, the sets of C^1- and of E^1-points and their dual role emerged as particularly important. However, we have yet to demonstrate the usefulness of these concepts by showing that each of this type of points determines the wedge—each in its own way. For this purpose we shall give a full account of a classical theorem due to Mazur which implies that *for each wedge with non-empty interior in a separable Banach space, the C^1-points form a dense G_δ in its boundary.* From this result we shall derive the fact that any wedge in a finite dimensional vector space is the closed convex hull of its E^1-points. This result again is classical and was established by Straszewicz. We will complement these results by additional information which is relevant for our later applications. In particular, we shall inspect some special classes of finite dimensional wedges and their geometric properties.

The Density Theorem

Mazur's theorem will be established through a sequence of lemmas. Of course, the Baire category theorem does play a central role, and the fact that convex functions are almost everywhere differentiable has, not unexpectedly, to be invoked.

From now on through the final proof of Mazur's Theorem, L shall be a topological vector space upon we shall impose the hypothesis of being a Banach space and of separability as needed. Furthermore, W will denote a closed convex subset with non-empty interior. Since we used the concepts of *support hyperplane* and of C^1-*points* only in the case of wedges, let us observe here, that both apply perfectly well to our set W: A *support hyperplane of W at $w \in W$* is any set of the form $\omega^{-1}(\omega(w))$ with a linear functional ω of L such that $\omega(x) \geq \omega(w)$ for all $x \in W$. Since the zero set of a discontinuous linear functional is dense but that of ω has a non-empty open set in its complement, such a functional is automatically continuous. A C^1-*point $w \in W$* is a point at which W has precisely *one* support hyperplane. Since the statement of the theorem that the set of C^1-points is a dense G_δ in the boundary ∂W is translation invariant we may and shall assume that

$$0 \in \operatorname{int} W.$$

We begin by recording the definition of the familiar Minkowski functional of W.

I.3.1.   **Definition.**   The *Minkowski functional of W* is defined by

$$p \colon L \to \mathbb{R}^+, \qquad p(x) = \inf\{r \in \mathbb{R}^+ : x \in r{\cdot}W\}.$$

I.3.2.   **Lemma.**   *The Minkowski functional satisfies*
   (i) $p(x) \geq 0$ *for all $x \in L$.*
   (ii) $p(x + y) \leq p(x) + p(y)$ *for all $x, y \in L$.*
   (iii) $p(r{\cdot}x) = rp(x)$ *for all $x \in L$ and all $r \in \mathbb{R}^+$.*

Proof.   The proof is standard functional analysis and is an exercise.   ∎

We shall say that a function $p \colon L \to \mathbb{R}$ is *sublinear* if it satisfies conditions (ii) and (iii) above. The function p would be a *seminorm* if, in addition to the conditions (i),(ii),and (iii) it satisfies
   (iii′) $p(-x) = p(x)$ *for all $x \in L$.*
The Minkowski functional is a seminorm if and only if W satisfies $W = -W$.

I.3.3.   **Lemma.**   *If L is a Banach space, then p satisfies a Lipschitz condition, that is*
   (iv) *There is a constant $c > 0$ such that $|p(u) - p(v)| \leq c\|u - v\|$ for all $u, v \in L$. In particular, p is continuous.*

Proof.   By Lemma I.3.2, $p(u) \leq p(u - v) + p(v)$, whence $|p(u) - p(v)| \leq \max\{p(u - v), p(v - u)\}$. Now W is a neighborhood of 0 by assumption, and thus there is a closed ε-ball around 0 contained in W. If we set $c = \varepsilon^{-1}$, then, since we are in a Banach space, the closed unit ball B is contained in $c{\cdot}W$. Hence for $x \neq 0$ in L we have $\frac{1}{\|x\|}{\cdot}x \in B \subseteq c{\cdot}W$, whence $p(\frac{1}{\|x\|}{\cdot}x) \leq c$ and thus $p(x) \leq c\|x\|$ for all non-zero and then for all $x \in L$. The claim is now an easy consequence.   ∎

I.3.4.   **Lemma.**   *For any two elements $x, y \in L$ the set*

$$(1) \qquad \{\frac{p(x + h{\cdot}y) - p(x)}{h} : h > 0\} = \{p(y + r{\cdot}x) - p(r{\cdot}x) : r > 0\}$$

has a greatest lower bound and this is equal to

$$(2) \qquad p_x(y) \overset{\text{def}}{=} \lim_{\substack{h \to 0 \\ h > 0}} \frac{p(x + h{\cdot}y) - p(x)}{h}.$$

Proof.   We note $\frac{1}{h}(p(x + h{\cdot}y) - p(x)) = p(y + \frac{1}{h}{\cdot}x) - p(\frac{1}{h}{\cdot}x)$ for $h > 0$, and this allows us to conclude the equality in (1). We claim that the function

$$t \mapsto p(x + t{\cdot}y) : \mathbb{R} \to \mathbb{R}$$

is convex. This claim follows readily from the subadditivity of p. This implies that the function

$$t \mapsto \frac{p(x + t{\cdot}y) - p(x)}{t} : \mathbb{R} \setminus \{0\} \to \mathbb{R}$$

is non-decreasing. In particular, the left hand set in (1) has a lower bound. The assertion follows. ∎

Obviously, $p_x(y)$ is the right-derivative of $t \mapsto p(x + t{\cdot}y)$, and thus the forward directional derivative of p at x in the direction y (although not normalized). The left derivative is $-p_x(-y)$, that is, the "backward" directional derivative of p at x in the direction y.

I.3.5. **Lemma.** (Properties of p_x) *For each $x \in L$, the function $p_x: L \to \mathbb{R}$ has the following properties:*

(i) $p_x \leq p$.

(ii) p_x *is sublinear.*

(iii) $0 \leq p_x(y) + p_x(-y)$ *for all $y \in L$.*

(iv) $p_{t{\cdot}x} = p_x$ *for all $t > 0$.*

(v) $p_x(t{\cdot}x) = tp(x)$ *for all $t \in \mathbb{R}$.*

(vi) *If L is a Banach space then p_x satisfies a Lipschitz condition with any Lipschitz constant c which works for p.*

Proof. (i) We note $p_x(y) \leq p(y + r{\cdot}x) - p(r{\cdot}x)$ for all $r > 0$ by Lemma I.3.4, and since p is continuous by Lemma I.3.3 the assertion follows for r tending to 0. (ii) For $u, v \in L$ we have $p_x(u + v) \leq p(u + v + r{\cdot}x) - p(r{\cdot}x)$ for all $r > 0$, and if we replace r by $r' + r''$ with positive r' and r'' we see $p(u + v) \leq p(u + v + (r' + r''){\cdot}x) - p((r' + r''){\cdot}x) = p((u + r'{\cdot}x) + (v + r''{\cdot}x)) - r'p(x) - r''p(x) \leq (p(u + r'{\cdot}x) - p(r'{\cdot}x)) + (p(v + r''{\cdot}x) - p(r''{\cdot}x))$ by sublinearity of p. Since $r' > 0$ and $r'' > 0$ were arbitrary, $p_x(u + v) \leq p_x(u) + p_x(v)$ follows. The proof of the positive homogeneity of p_x is simpler and is left as an exercise. (iii) Since $p_x(y)$ is the right derivative at 0 of the convex function $t \mapsto p(x + t{\cdot}y)$ and $-p_x(-y)$ is its left derivative, $-p_x(-y) \leq p_x(y)$ follows immediately. (iv) Let $t > 0$. Then $p_{t{\cdot}x}(y) = \inf\{p(y + r{\cdot}(t{\cdot}x)) - p(r{\cdot}(t{\cdot}x)) : r > 0\}$, and by making the parameter substitution $rt \mapsto s$, we recognize this infimum instantaneously as $p_x(y)$. (v) Now let t be an arbitrary real number. Then $p(x + h{\cdot}(t{\cdot}x)) = (1 + ht)p(x)$ for all h with $0 < h < 1/(1 + |t|)$ by Lemma I.3.2.iii. Thus

$$p_x(t{\cdot}x) = \lim_{\substack{h \to 0 \\ h > 0}} \frac{p(x + h{\cdot}(t{\cdot}x)) - p(x)}{h} = tp(x).$$

(vi) Since p_x is sublinear, the argument in the proof of Lemma I.3.3 shows $|p_x(u) - p_x(v)| \leq \max\{p_x(u - v), p_x(v - u)\}$. Now assume that L is a Banach space. Then by (i) above we have $p_x(u - v) \leq p(u - v) \leq c\|u - v\|$ with a Lipschitz constant c for p. The same upper bound applies to $p_x(v - u)$. We conclude $p_x(u) - p_x(v) \leq c\|u - v\|$. ∎

We are now ready to link the directional derivative p_x with the idea of a support hyperplane.

I.3.6.   Lemma.   *For any* $x \in \partial W$, *and for any linear functional* ω *on* L, *the following conditions are equivalent:*

(1) $\omega^{-1}(1)$ *is a support hyperplane of* W *at* x.

(2) $\omega \leq p$ *and* $\omega(x) = 1$.

(3) $\omega \leq p_x$.

Proof.   Since x is a point in the boundary of W we know $p(x) = 1$. If (1) is satisfied, then trivially, $\omega(x) = 1$, and $w \in W$ implies $\omega(w) \leq \omega(x) = 1$. Now let $y \in L$. If $p(y) = 0$, then $\mathbb{R}^+ \cdot y \subseteq W$ and thus $r\omega(y) = \omega(r \cdot y) \leq 1$, whence $\omega(y) \leq 0 = p(y)$. If, on the other side, $p(y) > 0$, then $p(y)^{-1} \cdot y \in W$, and thus $\frac{\omega(y)}{p(y)} = \omega(p(y)^{-1} \cdot y) \leq 1$, that is, $\omega(y) \leq p(y)$. Thus (2) is proved. Conversely, if (2) is satisfied, then $x \in \omega^{-1}(1)$, and if $w \in W$, then $\omega(w) \leq p(w)$; but by the definition of p we have $p(w) \leq 1$ as $w \in W$, and thus $\omega(W) \subseteq] - \infty, 1]$, which proves (1).

It remains to show that (2) and (3) are equivalent. Because of Lemma I.3.5.i, if (3) is satisfied, then $\omega \leq p_x \leq p$. Since $x \in W$ we have $\omega(x) \leq p(x) = 1$. But by Lemma I.3.5.v we know $p_x(-x) = -p(x) = -1$, whence $-\omega(x) = \omega(-x) \leq p_x(-x) = -1$. Thus $\omega(x) = 1$, and (2) is proved. Conversely, if (2) holds, then for all $h > 0$ we have $\omega(y) = \frac{1}{h}\big(\omega(x + h \cdot y) - \omega(x)\big) = \frac{1}{h}\big(\omega(x + h \cdot y) - p(x)\big) \leq \frac{1}{h}\big(p(x + h \cdot y) - p(x)\big)$. Passing to the limit as $h \searrow 0$ we find $\omega(y) \leq p_x(y)$ for arbitrary y, and this is (3)

Now we are closing in on a characterization of C^1 points in terms of directional derivatives.

I.3.7.   Lemma.   *Let* $x \in \partial W$, *and suppose that* D *is any dense subset of* L *whatsoever. Assume that* L *is a Banach space. Then the following statements are equivalent:*

(1) $x \in C^1(W)$.

(2) p_x *is linear.*

(3) $p_x(y) + p_x(-y) = 0$ *for all* $y \in L$.

(4) $p_x(y) + p_x(-y) = 0$ *for all* $y \in D$.

(5) $p_x(y) + p_x(-y) \leq 0$ *for all* $y \in D$.

Proof.   Let us quickly decide that (2) through (5) are equivalent: By Lemma I.3.5.ii, conditions (2) and (3) are equivalent. By Lemma I.3.5.vi, the function p_x is continuous, whence (3) and (4) are equivalent. By Lemma I.3.5.iii, conditions (4) and (5) are equivalent.

Suppose (2) holds. Let $\omega^{-1}(1)$ be a support hyperplane of W at x. Then by Lemma I.3.6, $\omega \leq p_x$, that is, $p_x - \omega \geq 0$, and if p_x is linear, this means that $p_x - \omega = 0$. Thus the support hyperplanes of W at x are unique, and this means $x \in C^1(W)$ by definition.

Finally, let us prove that not (2) implies not (1). Thus we assume that we find vectors $x, y \in L$ such that $-p_x(-y) < p_x(y)$. We must find at least two different support hyperplanes of W at x. Since $p_x(z)$ and $-p_x(-z)$ are the right-, respectively, left-derivatives at 0 of the convex function $t \mapsto p(x + t \cdot z)$, and since

$p(x) = 1$, we know that

$$(*) \qquad \begin{aligned} 1 + tp_x(+z) &\le p(x + t \cdot z) \quad \text{for all} \quad t \in \mathbb{R},\ z \in L, \\ 1 - tp_x(-z) &\le p(+t \cdot x) \quad \text{for all} \quad t \in \mathbb{R},\ z \in L. \end{aligned}$$

Now we choose $z = y \mp p_x(\pm y) \cdot x$. We observe $p_x(y) = p_x(z \pm p_x(\pm y) \cdot x) \le p_x(z) + p_x(\pm p_x(\pm y) \cdot x) = p_x(z) \pm p_x(\pm y)p(x) = p_x(z) \pm p_x(\pm y) \le p_x(z) + p_x(y)$ in view of Lemma I.3.5.iii and v. Similarly, we see that $p_x(-y) \le p_x(-z) + p_x(-y)$. Thus $p_x(z) \ge 0$. This together with $(*)$ above implies $1 \le p(x + t(y \mp p_x(\pm y) \cdot x)$, and from this we conclude

$$x + \mathbb{R} \cdot (y \mp p_x(\pm y) \cdot x) \cap \operatorname{int} W = \varnothing$$

since $w \in \operatorname{int} W$ implies $p(w) < 1$. Now the Theorem of Hahn and Banach yields two hyperplanes $T_\pm$ containing the two straight lines $g_\pm = x + \mathbb{R} \cdot (y \mp p_x(\pm y) \cdot x)$ but not meeting the interior $\operatorname{int} W$ of W. If we can ascertain that $T_+ \neq T_-$, we are finished, since these two hyperplanes are support hyperplanes of W at x. But since $p_x(-y) < p_x(y)$ we have $g_+ \neq g_-$. Suppose that $g_- \subseteq T_-$. Then we have $x + (y + p_x(-y) \cdot x) \in T_-$; but since $g_+ \subseteq T_-$, we also have $x - (y - p_x(y) \cdot x)$. Thus the linear subspace $T_- - x$ contains $(p_x(y) + p_x(-y)) \cdot x$. By our very assumption, $p_x(y) + p_x(-y) \neq 0$, and thus $x \in T_- - x$, that is $0 \in T$ follows. But since 0 is in the interior of W, which is disjoint from T_-, this is a contradiction, and our proof is finished. ∎

We remark, that the linearity of p_x means that the function p has directional derivatives at x in all directions.

Henceforth we fix a dense subset D of L; later this set will be assumed to be countable. If we set

$$G(y) = \{x \in L : p_x(y) + p_x(-y) \le 0\} \qquad \text{and} \qquad G = \bigcap_{y \in D} G(y),$$

then the preceding lemma says

$$(3) \qquad\qquad C^1(W) = \partial W \cap G.$$

From Lemma I.3.5.iv we know that all sets $G(y)$ and then also G are stable under multiplication by positive scalars. Now let $x \in \partial W$. Then $p(x) = 1$. If there is a sequence $x_n \in G$ with $x = \lim x_n$, then $\lim p(x_n) = 1$ by the continuity of p (see Lemma I.3.3); thus $\lim p(x_n)^{-1} \cdot x_n = x$ with $p(x_n)^{-1} \cdot x_n$ having p-value 1 by Lemma I.3.2.iii, hence being in ∂W. Thus we have

I.3.8.   Remark. $C^1(W)$ is a dense G_δ in ∂W if G is a dense G_δ in L. ∎

(We recall that a set in a topological space is called a G_δ if it is the intersection of a countable set of open sets.)

We now assume that D is countable and does not contain 0. Then $G = \bigcap_{y \in D} G(y)$ is a countable intersection. If each $G(y)$ is a G_δ, then G is a G_δ. If we can show that each $G(y)$ is a countable intersection of dense open sets, then G is a countable intersection of dense open sets and then by the *Baire category theorem*, G is dense in L, and it will also be a G_δ. Thus we have to show that for each $y \in D$

(4) $G(y)$ *is a countable intersection of dense open sets.*

I.3.9.   Lemma.   *For each $y \in D$, the following statements are equivalent:*
  (1) $p_x(y) + p_x(-y) \leq 0$.
  (2) *For each $m \in \mathbb{N}$ there is an $n \in \mathbb{N}$ such that $p(x + \frac{1}{n} \cdot y) + p(x - \frac{1}{n} \cdot y) < \frac{1}{mn}$.*

Proof.   We recall

$$p_x(y) + p_x(-y) = \lim_{h \searrow 0} \frac{1}{h} \Big(\big(p(x + h \cdot y) - p(x)\big) - \big(p(x - h \cdot y) - p(x)\big) \Big),$$

that is,

$$p_x(y) + p_x(-y) = \lim_{h \searrow 0} \frac{p(x + h \cdot y) - p(x - h \cdot y)}{h}.$$

Thus, condition (1) is tantamount to saying that for each $m \in \mathbb{N}$ there is a $\delta > 0$ such that $0 < h < \delta$ implies $\frac{p(x + h \cdot y) - p(x - h \cdot y)}{h} < \frac{1}{m}$. Since we certainly find an $n \in \mathbb{N}$ such that $\frac{1}{n} < \delta$, this certainly implies condition (2). On the other hand, the function $t \mapsto \frac{1}{t} \Big(\big(p(x + t \cdot y) - p(x)\big) + \big(p(x - t \cdot y) - p(x)\big) \Big) = \frac{1}{t} \big(p(x + t \cdot y) + p(x - t \cdot y)\big)$ is non-decreasing as we saw earlier. Thus (2) implies that $0 < h < \frac{1}{n}$ entails $\frac{p(x + h \cdot y) + p(x - h \cdot y)}{h} < n \big(p(x + \frac{1}{n} \cdot y) + p(x - \frac{1}{n} \cdot y)\big) < \frac{1}{m}$, and this is equivalent to condition (1) as we noted above.   ∎

Now we set $U_{m,n} = \{x \in L : p(x + \frac{1}{n} \cdot y) + p(x - \frac{1}{n} \cdot y < \frac{1}{mn}\}$ for $m, n \in \mathbb{N}$. Then $U_{m,n}$ is open since p is continuous. By the preceding Lemma I.3.9 we have

(5) $$G(y) = \bigcap_{m \in \mathbb{N}} \bigcup_{n \in \mathbb{N}} U_{m,n}.$$

In view of this condition (5) our desired conclusion (4) will be shown as soon as we can show that for each $m \in \mathbb{N}$,

(6) $$\bigcup_{n \in \mathbb{N}} U_{m,n} \quad \text{is dense in} \quad L.$$

We prove this claim by contradiction and thereby finish the proof of the density theorem. If condition (6) does not hold, then there is an $x \in L$ and a positive number r such that the open ball with radius r is disjoint from $U_{m,n}$ for all n. In other words, $\|y\| < r$ implies $x + y \notin U_{m,n}$ for all n. If we take $|t| < \frac{r}{\|y\|}$ (recalling that $0 \notin D$!), we note $x + t \cdot y \notin U_{m,n}$ for all n, that is, for each such t there is an $m \in N$ such that for all $n \in N$ we have $p(x + t \cdot y + \frac{1}{n} \cdot y) + p(x + t \cdot y - \frac{1}{n} \cdot y) \geq \frac{1}{mn}$. Consider the function

$$\kappa :] - \frac{r}{\|y\|}, \frac{r}{\|y\|} [\longrightarrow \mathbb{R}, \qquad \kappa(t) = p(x + t \cdot y).$$

If κ is differentiable at s, then for every $m \in \mathbb{N}$ there is an $n \in \mathbb{N}$ such that $\left(n\big(p(x+s\cdot y+\frac{1}{n}\cdot y)-p(x+s\cdot y)\big)-\kappa'(s)\right)-\left(-n\big(p(x+s\cdot y-\frac{1}{n}\cdot y)-p(x+s.y)\big)-\kappa'(s)\right) <$ $\frac{1}{m}$. We conclude now from what we saw before that κ' cannot be differentiable in any point of its domain. But as we already saw in the proof of Lemma I.3.4, the function κ is convex. But then it is differentiable with the possible exception of countably many points. (See for instance R.T. Rockafellar, Convex Analysis, Princeton University Press 1970, p.244, Theorem 25.3.) This contradiction proves the claim that $\bigcup_{n\in\mathbb{N}} U_{m,n}$ is dense. We have therefore proved the following theorem:

I.3.10. **Theorem.** (Mazur's Density Theorem) *Let W be a closed convex set with non-empty interior in a separable Banach space. Then $C^1(W)$ is a dense G_δ in the boundary ∂W.* ∎

The hypothesis of separability is necessary. If L is the space $L^\infty([0,1])$ with respect to Lebesgue measure, and if W is the unit ball in L, then $C^1(W) = \emptyset$.

We shall use the Density Theorem in the following form:

I.3.11. **Corollary.** (The Density Theorem for Wedges). *Let W be a wedge with non-empty interior in a separable Banach space. Then $C^1(W)$ is a dense G_δ in the boundary ∂W.* ∎

The Density Theorem for Wedges can be exploited in a variety of ways. The first observation we make is that it allows us to represent any weakly closed wedge in an economical way as an intersection of half-spaces.

I.3.12. **Proposition.** *Let W be a wedge with non-empty interior in a separable Banach space L. Let D be any dense subspace of $C^1(W)$ $\big($for instance, $D = C^1(W)\big)$. Then*

(i)
$$W = \bigcap_{x\in D} L_x,$$

and

(ii)
$$H(W) = \bigcap_{x\in D} T_x.$$

Proof. Since $T_x = H(L_x)$ by Proposition I.2.7.i and, since picking the edge commutes with forming intersections, (ii) is a consequence of (i); so we must prove (i).

We consider the natural duality between L and its topological dual. Clearly $W \subseteq \bigcap_{x\in D} L_x$, since each L_x contains W. For a proof of the reverse inclusion, let $y \in \bigcap_{x\in D} L_x$. This means that $y \in (\operatorname{op} x)^*$ for each $x \in D$, and this is tantamount to saying that

$$(\forall x \in D)(\forall \omega \in \operatorname{op} x)\langle \omega\, x\rangle \geq 0,$$

and since $\operatorname{op} x = x^\perp \cap W^*$ we can reformulate this as

$$(\forall x \in D, \omega \in W^*)\langle \omega, x\rangle = 0 \Rightarrow \langle \omega, y\rangle \geq 0.$$

Equivalently,

$$(\forall \omega \in W^*)\big((\forall x \in D)\langle \omega, x \rangle = 0\big) \Rightarrow \langle \omega, y \rangle \geq 0.$$

But we have assumed that D is dense in $C^1(W)$. From the Density Theorem I.3.11 we know that $C^1(W)$ is dense in the boundary ∂W. Thus the statements

$$(\forall x \in D)\langle \omega, x \rangle = 0$$

and

$$(\forall x \in \partial W)\langle \omega, x \rangle = 0$$

are equivalent. Therefore our original statement $y \in \bigcap_{x \in D} L_x$ is equivalent to

$$y \in \bigcap_{x \in \partial W} L_x.$$

We claim that this suffices to establish $y \in W$, as desired. If not, then let $w \in \operatorname{int} W$. The straight line segment connecting w and y contains a boundary point $x \in \partial W$ in its interior. Now we pick a non-zero $\omega \in x^\perp \cap W^* = \operatorname{op} x$, that is we select a support hyperplane $\omega^\perp$ of W at x. Then $\langle \omega, y \rangle < 0 < \langle \omega, w \rangle$ and thus $y \notin \omega^*$. But $\omega \in \operatorname{op} x$ implies $L_x = (\operatorname{op} x)^* \subseteq \omega^*$, whence $y \notin L_x$ which is the desired contradiction. ∎

At a later point we need a sharpening of the second part of the preceding proposition in which we allow the index set D over which the intersection is taken to be reduced even further, albeit in a controlled fashion.

In the proof of the following proposition, we shall use some information which we record in an exercise:

EI.3.1.   Exercise. Let W be a convex body in a topological vector space and w a point in the boundary ∂W. If T is a hyperplane containing w and $T \cap \partial W$ is a neighborhood of w in ∂W, then T is a support hyperplane of W. ∎

I.3.13.   Proposition. *Let W be a wedge with non-empty interior in a Banach space L. Let D be a dense subset of $C^1(W)$, and T a closed hyperplane of L. Then the following conclusions hold:*

(i) *If $T_x = T$ for all $x \in D \cap T$, then W is contained in one of the two half-spaces, say T^+, bounded by T, and*

(7)
$$W = T^+ \cap \bigcap_{x \in D \setminus T} L_x.$$

(ii) *If there is an $x \in D \cap T$ with $T_x \neq T$, then $D \setminus T$ is dense in $C^1(W)$, and*

(8)
$$W = \bigcap_{x \in D \setminus T} L_x.$$

(iii) *In any case,*

(9)
$$T \cap \bigcap_{x \in D \setminus T} T_x \subseteq H(W).$$

Proof.   (i) Set $D' = \{x \in D \setminus T : T_x = T\}$. Then $W = \bigcap_{x \in D} L_x$ by Proposition I.3.12(i), and thus $W \subseteq \bigcap_{x \in D'} L_x$. For $x \in D$, we have $T_x = T$, hence $L_x = T^+$ or $L_x = -T^+$. Since $\operatorname{int} W \neq \emptyset$, all L_x must be equal, say to T^+. Then Proposition I.3.12(i) implies (7).

(ii) We claim that $D \setminus T$ is still dense in $C^1(W)$; then Proposition I.3.12.i again will show that $W = \bigcap_{x \in D \setminus T} L_x$ and thus confirm the assertion in this case. In order to verify our claim, by the Density Theorem it suffices to show that $\partial W \cap T$ is nowhere dense in ∂W. Assume, by way of contradiction, that there is an open set U such that $\emptyset \neq T \cap U \subseteq \partial W$. Thus by Exercise EI.3.1, T is necessarily a support hyperplane of W. But we have an $x \in D \cap T$ such that $T \neq T_x$. Since $x \in C^1(W)$, the only support hyperplane of W at x is T_x, and thus T cannot be a support hyperplane of W at x. This is a contradiction which proves our claim.

(iii) is a consequence of (i) and (ii).   ∎

I.3.14.   Corollary.   *Let W be a wedge in a separable Banach space L. Suppose that V is a vector subspace not contained in $H(W)$, and suppose that T is a closed hyperplane containing V. Then there is an $x \in C^1(W)$ with $x \in T$ and $V \nsubseteq T_x$.*

Proof.   Suppose not. Then

$$V \subseteq T \cap \bigcap_{x \in C^1(W) \setminus T} T_x \subseteq H(W)$$

by Proposition I.3.13, contradiction.   ∎

The Theorem of Straszsewicz

We have observed in the previous section that there is a certain duality between the concepts of a C^1-point and that of an E^1-point. We expect therefore some sort of counterpart of the density theorem for C^1-points. Here it is:

I.3.15.   Proposition.   *Let W a wedge with non-empty interior in a separable Banach space L which is in duality with a topological vector space $\widehat{L}$. Suppose the following hypothesis:*

(E^1) $\qquad\qquad$ $\operatorname{algint}(\operatorname{op} \omega) \neq \emptyset$ $\qquad$ *for all* $\qquad$ $\omega \in E^1(W^*)$.

Then

$$W^* = \operatorname{cl}\Big(\sum_{\omega \in E^1(W^*)} \mathbb{R}^+ {\cdot} \omega \Big).$$

Proof.   We compute

$$\Big(\sum_{\omega \in E^1(W^*)} \mathbb{R}^+ {\cdot} \omega \Big)^* = \bigcap_{\omega \in E^1(W^*)} \omega^*$$

$$= \bigcap \{ L_x : x \in W, \operatorname{op} x = \mathbb{R}^+ {\cdot} \omega \quad \text{for some} \quad \omega \in E^1(W^*) \}$$

in view of the fact that $L_x = (\mathrm{op}\, x)^*$. But by Proposition I.2.25 we know that $x \in C^1(W)$ implies $\mathrm{op}\, x = \mathbb{R}^+ \cdot \omega$ for some $\omega \in E^1(W^*)$ and that $\omega \in E^1(W^*)$ implies $\mathrm{algint}(\mathrm{op}\, \omega) \subseteq C^1(W)$. Thus hypothesis (E^1) implies that the set of all $x \in W$ for which there is an $\omega \in E^1(W^*)$ such that $\mathrm{op}\, x = \mathbb{R}^+ \cdot \omega$ is exactly $C^1(W)$. As a consequence, $\left(\sum_{\omega \in E^1(W^*)} \mathbb{R}^+ \cdot \omega \right)^* = \bigcap_{x \in C^1(W)} L_x$. However, by Proposition I.3.12, this last expression equals W. Applying $(\bullet)^*$ to both sides of the resulting equation, we obtain the assertion of the proposition in view of the fact that the operations $(\bullet)^{**}$ and cl agree on convex sets. ∎

As a corollary we obtain the Theorem of Straszewicz:

I.3.16.   Theorem. (The Straszewicz Spanning Theorem) *If W is a pointed cone in a finite dimensional vector space, then*

$$W = \overline{\sum_{x \in E^1(W)} \mathbb{R}^+ \cdot x}.$$

Proof.   In a finite dimensional vector space the algebraic interior of a wedge is never empty. The theorem thus follows from the preceding proposition by duality. ∎

Consequences and Refinements

The Density Theorem yielded important consequences. One may consider the Density Theorem as a result on approximation, namely, the approximation of arbitrary boundary points in a wedge by C^1-points. We actually need a rather subtle refinement of this aspect of the density theorem in the case of finite dimensional wedges. This refinement and a first consequence will occupy us for the remainder of this section.

I.3.17.   Lemma. *Let $V \subseteq W$ be wedges in a topological vector space L (which is in duality with a topological vector space $\widehat{L}$). If V is a neighborhood of x in W, then $\mathrm{op}_V\, x = \mathrm{op}_W\, x$, that is, $x^\perp \cap V^* = x^\perp \cap W^*$.*

In particular, $\mathrm{op}\, x = x^\perp \cap W^*$ is an exposed face of V^*.

Proof.   Clearly, $x^\perp \cap W^* \subseteq x^\perp \cap V^*$. We have to prove the reverse containment. By duality, this means that we must show $(x^\perp \cap W^*)^* \subseteq (x^\perp \cap V^*)^*$. Now $(x^\perp \cap W^*)^* = \mathrm{cl}(W + \mathbb{R} \cdot x)$, so that we must show $\mathrm{cl}(W + \mathbb{R} \cdot x) \subseteq \mathrm{cl}(V + \mathbb{R} \cdot x)$, for which $W \subseteq V + \mathbb{R} \cdot x$ is certainly sufficient. Now let $w \in W$. Since V is a neighborhood of x in W, there is a real number $r > 0$ such that $x + r \cdot w \in V$. Then $w \in V - \frac{1}{r} \cdot x \subseteq V + \mathbb{R} \cdot x$. ∎

For the next lemma we recall that an element x of a wedge W is called an *extreme point of W* if $\mathbb{R}^+ \cdot x$ is a face of W. Thus every E^1-point is extreme, but the converse may be false, as simple examples show.

I.3.18.   Lemma.   *If V is a wedge in a topological vector space L in duality with $\widehat{L}$, then every extreme point of $\operatorname{op} x$ is an extreme point of V^*.*

Proof.   Let ε be an extreme point of $\operatorname{op} x$ and suppose that $\varepsilon = \alpha + \beta$ with $\alpha, \beta \in V^*$. Then $\alpha = \varepsilon - \beta \in (\operatorname{op} x - V^*) \cap V^* \subseteq \operatorname{cl}(\operatorname{op} x - V^*) \cap V^* = \operatorname{op} x$ by Corollary I.2.2 since $\operatorname{op} x$ is an exposed face by Proposition I.2.3. Similarly, $\beta \in \operatorname{op} x$. But since ε is an extreme point of $\operatorname{op} x$, it follows that $\alpha, \beta \in \mathbb{R}^+\cdot\varepsilon$, which we had to show.   ∎

In the next lemma we shall use a fact on compact convex sets in locally convex topological vector spaces. Here we recall that a point x of a convex set C in a vector space is called an *extreme point of C* if $\{x\}$ is a face of C.

Fact.   *If C is a compact convex set in a locally convex topological vector space and C is the closed convex hull of a subset S of C, then every extreme point of C is contained in S.*

I.3.19.   Lemma.   *If x is any point in a generating wedge V in a finite dimensional vector space L, and if ε is any extreme point of $\operatorname{op} x$, then $\varepsilon \in \overline{E^1(V^*)}$.*

Proof.   Since V is generating, V^* is pointed and thus has a compact base B by Proposition I.2.30. A point $x \in B$ is an extreme point of the compact convex set B if and only if it is an extreme point of the wedge V^*. By Straszsewicz's Spanning Theorem I.3.15, V^* is the closed convex hull of $E^1(V^*)$. Hence B is the closed convex hull of $B \cap E^1(V^*)$. Since ε is an extreme point of $\operatorname{op} x$, it is also an extreme point of V^* by Lemma I.3.18. Hence it is an extreme point of B. It then follows from the Fact quoted before this lemma that $\varepsilon \in \overline{E^1(V^*)}$

I.3.20.   Lemma.   *Suppose the following data are given in a finite dimensional vector space L:*

   (i)   *A generating wedge W and a non-zero point $x \in W$;*

   (ii)   *a pointed wedge V contained in W such that V is a neighborhood of x in W;*

   (iii)   *an affine hyperplane A containing x such that $A \cap V$ is a base for V;*

   (iv)   *an extreme point ε of $\operatorname{op} x$.*

   Then there exists a convergent sequence $(x_n, \omega_n) \in C^1(W) \times E^1(W^*)$ such that $\lim(x_n, \omega_n) = (y, \varepsilon)$ with some $y \in A \cap V$, and that $x_n \perp \omega_n$.

Proof.   Since $V \subseteq W$ is a neighborhood of $x \in W$ and W is generating, V is itself generating, and the wedges $W^* \subseteq V^*$ are pointed. In the first part of the proof we concentrate on V and its dual. Since V^* is pointed, we may pick a compact base B for V^* containing ε by Proposition I.2.30. The assertion of the lemma is such that our making V smaller, if necessary, will not invalidate the conclusion. Since in the affine space A, every point has a neighborhood basis of sets each of which is obtained as a finite intersection of closed half-spaces, it is no loss of generality to assume that V is the intersection of W and a finite number of closed half-spaces. Hence we assume now that

$$V = W \cap \mu_1^* \cap \ldots \mu_m^* \quad \text{with suitably chosen} \quad \mu_1, \ldots, \mu_m \in B \subseteq V^*.$$

By Lemma I.3.18, and by the choice of ε in (iv), ε is an extreme point of V^*. Thus by Lemma I.3.19, we know that $\varepsilon \in \overline{E^1(V^*)}$. Hence we find a sequence of elements $\omega_n \in \widehat{L}$ such that

$$\varepsilon = \lim \omega_n, \qquad \omega \in B \cap E^1(V^*).$$

But now the map $\sigma^B : C^1(V) \to B \cap E^1(V^*)$ of the Transgression Theorem I.2.35 is surjective; hence we find elements $x_n \in A \cap C^1(V)$ with $\sigma^B(x_n) = \omega_n$, that is, $\{\omega_n\} = B \cap x_n^\perp$. Since $V \cap A$ is compact, we may pick a convergent subsequence of the x_n and after renaming the sequences of the x_n and the ω_n assume that $(y, \varepsilon) = \lim(x_n, \omega_n)$.

We are not finished. We need sequences x_n and ω_n such that $x_n \in C^1(W)$ and $\omega_n \in E^1(W)$. For this purposes we shall show that infinitely many of the x_n miss the hyperplanes $\mu_1^\perp, \ldots, \mu_m^\perp$, whose union is the boundary of V in W. Then we shall find a subsequence of the x_n in the interior of V in W and then rename the sequences x_n and ω_n and assume that all x_n are in the interior of V in W. But then Lemma I.3.17 applies to all x_n and shows $x_n^\perp \cap V^* = x_n^\perp \cap W^*$ and thus $T_{x_n}(V) = (x_n^\perp \cap V^*)^\perp = (x_n^\perp \cap W^*)^\perp = T_{x_n}(W)$. Since $x_n \in C^1(V)$, we know that $T_{x_n}(V)$ is a hyperplane; hence $T_{x_n}(W)$ is a hyperplane, and that means that $x_n \in C^1(W)$ by Proposition I.2.17. Now $\omega_n \in x_n^\perp \cap V^* = x_n^\perp \cap W^* = \mathbb{R}^+ \cdot \omega_n$ by Proposition I.2.25, and so $\omega_n \in E^1(W^*)$. This will complete the proof of the lemma.

Thus it remains to show that the x_n are not eventually in any of the hyperplanes $\mu_k^\perp$, $k = 1, \ldots, m$. Suppose this claim were false. Then we would find at least one $k \in \{1, \ldots, m\}$ such that $\mu_k \in x_n^\perp \cap B$ for all but a finite number of the n. But $x_n^\perp \cap B = \{\omega_n\}$ by the original choice of the x_n. Thus we should have $\mu_k = \omega_n$ for all sufficiently large n. Then for these n we have $\langle \mu_k, x \rangle = \langle \omega_n, x \rangle = \langle \lim \omega_n, x \rangle = \langle \varepsilon, x \rangle = 0$. Yet this would mean that x is in the boundary of V in W whereas we hypothesized that V was a neighborhood of x in W. This contradiction completes the proof.   ∎

This somewhat delicate lemma now allows us to prove the approximation theorem we are after.

I.3.21.   Theorem. (The Approximation Theorem) *Let W be a generating wedge in a finite dimensional vector space, and let B be an arbitrary compact base of the dual W^* (which always exists). Then for every non-zero boundary point x of W and every extreme point ε of $\mathrm{op}\, x$ in B there is a sequence x_n of C^1-points of W such that $x = \lim x_n$ and $\varepsilon = \lim \sigma^B(x_n)$ with the transgression function σ^B for the base B.*

Proof.   We fix an affine hyperplane A through x with $0 \notin A$. Let U be an open neighborhood of x in the vector space L and U' an open neighborhood of ε in the dual $\widehat{L}$. Then we find ourselves a pointed wedge V which is a neighborhood of x in W such that $V \cap A \subseteq U$; this is readily seen to be possible. Now we apply Lemma I.3.20 and find a sequence $(x_n', \omega_n') \in C^1(W) \times E^1(W^*)$, $x_n' \in A \cap V \subseteq U$, $\lim \omega_n' = \varepsilon$ and $x_n \perp \omega_n$. Thus we pick an index n so that $\omega_n' \in U'$. Then $(x_n', \omega_n') \in U \times U'$. By Definition I.2.34(ii) of the transgression function we have $\omega_n' = \sigma^B(x_n')$. If U_n denotes a neighborhood basis for x in L and U_n' a neighborhood basis for ε in $\widehat{L}$, then this construction gives us an element $x_n \in U_n \cap C^1(W)$ such that $\sigma^B(x_n) \in U_n'$. This proves the theorem.   ∎

One can give this result an alternative, but equivalent formulation. Let us define the binary relation Ext from W to W^* by

$$\mathrm{Ext}_{\mathrm{B}} = \{(x, \varepsilon) \in W \times W^* : \varepsilon \quad \text{is an extreme point of} \quad B \cap \mathrm{op}\, x\}.$$

Let us denote with $G(\sigma^B)$ the graph $\{(x, \sigma^B(x)) : x \in C^1(W)\}$ considered as a subset of $W \times W^*$. Then we can formulate the approximation theorem as follows

I.3.22.   **Corollary.**   *Let W be a generating wedge in a finite dimensional vector space and B an arbitrary compact base of the dual wedge W^*. Then*

$$\mathrm{Ext}_{\mathrm{B}} \subseteq \overline{G(\sigma^B)}.$$

∎

The most important immediate application is a theorem on confining vector fields in a generating wedge. We shall use this result in an essential fashion in the Lie theory of wedges.

We recall that for a generating wedge W we have $L_x = L_x(W) = (\mathrm{op}\, x)^* = (x^\perp \cap W^*)^* = \overline{W + \mathbb{R} \cdot x}$.

I.3.23.   **Theorem.**   (The Confinement Theorem).   *Let W be a generating wedge in a finite dimensional vector space L and $f : W \to L$ a continuous function satisfying*

$$f(x) \in L_x \qquad \text{for all} \quad x \in C^1(W).$$

Then

$$f(x) \in L_x \qquad \text{for all} \quad x \in \partial W.$$

Proof.   Since $\mathrm{op}\, x$ is the closed convex hull of its extreme points, we have $L_x = \{\varepsilon : \varepsilon \text{ is an extreme point of } \mathrm{op}\, x\}^*$, and this shows

$$(*) \qquad y \in L_x \Longleftrightarrow \langle \varepsilon, y \rangle \geq 0 \quad \text{for all extreme points } \varepsilon \text{ of} \quad \mathrm{op}\, x.$$

If $x \varepsilon \partial W$, and if ε is an extreme point of $\mathrm{op}\, x$, then by the Approximation Theorem I.3.21 there is a sequence of points $x_n \in C^1(W)$ converging to x so that $\lim \sigma^B(x_n) = \varepsilon$ for a suitably chosen compact base B of W^* through ε. Now $f(x_n) \in L_{x_n} = (\mathrm{op}\, x_n)^*$ and $\sigma^B(x_n) \in \mathrm{op}\, x_n$ imply $\langle \sigma^B(x_n), f(x_n) \rangle \geq 0$. Thus $\langle \varepsilon, f(x) \rangle = \lim \langle \sigma^B(x_n), f(x_n) \rangle \geq 0$ by the continuity of f. Since ε was an arbitrary extreme point of $\mathrm{op}\, x$, condition $(*)$ now implies $f(x) \in L_x$. ∎

Figuratively speaking, a vector field cannot escape through the corners of the wedge; it is confined by its tangent *hyperplanes*.

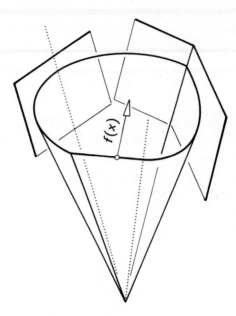

Figure 1

4. Special finite dimensional cones

This section is devoted to examples and illustrations of geometric properties of special classes of wedges in finite dimensional vector spaces, notably the classes of polyhedral wedges and of Lorentzian cones.

The previous sections provided some fairly powerful tools to work on the geometry of finite dimensional wedges. We saw indications in Proposition I.2.9, Corollary I.2.14, and Corollary I.2.19 that for many purposes it is sufficient to consider pointed cones. Although this will not be true for our later applications in all instances, as long as we wish to investigate special classes of wedges for their purely geometric properties, we may restrict our attention to pointed cones, and we will do so for the larger portion of this section in which we shall investigate what we mean by a cone being "round". Our discourse on "roundness" will become somewhat technical towards the end; this will be due to the fact that we prepare material for later applications which by its nature belongs into this chapter. However, the reader may decide to skip these portions of the section at the present time and return to them at a later time when they are needed.

Yet we begin with a particularly simple type of wedge for which the restriction to pointed specimens offers no advantage, namely wedges spanned by finitely many half-rays.

Polyhedral Wedges

In our first definition we repeat for the sake of completeness the definition of the concept of a half-space which we have used now and then in past sections.

I.4.1.   Definition.   (i) A wedge W in a topological vector space is called a *half space* if there is a non-zero continuous functional ω on L such that $W = \omega^* = \{x \in L : \langle \omega, x \rangle \geq 0\}$.

(ii) A wedge W is called *polyhedral* if it is the intersection of finitely many half-spaces.

Thus a wedge in a finite dimensional vector space is a half-space if and only if its dual is a half line $\mathbb{R}^+ \cdot \omega$, and it is polyhedral if there are non-zero elements

$\omega_1, \ldots, \omega_m \in \widehat{L}$ such that $W = \bigcap_{k=1}^m \omega_k^*$. Let us consider a number of equivalent characterizations for the property of being polyhedral.

I.4.2.   Proposition. *Let W be a generating wedge in a finite dimensional vector space. Then the following conditions are equivalent:*

(1) *W is polyhedral.*

(2) *There is a finite number of E^1- points $\varepsilon_1, \ldots, \varepsilon_p$ in W^* such that $W^* = \mathbb{R}^+ \cdot \varepsilon_1 + \cdots + \mathbb{R}^+ \cdot \varepsilon_p$.*

(3) *$E^1(W^*)$ is a finite union of half-lines.*

(4) *W has only finitely many tangent hyperplanes.*

(5) *The set of all half spaces $L_x(W)$, $x \in C^1(W)$ is finite.*

Proof.   (1)$\Rightarrow$(2): Since W is generating, W^* is pointed by Proposition I.1.7.i. Thus W^* has a compact base B by Proposition I.2.30. Since W is polyhedral, there are finitely many elements $\omega_1, \ldots, \omega_m \in B$ such that $W = \bigcap_{k=1}^m \omega_k^*$. By Proposition I.1.6.i, this means $W^* = \overline{\sum_{k=1}^m \mathbb{R}^+ \cdot \omega_k}$. In other words, B is the closed convex hull of the finite set $\{\omega_1, \ldots, \omega_m\}$. But then it is indeed the convex hull of this set, since the convex hull of any finite set is compact as the image under an affine map of a simplex (that is, the span of a finite set of linearly independent vectors in $\mathbb{R}^m$). Let $\{\varepsilon_1, \ldots, \varepsilon_p\}$ denote the subset of the ω_k which are extreme points of B. Indeed, after the **Fact** on compact convex sets which we used just before Lemma I.3.19 in the previous section, the set of extreme points must be a subset of the finite, hence closed set of the ω_k. Now B is the convex hull of $\{\varepsilon_1, \ldots, \varepsilon_p\}$. We shall finish the proof of (2) by remarking that each of the points ε_k is in fact an E^1-point. But we note that it easily possible to find a support hyperplane T of W^* in ε_k which will not contain any of the other finitely many ε_j. But then $\mathbb{R}^+ \cdot \varepsilon_k = W^* \cap T$, and by Proposition I.2.23 we know $W^* \cap T = E_{\varepsilon_k}$. The relation $\mathbb{R}^+ \cdot \varepsilon_k = E_{\varepsilon_k}$, however, says exactly that ε_k is an exposed point (see Definition I.2.1).

(2)$\Rightarrow$(3): By (2) we have $W^* = \mathbb{R}^+ \cdot \varepsilon_1 + \cdots + \mathbb{R}^+ \cdot \varepsilon_m$ with $\varepsilon_k \in E^1(W^*)$ for $k = 1, \ldots, m$. Thus $\mathbb{R}^+ \cdot \varepsilon_1 \cup \cdots \cup \mathbb{R}^+ \cdot \varepsilon_m \subseteq E^1(W^*)$. Conversely, let ω be an E^1-point of W^*. We may assume that $\omega \in B$ for a compact base B of W^* which we also may assume to contain all ε_k. Every E^1-point in B is, in particular, an extreme point; yet the extreme points of the finite polyhedron are exactly the exposed points ε_k. Thus $\omega = \varepsilon_j$ for some j and the reverse containment $E^1(W^*) \subseteq \mathbb{R}^+ \cdot \varepsilon_1 \cup \cdots \cup \mathbb{R}^+ \cdot \varepsilon_m$ is established.

(3)$\Rightarrow$(4): We recall that a tangent space T_x of W at a point x is a hyperplane if and only if $\operatorname{op} x \backslash \{0\} \subseteq E^1(W^*)$ by Proposition I.2.25. In other words, the tangent hyperplanes of W are exactly the sets $\omega^\perp$ with $0 \neq \omega \in E^1(W^*)$. The asserted implication follows.

(4)$\Rightarrow$(5): From Proposition I.3.12 we know that W is the intersection of the half-spaces L_x, $x \in C^1(W)$. By Proposition I.2.7.i, the tangent hyperplane T_x is the boundary $H(L_x)$ of the half space L_x. Thus by (4) there are only finitely many half spaces L_x, $x \in C^1(W)$.

(5)$\Rightarrow$(1): This is trivial.   ∎

I.4.3.   Proposition.   *A wedge in a finite dimensional vector space is polyhedral if and only if its lattice of exposed faces is finite.*

Proof.   We may assume that W is a generating wedge in a finite dimensional vector space. Accordingly, W^* is pointed and thus has a compact base. The wedges W and W^* have antiisomorphic lattices of exposed faces by Corollary I.2.4. If W is polyhedral, then B is a finite polyhedron, that is, the convex hull of a finite set by Proposition I.4.2. The lattice of faces of B is then finite; equivalently, the lattice of faces of W^* is finite. Every exposed face is a face, hence the lattices of exposed faces of both W^* and W are finite. Conversely, if the lattice of exposed faces of W is finite then so is that of W^* and then in particular the set of exposed half-lines is finite. The union of these half-lines is $E^1(W^*) \cup \{0\}$. The sum of these half-lines is closed, hence is W^* by the Spanning Theorem I.3.16. Hence Proposition I.4.2 shows that W is polyhedral.   ∎

I.4.4.   Corollary.   *A wedge in a finite dimensional vector space is polyhedral if and only if its dual is polyhedral.*

Proof.   Immediate from the preceding proposition in view of the fact that a wedge and its dual have antiisomorphic lattices of exposed faces after Corollary 2.4.   ∎

I.4.5.   Corollary.   *A wedge in a finite dimensional vector space is polyhedral if and only if its associated pointed cone is polyhedral.*

Proof.   Immediate from Proposition I.4.3 and Corollary I.2.11.   ∎

I.4.6.   Proposition.   *If W is a finite dimensional generating polyhedral wedge, and $\mathcal{T}$ the finite set of tangent hyperplanes of W, then $E^1(W)$ is a finite union of open half lines and*

$$C^1(W) = \bigcup_{T \in \mathcal{T}} \mathrm{algint}(T \cap W).$$

Proof.   Exercise.   ∎

The *hyperquadrants* $(\mathbb{R}^+)^n$ in $\mathbb{R}^n$ are the most commonly known polyhedral cones. They are, in a manner of speaking, the simplest proper generating cones to be found in finite dimensional vector spaces.

However, there are other important special types of generating cones in $\mathbb{R}^n$ whose geometric nature is, in a sense, opposite to that of the polyhedral wedges. They are the ones associated with quadratic forms.

Lorentzian Cones

We recall that a *quadratic form* q on a vector space L is a symmetric bilinear map $q : L \times L \to \mathbb{R}$. Instead of writing $q(x,x)$ when the arguments are equal we shall also write $q(x)$.

I.4.7.   Definition.   Let L be a finite dimensional vector space and q a quadratic form on L.

(i) We write $x \perp_q y$ if $q(x,y) = 0$. Accordingly, if M is any subset of L we set $M^{\perp_q} = \{x \in L : q(x,y) = 0 \quad \text{for all} \quad y \in M\}$. If $M, N \subseteq L$, then $M \perp_q N$ means $x \perp_q y$ for all $x \in M$ and $y \in N$.

(ii) q is said to be *non-degenerate* if $L^{\perp_q} = \{0\}$.

(iii) q is called *positive definite*, respectively, *negative definite* if $q(x) \geq 0$, respectively, $q(x) \leq 0$ for all $x \in L$ and $q(x) = 0$ implies $x = 0$. If q is positive definite we shall call the space (L,q) equipped with the quadratic form q a *Hilbert space* and normally write $q(x,y) = \langle x \mid y \rangle$. The form q is then called a *scalar product*.(Cf. Section I.1.1)

(iv) We shall say that (L,q) is an *orthogonal direct sum* of the vector subspaces L_1 and L_2 if $L_1 \perp_q L_2$ and $L = L_1 \oplus L_2$, algebraically. In this case we shall set $q_k = q|(L_k \times L_k)$, $k = 1,2$ and write $(L,q) = (L_1, q_1) \oplus (L_2, q_2)$.

(v) A quadratic form q will be called a *Lorentzian form* if (L,q) is the orthogonal direct sum $(L_1, q_1) \oplus (L_2, q_2)$ of a Hilbert space (L_1, q_1) and a 1-dimensional vector subspace L_2 with a negative definite form q_2. We will also say that (L,q) is a *Lorentzian* space.

Some remarks are in order. If q is non-degenerate, then L may be identified with its own dual via $x \mapsto q(x, \bullet) : L \to \mathrm{Hom}(L, \mathbb{R})$. Once this identification is made, the binary relation $\perp_q$ becomes identified with the annihilator relation $\perp$ which was used in the preceding sections in the context of duality. When no confusion is possible, we will therefore omit the index q. If $(L,q) = (L_1, q_1) \oplus (L_2, q_2)$ orthogonally, then for the elements $x = x_1 \oplus x_2$ and $y = y_1 \oplus y_2$, appropriately decomposed, we have $q(x,y) = q_1(x_1, y_1) + q_2(x_2, y_2)$, and q is non-degenerate if and only if both q_1 and q_2 are non-degenerate.

A quadratic form q is Lorentzian if we find a decomposition $L = H \oplus \mathbb{R} \cdot e$ and a scalar product on H such that $q(x' \oplus \xi \cdot e, y' \oplus \eta \cdot e) = \langle x' \mid y' \rangle - \xi \eta$. A 2-dimensional Lorentzian space is also referred to as a *hyperbolic plane*. It is clear that any Lorentzian space is the orthogonal direct sum of a Hilbert space and a hyperbolic plane. The decompositions of a Lorentzian space are, of course, not unique.

But let us assume that (L,q) is a Lorentzian space which we decompose orthogonally into a Hilbert space H and a one dimensional subspace $\mathbb{R} \cdot e$ such that $q(x' \oplus \xi \cdot e) = \langle x' \mid x' \rangle - \xi^2$. Then for $x = x' \oplus \xi \cdot e$ we have

$$q(x) \leq 0 \iff \langle x' \mid x' \rangle \leq \xi^2.$$

It follows, that the set $\{x \in L : q(x) \leq 0\}$ is the union of the two sets

$$L^+ \stackrel{\text{def}}{=} \{x \in L : q(x) \leq 0, \xi \geq 0\} \qquad \text{and} \qquad L^- \stackrel{\text{def}}{=} \{x \in L : q(x) \leq 0, \xi \leq 0\},$$

whose intersection is the set $\{0\}$, and that each of these sets is a convex "circular" cone, inasmuch as $L^+ = \mathbb{R}^+\cdot B$ with $B = \{x = x' \oplus e : \langle x' \mid x' \rangle \leq 1\}$ and $L^- = -L^+$. We shall therefore make the following definition:

I.4.8.   Definition. A wedge W in a finite dimensional vector space L is called a *Lorentzian cone* if it is generating and pointed and if there is a quadratic form q on $\mathbf{L}$ and a linear form $\omega \in W^*$ such that $W = \{x \in L : q(x) \leq 0, \omega(x) \geq 0\}$. We shall call the form q and the wedge W *associated*.

Notice that we did not postulate that the quadratic form be Lorentzian. This is in fact a consequence.

I.4.9.   Proposition. *For a wedge W in a finite dimensional vector space L, the following conditions are equivalent:*

(1) W *is a Lorentzian cone.*

(2) L *can be written as a direct sum $H \oplus \mathbb{R}\cdot e$ with a Hilbert space $(H, \langle \bullet \mid \bullet \rangle)$ such that $x = x' \oplus \xi\cdot e \in L$ is in W if and only if $\langle x' \mid x' \rangle \leq \xi^2$ and $0 \leq \xi$.*

Proof.   We observed before that (2) implies (1). In order to prove the reverse, we assume that W is Lorentzian. We may assume that ω is in the interior of W^* as W is pointed. Now we set $H = \omega^\perp$. Then $W \cap H = \{0\}$. Then $q(x) \geq 0$ for $x \in H$, and if $q(x) = 0$ for $x \in H$, then $x = 0$. Thus $\langle \bullet \mid \bullet \rangle = q|(H \times H)$ is a scalar product on H. As a consequence, $H^\perp \cap H = \{0\}$; but since H is a hyperplane of L, then the vector space $H^\perp$ is at least one dimensional, and since it meets H trivially, it must be exactly one dimensional. If $q(x) = 0$ for all $x \in H^\perp$, then $W = \{x \in L : q(x) \leq 0, \omega(x) \geq 0\}$ is necessarily a half line of $H^\perp$ in contradiction with the hypothesis that W is generating. Thus the restriction q' of q to $H^\perp \times H^\perp$ is non-degenerate, whence q is non-degenerate. If q' were positive definite, then q would be positive definite, since L is the orthogonal direct sum of H and $H^\perp$. But this would force the generating wedge W to be singleton by its very definition which is impossible. Thus q' is negative definite, since $H^\perp$ is one dimensional, and there is a vector e with $q(e) = -1$, $\omega(e) > 0$. Now everything falls into place: $q(x' \oplus \xi\cdot e) = \langle x' \mid x' \rangle - \xi^2$ and the remainder of statement (2) is now clear in view of the definition of W.   ∎

In dealing with Lorentzian forms, the following variation of the theme of the Cauchy–Schwarz inequality proves to be useful:

I.4.10.   Lemma. *Let (L, q) be a Lorentzian space and ω a linear form such that $q(u) \geq 0$ for all u in the hyperplane $\omega^\perp$. Then for elements $x, y \in L$, the relations*

$$x \neq 0, \qquad q(x) = 0, \qquad q(y) < 0, \qquad \omega(x)\omega(y) \geq 0$$

imply the relation

$$q(x, y) < 0.$$

Also, the relations

$$q(x) = 0, \qquad q(y) \leq 0 \qquad \omega(x)\omega(y) \geq 0$$

imply

$$q(x, y) \leq 0.$$

Proof. We write $L = H \oplus \mathbb{R} \cdot e$ with $H = \omega^\perp$ and $q(e) = -1$ so that $q(x' \oplus \xi \cdot e, y \oplus \eta \cdot e) = \langle x' \mid y' \rangle - \xi\eta$. Then $q(x) = 0$ means $\langle x' \mid x' \rangle = \xi^2$ and $q(y) < 0$ is equivalent to $\langle y' \mid y' \rangle < \eta^2$. The Cauchy–Schwarz inequality now shows $\langle x' \mid y' \rangle^2 \leq \langle x' \mid x' \rangle \langle y' \mid y' \rangle < \xi^2\eta^2$, since $\xi \neq 0$. Also $\xi\eta = \omega(x)\omega(y) \geq 0$. Thus, by extracting square roots, we conclude $\langle x' \mid y' \rangle < \xi\eta$, and this proves the first assertion of the lemma in view of $q(x, y) = \langle x' \mid y' \rangle - \xi\eta < 0$. The second is proved by a minute variation of the same proof. ∎

In the following proposition we identify L with its dual $\hat{L} = \mathrm{Hom}(L, \mathbb{R})$ via the assignment $x \mapsto q(x, \bullet)$. The dual pairing of L with itself is given by $\langle x, y \rangle = q(x, y)$.

I.4.11.   Proposition. *Let W be a Lorentzian cone, q an associated Lorentzian form, and ω a linear form as required in* Definition I.4.8. *Then the following conclusions hold:*

(i) $\mathrm{int}\, W = \{x \in L : q(x) < 0, \quad \omega(x) \geq 0\}$ *and* $\partial W = \{x \in L : q(x) = 0, \quad \omega(x) \geq 0\}$.

(ii) $W^* = -W$; *in particular, W and W^* are isomorphic.*

(iii) *If $x \in \partial W$, then*

    (a) $\mathrm{op}\, x = -\mathbb{R}^+ \cdot x$,

    (b) $L_x = \{y \in L : q(x, y) \leq 0\} = -x^*$,

    (c) $T_x = \{y \in L : q(x, y) = 0\} = x^\perp$,

    (d) $E_x = \mathbb{R}^+ \cdot x$.

(iv) $C^1(W) = E^1(W) = \partial W \setminus \{0\}$.

Proof. (i) is straightforward.

Before we proceed to prove (ii) we prove (iiib). From Remark I.1.10 we know that $L_x = \overline{W} + \mathbb{R} \cdot x$. Let $x \in \partial W \setminus \{0\}$, that is, assume $q(x) = 0$ and $\omega(x) > 0$. Now suppose $w \in W$. Then $q(w) \leq 0$ and $\omega(w) \geq 0$. So by Lemma I.4.10 we have $q(x, w) \leq 0$. For each $r \in \mathbb{R}$, this implies $q(x, w + r \cdot x) = q(x, w) + rq(x) \leq 0$. Hence $q(x, \overline{W} + \mathbb{R} \cdot x) \subseteq\ ]-\infty, 0]$. Conversely, suppose that $q(x, y) < 0$. Then $y = (y + r \cdot x) - r \cdot x$ for any $r \in \mathbb{R}$. Further, $q(y + r \cdot x) = q(y) + 2rq(x, y) + r^2 q(x) = q(y) + 2rq(x, y)$; this is majorized by 0 for all sufficiently large positive r since $q(x, y) < 0$. Finally, $\omega(y + r \cdot x) = \omega(y) + r\omega(x)$ will be greater than 0 for all sufficiently large r since $\omega(x) > 0$. Thus for large enough r we have $y + r \cdot x \in W$. Thus the open half space $\{y \in L : q(x, y) < 0\}$ is contained in $W + \mathbb{R} \cdot x$, and its closure is contained in $\overline{W} + \mathbb{R} \cdot x = L_x$. This completes the proof of (iiib).

Now we turn to the proof of (ii). From Proposition I.3.12 we know that $W = \bigcap_{x \in \partial W} L_x$; in other words, $y \in W$ if and only if $q(x, y) \leq 0$ for all $x \in \partial W$ if and only if $q(x, y) \leq 0$ for all $x \in W$ —as $W = (\partial W)^{**}$. But in view of our definitions and conventions, $y \in W^*$ if and only if $q(x, y) = \langle y, x \rangle \geq 0$ for all $x \in W$. It follows that $y \in W^*$ if and only if $-y \in W$. This proves (ii).

Next we prove (iii). First (c) and (d): Since T_x is the boundary hyperplane of the half space L_x by Proposition I.2.7.i, condition (b) implies (c). Then by the same result we have

$$E_x = T_x \cap W = \{y \in W : q(x, y) = 0, q(y) \leq 0, \quad \text{and} \quad \omega(y) \geq 0\}.$$

By Lemma I.4.10, $q(x) = q(y) = q(x, y) = 0$ implies $q(s \cdot x + t \cdot y, v \cdot x + v \cdot y) = 0$ for all $s, t, u, v \in \mathbb{R}$. But by the very definition of a Lorentzian form in I.4.7.v, it cannot induce the zero form on any 2-dimensional vector subspace. Hence y must be linearly dependent of x. Because of $\omega(y) \geq 0$ for $y \in E_x$, we conclude $E_x = \mathbb{R}^+ \cdot x$. This completes the proof of (d).

In order to finish the proof of (iii), we have to show (a): From Proposition I.2.7.ii we know $\operatorname{op} x = \operatorname{op} T_x = T_x^\perp \cap W^*$. But $T_x^\perp = x^{\perp\perp} = \mathbb{R} \cdot x$ and $W^* = -W$ by (i) above. Thus $\operatorname{op} x = -\mathbb{R}^+ \cdot x$ follows.

(iv) By (iiic), the tangent space T_x is a hyperplane for each $x \in \partial W \setminus \{0\}$; hence each of these x is a C^1-point by I.2.25. By (iiid), the exposed face E_x generated by x is $\mathbb{R}^+ \cdot x$ for each $x \in W \setminus \{0\}$. Hence each of these points is an E^1-point by Definition I.2.1. The proof of the proposition is now complete. ∎

It appears that our analysis of the polyhedral wedges as well as that of the Lorentzian cones is protracted even though we seem to be dealing with objects very close to our geometric intuition in 3-space. Yet we have to face up to the situation: In the preceding sections we have built up a considerable conceptual apparatus to deal with cones, notably those in finite dimensional vector spaces. This apparatus has to stand the test whether or not it applies to classical circumstances such as conics defined by quadratic forms. This requires verification, and the necessary arguments cannot simply be dismissed.

When we consider Lorentzian cones we perceive a geometrical and an algebraic aspect. The geometrical one manifests itself through the Lorentzian cones, the algebraic one through the Lorentzian forms. Each Lorentzian form determines two Lorentzian cones W and $-W$. Conversely, a Lorentzian cone determines its associated Lorentzian form uniquely only up to a non-zero scalar multiple. This is not entirely obvious. It is a consequence of the following lemma. We shall later have to refer to this lemma again. We shall prove a little more than we need. We call a quadratic form q *semidefinite*, if either $q(x) \geq 0$ for all $x \in L$ or $q(x) \leq 0$ for all $x \in L$.

I.4.12. Lemma. *Let q_1 and q_2 be two quadratic forms on a finite dimensional real vector space L and assume that they are not semidefinite and have the same zero-set $Z = \{x \in L : q_1(x) = 0\} = \{x \in L : q_2(x) = 0\}$. Then $q_1 = sq_2$ for some non-zero real number s.*

Proof. For $k = 1, 2$, the set $N_k = \{x \in Z : q_k(z, x) = 0 \text{ for all } z \in Z\}$ is a vector space and is geometrically characterized as the set of all $x \in Z$ such that for any $z \in Z$, the line segment joining x and z is also in Z. Thus we may conclude $N_1 = N_2$. Now both q_1 and q_2 induce on the quotient vector space

L/N_1 non-degenerate quadratic forms with the same zero-set. If these differ only by a non-zero scalar, then the same holds for q_1 and q_2. It is therefore no loss of generality to assume that q_1 and q_2 are non-degenerate. In particular, there is then an automorphism f of the vector space L such that $q_1(u, v) = q_2(f(u), v) = q_2(u, f(v))$. In order to prove the lemma, we have to show that f is a scalar multiple of the identity. We fix a non-zero element $x \in Z$. The tangent hyperplane of Z in x is $\{y \in L : q_1(x, y) = 0\}$ as well as $\{y \in L : q_2(x, y) = 0\}$. The linear forms $y \mapsto q_1(x, y)$ and $y \mapsto q_2(x, y) = q_1(f(x), y)$ therefore differ by a non-zero scalar $s(x)$, that is, we have a function $s : Z \setminus \{0\} \to \mathbb{R} \setminus \{0\}$ with $q_1(x, y) = s(x)q_2(x, y)$ for all $y \in L$. This means $f(x) = s(x) \cdot x$ for all $x \in Z$. If we fix an element y with $q_2(x, y) \neq 0$, then we have $s(z) = q_1(z, y)q_2(z, y)^{-1}$ for all $z \in Z$ in a neighborhood of x. Thus s is a continuous function from $Z \setminus \{0\}$ into the spectrum of f, which is discrete. Hence s is constant on the connected components of Z. If $\dim L = 1$, then all quadratic forms are semidefinite. If $\dim L = 2$, then we may assume that $q_1(x, y) = x_1 y_1 - x_2 y_2$ with respect to a suitable basis and that $q_2(x, y) = a x_1 y_1 + b(x_1 y_2 + x_2 y_1) + c x_2 y_2$, so that $x = (x_1, x_2)$ is in Z if and only if $x_1 = x_2$ or $x_1 = -x_2$. The substitution of $x = y = (1, 1)$ and of $x = y = (1, -1)$ into q_2 yields the equations $a + 2b + c = 0$ and $a - 2b + c = 0$, and these give $a = c$ and $b = 0$. The lemma follows directly in this case. We may therefore now assume that $\dim L \geq 3$. In this case each component Y of Z spans L, so that the relation $f(y) = t \cdot y$ for all $y \in Y$ with the constant value t of the function s on Y implies $f(u) = t \cdot u$ for all $u \in L$, and this is what we had to show.   ■

Round cones

Polyhedral wedges are nowhere round; Lorentzian wedges are about as round as a cone can be. These intuitive ideas need to be made precise. We discuss in the remainder of this section what local roundness should mean.

We have seen in Proposition I.4.11 that on a Lorentzian cone the C^1-points and the E^1-points agree. In particular, this implies that the E^1 points are dense in the boundary of a Lorentzian wedge. The dual of a Lorentzian wedge, we recall, is likewise Lorentzian. We shall take these remarks as a departure point for definitions of "roundness", first of a local one, then a global one.

We recall from Proposition I.2.8, that any wedge with exposed points is necessarily pointed, and that the dual of a pointed wedge is generating after Proposition I.1.7.

I.4.13.   **Definition.** Let W be a generating wedge in a finite dimensional vector space L. We say that

(i) W is *locally round at the boundary point* $x \in \partial W$ if

(a) $x \in C^1(W) \cap E^1(W)$, and

(b) $\overline{E^1(W^*)}$ is a neighborhood of any non-zero point of the half line $\mathrm{op}\, x$ in ∂W^*,

and we say that

(ii) W is *round* if the set $\{x \in \partial W : W$ is locally round at $x\}$ is dense in $C^1(W)$.

We notice that, by the Density Theorem I.3.11, W is round only if $E^1(W)$ is dense in ∂W. Observe that a round wedge is always a pointed generating cone.

We observe immediately that any Lorentzian cone is round, as it should be. Because of the somewhat technical nature of these definitions it is of interest to have stronger sufficient conditions which may be more easily handled.

I.4.14.   Proposition. *Suppose that W is a pointed generating cone such that $E^1(W)$ is a subset of $C^1(W)$. Then $C^1(W^*)$ is a subset of $E^1(W^*)$ and W is round.*

Proof.   Let B be a compact base for W, and B_* a compact base for W^*. Let $s : B \cap C^1(W) \to B_* \cap E^1(W^*)$ and $s_* : B_* \cap C^1(W^*) \to B \cap E^1(W)$ be the restrictions and corestrictions of the transgression maps of the Transgression Theorem I.2.35, respectively. By hypothesis, $E^1(W) \subseteq C^1(W)$. For each $\omega \in B_* \cap C^1(W^*)$ we have $s\big(s_*(\omega)\big) = \sigma^{B_*}\big(\sigma^B(\omega)\big) = \omega$ by the very definition of the transgression functions (see Theorem I.2.35). Hence $C^1(W^*) \cap B_* = s\big(s_*(C^1(W^*))\big) \subseteq \operatorname{im} s = E^1(W^*)$. This proves the first assertion. Since $C^1(W^*)$ is dense in ∂W^* by the Density Theorem, we conclude $\overline{E^1(W^*)} = \partial W^*$. Thus condition (ib) of Definition I.4.13 is satisfied for any x in the dense subset $E^1(W)$ of $C^1(W)$.   ∎

This shows, in particular, that every Lorentzian cone is indeed round in view of Proposition I.4.11.ii.d.

It is perhaps useful to record the following remark which is readily verified from the definition and the Density Theorem:

I.4.15.   Remark. Suppose that W is a pointed generating cone such that $C^1(W) \cap E^1(W)$ is a dense subset of $C^1(W)$ and $C^1(W^*) \cap E^1(W^*)$ is a dense subset of $C^1(W^*)$. Then W is round.   ∎

There are a few geometric ideas which relate to the idea of roundness; in our later applications in Chapter II, Section 5, we need to know these aspects of roundness.

I.4.16.   Definition. Let L be a finite dimensional vector space and W a wedge in L. For each point $x \in C^1(W)$ and each open neighborhood U of x in L we set

$$D(U) = \bigcap_{z \in C^1(W) \cap U} T_z.$$

Further, let $\mathcal{U}$ denote the filter basis of open neighborhoods of x in L. Then we set

$$D(x) = \bigcup_{U \in \mathcal{U}} D(U).$$

In a dual vein we define

$$E(U) = \sum_{z \in E^1(W) \cap U} \mathbb{R} \cdot z,$$

and

$$E(x) = \bigcap_{U \in \mathcal{U}} E(U).$$

I.4.17.   Lemma.   *Under the circumstances of* Definition I.4.16 *the following statements hold:*

(i) $\{D(U) : U \in \mathcal{U}\}$ *is an ascending and* $\{E(U) : U \in \mathcal{U}\}$ *a descending family of vector subspaces of* L. *For all* $U \in \mathcal{U}$ *we have* $D(U) \subseteq T_x$ *and* $x \in E(U)$.

(ii) *There is a neighborhood* $U_0 \in \mathcal{U}$ *such that for all* $U \in \mathcal{U}$, $U \subseteq U_0$, *the equalities* $D(U) = D(x)$ *and* $E(U) = E(x)$ *hold.*

(iii) $H(W) \subseteq D(x)$ *for all* $x \in C^1(W)$.

Proof.   (i) If $U \subseteq V$ in $\mathcal{U}$, then $D(U) \supseteq D(V)$ is clear from the definition. Also, $D(U) \subseteq T_x$ is obvious. Statement (ii) simply follows from the fact that all chains in the lattice of vector subspaces in a finite dimensional vector space have finite length, hence trivially satisfy the ascending chain condition. The assertions on E are proved similarly.

(iii) By Propositions I.2.8 or I.2.18, we have $H(W) \subseteq T_x$ for all $x \in W$. Hence (iii) follows.   ∎

The vector spaces $D(x)$ and $E(x)$ are new geometric invariants attached to a wedge and a point $x \in W$. They are related to each other via duality.

I.4.18.   Proposition.   *Let* $x \in C^1(W)$ *with a generating wedge* W *in* L, *and let* $\omega \in x^\perp \cap W^*$ *be such that* $E_\omega = x^\perp \cap W^*$. *Then*

$$D(x)^\perp \subseteq E(\omega)$$

and equality holds if the transgression function σ *of the* Transgression Theorem I.2.35 *is open at* x.

Proof.   There is an open neighborhood V_0 of ω in $\widehat{L}$ so that for all open neighborhoods V of ω in V_0 the vector space $E(\omega)$ is the span of all $\mathbb{R} \cdot \mu$ with $\mu \in E^1(W^*) \cap V$. Likewise, there is an open neighborhood U_0 of x in L so that for all neighborhoods U of x in U_0 we have $D(x) = \bigcap_{u \in C^1(W) \cap U} T_u$. Now let B be any compact base of W^* containing ω and let U be a neighborhood of x in U_0 which is so small that $U \subseteq U_0$ and that $\sigma^B(U) \subseteq V_0$. This is possible because of the continuity of σ^B. Now

$$D(x)^\perp = \Big(\bigcap_{u \in C^1(W) \cap U} T_u \Big)^\perp = \sum_{u \in C^1(W) \cap U} T_u^\perp.$$

(Note that the sum of any family of vector subspaces in a finite dimensional vector space is a vector subspace and thus is automatically closed.) But now for each $u \in C^1(W) \cap U$ we find a $\mu \in E^1(W^*) \cap V_0$ via $\mu = \sigma^B(u)$ such that $T_u^\perp = \mathbb{R} \cdot \mu$. Thus

$$D(x)^\perp \subseteq \sum_{\mu \in E^1(W^*) \cap V_0} \mathbb{R} \cdot \mu \;=\; E(\omega)$$

as asserted.

If σ is open at x, then the set of all $\mu \in E^1(W) \cap V_0$ for which there is a $u \in C^1(W) \cap U$ with $\langle \mu, u \rangle = 0$ is a neighborhood V of ω in V_0, and then $D(x)^\perp = \sum\{\mathbb{R} \cdot \mu : \text{ there is a } u \in C^1(W) \cap U \text{ with } \langle \mu, u \rangle = 0\} = E(\omega)$.   ∎

I.4.19. **Definition.** A wedge W is called *weakly round* if $D(x) = \{0\}$ for a dense set of C^1-points x in $C^1(W)$.

The following example illustrates these concepts in $L = \mathbb{R}^4$. It is a deceptive practice to form one's intuition on matters of convex cones on 3-dimensional examples. The interesting things happen from dimension 4 on upwards. We can represent a 4-dimensional pointed cone completely by depicting one of its 3-dimensional bases (see Proposition I.2.30). The following figure represents a compact base B of a 4-dimensional pointed cone W and a base B_* of its dual W^*. For the C^1-point x indicated in the picture, $D(x)$ is a 1-dimensional subspace. The dual picture shows a $\omega \in W^*$ with $T_x = \omega^\perp$ at which $\dim E(\omega) = 3$.

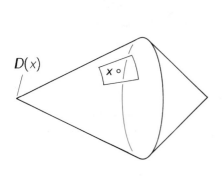
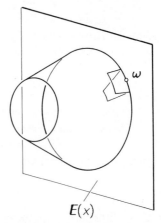

I.4.20. **Proposition.** *Let W be a generating wedge in a vector space L and let x be a boundary point of W. If W is locally round at x, then $D(x) = \{0\}$ and $E(\omega) = \widehat{L}$ for any $\omega \in E^1(W^*)$ with $\langle \omega, x \rangle = 0$.*

Proof. There is a neighborhood V of ω in $\widehat{L}$ such that $E(\omega) = E(V) = \widehat{L}$. If not, then

$$\sum_{\nu \in E^1(W^*) \cap V} \mathbb{R} \cdot \nu \subseteq H$$

for some hyperplane H of $\widehat{L}$ containing ω. But then $\overline{E^1(W^*)} \cap H$ is a neighborhood of ω in ∂W^*. By Exercise EI.3.1 preceding Proposition I.3.13, H is a support hyperplane of W^* at ω. Then $E_\omega = \mathbb{R}^+ \cdot \omega = H \cap W^*$ by Proposition I.2.23. This is a contradiction. By hypothesis x is a C^1-point as well as an E^1-point, and thus σ is open at x by the Transgression Theorem I.2.35. The assertion $D(x) = \{0\}$ then follows from Proposition I.4.18. ∎

In particular, each round wedge is weakly round. Notably, we have the following chain of implications: Lorentzian $\Longrightarrow$ round $\Longrightarrow$ weakly round.

More on quadratic forms and wedges

Later developments will depend on a closer inspection of the geometric situation of a vector space (L, q) equipped with a non-degenerate quadratic form q

and a wedge W such that the tangent hyperplanes T_x for all C^1-points x in some open set of the boundary ∂W are also tangent hyperplanes of the zero set $Z_q = \{y \in L : q(y) = 0\}$. The reader may wish to defer the reading of this material to the point where it will be needed in Chapter II, Section 6.

If q is a non-degenerate quadratic form on a vector space L, then L decomposes into an orthogonal direct sum of two Hilbert spaces $L_1 \oplus L_2$ such that $q(x_1 \oplus x_2, y_1 \oplus y_2) = \langle x_1 \mid y_1 \rangle - \langle x_2 \mid y_2 \rangle$. We shall fix a non-zero $y \in Z_q$; for our discussion it will be no restriction of generality to realize the relation $q(y) = 0$ by assuming that $\langle y_1 \mid y_1 \rangle = \langle y_2 \mid y_2 \rangle = 1$. We define $y' = y_1 - y_2$. Then $q(y') = \langle y_1 \mid y_1 \rangle - \langle -y_2 \mid -y_2 \rangle = 0$ and $q(y, y') = \langle y_1 \mid y_1 \rangle - \langle y_2 \mid -y_2 \rangle = 2 \neq 0$. Thus $\mathbb{R}{\cdot}y + \mathbb{R}{\cdot}y' = \mathbb{R}{\cdot}y_1 + \mathbb{R}{\cdot}y_2$ is a hyperbolic plane. Its orthogonal complement we write as the orthogonal direct sum $L'_1 \oplus L'_2$ where L'_j is the orthogonal complement of y_j in the Hilbert space L_j for $j = 1, 2$. After these specifications, every element $x \in L$ can be written uniquely as

$$(1) \quad x = (1 + k){\cdot}y + f{\cdot}y' \oplus h_1 \oplus h_2 \quad \text{with} \quad k, f \in \mathbb{R} \quad \text{and} \quad h_1 \in L'_1, \ h_2 \in L'_2,$$

the sum being orthogonal where indicated by "$\oplus$".

We shall now describe the points x in Z_q in a vicinity of y using the variables k, h_1, and h_2 as coordinates; these will uniquely determine f as a function of these coordinates.

I.4.21.   Lemma. *For $y \in Z_q$ as above, the following statements are equivalent:*

(A) $x \in Z_q$.

(B) *There are elements $h_j \in L'_j$, $j = 1, 2$ and real numbers k, $f = f(k, h_1, h_2)$ such that either*

(a) $\langle h_1 \mid h_1 \rangle = \langle h_2 \mid h_2 \rangle$ *and* $1 = -k$, *or*

(b) $x = (1 + k){\cdot}y + f{\cdot}y' \oplus h_1 \oplus h_2$ *with* $f(k, h_1, h_2) = -\frac{1}{4} \frac{\langle h_1 | h_1 \rangle - \langle h_2 | h_2 \rangle}{1 + k}$.

Proof.   First we write $h = h_1 \oplus h_2$ and recall $q(y) = q(y') = q(y, h) = q(y', h) = 0$, $q(y, y') = 2$. Then (A) means $q(x) = 0$, and this is equivalent to $0 = q(s{\cdot}y + s'{\cdot}y' \oplus h) = 4ss' + q(h)$, that is, to $ss' = -\frac{1}{4}q(h) = -\frac{1}{4}(\langle h_1 \mid h_1 \rangle - \langle h_2 \mid h_2 \rangle)$. If $s = 0$, then this means $q(h) = 0$ and we find ourselves in case (a). If $s \neq 0$ we set $k = s - 1$ and $f = s'$; and then $q(x) = 0$ is equivalent to

$$f = \frac{1}{4} \frac{\langle h_2 \mid h_2 \rangle - \langle h_1 \mid h_1 \rangle}{1 + k}.$$

This is case (b).   ∎

Our discussion will now proceed as follows: We shall assume

$$(2) \qquad\qquad\qquad \dim L_1 \geq 2 \quad \text{and} \quad \dim L_2 \geq 2.$$

This condition is equivalent to

$$(3) \qquad\qquad\qquad \dim L'_1 \geq 1 \quad \text{and} \quad \dim L'_2 \geq 1.$$

After further analysis of the structure of the quadric Z_q in the vicinity of y, we shall show that certain convexity assumptions stemming from the presence of wedges will lead to condition (3) and hence (2) to a contradiction; therefore these additional hypotheses will force the form q to be either Lorentzian or definite; the latter case will be ruled out if there is a vector of zero q-length.

Set

$$U = \{x \in L : x = (1 + k) \cdot y + f \cdot y' \oplus h_1 \oplus h_2 \quad \text{such that} \quad k > -1\}.$$

Then U is an open neighborhood of y in L. We set $U^+ = \{x \in U \cap Z_q : f > 0\}$ and $U^- = \{x \in U \cap Z_q : f < 0\}$. Then by Lemma I.4.21, for each $x \in U \cap Z_q$ we have

(4)
$$f > 0 \iff \langle h_2 \mid h_2 \rangle > \langle h_1 \mid h_1 \rangle,$$
$$f < 0 \iff \langle h_1 \mid h_1 \rangle > \langle h_2 \mid h_2 \rangle.$$

In view of Lemma I.4.21 and condition (3), U^+ and U^- are non-empty disjoint open subsets of Z_q such that $U^+ \cup U^-$ is dense in $U \cap Z_q$.

Furthermore, we claim that

(5) $\qquad$ If $\quad y + z \in U^+, \quad$ then $\quad y + r \cdot z \in U^+ \quad$ for $\quad 0 < r \le 1,$

and that a corresponding statement holds for U^- in place of U^+. Indeed, if $y + z \in U$ and we write $y + z = (1 + k) \cdot y + f \cdot y' \oplus h_1 \oplus h_2$ as we did before, then $z = k \cdot y + f \cdot y' \oplus h_1 \oplus h_2$, and if $k > -1$, then also $rk > -1$ for all $0 < r \le 1$. Also, for the same numbers r, the inequality $\langle h_2 \mid h_2 \rangle > \langle h_1 \mid h_1 \rangle$ implies $\langle r \cdot h_2 \mid r \cdot h_2 \rangle > \langle r \cdot h_1 \mid r \cdot h_1 \rangle$, and similarly for the reverse inequality. This establishes the claim.

As an immediate consequence of (5) and its counterpart for U^- we have

(6) $\qquad\qquad\qquad y \in \overline{U^+}, \quad$ and $\quad y \in \overline{U^-}.$

At this point we bring in a *generating* wedge V which is linked to the quadratic form q and the point $y \in Z_q$ by the following condition:

(7) $\qquad$ *A neighborhood of y in $E^1(V)$ is contained in Z_q.*

The example of the polyhedral wedges shows that the set of E^1 points may be very thin topologically. In order to utilize condition (7) above it is therefore necessary to secure the existence of sufficiently many E^1-points in the boundary of V. Therefore we require

(8) $\qquad\qquad \overline{E^1(V)}$ *is a neighborhood of y in ∂V.*

By Condition (7), a neighborhood $E^1_+(V)$ of y in $E^1(V)$ is contained in Z_q. By Condition (8) we know that $\overline{E^1(V)}$ is a neighborhood of y in ∂V. Thus $N \stackrel{\text{def}}{=} \overline{E^1_+(V)}$ is a neighborhood of y in the topological $(n - 1)$-manifold ∂V, where $n = \dim L$. Since $N \subseteq Z_q$ and Z_q is an $(n - 1)$-manifold, by invariance of domain, N is a neighborhood of y in Z_q. Then the hyperplane $T \stackrel{\text{def}}{=} y^{\perp_q}$

is a tangent hyperplane at y of V as well as of Z_q and therefore is of the form $T_y(V)$. But V is convex and thus is contained in one of the closed half-spaces H or $-H$ bounded by T. Let us assume that H is that half-space which contains y'. Then $(H \setminus T) \cap Z_q \cap U = U^+$ by the definition of U and U^+, and, accordingly, $(U \setminus H) \cap Z_q = U^-$. It now follows that one of the two sets $N \cap U^+$ or $N \cap U^-$ must be empty. But N is a neighborhood of y in Z_q and thus by (6) above, $y \in \overline{N \cap U^+}$ and $y \in \overline{N \cap U^-}$. This is a contradiction. Hence condition (3) must be violated under the present hypothesis on V.

The preceding analysis gives us the following result:

I.4.22.   Proposition. *Let L be a finite dimensional vector space with $\dim L \geq 2$ and with a non-degenerate quadratic form q. Let V be any generating wedge in L with an E^1-point y such that the following conditions are satisfied:*

(i) *Some neighborhood of y in $E^1(V)$ is contained in the zero-set Z_q of q.*

(ii) *$\overline{E^1(V)}$ is a neighborhood of y in ∂V.*

Then q is Lorentzian.

Proof.   We have seen that the hypotheses (i)=(7) and (ii)=(8) imply that $\dim L_1 \leq 1$ or $\dim L_2 \leq 1$. If $\dim L_1 = 0$ or $\dim L_2 = 0$, then q would be definite which is impossible since there is a non-zero vector y with $q(y) = 0$. Hence $\dim L_j = 1$ for at least one of the indices $j = 1, 2$, and this means that q is Lorentzian. ■

This result must necessarily appear technical as long as we do not know how the hypotheses (i) and (ii) of the preceding proposition arise. They will occur in the following situation:

Let (L, q) be a finite dimensional vector space equipped with a non-degenerate quadratic form, and identify L with its dual via q as we did earlier in this section. Recall that for any wedge W in L, the dual wedge W^* is now a wedge in L.

I.4.23.   Proposition. *Let L be a finite dimensional vector space and W a generating cone which is locally round at x. Suppose further that there is a non-degenerate quadratic form q such that the following condition is satisfied:*

For all C^1-points u of W sufficiently close to x, the tangent hyperplane $T_u(W)$ is also a tangent hyperplane of the zero-set Z_q of q at a point z of $\operatorname{op} u$.

Then q is Lorentzian.

Proof.   We shall apply the preceding Proposition I.4.22 with $V = W^*$ and with any non-zero point $y \in \operatorname{op} x = x^\perp \cap W^*$. Condition (ii) of I.4.22 is clear from the fact that W is locally round at x. We have to verify (i). Since $x \in C^1(W) \cap E^1(W)$, the transgression function σ is open at x by the Transgression Theorem I.2.35. Thus we find an open neighborhood U of x in $C^1(W)$ such that (a) all T_u with $u \in U$ are tangent hyperplanes of Z_q and that (b) there is an open neighborhood U' of y in $E^1(W^*)$ such that for each $z \in U'$ there is an $u \in U$ with $q(u, z) = \langle z, u \rangle = 0$. We claim that $U' \subseteq Z_q$; this will establish condition (i) of Proposition I.4.22 and finish the proof. But if z and u are as above, then $z \in E^1(W^*)$ and thus $\operatorname{op} u = \mathbb{R}^+ \cdot z$. By hypothesis, z is the point of contact of $T_u = z^\perp$ with Z_q. ■

5. The invariance of cones under flows

A flow is a one parameter semigroup of transformations on a manifold; we shall give a precise definition below. If it is differentiable, we can associate with each orbit its tangent vector at its origin; this gives us a vector field. Conversely, if we are given a vector field which satisfies a Lipschitz condition locally, then we can construct a flow— at least locally— whose orbits start off in the right direction.

If W is a wedge in a finite dimensional vector space L, and if $X: L \to L$ is a vector field on L satisfying a Lipschitz condition locally, then we have a local flow. In our later applications in a variety of contexts we have to know for which properties of the vector field X the wedge W is invariant under the associated flow. There is a more general result due to Bony and Brezis which describes such conditions accurately. In this section we shall present in a self contained manner a new proof of a somewhat more general result due to Lawson. The generalization will come in handily later, and the proof is not more difficult.

First we have to face a geometric definition of a tangent vector of an arbitrary subset at a point. This idea we shall immediately link with the concepts we used in the preceding sections in the context of wedges and convex sets. We shall then proceed to discuss the idea of measuring the distance of a point from a closed set S in a suitable fashion which will allow us to express conveniently the fact that a continuous piecewise differentiable curve starting in S remains trapped in S. We then discuss the Bony-Brezis-Lawson formalism and apply it in the special case that S is a wedge W. Special attention must be given to the case of linear flows on L, because it is here that most of our applications lie.

Subtangent vectors and tangent vectors

I.5.1. **Definition.** Let L be a topological vector space and $S \subseteq L$. A function $\alpha: D \to \overline{S}$ with $D \subseteq \mathbb{R}^+$, $\alpha(D \cap \,]0, \infty[) \subseteq S$ is said to be *right-differentiable at the origin* or, briefly, 0-*right-differentiable* provided the following conditions are satisfied:

  (i) 0 is a cluster point of positive numbers in D, that is, $0 \in \overline{D \cap \,]0, \infty[}$, and $0 \in D$.

(ii) The following limit exists in L:

$$(1) \qquad\qquad \lim_{\substack{h \searrow 0 \\ h \in D}} \frac{1}{h}\big(\alpha(h) - \alpha(0)\big).$$

The limit in (ii) will be denoted $\alpha'_+(0)$ and called the *right-derivative of α in 0* or, in short, the *0-right-derivative of α*.

A vector $y \in L$ will be called a *subtangent vector of S at x* for any point $x \in L$ if there is a *0-right differentiable function α* with $\alpha(0) = x$ and $y = \alpha'_+(0)$. It is called a *tangent vector of S at x* if both y and $-y$ are subtangent vectors.

The concept of a tangent vector is therefore settled through that of a subtangent vector. Of course, y is a tangent vector of S at x if and only if there is a function $\alpha \colon D \to S$ whose domain $D \subseteq \mathbb{R}$ clusters to 0 from the right and left which is differentiable at 0 such that $\alpha(0) = x$ and $\alpha'(0) = y$ with the *two sided* derivative of α at 0. The following remark is useful, because it tends to simplify geometrical considerations:

I.5.2.   Proposition. *For a subset S in a topological vector space L, the following two conditions are equivalent:*

(1) *y is a subtangent vector of S at x.*

(2) *There is a sequence of positive real numbers r_n and a sequence of elements $s_n \in S$ such that*

   (i) *$\lim r_n \cdot (s_n - x) = y$, and*

   (ii) *$\lim r_n^{-1} = 0$.*

Proof.   (1) $\Rightarrow$ (2): In the domain D of the function α whose 0-right derivative gives y we pick a sequence t_n of positive real numbers converging to 0, set $r_n = t_n^{-1}$ and $s_n = \alpha(t_n)$. Conditions (i) and (ii) are now immediate.

(2) $\Rightarrow$ (1): We define $D = \{0\} \cup \{r_n^{-1} : n = 1, \ldots\}$ and set $\alpha(0) = x$, $\alpha(r_n^{-1}) = s_n$. Then by (2), α has y as a 0-right-derivative.   ∎

Let us determine what the subtangent and tangent vectors of a wedge are.

I.5.3.   Proposition. *Let W be a wedge in a Banach space space L considered to be in duality with its topological dual. Then for $y \in L$ and $x \in W$, the following statements are equivalent:*

(1) *y is a subtangent vector of W at x.*

(2) *$y \in L_x(W) = (\operatorname{op} x)^* = \overline{W - \mathbb{R}^+ \cdot x}$.*

Proof.   First we recall from Proposition I.1.9 and Definition 1.10 that $L_x(W) = (\operatorname{op} x)^* = \operatorname{cl}(W - \mathbb{R}^+ \cdot x)$. Moreover, in a locally convex vector space, the closure of a convex set agrees with its weak closure in view of the Hahn Banach formalism. Thus $y \in L_x(W)$ simply means $y \in \overline{W - \mathbb{R} \cdot x}$.

(1) $\Rightarrow$ (2). We know $y = \lim r_n \cdot (w_n - x)$ and $r_n \to \infty$ by Proposition I.5.2. But since $r_n \cdot (w_n - x) \in W + \mathbb{R} \cdot x$, we know $y \in \overline{W - \mathbb{R}^+ \cdot x}$

(2) $\Rightarrow$ (1). Suppose that $y \in \overline{W - \mathbb{R} \cdot x}$. Hence we find elements $v_n \in W$ and positive real numbers r_n such that $y = \lim(v_n - r_n \cdot x)$. If the sequence of the r_n

is bounded, then after extracting a convergent subsequence and renaming we may assume that $r = \lim r_n$ exists. Then $y + r{\cdot}x = \lim v_n \in W$, whence $y = w - r{\cdot}x$ with $w \in W$. Now let $\alpha(t) = t{\cdot}(w - r{\cdot}x) + x = t{\cdot}w + (1 - tr){\cdot}x$. Then $\alpha(t) \in W$ for all $0 \le t \le r^{-1}$, $\alpha(0) = x$ and $\alpha'_+(0) = w - r{\cdot}x = y$. Thus y is a subtangent vector of W at x. On the other hand, if the sequence of the r_n is unbounded, we may assume without loss of generality that it grows beyond all bounds. Then $\lim r_n^{-1} = 0$, and if we set $w_n = r_n^{-1}{\cdot}v_n$, then $w_n \in W$ and $y = \lim r_n(w_n - x)$. This shows that y is a subtangent vector of W at x by Proposition I.5.2. ∎

I.5.4. **Corollary.** *Let W be a wedge in a Banach space L in duality with its topological dual. Let $x \in W$ and $y \in L$. Then the following conditions are equivalent:*

(1) *y is a tangent vector of W at x.*

(2) *$y \in T_x(W)$.*

Proof. Condition (1) means that both y and $-y$ are subtangent vectors of W at x. Condition (2) says $y \in L_x(W) \cap -L_x(W)$ by Proposition I.2.7.i. Thus (1) and (2) are equivalent by the preceding proposition. ∎

We leave it to the reader to contemplate these observations for the case of other infinite dimensional vector spaces. As long as we consider Banach spaces, the following definition is in accordance with the notation in the previous sections.

I.5.5. **Definition.** If S is a subset of a topological vector space L and $x \in \overline{S}$, then the set of all subtangent vectors of S at x will be denoted $L_x(S)$ and the set of all tangent vectors of S at x will be called $T_x(S)$.

I.5.6. **Proposition.** *If S is a subset in a Banach space, then $L_x(S) = L_x(\overline{S}) = \overline{L_x(S)}$ and $\mathbb{R}^+{\cdot}L_x(S) = L_x(S)$ for $x \in \overline{S}$.*

Proof. If $y \in L_x(\overline{S})$, then $y = \lim r_n(c_n - x)$ with $r_n \to \infty$ and $c_n \in \overline{S}$. For each n we select an element $s_n \in S$ with $\|c_n - s_n\| \le r_n^{-2}$. Then $\|r_n{\cdot}c_n - r_n{\cdot}s_n\| \le r_n\|c_n - s_n\| \le r_n r_n^{-2} = 1/r_n \to 0$. Hence $y = \lim r_n{\cdot}(c_n - x) = \lim r_n{\cdot}(s_n - x)$ and thus $y \in L_x(S)$. Since trivially $L_x(S) \subseteq L_x(\overline{S})$, this shows $L_x(S) = L_x(\overline{S})$. If $y \in \overline{L_x(S)}$, then $y = \lim y_n$ with $y_n \in L_x(S)$. Then $y_n = \lim_{k \to \infty} r_{nk}{\cdot}(s_{nk} - x)$ with $\lim_{k \to \infty} r_{nk}^{-1} = 0$, and with $s_{nk} \in S$. If now U is any open neighborhood of x and N any natural number, then we find a natural number n_0 such that $y_n \in U$ for $n \ge n_0$. It suffices to consider one of the elements $y_n = \lim_{k \to \infty} r_{nk}{\cdot}(s_{nk} - x)$ and find a natural number k so that $r_{nk} > N$ and $r_{nk}{\cdot}(s_{nk} - x) \in U$. This will allow us to extract from $\{(r_{nk}, s_{nk}) : n, k \in \mathbb{N}\}$ a subsequence $(n_j, s_j)_{j \in \mathbb{N}}$ with $r_j \to \infty$ and $y = \lim r_j{\cdot}(s_j - x)$. This proves the first assertion.

Finally suppose that $y \in L_x(S)$ and $r \in \mathbb{R}^+$. If $y = \lim r_n{\cdot}(s_n - x)$, with r_n growing beyond all bounds. Then $r{\cdot}y = \lim(rr_n){\cdot}(s_n - x)$, and if $r > 0$ this shows $r.y \in L_x(S)$. Since $x \in \overline{S}$, then $0 = \lim n{\cdot}(x - x)$ is in $L_x(\overline{S}) = L_x(S)$, whence $r.y \in L_x(S)$ if $r = 0$, too. ∎

The following proposition, although not needed in our later developments, sheds more light on the nature of subtangent vectors.

I.5.7. **Proposition.** *Let S be a closed subset in a finite dimensional vector space L. Suppose that $g: U \to \mathbb{R}$ is a function on an open neighborhood of $x \in S$ such that*

(i) $g(x) = 0$,

(ii) $g(u) \leq 0$ *for all $u \in U \cap S$*,

(iii) *g has a derivative $g'(x): L \to \mathbb{R}$ at x.*

Then

$$g'(x)(v) \leq 0 \quad \text{for all subtangent vectors} \quad v \in L_x(S).$$

Proof. A subtangent vector $v \in L_x(S)$ is of the form $v = \alpha'(0)$ with $\alpha(0) = x$, where α is a suitable 0-right-differentiable function with $\alpha(t) \in S$ according to Definition I.5.1. Then, by the chain rule, we have $g'(x)(v) = g'\big(\alpha(0)\big)\big(\alpha'(0)\big) = (g \circ \alpha)'(0) = \lim_{t \searrow 0} t^{-1}\big(g(\alpha(t)) - g(x)\big) = \lim_{t \searrow 0} g\big(\alpha(t)\big)/t \leq 0$ by hypothesis (ii). ∎

I.5.8. **Proposition.** *If S is a convex set in a Banach space, then $L_x(S)$ is a wedge for any $x \in \overline{S}$.*

Proof. Exercise.   ∎

A Lemma in Calculus I

In order to prove the central result on the invariance of closed sets under flows it will be convenient to have available a lemma which is a variation of the theme of calculus: A real valued function of a real variable is uniquely determined by its derivatives—if these exist in sufficient quantity.

I.5.9. **Lemma.** *Suppose that a continuous function $f: [a, b] \to \mathbb{R}^+$ with $f(a) = 0$ satisfies the following hypotheses:*

There is a positive number c, and for each $x \in [a, b]$ there exists a function $s_x: D_x \to \mathbb{R}$ with a subset $D_x \subseteq \mathbb{R}^+$ containing the non-isolated element $0 \in D_x$ such that s_x is continuous at 0 with $s_x(0) = 0$ and that the following estimate holds:

(1) $$f(x + h) \leq f(x) + h\big(cf(x) + s_x(h)\big) \quad \text{for all} \quad h \in D_x.$$

Then $f = 0$.

Proof. Condition (1) is equivalent to

(2) $$f(x + h) \leq (1 + ch)f(x) + hs_x(h) \quad \text{for all} \quad h \in D_x.$$

Since $1 + ch \leq e^{ch}$ for all $h \geq 0$, we conclude from this that

(3) $$f(x + h) \leq e^{ch} f(x) + hs_x(h) \quad \text{for all} \quad h \in D_x.$$

We now fix an arbitrary $\varepsilon > 0$ and define, for each $x \in [a, b]$ a subset $M(x)$ of $[a, x]$ depending also on ε by

$$(4) \qquad\qquad M(x) = \{t \in [a, x] : f(t) \le \varepsilon e^{ct}(t - a)\}.$$

On account of $a \in M(x)$ we know $M(x) \ne \emptyset$. If we set $m = \sup M(x)$ we therefore have $a \le m \le x$. Since f is continuous, $M(x)$ is closed in $[a, x]$ and thus $m \in M(x)$. We claim that $m = x$. In order to prove this claim we assume $m < x$ and derive a contradiction. Since 0 is not isolated in D_m and s_m is continuous at 0 with $s_m(0) = 0$, we find a positive $d \in D_m$ so that $m < m + d \le x$ and that

$$(5) \qquad\qquad s_m(d) \le \varepsilon e^{c(m+d)}.$$

Hence (3) yields

$$(6) \qquad\qquad f(m + d) \le e^{cd} f(m) + \varepsilon d e^{c(m+d)}.$$

Using $m \in M(x)$ we conclude from this and from (4) that

$$
\begin{aligned}
(7) \qquad f(m + d) &\le e^{cd}\big(\varepsilon e^{cm}(m - a)\big) + \varepsilon d e^{c(m+d)} \\
&= \varepsilon e^{c(m+d)}(m - a) + \varepsilon d e^{c(m+d)} \\
&= \varepsilon e^{c(m+d)}\big((m + d) - a\big).
\end{aligned}
$$

But this implies $m + d \in M(x)$ by (4) in contradiction to $m = \sup M(x)$.

Thus our claim $m = x$ is proved. In particular $x \in M(x)$. Because of $M(x) \subseteq M(b)$ for all $x \in [a, b]$ we thus find $x \in M(b)$ for all $x \in [a, b]$, that is,

$$(8) \qquad 0 \le f(x) \le \varepsilon e^{cx}(x - a) \le \varepsilon e^{cb}(b - a) \quad \text{for all} \quad x \in [a, b].$$

Since $\varepsilon > 0$ was arbitrary, $f = 0$ follows. $\qquad\blacksquare$

Flows, vector fields

After a number of more or less technical definitions which secure our terminology we shall arive at the main result of this section.

I.5.10. **Definition.** A *flow* on a manifold M is a transformation semigroup action

$$(r, x) \mapsto F_r(x) : \mathbb{R}^+ \times M \to M$$

with $F_0 = 1_M$, the identity map of M, and $F_r F_s = F_{r+s}$ for all $r, s \in \mathbb{R}^+$.

We will be concerned only with differentiable actions.

I.5.11.   Remark. If a flow F on M is differentiable, then it defines a vector field $X \colon M \to T(M)$ with

$$(9) \qquad\qquad X(m) = \frac{d}{dt} F_t(m)|_{t=0},$$

where $T(M)$ denotes the tangent bundle of the manifold.   ∎

The theory which we are about to present is basically local in character, and its results can be immediately generalized to the global situation. But because of the local character of our theory we capture its full import by restricting our attention to the case of manifolds M which are open subsets of a finite dimensional vector space L. In this case the tangent bundle is trivial and we may therefore interpret any vector field X on M as a function $X \colon M \to L$. Then (9) above simply reads

$$(10) \qquad\qquad X(m) = \frac{d}{dt} F_t(m)|_{t=0} = \lim_{t \searrow 0} \frac{1}{t} \cdot (F_t(m) - m).$$

I.5.12.   Remark. Suppose that M is an open subset of a finite dimensional vector space L. If F is a differentiable flow on M then the trajectories x defined by $x(t) = F_t(m)$ are the solutions of the initial value problem

$$(11) \qquad\qquad \dot{x}(t) = X(x(t)), \qquad x(0) = m.$$

Conversely, if a vector field X on M satisfies a local Lipschitz condition, then the Theorem of Picard and Lindelöf shows that the problem (11) has unique local solutions. If the maximal extensions of these solutions are defined for all $t \in \mathbb{R}^+$, then the definition $F_t(m) = x(t)$, where x is a solution of (11), defines a flow.   ∎

In general, however, the extensions of the local solutions of the initial value problem (11) will run into obstructions. This will actually be the case for some of the applications we envisage. It is therefore important that we define, right in the beginning, the concept of a *local* flow, even in the case that M is an open subset of L.

I.5.13.   Definition. A *local flow* on a smooth manifold M is a differentiable function

$$(t, m) \mapsto F_t(m) : U \to M,$$

where U is an open subset of $\mathbb{R}^+ \times M$ containing $\{0\} \times M$ such that the following conditions are satisfied:

(i) $F_0 \colon M \to M$ is the identity map of M,
(ii) If (t, m), $(s, F_t(m))$ and $(s+t, m) \in U$, then $F_{s+t}(m) = F_s F_t(m)$.
(iii) If $(t, m) \in U$ and $0 \le s \le t$, then $(s, m) \in U$.

A moment's contemplation of condition (iii) will convince us that it is a sort of fiberwise convexity condition on U which rules out the possible presence of unwanted "holes" or "folds" in U. Since we can, for the most part, pass from U to smaller open neighborhoods of $\{0\} \times M$, this is no real restriction of generality.

Definition (9) above associates with a local flow a vector field; and every locally Lipschitzian vector field defines a local flow; two such local flows will agree on their common domain. We shall say that any such flow is *generated by the vector field X*.

I.5.14. **Definition.** Suppose that we are given a closed subset S of M. We shall say that S is *invariant under a local flow* F if the orbits of elements in S stay in S for all time. In other words, S is invariant under F if $F_t(m) \in S$ for all $m \in S$ and all t with $(t, m) \in U$.

We also have to recall the following definition.

I.5.15. **Definition.** A vector field $X: M \to M$ for an open set $M \subseteq L$ is called *locally Lipschitzian* if for any fixed norm on L there is an open cover of M such that on each open set V of the cover there is a real number c such that $\|X(u) - X(v)\| \leq c\|u - v\|$ for all $u, v \in V$.

For the formulation of the main results it is important to recall the convention expressed in the following definition:

I.5.16. **Definition.** Let M be an open subset of a finite dimensional vector space L and S a closed subset of L contained in M. Let X be a locally Lipschitzian vector field on M. We shall call S *invariant under* X if S is invariant under any local flow on M generated by X. We shall call S *fully invariant under* X if S is invariant under X *and under* $-X$.

The set S is then invariant for X if and only if any solution $x: [0, T] \to M$ of the initial value problem (11) above which starts in S, that is satisfies $x(0) \in S$ stays in S. Likewise, S is fully invariant under X if and only if any solution $x: [-T, T] \to M$ with $T > 0$ of the initial value problem (11) which starts in S stays in S forward and backward in time. Now we are ready to formulate and prove a full characterization for the invariance of closed sets under a given vector field:

I.5.17. **Theorem.** (The Invariance Theorem for Vector Fields) *Let L be a finite dimensional vector space, S a closed subset, and M an open subset containing S. Let $X: M \to L$ be a locally Lipschitzian vector field on M. Then the following conditions are equivalent:*

(1) *S is invariant under X.*

(2) *$X(y) \in L_y(S)$ for all $y \in S$.*

Proof. $(1) \Rightarrow (2)$: Let $y \in S$ and let $x: [0, T] \to M$ be a trajectory given by $\dot{x}(t) = X(x(t))$, $x(0) = y$ with $T > 0$. By (1) we know that $x(t) \in S$ for all $t \in [0, T]$. Then $X(y) = \dot{x}(0) = \lim_{t \searrow 0} t^{-1}\big(x(t) - x(0)\big) \in L_y(S)$ by Definitions I.5.1 and 5.5.

$(2) \Rightarrow (1)$: Suppose that $x: [0, T] \to M$ is a trajectory given by $\dot{x}(t) = X\big(x(t)\big)$, $x(0) \in S$ with $T > 0$. We set $I = \{t \in [0, T] : x([0, t]) \subseteq S\}$. Then I is clearly an interval containing 0. Since x is continuous and S is closed, I is closed in $[0, T]$. We shall show that I is open; this will show that $I = [0, T]$ and thus prove $x([0, T]) \subseteq S$. This will finish the proof.

In order to prove our claim suppose that $a \in I$ and that $a < T$. We must show that there is a b with $a < b \leq T$ such that $x([a, b]) \subseteq S$. For this purpose we choose a closed neighborhood U of $x(a)$ such that

(a) $$\|X(u) - X(v)\| \leq c\|u - v\|$$

holds for $u, v \in U$. By the continuity of x we find a $b > a$, $b \leq T$ such that $x([a, b]) \subseteq U$.

Now we utilize $\dim L < \infty$ and the axiom of choice in defining a function

(min) $z: [a, b] \to S \cap U$ by $\|x(t) - z(t)\| = \min\{\|x(t) - s\| : s \in S \cap U\}$;

indeed if $r = \inf\{\|x(t) - s\| : s \in S\}$ and B is the closed ball around $x(t)$ with radius $2r + 1$, then B is compact and so the continuous function $s \mapsto \|x(t) - s\|: B \cap S \cap U \to \mathbb{R}^+$ attains its minimum.

By hypothesis (2) we have $X(z(t)) \in L_{z(t)}$. Hence by Proposition I.5.3 and Definition I.5.1, if $0 \leq t < T$ there is a set $D_t \subseteq R^+$ with the non-isolated element $0 \in D_t$ and with $t + D_t \subseteq [0, T]$, such that there is a function $\alpha_t: D_t \to \overline{S} = S$ with $\alpha_t(0) = z(t)$ and $X(z(t)) = \alpha'_+(0)$. This means that we have a function $\sigma_t: D_t \to L$ such that

(b) $\alpha_t(h) = z(t) + h \cdot X(z(t)) + h \cdot \sigma_t(h)$

and

$$\lim_{\substack{h \searrow 0 \\ h \in D_t}} \sigma_t(h) = 0.$$

Since x is differentiable in t with derivative $X(x(t))$ we know

(c) $x(t + h) = x(t) + h \cdot X(x(t)) + h.\rho_t(h)$ for $h \in D_t$

with a function $\rho: D_t \to L$ with

$$\lim_{\substack{h \searrow 0 \\ h \in D_t}} \rho_t(h) = 0.$$

Now $z(t + h)$ is an element of S with minimal distance from $x(t + h))$. Thus we conclude

(d) $\|x(t + h) - z(t + h)\| \leq \|x(t + h) - \alpha_t(h)\|$.

But (b) and (c) imply

$\|x(t + h) - \alpha_t(h)\|$

(e) $= \|x(t) + h \cdot X(x(t)) + h.\rho_t(h) - z(t) - h \cdot X(z(t)) - h \cdot \sigma_t(h)\|$

$\leq \|x(t) - z(t)\| + h\|X(x(t)) - X(z(t))\| + h\|\rho_t(h) - \sigma_t(h)\|$.

Now we set $s_t(h) = \|\rho_t(h) - \sigma_t(h)\|$ and note that

(f) $\|X(x(t)) - X(z(t))\| \leq c\|x(t) - z(t)\|$

by (a) above. We take (d),(e) and (f) together and find

(g) $\|x(t + h) - z(t + h)\| \leq \|x(t) - z(t)\| + hc\|x(t) - z(t)\| + hs_t(h)$.

We denote with $\text{dist}(z, Z)$ the distance in L of a point z from a set $Z \neq \emptyset$ given by $\inf\{\|z - z'\| : z' \in Z\}$. The function $z \mapsto \text{dist}(z, Z)$ is continuous for any non-empty subset Z. Hence the function $f: [a, b] \to \mathbb{R}^+$ defined by $f(t) = \text{dist}(x(t), S \cap U)$ is continuous. Since $a \in I$ we know that $f(a) \in S$, whence $f(a) = 0$. But $f(t) = \|x(t) - z(t)\|$, and thus (g) reads

$$f(t + h) = f(t) + h(cf(t) + s_t(h)) \quad \text{for all} \quad h \in D_t, \quad t \in [a, b].$$

Thus the hypotheses of Lemma I.5.9 are satisfied and we conclude that $f = 0$. This means $x(t) = z(t)$ and thus $f([a, b]) \subseteq S$. Since $f([0, a]) \subseteq S$ because of $a \in I$, this proves that $f([0, b]) \subseteq S$. Thus $b \in I$ and this is what we had to show.   ∎

A few remarks are in order. Firstly, except through condition (min), we have not used that L has finite dimension. Thus the theorem continues to hold in Banach spaces for those sets S for which to each $x \in L$ the distance $\text{dist}(x, S)$ is realized in the form $\text{dist}(x, s)$ with an $s \in S$.

Secondly, we should keep in mind that for the proof of the continuity of the function f we have *not* called upon any property of the function $z : [a, b] \to S$; in general this function will not be continuous. We used the fact that f was defined via the distance function of a variable point from a fixed set, and this function is always continuous.

Finally, while the formulation of the theorem is local, it is no serious problem to state a global version. We leave this as an exercise.

EI.5.1. **Exercise.** Formulate and prove the global version of the Invariance Theorem for Vector Fields for arbitrary closed subsets in arbitrary smooth manifolds.

However, we do need the version of the theorem which states a conclusion on the *full* invariance of a closed set under a vector field:

I.5.18. **Corollary.** *Under the general circumstances of Theorem I.5.17, the following conditions are equivalent:*

(1) S *is fully invariant under* X.

(2) $X(y) \in T_y(S)$ *for all* $y \in S$.

Proof. By Theorem I.5.17, condition (1) is equivalent to $X(y) \in L_y(S)$ and $-X(y) \in L_y(S)$. But this in turn is equivalent to (2) above since $T_y(S) = L_y(S) \cap -L_y(S)$ by Definitions I.5.1 and I.5.5. ∎

For the applications we shall make in the immediate future, this version of the invariance theorem will be completely adequate, and we could proceed to these applications right away. However, increased generality is readily available after a few definitions. We pause for these additional deliberations.

A vector field associates with the points of a manifold M a vector in the tangent space at that point. We shall encounter on many occasions that we have prescriptions, through which to each point m of a manifold there is associated a *whole set* of vectors in the tangent hyperplane $T_m(M)$ at this point. Since we will almost exclusively work with manifolds M which are open sets of L, we will, as before, identify $T_m(M)$ with L. A typical such assignment would be $x \mapsto L_x(S)$ which assigns, for a given closed set S in L, to a point $x \in L$ the empty set if $x \notin S$, else the set of subtangent vectors of the set S. Another example is the assignment of a vector subspace of $T_m(M)$ to the point m; these assignments are known under the name of *distributions*. We will adopt this terminology for the more general concept:

I.5.19. **Definition.** Let M be a smooth manifold. A *vector field distribution* on M is a function $m \mapsto \Xi(m)$ which assigns to each $m \in M$ a subset $\Xi(m) \subseteq T_m(M)$ in the tangent hyperplane of M at m. More specially, if M is an open subset of a finite dimensional vector space, then a *vector field distribution* is a function $\Xi : M \to 2^L$. If, in this situation, S is a closed subset of L contained in

M, then the vector field distribution $x \mapsto L_x(S)$ with $L_x(S) = \emptyset$ for $x \notin S$ is called the *subtangent distribution for S*.

I.5.20.   Definition. Let M be a smooth manifold and Ξ a vector field distribution on M. A trajectory $x : [0, T] \longrightarrow M$ is said to be *subordinate to Ξ* if

$$ \dot{x}(t) \in \Xi(x(t)) \tag{12} $$

for all $t \in [0, T[$ with the possible exception of some countable subset N.

Relations of the type of (12) are also called *differential inclusions*. They are clearly a generalization of the idea of ordinary differential equations.

We need to generalize the idea of a local Lipschitz condition to vector field distributions.

I.5.21.   Definition. A vector field distribution $\Xi \colon M \to 2^L$ on an open subset M of a finite dimensional vector space L is called *locally Lipschitzian* if there is an open cover of M such that for each open set U from the cover there is a non-negative constant C such that

$$ (\forall u, v \in U)(\forall x \in \Xi(u))(\exists y \in \Xi(v)) \| x - y \| \leq C \| x \| \| u - v \|. \tag{13} $$

It is a good exercise to contemplate the case of a locally Lipschitz vector field X for which we define an obvious vector field distribution by $\Xi(x) = \{X(x)\}$. Indeed, this distribution will be locally Lipschitzian in the sense of (13); and it would be even if the factor $\|x\|$ had been omitted from the right hand side of the inequality in (13). But then one remembers that a Lipschitz condition for vector field distributions must, in particular, remain meaningful for the case of a distribution of vector subspaces; this requires this factor, and we shall see later under what conditions we obtain Lipschitzian vector field distributions. For the moment we are satisfied with proving the following result:

I.5.22.   Theorem. *Let M be an open subset of a finite dimensional vector space L, and S a closed subset of L contained in M. Let $\Xi \colon M \to 2^L$ be a locally Lipschitzian vector field distribution satisfying*

$$ \Xi(s) \subseteq L_s(S) \qquad \text{for all} \quad s \in S. $$

Then any trajectory starting in S and being subordinate to Ξ stays in S, provided that its forward tangents are bounded locally.

Proof.   The proof is practically the same as that of Theorem I.5.17. We shall therefore restrict ourselves to some indications. Once we have reached the point in the proof to consider the element $a \in I$ we choose b so small that $K = \sup_{a \leq t \leq b} \| \dot{x}(t) \|$ exists and that $x([a, b])$ is contained in a closed neighborhood U on which (13) holds with a constant C. We choose z as in the proof of Theorem I.5.17. and pick, for each $t \in [a, b]$, a vector $z^*(t) \in \Xi(z(t))$ which will then satisfy

$$ \| \dot{x}(t) - z^*(t) \| \leq CK \| x(t) - z(t) \|. \tag{h*} $$

Since $z^*(t) \in \Xi\big(z(t)\big) \subseteq L_{z(t)}(S)$, instead of relation (b) in the proof of Theorem I.5.17 we work with relation

(b*)
$$\alpha_t(h) = z(t) + h \cdot z^*(t) + h \cdot \sigma_t(h).$$

Instead of (e) we now have

(e*)
$$\|x(t+h) - \alpha_t(h)\| = \|x(t) + h \cdot \dot{x}(t) + h \cdot \rho_t(h) - z(t) - h \cdot z^*(t) - h \cdot \sigma_t(h)\|$$
$$\leq \|x(t) - z(t)\| + h\|\dot{x}(t) - z^*(t)\| + h\|\rho_t(h) - \sigma_t(h)\|.$$

But then (h*) and (e*) yield (g) again with $c = CK$. The proof then proceeds as in that of Theorem I.5.17. ∎

The invariance of wedges and vector fields

We now apply the central result, the Invariance Theorem for Vector Fields I.5.17, to the special case that we have a vector field X on all of L and that the closed set S is in fact a wedge W. It is a very important point for our later applications, and a point which is easily overlooked in specializing the general result, that we can relax the condition that the vector field has to be in the subtangent wedge $L_x(W)$ for *all* $x \in W$. It suffices to know this for all C^1-points. It should be clear that it would suffice to have the vector field defined on an open neighborhood of a wedge.

I.5.23. **Theorem.** (The Invariance Theorem for Wedges and Vector Fields) *Let W be a generating wedge in a finite dimensional vector space L and a locally Lipschitzian vector field $X : L \to L$. Then the following conditions are equivalent:*

(1) *W is invariant under X.*

(2) *$X(w) \in \overline{W - \mathbb{R}^+ \cdot w}$ for all $w \in W$.*

(3) *$X(c) \in \overline{W - \mathbb{R}^+ \cdot c}$ for all $c \in C^1(W)$.*

Proof. The equivalence of (1) and (2) follows from the Invariance Theorem for Vector Fields I.5.17 and the fact that $L_w(W) = \overline{W - \mathbb{R} \cdot w}$ by Proposition I.5.3. It is trivial that (2) implies (3). For the implication $(3) \Rightarrow (2)$ we invoke the Confinement Theorem I.3.23. This theorem shows that $X(w) \in L_w(W)$ for all $w \in \partial W$, and since $L_w(W) = L$ for all $w \in \operatorname{int} W$, condition (2) follows. ∎

This is *the* essential application of the Confinement Theorem in our context.

We also remember, that the wedges $\overline{W - \mathbb{R}^+ \cdot c}$ occurring in condition (3) are all closed half-spaces bounded by the tangent hyperplanes $T_c(W)$.

I.5.24. **Corollary.** *Under the general circumstances of Theorem I.5.23, the following conditions are equivalent:*

(1) *W is fully invariant under X.*

(2) *$X(w) \in T_w(W)$ for all $w \in W$.*

(3) $X(c) \in T_c(W)$ *for all* $c \in C^1(W)$.

Proof. The equivalence of (1) and (2) is a special case of Corollary I.5.18. By the preceding Theorem I.5.23 and the definition of full invariance, condition (1) is equivalent to saying that $X(c) \in L_c(W)$ and $-X(c) \in L_c(W)$ for all $c \in C^1(W)$. But $T_c(W) = L_c(W) \cap -L_c(W)$ whence (1) and (3) are equivalent. ∎

This says that a wedge is fully invariant under a locally Lipschitzian vector field if and only if its vectors at each C^1-point of W are in the tangent hyperplane at that point.

Our concern will be almost exclusively with linear vector fields, that is, with vector fields which on L are simply given by any linear self-map $X: L \to L$, so that $X(x)$ happens to be the image of the vector x under X. For such a vector field, the associated flow is likewise linear. Indeed every such X generates the flow $F : \mathbb{R} \times L \longrightarrow L$ given by

$$F_t(x) = e^{t \cdot X} x = x + t \cdot X x + \frac{t}{2!} \cdot X^2 x + \cdots$$

Here, however, we have an additional aspect which is important to us in view of duality. Each linear self-map X of L defines an adjoint $\widehat{X}: \widehat{L} \to \widehat{L}$. Correspondingly, the flow F has an adjoint flow $\widehat{F}$ on $\widehat{L}$ given by $\widehat{F}_t = e^{t \cdot \widehat{X}}$. The following observation is then certainly not unexpected:

I.5.25.   **Proposition.** *For a linear self-map X of the finite dimensional vector space L and a wedge W in L, the following conditions are equivalent:*

(1) $e^{t \cdot X}(W) \subseteq W$ *for all* $t \in \mathbb{R}^+$.

(1') $e^{t \cdot \widehat{X}}(W^*) \subseteq W^*$ *for all* $t \in \mathbb{R}^+$.

Proof. In view of the fact that $e^{t \cdot \widehat{X}} = (e^{t \cdot X})\widehat{}$, this is a direct consequence of the following lemma. ∎

I.5.26.   **Lemma.** *Let V and W be wedges in a finite dimensional vector space L, and $\alpha: L \to L$ a linear map. Then $\alpha(V) \subseteq W$ if and only if $\widehat{\alpha}(W^*) \subseteq V^*$.*

Proof. The relation $\alpha(V) \subseteq W$ is equivalent to

$$(14) \qquad\qquad \langle \omega, \alpha(v) \rangle \geq 0 \quad \text{for all} \quad v \in V, \ \omega \in W^*$$

as $W = W^{**}$ by duality. But by the definition of the adjoint map, (14) is equivalent to

$$(15) \qquad\qquad \langle \widehat{\alpha}(\omega), v \rangle \geq 0 \quad \text{for all} \quad v \in V, \ \omega \in W^*.$$

Condition (15) means exactly $\widehat{\alpha}(W^*) \subseteq V^*$ in view of duality. ∎

Now we are ready for the linear invariance theorem:

I.5.27. **Theorem.** (The Linear Invariance Theorem for Wedges) *Let W be a generating wedge in a finite dimensional vector space L and $X: L \to L$ a linear map. Then the following conditions are equivalent:*

(1) $e^{t \cdot X} W \subseteq W$ *for all $t \in \mathbb{R}^+$ (respectively, for all $t \in \mathbb{R}$).*

(2) $X(w) \in L_w$ *(respectively, $X(w) \in T_w$) for all $w \in W$.*

(3) $X(c) \in L_c$ *(respectively, $X(c) \in T_c$) for all $c \in C^1(W)$.*

(4) $X(e) \in L_e$ *(respectively, $X(e) \in T_e$) for all $e \in E^1(W)$.*

Proof. The equivalence of (1),(2) and (3) is a special case of the Invariance Theorem for Wedges and Vector Fields I.5.23 and its Corollary I.5.24 applied to $W - W$. We still have to show the equivalence of (4) with these conditions. We invoke duality and Proposition I.5.25 from which we know that the first three conditions are equivalent to

(1′) $e^{t \cdot \widehat{X}} W^* \subseteq W^*$ for all $t \in \mathbb{R}^+$ (respectively, $t \in \mathbb{R}$).

But since we have already the equivalence of (1),(2) and (3) it follows that (1′) is equivalent to

($3'_a$) $\widehat{X}(\omega) \in L_\omega$ for all $\omega \in C^1(W^*)$,

respectively,

($3'_b$) $\widehat{X}(\omega) \in T_\omega$ for all $\omega \in C^1(W^*)$.

Now $\widehat{X}(\omega) \in L_\omega$ is equivalent to $X(L^*_\omega) \subseteq \omega^*$ by Lemma I.5.26. But $L^*_\omega = \mathrm{op}\,\omega$ in view of Proposition I.1.9 (and Definition I.1.10). If W is pointed, then W^* is generating and so, if ω is a C^1-point, then $\mathrm{op}\,\omega$ is a half line $\mathbb{R}^+ \cdot e$ with $e \in E^1(W)$ by Proposition I.2.25. By Proposition I.2.26, all E^1-points e of W are elements of some $\mathrm{op}\,\omega$ with $\omega \in C^1(W^*)$. Thus ($3'_a$) is equivalent to $X(e) \in \omega^*$ for all E^1-points e of W and all C^1-points $\omega \in \mathrm{op}\,e$. This means that for all $e \in E^1(W)$ we have

$$(16) \qquad X(e) \in \bigcap_{\omega \in \mathrm{op}\,e} \omega^* = \Big(\sum_{\omega \in \mathrm{op}\,e} \mathbb{R}^+ \cdot \omega \Big)^* = (\mathrm{op}\,e)^* = L_e.$$

Now the validity of ($3'_b$) is tantamount to the validity of ($3'_a$) for X and for $-X$ and hence to the validity of (16) for X and $-X$, for all $e \in E^1(W)$. This means $X(e) \in T_e$ for all $e \in E^1(W)$. Thus we have shown that (3′) is equivalent to (4) if W is pointed.

If W is arbitrary, then (1) implies that $X\big(H(W)\big) \subseteq H(W)$ as does (4) in view of Proposition I.2.7. Hence X induces a linear map $\widetilde{X}$ on $\widetilde{L} \overset{\text{def}}{=} L/H(W)$, where $\widetilde{W} \overset{\text{def}}{=} W/H(W)$ is the associated pointed cone. We write

$(\widetilde{1})$ $\qquad\qquad e^{t \cdot \widetilde{X}} \widetilde{W} \subseteq \widetilde{W}$ for all $t \in \mathbb{R}^+$, respectively, $t \in \mathbb{R}$,

$(\widetilde{4})$ $\qquad \widetilde{X}(\widetilde{e}) \in L_{\widetilde{e}}(\widetilde{W})$, respectively, $\widetilde{X}(\widetilde{e}) \in T_{\widetilde{e}}(\widetilde{W})$, for all $\widetilde{e} \in E^1(\widetilde{W})$.

We have seen above that $(\widetilde{1})$ and $(\widetilde{4})$ are equivalent. But (1) and $(\widetilde{1})$ are clearly equivalent and so are (4) and $(\widetilde{4})$. This completes the proof. ∎

I.5.28.   Corollary.   *Let W be a generating wedge in a finite dimensional vector space L. Then the set of all $X \in \mathrm{Hom}(L, L)$ such that*

$$e^{t \cdot X} W \subseteq W \quad \text{for all} \quad t \in \mathbb{R}^{+}, \quad \text{respectively,} \quad t \in \mathbb{R}$$

is a wedge, respectively vector subspace, of $\mathrm{Hom}(L, L)$.

Proof.   By Theorem I.5.27, the set in question is the set of all $X \in \mathrm{Hom}(L, L)$ such that $X(e) \in L_e$, respectively, $X(e) \in T_e$ for all $e \in E^1(W)$. The set of all $X \in \mathrm{Hom}(L, L)$ satisfying $X(e) \in L_e$ for any fixed $e \in E^1(W)$ is clearly a wedge. Similarly, the set of all $X \in \mathrm{Hom}(L, L)$ satisfying $X(e) \in T_e$ is a vector space. The assertion is now clear.   ∎

Problems for Chapter I

PI.1.   Problem.   *Prove or disprove the* Transgression Theorem I.2.35 *for infinite dimensional wedges.*   ∎

PI.2.   Problem.   *Give infinite dimensional versions of the* Invariance Theorem for Vector Fields I.5.17.   ∎

PI.3.   Problem.   *When is the sum of two wedges closed (see* Paragraphs I.2.31 through 33) *for the general context.*   ∎

PI.4.   Problem.   *Give an infinite dimensional version of the* Straszewicz Spanning Theorem I.3.15 and 16.   ∎

Notes for Chapter I

Section 1.   The material in Section 1 is standard duality of wedges. It primarily serves to fix the notation which we shall use in this book.

Section 2.   The calculus of faces of convex sets and cones is standard. However, we concentrate on a special type of faces, namely, the *exposed* faces. These fit into the scheme of duality. A general source for convex analysis is the book by Rockafellar [Rock70], but there are many other references (see e.g. Fenchel [Fe53], or Bonnesen and Fenchel [BoFe34]). The results in the Transgression Theorem I.2.35 first appeared in [HH85c]. Some of the other results of a more technical nature such as Propositions I.2.31 through 33 are provided here for later use.

Section 3.   The central results in this section are classical. Mazur's Density theorem I.3.10 was proved in 1933 (see [Ma33]) and the Straszswicz Spanning Theorem I.3.16 in 1935 (see [St35]). Nevertheless, the use we shall make of these results is based on some consequences of these results which are not immediate. Such results are the Approximation Theorem I.3.21 and the Confinement

Theorem I.3.23. They require the entire machinery developed up to this point. These results were published in [HH85c].

Section 4. The general theory of wedges is specialized to polyhedral cones in euclidean n-spane and cones which are defined by a Lorentzian form—such as the forward light cone in Minkowski space. The machinery built up in the earlier sections allows us to define in a precise and formal fashion what we mean when we say that a cone is locally round at a point or that it is locally round. These concepts allow us to establish sufficient conditions for quadratic forms related to cones to be in fact Lorentzian: See Propositions I.4.22 and 23.

Section 5. The concept of a subtangent vector of a set at a point is classical and we make no attempt to survey the literature. It enters the core result of this section, the Invariance Theorem for Vector Fields I.5.17. This is due to Bony [Bon69] and Brezis [Bre70]. See also [Re72]. We offer an alternative proof for the theorem which fits our conceptual frame work. It is due to C.Terp. The generalization in Theorem I.5.22 was given by Lawson [Law87b]; his proof is different and resembles the one given by Hörmander [Hö83] for Theorem I.5.17. The Invariance Theorem and the Confinement Theorem of Section 3 jointly yield the new Invariance Theorem for wedges and Vector Fields I.5.23. Together with its corollaries this will be applied extensively later. The presence of the equivalent condition (3) is the reason why this theorem is not a simple and direct consequence of the Bony–Brezis Theorem. Theorem I.5.23 is due to Hilgert and Hofmann [HH86c].

Chapter II

Wedges in Lie algebras

In the first chapter we studied the geometrical and analytical structure of wedges in a finite dimensional real vector space. In the current chapter we make the additional assumption that the vector space whose wedges we consider is, in addition, a Lie algebra. This new element brings us to the objective of this book which is the investigation of the infinitesimal, the local and the global theory of semigroups in a Lie group. Just as in conventional Lie theory the infinitesimal generators of a Lie group are the elements of a Lie algebra we shall find that the infinitesimal generators of a subsemigroup of a Lie group are the elements of a wedge in the Lie algebra. We shall have to see which wedges occur and in which way their structure is linked to the algebraic data of the Lie algebra. In the same sense in which one might say that the theory of real Lie algebras is the infinitesimal theory of Lie groups we could say of the material of this chapter that it presents the algebraic and geometric aspects of the infinitesimal theory of semigroups in Lie groups.

In the first section we derive the characteristic property of those wedges which arise as subtangent sets of (local) semigroups in Lie groups. We shall be brief here since the correspondence between local semigroups and their tangent wedges will be discussed in depth in Chapter IV. We do enough in this section to motivate the definition of what we shall call a Lie wedge. A Lie wedge is to a subsemigroup of a Lie group what a Lie subalgebra is to a subgroup. In the traditional scheme of things, a normal subgroup yields an ideal in the Lie algebra. The corresponding generalization of an ideal to wedges is the idea of an invariant wedge in a Lie algebra. It will be our objective in Chapter III to give an essentially complete theory of invariant cones in Lie algebras—not a minor undertaking, as we shall see. At this point however, we shall be satisfied with finding a characterization of Lie wedges and invariant wedges in terms of the Lie brackets plus the geometry of the wedges. This requires a good deal of the information which we prepared in Chapter I.

It turns out that, in contrast with local Lie groups, which are (locally) ruled by one parameter subgroups, local semigroups are not always ruled by local one parameter semigroups. Those wedges which are ruled by local one parameter semigroups give rise to a special class of Lie wedges which we shall call Lie semi-algebras. Section 2 is devoted to their introduction and characterization. All invariant wedges are Lie semialgebras. The class of almost abelian Lie algebras

gives a first supply of non-invariant Lie semialgebras; the subsequent sections will
unfold a rich theory and a much greater wealth of examples. As a decisive tool
for working with semialgebras we introduce their characteristic functions. Some of
its elementary properties are discussed here. The last part of the second section
dealing with analytic extension aspects has interesting consequences but may be
skipped during a first reading.

Section 3 deals with Lie algebras of dimension 2, 3, and 4 and their semialgebras. The purpose of this section is twofold: Firstly, we form a better intuition
from knowing as thoroughly as possible what the theory yields in low dimensions.
Secondly, in the general theory we shall have to fall back on precise information
coming out of our thorough knowledge of low dimensional subalgebras.

A deeper theory of Lie semialgebras follows in Sections 4, 5, and 6. For the
first time we observe here a remarkable link between the new semigroup oriented
Lie theory and classical Lie algebra theory; it will be corroborated in Chapter III
when we deal with invariant cones. The connection is implemented by the Cartan
subalgebras of a Lie algebra. It is then not surprising that we find the representation
theory of nilpotent and solvable Lie algebras to play a role. Lie semialgebras exhibit
a general tendency to be invariant. The first strong indication of this tendency is
the the following result in which Section 5 culminates: A tangent hyperplane of
any generating Lie semialgebra which is not invariant must be a subalgebra. The
inclination of semialgebras to be invariant is also evident in the situation of round
semialgebras and, notably, in the case of Lorentzian semialgebras which we shall
completely understand after Section 6.

Not every Lie algebra can contain Lie semialgebras with inner points. In
Section 7 we investigate which consequences on the structure of the containing Lie
algebra are implied by the presence of Lie semialgebras with inner points.

1. Lie wedges and invariant wedges in Lie algebras

The first step in this section is the definition of a Lie wedge. In the same
spirit in which the subalgebras of the Lie algebra $L = \mathbf{L}(G)$ of a Lie group G
generate the analytic subgroups of G, Lie wedges generate subsemigroups of G.
This will be made more precise in the later parts of this book. But we must
understand the basic principle underlying the definition of a Lie wedge now. This
requires a brief detour into Lie theory.

Suppose that $\exp: L \to G$ is the exponential function of a finite dimensional
Lie group. Let B be an open neighborhood of 0 which is symmetric and star-shaped
(that is, satisfies $-B = B$ and $[0,1]\cdot B = B$) in L such that for all $x, y \in B$ the

Campbell–Hausdorff series

$$x * y = x + y + \frac{1}{2}[x, y] + \cdots, \qquad x, y \in B$$

converges absolutely. It is convenient to assume that $B * B * B$ is defined. We shall call such neighborhoods *Baker–Campbell–Hausdorff neighborhoods* or briefly *C–H neighborhoods of L*. (See also Definition IV.1.1 below.) For $x, y \in B$ we have $\exp(x * y) = \exp x \exp y$; indeed this holds for all x and y sufficiently close to 0, and then throughout $B \times B$ by analytic extension as both functions

$$(x, y) \mapsto \exp(x * y) : B \times B \to G,$$
$$(x, y) \mapsto \exp x \exp y : B \times B \to G$$

are analytic.

   If a subset $M \subseteq G$ satisfies $1 \in M$ and $MM \subseteq M$, then $S \overset{\text{def}}{=} B \cap \exp^{-1} M$ satisfies

   (i) $0 \in S \subseteq B$,

   (ii) $x, y \in S$ and $x * y \in B$ imply $x * y \in S$.

   We shall say (and later repeat more formally) that any subset $S \subseteq B$ satisfying (i) and (ii) is a *local subsemigroup of L with respect to B*. If S is a local subsemigroup of L with respect to B, then so is $\overline{S} \cap B$. A key lemma which we shall prove later (see Proposition IV.1.21) and record here in advance exhibits an important property of the subtangents of a closed local semigroup S at 0.

II.1.1. **Lemma.** *For a local subsemigroup S of a finite dimensional Lie algebra L with respect to a C–H neighborhood B the following two conditions are equivalent for any $x \in L$:*

   (1) $x \in L_0(S)$.

   (2) $\mathbb{R}^+ \cdot x \cap B \subseteq \overline{S}$. ∎

   This shows immediately that $L_0(S)$ is stable under scalar multiplication by $\mathbb{R}^+$. From Proposition I.5.6 we know that $L_0(S) = L_0(\overline{S})$ is topologically closed. For any two vectors $x, y \in L$ we have

$$x + y = \lim_{n \to \infty} n(\frac{1}{n} \cdot x * \frac{1}{n} \cdot y).$$

   With the aid of Lemma II.1.1 above we see now at once that $L_0(S)$ is additively closed. (A formal proof will follow in the proof of Proposition IV.1.25.) We conclude that $L_0(S)$ *is a wedge*. But more is true:

II.1.2. **Proposition.** *Suppose that W is a wedge in a finite dimensional Lie algebra L and that W is the set of subtangents of a local semigroup of L with respect to some C–H neighborhood of L. Then*

(LW) $\qquad e^{\operatorname{ad} x} W = W \quad$ *for all* $\quad x \in H(W) = W \cap -W,$

where $(\operatorname{ad} x)(y) = [x, y]$.

Proof. Suppose that B is a C–H neighborhood of L and S a local semigroup of L with respect to B. Then $W = L_0(S)$ by hypothesis, and by Proposition I.5.6 and the fact that $\overline{S} \cap B$ is again a local subsemigroup of L with respect to B, we may assume that S is closed in B. Now let $x \in H(W)$. Then x and $-x$ are in W, and from Lemma II.1.1 we know that

$$(*) \qquad\qquad\qquad \mathbb{R} \cdot x \cap B \subseteq S.$$

Now let $y \in W$. Then $\mathbb{R}^+ \cdot y \cap B \subseteq S$. For all sufficiently small elements x' and y' we know $e^{\operatorname{ad} x'} y' = x' * y' * -x'$. We apply this piece of information with $x' = s \cdot x$ and $y' = t \cdot y$, and with s and t so small that $x' * y' * -x'$ is in B. Then $x' * y' * -x' \in S$, since S is a local semigroup with respect to B and since $x' \in B$ by $(*)$. Thus $t \cdot e^{\operatorname{ad} x'} y \in S$ for all sufficiently small $t \in \mathbb{R}^+$, and thus $e^{\operatorname{ad} x'} y \in W$. Consequently, $e^{s \cdot \operatorname{ad} x} W \subseteq W$ for all sufficiently small $s \in \mathbb{R}$. Hence $e^{s \cdot \operatorname{ad} x} W = W$ for all sufficiently small $s \in \mathbb{R}$. Since for all $T \in \operatorname{Hom}(L, L)$ we have $e^{s \cdot T} e^{s' \cdot T} = e^{(s+s') \cdot T}$, we conclude $e^{r \cdot \operatorname{ad} x} W = W$ for all $r \in \mathbb{R}$, and in particular

$$e^{\operatorname{ad} x} W = W.$$

This concludes the proof.   ∎

For a slightly more detailed organization of this proof see Theorem IV.1.27.

This justifies the following definition:

II.1.3.   Definition. A wedge W in a topological Lie algebra L is called a *Lie wedge* if it satisfies the condition

$$e^{\operatorname{ad} x} W = W \qquad \text{for all} \quad x \in H(W).$$

As is usual in this book, we shall encounter Lie wedges most frequently in finite dimensional Lie algebras. But we point out that the concept is certainly viable in the more general case of Dynkin algebras.

We observe that one of the results of Section 5 in Chapter I allows us to associate with a wedge W in a Lie algebra L of finite dimensions a vector space $\mathcal{L}(W, L)$ as follows:

II.1.4.   Lemma. *Consider a finite dimensional Lie algebra L and a wedge W in L. Set*

$$\mathcal{L}(W, L) = \{x \in L : e^{\operatorname{ad} x} W = W\}.$$

Then $\mathcal{L}(W, L)$ is a Lie subalgebra of L.

Proof. We know from Corollary I.5.28 that

$$V = \{T \in \operatorname{Hom}(L, L) : e^T W = W\}$$

is a vector subspace of $\operatorname{Hom}(L, L)$. The function $\operatorname{ad} \colon L \to \operatorname{Hom}(L, L)$ is certainly linear. Hence $\mathcal{L}(W, L) = (\operatorname{ad})^{-1}(V)$ is a vector space.

The group $\mathrm{Gl}(L)$ of all vector space automorphisms of L has as its Lie algebra the Lie algebra $\mathrm{gl}(L)$ of all vector space endomorphisms of L with the Lie bracket $[X, Y] = XY - YX$, and with the exponential function $X \mapsto e^X = 1 + X + \frac{1}{2!} \cdot X^2 + \cdots$. The set G of all $g \in \mathrm{Gl}(L)$ with $gW = W$ is a closed subgroup of $\mathrm{Gl}(L)$ and is, therefore, a Lie group. (Recall that a locally compact subgroup of any Lie group is a Lie group!) Its Lie algebra $\mathbf{L}(G)$ is the set of all $X \in \mathrm{gl}(L)$ such that the exponential function maps $\mathbb{R} \cdot X$ into G. Thus $\mathbf{L}(G) = V$: Indeed, $\mathbf{L}(G) \subseteq V$ is immediate from the definition of V, and the reverse follows since V is a vector space.

Now $\mathrm{ad} \colon L \to \mathrm{gl}(L)$ is in fact a homomorphism of Lie algebras by the Jacobi identity. Hence $\mathcal{L}(W, L) = (\mathrm{ad})^{-1} \mathbf{L}(G)$ is indeed a Lie algebra. ∎

II.1.5. **Definition.** We call $\mathcal{L}(W, L)$ the Lie subalgebra of L *determined by the wedge W.*

Now we can reformulate the definition of a Lie wedge:

II.1.6. **Proposition.** *A wedge W in a Lie algebra is a Lie wedge if and only if its edge $H(W)$ is contained in the Lie algebra $\mathcal{L}(W, L)$ determined by W.* ∎

The following observation is useful:

II.1.7. **Lemma.** *If A and B are vector subspaces of a finite dimensional Lie algebra (or, more generally, closed vector subspaces in a Dynkin algebra), then the following statements are equivalent:*

(1) $e^{\mathrm{ad}\, x} B \subseteq B$ *for all $x \in A$ (that is, $A \subseteq \mathcal{L}(B, L)$).*

(2) $[A, B] \subseteq B$ *(that is, A is in the normalizer of B).*

Proof. $(1) \Rightarrow (2)$: If $x \in A$ and $t \in \mathbb{R} \setminus \{0\}$, then $t^{-1} \cdot (e^{t \cdot \mathrm{ad}\, x} - 1_L) b \in B$ for all $b \in B$, since B is a vector space. (Note that this conclusion would break down if B were merely a wedge!) Passing to the limit for $t \to 0$ we obtain $(\mathrm{ad}\, x) b \in B$, that is $[x, b] \in B$. This proves (2).

$(2) \Rightarrow (1)$: This is a direct consequence of $e^X = 1_L + X + \frac{1}{2!} \cdot X^2 + \cdots$. ∎

II.1.8. **Corollary.** *If W is a Lie wedge, then the edge $H(W)$ is a subalgebra.*

Proof. If T is a vector space endomorphism of L and $TW \subseteq W$, then $T(H(W)) = T(W \cap -W) \subseteq TW \cap -TW = H(TW) \subseteq H(W)$. We apply this with $T = e^{\mathrm{ad}\, x}$ for $x \in H(W)$ and find $e^{\mathrm{ad}\, x} H(W) \subseteq H(W)$ for all $x \in H(W)$. By Lemma II.1.7, we have $[H(W), H(W)] \subseteq H(W)$, as asserted. ∎

One should not be led to the conclusion that every wedge W in a Lie algebra whose edge is a Lie algebra is also a Lie wedge. For example, every wedge with a 1-dimensional edge has a Lie algebra as edge, trivially. But if we consider the 3-dimensional simple Lie algebra $\mathrm{so}(3)$ with the basis e_j, $j \in \mathbb{Z}/3\mathbb{Z}$ such that $[e_j, e_{j+1}] = e_{j+2}$ for all j, then $W = \mathbb{R} \cdot e_0 + \mathbb{R}^+ \cdot e_1 + \mathbb{R}^+ \cdot e_2$ is not a Lie wedge since $e^{t \cdot \mathrm{ad}\, e_0} e_1 = (\cos t) \cdot e_1 + (\sin t) \cdot e_2$.

Since $H(W)$ is a Lie algebra, the Lie subalgebra $\mathrm{ad}\, H(W)$ of $\mathrm{gl}(L)$ generates an analytical subgroup of $\mathrm{Gl}(L)$, namely, the subgroup generated by $e^{\mathrm{ad}\, H(W)}$.

This subgroup leaves W invariant by Lemmas II.1.4 and II.1.7. We can view this statement as a "minimal" invariance requirement which a wedge W in a Lie algebra L has to satisfy in order to link up with the Lie algebra structure. A "maximal" one would be that W be invariant under *all* Lie algebra automorphisms.

II.1.9.   Definition.   A wedge W in a Lie algebra L is called *invariant* if

(INV) $e^{\operatorname{ad} x} W = W$ for all $x \in L$,

equivalently, if and only if $L \subseteq \mathcal{L}(W, L)$.

We say that W is *relatively invariant* if

(RINV) $e^{\operatorname{ad} x} W = W$ for all $x \in W$,

equivalently, if and only if $W \subseteq \mathcal{L}(W, L)$.

In view of Lemma II.1.4, W is relatively invariant if and only if $W - W \subseteq \mathcal{L}(W, L)$ if and only if the Lie algebra generated by W is in $\mathcal{L}(W, L)$. In other words, W is relatively invariant if and only if W is invariant in the Lie algebra generated by W. If W is generating, then there is no difference between invariance and relative invariance.

Obviously, there is a hierarchy

invariant $\Longrightarrow$ relatively invariant $\Longrightarrow$ Lie wedge.

II.1.10.   Proposition.   *If W is an invariant wedge in a finite dimensional Lie algebra L, then $H(W)$ and $W - W$ are ideals in L. The pointed cone $W/H(W)$ associated with W is invariant in $L/H(W)$, and the vector space $(W - W)/H(W)$ generated by it is an ideal of $L/H(W)$.*

Proof.   The first assertions follow from the validity of the relations $e^{\operatorname{ad} x} H(W) \subseteq H(W)$ and $e^{\operatorname{ad} x}(W - W) \subseteq W - W$ for all $x \in L$. The remainder is then straightforward.   ∎

In view of this proposition, in many cases *the study of invariant wedges reduces to the study of* pointed *invariant cones in Lie algebras*, and indeed often to that of *pointed* generating *invariant cones*.

Let us pause and look at the concepts we have considered. The data are the Lie algebra L and a wedge W in L. These data are purely algebraic and geometric. The definitions, however, do not fully reflect this character, because they rely on the analysis of convergent power series in the algebra of operators on L. The preparations we made in Chapter I, nevertheless, allow us to remedy this situation and give characterizations which refer only to the Lie bracket and the geometric data of the wedge W such as the tangent spaces. In fact the key theorem is an immediate consequence of earlier work in Section 5 of Chapter I.

II.1.11.   **Theorem.**   (The Theorem on Wedges in Lie Algebras) *If W is a wedge in a finite dimensional Lie algebra L, then for any element $y \in L$, the following statements are equivalent:*

(1) $y \in \mathcal{L}(W, L)$ (*that is, $e^{\operatorname{ad} y} W = W$*).

(2) $[x, y] \in T_x$ *for all $x \in W$.*

(3) $[x, y] \in T_x$ *for all $x \in C^1(W)$.*

(4) $[x, y] \in T_x$ *for all $x \in E^1(W)$.*

Proof.   This is straightforward from Theorem I.5.27 with $\operatorname{ad} y = X$.   ∎

This theorem now allows us to write down characterization theorems for Lie wedges and invariant wedges in terms of algebra and geometry exclusively.

First the Lie wedges!

II.1.12.   **Theorem.**   (The Characterization Theorem for Lie Wedges) *For a wedge W in a finite dimensional Lie algebra L, the following conditions are equivalent:*

(1) W *is a Lie wedge.*

(2) $[x, H(W)] \subseteq T_x$ *for all $x \in W$.*

(3) $[x, H(W)] \subseteq T_x$ *for all $x \in C^1(W)$.*

(4) $[x, H(W)] \subseteq T_x$ *for all $x \in E^1(W)$.*

Proof.   From Proposition II.1.6 we know that W is a Lie wedge if and only if $H(W) \subseteq \mathcal{L}(W, L)$. The theorem is then immediate from Theorem II.1.11 above.   ∎

In exactly the same vein, we have the following two theorems:

II.1.13.   **Theorem.**   (The Characterization Theorem for Relatively Invariant Wedges) *For a wedge W in a finite dimensional Lie algebra L, the following conditions are equivalent:*

(1) W *is relatively invariant.*

(2) $[x, W] \subseteq T_x$ *for all $x \in W$.*

(3) $[x, W] \subseteq T_x$ *for all $x \in C^1(W)$.*

(4) $[x, W] \subseteq T_x$ *for all $x \in E^1(W)$.*   ∎

II.1.14.   **Theorem.**   (The Characterization Theorem for Invariant Wedges) *For a wedge W in a finite dimensional Lie algebra L, the following conditions are equivalent:*

(1) W *is invariant.*

(2) $[x, L] \subseteq T_x$ *for all $x \in W$.*

(3) $[x, L] \subseteq T_x$ *for all $x \in C^1(W)$.*

(4) $[x, L] \subseteq T_x$ *for all $x \in E^1(W)$.*   ∎

From Definition II.1.3 as well as from the Characterization Theorem for Lie Wedges II.1.12 one observes at once that every pointed cone in any Lie algebra is automatically a Lie wedge. For easy reference we record:

II.1.15.   Corollary.   *Every pointed cone in a finite dimensional Lie algebra is a Lie wedge.*   ∎

Furthermore, the following remark is also immediate:

II.1.16.   Corollary.   *Let W be a vector space in a finite dimensional Lie algebra L. Then W is a Lie wedge if and only if W is a subalgebra, if and only if W is relatively invariant. Moreover, W is invariant if and only if W is an ideal of L.*

Proof.   If W is a vector space, then $W = H(W) = W - W$ and $T_x = W$ for all $x \in W$. The assertions are then immediate from Theorems II.1.12, 13, 14.   ∎

A general theory of Lie wedges is not available today. We know the Lie wedges in low dimensional algebras such as $sl(2, \mathbb{R})$ (see Section 3 below). The question whether every Lie wedge generates a local semigroup whose precise tangent wedge it is turns out to be difficult. It will be discussed in detail and answered affirmatively in Chapter IV.

The theory of invariant wedges is, except for certain details, in a satisfactory state and will be the subject of Chapter III.

The remainder of the present chapter—which will not be short by any means!—will be concerned with a special kind of Lie wedge which interpolates the hierarchy of wedges between Lie wedges and relatively invariant wedges. Let us briefly discuss now why there is a demand for such an interpolation.

Let us consider the 3-dimensional *Heisenberg algebra* $L = \mathbb{R} \cdot p \oplus \mathbb{R} \cdot q \oplus \mathbb{R} \cdot e$ with $[p, q] = e$ and $[h, e] = 0$ for all $h \in L$. We write $(\xi, \eta, v) = \xi \cdot p + \eta \cdot q + v \cdot e$ and note $(\nu, \mu, u) * (\xi, \eta, v) = (\nu + \xi, \mu + \eta, u + v + \frac{1}{2} \cdot (\nu\eta - \mu\xi))$. If we set $x = \nu \cdot p + \mu \cdot q$ and $y = \xi \cdot p + \eta \cdot q$, further $x \wedge y = \nu\eta - \mu\xi$, then $(x, u) * (y, v) = (x + y, u + v + \frac{1}{2}(x \wedge y))$. The wedge $W = \mathbb{R}^+ \cdot p + \mathbb{R}^+ \cdot q = \{(s, t, 0) : s, t \in \mathbb{R}^+\}$ is a Lie wedge by Corollary II.1.15. We compute $(s, 0, 0) * (0, s, 0) = (s, s, s^2/2)$. This means that no matter how small a positive s is chosen, this product is outside W. Indeed, as we shall see later in Chapter V, the semigroup S generated in the Lie group $(L, *)$ by W is $\{(\xi, \eta, v) : 0 \leq \xi, \eta, \quad v \leq \frac{1}{2}\xi\eta\}$. (Indeed: let q be the quadratic form on $\mathbb{R}^2$ given by $q(x, y) = \nu\eta + \mu\xi$, then $x \wedge y = \nu\eta - \mu\xi \leq q(x, y)$ since $2\mu\xi \geq 0$ as $\mu, \xi \geq 0$. Thus $u \leq \frac{1}{4}q(x, x)$ and $v \leq \frac{1}{4}q(y, y)$ imply $u + v + \frac{1}{2}(x \wedge y) \leq \frac{1}{4}(q(x, x) + q(y, y) + 2q(x, y)) = \frac{1}{4}q(x + y, x + y)$; similarly, $-\frac{1}{4}q(x + y, x + y) \leq u + v + \frac{1}{2}(x \wedge y)$. Moreover, $(s, 0, 0) * (0, t, 0) = (s, t, \frac{1}{2}st)$ and $(0, t, 0) * (s, 0, 0) = (s, t, -\frac{1}{2}st)$.)

It is a very prevalent phenomenon that a (local) semigroup S having a prescribed Lie wedge W as tangent object $\mathbf{L}(S)$ is not contained in W no matter how small the C-H-neighborhood B of reference may be chosen. This is in vivid contrast with the case of subalgebras and local Lie groups: Local Lie groups are always ruled by local one parameter semigroups, if they are chosen sufficiently small.

This calls for an investigation of those local semigroups which are ruled by local one parameter semigroups or which, equivalently, are locally divisible. A Lie wedge W generating such a local semigroup determines a sufficiently small C-H-neighborhood B in L such that $(W \cap B) * (W \cap B) \subseteq W$. Such a wedge will be called a Lie semialgebra. The investigation of these wedges is the objective of the remainder of this chapter.

2. Lie Semialgebras

This section is devoted to the development of the basic theory of Lie semi-algebras and their fundamental invariants and characterizations.

The analytic function g(X)

II.2.1.   **Definition.**   In the algebra $\mathbb{Q}[[X]]$ of all formal power series in one variable over the field $\mathbb{Q}$ of rationals we set

$$f(X) = \frac{e^{-X} - 1}{-X} = 1 - \frac{1}{2!}X + \frac{1}{3!}X^2 + \cdots + \frac{(-1)^n}{(n+1)!}X^n + \cdots,$$

and

$$g(X) = f(X)^{-1} = \frac{X}{1 - e^{-X}} = 1 + \frac{1}{2}X + \sum_{n=1}^{\infty} \frac{b_{2n}}{(2n)!}X^{2n}$$

with the Bernoulli numbers b_n. (Cf. [Bou61], Chap. VI, §1, n° 4 ff.)

II.2.2.   **Lemma.**   (i) *The series* $f(X)$ *defines an entire function* $z \mapsto f(z)$: $\mathbb{C} \to \mathbb{C}$ *whose zeros are of the first order and are in the points of* $P \overset{\text{def}}{=} 2\pi i\mathbb{Z} \setminus \{0\}$.

(ii) *The radius of convergence of the power series* $g(X)$ *in* $\mathbb{C}$ *is* 2π, *and the formula* $g(z) = f(z)^{-1}$ *defines an analytic function* $\mathbb{C} \setminus P \to \mathbb{C}$ *which extends the function defined by the power series* $g(X)$.

(iii) *On* $\mathbb{R}$ *both* f *and* g *are positive. Both map* $\mathbb{R}$ *diffeomorphically onto the positive half-line.*

(iv) *If* A *is a complex Banach algebra, then the function* $f: A \to A$ *defined by* $f(u) = \sum_{n=0}^{\infty} (n+1)!^{-1}(-u)^n$ *is analytic. If* $u \in A$ *is such that* $\operatorname{Spec} u$ *does not meet* P, *then* $g(u) \in A$ *is well defined by the analytic functional calculus (for instance by* $g(u) = \frac{1}{2\pi i} \int g(z)(z-u)^{-1}dz$, *the integral extended over a simply closed rectifiable curve bounding an open neighborhood of* $\operatorname{Spec} u$ *whose closure does not meet* P). *Moreover,* $g(u) = f(u)^{-1}$.

(v) *If* L *is a Banach Lie algebra (satisfying* $\|[x,y]\| \le \|x\| \cdot \|y\|$), *and if* D *is an open subset of* L *such that* $P \cap \operatorname{Spec} \operatorname{ad} u = \emptyset$ *for all* $u \in D$, *then for* *each $y \in L$ the function $x \mapsto f(\operatorname{ad} x)y \colon L \to L$ and the function $x \mapsto g(\operatorname{ad} x)y = f(\operatorname{ad} x)^{-1}y \colon D \to L$ are analytic. The function $g(\operatorname{ad} x) \colon L \to L$ is an automorphism of the vector space L for each $x \in D$.*

Proof.   (i) The function f given by

$$f(z) = \frac{e^{-z} - 1}{-z} = \sum_{n=0}^{\infty} \frac{(-z)^n}{(n+1)!}$$

is an entire function. Except for 0, at which f takes the value 1, the functions f and $z \mapsto e^{-z} - 1$ have the same zeros of the same order.

(ii) The function $g \colon \mathbb{C} \setminus P \to \mathbb{C}$ defined by $g(0) = 1$ and $g(z) = z(1 - e^{-z})^{-1} = f(z)^{-1}$ for $z \neq 0$ is holomorphic and has poles of the first order in the points of P. The power series expansion of g around 0 therefore converges on the largest open disc contained in the domain of holomorphy, and this disc has radius 2π.

(iii) We note $f'(0) = -1/2$ and

$$f'(z) = \frac{ze^{-z} + (e^{-z} - 1)}{z^2} = -\frac{e^{-z}(e^z - 1 - z)}{z^2} \quad \text{for} \quad z \neq 0.$$

If z is real non-zero, this number is negative. Hence f is strictly decreasing on $\mathbb{R}$. For positive z, the value of $f(z) = \frac{1 - e^{-z}}{z}$ is positive. Hence $f(z) > 0$ for all $z \in \mathbb{R}$ follows. Since $\lim_{z \to \pm\infty} f(z) = \left\{ \begin{smallmatrix} 0 \\ \infty \end{smallmatrix} \right.$ we deduce that f maps $\mathbb{R}$ diffeomorphically onto the positive half-line. Hence $g(z) = 1/f(z)$ is likewise positive and maps $\mathbb{R}$ diffeomorphically onto the positive half-line in a strictly increasing fashion.

(iv) The assertions of (iv) belong to the area of functional calculus in Banach algebras for which we refer to [Bou67], Chap. I, §4, n° 8.

In order to establish (v), we let A be the Banach algebra of all bounded operators on L and note that $\operatorname{ad} \colon L \to A$ is analytic, as is the function $T \mapsto Ty \colon A \to L$. Thus $x \mapsto g(\operatorname{ad} x)y \colon D \to L$ is analytic. Since $g(\operatorname{ad} x)$ has, for $x \in D$, the inverse $f(\operatorname{ad} x)$, the last assertion follows.   ∎

Now let us fix a Banach Lie algebra L and an open set $D \subseteq L$ such that $\operatorname{Spec}(\operatorname{ad} u)$ never meets P for any $u \in D$. We fix $y \in L$. Then $X \colon D \to L$, where $X(x) = g(\operatorname{ad} x)y$ is a locally Lipschitzian vector field (being even analytic by Lemma II.2.2.iv!). In Section 5 of Chapter I we already observed that such a vector field generates a local flow.

If now W is a wedge in L and L is finite dimensional, we can apply the results of Section 5 of Chapter I to determine when $W \cap U$ is invariant under this local flow, that is, when $W \cap U$ is invariant under X (see Definition I.5.16). For a better understanding of the following result, we recall from Proposition I.5.3 that the set of subtangent vectors of W at x is $L_x(W) = \overline{W - \mathbb{R}^+ \cdot x}$.

II.2.3.   Proposition. (The g-Invariance Theorem for Wedges in Lie algebras) *Let W be a generating wedge in a finite dimensional Lie algebra L. Let D be an* *open neighborhood of* 0 *in* L *such that* $x \in D$ *implies* $(\mathrm{Spec\,ad}\,x) \cap P = \emptyset$. *For any* $y \in L$ *consider the non-linear vector field*

$$X \colon D \to L, \qquad X(x) = g(\mathrm{ad}\,x)y.$$

Then the following statements are equivalent:

(1) $W \cap D$ *is invariant under* X.

(2) $g(\mathrm{ad}\,x)y \in L_x(W)$ *for all* $x \in W \cap D$.

(3) $g(\mathrm{ad}\,x)y \in L_x(W)$ *for all* $x \in C^1(W) \cap D$.

Proof.   Since D is open, we note that $x \in W \cap D$ implies $L_x(W \cap D) = L_x(W)$. Our theorem is then a direct consequence of the Invariance Theorem for Wedges and Flows I.5.23.   ∎

We note that we cannot invoke a statement on E^1-points here, since the vector field X is not linear and thus Theorem I.5.27 does not apply.

Invariance of vector fields under local translation

We owe the reader an explanation why the vector field X defined via g is so important. The key is found in the next lemmas.

II.2.4.   Lemma. *In the algebra* $\mathbb{Q}[[X, Y]]$ *of formal power series in two non-commuting variables over the field of rationals let* I *denote the ideal of all power series whose monomial summands have at least degree* 2 *in* Y. *Then the power series* $g(\mathrm{ad}\,X)Y = Y + \frac{1}{2}[X, Y] + \frac{1}{12}[X, [X, Y]] + \cdots$ *satisfies*

$$X * Y - g(\mathrm{ad}\,X)Y - X \in I.$$

Proof.   See e.g. N.Bourbaki, Groupes et algèbres de Lie, Chap.II, §6, Ex.3c.   ∎

II.2.5.   Lemma. *Let* B *be a C-H-neighborhood in a finite dimensional Lie algebra (or, more generally, a Dynkin algebra). Let* $u \in B$ *and define* $\lambda_u \colon B \to L$ *by* $\lambda_u(x) = u * x$. *Then the derivative* $\lambda_u'(0) \colon L \to L$ *of* λ_u *at* 0 *is* $g(\mathrm{ad}\,u)$.

Proof.   By Lemma II.2.4 we have $\lambda_u(x) = u * x = u + g(\mathrm{ad}\,u)x + \mathrm{o}(x)$ with

$$\lim_{x \to 0} \|x\|^{-1} \cdot \mathrm{o}(x) = 0.$$

∎

II.2.6.   Lemma. *Let* $x \in B$ *and* $y \in L$. *We set* $u(t) = x * t{\cdot}y$ *for all* $|t| < \varepsilon$ *for a suitably chosen* $\varepsilon > 0$. *Then* $u \colon]-\varepsilon, \varepsilon[\to L$ *is the unique solution (on its domain interval) of the initial value problem*

$$\dot{u}(t) = X\big(u(t)\big), \qquad u(0) = x,$$

where

$$X(z) = g(\mathrm{ad}\,z)y.$$

Proof.   We note $u(t + h) = x * (t + h){\cdot}y = x * t{\cdot}y * h{\cdot}y = u(t) * h.y = u(t) + h{\cdot}g\big(\mathrm{ad}\,u(t)\big)y + \mathrm{o}(h)$ by Lemma II.2.5. It follows that $\dot{u}(t) = g(\mathrm{ad}\,u(t))y = X\big(u(t)\big)$ and $u(0) = x * 0{\cdot}y = x$. Since X is analytic, hence locally Lipschitzian, the solution of the initial value problem is unique on its domain.   ∎

The solution $t \mapsto u(t)$ of the initial value problem $\dot{u}(t) = X\big(u(t)\big)$, $u(0) = x$ defines the local orbit of x of the local flow generated by X via $F_t x = u(t)$. Thus we have the following remark:

II.2.7.   Remark. If B is a C-H-neighborhood of 0 in L and the vector field $X: B \to L$ is defined by $X(b) = g(\operatorname{ad} b)y$, then the equation

$$F_t x = x * t{\cdot}y.$$

gives the flow generated by X.   ∎

Definition and characterization of Lie semialgebras

We formalize the definition of a Lie semialgebra:

II.2.8.   Definition. A *Lie semialgebra* is a wedge W in a Dynkin algebra L such that there is some C-H-neighborhood B of 0 in L satisfying

(LSA) $(W \cap B) * (W \cap B) \subseteq W.$

This definition depends on B. There is no guarantee at this point that for a larger C-H-neighborhood B' of L the condition

$$(W \cap B') * (W \cap B') \subseteq W$$

is still satisfied.

It is clear from the definition that $S = W \cap B$ is a local semigroup of L with respect to B (see Section 1 above). Also, obviously $L(S) = L_0(W \cap B) = W$. Hence *every Lie semialgebra is a Lie wedge.* We have observed at the end of Section 1 that the converse is not the case. If $L = \mathbb{R}{\cdot}e + \mathbb{R}{\cdot}f$ is the non-abelian 2-dimensional Lie algebra with $[e,f] = f$, then $W = \mathbb{R}{\cdot}e + \mathbb{R}^{+}{\cdot}f$ is a half-space and a Lie semialgebra (we will soon have the tools to verify this directly; in fact in Section 3 below we shall determine all Lie semialgebras of dimensions up to 3). However, W is not an invariant wedge, for the edge of an invariant wedge is an ideal by Proposition II.1.10 and $H(W) = \mathbb{R}{\cdot}e$ is not an ideal of L.

II.2.9.   Lemma. *Let W be a Lie semialgebra in L with respect to the C-H-neighborhood B. For $y \in W$ we choose $\varepsilon > 0$ so that $t{\cdot}y \in B$ for $|t| < \varepsilon$. Then for any $x \in W \cap B$ we have*

(HW) $x * t{\cdot}y \in W \qquad \text{for all} \quad 0 \le t < \varepsilon.$

Conversely, suppose that for a wedge W, we find a C-H-neighborhood B with the following property: For all $x \in W \cap B$ and $y \in W$ there is an $\varepsilon > 0$ such that for all $0 \le t < \varepsilon$ with $t{\cdot}y \in B$ Condition (HW) *is satisfied. Then W is a Lie semialgebra.*

Proof.   Even though the formulation of the lemma is longish and technical, its proof is immediate from the definitions.   ∎

We develop this one step further. The next lemma is crucial towards a characterization theorem for Lie semialgebras:

II.2.10. **Lemma.** *Let W be a wedge in a finite dimensional Lie algebra. Then W is a Lie semialgebra if and only if there is a C-H-neighborhood B such that for all $y \in W$ the closed set $W \cap B$ in B is invariant under the vector field X defined by $X(x) = g(\operatorname{ad} x)y$, $x \in B$,*

Proof. By Lemma II.2.9, W is a Lie semialgebra if and only if there is a C-H-neighborhood B such that $x * t \cdot y \in W$ for all $x \in W \cap B$ and all $y \in W$ for all sufficiently small t. This is equivalent to saying that the trajectories $t \mapsto F_t x = x * t \cdot y$ stay in $W \cap B$ in view of Remark II.2.7. But the local flow F_t is generated by X. Hence by Definition I.5.16, the condition is equivalent to saying that $W \cap B$ is invariant under the vector field X. ∎

This lemma opens up the possibility of applying the g-Invariance-Theorem for Wedges II.2.3. As an immediate consequence of Lemma II.2.10 above and that result we obtain

II.2.11. **Lemma.** *A wedge W in a finite dimensional Lie algebra L is a Lie semialgebra if and only if the following condition is satisfied:*

(1) $$g(\operatorname{ad} x)W \subseteq L_x(W) \qquad \text{for all} \quad x \in W \cap B.$$

If W is, in addition, generating, Condition (1) is equivalent to

(2) $$g(\operatorname{ad} x)W \subseteq L_x(W) \qquad \text{for all} \quad x \in C^1(W) \cap B.$$

∎

Now we recall once more from Proposition I.5.3. that $L_x(W) = \overline{W - \mathbb{R}^+ \cdot x}$. We use this and the remark $g(\operatorname{ad} x)x = x$ to see that

$$g(\operatorname{ad} x)L_x(W) \subseteq \overline{g(\operatorname{ad} x)(W - \mathbb{R}^+ \cdot x)} \subseteq \overline{g(\operatorname{ad} x)W + \mathbb{R} \cdot x}.$$

It follows that Condition (1) of Lemma II.2.11 is equivalent to the condition

(3) $$g(\operatorname{ad} x)L_x(W) \subseteq L_x(W) \qquad \text{for all} \quad x \in W \cap B.$$

Every vector space endomorphism preserving a wedge W preserves its edge. Hence condition (3) entails

(4) $$g(\operatorname{ad} x)T_x(W) \subseteq T_x(W) \qquad \text{for all} \quad x \in W \cap B.$$

Under these circumstances, the following lemma helps:

II.2.12. **Lemma.** *Let $h(z) = 1 + a_1 z + \cdots$ be a power series on $\mathbb{C}$ with a positive radius r of convergence and with $a_1 \neq 0$. Consider a Banach space E over $\mathbb{C}$ and V a closed vector subspace. If we denote with A the Banach algebra of all bounded operators on E, then the following conditions are equivalent for $T \in A$:*

(a) *$h(t \cdot T)V \subseteq V$ for all real t of a set I of real numbers having 0 as cluster point.*

(b) *$TV \subseteq V$.*

Proof. (b)$\Rightarrow$(a) is clear since $SV \subseteq V$ implies $h(S)v = v + a_1 Sv + \cdots \in V$ for all $v \in V$ and $\|S\| < r$, as V is closed. Conversely, since $t^{-1} \cdot (h(t \cdot T) - 1)v \to a_1 Tv$ for $t \to 0$ in I, if $v \in V$ and (a) is satisfied, then $t^{-1} \cdot (h(t \cdot T) - 1)v = t^{-1} \cdot h(t \cdot T)v - t^{-1} \cdot v \in V - V = V$. Hence $a_1 \cdot Tv \in V$ and thus, since $a_1 \neq 0$, also $Tv \in V$. ∎

We note that the implication (a) $\Rightarrow$ (b) breaks down if V is only a wedge rather than a vector space.

An application of Lemma II.2.12 with $h = g$ and $T = \operatorname{ad} x$, $V = T_x(W)$ shows that (4) implies

$$(5) \qquad\qquad [x, T_x] \subseteq T_x \qquad \text{for all} \quad x \in W.$$

We note here that the restriction on x to be in B can now be dropped since $\operatorname{ad} x$ is linear. If $x \in \operatorname{algint} W$, then $T_x = W - W$. Thus (5) implies, in particular,

$$[x, W] \subseteq [x, T_x] \subseteq T_x = W - W \quad \text{for} \quad x \in \operatorname{algint} W,$$

whence $[W, W] \subseteq W - W$. Therefore (5) entails

$$(6) \qquad [x, T_x] \subseteq T_x \quad \text{for all} \quad x \in C^1(W) \quad \text{and} \quad [W, W] \subseteq W - W.$$

This enables us to secure an important property of Lie semialgebras:

II.2.13.   Proposition. *If W is a Lie semialgebra in a finite dimensional Lie algebra, then $W - W$ is a Lie algebra.*

Proof.   Let $x \in \operatorname{algint} W$. Then $T_x(W) = W - W$. By (5) we have $[x, W - W] \subseteq W - W$ for all $x \in \operatorname{algint} W$, and since $\operatorname{algint} W$ spans $W - W$, the proposition follows.   ∎

This proposition allows us to restrict our attention to generating Lie semialgebras without losing generality. We should recall in the context of this proposition that it is not hard to find a Lie algebra with a pointed cone W such that $W - W$ is not a Lie algebra; we know from Corollary II.1.15 that every pointed cone is a Lie wedge. At the end of Section 1 we had an example in the Heisenberg algebra.

We now propose to show that Condition (6) implies Condition (3). Once this is shown we know that Conditions (1), (3), ..., (6) are equivalent.

So we assume (6). In view of $[W, W] \subseteq W - W$ we may and shall assume that $L = W - W$. We now apply Lemma II.2.12. If $g(\operatorname{ad} x)T_x \subseteq T_x$ for $x \in C^1(W) \cap B$, then T_x is a hyperplane, bounding the half-space $L_x = L_x(W)$. Let $w \in \operatorname{int} W$. Since $t \cdot x \in C^1(W) \cap B$ for $t \in [0, 1]$, and $T_{t \cdot x} = T_x$, we know that $g(t \cdot \operatorname{ad} x)T_x \subseteq T_x$ for all $t \in [0, 1]$. Thus $g(t \cdot \operatorname{ad} x)$ either preserves the two half-spaces bounded by T_x or else exchanges them. Now $t \mapsto g(t \cdot \operatorname{ad} x)w : [0, 1] \to L$ is a continuous curve joining w and $g(\operatorname{ad} x)W$. But this curve does not penetrate the hyperplane T_x since $g(\operatorname{ad} t \cdot x)$ is an automorphism by Lemma II.2.2.iv and T_x is $g(\operatorname{ad} t \cdot x)$-invariant. It follows for reasons of connectivity that $g(\operatorname{ad} x)$ leaves the half-space invariant which contains W. That half-space is exactly L_x and this establishes our claim.

Now we are ready for the crucial characterization theorem for Lie semi-algebras.

II.2.14.   **Theorem.**   (The Characterization Theorem for Lie Semialgebras) *For a wedge W and a Campbell–Hausdorff neighborhood B in a Lie algebra L, the following conditions are equivalent:*

    (1) W *is a Lie semialgebra.*

    (2) $[x, T_x] \subseteq T_x$ *for all $x \in W$.*

    (3) $g(\operatorname{ad} x)(W) \subseteq \overline{W - \mathbb{R}^+ \cdot x}$ *for all $x \in W \cap B$.*

If W is also generating, then these conditions are also equivalent to

    (4) $[x, T_x] \subseteq T_x$ *for all $x \in C^1(W)$.*

Proof.   This is just a summary of the preceding discussion.   ∎

    It is instructive to juxtapose the characterization theorems for Lie wedges, Lie semialgebras, relatively invariant wedges, and invariant wedges in the following

II.2.15.   **Scholium.**   (The Hierarchy of Wedges in a Lie Algebra) *A wedge in a finite dimensional Lie algebra L is a*

(A)	*Lie wedge if and only if*	$[x, H(W)] \subseteq T_x$,
(B)	*Lie semialgebra if an only if*	$[x, T_x] \subseteq T_x$,
(C)	*relatively invariant if and only if*	$[x, W] \subseteq T_x$,
(D)	*invariant wedge if and only if*	$[x, L] \subseteq T_x$,

for all $x \in W$ (respectively, for all $x \in C^1(W)$, if W is generating). There is a sequence of implications

Lie wedge $\Leftarrow$ Lie semialgebra $\Leftarrow$ relatively invariant wedge $\Leftarrow$ invariant wedge.   ∎

    In Cases (A),(C) and (D), we may test the characterizing condition also for all $x \in E^1(W)$. However, the exposed points have to be dropped as a tool for characterizing Lie semialgebras. This is inherent in the nature of things. Indeed suppose that W is a polyhedral wedge. Then for an exposed point x we have $T_x = \mathbb{R} \cdot x$ and $[x, T_x] = [x, \mathbb{R} \cdot x] = \{0\}$. However, if we recall the example of the Heisenberg algebra $L = \mathbb{R} \cdot p \oplus \mathbb{R} \cdot q \oplus \mathbb{R} \cdot e$ in the end of Section 1, then $W = \mathbb{R}^+ \cdot p + \mathbb{R} \cdot q$ is a polyhedral wedge, hence satisfies $[x, T_x] \subseteq T_x$ for all exposed points (trivially!), but W is only a Lie wedge, yet not a Lie semialgebra as we saw at the end of Section 1.

    The Characterization Theorem II.2.14 frees the definition of a Lie semialgebra from the special role of one given C-H-neighborhood.

II.2.16.   **Corollary.**   *For a wedge W in a finite dimensional Lie algebra, the following conditions are equivalent:*

    (1) W *is a Lie semialgebra.*

    (2) *For any C-H-neighborhood B of L the condition $(W \cap B) * (W \cap B) \subseteq W$ is satisfied.*

Proof.   Clearly, $(2) \Rightarrow (1)$. Now suppose, conversely, that (1) is satisfied and let B be an arbitrary C-H-neighborhood. We pick $x, y \in W \cap B$ so that $x * y \in B * B$. Now we must show that $x * y \in W$. For this purpose we consider the trajectory $u(t) = x * t \cdot y$, $t \in [0,1]$. Then $u(0) = x \in W$. The vector field $X: B * B \to L$, $X(b) = g(\operatorname{ad} b)y$, generates a local flow and $F_t x = u(t)$ is the orbit of x. By

Theorem II.2.14 and the Invariance Theorem for Wedges and Vector Fields I.5.23, $W \cap B * B$ is invariant under X. Hence the trajectory u stays in W. Thus $x * y = u(1) \in W$. ∎

We shall pursue the ideas around this corollary in greater depth under the heading "Analytic Extension Aspects" at the end of this section.

As a consequence of the preceding corollary, we can now show that the intersection of any family of Lie semialgebras is a Lie semialgebra:

II.2.17.   Corollary.   *Let $\{W_j : j \in J\}$ be a family of Lie semialgebras in a finite dimensional Lie algebra L. Then*

$$W = \bigcap_{j \in J} W_j$$

is a Lie semialgebra.

Proof.   We fix a C-H-neighborhood B of L. Then, by Corollary II.2.15,

$$(W_j \cap B) * (W_j \cap B) \subseteq W_j \qquad \text{for all} \quad j \in J.$$

As a consequence, since $W \subseteq W_j$ for all j,

$$(W \cap B) * (W \cap B) \subseteq W_j \qquad \text{for all} \quad j \in J.$$

But this entails

$$(W \cap B) * (W \cap B) \subseteq W,$$

whence W is a Lie semialgebra. ∎

Faces of Lie semialgebras

The information from Chapter I on the geometry of wedges is now used to provide insight into the facial structure of Lie semialgebras. We recall from Definition I.2.6 that $E_x = E_x(W)$ denotes the smallest exposed face of a wedge W containing the element x. By Proposition I.2.7, we have $E_x = T_x \cap W$. In view of Proposition I.2.23, *every exposed face of a finite dimensional wedge W is of the form E_x for some $x \in W$*. We recall furthermore from Lemma II.1.7, that for any vector subspace T of a finite dimensional Lie algebra L we have

$$\mathcal{L}(T, L) = \{x \in L : e^{\operatorname{ad} x} T = T\} = \{x \in L : [x, T] \subseteq T\}.$$

From Lemma II.1.4 we know already, that $\mathcal{L}(T, L)$ is in fact a Lie subalgebra of L. We make the following observation:

II.2.18. **Lemma.** $T \cap \mathcal{L}(T,L)$ *is a Lie algebra contained in T and T is a* $\mathcal{L}(T,L)$*-module, hence, in particular, a $T \cap \mathcal{L}(T,L)$-module.*

Proof. If $x \in \mathcal{L}(T,L)$ and $y \in T$, then $[x,y] \in [\mathcal{L}(T,L),T] \subseteq T$. In particular, if both x and y are in $T \cap \mathcal{L}(T,L)$, then $[x,y] \in T \cap \mathcal{L}(T,L)$ since $\mathcal{L}(T,L)$ is a Lie algebra. The lemma is now clear. ∎

II.2.19. **Lemma.** *Let W be a Lie semialgebra in a finite dimensional Lie algebra and let T be a tangent space of W at some point, that is, $T = T_z$. Then the following conclusions hold:*

(i) $\{x \in W : T_x = T\} \subseteq T \cap \mathcal{L}(T,L)$.

(ii) *The exposed face $F = W \cap T$ of W is contained in $T \cap \mathcal{L}(T,L)$, that is,* $[F,T] \subseteq T$.

(iii) *F is a Lie semialgebra, and $F - F$ is a Lie algebra.*

(iv) *T is an $F - F$-module.*

Proof. (i) is a consequence of the Characterization Theorem for Lie Semialgebras II.2.14 (2), after which $[x, T_x] \subseteq T_x$ for all $x \in W$.

(ii): Let $x \in \text{algint}\, F$. Then $T_x = T$ and $E_x = T \cap W = F$. Now (i) implies $x \in \mathcal{L}(T,L)$, and since the algebraic interior of F is dense in F we conclude $E_x = F \subseteq \mathcal{L}(T,L)$. But this means exactly $[F,T] \subseteq T$.

(iii): We have $F \subseteq T \cap \mathcal{L}(T,L)$ by (ii) above. Hence $F = W \cap T = W \cap (T \cap \mathcal{L}(T,L))$. But W is a Lie semialgebra by hypothesis and $T \cap \mathcal{L}(T,L)$ is a Lie algebra by Lemma II.2.18. Hence F is a Lie semialgebra by Corollary II.2.17. Then $F - F$ is a Lie algebra by Proposition II.2.13.

(iv): From (ii) we know $[F,T] \subseteq T$, and thus $[F - F, T] \subseteq T$, and this makes T an $F - F$-module. ∎

II.2.20. **Theorem.** *Let W be a semialgebra in a finite dimensional Lie algebra. Then every exposed face F of W is a Lie semialgebra and $F - F$ is a Lie subalgebra of L. If T is the tangent space of W in the face F (according to Remark I.1.10), then $[F - F, T] \subseteq T$, that is, T is an $F - F$-submodule of L.*

Proof. Lemma II.2.19 completely settles the proof of the theorem. ∎

II.2.21. **Remark.** If W is a wedge in a Lie algebra L and each *exposed* face F and the tangent space T of W in F satisfy $[F,T] \subseteq T$, then W is a Lie semialgebra.

Proof. Let $x \in W$. Then $[x, T_x] \subseteq [E_x, T_x] \subseteq T_x$ by hypothesis, and so W is a Lie semialgebra by the Characterization Theorem for Lie Semialgebras II.2.14. ∎

EII.2.1. **Exercise.** Discuss the situation for invariant wedges in place of Lie semialgebras.

In Chapter I, Section 4 we discussed polyhedral wedges in finite dimensional vector spaces. For a polyhedral wedge W we have $T_x = E_x - E_x$. By contrast, in a Lorentzian cone of dimension 3 or more, all tangent spaces T_x are hyperplanes, while all spaces $E_x - E_x$ are 1-dimensional. Thus, for polyhedral wedges, the following corollary is relevant:

II.2.22.   Corollary.   *Let W be a wedge in a finite dimensional Lie algebra such that $T_x = E_x - E_x$ for all $x \in W$. Then the following statements are equivalent:*

(1) *W is a Lie semialgebra.*

(2) *Every tangent space of W is a Lie subalgebra.*

Proof.   If (1) is satisfied, then $T_x = E_x - E_x$ is a Lie algebra by Theorem II.2.20. Conversely, if (2) is satisfied, then $[x, T_x] \subseteq [T_x, T_x] \subseteq T_x$, and thus W is a Lie semialgebra by the Characterization Theorem for Lie Semialgebras II.2.14.   ∎

The proof of the following corollary, which is modelled after the preceding one, is left to the reader as an exercise:

II.2.23.   Corollary.   *If W is a wedge with $T_x = E_x - E_x$ for all $x \in W$ in a Lie algebra, then the following statements are equivalent:*

(1) *W is an invariant wedge.*

(2) *Every tangent space of W is an ideal.*   ∎

EII.2.2.   Exercise.   Prove Corollary II.2.23.

Half-space Semialgebras

One of the most important special cases of a polyhedral wedge is that of a half-space. We record this special case separately.

II.2.24.   Corollary.   *Let W be a half-space in a finite dimensional Lie algebra. Then the following conditions are equivalent:*

(1) *W is a Lie semialgebra.*

(2) *The boundary hyperplane $H(W) = \partial W$ of W is a Lie algebra.*

Also, the following statements are equivalent:

(1) *W is an invariant wedge.*

(2) *The boundary hyperplane $H(W) = \partial W$ of W is an ideal.*

∎

Since we will have to deal with this type of semialgebra quite frequently in the further development of the theory we introduce a definition:

II.2.25.   Definition.   Any half-space in a finite dimensional Lie algebra is called a *half-space semialgebra* if its boundary hyperplane is a Lie subalgebra.

We observe that every hyperplane subalgebra H of a Lie algebra L determines two half-space semialgebras, namely, the two half-spaces bounded by H.

II.2.26.   Remark.   The intersection of any family of half-space semialgebras is a Lie semialgebra.

Proof.   This is an immediate consequence of Corollary II.2.17.   ∎

In very low dimensional Lie algebras, these results become particularly incisive because of the following simple remark:

II.2.27.   **Lemma.**   *Let F be a Lie subalgebra of a Lie algebra L and let T be a vector subspace with $F \subseteq T$ and $[F, T] \subseteq T$. If $\dim T \leq 1 + \dim F$, then T is a subalgebra.*

Proof.   Let $x \in T$ but $x \notin F$. Then $T = F + \mathbb{R} \cdot x$ and thus $[T, T] \subseteq [F + \mathbb{R} \cdot x, F + \mathbb{R} \cdot x] \subseteq [F, F] + [F, x] \subseteq [F, T] \subseteq T$. ∎

A consequence is the following proposition:

II.2.28.   **Proposition.**   *A wedge W in a Lie algebra of dimension not exceeding 3 is a Lie semialgebra if and only if all of its tangent spaces are Lie algebras. Every generating Lie semialgebra in such a Lie algebra is the intersection of half-space semialgebras.*

Proof.   Suppose that $\dim L \leq 3$ and that W is a wedge in L. Then for every $x \in W$, we have $\dim T_x \leq \dim(E_x - E_x) + 1$. Thus $[E_x, T_x] \subseteq T_x$ if and only if T_x is a subalgebra by the preceding Lemma II.2.27. Hence W is a Lie semialgebra precisely if all T_x are Lie algebras.

Suppose now that W is a generating Lie semialgebra. Then each half-space $\overline{W - \mathbb{R}^+ \cdot x}$ for $x \in C^1(W)$ is a half-space semialgebra by Corollary II.2.24 since T_x, its boundary hyperplane, is a Lie subalgebra by what we just saw. But W is the intersection of all half-spaces $L_x(W) = \overline{W - \mathbb{R} \cdot x}$ with $x \in C^1(W)$ by Proposition I.3.12. The proposition is proved. ∎

In the next section we shall classify all Lie semialgebras of dimension less than 4 (and some 4-dimensional types).

Almost abelian Lie algebras

Let us consider a Lie algebra L all of whose vector subspaces are subalgebras. What can we say about L? Certainly all abelian Lie algebras have this property. Perhaps a bit surprisingly, a Lie algebra with this property need not be abelian.

If N is an abelian ideal of L, then L/N acts on N via $(x + N) \cdot a = [x, a]$. If N is also a hyperplane, then this action is completely determined by that of one element $e + N$, $e \notin N$, since L/N is 1-dimensional. Let us make the following definition:

II.2.29.   **Definition.**   A Lie algebra L is called *almost abelian* if there is a hyperplane ideal N such that

(i) $[N, N] = \{0\}$,

(ii) there is a functional $\omega \in \widehat{L}$ such that $[x, n] = \langle \omega, x \rangle \cdot n$ for all $x \in L$ and $n \in N$.

We note at once that an almost abelian Lie algebra is abelian if and only if $\omega = 0$. If $\omega \neq 0$, then N is the nilradical of L.

The almost abelian Lie algebras are characterized in the following theorem.

II.2.30.   Theorem.   (Characterization Theorem of Almost Abelian Lie Algebras) *For a finite dimensional Lie algebra L, the following statements are equivalent:*

(1) L *is almost abelian.*

(2) *Every hyperplane in L is a subalgebra.*

(3) *Every vector space in L is a subalgebra.*

(4) *Every wedge W in L is a Lie semialgebra.*

(5) *Every half-space is a Lie semialgebra.*

Proof.   Since every vector space is an intersection of hyperplanes, statements (2) and (3) are equivalent. (3)$\Rightarrow$(4): Let W be a wedge and $x \in W$. Then T_x is a subalgebra by (3), whence $[x, T_x] \subseteq T_x$, and thus W is a Lie semialgebra by the Characterization Theorem for Lie Semialgebras II.2.14. Trivially, (4) implies (5), and (5) implies (2) by Corollary II.2.24. Thus (2) through (5) are equivalent.

(1)$\Rightarrow$ (2): Let H be a hyperplane in L. We must show that H is a subalgebra. If $H = N$, then H is even an ideal. If $H \neq N$, then there is an $h \in H \backslash N$ and $H = \mathbb{R} \cdot h \oplus (H \cap N)$. But then $[H, H] \subseteq [h, H \cap N] + [H \cap N, H \cap N] \subseteq H \cap N$ since $[h, x] = \langle \omega, h \rangle \cdot x$ for all $x \in H \cap N$ and N is abelian.

(3)$\Rightarrow$(1): We assume (3) and claim that L is solvable. For if not, then L would contain a subalgebra which is isomorphic to $\mathrm{sl}(2, \mathbb{R})$ or to $\mathrm{so}(3)$, and both of these algebras contain two dimensional vector spaces which are not subalgebras. Let N be the nilradical of L. We claim that N is abelian. Assume that this is not the case. Then the last term of the descending central series of N is non-zero and central. Thus there are two elements p and q such that $[p, q] \neq 0$ and $[p, q]$ is central. Then the vector space V spanned by p and q cannot contain any non-zero central element $s \cdot p + t \cdot q$; for $[p, s \cdot p + t \cdot q] = 0$ implies $t = 0$. Similarly we show $s = 0$. Therefore, V is not a subalgebra. This contradiction shows that N must be abelian.

If $N = L$, then L is abelian and we are finished. Suppose $N \neq L$. Then L is not abelian and $[L, L] \subseteq N$ (see N.Bourbaki, Groupes et algèbres de Lie, Chap. I, §6 n° 4, Prop.6). Let $e \in L \backslash N$. Then $[e, N] \neq \{0\}$ because otherwise $L = \mathbb{R} \cdot e + N$ would be abelian contrary to our assumption. Now $[e, n] = s \cdot n$ since N is an ideal and $\mathbb{R} \cdot e + \mathbb{R} \cdot n$ a subalgebra. Suppose that $[e, m] = t \cdot m$ for an $m \in N$. Then $[e, m + n] = r \cdot (m + n)$ with a suitable r on the one hand and $[e, m + n] = s \cdot m + t \cdot n$ on the other. If m and n are linearly independent, this implies $s = r = t$. This allows us to define a function $\omega \colon L \to \mathbb{R}$ by $[x, n] = \omega(x) \cdot n$ for $x \in L$ and $n \in N$. Clearly, ω is linear and $N \subseteq \ker \omega$. However, the relation $k \in \ker \omega$ implies that k is in the centralizer of N. We have shown above that N is abelian. Hence $N + \mathbb{R} \cdot K$ is an abelian ideal. Since N is the largest such ideal, $k \in N$. Hence $N = \ker \omega$ and N is a hyperplane. Thus L is almost abelian.   ∎

II.2.31.   Proposition.   *If W is a non-zero invariant wedge in a non-abelian almost abelian Lie algebra L with nilradical N, then W is L, or a half-space bounded by N, or an arbitrary wedge in N.*

Proof.   By Proposition II.1.10, the vector space $I = W - W$ spanned by W is an ideal. Hence it is L or a vector subspace of N. In the latter case, W is a wedge in N. Every wedge in N is invariant. If $I = L$, then W is generating.

Let $x \in W \setminus N$. Then $[x, L] \subseteq T_x$ by Theorem II.1.14. Now let $n \in N$. Then $\langle \omega, x \rangle \cdot n = [x, n] \in T_x$. Since L is non-abelian and $x \notin N$ we have $\langle \omega, x \rangle \neq 0$. Hence $n \in T_x$. Thus $N \subseteq T_x$ for all $x \in W \setminus N$. Then $N \subseteq H(W)$ by Proposition I.3.13. The claim now follows. ∎

We remark that every non-abelian almost abelian Lie algebra is, in particular, metabelian, that is, has an abelian commutator algebra. It is not hard to verify by explicit calculation that every almost abelian Lie algebra is exponential in the sense that the exponential function is a diffeomorphism onto one of the associated Lie groups:

Indeed, any $(n + 1)$-dimensional non-abelian almost abelian Lie algebra L is isomorphic to the Lie algebra of all n by n matrices of the form

$$\begin{pmatrix} t \cdot E_n & X \\ 0 & 0 \end{pmatrix}, \qquad X = \begin{pmatrix} x_1 \\ \vdots \\ x_n \end{pmatrix}, \qquad t, x_1, \ldots, x_n \in \mathbb{R}$$

under the usual Lie bracket $[A, B] = AB - BA$. The corresponding Lie group G is the Lie group of all matrices

$$\begin{pmatrix} t \cdot E_n & X \\ 0 & 1 \end{pmatrix}$$

with t and X as above. The exponential function $\exp \colon L \to G$ is then calculated via the exponential series and is given by

$$\exp \begin{pmatrix} t \cdot E_n & X \\ 0 & 0 \end{pmatrix} = \begin{pmatrix} e^t \cdot E_n & f(-t) \cdot X \\ 0 & 1 \end{pmatrix}$$

with $f(-t) = (e^t - 1)/t$ (see Definition II.2.1). Since f maps $\mathbb{R}$ diffeomorphically onto the positive half-line, the function $\exp$ is a diffeomorphism from L onto G.

Thus the Campbell–Hausdorff multiplication on any C-H-neighborhood of L has an analytic extension to a global group operation $(x, y) \mapsto x * y \colon L \times L \to L$. We shall show later in this section that every wedge W in L is a $*$-subsemigroup, whence $\exp W$ is a subsemigroup of G.

The characteristic function of a Lie algebra

In Chapter I, we discussed the geometry of wedges in finite dimensional vector spaces in terms of duality. One particular feature was the transgression function from the set of C^1-points of a generating wedge W onto the E^1-points of a fixed base of the dual cone W^*. We recall:

If W is generating in L, then W^* is pointed in L and admits (in many ways) a compact base $B = \{\omega \in W^* : \langle \omega, e \rangle = 1\}$ with a fixed element $e \in \operatorname{int} W$. Then $W^* = \mathbb{R}^+ \cdot B$, and the transgression function $\sigma^B : C^1(W) \to B \cap E^1(W^*)$ associates with each C^1-point x the unique functional $\sigma^B(x) \in B \cap x^\perp$. The transgression function is continuous by the Transgression Theorem I.2.35.

On the other hand, if L is a finite dimensional Lie algebra, then we have two canonical representations: The first one is the adjoint representation $\mathrm{ad}\colon L \to \mathrm{gl}(L)$ given by $(\mathrm{ad}\,x)y = [x,y]$, and the other is the dual representation, also called the *coadjoint representation*

$$\widehat{\mathrm{ad}}\colon L \to \mathrm{gl}(\widehat{L}), \qquad \widehat{\mathrm{ad}}(x) = -\widehat{\mathrm{ad}\,x},$$

that is

$$\langle \widehat{\mathrm{ad}}(x)\omega, y\rangle = -\langle \omega, [x,y]\rangle.$$

II.2.32.   Definition.   If L is a Lie algebra, we denote with $\widehat{x}\colon \widehat{L} \to \widehat{L}$ the linear map given by $\widehat{x} = \widehat{\mathrm{ad}}(x) = -\widehat{\mathrm{ad}\,x}$, that is, $\langle \widehat{x}(\omega), y\rangle = -\langle \omega, [x,y]\rangle$.

We call that with this notation we have $[x,y]^\wedge = \widehat{x}\widehat{y} - \widehat{y}\widehat{x}$.

In the following paragraphs we bring the concepts of the coadjoint representation and of the transgression function together. We have seen in Proposition II.2.13 that in dealing with Lie semialgebras we may restrict our attention to *generating* Lie semialgebras. This makes the formalism of the transgression function

$$\sigma^B \colon C^1(W) \to B \cap E^1(W^*), \qquad \{\sigma^B(x)\} = x^\perp \cap B$$

available to us.

If $x \in C^1(W)$, then $T_x = \sigma^B(x)^\perp$ and $L_x = \sigma^B(x)^*$. By the Characterization Theorem for Lie Semialgebras II.2.14, we know that W is a Lie semialgebra if and only if

$$(1) \qquad\qquad (\mathrm{ad}\,x)\sigma^B(x)^\perp \in \sigma^B(x)^\perp \text{ for all } x \in C^1(W).$$

Now let us consider a non-zero functional $\omega \in \widehat{L}$ and a linear self-map T of L. Then $T\omega^\perp \subseteq \omega^\perp$ holds if and only if $\langle \widehat{T}\omega, y\rangle = \langle \omega, Ty\rangle = 0$ for all $y \in \omega^\perp$. This, in turn, is equivalent to $\widehat{T}\omega \in \mathbb{R}\cdot\omega$, that is, to the statement that ω is an eigenvector of $\widehat{T}$. We apply this information with $T = \mathrm{ad}\,x$ and $\omega = \sigma^B(x)$ and find that Condition (1) above is equivalent to saying that for each $x \in C^1(W)$ there is a real number $\lambda(x)$ such that

$$\widehat{x}\big(\sigma^B(x)\big) = -\lambda(x)\cdot\sigma^B(x).$$

Here we introduce the minus sign in view of the coadjoint representation according to which $\widehat{x} = \widehat{\mathrm{ad}}(x) = -\widehat{\mathrm{ad}\,x}$, whence

$$\widehat{\mathrm{ad}\,x}\big(\sigma^B(x)\big) = \lambda(x)\cdot\sigma^B(x).$$

Since $\sigma^B(t\cdot x) = \sigma^B(x)$ for all $x \in C^1(W)$ and all $t > 0$, we also have $\lambda(t\cdot x) = \lambda(x)$ for these x and t. We notice also that the eigenvalue $\lambda(x)$ is independent of the choice of a particular base B of W.

II.2.33. **Definition.** Let W be a generating Lie semialgebra in a finite dimensional Lie algebra L. The *characteristic function* $\lambda\colon C^1(W) \to \mathbb{R}$ of W is the function which assigns to each C^1-point x of W that real number $\lambda(x)$ which satisfies

$$\widehat{x}\big(\sigma^B(x)\big) = -\lambda(x){\cdot}\sigma^B(x),$$

where $\sigma^B : C^1(W) \to B \cap E^1(W^*)$ is the transgression function for any base B of W (see Theorem I.2.35).

The following propositions show the basic properties of the characteristic function.

II.2.34. **Proposition.** *Let W be a generating Lie semialgebra in L and suppose $B = \{\omega \in W^* : \langle \omega, e \rangle = 1\}$ is a base of W^* for a suitable $e \in \operatorname{int} W$. Then*

(†) $$\lambda(x) = -\langle \widehat{x}\big(\sigma^B(x)\big), e \rangle = \langle \sigma^B(x), [x, e] \rangle \text{ for all } x \in C^1(W).$$

In particular, $\lambda\colon C^1(W) \to \mathbb{R}$ is continuous.

Proof. Condition (†) is straightforward to compute from the definitions in view of the fact that $\langle \sigma^B(x), e \rangle = 1$ for all C^1-points x of W. Since σ^B is continuous by the Transgression Theorem I.2.35, the proposition is proved. ∎

II.2.35. **Proposition.** (a) *Let x be an arbitrary boundary point of a generating Lie semialgebra W in L and ε an arbitrary non-zero extreme point of $\operatorname{op} x$. Then there is a real number $\lambda_{x,\varepsilon}$ such that*

$$[x, y] - \lambda_{x,\varepsilon}{\cdot}y \in \varepsilon^\perp \text{ for all } y \in L.$$

(b) *Suppose that the following limit exists:*

$$\lambda = \lim_{\substack{x' \to x \\ x' \in C^1(W)}} \lambda(x').$$

Then $\lambda_{x,\varepsilon} = \lambda$ for all extreme points ε of $\operatorname{op} x$ and

$$[x, y] - \lambda{\cdot}y \in T_x \text{ for all } y \in L.$$

Proof. We fix a base B of W as in Proposition II.2.34, apply the Approximation Theorem I.3.21, and find a sequence x_n of C^1-points of W such that $x = \lim x_n$ and $\varepsilon = \lim \sigma^B(x_n)$. Proposition II.2.34 yields $\lambda(x_n) = \langle \sigma^B(x_n), [x_n, e] \rangle$. The limit of the right hand side as $n \to \infty$ exists and equals $\langle \varepsilon, [x, e] \rangle$. We abbreviate this number by $\lambda_{x,\varepsilon}$ and have $\lambda_{x,\varepsilon} = \lim \lambda(x_n)$.

From Definition II.2.33 we know that for any element $y \in L$ we have $0 = \langle \widehat{x}\big(\sigma^B(x_n)\big) + \lambda(x_n){\cdot}\sigma^B(x_n), y \rangle = \langle \sigma^B(x_n), -[x_n, y] + \lambda(x_n){\cdot}y \rangle$. Passing to the limit as $n \to \infty$ we get

$$\langle \varepsilon, -[x, y] + \lambda_{x,\varepsilon}{\cdot}y \rangle = 0,$$

and this is the claim of Part (a) of the Proposition.

In order to prove Part (b) we simply observe that $\lambda = \lim \lambda(x_n) = \lambda_{x,\varepsilon}$, and thus $[x, y] + \lambda \cdot y \in \varepsilon^{\perp}$ for all extreme points ε of $\operatorname{op} x$. Since $\operatorname{op} x$ is the closed convex hull of the set $\operatorname{Ext}(\operatorname{op} x)$ of its extreme points, we have

$$\bigcap_{\varepsilon \in \operatorname{Ext}(\operatorname{op} x)} \varepsilon^{\perp} = \bigcap_{\omega \in \operatorname{op} x} \omega^{\perp} = (\operatorname{op} x)^{\perp}.$$

By Definition I.1.10 we have $T_x = (\operatorname{op} x)^{\perp}$. At this point we conclude $[x, y] - \lambda \cdot y \in T_x$, which is the assertion.   ∎

II.2.36.   Corollary.   (a) *If W is a generating Lie semialgebra in L, then for every C^1-point x of W we have*

(†) $[x, y] - \lambda(x) \cdot y \in T_x$ *for all $y \in L$.*

(b) *If the characteristic function of W extends continuously to a subset D of the boundary of W containing $C^1(W)$, then* (†) *above holds for all $x \in D$.*

Proof.   This corollary is an immediate consequence of the preceding proposition.   ∎

With the aid of the characteristic function we obtain a convenient criterium which allows us to single out the invariant wedges in the class of all Lie semialgebras.

II.2.37.   Proposition.   *A generating Lie semialgebra W in a Lie algebra L is an invariant wedge if and only if its characteristic function vanishes identically.*

Proof.   By the Characterization Theorem for Invariant Wedges II.1.14, W is invariant if and only if $[x, L] \subseteq T_x$ for all $x \in C^1(W)$. If we assume this condition to be satisfied and take it together with Corollary II.2.36.a, we observe $\lambda(x) \cdot y \in T_x$ for any C^1-point x and an arbitrary element y in L. In particular, we take any $y \in L \setminus T_x$; this choice forces $\lambda(x) = 0$. Thus λ vanishes identically. Conversely, if this is true, then $[x, y] \in T_x$ for all $x \in C^1(W)$ and all $y \in L$ by Corollary II.2.36.a again. By our initial remark, this is equivalent to the invariance of W.   ∎

Analytic Extension Aspects of Lie Semialgebras

In exponential Lie algebras (such as nilpotent or almost abelian ones) the Campbell-Hausdorff multiplication permits a global extension. Corollary II.2.16 and its proof contains the nucleus of an argument which should allow us to conclude that Lie semialgebras remain closed under an analytical extension of the Campbell-Hausdorff multiplication. In the following paragraphs we shall see that this is indeed the case. In particular we shall show that in any exponential Lie algebra, a Lie semialgebra is a subsemigroup with respect to the globally extended Campbell-Hausdorff multiplication.

In the following discussion let $\mathbb{I}$ denote the unit interval $[0.1]$. Let L be a Dynkin algebra and B a C-H-neighborhood in L. We say that *the Campbell-Hausdorff multiplication allows an analytic extension* m *to the pair* $(x, y) \in L \times L$ if there is an open set D in $L \times L$ containing $\mathbb{I} \cdot x \times \mathbb{I} \cdot y$ and an analytic function $M : D \to L$ such that for all $(u, v) \in D \cap (B \times B)$ one has $m(u, v) = u * v$, and that $m(\mathbb{I} \cdot x \times \mathbb{I} \cdot y) \times \{0\} \subseteq D$. Note that this definition is not entirely symmetric in the two arguments of the analytic function m.

Under these circumstances, we set $E = \{x \in L : (x, 0) \in D\}$ and define a vector field $X : E \to L$ by $X(z) = (D_2 m)(z, 0)(y)$, where $D_2 m : D \to \mathrm{Hom}(L, L)$ is the second partial derivative of m. This vector field is clearly analytic.

II.2.38.   Lemma.   *The function* $u : \mathbb{I} \to L$, $u(t) = m(x, t \cdot y)$ *is well defined and is the unique solution on its domain of the initial value problem*

$$\dot{u}(t) = X\big(u(t)\big), \qquad u(0).$$

Proof.   We consider the analytic functions $\varphi, \psi : \mathbb{I}^2 \to L$ given by

$$\varphi(s, t) = (D_2 m)(s \cdot x, t \cdot y)(y) \qquad \text{and} \quad \psi(s, t) = (D_2 m)(m(s \cdot x, t \cdot y), 0)(y).$$

Both are well-defined since $\mathbb{I} \cdot x \times \mathbb{I} \cdot y \subseteq D$. For all sufficiently small s and t we have $s \cdot x, t \cdot y \in B$, hence $m(s \cdot x, t \cdot y) = s \cdot x * t \cdot y$ and if s, t, h are sufficiently small, therefore, $m(s \cdot x, (t + h) \cdot y) = m\big(m(s \cdot x, t \cdot y), h \cdot y\big) = m(s \cdot x, t \cdot y) + h \cdot g\big(\mathrm{ad}\, m(s \cdot x, t \cdot y)\big)(y) + o(h)$, whence $g\big(\mathrm{ad}\, m(s \cdot x, t \cdot y)\big) = \frac{d}{dt} m(s \cdot x, t \cdot x)|_{r=t} = (D_2 m)(s \cdot x, t \cdot y)(y) = \varphi(s, t)$. On the other hand, from Lemma II.2.5, we know that $g(\mathrm{ad}\, u)(y) = \lambda'_u(0)(y) = (D_2 m)(u, 0)(y)$. Hence $g\big(\mathrm{ad}\, m(s \cdot x, t \cdot y)\big)(y) = \psi(s, t)$. Since φ and ψ are analytic on the connected set $\mathbb{I}^2$, it follows that $\varphi = \psi$. In particular we have $\dot{u}(t) = (D_2 m)(x, t \cdot y)(y) = \varphi(1, t) = \psi(1, t) = (D_2 m)\big(m(x, t \cdot y), 0\big)(y) = X\big(u(t)\big)$. Since the vector field X is analytic, solutions are unique. The assertion follows. ∎

II.2.39.   Lemma.   *Let W be a Lie semialgebra in a finite dimensional Lie algebra L. Suppose that the Campbell-Hausdorff multiplication allows an analytic extension $m : L \to L$ to (x, y). Set $E = \{u \in L : \mathbb{I} \cdot u \times \{0\} \subseteq D\}$. Suppose that $x \in$*

$E \cap W$ and $y \in W$. *Define the vector field* $X: E \to L$ *by* $X(u) = (D_2m)(u, 0)(y)$.
Then $X(x) \in L_x(W)$.

Proof. We first note that E is open in L. By the Confinement Theorem I.3.23,
it suffices to show the claim for $x \in C^1(W)$. We recall that $L_x = (\mathrm{op}\, x)^*$. Thus let
$0 \neq \omega \in \mathrm{op}\, x$. We must show that $\langle \omega, X(x) \rangle \geq 0$. We define $\alpha(t) = \langle \omega, X(t \cdot x) \rangle$ for
$t \in \mathbb{I}$, recalling that $\mathbb{I}.x \in E$. Now $X(u) = (D_2m)(u, 0)(y)$. For all t sufficiently
close to 0 we have $(D_2m)(t \cdot x, 0) = g(t \cdot \mathrm{ad}\, x)$ and thus, for sufficiently small t we
have
$$\alpha(t) = \langle \omega, g(t \cdot \mathrm{ad}\, x)(y) \rangle.$$

But $g(t.\,\mathrm{ad}\, x)\hat{} = g(t \cdot \widehat{\mathrm{ad}\, x}) = g(-t \cdot \hat{x})$. By Definition II.2.33 and since $\mathrm{op}\, x = \mathbb{R}^+ \cdot \omega$,
we have $-t \cdot \hat{x}(\omega) = \lambda(x) \cdot \omega$. Hence $g(-t \cdot \hat{x})\omega = g(t\lambda(x)) \cdot \omega$. Therefore,

$(\dagger)$ $$\alpha(t) = g\big(t\lambda(x)\big)\langle \omega, y \rangle$$

for all sufficiently small t. By the analyticity of α, equation $(\dagger)$ persists for all
$t \in \mathbb{I}$. Since $y \in W \subseteq L_x$ and $\omega \in L_x^*$, we have $\langle \omega, y \rangle \geq 0$. Now we take $t = 1$ and
observe that $\alpha(1) = g(\lambda)\langle \omega, y \rangle \geq 0$ by Lemma II.2.2.iii. This we had to show. ∎

   We summarize these results in the following theorem:

II.2.40.   **Theorem.**   (Closure Theorem for Lie semialgebras) *Let* W *be a Lie
semialgebra in a finite dimensional Lie algebra* L. *Suppose that* $x, y \in W$ *and that*
$m: D \to L$ *is an analytic function on an open subset* D *of* $L \times L$ *satisfying the
following conditions:*

  (i) $m(u, v) = u * v$ *for all* u, v *sufficiently close to 0.*
  (ii) $\mathbb{I} \cdot x \times \mathbb{I} \cdot y \subseteq D$.
  (iii) $m(\mathbb{I} \cdot x \times \mathbb{I} \cdot y) \times \{0\} \subseteq D$.
  (iv) $\mathbb{I} \cdot \big(D \cap (L \times \{0\})\big) \subseteq D$.
     Then $m(x, y) \in W$.

Proof.   We consider the trajectory $u(t) = m(x, t \cdot y), 0 \leq t \leq 1$. Then this curve
is the solution of the initial value problem
$$\dot{u}(t) = X\big(u(t)\big), \qquad u(0) = x$$

with $X(u) = (D_2m)(u, 0)(y)$ for all $u \in E$, because Lemma II.2.38 applies on
account of Conditions (i),(ii), and (iii). But by Condition (iv), Lemma II.2.39
applies and shows $X\big((u(t)\big) \in L_{u(t)}$. Hence $u(t)$ stays in W by Theorem I.5.17.
Thus $m(x, y) = u(1) \in W$. ∎

II.2.41.   **Corollary.**   *Let* $\exp: L \to G$ *be the exponential function of a finite
dimensional Lie group. Let* U *be an open neighborhood of 0 in* L *which satis-
fies* $\mathbb{I}.U \subseteq U$. *Suppose further that* $\exp|U : U \to V$ *is a diffeomorphism onto
an open subset of* G. *Set* $D = \{(x, y) \in L \times L : \exp x \exp y \in V\}$. *Then*
$m: D \to L$, $m(x, y) = (\exp|U)^{-1}(\exp x \exp y)$, *is an analytic extension of the
Campbell-Hausdorff multiplication of* L. *If* $(x, y) \in D$ *satisfies*

$(*)$ $$(\exp s \cdot x)(\exp t \cdot y) \in V \quad \text{for all} \quad 0 \leq s, t \leq 1,$$

then for any Lie semialgebra W with $x, y \in W$ the condition

$$m(x, y) \in W$$

is satisfied.

Proof. We verify the hypotheses of the Closure Theorem for Lie Semialgebras II. 2.40. Clearly (i) holds. By $(*)$, we have (ii). If $0 \leq s, t \leq 1$ and $z = m(s \cdot x, t \cdot y)$, then $\exp z = (\exp s \cdot x)(\exp t \cdot y)$, whence (iii). Finally, $(x, 0) \in D \cap (L \times \{0\})$ if and only if $\exp x = \exp x \exp 0 \in V$ if and only if $x \in U$. By hypothesis, $1 \cdot U \subseteq U$. Hence Theorem II.2.40 applies and proves the claim. ∎

If L is a nilpotent Lie algebra, then we may take $G = (L; *)$ and $\exp = \mathrm{id}_L$. This is a reminder that every nilpotent Lie algebra is exponential. We saw that all almost abelian Lie algebras are exponential. All exponential Lie algebras are necessarily solvable. The 3-dimensional Lie algebra of the group of motions of $\mathbb{R}^2$ is solvable but not exponential.

II.2.42. **Corollary.** *Let L be an exponential Lie algebra and $m : L \times L \to L$ the multiplication obtained by transporting the group multiplication to L. Then every Lie semialgebra W satisfies $m(W \times W) \subseteq W$, that is, W is a subsemigroup with respect to the multiplication m.* ∎

3. Low dimensional and special Lie semialgebras

There are two reasons why we should familiarize ourselves with special examples of the concepts which were introduced in the preceding sections: Firstly, we must form a good intuition of the geometry, the algebra, and the analysis of the situation in low dimensional Lie algebras; low dimensional vector spaces are close to our concrete geometrical intuition. Secondly, however, a complete classification of semialgebras in low dimensional Lie algebras will play a crucial role in our later development of a general theory of semialgebras. We shall see that general questions in the theory of Lie semialgebras can frequently be reduced to questions on low dimensional subalgebras. Remarkably, we have to go up to dimension 4 in some instances, and a complete knowledge of semialgebras in Lie algebras up to dimension 3 will be necessary. Furthermore, there is a class of solvable Lie algebras which support invariant Lorentzian cones, and we shall describe this family in this section, too.

We shall proceed as follows: Firstly, we consider all Lie algebras up to dimension 3 and completely describe the Lie semialgebras contained in them. Secondly, we shall define the family of what we shall call the standard Lorentzian solvable algebras (sometimes also referred to as the "oscillator algebras"), an infinite family of solvable Lie algebras. Thirdly we shall describe all Lie semialgebras in those 4-dimensional solvable Lie algebras which have a Heisenberg algebra as commutator algebra. The fourth step is an attempt to juxtapose these with the two 4-dimensional reductive Lie algebras and their invariant wedges. Finally, we intrduce another infinite one parameter family of solvable Lie algebras which is of considerable theoretical interest for the theory of Lie semialgebras as we shall find out in Section 5. The members of this family have solvable length 3; the nilpotent class of their nilradicals, however, surpasses all bounds.

$\dim L \leq 3$: **The solvable case**

II.3.1.   Lemma.   *A Lie algebra of dimension less than 3 is almost abelian. Any wedge is a Lie semialgebra in such an algebra.*

Proof.   The only Lie algebras of dimension 2 are the abelian one and the algebra of matrices

$$\begin{pmatrix} a & b \\ 0 & 0 \end{pmatrix}, a, b \in \mathbb{R}$$

This algebra is almost abelian, its hyperplane commutator algebra being the subalgebra of all matrices with $a = 0$. The remainder then follows from Theorem II.2.30. ∎

II.3.2.   Lemma. *A non-abelian solvable Lie algebra of dimension 3 is isomorphic to the Lie algebra of all 3×3 matrices of the form*

$$\begin{pmatrix} t \cdot A & \mathbf{u} \\ 0 & 0 \end{pmatrix}, \qquad A = \begin{pmatrix} a_{11} & a_{12} \\ a_{21} & a_{22} \end{pmatrix}, \qquad \mathbf{u} = \begin{pmatrix} u \\ v \end{pmatrix}, \qquad a_{jk}, u, v \in \mathbb{R}.$$

The matrix A may be assumed to be in real Jordan normal form. This allows for the following possibilities:

(i) $A = \begin{pmatrix} \lambda & \omega \\ -\omega & \lambda \end{pmatrix}, \lambda, \omega \in \mathbb{R}, \omega \neq 0,$

(ii) $A = \begin{pmatrix} \lambda & 1 \\ 0 & \lambda \end{pmatrix}, \lambda \in \mathbf{R},$

(iii) $A = \begin{pmatrix} \lambda_1 & 0 \\ 0 & \lambda_2 \end{pmatrix}, \lambda_1, \lambda_2 \in \mathbb{R}.$

Proof.   For a complete classification of the 3-dimensional sovable Lie algebras we refer e.g. to [Jacobson], p.11 ff. ∎

We observe that in case (iii) one of the two eigenvalues of A may be 0. In this case L is the direct sum of a 1-dimensional center and the 2-dimensional non-abelian Lie algebra. Case (i) arises when A has two complex conjugate eigenvalues $\lambda \pm i\omega$, and case (ii) includes the case $\lambda = 0$ which results in L being the Heisenberg algebra, the only non-abelian nilpotent algebra of dimension 3 with $[L, L] = Z(L)$, the center of L being 1-dimensional.

II.3.3.   Lemma. *Suppose that L is a Lie algebra, J an ideal of codimension 1, and H a hyperplane in L. We consider the following two statements:*

(1) $H \cap J$ *is an ideal.*

(2) H *is a subalgebra.*

Then (1) implies (2), and if $H \cap J$ is an ideal in J, then both are equivalent. This is the case, in particular, when J is abelian.

Proof.   (1)$\Rightarrow$(2): If $H = J$, then the assertion is trivial. Otherwise $H \cap J$ has codimension 2. If e is any element in $H \setminus (H \cap J)$, then $H = \mathbb{R} \cdot e + (H \cap J)$ is clearly a subalgebra as the sum of a subalgebra and an ideal.

(2)$\Rightarrow$(1): Here we assume that $H \cap J$ is an ideal in J. The case $H = J$ is once more trivial. If there is an $e \in H \setminus J$, then $L = \mathbb{R} \cdot e + J$, and $[e, H \cap J] \subseteq [e, H] \cap [e, J] \subseteq H \cap J$ by (2), and $[J, H \cap J] \subseteq H \cap J$, since $H \cap J$ is an ideal in J. The claim follows. ∎

We now proceed to characterize all possible generating Lie semialgebras in a 3-dimensional solvable Lie algebra. The case of lower dimensions is clear, since by Lemma II.3.1 above all Lie algebras are almost abelian in this case and any wedge in an almost abelian Lie algebra is a Lie semialgebra by Theorem II.2.30. Furthermore, in view of Proposition II.2.13, it is no loss of generality if we restrict our attention to generating Lie semialgebras.

In order to find all generating Lie semialgebras in the Lie algebras of dimension not exceeding 3 it suffices after Proposition II.2.28 to find all half-space semialgebras, because any Lie semialgebra in a Lie algebra of dimension ≤ 3 is an intersection of those.. But then, by Corollary II.2.24, in order to describe all half-space semialgebras it it sufficient to exhibit all hyperplane subalgebras. In any almost abelian Lie algebra, every hyperplane is a subalgebra and every wedge is a Lie semialgebra. Thus we assume henceforth that L is not almost abelian.

According to Lemma II.3.2 we may write $L = \mathbb{R}{\cdot}e + J$ with a 2-dimensional abelian ideal J such that $\operatorname{ad} e$ induces on J an endomorphism f which, with respect to a suitable basis in J, is represented by a matrix A of the three types indicated in Lemma II.3.2. The ideals I of L contained in J are exactly the eigenspaces of f. By Lemma II.3.3, a hyperplane H other than J is a subalgebra of L if and only if it meets J in one of these eigenspaces. We run through these cases:

Case i. The eigenvalues are not real; there is no non-trivial eigenspace. The ideal J is the only hyperplane subalgebra.

Case ii. There is exactly one 1-dimensional eigenspace I of f. The hyperplane subalgebras are exactly the ones containing I. If L is the Heisenberg algebra, then I is the center.

Case iii. Since we assume L not to be almost abelian, the two eigenvalues λ_1 and λ_2 are different. There are exactly 2 eigenspaces I_1 and I_2. The hyperplane subalgebras of L are exacly the hyperplanes containing I_1 or I_2.

Every hyperplane subalgebra bounds 2 half-space semialgebras; every Lie semialgebra is the intersection of half-space semialgebras in dimensions below 4. Hence we have the following result:

II.3.4.   Theorem. (First Classification Theorem of Low Dimensional Semialgebras) *If L is a solvable Lie algebra with $\dim L \leq 3$ which is not almost abelian, then L contains a 2-dimensional abelian ideal J and the following possibilities occur.*

(i) *J contains no 1-dimensional ideal,*

(ii) *J contains precisely one 1-dimensional ideal I,*

(iii) *J contains two different 1-dimensional ideals I_1 and I_2.*

If W is a generating Lie semialgebra in L, then it is polyhedral. In case (i) it is one of the two half space semialgebras bounded by J. In case (ii) it must contain I, and every wedge containing I is indeed a Lie semialgebra. In case (iii), W is the intersection of half-spaces containing I_1 or I_2 in their boundaries, and all intersections of this type are Lie semialgebras. Each Lie semialgebra arising in this fashion is the intersection of $0, 1, 2, 3$ or 4 half-space semialgebras.

The Lie semialgebra W is an invariant wedge if and only if W is a half-space semialgebra bounded by J.

Proof. Everything except the last assertion has been proved, and this follows from Corollary II.2.24. ∎

Notice that in case (iii) there are polyhedral semialgebras with four edges and four sides.

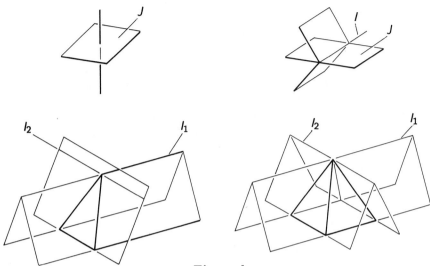

Figure 1

$\dim L = 3$: **The semisimple case**

There are two semisimple 3-dimensional real Lie algebras, namely, so(3) and sl(2, $\mathbb{R}$). We shall write sl(2) for the latter. Since so(3) has no 2-dimensional subalgebra, it has no Lie semialgebras except so(3) and the obvious 1- and 0-dimensional ones. We have to discuss sl(2), which is very interesting with respect to its Lie semialgebras, as it is in so many other regards.

The Cartan–Killing form $B(X, Y) = tr(\text{ad}\,X\,\text{ad}\,Y)$ is, up to scalar multiplication, the only invariant bilinear form on the simple Lie algebra sl(2). The bilinear form $b(X, Y) = \frac{1}{2}tr\,XY$ is also invariant and non-zero, and it is not hard to determine that $B = 8b$. (We recall that a bilinear form F on a Lie algebra L is called *invariant*, if $F(x, [y, z]) = F([x, y], z)$ for all $x, y, z \in L$.)

We consider the following particular elements of sl(2):

$$H = \begin{pmatrix} 1 & 0 \\ 0 & -1 \end{pmatrix}, \qquad P = \begin{pmatrix} 0 & 1 \\ 0 & 0 \end{pmatrix}, \qquad Q = \begin{pmatrix} 0 & 0 \\ 1 & 0 \end{pmatrix},$$

$$T = P + Q = \begin{pmatrix} 0 & 1 \\ 1 & 0 \end{pmatrix}, \qquad \text{and} \qquad U = P - Q = \begin{pmatrix} 0 & 1 \\ -1 & 0 \end{pmatrix}.$$

Then $\{H, P, Q\}$ is a basis satisfying the following multiplication table:

(1) $\qquad [H, P] = 2P, \quad [H, Q] = -2Q, \quad \text{and} \quad [P, Q] = H.$

Also, $\{H, T, U\}$ is a basis with the following multiplication table:

(2) $\qquad [H, T] = 2U, \quad [H, U] = 2T, \quad \text{and} \quad [U, T] = 2H.$

The Cartan–Killing form of sl(2) is calculated via the form b through the formulae

$$(3) \qquad b\left(\begin{pmatrix} h & p \\ q & -h \end{pmatrix}\right) = b(h{\cdot}H + p{\cdot}P + q{\cdot}Q) = h^2 + pq,$$

$$(4) \qquad b\left(\begin{pmatrix} h & t+u \\ t-u & -h \end{pmatrix}\right) = h^2 + t^2 - u^2.$$

In particular, we see from (4) that the Cartan–Killing form is Lorentzian, as it has signature $+\,+\,-$. Notice that (3) says $b(A) = -\det A$.

Nothing except computational convenience distinguishes any basis over another. However, the Cartan–Killing form determines a geometric object which is independent of any choice of basis and is invariant under all automorphisms of L, namely, the so-called *standard double cone* and its boundary, which we define as follows:

$$(5) \qquad \mathcal{W} = \{X : B(X) \le 0\}, \qquad \partial\mathcal{W} = \{X : B(X) = 0\}.$$

Clearly, $\partial\mathcal{W}$ is the boundary of $\mathcal{W}$, and the interior of $\mathcal{W}$ is given by $\operatorname{int}\mathcal{W} = \{X : B(X) < 0\}$. The double cone is the union of two cones $\mathcal{W}^+$ and $\mathcal{W}^- = -\mathcal{W}^+$ which we obtain from the two components of $\mathcal{W} \setminus \{0\}$ by reinserting 0. Let us arbitrarily decree that the one containing U is $\mathcal{W}^+$. The separation of $\mathcal{W}$ into two double cones is no longer canonical: If $\alpha : \mathrm{sl}(2) \to \mathrm{sl}(2)$ denotes the (outer) automorphism given by $\alpha(X) = TXT$, then α exchanges $\mathcal{W}^+$ and $\mathcal{W}^-$.

It is convenient to use the non-degenerate Cartan–Killing form B to identify the vector space dual of L with L itself under the isomorphism $x \mapsto B(x, \bullet) : L \to \widehat{L}$. Then for any subset S of sl(2) we have $S^\perp = \{X : B(X,Y) = 0 \text{ for all } Y \in S\}$. Then $S^{\perp\perp}$ is the vector subspace generated by S. If φ is any automorphism of sl(2), then $\varphi(S^\perp) = \varphi(S)^\perp$ since B is invariant under any automorphism. The following proposition is clearly relevant in our context:

II.3.5.   Proposition. *For a plane E in* sl(2) *the following statements are equivalent:*

 (1) *E is a subalgebra.*
 (2) *$E = X^\perp$ for some $X \ne 0$ with $B(X) = 0$.*
 (3) *$E = X^\perp$ for some $X \in E$.*
 (4) *$E^\perp \subseteq E$.*
 (5) *E is tangent to $\mathcal{W}$.*
 (6) *$E = [X, \mathrm{sl}(2)]$ for some $X \in E$.*

Furthermore, all such E are conjugate under the group $e^{\mathbb{R}\cdot \operatorname{ad} U}$.

Proof.   First we note that the the invariance of B implies $B([X,Y], X) = 0$, hence

$$(\dagger) \qquad \operatorname{im} \operatorname{ad} X = [X, \mathrm{sl}(2)] \subseteq X^\perp.$$

The equivalence of (2) through (5) is standard analytic geometry of Lorentzian forms. Indeed, the equivalence of (2) and (3) is clear from the definitions. If $E = X^\perp$, then $E^\perp = \mathbb{R} \cdot X$, and so (3) and (4) are the same. $(3) \Rightarrow (5)$: If $Y \in E \cap \partial W$, then also $E = Y^\perp$ and thus $E = X^\perp \cap Y^\perp = (\mathbb{R} \cdot X + \mathbb{R} \cdot Y)^\perp$, whence X and Y have to be linearly dependent. Hence (5) follows. $(5) \Rightarrow (3)$: By (5), $Y \in E$, and $B(Y) \leq 0$ implies $Y \in \mathbb{R} \cdot X$, where $X \in E \cap \partial W$, $X \neq 0$. If now Z is any element of E such that $B(X, Z) \neq 0$, then we set $t = \frac{B(Z)}{2B(X,Z)}$ and find $B(tX - Z) = 0$ and thus $Z \in \mathbb{R} \cdot X$ after the preceding. Hence $E \subseteq X^\perp$ and thus $E = X^\perp$ for reasons of dimension.

$(1) \Rightarrow (2)$: From (†) we have $[X, X^\perp] \subseteq X^\perp$. If $X^\perp$ is a subalgebra, then it follows that $X^\perp$ is an ideal in $X^\perp + \mathbb{R} \cdot X$. But since $sl(2)$ is simple, this happens only if $X \in X^\perp$. Now every plane E is of the form $X^\perp$ with some non-zero X. Hence (2) is established.

$(2) \Rightarrow (1)$: Let $Y \in X^\perp \setminus \mathbb{R} \cdot X$. Then $X^\perp = \mathbb{R} \cdot X + \mathbb{R} \cdot Y$, since $X^\perp$ is 2-dimensional. Now $[X, Y] \in [X, sl(2)] \subseteq X^\perp$, so that $[X^\perp, X^\perp] \subseteq X^\perp$. That is, $X^\perp$ is a subalgebra.

Let us next deal with the conjugacy of the planar algebras E. Firstly, $E = X^\perp$ with $X \in \partial W \setminus \{0\}$ for any plane algebra E by the preceding. Secondly, B is invariant, whence all inner automorphisms φ satisfy $\varphi(E) = \varphi(X^\perp) = \varphi(X)^\perp$. Therefore it suffices to show that the 1-dimensional subspaces of ∂W are permuted transitively by $e^{\mathbb{R} \cdot U}$. But with respect to the basis $\{H, T, U\}$ we have

$$e^{t \, \mathrm{ad} \, U} = \begin{pmatrix} \cos 2t & \sin 2t & 0 \\ -\sin 2t & \cos 2t & 0 \\ 0 & 0 & 1 \end{pmatrix}, \qquad t \in \mathbb{R}.$$

The vector $2P = T + U$ gets transformed into $\sin 2t \cdot H + \cos 2t \cdot T + U$. This proves the asserted conjugacy.

Finally we prove the equivalence of (6) with the other conditions. First $(3) \Rightarrow (6)$: The centralizer of an element X in a Lie algebra is precisely $\ker \mathrm{ad}\, X$. Since $sl(2)$ has no abelian subalgebras of dimension greater than 1, it follows that $\ker \mathrm{ad}\, X = \mathbb{R} \cdot X$ if $X \neq 0$. Then

(††) $$[X, sl(2)] = X^\perp$$

for dimensional reasons. But $(6) \Rightarrow (3)$ in view of (††). ∎

In particular, every planar subalgebra is conjugate to the non-abelian algebra

$$\mathbb{R} \cdot H + \mathbb{R} \cdot P = \left\{ \begin{pmatrix} r & s \\ 0 & -r \end{pmatrix} : \quad r, s \in \mathbb{R} \right\}.$$

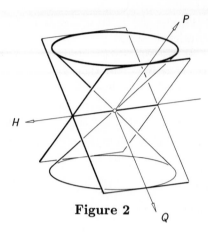

Figure 2

The level sets of the function $X \mapsto B(X)$ are the sets $\{h{\cdot}H + t{\cdot}T + u{\cdot}U : h^2 + t^2 - u^2 = \text{const}\}$. We see at once that they are hyperboloids if the constant on the right side is different from zero, and they agree with the boundary ∂W if the constant is zero.

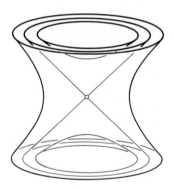

Figure 3

II.3.6.   Proposition. *The orbits in* $\mathrm{sl}(2)$ *of the adjoint action are the connected components of the non-zero level sets,* $\{0\}$*, and the two components of* $\partial W \setminus \{0\}$*. In particular, the adjoint group acts transitively on the set of one dimensional subspaces meeting the interior of* W*, on the set of one dimensional subspaces of* ∂W*, and on the remaining set of 1-dimensional subspaces.*

Proof.   Clearly, the level sets of B are invariant, since B is invariant. We have already observed that the adjoint group contains the rotation group around $\mathbb{R}{\cdot}U$ as axis. The orbit of U under $e^{\mathbb{R}\cdot \mathrm{ad}\,H}$ is $\{e^{2h}{\cdot}P - e^{-2h}{\cdot}Q : h \in \mathbb{R}\}$, and the orbit of T under this group is $\{e^{2h}{\cdot}P + e^{-2h}{\cdot}Q : h \in \mathbb{R}\}$. The orbit of P under the same group is $\{e^{2h}{\cdot}P : h \in \mathbb{R}\} = e^{\mathbb{R}}{\cdot}P$ and that of Q is $e^{\mathbb{R}}{\cdot}Q$. With these observations it is now straightforward to finish the proof of the claim.   ∎

We are now ready to determine the Lie semialgebras in $\mathrm{sl}(2)$. If W is a Lie semialgebra, then $W - W$ is a subalgebra by Proposition II.2.13. Since all

2-dimensional subalgebras are conjugate and isomorphic to the almost abelian non-abelian 2-dimensional algebra, whose Lie semialgebras we know to range through the set of all wedges by Theorem II.2.30, we may restrict our attention to generating Lie semialgebras. In order to have a convenient notation, for $B(X) = 0$ and $X \neq 0$ we set

$$X^* = \{Y : \quad B(X, Y) \geq 0\}.$$

Then X^* is a half space bounded by $X^\perp$, and in view of Corollary II.2.24 and Proposition II.3.5, X^* is a half-space semialgebra. Conversely, all half-space semialgebras are of the form X^* with a suitable $X \in \partial \mathcal{W}$.

II.3.7.   Theorem.   (Second Classification Theorem of Low Dimensional Semialgebras) *Let W be a generating Lie semialgebra in* $\mathrm{sl}(2)$. *Then the following conclusions hold:*

(i) *W is the intersection of a family of half-space semialgebras of the form X^*, $X \in \partial \mathcal{W} \setminus \{0\}$.*

(ii) *W is an invariant wedge if and only if $W = \mathcal{W}^+$ or $W = \mathcal{W}^-$.*

(iii) *If $\dim H(W) = 1$, then W is conjugate to $P^* \cap Q^* = \{h{\cdot}H + p{\cdot}P + q{\cdot}Q : (h, p, q) \in \mathbb{R} \times \mathbb{R}^+ \times \mathbb{R}^+\}$, or to $P^* \cap -Q^* = \{h{\cdot}H + p{\cdot}P + q{\cdot}Q : \quad (h, p, q) \in \mathbb{R} \times \mathbb{R}^+ \times \mathbb{R}^+\}$, or to the negative of the latter.*

(iv) *Every weakly round semialgebra in* $\mathrm{sl}(2)$ *is invariant.*

Proof.   (i) has already been proved; see in particular Proposition II.2.28.

(ii) The two wedges $\mathcal{W}^+$ and $\mathcal{W}^-$ are invariant; indeed $\mathcal{W}$ is invariant under all automorphisms, since B is invariant under all automorphisms. Hence the interior of $\mathcal{W}$ is invariant under all automorphisms and so each of the two connected components of $\mathrm{int}\,\mathcal{W}$ remain invariant under the identity component of the full automorphism group of $\mathrm{sl}(2)$. But the group of inner automorphisms, generated by $e^{\mathrm{ad}\,\mathrm{sl}(2)}$ is connected. Hence each of the components of $\mathrm{int}\,\mathcal{W}$ is invariant under all inner automorphisms and hence so are $\mathcal{W}^+$ and $\mathcal{W}^-$, their respective closures.

Now let W be an invariant non-zero proper wedge. Then $\partial W \setminus \{0\}$ is an invariant connected topological 2-manifold with 0 in its closure. Now Proposition II.3.6 implies $\partial W \setminus \{0\} = \partial \mathcal{W}^+ \setminus \{0\}$ or $\partial W \setminus \{0\} = \partial \mathcal{W}^- \setminus \{0\}$. Thus $X = \mathcal{W}^+$ or $W = \mathcal{W}^-$.

(iii) Let W be a generating Lie semialgebra in $\mathrm{sl}(2)$ with 1-dimensional edge $H(W)$. Then W must be the intersection of two half-space semialgebras X^* and Y^* by (i) above. If $0 \neq Z \in H(W) = X^\perp \cap Y^\perp$, then $B(Z) > 0$. Hence by Proposition II.3.6, a scalar multiple $r{\cdot}Z$ is conjugate to H. But then W is conjugate to one of the four Lie semialgebras having $\mathbb{R}{\cdot}H$ as an edge, namely, $-P^* \cap -Q^*$, $P^* \cap Q^*$, $-P^* \cap Q^*$, and $P^* \cap -Q^*$. The first two of these are conjugate under rotation. Thus assertion (iii) follows.

(iv) Firstly, suppose that $\mathcal{W}^+ \subseteq W$. If contrary to the claim of invariance of W we had $\mathcal{W}^+ \neq W$, then there is a $y_0 \in \partial W \setminus \mathcal{W}^+$. Since W is closed and $C^1(W)$ is dense in ∂W by the Density Theorem for Wedges I.3.11, then we find an element $y \in C^1(W) \setminus \mathcal{W}^+$. Now consider $x \in T_y(W) \cap \mathcal{W}^+$, $x \neq 0$. Because of $\mathcal{W}^+ \subseteq W$ we have $x \in T_y(W) \cap W = E_y(W)$ by Proposition I.2.7. As x and y are linearly independent, $\dim E_y(W) = 2$. Now for all $z \in \mathrm{algint}\, E_y$ we have $D(z) \neq \{0\}$ (see Definition I.4.16). But then W is not weakly round (see Definition I.4.19), contrary to our assumption.

The second case is that W is contained in some conjugate of $P^* \cap Q^*$ We will formulate below an Exercise to show that all these semialgebras are in fact polyhedral. But W is assumed to be weakly round, hence cannot be polyhedral. ∎

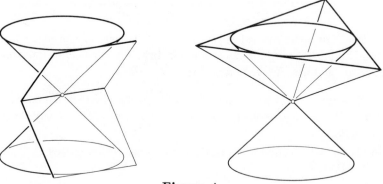

Figure 4

The Lie semialgebra $P^* \cap Q^* = \{h \cdot H + p \cdot P + q \cdot Q : \quad p, q \geq 0, \quad h \in \mathbb{R}\}$ deserves a special name, because it will turn out to be of significance in the global theory of subsemigroups of Lie groups as we shall see in Chapter V.

II.3.8.   Definition.   We write $\mathrm{sl}(2)^+ = \mathrm{sl}(2, \mathbb{R})^+ = P^* \cap Q^*$.

The choice of this notation is due to the fact that the exponential function $\exp : \mathrm{sl}(2) \to \mathrm{Sl}(2)$ will map $\mathrm{sl}(2)^+$ diffeomorphically onto the semigroup of all 2×2 matrices of determinant 1 all of whose entries are non-negative. This semigroup will be called $\mathrm{Sl}(2)^+$. Notice that $-Q^* = (-Q)^*$ is a conjugate of P^*.

According to Theorem II.3.7, there will be numerous pointed generating Lie semialgebras in $\mathrm{sl}(2)$. We observe how they fall into two disjoint classes:

Class 1 : Semialgebras which are the intersections of conjugates of $-P^*$ and their images under multiplication with -1. Since every conjugate of $-P^*$ contains $\mathcal{W}^+$, all of these semialgebras contain $\mathcal{W}^+$ or $\mathcal{W}^-$, respectively.

Class 2 : Semialgebras which are the subsemialgebras of some conjugate of $\mathrm{sl}(2)^+$

We will see later, that all pointed Lie semialgebras of Class 2 are mapped homeomorphically onto a subsemigroup of $\mathrm{Sl}(2)$, whereas the Lie semialgebras of Class 1 are not the tangent objects of any (global) semigroup in $\mathrm{Sl}(2)$. However, they are realized as tangent objects of subsemigroups in the universal covering group of $\mathrm{Sl}(2)$.

EII.1.   Exercise.   Show that all pointed Lie semialgebras of Class 2 are necessarily polyhedral and are, in fact, the intersection of 2, 3, or 4 half- space semialgebras.

Examples of Lorentzian cones

We have seen that the 3-dimensional Lie algebra sl(2) contains two invariant cones defined by a Lorentzian form which is invariant. An almost abelian Lie algebra has the property that every wedge is a Lie semialgebra. So, in particular, every cone consisting of one half of the double cone defined by any Lorentzian form is a Lie semialgebra. We shall now give some additional examples of invariant Lorentzian forms on Lie algebras and thus some further examples of invariant Lorentzian cones. At a later point we will prove that, by and large, the examples we have described exhaust the total supply of invariant Lorentzian forms in Lie algebras.

Firstly, we make the following simple observation: If L is the direct sum of two ideals I and J, and if there are invariant quadratic forms q_I and q_J on I and J, respectively, then the form $q = q_I \oplus q_J$ defined by $q(x \oplus y, x' \oplus y') = q_I(x, x') + q_J(y, y')$ is invariant on L, and I and J are orthogonal with respect to q. We shall also say that (L, q) is the orthogonal direct sum of (I, q_I) and (J, q_J). Of course, q will be non-degenerate if and only if both q_I and q_J are non-degenerate.

Let us make the convention that the term *Lorentzian form* will always refer to a non-degenerate form.

II.3.9. **Proposition.** *Every compact Lie algebra with non-trivial center supports an invariant Lorentzian form, hence contains invariant Lorentzian cones.*

Proof. We can write $L = I \oplus J$ with a one dimensional ideal I and a compact ideal J of codimension 1. Let q_I be any negative definite quadratic form on I; trivially, q_I is invariant. Let q_J be an arbitrary positive definite invariant quadratic form on J; such exist, since J is compact. Then $q = q_I \oplus q_J$ is the required Lorentzian invariant form. ∎

II.3.10. **Proposition.** *If L is a direct sum of two ideals I and J, and I is compact and J supports an invariant Lorentzian form , then L supports an invariant Lorentzian form.*

Proof. Just select a positive definite invariant quadratic form q_I on I by virtue of compactness and let q_J be an invariant Lorentzian form on J. Then $q = q_I \oplus q_J$ is an invariant Lorentzian form on L. ∎

In particular, the direct product sl(2) $\times L$ with a compact Lie algebra L as factor always supports an invariant Lorentzian form.

Now we shall construct another type of Lie algebra which supports invariant Lorentzian forms. The family will consist of even dimensional solvable Lie algebras.

II.3.11. **Proposition.** *Let V be a finite dimensional non-zero Hilbert space over $\mathbb{R}$ with an inner product $\langle \bullet \mid \bullet \rangle$ and with a skew symmetric vector space automorphism d (so that in fact $\langle dx \mid y \rangle + \langle x \mid dy \rangle = 0$. The existence of d will make V even dimensional!) On the vector space $L = \mathbb{R} \times V \times \mathbb{R}$ we define a bilinear multiplication through*

$$(6) \qquad [(u, x, v), (u', x', v')] = (0, u \cdot dx' - u' \cdot dx, \langle dx \mid x' \rangle),$$

and a quadratic form q by

$$(7) \qquad q((u, x, v), (u', x', v')) = uv' + u'v + \langle x \mid x' \rangle.$$

Then L *is a Lie algebra, and* q *an invariant Lorentzian form such that* L *is the orthogonal direct sum of the hyperbolic plane* $\mathbb{R} \times \{0\} \times \mathbb{R}$ *and the Hilbert space* $\{0\} \times V \times \{0\}$.

The algebra L *is solvable so that* $L' = \{0\} \times V \times \mathbb{R}$ *and* $L'' = \{0\} \times \{0\} \times \mathbb{R} = Z(L)$ *(the center of* L*). Further,* $[L, L'] = L'$*, so* L *is not nilpotent. The nilradical of* L *is* L'.

Proof.   i) First we show that L is a Lie algebra of the type described in the proposition. If one equips $V \times \mathbb{R}$ with the multiplication given by $[(x, v), (x', v')] = (0, \langle dx \mid x' \rangle)$, then it becomes a nilpotent algebra of Heisenberg type with commutator algebra and center $\{0\} \times \mathbb{R}$. The function D given by $D(x, v) = (dx, 0)$ is a derivation, since $D[(x, v), (x', v')] = D(0, \langle dx \mid x' \rangle) = (0, 0)$ on one hand and $[D(x, v), (x', v')] + [(x, v), D(x', v')] = (0, \langle d^2 x \mid x' \rangle) + (0, \langle dx \mid dx' \rangle) = (0, 0)$ in view of the antisymmetry of d, on the other. Now L is simply the semidirect product of this Heisenberg algebra with $\mathbb{R}$ with respect to the derivation D. Since d is an automorphism, $L' = \{0\} \times V \times \mathbb{R}$ follows and the assertion $L'' = \{0\} \times \{0\} \times \mathbb{R}$ are straightforward. The commutator algebra of any solvable Lie algebra is contained in its nilradical (see e.g. [Bou75], Chap. I, §5, n° 3, Cor. 1). Since L' is a hyperplane and L is not nilpotent, L' is the nilradical.

ii) Clearly, q is a Lorentzian form so that L is the orthogonal sum of the hyperbolic plane $\mathbb{R} \times \{0\} \times \mathbb{R}$ and the Hilbert space $\{0\} \times V \times \{0\}$. We must verify the invariance of q. We set $w_j = (u_j, x_j, v_j)$, $j = 1, 2, 3$ and calculate

$$q([w_1, w_2], w_3) = q((0, u_1 \cdot dx_2 - u_2 \cdot dx_1, \langle dx_1 \mid x_2 \rangle), (u_3, x_3, v_3))$$
$$= \langle dx_1 \mid x_2 \rangle u_3 + u_1 \langle dx_2 \mid x_3 \rangle - u_2 \langle dx_1 \mid x_3 \rangle,$$

and

$$q(w_1, [w_2, w_3]) = q((u_1, x_1, v_1), (0, u_2 \cdot dx_3 - u_3 \cdot dx_2, \langle dx_2 \mid x_3 \rangle))$$
$$= u_1 \langle dx_2 \mid x_3 \rangle + u_2 \langle x_1 \mid dx_3 \rangle - u_3 \langle x_1 \mid dx_2 \rangle.$$

Because of the antisymmetry of d, these two expressions are equal, as required. ∎

We shall say that a Lie algebra (L, q) together with an invariant quadratic form is *irreducible*, if it is not the orthogonal direct sum of two ideals. Later we shall show that all algebras L obtained in the preceding proposition are irreducible with respect to the form q.

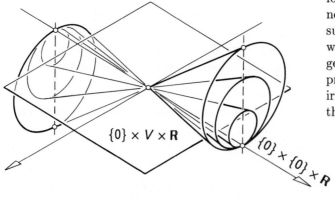

$\{0\} \times V \times \mathbb{R}$

$\{0\} \times \{0\} \times \mathbb{R}$

Figure 5

II.3.12.   Proposition.   *The Lie algebra* (L, q) *with the Lorentzian form constructed in* Proposition II.3.11 *is irreducible.*

Proof. First we show that every ideal I of L contains $Z(L) = \{0\} \times \{0\} \times \mathbb{R}$. Let $(u, x, v) = w \neq 0$ be an element of I. Then I contains all $(0, x', v')$ with $x' = u \cdot dx'' - u'' \cdot dx$ and $v' = \langle dx \mid x'' \rangle$ with $x'' \in V$ and $u'' \in \mathbb{R}$. If $u \neq 0$, then $\{0\} \times V \times \{0\} \subseteq I$, since d is non-singular. This will imply $L' = \{0\} \times V \times \mathbb{R} \subseteq I$. If $u = 0$, $x \neq 0$, then $\{0\} \times \mathbb{R} \cdot dx \times \mathbb{R} \subseteq I$ follows. If $u = 0$ and $x = 0$, then $w \in Z(L)$. In each case, $Z(L) \subseteq I$ is true. As a consequence, no non-zero proper ideal can have an orthogonal complement. Hence L is irreducible. ∎

In order to establish uniqueness results on (L, q) we need some preparation.

II.3.13. **Lemma.** *If L is as in* Proposition II.3.11 *and f is an L-module endomorphism of L (i.e., satisfies $f[x, y] = [x, f(y)]$), then there are two numbers r and s in $\mathbb{R}$ such that $f(u, x, v) = (ru, r \cdot x, rv + su)$.*

Proof. Since f must leave L' and L'' invariant, we know that f can be written in the following form:

$$f(u, x, v) = (ru, u \cdot x_0 + F(x), su + \langle x_1 \mid x \rangle + tv)$$

with suitable numbers $r, s, t \in \mathbb{R}$, vectors $x_0, x_1 \in V$, and a linear map $F : V \to V$. Now we calculate

$$f[(u, x, v), (u', x', v')] = f(0, u \cdot dx' - u' \cdot dx, \langle dx \mid x' \rangle)$$
$$= \big(0, u \cdot F(dx') - u' \cdot F(dx), u \langle x_1 \mid dx' \rangle - u' \langle x_1 \mid dx \rangle + t \langle dx \mid x' \rangle \big).$$

On the other hand,

$$[(u, x, v), f(u', x', v')] = [(u, x, v), (ru', u' \cdot x_0 + F(x'), su' + \langle x_1 \mid x' \rangle + tv')]$$
$$= (0, uu' \cdot dx_0 + u \cdot dF(x') - ru' \cdot dx, u' \langle dx \mid x_0 \rangle + \langle dx \mid F(x') \rangle).$$

We equate these expressions and consider the second and third components. We specialize variables:

(i) $u = 0$, $x' = 0$: $F(dx) = r \cdot dx$ for all x; in other words $F = r \cdot 1_V$, since d is non-singular.

(ii) $x = x' = 0$: $0 = uu' \cdot dx_0$, that is, $x_0 = 0$, since d is non-singular.

(iii) $x = 0$: $u \langle x_1 \mid dx' \rangle = 0$ for all $u \in \mathbb{R}$, $x' \in V$, that is, $x_1 = 0$, since d is non-singular.

(iv) $t \langle dx \mid x' \rangle = \langle dx \mid r \cdot x' \rangle$ for all $x, x' \in V$, in view of (i). Therefore $t = r$. This completes the proof of Lemma II.3.13. ∎

II.3.14. **Proposition.** *If b is any invariant quadratic form on the Lie algebra L of* Proposition II.3.11, *and if $V \neq 0$, then*

$$b((u, x, v), (u', x', v')) = ruv' + ru'v + suu' + r \langle x \mid x' \rangle$$

for some real numbers r and s. If b is in addition non-degenerate, then, up to a non-zero scalar multiple, the relation

$$b((u, x, v), (u', x', v')) = q((u, x, v), (u', x', v')) + suu'$$

holds with a suitable $s \in \mathbb{R}$.

Proof.   Consider a quadratic form b. Then there is a vector space automorphism f of L such that $b(w, w') = q(f(w), w') = q(w, f(w'))$ since q is non-degenerate and b is symmetric. If b is invariant, we calculate $q(w, f[w', w'']) = b(w, [w', w'']) = b([w, w'], w'') = q([w, w'], f(w'')) = q(w, [w', f(w'')])$ for all $w, w', w'' \in L$ in view of the invariance of q. Hence $f[w', w''] = [w', f(w'')]$ since q is non-degenerate. Hence f is an L-module endomorphism of L, and by Lemma II.3.13, there are real numbers r and s such that $f(u, x, v) = (ru, r{\cdot}x, rv + su)$. Thus

$$b((u, x, v), (u', x', v')) = ruv' + ru'v + suu' + r\langle x \mid x' \rangle.$$

If b is also non-degenerate, then f is non-singular, and thus $r \neq 0$. The assertion now follows readily.   ∎

A straightforward calculation shows, that the equation $g(u, x, v) = (u, x, v + \frac{s}{2}u)$ for $s \in \mathbb{R}$ defines an automorphism g of L. This allows us to state

II.3.15.   Corollary.   *Let L be as in* Proposition II.3.11 *and suppose that q and Q are two Lorentzian invariant forms on L. Then there is a non-zero $r \in \mathbb{R}$ and an $s \in \mathbb{R}$ such that $Q(w, w') = rq(g(w), g(w'))$ with $g(u, x, v) = (u, x, v + \frac{s}{2}u)$. Thus, up to scalar multiplication and up to (very simple) automorphisms, the Lorentzian form on L is unique.*

Proof.   From Proposition II.3.14 we know that $Q(w, w') = rq(w, f(w'))$ with $f(u, x, v) = (u, x, v + su)$. Define $g: L \to L$ by $g(u, x, v + \frac{s}{2}u)$. Now $g^2 = f$, and thus $q(w, f(w')) = q(w, g^2(w')) = q(g(w), g(w'))$ since g is q-symmetric.   ∎

EII.2.   Exercise.   If L is as in Proposition II.3.11 with $V \neq \{0\}$ and f is a self-map of L, then f is an automorphism of L if and only if there are real numbers r, s and p with $p > 0$, a vector x_0, and an orthogonal transformation T of V satisfying $Td = dT$ such that

$$f(u, x, v) = (u, r{\cdot}x_0 + p{\cdot}T(x), su - p\langle x_0 \mid T(x) \rangle + p^2 v).$$

After all this information about the properties and uniqueness of L and q the following definition appears justified:

II.3.16.   Definition.

  (i) A *Lorentzian Lie algebra* is a pair (L, q) with a Lie algebra L and an invariant non-degenerate Lorentzian form.

  (ii) If L and q are as in Proposition II.3.11 with $\dim V = 2n > 0$, then (L, q) is called the *standard solvable Lorentzian Lie algebra* A_{2n+2} *of dimension* $2n + 2$.

  (iii) If L is a compact Lie algebra with a non-degenerate Lorentzian form q according to Proposition II.3.9, we say that (L, q) is a *compact Lorentzian Lie algebra*.

(iv) We shall call $(\mathrm{sl}(2,\mathbb{R}), B)$ the *Lorentzian* $\mathrm{sl}(2)$-*algebra*.

At a later point we shall show that the standard solvable Lorentzian algebras and the Lorentzian $\mathrm{sl}(2)$-algebra are the only irreducible ones. Specifically, we shall prove, that a Lorentzian Lie algebra is either compact or the orthogonal direct sum of two ideals, one of which is an irreducible Lorentzian algebra (of one the types just mentioned), while the second is a compact Lie algebra with a positive definite quadratic form.

We should also point out that the Lie algebras underlying our standard Lorentzian Lie algebras of dimension 2n+2 are called oscillator algebras in quantum mechanics because they describe a system of a harmonic oscillator in n-dimensional euclidean space.

More on 4-dimensional solvable examples

The diversity of solvable Lie algebras increases drastically when we pass from dimension 3 to dimension 4. It would hardly be feasible to attempt a classification of these Lie algebras together with their possible Lie semialgebras. However, the standard Lorentzian solvable Lie algebra of dimension 4 has already been recognized as the receptacle of interesting Lie semialgebras, namely, invariant Lorentzian cones. The 4-dimensional standard Lorentzian Lie algebra has the (3-dimensional) Heisenberg algebra as commutator algebra. We shall therefore consider in this subsection those 4-dimensional Lie algebras whose commutator algebra is the Heisenberg algebra and determine the possible pointed generating Lie semialgebras in them. The principal result will reveal nothing new: Only the standard Lorentzian algebra is capable of supporting pointed generating Lie semialgebras, and these have to be the Lorentzian cones we discussed in the preceding subsection.

First let $H = V \times \mathbb{R}$ denote the Heisenberg algebra which, for the moment, is still given by an arbitrary finite dimensional Hilbert space V of even dimension and a skew symmetric operator d on V such that the multiplication on H is given by

$$[(x, v), (x', v')] = (0, \langle dx \mid x'\rangle).$$

The hyperplanes E of H can be described as follows: Firstly, there are all those which are of the form $V_1 \times \mathbb{R}$ with a hyperplane V_1 of V. Secondly, for each other hyperplane E there is a vector $z \in V$ such that $E = \{(x, v) \in H : \quad v = \langle z \mid x\rangle\}$.

For our purposes we need to know when a derivation D of H fixes a hyperplane E as a whole. This requires complete information on the derivations of H. Since a derivation respects the center, it is necessarily of the form $D(x, v) = (F(x), \langle x \mid y\rangle + rv)$ with a vector $y \in H$ and some real number r. Now note $D[(x, v), (x', v')] = D(0, \langle dx \mid x'\rangle) = (0, \langle r \cdot dx \mid x'\rangle)$ on one hand and $[D(x, v), (x', v')] + [(x, v), D(x', v')] = (0, \langle dF(x) \mid x'\rangle + \langle d(x) \mid F(x')\rangle) = (0, \langle (dF + F^*d)(x) \mid x'\rangle)$ (where F^* is the adjoint of F) on the other. Thus in order that D be a derivation it is necessary and sufficient that

$$(8) \qquad\qquad r \cdot d = dF + F^*d,$$

or, equivalently, that

$$(9) \qquad\qquad r \cdot 1 = dFd^{-1} + F^*.$$

Taking traces on both sides we obtain $rn = \operatorname{tr} dFd^{-l} + tr F^*$ with $n = \dim V$. But $\operatorname{tr} dFd^{-1} = \operatorname{tr} F = \operatorname{tr} F^*$, whence necessarily $r = \frac{2}{n} \operatorname{tr} F$. Thus D is a derivation of H exactly when

$$(10) \qquad\qquad \frac{2}{n} \operatorname{tr} F \cdot 1 = dFd^{-1} + F^*, \quad \text{and} \quad r = \frac{2}{n} \operatorname{tr} F.$$

Now we specialize to the case $n = \dim V = 2$ and introduce a basis relative to which d has the matrix

$$\begin{pmatrix} 0 & 1 \\ -1 & 0 \end{pmatrix}$$

and F the matrix

$$\begin{pmatrix} a_{11} & a_{12} \\ a_{21} & a_{22} \end{pmatrix}.$$

A rapid calculation shows that $dFd^{-1} + F^*$ has the form

$$\begin{pmatrix} a_{22} + a_{11} & 0 \\ 0 & a_{11} + a_{22} \end{pmatrix}.$$

Now $\operatorname{tr} F = a_{11} + a_{22}$, and thus according to (10) the map D is a derivation if and only if $r = \operatorname{tr} F$. Let us summarize this in the following statement:

II.3.17.   **Remark.**   If $H = \mathbb{R}^2 \times \mathbb{R}$ denotes the 3-dimensional Heisenberg algebra with the multiplication $[(x, v), (x', v')] = (0, \det(x, x'))$, then

(i) every pair $(F, y) \in gl(2, \mathbb{R}) \times \mathbb{R}^2$ defines a derivation $D = D_{(F,y)}$ by $D(x, v) = (F(x), \langle x \mid y \rangle + (\operatorname{tr} F)v)$,

(ii) the function $(F, y) \mapsto D_{(F,y)} : gl(2, \mathbb{R}) \times \mathbb{R}^2 \to \operatorname{Der}(H)$ is an isomorphism of Lie algebras, if the domain vector space is given the Lie multiplication $[(F, y), (F', y')] = ([F, F'], F(y') - F'(y))$

(iii) If there is a $z \in H$ with $y = (F - (\operatorname{tr} F) \cdot 1)^* z$, then $\alpha D_{(F,y)} \alpha^{-1} = D_{(F,0)}$ for the automorphism α of H with $\alpha(x, v) = (x, v - \langle z \mid x \rangle)$.

Proof.   (i) has been proved, and it is clear that the function in (ii) is bijective from what we said before. The claim that it is a morphism of Lie algebras can be safely left as an exercise to the reader. For a proof of (iii) it is easy to prove that α is an automorphism. The remaining verification is then a matter of straightforward calculation.   ■

Now let us suppose that H is the 3-dimensional Heisenberg algebra and E is a hyperplane in H. If $E = V_1 \times \mathbb{R}$ with a 1-dimensional vector subspace V_1 in $\mathbb{R}^2$, then we have $D(E) \subseteq E$ with $D = D_{(F,y)}$ if and only if $F(V_1) \subseteq V_1$, that is, if V_1 is spanned by an eigenvector. If $E = \{(x, v) \in H : v = \langle z \mid x \rangle\}$, then $D(E) \subseteq E$

if and only if $\big(F(x), \langle x \mid y \rangle + (\operatorname{tr} F)\langle z \mid x \rangle\big) = D(x, \langle z \mid x \rangle) = \big(F(x), \langle z \mid F(x) \rangle\big)$, that is, exactly when

$$(11) \qquad \qquad \langle z \mid F(x) \rangle = \langle x \mid y \rangle + (\operatorname{tr} F)\langle z \mid x \rangle.$$

This is equivalent to

$$(12) \qquad \qquad \langle z \mid \big(F - (\operatorname{tr} F){\cdot}\mathbf{1}\big)(x) \rangle = \langle y \mid x \rangle,$$

that is, to

$$(13) \qquad \qquad y = \big(F^* - (\operatorname{tr} F^*){\cdot}\mathbf{1}\big)(z).$$

From (13) it follows that y determines z uniquely if $\operatorname{tr} F$ is not an eigenvalue of F, since this is the exact condition for $F - (\operatorname{tr} F){\cdot}\mathbf{1}$ and, equivalently, $F^* - (\operatorname{tr} F^*){\cdot}\mathbf{1}$ to be invertible. Now let λ_1 and λ_2 be the eigenvalues of F in $\mathbf{C}$. Then $\lambda_1 = \operatorname{tr} F = \lambda_1 + \lambda_2$ if and only if $\lambda_2 = 0$. This happens if and only if F is singular. Let us summarize:

II.3.18. **Lemma.** *Suppose that $D = D_{(F,y)}$ is a derivation of the 3-dimensional Heisenberg algebra H with a non-singular endomorphism F. Suppose that E is a hyperplane of H with $D(E) \subseteq E$. Then the following cases occur:*

(a) $E = \mathbf{R}{\cdot}x + Z(H)$ *with an eigenvector $x \neq 0$ of F and $Z(H) = \{0\} \times \mathbf{R}$.*

(b) $E = \{(x, v) \in H : v = \langle y \mid \big(F - (\operatorname{tr} F){\cdot}\mathbf{1}\big)^{-1}(x) \rangle\}$. ∎

If we represent the non-singular endomorphisam F in real Jordan normal form, then like in Remark II.3.2, the matrix representing F can have one of the following forms:

$$(\mathrm{i}) \qquad \qquad \begin{pmatrix} \lambda & \omega \\ -\omega & \lambda \end{pmatrix}, \quad \lambda, \omega \in \mathbf{R}, \quad \omega \neq 0,$$

$$(\mathrm{ii}) \qquad \qquad \begin{pmatrix} \lambda & 1 \\ 0 & \lambda \end{pmatrix}, \quad \lambda \in \mathbf{R} \setminus \{0\},$$

$$(\mathrm{iii}) \qquad \qquad \begin{pmatrix} \lambda_1 & 0 \\ 0 & \lambda_2 \end{pmatrix}, \quad \lambda_1, \lambda_2 \in \mathbf{R} \setminus \{0\}.$$

(One should note, however, that the basis for which the real Jordan form of F is realized is not in general the basis for which d has an antidiagonal matrix.)

In case (i) there is no real eigenvalue, and the planes of type (a) above do not occur. The matrix of $F - (\operatorname{tr} F){\cdot}\mathbf{1}$ is

$$\begin{pmatrix} -\lambda & \omega \\ -\omega & -\lambda \end{pmatrix}$$

and its inverse is

$$(\lambda^2 + \omega^2)^{-1} \begin{pmatrix} -\lambda & -\omega \\ \omega & -\lambda \end{pmatrix}.$$

The plane E is uniquely determined by D.

In case (ii) there is precisely one 1-dimensional eigenspace, and

$$\left(F - (\operatorname{tr} F) \cdot \mathbf{1}\right)^{-1}$$

has the matrix

$$\begin{pmatrix} -\lambda^{-1} & -\lambda^{-2} \\ 0 & -\lambda^{-1} \end{pmatrix}.$$

The plane E is uniqely determined by D.

In case (iii) there are precisely two 1-dimensional eigenspaces, provided the two eigenvalues are different, and every 1-dimensional vector subspace of V is spanned by an eigenvector if the two eigenvalues agree. The matrix of

$$\left(F - (\operatorname{tr} F) \cdot \mathbf{1}\right)^{-1}$$

is

$$\begin{pmatrix} -\lambda_2^{-1} & 0 \\ 0 & -\lambda_1^{-1} \end{pmatrix}.$$

After these preparations we shall now consider 4-dimensional Lie algebras whose commutator algebra is isomorphic to the Heisenberg algebra H. Such a Lie algebra is then necessarily isomorphic to one of the form $L = \mathbb{R} \times V \times \mathbb{R}$ with $V = \mathbb{R}^2$ and $[(u, x, v), (u', x', v')] = \left(0, u \cdot F(x') - u' \cdot F(x), (\operatorname{tr} F)(uv' - u'v) + \det(x, x')\right)$ with a linear operator F on V and a vector $y \in V$, since L is a semidirect product of H and $\mathbb{R}$ with respect to some derivation $D = D_{(F,0)}$ in view of Remark II.3.17. In order for $\{0\} \times H = \{0\} \times V \times \mathbb{R}$ to be the commutator algebra L', it is necessary and sufficient that F be invertible. If we now set $w = (1, x, v)$, then $(\operatorname{ad} w) \mid L'$ induces on H a derivation $D_{(F,0)} + \operatorname{ad}(0, x, v) = D_{(F,0)} + D_{(0,d(x))}$ with $d(x_1, x_2) = (x_2, -x_1)$. In view of Remark II.3.17, we have

$$(14) \qquad\qquad \left(\operatorname{ad}(1, x, v)\right) \mid H = D_{(F, d(x))}$$

allowing for a mild abuse of language in identifying $\{0\} \times V \times \mathbb{R}$ with $V \times \mathbb{R}$ in the obvious way.

Let now T be a hyperplane of L different from L' and containing $w = (1, x, v)$. For T to be invariant under $\operatorname{ad} w$ it is necessary and sufficient that the plane E defined via $T \cap L' = \{0\} \times E$ be invariant under the derivation $D = D_{(F, d(x))}$.

Finally, suppose that W is a generating Lie semialgebra in L. If the interior of W meets L', then $W \cap L'$ is a generating Lie semialgebra in the Heisenberg algebra, and we know from Theorem II.3.4 that then $L'' = Z(L') \subseteq W \cap H$. Conversely, if $L'' \subseteq W$, then L/L'' is a 3-dimensional solvable Lie algebra such that W/L'' is a generating Lie semialgebra in a Lie algebra which is 3-dimensional and solvable. All of these were classified in Theorem II.3.4, and thus we know all of those W as the full inverse images of the Lie semialgebras in L/L'' under the quotient morphism $L \to L/L''$. Thus we shall assume that the interior of W and L' are disjoint. In addition we shall assume that L'' is not contained in W for the reason just given: In this case we know how to classify the possible W as pullbacks of the known wedges in 3-dimensioal algebras. Now W is contained in one of the two half-spaces bounded by L'. If W is a Lie semialgebra, then so is $-W$. Thus we shall restrict our attention to the case that W is in the half-space containing $(1, 0, 0)$. Let $w \in C^1(W)$, $w \notin L'$. We may assume that w is of the form $(1, x, v)$ with $(x, v) \in H$. Then T_w is a tangent hyperplane of the form $\mathbb{R} \cdot w + \{0\} \times E$ with a plane E in H. By Theorem II.2.14 we know that $[w, T_w] \subseteq T_w$ and thus if $D = D_{(F, dx)}$ is the derivation induced by $\operatorname{ad} w|L'$ on H, we have $D(E) \subseteq E$. We define the set K in H by $\{0\} \times K = (W \cap ((1, x, v) + L')) - (1, x, v)$. Then K is a closed convex subset of H and E is a tangent plane to it. There are two possibilities for E according to Lemma II.3.18. Either E contains $Z(H)$ and an eigenspace of F on V (type (a)), or else

$$E = \{(x', v') : v' = \langle dx \mid (F - (\operatorname{tr} F) \cdot \mathbf{1})^{-1} x' \rangle\},$$

that is,

(15) E is the graph of the function $\langle (F^* - (\operatorname{tr} F) \cdot \mathbf{1})^{-1} dx \mid \bullet \rangle : V \to \mathbb{R}$

(type (b)). If all tangent hyperplanes T_w arose from planes E of type (a), then W would have to contain L'', contrary to what we have assumed above. Thus at least one tangent hyperplane T_w originates from a tangentplane E of K of type (b). Now the closed convex set K is either above or below E. For the sake of the argument let us assume that it is above E. We now define the *support function* of K to be

(16) $\kappa : V \to \mathbb{R} \cup \{+\infty\}, \quad \kappa(y) = \min\{t : (y, t) \in K\}.$

Since K has interior points, there is an open subset U of V on which κ is finite. If K were below E, we would simply deal with a dual argument. We know that κ is a closed convex function which is, among other things, locally Lipschitz continuous on U and differentiable on a dense subset of U whose complement has Lebesgue measure 0. (For this sort of information see for instance Rockafellar, Convex Analysis, p.86 and p.246.) We further know that in all points, where κ is differentiable, its derivative is given by

(17) $\kappa'(x) = f(x) \quad \text{with} \quad f = (F^* - (\operatorname{tr} F) \cdot \mathbf{1})^{-1} d$

in view of (15). At this point we invoke a lemma on convex functions:

II.3.19.   Lemma. *Let* $\kappa : U \to \mathbb{R}$ *be a closed convex function on an open connected subset of* $\mathbb{R}^n$. *Suppose that there is a linear automorphism* f *of* $\mathbb{R}^n$ *such that* $\kappa'(x) = f(x)$ *wherever the derivative exists. Then*

   (i) f *is symmetric and positive definite,*

   (ii) $\kappa(x) = \frac{1}{2}\langle f(x) \mid x \rangle + c$ *for some constant* c.

Proof. We invoke Theorem 25.6 of Rockafellar, Convex Analysis, p.246. Since U is open, the normal cone $K(x)$ to U in x is $\{0\}$ for all $x \in U$. Further, the set $S(x)$ of all limits of sequences of the form $\kappa'(x_n)$ with κ differentiable at x_n and $x = \lim x_n$ is $\{f(x)\}$, since $\kappa'(x_n) = f(x_n)$. Thus $\partial f(x) = \{f(x)\}$ in the notation of Rockafellar. Then κ is differentiable on all of U by Theorem 25.1 on p.242 of Rockafellar. The assertions (i) and (ii) are now elementary calculus. ∎

We now apply this lemma to our situation and investigate under which circumstances $f = \bigl(F^* - (\operatorname{tr} F)\cdot\mathbf{1}\bigr)^{-1} d$ is symmetric and positive definite. From (3) and the hypothesis that f be symmetric we compute $F^* d - (trF)\cdot d = -dF = -(dF)^* = -F^* d^* = -F^*(-d) = F^* d$, whence $trF = 0$. In the three cases described after Lemma II.3.18 this has the following consequences: In case (i) we find $\lambda = 0$. Case (ii) is ruled out. In case (iii) we have $\lambda_2 = -\lambda_1$ and thus $\det F < 0$. But we also know that $f = F^* d$ is positive definite, whence $0 < \det f = \det F^* \det d = \det F \det d$. But $\det d = 1$, which yields $\det F > 0$, a contradiction. Thus case (iii) is ruled out, too. In case (i), after rescaling, we may assume that $\omega = 1$ and find that the standard Lorentzian solvable Lie algebra of dimension 4 is the only 4-dimensional Lie algebra whose commutator algebra is a Heisenberg algebra which is capable of supporting a generating Lie semialgebra whose interior does not meet L' and which does not contain L''. In case (i), the planes E of type (a) cannot occur. By Lemma II.3.19 we have $\kappa(x) = \frac{1}{2}\langle x \mid x \rangle + c$ which yields the Lorentzian cones we have discussed from Proposition II.3.11 through Definition II.3.16 This then gives us the following theorem:

II.3.20.   Theorem. *Let L be a 4-dimensional Lie algebra whose commutator algebra is the 3-dimensional Heisenberg algebra. Suppose that W is a Lie semialgebra satisfying the following conditions*

 (i) *$L' \cap \operatorname{int}(W) = \varnothing$,*

 (ii) *$L'' \not\subseteq W$, and*

 (iii) *W is not a half-space bounded by L'.*

Then (L, W) is the standard Lorentzian solvable Lie algebra of dimension 4. ∎

In particular, the Lie semialgebra W is one of the Lorentzian cones of which the standard Lorentzian solvable Lie algebras are full.

We shall see later (after Theorem II.4.13 below) that condition (ii) implies (i) in Theorem II.3.20. If $L'' \subseteq W$, then we shall have a classification of the possible cases through Lemma II.4.6 below.

The non-solvable 4-dimensional examples

It is very instructive to contrast the preceding class of examples with the class of non-solvable 4-dimensional Lie algebras. However, in contrast with the preceding developments in this section, not everything we shall say about these algebras can be treated in a self-contained fashion. The previous parts were proved completely at this point because many pieces of information accumulated in this process will be used in general developments in subsequent sections and chapters. For the non-solvable 4-dimensional Lie algebras, however, the situation is different. We can present certain aspects of these Lie algebras and their invariant cones at this point only if we borrow certain results from the full stock of information on Lie semialgebras and invariant cones which we will build up in the remainder of this chapter and in the next one.

A 4-dimensional Lie algebra accomodates at most 3-dimensional semisimple and hence simple algebras. Thus any Levi subalgebra S of L is isomorphic either to $so(3)$ or to $sl(2, \mathbb{R})$. The radical R of L is then necessarily 1-dimensional, and all 1-dimensional S-modules are necessarily trivial for any non-zero semisimple Lie algebra S. This means that L is reductive and thus either isomorphic to

(18) $$u(2) \cong \mathbb{R} \oplus so(3),$$

or to

(19) $$gl(2) \overset{\text{def}}{=} gl(2, \mathbb{R}) \cong \mathbb{R} \oplus sl(2).$$

The following observation is a tiny special case of a much more general theorem which we shall prove in Section 6 below.

II.3.21.   **Remark.**   Both $u(2)$ and $gl(2)$ support invariant Lorentzian forms.

Proof.   We begin by fixing a positive number t and consider the two cases separately.

(a) If

$$X = \begin{pmatrix} ix + iy & u + iv \\ -u + iv & ix - iy \end{pmatrix} \in u(2),$$

then $q_t(X) = -tx^2 + y^2 + u^2 + v^2$ defines an invariant Lorentzian quadratic form on $u(2)$. This may be verified directly; the idea, however, is to define q as the direct sum of a negative definite quadratic form on the abelian ideal $\mathbb{R}i{\cdot}1$ and a positive definite invariant quadratic form on the ideal $su(2)$.

(b) If

$$X = \begin{pmatrix} a & b \\ c & d \end{pmatrix} \in gl(2),$$

then $q_t(X) = \frac{t}{4}(a+d)^2 + \frac{1}{4}(a-d)^2 + bc$ defines an invariant Lorentzian form on $gl(2)$. Again this may be verified directly; the idea, however, is to define q as the direct sum of as positive definite quadratic form on the ideal $\mathbb{R}{\cdot}1$ and the quadratic form b on $sl(2)$ which we know from the discussion of the semisimple case in dimension 3 above. Observe that the projection of the matrix X into the space of scalar multiples of the identity matrix is $\frac{(a+d)}{2}{\cdot}1$ while the projection into $sl(2)$ is

$$\begin{pmatrix} \frac{(a-d)}{2} & b \\ c & \frac{(d-a)}{2} \end{pmatrix}$$

∎

Note that the Lorentzian forms are not uniquely determined up to scalar multiples, since we are free to choose $t > 0$ as we please.

With each Lorentzian invariant form q on L we obtain two invariant pointed generating cones in L. It is important to notice right away and once and for all that *invariant cones in a Lie algebra which is a direct sum of two ideals need not at all be a direct sum of invariant pointed generating cones in the two ideal summands.* Observe further that the Lorentzian cones in the two Lie algebras u(2) and gl(2) are of an entirely different type in relation to the structure of the Lie algebra: In u(2), the axis of rotational symmetry of the Lorentzian cone is the center, in gl(2) the axis of rotational symmetry is located inside $\mathcal{W}^{\pm}$ in the Levi subalgebra sl(2). In the former case, the projection of the invariant cone into the Levi algebra is the whole Levi algebra, while in the latter case this projection is one of the two possible invariant Lorentzian cones in the Levi subalgebra.

II.3.22.   Proposition. *The Lorentzian invariant cones in* u(2) *are the only generating Lie semialgebras different from* u(2) *and the half-spaces bounded by* su(2).

Proof.   Our first claim is that all generating Lie semialgebras are invariant. This is not obvious, and we shall later prove a more general result which will entail this claim (see Proposition II.6.15). So let us assume that W is a generating invariant cone which is not the whole space nor one of the two invariant half-spaces. The edge of W is an ideal by Proposition II.1.10. There are only two non-zero ideals, namely, the center and the commutator algebra. The latter is ruled out as edge. If the center is $H(W)$, then W is a direct sum of the center and a pointed invariant generating cone in su(2) by Proposition I.2.12, but such a cone does not exist. Yet su(2) does not contain any invariant non-zero proper wedges; this is, for instance a consequence of the fact that the group of inner automorphisms of su(2) operates transitively on the unit sphere with respect to an invariant scalar product on su(2). (In fact it follows from more general results in Chapter III below: See for instance Proposition III.2.2.). Thus W is pointed. Let E denote the hyperplane $\mathbf{1} + \mathrm{su}(2)$. We shall assume without loss of generality that W is on the same side of su(2), that is, that $W \cap E \neq \emptyset$. The group of inner automorphisms of $u(2)$ acts as the group of rotations on the three dimensional euclidean affine space E. The only fixed point is $\mathbf{1}$, and the convex compact invariant subset $W \cap E$ is a ball B around $\mathbf{1}$ with radius $r > 0$. But this means that $W = \mathbb{R}^{+} \cdot B$ is a Lorentzian cone.   ∎

After we discovered numerous generating Lie semialgebras in sl(2) in the earlier parts of this section we should not be surprised to find that gl(2) has a much greater variety of generating Lie semialgebras than u(2). Of course, whenever V is a generating Lie semialgebra in sl(2), then $\mathbb{R} \cdot \mathbf{1} \oplus V$ is a generating Lie semialgebra of gl(2). We exclude this case. We do not know whether under these circumstances, W is invariant. (Proposition II.6.15 below discusses related matters). At any rate we shall describe the invariant pointed generating cones in gl(2). First we shall establish a simple lemma:

II.3.23.   Lemma. *Suppose that $L = R \oplus S$ with $S = \mathrm{sl}(2)$ and with an abelian ideal R. If W is any invariant cone in L and if p denotes the projection of L into S, then $W \cap S = p(W)$.*

Proof. Obviously, $W \cap S \subseteq p(W)$. We must show the reverse inclusion. Hence take $w \in W$ and write $w = r \oplus s$ with $r \in R$ and $s \in S$. Our claim amounts to showing that $s \in W$. Let G denote the group of inner automorphisms of S. Then the orbit of w under the group of inner automorphisms of L is simply $r \oplus Gs$. If C denotes the closed convex hull of Gs in S, then $r \oplus C \subseteq W$ by invariance and convexity of W. Now we have explicitly described the G-orbits of $S = \mathrm{sl}(2)$ in Proposition II.3.6 above. We conclude from this description that $t \cdot s \in C$ for all $1 \leq t \in \mathbb{R}$. Hence $\frac{1}{n} \cdot r \oplus s = \frac{1}{n} \cdot (r \oplus n \cdot s) \in W$ for $n = 1, 2, \ldots$ If we note that the limit of this sequence for $n \to \infty$ is s and recall that W is closed, we find $s \in W$ as asserted. ∎

Now we are going to construct invariant pointed generating cones in $\mathrm{gl}(2)$. Let $H^{\pm}$ denote the half-space semialgebras of $\mathrm{gl}(2)$ bounded by $\mathrm{sl}(2)$ and containing $\pm \mathbf{1}$, respectively. Fix two positive real numbers $t^{\pm}$ and define two Lorenzian invariant forms $q_{t\pm}$ on $\mathrm{gl}(2)$ with these parameters as in the proof of Remark II.3.21 above. For each of the two forms let $W_{\pm}$ denote one of the invariant Lorentzian cones. Assume, in order to fix notation, that both of them intersect $\mathrm{sl}(2)$ in $\mathcal{W}^{+}$; after Lemma II.3.23 they have to intersect $\mathrm{sl}(2)$ in either $\mathcal{W}^{+}$ or $\mathcal{W}^{-}$.

II.3.24. **Proposition.** *The following subsets of* $\mathrm{gl}(2)$ *are invariant pointed generating cones:*

(i) $H^{+} \cap W_{+}$ *and* $H^{-} \cap W_{-}$,

(ii) $\mathbb{R}^{+} \cdot \mathbf{1} \oplus \mathcal{W}^{+}$ *and its reflection in* $\mathrm{sl}(2)$,

(iii) $(H^{+} \cap W_{+}) \cup (H^{-} \cap W_{-})$,

(iv) $(\mathbb{R}^{+} \cdot \mathbf{1} \oplus \mathcal{W}^{+}) \cup (H^{-} \cap W_{-})$ *and* $(-\mathbb{R}^{+} \cdot \mathbf{1} \oplus \mathcal{W}^{+}) \cup (H^{+} \cap W_{+})$.

These exhaust all possible types.

Proof. Since the intersection of two invariant wedges is invariant , the sets of type (i) are invariant cones. The sets of type (ii) are ostensibly invariant from their definition. Regarding sets of type (iii) and (iv) we note firstly that all of the sets in question are pointed generating cones, and secondly they are unions of invariant subsets, hence are themselves invariant.

The claim that these types exhaust all possibilities is much harder to establish. We defer the proof until Section 9 of Chapter III, where we shall have a complete set of tools to deal with problems like this in full generality. ∎

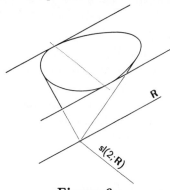

Figure 6

The variety of invariant wedges in gl(2) is quite remarkable. Only type (ii) is the type which adjusts to decomposition of the reductive algebra gl(2) into a direct sum of ideals. Precisely the type (iii) cones are round. The others are not even weakly round.

In the course of this discussion, we noted the following fact, which we summarize here for easy reference:

II.3.25.   Theorem.   *Every pointed invariant Lie semialgebra in a non-solvable Lie algebra of dimension 4 is either Lorentzian, the direct sum of a half line and a 3-dimensional Lorentzian cone, half of a Lorentzian cone, or the union of two halves of Lorentzian cones.*   ∎

Another special class of solvable Lie algebras

We now inspect a special class of solvable Lie algebras which is quite different from the class of standard solvable Lorentzian Lie algebras A_{2n+2} which we considered above. Let us begin by considering a real vector space L of dimension $n+2$ on which we fix a basis $x, y, z_1, \ldots, z_n$. We set $z_{n+1} = 0$ and define a bracket whose non-zero entries on the basis are given by

(20) $[x, y] = y + z_1, \quad [x, z_m] = m \cdot z_m, \quad [y, z_m] = z_{m+1} \quad$ for $\quad m = 1, \ldots, n.$

One verifies directly that this bracket makes L into a solvable Lie algebra.

II.3.26.   Definition.   We shall denote the $n+2$-dimensional solvable Lie algebra L by Γ_n. We say $\Gamma_{-1} = \mathbb{R}$.

II.3.27.   Lemma.   *We consider Γ_n for $n \geq 1$, that is, $\dim \Gamma_n \geq 3$.*

(i) *Γ_n has the following k-dimensional ideals: $J_k = \mathrm{span}\{z_{n+1-k}, \ldots, z_n\}$, $k = 1, \ldots, n$, $J_{n+1} = \mathbb{R} \cdot y + J_n$. Moreover, $J_{n+1} = \Gamma_n' = N = $ nilradical of Γ_n, $J_{n-k} = N^{[k]} = [N, N^{[k-1]}] = k$-th term of the descending central series $= n - k$-th term of the ascending central series for $k = 1, \ldots, n-1$. All J_k (including J_n!) are characteristic.*

(ii) *Set $d = (\mathrm{ad}\, x)|J_n$ and $D = (\mathrm{ad}\, y)|J_n$. Then $[d, D] = D$ and Γ_n/J_n is the 2-dimensional non-abelian algebra.*

(iii) *The commutator algebra N is metabelian and nilpotent of length n.*

(iv) *The proper non-zero ideals of Γ_n are the J_m, $m = 1, \ldots, n+1$. The non-abelian homomorphic images of Γ_n are isomorphic to Γ_m with $m \leq n$.*

(v) *The simply connected Lie group G_n for Γ_n is the semidirect product $(N, *) \rtimes \mathbb{R}$ with $(n, r)(n', r') = (n * e^{r \cdot \mathrm{ad}\, x} n', r + r')$ with the Campbell–Hausdorff multiplication $*$ on the nilpotent algebra N. The exponential function $\exp \colon \Gamma_n \to G_n$ is a diffeomorphism.*

(vi) *The Campbell–Hausdorff multiplication $* \colon \Gamma_n \times \Gamma_n \to \Gamma_n$ is globally defined by $u * v = \exp^{-1}(\exp u \exp v)$ and $(\Gamma_n, *) \cong G_n$ via $\exp$.*

(vii) *If W is a Lie semialgebra in Γ_n then $(W, *)$ is a subsemigroup of $(\Gamma_n, *)$. If I is an ideal of Γ_n, then $(W + I, *)$ is a semigroup and $\overline{W + I}$ is a semialgebra.*

Proof. All assertions through (vi) are straightforward from the definitions. The first part of (vii) follows from Corollary II.2.42. If I is an ideal, then $(I, *)$ is a closed normal subgroup of Γ_n, and $w * I \subseteq w + I$. But $w * I$ is a closed submanifold of $w + I$ of the same dimension, whence equality follows. Thus $W + I = W * I$ is a subsemigroup of $(\Gamma_n, *)$. Hence $\overline{W + I}$ is a subsemigroup, too, and then a semialgebra since it is a wedge. ∎

Note that Γ_0 is the non-abelian 2-dimensional algebra, and Γ_2 is a 4-dimensional Lie algebra whose commutator algebra is the 3-dimensional Heisenberg algebra.

II.3.28. **Proposition.** *Every generating Lie semialgebra of the Lie algebra Γ_n contains the unique n-dimensional ideal J_n.*

Proof. For $n = 0$, the proposition is trivial. For $n = 1$, the claim is true by the First Classification Theorem of Low Dimensional Semialgebras II.3.4, Case (ii). Assume then W is a counterexample to the proposition in Γ_n with minimal $n > 1$. Since Cartan algebras in solvable algebras are conjugate and dense, it is no loss of generality to assume that the Cartan algebra $\mathbb{R}.x$ satisfies $\mathbb{R}.x \cap \operatorname{int} W \neq \emptyset$, say, $x \in \operatorname{int} W$. If $J_1 \subseteq W$, then it is straightforward to verify that W/J_1 is a counterexample in $\Gamma_n/J_1 \cong \Gamma_{n-1}$ (see the next Proposition II.4.1 for a more general result). Since this contradicts the minimality of n, it follows that no counterexample in Γ_n contains J_1. We claim further that $\mathbb{R}.z_{n-1} \not\subseteq W$. Otherwise pick $r > 0$ so small that $x + ry \in \operatorname{int} W$. Since W is a Lie wedge,

$$e^{-t \cdot \operatorname{ad} z_{n-1}}(x + r \cdot y) - t(n-1) \cdot z_{n-1}$$
$$= x + r \cdot y + t(n-1) \cdot z_{n-1} + tr \cdot z_n - t(n-1) \cdot z_{n-1}$$
$$= x + r \cdot y + tr \cdot z_n \in W$$

for all $t \in \mathbb{R}$, and thus $J_1 = \mathbb{R}.z_n \subseteq W$. But we have just seen that this is impossible.

By Lemma II.3.27.vii, $\overline{W + \mathbb{R}.z_n} = \overline{W + J_1}$ is a Lie semialgebra. Now $W \cap A$ is a generating semialgebra in the subalgebra $A = \mathbb{R}.x + J_n$. Since $\mathbb{R}.z_{n-1} \not\subseteq W \cap A$, there exists a C^1-point $w = x + m \in \partial(W \cap A)$, $m \in J_n$, such that the tangent hyperplane T_w to $W \cap A$ at w in A does not contain $\mathbb{R}.z_{n-1}$. Since $\operatorname{ad} w|J_n$ has eigenvectors $z_1, \ldots, z_n$, it follows that the $(\operatorname{ad} w)$-invariant hyperplane $T_w \cap J_n$ of J_n is spanned by $\{z_1, \ldots, z_n\} \setminus \{z_{n-1}\}$. In particular, $J_1 \subseteq T_w$. Then T_w extends in Γ_n to a support hyperplane T of W, and the half-space H containing W also contains J_1, but not $\mathbb{R}.z_{n-1}$. Thus $\overline{W + J_1} \subseteq H$ does not contain $\mathbb{R}.z_{n-1}$. Again $\overline{W + J_1}/J_1$ yields a lower dimensional counterexample, a contradiction. ∎

4. Reducing Lie semialgebras, Cartan algebras

Our eventual goal is to classify Lie semialgebras as completely as possible. Any attempt at such a classification is greatly helped by certain reductions which we discuss now. We shall also describe a number of devices which we then shall use in a standard fashion time and again. First applications will in fact be given in this section.

We have observed in Theorem II.2.13 that for any Lie semialgebra W in a Lie algebra L the span $W - W$ is a Lie subalgebra. It is therefore reasonable to restrict one's attention to generating Lie semialgebras for which, we recall, $W - W = L$. Now we wish to investigate to what extent it is justified to pass to factor algebras. In general, if W is a wedge in a finite dimensional vector space L and if I is a vector subspace of L, the passage to the quotient L/I is problematical, since the set $W + I$ may fail to be closed, whence the image $(W + I)/I$ of W in the factor space may fail to be a wedge. As a general rule, passage to a factor space is only indicated when $I \subseteq W$.

Proposition I.2.32, however, gives us a little more leeway. Namely, if W is a wedge and I a vector subspace in a finite dimensional vector space L, and if $W \cap I$ is a vector space, then $W + I$ is a wedge. This applies, in particular to the case that $W \cap I = \{0\}$. Now the question arises whether a wedge $W + I$ in a finite dimensional Lie algebra is in fact a Lie semialgebra whenever W is a Lie semialgebra and I is an ideal. This would generalize the familiar fact that for a Lie subalgebra W and an ideal I, the sum $W + I$ is always a subalgebra. The following proposition gives an affirmative answer:

II.4.1.   Proposition.   *Let W be a Lie semialgebra and I an ideal in a finite dimensional Lie algebra L. If the sum $W + I$ is closed, then it is a Lie semialgebra, and $(W + I)/I$ is a Lie semialgebra in the quotient algebra L/I.*

If on the other hand, V is a Lie semialgebra in L/I, then its full inverse image W in L is a Lie semialgebra containing I.

Proof.   We consider a Campbell–Hausdorff neighborhood B of 0 in L and the Campbell–Hausdorff multiplication $* : B \times B \to L$. We pick a second Campbell–Hausdorf neighborhood B' so small that $B' * \cdots * B'$ (6 factors) is contained in B. In the local theory of Lie groups one verifies that for any Campbell–Hausdorff neighborhood C with $C * C \subseteq B'$ we have

$$(1) \qquad \big(x * (I \cap B')\big) \cap C = (x + I) \cap C \qquad \text{for each} \qquad x \in C.$$

Now we exploit Proposition I.2.33 and find a neighborhood U of 0 in $W + I$ contained in C so that $U = S_W + S_I$ with subsets $S_W \subseteq W$ and $S_I \subseteq I$. Finally, we choose a last Campbell–Hausdorff neighborhood B'' so that $B'' \subseteq C$ and $B'' \cap (W + I) \subseteq U$.

Now suppose that $u, u' \in (W + I) \cap B''$. We must show that $u * u' \in W + I$; this will show that $W + I$ is a Lie semialgebra. Since $B'' \cap (W + I) \subseteq U$ we have $u = w + j$ and $u' = w' + j'$ with $w, w' \in W \cap U \subseteq W \cap C$ and $j, j' \in I \cap U \subseteq I \cap C$. By (1) above we can write $u = w * i$ and $u' = w' * i'$ with $i, i' \in I \cap B'$. Now we calculate $u * u' = w * w' * (-w') * i * w' * i'$. But $w * w' \in (W \cap B') * (W \cap B') \subseteq W \cap B$, and $(-w') * i * w' * i'$ is in $B' * B' * B' * B' \subseteq B$, but also equals $e^{-\operatorname{ad} w'} i * i'$, hence is also in I. Thus $u * u' \in (W \cap B) * (I \cap B) \subseteq W + I$. We have shown now that $W + I$ is a Lie semialgebra.

Next we show that $(W + I)/I$ is a Lie semialgebra in L/I. Since $W + I$ is a Lie semialgebra, by changing notation, if necessary, we may assume that $I \subseteq W$. We denote with $p: L \to L/I$ the quotient morphism and let B and B' be as before. We set $D = p(B')$. Now take two elements $X, X' \in D \cap W/I$. Then we find elements $b, b' \in B'$ and $w, w' \in W$ such that $X = p(b) = p(w)$ and $X' = p(b') = p(w')$. Now the elements $i = w - b$ and $i' = w' - b'$ are in I and we have $b = w - i$, $b' = w' - i'$ both contained in $W \cap B'$, since $I \subseteq W$. Thus $b * b' \in W$, because W is a Lie semialgebra. Therefore $X * Y = p(b * b') \in p(W) = W/I$.

Finally suppose that V is a Lie semialgebra in L/I. We claim that $p^{-1}(V)$ is a Lie semialgebra. Let $w, w' \in p^{-1}(V) \cap B'$. Then $p(w), p(w') \in V \cap D$, and thus $p(w) * p(w') \in (V \cap D) * (V \cap D) \subseteq V$, since V is a Lie semialgebra. Thus $w * w' \in W$ which we wanted to show. ∎

The following is an unresolved question:

Problem. *If W is a Lie semialgebra and I an ideal in a finite dimensional Lie algebra, is $\overline{W + I}$ a Lie semialgebra?*

If we consider the Lie algebra L as an extension of I by L/I then any Lie semialgebra V in L/I gives rise to an "extension" of V by I, namely, $p^{-1}(V)$. We interpret this to mean that, for the purposes of classification of Lie semialgebras, we might just as well factor the largest ideal of L contained in a given Lie semialgebra W and work with W/I in L/I. Notice that such a largest ideal exists always, since the sum of any family of ideals is an ideal. These comments are to motivate the following definition:

II.4.2.   Definition. A Lie semialgebra W in a Lie algebra L will be called *reduced*, if it is generating (!) and does not contain any non-zero ideals of L. If W is a Lie semialgebra in a Lie algebra L, and if I_W denotes the largest ideal of $W - W$ contained in W, then the Lie semialgebra W/I_W is called the *reduction of W in* $(W - W)/I_W$.

Naturally, the reduction of W is reduced in $(W - W)/I_W$

The following result shows that the presence of reduced Lie semialgebras always implies the existence of pointed generating Lie semialgebras.

II.4.3.   Theorem. (The Pointing Procedure) *Let W be a reduced Lie semialgebra in L. Then for each $x \in \operatorname{int} W$ there is at least one pointed Lie semialgebra contained in W and containing x in its interior.*

This theorem can also be expressed in the following fashion:

II.4.4.   Corollary.   *For any reduced Lie semialgebra W in L we have*

$$\operatorname{int} W \subseteq \bigcup \{\operatorname{int} V : V \text{ is a pointed generating wedge contained in } W\}$$

■

Theorem II.4.3 is a consequence of the following lemma:

II.4.5.   Lemma.   *If W is a generating Lie semialgebra and $x \in \operatorname{int} W$, then there is a wedge W' containing x in its interior such that $W' \subseteq W$ and such that $H(W')$ is an ideal of L.*

Proof.   Let U_1 be an open symmetric neighborhood of 0 in L and V_1 an open neighborhood of x in $\operatorname{int} W$ such that $e^{\operatorname{ad} U_1} V_1 \subseteq W$. Then $V_1 \subseteq e^{\operatorname{ad} u} W$ for all $u \in U_1$ by the symmetry of U_1. If, for any subset S of L, we set $W(S) = \bigcap \{e^{\operatorname{ad} x} W : x \in S\}$, then we have $V_1 \subseteq W(U_1)$. Each $W(U)$ is a Lie semialgebra by Corollary II.2.17, and if $V \subseteq U$ then $W(V) \supseteq W(U)$. Since the lattice of vector subspaces of L satisfies the ascending chain condition, there is an open neighborhood U_2 of 0 contained in U_1 such that for all open neighborhoods U of 0 in U_2, the vector spaces $H(W(U))$ have one and the same dimension. Let us now find a convex Campbell–Hausdorff neighborhood U so small that $U * U \subseteq U_2$ and set $W' = W(U)$. Next consider any $u \in U$. Then $e^{\operatorname{ad} u} H(W') = H(e^{\operatorname{ad} u} W(U)) = H(e^{\operatorname{ad} u} \bigcap \{e^{\operatorname{ad} v} W : v \in U\}) = H(\bigcap \{e^{\operatorname{ad} u} e^{\operatorname{ad} v} W : v \in U\}) = H(\bigcap \{e^{\operatorname{ad}(u * v)} W : v \in U\}) \subseteq H(\bigcap \{e^{\operatorname{ad} w} W : w \in U_2\}) = H(W(U_2)) = H(W(U))$ (by the choice of U_2!) $= H(W')$. Thus for all $u \in U$ and all real t with $|t| \leq 1$ we have $e^{t \cdot \operatorname{ad} u} H(W') \subseteq H(W')$. Differentiating with respect to t at 0 then yields $[u, H(W')] = (\operatorname{ad} u)(H(W')) \subseteq H(W')$ for all $u \in U$. Thus the normalizer of $H(W')$ in L contains a neighborhood U of 0 in L, hence contains all of L. Thus $H(W')$ is an ideal of L. Since $V_1 \subseteq W(U) = W'$, we know that $x \in \operatorname{int} W'$. The relation $W' = W(U) \subseteq W$ is trivial. The lemma is proved.   ■

Before we can apply the pointing procedure we have to discuss another device which we will apply time and again. It is in this procedure our knowledge of Lie semialgebras in low dimensions which we have accumulated in the previous section comes to bear fruit.

II.4.6.   Lemma.   (The Standard Testing Device) *Let L be a Lie algebra with a generating Lie semialgebra W. Let J be a subalgebra and x an element in the normalizer of J in L. Suppose that one of the following hypotheses are satisfied:*

(a) $x \in \operatorname{int} W,$

(b) $J \cap \operatorname{int} W \neq \emptyset.$

Then $A = J + \mathbb{R} \cdot x$ is a subalgebra, and $W \cap A$ is a generating Lie semialgebra of A.

If J is an ideal to begin with, then any $x \in \operatorname{int} W$ is in the normalizer of J.

If $\dim J = 2$ and $x \notin J$, then $\dim A = 3$ and A is a 3-*dimensional solvable Lie algebra with a generating Lie semialgebra* $W \cap A$ *to which we can apply the* First Classification Theorem of Low Dimensional Lie Semialgebras II.3.4. *In fact, if J is already the 2-dimensional abelian ideal of that theorem, then the following is a complete list of the possible cases:*

   (0) *A is almost abelian.*

   (i) *J contains no 1-dimensional ideal of A. Then W contains J and $A \cap W$ is one of the two half-spaces of A bounded by J.*

   (ii) *J contains precisely one 1-dimensional ideal I. Then $I \subseteq W$. Thus $W \cap A$ is polyhedral with I in its edge. Indeed W is a half-space or the intersection of two half-spaces of A bounded by J.*

   (iii) *J contains precisely two 1-dimensional ideals I_1 and I_2. Here $A \cap W$ is a polyhedral semialgebra which is the intersection of at most four half-space semialgebras in A containing either I_1 or I_2.*
   The following conclusions hold:

   (C1) *If $\dim\big(H(W) \cap J\big) \leq 1$, then case (i) is ruled out.*

   (C2) *If $\dim\big(H(W) \cap J\big) = 0$, then cases (i) and (ii) are ruled out.*

Proof.   Since x is in the normalizer of J in L, the vector space $A = J + \mathbb{R}{\cdot}x$ is a subalgebra. Since the intersection of any family of Lie semialgebras is a Lie semialgebra by Corollary II.2.17, the intersection $W \cap A$ is a Lie semialgebra. Either one of the hypotheses (a) or (b) will guarantee that the interior of $W \cap A$ in A is not empty. Hence $W \cap A$ is generating in A. The remainder of the lemma is then a consequence of Theorem II.3.4.   ∎

As a first application of the standard testing device we prove the following still rather technical lemma:

II.4.7.   Lemma.   *Let L be a Lie algebra with a 2-dimensional abelian subalgebra J and a generating Lie semialgebra W. Suppose that $x \in W$ is in the normalizer of J but not in J. Assume that*

   (i) *$(x + J) \cap \operatorname{int} W \neq \varnothing$.*

   (ii) *$(\operatorname{ad} x)|J$ is not a scalar multiple of 1_J.*
   If we now set $A = J + \mathbb{R}{\cdot}x$, then we have the following conclusions:

   (I) *$(E_x \cap A) - (E_x \cap A) = T_x \cap A$. In particular, $T_x \cap A$ is a Lie algebra.*

   (II) *$T_x \cap J$ is an eigenspace of $(\operatorname{ad} x)|J$.*

Proof.   By (i), the wedge $W \cap A$ is a generating Lie semialgebra in A. We can apply the standard testing device II.4.6 to the 3-dimensional algebra A. By hypothesis (ii), A is not almost abelian. Lemma II.4.6 says that the wedge $W \cap A$ is polyhedral. Hence we can conclude $E_x(W \cap A) - E_x(W \cap A) = T_x(W \cap A)$, where the notation indicates that the exposed face E_x generated by x is taken in A, as is the tangent hyperplane T_x. From Proposition I.2.18, we know $E_x(W \cap A) = E_x(W) \cap A$ and $T_x(W \cap A) = T_x(W) \cap A$. With Corollary II.2.22 this is conclusion (I). But now Lemma II.3.3 proves assertion (II).   ∎

II.4.8.   Remark.   Condition (i) of Lemma II.4.7 is implied by the following conditions (i′) $x \in C^1(W)$ and $J \not\subseteq T_x$.

Proof.   If (i) is violated, then $x + J$ is contained in some support hyperplane T of W at x by the Theorem of Hahn and Banach. But $T = T_x$ since $x \in C^1(W)$. Thus $J \subseteq T_x - x = T_x$, contradicting (i′).   ■

The next result is the first in the current line which has independent interest. Let us recall that a *minimal ideal* in a Lie algebra is a non-zero proper ideal in which $\{0\}$ is the only ideal of L.

II.4.9.   Theorem. *Let J be a minimal ideal of L and suppose that* $\dim J = 2$. *If W is a generating Lie semialgebra, then $J \subseteq W$.*

Proof.   Since the commutator algebra $[J, J]$ is a characteristic ideal of J, it is an ideal of L. Hence J must be abelian, since all 2-dimensional algebras are solvable, which rules out $[J, J] = J$. Let $\rho \colon L \to gl(J)$ be the representation obtained by restricting the adjoint representation. Let $Q = \rho^{-1}(\mathbb{R} \cdot \mathbf{1}_J)$. Then Q is an ideal which is proper, because otherwise all 1-dimensional vector subspaces of J would be ideals, in contradiction to the minimality of J. Let T be any hyperplane containing Q. We finish our proof via an argument by contradiction. We start by assuming that $J \not\subseteq H(W)$. We apply Corollary I.3.14 with $V = J$. In this way we now find an $x \in C^1(W) \setminus T$ with $J \not\subseteq T_x$. Then $L = T_x + J$. Since $x \notin T$ we know that $x \notin Q$ since $Q \subseteq T$. But then $\rho(x)$ is not a scalar multiple of $\mathbf{1}_J$. Now Lemma II.4.7 applies and yields that $T_x \cap A = (E_x \cap A) - (E_x \cap A)$ is a subalgebra. Then $[T_x \cap J, L] = [T_x \cap J, T_x + J] \subseteq T_x \cap J$. Thus $T_x \cap J$ is a 1-dimensional ideal of L, a contradiction to our hypothesis on J.   ■

II.4.10.   Corollary. *A Lie algebra which supports a reduced Lie semialgebra cannot have any minimal ideals of dimension 2.*

Proof.   Any minimal ideal of dimension 2 is contained in any generating Lie semialgebra by Theorem II.4.9, hence is contained, in particular, in any reduced semialgebra. But this is impossible.   ■

We have seen in Theorem II.4.9 that a generating Lie semialgebra W of a Lie algebra L necessarily contains the ideal spanned by all two dimensional minimal ideals. A very important variant of this line of thought attempts to detect situations in which the commutator algebra $[L, L]$ has to be contained in W. This is a very reasonable issue: *If* a wedge W in a Lie algebra contains $[L, L]$, then it is invariant, since $e^{\mathrm{ad}\, x} y \equiv y$ mod $[L, L]$. Hence, a fortiori, it is a Lie semialgebra. From a Lie semialgebra point of view, we may therefore consider such wedges as trivial Lie algebras. Thus we make the following definition:

II.4.11.   Definition. A Lie semialgebra W in a Lie algebra is called *trivial* if and only if it contains the commutator algebra.

II.4.12.   Proposition. *Suppose that W is a generating Lie semialgebra in L. Then the following statements are equivalent:*

   (1) *W is trivial in L.*

   (2) *The reduction W/I_W is trivial in L/I_W.*

(3) L/I_W is abelian and W/I_W is pointed.

(4) L/I_W is abelian.

Proof. Firstly suppose that W is trivial. Then the maximal ideal I_W of L in W contains $[L, L]$ and W/I_W is a wedge in the abelian Lie algebra L/I_W, hence is trivial. Next, suppose that W/I_W is trivial in L/I_W. Then $(L/I_W)' \subseteq W/I_W$. Now $H(W/I_W)$ contains $(L/I_W)'$, hence is an ideal and thus a point, since W/I_W is reduced. Thus W/I_W is pointed. $(3) \Rightarrow (4)$ is trivial. Finally assume (4). Then $(L/I_W)'$ is singleton. Since $(L/I_W)' = (L' + I_W)/I_W$, this means $L' + I_W \subseteq I_W \subseteq W$. Hence W is trivial. ∎

Results which assert that generating Lie semialgebras in a Lie algebra have to be trivial we shall call *triviality theorems*. Here is one:

II.4.13. **Theorem.** (The First Triviality Theorem—the Nilpotency Theorem) *Every generating Lie semialgebra in a nilpotent Lie algebra is trivial.*

Proof. By Proposition II.4.12 we may assume that W is reduced. Then the theorem will be proved if we can show that L is abelian. By the pointing procedure II.4.3 it suffices to prove that a nilpotent Lie algebra L is abelian if it contains a pointed generating Lie semialgebra W. Suppose that this is not the case and consider a counterexample to the claim such that $\dim L$ is minimal. Let $L^{[0]}, \dots, L^{[m+1]} = [L, L^{[m]}], \dots$ be the descending central series. Let n be the largest natural number so that $L^{[n]} \neq \{0\}$. Then $n > 0$ since L is not abelian. Then any non-zero element in $L^{[n]}$ is of the form $\sum_{k=1}^{p} [x_k, y_k]$ with $x_k \in L$ and $y_k \in L^{[n-1]}$. At least one of the brackets in the sum has to be non-zero. Thus we find an element $z_0 = [x_0, y] \neq 0$ with $z_0 \in L^{[n]}$, further $x_0 \in L$ and $y \in L^{[n-1]}$. Since W is generating, there are elements $x, x' \in \operatorname{int} W$ such that $x_0 = x - x'$. Hence $z_0 = z - z'$ with $z = [x, y]$ and $z' = [x', y]$. One of these elements is non-zero. Assume $z \neq 0$. We set $A = \mathbb{R} \cdot x + \mathbb{R} \cdot y + \mathbb{R} \cdot z$. Then A is a Heisenberg algebra, hence is non-abelian. But it contains a pointed generating Lie semialgebra $W \cap A$, since $x \in \operatorname{int} W$. Hence A is a counterexample to the claim, and by the minimality of L we have $L = A$. The standard testing device II.4.6 applies with $J = \mathbb{R} \cdot y + \mathbb{R} \cdot z$ and shows that this situation is not possible by conclusion (C2). This contradiction proves the claim and thus establishes the theorem. ∎

We observe that the proof can be reorganized into a proof by traditional induction which does not use the pointing device.

II.4.14. **Corollary.** *Let N be any nilpotent subalgebra of a Lie algebra L with a generating Lie semialgebra W. Then $N \cap \operatorname{int} W \neq \emptyset$ implies $[N, N] \subseteq W$. This applies, in particular, to the nilradical of L and, a fortiori, to $[L, R]$ in place of N, where R denotes the radical of L.* ∎

II.4.15. **Corollary.** *Let L be a Lie algebra with a reduced Lie semialgebra W. Then any nilpotent subalgebra N meeting $\operatorname{int} W$ is necessarily abelian.*

Proof. Let $x \in N \cap \operatorname{int} W$. By the pointing procedure II.4.3, there is a pointed generating Lie semialgebra W' with $x \in \operatorname{int} W'$. Then Corollary II.4.14 shows that $[N, N] \subseteq H(W') = \{0\}$. Hence N is abelian. ∎

This is important. Nilpotent subalgebras are quite prevalent in Lie algebras due to the fact that every regular element of L is contained in one such and the set of regular elements is dense. In fact, if x is regular, then the nilspace H of $\operatorname{ad} x$, that is, the space of all $y \in L$ such that for some natural number n we have $(\operatorname{ad} x)^n y = 0$, is the *Cartan algebra* associated with x. In this book we shall use freely a variety of standard pieces of information on Cartan algebras; a good source of reference is [Bou75], Chap.VII et VIII. The previous results now lead us to the following theorem.

II.4.16.   Theorem.   (The First Cartan Algebra Theorem) *In a Lie algebra supporting a reduced generating Lie semialgebra, any Cartan algebra is necessarily abelian.*

Proof.   Let x be a regular element in $\operatorname{int} W$. Since the set of regular elements is dense and W is generating, such an element exists. Now let H be the nilspace of $\operatorname{ad} x$. This is a nilpotent subalgebra meeting $\operatorname{int} W$. Hence by Corollary II.4.15, H is abelian. Now consider the complexification $L_{\mathbb{C}} = \mathbb{C} \otimes L$ of L. Then $\mathbb{C} \otimes H$ is a Cartan subalgebra of $L_{\mathbb{C}}$. Likewise, for any other Cartan subalgebra K of L, the complexification $\mathbb{C} \otimes K$ is a Cartan subalgebra of $L_{\mathbb{C}}$. However, in a complex Lie algebra, all Cartan subalgebras are conjugate. Since H is abelian, so is $\mathbb{C} \otimes H$. Consequently the conjugate $\mathbb{C} \otimes K$ is likewise abelian, and this implies the commutativity of K.   ∎

II.4.17.   Corollary.   *If L is an arbitrary Lie algebra with a generating Lie semialgebra W and if H is an arbitrary Cartan subalgebra of L, then $[H, H] \subseteq W$.*

Proof.   Let I_W be the largest ideal of L contained in W. Then $(H + I_W)/I_W$ is a Cartan algebra in the Lie algebra L/I_W supporting the reduced Lie semialgebra W/I_W. Hence it is abelian by Theorem II.4.16. This means that $[H, H] \subseteq I_W \subseteq W$.   ∎

A good deal can be said on solvable subalgebras once we have all of this information. The first step provides a technical tool for dealing with certain adjoint representations which we shall have occasion to use quite frequently.

II.4.18.   Lemma.   *Let I be an abelian ideal of a Lie algebra L with a generating Lie semialgebra W. Denote $\rho: L \to gl(I)$ the representation obtained from the adjoint representation by restriction via $\rho(x) = (\operatorname{ad} x)|I$. If $\dim(H(W) \cap I) \leq 1$, then $\operatorname{Spec} \rho(x) \subseteq \mathbb{R}$ for all $x \in \operatorname{int} W$.*

Proof.   We pick $x \in \operatorname{int} W$ and consider a minimal $\rho(x)$-invariant subspace J of I. If $\dim J = 1$, then J is an eigenspace for some real eigenvalue. If $\dim J = 2$, then we apply the Standard Testing Device II.4.6 to $A = J + \mathbb{R} \cdot x$ and notice that we can apply conclusion (C1) to rule out our present situation because we are exactly in case (i). This procedure shows that all minimal $\rho(x)$-invariant subspaces of I are 1-dimensional. Thus all eigenvalues of $\rho(x)$ are real.   ∎

Notice that abelian ideals always exist in any Lie algebra which is not semisimple.

For the proof of the following result we need a general lemma.

II.4.19. **Lemma.** *Let L be a finite dimensional real Lie algebra with an abelian ideal I and denote with $\rho\colon L \to \mathrm{gl}(I)$ the representation given by $\rho(p) = (\operatorname{ad} p)|I$. Let S denote a non-empty open subset of L such that every element of $\rho(S)$ is diagonalizable and suppose that $\rho(L)$ is solvable. Then L is simultaneously diagonalizable or else there exist linearly independent elements $x, y, z_1, \ldots, z_n \in L$, $n \geq 1$ such that a non-zero scalar multiple of x is in S, that $[y, z_m] = z_{m+1}$ for $m = 1, \ldots, n$ with $z_{m+1} = 0$, and that one of the two cases occurs for the span A of $x, y, z_1, \ldots, z_n$:*

(i) *$[x, y] = y + z_1$, $[x, z_m] = m \cdot z_m$, and $A \cong \Gamma_n$ (see Definition II.3.26).*

(ii) *$n = 2$ and there is a real number $\lambda \neq 0$ such that $[x, y] = y$, $[x, z_1] = \lambda \cdot z_1$ and $[x, z_2] = (\lambda + 1) \cdot z_2$. The commutator algebra A' is spanned by y, z_1, z_2, and is the Heisenberg algebra.*

Proof. Let $I = V_n \supseteq V_{n-1} \supseteq \ldots \supseteq V_0 = \{0\}$ be a Jordan–Hölder series of $\rho(L)$-submodules. If ρ_m is the representation induced by ρ on the simple module V_m/V_{m-1} for $m = 1, \ldots, n$, then $\rho_m(S)$ consists of diagonalizable elements, and since $\rho_m(L)$ is abelian, $\rho_m(S)$ is simultaneously diagonalizable. Since S spans L, $\rho_m(L)$ is diagonalizable. That is, $\dim V_m/V_{m-1} = 1$ for $m = 1, \ldots, n$. Thus $\rho(L)$ is triagonalizable, and the ρ_m may be considered as the weights of L of ρ. Now we assume that L is not diagonalizable, hence not abelian. If $s \in S$ and $\rho_m(s) = 0$ for all m, then $s = 0$ because $\rho(s)$ is diagonalizable. Hence there is a non-zero weight ρ_m and L cannot be nilpotent. Since S is open in L, there is a regular element $s \in S$. The Cartan algebra generated by $\rho(s)$ in $\rho(L)$ has real non-zero roots since $\rho(L)$ is triangular and not nilpotent. Let ω be one of them. There is a smallest positive natural number n and a root vector φ for ω such that $\rho(y') \overset{\mathrm{def}}{=} \{\operatorname{ad}\rho(s) - \omega(\rho(s)) \cdot \mathbf{1}\}^n \varphi \neq 0$ and $\{\operatorname{ad}\rho(s) - \omega(\rho(s)) \cdot \mathbf{1}\}\rho(y') = 0$. Hence $\rho([s, y']) = [\rho(s), \rho(y')] = r \cdot \rho(y')$ with $r = \omega(\rho(s)) \neq 0$. Set $x = r^{-1} \cdot s$. Then $\tau = (\operatorname{ad} x)|I$ is diagonalizable and satisfies $[\tau, \rho(y')] = \rho(y')$. Now $\rho(y') = [x, y'] = y' + \sum_{\lambda \in \mathbb{R}} v_\lambda$ with $v_\lambda \in I$ such that $\tau(v_\lambda) = \lambda \cdot v_\lambda$ for all $\lambda \in \mathbb{R}$. If we set $y = y' + \sum_{\lambda \in \mathbb{R} \setminus \{1\}} (1 - \lambda)^{-1} \cdot v_\lambda$, then $[x, y] = y' + \sum_{\lambda \in \mathbb{R} \setminus \{1\}} (v_\lambda + \frac{\lambda}{1 - \lambda} \cdot v_1) + v_1 = y + v_1$. There are two cases: Case (i). $v_1 \neq 0$. Then we set $z_1 = v_1$, $z_k = \rho(y)^k(z_1)$ for $k = 1, 2, \ldots$ The relation $[\tau, \rho(y)] = \rho(y)$ shows $\tau(z_k) = k \cdot z_k$. Hence $z_k \neq 0$ for $k = 1, \ldots, n$ implies that these z_k are linearly independent. Hence there is a smallest n such that $z_{n+1} = \rho^{n+1}(y) = 0$. The elements $x, y, z_1, \ldots, z_n$ satisfy the requirements. Case (ii). $v_1 = 0$. The relation $[\tau, \rho(y)] = \rho(y)$ shows that the eigenspaces V_λ of τ satisfy $\rho(y)(V_\lambda) \subseteq V_{\lambda+1}$. Since $\rho(y) \neq 0$, there is an eigenvalue λ of τ and a vector $z_1 \in V_\lambda$ such that $0 \neq z_2 = \rho(y)(z_1)$ and $\rho(y)(z_2) = 0$. If $\lambda \neq 0$, then x, y, z_1, z_2 satisfy the requirements. If $\lambda = 0$, we find an $r > 0$ so that a scalar multiple of $\overline{x} = x + r \cdot z_1$ is in S. We set $\overline{y} = y$ and $\overline{z}_1 = -r \cdot z_2$. Then $[\overline{x}, \overline{y}] = [x + r \cdot z_1, y] = y - r \cdot z_2 = \overline{y} + \overline{z}_1$, $[\overline{x}, \overline{z}_1] = [x + r \cdot z_1, -r \cdot z_2] = -r \cdot z_2 = \overline{z}_1$, and $[\overline{y}, \overline{z}_1] = [y, -r \cdot z_2] = 0$. Hence the triple $\overline{x}, \overline{y}, \overline{z}_1$ is an instance of Case (i) with $n = 1$. ∎

II.4.20. **Theorem.** *Let W be a reduced generating Lie semialgebra in a Lie algebra L. Suppose that I is an abelian ideal and $\rho\colon L \to \mathrm{gl}(I)$ the representation obtained from the adjoint representation by restriction. Then $\rho(x)$ is diagonalizable for all $x \in \operatorname{int} W$. If $\rho(L)$ is solvable, then $\rho(L)$ is simultaneously diagonalizable.*

Proof. Let $x \in \operatorname{int} W$. By the pointing procedure II.4.3 we find a pointed and generating Lie semialgebra W' with x in its interior. By Lemma II.4.18, the morphism $\rho(x)$ has real spectrum. Suppose that $\rho(x)$ is not diagonalizable. Then there is at least one eigenvalue λ of $\rho(x)$ and an eigenvector e such that for another vector e' we have $\rho(x)(e') = \lambda \cdot e' + e$. Then the algebra $J + \mathbb{R} \cdot x$, where J is the span of e and e', is of type (ii) in the list of the Standard Testing Device II.4.6. But then $I = \mathbb{R} \cdot e$ would have to be contained in W' contrary to the fact that W' is pointed.

Now suppose that $\rho(L)$ is solvable. Then $\rho(L)$ is a solvable Lie algebra of endomorphisms of I, and $S = \rho(\operatorname{int} W')$ is an open subset all of whose members are diagonalizable. If $\rho(L)$ is not diagonalizable, then by Lemma II.4.19, there is a subalgebra A meeting the interior of W' and $A \cong \Gamma_n$ or $\dim A = 4$ and A' is a Heisenberg algebra. The first case is impossible by Proposition II.3.28. The second case is impossible by Theorem II.3.20. ∎

There are some remarkable consequences of these results. Before we encounter the first ones in this section, we need some purely Lie algebra theoretical lemmas which have some independent interest.

II.4.21.   Lemma. *Suppose that L is a finite dimensional real Lie algebra with a nilpotent subalgebra H which is not contained in a properly larger solvable subalgebra. Then $\operatorname{ad} h$ has purely imaginary spectrum for all $h \in H$.*

Proof. Let $L_{\mathbb{C}}$ be the complexification and identify L with its image $1 \otimes L$ in $L_{\mathbb{C}}$. Then $L_{\mathbb{C}} = L + iL$, and $H_{\mathbb{C}} = H + iH$ is the complexification of H. We consider the primary (or Jordan) decomposition of $L_{\mathbb{C}}$ with respect to $\operatorname{ad} h$:

$$L_{\mathbb{C}} = L^0 \oplus \bigoplus_{\lambda \neq 0} L^\lambda,$$

where the sum extends over the eigenvalues λ of $\operatorname{ad} h$. Since H is nilpotent, we have $H \subseteq L^0$. Now set $I = \bigoplus_{\operatorname{Re} \lambda > 0} L^\lambda$ and $A = H + I$. Since $[L^\lambda, L^\mu] \subseteq L^{\lambda + \mu}$, it follows that I is a nilpotent ideal of A. Since A/I is isomorphic to H and hence also nilpotent, it follows that A is solvable. Therefore, $A \cap L = H$ since H is not contained properly in a solvable subalgebra. From this it follows that $I \cap L = \{0\}$. We claim that this implies $I = \{0\}$: Indeed if $x + iy$ is an eigenvector for λ with $\operatorname{Re} \lambda > 0$, then $x - iy$ is an eigenvector for $\overline{\lambda}$, and thus $2x = (x + iy) + (x - iy) \in I \cap L = \{0\}$. Similarly, we see that $y = 0$. This proves the claim and shows that there are no eigenvalues with positive real parts. But analogously we show that there are no eigenvalues with negative real parts. Therefore the lemma is proved. ∎

II.4.22.   Lemma. *Let L be as in the preceding lemma and assume in addition that L is a Lie algebra of linear transformations of a finite dimensional real vector space such that H is diagonalizable. Then $L = H$.*

Proof. We pick a basis of V with respect to which the matrix of h is diagonal for each $h \in H$. Let h_{jj} denote j-th diagonal entry of the matrix of h and let $e_{jk} = (\delta_{pj} \delta_{kq})_{p,q=1,\ldots n}$ denote the matrix having coefficient 1 in the j-th row and the k-th column and zeroes elsewhere. Then, in the Lie algebra $\operatorname{gl}(V)$ we have

$$(\operatorname{ad}_{\operatorname{gl}(V)} h)(e_{jk}) = (h_{jj} - h_{kk}) \cdot e_{jk}, \qquad j, k = 1, \ldots, n.$$

Thus, if $\operatorname{Spec} T$ denotes the spectrum of an endomorphism T of V we have

$$\operatorname{Spec}(\operatorname{ad}_{\operatorname{gl}(V)} h) = \{h_{jj} - h_{kk} : j, k = 1, \ldots, n\}.$$

Since $\operatorname{Spec} \operatorname{ad}_L h \subseteq \operatorname{Spec} \operatorname{ad}_{\operatorname{gl}(V)} h$ we know that $\operatorname{Spec} \operatorname{ad}_L h \subseteq \mathbb{R}$. From the preceding Lemma II.4.21 we have $\operatorname{Spec} \operatorname{ad}_L h \subseteq i\mathbb{R}$. It follows that all eigenvalues of $\operatorname{ad} h$ on L are zero for all $h \in H$. But then all $\operatorname{ad} h$, $h \in H$ are nilpotent (for instance by the Theorem of Cayley and Hamilton or in view of the Jordan decomposition over $\mathbb{C}$). Let us now assume, contrary to the assertion of the lemma, that $H \neq L$. Then, if $h_{L/H}$ denotes the vector space endomorphism of L/H induced by $\operatorname{ad} h$, we know that all $h_{L/H}$, $h \in H$ are nilpotent. By Engel's Theorem there is a non-zero vector $x + H$ of L/H such that $[H, x + H] \subseteq H$. Then $\mathbb{R} \cdot x \oplus H$ is a solvable subalgebra properly containing H. Since this is impossible the assumption $L \neq H$ is refuted. ∎

We summarize the information contained in the preceding two lemmas as follows:

II.4.23. **Lemma.** *Let $\pi: L \to \operatorname{gl}(n, \mathbb{R})$ be a finite dimensional representation of a real Lie algebra such that L contains a nilpotent subalgebra H with the property that $\pi(H)$ is not contained in any larger solvable subalgebra of $\pi(L)$ and that $\pi(H)$ is diagonalizable. Then $\pi(L) = \pi(H)$. In particular, $\pi(L)$ is abelian and $[L, L] \subseteq \ker \pi$.*

Proof. We know that the homomorphic image $\pi(H)$ of H is a nilpotent subalgebra of $\pi(L)$. Then Lemma II.4.22 applies and shows that $\pi(L) = \pi(H)$. The remainder is clear, since $\pi(H)$ is diagonalizable, hence abelian. ∎

Now we apply this information to prove the following theorem:

II.4.24. **Theorem.** *Let L be a Lie algebra which contains a reduced generating Lie semialgebra. Suppose that I is an abelian ideal and $\rho: L \to \operatorname{gl}(I)$ the representation given by $\rho(x) = (\operatorname{ad} x)|I$. Then $\rho(L)$ is (simultaneously) diagonalizable.*

Proof. Pick any regular element $x \in \operatorname{int} W$; as usual this is possible since W is generating and the set of regular elements is dense. By the pointing procedure II.4.3, there is a pointed generating Lie semialgebra W' containing x in its interior. We define H to be the nilspace of $\operatorname{ad} x$, that is the Cartan algebra associated with x. Suppose that P is any subalgebra of L containing H such that $\rho(P)$ is solvable. By Theorem II.4.20, the algebra $\rho(P)$ is diagonalizable. Thus it is abelian. But $\rho(H)$ is a Cartan algebra of $\rho(L)$ and thus is its own normalizer. This implies $\rho(P) = \rho(H)$. Now the preceding Lemma II.4.23 applies and shows that $\rho(L)$ is diagonalizable. ∎

The following remarks are then immediate consequences:

II.4.25. **Remark.** Under the circumstances of Theorem II.4.24, $[L, L] \subseteq \ker \rho$, and $[L, L]$ is in the centralizer of I in L. ∎

The importance of Theorem II.4.24 lies in the fact that every Lie algebra which is not semisimple contains non-zero abelian ideals. In our further study of Lie semialgebras in the subsequent sections we will make heavy use of this result.

5. The base ideal and Lie semialgebras

In any Lie algebra one has numerous characteristic ideals such as the radical, the nilradical, the ideals of the commutator series or of the descending or ascending central series. We need another characteristic ideal which we shall call the base ideal and which we shall introduce and discuss in general terms in the beginning of this section. Later we shall introduce a canonical way of associating a function with a Lie semialgebra which we shall call the characteristic function of the Lie semialgebra. We shall then bring the two lines of argument together in an analysis of the characteristic function in its finer aspects. The section will culminate in a theorem which says that the tangent hyperplane T_x of a generating Lie semialgebra at x is necessarily a subalgebra if the characteristic function does not vanish in x.

The base ideal

II.5.1.   Definition. If L is a Lie algebra, we shall call its *base ideal* $M(L)$ the span of all 1-dimensional ideals of L. It is, of course, understood that the base ideal is $\{0\}$ if there are no 1-dimensional ideals.

If I and J are 1-dimensional ideals, then $[I, J] \subseteq I \cap J = \{0\}$ if $I \neq J$. It follows that $M(L)$ is abelian.

If L is semisimple, then $M(L) = \{0\}$. The base ideal $M(L)$ will always contain the center $Z(L)$. If L is almost abelian, then either $M(L) = L$ (if L is abelian) or $M(L) = [L, L]$.

II.5.2.   Definition. For each linear form $\alpha: L \to \mathbb{R}$ of L we define $M_\alpha(L) = \{y \in L : [x, y] = \alpha(x) \cdot y \quad \text{for all} \quad x \in L\}$. We call α a *base root* if $M_\alpha(L) \neq \{0\}$; the (finite!) set of base roots will be denoted $B(L)$. For a base root α the vector space $M_\alpha(L)$ is called the *base root space* of α.

The zero form is a base root if and only if the center $Z(L)$ is non-zero, and then $M_0(L) = Z(L)$. Let us record:

II.5.3. **Remark.** The center $Z(L)$ of any Lie algebra is the component $M_0(L)$ of the base ideal. ∎

By its very definition, each base root is a Lie algebra homomorphism $\alpha\colon L \to \mathbb{R}$. In particular, $M(L) \subseteq L' + Z(L) \subseteq \ker \alpha$.

Recall that an ideal is characteristic if it is invariant under all derivations.

II.5.4. **Proposition.** *Each base root space $M_\alpha(L)$, $\alpha \in B(L)$ is a characteristic ideal, and $M(L)$ is the direct sum of all base root spaces.*

Proof. We let the automorphism group $\operatorname{Aut} L$ of L act on the right of $B(L)$ via $\alpha \cdot \varphi = \alpha \circ \varphi$ for $\varphi \in \operatorname{Aut} L$. Since $B(L)$ is discrete and the action is continuous, the identity component $\operatorname{Aut}_0 L$ acts trivially. It follows that each automorphism from $\operatorname{Aut}_0 L$ leaves $B(L)$ pointwise fixed. But the Lie algebra $L(\operatorname{Aut}_0 L) = L(\operatorname{Aut} L)$ is the derivation algebra $\operatorname{Der} L$ of L. Thus each base root is fixed under e^D for $D \in \operatorname{Der} L$. Now consider $y \in M_\alpha(L)$. Then $[x, y] = \alpha(x) \cdot y$ for all $x \in L$. If φ is an automorphism of L then we derive $[\varphi(x), \varphi(y)] = \alpha(x) \cdot \varphi(y) = (\alpha \circ \varphi^{-1})(\varphi(x)) \cdot \varphi(y)$. If we set $\varphi = e^D$ and $\varphi(x) = u$, by the preceding observations we obtain from this

$$[u, e^D y] = \alpha(u) \cdot e^D y \text{ for all } u \in L,\, D \in \operatorname{Der} L,\, y \in M_\alpha(L).$$

This means that

$$e^{t \cdot D} M_\alpha(L) \subseteq M_\alpha(L) \text{ for all } t \in \mathbb{R},\, D \in \operatorname{Der} L.$$

Differentiation with respect to t yields $D\big(M_\alpha(L)\big) \subseteq M_\alpha(L)$ for all derivations D of L and this is the asserted claim that $M_\alpha(L)$ is a characteristic ideal.

Now we show that $M(L)$ is the direct sum of the base root spaces $M_\alpha(L)$ as α ranges through $B(L)$. Since every 1-dimensional ideal is contained in one of the base root spaces, $M(L)$ is contained in $\sum_{\alpha \in B(L)} M_\alpha(L)$. The reverse inclusion is obvious. Hence we have equality and we must show that the sum is direct. For a proof we may refer to [Bou75], Ch.VII, §1, n°1, Proposition 3. The proof is complete. ∎

II.5.5. **Corollary.** *If we denote with $\sigma\colon L \to gl(M(L))$ the representation given by $\sigma(x) = (\operatorname{ad} x)|M(L)$ and with $p_\alpha\colon M(L) \to M(L)$ the projection onto the direct summand $M_\alpha(L)$, then $\sigma(x) = \sum_{\alpha \in B(L)} \alpha(x) \cdot p_\alpha$.* ∎

We shall maintain the notation for the representation σ induced on $M(L)$ by the adjoint representation.

Special metabelian Lie algebras

Let us generalize the concept of an almost abelian algebra by constructing a class of examples of metabelian Lie algebras. We recall that a Lie algebra is called metabelian if $[L, L]$ is abelian.

II.5.6. **Proposition.** *Suppose that $V = \bigoplus_{k=1}^n V_k$ is a direct sum of finite dimensional real vector spaces and that $\lambda_1, \ldots, \lambda_n$, $\lambda_j \neq 0$ for all $j = 1, \ldots, n$ are different non-zero real numbers. Set $L = \mathbb{R} \times V$ and define a bracket $[(u, x_j), (u', x'_k)] = (0, u\lambda_k \cdot x'_k - u'\lambda_j \cdot x_j)$ for $u \in \mathbb{R}$ and $x_j \in V_j$, $x_k \in V_k$. Then L is a metabelian Lie algebra satisfying the following conditions:*

(i) $M(L) = \{0\} \times V$.

(ii) $[L, L] = \{0\} \times \bigoplus\{V_j : \lambda_j \neq 0\}$.

(iii) *If all λ_j are non-zero, set $H = \mathbb{R} \times \{0\}$; if $\lambda_k = 0$, set $H = \mathbb{R} \times V_k$. Then H is a Cartan subalgebra of L.*

Proof. The Lie algebra L we construct is the semidirect product of the abelian Lie algebra V and the abelian Lie algebra $\mathbb{R}$ which acts on V so that $r \in \mathbb{R}$ operates as the derivation $x_1, \ldots, x_n \mapsto r\lambda_1 \cdot x_1, \ldots, r\lambda_n \cdot x_n$. This remark makes everything straightforward with the possible exception of property (iii). Now if all λ_j are non-zero, then the abelian subalgebra $H = \mathbb{R} \times \{0\}$ is necessarily its own normalizer, hence is a Cartan subalgebra. If, on the other hand, $\lambda_1 = 0$, say, then the abelian subalgebra $H = \mathbb{R} \times V_1$ is its own normalizer, hence is a Cartan algebra. ∎

It should be clear that L is abelian if and only if $n = 1$ and $\lambda_1 = 0$ and almost abelian non-abelian if $n = 1$ and $\lambda_1 \neq 0$. If L is non-abelian, then $M(L)$ is its nilradical.

II.5.7. **Definition.** A Lie algebra L which is isomorphic to one of the algebras $\mathbb{R} \times V$ just constructed in Proposition II.5.6 will be called a *special metabelian Lie algebra*.

Having understood the example of a special metabelian algebra we can easily see the next step of generality: If E is a finite dimensional vector space and $V = \bigoplus_{k=1}^n V_k$ a direct sum of finite dimensional vector spaces, and if further $\alpha_1, \ldots, \alpha_n$ are different linear forms on E, then the vector space $L = E \times V$ becomes a metabelian Lie algebra with respect to the bracket $[(u, x), (u', x')] = (0, \sum_{j,k=1}^n (\alpha_k(u) \cdot x'_k - \alpha_j(u') \cdot x_j))$. The linear functionals α_k may be considered as functionals on $E \times \{0\} \subseteq L$ and as such they extend trivially to functionals on L with $\{0\} \times V$ in their kernels. If $\bigcap_{k=1}^n \ker \alpha_k = \{0\}$, then these extensions form the base roots of L and $M(L) = \{0\} \times V$, the root spaces being $\{0\} \times V_k$. If none of the α_k is zero, then $E \times \{0\}$ is a Cartan subalgebra of L. If $\alpha_1 = 0$, then $E \times V_1$ is a Cartan subalgebra. If x is any element of E not contained in any of the kernels of the α_k, then the subalgebra $\mathbb{R} \cdot x + \{0\} \times V$ is a special metabelian algebra. The point of the brief discussion of the more general type of metabelian algebra is that they occur often quite naturally in a typical way in any Lie algebra which has a non-zero base ideal. This is exemplified by the following proposition.

II.5.8. **Proposition.** *Let L be a Lie algebra and H a Cartan subalgebra. Set $L_H = H + M(L)$. Then L_H is a solvable subalgebra with H as a Cartan subalgebra and $M(L_H) = M(L) + M_0(L_H)$. The function $\alpha \mapsto \alpha' = \alpha|L_H: B(L) \to B(L_H)$ is injective; if it fails to be surjective, then the only base root of L_H which may not be in the image is 0. For any non-zero base root α of L we have $M_{\alpha'}(L_H) = M_\alpha(L)$. Moreover, $H \cap M(L) = M_0(L) = Z(L)$ and if H is abelian then L_H is metabelian.*

Proof.   Since $M(L)$ is an abelian ideal of L and H a nilpotent subalgebra, clearly L_H is a solvable subalgebra. Every 1-dimensional ideal of L is in $M(L)$, hence in L_H and is a 1-dimensional ideal of the subalgebra of L_H. Hence $M(L) \subseteq M(L_H)$. Similarly, we see that $M_\alpha(L) \subseteq M_{\alpha'}(L_H)$ with $\alpha' = \alpha|L_H$. The homomorphic image $\sigma(H)$ of H in $\sigma(L)$ is a Cartan subalgebra of $\sigma(L)$. Since $\sigma(L) = \sum_{\alpha \in B(L)} \alpha(L) \cdot p_\alpha$ is abelian, we have $\sigma(H) = \sigma(L)$. Thus

(1) $$(\forall x \in L)(\exists h \in H)\big(\forall \alpha \in B(L)\big)\alpha(x) = \alpha(h).$$

Now let α and β be two different base roots of L. This implies the existence of an $x \in L$ such that $\alpha(x) \neq \beta(x)$. By (1) above there is an $h \in H$ such that $\alpha'(h) = \alpha(x) \neq \beta(x) = \beta'(h)$. Thus $\alpha \mapsto \alpha' : B(L) \to B(L_H)$ is injective.

By Remark II.5.3 we know $Z(L_H) = M_0(L_H)$. Since H is also a Cartan algebra of L_H we have $Z(L_H) \subseteq H$. Conversely, $[H, H \cap M(L_H)] \subseteq H \cap M(L_H)$ and $[M(L_H), H \cap M(L_H)] = 0$, as $M(L_H)$ is abelian. Thus $H \cap M(L_H)$ is an ideal of L_H, thus is an L_H-invariant subspace of $M(L_H)$, hence is of the form $\sum_{\beta' \in B(L_H)} H \cap M_{\beta'}(L_H)$. Now let $h \in H \cap M_{\beta'}(L_H)$. Then for all $k \in H$ we have $[k, h] = \beta'(k) \cdot h$. If $\beta' \neq 0$, then we find a $k \in H$ with $\beta'(k) \neq 0$, and if we also had $h \neq 0$, then $\mathbb{R} \cdot k + \mathbb{R} \cdot h$ would be the non-abelian 2-dimensional solvable Lie algebra which cannot be a subalgebra of the nilpotent algebra H. We conclude $H \cap M(L_H) \subseteq M_0(L_H)$. Hence we have $H \cap M(L_H) = M_0(L_H)$. For $\alpha \in B(L)$ and $\alpha \neq 0$ we conclude $M_{\alpha'}(L_H) = M_\alpha(L)$, since $M_{\alpha'}(L_H) \cap H \subseteq M_0(L_H) \cap M_{\alpha'}(L_H) = \{0\}$, whence $M_{\alpha'}(L_H) \subseteq M(L) = \sum_{\gamma \in B(L)} M_\gamma(L)$, and thus $M_{\alpha'}(L_H) \subseteq M_\alpha(L)$. Since $L_H = H + M(L)$ and $M(L) \subseteq M(L_H)$, the relation $H \cap M(L_H) = M_0(L_H)$ also implies $M(L_H) \subseteq M(L) + M_0(L_H) \subseteq M(L_H)$ and $H \cap M(L) = M(L) \cap M_0(L_H) = M_0(L_H)$. If H is abelian, then $[L_H, L_H] \subseteq M(L)$ and thus L_H is metabelian. ∎

We remark in passing that the component $M_0(L_H) = Z(L_H)$ which somewhat disturbed the descending from L to L_H is simply $H \cap \ker \sigma$.

The preceding proposition permits us to conclude that in any Lie algebra with a base ideal, one finds a special metabelian subalgebra with the same base ideal.

II.5.9.   Corollary.   *Let L be an arbitrary Lie algebra. Then for any regular element $x \in L$ the subalgebra $L^{(x)} \stackrel{\text{def}}{=} \mathbb{R} \cdot x + M(L)$ is a special metabelian subalgebra with $M(L^{(x)}) = M(L)$. Moreover, the function $\alpha \mapsto \alpha' = \alpha|L^{(x)} : B(L) \to B\big(L^{(x)}\big)$ is a bijection so that $M_{\alpha'}(L^{(x)}) = M_\alpha(L)$.*

Proof.   We claim that $\alpha \in B(L)$ and $\alpha(x) = 0$ implies $\alpha = 0$. Indeed, if $\alpha(x) = 0$, then $[x, M_\alpha(L)] = \alpha(x) \cdot M_\alpha(L) = \{0\}$. Thus $M_\alpha(L)$ is in the nilspace of $\operatorname{ad} x$, which is exactly the Cartan subalgebra H associated with x. Hence $M_\alpha(L) \subseteq H$. By Proposition II.5.8 this implies $\alpha = 0$. Thus every base root $\alpha \in B(L)$ produces a base root $\alpha' = \alpha|L^{(x)} \in B(L^{(x)})$. Let us suppose that for base roots α and β of L we have $\alpha(x) = \beta(x)$. Then the nilspaces of $\operatorname{ad} x - \alpha(x) \cdot \mathbf{1}$ and $\operatorname{ad} x - \beta(x) \cdot \mathbf{1}$ agree. These nilspaces are the root spaces of H for $\alpha|H$ and $\beta|H$, which therefore coincide. Since M_α and M_β, respectively, are the intersections of these root spaces with $M(L)$, we conclude $M_\alpha(L) = M_\beta(L)$ and thus $\alpha = \beta$ by Proposition II.5.4 above. Thus the $M_\alpha(L)$ are the Jordan components for

$\operatorname{ad} x | M(L)$ for the *different* eigenvalues $\alpha(x)$, $\alpha \in B(L)$. The assertion is now clear.   ∎

At this point we leave the general theory of the base ideal and return to the presence of Lie semialgebras in a Lie algebra. The principal results of the previous sections now allow us to draw some important conclusions very quickly.

Base ideals and Lie semialgebras

II.5.10.   Theorem. *Let L be a Lie algebra which contains a reduced Lie semialgebra. Then the base ideal $M(L)$ is the sum of all abelian ideals and thus is the unique largest abelian ideal.*

Proof.   By Theorem II.4.24, for any abelian ideal I of L, the representation $\rho : L \to gl(I)$ given by $\rho(x) = (\operatorname{ad} x)|I$ gives rise to a *diagonalizable* Lie algebra $\rho(L)$. But by the definition of the base ideal this implies $I \subseteq M(L)$. Since $M(L)$ is itself an abelian ideal, the claim is proved.   ∎

II.5.11.   Corollary. *Every Lie algebra which supports a reduced Lie semialgebra and which is not semisimple has a non-zero base ideal.*

Proof.   If R denotes the radical of L then $R \neq \{0\}$ since L is not semisimple. Let $R^{(m)}$ denote the commutator series of R. If $R^{(n+1)} = \{0\}$ while $R^{(n)} \neq \{0\}$, then $R^{(n)}$ is a non-zero characteristic abelian ideal of R. Hence it is a non-zero abelian ideal of L and then by Theorem II.5.10 is contained in $M(L)$.   ∎

At this point we shall relate the structure of the space of C^1-points with the base ideal. In fact, we shall decompose the set of C^1-points according to the presence of base roots in L.

II.5.12.   Definition. Let L be a Lie algebra and W a generating Lie semialgebra. For each base root $\alpha \in B(L)$ we set

$$(2) \qquad\qquad C_\alpha^1(W) = \{x \in C^1(W) : M_\alpha(L) \not\subseteq T_x\},$$

and

$$(3) \qquad\qquad C_*^1(W) = \{x \in C^1(W) : M(L) \subseteq T_x\}.$$

Finally we set

$$(4) \qquad\qquad C_B^1(W) = C^1(W) \setminus C_*^1(W).$$

II.5.13. **Lemma.** *Given L and W, the following conclusions hold:*

(i) $C_B^1(W) = \bigcup_{\alpha \in B(L)} C_\alpha^1(W)$.

(ii) $C^1(W)$ *is the disjoint union of* $C_B^1(W)$ *and* $C_*^1(W)$.

(iii) *Each* $C_\alpha^1(W)$ *is open in* $C^1(W)$ *while* $C_*^1(W)$ *is closed.*

(iv) *If* $M_\alpha(L) \cap \operatorname{int} W \neq \emptyset$, *then* $C_\alpha^1(W) = C^1(W)$.

Proof. (i) By definition, $C_B^1(W)$ is the set of all $x \in C^1(W)$ such that $M(L) \not\subseteq T_x$. By Proposition II.5.4, $M(L)$ is the direct sum of the base root spaces $M_\alpha(L)$. Hence $M(L) \not\subseteq T_x$ if and only if for some α we have $M_\alpha(L) \not\subseteq T_x$. This proves (i).

(ii) is trivial.

(iii) The relation $M_\alpha(L) \subseteq T_x$ is equivalent to $T_x^\perp \subseteq M_\alpha(L)^\perp$. This statement is equivalent to $\left(T_x^\perp \cap (W^* \setminus \{0\})\right) \subseteq M_\alpha(L)^\perp$, and if we use the notation of Definition I.2.34, we recognize this as equivalent to $\sigma(x) \subseteq \Pi\left(M_\alpha(L)^\perp \cap (W^* \setminus \{0\})\right)$. Since the set on the right hand side is closed in $\Pi(W^*)$, and since σ is continuous by the Transgression Theorem I.2.35, the set $\{x \in C^1(W) : M_\alpha(L) \subseteq T_x\}$ is closed. Upon passing to complements we arrive at the first assertion; the last is a consequence of (ii) and (i).

(iv) Let $w \in M_\alpha(L) \cap \operatorname{int}(W)$, then, for any $x \in C^1(W)$, we have $w \notin T_x$. Hence $x \in C_\alpha^1(W)$. ∎

There is a feeling that the sets $C_\alpha^1(W)$ should be disjoint. A moments contemplation will convince the reader, however, that there is no obvious reason to expect them to be disjoint in general. We shall nevertheless prove, with some effort, that this is the case.

Let us first look at the situation of special metabelian Lie algebras.

II.5.14. **Lemma.** *Let T be a hyperplane in a special metabelian Lie algebra L. Then the following conditions are equivalent*

(1) *If* $T \not\subseteq M(L)$, *then there is a* $t \in T \setminus M(L)$ *such that* $[t, T] \subseteq T$.

(2) *T is a subalgebra.*

(3) *T ∩ M(L) is an ideal.*

(4) *There is one base root* $\alpha \in B(L)$ *such that* $T \cap M(L) = \left(T \cap M_\alpha(L)\right) \oplus \bigoplus_{\alpha \neq \beta \in B(L)} M_\beta(L)$.

Proof. By Lemma II.3.3, conditions (2) and (3) are equivalent. Further, (4) implies (3) in an obvious fashion, since each vector subspace of any $M_\alpha(L)$ is an ideal. Conversely, if (3) is satisfied, then for any $x \in L$ such that $L = M(L) \oplus \mathbb{R} \cdot x$, the space $T \cap M(L)$ is an ad x-invariant subspace. If $T = M(L)$, then (4) is trivially true. If $T \neq M(L)$, we may assume $T \cap M(L) = M(T) = \bigoplus_{\alpha \in B(T)} M_\alpha(T) = \sum_{\alpha \in B(L)} M_\alpha(T)$ with the proviso that $M_\alpha(T)$ could be zero for some base root α of L. Now $M_\alpha(T) = T \cap M_\alpha(L)$ is either $M_\alpha(L)$ or a hyperplane in $M_\alpha(L)$, and the latter case occurs in precisely one case for dimensional reasons. Thus (4) holds. Clearly (2) implies (1). There remains a proof of the implication (1) ⇒ (3): If $T \subseteq M(L)$, then T is a subalgebra since $M(L)$ is abelian. Suppose now that $T \not\subseteq M(L)$. We write $L = \mathbb{R} \cdot t \oplus M(L)$ with a suitable element $t \in T \setminus M(L)$. Now $[t, T] = [t, \mathbb{R} \cdot t + (M(L) \cap T)] = [t, M(L) \cap T]$. Now this set is in $M(L) \cap T$ if and only if $M(L) \cap T$ is an ad t-invariant subspace of $M(L)$. This, however, is equivalent to statement (3). ∎

II.5.15. **Proposition.** *Let W be a generating Lie semialgebra in a special metabelian Lie algebra. Then W is the intersection of half-space Lie semialgebras whose boundaries T intersect $M(L)$ in subspaces of the form*

$$\bigoplus_{\alpha \in B(L)} N_\alpha(L) \quad with \quad N_\alpha(L) = M_\alpha(L)$$

with possible exception of one base root for which

$$N_\alpha(L) \quad is \ a \ hyperplane \ in \quad M_\alpha(L).$$

Conversely, the intersection of any family of such half-space Lie semialgebras is a Lie semialgebra.

Proof. Let $x \in C^1(W)$. If $T_x = M(L)$ or if $x \in C^1(W) \setminus M(L)$, then T_x is a subalgebra of the desired form by Lemma II.5.14, and Corollary II.2.24 shows that $\overline{W + \mathbb{R} \cdot x}$ is a Lie semialgebra bounded by T_x. But then Proposition I.3.13 applied with $T = M(L)$ and $D = C^1(W)$ shows that W is the intersection of half-space Lie semialgebras $L_x(W) = \overline{W + \mathbb{R} \cdot x}$. Remark II.2.26 implies the converse. ∎

Since we can easily construct special metabelian Lie algebras after Definition II.5.7, we also find a large supply of Lie semialgebras in them by the preceding proposition. In particular, it is clear that we can construct polyhedral Lie semialgebras in this way whose set of C^1-points has as many pieces of the form $C^1_\alpha(W)$ as we like.

II.5.16. **Corollary.** *If W is a generating Lie semialgebra in a special metabelian Lie algebra, then the $C^1_\alpha(W)$ are disjoint and non-empty.*

Proof. It follows from Proposition II.5.15 that each hyperplane T_x of a C^1-point x can fail to contain at most 1 of the base root spaces $M_\alpha(L)$. According to the definition of $C^1_\alpha(W)$, this proves the assertion. ∎

The standard testing device now allows us to conclude more:

II.5.17. **Proposition.** *If $W \neq L$ is a generating Lie semialgebra in a Lie algebra L, then the spaces $C^1_\alpha(W)$ are disjoint.*

Proof. We assume that we have two different base roots α and β such that $C^1_\alpha(W) \cap C^1_\beta(W) \neq \emptyset$. Thus there is an element $y \in C^1_\alpha(W) \cap C^1_\beta(W)$. Hence there exists an element $u \in M_\alpha(L) \setminus T_y$. We set $J = \mathbb{R} \cdot u$. The set $\mathrm{reg}(L)$ of regular points of L is open and dense. Thus the set $U \overset{\mathrm{def}}{=} \mathrm{reg}(L) + J$ is open and dense in L, and it satisfies $U + J = U$ and is stable under non-zero scalar multiplication. The complement $K = L \setminus U$ is closed, nowhere dense, stable under non-zero scalar multiplication and satisfies $K + J = K$. Let $p_u : L \to T_y$ be the projection along u, which is well defined since $u \notin T_y$. The set $C^1_\alpha(W) \cap C^1_\beta(W)$ is an open neighborhood of y in $C^1(W)$. Hence there is an open neighborhood V of y in the boundary ∂W such that $V^1 \overset{\mathrm{def}}{=} V \cap C^1(W)$ is contained in $C^1_\alpha(W) \cap C^1_\beta(W)$. We may assume that V was picked so small, that p_u maps V homeomorphically onto an open neighborhood of y in T_y. By the Density Theorem for Wedges I.3.11, the space $C^1(W)$ is a dense subset of ∂W. Hence $p_u(V^1)$ is a dense subset of $p_u(V)$. Thus $p_u(V^1) + \mathbb{R} \cdot u$ is dense in the open subset $p_u(V) + \mathbb{R} \cdot u$ of L; on the other hand, $p_u(V^1) + \mathbb{R} \cdot u = V^1 + \mathbb{R} \cdot u \subseteq C_\alpha^1(W) \cap C_\beta^1(W) + J$. We conclude from this that

$$(5) \qquad\qquad C_\alpha^1(W) \cap C_\beta^1(W) \not\subseteq K,$$

for otherwise the set $C_\alpha^1(W) \cap C_\beta^1(W) + J$ which, as we just saw, is not nowhere dense, would be contained in the nowhere dense subset K. Thus

$$(6) \qquad\qquad \text{there is a} \quad y' \in C_\alpha^1(W) \cap C_\beta^1(W) \cap U,$$

which means that we find a regular element x of L and an element $j \in J$ such that $y' = x + j$ is a C^1-point of W with $M_\alpha(L) \not\subseteq T_{y'}$ and $M_\beta(L) \not\subseteq T_{y'}$. Now we consider the subalgebra $L^{(x)} \overset{\text{def}}{=} \mathbb{R} \cdot x + M(L)$. This is a special metabelian subalgebra whose base ideal agrees with $M(L)$ by Corollary II.5.9 and which contains the point y'. Since $M(L) \not\subseteq T_{y'}$ we know that $L^{(x)} \cap \operatorname{int} W \neq \emptyset$ by Remark II.4.8. Then y' is a C^1-point of the generating wedge $W^{(x)} \overset{\text{def}}{=} W \cap L^{(x)}$ of $L^{(x)}$, and by Corollary I.1.12, it satisfies

$$(7) \qquad M_\alpha(L^{(x)}) \not\subseteq T_{y'}(W^{(x)}) \quad \text{and} \quad M_\beta(L^{(x)}) \not\subseteq T_{y'}(W^{(x)}).$$

By Corollary II.5.9, $\alpha|L^{(x)} \neq \beta|L^{(x)}$. However, this means $y' \in C_\alpha^1(W^{(x)}) \cap C_\beta^1(W^{(x)})$ which is impossible by Corollary II.5.16. This contradiction proves the claim. ∎

We are now prepared for the fine structure theorem for C^1. The notation will be that of Definition II.5.12.

II.5.18. **Theorem.** (Fine Structure Theorem for C^1) *Let W be a generating Lie semialgebra in a finite dimensional vector space L. Then the following conclusions hold:*

(i) *$C^1(W)$ is the disjoint union of the open subsets $C_\alpha^1(W)$, $\alpha \in B(L)$, and the closed subset $C_*^1(W)$. In particular, the $C_\alpha^1(W)$ are open-closed in $C_B^1(W)$.*

(ii) *If $C_\alpha^1(W)$ is empty, then $M_\alpha(L) \subseteq H(W) \subseteq W$.*

(iii) *If W is reduced, all $C_\alpha^1(W)$ are non-empty.*

Proof. (i) The set $C_B^1(W)$ is the disjoint union of the open subsets $C_\alpha^1(W)$, $\alpha \in B(L)$ by Lemma II.5.13(iii) and Proposition II.5.17. The assertion is then clear.

(ii) If $C_\alpha^1(W) = \emptyset$, then $M_\alpha(L) \subseteq T_x$ for all $x \in C^1(W)$. Thus Proposition I.3.12 implies the claim.

(iii) This is a consequence of (ii), since every $M_\alpha(L)$ is an ideal by Proposition II.5.4. ∎

This information is particularly useful, if $C_B^1(W)$ is dense in $C^1(W)$ and consists of one $C_\alpha^1(W)$ only. The latter is certainly the case after Proposition II.5.17, if $C_B^1(W)$ happens to be connected.

II.5.19.   Corollary.   *Let W be a generating Lie semialgebra in a Lie algebra L. If $C^1_\alpha(W)$ is dense in $C^1(W)$, then*

$$\bigoplus_{\alpha \neq \beta \in B(L)} M_\beta(L) \subseteq H(W) \subseteq W.$$

In particular, if W is reduced, then

$$M(L) = M_\alpha(L).$$

Proof.   By Lemma II.5.13(iii), all $C^1_\beta(W)$ are open in $C^1(W)$. From Proposition II.5.17 we know that they are all disjoint. Hence $\alpha \neq \beta \in B(L)$ implies $C^1_\beta(W) = \emptyset$, so that $M_\beta(L) \subseteq H(W)$ by Theorem II.5.18(ii).   ∎

Now we bring the preceding discussions together with the characteristic function $\lambda \colon C^1(W) \to \mathbb{R}^+$ of a Lie semialgebra W.

II.5.20.   Theorem.   *Let α be a base root of a Lie algebra with a generating Lie semialgebra W. Then α and λ agree on the set $\overline{C^1_\alpha(W)}$, where the closure is taken in $C^1(W)$.*

Proof.   Let $x \in C^1(W)$ and $y \in M_\alpha(L)$. Then $[x, y] = \alpha(x) \cdot y$ by the definition of $M_\alpha(L)$. On the other hand, $[x, y] = \lambda(x) \cdot y + t(x, y)$ with $t(x, y) \in T_x$ by the definition of the characteristic function. Thus we observe

$$(8) \qquad\qquad \big(\alpha(x) - \lambda(x)\big) \cdot y = t(x, y) \in T_x \cap \mathbb{R} \cdot y.$$

Since $M_\alpha(L) \not\subseteq T_x$ by the definition of $C^1_\alpha(W)$ we actually find a $y \in M_\alpha(L) \setminus T_x$. In this case we must necessarily have $t(x, y) = 0$ by (8). Since $y \neq 0$, this means $\alpha(x) = \lambda(x)$. Since both $\alpha|C^1(W)$ and λ are continuous (see Proposition II.2.34), the assertion follows.   ∎

This theorem says in effect, that on the portion $C^1_B(W)$ of the C^1-boundary of a Lie semialgebra, the characteristic function is explicitly identifiable as being pieced together from restrictions of the base roots $\alpha|C^1_\alpha(W)$. If it should happen that $M(L) \cap \operatorname{int} W \neq \emptyset$, then no tangent hyperplane can contain $M(L)$ and thus $C^1_*(W) = \emptyset$; in this case $C^1(W) = C^1_B(W)$ and the structure of the characteristic function is completely described in terms of the base roots.

The following corollary is a variant of Theorem II.5.18 in the special case of invariant wedges. Recall in the following that an invariant wedge is pointed if and only if it is reduced.

II.5.21.   Corollary.   *Let W be an invariant generating wedge in a finite dimensional Lie algebra L. Then the following conditions hold:*

(i) $\bigoplus_{0 \neq \alpha \in B(L)} M_\alpha(L) \subseteq H(W) \subseteq W$.

(ii) *If W is pointed, then $M(L) = Z(L)$. In particular, every abelian ideal is central.*

Proof.   (i) We know from Proposition II.2.37 that W is invariant if and only if its characteristic function vanishes identically. By Theorem II.5.20, this means that $C^1_\alpha(W) = \emptyset$ for $0 \neq \alpha \in B(L)$. Then (i) follows from Theorem II.5.18 above.

(ii) If W is pointed, then $M_\alpha(L) = \{0\}$ for $0 \neq \alpha \in B(L)$. Hence $M(L) = M_0(L) = Z(L)$.   ∎

We recall at this point the concept of a weakly round cone from Definition I.4.19. For such cones, the consequences are considerable.

II.5.22. **Theorem.** *Let L be a finite dimensional Lie algebra which is not semisimple and let W be a weakly round Lie semialgebra. Then there is a base root α such that the characteristic function of W is equal to $\alpha|C^1(W)$ and that $M(L) = M_\alpha(L)$.*

Proof. Since a weakly round cone is pointed in view of Lemma I.4.17, Corollary II.5.11 applies to show $M(L) \neq \{0\}$. Let $\alpha \in B(L)$ be an arbitrary base root. We shall show that $C_\alpha^1(W)$ is dense in $C^1(W)$. Once this is shown, Theorem II.5.20 will prove $\lambda = \alpha|C^1(W)$, and then Corollary II.5.19 will do the rest. In order to prove our claim we pick an arbitrary non-zero element $y \in M_\alpha(L)$ and set $S = \{x \in C^1(W) : y \notin T_x\}$. Since $y \in M_\alpha(L)$ it suffices to show that S is dense in $C^1(W)$. For a proof of this claim let x be an arbitrary C^1-point of W and let U be any open neighborhood of x. Since W is weakly round, there is an element $x_0 \in U \cap C^1(W)$ such that $D(x_0) = \{0\}$. By Lemma I.4.17, there is an open neighborhood U' of x_0 in U such that $D(x_0) = \bigcap_{u \in U' \cap C^1(W)} T_u$. Hence we find an $x' \in U' \cap C^1(W) \subseteq U \cap C^1(W)$ such that $y \notin T_{x'}$. Thus $x' \in S \cap U$. This shows that S is dense in $C^1(W)$; the proof is complete. ∎

This theorem will render particularly good service in such tasks as the classification of Lorentzian Lie semialgebras. The theorem itself suggests that we look more intensely at Lie algebras with a base ideal and a Lie semialgebra whose characteristic function agrees with a base root restricted to the set of C^1-points. Before we do this, however, we need extra information on how the presence of Lie semialgebras influences nilpotent ideals.

Nilpotent ideals

Theorem II.4.24 strongly limits the nilpotent ideals appearing in a Lie algebra with a reduced Lie semialgebra.

First we recall some notation. If N is a Lie algebra, then its *descending central series* is the sequence of ideals $N^{[0]} = N$, $N^{[k+1]} = [N, N^{[k]}]$, $k = 0, 1, \ldots$ We say that N is *nilpotent of class $m + 1$* if $N^{[m]} \neq \{0\}$, $N^{[m+1]} = \{0\}$. Thus N is of class one if and only if it is abelian.

II.5.23. **Proposition.** (a) *Suppose that N is a nilpotent Lie algebra of class $m + 1 \geq 3$. Then $N^{[m-1]}$ is abelian.*

(b) *If N is a nilpotent ideal of a Lie algebra L which contains a reduced Lie semialgebra, then $N^{[2]}$ $(= [N, N']) = \{0\}$, that is, the class of nilpotency of N is at most 2.*

Proof. (a) If $m \geq 2$, then

$$[N^{[m-1]}, N^{[m-1]}] \subseteq [N^{[1]}, N^{[m-1]}] = [[N, N], N^{[m-1]}]$$
$$\subseteq [[N, N^{[m-1]}], N] + [N, [N, N^{[m-1]}]] \subseteq N^{[m+1]} = \{0\}.$$

Thus the ideal $N^{[m-1]}$ is abelian.

(b) Now we assume that N is a nilpotent ideal in a Lie algebra L containing a reduced Lie semialgebra. We claim that the nilpotency class $m+1$ of N is at most 2. For a proof by contradiction we assume $m \geq 2$. Then $N^{[m-1]}$ is abelian by (a) above. For each $x \in N$, $(\operatorname{ad} x)|N^{[m-1]}$ is nilpotent. But it is also diagonalizable by Theorem II.4.24. Thus $(\operatorname{ad} x)|N^{[m-1]}$ is the zero mapping. This yields $N^{[m]} = [N, N^{[m-1]}] = \{0\}$, a contradiction to $N^{[m]} \neq \{0\}$. We conclude that $m < 2$, i.e., $N^{[2]} = \{0\}$.   ∎

The derivation of Theorem II.3.20 required substantial effort. We use this theorem to extend the Standard Testing Device II.4.6.

II.5.24.   Lemma. (The Second Testing Device) *Let L be a Lie algebra containing a reduced Lie semialgebra W, let N be a nilpotent subalgebra, and let $h \in \operatorname{int}(W) \setminus N$ be in the normalizer of N. Then $A = N + \mathbb{R} \cdot h$ is a solvable subalgebra containing a reduced Lie semialgebra with h in its interior.*

(i) *If for $x, y \in N$, $y + ix$ is an eigenvector for the eigenvalue $\zeta = a + bi$, $b \neq 0$ for the extension of $\operatorname{ad} h$ to the complexification $L_{\mathbb{C}} = L + iL$, then $a = 0$, $z = [x, y] \neq 0$, and $[h, z] = [x, z] = [y, z] = 0$. Hence the span of $\{h, x, y, z\}$ is the standard Lorentzian solvable Lie algebra of dimension 4. Furthermore, if N is an ideal and h is a regular element, then $z \in Z(L)$.*

(ii) *If for $x \in N$, $(\operatorname{ad} h - \lambda \cdot \mathbf{1})^n(x) = 0$ for some $\lambda \in \mathbb{R}$ and positive integer n, then $[h, x] = \lambda \cdot x$.*

(iii) *If for $x, y \in N$, $[h, x] = \lambda \cdot x$ and $[h, y] = \mu \cdot y$ for some $\lambda, \mu \in \mathbb{R}$, then $[x, y] = 0$. Hence the span of $\{h, x, y\}$ is almost abelian or an algebra of type* (iii) *in the* Classification Theorem of Low Dimensional Semialgebras III.3.4.

Proof.   The subalgebra A is solvable since $[A, A] \subseteq N$ and N is nilpotent. By the pointing procedure II.4.3, we can find a pointed Lie semialgebra containing h in its interior, and we henceforth rename W to be this wedge. Then for any subalgebra M containing h, $W \cap M$ will be a pointed generating Lie semialgebra in M.

(i) We have in $L_{\mathbb{C}} = L + iL$

$$[h, y] + i[h, x] = \operatorname{ad} h(y + ix) = (a + bi)(y + ix) = (-b \cdot x + a \cdot y) + i(a \cdot x + b \cdot y).$$

Hence $[h, x] = a \cdot x + b \cdot y$ and $[h, y] = -b \cdot x + a \cdot y$. Then

$$[h, z] = [[h, x], y] + [x, [h, y]] = a \cdot [x, y] + a \cdot [x, y] = 2a \cdot z.$$

We show first that $z \neq 0$. For if $z = 0$, then $\{h, x, y\}$ spans a 3-dimensional algebra for which the span of $\{x, y\}$ is a minimal 2-dimensional ideal. But this is impossible according to the Standard Testing Device II.4.6.

We have $0 \neq z = [x, y] \in [N, N] = N'$. By Proposition II.5.23, $[x, z] = 0 = [y, z]$. Also z is linearly independent from $\{x, y\}$ since z and $y + ix$ belong to different eigenspaces of $\operatorname{ad} h$. Thus $\{h, x, y, z\}$ spans a 4-dimensional subalgebra M with commutator algebra the Heisenberg algebra spanned by $\{x, y, z\}$. By the First Triviality Theorem II.4.13, $[M, M] \cap \operatorname{int}(W) = \emptyset$. Since $W \cap M$ is a pointed generating Lie semialgebra in M, it follows from Theorem II.3.20 that M must be the standard Lorentzian algebra of dimension 4. But if this is the case, then $M'' = \mathbb{R} \cdot z$ is the center. Thus $[h, z] = 0$, which means that $a = 0$.

Suppose now that N is an ideal and h is a regular element. Then $H = \{v : (\operatorname{ad} h)^n(v) = 0$ for some $n\}$ is a Cartan subalgebra. Since H is abelian by the First Cartan Algebra Theorem II.4.16, it follows that $H = \ker(\operatorname{ad} h)$. Note that $z \in H$ and $z \in N'$, an abelian ideal by Proposition II.5.23.

Consider the homomorphism $\rho : L \to \operatorname{gl}(N')$ of Theorem II.4.24. By that theorem $\rho(L)$ is commutative. Since $\rho(H)$ must be a Cartan subalgebra of $\rho(L)$, it follows that $\rho(H) = \rho(L)$. Thus for $u \in L$, there exists $g \in H$ with $\rho(g) = \rho(u)$. Then

$$[u, z] = \rho(u)(z) = \rho(g)(z) = [g, z] = 0$$

since $g, z \in H$. Thus $z \in Z(L)$.

(ii) If (ii) fails, then $\ker(\operatorname{ad} h - \lambda \cdot \mathbf{1})^2 | N \neq \ker(\operatorname{ad} h - \lambda \cdot \mathbf{1}) | N$. So there exist $u, v \in N$ such that $[h, u] = \lambda \cdot u + v$, $[h, v] = \lambda \cdot v$, $v \neq 0$. Let $z = [u, v]$. If z were 0, then $\{h, u, v\}$ would span a Lie algebra of type (ii) in the Standard Testing Device II.4.6, but it is shown there that this is impossible (since W is pointed). Hence $z \neq 0$. We also have

$$[h, z] = \big[[h, u], v\big] + \big[u, [h, v]\big] = [\lambda \cdot u + v, v] + [u, \lambda \cdot v] = 2\lambda \cdot z.$$

Since $[N, N'] = \{0\}$ by Proposition II.5.23, we have $[u, z] = 0 = [v, z]$. It follows that the span J of $\{v, z\}$ is an abelian ideal in the Lie algebra D, the span of $\{h, u, v, z\}$. Thus $(\operatorname{ad} u) | J$ is diagonalizable by Theorem II.4.20 (since $W \cap D$ is a pointed generating Lie semialgebra in D). But this is patently not the case, since $(\operatorname{ad} u) | J$ is nilpotent and non-zero.

(iii) Suppose $z = [x, y] \neq 0$. As previously, one verifies that $[h, z] = (\lambda + \mu)z$ and that $[x, z] = 0 = [y, z]$ (since $z \in N'$). The span J of $\{y, z\}$ is again an abelian ideal of D, the span of $\{h, x, y, z\}$, but $(\operatorname{ad} x) | J$ is nilpotent and non-zero, in contradiction to Theorem II.4.20. ■

We return now to the class of Lie algebras appearing in Proposition II.5.8 and the comments preceding it.

$$[h, z] = \big[[h, x], y\big] + \big[x, [h, y]\big] = a \cdot [x, y] + a \cdot [x, y] = 2a \cdot z.$$

II.5.25. Definition. A Lie algebra L is called *basic metabelian* if $L = H + M(L)$, where $H \neq L$ is an abelian Cartan subalgebra and $M(L)$ is the base ideal.

II.5.26. Proposition. *Let L be a solvable Lie algebra with an abelian Cartan algebra $H \neq L$. Then L is basic metabelian if and only if $M(L)$ is equal to the nilradical (the largest nilpotent ideal) of L.*

Proof. Let $N = M(L)$ be the nilradical. Since $[L, L] \subseteq N$ as a consequence of Lie's Theorem, L/N is abelian. Thus the image of H in L/N must be all of L/N, since the image of Cartan algebra is a Cartan algebra. It follows that $L = H + N = H + M(L)$.

Conversely, suppose that $L = H + M(L)$. Since the nilradical N contains the abelian ideal $M(L)$, either $N = M(L)$ or $(N \cap H) \setminus M(L) \neq \emptyset$. But if $x \in (N \cap H) \setminus M(L)$, then $[x, H] = \{0\}$ since H is abelian, and $(\operatorname{ad} x) | M(L) = 0$, since it is diagonalizable and nilpotent. Thus $x \in Z(L) \subseteq M(L)$, a contradiction. Hence $N = M(L)$. ■

Problem. Characterize the Lie semialgebras in a basic metabelian Lie algebra.

II.5.27.   Theorem. *Let L be a Lie algebra containing a reduced Lie semi-algebra and let R and N denote the radical and the nilradical of L, respectively. Suppose that $R \neq \{0\}$ and $N \neq L$. Then the following conditions are equivalent:*

(1) $M(L) = N$.

(1') $[N, N] = \{0\}$.

(2) $[N, N] \cap Z(L) = \{0\}$.

(3) $(\operatorname{ad} x)|R$ *has real spectrum for all $x \in L$.*

(4) $(\operatorname{ad} h)|N$ *has real spectrum for some $h \in \operatorname{int}(W) \setminus N$.*

Proof. $(1) \Rightarrow (1')$: If $M(L) = N$, then $[N, N] = \{0\}$, since $M(L)$ is abelian.

$(1') \Rightarrow (2)$ is trivial.

$(2) \Rightarrow (4)$: This follows at once from the Second Testing Device II.5.24(i) for any regular element $h \in \operatorname{int}(W) \setminus N$.

$(4) \Rightarrow (1)$: It follows from the fact that $(\operatorname{ad} h)|N$ has real spectrum, from the Second Testing Device II.5.24(ii), and from the Jordan decomposition that N has a basis $\{v_1, \ldots, v_n\}$ consisting of eigenvectors for $\operatorname{ad} h$. Then $[v_i, v_j] = 0$ for $1 \leq i, j \leq n$ by the Second Testing Device II.5.24(iii). Thus N is abelian, and hence $N \subseteq M(L)$ by Theorem II.5.10. Since always $M(L) \subseteq N$, we have that $N = M(L)$.

$(1) \Rightarrow (3)$: For $x \in L$, $[x, R] \subseteq [L, L] \cap R$. The latter is a nilpotent ideal (see [Bou75], Chap. 1, §5, n° 3, Theorem 1), hence is contained in N. The linear operator induced by $\operatorname{ad} x$ on R/N is zero. Also, $(\operatorname{ad} x)|N$ has real spectrum, since $N = M(L)$. Thus $(\operatorname{ad} x)|R$ has real spectrum.

$(3) \Rightarrow (4)$: Immediate.   ∎

II.5.28.   Proposition. *Let L be a Lie algebra containing a reduced Lie semi-algebra W. For $w \in W$, the spectrum of $\operatorname{ad} w$ is contained in $\mathbb{R} \cup i\mathbb{R}$.*

Proof.   Let $h \in \operatorname{int}(W)$. Let $L_{\mathbb{C}} = L + iL$, and let $L + iL = \bigoplus L^{a+bi}$ be the primary decomposition of $L_{\mathbb{C}}$ over all $a + bi$ in the spectrum of $\operatorname{ad} h$. For a proof by contradiction we assume that there is an element $y + ix \in L^{a+bi}$ with $a \neq 0 \neq b$. If we take $|a|$ maximal with respect to this property, we have $L^{3a+bi} = \{0\}$. We calculate $[h, y] + i[h, x] = \operatorname{ad} h(y + ix) = (a + bi)(y + ix) = (-b{\cdot}x + a{\cdot}y) + i(a{\cdot}x + b{\cdot}y)$. Hence $[h, x] = a{\cdot}x + b{\cdot}y$ and $[h, y] = -b{\cdot}x + a{\cdot}y$. Now we set $z = [x, y]$. Then $[h, z] = [[h, x], y] + [x, [h, y]] = a{\cdot}[x, y] + a{\cdot}[x, y] = 2a{\cdot}z$. Therefore $[y + ix, z] \in L^{3a+bi} = \{0\}$. This shows $[y, z] = [x, z] = 0$. Hence $\{x, y, z\}$ spans a nilpotent algebra which is normalized by h. Thus the Second Testing Device II.5.24 applies and shows that $b = 0$. This is the desired contradiction which shows that the spectrum of $\operatorname{ad} h$ is contained in $\mathbb{R} \cup i\mathbb{R}$.

Now let $w \in W$. Then $w = \lim h_n$, $h_n \in \operatorname{int} W$, and hence the linear operator $\operatorname{ad} w$ is the limit of the $\operatorname{ad} h_n$. We have just seen that the spectrum of each $\operatorname{ad} h_n$ is contained in the closed set $\mathbb{R} \cup i\mathbb{R}$. Since the values of the spectrum vary continuously (see e.g. [Bou67], Chap. I, §4, n° 8, Proposition 10(i)), the same conclusion follows for the spectrum of $\operatorname{ad} w$.   ∎

Problem. Let L be a Lie algebra which admits a reduced Lie semialgebra and let N be a nilpotent ideal. Is $[N, N]$ contained in the center $Z(L)$?

II.5.29. **Proposition.** *Let W be a reduced Lie semialgebra in a Lie algebra L, let $h \in \mathrm{int}(W)$, and consider $\mathrm{ad}\, h \colon L \to L$.*

(i) For each real eigenvalue $\lambda \neq 0$ of $\mathrm{ad}\, h$, the generalized eigenspace L^λ is equal to the eigenspace L_λ (i.e., the algebraic and geometric multiplicities of λ agree).

(ii) If $\lambda, \mu > 0$ or $\lambda, \mu < 0$, then $[L_\lambda, L_\mu] = \{0\}$.

(iii) If, additionally, h is regular, then $[L_\lambda, L_{-\mu}] = \{0\}$ for $\lambda, \mu > 0$, $\lambda \neq \mu$.

Proof. (i) The set $N = \bigoplus \{L^\lambda \colon \lambda > 0\}$ is the nilradical of $A = \mathbb{R} \cdot h + N$. Also $(\mathrm{ad}\, h)|N$ has real spectrum, and there exists a reduced Lie semialgebra in A containing h in its interior. Hence by Theorem II.5.27, $M(A) = N$. It follows that each L^λ is a sum of all base root spaces $M_\alpha(A)$ such that $\alpha(h) = \lambda$; in particular $L_\lambda = L^\lambda$. A similar proof holds for $\lambda < 0$.

(ii) For $\lambda, \mu > 0$, $[L_\lambda, L_\mu] = \{0\}$ since $M(A)$ is abelian, and similarly for $\lambda, \mu < 0$.

(iii) Suppose $[L_\lambda, L_{-\mu}] \neq \{0\}$ for $\lambda, \mu > 0$, $\lambda \neq \mu$; we suppose that μ and λ are chosen as small as possible for this to happen. We assume without loss of generality that $0 < \mu < \lambda$ (the other case being analogous). Let $H = \{k \colon (\mathrm{ad}\, h)^n(k) = 0 \text{ for some } n\}$. Since h is regular, H is a Cartan algebra, and hence abelian by the First Cartan Algebra Theorem II.4.16. Hence each $\mathrm{ad}\, k$ preserves L_ν for $k \in H$. It follows from this observation and part (ii) that $L_{-\mu}$ is an abelian ideal of the subalgebra $M = H + \bigoplus_{\nu < 0} L_\nu$. Since M also supports a reduced Lie semialgebra, it follows from Theorem II.4.24 that $\{(\mathrm{ad}\, x)|L_{-\mu} \colon x \in M\}$ is simultaneously diagonalizable. Hence we can pick a basis $\{y_1, \ldots, y_n\}$ of $L_{-\mu}$ such that each y_i is an eigenvector for $\mathrm{ad}\, k$ for each $k \in H$. Since $[L_\lambda, L_{-\mu}] \neq \{0\}$, there exist $x \in L_\lambda$ and $y = y_i \in L_{-\mu}$ such that $z = [x, y] \neq 0$.

We consider first the easier case to dispose of, namely $\lambda \neq 2\mu$. In this case $\lambda - \mu > 0$ and $\lambda - \mu \neq \mu$, so $[L_{\lambda - \mu}, L_{-\mu}] = \{0\}$ by the minimal choice of λ and μ. Thus since $z = [x, y] \in L_{\lambda - \mu}$, we have $[z, y] = 0$. By part (ii), $[x, z] = 0$. Thus $A = \mathbb{R} \cdot h + \mathbb{R} \cdot x + \mathbb{R} \cdot y + \mathbb{R} \cdot z$ is a Lie algebra supporting a reduced Lie semialgebra and $N = \mathbb{R} \cdot x + \mathbb{R} \cdot y + \mathbb{R} \cdot z$ is a nilpotent ideal of A. By the Second Testing Device II.5.24(iii), $[x, y] = 0$, a contradiction.

We consider the alternate case that $\lambda = 2\mu$. In this case $z \in L_\mu$. Then again $[x, z] = 0$ by part (ii). If $[z, y] = 0$, then we obtain a contradiction precisely as in the preceding paragraph. Thus we assume that $[z, y] = k \neq 0$. We have $k \in [L_\mu, L_{-\mu}] \subseteq L_0 = H$, since h is regular. Then y is an eigenvector for $\mathrm{ad}\, k$ by choice of y, i.e., $[k, y] = \gamma \cdot y$ for some $\gamma \in \mathbb{R}$. Furthermore, $[k, x] = [[z, y], x] = [[z, x], y] + [z, [y, x]] = 0$. Also $[h, k] = 0$ since $h, k \in H$. We compute

$$[x, k] = [x, [z, y]] = [[x, z], y] + [z, [x, y]] = [0, y] + [z, z] = 0$$

and

$$[k, z] = [k, [x, y]] = [[k, x], y] + [x, [k, y]] = 0 + \gamma \cdot [x, y] = \gamma \cdot z.$$

It follows that A, the span of $\{h, x, y, z, k\}$, is a subalgebra supporting a reduced Lie semialgebra. The span N of $\{x, y, z, k\}$ is a nilpotent ideal of A since

$$0 = [k, k] = [k, [z, y]] = [[k, z], y] + [z, [k, y]] = 2\gamma \cdot [z, y] = 2\gamma \cdot k$$

implies $\gamma = 0$. But again the Second Testing Device II.5.24(iii) yields the contradiction $[x, y] = 0$. ∎

Base ideals and Cartan algebras

II.5.30.   Theorem. (The Second Cartan Algebra Theorem) *Let W be a reduced generating Lie semialgebra in L and assume that its characteristic function is $\alpha | C^1(W)$ for some $\alpha \in \tilde{L}$. Then either W is an invariant wedge or all Cartan subalgebras of L are 1-dimensional.*

Proof.   We assume that W is not invariant. By Corollary II.2.37 this means that the characteristic function does not vanish identically. Our hypothesis on the characteristic function implies that $\alpha \neq 0$. We must show that all Cartan algebras are 1-dimensional. For this purpose it suffices to find *one* Cartan algebra which is 1-dimensional. (Indeed, if H is a 1-dimensional Cartan algebra, then its complexification H_C has complex dimension 1. As all Cartan algebras in the complexification L_C are conjugate, they all have complex dimension 1, and if K is any Cartan subalgebra of L, then K_C is one of these, whence $\dim K = 1$.) By hypothesis, the Lie semialgebra W is generating, and thus $\operatorname{int} W \neq \emptyset$. The set of regular elements of L is dense. Hence we find a regular element $x \in \operatorname{int} W$. By the Pointing Procedure II.4.3, there is a pointed generating Lie semialgebra W' containing x in its interior. The nilspace H of $\operatorname{ad} x$ is the Cartan algebra associated with x. We now suppose that $\dim H > 1$, derive a contradiction and thereby complete the proof. Since $H \cap W'$ cannot be a half line, the boundary $\partial(H \cap W')$ spans H. If $y \in \partial(H \cap W')$, then by Corollary II.2.36(b), for all $h \in H$ we observe $[y, h] - \alpha(y) \cdot h \in T_y(W') \cap H$. The First Cartan Algebra Theorem II.4.16 tells us that H is abelian. Hence $[y, x] = 0$. Thus $-\alpha(y) \cdot x \in T_y(W')$ for all $y \in \partial(H \cap W')$. Since $x \in \operatorname{int} W'$, this implies $\alpha(y) = 0$ for all $y \in \partial(H \cap W')$. As $\partial(H \cap W')$ spans H we conclude that $H \subseteq \ker \alpha$. In particular, $\alpha(x) = 0$. But the regular points fill out a dense subset of $\operatorname{int} W$, whence we conclude $\alpha(W) = \{0\}$. Since $L = W - W$, we find $\alpha = 0$, the desired contradiction.   ∎

The question is now what we know about Lie algebras with 1-dimensional Cartan algebras in general and what the additional hypothesis of having a reduced Lie semialgebra entails in addition.

II.5.31.   Lemma. *Let a Lie algebra L have a 1-dimensional Cartan algebra, and let R denote its radical. Then either $L = R$ or $L/R \cong \operatorname{sl}(2, \mathbb{R})$ or $L/R \cong \operatorname{so}(3)$. If L is solvable, then the commutator algebra L' has codimension 1.*

Proof.   The Cartan algebras of L/R are at most one dimensional, being homomorphic images of the Cartan algebras of L. If the Lie algebra L/R is non-zero, then it is a semisimple algebra with 1-dimensional Cartan algebras, leaving only the two real forms of $\operatorname{sl}(2, \mathbb{C})$ we have listed. Now suppose that L is solvable and let H be a 1-dimensional Cartan algebra. The commutator algebra L' is now nilpotent. In a nilpotent Lie algebra, every proper subalgebra is properly contained in its normalizer. Thus $H \not\subseteq L'$. Now find a hyperplane I containing L' not containing H. Then $L = I \oplus H$ and I is an ideal. Let x be a non-zero element of H. Since x

spans H and H is its own normalizer, $(\operatorname{ad} x|I)$ cannot have the eigenvalue 0. This implies that $(\operatorname{ad} x|I)$ is bijective, and so $I = [x, I] \subseteq L'$. Hence $I = L'$ and thus L' is a hyperplane. ∎

II.5.32. **Theorem.** (The Rank 1 Structure Theorem) *Let W be a reduced Lie semialgebra in a Lie algebra L with a 1-dimensional Cartan algebra. Then one and only one of the following three situations is possible:*

(i) *$L = H$ and W is one of the two closed half lines.*

(ii) *Up to isomorphism, $L = \operatorname{sl}(2, \mathbb{R})$, and W is one of the semialgebras described in the* Second Classification Theorem II.3.7.

(iii) *L is a special metabelian Lie algebra, and W is one of the Lie semialgebras described in* Proposition II.5.15.

Proof. Let H be a 1-dimensional Cartan algebra of L, and assume $L \neq H$. If L is semisimple, then by Lemma II.5.31, we have $L \cong \operatorname{sl}(2, \mathbb{R})$, so(3). But so(3) has no proper generating Lie semialgebras since it has no 2-dimensional subalgebras. Hence we are in situation (ii). So we may assume from here on out that L is not semisimple. By Corollary II.5.11 we know that $M(L) \neq \{0\}$. If, as in earlier parts of this section, $\sigma: L \to \operatorname{gl}(M(L))$ is the representation given by $\sigma(x) = (\operatorname{ad} x|M(L))$, then $\sigma(H)$ is a Cartan algebra of $\sigma(L)$. Since $\sigma(L)$ is abelian, we conclude $\sigma(H) = \sigma(L)$, that is, $L = H + \ker \sigma$. We claim that no base root is zero; for otherwise $M_0(L)$ would be a non-zero center of L hence would be contained in H. Since $\dim H = 1$ we would have $H = M_0(L)$, and this would mean $L = H$ since H is its own normalizer. But we ruled out this case in the beginning. In particular we have shown that $\sigma \neq 0$ and thus $N \overset{\text{def}}{=} \ker \sigma \neq L$. Thus L is the semidirect product of N and H. We now claim that this implies the solvability of L. Let $R(N)$ be the radical of N. Then $N/R(N)$ is a semisimple ideal of $L/R(N)$. Thus $L/R(N)$ is the direct vector space sum of the Cartan algebra $\bigl(H + R(N)\bigr)/R(N)$, necessarily of dimension one, and the semisimple ideal $N/R(N)$. Then $L/R(N)$ must have a one dimensional radical and thus is reductive. If $N/R(N)$ is non-zero then the rank of $L/R(N)$ exceeds one, and thus $\operatorname{rank}(L) > 1$ in contradiction with the hypothesis. Therefore $N = R(N)$. Then N is solvable and thus $L = H + N$ is solvable, too.

Since L is the semidirect product of the ideal N and the 1-dimensional subalgebra H, we conclude that L/N is abelian, whence $L' \subseteq N$. But since H is a Cartan algebra, we have $L = H + L'$; thus $N = L'$. Hence L' is also the maximal nilpotent ideal and is, therefore, the nilradical of L. By the definition of N, the base ideal $M(L)$ is the center of N. The theorem will be proved if we can show that $M(L) = N$. But this follows from Theorem II.5.27, since we have seen $Z(L) = \{0\}$. ∎

We amalgamate this information with other insights we gained before and find the following consequences:

II.5.33. **Corollary.** *Let W be a reduced generating Lie semialgebra in a Lie algebra whose characteristic function agrees with some base root α restricted to $C^1(W)$. Then either W is invariant, or L is almost abelian and W is an arbitrary wedge in L.*

Proof.   By the Second Cartan Algebra Theorem II.5.30, W is invariant or all Cartan algebras are 1-dimensional. In the latter case by Theorem II.5.32, L is 1-dimensional or sl(2) or a special metabelian algebra. In the case of sl(2) there are no base roots. If the last case occurs, then by Corollary II.5.19, $M(L) = M_\alpha(L)$ for a suitable base root, and this means that L is almost abelian.   ∎

II.5.34.   Corollary.   *Let W be a weakly round generating Lie semialgebra in a Lie algebra which is not semisimple. Then W is invariant or L is almost abelian and W is an arbitrary weakly round wedge.*

Proof.   Since L is not semisimple, Theorem II.5.22 applies and shows that the characteristic function of W agrees with a base root on all of $C^1(W)$. Hence Corollary II.5.33 applies.   ∎

EII.5.1   Exercise. Use the Second Classification Theorem for Low Dimensional Lie Semialgebras II.3.7 to verify that sl(2) contains only two weakly round Lie semialgebras.

II.5.35.   Corollary.   *If W is a Lorentzian Lie semialgebra in a Lie algebra which is not semisimple, then W is invariant or L is almost abelian.*

Proof.   The proof is immediate from the preceding results.   ∎

Tangent hyperplane subalgebras

We have seen in Section II.3 that tangent hyperplanes at C^1-points of Lie semialgebras are subalgebras if the algebra is 3-dimensional. This property persists in almost abelian and special metabelian algebras (see Proposition II.5.15). We show in this section that the tangent hyperplane at any C^1-point p of a generating Lie semialgebra is a subalgebra if the characteristic function λ is non-zero at p. The result is not an easy one and depends heavily on much of the machinery developed earlier in this chapter. Since the condition of having hyperplane subalgebras is rather restrictive, this result sheds light on the tendency of Lie semialgebras to be invariant, which by Proposition II.2.37 happens precisely when the characteristic function vanishes.

Let W be a generating Lie semialgebra in a Lie algebra L. We consider now an alternate way of obtaining the value $\lambda(x)$ of the characteristic function for a fixed $x \in C^1(W)$. We complexify L to obtain the complex Lie algebra $L_\mathbb{C} = L \otimes \mathbb{C} = L \oplus iL$. For the transformation $\mathrm{ad}\, x$ on $L_\mathbb{C}$, one obtains a Jordan (or primary) decomposition of $L_\mathbb{C}$ into a finite direct sum of generalized eigenspaces $L_\mathbb{C} = \sum L^\lambda$ indexed by the eigenvalues of $\mathrm{ad}\, x$, where L^λ is the largest subspace satisfying $(\mathrm{ad}\, x - \lambda I)^n L^\lambda = 0$ for some n (alternately this decomposition may be viewed as the root decomposition of $L_\mathbb{C}$ with respect to the one-dimensional subalgebra containing x).

Let T_x denote the tangent hyperplane to W at x. Since W is a Lie semialgebra, $[x, T_x] \subseteq T_x$. Note that for $T_\mathbb{C} = T_x + iT_x$, $[x, T_\mathbb{C}] \subseteq T_\mathbb{C}$. The set of eigenvalues for $\operatorname{ad} x|T_{\mathbf{C}}$ is contained in the set of eigenvalues for $\operatorname{ad} x$, and $T_{\mathbf{C}} = \bigoplus T^\mu$, where $T^\mu \subseteq L^\mu$. Since $T_{\mathbf{C}}$ has (complex) codimension 1 in $L_{\mathbf{C}}$, it must be the case that $L^\mu = T^\mu$ for all but one μ, and in the exceptional case $\mu = \nu$, T^ν has codimension 1 in L^ν, or else L^ν has dimension 1 and ν is not an eigenvalue for $\operatorname{ad} x|T_{\mathbf{C}}$.

II.5.36. **Lemma.** *If $\mu \neq \lambda(x)$, then $L^\mu = T^\mu$; $L^{\lambda(x)} \cap T_{\mathbf{C}}$ has codimension 1 in $L^{\lambda(x)}$.*

Proof. The proof follows from the remarks before the lemma if we show that $\nu = \lambda(x)$ is the exceptional case.

Suppose $z \notin T_{\mathbf{C}}$ and ζ is a complex number such that $[x, z] - \zeta z \in T_{\mathbf{C}}$. It follows from Corollary II.2.36 that $[x, z] - \lambda(x) z \in T_{\mathbf{C}}$. Subtracting, we obtain $(\lambda(x) - \zeta) z \in T_{\mathbf{C}}$. Since $z \notin T_{\mathbf{C}}$, it follows that $\lambda(x) = \zeta$.

Now suppose $(\operatorname{ad} x - \nu I)^n(z) = 0$ for some $n \geq 1$. Then there exists j, $0 \leq j < n$, such that $z_j = (\operatorname{ad} x - \nu I)^j(z) \notin T_{\mathbf{C}}$, but $(\operatorname{ad} x - \nu I)(z_j) = (\operatorname{ad} x - \nu I)^{j+1}(z) \in T_{\mathbf{C}}$. Then $[x, z_j] - \nu z_j \in T_{\mathbf{C}}$, so by the preceding paragraph $\nu = \lambda(x)$. $\blacksquare$

Now $T_{\mathbf{C}}$ is a subalgebra of $L_{\mathbf{C}}$ if and only if T_x is a subalgebra of L. Hence if T_x is not a subalgebra of L, then $T_{\mathbf{C}}$ is not a subalgebra of $L_{\mathbf{C}}$. It follows that there must exist eigenvalues η and ζ of $\operatorname{ad} x$ such that $[T^\eta, T^\zeta] \not\subseteq T_{\mathbf{C}}$. Now $[T^\eta, T^\zeta] \subseteq L^{\eta+\zeta}$; by the preceding Lemma II.5.36 it must be the case that $\lambda(x) = \eta + \zeta$. If the Lie semialgebra W is reduced, then by Propostion II.5.28 the spectrum of $\operatorname{ad} x$ is contained in $\mathbb{R} \cup i\mathbb{R}$. Hence if $\lambda(x) \neq 0$, then it must be the case that $\eta, \zeta \in \mathbb{R}$. But for a real eigenvalue η, we have

$$T^\eta = \{u + iv : u, v \in T_x, (\operatorname{ad} x - \eta \cdot \mathbf{1})^n(u) = 0 = (\operatorname{ad} x - \eta \cdot \mathbf{1})^n(v) \quad \text{for some} \quad n\}.$$

Hence it must be the case that there exist $y \in T_x \cap T^\eta$ and $z \in T_x \cap T^\zeta$ such that $[y, z] \notin T_x$, but $[y, z]$ is in the generalized eigenspace for $\eta + \zeta = \lambda(x)$ in L. We have thus derived the following lemma.

II.5.37. **Lemma.** *Let W be a reduced Lie semialgebra in L and let $x \in C^1(W)$. If $\lambda(x) \neq 0$ and T_x is not a subalgebra, then there exist real eigenvalues μ and ν and generalized eigenvectors y and z in T_x for μ and ν, respectively, such that $\mu + \nu = \lambda(x)$ and $[y, z] \notin T_x$.* $\blacksquare$

We need some additional lemmas before embarking on the proof of the theorem.

II.5.38. **Lemma.** *Let W be a generating Lie semialgebra in a Lie algebra L. Then in $C^1(W)$, the set $\{p \in C^1(W) : T_p$ is a subalgebra$\}$ is closed. Specifically, if a C^1-point p is the limit of C^1-points p_n such that T_{p_n} is a subalgebra for each n, then T_p is a subalgebra.*

Proof. Let B be a compact base of the pointed dual cone W^*. By the Transgression Theorem I.2.35, $p_n^* \overset{\text{def}}{=} \sigma^B(p_n)$, $\{p_n^*\} = B \cap p_n^\perp$ converges to $p^* \overset{\text{def}}{=} \sigma^B(p)$, $\{p_n^*\} = B \cap p^\perp$. Pick $v \in L$ such that $p^*(v) = 1$. Now let $x, y \in T_p = \ker(p^*)$; then $p^*(x) = p^*(y) = 0$. If $x_n = x - \big(p_n^*(x)/p_n^*(v)\big) \cdot v$ and $y_n =$

$y - \left(p_n^*(y)/p_n^*(v)\right)\cdot v$, then one verifies directly that $x_n, y_n \in T_{p_n}$ (since $p_n^*(x_n) = p_n^*(y_n) = 0$), that $x_n \to x$ and that $y_n \to y$. Thus $[x_n, y_n] \to [x, y]$. Since $[x_n, y_n] \in T_{p_n}$, we have $p_n^*([x_n, y_n]) = 0$. It follows that $p^*([x, y]) = 0$, i.e., $[x, y] \in T_p$. ∎

II.5.39.   Definition.   Let L be a real Lie algebra. For an element $x \in L$ we write $\#(x)$ for the number of non-zero real eigenvalues of $\operatorname{ad} x$.

An element $x \in L$ is called *strongly regular* if x is a regular element and if there exists a neighborhood N of x such that $\#(y) \leq \#(x)$ for all $y \in N$.

II.5.40.   Lemma.   *Let W be a reduced Lie semialgebra in L. The set of strongly regular elements in $\operatorname{int} W$ is open and dense in W. Furthermore, if $x \in \operatorname{int} W$ is strongly regular, then for each real eigenvalue $\lambda \neq 0$ of $\operatorname{ad} x$, there is a unique real root α on the Cartan algebra $H = \ker(\operatorname{ad} x)$ such that $\alpha(x) = \lambda$ and $L_\lambda \subseteq L^\alpha$.*

Proof.   Let $U \neq \varnothing$ be an open set in $\operatorname{int}(W)$. Pick a regular element $x \in U$ such that $\#(x) = \max\{\#(y) : y \in U\}$. Let ε be half of the minimal distance between any distinct pair of the real eigenvalues of $\operatorname{ad} x$. The set of all regular elements of L is open, and the set of elements $y \in L$ such that $\operatorname{Spec} \operatorname{ad} y$ is in an ε-neighborhood of $\operatorname{Spec} \operatorname{ad} x$ and that $\operatorname{Spec} \operatorname{ad} x$ is an ε-neighborhood of $\operatorname{Spec} \operatorname{ad} y$ is open (see [Os73], Appendix K). Hence we find an open neighborhood $V \subseteq U$ of x such that V consists entirely of regular elements and such that each member of the spectrum of $\operatorname{ad} y$ is within ε of some member of the spectrum of $\operatorname{ad} x$ and vice versa.

Let $y \in V$, and let μ be a non-zero real eigenvalue of $\operatorname{ad} x$. The ε neighborhood around μ in the complex plane misses the imaginary axis since 0 is an eigenvalue of $\operatorname{ad} x$. It follows from Proposition II.5.28 and the choice of V that $\operatorname{ad} y$ must have a real eigenvalue within ε of μ. By the choice of ε, this implies $\#(y) \geq \#(x)$. Considering the choice of x we conclude $\#(x) = \#(y)$. Hence y is also strongly regular.

Let H be the nilspace of $\operatorname{ad} x$; it is standard that H is a Cartan subalgebra. By the First Cartan Algebra Theorem II.4.16, H is abelian, and hence is the kernel of $\operatorname{ad} x$. Consider $A = H + N^+$ for $N^+ = \bigoplus\{L_\mu : \mu > 0\}$, where L_μ is the eigenspace for the eigenvalue μ of $\operatorname{ad} x$. Then N^+ is a nilpotent ideal of A, and hence A is solvable. Since $(\operatorname{ad} x)|N^+$ has real spectrum, by Theorem II.5.27 we have $N^+ = M(A)$. By Proposition II.5.4, $M(A)$ is the direct sum of its base root spaces $M_\alpha(A)$. If the base root α is restricted to H, we then obtain a (real) root for the Cartan algebra H. Thus N^+ is contained in the direct sum $\bigoplus L^\alpha$ taken over all real roots α of H. If N^- is defined to be the sum of the eigenspaces of $\operatorname{ad} x$ for negative real eigenvalues, then a similar argument establishes that $N^- \subseteq \bigoplus L^\alpha$. Hence each eigenspace for every non-zero real eigenvalue of $\operatorname{ad} x$ is contained in $\bigoplus L^\alpha$. For any $y \in L_\lambda$ we write $y = \sum y_\alpha$ with $y_\alpha \in L^\alpha$, and compute $\sum \lambda \cdot y_\alpha = [x, y] = \sum \alpha(x) \cdot y_\alpha$. This means that $y_\alpha = 0$ whenever $\alpha(x) \neq \lambda$. Thus we see that

$$(\dagger) \qquad\qquad L_\lambda = \bigoplus\{L_\lambda \cap L^\alpha : \alpha(x) = \lambda\}.$$

This shows that we have at least *one* real root α such that $\alpha(x) = \lambda$.

In order to show the uniqueness of such an α we pick $h \in H \cap V$ such that h is not in $\ker(\alpha - \beta)$ for any two distinct real roots α and β of H and h is also not in $\ker(\alpha)$ for any real root α; this is possible since there are only finitely many hyperplanes in H to be avoided and $H \cap V$ is open in H. Then $\#(h)$ is not less than the number r of distinct real roots. Since $\#(x) = \max\{\#(y) : y \in U\} \geq \#(h)$, it follows that $\#(x) \geq r$. Since by the preceding paragraph we can represent any non-zero real eigenvalue of $\operatorname{ad} x$ as some $\alpha(x)$, it follows that $\alpha \neq \beta$ must imply $\alpha(x) \neq \beta(x)$. This shows that there is at most one real root α with $\lambda = \alpha(x)$, and (†) implies that, for this unique α, we have

$$(\dagger\dagger) \qquad\qquad L_\lambda = L^\alpha.$$

This completes the proof. ∎

II.5.41. **Theorem.** (The Tangent Hyperplane Subalgebra Theorem) *Let W be a generating Lie semialgebra in a Lie algebra L. If $p \in C^1(W)$ satisfies $\lambda(p) \neq 0$, then the tangent hyperplane T_p is a subalgebra.*

Proof. Let I_W denote the largest ideal contained in W. The image of W in L/I_W is a Lie semialgebra by Proposition II.4.1, and as in Proposition I.2.9 the image of p is a C^1-point of the quotient wedge with unique tangent hyperplane the image of T_p. It is a direct verification that the characteristic function takes on the same value at p and at the image of p in L/I_W. If we show the image of T_p is a subalgebra of L/I_W, then the inverse image is T_p (see Proposition I.2.9), and hence T_p will be a subalgebra of L. Thus it suffices to prove the theorem for reduced Lie semialgebras.

Let W be a reduced Lie semialgebra in L, and let $p \in C^1(W)$ such that $\lambda(p) \neq 0$. Let $B_\varepsilon(p)$ be an open ball of radius ε around p such that the characteristic function is bounded away from 0 on the set of C^1-points in $B_\varepsilon(p)$ (this is possible since by Proposition II.2.34 the characteristic function is continuous on the C^1-points). Since ε can be chosen arbitrarily small, to show that T_p is a subalgebra, it suffices by Lemma II.5.38 to show that there exists a C^1-point in $B_\varepsilon(p)$ for which the tangent hyperplane is a subalgebra. This will be our method of proof.

Let $x \in L \setminus T_p$. If there existed $\delta > 0$ such that $e^{t \operatorname{ad} x}(p) \in W$ for $-\delta < t < \delta$, then evaluating the derivative of $t \mapsto e^{t \operatorname{ad} x}(p)$ at 0, we would obtain $[x, p] \in T_p$. By Corollary II.2.36, we have $[p, x] - \lambda(p) \cdot x \in T_p$. We conclude $\lambda(p) \cdot x = -[x, p] - ([p, x] - \lambda(p) \cdot x) \in T_p$, a contradiction since $x \notin T_p$ and $\lambda(p) \neq 0$.

Pick an open set U containing p and $\delta > 0$ such that $e^{t \operatorname{ad} x}(u) \in B_\varepsilon(p)$ for $u \in U$ and $|t| < \delta$. By the preceding paragraph $e^{t \operatorname{ad} x}(p) \notin W$ for some $t \in]-\delta, \delta[$. Then there exists a strongly regular element $h \in U \cap \operatorname{int}(W)$ such that $e^{t \operatorname{ad} x}(h) \notin W$, since by Lemma II.5.40 the strongly regular elements are dense in the interior of W.

We have $e^{0 \cdot \operatorname{ad} x}(h) = h \in \operatorname{int}(W)$ and $e^{t \cdot \operatorname{ad} x}(h) \notin W$. Since the interval from 0 to t is connected, there exists s between 0 and t such that $e^{s \cdot \operatorname{ad} x}(h) \in \partial W$. By Lemma II.5.40, we can pick an open set V of strongly regular elements such that $h \in V \subseteq U$. Since $e^{s \cdot \operatorname{ad} x}(V)$ is an open set meeting $\partial(W)$ and since $C^1(W)$ is dense in $\partial(W)$, we conclude that $p' = e^{s \cdot \operatorname{ad} x}(k) \in C^1(W)$ for some $k \in V$. Also by choice of U and $B_\varepsilon(p)$, $\lambda(p') \neq 0$ and $p' \in B_\varepsilon(p)$. If we now apply the Lie algebra automorphism $e^{-s \cdot \operatorname{ad} x}$, we conclude that k is a C^1-point for the reduced Lie semialgebra $W' = e^{-s \cdot \operatorname{ad} x}(W)$ and that the characteristic function for W' evaluated at k agrees with the characteristic function for W evaluated at p', and hence is non-zero. Furthermore, the tangent hyperplane $T_{p'}(W)$ to W at p' will be a subalgebra if and only if the tangent hyperplane $T_k(W')$ to W' at k is a subalgebra, since the automorphism must carry the tangent hyperplane to W at p' to the tangent hyperplane to W' at k. Thus to complete the proof it suffices to show that $T_k = T_k(W')$ is a subalgebra. But in working with T_k, we can use the fact that k is simultaneously a C^1-point of W' and a strongly regular interior point of W.

Now we assume that T_k is not a subalgebra and we shall derive a contradiction. Then by Lemma II.5.37 there exist real eigenvalues μ and ν for $\operatorname{ad} k$ and generalized eigenvalues y and z in T_k for μ and ν respectively such that $[y, z] \notin T_k$ and $\mu + \nu = \lambda(k)$. But it follows from Proposition II.5.29 that $L^\mu = L_\mu$ and $L^\nu = L_\nu$, and then parts (ii) and (iii) of that proposition imply $[y, z] = 0$ if μ and ν are both non-zero (the case that one is the negative of the other is ruled out since $\lambda(k) \neq 0$). Hence we must have $\mu = 0$ and $\nu = \lambda(k)$ (or vice-versa). Set $H = \{k' \in L : (\operatorname{ad} k)^n(k') = 0 \text{ for some } n\}$; then H is a Cartan subalgebra. By the First Cartan Algebra Theorem II.4.16, H is abelian and so $H = \ker \operatorname{ad} k$. But then $y \in H$ and $z \in L^\nu = L_\nu$. By Lemma II.5.40, there is a unique real root α with respect to H such that $\alpha(k) = \nu$ and $L_\nu = L^\alpha$. Then $z \in L^\alpha$ and thus $[y, z] = \alpha(y) \cdot z$. Since $z \in T_k$ we have found that $[y, z] \in T_k$. This contradiction completes the proof. ∎

6. Lorentzian Lie semialgebras

We recall: A quadratic form on a vector space L is a symmetric bilinear form $q: L \times L \to \mathbb{R}$, and often we write $q(x)$ instead of $q(x,x)$. If L is a Lie algebra, q is called *invariant* if $q(x, [y, z]) = q([x, y], z)$ for all $x, y, z \in L$. The most common examples of Lie algebras equipped with an invariant non-degenerate quadratic form are the semisimple algebras with their Cartan–Killing form, or compact Lie algebras with some positive definite invariant scalar product. We have seen a whole series of solvable Lie algebras in Section 3 which carry invariant Lorentzian forms, and all theses examples are by no means exhaustive. Invariant quadratic forms on Lie algebras have been investigated in recent years by various authors (see e.g. Hofmann and Keith, Medina and Revoy, Vinberg and his school). We shall consider in this section Lie algebras L equipped with some invariant non-degenerate bilinear form and study Lie semialgebras in them. This will eventually lead to a complete classification of all Lorentzian Lie semialgebras. On the way we shall show that Lie semialgebras satisfying certain roundness conditions either force the containing Lie algebra to be almost abelian or else are invariant, in which case the theory of Chapter III takes over.

We shall, in this section, identify the dual $\widehat{L}$ of L canonically with L via the isomorphism $x \mapsto q(x, \bullet): L \to \widehat{L}$. In particular, for any subset $X \subseteq L$ we set $X^{\perp} = \{x \in L : q(x, u) = 0 \text{ for all } u \in X\}$. Also the dual W^* of a cone $W \subseteq L$ is a cone in L under our identification: In fact we have $W^* = \{x \in L : q(x, u) \geq 0 \quad \text{for all} \quad u \in W\}$. The *zero set* of q is the set $Z_q = \{z \in L : q(z) = 0\}$. This set is a quadric in the vector space L stable under scalar multiplication. (Of course, if q is positive or negative definite, then $Z_q = \{0\}$!) Any hyperplane T of L is of the form $y^{\perp}$ for some non-zero vector $y \in L$. It is a tangent hyperplane of Z_q in y if and only if $q(y) = 0$ and $T = y^{\perp}$.

EII.6.1.   **Exercise.**   Prove the statement asserted in the preceding paragraph.

Lie semialgebras in Lie algebras with invariant quadratic form

We begin our discussion in earnest by identifying those pairs (x, T) consisting of a point x and a hyperplane T which qualify as being the tangent hyperplanes

T_x of a generating Lie semialgebra in a C^1-point x.

II.6.1.   Definition.   A pair (x, T) consisting of a hyperplane T of a Lie algebra L and a point $x \in T$ is called *good* if

(1) $$[x, T] \subseteq T.$$

It will be called *excellent* if even

(2) $$[x, L] \subseteq T.$$

Under the present circumstances, with a non-degenerate invariant quadratic form around, good pairs can be expressed in terms of 2-dimensional subalgebras:

II.6.2.   Theorem.   *Let L be a finite dimensional Lie algebra with an invariant quadratic form q. For a non-zero element $x \in L$ and a hyperplane T (which need not necessarily contain x) the following statements are equivalent:*

(A) (x, T) *is a good pair.*

(B) *There is a non-zero $y \in L$ such that $[x, y] \in \mathbb{R} \cdot y$, and $T = y^\perp$.*

Moreover, if, in condition (B) the relation $[x, y] = t \cdot y$ holds with a non-zero $t \in \mathbb{R}$, then $q(x, y) = q(y) = 0$ and T is tangent to Z_q in the point y. In particular, $x \in T$.

Proof.   (A)$\Rightarrow$(B): We know that $T^\perp$ is a 1-dimensional subspace, hence is of the form $\mathbb{R} \cdot y$ with some non-zero $y \in L$. For each $t \in T$ we have $[x, t] \in T$ by (A). Hence $0 = q([x, t], y) = q(x, [t, y]) = -q([x, y], t)$ for all $t \in T$ by invariance. Thus $[x, y] \in T^\perp = \mathbb{R} \cdot y$, as asserted. (B)$\Rightarrow$(A): We have just seen that $q([x, t], y) = -q([x, y], t)$ for all $t \in T$. By (B) we know $[x, y] \in \mathbb{R} \cdot y$. Hence $q([x, t], y) \in -q(\mathbb{R} \cdot y, t) = \{0\}$ for all $t \in T$, whence $[x, T] \subseteq y^\perp = T$.

Now suppose that there is a $t \in \mathbb{R}$ with $[x, y] = t \cdot y$ and $t \neq 0$. Then $q(x, y) = t^{-1} q(x, [x, y]) = t^{-1} q([x, x], y) = 0$. Let $s \in \mathbb{R}$ be non-zero. Then $[s \cdot x + y, y] = st \cdot y$. But then, setting $x' = s \cdot x + y$ and observing $[x', y] = st \cdot y$ by what we just saw, we conclude $q(s \cdot x + y, y) = q(x', y) = 0$. Since this equation holds for all $s \neq 0$, by continuity we infer $q(y, y) = 0$. Since $T = y^\perp$ and $q(y) = 0$, the hyperplane T is tangent to Z_q in y.   ∎

II.6.3.   Corollary.   *Under the general circumstances of* Theorem II.6.2, *the following conditions are also equivalent:*

(a) (x, T) *is an excellent pair.*

(b) *There is a non-zero $y \in L$ such that $[x, y] = 0$ and $T = y^\perp$.*

Proof.   (a)$\Rightarrow$(b): $0 = q([x, t], y) = -q([x, y], t)$ for all $t \in L$ this time. Hence $[x, y] \in L^\perp = \{0\}$ since q is non-degenerate.

(b)$\Rightarrow$(a): $[x, y] = 0$ implies, in the same equation, the relation $q([x, t], y) = 0$ for all $t \in L$. Thus $[x, L] \subseteq y^\perp = T$.   ∎

From Corollary II.6.3 we observe, in particular, that $(x, x^\perp)$ is always an excellent pair for $x \neq 0$.

If W is a Lie semialgebra in L and x is a C^1-point of W, then (x, T_x) is a good pair with which a non-negative real number $\lambda(x)$ is canonically associated. This number links up with the data of Theorem II.6.2 as follows:

II.6.4. **Proposition.** *Let W be a generating Lie semialgebra in a Lie algebra L with a non-degenerate invariant quadratic form. For each C^1-point x of W we choose a $y \in E^1(W^*)$ in such a fashion that $y^\perp = T_x$. Then $[x,y] = -\lambda(x) \cdot y$.*

Proof. By Corollary II.2.36, we find that for all $u \in L$ we have $[x,u] = \lambda(x) \cdot u + t(x,u)$ with a suitable $t(x,u) \in T_x$. Then $q(t(x,u),y) = 0$ and we derive $q([x,y],u) = -q(y,[x,u]) = -q(y,\lambda(x) \cdot u) = -q(\lambda(x) \cdot y, u)$ for all $u \in L$. As q is non-degenerate, the claim follows. ∎

II.6.5. **Corollary.** *Suppose that $W \neq L$ is a generating Lie semialgebra in a Lie algebra L with an invariant non-degenerate quadratic form. Then the following conclusions hold:*

(i) *If L does not contain any 2-dimensional non-abelian subalgebras, then W is necessarily invariant.*

(ii) *If L is a compact Lie algebra, then W is invariant.*

(iii) *If L is a standard solvable Lorentzian Lie algebra, then W is invariant.*

(iv) *If L is a direct sum of a compact ideal and an ideal which is a standard solvable Lorentzian Lie algebra, then W is invariant.*

Proof. (i) We know that W is invariant if $\lambda(x) = 0$ for all $x \in C^1(W)$ by Corollary II.2.37. This condition is satisfied if L does not contain a copy the 2-dimensional non-abelian algebra by the preceding Proposition II.6.4.

(ii) Every solvable subalgebra in a compact Lie algebra is abelian. Then claim then follows from (i).

(iii) If L is a standard solvable Lorentzian algebra $\mathbb{R} \times V \times \mathbb{R}$, then $L/Z(L)$ is a metabelian algebra without non-zero base ideal. It is a very immediate consequence, that this quotient algebra cannot contain a copy of the non-abelian 2-dimensional algebra. Hence if L were to contain a copy of this algebra, it would have to be contained in $L' = \{0\} \times V \times \mathbb{R}$ or else contain $Z(L)$. The first case is ruled out by the fact that L' is nilpotent, and no nilpotent algebra can contain the 2-dimensional non-abelian algebra, and the second case is ruled out since any 2-dimensional algebra containing a non-zero central algebra must be abelian. The claim now follows from (i).

(iv) The direct product of two Lie algebras neither of which contains a copy of the 2-dimensional non-abelian algebra does not contain one either. Thus the previous results secure the claim. ∎

The preceding result could be extended somewhat. One knows other solvable Lie algebras which support non-degenerate invariant quadratic forms (see Hofmann and Keith, Medina and Revoy). However, for nilpotent Lie algebras we already have a better result in the form of Theorem II.4.13 and for solvable algebras our later results in this section for most practical purposes reduce us to the case of the Lorentzian solvable algebras.

The following definition facilitates the formulation of the subsequent results.

II.6.6. **Definition.** If W is a generating Lie semialgebra in a Lie algebra L, we set

(i) $C^1_+(W) = \{x \in C^1(W) : \lambda(x) > 0\}$,

(ii) $E^1_+(W^*) = \{\omega \in W^* :$ there is an $x \in W$ such that $\langle \omega, x \rangle = 0$ and $x \in C^1_+(W)\}$.

Recall our convention on identifying W^* with a wedge in L in the presence of a non-degenerate invariant bilinear form; accordingly, $E^1_+(W^*)$ is identified with a subset of L. Indeed, we can write $E^1_+(W^*) = \bigcap \{x^\perp \cap W^* : x \in C^1_+(W)\}$.

II.6.7.   Lemma. *If W is a generating Lie semialgebra in a Lie algebra L, then the following conclusions hold:*

(i) *$C^1_+(W)$ is open in $C^1(W)$.*

(ii) *If $x \in C^1_+(W) \cap E^1(W)$, and $\omega \in E^1(W^*)$ satisfy $\langle \omega \mid x \rangle = 0$, then $E^1_+(W^*)$ is a neighborhood of $\omega \in E^1(W^*)$.*

Proof.   (i) This follows from the continuity of the characteristic function λ according to Proposition II.2.34 in view of the definition of $C^1_+(W)$.

(ii) We recall the transgression function $\sigma : C^1(W) \to \Pi(E^1(W^*))$ of Theorem I.2.35. Clearly, $E^1_+(W^*)$ is the full inverse image in $E^1(W^*)$ of $\sigma(C^1_+(W))$ under the coset map. The claim is therefore a consequence of the Transgression Theorem I.2.35.iii.   ■

II.6.8.   Corollary.   *Let L be a Lie algebra with a non-degenerate invariant quadratic form and W a generating Lie semialgebra in L. Then for $x \in C^1_+(W)$ and $y \in E^1_+(W^*)$ with $q(x, y) = 0$, the tangent hyperplane T_x of W in x is the tangent hyperplane $y^\perp$ of Z_q in y. In particular, $q(y) = 0$. In other words, $E^1_+(W^*) \subseteq Z_q$.*

If $x \in C^1_+(W) \cap E^1(W)$, then $\overline{E^1_+(W^*)}$ is a neighborhood of y in $\overline{E^1(W^*)}$.

Proof.   For any pair of points x, y as described in the corollary, we have $\lambda(x) \neq 0$, and so Theorem II.6.2 shows that $q(y) = 0$, and that $y^\perp$ is the tangent hyperplane of Z_q in y, and from the Transgression Theorem I.2.35 we know that $y^\perp = T_x$.

Now suppose that $x \in C^1_+(W) \cap E^1(W)$. Then by Lemma II.6.7, the set $E^1_+(W^*)$ is a neighborhood of y in $E^1(W^*)$. The last assertion then follows.   ■

II.6.9.   Corollary.   *If W is a generating semialgebra in L and L has a non-degenerate invariant quadratic form q, then W is the intersection of half-spaces $L_x(W) = \overline{W - \mathbb{R} \cdot x}$ whose boundaries T_x either*

(i) *are tangent hyperplanes of Z_q, or*

(ii) *satisfy $[x, L] \subseteq T_x$.*

Proof.   This corollary is clear from the preceding results.   ■

If we make the relatively mild assumption on W amounting to the requirement that W be locally round at one point in the sense of Definition I.4.13, then the preceding corollary can be sharpened significantly. In fact we shall make precise the contention that in the presence of invariant non-degenerate quadratic forms, Lie semialgebras have a strong tendency to be invariant. For this purpose we bring the results of Chapter I, Section 4 to bear and formulate:

II.6.10.   Theorem.   *Let L be a finite dimensional Lie algebra with an invariant non-degenerate quadratic form q. Suppose that W is a generating Lie semialgebra which has a C^1-point x with the following properties:*

(i) $\lambda(x) > 0$,

(ii) W *is locally round at x.*

Then q is Lorentzian.

Proof.   By Corollary II.6.9, the hypotheses of Proposition I.4.23 are satisfied. This proposition proves the theorem.   ∎

Lorentzian Lie algebras

Now we are motivated to determine all Lie algebras (L, q) equipped with an invariant Lorentzian form. These are the *Lorentzian Lie-algebras* of the headline. Some simple reductions at the beginning will facilitate this task.

II.6.11.   Lemma.   *If $(L, q) = (I, q_I) \oplus (J, q_J)$ is an orthogonal direct sum of ideals, and K is an ideal of I, then K is an ideal of L.*

Proof.   Note that $[L, K] = [I + J, K] \subseteq [I, K] + [J, K] \subseteq K + \{0\}$ since orthogonal ideals annihilate each other.   ∎

II.6.12.   Lemma.   *Every Lie algebra (L, q) with an invariant non-degenerate quadratic form decomposes into an orthogonal direct sum $(L, q) = \bigoplus_{j=1}^{n}(L_j, q_j)$ of irreducible ideals, that is, ideals which cannot be further decomposed into an orthogonal sum of ideals, and each of the restrictions $q_j = q|(L_j \times L_j)$ is non-degenerate.*

Proof.   Since L is finite dimensional we can write L as an orthogonal direct sum of ideals none of which can be further decomposed into a direct sum of ideals of L. Then each of the summands is irreducible by Lemma II.6.11.   ∎

II.6.13.   Remark.   *Let (L, q) be a Lorentzian Lie algebra. Then (L, q) is the direct orthogonal sum of a compact ideal (J, q_J) with a positive definite quadratic form q_J and an irreducible ideal (I, q_I) such that either*

(i) I *is 1-dimensional and q_I negative definite, or*

(ii) (I, q_I) *is Lorentzian and irreducible.*

Proof.   Let $(L, q) = \bigoplus_{j=1}^{n}(L_j, q_j)$ be an orthogonal direct decomposition into irreducible ideals. Since $q = q_1 \oplus \cdots \oplus q_n$ is a Lorentzian form decomposed orthogonally, n-1 of the summands have to be positive definite, say, $q_2, \ldots, q_n$. Then q_1 is either negative definite, in which case $\dim I_1 = 1$ follows, or else is Lorentzian. Set $I = L_1$ and $q_I = q_1$, moreover $J = L_2 \oplus \cdots \oplus L_n$ and $q_J = q_2 \oplus \cdots \oplus q_n$.   ∎

In in view of this remark, it suffices to determine all *irreducible* Lorentzian Lie algebras.

Irreducible Lorentzian Lie algebras

In classifying irreducible Lorentzian Lie algebras (L, q) we proceed by considering cases.

Case 1. *L has a hyperplane ideal I.*

The annihilator $I^\perp$ is a 1-dimensional ideal. Since $[L, L] \subseteq I$ as $L/I \cong \mathbb{R}$ is abelian, we have $q([L, I^\perp], L) \subseteq q(I^\perp, [L, L]) \subseteq q(I^\perp, I) = \{0\}$, and thus $[L, I^\perp] = \{0\}$ since q is non-degenerate. Hence $I^\perp$ is central. Further, $I^\perp \subseteq I$, for otherwise $L = I \oplus I^\perp$ in contradiction with irreducibility. Since q is non-degenerate, we find a $y \in Z_q \setminus I$. Then $H \overset{\text{def}}{=} \mathbb{R} \cdot y + I^\perp$ is an abelian subalgebra. Since $y \notin I = (I^\perp)^\perp$, the restriction $q_H \overset{\text{def}}{=} q|(H \times H)$ is not zero. Hence (H, q_H) is a hyperbolic plane. Also, $H^\perp \subseteq (I^\perp)^\perp = I$, whence $H^\perp \cap H \subseteq I \cap H = I^\perp$. If we choose a z with $I^\perp = \mathbb{R} \cdot z$, then for any $h \in H^\perp \cap H$, we have $h = r \cdot z$ for some $r \in \mathbb{R}$, and $0 = q(h, y) = rq(z, y)$. Since $q(y, y) = q(z, z) = 0$ and $q_H \neq 0$, we conclude $q(h, y) \neq 0$ and thus $r = 0$, that is, $h = 0$. Hence $H^\perp \cap H = \{0\}$, and thus $(H^\perp, q_\perp)$, $q_\perp = q|(H^\perp \times H^\perp))$, is a Hilbert space. Now q induces on the factor algebra $I/I^\perp$ an invariant quadratic form Q via $Q(x + I^\perp, x' + I^\perp) = q(x, x')$. We observe $I = H^\perp + I^\perp$, so that $(I/I^\perp, Q)$ is isomorphic to $(H^\perp, q_\perp)$ as a space equipped with a quadratic form. Thus Q is positive definite, and thus $I/I^\perp$ is a compact Lie algebra K. But then $K = Z \oplus S$ with the center Z and the semisimple commutator algebra $K' = S$ as characteristic ideals. We write $\widetilde{Z}$ for the full inverse image of Z in I. Since $I^\perp$ is central in L and Z is abelian, $[\widetilde{Z}, \widetilde{Z}] \subseteq I^\perp \subseteq Z(L)$, whence $\widetilde{Z}$ is nilpotent (and in fact a Heisenberg algebra). By picking a Levi complement for $\widetilde{Z}$ in I we may write $I = S + \widetilde{Z}$ and observe $[S, \widetilde{Z}] \subseteq I^\perp$. Once more we use that $I^\perp$ is central and conclude that the representation of S on $\widetilde{Z}$ is nilpotent, hence must be zero since S is compact semisimple. Hence $[S, \widetilde{Z}] = \{0\}$ and I is the direct sum of two characteristic ideals $S \oplus \widetilde{Z}$. In particular, S and $\widetilde{Z}$ are ideals of L. Moreover, since $S = [S, S]$, we know $q(S, \widetilde{Z}) = q([S, S], \widetilde{Z}) = q(S, [S, \widetilde{Z}]) = \{0\}$, that is, $S \perp \widetilde{Z}$. Since S and $\widetilde{Z}$ are ideals of L, they are stable under $\operatorname{ad} y$. Since S is semisimple, the derivation $\operatorname{ad} y|S$ is inner. Hence there is an $s \in S$ such that $[y, x] = [s, x]$ for all $x \in S$. Now we set $T = \mathbb{R} \cdot (y - s) + \widetilde{Z}$. Then $[T, T] \subseteq [y - s, \widetilde{Z}] + [\widetilde{Z}, \widetilde{Z}] \subseteq \widetilde{Z} \subseteq T$, and $[T, S] \subseteq [y - s, S] + [\widetilde{Z}, S] = \{0\}$. We note $S + T = \mathbb{R} \cdot (y - s) + (S + \widetilde{Z}) = \mathbb{R} \cdot (y - s) + I = \mathbb{R} \cdot y + I = L$ in view of $s \in I$. Furthermore, the sum of vector space $L = \mathbb{R} \cdot (y - s) + S + \widetilde{Z}$ is direct, hence $S \cap T = S \cap (\mathbb{R} \cdot (y - s) + \widetilde{Z}) = \{0\}$. Thus $L = S \oplus T$ is a direct sum of ideals. Since S is semisimple, $[S, S] = S$, whence $q(S, T) = q([S, S], T) = q(S, [S, T]) = q(S, \{0\}) = \{0\}$. Hence L is the orthogonal direct sum of the two ideals S and T, and $T \neq \{0\}$. Since we assume (L, q) to be irreducible, this implies $S = \{0\}$.

We now know $L = \mathbb{R} \cdot y + H^\perp + \mathbb{R} \cdot z$ with $H = \mathbb{R} \cdot y + \mathbb{R} \cdot z$.

At this point we observe $[y, H] = [y, \mathbb{R} \cdot y + I^\perp] = \{0\}$. Thus $q([y, H^\perp], H) = q([y, H], H^\perp) = q(\{0\}, H^\perp) = \{0\}$, whence $[y, H^\perp] \subseteq H^\perp$. We may and will now assume that the elements y and z are normalized in such a fashion that $q(y,z) = 1$. If we consider any element $x \in I^\perp$, then $x = t \cdot z$ for some $t \in \mathbb{R}$, and thus $q(x,y) \cdot z = q(t \cdot z, y) \cdot z = tq(z,y) \cdot z = t \cdot z = x$. Since $I/I^\perp$ is abelian, $[I,I] \subseteq I^\perp$. Hence we can apply this argument to an element $x = [a,b]$ with $a, b \in H^\perp$ and find

$$(*) \qquad [a,b] = q([a,b],y) \cdot z = q([y,a],b) \cdot z.$$

We now denote with $(V, \langle \bullet | \bullet \rangle)$ the Hilbert space $(H^\perp, q_\perp)$. Recalling the set-up of Proposition II.3.11, we consider on $\mathbb{R} \times V \times \mathbb{R}$ the componentwise addition and scalar multiplication and the Lie algebra bracket $[(u,x,v),(u',x',v')] = (0, u \cdot dx' + u' \cdot dx, \langle dx \mid x' \rangle)$ with $dx = [y,x]$ and introduce a quadratic form

$$\bar{q}((u,x,v),(u',x',v')) = uv' + u'v + \langle x \mid x' \rangle.$$

We claim that d is a vector space automorphism of V. We must show its injectivity. For a proof of this assertion, let $a \in \ker d$. Then a is in the centralizer of y in $H^\perp$. If $b \in H^\perp$, then $0 = q([y,a],b) \cdot z = [a,b]$ by $(*)$ above. So a centralizes $H^\perp$, but also $I^\perp$. Hence a is central in $\mathbb{R} \cdot y + H^\perp + I^\perp = L$. We recall that (L,q) is the orthogonal direct sum of H and the Hilbert space $(H^\perp, q_\perp)$. Thus every vector subspace J of $H^\perp$ satisfies $J \cap J^\perp = \{0\}$. The irreducibility of (L,q) then shows that $H^\perp$ cannot contain non-zero ideals of L. It follows that $Z(L) \cap H^\perp = \{0\}$. This implies $a = 0$ as asserted.

If we now define $f : \mathbb{R} \times V \times \mathbb{R} \to L$ by $f(u,x,v) = u \cdot y + x + v \cdot z$, then f is readily checked to be an isomorphism of Lie algebras, and since further $q(u \cdot y + x + v \cdot z, u' \cdot y + x' + v' \cdot z) = uv' q(y,z) + vu' q(z,y) + q(x,x')$, we see that $\bar{q}(w,w') = q(f(w), f(w'))$. Thus (L,q) is isomorphic to a standard Lorentzian Lie algebra. (See Figure 5 of Section 3 above.)

Case 2. *L does not have a hyperplane ideal.*

There is always the posssibility that $\dim L = 1$. In the following arguments we shall assume $\dim L > 0$.

First we claim that L is semisimple. For if not, then we remember that the set $\{x \in L : q(x) \leq 0\}$ is the union of two invariant Lorentzian cones. In particular, L supports a pointed Lie semialgebra. By Corollary II.5.11, the base ideal of L is non-zero. Hence L contains at least one 1-dimensional ideal I. Then $I^\perp$ is a hyperplane ideal which we assume not to exist in the present case. If L were not simple, then $L = I \oplus J$ with two non-zero semisimple ideals I and J. Then $q(I,J) = q([I,I],J) = q(I,[I,J]) = q(I,\{0\}) = \{0\}$. Thus the sum is orthogonal. But this is a contradiction to the irreducibility of (L,q). Thus L is simple.

First we prove that a simple Lie algebra L on which the Cartan Killing form B is a scalar multiple of a Lorentzian form is necessarily $\mathrm{sl}(2,\mathbb{R})$. Indeed, if K is a maximal compactly embedded Lie subalgebra of L, then $\dim K$ is the number of minus signs in any diagonal representation of B. (See for instance [Ti67].) If for some non-zero $r \in \mathbb{R}$ the quadratic form rB is Lorentzian, then either $\dim K = \dim L - 1$ if $r < 0$ or $\dim K = 1$ if $r > 0$. The only real simple Lie algebra having a subalgebra of codimension 1 is $\mathrm{sl}(2,\mathbb{R})$ (see for instance J.Tits, loc. cit.); but the subalgebras of dimension 2 are not compact. Thus there remains the case $r > 0$. However, in this case, L has a 1-dimensional maximal compact subalgebra. But $sl(2,\mathbb{R})$ is the only simple real Lie algebra with this property (as an inspection of the tables of simple Lie algebras shows).

Next we show that an invariant Lorentzian form on a real simple Lie algebra is a scalar multiple of the Cartan Killing form. As a first step in the proof of this claim we show that L cannot be the underlying real Lie algebra of a complex Lie algebra. (We will show later that any generating Lie semialgebra in a complex Lie algebra must contain the commutator algebra; this would prove this assertion at once. But we give a direct proof here.) Suppose that L is a complex Lie algebra. Pick a regular element w with $q(w) < 0$. Let L^0 be the nilspace of w and $L = L^0 \oplus \bigoplus_{\alpha \in \Omega^+}(L^\alpha + L^{-\alpha})$ the Cartan decomposition with respect to the Cartan algebra L^0, and with a suitable set Ω^+ of positive roots. Then $B \stackrel{\text{def}}{=} L^0 \oplus \bigoplus_{\alpha \in \Omega^+} L^\alpha$ is a Borel subalgebra, and $q_B = q|(B \times B)$ is a Lorentzian form on B since $w \in B$. (We may consider this argument as a version of our standard testing device: Indeed w is in the interior of one of the two invariant Lorentzian cones defined by q, say W. Then $W \cap B$ is a Lorentzian cone on B defined by q_B.) But now B is solvable, and by Remark II.6.13, B is the orthogonal direct sum of a compact solvable, hence abelian ideal (J, q_J) and an ideal (I, q_I) which is either 1-dimensional with a negative definite q_I, or is irreducible solvable Lorentzian. If I is abelian, then B is abelian. Hence $\Omega^+ = \emptyset$ and so $L = L^0$ is abelian. This is a contradiction to the assumptions that L does not have a hyperplane ideal unless $\dim L = 1$. Since we assume $\dim L > 1$, this possibility is ruled out. If, however, I is irreducible Lorentzian, then it is of the form A_{2n} by Case 1 discussed beforehand. It follows that $L^0 = A_{2n}^0 \oplus J$, where A_{2n}^0 is the 2-dimensional Cartan algebra of A_{2n} isomorphic to $\mathbb{R} \times \{0\} \times \mathbb{R}$, and it further follows that there is just one root α in Ω^+ so that $L^\alpha \cong \{0\} \times V \times \{0\}$ in A_{2n}. But by Section II.3 this root takes on purely imaginary values only, which is impossible in a complex Lie algebra.

As a second step we now note that the complexification $L_\mathbb{C}$ of L is simple, since L does not carry a complex structure (see for instance [Ti67]). The natural extension $q_\mathbb{C}$ of q to $L_\mathbb{C}$ is a complex valued quadratic non-degenerate form on $L_\mathbb{C}$. If $B_\mathbb{C}$ denotes the Cartan–Killing form of $L_\mathbb{C}$, then there is a unique vector space automorphism f of $L_\mathbb{C}$ such that $q_\mathbb{C}\big(f(x), y\big) = q_\mathbb{C}\big(x, f(y)\big) = B_\mathbb{C}(x, y)$, and since both $q_\mathbb{C}$ and $B_\mathbb{C}$ take real values on L, we also know that f is the complexification of a real vector space automorphism g. A quick calculation using the invariance of both $q_\mathbb{C}$ and $B_\mathbb{C}$ shows $f([x,y]) = [f(x), y]$. (Indeed $q_\mathbb{C}(f([x,y], z) = B_\mathbb{C}([x,y], z) = B_\mathbb{C}(x, [y,z]) = q_\mathbb{C}(f(x), [y,z]) = q_\mathbb{C}([f(x),y], z)$. Thus f is an automorphism of the adjoint $L_\mathbb{C}$-module $L_\mathbb{C}$, which is simple. Hence by Schur's Lemma, $f = (a + ib)\cdot\mathbf{1}_{L_\mathbb{C}}$. But since f is the complexification of a real vector space automorphism, $b = 0$. Then we conclude $aq = B$, and this proves our claim that the Cartan Killing form B of L is a scalar multiple of q.

Taking the two pieces of information together we obtain a proof that, up to rescaling of q the Lorentzian Lie algebra (L, q) must be isomorphic to the Lorentzian Lie algebra $(sl(2,\mathbb{R}), B)$ as soon as it is irreducible and has no hyperplane ideal.

Our classification of Lorentzian Lie algebras is now complete. We have shown the following theorem:

II.6.14.   **Theorem.** (Classification of Lorentzian Lie algebras) *Let (L, q) be a*

Lorentzian Lie algebra. Then three mutually exclusive cases arise:

(1) *L is compact and is an orthogonal direct sum $I \oplus K$ of two ideals with I being one dimensional and central such that $q|(I \times I)$ is negative definite and $q|(K \times K)$ is positive definite.*

(2) *L is an orthogonal direct sum $A_{2n} \oplus K$ of a standard Lorentzian solvable algebra of dimension $2n$ and a compact ideal.*

(3) *L is a unique orthogonal direct sum $\mathrm{sl}(2, \mathbb{R}) \oplus K$ with a compact ideal K and where the restriction to the first summand is a positive scalar multiple of the Cartan–Killing form.*

Proof.   If (L, q) is not of type (1), then it is the orthogonal direct sum of an irreducible Lorentzian Lie algebra and a compact Lie algebra with a positive quadratic form by Remark II.6.13. The classification then shows that we have to have type (2) or type (3). The uniqueness of the decomposition in type (3) follows from the fact, that in the decomposition $\mathrm{sl}(2, \mathbb{R}) \oplus K$ both $\mathrm{sl}(2, \mathbb{R})$ and K are characteristic ideals which are not isomorphic.   ∎

The explicit information given in the preceding structure theorem allows us to say a few things about the possible generating Lie semialgebras in Lorentzian Lie algebras.

II.6.15.   Proposition.   *Let (L, q) be a Lorentzian Lie algebra and W a generating Lie semialgebra. Then W is invariant, or else L contains a copy of $\mathrm{sl}(2, \mathbb{R})$, hence is an orthogonal direct sum of the form $\mathrm{sl}(2, \mathbb{R}) \oplus K$ with a compact ideal K.*

If $x \in C^1(W)$ and $\lambda(x) \neq 0$, then T_x is a subalgebra containing K.

Proof.   We claim that a Lorentzian Lie algebra of type (1) or (2) according to the Classification Theorem II.6.14 cannot contain any non-abelian 2-dimensional Lie algebras. Indeed for compact Lorentzian and standard solvable Lorentzian algebras this was observed in Corollary II.6.5. But if S is a non-abelian 2-dimensional subalgebra contained in a direct sum of two ideals $I \oplus J$ such that J is compact, then the projection of the solvable algebra S into J must necessarily be abelian, hence can be at most 1-dimensional. It follows that the projection into I must be faithful. Thus I must contain a 2-dimensional non-abelian algebra. Our claim then follows from Theorem II.6.14. But if L does not contain any non-abelian 2-dimensional subalgebras, then W must be invariant by Corollary II.6.5. Thus if W is not invariant, then necessarily L is of type (3) in Theorem II.6.14.

Now suppose that $L = \mathrm{sl}(2, \mathbb{R}) \oplus K$ with a compact ideal K. Consider $x \in C^1(W)$ and suppose that $\lambda(x) \neq 0$. Then $T_x = y^\perp$ for an element $y \in x^\perp \cap W^*$ and $[x, y] = -\lambda(x) \cdot y$ by Corollary II.6.4. Thus $S = \mathbb{R} \cdot x + \mathbb{R} \cdot y$ is a non-abelian 2-dimensional subalgebra of L with commutator $\mathbb{R} \cdot y$. The projection into the compact Lie algebra K must annihilate this commutator algebra; hence $y \in \mathrm{sl}(2, \mathbb{R})$. The projection E of S into $\mathrm{sl}(2, \mathbb{R})$ is faithful as we saw in the previous paragraph. Now E is one of the 2-dimensional subalgebras of $\mathrm{sl}(2)$ well known to us from Proposition II.3.5. If x_1 and x_2 are the projections of x into the two orthogonal summands, then $x = x_1 \oplus x_2$ and $E = \mathbb{R} \cdot x_1 + \mathbb{R} \cdot y$. From Proposition II.3.5 we recall $y^\perp \cap \mathrm{sl}(2) = E$. Furthermore, $K \subseteq y^\perp$ since the sum decomposition of L is orthogonal. Hence

$$T_x = y^\perp = E \oplus K,$$

and this completes the proof of the proposition.   ∎

Recall that for a wedge W in a vector space L we have $L_x(W) = \overline{W - \mathbb{R} \cdot x}$ and that for $x \in C^1(W)$, this set is that closed half-space bounded by T_x which contains W.

II.6.16.   Corollary.   *Suppose that $L = \mathrm{sl}(2, \mathbb{R}) \oplus K$ with a compact Lie algebra K, and that W is a generating Lie semialgebra of L. Set*

$$W_+ \stackrel{\text{def}}{=} \bigcap_{x \in C_+^1(W)} L_x(W).$$

Then W_+ is a Lie semialgebra containing $\overline{W + K}$.

Proof.   Let $x \in C_+^1(W)$. Then T_x is a subalgebra by Proposition II.6.15. Hence $L_x(W)$ is a half space semialgebra by Corollary II.2.24. Hence W_+ is a Lie semialgebra as the intersection of Lie semialgebras by Corollary II.2.17. Since every $L_x(W)$ with $x \in W$ contains W, clearly W_+ contains W. Moreover, by Proposition II.6.15, we have $K \subseteq T_x \subseteq L_x(W)$ for each $x \in C_+^1(W)$. Hence $K \subseteq W_+$. Thus, since W_+ is obviously closed as the intersection of closed half-spaces, $\overline{W + K} \subseteq W_+$.   ∎

It follows from these results that, in type (3) Lorentzian Lie algebras, the generating semialgebras W for which $C_+^1(W)$ is dense in $C^1(W)$ are the full pullbacks under the projection $L \to \mathrm{sl}(2)$ of the Lie semialgebras of $\mathrm{sl}(2)$ familiar to us after the second classification theorem of low dimensional Lie semialgebras II.3.7. All Lie semialgebras in the summand K are invariant by Corollary II.6.5. The invariant wedges in a compact Lie algebra we shall completely describe in Chapter III. In order that there be pointed generating ones, K has to have a non-zero center. If such a center exists, then we can now easily construct generating pointed Lie semialgebras in L which are not invariant, using non-invariant semialgebras in $\mathrm{sl}(2)$. If, however, K is semisimple, then K does not contain any proper invariant generating cone. Then the projection of W into K cannot be contained in any closed half-space, and then, being a set which is closed under addition and non-negative scalar multiplication, must be all of K. An example of such a W is given by either of the two invariant Lorentzian cones defined by q.

We shall now show that round Lie semialgebras in a Lorentzian Lie algebra are necessarily invariant. For this purpose we formulate a proposition.

II.6.17.   Proposition.   *If W is a generating Lie semialgebra in $L = \mathrm{sl}(2, \mathbb{R}) \oplus K$ with a compact ideal K, and if $x \in C_+^1(W)$, then $K \subseteq D(x)$, where $D(x)$ is the vector space defined in Definition I.4.16. In particular, if $K \neq \{0\}$, then W is not round.*

Proof.   Let $x \in C_+^1(W)$. Since $C_+^1(W)$ is open in $C^1(W)$ by Lemma II.6.7.i, there is an open neighborhood V of x in L such that $V \cap C^1(W) \subseteq C_+^1(W)$. Now let $U \subseteq V$ be an open neighborhood of x in L which is so small that $D(U) = D(x)$ in view of Lemma I.4.17. Then for $v \in U \cap C^1(W)$ we have $v \in C_+^1(W)$ and thus $K \subseteq T_v$ by Proposition II.6.15. Hence

$$K \subseteq \bigcap_{u \in C^1(W) \cap U} T_u = D(U) = D(x).$$

Since $C^1_+(W)$ is open and non-empty in $C^1(W)$, the wedge W is not locally round by Proposition I.4.20 for $K \neq \{0\}$. ∎

As a consequence of this proposition we have shown, in the case of a Lorentzian Lie algebra, that a generating and round Lie semialgebra is necessarily invariant. Indeed, of the Lorentzian Lie algebras, only type (3) Lie algebras can contain non-invariant generating Lie semialgebras; so consider $L = \mathrm{sl}(2) \oplus K$. If $K = \{0\}$, then the assertion follows from Theorem II.3.7, and if $K \neq \{0\}$, then it follows from the preceding proposition.

A full classification of all Lie semialgebras in Lorentzian Lie algebras is an open research problem.

Our previous results, taken together, yield the following information:

II.6.18. **Theorem.** *Let L be a Lie algebra with a non-degenerate invariant quadratic form q and let W be a generating Lie semialgebra which is locally round at some boundary point. Then q is Lorentzian and (L, q) is one of the three types described in* Theorem II.6.14. *Either W is invariant, or else W is a non-invariant Lie semialgebra in a Lorentzian Lie algebra of type $\mathrm{sl}(2) \oplus K$ with a compact ideal K. In this case, W cannot be round, and for every C^1-point x of W with $[x, L] \not\subseteq T_x$ the hyperplane T_x is a subalgebra, and the half-space $L_x(W)$ is a half-space semialgebra.* ∎

We remind the reader again, that this theorem applies, in particular, to semisimple Lie algebras, whose Cartan–Killing form is non-degenerate invariant. The theorem thus gives the curious conclusion that, for a generating Lie semialgebra which is very round in at least one point, every tangent hyperplane T either satisfies $[T \cap W, L] \subseteq T$ or else is a subalgebra. The same conclusion is true for polyhedral Lie semialgebras. Whenever subalgebras of codimension 1 are present, the Lie algebra L has to be very special: If T is a subalgebra of codimension 1 in a Lie algebra L, and if I is the largest ideal of L contained in T, then $L/I \cong \mathbb{R}$ or L/I is the 2-dimensional non-abelian algebra or $L/I \cong \mathrm{sl}(2, \mathbb{R})$ by a theorem of Lie's (reproved several times). Thus in order to find a non-invariant Lie semialgebra in, let us say, a simple Lie algebra different from $\mathrm{sl}(2, \mathbb{R})$, it cannot be very round in any boundary point, nor can it be very flat (in the sense that $E_x = T_x \cap W$ is an $(n-1)$-dimensional wedge).

Lorentzian Lie semialgebras

We now complete the classification of Lorentzian Lie semialgebras. With all the information we have gathered up to now, this is no longer complicated.

II.6.19. **Theorem.** *Let L be a finite dimensional Lie algebra and W a weakly round generating Lie semialgebra which is locally round in at least one boundary point. Then the following possibilities occur:*

(i) *L is almost abelian and W an arbitrary wedge (given the specified geomet-rical properties).*

(ii) *W is invariant.*

Proof.   Recall from Lemma I.4.17 and Definition I.4.19 that every weakly round wedge is pointed. If L is not semisimple, then Corollary II.5.34 applies and proves the claim. However, if L is semisimple, then the Cartan–Killing form is non-degenerate, and thus Theorem II.6.18 applies and shows that W is invariant.   ∎

In a later chapter, we shall investigate invariant wedges in Lie algebras and their structure theory. At this point we formulate the conclusive result for Lorentzian Lie semialgebras. Lorentzian wedges are round, so Theorem II.6.19 applies. What remains in this case is the classification of possible Lorentzian Lie semialgebras.

II.6.20.   Lemma.   *Let W be a Lorentzian Lie semialgebra in a Lie algebra L and let q denote an associated quadratic form (see* Definition I.4.8*). Then*

$$\lambda(x)q(x,y) = q([x,y],x) \text{ for all } (x,y) \in C^1(W) \times L.$$

Proof.   Let $x \in C^1(W)$. Then $u \in T_x$ if and only if $q(x,u) = 0$. Hence Condition (†) in Corollary II.2.36 is equivalent to

$$q([x,y] - \lambda(x){\cdot}y, x) = 0 \text{ for all } (x,y) \in C^1(W) \times L.$$

But this establishes the lemma.   ∎

II.6.21.   Lemma.   *Let W be a Lorentzian Lie semialgebra in a Lie algebra L. Suppose that q is an associated quadratic form, and Z_q its zero set. Then the following conditions are equivalent:*

(1) *There is an $e \in L$ with $q(e,e) < 0$ and $q([x,e],x) = 0$ for all $x \in Z_q$.*

(2) *W is an invariant wedge.*

(3) *$q([x,y],x) = 0$ for all $(x,y) \in Z_q \times L$.*

(4) *$q([x,y],x) = 0$ for all $(x,y) \in L \times L$.*

(5) *q is invariant.*

Proof.   (1) $\Rightarrow$ (2): If $x \in C^1(W)$, then $q(x,x) = 0$ and $x \neq 0$, whence $q(e,e) < 0$ implies $q(x,e) < 0$ by Lemma I.4.10. But now Lemma II.6.20 implies $\lambda(x) = 0$. Hence W is invariant by Proposition II.2.37. (2) $\Rightarrow$ (3): If the wedge W is invariant, then its characteristic function vanishes, and so $q([x,y],x) = 0$ for all $(x,y) \in Z_q \times L$ by Lemma II.6.20, as $Z_q = C^1(W) \cup \{0\} \cup -C^1(W)$. (3) $\Rightarrow$ (1): Trivial. (4) $\Leftrightarrow$ (5): Trivially, (5) implies (4), and if we know (4) then (5) follows by polarization: $0 = q([x+z,y], x+z) = q([x,y],x) + q([z,y],x) + q([x,y],z) + q([z,y],z) = q([z,y],x) + q([x,y],z)$. (2) $\Rightarrow$ (4): We take an arbitrary element $y \in L$ and consider the two quadratic forms defined by $q_1(x) = q(e^{\mathrm{ad}\,y}x)$ and $q_2 = q$. If $q_2(x) = 0$, then x or $-x$ is on the boundary of W since q and W are associated. But since W is invariant by (2), and so $\pm e^{\mathrm{ad}\,y}x$, respectively, is on the boundary of W. Hence $q_1(x) = 0$. Similarly, $q_1(x) = 0$ implies $q_2(x) = 0$. Thus q_1 and q_2

have the same zero-set. Hence, by Lemma I.4.12, they differ by a scalar multiple. Thus for any $t \in \mathbb{R}$ and any $y \in L$ we find a non-zero real number $s(t, y)$ such that

$$(*) \qquad q(e^{t \cdot \operatorname{ad} y} x) = s(t, y) \cdot q(x) \quad \text{for} \quad (t, x, y) \in \mathbb{R} \times L \times L.$$

Fixing an x with $q(x) \neq 0$ we notice that s is an analytic function of (t, y). Thus we may differentiate both sides of $(*)$ with respect to t, thus obtaining

$$2q(e^{t \cdot \operatorname{ad} y}[y, x], e^{t \cdot \operatorname{ad} y} x) = s_t(t, y) q(x) \quad \text{for} \quad (t, x, y) \in \mathbb{R} \times L \times L.$$

If we abbreviate $s_t(0, y)$ by $s(y)$, then upon letting $t = 0$ we obtain

$$2q([y, x], x) = s(y) q(x) \quad \text{for} \quad (x, y) \in L \times L.$$

Taking again an x with $q(x) \neq 0$, and setting $y = x$ we see $s(x) = 0$ for $x \in L \setminus Z_q$. But since s is continuous and $L \setminus Z_q$ is dense in L, we conclused that s vanishes identically. But his means that $q([x, y], x) = 0$ for all $(x, y) \in L \times L$. Since, trivially, (4) implies (3), the lemma is proved. ∎

This lemma allows us to conclude that any Lie algebra containing an invariant Lorentzian cone is Lorentzian with respect to some invariant Lorentzian form, and indeed one that is associated with the given cone. Theorem II.6.14 then gives us a full classification of the possible Lorentzian Lie algebras (L, q) and thus of the possible Lorentzian invariant cones associated with q.

II.6.22. **Theorem.** (The Classification Theorem of Lorentzian Lie Semialgebras) *Let W be a Lorentzian Lie semialgebra in a finite dimensional Lie algebra L. Let q be any Lorentzian form associated with W. Then one of the four following cases occurs:*

   (i) *L is almost abelian and W is an arbitrary Lorentzian cone in L.*

   (ii) *L is a compact not semisimple Lie algebra and is an orthogonal direct sum $Z \oplus K$ with a 1-dimensional central ideal Z such that $q|(Z \times Z)$ is negative definite and a compact ideal K such that $q|(K \times K)$ is positive definite.*

   (iii) *(L, q) is an orthogonal direct sum of a standard solvable Lorentzian Lie algebra and a compact ideal on which q is positive definite.*

   (iv) *$L \cong (\mathrm{sl}(2, \mathbb{R}), B) \oplus (K, q)$, where B is a scalar multiple of the Cartan–Killing form, and (K, q) is a compact ideal on which q is positive definite.*

Proof. By Theorem II.6.19 we have either case (i), or else W is invariant. In this case, by Lemma II.6.21 above, q is invariant. Thus (L, q) is a Lorentzian Lie algebra, and Theorem II.6.14 applies and proves the remainder. ∎

Let us summarize what we have achieved in this section: We have certainly arrived at a complete classification of Lorentzian Lie semialgebras. But on our way to get there we have accumulated considerable information applying to much more general Lie semialgebras. In the context of Lie algebras allowing a non-degenerate invariant quadratic form (so for instance all semisimple algebras) we have seen that roundness hypotheses on the semialgebra forces the algebra to be Lorentzian, in which case at least the algebras can be completely characterized while the information on the semialgebras involved, even though rather substantial, remains incomplete in the case that L is a direct sum of sl(2) and a non-zero compact ideal. We have seen in general that in the class of Lie algebras allowing an invariant quadratic form, Lie semialgebras seem to have an urge to be invariant in the sense that a tangent hyperplane T either satisfies $[T \cap W, L] \subseteq T$ or else must be a subalgebra; and subalgebras of codimension 1 are rare. What is still missing after this and the preceding section is an understanding of non-invariant Lie semialgebras in semisimple Lie algebras which do not contain sl(2, $\mathbb{R}$) as a factor. We shall have a virtually conclusive theory of the invariant ones in Chapter III.

7. Lie algebras with Lie semialgebras

The presence of generating Lie semialgebras—notably of reduced ones—restricts the structure of the surrounding Lie algebra. Conversely, restrictions on the surrounding Lie algebra restrict the possible generating Lie semialgebras contained in it. We have observed much of this interplay in the previous sections. In the previous sections we have accumulated tools for the investigation of the structure of both the Lie algebra and the Lie semialgebra contained in it; in this section we shall apply these tools in order to accumulate more information on the structure theory of the pairs (L, W) consisting of a finite dimensional real Lie algebra L and a generating Lie semialgebra W which, in most instances, we shall assume to be reduced with no loss to the generality of our statements.

One of the guiding thoughts in this section is the idea to force triviality of a Lie semialgebra in the sense of Definition II.4.11. We recall: A wedge W in a Lie algebra L is called *trivial* if it contains the commutator algebra $L' \overset{\text{def}}{=} [L, L]$. Every such wedge is invariant. Trivial wedges therefore extend our hierarchy of wedges of interest in a Lie algebra as follows:

$$\text{trivial} \Rightarrow \text{invariant} \Rightarrow \text{semialgebra} \Rightarrow \text{Lie wedge}$$

We shall call *invariance theorem* any result which from suitable hypotheses allows the conclusion that a Lie semialgebra is invariant, whereas we shall call *triviality theorem* any proposition which concludes the triviality of a Lie semialgebra. We have seen examples of both types of results, and some of these earlier examples we shall repeat in this section for a more complete listing in one place. We shall begin with invariance theorems, move on to triviality theorems, and then apply both for a number of structure theorems on (L, W).

Invariance Theorems

We recall from Section II.5, that the center $Z(L)$ of a Lie algebra L is the base root space $M_0(L)$ for the base root 0 (see Remark II.5.3).

II.7.1.   Theorem.   (The First Invariance Theorem) *If $Z(L)$ meets the interior of a Lie semialgebra W of L, then W is invariant.*

Proof.   By Lemma II.5.13(iv), if $M_0(L) \cap \operatorname{int} W \neq \emptyset$, then $C_0^1(W) = C^1(W)$. This means that the characteristic function λ vanishes on all of $C^1(W)$, and thus W is invariant by Proposition II.2.37.   ∎

A reformulation of the Second Cartan Algebra Theorem II.5.30 with a slightly modified emphasis yields directly the following result:

II.7.2.   Theorem. (The Second Invariance Theorem) *Let W be a generating Lie semialgebra in the Lie algebra L which is not semisimple and has rank at least two. If W is reduced and its characteristic function is the restriction of a base root—in particular if W is weakly round—then W is invariant.*

Proof.   Firstly we note that a weakly round Lie semialgebra is pointed, hence reduced. Then by Theorem II.5.22, its characteristic function is the restriction of a base root.

If, however, W is reduced and its characteristic function is the restriction of a base root, then the theorem is just a reformulation of the Second Cartan Algebra Theorem II.5.30.   ∎

In II.6.5 we showed the following invariance theorem:

II.7.3.   Proposition. (The Third Invariance Theorem) *Suppose that L is the direct sum of two ideals of which the first one is either zero or a standard Lorentzian solvable Lie algebra and of which the second is compact. Then any generating Lie semialgebra in L is invariant.*

In particular, every generating Lie semialgebra in a compact Lie algebra is invariant.   ∎

II.7.4.   Theorem. (The Fourth Invariance Theorem) *Suppose that L is a Lie algebra without any hyperplane subalgebra. In particular, this is true for any semisimple Lie algebra without an $\mathrm{sl}(2, \mathbb{R})$-factor. Then any generating Lie semialgebra is invariant.*

Proof.   This follows directly from the Tangent Hyperplane Subalgebra Theorem II.5.41 in view of Proposition II.2.37.   ∎

Triviality theorems

For the sake of completeness we record once more the important First Trivality or Nilpotency Theorem II.4.13:

II.7.5.   Theorem. (The First Triviality Theorem) *Every generating Lie semialgebra in a nilpotent Lie algebra is trivial.*   ∎

The second triviality theorem uses the Nilpotency Theorem but is entirely different in its general character. It will yield another relatively wide class of real Lie algebras in which any generating Lie semialgebra must be trivial.

II.7.6. **Theorem.** (The Second Triviality Theorem) *Let L be a finite dimensional real Lie algebra which is the underlying Lie algebra of a complex Lie algebra. Then every Lie semialgebra in L is trivial.*

Proof. The proof proceeds in two steps.

Step 1: The claim is true for any complex almost abelian Lie algebra, that is any complex Lie algebra L containing a complex abelian hyperplane ideal J such that there is a vector $u \in L \setminus J$ and a complex number γ such that $[u, x] = \gamma \cdot x$ for all $x \in J$. Step 2: The claim is true for any complex Lie algebra L.

Let us first assume that we have the information in step 1 and prove step 2. For this purpose let W be a generating Lie semialgebra of L; we must show that $L' \subseteq W$. For this purpose it is no restriction of generality to assume that W is reduced. (See Proposition II.4.12.) Now we have to show that L is abelian. By the pointing procedure II.4.3 this claim will be established if it can be established for pointed generating Lie semialgebras. We will therefore assume now that W is pointed and generating. We select a regular element $x \in \operatorname{int} W$. Then the nilspace H of $\operatorname{ad} x$ is a Cartan algebra. For any complex linear functional α of H we write

$$(1) \qquad L^\alpha = \{y \in L : (\exists n \in \mathbb{N})\, \big(\operatorname{ad} h - \alpha(h) \cdot \mathbf{1}\big)^n y = 0\}.$$

Now we use the hypothesis that L is a complex Lie algebra in order to conclude

$$(2) \qquad L = H \oplus \bigoplus_{\alpha \in \Omega} L^\alpha,$$

where Ω is the set of *roots* of H, that is, the set of linear functionals $\alpha \in \widehat{L}$ for which $L^\alpha \neq \{0\}$. We record that for all functionals $\alpha, \beta \in \widehat{L}$ we have

$$(3) \qquad [L^\alpha, L^\beta] \subseteq L^{\alpha+\beta}.$$

Our standard reference for these matters is [Bou75]. From the first Cartan Algebra Theorem II.4.17 we know that H is abelian, and we have to show that $L = H$, in other words, that there are no non-zero roots. So by way of contradiction let us assume that α is a non-zero root. Then there is a natural number $m \in \mathbb{N}$ such that $m\alpha$ is a root while no $n\alpha$ is a root for $n > m$. Then $L^{2m\alpha} = \{0\}$, and from (3) above we know

$$(4) \qquad [\mathbb{C} \cdot x, L^\alpha] \subseteq L^\alpha \quad \text{and} \quad [L^\alpha, L^\alpha] = \{0\}.$$

Thus $A \stackrel{\text{def}}{=} \mathbb{C} \cdot x \oplus L^\alpha$ is a subalgebra, and, in view of Theorem II.5.10, it is in fact a complex almost abelian Lie algebra (with $J = L^\alpha, u = x$, and $\gamma = m\alpha(x)$. Moreover, since $x \in \operatorname{int} W$, the wedge $W \cap A$ is generating and pointed in A, and so the standard testing device allows us to apply step 1 to A and conclude that A is abelian. This means $m\alpha(x) = 0$, that is, $\alpha(x) = 0$. Since x is regular, this implies $\alpha = 0$ which is contrary to our assumption that α is a non-zero root. The proof of step 2 is therefore complete.

It remains for us to prove step 1. We assume that $L = \mathbb{C} \cdot u \oplus J$ is an almost abelian complex Lie algebra with

$$(5) \qquad [u, x] = \gamma \cdot x \quad \text{for all} \quad x \in J,$$

for some $u \notin J$, and that W is a generating Lie semialgebra. If L is abelian, there is nothing to prove, hence we may assume that L is non-abelian. This means $\gamma \neq 0$. We have to show that $L' = J$ is in W. Since W is generating, hence not contained in J, we may assume that u is in the interior of W, as u may be replaced by any vector in $u + J$. Now $B \overset{\text{def}}{=} \mathbb{C} \cdot u \oplus \mathbb{C} \cdot v$ is a subalgebra for any $0 \neq v \in J$, and $W \cap B$ is a generating Lie semialgebra of B. It suffices to show that $v \in W$. Thus it remains to show that *any Lie semialgebra in the real Lie algebra underlying the complex 2-dimensional non-abelian Lie algebra is trivial*. Since $u \in \operatorname{int} W$ we find a non-real complex number c close by 1 such that $c \cdot u \in \operatorname{int} W$. We apply the standard testing device II.4.6 with J as the complex 1-dimensional commutator algebra and $x = c \cdot u$. Then $A = \mathbb{R} \cdot x + J$ is a real 3-dimensional Lie algebra with a generating Lie semialgebra $W \cap A$. Since c is not real, we have case (i), and Lemma II.4.6 now implies that $J \subseteq W$ which we had to show.   ∎

We should observe that the second triviality theorem implies for instance that the real 6-dimensional Lie algebra $\operatorname{sl}(2, \mathbb{C})$ does not contain any proper Lie semialgebras while the Lie algebra $\operatorname{sl}(2, \mathbb{R})$ is full of them.

In the first and second triviality theorem we considered Lie semialgebras in general and added hypotheses on the Lie algebra to conclude triviality. In the following triviality results we generally assume that the wedge in question is invariant; we wish to explore which additional hypotheses force triviality.

II.7.7.   Lemma.   *Let L a Lie algebra such that $L' = [x, L]$ for all $x \in L$ outside some hyperplane H containing L'. Then every generating invariant wedge W is trivial.*

Proof.   For all $x \in C^1(W) \setminus H$ we note $[L, L] = [x, L] \subseteq T_x$, and by hypothesis we have $L' \subseteq H$. Now Proposition I.3.13 shows that $L' \subseteq H(W)$.   ∎

Let us remark in passing that a hyperplane H is a Lie algebra L is an ideal if and only if it contains L'.

II.7.8.   Theorem.   (The Third Triviality Theorem) *Let W be an invariant generating wedge in a solvable Lie algebra whose nilradical we denote with N. If $N \cap \operatorname{int} W \neq \varnothing$, then W is trivial.*

Proof.   By Proposition II.4.12, the wedge W is trivial if and only if its reduction W/I_W in L/I_W is trivial. Note that $(N + I_W)/I_W$ is contained in the nilradical of L/I_W. Thus if N meets $\operatorname{int} W$, then the nilradical of L/I_W meets $\operatorname{int}(W/I_W)$. Hence we may assume that W is reduced. The commutator algebra $[N, N]$ is a characteristic ideal of N, hence an ideal of L. Since $N \cap \operatorname{int} W \neq \varnothing$, the Lie semialgebra $N \cap W$ is generating in N, hence is trivial by the First Triviality Theorem II.7.5. Thus $N' \subseteq W$, whence $N' = \{0\}$ as W is reduced. Thus we know that N is abelian. As L' is contained in N, we have to show that $L = N$. Using now the invariance of W, from Corollary II.5.21(ii), we infer that $M(L) = Z(L)$. By Theorem II.5.10, we have $N \subseteq M(L)$. Hence $L' \subseteq Z(L)$, whence L is nilpotent, hence equal to N.   ∎

Such simple examples as any non-abelian almost abelian Lie algebra L with an arbitrary generating wedge whose interior meets $L' = N$ show that the previous triviality theorem does not hold for Lie semialgebras in place of invariant wedges.

II.7.9. **Corollary.** *Let W be an invariant generating wedge in a Lie algebra L. Let R denote the radical and N the nilradical of L. If $N \cap \operatorname{int} W \neq \emptyset$, then $[L, R] \subseteq W$.*

Proof. We apply the standard testing device. Let x be an arbitrary element of L. Then $\mathbb{R} \cdot x + R$ is a solvable subalgebra A of L. The nilradical N_A of A contains N. Hence N_A meets the interior of $W \cap A$. Thus the Third Triviality Theorem II.7.8 applies to A and shows that $W \cap A$ is trivial, that is, that $[A, A] \subseteq W$. But $[A, A] = [x, R] + [R, R]$. In particular, $[x, R] \subseteq W$, and since x was arbitrary, $[L, R] \subseteq W$ follows. ∎

Recall that a Lie algebra is *reductive* if it is a direct sum of an abelian and of a semisimple ideal.

II.7.10. **Corollary.** *If a Lie algebra contains a generating invariant pointed cone whose interior meets the nilradical, then it is reductive.*

Proof. From Corollary II.7.9 above we have $[L, R] \subseteq H(W) = \{0\}$. Hence R is central, which we had to show. ∎

More can be said under these circumstances. In fact, it turns out that L is compact. In order to establish this additional information, we have to wait until we have more information on invariant cones which we shall provide in III.2.3 ff.

The previous triviality results dealt with invariant wedges whose interior meets the nilradical. The next result deals with the case that a generating invariant wedge misses the commutator algebra completely. This case is in many respects opposite to the previous ones.

II.7.11. **Theorem.** (The Fourth Triviality Theorem) *Let L be a solvable Lie algebra and W a generating invariant wedge in L with $W \cap L' = H(W) \cap L'$. Then W is trivial.*

Proof. Since W is invariant, $H(W)$ is an ideal, and

$$\bigl(L/H(W)\bigr)' = \bigl(L' + H(W)\bigr)/H(W).$$

Then $W/H(W) \cap \bigl(L/H(W)\bigr)' = \bigl(W \cap L' + H(W)\bigr)/H(W)$. If $w = x + h \in W$ with $x \in L'$ and $h \in H(W)$, then $w - h = x \in W \cap L' = H(W) \cap L'$ by hypothesis, and thus $w \in H(W) + h = H(W)$. Hence we have $W \cap L' + H(W) = H(W)$. Upon passing to the factor algebra, we may therefore assume that W is an invariant generated pointed cone with $W \cap L' = \{0\}$. We have to show that L is abelian.

We prove this assertion by induction with respect to the dimension of L. For this purpose we assume that L is a counterexample to the claim with minimal dimension. We claim that $L'' = \{0\}$, that is, that L' is abelian. By Proposition I.2.32, the wedge $W + L''$ is closed and invariant, since $W \cap L'' = \{0\}$. The intersection of the pointed invariant generating cone $(W + L'')/L''$ with L'/L'' is trivial: Indeed, if $x = w + y \in L'$ with $w \in W$ and $y \in L''$, then $w = x - y \in L' \cap W = \{0\}$, and thus $x = y \in L''$, whence $L' \cap (W + L'') = L''$. Thus, if L'' were non-zero, then L/L'' could not be a counterexample to the theorem by the minimality of $\dim L$. Hence L/L'' would be abelian. This would mean

$L' = L'' \neq \{0\}$ which is incompatible with the assumption that L be solvable. This contradiction proves the claim that L' is abelian.

Now let x be any element in $\operatorname{int} W$ and consider the algebra $A = \mathbb{R}{\cdot}x + L'$. Then A and the invariant generating pointed cone $W \cap A$ satisfy the hypotheses of the theorem. If A is different from L, then A is abelian by the minimality of the counterexample L. This means $[x, L'] = \{0\}$. If this holds for all $x \in \operatorname{int} W$, then $\operatorname{int} W$ centralizes L', and since W is generating, all of L centralizes L'. Hence L is nilpotent, and thus by the First Triviality Theorem II.7.5, L is abelian and thus cannot be a counterexample to the theorem. Hence there is at least one $x \in \operatorname{int} W$ such that $L = \mathbb{R}{\cdot}x + L'$. Since L is not abelian, L' cannot be central. Moreover, L is special metabelian with $L' = M(L)$. Since W is pointed, the invariance of W entails $M(L) = Z(L)$ by Corollary II.5.21(ii). This contradiction proves the theorem. ∎

We summarize the information derived in Theorem II.7.8 and Theorem II.7.11:

II.7.12.   Corollary.   *Let W be a generating invariant wedge in a solvable Lie algebra. Then W is trivial if at least one of the following conditions is satisfied:*

(i) *The nilradical of L meets W in its interior.*

(ii) *The commutator algebra of L meets W in its edge.*

For pointed cones W, condition (ii) *is satisfied if $W \cap L' = \{0\}$.* ∎

We should bear in mind the example of the standard Lorentzian solvable Lie algebras and their invariant Lorentzian cones as an illustration of how sharp these results are. These cones fail to be trivial by a long shot. Also the special metabelian Lie algebras and their various Lie semialgebras illustrate the point, that without invariance or the presence of a center, non-trivial Lie semialgebras occur copiously even in relatively uncomplicated solvable Lie algebras.

II.7.13.   Corollary.   *Let L be a solvable Lie algebra and W a generating invariant wedge different from L. Then W has a support hyperplane which is an ideal.*

Proof.   Every support hyperplane contains $H(W)$. Thus, if W is trivial, every support hyperplane contains L', hence is an ideal. Suppose now that W is not trivial. Then, by Corollary II.7.12, there is an $x \in W \setminus H(W)$ with $x \in L'$ and $L' \cap \operatorname{int} W = \varnothing$. By the Theorem of Hahn and Banach, this implies the existence of an $\omega \in L'^{\perp} \cap W^*$. In particular $\langle \omega \mid x \rangle = 0$. Then $T \overset{\text{def}}{=} \omega^{\perp}$ is a support hyperplane of W at x containing L'. Hence it is an ideal. ∎

Lie semialgebras forcing structure theorems

The hypothesis that the interior of a Lie semialgebra meets the nilradical of the algebra has strong consequences for the structure of the algebra. We need a lemma:

II.7.14.   Lemma. *Let N be an ideal of L containing L' and suppose that the Cartan algebras of L are abelian. Then L is a semidirect sum of N and an (abelian) subalgebra V.*

Proof.   Let H be a Cartan algebra of L. Since $(H+N)/N$ is a Cartan algebra of L/N and L/N is abelian, it follows that $N + H = L$. Let V be a vector space complement for $N \cap H$ in H. Then L is the direct sum of the vector spaces N and V. But N is an ideal, and V is a subalgebra, since H is abelian.   ∎

II.7.15.   Lemma. *Let L be a solvable Lie algebra and W a reduced generating Lie semialgebra such that the nilradical N meets $\operatorname{int} W$. Then N is abelian and L is the semidirect sum of N and an abelian algebra.*

Proof.   By the First Cartan Algebra Theorem II.4.16, any Cartan algebra H is abelian. The First Triviality Theorem II.7.5 and the fact that W is reduced imply that N is abelian. The lemma then follows from Lemma II.7.14 above.   ∎

II.7.16.   Theorem. *Let L be a Lie algebra with nilradical N and with a reduced Lie semialgebra whose interior meets N. Then the following conclusions hold:*

(i) *N is abelian.*

(ii) *The radical R is the semidirect sum of N and an abelian subalgebra V.*

(iii) *L is the direct sum of R and a semisimple ideal S.*

(iv) *As direct sum of vector spaces, $L = N \oplus V \oplus S$.*

Proof.   Claims (i) and (ii) follow from Lemma II.7.15 above via the standard testing device. Since N is abelian, the representation $\rho\colon L \to \operatorname{gl}(N)$ given by $\rho(x) = \operatorname{ad} x|N$ is diagonalizable by Theorem II.4.24. The kernel K of ρ is the centralizer of N, and by the definition of the nilradical, $K \cap R = N$. It was observed in Remark II.4.25, that $L' \subseteq K$. Let S be any Levi complement for R in L. Then $S \subseteq K$ and thus $[N, S] = \{0\}$, and we conclude $K = N \oplus S$. But then S is a characteristic ideal of the reductive ideal K and hence is an ideal in L. Thus $L = R \oplus S$, a direct sum of ideals. This proves (iii). Conclusion (iv) is an immediate consequence of the preceding results.   ∎

Problems for Chapter II

PII.1.   Problem.   (a) *Classify all Lie semialgebras.*

     (b) *Classify all Lie semialgebras in Lorentzian Lie algebras (in solvable, semisimple, other types of special) Lie algebras.*

PII.2.   Problem.   *Describe all basic metabelian Lie algebras and all Lie semialgebras they might contain.*   ∎

PII.3.   **Problem.**   *Give a very explicit description of all hyperplane subalgebras and of all half-space semialgebras in a finite dimensional Lie algebra. Cf.* II.2.24, II.5.41, IV.1.34, VI.5.2. ∎

PII.4.   **Problem.**   (a) *Are homomorphic images of Lie semialgebras (or the closures of these images) Lie semialgebras?*

(b) *Suppose that W is a Lie semialgebra and I an ideal in a finite dimensional Lie algebra. Is $\overline{W + I}$ a Lie semialgebra?*

PII.5.   **Problem.**   *What are the infinite dimensional versions of* Theorems II.1.11,14, *and* II.2.14 *—if they exist?* ∎

PII.6.   **Problem.**   *Let W be a reduced generating Lie semialgebra in a Lie algebra L. Set $C_+^1(W) = \{x \in C^1(W) : \lambda(x) \neq 0\}$ and $C_0^1(W) = \{x \in C^1(W) : \lambda(x) = 0\}$ and define $W_+ = \bigcap_{x \in C_+^1(W)} L_x(W)$ and $W_0 = \bigcap_{x \in C_0^1(W)} L_x(W)$. How do W_+ and W_0 relate to W? Compute the characteristic functions of W_+ and W_0. Cf.* II.2.37, II.6.6. ∎

PII.7.   **Problem.**   *Suppose that L is a finite dimensional Lie algebra and N a nilpotent ideal. Does the existence of a reduced generating Lie semialgebra in L imply $[N,N] \subseteq Z(L)$? (Cf.* II.5.29.) ∎

PII.8.   **Problem.**   *Let $M \subseteq L$. Give a good internal description of the smallest Lie wedge in L containing M.* ∎

Notes for Chapter II

Section 1.   The fact that the set of subtangent vectors of a semigroup at the identity of a Lie group is a wedge has been noted repeatedly (see e.g. Loewner [Loe64], Brockett [Bro73]), and in some sense, dates back to Sophus Lie. The result in Proposition II.1.2 that this wedge in fact satisfies condition (LW) was independently observed by Ol'shanskiĭ [Ol81] and Hofmann and Lawson [HL81]. The Theorem on Wedges in Lie algebras II.1.11 can be found in Hilgert and Hofmann [HH86c]. This applies also to the Characterization Theorem of Wedges II.1.14, while the results in Theorems II.1.12 and II.1.13 are formulated here for the first time. The first treatment of invariant wedges in Lie algebras in print was given by Vinberg [Vi80].

Section 2.   All results up to the topic of almost abelian Lie algebras (Definition II.2.29) are from [HH85c]. Almost abelian Lie algebras and their Lie semialgebras were discussed by Hofmann and Lawson in [HL81] and by Graham and deVun in [GdeV88]. The concept of the characteristic function of a Lie semialgebra was introduced by Hilgert and Hofmann in [HH85d]. The analytic extension aspects of Lie semialgebras leading to the Closure Theorem for Lie Semialgebras II.2.40 are new. This material was used in [Hi86c].

Section 3.   The results in the First Classification Theorem of Low Dimensional Semialgebras II.3.4 are due to Hilgert and Hofmann [HH85c] and Lawson and Ruppert (Seminar Notes 1986), see also [Law87a]. The Second Classification Theorem of Low Dimensional Semialgebras II.3.7 is due to

Hilgert and Hofmann [HH85b] and [HH85c] (except for some additions). Lorentzian Semialgebras (see Sections II.3.9 through II.3.16) were discussed by Hilgert and Hofmann in [HH85d] and by Levichev in [Le86]. For related material see also Medina and Revoy [MR83], [MR84]. The material leading up to Theorem II.3.20 and in the remainder of the section is new. An alternate proof of II.3.20 appears in [Law87b].

Section 4 and 5. The contents of Sections 4 and 5 are new. The Tangent Hyperplane Theorem II.5.4 was first presented in unpublished notes by Lawson and is published here for the first time.

Section 6. The results of this section generalize and reorganize the results of Lorentzian Lie semialgebras by Hilgert and Hofmann [HH85d], Levichev [Le86] and [Le87], and Lawson [Law87b].

Section 7. The First Triviality Theorem II.7.5 is due to Hofmann and Lawson [HL81] and the Second Triviality Theorem II.7.6 to Hilgert and Hofmann [HH85a]. The rest of this section is published here for the first time.

Chapter III

Invariant cones

In the hierarchy of wedges in a Lie algebra, invariant wedges are the most special (see Scholium II.2.15). It is, therefore, reasonable to expect the richest and most explicit theory for this class. We recall that a wedge W in a finite dimensional Lie algebra W is *invariant* according to Definition II.1.9 if and only if it satisfies

$$e^{\operatorname{ad} x}W = W \qquad \text{for all } x \in L.$$

The Characterization Theorem for Invariant Wedges II.1.14 gives a precise description of invariant wedges in terms of the geometry of W through the reference to tangent spaces T_x, C^1-points, and E^1-points of W, and in terms of the Lie algebra structure through the invoking of the Lie bracket. We have seen a broad range of sufficient conditions which force a Lie semialgebra to be invariant (see the Invariance Theorems II.7.1 through 4). By Proposition II.1.10 we know that we are not losing any generality if we restrict the study of invariant wedges to generating pointed invariant cones.

The purpose of this chapter is to develop a complete theory of invariant generating cones in a finite dimensional Lie algebra. A certain amount of the background theory will have to deal with cones in finite dimensional vector spaces which are invariant under certain groups of linear automorphisms. The first sections deal with this general situation. They culminate in a version of the theory of Perron and Frobenius suitable for cones and linear groups and semigroups. After that we shall turn to the question of invariant generating cones in Lie algebras and find that we need a considerable amount of Lie algebra and Lie group theory in order to deal with the issue at the level of generality which we aspire. Some of the proofs we provide in one of our appendices so that the treatment is self contained. Not everything we need appears to be available in the existing literature. In the final phases of the proof of the general characterization theorem for invariant cones, we must call on a variety of results from the theory of semisimple Lie algebras for which we refer to the appropriate texts.

1. The automorphism group of a wedge

In this section we fix a finite dimensional real vector space L and a wedge W in L. The groups and semigroups which we consider are all contained in the group $\mathrm{Gl}(L)$ of all vector space automorphisms of L. The Lie algebra $\mathbf{L}\big(\mathrm{Gl}(L)\big)$ of this group is $\mathrm{gl}(L)$, the Lie algebra of all vector space endomorphisms $X : L \to L$ with the Lie bracket $[X, Y] = XY - YX$. The exponential function is the one given by the exponential series

$$\exp : \mathrm{gl}(L) \to \mathrm{Gl}(L), \qquad \exp X = \sum_{n=0}^{\infty} \frac{1}{n!} X^n.$$

For any subset S of $\mathrm{Gl}(L)$, we shall set for the purposes of this section

$$\mathbf{L}(S) = \{X \in \mathrm{gl}(L) : \exp t{\cdot}X \in S \quad \text{for all } t \geq 0\}.$$

If S is a closed subgroup, then $\mathbf{L}(S)$ is precisely its Lie algebra and its exponential function is the restriction of $\exp$ to $\mathbf{L}(S)$. If S is a closed semigroup, we let B denote a Campbell–Hausdorff neighborhood of $\mathrm{gl}(L)$ which is mapped diffeomorphically onto an open neighborhood U of $\mathbf{1}$ via $\exp|B : B \to U$. Then $(\exp|B)^{-1}(U \cap S)$ is a closed local semigroup in $\mathrm{gl}(L)$ with respect to B. (See the beginning of Section 1 of Chapter II.) In view of Lemma II.1.1 we know that $\mathbf{L}(S) = L_0\big((\exp|B)^{-1}S\big)$. This remark is relevant because it allows us to conclude that $\mathbf{L}(S)$ is a Lie wedge if S is a semigroup.

III.1.1. Definition.   (i) We define

$$\mathrm{Aut}(W, L) = \{g \in \mathrm{Gl}(L) : gW = W\}$$

and call this group *the automorphism group of the pair* (L, W). If W is generating, we shall simply write $\mathrm{Aut}(W)$ instead of $\mathrm{Aut}(W, L)$ and call $\mathrm{Aut}(W)$ *the automorphism group of the wedge* W.
   (ii) We define

$$\mathrm{End}(W, L) = \{g \in Gl(L) : gW \subseteq W\}$$

and call this semigroup *the semigroup of regular endomorphisms of the pair* (L, W). If W is generating, we write $\mathrm{End}(W)$ instead of $\mathrm{End}(W, L)$ and call $\mathrm{End}(W)$ the *regular endomorphism semigroup of* W.

Clearly, $\text{Aut}(W, L)$ is a closed subgroup of $\text{Gl}(L)$ and is, therefore, a Lie group with a well defined Lie subalgebra $\mathbf{L}\big(\text{Aut}(W, L)\big)$ of $\text{gl}(L)$. One of the tasks of this section is the explicit description of this Lie algebra.

We now recall from the material in Proposition I.2.9 through Corollary I.2.14 how a wedge W is determined by its edge $H(W) = W \cap -W$ and its associated pointed wedge $W/H(W)$. We should like to discuss now how these invariants of a wedge are reflected in its automorphism group and its regular endomorphism semigroup. First we note that every endomorphism g of the vector space L with $gW \subseteq W$ preserves the edge, whence $g|H(W) : H(W) \to H(W)$ is a well defined endomorphism of the vector space $H(W)$. Accordingly, there is a well defined endomorphism $\bar{g} : L/H(W) \to L/H(W)$ given by $\bar{g}(x + H(W)) = g(x) + H(W)$.

III.1.2. **Definition.** For a wedge W in a finite dimensional vector space L we define

$$\alpha \colon \text{End}(W, L) \to \text{Aut}\, H(W) \times \text{End}\big(W/H(W), L/H(W)\big), \quad \alpha(g) = (g|H(W), \bar{g}).$$

By a slight abuse of language we shall denote the restriction and corestriction

$$\alpha \colon \text{Aut}(W, L) \to \text{Aut}\, H(W) \times \text{Aut}\big(W/H(W), L/H(W)\big)$$

with the same letter α.

Clearly, α is a morphism of semigroups (respectively, groups). We have $\alpha(g) = 1$ if and only if g fixes each element of $H(W)$ and satisfies $g(x) - x \in H(W)$ for all $x \in L$. We shall call these vector space automorphisms of L *transvections along* $H(W)$ and denote their totality with $T(H(W), L) \overset{\text{def}}{=} \ker \alpha$. With this notation we have the following decomposition statements:

III.1.3. **Proposition.** *For any wedge W in a finite dimensional vector space L there are splitting exact sequences*

$$1 \to T(H(W), L) \overset{\text{inc}}{\to} \text{Aut}(W, L) \overset{\alpha}{\to} \text{Aut}\, H(W) \times \text{Aut}\big(W/H(W), L/H(W)\big) \to 1,$$

$$1 \to T(H(W), L) \overset{\text{inc}}{\to} \text{End}(W, L) \overset{\alpha}{\to} \text{Aut}\, H(W) \times \text{End}\big(W/H(W), L/H(W)\big) \to 1.$$

In particular, the automorphism group of the pair (L, W) is the semidirect product of the normal subgroup of transvections along $H(W)$ with a subgroup which is isomorphic to the direct product of $\text{Gl}\big(H(W)\big)$ with the automorphism of the pair $\big(W/H(W), L/H(W)\big)$ which represents the associated pointed cone of W.

Proof. We decompose L in the form $L = H(W) \oplus V$ with a vector space complement V and we apply Proposition I.2.12 in order to represent W in the form $W = H(W) \oplus (W \cap V)$. We set

$$S = \{g \in \text{End}(W, L) : gV = V\}.$$

Then S is a closed subsemigroup of $\operatorname{End}(W, L)$, and $G = S \cap \operatorname{Aut}(W, L)$ is a closed subgroup, namely, the group of invertible elements of S. The restriction

$$\alpha|S : S \to \operatorname{Aut} H(W) \times \operatorname{End}\big(W/H(W), L/H(W)\big)$$

is a isomorphism of topological semigroups. For if $p\colon V \to L/H$ is the isomorphism obtained by restriction the quotient morphism, then p maps the pair $(V \cap W, V)$ isomorphically onto the pair $\big(W/H(W), L/H(W)\big)$ according to Corollary I.2.13. The inverse of α is then given by

$$\alpha^{-1}(f, \varphi)(h \oplus v) = f(h) + p^{-1}\varphi p(v)$$

where $f \in \operatorname{Aut}(H(W), \varphi \in \operatorname{Aut} L/H(W), h \in H(W)$, and $v \in V$. Thus $(\alpha|S)^{-1}$ followed by the inclusion of S into $\operatorname{End}(W, L)$ is a splitting right inverse for α in the first exact sequence.

Quite analogously,

$$\alpha|G : G \to \operatorname{Aut} H(W) \times \operatorname{Aut}\big(W/H(W), L/H(W)\big)$$

is an isomorphism providing the required splitting of the second exact sequence. ∎

This proposition allows us to decompose the groups and semigroups of automorphisms or regular endomorphisms of a wedge in such a fashion that there will be no real loss in generality if we restrict our attention to pointed cones.

In the breakdown of the automorphism group of a wedge into its constituents, the groups $\operatorname{Aut} H(W) = \operatorname{Gl}\big(H(W)\big)$ and $T\big(H(W), L\big)$ belong to standard linear algebra, while $\operatorname{Aut}\big(W/H(W), L/H(W)\big)$ is the automorphism group of a *pointed* cone. However, the automorphism group of a pointed cone breaks down even further.

III.1.4.   Definition.   We shall call an element $g \in \operatorname{End}(W, L)$ *special*, if $\det g = 1$, and we denote $\operatorname{SEnd}(W, L)$ the semigroup of all special regular endomorphisms of the pair (W, L). Accordingly we set $\operatorname{SAut}(W, L) = \operatorname{Aut}(W, L) \cap \operatorname{SEnd}(W, L)$. If W is generating, we again write $\operatorname{SEnd}(W)$ and $\operatorname{SAut}(W)$ without reference to L.

By the very definition of a wedge (see Definition I.1.1), each positive homothety $r \cdot 1_L$ with $r > 0$ is in $\operatorname{Aut}(W, L)$. Thus every $g \in \operatorname{End}(W, L)$ decomposes uniquely into a product

$$g = \big((\det g) \cdot 1_L\big)\big((\det g)^{-1} \cdot g\big)$$

of a homothety and a special linear transformation from $\operatorname{Sl}(L)$. Both factors are in $\operatorname{End}(W, L)$ if an only if $\det g > 0$. This situation is illustrated in the following example: Let $L = \mathbb{R}^2$, $W = (\mathbb{R}^+)^2$, and

$$g = \begin{pmatrix} 0 & 1 \\ 1 & 0 \end{pmatrix}.$$

Then $g \in \operatorname{Aut} W$, but $\det g = -1 < 0$. This automorphism cannot be decomposed *inside* $\operatorname{Aut} W$ into a product of a homothety and a special automorphism.

This suggests that we single out the subgroup $\operatorname{Gl}^+(L)$ of all $\operatorname{Gl}(L)$ of index 2 consisting of all automorphisms with positive determinant and write

$$\operatorname{End}^+(W, L) = \operatorname{End}(W, L) \cap \operatorname{Gl}^+(L), \text{ and } \operatorname{Aut}^+(W, L) = \operatorname{Aut}(W, L) \cap \operatorname{Gl}^+(L).$$

For the following straightforward proposition, which we record for easy reference, we recall $\mathbb{P} = \mathbb{R}^+ \setminus \{0\} = \operatorname{Aut}(\mathbb{R}^+, \mathbb{R})$.

III.1.5.   Proposition.   *The following product decompositions are direct:*

$$\text{End}^+(W, L) = (\mathbb{P} \cdot \mathbf{1}_L)\,\text{SEnd}(W, L) \cong \mathbb{P} \times \text{SEnd}(W, L),$$

$$\text{Aut}^+(W, L) = (\mathbb{P} \cdot \mathbf{1}_L)\,\text{SAut}(W, L) \cong \mathbb{P} \times \text{SAut}(W, L).$$

The index of $\text{Aut}^+(W, L)$ *in* $\text{Aut}(W, L)$ *is 1 or 2, and* $\text{Aut}^+(W, L)$ *is open in* $\text{Aut}(W, L)$, *while* $\text{End}^+(W, L)$ *is open in* $\text{End}(W, L)$ ∎

We recall that $\mathbf{L}\big(\text{Sl}(L)\big) = \text{sl}(L) = \{X \in \text{gl}(L) : \text{tr}\,X = 0\}$. As an immediate consequence of the preceding proposition and this remark, we record a conclusion on the tangent objects:

III.1.6.   Corollary.   *On the level of the Lie algebra* $\text{gl}(L)$ *the following conclusions hold:*

$$\mathbf{L}\big(\text{End}(W, L)\big) = \mathbf{L}\big(\text{End}^+(W, L)\big) \cong \mathbb{R} \oplus \mathbf{L}\big(\text{SEnd}(W, L)\big),$$

$$\mathbf{L}\big(\text{Aut}(W, L)\big) = \mathbf{L}\big(\text{Aut}^+(W, L)\big) \cong \mathbb{R} \oplus \mathbf{L}\big(\text{SAut}(W, L)\big).$$

Moreover,

$$\mathbf{L}\big(\text{SEnd}(W, L)\big) = \{X \in \mathbf{L}\big(\text{End}(W, L)\big) : \text{tr}\,X = 0\},$$

$$\mathbf{L}\big(\text{SAut}(W, L)\big) = \{X \in \mathbf{L}\big(\text{Aut}(W, L)\big) : \text{tr}\,X = 0\}.$$

∎

For the following remarks in which we attempt to illustrate the situation, let us assume that W is pointed and generating. How can we visualize the transformations of $\text{SAut}(W)$?

If ω is an arbitrary element of the interior of W^*, then $B = W \cap \omega^{-1}(1)$ is a compact base of W (see Proposition I.2.28 and Definition I.2.29). The point set $K = \{w \in W : 0 \leq \langle \omega, w \rangle \leq 1\}$ is a pyramid with base B and vertex 0. We fix a scalar product on L and thus assign a volume $M(K)$ to K and an $n - 1$-dimensional measure $m(B)$ to B where $n = \dim L$. If $d(\omega)$ denotes the distance of $\omega^{-1}(1)$ from the origin 0, we have

$$M(K) = \frac{1}{n}d(\omega)m(B).$$

Now every $g \in \text{SAut}(W)$ preserves volume, whence $d(\widehat{g}\omega)m(gB) = nM(gK) = nM(K) = d(\omega)m(B)$ or, equivalently,

$$d(\widehat{g}\omega) : d(\omega) = m(B) : m(gB).$$

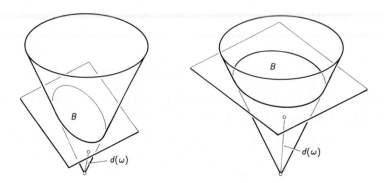

Figure 1

However, the transitivity of the automorphism group may be severely defective. There are three dimensional polyhedral cones in $\mathbb{R}^3$ for which $\mathrm{SAut}(W) = \{\mathbf{1}_L\}$. The most homogeneous cones are the Lorentzian ones which we have already treated extensively in the context of Lie semialgebras. We list a few examples. The verification of the details we leave as an exercise.

1.   Example. For each $n \geq 3$ there are generating wedges W in $L = \mathbb{R}^n$ with $\mathrm{SAut}\, W = \{\mathbf{1}_L\}$. One may chose W polyhedral.

2.   Example. For each $n \in \mathbb{N}$ we let $L = \mathbb{R}^n$ and $W = (\mathbb{R}^+)^n$. Then W is pointed and generating. The semigroup $\mathrm{End}\, W$ may be identified with the semigroup of all invertible $n \times n$ matrices with non-negative entries, while $\mathrm{Aut}\, W$ may be identified with the group of all monomial matrices with non-negative entries. The identity component $\mathrm{Aut}_0\, W$ thereby corresponds to the group of diagonal matrices with positive diagonal entries. Further, $\mathrm{Aut}\, W$ is the semidirect product of this group with the group of all $n \times n$ permutation matrices. (The latter, of course, is isomorphic to the symmetric group on n elements.)

3.   Example. For each $n \geq 2$, let $L = \mathbb{R}^n$ and let q denote the standard Lorentzian form given by $q(x) = x_1^2 + \cdots + x_{n-1}^2 - x_n^2$. Let $\mathrm{O}(n-1,1)$ denote the Lorentz group of all $g \in \mathrm{Gl}(n)$ leaving q invariant. If W is the associated Lorentzian cone given by $x_n \geq 0$ (see Definition I.4.8), and if $\mathrm{O}_0(n-1,1)$ denotes the identity component of the Lorentz group, then $\mathrm{SO}(n-1,1) \subseteq \mathrm{O}_0(n-1,1) = \mathrm{Aut}\, W$. The group $\mathrm{Aut}\, W$ operates transitively on the interior of W and doubly transitive on ∂W.

All endomorphisms of a pair (W, L) so far were assumed to be regular, that is *invertible* in $\mathrm{Hom}(L, L)$. But arbitrary, possibly singular endomorphisms also play a role, although no central one in this book.

III.1.7.   Definition. Let W be a wedge in a finite dimensional vector space L. We say that a vector space endomorphism $g \in \mathrm{Hom}(L, L)$ is an *endomorphism of the pair* (W, L) if $gW \subseteq W$. The set of all those endomorphisms will be denoted $\mathrm{Hom}_L\, W$.

We denote with $M_n(\mathbb{R})$ the algebra of all real $n \times n$-matrices. Recall that $\mathrm{Gl}(n)$ is an open subset of $M_n(\mathbb{R})$. The proof of the following proposition is then straightforward.

III.1.8.   Proposition. *For a wedge W in a finite dimensional vector space L the set $\mathrm{Hom}_L W$ is a multiplicative semigroup (under composition of maps) and a wedge in $\mathrm{Hom}(L, L) \cong M_n(\mathbb{R})$. The group of elements which are invertible in $\mathrm{Hom}_L W$ is $\mathrm{Aut}(W, L)$; the subsemigroup $\mathrm{End}(W, L)$ of endomorphisms invertible in $\mathrm{Hom}(L, L)$ is open in $\mathrm{Hom}_L W$.* ∎

We shall say that $\mathrm{Hom}_L W$ is an *associative semialgebra*. Associative semialgebras differ from associative algebras only insofar as addition need not be invertible and scalar multiplication is only defined for non-negative scalars.

The Lie algebra of the automorphism group of a wedge

We conclude the section by determining the Lie algebra of $\mathrm{Aut}(W, L)$. As we observed before, as a closed subgroup of the Lie group $\mathrm{Gl}(L)$, the group $\mathrm{Aut}(W, L)$ is a Lie group. But the situation here is different from many otherwise comparable ones. Lie groups are frequently defined as automorphism groups of a given mathematical structure. As a rule, their very definition allows an immediate determination of its Lie algebra. One of the most familiar examples is the group $\mathrm{O}(q) = \{g \in \mathrm{Gl}(L) : q(gx) = q(x) \text{ for all } x \in L\}$ of all automorphisms respecting a given quadratic form q on L. A routine differentiation process for the function $t \mapsto q(e^{t \cdot X} x)$ for $X \in \mathrm{gl}(L)$ shows that $\mathbf{L}\big(\mathrm{O}(q)\big) = \{X \in \mathrm{gl}(q) : q(Xx, y) + q(x, Xy) = 0\}$. Yet in the case of $\mathrm{Aut}(W, L)$ it is not a priori clear how $\mathbf{L}\big(\mathrm{Aut}(W, L)\big)$ should be characterized and how the structure of W itself ought to be expressed in such a characterization. However, the suitable tool has already been forged in the form of the Linear Invariance Theorem for Wedges I.5.27. All that is needed is a recasting of that theorem in the language of the present section. We must remember the set of subtangent vectors of a wedge at a point $x \in W$: By Proposition I.5.3 this set is $L_x = L_x(W) = \overline{W - \mathbb{R}^+ \cdot x}$.

III.1.9.   Theorem. *Let W be a wedge in a finite dimensional vector space L. Then, for an endomorphism X of L, the following statements are equivalent:*

(1) $X \in \mathbf{L}\big(\mathrm{End}(W, L)\big)$.

(2) $X(w) \in L_w$ *for all* $w \in W$.

(3) $X(c) \in L_c$ *for all* $c \in C^1(W)$.

(4) $X(e) \in L_e$ *for all* $e \in E^1(W)$.

(5) $\widehat{X}(\omega) \in L_\omega(W^*)$ *for all* $\omega \in W^*$.

Proof.   In view of Definition III.1.1, this is just a reformulation of Proposition I.5.25 and Theorem I.5.27. ∎

The proof of the following theorem is exactly the same:

III.1.10.   **Theorem.**   *Let W be a wedge in a finite dimensional vector space L. The for an endomorphism X of L, the following statements are equivalent:*

(1) $X \in \mathbf{L}\big(\mathrm{Aut}(W, L)\big)$.

(2) $X(w) \in T_w$ *for all* $w \in W$.

(3) $X(c) \in T_c$ *for all* $c \in C^1(W)$.

(4) $X(e) \in T_e$ *for all* $e \in E^1(W)$.

(5) $X(\omega) \in T_\omega$ *for all* $\omega \in W^*$.   ∎

The special case of a Lie algebra L

If, in particular, L is a finite dimensional Lie algebra, then we have inside the Lie group $\mathrm{Gl}(L)$ several analytic subgroups. Firstly, the group $\mathrm{Aut}\, L$ of all Lie algebra automorphisms is closed in $\mathrm{Gl}(L)$ and is therefore a Lie group. Its Lie algebra is $\mathrm{Der}\, L$, the Lie subalgebra in $\mathrm{gl}(L)$ of all derivations, that is, linear self-maps $D: L \to L$ which satisfy $D[x, y] = \big([Dx, y] + [x, Dy]\big)$ for all $x, y \in L$. Then $\mathrm{Aut}(W, L) \cap \mathrm{Aut}\, L$ is the Lie group of automorphisms of the pair (W, L) which, in addition, are Lie algebra automorphisms. Its Lie algebra is $\mathbf{L}\big(\mathrm{Aut}(W, L)\big) \cap \mathrm{Der}\, L$, that is, the Lie algebra of all derivations X of L satisfying the conditions of Theorem III.1.10 above.

More delicate is the situation of the so-called inner automorphisms of L. Firstly, the definition of an inner derivation is not problematic: We recall that a derivation D is called *inner* if there is an $x \in L$ such that $D = \mathrm{ad}\, x$, that is $Dy = [x, y]$. The set $\mathrm{ad}\, L$ of all inner derivations is an ideal of $\mathrm{Der}\, L$, and it is isomorphic to $L/Z(L)$ in view of the exact sequence

$$0 \to Z(L) \overset{\mathrm{inc}}{\to} L \overset{\mathrm{ad}}{\to} \mathrm{Der}\, L \overset{\mathrm{quot}}{\longrightarrow} \mathrm{out}\, L \to 0,$$

where the cokernel $\mathrm{out}\, L$ of ad is simply the factor algebra $(\mathrm{Der}\, L)/(\mathrm{ad}\, L)$, also called the *outer automorphism algebra*.

Now it is a fact from the foundations of Lie group theory, that every subalgebra A of the Lie algebra $\mathbf{L}(G)$ of a Lie group G generates a unique analytic subgroup H of G such that $\mathbf{L}(H) = \{X \in \mathbf{L}(G) : \exp \mathbb{R}.X \subseteq H\}$ is exactly A. Moreover, there is a unique Lie group topology on H making it into a Lie group $\dot{H}$ with $\mathbf{L}(\dot{H}) = A$ so that its exponential function is obtained from that of G by restriction: $\exp_{\dot{H}} : A \to \dot{H}$ is $\exp_G |A$. The following diagram illustrates the situation:

$$
\begin{array}{ccc}
A & \overset{\mathrm{inc}}{\longrightarrow} & \mathbf{L}(G) \\
{\scriptstyle \exp_{\dot{H}}} \downarrow & & \downarrow {\scriptstyle \exp_G} \\
\dot{H} & \underset{\mathrm{inc}}{\longrightarrow} & G.
\end{array}
$$

We apply this to the Lie group $G = \mathrm{Aut}\, L$ with the Lie algebra $\mathbf{L}(G) = \mathrm{Der}\, L$ and the subalgebra $A = \mathrm{ad}\, L$. The preceding reasoning provides us with an analytic subgroup $\operatorname{Inn} L$ of $\operatorname{Aut} L$ with Lie algebra $\mathbf{L}(\operatorname{Inn} L) = \operatorname{ad} L$ and with an internal Lie group topology relative to which it becomes a Lie group with exponential function

$$\operatorname{ad} x \mapsto e^{\operatorname{ad} x} : \operatorname{ad} L \longrightarrow (\operatorname{Inn} L)^{\cdot}.$$

III.1.11. Definition. The group $\operatorname{Inn} L$ in the automorphism group of a Lie algebra L is called the group of inner automorphisms, and its members are inner automorphisms. The closure in $\operatorname{Aut} L$ of $\operatorname{Inn} L$ will be written $\operatorname{INN} L$.

Unfortunately, the group of inner automorphisms need not be closed in $\operatorname{Aut} L$, which is itself a closed subgroup of $\operatorname{Gl}(L)$. As every analytic subgroup H of a Lie group G is generated by $\exp \mathbf{L}(H)$, the group of inner automorphisms is generated by $e^{\operatorname{ad} L}$. Thus every inner automorphism is a finite product of elements of the form $e^{\operatorname{ad} x}$. The elements of $\operatorname{INN} L$ are limits in $\operatorname{Hom}(L, L) \cong M_n(\mathbb{R})$, $n = \dim L$, of sequences of inner automorphisms. We shall describe in the Appendix an example of a 4-dimensional metabelian Lie algebra L for which the analytic group $\operatorname{Inn} L$ is not closed.

With this terminology, we can rephrase the definition of invariance of a wedge in a Lie algebra as given in Definition II.1.9. Firstly, the Lie algebra $\mathcal{L}(W, L)$ which we found in II.1.4 is none other than

$$\mathcal{L}(W, L) = (\operatorname{ad})^{-1} \mathbf{L}\big(\operatorname{Aut}(W, L)\big).$$

The invariance condition Definition I.1.9 is rephrased as follows:

III.1.12. Remark. A wedge in a Lie algebra L is invariant if and only if it is invariant under all inner automorphisms if and only if it is invariant under the Lie subgroup $\operatorname{INN} L$ of $\operatorname{Gl}(L)$. ∎

2. Compact groups of automorphisms of a wedge

The set-up is simple: We consider a compact subgroup G of the Lie group $\text{Aut}(W, L)$ for a wedge W in a finite dimensional vector space L. Our first result is a structure theorem for W in terms of invariants determined by the compact group G.

III.2.1.   Theorem.   (The First Theorem on Compact Automorphism Groups of Cones) *Let W be a wedge in a finite dimensional vector space L. Then there exist*

(a) *a vector $w \in W$ fixed under G,*

(b) *a vector subspace $E \subseteq L$ invariant under G,*

(c) *a neighborhood C of 0 in $E \cap (W - W)$ invariant under G,*

such that C is compact and convex and that

(i)
$$L = H(W) \oplus E \oplus \mathbb{R}{\cdot}w,$$

and

(ii)
$$W = H(W) \oplus \mathbb{R}^+{\cdot}(C + w).$$

In particular, if $W \neq H(W)$, then there is a non-zero fixed vector w for G in the algebraic interior $\text{algint}\, W$ *of W (see* Definition I.2.20). *If W is a pointed cone, there is a G-invariant compact base of W.*

Proof.   By Weyl's trick, we may assume that L carries a G-invariant inner product. (Indeed, if $\langle \bullet \mid \bullet \rangle$ is an arbitrary inner product of L, we set $(x \mid y) = \int \langle gx \mid gy \rangle dg$ with normalized Haar measure dx on G. Then $(\bullet \mid \bullet)$ is an invariant inner product for L.) As W is G-invariant, the edge $H(W)$ and the vector space $W - W$ spanned by W are invariant, too. Let V denote the orthogonal complement of $H(W)$ in $W - W$ and V' the orthogonal complement of $W - W$ in L. Then L is an orthogonal direct sum of G-invariant vector spaces

$$L = H(W) \oplus V \oplus V', \quad W - W = H(W) \oplus V,$$

und W decomposes accordingly into an orthogonal direct sum

$$W = H(W) \oplus (W \cap V), \quad (W \cap V) - (W \cap V) = V$$

of the edge and a pointed G-invariant cone generating V. (See Proposition I.2.12.) It is then clear that our theorem is true as soon as it is proved for non-zero pointed generating invariant cones. This is what we shall assume for the remainder of the proof.

Let w_0 be an arbitrary point in algint W. Then $w = \int g w_0 dg$, the barycenter of the orbit $G w_0 \subseteq W$, is in the convex closure of the orbit, hence is in W. By the invariance of Haar measure, w is a G-fixed point. However, more is true: Since W is finite dimensional, algint $W = \text{int}(W \cap V)$ with topological interior taken in V by Lemma I.2.21. Since every $g \in G$ is a homeomorphism of $W \cap V$, we have $G w_0 \subseteq \text{int}(W \cap V)$. Since the orbit $G w_0$ is compact, its closed convex hull is still contained in $\text{int}(W \cap V)$, since it is contained in the open half-space bounded by any support hyperplane of W. It follows that $w \in \text{int } W = \text{algint } W$. If we let F denote the vector space of all G-fixed points, we have shown that $F \cap \text{int } W \neq \emptyset$.

Now we consider the dual $W^* = \{x \in L : (w \mid x) \geq 0 \text{ for all } w \in W\}$ of W obtained by identifying $\widehat{L}$ with L as we did in the last paragraphs of Section 1 of Chapter I. By the choice of the invariant inner product, the action of G is by orthogonal transformations. Hence W^* is invariant under G. Since W is pointed and generating, W^* is generating and pointed by Proposition I.1.7. Hence we find an element $w^* \in \text{int } W^*$ which is fixed under G. We may assume that, after appropriate normalization, $(w^* \mid w) = 1$ with a fixed vector $w \in F \cap \text{int } W$. Now $E = (w^*)^\perp$ and $E + w = \{x \in L : (w^* \mid x) = 1\}$ are G-invariant hyperplanes. Furthermore, $C \stackrel{\text{def}}{=} E \cap (w - W)$ is a compact convex invariant neighborhood of 0 in E such that $w + C = W \cap (E + w)$ is a compact base of W. In particular, $W = \mathbb{R}^+ \cdot (C + w)$. Since $L = E \oplus \mathbb{R} \cdot w$ is an orthogonal decomposition, the theorem is proved for pointed generating cones, and then for arbitrary wedges W after our initial observations. ∎

The First Theorem on Compact Automorphism Groups of Cones III.2.1 has an immediate consequence for compact Lie algebras.

III.2.2. **Proposition.** *Let L be a semisimple compact Lie algebra. Then*

(i) *every invariant wedge W in L is a vector space, and*

(ii) *L is the only generating Lie semialgebra of L.*

Proof. (i) Let W be an invariant wedge in L and assume that $W \neq H(W)$. By Theorem III.2.1, the compact group $\text{INN } L$ has a non-zero fixed point x in W. Such an element is central in L. But L is semisimple and thus has zero center. This contradiction shows $W = H(W)$.

(ii) Every generating Lie semialgebra W of L is invariant by Corollary II.6.5(ii). Then the assertion follows at once from (i) above. ∎

In the proof of Theorem III.2.1, we saw the rudiments of the following lemma whose explicit formulation will be convenient in the subsequent arguments.

III.2.3. **Lemma.** *Let L be a finite dimensional vector space and S a subsemigroup of $\text{Hom}(L, L)$. Let $\widehat{S}$ denote the semigroup of adjoint endomorphisms $\widehat{g}$ of $\widehat{L}$ with $g \in S$. We let x be a non-zero vector in L and set $T = x^\perp \subseteq \widehat{L}$. Then we have the following conclusions:*

(i) $\mathbb{R}{\cdot}x$ is S-invariant if and only if T is $\widehat{S}$-invariant.

(ii) If $Sx \subseteq \mathbb{R}{\cdot}x$, then there is a semigroup homomorphism $\chi\colon S \to \mathbb{R}$ into the multiplicative semigroup of real numbers such that $gx = \chi(g){\cdot}x$ and $\widehat{g}\omega \in \chi(g){\cdot}\omega + T$.

(iii) x is a fixed vector for S if and only if all cosets $\omega + T$ are $\widehat{S}$-invariant.

Proof.   The proofs are straightforward. In fact, (i) is just a special case of the more general situation of a vector subspace V of L which is S-invariant if and only if its annihilator $V^{\perp}$ in $\widehat{L}$ is $\widehat{S}$-invariant. (ii) follows readily from (i), and (iii) is an immediate consequence of (ii).   ∎

We shall now prove a sort of converse of the First Theorem on Compact Automorphism Groups of Cones.

III.2.4.   Theorem.   (Second Theorem on Compact Automorphism Groups of Cones) *Suppose that W is a pointed generating cone in a finite dimensional vector space L and that G is a subgroup of $\operatorname{Aut} W$. Then the following conditions are equivalent:*

(1) *G has a fixed point in the interior of W.*

(2) *G leaves a suitable compact base B of W invariant.*

(3) *The closure of G in $\operatorname{Aut} W$ is compact.*

Proof.   From Theorem III.2.1 we know that (3) implies (1) and (2). Suppose (2); we shall show (3). The convex hull C of $B \cup -B$ is a compact connected neighborhood of 0, for if $b \in B \cap \operatorname{int} W$, then $b/2$ is in the interior of C, whence also $-b/2$ is in the interior of C, and thus 0 is in the interior of C. Further, C is G-invariant, and thus G is a subgroup of the group of isometries relative to the norm on L which has C as unit ball. This group is compact as a closed and bounded subset of $\operatorname{Hom}(L, L)$. Hence (3) follows. It remains to show that (1) implies (3): Let $x \in \operatorname{int} W$ be a G-fixed point. The dual wedge W^{*} is generating and pointed by Proposition I.1.7. By Lemma III.2.3 above, the group $\widehat{G}$ of adjoints of elements of G leaves the hyperplane $T = x^{\perp}$ and all of its cosets invariant. Since x is an inner point of W, we have $T \cap W^{*} = \{0\}$, and thus if $\omega \in \operatorname{int} W^{*}$, then $(T + \omega) \cap W^{*}$ is an invariant compact base of W^{*}. The equivalence of (2) and (3) then shows that $\widehat{G}$ is relatively compact in $\operatorname{Gl}(\widehat{L})$. Since $g \mapsto \widehat{g} \colon \operatorname{Gl}(L) \to \operatorname{Gl}(\widehat{L})$ is a homeomorphism mapping G to $\widehat{G}$ we conclude that G is relatively compact in $\operatorname{Gl}(L)$. Since $\operatorname{Aut} W$ is closed in $\operatorname{Gl}(L)$, condition (3) is proved.   ∎

We apply this result to obtain further information on Lie algebras accommodating invariant generating pointed cones. (We promised such results after Corollary II.7.10.)

III.2.5.   Lemma.   *Let W be a generating invariant pointed cone in a Lie algebra L such that $Z(L) \cap \operatorname{int} W \neq \emptyset$. Then L is compact.*

Proof.   Let $\operatorname{Inn} L$ denote the group generated in $\operatorname{Gl}(L)$ by all $e^{\operatorname{ad} x}$ with $x \in L$, and let Γ denote its closure in $\operatorname{Gl}(L)$. Then $\Gamma \subseteq \operatorname{Aut} W$. If $Z(L) \cap \operatorname{int} W$ contains an element z, then it is a fixed point of Γ. Then Theorem III.2.4 implies that Γ is compact. Now we have an exact sequence of vector spaces

$$0 \to Z(L) \overset{\text{inc}}{\to} L \overset{\text{ad}}{\to} L(\Gamma).$$

Now $L(\Gamma)$ is a compact Lie algebra as the Lie algebra of a compact group. Consequently, the subalgebra ad L is a compact Lie algebra and thus $L/Z(L) \cong \text{ad}\,L$ is a compact Lie algebra. By Corollary II.7.10, L is reductive. Hence it follows that L itself must be a compact Lie algebra. ∎

III.2.6. **Proposition.** *Let L contain a generating invariant pointed cone whose interior meets the nilradical. Then L is compact.*

Proof. Let N be the nilradical and x an arbitrary element of L. Then $A = \mathbb{R}{\cdot}x + N$ is a subalgebra with a generating invariant pointed cone $A \cap W$. By the Third Triviality Theorem II.7.8, the algebra A is abelian. Thus $[x, N] = \{0\}$. Hence N is central and the preceding Lemma III.2.5 applies. ∎

III.2.7. **Corollary.** *If W is a generating Lie semialgebra in a Lie algebra L such that $Z(L) \cap \text{int}\,W \neq \varnothing$. Then the following conclusions hold:*

    (i) *W is invariant.*

    (ii) *$[L, R] \subseteq W$.*

    (iii) *$L/H(W)$ is compact.*

    (iv) *If L is solvable, then W is trivial.*

Proof. By the First Invariance Theorem II.7.1, W is invariant. Hence Corollary II.7.9 shows (ii). Since W is invariant, $H(W)$ is an ideal. Then $W/H(W)$ is a generating invariant pointed cone in $L/H(W)$. Since $\big(Z(L)H(W)\big)/H(W)$ is central in $L/H(W)$, the center of $L/H(W)$ meets the interior of $W/H(W)$. Hence Proposition III.2.6 applies and proves (iii). Finally, (iv) is a consequence of (ii). ∎

If a locally compact group is compact modulo its identity component, it is known to possess maximal compact subgroups. The additive group of the field of p-adic rationals is an example of a locally compact group which is not compact but is a union of its compact subgroups. (Discrete examples are furnished by all infinite abelian torsion groups.) Since the connectivity structure of the automorphism group of a cone depends strongly on the structure of the cone, it is not a priori clear that the general fact on locally compact groups applies. The existence of maximal compact subgroups is nevertheless secured by the following corollary in which the general fact is invoked for $\text{Gl}(L)$.

III.2.8. **Corollary.** *Let W be a pointed generating cone in a finite dimensional vector space L. Then $\text{Aut}\,W$ has maximal compact subgroups, and for every such group K there is a $w \in \text{int}\,W$ such that*

(†) $$K = (\text{Aut}\,W)_w = \{g \in \text{Aut}\,W : gw = w\}.$$

Two such subgroups $(\text{Aut}\,W)_v$ and $(\text{Aut}\,W)_w$ are conjugate under any element $h \in \text{Aut}\,W$ which satisfies $v = hw$.

Proof. Let $\mathcal{C}$ denote the set of all compact subgroups of $\operatorname{Aut} W$ partially ordered by $\subseteq$. For $C \in \mathcal{C}$ the fixed point set $F(C) = \{x \in L : Cx = \{x\}\}$ is a vector subspace of L. If $\mathcal{D} \subseteq \mathcal{C}$ is upward directed, then $\{F(D) : D \in \mathcal{D}\}$ is a filterbasis of finite dimensional vector spaces. Hence there is a $D' \in \mathcal{D}$ such that $F(D') = \bigcap_{D \in \mathcal{D}} F(D)$. By Theorem III.2.4 there is a $w \in F(D') \cap \operatorname{int} W$. Then $w \in F(D)$, that is, $D \subseteq (\operatorname{Aut} W)_w$ for all $D \in \mathcal{D}$. But $(\operatorname{Aut} W)_w \in \mathcal{C}$ by Theorem III.2.4. Hence $\mathcal{C}$ is inductive, and thus has maximal elements K. If we take $\mathcal{D} = \{\mathcal{K}\}$ in the preceding argument then the relation $K \subseteq (\operatorname{Aut} W)_w$ and the maximality of K give us ($\dagger$).

If G is a group acting on a set M and $G_w = \{g \in G : gw = w\}$ is the stability group at w, then $g \in hG_wh^{-1}$ if and only if $h^{-1}ghw = w$, that is, if and only if $g \in G_{hw}$. This proves the last assertion. ∎

Applications to Lie algebras with invariant cones

We shall now apply these results to the special case that L is a finite dimensional Lie algebra with an invariant pointed generating cone W.

Firstly we expand our terminology on the closure $\operatorname{INN} L$ of the group of inner automorphisms.

III.2.9.   Definition. If K is a subalgebra of L, we write $\operatorname{Inn}_L K$ for the analytic subgroup of $\operatorname{INN} L$ whose Lie algebra is exactly $\operatorname{ad} K$. In other words, $\operatorname{Inn}_L K$ is the subgroup algebraically generated by $e^{\operatorname{ad} K}$. The closure of $\operatorname{Inn}_L K$ in $\operatorname{INN} L$ will be denoted $\operatorname{INN}_L K$.

We observe that $\operatorname{Inn}_L L = \operatorname{Inn} L$ and $\operatorname{INN}_L L = \operatorname{INN} L$.

III.2.10.   Definition. Let L be a finite dimensional Lie algebra. A Lie subalgebra K is said to be *compactly embedded* (in L) if $\operatorname{INN}_L K$ is compact. It is traditional to call a Lie algebra L *compact* if, in this terminology, it is compactly embedded in itself.

We shall say that an element $x \in L$ is *compact* if $\mathbb{R} \cdot x$ is a compactly embedded subalgebra, that is, if

$$\operatorname{INN}_L \mathbb{R} \cdot x = \overline{e^{\mathbb{R} \cdot \operatorname{ad} x}}$$

is compact in $\operatorname{INN} L$.

The set of all compact elements of a Lie algebra L will be written $\operatorname{comp} L$.

We observe that we have defined

$$\operatorname{comp} L = \{x \in L : \overline{e^{\mathbb{R} \cdot \operatorname{ad} x}} \text{ is compact in } \operatorname{INN} L\}.$$

If K is a compactly embedded subalgebra, then $K \subseteq \operatorname{comp} L$ is clear. It is less obvious, but nevertheless correct that, conversely, $K \subseteq \operatorname{comp} L$ for a subalgebra K implies that K is compactly embedded. (See Corollary A.2.21.)

The following remarks are mere linear algebra. If L is a finite dimensional vector space and T an endomorphism, then T extends to a unique endomorphism $T_{\mathbb{C}} = \mathbf{1} \otimes T$ of the complexification $L_{\mathbb{C}} = \mathbb{C} \otimes L$. We note that $e^{T_{\mathbb{C}}} = (e^T)_{\mathbb{C}} = \mathbf{1} \otimes e^T$. If L is a real Lie algebra and $x \in L$, then we consider L identified with a real subalgebra of $L_{\mathbb{C}}$. Then $x \in L_{\mathbb{C}}$ and $\operatorname{ad}_{L_{\mathbb{C}}} x = (\operatorname{ad}_L x)_{\mathbb{C}}$. It follows that *an element x in a finite dimensional real Lie algebra L is compact if and only if it is compact in (the real Lie algebra underlying) $L_{\mathbb{C}}$*. The following remark is now an immediate consequence of the Jordan decomposition of $\operatorname{ad}_{L_{\mathbb{C}}} x$ and the definition of a compact element:

III.2.11. **Proposition.** *In a finite dimensional Lie algebra L we have*

$$\operatorname{comp} L = \{x \in L : \operatorname{ad} x \text{ is semisimple and } \operatorname{Spec} \operatorname{ad} x \subseteq i \cdot \mathbb{R}\}. \qquad \blacksquare$$

To illustrate these concepts we recall the example $L = \operatorname{sl}(2, \mathbb{R})$ from Section 3 of Chapter II. Here $\operatorname{comp} L = \{X \in \operatorname{sl}(2, \mathbb{R}) : B(X) < 0\} = \operatorname{int} \mathcal{W}^+ \cup \operatorname{int} \mathcal{W}^-$ is the interior of the double cone bounded by the zero-set of the Lorentzian Cartan–Killing form.

If $L = A_2$ is the standard Lorentzian 4-dimensional solvable algebra (the harmonic oscillator algebra), then $\operatorname{comp} L = (L \setminus [L, L]) \cup [L, [L, L]]$ contains the interiors of the two half-spaces bounded by the hyperplane ideal L' plus the center $Z(L) = [L, [L, L]]$, which is a line inside the 3-space L'.

The following theorem is more sophisticated than the preceding proposition. We shall give a proof of this purely Lie-theoretical fact in the Appendix. We must recall here that every Cartan algebra H in a finite dimensional real Lie algebra L is the generalized null-space $L^0(x) = \{y \in L : (\exists n \in \mathbb{N})(\operatorname{ad} x)^n y = 0\}$ of $\operatorname{ad} x$ for some regular element $x \in L$. (See [Bou75], Chap. VII, §2, n° 3, Théorème 1.) The gist of the next theorem is the fact that a regular point determines a *compactly embedded* Cartan algebra if and only if it is an *inner* point of the set of compact elements.

III.2.12. **Theorem.** (The Third Cartan Algebra Theorem) *Let H be a Cartan algebra of a finite dimensional real Lie algebra L and let x be any element of L. Then the following conditions are equivalent:*

(1) $x \in \operatorname{int}(\operatorname{comp} L)$.

(2) $\ker(\operatorname{ad} x) = \bigcup \{K : K \in \mathcal{H}, x \in K\}$, *where $\mathcal{H}$ is the set of all compactly embedded Cartan algebras.*

(3) $\ker(\operatorname{ad} x) \subseteq \operatorname{comp} L$.

If further $H = L^0(x)$, then these conditions are also equivalent to

(4) H *is a compactly embedded Cartan algebra.*

Moreover, all compactly embedded Cartan algebras of L are conjugate under inner automorphisms (that is, under automorphisms of $\operatorname{Inn} L$). All compactly embedded Cartan algebras are abelian. Hence if there is at least one of them, then all Cartan algebras are abelian.

Proof. See Theorem A.2.25, Theorem A.2.27, and Proposition A.2.28. $\qquad \blacksquare$

Since a compactly embedded Cartan algebra $H = L^0(x)$ with a regular $x \in L$ is abelian, it follows that $H = \ker(\operatorname{ad} x)$.

III.2.13.   Corollary.   *In any finite dimensional Lie algebra L the following inclusions hold:*

$$\text{int}(\text{comp}\,L) \subseteq \bigcup_{H \in \mathcal{H}} H \subseteq \text{comp}\,L,$$

where $\mathcal{H}$ is the conjugacy class of all compactly embedded Cartan algebras of L.

Moreover, all regular elements of the middle set are contained in the leftmost set. In particular, $\text{int}(\text{comp}\,L)$ is dense in the middle set.

Proof.   The containments are clear from the preceding Theorem III.2.12 and the definition of a compactly embedded algebra. If x is regular and $H = L^0(x)$ is compactly embedded, then $H = \ker(\text{ad}\,x)$ and Theorem III.2.7 implies $x \in \text{int}(\text{comp}\,L)$. For the last assertion, recall that the set of regular elements of L contained in any Cartan algebra H is dense in H.   ∎

Now we bring everything to bear on invariant cones in Lie algebras.

III.2.14.   Theorem.   (The Fourth Cartan Algebra Theorem) *Let W be an invariant pointed generating cone in a finite dimensional real Lie algebra L. Let $\mathcal{H}$ denote the conjugacy class of all compactly embedded Cartan algebras. Then the following conclusions hold:*

  (i) $\text{int}\,W \subseteq \text{comp}\,L$.

  (ii) *$\mathcal{H} \neq \emptyset$, and all Cartan algebras of L are abelian.*

  (iii) *If H is any compactly embedded Cartan algebra of L, then $H \cap \text{int}\,W \neq \emptyset$.*

Proof.   (i) Let $x \in \text{int}\,W$. Define $G = \text{INN}_L \mathbb{R}\cdot x$. Then G is a subgroup of $\text{INN}\,L$, and G has x as a fixed vector since $e^{t\cdot\text{ad}\,x}x = x + [t\cdot x, x] + \cdots = x$ for all $t \in \mathbb{R}$. Then by the Second Theorem on Compact Automorphism Groups of Cones III.2.4, G is compact. Hence $x \in \text{int}(\text{comp}\,L)$ by Definition III.2.10.

(ii) By Corollary III.2.13, $\text{int}\,\text{comp}\,L$ is contained in the union of the compactly embedded Cartan algebras. Therefore $\mathcal{H} \neq \emptyset$. It follows then from the Third Cartan Algebra Theorem III.2.12 that all Cartan algebras of L are abelian.

(iii) Consider a compactly embedded Cartan algebra H of L. By Corollary III.2.13 and (i) above we find a compactly embedded Cartan algebra H_1 with $H_1 \cap \text{int}\,W \neq \emptyset$. By the Third Cartan Algebra Theorem III.2.12 there is an inner automorphism $\gamma \in \text{Inn}\,L$ with $\gamma(H) = H_1$. Hence $\emptyset \neq H_1 \cap \text{int}\,W = \gamma(H) \cap \gamma(\text{int}\,W)$ (by the invariance of W!) $= \gamma(H \cap \text{int}\,W)$ because γ is bijective. This proves (iii).   ∎

This theorem contains, in particular, the conclusion that all Cartan algebras in a Lie algebra L have to be abelian if it supports an invariant pointed generating cone. This information is already available: By the First Cartan Algebra Theorem II.4.16, Cartan algebras have to be abelian whenever L supports a reduced generating Lie semialgebra. By Scholium II.2.15, every invariant wedge is a Lie semialgebra. An invariant wedge is reduced if and only if it is pointed by Proposition II.1.10.

With all of this information at our hands, we can take a first important step towards a classification of invariant cones in Lie algebras. In fact we shall see presently, that invariant pointed generating cones are uniquely determined by their intersections with compactly embedded Cartan algebras— these always exist in the presence of such cones! Cartan algebras are very thin slices in a Lie algebra in general, and the way they slice pointed generating invariant cones determines these uniquely!

III.2.15.   Theorem. (Uniqueness Theorem for Invariant Cones) *Let W be an invariant pointed generating cone in a finite dimensional real Lie algebra L. Let H be any compactly embedded Cartan algebra. Then*

$$\operatorname{int} W = (\operatorname{Inn} L)\operatorname{algint}(H \cap W).$$

In particular, if H_1 and H_2 are compactly embedded Cartan algebras and W_1 and W_2 are invariant pointed generating cones of L such that $H_1 \cap W_1$ is conjugate to $H_2 \cap W_2$ under an inner automorphism, then $W_1 = W_2$.

Proof.   First we claim that

$(*)$ $\qquad\qquad\qquad\qquad \operatorname{algint}(H \cap W) = H \cap \operatorname{int} W.$

The left side clearly contains the right side. We must show the reverse containment. Let $h \in H\cap W$ and fix some element $w \in H\cap\operatorname{int} W$. Then $t\cdot h+(1-t)\cdot w \in H\cap\operatorname{int} W$ for $0 \le t < 1$, and since all points of $\operatorname{algint}(H \cap W)$ are of the form $t\cdot h + (1 - t)\cdot w$ with a fixed $w \in H\cap\operatorname{int} W$ and suitable elements $h \in H\cap W$, $0 \le t < 1$, the claim $(*)$ follows.

By Corollary III.2.13,

$$\operatorname{int} W = \bigcup_{K\in\mathcal{H}} K \cap \operatorname{int} W,$$

where, as before, $\mathcal{H}$ denotes the class of compactly embedded Cartan subalgebras of L. Hence for each $K \in \mathcal{H}$ we find a $\gamma \in \operatorname{Inn} L$ such that $K \cap \operatorname{int} W = \gamma(H) \cap \gamma(\operatorname{int} W)$ (by the invariance of W!) $= \gamma(H \cap \operatorname{int} W) = \gamma\big(\operatorname{algint}(H \cap W)\big)$ by Claim $(*)$. We have now shown

$(**)$ $\qquad\qquad\qquad \operatorname{int} W = (\operatorname{Inn} L)\operatorname{algint}(H \cap W).$

Now suppose that W_1, W_2, H_1 and H_2 satisfy the hypotheses of the theorem. By $(**)$ we have $\operatorname{int} W_n = (\operatorname{Inn} L)\operatorname{algint}(H_n \cap W_n)$ for $n = 1, 2$. Since $H_1 \cap W_1$ and $H_2 \cap W_2$ are assumed to be conjugate under an inner automorphism, the algebraic interiors of these wedges are likewise conjugate under some inner automorphism. It follows that $\operatorname{int} W_1 = \operatorname{int} W_2$, whence $W_1 = W_2$.   ∎

This theorem reduces, in principle, the classification of invariant pointed generating cones in a Lie algebra to the study of all possible pointed generating cones in a fixed compactly embedded Cartan algebra which are the traces of invariant cones in the Lie algebra. An abelian Cartan algebra, however, is simply a vector space; therefore we shall have to find additional structural elements attached to it which permit us to determine when a cone is such a trace. Such a structure is provided by the roots of the algebra with respect to the given compactly embedded Cartan algebra. This topic will be addressed in later sections.

Minimal and maximal invariant cones

For a further immediate application of the theorems on compact auto-morphism groups let us now consider a finite dimensional vector space L with a non-degenerate quadratic (that is, bilinear symmetric) form $q: L \times L \to \mathbb{R}$ (compare Definition I.4.7). We shall write

$$\mathrm{O}(q) = \{g \in \mathrm{Aut}\, L : q(gx, gy) = q(x, y) \text{ for all } x, y \in L\}.$$

Then $\mathrm{O}(q)$ is a closed subgroup of $\mathrm{Aut}\, L$ and thus a Lie group. We may and shall identify L with its dual $\widehat{L}$ via the isomorphism

$$x \mapsto \big(y \mapsto q(x, y)\big): L \to \widehat{L}.$$

Under this identification, the dual W^* of a wedge $W \subseteq L$ is given by

$$W^* = \{x \in L : q(x, w) \geq 0 \text{ for all } w \in W\}.$$

It is an immediate consequence of the definition of the group $\mathrm{O}(q)$ that *for any subgroup G of $\mathrm{O}(q)$ a wedge W in L is G-invariant if and only if its dual W^* is G-invariant.*

Now let G be a closed subgroup of $\mathrm{O}(q)$ such that G/G_0 is compact where G_0 denotes the identity component of G. Then G has a maximal compact subgroup K and any compact subgroup is contained in some conjugate of K. Moreover, there is a "manifold factor" E, that is, a subset diffeomorphic to $\mathbb{R}^n$ such that $(e, k) \mapsto ek: E \times K \to G$ is a diffeomorphism (see for instance [Hoch65], Theorem 3.1, p.180). If $x \in L$ satisfies $Kx = \{x\}$, then $Gx = Ex$ for any subset $E \subseteq G$ whatsoever as long as it satisfies $G = EK$.

III.2.16.   Remark.   The following statements on a wedge W are equivalent:

(1) W is a minimal G-invariant wedge which is not a vector space.

(2) W^* is a maximal G-invariant wedge which is not a vector space.

(3) W is not a vector space and is the closed convex hull of $\mathbb{R}^+ \cdot Ex = E(\mathbb{R}^+ \cdot x)$ for all K-fixed vectors $x \in W$.

(4) W^* is not a vector space and $W^* = \{w \in L : q(w, ex) \geq 0 \text{ for all } e \in E\}$ for all K-fixed vectors $x \in W$.

Moreover, if these conditions are satisfied, there exist non-zero K-fixed vectors.

Proof.   The equivalence of (1) and (2) and that of (3) and (4) is a simple consequence of duality. We prove the equivalence of (1) and (3). Suppose (1). By the minimality of W, for each $x \in W$ the wedge W is the closed convex hull of $\mathbb{R}^+ \cdot Gx$. If x is a K-fixed vector, then $Gx = Ex$ in view of the preceding remarks. Hence (3) follows. Conversely, suppose that (3) holds. Let W_0 be a G-invariant wedge contained in W and suppose that W_0 is not a vector space. By the First Theorem on Compact Automorphism Groups of Cones III.2.1, there is a K-fixed vector in $\mathrm{algint}\, W_0$. Condition (3) then implies $W \subseteq W_0$ and thus $W = W_0$. Hence W is G-invariant and is minimal with respect to this property. In passing we have also proved the last assertion.   ∎

Under particularly opportune circumstances, this set-up allows us to conclude that all G-invariant wedges of L are, up to sign, sandwiched between a unique maximal and a unique minimal G-invariant wedge.

III.2.17. **Proposition.** *Let L be a finite dimensional vector space with a non-degenerate quadratic form q and a vector $z \in L$ with $q(z) > 0$. Suppose that G is a closed subgroup of $O(q)$ and any decomposition $G = EK$ with a maximal compact subgroup K and a subset E. If $\mathbb{R}.z$ is the precise fixed point set of K, there are G-invariant wedges $W^{\min}$ and $W^{\max}$ such that for each G-invariant wedge W there is a $\sigma \in \{1, -1\}$ with*

$$(1) \qquad\qquad W^{\min} \subseteq \sigma \cdot W \subseteq W^{\max}.$$

Moreover,

$$(2) \qquad\qquad W^{\min} = \text{closed convex hull of } \mathbb{R}^+ \cdot Ez,$$

$$(3) \qquad\qquad W^{\max} = \{w \in L : q(w, ez) \geq 0 \text{ for all } e \in E\}.$$

If there exist pointed G-invariant wedges, then $W^{\min}$ is pointed and $W^{\max}$ is generating.

Proof. We define $W^{\min}$ and $W^{\max}$ via (2) and (3). Then these two wedges are duals of each other. Let W be a G-invariant wedge which is not a vector space. Then by the First Theorem on Compact Automorphism Groups of Cones III.2.1, there is a K-fixed vector $z' \in \operatorname{algint} W$. In particular, $z' \neq 0$. As $\mathbb{R} \cdot z$ is the precise fixed point vector space of K, there is a non-zero real number r with $z' = r \cdot z$. We set $\sigma = r/|r|$. Then $Gz = Ez \subseteq \sigma W$, whence $W^{\min} \subseteq \sigma \cdot W$. Since q is G-invariant, W^* is G-invariant. Since W is not a vector space, also W^* cannot be a vector space. Hence, by what we just saw, we have $W^{\min} \subseteq (\sigma \cdot W)^*$. Hence by duality, according to Proposition I.1.4, we find $\sigma \cdot W = (\sigma \cdot W)^{**} \subseteq (W^{\min})^* = W^{\max}$. If we assume that pointed non-singleton G-invariant wedges exist, then $W^{\min}$, being contained in all G-invariant pointed wedges up to sign, is itself pointed. So by duality, $W^{\max}$ is generating. ∎

This applies to certain Lie algebras in the following form:

III.2.18. **Proposition.** (The Minimax Theorem for Invariant Cones) *Let L denote a finite dimensional real Lie algebra and let K denote a maximal compactly embedded subalgebra of L. Suppose that the following hypotheses are satisfied:*

(i) *L supports a pointed invariant wedge.*

(ii) *L supports a non-degenerate invariant quadratic form q.*

(iii) *K contains an element z such that K is the centralizer $\ker \operatorname{ad} z$ of z in L.*

(iv) *$q(z) > 0$.*

Then there exists a pointed invariant wedge $W^{\min}$ and a generating invariant wedge $W^{\max}$ such that for each invariant wedge W, which is not a vector space, there is a $\sigma \in \{1, -1\}$ such that

$$W^{\min} \subseteq \sigma \cdot W \subseteq W^{\max}.$$

Moreover, for every subset $E \subset \mathrm{INN}\, L$ *such that* $\mathrm{INN}\, L = E(\mathrm{INN}_L\, K)$ *one has*

$$W^{\min} = closed\ convex\ hull\ of\quad \mathbb{R}^+ \cdot E(z),$$

$$W^{\max} = \{w \in L : q\big(w, e(z)\big) \geq 0\ for\ all\ e \in E\}.$$

Proof.   This is an immediate consequence of the preceding Proposition III.2.17 if we set $G = \mathrm{INN}\, L$ and observe that $\mathrm{INN}_L\, K$ is a maximal compact subgroup of G (see Theorem A.2.11). ∎

3. Frobenius–Perron theory for wedges

The classical Frobenius–Perron theory deals with the eigenvectors and eigenvalues of an $n \times n$ matrix with non-negative or with positive entries. Every such matrix corresponds to an endomorphism $g \in \mathrm{Hom}_L W$ with $L = \mathbb{R}^n$ and $W = (\mathbb{R}^+)^n$. (see Definition III.1.7). If we consider the semigroup $S = \{\mathbf{1}, g, g^2, \ldots\} \subseteq \mathrm{Hom}_L W$, then we are, in the search for an eigenvector w of g in W, looking in fact for a ray $\mathbb{R}^+ \cdot w$ in W which is invariant under S, and the search for a corresponding eigenvalue λ amounts to finding semigroup homomorphism $\chi \colon S \to \mathbb{R}^+$ into the multiplicative semigroup of non-negative real numbers such that $gw = \chi(g) \cdot w$. In the special case before us we simply have $\chi(g^n) = \lambda^n$.

Now we consider the general situation. Let W be a wedge in a finite dimensional vector space L and S a subsemigroup of the multiplicative semigroup of the semialgebra $\mathrm{Hom}_L W$. *We shall be looking for sufficient conditions under which there is a non-zero ray $R = \mathbb{R}^+ \cdot w \subseteq W$ which is invariant under S.* The example $L = \mathrm{sl}(2, \mathbb{R})$ and $W = \mathcal{W}^+$ (see Theorem II.3.7 and Exercise 3 following Corollary III.1.6) shows that such an invariant ray may not exist for large enough S. On the other hand, the results of the previous section, notably Theorems III.2.1 and 4 illustrate well that under suitable circumstances, here if S is a compact group of automorphisms, very sharp conclusions may be drawn. The methods, however, differ according to the circumstances. In the case of compact groups S, the appropriate tool was the normalized Haar measure on S.

The case of abelian semigroups

III.3.1.   Theorem.   (The Frobenius–Perron Theorem for Abelian Semigroups) *Suppose that W is a wedge in a finite dimensional vector space L satisfying $W \neq H(W)$. If S is any abelian subsemigroup of $\mathrm{Hom}_L W$ then there is a non-zero $w \in W$ such that $Sw \subseteq \mathbb{R}^+ \cdot w$.*

Proof.   We proceed in several steps. Claim 1: Every $g \in \mathrm{Hom}_L W$ has a non-zero eigenvector in W. We prove this claim: If W contains a $w \neq 0$ with $gw = 0$ then w is the eigenvector we want. If $gw \neq 0$ for all $w \in W \setminus \{0\}$, then g induces a continuous selfmap $\bar{g}$ of the semiprojective space $\Pi(W)$ of W. (See Definition I.2.29.) Since $W \neq H(W)$ this space is a compact $n-1$-cell with $n = \dim(W - W)$ by Proposition I.2.30. Hence by the Fixed Point Theorem of Brouwer, $\bar{g}$ has at least one fixed point in $\Pi(W)$, and this means exactly that there is a non-zero w in W with $gw \in \mathbb{R}^+ \cdot w$. Thus Claim 1 is established.

For the purposes of this proof only we shall call an S-invariant subwedge W' of W *irreducible* if it is non-zero and does not contain any non-zero S-invariant subwedge properly. Claim 2: W contains irreducible subwedges. We shall prove this claim by observing that the set of all non-zero S-invariant subwedges of W is inductive with respect to $\supseteq$. Indeed if $\{W_j : j \in J\}$ is a descending tower of non-zero S-invariant subwedges, then $V \overset{\text{def}}{=} \bigcap_{j \in J} W_j$ is certainly an S-invariant subwedge of W. However, by the compactness of all ΠW_j the intersection $\bigcap_{j \in J} \Pi W_j$ is not empty, whence V is non-zero. This shows the asserted inductivity. Now Zorn's Lemma applies and shows that W contains a minimal non-zero S-invariant subwedge W'. By minimality, W' is irreducible, and so Claim 2 is proved.

The proof of the theorem will be complete if we establish Claim 3: Every irreducible wedge is 1-dimensional. For the proof of this claim we assume that W is irreducible and show $\dim W = 1$. Now let $g \in S$. By Claim 1, g has a non-zero eigenvector for the eigenvalue λ in W. If L_λ denotes the eigenspace of g for the eigenvalue λ, then $W \cap L_\lambda$ is non-zero by what we just saw. Let $x \in L_\lambda$ and $h \in S$. Since S is abelian, we have $g(hx) = h(gx) = h(\lambda \cdot x) = \lambda \cdot (hx)$, that is, $hx \in L_\lambda$. Thus L_λ is S-invariant, and so $W \cap L_\lambda$ is a non-zero S-invariant subwedge of W and thus agrees with W by irreducibility of W. Thus $W \subseteq L_\lambda$. Hence $g|(W - W)$ is a scalar multiplication. Since $g \in S$ was arbitrary, every subwedge of W is invariant. The irreducibility of W now implies $\dim W = 1$, as asserted. ∎

The crucial tool in this proof was the Brouwer Fixed Point Theorem. Also it was essential that the semiprojective space $\Pi(W)$ of W was compact. The theorem has a straightforward generalization to any locally compact wedge W in any locally convex topological vector space.

EIII.3.1.   **Exercise.** If $W \neq H(W)$ is a locally compact wedge in a locally convex topological vector space L and if S is a commutative semigroup of continuous endomorphisms of L preserving W. Then there is a non-zero $w \in W$ such that $Sw \subseteq \mathbb{R}^+ \cdot w$. ∎

In contrast with the case the S is a compact automorphism group of W we cannot guarantee that the common eigenvector w is in the interior of W even if W is pointed and generating. For an example consider $L = \mathbb{R}^2$, $W = (\mathbb{R}^+)^2$ and

$$S = \{ \begin{pmatrix} s & 0 \\ 0 & t \end{pmatrix} : 0 \leq s, t \}.$$

Another instructive example involving this time a half-space W is $L = \mathbb{R}^2$, $W = \mathbb{R}^+ \times \mathbb{R}$, and $S = \{\mathbf{1}, g, g^2, \ldots\}$ with

$$g = \begin{pmatrix} 1 & 1 \\ 0 & 1 \end{pmatrix}.$$

The case of solvable groups

The desire to generalize the previous result for abelian semigroups at least to solvable subgroups of $\operatorname{Aut} W$ is natural. However, there are natural boundaries to such generalizations. This is illustrated by the following examples:

1. **Example.** Let $L = \mathbb{R}^3$, $W = \mathbb{R} \times \mathbb{R} \times \mathbb{R}^+$ and

$$G = \left\{ \begin{pmatrix} \cos t & -\sin t & r \\ \sin t & \cos t & s \\ 0 & 0 & 1 \end{pmatrix} : r, s, t \in \mathbb{R} \right\}.$$

Then G is a three dimensional connected solvable (in fact metabelian) Lie group in $\operatorname{Aut} W$. In fact, G is isomorphic to the group of euclidean motions of $\mathbb{R}^2$. Also, W is a half space. But G has no non-zero common eigenvector.

2. **Example.** Let $L = \mathbb{R}^2$, $W = (\mathbb{R}^+)^2$, and

$$G = \left\{ \begin{pmatrix} s & 0 \\ 0 & t \end{pmatrix} : s, t > 0 \right\} \cup \left\{ \begin{pmatrix} 0 & s \\ t & 0 \end{pmatrix} : s, t > 0 \right\}.$$

Then G is a two dimensional Lie group whose identity component G_0 is isomorphic to $\mathbb{R}^2$ and has index 2 in G. The group G is solvable and in fact metabelian. The cone W is pointed and generating. But G does not have a common eigenvector in W.

3. **Example.** Let $\mathbb{T} = \mathbb{R}/\mathbb{Z}$ denote the circle group and let $C(\mathbb{T}, \mathbb{R})$ be the Banach space of all continuous real valued function on $\mathbb{T}$ with the supremum norm. We set $L = M(\mathbb{T}) = C(\mathbb{T}, \mathbb{R})\hat{\ }$, its topological dual, equipped with the weak $*$-topology, that is, the topology of pointwise convergence of linear functionals. In other words, L is the space of Radon measures on $\mathbb{T}$. If $W = M^+(\mathbb{T})$ denotes the set of all non-negative measures. Then W is a locally compact pointed cone in L with the space $P(\mathbb{T})$ of probability measures on $\mathbb{T}$ as a compact base.

We let $\mathbb{T}$ act on $C(\mathbb{T}, \mathbb{R})$ on the right by $f_t(x) = f(t + x)$ and form the semidirect product $G = \mathbb{T} \ltimes C(\mathbb{T}, \mathbb{R})$ with the multiplication $(s, f)(t, g) = (st, f_t + g)$ and let G act on L via $\big((t, f), \mu\big) \mapsto \nu$ with $\int h(x) d\nu(x) = \int e^{f(x)} h(t+x) d\mu(x)$. This action is jointly continuous on L and leaves W invariant. Hence we may identify G with a subgroup of $\operatorname{Hom}_L W$. But no one dimensional ray in W remains invariant under this action. The group G is an infinite dimensional metabelian connected Lie group.

The proofs of the assertions pertaining to Examples 1 and 2—which are of main interest for our purposes here— are straightforward. The proofs of the claims concerning Example 3 require more careful consideration.

EIII.3.2. **Exercise.** Verify the properties of Example 3 above.

In the presence of these examples it is perhaps a little surprising that anything at all can be said in the solvable case.

First we need a lemma which requires a little background discussion.

We fix a subgroup H of $\operatorname{Gl}(L)$ and consider the set $M = \operatorname{Hom}(H, \mathbb{R}^\times)$ of all homomorphisms from H into the multiplicative semigroup $\mathbb{R}^\times$ of real numbers. For each $\chi \in M$ we set

(†) $$L_\chi = \{x \in L : hx = \chi(h) \cdot x \text{ for all } h \in H\}.$$

We note that L_χ will be $\{0\}$ in most cases. Now let G be any subset of $\mathrm{Gl}(L)$ which is in the normalizer of H in $\mathrm{Gl}(L)$. Next we fix an arbitrary $g \in G$. Then $g^{-1}hg \in H$ for all $h \in H$, and for each $\chi \in M$, the function $g\cdot\chi\colon H \to \mathbb{R}^\times$ given by $(g\cdot\chi)(h) = \chi(g^{-1}hg)$ is again in M. Thus G operates on M on the left, and if G is a group or semigroup, then this action is a group or semigroup action. We claim that

$$(\dagger\dagger) \qquad\qquad gL_\chi = L_{g\cdot\chi}.$$

Indeed $x \in L_\chi$ precisely when $hx = \chi(h)\cdot x$ for all $h \in H$ by $(\dagger)$. This holds if and only if $(ghg^{-1})(gx) = \chi(h)\cdot gx$ for all $h \in H$. Thus $x \in L_\chi$ is equivalent to $h(gx) = \chi(g^{-1}hg)\cdot gx = (g\cdot\chi)(h)\cdot gx$ for all $h \in H$. This means $gx \in L_{g\cdot\chi}$, and $(\dagger\dagger)$ is proved.

Now we let $R \subseteq M$ denote the set of all $\chi \in M$ for which $L_\chi \neq \{0\}$. This is a finite set since L is finite dimensional and the sum of the L_χ is necessarily direct (see [Bou75], Chap.VII, §1, n° 1, Proposition 3). The action of G permutes R hence induces an action on the finite set of vector spaces $E = \{L_\chi : \chi \in R\}$. For each permutation $\sigma \in S(E)$, where $S(E)$ denotes the full symmetric group on the finite set E, we set

$$(*) \qquad\qquad \Gamma_\sigma = \{g \in \mathrm{Gl}(L) : gL_\chi = \sigma(L_\chi) \text{ for all } \chi \in \mathbb{R}\}.$$

The action of $\mathrm{Gl}(L)$ on the Grassmann variety of all vector subspaces of L is algebraic and thus the stability subgroup $\mathrm{Gl}(L)_\chi = \{g \in \mathrm{Gl}(L) : gL_\chi = L_\chi\}$ of this action at each L_χ is an algebraic subvariety of $\mathrm{Gl}(L)$. For an arbitrary pair $\chi, \chi' \in R$, the set $\{g \in \mathrm{Gl}(L) : gL_\chi = L_{\chi'}\}$ is either empty, or else, if it contains an element g_0, is a coset $g_0\,\mathrm{Gl}(L)_\chi$ of such a stability group and is, therefore likewise an algebraic subvariety of $\mathrm{Gl}(L)$. Since Γ_σ is a finite intersection of such varieties, it follows that Γ_σ *is an algebraic subvariety of* $\mathrm{Gl}(L)$.

The action of G on E allows us the conclusion that G is contained in the union U of the family $\{\Gamma_\sigma : \sigma \in S(E)\}$ of algebraic subvarieties of $\mathrm{Gl}(L)$. Two members of this family either agree or are disjoint. Hence each connected component of the algebraic variety U *with respect to the Zariski topology* of $\mathrm{Gl}(L)$ is necessarily contained in one of the varieties Γ_σ, since each of them is closed by the very definition of the Zariski topology, but is also open as the complement of the finitely many closed subsets $\Gamma_{\sigma'} \neq \Gamma_\sigma$ of U. Therefore, if $G \subseteq \mathrm{Gl}(L)$ *is assumed to be connected with respect to the Zariski topology* of $\mathrm{Gl}(L)$, *then there exists a permutation* σ *of* E *such that* $G \subseteq \Gamma_\sigma$. This allows us to establish the following lemma:

III.3.2.   **Lemma.** *Suppose that* H *is a subgroup of* $\mathrm{Gl}(L)$ *for a finite dimensional vector space* L. *Let* G *be a subset of the normalizer of* H *is* $\mathrm{Gl}(L)$ *which contains the identity* $\mathbf{1}$ *and is connected with respect to the Zariski topology of* $\mathrm{Gl}(L)$. *Then* G *leaves all eigenspaces* L_χ, $\chi \in \mathrm{Hom}(H, \mathbb{R}^\times)$ $\big($*defined by* $(\dagger)\big)$ *of* H *invariant.*

Proof.   From our preceding discussion we know that there is a permutation σ of the set E such that $G \subseteq \Gamma_\sigma$. Thus for each $\chi \in R$ we have $gL_\chi = \sigma(L_\chi)$ for all $g \in G$. But $\mathbf{1} \in G$, whence $\sigma(L_\chi) = \mathbf{1}(L_\chi) = L_\chi$. Thus $gL_\chi = L_\chi$ for all $g \in G$. This is what we had to show.   ∎

One might wonder why we bother with the Zariski topology if a result like the one expressed in the preceding lemma seems intuitively much more natural in the case of the usual topology which is much finer than the Zariski topology. The reason is that relatively to a coarse topology, connected components are large and that, as a consequence, certain special subgroups of algebraic groups are known to be connected with respect to the Zariski topology while they may fail to be connected relative to some Lie group topology. We record two such examples because they will be applied in the context of cones:

III.3.3. **Proposition.** *Let A be an algebraic group equipped with the Zariski topology and G a subgroup. Then G is connected in the following two special cases:*

(i) *G is the centralizer of some torus T in A.*

(ii) *G is a parabolic subgroup of A.*

Proof. For these results and the theory of algebraic groups we refer to sources on algebraic groups such as for instance [Hu75], notably pages 140 and 143. ∎

If A is a subgroup of $\mathrm{Gl}(L)$ and therefore also has a natural Lie group topology, G need be connected in neither of the two cases relative to the Lie group topology.

We now exploit Lemma III.3.2 for our purposes and formulate a lemma which serves as the basis for a proof by induction of our final result:

III.3.4. **Lemma.** *Let W be a pointed cone in a finite dimensional real vector space L and let H be a subgroup of $\mathrm{Aut}(W, L)$ which has a common non-zero eigenvector in W. If G is a semigroup in $\mathrm{End}(W, L)$ which normalizes H, which is connected in the Zariski topology and which is either abelian or else is a compact group (with respect to the ordinary topology induced from that of $\mathrm{Gl}(L)$), then GH has a non-zero common eigenvector in W.*

Proof. Since H has a common eigenvector in W, there exists a $\chi \in \mathrm{Hom}(H, \mathbb{R}^{\times})$ such that $W \cap L_{\chi} \neq \{0\}$. By Lemma III.3.2, $GL_{\chi} = L_{\chi}$, and thus G leaves $W \cap L_{\chi}$ invariant because $G \subseteq \mathrm{End}(W, L)$. Since $H(W \cap L_{\chi}) = \{0\}$, then Theorem III.3.1 applies if G is abelian and Theorem III.2.1 if G is compact. At any rate we have a non-zero $w \in W \cap L_{\chi}$ with $Gw \subseteq \mathbb{R}^{+} \cdot w$. But since $hw = \chi(h) \cdot w$ for all $h \in H$ as $w \in L_{\chi}$ we have $GHw \subseteq \mathbb{R}^{+} \cdot w$. ∎

It is instructive to notice that for a wedge in general the proof may break down if $W \cap L_{\chi} \subseteq H(W)$, since the Theorem III.3.1 is not applicable to the vector space $W \cap L_{\chi}$.

With the aid of this lemma we can now proceed to prove the following result on solvable automorphism groups of pointed cones which is, as the examples above have shown, rather sharp.

III.3.5. **Theorem.** (The Frobenius–Perron Theorem for Solvable Groups) *Suppose that W is a proper cone in a finite dimensional real vector space L and that S is a solvable subgroup of $\mathrm{Aut}(W, L)$. Suppose further that at least one of the two following conditions is satisfied:*

(i) *S is arcwise connected.*

(ii) *S is an algebraic subgroup of* Gl(L) *and is connected with respect to the Zariski topology.*

Then S has a non-zero common eigenvector in W . Moreover, there is an S - invariant hyperplane T of W − W with T ∩ algint $W = \emptyset$.

Proof.   In Case (i), S is an analytic subgroup by Yamabe's Theorem (see [Bou75], Chap. III, §8, Ex.4). Hence in both cases we can speak in a meaningful way about dimension and we can embark on a proof by induction with respect to dimension. If the theorem is false, then there exists a counterexample for which S has minimal dimension. The algebraic commutator group S' is arcwise connected in case (i) and is Zariski connected in case (ii) (see for instance [Hu75], page 110). Let H be a normal subgroup containing S' which is maximal with respect to being arcwise connected in case (i), with respect to being Zariski connected in case (ii), and with respect to having a dimension smaller than that of S. Then H cannot be a counterexample, and thus H has a non-zero common eigenvector in W.

We now claim that in both cases there is an abelian subgroup G of S which, as H is normal, is automatically in the normalizer of H, which satisfies $S = GH$ and is, in case (i), arcwise connected and, in case (ii), Zariski connected. If we have G, the we apply Lemma III.3.4 and find that S itself has a common eigenvector, contrary to our assumption that S furnishes a counterexample to our theorem. This contradiction will then finish the proof.

In both cases we consider the Lie algebra $\mathbf{L}(S)$ of S. We may assume that $\dim \mathbf{L}(H) = \dim \mathbf{L}(S) - 1$, because in case (ii) we may invoke the fact that S/S' is a Zariski connected abelian algebraic group, whose structure we know (see for instance [Bor69], page 156). Now we let V be a 1-dimensional subalgebra of $\mathbf{L}(S)$ such that $\mathbf{L}(S) = \mathbf{L}(H) \oplus V$. In case (i) we may take $G = \exp V$. In case (ii) we consider the algebraic Lie algebra $\alpha(V)$ generated by V (see for instance [Bor69], page 195). If G is the algebraic Zariski connected subgroup with $\mathbf{L}(G) = \alpha(V)$, then G has the required properties. Thus the proof for the existence of the non-zero eigenvector is finished.

Finally, assume that W is generating. Then W^* is pointed and generating. The adjoint action of G on $\widehat{L}$ has a common non-zero eigenvector $\omega \in W^*$. Let $T = \omega^\perp$. Then $T \cap$ algint $W = \emptyset$, and T is S-invariant by Lemma III.2.3(ii). ∎

In Case (ii), this theorem is a real version of the Kolchin-Lie Theorem for algebraic groups over an algebraically closed field. (See for instance [Bor69], Corollary 10.5 on page 243.)

III.3.6.   Corollary.   *Let W be a pointed cone in a finite dimensional vector space L , let further S be a solvable subgroup of* Aut(W) *which is arcwise connected or algebraic and Zariski connected, and let G be a compact connected subgroup of* Aut(W, L) *which normalizes S . Then GS has a common non-zero eigenvector in W .*

Proof.   By the Frobenius–Perron Theorem for Solvable Groups III.3.5, S has a common non-zero eigenvector. An application of Lemma III.3.4 then proves the corollary. ∎

III.3.7.   Corollary.   *Let G be any connected Lie subgroup of* Aut(W, L) *for a pointed cone W in a finite dimensional real vector space L . If G is compact* *modulo its radical, then G has a common eigenvector in W .*

Proof.   Let R be the radical of G. Then R is a connected solvable normal Lie subgroup of G. If K is any semisimple Levi complement for R in G, then K is a compact semisimple Lie group since G/R is compact. The preceding corollary then proves the claim.   ∎

III.3.8.   Corollary.   *Let W be a pointed cone in a finite dimensional real vector space L. If G is an algebraic subgroup of* $\mathrm{Aut}(W, L)$ *and P is a parabolic subgroup of G (see for instance* [Bo69] *pages following* 261*), then P has a non-zero common eigenvector in W .*

Proof.   By Proposition III.3.3, the parabolic subgroup P is connected. Then the Frobenius–Perron Theorem for Solvable Groups III.3.45 proves the claim.   ∎

The reader should review the three examples displayed after Theorem III.3.1 and become aware once more that these results leave no perceptible room for generalizations. It should be clear by now that in the Frobenius–Perron theory for solvable groups on pointed cones the principal tools, apart from standard Lie group theory, are the Brouwer Fixed Point Theorem and the methods which algebraic group theory provides for those parts involving hypotheses on connectivity in the Zariski topology. The importance of this aspect, notably in the form of Corollary III.3.8, for the Frobenius–Perron theory of semisimple Lie groups we shall see in the next section.

4. The theorems of Kostant and Vinberg

The example of the vector space $L = \mathrm{sl}(2, \mathbb{R})$ with the Lorentzian cone $W = W^+$ and the group of W-automorphisms $\mathrm{Inn}\, L \cong \mathrm{PSL}(2, \mathbb{R})$ shows that a semisimple Lie subgroup of $\mathrm{Aut}(W, L)$ does not in general allow a non-zero common eigenvector in W.

Indeed, if w is a non-zero vector in a finite dimensional real vector space L and G is a subgroup of $\mathrm{Gl}(L)$ such that $Gw \subseteq \mathbb{R}.w$ and if the commutator subgroup G' of G is dense in G—which is certainly the case if G is a semisimple connected Lie group— then the homomorphism $\chi \colon G \to \mathbb{R}^\times$ defined by $gw = \chi(g){\cdot}w$ is necessarily constant since $\chi(G) \subseteq \mathbb{R}^\times$ is commutative. Hence it must be singleton because of $\overline{G'} = G$. Therefore, w is a G-fixed point. Thus a common eigenvector of a semisimple Lie group is necessarily a fixed vector. However, if W is a pointed cone in L and a Lie group $G \subseteq \mathrm{Aut}(W, L)$ has a non-zero fixed vector in W, then G is compact by the Second Theorem on Compact Automorphism Groups of Cones III.2.4. Hence a *non-compact semisimple Lie subgroup G of $\mathrm{Aut}(W, L)$ for a pointed cone L cannot have a common eigenvector.* As a consequence, there is no room for a Perron–Frobenius theory for semisimple non-compact Lie groups in the strict sense of the word.

There is a reasonable and important substitute for such a theory. It deals with the problem of finding necessary and sufficient conditions for the existence of an invariant pointed cone under the action of a given semisimple linear Lie group G on L. In one way or another, such conditions will have to be expressed in terms of structural entities which come along with the semisimple Lie group G.

One such class of entities is the class of maximal compact subgroups K and another one is the class of parabolic subgroups P.

III.4.1.   Proposition. *Let W be a pointed cone in a finite dimensional vector space L. If G is a semisimple connected Lie subgroup of $\mathrm{Aut}(W, L)$, and K a maximal compact and P a parabolic subgroup of G, then K has a common fixed point in the algebraic interior of W and P has a non-zero common eigenvector in W.*

Proof.   The first assertion is a consequence of Theorem III.2.1. The second assertion is just rephrasing Corollary III.3.8 above in view of the fact that G is algebraic (see for instance [Hoch71], page 89).   ■

After this proposition, the existence of common eigenvectors for K and P are certainly necessary conditions for the existence of a pointed invariant cone for a given semisimple linear Lie group G. It is remarkable, that these conditions are, for all practical purposes, sufficient as we shall see presently. First we need some preparation.

III.4.2.   Proposition.   *If L is a finite dimensional real vector space and G a connected semisimple analytic subgroup of $\mathrm{Gl}(L)$, then the following statements are equivalent:*

(1) *There is a G-invariant non-zero pointed cone W in L.*

(2) *There is a G-invariant generating wedge W in L with $W \neq L$.*

(3) *There is a G-invariant wedge W in L with $W \neq H(W)$.*

(4) *There is a decomposition $L = L_1 \oplus \cdots \oplus L_n$ into a direct sum of simple G-submodules such that at least one of the simple components contains a G-invariant non-zero pointed cone.*

Proof.   We preface the proof with the remark, that L is a semisimple G-module due to the semisimplicity of G. Indeed let $\mathfrak{g} = \mathbf{L}(G) \subseteq \mathrm{gl}(L)$ be the Lie algebra of G. Then a vector subspace of L is a G-module if and only if it is a $\mathfrak{g}$-module. By Weyl's Theorem, every $\mathfrak{g}$-module is semisimple. Thus every G-submodule of L has a module complement.

Second remark: If W is a G-invariant wedge, then $H(W)$ and $W - W$ are submodules. Thus we find submodules V and V' of L such that $L = (W - W) \oplus V'$ and that $W - W = H(W) \oplus V$. By Proposition I.2.12, $W \cap V$ is a pointed cone generating V, and since W and V are G-invariant, so is $W \cap V$. Furthermore, $W \oplus V'$ is an invariant generating wedge.

From the second remark is clear that the first three conditions are equivalent. Trivially, (4) implies (1).

We must show that (3) implies (4). Using the semisimplicity of the G-module L we decompose the submodules $H(W)$, V, and V' in direct sums of simple modules. Thus we obtain a sequence of submodules $\{L_m : m = 1, \ldots n\}$ such that $H(W) = L_1 \oplus \cdots \oplus L_j$, $V = L_{j+1} \oplus \cdots \oplus L_k$, and $V' = L_{k+1} \oplus \cdots \oplus L_n$ with $1 \leq j \leq k \leq n$. There is no loss in generality if we concentrate on V and $V \cap W$. We may therefore assume that W is pointed and generating in L, that is, $j = 0$ and $k = n$. If there is an index m with $W \cap L_m \neq \{0\}$, then we are finished since we have found an invariant pointed non-zero cone. This is certainly the case if L is simple. Now assume $n > 1$. Assume now that $W \cap L_m = \{0\}$ for $m = 1, \ldots, n$. Then the projection $L \to L_1 \oplus \cdots \oplus L_{n-1}$ maps W onto a pointed invariant cone in a module with fewer simple direct summands. The claim now follows by induction.   ∎

One should guard oneself against assuming that an invariant wedge W in a direct sum $L = L_1 \oplus \cdots \oplus L_n$ of simple G-modules decomposes accordingly into a direct sum $W = W_1 \oplus \cdots \oplus W_1$ of invariant wedges. Let $L = L_1 \oplus L_2$ with $L_1 = \mathrm{sl}(2, \mathbb{R})$ and $L_2 = \mathrm{so}(3)$ and let $G = \mathrm{Inn}\, L$. Then the quadratic form $q = B_1 \oplus (-B_2)$ (with the Cartan–Killing forms B_n on L_n, $n = 1, 2$) is Lorentzian and invariant (see the Classification Theorem of Lorentzian Lie algebras II.6.14) and hence yields an invariant Lorentzian cone W (see the Classification Theorem of Lorentzian Lie semialgebras II.6.22). But $W \cap L_1$ is one of the two standard Lorentzian invariant cones $\mathcal{W}^+$ or $\mathcal{W}^-$ in $\mathrm{sl}(2, \mathbb{R})$ discussed in Section 3 of Chapter 2, while $W \cap L_2 = \{0\}$. Hence $W \neq (W \cap L_1) \oplus (W \cap L_2)$.

III.4.3.   **Corollary.**   *Let G be a connected semisimple analytic subgroup of the group of vector space automorphisms of a finite dimensional real vector space L and suppose that W is a G-invariant wedge which is not a vector space. Then in W there exists a fixed vector for K and a common eigenvector for P, and there is a K-invariant hyperplane T with invariant cosets, and there is a P-invariant hyperplane H, neither of which meets the algebraic interior of W.*

Proof.   Again we write L as a direct sum of submodules $H(W) \oplus V \oplus V'$ with $W = H(W) \oplus (V \cap W) \oplus \{0\}$. When we apply Proposition III.4.1 to V and $V \cap W$ we immediately find the asserted fixed vector of K and common eigenvector of P. If we denote with $\widehat{G}$ the group of adjoint transformations $\widehat{g}$ on $\widehat{L}$, then $g \mapsto \widehat{g}^{-1} : G \to \widehat{G}$ is an isomorphism. Since W is G-invariant, W^* is $\widehat{G}$-invariant by Lemma I.5.26. The dual vector space $\widehat{L}$ is the direct sum of $\widehat{G}$-submodules which are isomorphic naturally to the duals of $H(W)$, V, and V' respectively. If, with this identification, we write $\widehat{L} = H(W)^{\widehat{\ }} \oplus \widehat{V} \oplus \widehat{V'}$, then we have $W^* = \{0\} \oplus (V \cap W)^{\widehat{\ }} \oplus \widehat{V'}$. Now Proposition III.4.1 applies to the $\widehat{G}$-module $\widehat{L}$ with the pointed invariant non-zero cone $(V \cap W)^{\widehat{\ }}$ and gives us a $\widehat{K}$-fixed vector κ and a common P-eigenvector π in $(V \cap W)^{\widehat{\ }}$. Then, in view of Lemma III.2.3(i), $\kappa^\perp$ and $\pi^\perp$ are K-, respectively, P-invariant hyperplanes. Also, $\kappa^\perp$ is the direct sum of $H(W)$, the annihilator of κ in V, which does not meet the algebraic interior of $V \cap W$, and V'. Hence it does not meet the algebraic interior of W. As a similar statement holds for $\pi^\perp$ the hyperplanes $\kappa^\perp$ and $\pi^\perp$ are the ones whose existence we asserted.   ∎

III.4.4.   **Proposition.**   *Let L be a finite dimensional real vector space and W a wedge with $W \neq H(W)$. If G is a semisimple analytic subgroup of $\operatorname{Aut}(W, L)$ and P a parabolic subgroup of G, then there exists a simple G submodule S of L and non-zero elements $w \in S$, $\omega \in \widehat{L}$ such that*

$$Pw \subseteq \mathbb{R} \cdot w, \qquad \widehat{P}\omega \subseteq \mathbb{R} \cdot \omega \qquad and \qquad \omega(S) \neq \{0\}.$$

Proof.   As in the proof of Corollary III.4.3 we decompose L into a direct sub $H(W) \oplus V \oplus V'$ of submodules with $W = H(W) \oplus (V \cap W) \oplus \{0\}$. Then it suffices to consider V and $V \cap W$. Thus we may assume that W is pointed and generating.

We now prove the claim by induction with respect to the number of simple summands of L. If there is a simple submodule S with $S \cap W \neq \{0\}$, then we apply Corollary III.4.3. to S and $S \cap W$; this gives us the desired element w and a non-zero linear form ω_S on S with $\widehat{P}\omega_S \subseteq \mathbb{R} \cdot \omega_S$. If we extend ω_S to a linear form $\omega \in \widehat{L}$ by defining it to be zero on a submodule complementary to S, then ω is the required linear form. In particular, this gets the induction started.

Now assume that there is no simple submodule meeting W non-trivially. We write $L = L_1 \oplus \ldots \oplus L_n$ with simple submodules L_k, $k = 1, \ldots, n$, $n < 1$. If p denotes the projection onto the first $n-1$ summands along L_n, then our assumption guarantees that $p(W)$ is a pointed generating invariant cone in $L' = L_1 \oplus \ldots \oplus L_n$. The induction hypotheis applies and gives us a simple submodule $S \subseteq L'$, a non-zero vector $w \in S$ with $Pw \subseteq \mathbb{R} \cdot w$ and a linear form ω' on L' whioch does not vanish on S and is a common eigenvector for the adjoint action of P. If we extend ω_S to a linear form on L by declaring it to be zero on L_{n-1}, then ω has the required properties.   ∎

Let us assume momentarily that G is an arbitrary analytic subgroup of $\mathrm{Gl}(L)$ for the real vector space L. The Lie algebra $\mathfrak{g} = \mathbf{L}(G) \subseteq \mathfrak{gl}(L)$ allows a complexification $\mathfrak{g}_C = \mathbb{C} \otimes \mathfrak{g}$ operating on the complexification $L_C = \mathbb{C} \otimes L$. The complex Lie group $\mathrm{Gl}(L_C)$ contains a complex analytic group G_C, namely, the one whose Lie algebra is exactly $\mathbf{L}(G_C) = \mathfrak{g}_C$. As usual, we identify L with a real vector subspace of L_C so that $L_C = L + i \cdot L$.

In the following discussion, we shall use standard information on semisimple Lie groups and semisimple Lie algebras such as the Cartan decomposition, the Iwasawa decomposition, and the Bruhat decomposition. For these crucial concepts, we refer to other sources (see e.g. [He78], page 182, page 403).

III.4.5.   Theorem.   (The Invariant Cone Theorem of Kostant and Vinberg) *Let L be a finite dimensional real vector space and G a semisimple Lie subgroup of $\mathrm{Gl}(L)$. Let $G = KAN$ be an Iwasawa decomposition of G and let P be the normalizer of the Borel subgroup AN in G. Then P is a parabolic subgroup of G and the following conditions are equivalent:*

(1) *K has a non-zero fixed point.*

(2) *There is a simple submodule S in L such that each of the following conditions holds:*

    (a) *P has a common eigenvector in S.*

    (b) *There is a P-invariant hyperplane in S.*

(3a) *There is a G-invariant pointed cone $W \neq \{0\}$ in L.*

(3b) *There is a G-invariant generating wedge $W \neq L$ in L.*

(3c) *There is a G-invariant wedge in L which is not a vector space.*

Proof.   For the fact that P is a parabolic subgroup of G see [Wa72], p.45 ff. Proposition III.4.2 tells us that (3a),(3b), and (3c) are equivalent. Proposition III.4.3 shows that (3b) implies (1), and Proposition III.4.4 shows that (3c) implies (2).

The principal task of the proof which now remains is to show that (1) implies (3b) and that (2) implies (3a). In either case we have a common non-zero eigenvector $w \in L$. If this eigenvector is to be contained in some G-invariant wedge W_0, then the entire orbit Gw is contained in W_0 and so is the wedge generated by this orbit. Hence we are going to consider the smallest wedge W containing the orbit Gw, and we have to show that W is not a vector space.

Let us now assume (1). Let K be a maximal compact subgroup of K and $\mathbf{k} = \mathbf{L}(K)$ its Lie algebra inside the Lie algebra $\mathfrak{g} = \mathbf{L}(G)$. Let $\mathfrak{g} = \mathfrak{k} \oplus \mathfrak{p}$ be a Cartan decomposition of $\mathfrak{g}$. We claim that there is a scalar product on L such that all operators $g \in e^{\mathfrak{p}}$ are symmetric with respect to this scalar product.

For a proof of this claim, we need a brief detour into the complexification of $\mathfrak{g}$ and G. In fact, $\mathfrak{g}_u = \mathfrak{k} \oplus i \cdot \mathfrak{p}$ is a compact real form of the complexification $\mathfrak{g}_C$. The subgroup G_u with $\mathbf{L}(G_u) = \mathfrak{g}_u$ is compact. By Weyl's unitary trick we may assume that there is a complex scalar product $(\bullet \mid \bullet)$ on L_C with $(gx \mid gy) = (x \mid y)$ for all $g \in G_u$. (See our remarks in the proof of Theorem III.2.1). Then all $Y \in \mathfrak{g}_u$ are skew hermitian with respect to $(\bullet \mid \bullet)$: Indeed the constant derivative of $t \mapsto (e^{t \cdot Y} x \mid e^{t \cdot Y} y) = (x \mid y)$ at 0 gives exactly $(Y(x) \mid y) + (x \mid Y(y)) = 0$ for all $x, y \in L_C$, that is, $Y^* = -Y$ with the adjoint operator Y^* of Y with respect to our scalar product. If now $X \in \mathfrak{p}$, then $i \cdot X \in \mathfrak{g}_u$, whence X is hermitian. Hence all elements of $e^{\mathfrak{p}}$ are hermitian. This proves our claim and signals the end of the detour into the complex domain.

Now hypothesis (1) grants us a non-zero K-fixed vector $w \in L$. Every element $g \in G$ has a polar decomposition $g = e^X k$ with $k \in K$ and $X \in \mathfrak{p}$. Then we compute $(gw \mid w) = (e^X kw \mid w) = (e^X w \mid w)$. Now we set $Z = \frac{1}{2} \cdot X$ and obtain a hermitian operator $T = e^Z$ with $T^* T = T^2 = e^X$. Then $(gw \mid w) = (T^* Tw \mid w) = (Tw \mid Tw) > 0$. The function $u \mapsto (u \mid w) : L \to \mathbb{R}$ is linear. Thus if u is any element of the closed convex hull C of the orbit Gw, we may conclude $(u \mid w) \geq 0$. The smallest wedge W containing Gw is $\overline{\mathbb{R}^+ \cdot C}$. Hence $(W \mid w) \subseteq \mathbb{R}^+$. Thus W is not a vector space and (3b) is proved.

Now we assume (2) and prove (3a). Let w be a non-zero common eigenvector of P in S. Let $L = S \oplus L'$ be a module decomposition. It will suffice now to show that $Gw + L'$ is contained in some closed half space, for then the closed wedge generated by Gw cannot equal S and thus will be the desired invariant cone.

We note next that P is connected with respect to the Zariski topology, because G is algebraic and thus Proposition III.3.3 applies. Since the homomorphism $\chi : P \to \mathbb{R}$ given by $gw = \chi(g) \cdot w$ is algebraic, we know $P \cdot w \subseteq \mathbb{R}^+ \cdot w$.

According to the Bruhat decomposition of G (see for instance [He78]) we find an element $g \in G$ such that $G = \overline{gNg^{-1}P}$. Since $N \subseteq P$, then $G \subseteq \overline{gPg^{-1}P}$. Now $Gw \subseteq (\overline{gPg^{-1}P})w \subseteq \overline{gPg^{-1}Pw} \subseteq \overline{\mathbb{R}^+ \cdot gPg^{-1}w}$.

Next let H_S be a P-invariant hyperplane in S guaranteed by (2a). Set $H = H_S \oplus L'$. Then $\widetilde{H} \overset{\text{def}}{=} gH$ is a hyperplane invariant under $\widetilde{P} \overset{\text{def}}{=} gPg^{-1}$. Let ω be a non-zero functional on L with $\widetilde{H} = \omega^\perp$. Then ω is a joint eigenvector of $\widetilde{P}^{\wedge}$ and since $S \not\subseteq \widetilde{H}$ we know that ω is non-zero. From $\widetilde{P}^{\wedge}\omega \subseteq \mathbb{R} \cdot \omega$, we conclude once again $\widetilde{P}^{\wedge}\omega \subseteq \mathbb{R}^+ \cdot \omega$ as we did above, using the Zariski connectedness of P. This means that each of the two half-spaces bounded by $\widetilde{H}$ is preserved by $\widetilde{P}$. Now w is contained in at least one of them, say H^+. Then $gPg^{-1}w \subseteq \widetilde{P}H^+ \subseteq H^+$ by the $\widetilde{P}$-invariance of H^+. Hence $Gw \subseteq \mathbb{R}^+ \cdot H^+ = H^+$. Thus $Gw + L_2 \subseteq H^+ + \widetilde{H} = H^+$ and this is all we need in order to complete our proof.   ∎

We note that the proof also shows that with the common eigenvector w of P in S mentioned in (2a) we have

(2aa) $\overline{\text{conv}}\left(\mathbb{R}^+ \cdot (Gw)\right) \neq S$

where $\overline{\text{conv}}$ denotes the formation of the closed convex hull.

III.4.6.   Remark. If, in addition to the conditions of Theorem III.4.5, L is simple and the $\widehat{G}$-module $\widehat{L}$ is isomorphic to the G-module L, then condition (2) in Theorem III.4.5 is equivalent to

(2a) P has a common eigenvector.

Proof.   If (2a) is satisfied, then under the present conditions, $\widehat{P}$ has a common eigenvector on $\widehat{L}$, and this is exactly Condition III.4.5(2)(b).   ∎

Application to Lie algebras with invariant cones

The situation which is of particular interest to us is the particular case that L is a finite dimensional real Lie algebra and $G = \text{Inn } L$. If L is semisimple, then G is algebraic, hence, in particular, a Lie group and thus we have $G = \text{Inn } L = \text{INN } L$. If now W is a wedge in L, it is invariant if and only if it is G-invariant. This observation permits us to translate the Invariant Cone Theorem of Kostant and Vinberg III.4.5 into this frame work.

III.4.7.   Theorem.   (The Existence Theorem of Invariant Cones in Semisimple Lie algebras) *Let $\mathfrak{g}$ denote a finite dimensional real semisimple Lie algebra. Let $\mathfrak{k}$ denote a maximal compact subalgebra and $\mathfrak{p}$ a minimal parabolic subalgebra (obtained as normalizer of $\mathfrak{a} \oplus \mathfrak{n}$ where $\mathfrak{g} = \mathfrak{k} \oplus \mathfrak{a} \oplus \mathfrak{n}$ is an Iwasawa decomposition of $\mathfrak{g}$). Then the following conditions are equivalent:*

(1) *The center of $\mathfrak{k}$ is non-zero.*

(2) *There is a 1-dimensional vector subspace $\mathfrak{f}$ in $\mathfrak{g}$ with $[\mathfrak{p}, \mathfrak{f}] \subseteq \mathfrak{f}$.*

(3a) *$\mathfrak{g}$ possesses an invariant non-zero pointed cone.*

(3b) *$\mathfrak{g}$ possesses a generating invariant proper wedge.*

(3c) *$\mathfrak{g}$ possesses an invariant wedge which is not a vector space.*

Proof.   We apply Theorem III.4.5. If K is the analytic subgroup with $\mathbf{L}(K) = \mathfrak{k}$, then a vector $x \in \mathfrak{g}$ is a K-fixed vector if an only if it is in the centralizer of $\mathfrak{k}$. For every such vector, $\mathbb{R} \cdot x + \mathfrak{k}$ is a compact algebra, and thus $x \in \mathfrak{k}$ is a consequence of the compactness of $\mathfrak{k}$. It follows that Condition (1) above corresponds exactly to Condition III.4.5(1).

Next we consider Condition (2). A one dimensional vector subspace $\mathfrak{f}$ satisfies $[\mathfrak{p}, \mathfrak{f}] \subseteq \mathfrak{f}$ if and only if $\mathfrak{f}$ is invariant under the action of the analytic group P generated by $\mathfrak{p}$. Thus this condition is equivalent to Condition III.4.5(2)(a). But in our present situation, the hypothesis of Remark III.4.6 is satisfied for each simple summand of L, since the Cartan–Killing form on $\mathfrak{g}$ is a non-degenerate invariant bilinear form which allows us to identify the adjoint module with its dual. Hence condition (2) above corresponds to condition III.4.5(2). Clearly, Conditions (3a),(3b), and (3c) above correspond to the analogous conditions in Theorem III.4.5. Thus the theorem is just a corollary of Theorem III.4.5.   ■

This theorem fails for solvable Lie algebras: The Lie algebra $\mathfrak{g}$ of the group of motions of the euclidean plane is a counterexample; this algebra is of the type (i) of Lemma II.3.2 with $\lambda = 0$ and $\omega = 0$. Let $\mathfrak{k}$ be any 1-dimensional vector space not contained in $J = [\mathfrak{g}, \mathfrak{g}]$. Then $\mathfrak{k}$ is a maximal compact and maximal compactly embedded subalgebra which is abelian hence satisfies the condition (1) of Theorem III.4.7. The ideal J is likewise abelian, but is not compactly embedded. The only invariant wedges in the algebra $\mathfrak{g}$ which are different from $\{0\}$ and $\mathfrak{g}$ are the two half-spaces bounded by $\mathfrak{g}'$ according to the First Classification Theorem of Low Dimensional Semialgebras II.3.4. The condition (3a) is violated; conditions

(3b) and (3c) hold. It is not particularly reasonable to ask for a commentary on an analog of Condition III.4.7(2) in our present situation; if the parabolic subalgebra $\mathfrak{p}$ of Theorem III.4.7 would have to be compared with anything in the algebra $\mathfrak{g}$ in the present situation it would have to be all of $\mathfrak{g}$. There is no 1-dimensional ideal in $\mathfrak{g}$, hence the analog of Condition III.4.7(2) fails; since there is a hyperplane ideal J, condition III.4.5(2)(a) is satisfied.

We observe one consequence which results from The Existence Theorem III.4.7 and the Fourth Cartan Algebra Theorem:

III.4.8.   Corollary.   *Let L be a finite dimensional simple real Lie algebra and K a maximal compact subalgebra. Then L contains a pointed generating invariant cone if and only if K is not semisimple. In particular, if K is not semisimple, then L has a compactly embedded Cartan subalgebra.*

Proof.   Since a compact Lie algebra is reductive, the first part of the corollary is a consequence of Theorem III.4.7. The second assertion, however, follows from Theorem III.2.14.   ■

5. The reconstruction of invariant cones

We have seen in the Third Cartan Algebra Theorem III.2.12 that any Lie algebra supporting an invariant pointed generating cone contains compactly embedded Cartan algebras. In the Uniqueness Theorem for Invariant Cones III.2.15 we observed that such a cone W is uniquely determined by its intersection with a fixed compactly embedded Cartan algebra H. A classification theory for invariant pointed generating cones in L in terms of H consists of two steps. Firstly, given the cone W in L we want to know how we can reconstruct W from $W \cap H$. We shall give at least one answer to this question in the following paragraphs. A second step which we shall take later will have to determine accurately which cones in H can actually arise as intersections $H \cap W$ with invariant generating cones W.

The orthogonal projection onto a compactly embedded Cartan algebra

We consider a finite dimensional real Lie algebra L and assume that it contains a compactly embedded Cartan algebra H. To say that the Cartan subalgebra H is compactly embedded is tantamount to saying that the group $\mathrm{INN}_L H$ (see Definition III.2.9) is a torus subgroup of $\mathrm{Aut}(L)$. We shall prove the following result in the Appendix because it belongs to the domain of pure Lie algebra and Lie group theory.

III.5.1. **Proposition.** *Let H be a compactly embedded Cartan algebra in a finite dimensional real Lie algebra L. Then $\mathrm{INN}_L H$ is its own centralizer in $\mathrm{INN}\, L$. The vector space H is the precise fixed point set of $\mathrm{INN}_L H$ on L.*

Proof. See Corollary A.2.36. ∎

Let us abbreviate the torus $\mathrm{INN}_L H$ with T. Then $T \subseteq \mathrm{Hom}(L, L)$. In the following we recall some basic facts on the representation theory of compact groups on finite dimensional vector spaces. If G is a compact group and $\pi: G \to \mathrm{Gl}(L)$ is a representation of G on a finite dimensional vector space, then the operator $P = \int_G \pi(g)dg$ with normalized Haar measure on G is a G-equivariant projection onto the space of fixed vectors F of L under $\pi(G)$. That is, $P^2 = P$, $P\big(\pi(g)(x)\big) = \pi(g)P(x) = P(x)$ for all $g \in G$. Moreover, L is the direct sum of F

and $F^+ \stackrel{\text{def}}{=} \ker P$, and F^+ is the span of all vectors $\pi(g)(x) - x$, $g \in G$, $x \in X$. On the G-module F^+, the vector 0 is the only fixed vector.

All of these facts are standard, and they are also readily verified. We apply this background information to the special situation that L is our given Lie algebra, G is the torus T.

III.5.2.   Definition.   We define the vector space endomorphism $p: L \to L$ by $p = \int_T t\, dt$ with normalized Haar measure dt on the torus T. In other words, for all $x \in L$ we have $p(x) = \int_T t(x)\, dt$.

III.5.3.   Proposition.   *Let H be a compactly embedded Cartan algebra in the finite dimensional Lie algebra L and let p be as in* Definition III.5.2. *Then $p: L \to L$ is T-equivariant projection onto the T-fixed point set H, that is, $p(tx) = tp(x) = p(x)$ for all $t \in T$. The Lie algebra L is the direct sum of H and the H-submodule $H^+ \stackrel{\text{def}}{=} \ker p$. In other words,*

$$L = H \oplus H^+ \text{ with } [H, H^+] \subseteq H^+, \text{ and } Z(H, L) = H,$$

where $Z(H, L)$ denotes the centralizer of H in L. With respect to any scalar product which is invariant under the action of T, the direct sum is orthogonal.

Proof.   The proof is clear from the preceding remarks and Proposition III.5.1. ∎

We note that in this proof we did not use the hard part of Proposition III.5.1, that is the information that T is its own centralizer in $\text{INN}\, L$ but only the straightforward fact that H is its own centralizer in L and thus the precise set of fixed vectors of T in L.

At this point we assume that W is a pointed generating invariant cone in L and H a fixed compactly embedded Cartan algebra. The set $p(W)$ is stable under addition and non-negative scalar multiplication in H, but it is not a priori clear whether it is even a wedge, that is, whether it is topologically closed. However, the following proposition sheds light onto this situation:

III.5.4.   Proposition.   *For an invariant wedge W and a compactly embedded Cartan algebra H the following equation holds:*

$$p(W) = H \cap W.$$

Proof.   Since p is a projection onto H, the inclusion $H \cap W \subseteq p(W)$ is trivial. Now let $w \in W$. Then $p(w) \in H$ by the definition of p, and since $gw \in W$ for all $g \in \text{INN}\, L$, in particular for all $t \in T$, then $p(w) = \int_T t(w)\, dw \in W$ as W is closed and convex. Hence $p(w) \in H \cap W$.   ∎

III.5.5.   Proposition.   *If, in addition to the conditions of* Proposition III.5.4, $W \neq L$, *then*

$$(\text{INN}\, L)(H^+) \cap \text{int}\, W = \emptyset.$$

Proof.   Since W is invariant, it suffices to show that

$$H^+ \cap \text{int}\, W = \emptyset.$$

Therefore, we assume the contrary of this condition and consider an element $x \in H^+ \cap \operatorname{int} W$. Then $0 = p(x) \in p(\operatorname{int} W) = \operatorname{int}_H(p(W))$ (since p is an open map!) $= \operatorname{int}_H(H \cap W)$ in view of the preceding Proposition III.5.4. But if 0 is an inner point of the wedge $H \cap W$ in H, then $H \cap W = H$, that is, $H \subseteq W \cap -W$. The edge of of the invariant wedge W is an ideal of L which must be proper, since we assumed $W \neq L$. However, no Cartan algebra can be contained in a proper ideal, since its image in the factor algebra is a Cartan algebra. This is a contradiction which proves the proposition. ∎

Facts on compactly embedded Cartan algebras

The following discourse requires that we first gather more information about Lie algebras with compactly embedded Cartan algebras. Details will be proved in the Appendix.

III.5.6.   Proposition. *Suppose that H is a compactly embedded Cartan algebra of a finite dimensional Lie algebra L. Then the following conclusions hold:*

  (i) *There is a unique maximal compactly embedded subalgebra $K(H)$ containing H.*

 (ii) *A subalgebra K of L is a maximal compactly embedded subalgebra of L if and only if $\operatorname{INN}_L K$ is a maximal compact subgroup of $\operatorname{INN} L$.*

(iii) *The normalizer $N(H) = \{g \in \operatorname{INN} L : g(\operatorname{INN}_L H)g^{-1} = \operatorname{INN}_L H\}$ of the maximal torus $\operatorname{INN}_L H$ in $\operatorname{INN} L$ is contained in the maximal compact subgroup $\operatorname{INN}_L K(H)$ of $\operatorname{INN} L$.*

 (iv) *$N(H)/(\operatorname{INN}_L H)$ is finite.*

Proof.   See Theorem A2.11 and Theorem A2.40. ∎

These results allow us to define, for each Lie algebra L with a compactly embedded Cartan algebra H, a *Weyl group* as follows:

III.5.7.   Definition. If H is a compactly embedded Cartan subalgebra of a finite dimensional Lie algebra L, then the finite group $N(H)/\operatorname{INN}_L H$ is called the *Weyl group of L* and is denoted $\mathcal{W}(H, L)$, or, if no confusion is possible, with $\mathcal{W}$. This group operates on H as follows: For $\nu \in \mathcal{W}(H, L)$, say $\nu = n(\operatorname{INN}_L H)$ with $n \in N(H) \subseteq \operatorname{Gl}(L)$ and for $h \in H$ we have

$$\nu \cdot h = n(h) \qquad \text{and} \qquad e^{\operatorname{ad} \nu \cdot h} = n e^{\operatorname{ad} h} n^{-1}.$$

Thus we have to consider the Weyl group as an important invariant of the pair (L, H). Whenever H is a compactly embedded Cartan algebra, there is a finite Weyl group W of linear self maps of H. This Weyl group is simply the classical Weyl group as defined for the compact group $\operatorname{INN}_L K(H)$ with respect to the maximal torus $\operatorname{INN}_L H$.

The trace of an invariant cone on a Cartan algebra

With this background we now return to the task at hand.

III.5.8.   Lemma. *If W is an invariant wedge in a finite dimensional Lie algebra L with a compactly embedded Cartan algebra H, then the wedge $H \cap W$ in H is invariant under the action of the Weyl group $\mathcal{W}(H, L)$.*

Proof.   The proof is immediate. If $w \in H \cap W$, and if $\nu = n(\text{INN}_L H)$ is an element of the Weyl group, then $\nu \cdot w = n(w) \in H$ since n is in the normalizer $N(H)$. But also $n(w) \in W$ since W is invariant. Hence $\nu \cdot w \in H \cap W$.   ∎

III.5.9.   Lemma. *If, in addition to the conditions of the preceding lemma, W is pointed and generating, and if $Z\big(K(H)\big)$ denotes the center of the unique maximal compactly embedded subalgebra $K(H)$ containing H, then $Z\big(K(H)\big) \cap \operatorname{int} W \neq \emptyset$.*

Proof.   The group $\text{INN}_L K(H)$ is a compact automorphism group of the pointed generating cone W. Hence, by Theorem III.2.1, it has a fixed point x in the interior of W. Then $e^{\operatorname{ad} k} x = x$ for all $k \in K(H)$, and this implies $[k, x] = 0$ for all $k \in K(H)$ (via differentiation). Hence $x \in Z\big(K(H)\big)$. But since H is a Cartan algebra of $K(H)$ we have $Z\big(K(H)\big) \subseteq H$. Thus $x \in H \cap \operatorname{int} W \cap Z\big(K(H)\big) = \operatorname{algint}(H \cap W) \cap Z\big(K(H)\big)$ in view of Condition $(*)$ in the proof of Theorem III.2.15.   ∎

We summarize the essential features of the preceding discourse in the following propositon:

III.5.10.   Proposition. *Let W be an invariant pointed generating cone in a finite dimensional Lie algebra L. Fix a compactly embedded Cartan algebra H and set $C = H \cap W$. We write $L = H \oplus H^+$ according to Proposition III.5.3 and recall the projection p from L onto H along H^+. Then the following conclusions hold:*

  (i) $C = p(W)$.

  (ii) C *is invariant under the Weyl group* $\mathcal{W}(H, L)$.

  (iii) $(\operatorname{algint} W) \cap (\text{INN } L)(H^+) = \emptyset$.

  (iv) $(\operatorname{algint} W) \cap Z\big(K(H)\big) \neq \emptyset$.   ∎

Reconstructing cones

With this proposition we have at least some idea what the trace of an invariant cone on a Cartan algebra is like. Eventually we expect to formulate a complete set of conditions which characterizes such traces fully. For the moment we shall address the simpler question by what devices we can reconstruct an invariant cone W when we are only given its trace $C = H \cap W$ on a compactly embedded Cartan algebra H. For this purpose we consider two canonical constructions:

III.5.11.   Definition.   For a wedge C in a compactly embedded Cartan algebra H of a Lie algebra L we write

(i) $C^\star$ for the closed convex hull of $(\operatorname{Inn} L)(C)$ (which is the same as the closed convex hull of $(\operatorname{INN} L)(C)$), and

(ii) $\widetilde{C} = \bigcap_{g \in \operatorname{Inn} L} gp^{-1}(C)$.

First we make some statements on $\widetilde{C}$:

III.5.12.   Proposition.   *Let C be a wedge in a compactly embedded Cartan algebra H of a Lie algebra L and let p again be the projection onto H along H^+. Then the following conlusions hold:*

(i) $\widetilde{C} = \{x \in L : p((\operatorname{Inn} L)x) \subseteq C\}$.

(ii) *$\widetilde{C}$ is an invariant wedge in L whose edge is the largest ideal of L contained in H^+ as soon as C is pointed.*

(iii) $H \cap \widetilde{C} = p(\widetilde{C}) = \{c \in C : p((\operatorname{Inn} L)c) \subseteq C\} \subseteq C$.

Proof.   (i) By the definition of $\widetilde{C}$ we have $x \in \widetilde{C}$ if and only if, for each inner automorphism g, the relation $x \in g^{-1}p^{-1}(C)$, that is, the relation $p(gx) \in C$ holds.

(ii) All sets $gp^{-1}(C)$ are wedges as g ranges through $\operatorname{Inn} L$. Thus the intersection of all of these sets is a wedge. Since each inner automorphism g permutes these sets, their intersection is invariant. The edge of an invariant wedge is always an ideal. The edge of $\widetilde{C}$ is obviously contained in the edge of the wedge $p^{-1}(C)$, and if C is pointed, this edge is $p^{-1}(0) = H^+$. Conversely, if I is any ideal of L contained in H^+, then $gI \subseteq gp^{-1}(0) \subseteq gp^{-1}(C)$, and thus I is contained in $\widetilde{C}$. Hence the edge of $\widetilde{C}$ is the largest ideal contained in H^+.

(iii) The first equality follows from Proposition III.5.4. If $c \in H \cap \widetilde{C}$, then, in particular, $c \in H \cap p^{-1}(C) = C$, and then the rest follows from (i) above.   ∎

It is now clear that the construction $C \mapsto \widetilde{C}$ will always produce an invariant wedge. If C is pointed, and H^+ does not contain any non-zero ideals, then $\widetilde{C}$ is pointed. The hypothesis that H^+ does not contain non-zero ideals is not grave: If I should be the largest such ideal, then we may pass to L/I without losing much generality for our purposes and continue our arguments in the factor algebra. More serious is the question whether $\widetilde{C}$ is non-zero—let alone generating. In general, $\widetilde{C}$ will indeed be $\{0\}$. The situation is somehow the reverse one with $C^\star$, and this is why we consider both constructions.

In the Appendix, we shall present the following Lemma

III.5.13.   Lemma.   *Let K be any compactly embedded subalgebra in a Lie algebra L, and suppose $k \in K$. Then the function*

$$(u, v) \mapsto e^{\operatorname{ad} u} v : L \times K \longrightarrow L$$

is open at the point $(0, k)$ if $\ker(\operatorname{ad} k) \subseteq K$.

Proof.   See Lemma A.2.24(iii).   ∎

III.5.14.   Proposition. *Let C be a wedge in a compactly embedded Cartan algebra H of a Lie algebra L. Then the following conclusions hold:*

(i) *$C^\star$ is the smallest invariant wedge of L containing C.*

(ii) *If C is generating in H, then $(\operatorname{Inn} L)(C)$ contains inner points of L.*

(iii) *If C is generating in H, then $C^\star$ is generating in L.*

Proof.   The proof of (i) is clear.

(ii) Suppose that C is generating in H, that is, has inner points of H and let c be a regular point of L in the H-interior of C; since the set of regular points of a Cartan algebra is dense, such a c exists. As c is regular and H is abelian by Theorem III.2.12, $H = \ker(\operatorname{ad} c)$. Now we can apply the preceding Lemma III.5.13 and we conclude that c is an inner point of $(\operatorname{Inn} L)(C)$ in L.

(iii) is now immediate from (ii).   ∎

We know that $C^\star$ is an invariant wedge of L which is generating as soon as C is generating in H. The problem here is whether $C^\star$ is in fact different from L—let alone pointed. In general we can have $C^\star = L$.

The information we have, however, suffices now to arrive at a satisfactory statement *if we know that C is the trace of an invariant wedge of L.*

III.5.15.   Theorem. (The Reconstruction Theorem of Invariant Cones) *Let H be a compactly embedded Cartan algebra in a real Lie algebra L, and let C be a pointed generating cone in H. Then the following statements are equivalent:*

(1) *There exists an invariant pointed cone W in L such that $C = H \cap W$.*

(2) *$C = H \cap \widetilde{C}$ and H^+ contains no non-zero ideal of L.*

(3) *$p\big((\operatorname{Inn} L)(C)\big) \subseteq C$, (that is, each conjugacy class of an element $c \in C$ projects into C under p), and H^+ contains no non-zero ideal of L.*

Moreover, if these conditions are satisfied, then $W = \widetilde{C} = C^\star$.

Proof.   (1)$\Rightarrow$(2): We assume that $C = H \cap W$ for some invariant pointed generating cone W in L. Then for each inner automorphism g we have $W = gW \subseteq gp^{-1}p(W) = gp^{-1}(H \cap W) = gp^{-1}(C)$ in view of Proposition III.5.10.i. Hence $W \subseteq \widetilde{C}$ by the definition of $\widetilde{C}$. Using the same for W and $\widetilde{C}$ we observe $C = p(W) \subseteq p(\widetilde{C}) = H \cap \widetilde{C} \subseteq H \cap C$ considering Proposition III.5.12.iii. Therefore, $C = H \cap \widetilde{C}$.

We must show that H^+ contains no non-zero ideal. Suppose that $I \neq \{0\}$ is an ideal of L contained in H^+. The torus $T = \operatorname{INN}_L H$ acts on H^+ without non-zero fixed point as H is the exact set of T-fixed points by Proposition III.5.1. Hence T acts without non-zero fixed points on I. Now let J be a minimal non-zero T-invariant vector subspace of I. Then $\dim J = 2$, because $\dim J = 1$ is only possible if J consists of T-fixed points as T is connected. Now J is a simple H-module under the adjoint action of H on H^+ (see Proposition III.5.3). Since T is a group of automorphisms of L, the vector space $[J, J]$ is a T-invariant subspace, too. Also $[J, J] \subseteq [I, I] \subseteq I \subseteq H^+$. But $\dim[J, J] \leq 1$, because of $\dim J = 2$ implying $\dim \bigwedge^2 J = 1$. Thus again $[J, J]$ consists of T-fixed points, but 0 is the only T-fixed point of H^+. Hence $[J, J] = \{0\}$, that is, J is abelian. Now J is an abelian ideal of the subalgebra $H \oplus J$. The representation $\rho: H \to \operatorname{gl}(J)$ given by $\rho(x) = \operatorname{ad} x|J$ has the property $\operatorname{Spec} \rho(x) \subseteq i \cdot \mathbb{R}$ by Proposition III.2.11, since

H is compactly embedded. Thus $\rho(H)$ is isomorphic to an abelian subalgebra of $\mathrm{gl}(2,\mathbb{R}) \cong \mathbb{R} \oplus \mathrm{sl}(2,\mathbb{R})$ in which each element has purely imaginary spectrum. Such a subalgebra is at most 1-dimensional. Since ρ is non-zero, $H_0 \stackrel{\text{def}}{=} \ker \rho$ is a hyperplane of H. But C is generating in H, so we find an element $h \in \mathrm{algint}\, C \backslash H_0$. Since the spectrum of $\rho(h)$ is purely imaginary, there are two basis vectors u and v of J such that $\rho(h)(u) = v$ and $\rho(h)(v) = -u$. Thus the subalgebra $A = \mathbb{R}\cdot h + \mathbb{R}\cdot u + \mathbb{R}\cdot v$ is defined by the identities $[h,u] = v$, $[h,v] = -u$, and $[u,v] = 0$. Hence A is the Lie algebra of the group of motions of the euclidean plane. Moreover $W \cap A$ is a pointed invariant cone in A, and because of $h \in \mathrm{int}\, W$ the cone $A \cap W$ is generating in A. By the First Classification Theorem of Low dimensional Semialgebras II.3.4, such a situation is impossible. This contradiction shows that I cannot exist as assumed. (Note that the last part of the proof is an application of the Standard Testing Device II.4.7!)

(2)$\Rightarrow$(3): Consider an inner automorphism g of L and and element c of C. From (2) we know $c \in \widetilde{C}$ and thus $c \in \widetilde{C} \subseteq g^{-1}p(C)$, whence $p(gc) \in C$ which was to be shown.

(3)$\Rightarrow$(1): We define $W = \widetilde{C}$ and know right away that W is an invariant wedge. By (3) we have $(\mathrm{Inn}\, L)(C) \subseteq p^{-1}(C)$ and thus $C \subseteq \widetilde{C} = W$ by Proposition III.5.12.i. Because of Proposition III.5.12.iii, we conclude $C = H \cap W$. By Proposition III.5.12.ii, the wedge W is pointed, since its edge is an ideal contained in H^+, hence must be zero in view of hypothesis (3). Finally we must show that W contains inner points. But by Proposition III.5.14.iii, $C^\star$ has inner points. But W is invariant and contains C, hence also $C^\star$ by Proposition III.5.14.i. Hence W has interior points. The equivalence of the conditions (1),(2) and (3) is now established.

Under these conditions we have seen $C^\star \subseteq \widetilde{C} = W$. In particular, this implies that $C^\star$ is pointed. But $C^\star$ is generating as we saw, and $H \cap C^\star = C = H \cap W$. Hence the Uniqueness Theorem for Invariant Cones III.2.15 shows $C^\star = W$. The proof of the theorem is now complete. ∎

In Proposition III.5.6 we saw that every compactly embedded Cartan algebra H in L uniquely determines a maximal compactly embedded subalgebra $K(H)$ containing H. If, however, the presence of a compactly embedded Cartan algebra is due to the existence of an invariant pointed generating cone W in L according to the Fourth Cartan Algebra Theorem II.2.14, then further information becomes available rather quickly.

Quite generally, as a compact Lie algebra, $K(H)$ is the direct sum of its center $Z(K(H))$ and its (semisimple!) commutator algebra $K(H)'$. Naturally, $K(H)$ is contained in the centralizer $Z(Z(K(H)), L)$ of $Z(K(H))$ in L. But in our special situation we have the following information:

III.5.16. **Theorem.** *Let W be an invariant pointed generating cone in a finite dimensional real Lie algebra L and let H denote a compactly embedded Cartan algebra. Then the following statements are true:*

(i) $Z(K(H)) \cap \mathrm{int}\, W \neq \emptyset$.

(ii) $K(H)' \cap W = \{0\}$.

(iii) $Z(Z(K(H)), L) = K(H)$.

Proof.   Statement (i) is just Lemma III.5.9.

(ii) follows directly from Proposition III.2.2.

(iii) Let us write $Z = Z\Big(Z\big(K(H)\big), L\Big)$. The group $\mathrm{INN}_L\, Z$ leaves W invariant and fixes at least one vector $x \in \mathrm{int}\, W$ according to (i) above. Thus $\mathrm{INN}_L\, Z$ is compact by the Second Theorem on Compact Automorphism Groups of Cones III.2.4. Then Z is a compactly embedded subalgebra. Since $K(H) \subseteq Z$ by the remarks preceding the proposition, then the maximality of $K(H)$ implies $K(H) = Z$. ∎

6. Cartan algebras and invariant cones

In this section we deal with finite dimensional real Lie algebras L possessing compactly embedded Cartan algebras H. As usual, we will denote with $L_{\mathbb{C}}$ the complexification of L and with $H_{\mathbb{C}}$ the corresponding complexification of the Cartan algebra H. Then $H_{\mathbb{C}}$ is a Cartan algebra of $L_{\mathbb{C}}$, and all Cartan algebras of $L_{\mathbb{C}}$ are conjugate to $H_{\mathbb{C}}$. Since H is abelian by the Third Cartan Algebra Theorem III.2.9, then $H_{\mathbb{C}}$ is abelian and therefore is its own centralizer in $L_{\mathbb{C}}$. We recall from Section 5, that L is the direct sum of H and a unique H-module complement H^+. It is then clear that $L_{\mathbb{C}}$ is the direct sum of $H_{\mathbb{C}}$ and the $H_{\mathbb{C}}$-module complement $H_{\mathbb{C}}^+$. The H-modules H^+ and $H_{\mathbb{C}}^+$ will be decomposed in their unique isotypic components. This decomposition will play a crucial role in this chapter and will be exploited for the structure theory of Lie algebras containing invariant pointed generating cones.

Roots and root decompositions

III.6.1.   Definition.   For any linear functional $\lambda \in \widehat{L_{\mathbb{C}}}$ we define $L_{\mathbb{C}}^\lambda = \{u \in L_{\mathbb{C}} : [h, u] = \lambda(h) \cdot u \quad \text{for all } h \in H\}$. We say that λ is a *root of* $L_{\mathbb{C}}$ *with respect to* $H_{\mathbb{C}}$ if $L_{\mathbb{C}}^\lambda \neq \{0\}$. The set of all roots with respect to $H_{\mathbb{C}}$ will be denoted Λ.

III.6.2.   Lemma.   *If H is compactly embedded, then the following conclusions hold:*

(i) *The Cartan algebra H acts semisimply on L and the complex Cartan algebra $H_{\mathbb{C}}$ acts semisimply on the complexification $L_{\mathbb{C}}$.*

(ii) *$u \in L_{\mathbb{C}}^\lambda$ if and only if for each $h \in L_{\mathbb{C}}$ there is a natural number n such that $(\operatorname{ad} h - \lambda \cdot \mathbf{1})^n(u) = 0$.*

(iii) *$L_{\mathbb{C}}^0 = H_{\mathbb{C}}$.*

Proof.   (i) If H is compactly embedded in L, then L is a T-module for the torus $T = \mathrm{INN}_L H$. Hence the complexification $L_{\mathbb{C}}$ is likewise a T-module under the extended action and as such every submodule has a complement by the compactness of T. Hence $L_{\mathbb{C}}$ is a semisimple T-module and thus a semisimple H-module.

Since every H-submodule of $L_{\mathbb{C}}$ is also an $H_{\mathbb{C}}$-submodule and vice versa, $H_{\mathbb{C}}$ acts semisimply on $L_{\mathbb{C}}$ under the adjoint action.

(ii) is a consequence of (i).

(iii) is equivalent to the statement that $H_{\mathbb{C}}$ is its own centralizer in $L_{\mathbb{C}}$. ∎

We shall frequently invoke the *root decomposition of $L_{\mathbb{C}}$ with respect to $H_{\mathbb{C}}$* which we recall in the following:

III.6.3.   Proposition.   (The Root Decomposition) *If L is a finite dimensional real Lie algebra with a compactly embedded Cartan algebra H, and if H^+ denotes the unique H-module complement of H, then*

$$(1) \qquad\qquad L_{\mathbb{C}} = H_{\mathbb{C}} + H_{\mathbb{C}}^+ \qquad with \qquad H_{\mathbb{C}}^+ = \bigoplus_{0 \neq \lambda \in \Lambda} L_{\mathbb{C}}^\lambda,$$

$$(2) \qquad\qquad [L_{\mathbb{C}}^\lambda, L_{\mathbb{C}}^\mu] \subseteq L_{\mathbb{C}}^{\lambda+\mu} \qquad for\ all\ \ \lambda, \mu \in \Lambda.$$

Proof.   In view of the decomposition $L = H \oplus H^+$ and Lemma III.6.2 above, these conclusions are standard Lie algebra theory for which we may safely refer to [Bou75], Chap VII.   ∎

A **g**-module for any Lie algebra **g** is called isotypic if it is the direct sum of isomorphic simple modules. Every semisimple **g**-module decomposes into a unique direct sum of isotypic components, that is, maximal isotypic submodules. This holds irrespective of the ground field. The isotypic components of the $H_{\mathbb{C}}$-module $L_{\mathbb{C}}$ are the root spaces $L_{\mathbb{C}}^\lambda$, $\lambda \in \Lambda$. But also, the H-module L decomposes into isotypic components over $\mathbb{R}$. The complexification $V_{\mathbb{C}}$ of each isotypic component V of L is a complex submodule, but not necessarily an isotypic one, since the complexification of a simple module is semisimple but not necessarily isotypic. It is clear that, for a thorough understanding of our present subject we need a careful analysis of the situation. Of course, H is the zero isotypic component of L, and $H_{\mathbb{C}}$ is the zero isotypic component of $L_{\mathbb{C}}$. It therefore suffices to concentrate on H^+ and its complexification $H_{\mathbb{C}}^+$.

Let V be any isotypic component of the H-module H^+. Then V is a direct sum of irreducible submodules. Suppose that M is an irreducible module to which they are all isomorphic. Then V is isomorphic to the tensor product over $\mathbb{R}$ of two modules $M \otimes V_0$ where V_0 is a trivial, that is, 0-module whose dimension gives the number of simple summands of V. In other words, the module action on $M \otimes V_0$ is $h \bullet (m \otimes v) = (h \bullet m) \otimes v$. The complexification $V_{\mathbb{C}}$ is isomorphic to $\mathbb{C} \otimes M \otimes V_0$ (all tensor products over $\mathbb{R}$!) which is a direct sum of $\dim V_0$ complex modules $M_{\mathbb{C}}$.

It therefore suffices to understand the complexification process for *simple* modules. Thus let M be a simple non-trivial real H-module such that each operator $m \mapsto h \bullet m$ has purely imaginary spectrum. Then M is the underlying real vector space of a 1-dimensional complex module on which H acts via purely imaginary scalar multiplications. In other words, $\dim M = 2$ and there is a real vector space automorphism I on M such that $I^2 = -\mathbf{1}_M$ and a non-zero linear form $\omega \in \widehat{H}$ such that

$$h \bullet m = \omega(h) \cdot Im.$$

The choice of I and ω is not unique; we can replace I by $-I$ and, simultaneously, ω by $-\omega$ without affecting the description of the H-action on M; but these two choices are the only possible ones.

The complexification $M_{\mathbb{C}} = \mathbb{C} \otimes M$ has 2 complex, that is, 4 real dimensions and, as a real vector space, carries two different complex structures: One given by the multiplication with i via $i \cdot (c \otimes m) = ic \otimes m$ and the other by the extension of I via $I(c \otimes m) = c \otimes Im$. The H-module structure of the complexification $M_{\mathbb{C}}$ is given by

$$h \bullet (c \otimes m) = c \otimes (h \bullet m) = c \otimes \omega(h) \cdot Im = \omega(h) \cdot I(c \otimes m).$$

However, on the complex 2-dimensional vector space $M_{\mathbb{C}}$, the automorphism I with the two eigenvalues i and $-i$ is diagonalizable. That is, $M_{\mathbb{C}} = M^+ \oplus M^-$ with two one dimensional complex subspaces $M^{\pm}$ such that $m^{\pm} \in M^{\pm}$ implies $Im^{\pm} = \pm i \cdot m^{\pm}$. In fact, if e is an non-zero vector in M (so that $M = \mathbb{R} \cdot e + \mathbb{R} \cdot Ie$), then

$$M^+ = \mathbb{C} \cdot (1 \otimes e - i \otimes Ie) \quad \text{and} \quad M^- = \mathbb{C} \cdot (1 \otimes e + i \otimes Ie).$$

Thus $M_{\mathbb{C}}$ is the direct sum of the two non-isomorphic simple complex H-modules M^+ and M^- such that

$$h \bullet m^{\pm} = \pm \omega(h) i \cdot m^{\pm}$$

with $m^{\pm} \in M^{\pm}$.

As simple *real* H-modules, M^+ and M^- are isomorphic. Indeed, the real vector space underlying $M_{\mathbb{C}}$ carries an involution defined by

$$\overline{c \otimes m} = \bar{c} \otimes m.$$

We note

$$\overline{h \bullet m^{\pm}} = \overline{\pm \omega(h) i \cdot m^{\pm}} = \mp \omega(h) i \cdot \overline{m^{\pm}},$$

which shows that $m \mapsto \overline{m}$ is the isomorphism between M^+ and M^-. The function

$$x \mapsto \frac{1}{2} \cdot (x + \bar{x})$$

is a real vector space isomorphism from each of M^+ and M^- onto M.

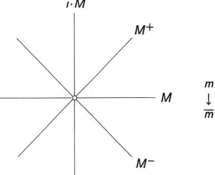

Figure 1

The H-module action on the complex module $M_{\mathbb{C}}$ extends to an $H_{\mathbb{C}}$-module action in the obvious fashion so that $(c \otimes h) \bullet u = c \cdot (h \bullet u)$ for $h \in H$, $u \in M_{\mathbb{C}}$. Also, the linear functional ω on H yields a complex functional on $H_{\mathbb{C}}$ via $\lambda(c \otimes h) = c\omega(h)i$. Then $M^{\pm}$ is a simple $H_{\mathbb{C}}$-module satisfying $h \bullet u = \pm \lambda(h) \cdot u$.

What was said on the simple module M and its complexification extends at once to any isotypic module $M \otimes V_0$ and its complexification $M_{\mathbb{C}} \otimes V_0$. This allows us to summarize our observations in the following lemma:

III.6.4.   Lemma.   *Let V be an isotypic H-module for the abelian Lie algebra H for which all operators $v \mapsto h \bullet v$ have purely imaginary spectrum. Then there is a (complex) linear functional $\lambda \in \widehat{H}_\mathbb{C}$ and a direct decomposition of the complex $H_\mathbb{C}$-module $V_\mathbb{C}$ into a direct sum of complex isotypic $H_\mathbb{C}$-modules*

$$V_\mathbb{C} = V_\mathbb{C}^\lambda \oplus V_\mathbb{C}^{-\lambda}, \quad \text{where } h \bullet v^\pm = \lambda(h) \cdot v^\pm \text{ for all } h \in H_\mathbb{C}, \quad v^\pm \in V_\mathbb{C}^{\pm\lambda}.$$

The restriction of λ to H is purely imaginary. The underlying real H-module $V_\mathbb{C}|_R$ of $V_\mathbb{C}$ is isotypic and the complex conjugation $c \otimes v \mapsto \bar{c} \otimes v$ of $V_\mathbb{C}$ establishes an involutive isomorphism between the real H-modules $V_\mathbb{C}^\lambda|_R$ and $V_\mathbb{C}^{-\lambda}|_R$. The function $x \mapsto \frac{1}{2} \cdot (x + \bar{x})$ is a real vector space isomorphism from each of $V_\mathbb{C}^{\pm\lambda}$ onto V.

The vector space $V_\mathbb{C}$ carries a second complex structure compatible with the first one: That is, there is a complex vector space automorphism $I: V_\mathbb{C} \to V_\mathbb{C}$ with $I^2 = -\mathbf{1}_{V_\mathbb{C}}$. The two spaces $V_\mathbb{C}^{\pm\lambda}$ are the eigenspaces of I for the eigenvalues $\pm i$, respectively. The real H-submodule V is I-invariant and the action of H on V is given by

$$h \bullet v = -i\lambda(h) \cdot Iv, \text{ for all } h \in H, v \in V \text{ with } \lambda(h) \in i \cdot \mathbb{R}.$$

∎

We note that also $-I$ is a complex structure on V and that for each $h \in H$ and $v \in V$ we can also write $h \bullet v = i\lambda(h) \cdot (-I)v$.

This lemma we apply immediately to the Root Decomposition III.6.3. The restriction $\lambda \mapsto -i\lambda|H$ maps the set Λ of roots of $L_\mathbb{C}$ with respect to $L_\mathbb{C}$ bijectively onto a set Ω of real linear functionals on H, called *real roots of L with respect to H*. We write

$$L^\omega = L \cap (L_\mathbb{C}^\lambda \oplus L_\mathbb{C}^{-\lambda}).$$

Note that $L^{-\omega} = L^\omega$. Any choice of a closed half space E in $\widehat{H}$ whose boundary meets the finite set Ω only in 0 allows us to represent Ω as a union $\Omega = \Omega^+ \cup -\Omega^+$ where $\Omega^+ = E \cap \Omega$. We shall call Ω^+ a *set of positive roots*.

III.6.5.   Theorem.   (The Real Root Decomposition) *Let L be a finite dimensional real Lie algebra with a compactly embedded Cartan algebra. For each choice of a set Ω^+ of positive roots there is a unique complex structure $I: H^+ \to H^+$ with $I^2 = -\mathbf{1}_{H^+}$ and a direct decomposition of L into isotypic H-submodules under the adjoint action*

$$(1) \qquad\qquad L = H \oplus H^+, \qquad H^+ = \bigoplus_{0 \neq \omega \in \Omega^+} L^\omega,$$

where the action is described by

$$(2) \qquad\qquad [h, x] = \omega(h) \cdot Ix \text{ for all } h \in H, \quad x \in L^\omega.$$

The complexification $L_\mathbb{C}^\omega$ is $L_\mathbb{C}^\lambda \oplus L_\mathbb{C}^{-\lambda}$, where λ is the canonical complex extension of $i\omega$, and where

$$(3) \qquad\qquad L^\omega = L \cap (L_\mathbb{C}^\lambda \oplus L_\mathbb{C}^{-\lambda})$$

For $x \in H^+$, set

(4)
$$x^\pm = x \mp i \cdot Ix.$$

Then

(5)
$$x = \frac{1}{2} \cdot (x^+ + x^-) \qquad and \qquad Ix = \frac{i}{2} \cdot (x^+ - x^-),$$

notably, $\mathbb{C} \cdot x + \mathbb{C} \cdot Ix = \mathbb{C} \cdot x^+ + \mathbb{C} \cdot x^-$. *Further,*

(6)
$$2i \cdot [x, Ix] = [x^+, x^-] \in H_\mathbb{C},$$

and if $x \in L^\omega$, then

(7)
$$[h, x^\pm] = \pm \lambda(h) \cdot x^\pm \qquad for\ all \qquad h \in H_\mathbb{C}.$$

If ω' is another real root and λ' the canonical complex extension of $i\omega'$, then for $p \in L^\omega$, $p' \in L^{\omega'}$, the elements $x = [p, p'] - [Ip, Ip']$ and $y = [p, Ip'] + [Ip, p']$ are real root vectors in $L^{\omega + \omega'}$ and

(8)
$$[p^+, p'^+] = x^+ = x - i \cdot y \qquad and \qquad [p^-, p'^-] = x^- = x + i \cdot y.$$

Proof. The first assertions through (3) are consequences of the preceding discussions. Straightforward calculations show that (4) entails (5) and (7). Now $[x, Ix] = [\frac{1}{2} \cdot (x^+ + x^-), \frac{i}{2} \cdot (x^+ - x^-)]$ at once implies (6). If we now consider $p \in L^\omega$ and $p' \in L^{\omega'}$, we set $x = [p, p'] - [Ip, Ip']$ and $y = [Ip, p'] + [p, Ip']$. From (4) we compute $[p^+, p'^+] = [p - i \cdot Ip, p' - i \cdot Ip'] = ([p, p'] - [Ip.Ip']) - i \cdot ([p, Ip'] + [Ip, p']) = x - i \cdot y$. Similarly we calculate $[p^-, p'^-] = x + i \cdot y$. We abbreviate $[p^+, p'^+] = x - i \cdot y$ with z an $[p^-, p'^-] = x + i \cdot y$ with $\overline{z}$. We define $\mu = \lambda + \lambda'$ and observe $[h, z] = \mu(h) \cdot z$ and $[h, \overline{z}] = -\mu(h) \cdot \overline{z}$ for all $h \in H_\mathbb{C}$. Then $[h, x] = [h, \frac{1}{2} \cdot (z + \overline{z})] = \frac{\mu(h)}{2} \cdot (z - \overline{z}) = (-i\mu)(h) \cdot \frac{i}{2} \cdot (z - \overline{z}) = (-i\mu)(h) \cdot y$. Similarly, $[h, y] = [h, \frac{i}{2} \cdot (z - \overline{z})] = \frac{i\mu(h)}{2} \cdot (z + \overline{z}) = -(-i\mu)(h) \cdot \frac{1}{2} \cdot (z + \overline{z}) = -(-i\mu)(h) \cdot x$. But $x, y \in L$ and $\mathbb{R} \cdot x + \mathbb{R} \cdot y$ is now either $\{0\}$ or and irreducible H-module with the real root $-i\mu | H = -i(\lambda + \lambda') | H = \omega + \omega'$. Hence, by the definition of I, the preceding calculations show that $y = Ix$ and x is a real root vector in $L^{\omega + \omega'}$. Then, finally, $[p^+, p'^+] = x - i \cdot Ix = x^+$ and $[p^-, p'^-] = x + i \cdot Ix = x^-$ according to (4). ∎

EIII.6.1. **Exercise.** In the notation of Theorem III.6.5, set $a = [p, p'] + [Ip, Ip']$ and $b = [p, Ip'] - [Ip, p']$ and show that a and b are root vectors in $L^{\omega - \omega'}$ and that $[p^+, p'^-] = a + i \cdot b$, $[p^-, p'^+] = a - i \cdot b$. Conclude, in particular, that $[p, p'] \in L^{\omega + \omega'} + L^{\omega - \omega'}$.

Among other things, this theorem enables us to define a certain quadratic function from $H_\mathbb{C}^+$ to $H_\mathbb{C}$ which will play a useful role in the sequel.

III.6.6. **Definition.** For a fixed selection Ω^+ of positive roots, the second complex structure $I: H_\mathbb{C}^+ \to H_\mathbb{C}^+$ permits us to define a bilinear function

$$Q : H_\mathbb{C}^+ \times H_\mathbb{C}^+ \to H_\mathbb{C} \qquad by\ Q(x, y) = -p_\mathbb{C}([x, Iy]),$$

where $p_\mathbb{C}$ is the projection of $L_\mathbb{C}$ onto $H_\mathbb{C}$ along $H_\mathbb{C}^+$ extending the projection p.

As is usual, we shall abbreviate $Q(x, x)$ with $Q(x)$. We note that, in particular, for $x \in L^\omega$ we have $Q(x) = -[x, Ix] = [Ix, x]$.

For the following lemma we set $\Lambda^+ = \{\lambda \in \Lambda : -i\lambda \in \Omega^+\}$.

III.6.7.   Lemma.   *The bilinear map Q is symmetric and is calculated as fol-lows: If $x = \sum_{0 \neq \lambda \in \Lambda} x_\lambda$ and $y = \sum_{0 \neq \lambda \in \Lambda} y_\lambda$ according to the root decomposition, then*

$$Q(x, y) = i \cdot \sum_{0 \neq \lambda \in \Lambda^+} ([x_\lambda, y_{-\lambda}] - [x_{-\lambda}, y_\lambda]).$$

If $x, y \in H^+$ and the decompositions in terms of real root vectors are

$$x = \sum_{0 \neq \omega \in \Omega^+} x_\omega \quad and \quad y = \sum_{0 \neq \omega \in \Omega^+} y_\omega,$$

then

$$Q(x, y) = \frac{i}{4} \cdot \sum_{0 \neq \omega \in \Omega^+} ([x_\omega^+, y_\omega^-] - [x_\omega^-, y_\omega^+]).$$

Proof.   Definition III.6.6 implies the symmetry of Q. From the bilinearity of Q we conclude that $Q(x, y)$ is the sum of all $Q(x_\lambda, y_\mu)$ where λ and μ range through the non-zero roots. From the definition of the space $L_\mathbb{C}^\mu$ we know that it is an eigenspace of I for the eigenvalue i if $\mu \in \Lambda^+$ and for the eigenvalue $-i$, otherwise. Thus $[x_\lambda, Iy_\mu] = \pm i[x_\lambda, y_\mu] \in L_\mathbb{C}^{\lambda+\mu}$ by the Root Decomposition III.6.3, whence $Q(x_\lambda, y_\mu) = -p_\mathbb{C}([x_\lambda, Iy_\mu]) = 0$ whenever $\lambda + \mu \neq 0$. If $\lambda \in \Lambda^+$, then $Q(x_\lambda, y_{-\lambda}) = -p_\mathbb{C}([x_\lambda, Iy_{-\lambda}]) = i[x_\lambda, y_{-\lambda}]$. Similarly, $Q(x_{-\lambda}, y_\lambda) = -i[x_{-\lambda}, y_\lambda]$. All of this taken together proves the first assertion of the lemma. The second follows from this and Theorem II.6.5(5) and (7).   ∎

We recall the Cartan–Killing form $B : L \times L \longrightarrow \mathbb{R}$ given by $B(x, y) = tr(\text{ad } x)(\text{ad } y)$. Let $L^{\perp_B}$ denote the radical $\{x \in L : B(x, y) = 0 \text{ for all } y \in L\}$ of the Cartan–Killing form. Now we shall deal with the Cartan–Killing form in our present situation.

III.6.8.   Proposition.   *Let H be a compactly embedded Cartan algebra in a finite dimensional Lie algebra L. We select a set Ω^+ of positive roots. Then the following conclusions hold:*

(i) *For all $h, k \in H$, one has*

$$B(h, k) = - \sum_{\omega \in \Omega^+} \omega(h)\omega(k) \dim L^\omega.$$

(ii) *$H \cap L^{\perp_B} = Z(L)$, where $Z(L)$ is the center of L.*

(iii) *If q is any invariant symmetric bilinear map defined on $L \times L$, then $q(x, Ix) = 0$ and $q(Ix) = q(x)$ for $x \in L^\omega$.*

Proof.   (i) We compute $B(h, k) = tr(\text{ad } h)(\text{ad } k) = \sum_{\omega \in \Omega^+} tr(\text{ad } h|L^\omega)(\text{ad } k|L^\omega)$. But for $x \in L^\omega$ we note $(\text{ad } h)(\text{ad } k)x = [h, [k, x]] = \omega(k) \cdot [h, Ix] = -\omega(h)\omega(k)x$. Claim (i) follows.

(ii) Let $h \in H \cap L^{\perp_B}$. This implies, in particular, $B(h) = 0$, which means that $\omega(h) = 0$ for all $\omega \in \Omega^+$, and this in turn means $[h, x] = 0$ for all $x \in L$, that is, $h \in Z(L)$. Conversely, if $h \in Z(L)$, then $\text{ad } h = 0$ and thus $B(h, x) = 0$ for all $x \in L$, that is, $x \in L^{\perp_B}$. This proves (ii).

(iii) There is an $h \in H$ such that $Ix = \omega(h)^{-1} \cdot [h, x]$ and $[Ix, h] = \omega(h) \cdot x$ by (2) and $I^2 = -1_{H^+}$. Hence $q(x, Ix) = \omega(h)^{-1} q(x, [h, x]) = 0$ by the invariance of q. Also, again by invariance, $q(Ix) = \omega(h)^{-1} q(Ix, [h, x]) = \omega(h)^{-1} q([Ix, h], x) = q(x)$.   ∎

This proposition has an important consequence which gives an insight into the structure of the nilradical.

III.6.9.   Theorem. *Suppose that H is a compactly embedded Cartan algebra of L. If R denotes the radical of L and N the nilradical, then*

(1)
$$H \cap [L, R] \subseteq Z(L) = H \cap N,$$

(2)
$$N = Z(L) \oplus (N \cap H^+).$$

(3)
$$N \cap H^+ = \bigoplus_{0 \neq \omega \in \Omega^+} N^\omega,$$

where $N^\omega = N \cap L^\omega$.

Proof.   It is a general fact that $L^{\perp_B} \supseteq N \supseteq [R, L]$ (see [Bou75], Chap.I, § 5, n° 3, Théorème 1, and n° 5, Remarque). Also $Z(L) \subseteq N$. Hence, from Proposition III.6.8.ii we conclude $Z(L) \subseteq H \cap Z(L) \subseteq H \cap N \subseteq H \cap L^{\perp_B} = Z(L)$. This proves (1). In order to understand (2) and (3) it suffices to realize that *every* ideal I of L is an H-module, hence decomposes uniquely into isotypic components according to

$$I = (I \cap H) \oplus \bigoplus_{0 \neq \omega \in \Omega^+} (I \cap L^\omega).$$

■

III.6.10.   Corollary. *If L is a finite dimensional Lie algebra with a compactly embedded Cartan algebra H, then $L/Z(L)$ is centerfree with the compactly embedded Cartan algebra $H/Z(L)$.*

If L is centerfree, then we have the following conclusions:

   (i) *The nilradical is contained in H^+.*

   (ii) *The Cartan–Killing form is negative definite on H. The dual $\widehat{H}$ of H can be identified with H under the isomorphism $\alpha \mapsto h_\alpha$, where $B(h_\alpha, h) = \alpha(h)$ for all $h \in H$.*

   (iii) *For $x = \sum_{\omega \in \Omega^+} x_\omega \in H^+$ we have $Q(x) = -\sum_{\omega \in \Omega^+} B(x_\omega) \cdot h_\omega$.*

Proof.   Let $Z_1 = Z(L)$ and define $Z_{n+1} \supseteq Z_n$ so that $Z_{n+1}/Z_n = Z(L/Z_n)$ and $Z_\infty = \bigcup_{n=1}^\infty Z_n$. Then Z_∞ is called the hypercenter of L and is contained in every Cartan algebra, as a proof by induction shows readily (see [Bou75], Chap.VII). Since the hypercenter is a nilpotent ideal, we have $Z_\infty \subseteq H \cap N$. By Theorem III.6.9(1), $H \cap N = Z(L) \subseteq Z_\infty$. Hence the center and hypercenter agree, that is, $L/Z(L)$ is center free. The image $H/Z(L)$ of H in the factor algebra is again a compactly embedded Cartan algebra.

   Now let us assume that L is centerfree. Then (i) follows from Theorem III.6.9 (2), and (ii) is immediate from Proposition III.6.8. We shall now prove (iii): We let $h \in H$ be arbitrary and calculate $B(h, Q(x))$. We write $x = \sum_{0 \neq \lambda \in \Lambda} x_\lambda$ in the complex root decomposition and have $Q(x) = -p([x, Ix]) =$

$2i \cdot \sum_{0 \neq \lambda \in \Lambda^+} [x_\lambda, x_{-\lambda}]$ by Lemma III.6.7. Now $B(h, [x_\lambda, x_{-\lambda}]) = B([h, x_\lambda], x_{-\lambda}) = \lambda(h) B(x_\lambda, x_{-\lambda})$. But for $z \in L^\mu$ we have $(\operatorname{ad} x_\lambda)^2 z \in L^{2\lambda + \mu}$ which allows us to conclude that $B(x_\lambda) = 0$ and, similary, $B(x_{-\lambda}) = 0$. Thus, if $x_\omega = x_\lambda + x_{-\lambda}$ with $\lambda = i\omega$, then $B(x_\omega) = B(x_\lambda) + 2B(x_\lambda, x_{-\lambda}) + B(x_{-\lambda}) = 2B(x_\lambda, x_{-\lambda})$. Hence $B(h, Q(x))$ equals $2i \cdot \sum_{0 \neq \lambda \in \Lambda^+} B(h, [x_\lambda, x_{-\lambda}]) = \sum_{0 \neq \lambda \in \Lambda^+} (i\lambda)(h) B(x_\omega) = -\sum_{0 \neq \omega \in \Omega^+} \omega(h) B(x_\omega)$, and using $\omega(h) = B(h_\omega, h)$ by the definition of h_ω, we find this expression equal to $B(-\sum_{0 \neq \omega \in \Omega^+} B(x_\omega) \cdot h_\omega, h)$ for all $h \in H$. Since B is non-degenerate on H by (ii), the assertion follows. $\blacksquare$

The test subalgebras

In Chapter II we saw numerous occasions where the Standard Testing Device II.4.7 converted detailed information on small subalgebras into powerful information. We shall now investigate how the root decomposition with respect to a compactly embedded Cartan algebra gives rise to a wealth of identifiable small subalgebras, which we shall call *test subalgebras*.

III.6.11.   Lemma. *Let H be a compactly embedded Cartan algebra of L and Ω^+ a set of positive roots. Then for each $0 \neq x \in L^\omega$, with $0 \neq \omega \in \Omega^+$, the vector space*

$$H(x) \stackrel{\text{def}}{=} H \oplus (\mathbb{R}.x + \mathbb{R} \cdot Ix)$$

is a subalgebra with compactly embedded Cartan algebra H. The center $Z(H(x))$ is $\ker \omega$. The vector space

$$\langle x \rangle = \mathbb{R} \cdot x + \mathbb{R} \cdot Ix + \mathbb{R}.Q(x)$$

is an ideal of $H(x)$.

Proof.   We know $[x, Ix] = -Q(x) \in H$, $[h, x] = \omega(h) \cdot Ix$, and $[h, Ix] = \omega(h) \cdot x$ by the definition of the root decomposition. Hence $H(x)$ is a subalgebra. An element $y = h + r \cdot x + s \cdot Ix$ with $h \in H$ and $r, s \in \mathbb{R}$ is central if and only if $[y, z] = 0$ for all $z \in H(x)$, and that is the case only if $[y, x] = [y, Ix] = 0$. But $[y, x] = \omega(h) \cdot Ix + s \cdot Q(x)$ and $[y, Ix] = -\omega(h) \cdot x - r \cdot Q(x)$. It follows that y is central only if $\omega(h) = 0$ and $r = s = 0$, that is, $y \in \ker \omega$. Conversely, if $y \in \ker \omega$, then $[y, x] = 0$ for all $z \in H(x)$. It is clear that H is a compactly embedded Cartan subalgebra of $H(x)$.

Since $[x, Ix] = -Q(x) \in H$, $[h, x] = \omega(h) \cdot Ix$, and $[h, Ix] = -\omega(h) \cdot x$, it follows that $\langle x \rangle$ is an ideal of $H(x)$. $\blacksquare$

We shall completely describe the structure of $H(x)$. We begin with $\langle x \rangle$:

III.6.12.   Proposition. *Suppose that H is a compactly embedded Cartan algebra of a finite dimensional real Lie algebra L and that ω is a positive non-zero* root with respect to H. Then for every non-zero $x \in L^\omega$ there are the following cases

$$\text{Case S}_- : \omega\big(Q(x)\big) < 0,$$
$$\text{Case S}_+ : \omega\big(Q(x)\big) > 0,$$
$$\text{Case N} : \omega\big(Q(x)\big) = 0 \quad and \quad Q(x) \neq 0,$$
$$\text{Case A} : Q(x) = 0.$$

In Case S_-, $\langle x \rangle \cong so(3) \cong su(2)$. In Case S_+, $\langle x \rangle \cong so(2,1) \cong sl(2,\mathbb{R})$. In Case N, $\langle x \rangle$ is the 3-dimensional Heisenberg algebra. In Case A, $\langle x \rangle$ is the 2-dimensional abelian Lie algebra.

Proof. Case S_-: We set

$$e_1 = -\omega\big(Q(x)\big)^{-1} \cdot Q(x), \quad e_2 = \frac{1}{\sqrt{-\omega\big(Q(x)\big)}} \cdot x, \quad e_3 = \frac{1}{\sqrt{-\omega\big(Q(x)\big)}} \cdot Ix.$$

Then $[e_j, e_{j+1}] = e_{j+2}$ for $j = 1, 2, 3 \pmod 3$. Thus $\langle x \rangle \cong so(3)$

Case S_+: We set

$$u = -2\omega\big(Q(x)\big)^{-1} \cdot Q(x),$$
$$t = \sqrt{\frac{2}{\omega\big(Q(x)\big)}} \cdot (x + Ix),$$
$$h = \sqrt{\frac{2}{\omega\big(Q(x)\big)}} \cdot (x - Ix).$$

Then $[u,t] = 2h$ in view of $[Q(x), x + Ix] = \omega\big(Q(x)\big) \cdot (Ix - x)$, and $[u,h] = -2t$ by a similar argument. Further, using $[x + Ix, x - Ix] = -2[x, Ix] = 2Q(x)$, we compute $[h,t] = 2u$. Thus $\langle x \rangle \cong sl(2,\mathbb{R})$ in view of relations (2) in the proof of Theorem II.3.4.

Case N: We set $p = x$, $q = Ix$, and $z = Q(x)$. Then $[p,q] = z$ and $[p,z] = [q,z] = 0$. Thus $\langle x \rangle$ is the 3-dimensional Heisenberg algebra.

Case A: Here $\langle x \rangle = \mathbb{R} \cdot x \oplus \mathbb{R} \cdot Ix$ with $[x, Ix] = 0$. The assertion is clear. ∎

III.6.13. **Definition.** We shall say that $\langle x \rangle$ is of compact simple type, non-compact simple type, nilpotent type, and abelian type if we are in Case S_-, S_+, N, and A, respectively.

III.6.14. **Proposition.** If H is a compactly embedded Cartan algebra and ω a non-zero positive root with respect to H, then for every $x \in L^\omega$ for which $\langle x \rangle$ is of simple type, the algebra $H(x)$ is reductive, that is

$$H(x) = Z\big(H(x)\big) \oplus \langle x \rangle.$$

Proof. We know that $Z\big(H(x)\big) = \ker \omega$ is a hyperplane of H by Lemma III.6.11. In the simple type, $Q(x) \notin \ker \omega$. Thus $H(x)$ decomposes into a direct sum $\ker \omega \oplus \mathbb{R} \cdot Q(x) \oplus \mathbb{R} \cdot x \oplus \mathbb{R} \cdot Ix$ and the assertion follows. ∎

III.6.15.   Proposition.   *If H is a compactly embedded Cartan algebra and ω a non-zero positive root with respect to H, and x a non-zero element in L^ω, and if h is any element in H with $\omega(h) = 1$, then the following two cases occur:*

(i) *If $\langle x \rangle$ is of nilpotent type, then there is a central subalgebra $C \subseteq \ker \omega$ such that*

$$H(x) = C \oplus A(x) \ \text{where} \ A(x) = \mathbb{R}{\cdot}h + \langle x \rangle$$

and where $A(x) \cong A_4$ is the 4-dimensional standard solvable Lorentzian (or harmonic oscillator) algebra.

(ii) *If $\langle x \rangle$ is of abelian type, then*

$$H(x) = Z\big(H(x)\big) \oplus A(x) \ \text{where} \ A(x) = \mathbb{R}{\cdot}h + \langle x \rangle$$

and where $A(x)$ is the 3-dimensional Lie algebra of the group of euclidean motions of the plane.

Proof.   The vector space H splits into a direct sum of $\mathbb{R}{\cdot}h$ and $\ker \omega = Z\big(H(x)\big)$. In any case we shall set $A(x) = \mathbb{R}{\cdot}h + \langle x \rangle$ and know that $A(x)$ is a subalgebra and indeed an ideal of $H(x)$. If $\langle x \rangle$ is of nilpotent type, then $0 \neq Q(x) \in \ker \omega$ by definition and we can decompose $\ker \omega$ into a direct sum of $\mathbb{R}{\cdot}Q(x)$ and some vector space complement C, and $H(x)$ is the direct sum of the ideals C and $A(x)$. The 4-dimensional algebra $A(x)$ is characterized by the relations $[h, x] = Ix, [h, Ix] = -x, [x, Ix] = Q(x)$, and $[Q(x), u] = 0$ for all $u \in A(x)$. Then $A(x) \cong A_4$ by Definition II.3.14. If $\langle x \rangle$, is of abelian type, then $H(x) = \ker \omega \oplus A(x)$, and $A(x)$ is characterized by the relations $[h, x] = Ix$, $[h, Ix] = -x$ and $[x, Ix] = 0$, whence $A(x)$ is the Lie algebra of the group of motions of the euclidean plane. (See Theorem II.3.4, Type (i).)   ∎

III.6.16.   Corollary.   *Under the circumstances of Propositions III.6.14 and 15, the following conditions are equivalent:*

(1) $H(x)$ *is a compact Lie algebra.*

(2) $\langle x \rangle$ *is of simple compact type (that is, is isomorphic to $\mathrm{so}(3)$).*

(3) $x \in K(H)$ *(see Proposition III.5.6).*

(4) $x \in \operatorname{comp} L$ *(see Definition III.2.7).*

Proof.   The equivalence of conditions (1) and (2) follows from the explicit information given in Propositions III.6.12, and 14. By Proposition III.5.6 H is contained in a *unique* maximal compactly embedded subalgebra $K(H)$; hence (1) implies (3). The implication $(3) \Rightarrow (4)$ is clear from the definitions. If (4) is satisfied then $x \in \operatorname{comp} H(x)$ we recall Proposition III.2.8 and inspect Propositions III.6.12 and 14 and conclude (1).   ∎

III.6.17.   Corollary.   *Under the circumstances of Propositions III.6.14 and 15, the following conditions are equivalent:*

(1) $H(x)$ *is a metabelian algebra.*

(2) $\langle x \rangle$ *is of abelian type, that is, is isomorphic to $\mathbb{R}^2$.*

(3) $Q(x) = 0$.

(4) *x is contained in some subalgebra I of L which is an H-module and is contained in H⁺.*

Proof. Lemma III.6.11 and Propositions III.6.12, and 14 give the equivalence of the first three conditions. If (2) is satisfied, then $I = \langle x \rangle$ is the required algebra. Conversely, if (4) is satisfied, then $Ix = \omega(h) \cdot Ix = [h, x] \in I$ and then $Q(x) = -[x, Ix] \in I \subseteq H^+$. But also $Q(x) \in H$, whence $Q(x) = 0$. This proves (3).   ∎

We shall now apply the Standard Testing Device in a way which by now is familiar to us. We assume that L is a finite dimensional real Lie algebra with an invariant pointed generating cone W. We fix a compactly embedded Cartan algebra H according to The Fourth Cartan Algebra Theorem III.2.11. Since $H \cap W$ is generating in H by this theorem, we may, for a given non-zero positive root ω choose an element $h \in (H \cap W)$ with the additional property that $\omega(h) = 1$. If x is any non-zero element in the root space L^ω, then $A(x) = \mathbb{R} \cdot h + \langle x \rangle$ is an at most 4-dimensional algebra meeting the interior of W. Thus $W \cap A(x)$ is an invariant pointed generating wedge in $A(x)$. However, if $\langle x \rangle$ is of the abelian type, then $A(x)$ is the euclidean motion algebra of dimension 3. According to the Standard Testing Device II.4.7.C2, this algebra does not contain any pointed generating Lie semialgebras, let alone invariant cones. Thus the abelian type is impossible and we have the following result

III.6.18.   Theorem. *If W is a pointed generating invariant cone in a finite dimensional real Lie algebra L and H is a compactly embedded Cartan algebra and $Q: H^+ \to H$ the quadratic function given by $Q(x) = -p([x, Ix])$, then $Q(x) = 0$ and $x \in L^\omega$ imply $x = 0$.*   ∎

From this theorem and the preceding Corollary III.6.17 we recover one aspect of the Reconstruction Theorem of Invariant Cones III.5.15: If L has a pointed generating invariant cone and H is any compactly embedded Cartan algebra, then H^+ cannot contain a non-zero subalgebra I with $[H, I] \subseteq I$, in particular no non-zero ideal. In fact, the relevant portion of the proof of Theorem III.5.15 anticipated the arguments of the present section.

Lie algebras with cone potential

In view of Theorem III.6.18 above, the following purely Lie algebraic definition appears useful:

III.6.19.   Definition. We say that a Lie algebra L *has cone potential* if L is a finite dimensional real Lie algebra with a compactly embedded Cartan algebra H such that $0 \neq x \in L^\omega$ for any non-zero positive root ω implies $Q(x) \neq 0$.

By Theorem III.6.18, every Lie algebra which actually contains an invariant generating pointed cone has cone potential.

III.6.20.   Proposition. *If a Lie algebra L has cone potential, then every abelian ideal is central. Thus the center of L is the unique largest abelian ideal. In particular, the center $Z(N)$ of the nilradical N and the last term of the commutator series of the radical are central in L.*

Proof.   Suppose that I is an abelian ideal of L. The H-module $I \cap H^+$ decomposes into isotypic components $I \cap L^\omega$. Also, since I is abelian, $I \cap H^+$ is an algebra. We claim $I \cap H^+ = \{0\}$; for if not, then there is some $0 \neq x \in I \cap L^\omega$, and then $\langle x \rangle$ is of abelian type by Corollary III.6.17. But exactly that is excluded if L has cone potential. Thus the only possibly non-zero isotypic component of I is $I \cap L^0 = I \cap H$. Thus $I \subseteq H$. As H^+ is an H-module and I is an ideal, we conclude $[I, H^+] \subseteq I \cap H^+ = \{0\}$, and thus $[I, L] = \{0\}$, since H is abelian. We have shown that I is central. The remainder is a simple consequence of this fact.■

III.6.21.   Proposition. *In any Lie algebra with cone potential, every non-zero ideal meets every compactly embedded Cartan algebra non-trivially.*

Proof.   Let L be a Lie algebra with cone potential and H a compactly embedded Cartan algebra. Let I be a non-zero ideal. Then it is an H-module under the adjoint action and thus has the unique decomposition

$$(*) \qquad\qquad I = (I \cap H) \bigoplus_{0 \neq \omega \in \Omega^+} (I \cap L^\omega)$$

into its isotypic components. Suppose that $0 \neq x \in I \cap L^\omega$. Since L has cone potential, the element $[x, Ix] \in H$ is non-zero. On the other hand, since I is an ideal, $[x, Ix] \in I$. Hence $H \cap I \neq \{0\}$. However, if $I \cap L^\omega = \{0\}$ for all non-zero positive roots, then $I = I \cap H$ by $(*)$. Since I is non-zero, the claim follows in this case, too.   ■

We shall now show that cone potential implies a significant restriction on the structure of the radical. In fact we shall show that the nilpotent class of the nilradical cannot exceed 2. We begin with a crucial result.

III.6.22.   Theorem. *Let L be a finite dimensional Lie algebra with cone potential, and let N denote its nilradical. Then its center $Z(N)$ is the center $Z = Z(L)$ of L and N/Z is abelian. In particular, the nilpotency class of N is at most 2.*

Proof.   The descending central series was defined inductively by $N^{[0]} = N$ and $N^{[k+1]} = [N, N^{[k]}]$ for $k = 0, 1, \ldots$ If N has class $m + 1$, then m denotes the largest integer for which $N^{[m]} \neq \{0\}$. We claim that $m < 2$. Assume $m \geq 2$. Then Proposition II.5.23(a) implies that $N^{[m-1]}$ is abelian. Now by Proposition III.6.20, $N^{[m-1]}$ is central. Then $N^{[m]} = [N, N^{[m-1]}] = \{0\}$, a contradiction. Thus $m \leq 1$, that is, $[N, N'] = N^{[2]} = \{0\}$. Hence N' is an abelian ideal and thus is central again by Proposition III.6.20. Thus $[N, N] = N' \subseteq Z$ and so N/Z is abelian.   ■

This gives a general structure theorem for Lie algebras with cone potential.

III.6.23.   Theorem. (Structure Theorem for Lie Algebras with Cone Potential) *Let L be a Lie algebra with cone potential, N its nilradical and Z its center.*

Then Z is the center of N and N/Z is abelian. If H is a compactly embedded Cartan algebra and Ω^+ a set of positive roots of L with respect to H, and if we set $N^\omega = N \cap L^\omega$, then for $\omega, \omega' \in \Omega^+$ we have

$$[N^\omega, N^{\omega'}] \begin{cases} \neq \{0\}, & \text{if } \omega = \omega' \neq 0; \\ = \{0\}, & \text{if } \omega \neq \omega' \text{ or } \omega = \omega' = 0. \end{cases}$$

Finally, if R is the radical of L, then $R \cap L^\omega = N^\omega$ for $0 \neq \omega \in \Omega^+$.

Proof. The center $Z(N)$ of N is an abelian ideal of L as it is characteristic. Hence by Proposition III.6.20, it is central in L, and $Z = Z(N)$ follows. The subalgebra $L_1 = H + N$ has H as compactly embedded Cartan algebra. Let N_1 be its nilradical. Clearly $N \subseteq N_1$. Since L_1 is solvable and has cone potential, Theorem III.6.22 applies to L_1 and shows that its center Z_1 is the center of N_1 and that N_1/Z_1 is abelian. This means that $[N, N] \subseteq [N_1, N_1] \subseteq Z_1$. But since H is a Cartan subalgebra of L_1, we know $Z_1 \subseteq H$, hence $[N, N] \subseteq H \cap [L, R]$ with the radical R of L, hence $[N, N]$ is contained in Z by Theorem III.6.9(1). This shows that N/Z is abelian as asserted.

Now let ω and ω' be two non-zero positive roots. If they agree, then for every non-zero $x \in N^\omega$ we have $[x, Ix] \neq \{0\}$, since L has cone potential. Next suppose $\omega \neq \omega'$. Let $p \in N^\omega$ and $p' \in N^{\omega'}$. We must show $[p, p'] = 0$, and we will assume that both p and p' are non-zero. We use the complexification $L_{\mathbb{C}}$ and refer to the Real Root Decomposition III.6.5. It suffices to show that $[p^\pm, p'^\pm] = 0$, where $p^\pm$ and $p'^\pm$ are root vectors for the complex roots $\pm\lambda$, respectively, $\pm\lambda'$ which are the canonical complex extensions of $i\omega$ and $i\omega'$, respectively. Since $\omega \neq \omega'$ we know

(1) $$\{\lambda, -\lambda\} \cap \{\lambda', -\lambda'\} = \emptyset.$$

Now

$$[p^\pm, p'^\pm] \in L_{\mathbb{C}}^{\pm\lambda\pm\lambda'} \cap [N_{\mathbb{C}}, N_{\mathbb{C}}].$$

By Theorem III.6.22, it follows that $[N_{\mathbb{C}}, N_{\mathbb{C}}] \subseteq H_{\mathbb{C}} = L_{\mathbb{C}}^0$. But by (1) we know that $L_{\mathbb{C}}^{\pm\lambda\pm\lambda'} \cap L_{\mathbb{C}}^0 = \{0\}$. Hence $[p^\pm, q^\pm] = 0$, as we had to show. Finally, let $x \in R \cap L^\omega$, $0 \neq \omega \in \Omega^+$. Find an $h \in H$ with $\omega(h) \neq 0$. Then $x = \omega(h)^{-1} \cdot [Ix, h] \in [L, R] \subseteq N$. This completes the proof of the theorem. ∎

In reference to the developments of Section 5 of Chapter II, it now emerges that *in a Lie algebra with cone potential, the center and the base ideal are one and the same thing.*

III.6.24. **Corollary.** *If R is a solvable Lie algebra with cone potential, N its nilradical and Z its center, and if H is a compactly embedded Cartan algebra, then*

$$R = H \oplus H^+, \qquad N = Z \oplus H^+, \qquad N/Z \text{ abelian}.$$

Moreover,

(i) $R' = [R, N] = N' \oplus H^+$, *and*

(ii) $R'' = N' = [H^+, H^+] \subseteq Z = H \cap N$.

In particular, if R is non-abelian, the solvable length of R is 3.

Proof. For the first assertion, we refer to Theorem III.6.5, Theorem III.6.9 and Theorem III.6.23. Recall that $R = H + N$, since $R' \subseteq N$, whence R/N is abelian and $(H + N)/N$ is the Cartan algebra of R/N.

In order to prove the remainder, note first that $R' = [H + N, H + N] = [H, N] + N'$. From the commutativity of N/Z we know that $N' \subseteq Z \subseteq H$. Further we have $H^+ \subseteq N$, hence $H^+ = [H, H^+] \subseteq [H, N] = [H, Z \oplus H^+] = [H, H^+]$, that is, $H^+ = [H, N]$. This finishes the proof of (i). Now $R'' = [N' + H^+, N' + H^+] = [H^+, H^+] = [Z + H^+, Z + H^+] = [N, N]$.   ∎

This applies, in particular, to the radical R of any Lie algebra L with cone potential, since for any ideal I in a Lie algebra with cone potential, the subalgebra $H + I$ has cone potential, where H is a compactly embedded Cartan algebra of L.

Note, however, that the nilradical of $H + R$ may be larger than the nilradical of L.

III.6.25.   Corollary.   *If L is a Lie algebra with cone potential and R its radical, then R has cone potential and thus* Corollary III.6.24 *applies to R.*   ∎

III.6.26.   Corollary.   *A nilpotent ideal in any Lie algebra with cone potential is either central or has nilpotent class 2.*

Proof. If I is an abelian ideal, then it is central by Proposition II.6.20. If I is not abelian, then I has at least nilpotent class 2. But since $I \subseteq N$, where N denotes the nilradical, and since N has class 2, then I has class 2.   ∎

In view of Theorem III.6.18 and Definition III.6.19, it need not be emphasized that the preceding results apply to Lie algebras supporting invariant generating pointed cones, but we record this remark for easy reference:

III.6.27.   Remark.   The conclusions of paragraphs III.6.20 through 26 are valid for any finite dimensional real Lie algebra containing invariant pointed generating cones.   ∎

Mixed Lie algebras with compactly embedded Cartan algebras

In the preceding results we provided some information on the relation of H and the nilradical N. We now turn to the radical R and find suitable Levi complements S which are appropriately related to H.

III.6.28.   Theorem.   *Let L be a finite dimensional real Lie algebra with a compactly embedded Cartan algebra H. Let R denote the radical of L. Then there exists a Levi complement S such that the following conditions are satisfied:*

   (i) $H = (H \cap R) \oplus (H \cap S)$, *and* $H \cap S$ *is a compactly embedded Cartan subalgebra of S.*

(ii) $H \subseteq Z(S,R) \oplus S$, where $Z(S,R) = \{x \in R : [x,S] = \{0\}\}$.

(iii) $[H,S] \subseteq S$ and $H + S = (H \cap R) \oplus S$ is a reductive subalgebra L_1 with the compactly embedded Cartan algebra H.

(iv) L is a semisimple L_1-module and $Z(S,R)$ is an L_1-submodule. In particular, there is a direct L_1-module decomposition $L = S \oplus Z(S,R) \oplus M_1$ with $R = Z(S,R) \oplus M_1$. Moreover, there is an L_1-module decomposition $Z(S,R) = (H \cap R) \oplus M_2$.

(v) Any real root decomposition (*see* Theorem III.6.5) and the Levi decomposition $L = R \oplus S$ are compatible, that is for each $\omega \in \Omega^+$,

$$L^\omega = (L^\omega \cap R) \oplus (L^\omega \cap S) \text{ and } R^\omega = L^\omega \cap R, \quad S^\omega = L^\omega \cap S,$$

where R^ω and S^ω denote the isotypic components of the H-modules R and S indexed by $\omega \in \Omega^+$ in an obvious way.

Proof.   (i) The abelian subalgebra $H \cap R$ defines a Fitting decomposition of L as a direct sum of the Fitting 0-component L_0 and a complementary Fitting 1-component (see for instance [Bou75], Chap.VII, § 1, n° 1, Corollaire 2). Then L_0 is the centralizer of $H \cap R$ in L and thus is a subalgebra of L with Cartan subalgebra H and $L = L_0 + R$ (see for instance loc.cit., § 3, Ex.11,g, p.62). Now we find a Levi complement S for the radical $R \cap L_0$ in L_0 such that $H = (H \cap R) \oplus (H \cap S)$ (see loc.cit. Ex.11,b, p.61). Because of $L = L_0 + R$, the Levi complement of $L_0 \cap R$ in L_0 is also a Levi complement of R in L. Thus $L = R \oplus S$ in such a fashion that $H = (H \cap R) \oplus (H \cap S)$. Since the intersection $H \cap S$ is the homomorphic image of H under the projection of L onto S along R, it it is a compactly embedded Cartan algebra of S.

(ii) and (iii) Since H is compactly embedded, $\operatorname{ad} x$ is semisimple for $x \in H$. Hence the Fitting 0-component L_0 is exactly the centralizer $Z(H \cap R, L)$ of $H \cap R$ in L. Now $S \subseteq L_0$ implies $[H \cap R, S] = \{0\}$. Hence $H + S = (H \cap R) \oplus S$ is a reductive algebra. The remainder of statements (ii) and (iii) is clear after these observations.

In order to prove (iv) we observe that L is a semisimple S-module by Weyl's Theorem and a semisimple $H \cap R$-module since H is compactly embedded. Since $[S, H \cap R] = \{0\}$ by (iii), L is a semisimple L_1-module. Since S is an L_1-submodule by (iii) and R is an L_1 submodule as an ideal, $L = R \oplus S$ is an L_1-module decomposition. We claim that $Z(S,R)$ is an L_1-module: To prove this we must show that $[H, Z(S,R)] \subseteq Z(S,R)$. Thus let $h \in H$ and $z \in Z(S,R)$. Take any $s \in S$. Then $[s, [h, z]] = [[s,h], z] + [h, [s, z]]$. The second summand is zero since $[s, z] = 0$ by the definition of the centralizer. From (iii) above we know that $[s, h] \in S$; hence the first summand vanishes for the same reason. This proves the claim. The existence of M_1 is just a consequence of the semisimplicity of the L_1-module R. For the rest it suffices to remark that $H \cap R$ is an L_1-module.

(v) Since $[H, S] \subseteq S$, we know that S is an H-module under the adjoint action, and, as an ideal, R is an H-module, regardless. From the definition of isotypic components of H-modules, we have $R^\omega = R \cap L^\omega$ and $S^\omega = S \cap L^\omega$ with an obvious understanding of the notation for the isotypic components of R and S. Since $L = R \oplus S$ is an H-module decomposition, we have $L^\omega = R^\omega \oplus S^\omega$. With this remark, the proposition is proved.   ∎

In particular, if we have the information that the centralizer of some Levi complement (and hence all Levi complements) is zero, then any compactly embedded Cartan algebra meets the radical trivially and is contained in some Levi complement. No mixed Lie algebra with cone potential can have this property by Theorem III.6.21. This gives the following remark:

III.6.29.   Corollary.   *If L is a Lie algebra with cone potential which is neither semisimple nor solvable, then the centralizer of any Levi subalgebra in the radical is non-zero.* ∎

In the proof of the following corollary, we need to recall the following fact:

III.6.30.   Lemma.   *If $L = R \oplus S$ is the Levi decomposition of a finite dimensional Lie algebra and I is any ideal of L, then $I = (I \cap R) \oplus (I \cap S)$.*

Proof.   We may factor the ideal $R \cap I$ and henceforth assume that $R \cap I = \{0\}$. Then $S_1 = S \cap (R + I)$ is a Levi complement for R in $R + I$. But I is a unique Levi complement for R in $R \oplus I$, since $[I, R] = \{0\}$ and thus $I = S_1$. Hence $I \subseteq S$ which we had to show. ∎

III.6.31.   Corollary.   *If H is a compactly embedded Cartan algebra of L, then every ideal of L which meets H trivially is contained in R.*

Proof.   Let I be an ideal of L with $I \cap H = \{0\}$. We choose S as in Theorem III.6.28 and invoke Lemma III.6.30 to show that $I = (R \cap I) \oplus (S \cap I)$. We have to show that $S \cap I = \{0\}$. But in a semisimple Lie algebra, any non-zero ideal is a direct summand and meets any of its Cartan algebras. Hence, if $S \cap I$ is non-zero, then so is its intersection with $H \cap S$, hence with H. ∎

Compact and non-compact roots in quasihermitian Lie algebras

We recall that a compactly embedded Cartan algebra H of a Lie algebra L determines a unique maximal compactly embedded subalgebra $K(H)$ containing H by Proposition III.5.6.

III.6.32.   Lemma.   *Let H be a compactly embedded Cartan algebra in L and $\omega \in \Omega^+$ a non-zero positive root. Then $K(H) \cap L^\omega = K(H)^\omega = L^\omega \cap \operatorname{comp} L$.*

Proof.   Let us note first that $K(H)$ is an H-submodule, whence $K(H)^\omega = L^\omega \cap K(H)$ for any root ω. The remainder then follows from Corollary III.6.16. ∎

It is not clear that L^ω has to be contained in $K(H)$ if it contains non-zero elements of $K(H)$. It is exactly this property which we shall establish in those algebras which support invariant pointed generating cones. The key is Theorem III.5.16. It motivates the following definition:

III.6.33.   Definition.   A Lie algebra L will be called *quasihermitian* if it is a finite dimensional real Lie algebra with a compactly embedded Cartan algebra H such that
$$Z_H \cap \operatorname{int}(\operatorname{comp} L) \neq \varnothing$$
for $Z_H \overset{\text{def}}{=} Z\big(K(H)\big)$.

The terminology is meant to be reminiscent of the context of hermitian symmetric spaces, where this phenomenon arises.

III.6.34.   Proposition.   *An element z in a Lie algebra L is in*
$$Z_H \cap \operatorname{int}(\operatorname{comp} L)$$
if and only if its centralizer $\ker(\operatorname{ad} z)$ *in L is exactly* $K(H)$.

Proof.   Suppose $z \in Z_H \cap \operatorname{int}(\operatorname{comp} L)$. Then $\ker(\operatorname{ad} z)$ is a compactly embedded subalgebra by Theorem III.2.12, and it contains H. Hence $\ker(\operatorname{ad} z) \subseteq K(H)$, by Proposition III.5.6. The reverse inclusion is obvious from the definition of Z_H. Conversely, suppose $\ker(\operatorname{ad} z) = K(H)$. Then $z \in Z_H$ since $[z, K(H)] = \{0\}$, and $\ker(\operatorname{ad} z)$ is a compactly embedded subalgebra. Then z is in the interior of $\operatorname{comp} L$ by Theorem III.2.12.   ∎

III.6.35.   Proposition.   *Let L be a quasihermitian Lie algebra. Suppose that ω is a non-zero root according to Theorem III.6.5. Then the following conditions are equivalent:*

(1) $\langle \omega, Z_H \cap \operatorname{int}(\operatorname{comp} L) \rangle = \{0\}$.

(2) *There is a non-zero element z in $Z_H \cap \operatorname{int}(\operatorname{comp} L)$ such that $\omega(z) = 0$.*

(3) $L^\omega \subseteq K(H)$.

(4) $L^\omega \cap K(H) \neq \{0\}$.

(5) $\omega \in Z_H^\perp$, *where $S^\perp$ denotes the annihilator in $\widehat{H}$ of any subset S of H.*

Proof.   Trivially, (1) implies (2). Suppose (2) and let $x \in L^\omega$. Then $0 = \omega(z) \cdot Ix = [z, x] = (\operatorname{ad} z)(x)$ in view of (2). Now $x \in \ker(\operatorname{ad} z) = K(H)$ by Proposition III.6.34. This proves (3). Trivially, (3) implies (4). Suppose (4): Consider a non-zero element $x \in L^\omega \cap K(H)$ and an arbitrary element $z \in Z_H$. We claim $\omega(z) = 0$. Since L is quasihermitian, the open subset $Z_H \cap \operatorname{int} \operatorname{comp} L$ of Z_H is non-empty. Hence it generates Z_H. Hence z is a linear combination of elements from this set. For the purposes of our proof it is therefore no loss of generality to assume that z is itself in this set. By Proposition III.6.34 we now know $\ker(\operatorname{ad} z) = K(H)$. Thus $0 = [z, x] = \omega(z) \cdot Ix$. Since x and thus Ix is non-zero, we conclude $\omega(z) = 0$. This proves our claim and shows (5). Trivially, (5) implies (1) and so our proposition is proved.   ∎

This calls for a definition:

III.6.36.   Definition.   Let Ω denote the set $\{i\lambda : \lambda \in \Lambda\}$ of all real roots of a quasihermitean Lie algebra with respect to a compactly embedded Cartan algebra H. We call a root $\omega \in \Omega$ *compact* if it is zero or satisfies the equivalent conditions of Proposition III.6.35. All other real roots are called *non-compact*. The set of compact roots $\Omega \cap Z_H^\perp$ will be denoted Ω_k, its complement will be written Ω_p. Accordingly, once a selection of a set Ω^+ of positive roots has been made, we write $\Omega_k^+ = \Omega_k \cap \Omega^+$ for the set of positive compact roots and $\Omega_p^+ = \Omega_p \cap \Omega^+$ for the set of positive non-compact roots. Finally we shall set

$$P(H) = \bigoplus_{\omega \in \Omega_p^+} L^\omega$$

III.6.37.   Lemma. *For any choice of a vector $z \in Z_H \cap \mathrm{int}(\mathrm{comp}\, L)$ there is a choice of a set Ω^+ of positive roots such that $\omega(z) > 0$ for all non-compact roots $\omega \in \Omega^+$.*

Proof.   We know that $\Omega_k \subseteq Z_H^\perp \subseteq z^\perp$ and $z^\perp \cap \Omega_p = \{0\}$ by Proposition III.6.35. Set $\Omega_p^+ = z^* \cap \Omega_p$ and select a hyperplane $h^\perp$ in $\hat{H}$ in such a fashion that it does not contain any non-zero compact roots and that $\Omega_p^+ \subseteq h^*$. Then $\Omega^+ = \Omega \cap h^*$ is the required selection of positive roots.   ∎

III.6.38.   Theorem. *Let L be a quasihermitian Lie algebra and fix a compactly embedded Cartan algebra H. Then the following conclusions hold:*

(i) $K(H) = H \oplus \bigoplus_{0 \neq \omega \in \Omega_k^+} L^\omega$.

(ii) $L = K(H) \oplus P(H)$ *and* $[K(H), P(H)] \subseteq P(H)$.

(iii) *If I is an ideal of L with $H \cap I = \{0\}$ then $I \subseteq P(H)$. In particular, among all ideals which meet H trivially, there is a unique largest one, and it is contained in the radical R and in $P(H)$.*

(iv) *For any choice of a set Ω^+ of positive roots such that $\omega(z) > 0$ for some $z \in Z_H \cap \mathrm{int}(\mathrm{comp}\, L)$ and all $\omega \in \Omega_p^+$ the vector space $P(H)$, endowed with the complex structure given by $I|P(H)$, is a complex $K(H)$-module, that is $[k, Ip] = I[k, p]$ for $k \in K(H)$ and $p \in P(H)$.*

(v) $R \subseteq H \oplus P(H)$.

(vi) *If L is semisimple, then $L = K(H) \oplus P(H)$ is a Cartan decomposition.*

Proof.   (i) is a reformulation of Proposition III.6.35 in view of Definition III.6.36.

(ii) Firstly, Theorem III.6.5, Proposition III.6.35 and Definition III.6.36 imply $L = K(H) \oplus P(H)$. Secondly, the compact group $G = \mathrm{INN}_L\, K(H)$ acts on L and, by Weyl's Unitary Trick, makes L into a semisimple G-module and, therefore, $K(H)$-module. Thus the submodule $K(H)$ allows a complementary $K(H)$-submodule P; that is $L = K(H) \oplus P$ with $[K(H), P] \subseteq P$. Then P is, in particular, an H-module and thus decomposes into isotypic components

$$P = \bigoplus_{0 \neq \omega \in \Omega^+} (P \cap L^\omega).$$

If $\omega \in \Omega_k^+$, then $L^\omega \subseteq K(H)$ by (i) above. Hence $P \cap L^\omega = \{0\}$. If, on the other hand, $\omega \in \Omega_p^+$, then $P \cap L^\omega \subseteq P(H)$. Hence $P \subseteq P(H)$. Then $K(H) \oplus P(H) = L = K(H) \oplus P$ implies $P = P(H)$, and thus (ii) is proved.

(iii) As an ideal, I is an H-module, whence $I = \bigoplus_{0 \neq \omega \in \Omega^+} (I \cap L^\omega)$ since $H \cap I = \{0\}$. Now assume that $0 \neq x \in I \cap L^\omega$ for some compact root ω. Then

$x \in K(H)$ by (i) and thus $\langle x \rangle$ is of simple compact type by Corollary III.6.16. Hence $0 \neq [x, Ix] \in H \cap I$ since I is an ideal. This contradicts $I \cap H = \{0\}$. Thus $I \cap L^\omega = \{0\}$ for all $\omega \in \Omega_k^+$. But this shows $I = \bigoplus_{\omega \in \Omega_p^+} (I \cap L^\omega) \subseteq P(H)$.

If $\mathcal{J}$ is the collection of all ideals I with $I \cap H = \{0\}$, then $\bigcup \mathcal{J} \subseteq P(H)$. Thus $J = \sum \mathcal{J}$ is still contained in $P(H)$ and so meets H trivially. Clearly, J is the unique largest element of $\mathcal{J}$. The relation $J \subseteq R$ follows from Corollary III.6.31.

(iv) We fix an element $0 \neq z \in Z_H \cap \text{int comp } L$. After Lemma III.6.37 we may assume that for our choice Ω^+ of positive roots we have $\omega(z) > 0$ for all non-compact roots ω. Since $\text{ad } z|L^\omega = \omega(z) \cdot I|L^\omega$ the operator $\varphi = \text{ad } z|P(H) = \bigoplus_{\omega \in \Omega_p^+} \omega(z) \cdot I|L^\omega$ is an automorphism of $P(H)$. Now let $k \in K(H)$. Since z is central in $K(H)$, the operator $\psi = \text{ad } k|P(H)$ commutes with φ. Then the operators $\sqrt{-\varphi^2} = \bigoplus_{\omega \in \Omega_p^+} \omega(z) \cdot \mathbf{1}_{L^\omega}$ and ψ also commute, as do $(-\varphi^2)^{-1/2}\varphi = I|P(H)$ and $\psi = \text{ad } k|P(H)$. But this means $I[k, x] = [k, Ix]$ for all $x \in P(H)$. This finishes the proof.

(v) We have $R = (R \cap K(H)) \oplus (R \cap P(H))$. But since any solvable compact Lie algebra is abelian, $R \cap K(H)$ is an abelian ideal of $K(H)$, hence is central in the compact Lie algebra $K(H)$ and thus is contained in any Cartan algebra. Hence $R \cap K(H) \subseteq H$.

(vi) If L is semisimple, then there exists a Cartan decomposition $L = K(H) \oplus P$ (see for instance [He78], III.7.). Then $[K(H), P] \subseteq P$, and so P is $K(H)$- and in particular an H-module. We have the following isomorphisms of H-modules: $P \cong L/K(H) \cong P(H)$. Since

$$L = H \oplus \bigoplus_{\omega \in \Omega_k^+} L^\omega \oplus \bigoplus_{\omega \in \Omega_p^+} L^\omega$$

is the unique decomposition of the H-module L into its isotypic components and $K(H) = \bigoplus_{\omega \in \Omega_k^+} L^\omega$ and $\bigoplus_{\omega \in \Omega_p^+} L^\omega$, we conclude $P = P(H)$. ∎

From Theorem II.5.16 and Definition III.6.33, the following remark is clear:

Remark. The information contained in Theorem III.6.38 applies, in particular, to all Lie algebras containing an invariant pointed generating cone. ∎

Constructing invariant cones: Reduction to the reductive case

We start the last subsection with technical remarks. By Definition III.6.6, the quadratic function Q depended on the complex structure I on H^+ and thus, after Theorem III.6.5, on the choice of a set Ω^+ of positive real roots. It is important to know whether the function $(h, x) \mapsto \omega(h) \cdot Q(x)$ depends on this choice. We recall that for $h \in H$, any positive real root ω, $x \in L^\omega$, and for any choice Ω^+ with $\omega \in \Omega^+$, we have

(1) $$[[h, x], x] = [\omega(h) \cdot Ix, x] = \omega(h) \cdot Q(x) \in H.$$

In particular we record explicitly that *the operator* $(\operatorname{ad} x)^2$ *maps the Cartan algebra H into itself whenever x is any root vector* $x \in L^\omega$, $\omega \in \Omega$. From (1) above it follows at once that $(\operatorname{ad} x)^2|H$ is a rank one operator on H.

III.6.39.   Definition.   Suppose that L is a real Lie algebra with a compactly embedded Cartan algebra H. For any non-zero real root $\omega \in \Omega$ we define a function $Q_\omega \colon H \times L^\omega \to H$ by

$$(2) \qquad\qquad Q_\omega(h, x) = (\operatorname{ad} x)^2(h) = \big[x, [x, h]\big] = \big[[h, x], x\big].$$

The function Q_ω is clearly linear in the first argument and quadratic in the second. Since it is just the restriction and corestriction of the function $(h, x) \mapsto (\operatorname{ad} x)^2(h) : L \times L \to L$ to $H \times L^\omega$ it does not depend on the choice of Ω^+. Also, (1) implies

$$(3) \qquad\qquad (\operatorname{ad} x)^2(h) = Q_\omega(h, x) = \omega(h){\cdot}Q(x).$$

We see the emphasis on the rank one operators $(\operatorname{ad} x)^2|H$.

We shall now reduce the problem of constructing invariant cones from cones in Cartan algebras to the same problem in reductive algebras. The key is the following basic lemma. It creates a link to the Reconstruction Theorem of Invariant Cones III.5.15.

III.6.40.   Lemma.   *Let L be a quasihermitian Lie algebra with cone potential, H a compactly embedded Cartan subalgebra, R the radical, N the nilradical, and S a Levi complement according to* Proposition III.6.31. *Denote with L_1 the reductive Lie algebra $H + S = (R \cap H) \oplus S$ and suppose that L_1 contains an invariant pointed generating cone W_1. We set*

$$C = W_1 \cap H = p(W_1),$$

where $p \colon L \to L$ is the projection onto H along H^+ according to Definition III.5.2. *We postulate the following*

Hypothesis: *For each non-compact root $\omega \in \Omega_p$ and each root vector $x \in L^\omega \cap N$, the following condition is satisfied:*

$$(\operatorname{ad} x)^2(C) = Q_\omega(C \times \{x\}) \subseteq C.$$

Conclusion: $p\big((\operatorname{INN} L)C\big) \subseteq C$.

Proof.   We select a set Ω^+ of positive roots. Under the adjoint action, L is a semisimple H-, hence $H \cap R$-module, as H is compactly embedded. By Weyl's Theorem, L is a semisimple S-module. But $H \cap R$ and S commute. Hence L is a semisimple L_1-module, and R is an L_1-submodule in which $H \cap R$ is an L_1-submodule. Hence we find a complementary L_1-submodule M in R such that

$$(4) \qquad\qquad R = (H \cap R) \oplus M.$$

In particular, M is an H-module. In the subalgebra $H + R = H \oplus M$, we have a compactly embedded Cartan subalgebra H and thus a canonical H-module decomposition $H \oplus (H^+ \cap R)$. It follows from (1) that $M = H^+ \cap R = (H \cap R)^+$. We invoke Corollary III.6.24 and observe that the nilradical N satisfies

$$(5) \qquad N = Z + M, \qquad \text{and } [M, M] \subseteq Z.$$

The morphism $x \mapsto \operatorname{ad} x \colon L \to \operatorname{ad} L$ maps L onto $\operatorname{ad} L \subseteq \operatorname{gl}(L)$ and has kernel Z. Thus

$$(6) \qquad L/Z \cong \operatorname{ad} L = \operatorname{ad} M \oplus \operatorname{ad}(H \cap R) \oplus \operatorname{ad} S = \operatorname{ad} M \oplus \operatorname{ad} L_1$$

where $\operatorname{ad} M$ is the abelian nilradical of $\operatorname{ad} L$. It follows from (6) that the analytic subgroup $\operatorname{Inn} L$ of $\operatorname{Gl}(L)$ is a product of the abelian normal subgroup $e^{\operatorname{ad} M}$ and the reductive group $\operatorname{Inn}_L L_1$:

$$(7) \qquad \operatorname{Inn} L = e^{\operatorname{ad} M} \cdot \operatorname{Inn}_L L_1.$$

Now let c be an arbitrary element of $C \subseteq H$, and m an arbitrary element of M. We note that $H \subseteq L_1$ and thus

$$(8) \qquad [m, c] \in [M, H] \subseteq [L_1, M] \subseteq M \subseteq H^+ = \ker p,$$

since M is an L_1-submodule. Further

$$(9) \qquad [m, [m, c]] \in [M, M] \subseteq Z$$

by (5). It now follows from (9) that

$$(10) \qquad e^{\operatorname{ad} m} c = c + [m, c] + \frac{1}{2}[m, [m, c]].$$

As a next step, we calculate $[m, [m, c]]$: We recall Theorem III.6.38(v) and consider the root decomposition $M = \bigoplus_{\omega \in \Omega_p^+} (M \cap L^\omega)$. Hence $m = \sum_{\omega \in \Omega_p^+} m_\omega$ with $m_\omega \in M \cap L^\omega$. Then $[c, m] = \sum_{\omega \in \Omega_p^+} [c, m_\omega] = \sum_{\omega \in \Omega_p^+} \omega(c) \cdot Im_\omega$. In view of (5) and the definition of $Q(m_\omega)$ we have $-p([m_\omega, Im_\omega]) = -[m_\omega, Im_\omega]$. For two different elements $\omega, \omega' \in \Omega_p^+$, by Theorem III.6.23, we have $[m_\omega, Im_{\omega'}] \in [N^\omega, N^{\omega'}] = \{0\}$, considering (5). Therefore,

$$(11) \qquad \begin{aligned} [m, [m, c]] &= - \sum_{\omega \in \Omega_p^+} \omega(c) \cdot [m_\omega, Im_\omega] = \sum_{\omega \in \Omega_p^+} \omega(c) \cdot Q(m_\omega) \\ &= \sum_{\omega \in \Omega_p^+} Q_\omega(c, m_\omega). \end{aligned}$$

By our *Hypothesis*, we conclude from (11) that

$$(12) \qquad [m, [m, c]] \in C.$$

Now let $g \in \operatorname{Inn} L$. Then by (7) there is a $g_1 \in \operatorname{Inn}_L L_1$ and an $m \in M$ such that $g = e^{\operatorname{ad} m} g_1$. If $c \in C$, then $m' = g_1^{-1} m \in M$, since M is an L_1-module. Then $[m, g_1 c] = g_1[g_1^{-1} m, c] = g_1[m', c] \in g_1 M \subseteq M \subseteq \ker p$ by (8). Thus

(13) $\qquad\qquad p([m, g_1 c]) = 0$ for all $m \in M, g_1 \in \operatorname{Inn}_L L_1$ and $c \in C$.

Further $[m, [m, g_1 c]] = g_1[g_1^{-1} m[g_1^{-1} m, c]]$. But $[m', [m', c]] \in Z$ by (9), whence $g_1[m', [m', c]] = [m', [m', c]] \in C \cap Z$ by (12) and (9). Thus

(14) $\qquad\qquad [m, [m, g_1 c]] \in C$ for all $m \in M, g_1 \in \operatorname{Inn}_L L_1$ and $c \in C \cap Z$.

Further we note that $g_1 c \in W_1$, since W_1 is invariant in L_1 and $C \subseteq W_1$. Hence

(15) $\qquad\qquad p(g_1 c) \in C \quad$ for all $g_1 \in \operatorname{Inn}_L L_1$.

Now we finally compute

$$p(gc) = p(e^{\operatorname{ad} m} g_1 c)$$
$$= p(g_1 c + [m, g_1 c] + \frac{1}{2} \cdot [m, [m, g_1 c]])$$
$$= p(g_1 c) + [m, [m, g_1 c]] \in C$$

using (13),(14), and (15). For reasons of continuity, this implies

(16) $\qquad\qquad p((\operatorname{INN} L) C) \subseteq C,$

which is what we had to show. $\qquad\qquad\qquad\qquad\qquad\qquad\qquad\qquad\qquad\blacksquare$

This gives the following result:

III.6.41. **Theorem.** (The Reduction Theorem) *Let L be a quasihermitian Lie algebra with cone potential and let H be a compactly embedded Cartan algebra containing a pointed generating cone C. Pick a Levi subalgebra S with $[H, S] \subseteq S$ which exists after* Proposition III.6.28. *Assume the following hypotheses:*

(i) *C is the trace of an invariant cone in the reductive Lie subalgebra $H + S$.*

(ii) *For each non-compact positive root $\omega \in \Omega_p^+$ and each root vector $x \in L^\omega \cap N$ the following relation holds:*

$$(\operatorname{ad} x)^2(C) \subseteq C.$$

Then there is a unique invariant pointed generating cone W in L with $C = H \cap W$. Moreover, $W_1 = W \cap L_1$.

Proof. Since L has cone potential, the vector space H^+ does not contain any non-zero ideal of L by Proposition III.6.21. Thus the existence and uniqueness of W follows from the Reconstruction Theorem of Invariant Cones III.5.15 and the preceding Lemma III.6.40. The Uniqueness Theorem for Invariant Cones III.2.12 applied to the two invariant pointed generating cones W_1 and $W \cap L_1$ of L_1 shows that $W_1 = W \cap L_1$. $\qquad\qquad\qquad\qquad\qquad\qquad\qquad\qquad\qquad\blacksquare$

This theorem is particularly simple for solvable Lie algebras L. In this case $S = \{0\}$, $L_1 = H$ and each pointed generating cone C in H is trivially the trace of an invariant pointed generating cone of L_1. Notice further that, in the solvable case, all positive roots are non-compact and that all root vectors for non-zero positive roots are in the nilradical. Thus from the Reduction Theorem III.6.41 we obtain immediately the following corollary:

III.6.42.   Corollary.   *Let L be a quasihermitian solvable Lie algebra with cone potential and suppose that C is a pointed generating cone of a compactly embedded Cartan subalgebra H. Assume that for each positive root ω and each root vector $x \in L^\omega$ the following condition is satisfied:*

$$(\operatorname{ad} x)^2(C) \subseteq C.$$

Then there is a unique invariant pointed generating cone W in L such that $C = W \cap H$.   ∎

We shall see in the next section that, for solvable Lie algebras this result is rather conclusive insofar as the hypothethis on the roots and root vectors is also necessary for the existence of W.

7. Orbits and orbit projections

In the Reduction Theorem III.6.41 we encountered certain sufficient conditions which allowed us to secure the characterization of invariant cones W in a Lie algebra L through cones C in a compactly embedded Cartan algebra H. In particular, we postulated that for each non-compact positive root $\omega \in \Omega_p^+$ and each root vector $x \in L^\omega$ in the nilradical we had

$$\omega(C) \cdot Q(x) \subseteq C.$$

We shall convince ourselves in this section that this condition is generally necessary. This will eventually allow us to give necessary and sufficient conditions on C to generate invariant cones W with $C = W \cap H$. The Reconstruction Theorem III.5.15 challenges us to find such conditions.

Orbits generated by root vectors

In this section we shall once more concentrate on a finite dimensional Lie algebra L with a compactly embedded Cartan algebra H and a fixed set of positive roots Ω^+ and, accordingly, a complex structure $I \colon H^+ \to H^+$ according to Theorem III.6.5. We shall consider non-zero elements $h \in H$ and root vectors $x \in L^\omega$, $\omega \in \Omega^+$.

We recall from Theorem III.6.5 that

$$(1) \qquad\qquad [h, x] = \omega(h) \cdot Ix.$$

Consequently,

$$(2) \qquad\qquad [[h, x], x] = -\omega(h) \cdot [x, Ix] = \omega(h) \cdot Q(x) \in H.$$

III.7.1. **Definition.** The function $\theta \colon L^\omega \to \mathbb{R}$ is given by

$$(3) \qquad\qquad \theta(x) = \omega(Q(x)) = -\omega([x, Ix]).$$

As a consequence of this definition, (3) implies

$$(4) \qquad\qquad \omega([[h, x], x]) = \omega(h)\theta(x).$$

III.7.2. **Lemma.** *Under the general assumptions in this section,*

$$
(5) \qquad (\operatorname{ad} x)^m (h) = \begin{cases} h & \text{for } m=0, \\ \omega(h)\theta(x)^n \cdot (-Ix) & \text{for } m = 2n + 1, \\ \omega(h)\theta(x)^n \cdot Q(x) & \text{for } m = 2(n+1), \end{cases}
$$

and

$$
(6) \qquad (\operatorname{ad} x)^m (Ix) = \begin{cases} Ix & \text{for } m = 0, \\ -\theta(x)^n \cdot Q(x) & \text{for } m = 2n + 1, \\ +\theta(x)^{n+1} \cdot Ix & \text{for } m = 2(n+1), \end{cases}
$$

where $n = 0, 1, 2, \ldots$.

Proof. It is clear that we have to verify this lemma by induction with respect to m. We prove the first assertion (5); the proof of (6) follows the same line and is left as an exercise to the reader. For $m=0$ there is nothing to prove. Now suppose that the lemma is proved up to an even m; we must show its validity for $m + 1$. If $m = 0$, then (1) above proves the claim. If $m = 2(n+1)$ with $n \geq 0$ then we calculate

$$
\begin{aligned}
(\operatorname{ad} x)^{(m+1)}(h) &= [x, \omega(h)\theta(x)^n \cdot Q(x)] \\
&= \omega(h)\theta(x)^n \cdot \big(-\omega(Q(x))\big) \cdot Ix \\
&= \omega(h)\theta(x)^{(n+1)} \cdot (-Ix),
\end{aligned}
$$

which proves the claim in this case.

Now suppose that $m = 2n + 1$ with $n \geq 0$. Then $(\operatorname{ad} x)^{(m+1)}(h) = [x, \omega(h)\theta^n(x) \cdot (-Ix)] = \omega(h)\theta(x)^n \cdot [x, -Ix] = \omega(h)\theta(x)^n \cdot Q(x)$, and, since $m + 1 = 2(n+1)$, this is what we had to show. ∎

For a simple representation of the subsequent result, we need a brief notation for some standard powerseries.

III.7.3. **Definition.** In the ring $\mathbb{Q}[[X]]$ of all power series with rational coefficients we set

$$
(7) \qquad
\begin{aligned}
c(X) &= \sum_{n=0}^{\infty} \frac{1}{(2(n+1))!} X^n \\
&= \frac{1}{2!} + \frac{1}{4!}X + \frac{1}{6!}X^2 + \cdots,
\end{aligned}
$$

$$
(8) \qquad
\begin{aligned}
s(X) &= \sum_{n=0}^{\infty} \frac{1}{(2n+1)!} X^n \cdots \\
&= 1 + \frac{1}{3!}X + \frac{1}{5!}X^2 + \cdots.
\end{aligned}
$$

We observe:

III.7.4.   **Lemma.**   *In* $\mathbb{Q}[[X]]$, *the following relations hold:*

$$(9) \qquad\qquad \cosh X = 1 + X^2 c(X^2),$$

$$(10) \qquad\qquad \sinh X = X s(X^2).$$

For real variables $t \in \mathbb{R}$ *we record:*

$$(11) \qquad\qquad c(t) = \begin{cases} \frac{\cosh\sqrt{t}-1}{t} & \text{*for* } t > 0; \\ \frac{\cos\sqrt{|t|}-1}{t}, & \text{*for* } t < 0. \end{cases}$$

$$(12) \qquad\qquad s(t) = \begin{cases} \frac{\sinh\sqrt{t}}{\sqrt{t}} & \text{*for* } t > 0; \\ \frac{\sin\sqrt{|t|}}{\sqrt{|t|}} & \text{*for* } t < 0. \end{cases}$$

Proof.   Straightforward from the definitions of cosh and sinh, cos and sin!   ∎

III.7.5.   **Theorem.**   (The Hyperbolic Sine-Cosine Theorem) *In any finite dimensional Lie algebra* L *with a compactly embedded Cartan algebra* H, *we have for all* $y \in L$, *all* $h \in H$ *and each root vector* $x \in L^\omega$, $\omega \in \Omega^+$, *the relations*

$$(12) \qquad\qquad e^{\operatorname{ad} x} y = \cosh(\operatorname{ad} x)y + \sinh(\operatorname{ad} x)y,$$

where

$$(13) \qquad \begin{aligned} \cosh(\operatorname{ad} x)h &= h + \omega(h)c\big(\theta(x)\big)\cdot Q(x), \\ \cosh(\operatorname{ad} x)Ix &= Ix + \theta(x)c\big(\theta(x)\big)\cdot Ix, \end{aligned}$$

and

$$(14) \qquad \begin{aligned} \sinh(\operatorname{ad} x)h &= -\omega(h)s\big(\theta(x)\big)\cdot Ix, \\ \sinh(\operatorname{ad} x)Ix &= -s\big(\theta(x)\big)\cdot Q(x). \end{aligned}$$

In particular, if p *is the canonical projection onto* H *of Definition III.5.2, then*

$$(15) \qquad \begin{aligned} p(e^{\operatorname{ad} x}h) &= \cosh(\operatorname{ad} x)h = h + \omega(h)c\big(\theta(x)\big)\cdot Q(x), \\ p(e^{\operatorname{ad} x}Ix) &= \sinh(\operatorname{ad} x)(Ix) = -s\big(\theta(x)\big)\cdot Q(x). \end{aligned}$$

Proof.   Condition (12) is simply an application of the power series formula $e^X = \cosh X + \sinh X$. Formulae (13) and (14) arise directly from Lemma III.7.2 in view of Definition III.7.3.   ∎

Naturally, the orbits $e^{\mathbb{R}\cdot\operatorname{ad} x}h$ and $e^{\mathbb{R}\cdot\operatorname{ad} x}Ix$ are contained in the test algebra $\mathbb{R}.h + \langle x \rangle$ (see Lemma III.6.11 and the subsequent discussion). We shall be primarily concerned with the projection of the h-orbit under $e^{\mathbb{R}\cdot\operatorname{ad} x}$ into H. It therefore remains to analyse the formulae of Theorem III.7.5 in terms of the different possible types of the test algebras $\langle x \rangle$ of Proposition III.6.12.

The definitions then immediately yield the following conclusions. Firstly, if $\theta(x) = 0$, we get:

III.7.6. **Corollary.** *If, under the circumstances of* Theorem III.7.5, *the test algebra* $\langle x \rangle$ *is of nilpotent or abelian type, then*

(16)
$$e^{\operatorname{ad} x} h = h + \frac{1}{2}\omega(h).Q(x) - \omega(h){\cdot}Ix.$$

∎

If $\theta(x) < 0$, we get:

III.7.7. **Corollary.** *If, under the conditions of* Theorem III.7.5, *the test algebra* $\langle x \rangle$ *is of compact simple type, then*

(17)
$$e^{\operatorname{ad} x} h = h + \omega(h)\theta(x)^{-1}(\cos\sqrt{|\theta(x)|} - 1){\cdot}Q(x)$$
$$- \omega(h)\sqrt{|\theta(x)|}^{-1}\sin\sqrt{|\theta(x)|}{\cdot}Ix.$$

∎

If $\theta(x) > 0$, we get:

III.7.8. **Corollary.** *If, under the circumstances of* Theorem III.7.5, *the test algebra* $\langle x \rangle$ *is of non-compact simple type, then*

(18)
$$e^{\operatorname{ad} x} h = h + \omega(h)\theta(x)^{-1}(\cosh\sqrt{\theta(x)} - 1){\cdot}Q(x)$$
$$- \omega(h)\sqrt{\theta(x)}^{-1}\sinh\sqrt{\theta(x)}{\cdot}Ix,$$

and

(19)
$$e^{\operatorname{ad} x} Ix = \cosh\sqrt{\theta(x)}{\cdot}Ix + \sqrt{\theta(x)}^{-1}\sinh\sqrt{\theta(x)}{\cdot}Q(x).$$

∎

We observe that for all real numbers $t \in \mathbb{R}$ we have $Q(t{\cdot}x) = -[t{\cdot}x, I(t{\cdot}x)] = t^2{\cdot}Q(x)$. This allows us to conclude at once from the preceding corollaries the following result on orbit projections:

III.7.9. **Proposition.** *Under the conditions of* Theorem III.7.5, *assume that* $\omega(h) \neq 0$ *and* $Q(x) \neq 0$. *Then*

(19)
$$p(e^{\mathbb{R}{\cdot}\operatorname{ad} x} h) = h + \begin{cases} \omega(h)\theta(x)^{-1}[0,2]{\cdot}Q(x), & \textit{if } \theta(x) < 0, \\ \omega(h)\mathbb{R}^+{\cdot}Q(x), & \textit{if } \theta(x) \geq 0, \end{cases}$$

∎

Now let us apply this result to formulate necessary conditions for a cone C in a compactly embedded Cartan algebra H to be the trace of an invariant pointed generating cone W.

III.7.10.   Proposition. *Let W be an invariant pointed generating cone in a finite dimensional Lie algebra L. If $C = H \cap W$ with a compactly embedded Cartan algebra H, then for all non-compact roots $\omega \in \Omega_p$ and all root vectors $x \in L^\omega$ we have*

$$(20) \qquad\qquad\qquad (\operatorname{ad} x)^2(C) \subseteq C.$$

Proof.   From Proposition III.5.4 and Proposition III.7.9 above we know that for each $c \in C$ and each root vector $x \in L^\omega$ for which $\theta(x) \geq 0$, we have $c + \mathbb{R}^+ \omega(c) \cdot Q(x) \subseteq C$. Then for all $t > 0$, in particular, $t^{-1} \cdot c + \omega(c) \cdot Q(x) = t^{-1}\big(c + t\omega(c) \cdot Q(x)\big) \in C$. If we let t tend to $+\infty$ we obtain $(\operatorname{ad} x)^2(c) = \omega(c) \cdot Q(x) \in C$. However, in view of Theorem III.6.38, $\theta(x) \geq 0$ happens exactly for the non-compact roots, since $\theta(x) < 0$ occurs exactly for $x \in K(H)$ by Corollary III.6.16.■

For solvable Lie algebras we now have in fact a necessary and sufficient condition for a cone in a compactly embedded Cartan algebra to be the trace of an invariant pointed generating cone. Combining Corollary III.6.42 with Proposition III.7.10 above we obtain at once the following result which is a model for the general type of characterization theorem we are looking for:

III.7.11.   Corollary.   *Let L be a solvable Lie algebra with a compactly embedded Cartan algebra H and a pointed generating cone C in H. Then the following conditions are equivalent:*

(1) *There is an invariant pointed generating cone W in L such that $C = W \cap H$.*

(2) *L is quasihermitian and has cone potential, and for each positive root ω and each root vector $x \in L^\omega$ the following relation holds:*

$$(\operatorname{ad}(x)^2)(C) \subseteq C.$$

■

8. Kostant's Convexity Theorem

In this section we deal with compact groups and shall present the result that the projection of an orbit of an element in a Cartan algebra of a compact Lie algebra under the adjoint group in this Cartan algebra is none other than the closed convex hull of the orbit of this element under the Weyl group. In many respects this is a remarkable result, since it exhibits in a very natural setting one of those circumstances where the projection of a real analytic manifold in some euclidean space into a suitable vector subspace is a convex polyhedron.

III.8.1.   Lemma. *Let G denote a finite dimensional real Lie group and let $B: \mathfrak{g} \times \mathfrak{g} \to \mathbb{R}$ be an invariant symmetric bilinear form on its Lie algebra $\mathfrak{g}$. Fix two elements $X, Y \in \mathfrak{g}$ in the Lie algebra of G. If K is any subgroup of G with Lie algebra $\mathfrak{k}$ and we define the* Hunt function

$$\varphi: K \to \mathbb{R} \quad by \quad \varphi(g) = B\big(X, \mathrm{Ad}(g)Y\big),$$

the the following statements are equivalent:

(1) *g is a critical point of φ.*

(2) *$[X, \mathrm{Ad}(g)Y] \in \mathfrak{k}^{\perp}$, where $\mathfrak{k}^{\perp} = \{Z \in \mathfrak{g}: B(Z, Z') = 0 \text{ for all } Z' \in \mathfrak{k}\}$.*

In particular, if $K = G$ and B is non-degenerate, then (1) is equivalent to

(2′) *$[X, \mathrm{Ad}(g)Y] = 0$.*

Proof.   The point g is a critical point of φ if and only if $\mathbf{1}$ is a critical point of $\varphi_g \stackrel{\mathrm{def}}{=} \varphi \circ \lambda_g$ where $\lambda_g(x) = gx$ on K. This means $d\varphi_g(\mathbf{1}) = 0$, and this is equivalent to $0 = d\varphi_g(\mathbf{1})(Z) = \frac{d}{dt}|_{t=0}\varphi_g(\exp t \cdot Z)$ for all $Z \in \mathfrak{k}$. Now $\varphi_g(\exp t \cdot Z) = B\big(X, \mathrm{Ad}(g \exp t \cdot Z)Y\big) = B\big(\mathrm{Ad}(g)^{-1}(X), e^{t \cdot \mathrm{ad}\, Z}Y\big)$ in view of the invariance of B under $\mathrm{Ad}(G)$, and in view of $\mathrm{Ad}(\exp t \cdot Z) = e^{t \cdot \mathrm{ad}\, Z}$. If we differentiate this expression with respect to t and evaluate for $t = 0$ we obtain $0 = d\varphi_g(\mathbf{1})(Z) = B\big(\mathrm{Ad}(g)^{-1}X, (\mathrm{ad}\, Z)Y\big) = -B([\mathrm{Ad}(g)^{-1}(X), Y], Z)$ by the invariance of B. Thus g is critical for φ exactly when $[\mathrm{Ad}(g)^{-1}X, Y] \in \mathfrak{k}^{\perp}$. As $\mathrm{Ad}(g)$ is an automorphism of $\mathfrak{k}$, this is exactly condition (2).   ∎

Now let G be compact Lie group and fix an invariant positive definite scalar product $\langle \bullet \mid \bullet \rangle$ on its Lie algebra $\mathfrak{g}$. This scalar product shall play the role of B in the preceding lemma. We select a maximal torus T in G. Its Lie algebra $\mathfrak{t}$ is a Cartan algebra of $\mathfrak{g}$. We let N denote the normalizer of T in G and write $\mathcal{W}(G)$ for the Weyl group N/T which operates on $\mathfrak{t}$ in the obvious fashion. Recall that the identity component of N is T and that, as a consequence, N/T is finite. For any $X \in \mathfrak{t}$ we denote with G^X the centralizer $\{g \in G: \mathrm{Ad}(g)(X) = X\}$ of X in G. Since G^X is exactly the centralizer of $\exp \mathbb{R}\cdot X$ and then also of the torus $\overline{\exp \mathbb{R}\cdot X}$ we know that G^X *is connected* (see for instance [Bou82], Chap.9, §2, N° 3, Corollaire 5). We consider $X, Y \in \mathfrak{t}$. If $g = xny$ with $x \in G^X$, $n \in N$ and $y \in G^Y$, then we first note that $\mathrm{Ad}(n)Y \in \mathfrak{t}$ since N is the normalizer of T and that, as a consequence, $[X, \mathrm{Ad}(n)Y] = 0$ as $\mathfrak{t}$ is abelian. Then we compute $[X, \mathrm{Ad}(g)(Y)] = [X, \mathrm{Ad}(x)\,\mathrm{Ad}(n)\,\mathrm{Ad}(y)(Y)] = \mathrm{Ad}(x)[\mathrm{Ad}(x)^{-1}(X), \mathrm{Ad}(n)(Y)] = \mathrm{Ad}(x)[X, \mathrm{Ad}(n)(Y)] = 0$ by the choice of x and y. Hence all points of $G^X N G^Y$ are critical for the function $\varphi: G \to R$ given by $\varphi(g) = \langle X \mid \mathrm{Ad}(g)(Y)\rangle$ by Lemma III.8.1. However, this set is the precise set of critical points:

III.8.2.   **Lemma.**   *A point $g \in G$ is φ-critical if and only if it is an element of $G^X N G^Y$, and the groups G^X and G^Y are connected.*

Proof.   After the preceding remarks and Lemma III.8.1 we have to show that for every element $g \in G$ with $[X, \mathrm{Ad}(g)(Y)] = 0$ we find elements $x \in G^X$, $n \in N$, and $y \in G^Y$ such that $g = xny$. If $\mathfrak{z}$ denotes the centralizer of X in $\mathfrak{g}$, then $\mathfrak{z} = \mathbf{L}(G^X)$ and $\mathfrak{t}$ is a Cartan subalgebra of $\mathfrak{z}$ and thus the G^X-conjugates of $\mathfrak{t}$ cover $\mathfrak{z}$. Hence there is an $x \in G^X$ such that $\mathrm{Ad}(g)(Y) \in \mathrm{Ad}(x)(\mathfrak{t})$. Thus $\mathrm{Ad}(x^{-1}g)(Y)$ is contained in $\mathfrak{t}$ on one hand, and in the orbit $\mathrm{Ad}(G)(Y)$ on the other. Now we claim that

$$(3) \qquad\qquad \mathfrak{t} \cap \mathrm{Ad}(G)(Y) = \mathcal{W}(G)(Y) = \mathrm{Ad}(N)(Y).$$

If this claim is proved, we are finished, for then we find an $n \in N$ such that $\mathrm{Ad}(x^{-1}g)(Y) = \mathrm{Ad}(n)(Y)$. If we now set $y = n^{-1}x^{-1}g$, then $\mathrm{Ad}(y)(Y) = Y$, that is $y \in G^Y$, and g is of the desired form xny. It therefore remains to show claim (3). For this purpose we look at a $Z' \in \mathfrak{t}$ with $Z' = \mathrm{Ad}(k)(Y)$ for some $k \in G$ and consider a regular element X' of $\mathfrak{t}$ and a regular element Y' of $\mathrm{Ad}(k)(\mathfrak{t})$. Then $[X', Z'] = 0$ and $[Y', Z'] = 0$.

The function $\psi: G^{Z'} \to \mathbb{R}$ given by $\psi(g) = \langle X' \mid \mathrm{Ad}(g)(Y')\rangle$ attains a minimum on $G^{Z'}$, say, in m. Then

$$(4) \qquad\qquad\qquad [X', \mathrm{Ad}(m)Y'] \in (G^{Z'})^{\perp}$$

by Lemma III.8.1. On the other hand, since $[X', Z'] = 0$ and $[\mathrm{Ad}(m)Y', Z'] = \mathrm{Ad}(m)[Y', \mathrm{Ad}(m)^{-1}Z'] = \mathrm{Ad}(m)[Y', Z'] = 0$, we have

$$(5) \qquad\qquad\qquad [X', \mathrm{Ad}(m)Y'] \in G^{Z'}.$$

From (4) and (5) we conclude

$$(6) \qquad\qquad\qquad [X', \mathrm{Ad}(m)Y'] = 0.$$

Since X' and Y' and thus also $\mathrm{Ad}(m)Y'$ are regular, their centralizers $Z(X', \mathfrak{g})$ and $Z(\mathrm{Ad}(m)Y', \mathfrak{g})$ in $\mathfrak{g}$, respectively, are Cartan algebras, and by (6) we conclude that they are equal, since two commuting regular elements generate the same Cartan algebra. But we have $\mathfrak{t} = Z(X', \mathfrak{g})$ and $\mathrm{Ad}(k)\mathfrak{t} = Z(Y', \mathfrak{g})$, whence $\mathrm{Ad}(m^{-1}k)\mathfrak{t} = \mathrm{Ad}(m)^{-1}Z(Y', \mathfrak{g}) = Z(\mathrm{Ad}(m)Y', \mathfrak{g}) = Z(X', \mathfrak{g}) = \mathfrak{t}$. Thus $n = m^{-1}k \in N$ as N is the normalizer of T and of $\mathfrak{t}$ in G. Thus $Z' = \mathrm{Ad}(m)^{-1}Z' = \mathrm{Ad}(m)^{-1}\mathrm{Ad}(k)Y = \mathrm{Ad}(n)Y$. This completes the proof of the containment $\mathfrak{t} \cap \mathrm{Ad}(G)Y \subseteq \mathrm{Ad}(N)Y$. Since the reverse containment is clear, condition (3) is proved.   ∎

III.8.3. **Lemma.** *If G is a compact connected Lie group and $\varphi: G \to \mathbb{R}$ is given by $\varphi(g) = \langle X \mid \mathrm{Ad}(g)Y \rangle$ for elements X and Y in the Cartan algebra $\mathfrak{t}$, then the following statements are equivalent:*

(7) *φ has a local minimum in g.*

(8) *$g = xny$ for elements $x \in G^X$, $n \in N$, and $y \in G^Y$, and there is a Weyl chamber $C \subseteq \mathfrak{t}$ such that $X, \mathrm{Ad}(n)Y \in \overline{C}$.*

(9) *φ has a global minimum in g.*

Moreover, if these conditions are satisfied, then $\varphi(g) = \varphi(n)$

Proof. (7)$\Rightarrow$(8): From Lemma III.8.2 we know $g = xny$ with $x \in G^X$, $n \in N$ and $y \in G^Y$. We have

$$(*) \qquad \varphi(x'g'y') = \varphi(g') \text{ for all } x' \in G^X, g' \in G \text{ and } y' \in G^Y,$$

and thus φ also takes a local minimum in n. For each $Z \in \mathfrak{g}$ we define

$$\begin{aligned}
\gamma(t) &= \varphi(n \exp t{\cdot}Z) = \langle \mathrm{Ad}(n)^{-1}X \mid \mathrm{Ad}(\exp t{\cdot}Z)Y \rangle \\
&= \langle \mathrm{Ad}(n)^{-1}X \mid e^{t.\,\mathrm{ad}\,Z}Y \rangle \\
&= \sum_{j=0}^{\infty} \frac{t^n}{n!} \langle \mathrm{Ad}(n)^{-1}X \mid (\mathrm{ad}\,Z)^n Y \rangle.
\end{aligned}$$

It follows that

$$\begin{aligned}
\gamma''(0) &= \langle \mathrm{Ad}(n)^{-1}X \mid [Z, [Z, Y]] \rangle \\
&= \langle X \mid [Z', [Z', \mathrm{Ad}(n)Y]] \rangle
\end{aligned}$$

with $Z' = \mathrm{Ad}(n)Z \in \mathfrak{g}$, and thus, since φ attains a local minimum in n and $\mathrm{Ad}(n)$ is an automorphism of $\mathfrak{g}$, we see that

$$(10) \qquad \langle X \mid [Z, [Z, \mathrm{Ad}(n)Y]] \rangle \geq 0 \text{ for all } Z \in \mathfrak{g}.$$

Now we apply the Real Root Decomposition Theorem III.6.5. to $\mathfrak{g}$ and find

$$(11) \qquad \mathfrak{g} = \mathfrak{t} \oplus \mathfrak{t}^+, \qquad \mathfrak{t}^+ = \bigoplus_{0 \neq \omega \in \Omega^+} \mathfrak{g}^\omega$$

for a suitable choice Ω^+ of positive real roots and with a suitable complex structure $I: \mathfrak{t}^+ \to \mathfrak{t}^+$ so that

$$(12) \qquad [h, Z] = \omega(h){\cdot}IZ \text{ for all } h \in \mathfrak{t}, Z \in \mathfrak{g}^\omega.$$

Then $Z \in \mathfrak{g}^\omega$ implies $[Z, \mathrm{Ad}(n)Y] = -\omega(\mathrm{Ad}(n)Y){\cdot}IZ$ and $[Z, [Z, \mathrm{Ad}(n)Y]] = -\omega(\mathrm{Ad}(n)Y){\cdot}[Z, IZ] = \omega(\mathrm{Ad}(n)Y)Q(Z)$ in the terminology of Definition III.6.6. Now $\langle X \mid [Z, [Z, \mathrm{Ad}(n)Y]] \rangle = \omega(\mathrm{Ad}(n)Y)\langle X \mid Q(Z) \rangle$. On the other hand, $\langle X \mid Q(Z) \rangle = \langle X \mid [IZ, Z] \rangle = \langle [X, IZ] \mid Z \rangle = \omega(X)\langle IZ \mid Z \rangle$. Thus (10) gives

$$(13) \qquad \omega(X)\omega(\mathrm{Ad}(n)Y)\langle IZ \mid Z \rangle \geq 0 \text{ for all } Z \in \mathfrak{g}.$$

We claim that $\langle IZ \mid Z \rangle > 0$ if $Z \neq 0$. Indeed $4\langle IZ \mid Z \rangle = \langle U \mid U \rangle - \langle V \mid V \rangle$ with $U = IZ + Z$ and $V = IZ - Z = IU$. But $\langle IU \mid IU \rangle = \omega(h)^{-1}\langle IU \mid [h,U] \rangle$ with a suitable $h \in \mathfrak{t}$ by (12). Since the scalar product is invariant we have $\langle IU \mid [h,U] \rangle = \langle [IU,h] \mid U \rangle = -\omega(h)\langle U \mid U \rangle$ in view of (12) and $I^2 = -1$ on $\mathfrak{t}^+$. It follows that

$$2\langle IZ \mid Z \rangle = \langle U \mid U \rangle = \|Z + IZ\|^2,$$

which proves the claim. Thus (13) yields

(14) $\omega(X)\omega(\mathrm{Ad}(n)Y) \geq 0$ for any $X, Y \in \mathfrak{t}$.

But then the elements X and $\mathrm{Ad}(n)Y$ are on the same side of the hyperplane $\omega^{-1}(0)$ for all $\omega \in \Omega^+$, and this means that these elements are in the same closed Weyl chamber. This proves (8).

The implication $(8) \Rightarrow (7)$ is clear from Lemma III.8.2. Hence (7) and (8) are equivalent. The implication $(9) \Rightarrow (7)$ is trivial. Therefore it remains to show

$(8) \Rightarrow (9)$: Since G is compact, φ attains a minimum on G, say in g'. In particular, φ has a local minimum in g', and thus by the equivalence of (7) and (8) we find elements $x' \in G^X$, $n \in N$, and $y' \in G^Y$ as well as a Weyl chamber C' such that $g' = x'n'y'$ and $X, \mathrm{Ad}(n)Y \in \overline{C'}$. We finish the proof of the lemma by showing that $\varphi(g) = \varphi(g')$. Since the Weyl group operates transitively on the Weyl chambers, there is a $w \in N$ such that $C' = \mathrm{Ad}(w)C$. Since $X \in \overline{C} \cap \overline{C'}$, we have $X, \mathrm{Ad}(w)X \in \overline{\mathrm{Ad}(w)C} = \overline{C'}$. Since in a compact *connected* Lie group each orbit of the action of the Weyl group on $\mathfrak{t}$ meets $\overline{C'}$ exactly once, we may conclude $\mathrm{Ad}(w)X = X$, that is, $w \in G^X \cap N$. Since $\mathrm{Ad}(n')Y \in \overline{C'} = \mathrm{Ad}(w)\overline{C}$ we note $\mathrm{Ad}(w^{-1}n')Y, \mathrm{Ad}(n)Y \in \overline{C}$. Once again, since the orbit of Y under the Weyl group meets $\overline{C}$ exactly once, we conclude $\mathrm{Ad}(w^{-1}n')Y = \mathrm{Ad}(n)Y$ and thus $n^{-1}w^{-1}n' \in G^Y \cap N$, whence $n' = wny''$ with $y'' \in G^Y$ and $w \in G^X$. In view of $(*)$ above we now observe $\varphi(g') = \varphi(n') = \varphi(n) = \varphi(g)$ which we had to show. ∎

At this point we have very pertinent information on the nature of the function

$$g \mapsto \langle X \mid \mathrm{Ad}(g)Y \rangle : G \to \mathbb{R} \text{ for } X, Y \in \mathfrak{t}.$$

If $p : \mathfrak{g} \to \mathfrak{t}$ denotes the orthogonal projection with kernel $\mathfrak{t}^+$, then $p(X) = X$ and the orthogonality of p implies

(15) $\langle X \mid p(\mathrm{Ad}(g)Y) \rangle = \langle X \mid \mathrm{Ad}(g)Y \rangle.$

We now want to compare the compact sets $A = p(\mathrm{Ad}(G)Y)$ and the convex hull B of $\mathcal{W}(G)Y$ in the real Hilbert space $\mathfrak{t}$. Since G is connected, for each linear functional f of $\mathfrak{t}$ the image $f(A)$ is a compact interval, and from Lemma III.8.3 we know that $f(A) = f(B)$ since the extrema of $\langle X \mid A \rangle$ and of the set $\langle X \mid \mathrm{Ad}(N)Y \rangle = \langle X \mid B \rangle$ are the same. This shows that the closed convex hull of A is necessarily B in view of the elementary Hahn-Banach Theorem on finite dimensional spaces and the fact that the convex hull of the finite set $\mathcal{W}(G)Y$ is closed. However, it is our objective to show that A equals B, that is, that A is already convex. This is now a geometric task.

Several reductions will simplify matters. Firstly we claim that if A is convex for all regular elements Y in $\mathfrak{t}$, then it is convex for all Y in $\mathfrak{t}$. Indeed, every $Y \in \mathfrak{t}$

is a limit $\lim Y_n$ of regular elements. The orbits $\mathrm{Ad}(G)Y_n$ converge in the space of compact subsets of $\mathfrak{g}$ with the Hausdorff metric topology to the orbit $\mathrm{Ad}(G)Y$. Hence the projections $p(\mathrm{Ad}(G)Y_n)$ converge to $p(\mathrm{Ad}(G)Y)$ in the space of compact subsets of $\mathfrak{t}$. But in this space the subspace of compact *convex* subsets is closed. Thus $p(\mathrm{Ad}(G)Y)$ is convex if all $p(\mathrm{Ad}(G)Y_n)$ are convex which proves the claim. Let us assume henceforth that Y is regular. This implies that its centralizer is exactly the maximal torus T, that is,

$$(16) \qquad G^Y = T \subseteq N \text{ and } NG^Y = N.$$

Next we note that we may assume that $\mathfrak{g}$ is semisimple, since otherwise $\mathfrak{g}$ is the direct sum $\mathfrak{z} \oplus \mathfrak{g}'$ of the center $\mathfrak{z}$ and the semisimple commutator algebra $\mathfrak{g}'$ so that $\mathfrak{z} \subseteq \mathfrak{t}$; it is then a very immediate observation that the entire matter is decided on the projection into $\mathfrak{g}'$. Therefore $\mathfrak{g}$ is semisimple from here on. Now 0 is the only central element and

$$(17) \qquad G^X \neq G \text{ for each non-zero } X \in T.$$

We consider the function

$$(18) \qquad F: G \to \mathfrak{t}, \qquad F(g) = p(\mathrm{Ad}(g)Y).$$

We say that F is *locally surjective at b* if arbitrarily small neighborhoods of b in G are mapped onto neighborhoods of $F(b)$ in $\mathfrak{t}$. By the Implicit Function Theorem, this is the case exactly when the linear map $d(F \circ \lambda_b)(\mathbf{1}): \mathfrak{g} \to \mathfrak{t}$ is surjective. This fails exactly when we find a non-zero X in the orthogonal complement of $d(F \circ \lambda_b)(\mathbf{1})(\mathfrak{g})$ in $\mathfrak{t}$. Then the derivative of the function φ given by $\varphi(g) = \langle X \mid \mathrm{Ad}(g)Y \rangle = \langle X \mid F(g) \rangle$ satisfies $d(\varphi \circ \lambda_b)(\mathbf{1}) = \langle X \mid d(F \circ \lambda_b)(\mathbf{1})(Y) \rangle = 0$ by the choice of X. This means that b is a critical point of φ and thus by Lemma III.8.2 we know that $b \in G^X NG^Y = G^X N$ in view of (16). Thus F *fails to be locally surjective exactly on the set*

$$(19) \qquad \Sigma_Y = \bigcup_{0 \neq X \in \mathfrak{t}} G^X N.$$

We recall that for any $X \in \mathfrak{t}$ the centralizer $\mathbf{L}(G^X)$ of X in $\mathfrak{g}$ in the root space decomposition (11) is exactly

$$(20) \qquad \mathbf{L}(G^X) = \bigoplus_{\substack{\omega \in \Omega^+, \\ \omega(X) = \{0\}}} \mathfrak{g}^\omega.$$

There are then only finitely many of these, that is, there is a finite sequence $X_1, \ldots, X_m$ of non-zero elements in $\mathfrak{t}$ such that

$$(21) \qquad \{G^X : 0 \neq X \in \mathfrak{t}\} = \{G^{X_1}, \ldots, G^{X_m}\},$$

since G^X is connected as we noted in Lemma III.8.2 and therefore is uniquely determined by $\mathbf{L}(G^X)$.

We also find finitely many elements $w_1, \ldots, w_q$ in N such that

(22) $$N = Tw_1 \cup \cdots \cup Tw_q,$$

where we may take for q the order of the Weyl group. Now from (19), (21), and (22) we may describe the singular set of points on which F is not locally surjective as

(23) $$\Sigma_Y = \bigcup_{\substack{j=1,\ldots,m \\ k=1,\ldots,q}} G^{X_j} w_k,$$

since $T \subseteq G^X$ for all $X \in \mathfrak{t}$.

It is now evident that this singular set is a finite union of compact submanifolds of G all of which have lower dimension by (17). Thus Σ_Y is a nowhere dense closed subset of G; in other words, F is *locally surjective on the dense open subset* $G \setminus \Sigma_Y$. In particular, A has dense interior. At this point it is reasonable to consider a boundary point $F(b)$ of $A = F(G)$ in $\mathfrak{t}$. Then $F \circ \lambda_b \colon G \to \mathfrak{t}$ maps $\mathbf{1}$ to a boundary point of A, hence cannot be locally surjective at $\mathbf{1}$. Thus $b \in \Sigma_Y$. In other words, if ∂A denotes the boundary of A in $\mathfrak{t}$, then

(24) $$\partial A \subseteq \Delta A \overset{\mathrm{def}}{=} \bigcup_{\substack{j=1,\ldots,m, \\ k=1,\ldots,q}} p(\mathrm{Ad}(G^{X_j}) \, \mathrm{Ad}(w_k) Y).$$

We now apply induction with respect to the dimension of G. (There is nothing to show if $\dim G = 0$!) By (17), the dimension of G^X is lower than that of G and T is contained in G^X. Also Y is regular in the Cartan algebra $\mathfrak{t}$ of $\mathbf{L}(G^X)$. Hence we know by induction hypothesis that $p(\mathrm{Ad}(G^{X_j}) \, \mathrm{Ad}(w_k) Y)$ is convex for all $j = 1, \ldots, m$ and $k = 1, \ldots, q$. Thus ΔA is a finite union of compact convex nowhere dense sets, and is, in particular, contained in a finite union of affine subvarieties $A_1, \ldots, A_r$ of $\mathfrak{t}$ of positive codimension. Then $\mathfrak{t} \setminus (A_1 \cup \cdots \cup A_r)$ is a finite union of open convex sets C_i. none of the C_j meets ∂A; hence by virtue of its connectivity, each C_j is either contained in the interior $\mathrm{int}\, A$ or the complement $\mathfrak{t} \setminus A$. It follows that A is the union of a finite family of closed convex sets $\overline{C_j}$. We claim now that A must actually be convex. Assume on the contrary that this is not the case. Then the interior of A cannot be convex either. Thus there are two points $a_1 = F(g_1)$ and $a_2 = F(g_2)$ in $\mathrm{int}\, A$ such that the straight line segment connecting the two points contains a point outside A. But then this segment necessarily contains at least one boundary point $t \cdot a_1 + (1 - t) \cdot a_2 \in \partial A$, $t \in [0, 1]$. Since A is a finite union of finite convex polyhedra, the boundary ∂A is piecewise affine. Hence the points $a \in \partial A$ for which there is an affine hyperplane T_a such that $T_a \cap \partial A$ is a neighborhood of a in ∂A is dense in ∂A. Thus, by moving $a_1 \in \mathrm{int}\, A$ a little we may assume that $a = t \cdot a_1 + (1 - t) \cdot a_2$ is such a point. Let $X \in \mathfrak{t}$ be a non-zero vector orthogonal to the hyperplane $T_a - a$. Then we consider φ defined by $\varphi(g) = \langle X \mid \mathrm{Ad}(g) Y \rangle = \langle X \mid p(\mathrm{Ad}(g) Y \rangle$ and notice that in the point $b \in G$ with $F(b) = p(\mathrm{Ad}(g) Y) = a$, the function φ attains a local extremum since there is an open neighborhood U of a in $\mathfrak{t}$ such that $U \cap F(G)$ equals $U \cap H$ where H is one of the closed half spaces bounded by the affine hyperplane T_a. On the other hand, since the value of $\varphi(b) = \langle X \mid a \rangle$ is between the values $\varphi(g_1) = \langle X \mid a_1 \rangle$ and $\varphi(g_2) = \langle X \mid a_2 \rangle$ as T_a disconnects the line segment from a_1 to a_2, the function φ does not attain an extreme value at b. This contradicts Lemma III.8.3. The contradiction finally establishes the convexity of A.

We have now completed the proof of the following main result of this section:

III.8.4.   Theorem.   (The Convexity Theorem) *Let G denote a compact connected Lie group and $\mathfrak{t}$ a Cartan algebra of its Lie algebra $\mathfrak{g}$ of G. Let $p: \mathfrak{g} \to \mathfrak{t}$ denote the orthogonal projection (with respect to an invariant scalar product on $\mathfrak{g}$). If $\mathcal{W} = \mathcal{W}(G)$ denotes the Weyl group acting on $\mathfrak{t}$ and Y is any element in $\mathfrak{t}$ then $p(\mathrm{Ad}(G)Y)$ is the convex hull of $\mathcal{W}(Y)$.*   ∎

III.8.5.   Corollary.   *Let L be a compact Lie algebra and H a Cartan subalgebra and $p: L \to H$ the projection of L onto H with kernel H^+ in the terminology of* Theorem III.6.5. *Let $\mathcal{W} = \mathcal{W}(H, L)$ be the Weyl group of L in the sense of* Definition III.5.7 *acting on H. Then for every element Y of H, the set $p((\mathrm{INN}\,L)(Y))$ (see* Definition III.1.11*) is the convex closure of $\mathcal{W}(Y)$.*

Proof.   Let Z denote the center of L. Then $L = Z \oplus L'$ and $Z \subseteq H$. Accordingly, we write $Y = Y_Z + Y'$ with $Y_Z \in Z$ and $Y' \in L'$. Write $G = \mathrm{INN}\,L$. Then $G(Y) = Y_Z + G(Y')$. Moreover, G is the closure in $\mathrm{Gl}(L)$ of the group $e^{\mathrm{ad}\,L} = e^{\mathrm{ad}\,L'}$. Thus it suffices to show that $p(e^{\mathrm{ad}\,L'}Y')$ is the convex hull of $\mathcal{W}Y'$ in $H \cap L'$. Since the restriction of $\mathcal{W}$ to $H \cap L'$ is the Weyl group of L', it is no loss of generality to assume that L is semisimple. Since L is now compact and semisimple, $\mathrm{ad}: L \to \mathfrak{g} = \mathbf{L}(G)$ with $G = \mathrm{INN}\,L = e^{\mathrm{ad}\,L}$ is an isomorphism. Moreover, $e^{\mathrm{Ad}(g)X} = ge^Xg^{-1} = e^{gXg^{-1}}$ for any $X \in \mathrm{gl}(L)$ and $g \in G$. In particular, if $X = \mathrm{ad}\,x$ for $x \in L$, then for any $y \in L$ we have $gXg^{-1}(y) = g[x, g^{-1}(y)] = [g(x), y] = (\mathrm{ad}\,g(x))(y)$ which shows that $\mathrm{ad}: L \to \mathfrak{g}$ is G-equivariant for the natural action of $G = \mathrm{INN}\,L$ on L and the adjoint action of G on its Lie algebra $\mathfrak{g}$. The isomorphism ad maps H isomorphically onto a Cartan algebra $\mathfrak{t}$ of $\mathfrak{g}$. Correspondingly, the complementary subspace H^+ goes to the complementary subspace $\mathfrak{t}^+$ according to Theorem III.6.5. If $p_{\mathfrak{g}}: \mathfrak{g} \to \mathfrak{t}$ is the projection with kernel $\mathfrak{t}^+$, then this implies $(\mathrm{ad}\,|H) \circ p = p_{\mathfrak{g}} \circ \mathrm{ad}$, and thus for each $Y \in H$ we have $\mathrm{ad}\big(p(G(Y))\big) = p_{\mathfrak{g}}\big(\mathrm{ad}(G(Y))\big) = p_{\mathfrak{g}}(\mathrm{Ad}(G)\,\mathrm{ad}\,Y)$. By the Convexity Theorem III.8.4, if we denote with "conv" the formation of the convex hull, the last term equals $\mathrm{conv}\,\mathcal{W}_{\mathfrak{g}}(\mathrm{ad}\,Y) = \mathrm{ad}(\mathrm{conv}\,\mathcal{W}(Y))$ since ad being G-equivariant also respects the actions of the Weyl groups. Since ad is an isomorphism, the corollary follows.   ∎

The convexity theorems have an immediate application to the characterization of invariant wedges in compact Lie algebras. We are referring to the setting of Section 5, notably to the circumstances of Proposition III.5.10 and Theorem III.5.15.

III.8.6.   Corollary.   *Suppose that L is a compact Lie algebra and H a Cartan subalgebra with a wedge C. Then the following statements are equivalent:*

(1) *There is an invariant wedge W in L such that $C = H \cap W$.*

(2) *C is invariant under the Weyl group.*

Proof.   The implication $(1) \Rightarrow (2)$ was already observed in Proposition III.5.10. So we assume (2) and prove (1). We define W as $\widetilde{C}$ as in Definition III.5.11(ii). Then W is an invariant wedge satisfying $H \cap W \subseteq C$ by Proposition III.5.12(ii) and (iii). But if $c \in C$, then $p((\text{INN } L)c) = \text{conv}(\mathcal{W}(c)) \subseteq C$ by Corollary III.8.5 and Condition (2) above. Then $c \in H \cap W$ in view of Proposition III.5.12(i) by the choice of W as $\widetilde{C} = \{x \in L : p((\text{Inn } L)x) \subseteq C\}$. Hence $C = H \cap W$ as asserted. Thus (1) and (2) are equivalent.   ∎

Note that if C is generating in H, then W is generating in L, since W contains $C^\star$, the closed convex hull of $(\text{Inn } L)(C)$ by Proposition III.5.14(i), and since $C^\star$ is generating in L by Proposition III.5.14(iii). If $C \neq H$, then L has a non-trivial center in view of Theorem III.4.7. Hence L is quasihermitian (see Definition III.6.36). The complement H^+ does not contain any non-zero ideal by Theorem III.6.38(iii). If C is pointed, then W is unique and equals $\widetilde{C}$ and $C^\star$ by the Reconstruction Theorem III.5.15.

In Corollary III.7.11 the solvable case was treated; there and in the preceding Corollary III.8.6 necessary and sufficient conditions were given on a pointed generating cone in a compactly embedded Cartan algebra to be the trace of an invariant pointed generating cone, but these conditions were quite different. It will be the task of the next sections to show that in the general case the two conditions together will be the necessary and sufficient criterion for C to be the trace of an invariant cone.

9. Invariant cones in reductive Lie algebras

At the present stage of information on the characterization problem for invariant cones in Lie algebras it remains to answer the following question:

Given a compactly embedded Cartan algebra H in a finite dimensional Lie algebra and a generating pointed cone C in H, what are the necessary and sufficient conditions for the existence of an invariant cone W in L with $C = W \cap H$?

We saw in the Reduction Theorem III.6.44 that the question reduces to the case of reductive Lie algebras. In that theorem, we encountered the following condition which is certainly necessary after Proposition III.7.10:

(I_0) $(\mathrm{ad}\, x)^2(C) \subseteq C$ for each non-compact root ω and each root vector $x \in L^\omega$.

For a better understanding of Condition (I_0) we recall that for any Lie algebra L with a compactly embedded Lie algebra H the linear operator $(\mathrm{ad}\, x)^2$ maps H into itself as soon as x is any root vector $x \in L^\omega$. The rank one operators $(\mathrm{ad}\, x)^2|H$ with $x \in L^\omega$ for all non-compact real roots ω are contained in a smallest wedge $\mathcal{C}(H, L)$ in the vector space $\mathrm{Hom}(H, H)$ of all vector space endomorphisms of H. Since any set of rank one operators together with the zero operator from a multiplicative semigroup, the wedge $\mathcal{C}(H, L)$ is a multiplicative semigroup and thus in fact an associative semialgebra. Since C is a cone, Condition (I_0) can now be rewritten in the equivalent form

(I) C is invariant under the semigroup $\mathcal{C}(H, L)$.

In the case of a solvable algebra, according to Proposition III.7.11, Condition (I) is also sufficient (providing L satisfies the inevitable conditions of having cone potential and being quasihermitian). Since Proposition III.5.10 we know that the following condition is also necessary:

(II) C is invariant under the Weyl group $\mathcal{W}(H, L)$.

In the case of a compact Lie algebra, this condition was also sufficient by Corollary III.8.6.

It is the objective of this section to show that conditions (I) and (II) are sufficient in the case of reductive algebras. With the information we have this will yield the main theorem that they are necessary and sufficient in general.

The arguments are protracted. The proof will be divided into two portions: As a first step we consider a suitable set Ω_p^+ of positive non-compact roots and the complement in H of the union of hyperplanes $\Sigma = \bigcup_{\omega \in \Omega_p^+} \omega^{-1}(0)$. We let $p: L \to H$ as before denote the projection with kernel H^+ and show that a suitable component C' of $H \setminus \Sigma$ meeting $Z_H \cap \mathrm{int}(\mathrm{comp}\, L)$ satisfies

$$(1) \qquad\qquad p\big((\mathrm{INN}\, L)(C')\big) \subseteq C'.$$

For this purpose we shall exhibit an invariant wedge W_0 with $C' = H \cap W_0$. The Reconstruction Theorem III.5.15 then gives us the tools to conclude (1). In the second part of the proof we shall use Kostant's Convexity Theorem in the form of Corollary III.8.5 in order to show that a cone C in H satisfying (I) and (II) also satisfies

$$(2) \qquad\qquad p\big((\mathrm{INN}\, L)(C)\big) \subseteq C.$$

The Reconstruction Theorem will then allow us to draw the desired conclusion.

In the entire section it is a standard hypothesis that L is a quasihermitian and reductive Lie algebra. (See Definition III.6.19 and Definition III.6.33.) Such an algebra automatically has cone potential as follows from the basic structure theory of semisimple complex Lie algebras and Lemma III.6.7. *We fix a compactly embedded Cartan algebra H, a choice of a non-zero $z \in H$ such that $\ker \mathrm{ad}\, z = K(H)$, and a choice of a set Ω^+ of positive roots such that $\omega(z) > 0$ for all non-compact roots.* (See Proposition III.6.34 and Lemma III.6.37.) For a clear understanding of Condition (I) we recall from Definition III.6.39 that, once a choice of positive roots is made, we have

$$(\dagger) \quad (\mathrm{ad}\, x)^2(C) = Q_\omega(C \times \{x\}) = \omega(C){\cdot}Q(x) \quad \text{for all} \quad \omega \in \Omega \quad \text{and} \quad x \in L^\omega.$$

Decomposing the Lie algebra

In order to avoid the technical problems arising from the fact that for a reductive quasihermitean algebra the closures of the components of $H \setminus \Sigma$ are, it is true, generating wedges, but not necessarily pointed cones, we decompose the algebra.

III.9.1.   Lemma.   *There is a decomposition*

$$(3) \qquad\qquad L = L_0 \oplus L_1 \oplus \cdots \oplus L_n,$$

such that L_0 is a compact and L_j is a simple and quasihermitian ideal for $j = 1, \ldots, n$ which is unique up to the order of summands.

Proof.   Any reductive algebra L has a decomposition (3) into a compact direct summand L_0 and simple non-compact direct summands L_j, and this direct sum decomposition is unique up to reordering of summands. We have to show that the summands L_j are quasihermitean. Let $K = K(H)$ be the maximal compactly embedded subalgebra containing H according to Proposition III.5.6. Then $H_j = H \cap L_j$ is a compactly embedded Cartan algebra of L_j and $K_j = K \cap L_j$ is the corresponding maximal compactly embedded subalgebra $K(L_j)$ for $j = 0, \ldots, n$. Now let $z \in H \cap \operatorname{int}(\operatorname{comp} L)$ be an element whose centralizer in L is $K(L)$ which exists after Definition III.6.34. We write $z = \sum_{j=0}^n z_j$ with $z_j \in L_j$. Then the centralizer of z_j in L_j is K_j. For $j \geq 1$ this implies $z_j \in \operatorname{int}(\operatorname{comp} L_j)$ by Proposition III.6.34. Thus L_j is quasihermitian. ∎

Now let $\omega \in \Omega_p^+$ be a non-compact positive root. Then by Lemma III.9.1 we can write $\omega = \sum_{j=0}^n \omega_j$ with linear functionals ω_j on H vanishing on H_k for $k = 0, \ldots, n$ with $k \neq j$. Then $\omega_j | H_j$ is either 0 or a non-compact positive root of L_j with respect to H_j. In fact, ω_j is non-zero exactly for one $j \in \{1, \ldots, n\}$. If we set $\Omega_p^+(j) = \{\omega_j : \omega \in \Omega_p^+, \omega_j \neq 0\}$ then we find a disjoint union

$$(4) \qquad \Omega_p^+ = \bigcup_{j=1}^n \Omega_p^+(j),$$

such that

$$\{\omega_j | H_j : \omega_j \in \Omega_p^+(j)\}$$

is a selection of a set of positive non-compact roots of L_j with respect to H_j. The hyperplane $\omega_j^{-1}(0)$ contains

$$\sum_{\substack{k=0,\ldots,n \\ k \neq j}} H_k,$$

for $j = 1, \ldots, n$, in other words,

$$(5) \qquad \omega_j^{-1}(0) = H_0 \oplus \cdots \oplus H_{j-1} \oplus (\omega_j | H_j)^{-1}(0) \oplus H_{j+1} \cdots \oplus H_n.$$

Now let C_j' be a component of $H_j \setminus \Sigma_j$ where Σ_j is the union of the finitely many kernels of $\omega_j | H_j$ with $\omega_j \in \Omega_p^+(j)$. Then it follows from (5) that

$$(6) \qquad C' = H_0 \oplus C_1' \oplus \cdots \oplus C_n'$$

is a component of $H \setminus \Sigma$, and all components of this open subset of H are obtained as in (6).

In order to show that $p\big((\operatorname{Inn} L)(C')\big) \subseteq C'$ it suffices by (6) to show that $p_j\big((\operatorname{INN} L_j)(C_j')\big) \subseteq C_j'$ for $j = 1, \ldots, n$.

Thus, until further notice, we shall assume that L *is a simple quasihermitian Lie algebra*. Therefore L is hermitian, that is, in addition to being quasihermitian, $\dim Z\big(K(H)\big) = 1$. (See for instance [He78], Chapter VIII, Proposition 6.2.) Under these circumstances, $\mathbb{R} \cdot z = Z\big(K(H)\big)$. We recall $\omega(z) > 0$ for all $\omega \in \Omega_p^+$. Let C_z' denote that component of $H \setminus \Sigma$ which contains z. In view of $\omega(z) > 0$ this means that

$$(7) \qquad C' = C_z' = \{h \in H : \omega(h) > 0 \text{ for all } \omega \in \Omega_p^+\}.$$

We claim that $\overline{C_z'}$ is the trace in H of an invariant cone in L.

Invariant cones in hermitian simple Lie algebras

We note first that by the Existence Theorem of Invariant Cones in Semi-simple Lie algebras III.4.7 there is a pointed invariant cone. The Cartan–Killing form B on L is non-degenerate, invariant, and negative definite on $K(H)$, hence $B(z, z)$ and the Minimax Theorem for Invariant Cones in Lie algebras III.2.18 is available with $q = -B$ and shows that we have invariant cones $W^{\min}$ and $W^{\max}$ which are unique up to sign such that for every invariant cone W which is not a vector space we find a $\sigma \in \{1, -1\}$ such that

$$(8) \qquad\qquad W^{\min} \subseteq \sigma \cdot W \subseteq W^{\max}.$$

Since L is simple and the edge of an invariant wedge as well as the vector space spanned by it are ideals by Proposition II.1.10, these wedges are necessarily pointed and generating. For easy reference we recall that

$$(9) \qquad\qquad W^{\min} = \text{closed convex hull of } (\text{INN } L)(\mathbb{R}^+ \cdot z),$$

and

$$(10) \qquad W^{\max} = (W^{\min})^* = \{x \in L : B(x, y) \le 0 \text{ for all } y \in W^{\min}\}$$
$$= \{x \in L : B(x, \gamma x) \le 0 \text{ for all } \gamma \in \text{INN } L\}$$
$$= \{x \in L : B(x, \gamma x) \le 0 \text{ for all } \gamma \in \text{Inn } L\}.$$

Tracing the maximal invariant wedge

We shall now work towards a proof that $W^{\max} \cap H = \overline{C'}$ which will then take care of our claim.

First we show

$$(11) \qquad\qquad W^{\max} \cap H \subseteq \overline{C'}.$$

We recall that $H \cap \text{int}(W^{\max}) = \text{int}_H(H \cap W^{\max})$ by Proposition III.5.4. Since the algebraic interior of a wedge is dense in the wedge it will therefore suffice if we show

$$(12) \qquad\qquad H \cap \text{int}(W^{\max}) \subseteq C'.$$

Let $h \in H \cap \text{int}(W^{\max})$. Since $z \in W^{\max}$, we even have $h + \mathbb{R}^+ \cdot z \subseteq \text{int}(W^{\max})$. If $\omega(h) > 0$, for all $\omega \in \Omega_p^+$, then $h \in C'$ by (7). If there were an $\omega \in \Omega_p^+$ such that $\omega(h) \le 0$ then $t \mapsto \omega(h + t \cdot z) = \omega(h) + t\omega(z)$, taking a non-positive value and being unbounded above because of $\omega(z) > 0$, must attain the value 0. That is, there is an $r \geq 0$ such that $\omega(h + r \cdot z) = 0$. By Theorem III.6.5(2), this implies $[h + r \cdot z, x] = 0$ for all $x \in L^\omega$. For these x we are in Case S_+ of Proposition III.6.12, whence $\theta(x) = \omega(Q(x)) > 0$ according to Definition III.7.1. For any $x \in L^\omega$ with $\theta(x) = 1$ and any $t \in R$ we now have $[h + r \cdot z + t \cdot x, Ix + Q(x)] = t \cdot [x, Ix + Q(x)] = t \cdot (-Q(x) - \omega(Q(x)) \cdot Ix) = -t \cdot (Ix + Q(x))$. Thus $\mathrm{ad}(h + r \cdot z + t \cdot x)$ has the eigenvalue $-t$, whence $h + r \cdot z + t \cdot x \notin \mathrm{comp}\, L$ for all $0 \neq t \in \mathbb{R}$ by Proposition III.2.11. However, $\mathrm{int}\, W^{\max} \subseteq \mathrm{comp}\, L$ by the Fourth Cartan Algebra Theorem III.2.14. It follows that $h + r \cdot z + t \cdot x \notin \mathrm{int}\, W^{\max}$ for all $0 \neq t \in \mathbb{R}$, and this obviously contradicts $h + r \cdot z \in \mathrm{int}\, W^{\max}$. This contradiction proves (12) and thus (11).

The second step is a proof of the converse of (11):

$$(13) \qquad\qquad C' \subseteq W^{\max}.$$

We begin by proving the following lemma:

III.9.2. **Lemma.** *The following condition is sufficient for (13):*

(ω) *There is an $\omega \in \Omega_p^+$ and a root vector $x \in L^\omega$ such that with $v = x + Q(x)$ the cone $W^{\max}$ can be described as follows:*

$$W^{\max} = \{x \in L : B(x, \kappa v) \leq 0 \text{ for all } \kappa \in \mathrm{INN}_L\, K(H)\}.$$

Proof. The invariance of B implies $B(x, \gamma v) = B(\gamma^{-1} x, v)$ for all $\gamma \in \mathrm{INN}\, L$. Now let $h \in C'$; we must show that $B\big(\kappa h, (x + Q(x))\big) \leq 0$ for all $\kappa \in \mathrm{INN}_L\, K(H)$. The compact Lie algebra $K(H)$ contains the conjugates κH with $\kappa \in \mathrm{INN}_L\, K(H)$, and we have $x \in L^\omega \subseteq P(H)$ with $P(H)$ as in Theorem III.6.38. By Theorem III.6.38(vi), $L = K(H) \oplus P(H)$ is a Cartan decomposition since L is simple. In particular, $K(H)$ and $P(H)$ are orthogonal with respect to B. Hence $B(\kappa h, x) = 0$ and our task reduces to showing that

$$B\big(\kappa^{-1} h, Q(x)\big) = B\big(h, \kappa Q(x)\big) \leq 0 \text{ for all } \kappa \in \mathrm{INN}_L\, K(H).$$

Now $\kappa Q(x) = \kappa[Ix, x] = [I\kappa x, \kappa x]$ since $P(L)$ is a complex $K(H)$- hence also $\mathrm{INN}_L\, K(L)$-module for the complex structure given by I in view of Theorem III.6.38(iv) since $\omega(z) > 0$ for all $\omega \in \Omega_p^+$. We write $\kappa x = \sum_{\mu \in \Omega_p^+} y_\mu$ with $y_\mu \in L^\mu$ and compute

$$B\big(h, \kappa Q(x)\big) = B(h, [I\kappa x, \kappa x]) = B([h, I\kappa x], \kappa x)$$
$$= -\sum_{\mu, \nu \in \Omega_p^+} \mu(h) B(y_\mu, y_\nu).$$

Now we use that, firstly, the L^μ are B-orthogonal and secondly, that $B|\big(P(H) \times P(H)\big)$ is positive definite in view of the fact that $K(H) \oplus P(H)$ is a Cartan decomposition of L. Thirdly we observe that $h \in C'$ implies $\mu(h) > 0$. With all of this we conclude $B\big(h, \kappa Q(x)\big) = -\sum_{\mu \in \Omega_p^+} \mu(h) B(y_\mu, y_\mu) \leq 0$, which proves the lemma. $\blacksquare$

Maximal real positive roots

The next step is a verification of the sufficient condition of Lemma III.9.2. In this subsection we assume that L is a reductive Lie algebra. At this point we shall embark on a modification of a traditional procedure (see [He78], Chapter VIII, Propositions 11,12) and once more enlarge our scope by considering the complexification $L_{\mathbb{C}} = L \oplus i \cdot L$ of L; we shall maintain the notation of Proposition III.6.3 and Theorem III.6.5. In particular, $\Omega = -i \cdot \Lambda | H$, where Λ is the set of roots. Accordingly we shall now write Λ_p^+ for the natural extension to $H_{\mathbb{C}}$ of the members of $i \cdot \Omega_p^+$. Let $M \subseteq \Lambda_p^+$ be an arbitrary subset and μ a maximal element in M. Let us set

$$(14) \qquad P_{\mathbb{C}}(M) = \sum_{\lambda \in M} (L_{\mathbb{C}}^\lambda + L_{\mathbb{C}}^{-\lambda}).$$

and

$$(15) \qquad M(\mu) = \{\lambda \in M : \lambda - \mu \notin \Lambda, \lambda + \mu \notin \Lambda, \lambda \neq \mu\},$$

We recall that for subsets X and Y of a Lie algebra we denote with $Z(X,Y)$ the centralizer of X in Y. We fix non-zero elements $x_\lambda \in L_{\mathbb{C}}^\lambda$ for $\lambda \in M$. In particular, $[x_\mu, x_{-\mu}]$ is a non-zero vector in $H_{\mathbb{C}}$.

For the following we define

$$\Lambda^+ \overset{\text{def}}{=} \{\lambda \in \Lambda : -i\lambda | H \in \Omega^+\}.$$

III.9.3.   Lemma.   (i) If $\lambda \in M' = M \setminus \{\mu\}$, then $\lambda - \mu \notin \Lambda^+$.
(ii) If $y = \sum_{\lambda \in M'} (c_\lambda \cdot x_\lambda + c_{-\lambda} \cdot x_{-\lambda})$, then

$$[y, x_\mu + x_{-\mu}] = \sum_{\lambda \in M'} (c_\lambda \cdot [x_\lambda, x_{-\mu}] + c_{-\lambda} [x_{-\lambda}, x_\mu]).$$

(iii) $Z(x_\mu + x_{-\mu}, P_{\mathbb{C}}(M)) = \mathbb{C} \cdot (x_\mu + x_{-\mu}) + P_{\mathbb{C}}(M(\mu))$.
(iv) If ω and ρ are two real roots in Ω_p^+ such that $\omega = -i \cdot \mu | H$ and $\rho = -i \cdot \lambda | H$ with $\lambda \in M(\mu)$, then the two functionals $\omega \pm \rho$ do not belong to Ω.

Proof.   (i) Let $\lambda \in M'$ and set $\delta = \lambda - \mu$. Then we have $\lambda = \mu + \delta$. Since $\lambda \in M'$ we know that $\delta \neq 0$. Thus, if $\delta \in \Lambda^+$ then $\lambda > \mu$ which contradicts the maximality of μ in M. Hence $\lambda - \mu \notin \Lambda^+$

(ii) By our choice of Ω^+ all elements $\lambda' \in \Lambda_k^+$ which are characterized by $-i \cdot \lambda' | H \in \Omega_k^+$, that is, by $\lambda'(z) = 0$, while the elements $\lambda' \in \Lambda_p^+$ are characterized by $-i\lambda'(z) > 0$. Also, for two positive roots λ_1 and λ_2 in Λ^+ the sum $\lambda_1 + \lambda_2$ is either a positive root or no root at all. These remarks allow us to compute

$[x_\lambda, x_\mu] \in L_{\mathbb{C}}^{\lambda+\mu} \cap [P(H)_{\mathbb{C}}, P(H)_{\mathbb{C}}] \subseteq L_{\mathbb{C}}^{\lambda+\mu} \cap K(H)_{\mathbb{C}} = \{0\}$. Thus $[x_\lambda, x_\mu] = 0$. Similarly we see that $[x_{-\lambda}, x_{-\mu}] = 0$. Thus, evaluating $[y, x_\mu + x_{-\mu}]$ we obtain

$$[y, x_\mu + x_{-\mu}] = \sum_{\lambda \in M'} (c_\lambda \cdot [x_\lambda, x_\mu] + c_{-\lambda} \cdot [x_{-\lambda}, x_\mu] + c_\lambda \cdot [x_\lambda, x_{-\mu}] + c_{-\lambda} \cdot [x_{-\lambda}, x_{-\mu}])$$

$$= \sum_{\lambda \in M'} (c_\lambda \cdot [x_\lambda, x_{-\mu}] + c_{-\lambda} \cdot [x_{-\lambda}, x_\mu]).$$

This proves (ii).

(iii) Let $x \in P_{\mathbb{C}}(M)$ be arbitrary. Then we can write $x = c_\mu \cdot x_\mu + c_{-\mu} \cdot x_{-\mu} + y$ with $y = \sum_{\lambda \in M'} (c_\lambda \cdot x_\lambda + c_{-\lambda} \cdot x_{-\lambda})$. The $H_{\mathbb{C}}$-component of $[x, x_\mu + x_{-\mu}]$ is $p([x, x_\mu + x_{-\mu}]) = [c_\mu \cdot x_\mu + c_{-\mu} \cdot x_{-\mu}, x_\mu + x_{-\mu}] = (c_\mu - c_{-\mu}) \cdot [x_\mu, x_{-\mu}]$.

Now we prove (iii) by first showing that the left side is contained in the right side. We thus assume that x commutes with $x_\mu + x_{-\mu}$ and first obtain $c_\mu = c_{-\mu}$. Then we also have $[y, x_\mu + x_{-\mu}] = 0$ and (ii) yields

$$(16) \qquad \sum_{\lambda \in M'} (c_\lambda \cdot [x_\lambda, x_{-\mu}] + c_{-\lambda} \cdot [x_{-\lambda}, x_\mu]) = 0.$$

We now claim that

$$(17) \qquad c_\lambda \cdot [x_\lambda, x_{-\mu}] = 0.$$

By way of contradiction we assume that for some $\lambda \in M'$ we had $c_\lambda \cdot [x_\lambda, x_{-\mu}] \neq 0$. Since $[x_\lambda, x_{-\mu}] \in L_{\mathbb{C}}^{\lambda-\mu}$, then $\lambda - \mu$ is a root. Now (16) gives us exactly one root $\lambda' \in M'$ such that $\lambda - \mu = -\lambda' + \mu$ and $c_\lambda \cdot [x_\lambda, x_{-\mu}] + c_{-\lambda'} \cdot [x_{-\lambda'}, x_\mu] = 0$. Then (i) above implies that $\lambda - \mu \in \Lambda^-$ and thus $\lambda' - \mu = \mu - \lambda \in \Lambda^+$ which contradicts (i) and thus proves our claim. Similarly we show that

$$(18) \qquad c_{-\lambda} \cdot [x_{-\lambda}, x_\mu] = 0.$$

Therefore, if $c_\lambda \neq 0$ for $\lambda \in M'$, then $[x_\lambda, x_{-\mu}] = 0$ by (17), whence $\lambda - \mu \notin \Lambda$ in view of the structure of reductive complex Lie algebras according to which $L_{\mathbb{C}}^{\alpha+\beta} = [L_{\mathbb{C}}^\alpha, L_{\mathbb{C}}^\beta]$ if α, β, and $\alpha + \beta$ are positive roots. Likewise $c_{-\lambda} \neq 0$ for $\lambda \in M'$ implies $-\lambda + \mu \notin \Lambda$ by (18). By the definition of $M(\mu)$, this means $y \in P_{\mathbb{C}}(M(\mu))$.

Next we prove that the right side of (iii) is contained in the left. We observe that $M(\mu) \subseteq M'$. If $\lambda \in M(\mu)$, then $[x_\lambda, x_{-\mu}], [x_{-\lambda}, x_\mu] = 0$ since by the definition of $M(\mu)$, neither $\lambda - \mu$ nor $-\lambda + \mu$ is a root. But then (14) forces $[P_{\mathbb{C}}(M(\mu)), x_\mu + x_{-\mu}]$ to be zero which is what we have to prove.

For a proof of (iv), we consider the Real Root Decomposition III.6.5, according to which $L^{\omega \pm \rho} \subseteq L_{\mathbb{C}}^{\mu \pm \lambda} \oplus L_{\mathbb{C}}^{-(\mu \pm \lambda)}$. However, by the definition of $M(\mu)$ and since $\lambda \in M(\mu)$, the right hand side is zero, and thus the real functionals $\omega \pm \rho$ cannot be real roots. ∎

III.9.4.   **Lemma.**   *Let ω be any maximal positive non-compact root, and let $x_\nu \in L^\nu$ be arbitrary non-zero elements for $\nu \in \Omega_p^+$. Then there is a set $\Pi \subseteq \Omega_p^+$ such that the following conditions hold:*

(i) $\omega \in \Pi$.

(ii) *The set $A_0 = \sum_{\nu \in \Pi} \mathbb{R} \cdot x_\nu$ is a maximal abelian subalgebra in $P(H)$.*

(iii) *If $\nu, \nu' \in \Pi$ and $\nu \neq \nu'$, then $\nu + \nu', \nu - \nu' \notin \Omega$.*

(iv) *If $J : P(H) \to P(H)$ is a vector space automorphism with $J(L^\nu) \subseteq L^\nu$ for all $\nu \in \Omega_p^+$, then $J(A_0)$ is a maximal abelian subspace of $P(H)$. In particular, $A = IA_0$ is a maximal abelian subspace of $P(H)$.*

Proof.   We fix arbitrary non-zero vectors $x_\nu \in L^\nu$ for all $\nu \in \Omega_p^+$. According to Theorem III.6.5(4) we set $x_\nu^\pm = x_\nu \mp i \cdot I x_\nu$. If λ is that complex root for which $\nu = -i \cdot \lambda | H$ we shall write $x_\lambda = x_\nu^+ \in L_\mathbb{C}^\lambda$ and $x_{-\lambda} = x_\nu^- \in L_\mathbb{C}^{-\lambda}$, and we have $x_\nu = \frac{1}{2} \cdot (x_\lambda + x_{-\lambda})$. We let λ_0 be that maximal root in Λ_p^+ for which $\omega = -i \cdot \lambda_0 | H$ with the given maximal non-compact positive root ω. Then we define $M_0 = \Lambda_p^+$ and construct inductively sets $M_0 \supseteq M_1 \supseteq \cdots \supseteq M_{k+1} = \emptyset$ by picking for a natural number n such that $M_n \neq \emptyset$ a maximal element $\lambda_n \in M_n$ and by setting $M_{n+1} = M_n(\lambda_n)$. There is a first number k such that $M_{k+1} = \emptyset$. We define $\Pi = \{\nu_n = -i \cdot \lambda_n | H : n = 0, \ldots, k\}$ and set

$$(19) \qquad\qquad A_0 = \sum_{\nu \in \Pi} \mathbb{R} \cdot x_\nu \subseteq P(H).$$

Note that $[A_0, A_0] = \{0\}$ by Lemma III.9.3 above and that, trivially, (i) is satisfied. Condition (iii) is a consequence of Lemma III.9.3(iv) in view of the inductive definition of Π.

We still have to show that A_0 is maximal abelian in $P(H)$. Assume by way of contradiction that we have an element $x \in P(H) \setminus A_0$ commuting with each element of A_0. As the sequence $P_\mathbb{C}(M_0) = P(H)_\mathbb{C} \supset \cdots \supset P_\mathbb{C}(M_n) + (A_0)_\mathbb{C} \supset \cdots$ descends to $(A_0)_\mathbb{C}$, we find a largest $m \in \{0, \ldots, k\}$ such that $x \in (P_\mathbb{C}(M_m) + (A_0)_\mathbb{C}) \setminus (P_\mathbb{C}(M_{m+1}) + (A_0)_\mathbb{C})$. Write $x = y + a$ with $y \in P_\mathbb{C}(M_m)$ and $a \in (A_0)_\mathbb{C}$. Then $[y, x_{\lambda_m} + x_{-\lambda_m}] = [x - a, x_{\lambda_m} + x_{-\lambda_m}] = [x - a, 2x_{\nu_m}] = 0$ since x commutes with all of $(A_0)_\mathbb{C}$ and a is a member of the commutative algebra $A_\mathbb{C}$. Now Lemma III.9.3 applies and shows $y \in \mathbb{C} \cdot x_{\nu_m} + P_\mathbb{C}(M_{m+1})$. But $a + \mathbb{C} \cdot x_{\nu_m} \subseteq (A_0)_\mathbb{C}$, whence $x \in P_\mathbb{C}(M_{m+1}) + (A_0)_\mathbb{C}$ contrary to the choice of m. This contradiction proves the claim that A_0 is maximal abelian in $P(H)$.

If J is as stated in (iv), then the elements $x_\nu' = J(x_\nu)$ satisfy the hypotheses of the lemma and its conclusions apply to $J(A_0) = \sum_{\nu \in \Pi} \mathbb{R} \cdot x_\nu'$.   ∎

III.9.5.   **Lemma.**   *We consider arbitrary non-zero elements $x_\nu \in L^\nu$ for all $\nu \in \Pi$ but normalized in such a fashion that $\theta(x_\nu) = \nu(Q(x_\nu)) = 1$. Set $x = -\frac{i\pi}{2} \cdot \sum_{\nu \in \Pi} x_\nu$ and define the automorphism α of $L_\mathbb{C}$ by*

$$\alpha = e^{\operatorname{ad} x} = \cos\left(-\frac{\pi}{2} \cdot \sum_{\nu \in \Pi} \operatorname{ad}(x_\nu)\right) + i \cdot \sin\left(-\frac{\pi}{2} \cdot \sum_{\nu \in \Pi} \operatorname{ad}(x_\nu)\right).$$

Then

(i) $\alpha(x_\nu) = x_\nu$,

(ii) $\alpha(Ix_\nu) = i\cdot Q(x_\nu)$,

(iii) $\alpha\big(Q(x_\nu)\big) = i\cdot Ix_\nu$.

Proof. For each $\nu \in \Pi$ we set $\alpha_\nu = e^{-\frac{i\pi}{2}\cdot \operatorname{ad} x_\nu}$. Since $A = IA_0$ is abelian, the α_ν commute with each other and α is the composition of all of them. Also all α_ν and hence α fix all elements of A which already proves (i). Since $\theta(x_\nu) = \nu\big(Q(x_\nu)\big) = 1$, by Lemma III.7.2 we have

$$\alpha_\nu(Ix_\nu) = \cos(-\frac{\pi}{2}\cdot \operatorname{ad} x_\nu)(Ix_\nu) + i\cdot \sin(-\frac{\pi}{2}\cdot \operatorname{ad} x_\nu)(Ix_\nu)$$
$$= \cos(-\frac{\pi}{2})\cdot Ix_\nu + i\cdot \sin(-\frac{\pi}{2})\cdot\big(-Q(x_\nu)\big)$$
$$= i\cdot Q(x).$$

If $\nu, \nu' \in \Pi$, $\nu \neq \nu'$, then $[x_\nu, Ix_{\nu'}] = \frac{1}{2}\cdot([x_\nu, Ix_{\nu'}] + [Ix_\nu, x_{\nu'}]) + \frac{1}{2}\cdot([x_\nu, Ix_{\nu'}] - [Ix_\nu, x_{\nu'}]) \in L^{\nu+\nu'} + L^{\nu-\nu'} = \{0\}$ by the Real Root Decomposition Theorem III.6.5 (and the Exercise following it) and Lemma III.9.4. Hence $\alpha_\nu(Ix_{\nu'}) = e^{-\frac{i\pi}{2}\cdot \operatorname{ad} x_\nu}Ix_{\nu'} = Ix_{\nu'}$. Thus

$$(20) \qquad \alpha_\nu(Ix_{\nu'}) = \begin{cases} i\cdot Q(x_\nu), & \text{if } \nu' = \nu; \\ Ix_{\nu'}, & \text{otherwise.} \end{cases}$$

Again by the definition of α and by Lemma III.7.2(5),

$$\alpha_\nu\big(Q(x_\nu)\big) = \cos(-\frac{\pi}{2}\cdot \operatorname{ad} x_\nu)\big(Q(x_\nu)\big) + i\cdot \sin(-\frac{\pi}{2}\cdot \operatorname{ad} x_\nu)\big(Q(x_\nu)\big)$$
$$= \cos(-\frac{\pi}{2})\cdot Q(x_\nu) + i\cdot \sin(-\frac{\pi}{2})\cdot(-Ix_\nu) = i\cdot Ix_\nu.$$

If $x \in L^\nu$ and $x' \in L^{\nu'}$ for $\nu \neq \nu'$, then

$$[x, Q(x')] = [x, [Ix', x']]$$
$$= [[x, Ix'], x'] + [Ix', [x, x']],$$

and by Lemma III.9.4(iii), each of these summands is zero. Thus $0 = [x, Q(x')] = -\nu\big(Q(x')\big)\cdot x$, and so

$$(21) \qquad \operatorname{ad} x_\nu\big(Q(x_{\nu'})\big) = 0 \text{ and } \nu\big(Q(x')\big) = 0 \text{ for all } x' \in L^{\nu'}, \nu \neq \nu',$$

and thus $\alpha_\nu\big(Q(x_{\nu'})\big) = e^{-\frac{i\pi}{2}\cdot \operatorname{ad} x_\nu}Q(x_{\nu'}) = Q(x_{\nu'})$. Hence

$$(22) \qquad \alpha_\nu\big(Q(x_{\nu'})\big) = \begin{cases} i\cdot Ix_\nu, & \text{if } \nu' = \nu; \\ Q(x_{\nu'}), & \text{otherwise.} \end{cases}$$

Since α is the composition of all of the α_ν, $\nu \in \Pi$, equation (ii) follows from (20): Since the α_ν commute, we compute $\alpha(Ix_\nu)$ by first applying all $\alpha_{\nu'}$ with $\nu' \neq \nu$; these all leave Ix_ν fixed by (20). Finally we apply α_ν which produces $i\cdot Q(x_\nu)$ as asserted in (ii). Similarly, we prove (iii) with the aid of (22). ∎

III.9.6.   **Lemma.** $\alpha(H_{\mathbb{C}})$ *is a Cartan algebra* $\widetilde{H}_{\mathbb{C}}$ *of* $L_{\mathbb{C}}$ *containing* $A_{\mathbb{C}}$ *, where* $A = IA_0$ *is as in Lemma III.9.5.*

Proof.   The claim is tantamount to $\alpha^{-1}(A_{\mathbb{C}}) \subseteq H_{\mathbb{C}}$. Now $\alpha^{-1}(Ix_\nu) = -i \cdot Q(x_\nu) \in i \cdot H \subseteq H_{\mathbb{C}}$. Then $\alpha^{-1}(A) \subseteq H_{\mathbb{C}}$ since α is an automorphism of $L_{\mathbb{C}}$, and the assertion follows.   ∎

We are now on our way to construct an Iwasawa decomposition of L with respect to the Cartan decomposition $K(H) \oplus P(H)$ and the maximal abelian subspace A of $P(H)$. We will first determine the restricted roots of the pair (L, A).

III.9.7.   **Lemma.**   *Let R denote the set of restricted roots of L with respect to A. Then the function $\lambda \mapsto \lambda \circ \alpha^{-1} \colon \Lambda \to R$ is well-defined and surjective. Specifically,*

$$(23) \qquad [a, \alpha(x)] = (\lambda \circ \alpha^{-1})(a) \cdot \alpha(x) \text{ for all } a \in A, \, x \in L_{\mathbb{C}}^\lambda.$$

Proof.   Since by Lemma III.9.6, the function $\alpha \colon L_{\mathbb{C}} \to L_{\mathbb{C}}$ is a Lie algebra automorphism carrying the Cartan algebra $H_{\mathbb{C}}$ to the Cartan algebra $\widetilde{H}_{\mathbb{C}} \stackrel{\text{def}}{=} \alpha(H_{\mathbb{C}})$, it follows that $\{\lambda \circ \alpha^{-1} : \lambda \in \Lambda\}$ is the set of roots with respect to $\widetilde{H}_{\mathbb{C}}$. Since $A \subseteq A_{\mathbb{C}} \subseteq \widetilde{H}_{\mathbb{C}}$ by Lemma III.9.6, the function $\lambda \circ \alpha^{-1}|A$ is a root on A, and its root space contains that of $\lambda \circ \alpha^{-1}$. Since the sum of the root spaces for the roots of $\widetilde{H}_{\mathbb{C}}$ is $L_{\mathbb{C}}$, the same is true for their restrictions to A. Since the sum of *all* root spaces with respect to A is direct (see [Bou75], Chap. VII, §1, n° 1, Proposition 3), it follows that the restrictions $\lambda \circ \alpha^{-1}|A$ exhaust the roots of A. ∎

Now for each $\rho \in R$ we set

$$\mathcal{R}_\rho = \{\lambda \in \Lambda : \rho = \lambda \circ \alpha^{-1}\}.$$

These sets measure the degree of non-injectivity of the function considered in the preceding lemma. We have an ordering of Λ which we may and shall henceforth assume to be a total order. We write

$$\lambda_\rho = \max \mathcal{R}_\rho$$

and now define an ordering of R via

$$(24) \qquad \rho \leq \rho' \quad \Longleftrightarrow \quad \lambda_\rho \leq \lambda_{\rho'} \text{ for all } \rho, \rho' \in R.$$

By this definition, the total order induced by that of Λ on $\{\lambda_\rho : \rho \in R\}$ is faithfully tranported to R. We claim that this order is compatible with addition. In fact, if ρ, ρ', and $\rho + \rho'$ are elements of R, then $\mathcal{R}_\rho + \mathcal{R}_{\rho'} \subseteq \mathcal{R}_{\rho+\rho'}$, and so $\lambda_\rho + \lambda_{\rho'} \leq \lambda_{\rho+\rho'}$. Since $0 \in \mathcal{R}_0$, the relation $0 \leq \rho$ holds if and only if $0 \leq \lambda_\rho$. We conclude that $\rho \geq 0$ implies $\rho + \rho' \geq \rho'$ for all $\rho' \in R$. In particular we note that $0 \leq \rho$ implies $-\rho \leq 0$. Since Λ is totally ordered, we conclude from this that we have a disjoint union $R = R^+ \cup R^-$ of the set of positive and that of negative roots. If $\omega \in \Omega_p^+$ is a maximal positive real root and λ is that maximal positive complex root which satisfies $\omega = -i \cdot \lambda|H$, then $\rho = \lambda \circ \alpha^{-1}$ is the largest element of R^+. We keep this ω, λ and ρ and choose an arbitrary non-zero $x_\omega \in L^\omega$, but normalized in such a fashion that $\omega\big(Q(x_\omega)\big) = 1$ and find that $x_\omega^+ = x_\omega - i \cdot I x_\omega \in L_\mathbb{C}^\lambda$, implying $\alpha(x_\omega^+) = \alpha(x_\omega) - i \cdot \alpha(I x_\omega) = x_\omega - i^2 \cdot Q(x_\omega) = x_\omega + Q(x_\omega)$. Therefore (23) implies $[a, x_\omega + Q(x_\omega)] = [a, \alpha(x_\omega^+)] = \lambda\big(\alpha^{-1}(a)\big)\alpha(x_\omega^+) = \rho(a)\big(x_\omega + Q(x_\omega)\big)$. Thus

$$(25) \qquad\qquad y_\rho \overset{\text{def}}{=} x_\omega + Q(x_\omega)$$

is a root vector for the maximal root ρ.

A suitable Iwasawa decomposition

We continue with the assumption that L is a reductive Lie algebra and create an Iwasawa decomposition of L. Recall from Theorem III.6.37(vi) that $L = K(H) \oplus P(H)$ is a Cartan decomposition. The vector space $A = \sum_{\nu \in \Pi} \mathbb{R} \cdot I x_\nu$ is a maximal abelian subspace of $P(H)$ by Lemma III.9.4(ii). If R^+ is the set of positive restricted roots with respect to A as in Lemma III.9.7 we may write the Iwasawa decomposition

$$L = K(H) \oplus A \oplus N \text{ with } N = \sum_{\sigma \in R^+} L^\sigma, \text{ and } A = \sum_{\nu \in \Pi} \mathbb{R} \cdot I x_\nu.$$

Then $[N, y_\rho] \subseteq \sum_{\sigma \in R^+} [L^\sigma, L^\rho] = \{0\}$ by the maximality of ρ. It follows that

$$(26) \qquad\qquad (\text{INN } L)(y_\rho) = \big(\text{INN}_L K(H)\big) e^{\text{ad } A}(y_\rho).$$

III.9.8. **Lemma.** $e^{\text{ad } A} y_\rho \subseteq \mathbb{R}^+ \cdot y_\rho$.

Proof. We recall $\omega\big(Q(x_\omega)\big) = 1$; then Corollary III.7.8(18) and (19) yield

$$e^{t \cdot \text{ad } I x_\omega}\big(Q(x_\omega)\big) = Q(x_\omega) + (\cosh t - 1) \cdot Q(x_\omega) + \sinh t \cdot x_\omega = \cosh t \cdot Q(x_\omega) + \sinh t \cdot x_\omega$$

and

$$e^{t \cdot \text{ad } I x_\omega}(x_\omega) = \cosh t \cdot x_\omega + \sinh t \cdot Q(x_\omega),$$

whence $y_\rho = x_\omega + Q(x_\omega)$ yields

$$(27) \qquad\qquad e^{t \cdot \text{ad } I x_\omega} y_\rho = (\cosh t + \sinh t) \cdot y_\rho = e^t \cdot y_\rho.$$

On the other hand, if $\nu \neq \omega$, then $\text{ad } x_\nu\big(Q(x_\omega)\big) = 0$ by (21). Hence

$$e^{t \cdot \text{ad } x_\nu} Q(x_\omega) = Q(x_\omega).$$

Since $[L^\nu, L^\omega] = \{0\}$ for $\nu \neq \omega$ in Π we trivially have $e^{t \cdot \text{ad } x_\nu} x_\omega = x_\omega$. Consequently,

$$(28)' \qquad\qquad e^{t \cdot \text{ad } x_\nu} y_\rho = y_\rho \text{ for all } \omega \neq \nu \in \Pi.$$

Now (27) and (28) prove the lemma. ∎

III.9.9.   Lemma.   (i) $y_\rho \in W^{\min}$.

   (ii) $W^{\min} =$ closed convex hull of $(\operatorname{INN} L)(\mathbb{R}^+ \cdot y_\rho)$

   (iii) $W^{\max} = \{x \in L: B(x, \gamma y_\rho) \le 0 \ for \ all \ \gamma \in \operatorname{INN}_L K(H)\}$.

Proof.   (i) From Corollary III.7.8(18) and $\omega(Q(x_\omega)) = 1$ we recall

$$(29) \qquad e^{t \operatorname{ad} I x_\omega} z = z + \omega(z)\big((\cosh t - 1)\cdot Q(x_\omega) + \sinh t \cdot x_\omega\big).$$

For $t \ne 0$ we divide (29) by $\sinh t$ and let t tend to $+\infty$. The right hand side then tends to $\omega(z)\cdot(Q(x_\omega) + x_\omega) = \omega(z)\cdot y_\rho$. But $\omega(z) > 0$. The left hand side stays in the closed convex hull of $(\operatorname{INN} L)(\mathbb{R}^+ \cdot z)$ which is none other than $W^{\min}$ by definition.

   (ii) The closed convex hull of $(\operatorname{INN} L)(\mathbb{R}^+ \cdot y_\rho)$ is an invariant wedge, and is contained in $W^{\min}$ by (i). Hence (ii) follows by the Minimax Theorem for Invariant Cones III.2.18.

   (iii) We recall from (10) and (ii) above that

$$W^{\max} = \{x \in L: B(x, \gamma y_\rho) \le 0 \ \text{with} \ \gamma \in \operatorname{INN} L\}.$$

Then (26) and Lemma III.9.8 imply that

$$(\operatorname{INN} L)(\mathbb{R}^+ \cdot y_\rho) \qquad \text{and} \qquad \big(\operatorname{INN}_L K(H)\big)(\mathbb{R}^+ \cdot y_\rho)$$

have one and the same closed convex hull. Then (iii) follows.   ∎

   We now come to the conclusion of the first part of this section.

III.9.10.   Theorem.   *Let L be a reductive quasihermitian Lie algebra and H a compactly embedded Cartan algebra. Let $\Sigma = \{h \in H: (\exists \nu \in \Omega_p)\nu(h) = 0\}$. Let C' be any component of $H \setminus \Sigma$. If an element $z \in Z_H \cap \operatorname{int}(\operatorname{comp} L)$ is contained in C', then $C' = \{h \in H: (\forall \nu \in \Omega_p)\nu(z)\nu(h) > 0\}$ and with*

$$C = \overline{C'} = \{h \in H: (\forall \nu \in \Omega_p)\nu(z)\nu(h) \ge 0\}$$

the following relations hold:

$$(30) \qquad\qquad p\big((\operatorname{INN} L)(C')\big) \subseteq C',$$

and

$$(31) \qquad\qquad p\big((\operatorname{INN} L)(C)\big) \subseteq C.$$

Proof.   We introduce an order on the set Ω of real roots according to Lemma III.6.37 such that $\omega(z) > 0$ for all $\omega \in \Omega_p^+$. Then the preceding results of the section apply. We have observed in Lemma III.9.1 and its consequences that it suffices to prove the theorem for simple hermitian symmetric algebras. In (11) we saw that $H \cap W^{\max} \subseteq C$. With the element $y_\rho = x_\omega + Q(x_\omega)$ we have $W^{\max} = \{x \in L: B(x, \kappa y_\rho) \le 0 \ \text{for all} \ \kappa \in \operatorname{INN}_L K(H)\}$ by Lemma III.9.9(iii), so that Lemma III.9.2 applies with $\omega(z) > 0$ and yields $C \subseteq W^{\max}$. Hence $C = H \cap W^{\max}$. But $W^{\max}$ is a pointed generating invariant cone in L. So the Reconstruction Theorem III.5.15 shows that $p\big((\operatorname{INN} L)(C)\big) \subseteq C$.

   In order to show that $p\big(\operatorname{INN} L)(C')\big) \subseteq C'$, we note from the Uniqueness Theorem for Invariant Cones III.2.15 that $\operatorname{int}(W^{\max}) = (\operatorname{INN} L)(C')$ and

$$p\big(\operatorname{int}(W^{\max})\big) \subseteq H \cap \operatorname{int}(W^{\max}) \subseteq \operatorname{algint}(H \cap W^{\max}) = C'.$$

                                                                                ∎

Exploiting sufficient conditions

We are now dealing with an arbitrary pointed generating cone C in H and assume that it satisfies

(I) $$(\operatorname{ad} x)^2(C) \subseteq C \quad \text{for all} \quad \omega \in \Omega_p, \, x \in L^\omega,$$

that is,

(I') $$\omega(C)Q(x_\omega) \subseteq C \quad \text{for all} \quad x_\omega \in L^\omega, \omega \in \Omega_p^+,$$

and

(II) $$\mathcal{W}(H, L)(C) \subseteq C,$$

with the Weyl group $\mathcal{W}(H, L)$ as defined in Definition III.5.7.

III.9.11.   Lemma. *Let L be a quasihermitian Lie algebra with cone potential and let H be a compactly embedded Cartan algebra. If C is a pointed generating cone in H satisfying* (I), *then*

  (i) $\operatorname{algint}(C)$ *is contained in a connected component of the set $H \setminus \Sigma$, where $\Sigma = \{h \in H : (\exists \omega \in \Omega_p^+)\langle \omega, h \rangle = 0\}$, and*

  (ii) $\operatorname{algint} C \subseteq \operatorname{int}(\operatorname{comp} L)$.

Proof.   (i) Since $\operatorname{algint}(C)$ is connected, it suffices to show that $\operatorname{algint}(C) \cap \Sigma = \emptyset$. But if we had $\langle \omega, x \rangle = 0$ for some $x \in \operatorname{algint}(C)$ and some $\omega \in \Omega_p^+$, then there would exist elements $u, v \in C$ such that $\langle \omega, u \rangle < 0 < \langle \omega, v \rangle$. Thus for any element $0 \neq x_\omega \in L^\omega$ we would conclude from (I) that both $Q(x_\omega)$ and $-Q(x_\omega)$ would be contained in C, and this would contradict the fact that C is pointed. This proves (i).

(ii) If $x \in \operatorname{algint} C$ then by (i) above, no root $\omega \in \Omega_p^+$ vanishes on x. But $\ker \operatorname{ad} x = H \oplus \sum_{\omega \in \Omega^+, \langle \omega, x \rangle = 0} L^\omega$ by the Real Root Decomposition III.6.5(2). Thus $K(H) = \sum_{\omega \in \Omega_k^+} L^\omega$, $P(H) = \sum_{\omega \in \Omega_p^+} L^\omega$ and $L = K(H) \oplus P(H)$ implies that $\ker \operatorname{ad} x \subseteq K(H)$. Then $x \in \operatorname{int}(\operatorname{comp} L)$ by the Third Cartan Algebra Theorem III.2.12 and (ii) is proved. ∎

III.9.12.   Lemma. *If L is a Lie algebra with a compactly embedded Cartan algebra H, and if Z_H denotes the center of $K(H)$, then for each $x \in H$ one has*

(32) $$Z_H \cap \operatorname{conv} \mathcal{W}(H, L)(x) \neq \emptyset.$$

Proof.   Let G denote the compact group $\operatorname{INN}_L K(H)$ and $d\kappa$ normalized Haar measure on G. We define the averaging operator $\Delta : L \to L$ by

$$\Delta(x) = \int_G \kappa(x) d\kappa.$$

Then Δ is a projection (that is, idempotent mapping) L onto the G-fixed points of L. Then $\Delta(L) = Z\big(K(H), L\big)$ (compare Lemma II.1.7). Since $K(H)$ is a maximal compact subalgebra of L by Proposition III.5.6(ii), $Z\big(K(H), L\big) \subseteq K(H)$, whence $\Delta(L) = Z\big(K(H), L\big)$ is simply the center Z_H of $K(H)$. Recall that $\Delta(x)$, as the barycenter of the orbit $\big(\mathrm{INN}_L K(H)\big)(x)$, is contained in the closed convex hull of the orbit. If $p: L \to H$ is the usual projection onto H with kernel H^+, then p induces on $K(H)$ the usual orthogonal projection onto H. Thus for any $x \in H$, the Convexity Theorem in the form of Corollary III.8.5 shows that $\Delta(x) = p\big(\Delta(x)\big) \in$
$$p\Big(\overline{\mathrm{conv}}\big(\mathrm{INN}_L(K(H))(x)\big)\Big) = \overline{\mathrm{conv}}\Big(p\{(\mathrm{INN}_L K(H))(x)\}\Big) = \mathrm{conv}\mathcal{W}(H, L)(x) \text{ and}$$
this proves the lemma.   ∎

III.9.13.   Lemma.   *In addition to the hypotheses of* Lemma III.9.11 *assume that also* (II) *is satisfied. Then there is an element* $z \in C \cap Z_H \cap \mathrm{int}(\mathrm{comp}\, L)$ *where* Z_H *denotes the center of* $K(H)$.

Proof.   By Lemma III.9.11(ii) it suffices to show that $Z_H \cap \mathrm{algint}\, C \neq \emptyset$. Now let $x \in \mathrm{algint}\, C$. Then (II) implies that $\mathcal{W}(H, L)(x) \subseteq \mathrm{algint}\, C$ since every invertible endomorphism of C maps $\mathrm{algint}\, C$ into itself. But $\mathrm{algint}\, C$ is convex, hence

$$\mathrm{conv}\mathcal{W}(H, L)(x) \subseteq \mathrm{algint}\, C.$$

Then Lemma III.9.12 proves the assertion.   ∎

At this point we fix for the entire remainder of the argument an element $z \in Z_H \cap \mathrm{algint}\, C \subseteq \mathrm{int}(\mathrm{comp}\, L)$ and assume according to Lemma III.6.37 that a choice of a set Ω^+ of positive roots has been made in such a fashion that

$$\langle \omega, z \rangle > 0 \quad \text{for all} \quad \omega \in \Omega_p^+.$$

III.9.14.   Lemma.   *With these conventions, we can replace condition* (I) *by*

(I'') $Q(x_\omega) \in C$ *for all* $x \in L^\omega, \quad \omega \in \Omega_p^+$ *and* $\mathrm{algint}\, C \subseteq H \setminus \Sigma$.

Proof.   By Lemma III.9.11, condition (I) implies $\mathrm{algint}\, C \subseteq H \setminus \Sigma$ and thus $\omega(\mathrm{algint}\, C) \subseteq\,]0, +\infty[$ or $\omega(\mathrm{algint}\, C) \subseteq\,]-\infty, 0[$ for all $\omega \in \Omega_p^+$. But $z \in \mathrm{algint}\, C$ and $\omega(z) > 0$ for all $\omega \in \Omega_p^+$. Thus $\omega(C) \subseteq\,]0, \infty[$, and thus (I) implies (I'). Conversely, if (I') is satisfied, then all $\langle \omega, c \rangle$ with $c \in \mathrm{algint}\, C$ have the same sign; since z is one of these, $\omega(C) \subseteq \mathbb{R}^+$. But then (I) clearly follows.   ∎

We need further concepts which all depend our previous choices. Indeed we set

(33) $C_1 = \{h \in H : \omega(h) \geq 0 \text{ for all } \omega \in \Omega_p^+\},$

and

(34) $C_2 = $ closed convex hull of $\{\omega(c) \cdot Q(x_\omega) : c \in C, x_\omega \in L^\omega, \omega \in \Omega_p^+\}.$

We now return to the case that L is a *reductive* quasihermitian algebra and thus satisfies $L = Z(L) \oplus [L, L]$. If we write $K'(H) = K(H) \cap [L, L]$, then $[L, L] = K'(L) \oplus P(H)$ is a Cartan decomposition of $[L, L]$ by Theorem III.6.38(vi). If θ is the corresponding Cartan involution, then $\langle x \mid y \rangle' = -B(x, \theta y)$ with the Cartan–Killing form B defines a positive definite scalar product on $[L, L]$. We choose an arbitrary scalar product $\langle \bullet \mid \bullet \rangle_Z$ on $Z(L)$ and consider L as the orthogonal direct sum of the two real Hilbert spaces $(Z(L), \langle \bullet \mid \bullet \rangle_Z)$ and $([L, L], \langle \bullet \mid \bullet \rangle')$, that is for $x = x_Z + x'$ and $y = y_Z + y'$ in $L = Z(L) \oplus [L, L]$, we set

$$\langle x \mid y \rangle = \langle x_Z \mid y_Z \rangle_Z + \langle x' \mid y' \rangle'.$$

The norm defined by this scalar product is

$$\|x\| = \sqrt{\langle x \mid x \rangle}.$$

Now we have the following information on the wedges C_1 and C_2:

III.9.15. **Lemma.** *If L is reductive, the following conclusions hold:*

(i) *C_1 and C_2 are the duals of each other with respect to the restriction to $H \times H$ of the scalar product $\langle \bullet \mid \bullet \rangle$.*

(ii) *For any selection of non-zero elements $x_\omega \in L^\omega$, $\omega \in \Omega_p^+$ one has*

$$C_2 = \sum_{\omega \in \Omega_p^+} \mathbb{R}^+ \cdot Q(x_\omega).$$

(iii) *Both C_1 and C_2 are polyhedral.*

(iv) *The edge $C_1 \cap -C_1$ of C_1 is $Z(L)$, and the span $C_2 - C_2$ of C_2 is $H \cap [L, L]$.*

(v) *Both C_1 and C_2 are invariant under the action of the Weyl group $\mathcal{W}(H, L)$.*

Proof. We recall the root decomposition of the complexification

$$L_{\mathbb{C}} = H_{\mathbb{C}} \oplus \bigoplus_{\lambda \in \Lambda} L_{\mathbb{C}}^\lambda,$$

and that each real root $\omega \in \Omega$ is the restriction $-i\lambda|H$ of a unique complex root $\lambda \in \Lambda$. All real root spaces L^ω are contained in $[L, L]$; all complex root spaces $L_{\mathbb{C}}^\lambda$ are contained in $[L_{\mathbb{C}}, L_{\mathbb{C}}] = [L, L] + i \cdot [L, L]$. The theory of complex semisimple Lie algebras gives us elements $h_\lambda \in H_{\mathbb{C}}$ such that $B_{\mathbb{C}}(h, h_\lambda) = \lambda(h)$ for all $h \in H_{\mathbb{C}}$ and $\lambda(h_\lambda) = 2$ with the Cartan–Killing form $B_{\mathbb{C}}$ on $[L_{\mathbb{C}}, L_{\mathbb{C}}]$. For all $\lambda \in \Lambda$ we have

(35)
$$\dim_{\mathbb{C}} L_{\mathbb{C}}^\lambda = 1$$

and $[L_{\mathbb{C}}^\lambda, L_{\mathbb{C}}^{-\lambda}] = \mathbb{C} \cdot h_\lambda$. From the Real Root Decomposition III.6.5 we conclude that

(36)
$$[L^\omega, L^\omega] = L \cap \mathbb{C} \cdot h_\lambda.$$

If x_ω is any non-zero element of L^ω for $\omega \in \Omega$, then from Theorem III.6.5(3) and (35) we know $L^\omega = \mathbb{R} \cdot x_\omega + \mathbb{R} \cdot I x_\omega$ and thus

$$(37) \qquad\qquad [L^\omega, L^\omega] = \mathbb{R} \cdot Q(x_\omega).$$

Let us normalize x_ω so that $\omega\big(Q(x_\omega)\big) = 2$. Then by (36) and (37), there exists a complex number c such that $Q(x_\omega) = c \cdot h_\lambda$. Now $B_\mathbb{C}(c \cdot h_\lambda, c \cdot h_\lambda) = c^2 \lambda(h_\lambda) = 2c^2$ on one hand and $B_\mathbb{C}(c \cdot h_\lambda, c \cdot h_\lambda) = c\lambda(c \cdot h_\lambda) = c\lambda\big(Q(x_\omega)\big) = ci\omega\big(Q(x_\omega)\big) = 2ci$ on the other. Hence $c = i$. This shows that $B\big(h, Q(x_\omega)\big) = B(h, i \cdot h_\lambda) = iB_\mathbb{C}(h, h_\lambda) = i\lambda(h) = -\omega(h)$ for all $h \in H \cap [L, L]$. Hence

$$(38) \quad \langle h \mid Q(x_\omega) \rangle = -B\big(h', Q(x_\omega)\big) = \omega(h') = \omega(h) \text{ for all } h = h_Z + h' \in H, \omega \in \Omega.$$

The fact that C_1 and C_2 are dual with respect to $\langle \bullet \mid \bullet \rangle$ now follows from the fact that, in view of (38) and of $\omega(c) \geq 0$ for all $c \in C$ and $\omega \in \Omega_p^+$, for all $h \in H$ each in the following sequence of statements is equivalent to the next:

$$\langle h \mid \omega(c)Q(x_\omega) \rangle \geq 0 \text{ for all } c \in C, \omega \in \Omega_p^+,$$
$$\omega(c)\omega(h) \geq 0 \text{ for all } c \in C, \omega \in \Omega_p^+,$$
$$\omega(h) \geq 0 \text{ for all } \omega \in \Omega_p^+.$$

This proves (i).

(ii) By our choice of Ω^+ we have $\omega(c) \geq 0$ for all $c \in C$, $\omega \in \Omega_p^+$. Because of $L^\omega = \mathbb{R} \cdot x_\omega + \mathbb{R} \cdot I x_\omega$ and the definition $Q(x) = [Ix, x]$ for root vectors x we note $\omega(C) \cdot Q(L^\omega) = \mathbb{R}^+ \cdot Q(x_\omega)$. Since Ω_p^+ is finite, $\sum_{\omega \in \Omega_p^+} \mathbb{R}^+ \cdot Q(x_\omega)$ is a wedge and (ii) follows.

(iii) By (ii), C_2 is polyhedral in view of Proposition I.4.2 and Corollary I.4.4. This last source also guarantees that C_1 is polyhedral.

(iv) The theory of semisimple complex Lie algebras yields that the h_λ span the Cartan algebra $H_\mathbb{C} \cap [L_\mathbb{C}, L_\mathbb{C}] = (H \cap [L, L])_\mathbb{C}$ of $[L_\mathbb{C}, L_\mathbb{C}]$. Thus the $Q(x_\omega) = i \cdot h_\lambda$ span $H \cap [L, L]$. Hence $C_2 - C_2 = H \cap [L, L]$. Then Proposition I.1.7 shows in view of (i) that $C_1 \cap -C_1 = (C_2 - C_2)^\perp = (H \cap [L, L])^\perp = Z(L)$.

(v) Let $n \in N(H)$, where $N(H)$ is the normalizer of H in $\mathrm{INN}\, L$ (see Proposition III.5.6). Since

$$N(H) \subseteq \mathrm{INN}_L(K)$$

by Proposition III.5.6(iii), then by Theorem III.6.38(iv), we have $n(Ix_\omega) = In(x_\omega)$ for $\omega \in \Omega_p^+$. Therefore $[h, n(x_\omega)] = n([n^{-1}(h), x_\omega]) = n\big(\omega(n^{-1}(h)) \cdot Ix_\omega\big) = \omega\big(n^{-1}(h)\big) \cdot I\big(n(x_\omega)\big)$. Thus $\omega \circ n^{-1} \in \Omega_p^+$ and $n(x_\omega) \in L^{\omega \circ n^{-1}}$. Moreover,

$$n\big(Q(x_\omega)\big) = [n(Ix_\omega), n(x_\omega)] = [In(x_\omega), n(x_\omega)]$$
$$= Q\big(n(x_\omega)\big).$$

It follows that the Weyl group $\mathcal{W}(H, L) = N(H)/\mathrm{INN}_L H$ permutes the half-rays $\mathbb{R}^+ \cdot Q(x_\omega)$, $\omega \in \Omega_p^+$. Then C_2 is invariant under the action of the Weyl group by (ii). By duality (i), also C_1 is invariant under the Weyl group. ∎

The descent procedure

The following lemma is a crucial step towards the proof of the main theorem. We recall the projection $p: L \to H$ with kernel H^+ and now also consider the projection $q: L \to P(H)$ with kernel $K(H)$.

III.9.16. **Lemma.** *Let $h_0 \in$ algint C_1 and let x be in the orbit $(\text{INN } L)(h_0)$. If $x \notin K(H)$ then there is a y in the same orbit such that*

(i) $\|q(y)\| < \|q(x)\|$,

(ii) $p(x) \in \text{conv} \mathcal{W}(H, L)p(y) + C_2$.

Proof. We write $x = \gamma(h_0) = x_K + x_P$ with $\gamma \in \text{INN } L$, $x_K \in K(H)$ and $x_P \in P(H)$. We consider two cases.

Case 1. The first case assumes that $x_K \in H$. Then $x = x_K + \sum_{\omega \in \Omega_p^+} x_\omega$, and $x \notin K(H)$. Thus there is at least one $\omega_0 \in \Omega_p^+$ such that $x_{\omega_0} \neq 0$. Theorem III.9.10 and $\omega_0(z) > 0$ imply $\omega_0(x_K) = \omega_0(p(x)) = \omega_0(p(\gamma h_0)) > 0$. Therefore we may consider the element $u = -(\omega_0(x_K))^{-1} \cdot I x_{\omega_0}$ and calculate $[u, x] = -[x_K, u] + [u, \sum_{\omega \in \Omega_p^+} x_\omega] = \omega_0(x_K)^{-1} \cdot [x_K, I x_{\omega_0}] - \omega_0(x_K)^{-1} \cdot \sum_{\omega \in \Omega_p^+} [I x_{\omega_0}, x_\omega] = I^2 x_{\omega_0} - \omega_0(x_K)^{-1} \cdot Q(x_{\omega_0}) - \omega_0(x_K)^{-1} \cdot \sum_{\omega \in \Omega_p^+, \omega \neq \omega_0} [I x_{\omega_0}, x_\omega]$, and if we set $r = \omega_0(x_K)^{-1} \cdot \sum_{\omega \in \Omega_p^+, \omega \neq \omega_0} [x_{\omega_0}, x_\omega]$, then $p(r) = 0$, but also $q(r) = 0$ since $[L^\omega, L^\nu] \subseteq [P(H), P(H)] \subseteq K(H)$ for $\nu, \omega \in \Omega_p^+$ as $K(H) \oplus P(H)$ is a Cartan decomposition, and we record

$$(39) \qquad [u, x] = -x_{\omega_0} - \omega_0(x_K)^{-1} \cdot Q(x_{\omega_0}) - r.$$

Now we set $y(t) = e^{t \cdot \text{ad} \, u} x = x + t \cdot [u, x] + O(t^2)$ and obtain the formula

$$(40) \quad y(t) = x_K + \sum_{\omega \in \Omega_p^+, \omega \neq \omega_0} x_\omega + (1 - t) \cdot x_{\omega_0} - \omega_0(x_K)^{-1} t \cdot Q(x_{\omega_0}) - t \cdot r + O(t^2).$$

Then

$$q(y(t)) = \sum_{\omega \in \Omega_p^+, \omega \neq \omega_0} x_\omega + (1 - t) \cdot x_{\omega_0} + O(t^2).$$

In the following we need an estimate in Hilbert space. Suppose that $u \perp v$ in a Hilbert space and that $O(t^2)$ denotes a function defined for all sufficiently small real numbers such that $t^{-2} \cdot O(t^2)$ remains bounded as t tends to 0. Then a simple calculation shows that there is an $1 \geq \varepsilon > 0$ such that for all $t \in]0, \varepsilon[$ we have

$$\|u + (1 - t) \cdot v + O(t^2)\|^2 \leq \|u + v\|^2.$$

Therefore we find an $\varepsilon \in]0, 1]$ such that for all $t \in]0, \varepsilon]$ we have

$$(41) \qquad \|q(y(t))\|^2 = \|(\sum_{\omega \in \omega_p^+, \omega \neq \omega_0} x_\omega) + (1 - t) \cdot x_{\omega_0} + O(t^2)\|^2$$

$$\leq \|(\sum_{\substack{\omega \in \Omega_p^+ \\ \omega \neq \omega_0}} x_\omega) + x_{\omega_0}\|^2$$

$$\leq \|\sum_{\omega \in \Omega_p^+} x_\omega\|^2$$

$$= \|q(x)\|^2,$$

since the sum of the L^ω is orthogonal. Now we claim that we find a $0 < t < \varepsilon$ such that

(42) $$p(x) - p\big(y(t)\big) \in C_2 = C_1^*.$$

Since C_1 is polyhedral by Lemma III.9.15(iii) we find finitely many vectors

$$v_1, \ldots, v_s \in C_1 \cap [L, L]$$

such that $C_1 = \mathbb{R}^+ {\cdot} v_1 + \cdots + \mathbb{R}^+ {\cdot} v_s + Z(L)$. Since x and $y(t)$ are in the INN L-orbit of h_0 they have the same $Z(L)$-component in $L = Z(L) \oplus [L, L]$. Therefore, (42) is equivalent to

(43) $$\langle p(x) - p\big(y(t)\big) \mid v_j \rangle \geq 0 \text{ for all } j = 1, \ldots, s.$$

In dealing with $y(t)$ we need information about $\omega_0(x_K)^{-1}{\cdot}Q(x_{\omega_0})$. By Lemma III.9.15(ii) we have $Q(x_{\omega_0}) \in C_2 = C_1^*$; hence $\langle Q(x_{\omega_0}) \mid v_j \rangle \geq 0$ for all $j = 1, \ldots, s$. The indices j fall into two classes:

Class 1: $\langle Q(x_{\omega_0}) \mid v_j \rangle > 0$. Since the root spaces L^ω are orthogonal to H, then (40) allows us to compute $\langle p(x) - p\big(y(t)\big), v_j \rangle = t\omega_0(x_K)^{-1}\langle Q(x_{\omega_0}) \mid v_j \rangle + O(t^2)$. Thus there is a $t_j \in {]}0, \varepsilon{[}$ such that $\langle p(x) - p\big(y(t)\big) \mid v_j \rangle \geq 0$ for all $t \in [0, t_j]$. Let τ be the minimum of the t_j with j in class 1. Then $0 < \tau \leq \varepsilon$ and for $t \in [0, \tau]$ we have

(44) $$\langle p(x) - p\big(y(t)\big) \mid v_j \rangle \geq 0 \text{ for all } j \text{ in class 1.}$$

Class 2: $\langle Q(x_{\omega_0}) \mid v_j \rangle = 0$. In this case we calculate $0 = \langle Q(x_{\omega_0}) \mid v_j \rangle = -B\big(Q(x_{\omega_0}), v_j\big) = -B([Ix_{\omega_0}, x_{\omega_0}], v_j) = B([x_{\omega_0}, Ix_{\omega_0}], v_j) = B(x_{\omega_0}, [Ix_{\omega_0}, v_j]) = -B(x_{\omega_0}, [v_j, Ix_{\omega_0}]) = \omega_0(v_j)B(x_{\omega_0}, x_{\omega_0}) = -\omega_0(v_j)\|x_{\omega_0}\|^2$. Since $x_{\omega_0} \neq 0$ this shows that $\omega_0(v_j) = 0$. But then we have $[u, v_j] = -\omega_0(x_K)^{-1}{\cdot}[Ix_{\omega_0}, v_j] = \omega_0(x_K)^{-1}\omega_0(v_j)I^2x_{\omega_0} = 0$. Thus, using the fact that the operator $e^{t \, \text{ad} \, u}$ on L fixes v_j and is symmetric with respect to $\langle \bullet \mid \bullet \rangle$ on account of $u \in P(H)$, we obtain $\langle p(x) - p\big(y(t)\big), v_j \rangle = \langle p\big(x - y(t)\big) \mid v_j \rangle = \langle x - y(t) \mid v_j \rangle = \langle x \mid v_j \rangle - \langle e^{t \cdot \text{ad} \, u}x \mid v_j \rangle = \langle x \mid v_j \rangle - \langle x \mid e^{t \, \text{ad} \, u}v_j \rangle = \langle x \mid v_j \rangle - \langle x \mid v_j \rangle = 0$. Thus (44) holds in fact for all $j = 1, \ldots, s$. Thus (43) is proved and the lemma is established for **Case 1**.

Case 2. Now we no longer assume that x_K is in H. The Cartan algebras of the compact Lie algebra $K(H)$ are all conjugate and their union is $K(H)$. Hence there is a $\kappa \in \text{INN}_L K(L)$ such that $\kappa(x_K) \in H$. Now $[K(H), P(H)] \subseteq P(H)$ implies $\kappa(x_P) \in P(H)$. Thus **Case 1** applies to the element $\kappa(x) = \kappa(x_K) + \kappa(x_P)$. Thus there is an element $y^* \in (\text{INN} \, L)(h_0)$ such that

$$\|q(y^\star)\| < \|q(\kappa(x))\| = \|\kappa\big(q(x)\big)\| = \|q(x)\|,$$

and

$$p\big(\kappa(x)\big) \in p(y^\star) + C_2,$$

since κ respects the norm and the subspaces $K(H)$ and $P(H)$. But now $p(x) = p(x_K) = p(\kappa^{-1}\kappa x_K) \in p\big((\mathrm{INN}_L K(H))\kappa(x_K)\big)$. Since $\kappa(x_K) \in H$ we note $\kappa(x_K) = p\big(\kappa(x_K)\big) = p\big(\kappa(x)\big)$ since $\kappa(x_P) \in \ker p$. Now let us apply the Convexity Theorem in the form of Corollary III.8.5, write $\mathcal{W}$ for the Weyl group $\mathcal{W}(H, L)$, and deduce $p\big((\mathrm{INN}_L K(H))\kappa(x_K)\big) = \mathrm{conv}\mathcal{W}\kappa(x_K) = \mathrm{conv}\mathcal{W}p(\kappa x) \subseteq \mathrm{conv}\mathcal{W}\big(p(y^\star) + C_2\big) \subseteq \mathrm{conv}\mathcal{W}p(y^\star) + C_2$, since C_2 is convex and invariant under the Weyl group. Thus $y^\star$ is the element we are looking for. ∎

III.9.17. **Proposition.** (The Orbit Projection Theorem) *Let L be a reductive quasihermitean Lie algebra, fix a compactly embedded Cartan algebra H and an element $z \in Z_H \cap \mathrm{int}(\mathrm{comp}\, L)$. Select a set Ω^+ of positive real roots such that $\omega(z) > 0$ for all $\omega \in \Omega_p^+$. Set*

$$C_1 = \{h \in H : \omega(h) \geq 0 \text{ for all } \omega \in \Omega_p^+\},$$

and

$$C_2 = \text{ closed convex hull of } \{\omega(c)\cdot Q(x) : c \in C, \quad x \in L^\omega, \quad \omega \in \Omega_p^+\}.$$

Then for each $h \in \mathrm{algint}\, C_1$ the following condition holds:

$$(45) \qquad p\big((\mathrm{INN}\, L)(h)\big) \subseteq \big(\mathrm{conv}\mathcal{W}(H, L)(h)\big) + C_2.$$

Proof. We define a binary relation $\prec$ on L as follows:

$$x \prec y \iff \begin{cases} \text{(i)} & \|q(y)\| \leq \|q(x)\|, \text{ and} \\ \text{(ii)} & p(x) \in \big(\mathrm{conv}\mathcal{W}p(y)\big) + C_2. \end{cases}$$

We set $\uparrow x = \{y \in L : x \prec y\}$ and abbreviate the orbit $(\mathrm{INN}\, L)(h)$ by B. Since the element h is in $\mathrm{comp}\, L$, it is semisimple, and so by a Theorem of Borel and Harish–Chandra, B is closed. (See for instance [Wa72], Proposition 1.3.5.5.). We claim that $\uparrow x$ is closed, too. Thus let $y = \lim y_n$ with $x \prec y_n$. Then the relation $\|p(y)\| \leq \|p(x)\|$ is clear. Further, there are elements $q_n \in C_2$ such that $p(x) \in q_n + \mathrm{conv}\,\mathcal{W}p(y_n)$ for all n. The set $-p(x) + \bigcup_{n=1}^\infty \mathrm{conv}\,\mathcal{W}p(y_n)$ is relatively compact and contains all q_n. Hence the sequence of the q_n has a cluster point q which will satisfy $p(x) \in q + \mathrm{conv}\,\mathcal{W}p(y)$, as one verifies readily. Hence $x \prec y$ and thus $\uparrow x$ is closed. The group $\mathrm{INN}\, L$ has a polar decomposition as $\mathrm{INN}_L K(H)\cdot e^{\mathrm{ad}\, P(H)}$. Thus we can write $\gamma \in \mathrm{INN}\, L$ in the form $\kappa e^{\mathrm{ad}\, s}$ with $\kappa \in \mathrm{INN}_L K(H)$ and $s \in P(H)$, and then $\|\gamma y\| = \|e^{\mathrm{ad}\, s}y\|$ for any y in L, since κ is an isometry. Now $e^{\mathrm{ad}\, s}$ is symmetric for all $s \in P(H)$; in particular, for each such s all operators $e^{t\cdot \mathrm{ad}\, s}$ are in the vector space of all symmetric operators, hence so is the derivative $\mathrm{ad}\, s$ at 0. Hence L is the direct sum of the eigenspaces L_λ for the eigenvalues λ. If $x \in L_\lambda$, then $x = x_K + x_P$ with $x_K \in K(H)$ and $x_P \in P(H)$. From $[P(H), K(H)] \subseteq P(H)$ and $[P(H), P(H)] \subseteq K(H)$, we conclude that $\lambda\cdot x_K + \lambda\cdot x_P = \lambda\cdot x = [s, x] = [s, x]_K + [s, x]_P$, hence $[s, x_P] = [s, x]_K = \lambda\cdot x_K$ and

$[s, x_K] = [s, x]_P = \lambda \cdot x_P$. If we set $\widetilde{x} = x_K - x_P$ then $[s, \widetilde{x}] = \lambda \cdot (x_P - x_K) = -\lambda \cdot \widetilde{x}$, that is, $\widetilde{x} \in L_{-\lambda}$. By the symmetry of $\mathrm{ad}\, s$, we find $\lambda \|x_P\|^2 = \langle \lambda \cdot x_P \mid x_P \rangle = \langle [s, x_K] \mid x_P \rangle = \langle x_K \mid [s, x_P] \rangle = \langle x_K \mid \lambda \cdot x_K \rangle = \lambda \|x_K\|^2$. If $\lambda \neq 0$ this implies $\|x_K\| = \|x_P\|$, and since $K(H)$ and $P(H)$ are orthogonal, $\langle x_K \mid x_P \rangle = 0$, whence $\|\widetilde{x}\|^2 = \|x\|^2 = \|x_K\|^2 + \|x_P\|^2 = 2\|x_P\|^2$. The map $x \mapsto \widetilde{x} \colon L_\lambda \to L_{-\lambda}$ is an involutive isomorphism with an obvious extension to an involution of L. Since 0 is an eigenvalue in view of $(\mathrm{ad}\, s)(s) = 0$, the set of eigenvalues of $\mathrm{ad}\, s$ is a disjoint union $\{0\} \dot\cup J \dot\cup (-J)$, and L is the orthogonal direct sum of L_0 and the spaces $L_\lambda \oplus L_{-\lambda}$, $\lambda \in J$ which are invariant under $\sim$ and the projections into $K(H)$ and $P(H)$. If $x \in L$ then $x = x_0 + \sum_{\lambda \in J}(x_\lambda + \widetilde{y_\lambda})$ with unique elements $x_0 \in L_0$ and $x_\lambda, y_\lambda \in L_\lambda$. Then $\|x\|^2 = \|(x_0)_K\|^2 + \|(x_0)_P\|^2 + 2\sum_{\lambda \in J} \|(x_\lambda + \widetilde{y_\lambda})_P\|^2 = \|(x_0)_K\|^2 - \|(x_0)_P\|^2 + 2\|x_P\|^2$. If $x_0 \in K(H)$ we obtain $\|x\|^2 = \|(x_0)_K\|^2 + 2\|x_P\|^2$. Applying $e^{\mathrm{ad}\, s}$ we find $e^{\mathrm{ad}\, s} x = x_0 + \sum_{\lambda \in J}(e^\lambda \cdot x_\lambda + e^{-\lambda} \cdot \widetilde{y_\lambda})$. If $k \in K(H)$, then $k_0 \in K(H)$ and $\|e^{\mathrm{ad}\, s} k\|^2 = \|k_0\|^2 + 2\|(e^{\mathrm{ad}\, s} k)_P\|^2 \leq \|k\| + 2\|(e^{\mathrm{ad}\, s} k)_P\|^2$. Thus we have

$$(46) \qquad \|\gamma(k)\|^2 \leq 2\|\big(\gamma(k)\big)_P\|^2 + \|k\|^2 \text{ for all } \gamma \in \mathrm{INN}\, L, k \in K(H).$$

Now we note that for any $y \in (\mathrm{INN}\, L)(h) \cap \uparrow x$ we have $y = \gamma(h)$ for some $\gamma \in \mathrm{INN}\, L$, and thus (46) implies that $\|y\|^2 = \|\gamma(h)\|^2 \leq 2\|\big(\gamma(h)\big)_P\|^2 + \|h\|^2 \leq 2\|x_P\|^2 + \|h\|^2$ since $\|y_P\| = \|q(y)\| \leq \|q(x)\| = \|x_P\|$ as $x \prec y$. This shows that the set

$$(\mathrm{INN}\, L)(h) \cap \uparrow x$$

is bounded and thus compact, because we saw before that it is closed.

Thus we may conclude that the continuous function

$$y \mapsto \|q(y)\| \colon (\mathrm{INN}\, L)(h) \cap \uparrow x \to \mathbb{R}$$

attains its minimum at some point y_0.

We claim that $y_0 \in K(H)$. For a proof we have to show that $q(y_0) = 0$. By way of contradiction let us assume that this were not the case. Now y_0 satisfies the hypotheses of Lemma III.9.16. Thus we find a $y_1 \in (\mathrm{INN}\, L)(h)$ such that $\|q(y_1)\| < \|q(y_0)\|$ and $p(y_0) \in \big(\mathrm{conv} \mathcal{W} p(y_1)\big) + C_2$, in particular, $y_0 \prec y_1$. The desired contradiction will result if we can show that $y_1 \in (\mathrm{INN}\, L)(h) \cap \uparrow x$. This, however, is certainly the case if the relation $\prec$ is transitive. So this is what we will now show. Thus suppose that $a, b, c \in L$ with $a \prec b$ and $b \prec c$. Obviously $\|p(c)\| \leq \|p(b)\| \leq \|p(a)\|$ and $p(a) \in \mathrm{conv} \mathcal{W} p(b) + C_2 \subseteq \mathrm{conv} \mathcal{W}(\mathrm{conv} \mathcal{W} p(c) + C_2) + C_2 = \mathrm{conv} \mathcal{W} p(c) + C_2$ and this indeed shows $a \prec c$.

Since $K(H)$ is covered by the conjugates of the Cartan algebra H, we find a $\gamma \in \mathrm{INN}_L K(H)$ such that $\gamma(y_0) \in H$. Since $y_0 \in (\mathrm{INN}\, L)(h)$, there is also a $\gamma' \in \mathrm{INN}\, L$ such that $y_0 = \gamma'(h)$. Hence $\gamma\gamma'(h) \in H$. But Theorem III.9.10 now shows that $\gamma\gamma'(h) \in \mathrm{algint}\, C_1$. Moreover, $\gamma\gamma'$ maps the centralizer $Z(h, L)$ onto the centralizer $Z(\gamma\gamma'(h), L)$. Both of these centralizers are contained in $K(H)$, since h, $\gamma\gamma'(h) \in \mathrm{algint}(C_1)$. Thus $\gamma\gamma'$ maps H onto another Cartan algebra H' of $K(H)$. Once more we find an element $\gamma'' \in \mathrm{INN}_L K(H)$ such that $H = \gamma''(H')$. It now follows that $\gamma''\gamma\gamma' \in N(H)$ by the definition of this normalizer (see Proposition III.5.6(iii)). Let $w = \gamma''\gamma\gamma' Z(H) \in \mathcal{W}$ be the corresponding Weyl group element.

We use Kostant's Convexity Theorem in the form of Corollary III.8.5 once more and calculate

$$p(x) \in \operatorname{conv}\mathcal{W}p(y_0) + C_2 = \operatorname{conv}\mathcal{W}p\big(\gamma^{-1}(\gamma(y_0))\big) + C_2$$
$$\subseteq \operatorname{conv}\mathcal{W}p\big(\operatorname{INN}_L K(H)\gamma(y_0)\big) + C_2$$
$$= \operatorname{conv}\mathcal{W}p\Big(\big(\operatorname{INN}_L K(H)\gamma''\big)\gamma(\gamma'h)\Big) + C_2$$
$$= \operatorname{conv}\mathcal{W}p\big(\operatorname{INN}_L K(H)w(h)\big) + C_2$$
$$= \operatorname{conv}\mathcal{W}p\big(\operatorname{INN}_L K(H)(h)\big) + C_2$$
$$= \operatorname{conv}\mathcal{W}\big(\operatorname{conv}\mathcal{W}h\big) + C_2$$
$$= \operatorname{conv}\mathcal{W}h + C_2.$$

This completes the proof of the proposition. ∎

At this point we have all the ingredients to prove the final major result of this chapter, the Classification Theorem for Invariant Cones. Recall that any Lie algebra supporting a pointed generating invariant cone must be quasihermitian with cone potential by Theorems III.5.16. and III.6.18. Also recall, for the record, that the operator $(\operatorname{ad} x)^2$ maps H into itself for any root vector $x \in L^\omega$, $\omega \in \Omega$ and that we define the wedge $\mathcal{C}(H, L)$ and multiplicative semigroup in $\operatorname{Hom}(H, H)$ to be the wedge generated by all rank one operators $(\operatorname{ad} x)^2 | H$ where x ranges through the vectors $x \in L^\omega$, $\omega \in \Omega_p$, where, for any quasihermitian Lie algebra, Ω_p, according to Definition III.6.36, denotes the set of all *non-compact* real roots.

III.9.18. **Theorem.** (The Classification Theorem for Invariant Cones) *Let L be a quasihermitian Lie algebra with cone potential and let H be any compactly embedded Cartan algebra. Suppose that C is any pointed generating cone in H. There is an invariant pointed generating cone W in L such that $C = H \cap W$ if and only if*

(I) $$\mathcal{C}(H, L)(C) \subseteq C,$$

(II) $$\mathcal{W}(H, L)(C) \subseteq C.$$

Proof. The necessity was established; it remains to show that (I) and (II) are sufficient. By the Reduction Theorem III.6.44 it suffices to prove the theorem for reductive Lie algebras. By the Reconstruction Theorem of Invariant Cones III.5.15, we only have to show that $p\big(\operatorname{INN} L)(C)\big) \subseteq C$. If $h \in \operatorname{algint}(C)$ we know from the preceding Proposition III.9.17 that $p\big((\operatorname{INN} L)(h)\big) \subseteq \operatorname{conv}\mathcal{W}h + C_2$. For the particular choice Ω^+ of positive real roots which we made in the introduction this section, we have

(†) $$(\operatorname{ad} x_\omega)^2(C) = Q_\omega(C \times \{x_\omega\}) = \omega(C) \cdot Q(x_\omega)$$

for all non-compact positive real roots $\omega \in \Omega_p^+$ and all root vectors $x \in L^\omega$. But now (II) implies that $\operatorname{conv}\mathcal{W}h \subseteq C$, and (†) implies that $C_2 \subseteq C$, so that we have have $p\big((\operatorname{INN} L)(\operatorname{algint} C)\big) \subseteq C$. Since $\operatorname{algint} C$ is dense in C, the claim follows by continuity. ∎

Problems for Chapter III

PIII.1.   Problem. (a) *Determine all Lie algebras which admit compactly embedded Cartan algebras. Cf.* [Ti67] *for the case of simple Lie algebras.*

(b) *Determine all Lie algebras with cone potential.*

(c) *Determine all quasihermitian Lie algebras.*   ■

PIII.2.   Problem. *Suppose that G is a subgroup of* Aut(L) *for a Lie algebra L. Determine the G-invariant cones in L. In particular, determine invariant cones which are not necessarily generating.*   ■

PIII.3.   Problem. *Characterize all Lie algebras admitting invariant pointed generating cones.*   ■

PIII.4.   Problem. *Establish a purely geometric theory for the euclidean spaces H equipped with a root system such it arises in a quasihermitian Lie algebra with cone potential, and for the cones C satisfying* Conditions (I) *and* (II) *of Theorem III.9.18.*   ■

Notes for Chapter III

Section 1. The content of this section is largely background material. Theorem III.1.9 was published in [HH86c].

Section 2. The First Theorem on compact Automorphism Groups of Cones is published in [HH86c]. The Second Theorem of Compact Automorphism Groups of Cones appears in [HH86d]. The idea of compactly embedded Cartan subalgebras in Definition III.2.10 was introduced in [HH86d]. In the special case of semisimple Lie algebras, the definition is standard (see Helgason [He78]). The Third Cartan Algebra Theorem III.2.12 appears in [HH86d] as do the Fourth Cartan Algebra Theorem III.2.14 and the Uniqueness Theorem III.2.15. Minimal and maximal invariant cones (see III.2.16 through 18) were considered for simple Lie algebras in the works of Vinberg [Vi80], Paneitz [Pa81], [Pa84] and Ol'shanskiĭ [Ol81], [Ol82a], [Ol82b]. In these papers it was first remarked that Cartan subalgebras play an important role.

Section 3. There is a well-known classical theory of eigenvectors and eigenvalues of non-negative matrices, dating back to the works of Frobenius and Perron. The Frobenius–Perron Theorem for solvable groups is based on the work of Vinberg [Vi80]. The presentation here follows the lines of the thesis of Terp [Te84].

Section 4. Vinberg's work [Vi80] and a result credited to Kostant (see [Se76]) constitute the central theme of this section. We unified and generalized their results somewhat in the Invariant

Cones Theorem of Kostant and Vinberg III.4.5, and we utilized some input from the thesis of Spindler [Spi84].

Section 5. The results in this section were presented in [HH86d]. The idea of studying invariant cones in simple Lie algebras is due to Vinberg [Vi80].

Section 6. Root decompositions are basic for Lie algebra theory. We develop here a variation of the classical repertoire by considering a compactly embedded Cartan algebra and by studying its real root decomposition. These developments, among other things, allow us to define the concept of a Lie algebra with cone potential (Definition III.6.19). The Structure Theorem for Lie Algebras with Cone Potential III.6.23 is new as is Theorem III.6.28. The idea of quasihermitian Lie algebras and the definition of compact and non-compact roots are standard in the semisimple case (see e.g. Helgason [He78]). Our Theorem III.6.38 is a structure theorem for *quasihermitean* Lie algebras. The Reduction Theorem III.6.41 is new. These results were announced in [HH88].

Section 7. The basic computations here are straightforward from the root decomposition and are used later. Corollary II.7.11 gives, for the first time, the characterization of invariant cones in solvable Lie algebras.

Section 8. There are now many proofs of Kostant's Convexity Theorem (see Theorem III.8.4) and of its variants in the literature. Kostant proved the theorem in [Ko73]; Atiyah gave a proof via Morse Theory [At82]. Related results are in Guillemin and Sternberg [GS82] and Kac and Peterson [KP84]. Our presentation is modelled after the proof given by Heckmann in his dissertation [Heck80].

Section 9. In Section 9 we deal with invariant cones in reductive Lie algebras. Invariant cones in simple Lie algebras were classified by Ol'shanskiĭ [Ol81], Paneitz [Pa84], and Kumaresan and Ranjan [KR82]. Our approach is an adaptation of the work of Kumaresan and Ranjan.

Chapter IV

The Local Lie theory of semigroups

At this juncture, the book changes its direction away from linear algebra and convexity towards analysis. Yet we shall see in which sense the preceding chapters serve as the foundations of an infinitesimal theory of a Lie theory of semigroups.

The purpose of this chapter is to develop the basic Lie theory for local subsemigroups of Lie groups and their tangent objects in the corresponding Lie algebra. There are several good motivations for considering the local theory. First of all, unlike the group case, there are infinitesimally generated finite dimensional local semigroups which are not locally isomorphic to any global subsemigroup of a Lie group. Thus the local and global theories diverge. Secondly, the local theory emerges in a natural way in several contexts. In geometric control theory there arises from a given set of vector fields a pseudogroup of local diffeomorphisms. In considering such questions as local attainability sets, local subsemigroups of this pseudogroup of local diffeomorphisms arise naturally. In another context in the study of differentiable and cancellable semigroups one frequently obtains theorems that guarantee that the semigroup can be embedded locally into a Lie group, but not necessarily globally. Thus in order to apply Lie machinery, one is restricted to the local setting. Thirdly, since in the Lie group case the exponential mapping is locally one-to-one, the group multiplication can be pulled back locally to a neighborhood of 0 in the Lie algebra, and the study of local semigroups can be carried out entirely in the Lie algebra with this induced multiplication. The Baker–Campbell–Hausdorff formula then gives an explicit internal characterization of this multiplication in terms of the Lie algebra structure alone, independent of the group from which it arose. Thus the machinery of Lie algebras can be brought to bear directly in the development of the local theory.

We work within the framework of a Dynkin algebra, i.e., a Lie algebra over the real or complex numbers admitting a complete norm for which the Lie bracket operation is continuous. These include, in particular, the finite dimensional real and complex Lie algebras, the focal point of our attention. For a Dynkin algebra the Baker–Campbell–Hausdorff multipication $*$ is given by the series expansion

$$x * y = x + y + \frac{1}{2}[x, y] + \dots$$

where the omitted terms consist of higher order Lie brackets of x and y. The Baker–Campbell–Hausdorff series converges on small neighborhoods of 0, and the resulting Baker–Campbell–Hausdorff multiplication ∗ endows the neighborhoods on which it converges with the structure of a local group.

We develop the local Lie theory of semigroups entirely in the context of the Baker–Campbell–Hausdorff multiplication on a neighborhood of 0 in a Lie algebra. However, as remarked in the preceding, this constitutes no real loss in generality for the investigation of the local theory in general Lie groups since the exponential mapping is locally a topological isomorphism from a neighborhood of 0 in the Lie algebra equipped with the multiplication given by the Baker–Campbell–Hausdorff series to a neighborhood of 1 in the Lie group. In our approach the theory also remains valid for infinite dimensional Dynkin algebras, which may have no global group associated with them.

Throughout our development we employ freely fundamental properties of Lie algebras and the Baker–Campbell–Hausdorff multiplication. These have been summarized in Appendix 1. The reader might prefer to adopt the viewpoint that we are working inside a Dynkin algebra (or finite dimensional Lie algebra) already equipped with a local binary operation ∗ satisfying the local group properties (i)–(iv) of Proposition A.1.5 and relating to the Lie algebra structure by the rules (A1)-(A9).

1. Local semigroups

In this section we work within a Dynkin algebra $\mathfrak{g}$.

IV.1.1.   Definition. An open neighborhood B of 0 is a *Baker–Campbell–Hausdorff neighborhood* (or *C-H-neighborhood*) if $x * y$ is defined and continuous on $\overline{B} \times \overline{B}$, all triple products are defined and associative, and B is symmetric (that is, $B = -B$) and star-shaped (that is, $r{\cdot}B \subseteq B$ for $0 \leq r \leq 1$).

In the following B will denote a Baker–Campbell–Hausdorff neighborhood.

There is no completely obvious approach to defining local semigroups. One that lends itself well to our purposes is the notion of a local semigroup (always assumed to be with identity) with respect to some fixed neighborhood of the identity.

IV.1.2.   Definition. We say $S \subseteq B$ is a *local semigroup with respect to* B if $0 \in S$ and $(S * S) \cap B \subseteq S$. A subset I of S is a *local left ideal* (respectively *local right ideal*) if $(S * I) \cap B \subseteq I$ (resp. $(I * S) \cap B \subseteq S$). A *local ideal* is simultaneously a local left and a local right ideal. A subset G of B is a *local group with respect to* B if G is a local semigroup with respect to B and $G = -G$.

IV.1.3. **Remark.** (i) If B_1 is a C-H-neighborhood contained in B and if S is a local semigroup with respect to B, then $S \cap B_1$ is a local semigroup with respect to B_1.

(ii) If S is a local semigroup with respect to B, then $\overline{S}^B = \overline{S} \cap B$ is also a local semigroup with respect to B.

Proof. (i) Trivial.

(ii) Suppose $x, y \in \overline{S}^B$. Then $x = \lim x_n$, $y = \lim y_n$ where $x_n, y_n \in S$ for all n. Then $x * y = \lim x_n * y_n$. If $x * y \in B$, then eventually $x_n * y_n \in B$ and hence $x_n * y_n \in S$. Thus $x * y \in \overline{S}^B$. Thus $\overline{S}^B$ is a local semigroup with respect to B. ∎

IV.1.4. **Proposition.** (i) *If S is a local semigroup with respect to a C-H-neighborhood B, then $H(S) = S \cap -S$, the set of units of S, is the unique maximal local subgroup with respect to B contained in S.*

(ii) *The complement I of $H(S)$ in S is the largest member in the set of local left ideals, the set of local right ideals, and the set of local ideals of S. Hence if a finite product of members of S is a unit of S, then so is each factor.*

Proof. (i) Suppose $x, y \in H(S)$. Then $x * y \in B$ implies $x * y \in S$ since S is a local semigroup with respect to B. Also $-(x * y) = (-y) * (-x)$ is in B (since $-B = B$) and thus is in S since $-y, -x \in S$. Thus $x * y \in H(S)$, and hence $H(S)$ is a local semigroup with respect to B. Clearly $-H(S) = H(S)$; thus $H(S)$ is a local group with respect to B.

If G is a local group with respect to B which is contained in S, then $G = -G \cap G \subseteq -S \cap S = H(S)$.

(ii) Suppose $x, y \in S$ and $x * y \in H(S)$. Then $-y * (-x) = -(x * y) \in H(S)$ and $-x \in B$. It follows that $-x = y * (-y) * (-x) \in (S * S) \cap B = S$. Thus $x \in H(S)$. Similarly $y \in H(S)$. It follows that if $x \in S$, $y \in I$, and $x * y \in B$ (resp. $y * x \in B$), then $x * y \in I$ (resp. $y * x \in I$), that is, I is a local ideal of S. If J is any local left ideal of S and $y \in J \cap H(S)$, then $0 = (-y) * y \in S * J \cap B \subseteq J$. Then $S = S * 0 \subseteq J$. Thus no proper local left (or, similarly, right) ideal of S meets the group of units, that is, I contains every proper local left and right ideal. The last statement follows (by induction) from the fact that I is a local ideal. ∎

We shall call the complement of $H(S)$ in S *the maximal local ideal of S*.

IV.1.5. **Definition.** Let $X \subseteq \mathfrak{g}$. An element $b \in B$ is *X-generated with respect to B* if $b \in \bigcup_n X^n$, where X^n is defined inductively by

$$X^1 = (B \cap X) \cup \{0\}, \quad X^n = (X^{n-1} * X^{n-1}) \cap B.$$

Let $\langle X \rangle_B$ (or $\langle X \rangle$ if B is understood) denote all elements of B that are X-generated with respect to B.

IV.1.6. **Proposition.** *The set $\langle X \rangle_B$ of all X-generated elements is the smallest local semigroup with respect to B containing $X \cap B$.*

Proof. Suppose $a \in X^m$, $b \in X^n$, where $m \leq n$. Since $0 \in X^k$ for all k, $a \in X^n$. If $a * b \in B$, then $a * b \in X^{n+1} \subseteq \langle X \rangle_B$. Thus $\langle X \rangle_B$ is a local semigroup.

If S is a local semigroup with respect to B containing $X \cap B$, then an easy induction yields $X^n \subseteq S$ for all n. Thus $\langle X \rangle_B$ is the smallest local semigroup containing $X \cap B$. ∎

IV.1.7.   Corollary.   *If $A \subseteq B$, then $G_B(A) = \langle A \cup -A \rangle_B$ is the smallest local group with respect to B containing A.*

Proof.   Clearly any local group with respect to B containing A contains $A \cup -A$ and hence $\langle A \cup -A \rangle_B$ by Proposition IV.1.6. Since $x \mapsto -x$ is an anti-isomorphism on B with respect to the Baker–Campbell–Hausdorff multiplication $*$ and since under this mapping $A \cup -A$ is carried to itself, it follows that $\langle A \cup -A \rangle_B$ is carried to itself. (Alternately one shows easily by induction that $\langle A \cup -A \rangle_B$ is closed under $x \mapsto -x$.) Thus it is a local group. ∎

Germs and local properties

Frequently we would like to consider two local semigroups with respect to different neighborhoods equivalent if their restrictions to a yet smaller neighborhood are equal. We develop now a convenient formal setting in which to consider such classes.

IV.1.8.   Definition.   We define an equivalence relation on the subsets of $\mathfrak{g}$ by saying that two subsets S and T *belong to the same germ* if there exists an open set U containing 0 such that $S \cap U = T \cap U$. A *germ* is an equivalence class of sets under this relation. The equivalence class of S is denoted by $\mathrm{Germ}(S)$.

The proof of the following proposition is straightforward from the definition and is hence omitted.

IV.1.9.   Proposition.   (i) *The intersection of two members of a germ is again in the germ, and hence the members of a germ form a filter base of sets if the empty set is not in the germ. If two germs give rise to the same filter, then the two germs are equal.*

(ii) *The relation of belonging to the same germ is preserved by finite unions, finite intersections, complementation, and set difference. Hence the standard set theoretic operations of union, intersection, complementation, and set difference extend in a natural way to germs.*

(iii) *We have*

$$\mathrm{Germ}(\overline{S}) = \mathrm{Germ}(\overline{T}) \qquad and \qquad \mathrm{Germ}(\mathrm{int}\, S) = \mathrm{Germ}(\mathrm{int}\, T)$$

whenever $\mathrm{Germ}(S) = \mathrm{Germ}(T)$; thus to each germ may be associated the closure and interior of that germ. ∎

We also formalize what it means for a set to have some property locally, that is, in some neighborhood of 0.

1. *Local semigroups*

IV.1.10. **Definition.** A *local property* $\mathcal{P}$ is a collection of ordered pairs (A, U) where $A \subseteq U$ and U is a neighborhood of 0. The local property $\mathcal{P}$ is said to be *hereditary* if for any $(A, U) \in \mathcal{P}$ and any C-H-neighborhood B, also $(A \cap B, U \cap B)$ is in $\mathcal{P}$. A subset S of $\mathfrak{g}$ containing 0 is said to be *locally $\mathcal{P}$* or to have $\mathcal{P}$ *locally* if for every open set W containing 0 there exists a neighborhood U of 0 such that $U \subseteq W$ and $(S \cap U, U)$ is in $\mathcal{P}$.

Following are some specific examples of local properties.

(i) The *local semigroup property* consists of all pairs (T, B) such that B is a C-H-neighborhood and T is a local semigroup with respect to B. By Remark IV.1.3 this property is hereditary. A set S is called a *local semigroup* if S has the local semigroup property locally.

(ii) The *divisibility property* consists of all pairs (A, U) such that $x \in A$ implies that $\frac{1}{n} \cdot x \in A$ for all positive integers n. This property is hereditary since a Baker–Campbell–Hausdorff neighborhood is star-shaped. This property gives rise to the notion of a subset S which is *locally divisible*.

(iii) The *locally closed* property consists of all pairs (A, U) such that A is closed in U. This property is hereditary and gives rise to the notion of a *locally closed* set.

IV.1.11. **Proposition.** (i) *Given a local property $\mathcal{P}$ and two members of a germ, then one member of the germ has property $\mathcal{P}$ locally if and only if the other does.*

(ii) *If the local property $\mathcal{P}$ is hereditary, then a subset S has the property locally if and only if there exists a C-H-neighborhood B_1 such that $(S \cap B_1, B_1)$ is in $\mathcal{P}$ if and only if $(S \cap B, B)$ is in $\mathcal{P}$ for all C-H-neighborhoods B contained in B_1.*

Proof. (i) Suppose S and T are in the same germ and $S \cap U = T \cap U$. Pick a neighborhood V of 0 contained in U such that $(S \cap V, V)$ is in $\mathcal{P}$. Since $S \cap V = T \cap V$, it follows that T has $\mathcal{P}$ locally if S does. The argument is clearly symmetric.

(ii) Suppose S has $\mathcal{P}$ locally. Then there exists an open set U containing 0 such that $(S \cap U, U)$ is in $\mathcal{P}$. Let B_1 be a fixed C-H-neighborhood contained in U and let B be a C-H-neighborhood contained in B_1. By the hereditary condition, $(S \cap B, B)$ is in $\mathcal{P}$. Conversely, suppose $(S \cap B_1, B_1)$ is in $\mathcal{P}$ for some C-H-neighborhood B_1. If U is any open set containing 0, pick a C-H-neighborhood B contained in $B_1 \cap U$. Again by the hereditary condition $(S \cap B, B)$ is in $\mathcal{P}$. Thus S has $\mathcal{P}$ locally. ∎

By part (i) of Proposition IV.1.11, we may say that the *germ has the property $\mathcal{P}$ locally*.

If $(S \cap B, B) \in \mathcal{P}$ for all C-H-neighborhoods contained in some neighborhood B_1 of 0 we say that $(S \cap B, B)$ *has the local property $\mathcal{P}$ for all B sufficiently small*.

Note that Proposition IV.1.11 holds, in particular, for the properties listed in the examples (i),(ii), and (iii) preceding it.

The following proposition gives basic equivalences for the fundamental notion of a local semigroup.

IV.1.12.   Proposition.   *The following statements are equivalent for $S \subseteq \mathfrak{g}$:*

  (1) *S is a local semigroup.*

  (2) *There exists a C-H-neighborhood B and a local semigroup T with respect to B such that $T = S \cap B$.*

  (3) *S is in the same germ as a local semigroup with respect to some C-H-neighborhood.*

  (4) *There exists a C-H-neighborhood B such that $(S \cap B) * (S \cap B) \subseteq S$.*

Proof.   The equivalence of (1) and (2) is immediate from part (ii) of Proposition IV.1.11, that (2) implies (3) is obvious, and (2) follows immediately from (4) by taking $T = S \cap B$. Thus we need only show that (3) implies (4). Suppose T is a local semigroup with respect to a C-H-neighborhood B and B_1 is a smaller C-H-neighborhood such that $T \cap B_1 = S \cap B_1$. Pick a C-H-neighborhood B_2 such that $B_2 * B_2 \subseteq B_1$. Then $(S \cap B_2) * (S \cap B_2) \subseteq T * T \cap B_1 \subseteq T \cap B_1$ (since T is a local semigroup with respect to B). But $T \cap B_1 \subseteq S$. This completes the proof.   ∎

    The next proposition gives two important germs associated with any given local semigroup germ.

IV.1.13.   Proposition.   *There is associated with the germ of each local semigroup a maximal local subgroup germ which contains the maximal local group of any member of the germ that is a local semigroup with respect to some C-H-neighborhood and there is a germ of locally closed local semigroups containing the closures of members of the original germ.*

Proof.   Let S be a local semigroup. Pick a C-H-neighborhood B such that $S \cap B$ is a local semigroup with respect to B. We show that the germ of $H(S \cap B)$ is independent of the choice of B and S. Suppose that T and B_1 are other choices. Let $B_2 \subseteq B \cap B_1$ be chosen so that $S \cap B_2 = T \cap B_2$. Then it follows from Proposition IV.1.4 that

$$H(S \cap B) \cap B_2 \;=\; H(S \cap B_2) \;=\; H(T \cap B_2) \;=\; H(T \cap B_1) \cap B_2.$$

Hence the same germ is generated in both cases.

    One verifies directly (see (iii) of Proposition IV.1.9) that given any germ, the closures of members of the germ all belong to a common germ. It follows from Remark IV.1.3 that this germ is a local semigroup germ.   ∎

    The following remark is immediate, since one characterization of a Lie semialgebra is that it is a wedge which is also a local semigroup (see Definition II.2.8).

IV.1.14.   Remark.   A Lie semialgebra is a closed locally divisible local semigroup.   ∎

    A converse to this remark will be established in Proposition IV.1.31 below.

The tangent set at 0

In this section we first recall the definition of a *subtangent vector* and a *tangent vector of a subset S* in a topological vector space L at a point $x \in L$ as they were introduced and discussed in Definition I.5.1 through Proposition I.5.7.

In our present setting, S will be a local semigroup in a Dynkin algebra $\mathfrak{g}$, and we are interested in the subtangent vectors at 0. These circumstances call for a special notation.

IV.1.15. **Definition.** Let S be a subset of a Banach space L. We write the set $L_0(S)$ of subtangent vectors of S at 0 briefly as $\mathbf{L}(S)$. Notably,

$$\mathbf{L}(S) = \{x \in E:\ x = \lim r_n \cdot y_n,\ y_n \in S,\ \lim r_n = \infty\}$$
$$= \{x \in E:\ x = \lim r_n \cdot y_n,\ r_n \geq 0,\ y_n \in S,\ \lim y_n = 0\}.$$

We shall call $\mathbf{L}(S)$ the *tangent set of S*.

Note that $x = \lim r_n \cdot y_n$ implies $\|x\| = \lim r_n \|y_n\|$; hence $\lim r_n = \infty$ if and only if $\lim y_n = 0$. Thus the last two sets in the preceding definition are indeed equal.

From Proposition I.5.6 we recall

IV.1.16. **Remark.** For any subset S in a Banach space L we have

$$\mathbf{L}(S) = \mathbf{L}(\overline{S}) = \overline{\mathbf{L}(S)} = \mathbb{R}^+ \cdot \mathbf{L}(S).$$

∎

The following observation will be useful in the present context:

IV.1.17. **Remark.** (i) If U is a neighborhood of 0, then $\mathbf{L}(S) = \mathbf{L}(S \cap U)$.

(ii) If two sets S and T belong to the same germ, then $\mathbf{L}(S) = \mathbf{L}(T)$.

Proof. (i) Suppose $x = \lim r_n \cdot y_n$, $y_n \in S$, $\lim y_n = 0$. Then $y_n \in U$ for large n. Hence $x \in \mathbf{L}(S \cap U)$. The reverse containment is immediate.

(ii) By part (i) $\mathbf{L}(S) = \mathbf{L}(S \cap U) = \mathbf{L}(T \cap U) = \mathbf{L}(T)$. ∎

By the preceding Remark IV.1.17, we may speak of a well-defined *tangent set of a germ*.

For the following lemma we recall from Definition I.5.1 the idea of a function $\alpha: D \to L$ with values in a Banach space L which is *right differentiable* at the origin.

IV.1.18.   **Lemma.**   *Let* $f : E \to F$ *be a function between Banach spaces which is differentiable at* y *with* $f(y) = z$. *If* $\alpha : D \to E$ *is right-differentiable at the origin with* $\alpha(0) = y$, *then* $f\alpha$ *is right-differentiable at the origin and* $(f\alpha)'_+(0) = df(y)(\alpha'_+(0))$.

   Alternately, if $x = \lim r_n(y_n - y)$, *then* $df(y)(x) = \lim r_n(f(y_n) - z)$.

Proof.   To say f is differentiable at y means that there exists a continuous linear operator $df(y)$ from E into F such that $f(y + h) - f(y) = df(y)(h) + o(h)$, where $\lim_{h \to 0}(o(h)/\|h\|) = 0$. Let α be as indicated in the lemma and let $x = \alpha'_+(0)$. In order to see that $f\alpha$ is right-differentiable at the origin we compute

$$\lim_{t \to 0^+} \left\| \frac{f(\alpha(t)) - z}{t} - df(y)(x) \right\| = \left\| \lim_{t \to 0^+} \frac{f(\alpha(t)) - z}{t} - \left(df(y)\left(\lim_{t \to 0^+} \frac{\alpha(t) - y}{t}\right)\right) \right\|$$

$$= \left\| \lim_{t \to 0^+} \frac{f(\alpha(t)) - f(y) - df(y)(\alpha(t) - y)}{t} \right\| = \lim_{t \to 0^+} \left\| \frac{o(\alpha(t) - y)}{t} \right\|.$$

The latter limit is 0 since for $\alpha(t) \neq y$, we have

$$\frac{o(\alpha(t) - y)}{t} = \frac{\|\alpha(t) - y\|}{t} \frac{o(\alpha(t) - y)}{\|\alpha(t) - y\|},$$

which approaches $\|\alpha'_+(0)\| \cdot 0 = 0$. Hence $(f\alpha)'_+(0)$ exists and equals $df(y)(x)$.

   The alternate version follows directly by employing the equivalence of Proposition I.5.2.   ∎

   The next result is an immediate consequence of the preceding lemma.

IV.1.19.   **Proposition.**   *Let* $f : E \to F$ *be a function between Banach spaces which is differentiable at* y *with* $f(y) = z$, *and let* $S \subseteq E$. *Then* $df(y)(L_y(S)) \subseteq L_z(f(S))$. *In particular, if* $f(0) = 0$ *and* f *is differentiable at* 0, *then*

$$df(0)(\mathbf{L}(S)) \subseteq \mathbf{L}(f(S)).$$

                                                                      ∎

   It is frequently desirable to adjust slightly the set of vectors giving rise to a subtangent vector. The next proposition gives certain conditions that allow such a replacement.

IV.1.20.   **Lemma.**   (The Replacement Lemma) *Suppose* $x = \lim r_n \cdot x_n$, *where* $\lim r_n = \infty$. *Then also* $x = \lim r_n \cdot y_n$ *if either of the following conditions hold for some* $M > 0$:

   (i) $\|x_n - y_n\| \leq M\|x_n\|\,\|z_n\|$ *where* $\lim z_n = 0$;

   (ii) $\|x_n - y_n\| \leq M\|y_n\|^2$ *and* $\lim \sup \|y_n\| < 1/M$.

Proof.   (i) We have $\|r_n \cdot x_n - r_n \cdot y_n\| = r_n\|x_n - y_n\| \leq Mr_n\|x_n\|\,\|z_n\| \to M\|x\|\|0 = 0$. Thus $\lim r_n \cdot y_n = x$.

   (ii) We first show $\lim y_n = 0$. If not, then $\lim \sup \|y_n\| = r$, where $0 < r < 1/M$. Then $r = \lim \sup \|y_n\| = \lim \sup \|x_n - y_n\|$ (since $\lim x_n = 0$!)

$\leq \lim \sup M \|y_n\|^2 = Mr^2 < r$ (since $r < 1/M!$). The contradiction $r < r$ implies $\lim y_n = 0$.

Pick a positive integer N so that $M \|y_n\| < 1/2$ for $N \leq n$. We then have

$$\|y_n\| \leq \|x_n - y_n\| + \|x_n\| \leq M \|y_n\|^2 + \|x_n\| \leq (1/2)\|y_n\| + \|x_n\|.$$

We conclude that $(1/2)\|y_n\| \leq \|x_n\|$, that is, $\|y_n\| \leq 2\|x_n\|$. Thus for $n > N$

$$\|r_n \cdot x_n - r_n \cdot y_n\| \leq r_n M \|y_n\|^2 \leq 2M \, r_n \|x_n\| \, \|y_n\| \to 2M \|x\| 0 = 0.$$

Hence $\lim r_n \cdot x_n = \lim r_n \cdot y_n$. ∎

The tangent wedge of a local semigroup

Now we turn to Lie algebras.

The Lie algebra $\mathfrak{g}$ of a Lie group G is a most important tool in the study of a Lie group since it allows many statements about the structure of G to be translated into algebraic language in $\mathfrak{g}$. In this subsection we associate with each local semigroup in a Dynkin algebra a tangent wedge and develop the basic properties of this association. The goal is to develop foundations of a "Lie algebraic" machinery suitable to deal with (local) semigroups and that can later be applied in the global setting.

In the remainder of this subsection we work within the following setting: Let $\mathfrak{g}$ be a Dynkin algebra with complete norm $\|\cdot\|$ satisfying $\| [x,y] \| \leq M \|x\| \, \|y\|$ for all $x, y \in \mathfrak{g}$. Let B denote a Baker–Campbell–Hausdorff neighborhood.

The following proposition gives important alternate characterizations of the tangent set for the case that S is a local semigroup which we will use frequently:

IV.1.21.   Proposition.   *Let S be a local semigroup with respect to B. Then*

$$\mathbf{L}(S) = \{x \in \mathfrak{g} : x = \lim n \cdot x_n, x_n \in S\} = \{x \in \mathfrak{g} : \mathbb{R}^+ \cdot x \cap B \subseteq \overline{S}^B\}.$$

Proof.   Clearly $\mathbf{L}(S)$ contains the first set. We finish the proof by showing that $\mathbf{L}(S)$ is contained in the second set and the second set is contained in the first set.

Let $x = \lim r_n \cdot x_n$ with $\lim r_n = \infty$. Then $\|x\| = \lim r_n \|x_n\|$ implies that $\lim x_n = 0$. Let $r > 0$ such that $rx \in B$. Let $m_n = [rr_n]$, the greatest integer less than or equal to rr_n. Then $r \cdot x = \lim rr_n \cdot x_n = \lim(m_n \cdot x_n + \varepsilon_n \cdot x_n)$, where $0 \leq \varepsilon_n < 1$. Since $\lim x_n = 0$, $r \cdot x = \lim m_n \cdot x_n$. Since $r \cdot x \in B$, we conclude that $m_n \cdot x_n \in B$ for all n large enough. Hence $j \cdot x_n \in B$ for $j = 1, \ldots, m_n$ since B is full. Since S is a local semigroup with respect to B and $x_n \in S$, we conclude that $m_n \cdot x_n \in S$. Thus $r \cdot x \in \overline{S}^B$, and hence $\mathbf{L}(S)$ is contained in the second set.

Suppose $\mathbb{R}^+ \cdot x \cap B \subseteq \overline{S}^B$. For each positive integer n large enough, pick $x_n \in S$ such that $\|\frac{1}{n} \cdot x - x_n\| < n^{-2}$. Then $x = \lim n \cdot x_n$, and hence the second set is contained in the first. ∎

IV.1.22.   Corollary.   *Suppose that S is a local semigroup with respect to B.*
Then $\mathbf{L}(S) \cap B \subseteq \overline{S}^B$.

Proof.   For $x \in \mathbf{L}(S) \cap B$, $x = 1 \cdot x \in \mathbb{R}^+ \cdot x \cap B \subseteq \overline{S}^B$.   ∎

The preceding results motivate the introduction of an important special subclass of local semigroups.

IV.1.23.   Definition.   A local semigroup S with respect to B is said to be *full* if $\mathbf{L}(S) \cap B \subseteq S$.

IV.1.24.   Proposition.   (i) *A local subsemigroup S with respect to B is full if and only if*

$$\mathbf{L}(S) = \{x \in \mathfrak{g} : \mathbb{R}^+ \cdot x \cap B \subseteq S\}.$$

(ii) *If the local semigroup S is closed in B, then S is full.*

Proof.   (i) Suppose $\mathbf{L}(S)$ is full. Then for $x \in \mathbf{L}(S)$, $\mathbb{R}^+ \cdot x \cap B \subseteq \mathbf{L}(S) \cap B \subseteq S$, and containment in the other direction holds by Proposition IV.1.21.

Assume now the converse hypothesis. If $x \in \mathbf{L}(S) \cap B$, then $x = 1 \cdot x \in \mathbb{R}^+ \cdot x \cap B \subseteq S$.

(ii) This part is a restatement of Corollary IV.1.22.   ∎

The definition of $\mathbf{L}(S)$ given in the first part of Proposition IV.1.24 for full local semigroups is frequently most useful in actually computing the tangent set for specific examples. Hence it would be a logical candidate for defining the tangent set of a local semigroup in general. However, we have opted for the approach of Definition IV.1.15 since it exhibits pleasing theoretical properties. It is, all the same, agreeable that the two approaches coincide in the (locally) closed case.

We turn now to the consideration of the structure of the tangent set of a local semigroup. We recall the concept of a wedge and its edge from Definition I.1.1.

IV.1.25.   Proposition.   *Let S be a local semigroup with respect to B. Then $\mathbf{L}(S)$ is a wedge, and $\mathbf{L}(S) = \mathbf{L}(\overline{S}^B)$.*

Proof.   That $\mathbf{L}(S)$ is closed, closed under multiplication by non-negative scalars, and that $\mathbf{L}(S) = \mathbf{L}(\overline{S}) = \mathbf{L}(\overline{S}^B)$ follows from Remark IV.1.16. Suppose $x, y \in \mathbf{L}(S)$. Then $x = \lim n \cdot x_n$ and $y = \lim n \cdot y_n$ where $\lim x_n = 0 = \lim y_n$ and $x_n, y_n \in S$ for all n. By the Trotter Product Formula (see (A3) of Appendix 1) $x + y = \lim n \cdot (x_n * y_n)$, and eventually $x_n * y_n \in B$ and hence in S. Thus $x + y \in \mathbf{L}(S)$.   ∎

In view of this proposition we shall also call the tangent set $\mathbf{L}(S)$ the *tangent wedge of S*.

We saw that the edge of a wedge is an important invariant of the geometry of the wedge. Now we have the following relationships between the maximal local group of a local semigroup and the edge of the tangent wedge.

IV.1.26. **Proposition.** *Let S be a local semigroup with respect to B. Then* $\mathbf{L}(H(S)) \subseteq H(\mathbf{L}(S))$, *and if S is full (which is the case for $S = \overline{S}^B$), then*

$$\mathbf{L}(H(S)) = H(\mathbf{L}(S)) = H(\mathbf{L}(\overline{S}^B)) = \mathbf{L}(H(\overline{S}^B)).$$

Proof. Note (from the definition or from Proposition IV.1.19) that $\mathbf{L}(-S) = -\mathbf{L}(S)$. Since $H(S) = S \cap -S$,

$$\mathbf{L}(H(S)) \subseteq \mathbf{L}(S) \cap \mathbf{L}(-S) = \mathbf{L}(S) \cap (-\mathbf{L}(S)) = H(\mathbf{L}(S)).$$

The middle equality in the string of equalities in the proposition follows from the previous proposition. Suppose $x \in \mathbf{L}(S) \cap (-\mathbf{L}(S))$. Then $\mathbb{R}^+ \cdot x \cap B \subseteq \overline{S}^B$ and $\mathbb{R}^+ \cdot (-x) \cap B \subseteq \overline{S}^B$. If $r \cdot x \in B$, then $r \cdot x \in \mathbf{L}(S) \cap B \subseteq S$ if S is full. Also $r \cdot x \in B$ implies that $r \cdot (-x) \in B$; thus $r \cdot (-x) \in B \cap \mathbf{L}(S) \subseteq S$. Since $r \cdot x, -r \cdot x \in S$, we conclude that $r \cdot x \in H(S)$ if $r \cdot x \in B$. Thus $\mathbb{R}^+ \cdot x \cap B \subseteq H(S)$, that is, $x \in \mathbf{L}(H(S))$ by Proposition IV.1.21. This establishes containment in the other direction, and we conclude that $\mathbf{L}(H(S)) = H(\mathbf{L}(S))$. This equality applies also to $\overline{S}^B$, which completes the proof. ∎

We recall from Definition II.1.3 that a wedge W is called a *Lie wedge* if it is invariant with respect to the action of $\mathrm{INN}_L H(W)$, in other words that $e^{\mathrm{ad}\, x} W = W$ for all $x \in H(W)$. The following important result was already displayed in Proposition II.1.2 in order to motivate the concept of a Lie wedge. Because of its significance we present here once more a full proof in a slightly different organization.

IV.1.27. **Theorem.** (The Lie Wedge Theorem) *The tangent wedge of a local semigroup is a Lie wedge.*

Proof. That the tangent set is a wedge follows from Remark IV.1.17 and Proposition IV.1.25. Let S denote the local semigroup, pick a C-H-neighborhood B such that $S \cap B$ is a local semigroup with respect to B, and pick a C-H-neighborhod U with $U * U * U \subseteq B$. Let $T = \overline{S \cap B}^B$, and pick $x \in H(T) \cap U$. Define $\alpha : U \to B$ by $\alpha(y) = x * y * (-x)$. Since T is a local semigroup with respect to B (Remark IV.1.3), $\alpha(T \cap U) \subseteq T$. By Formula (A6) of Appendix 1, $x * y * (-x) = e^{\mathrm{ad}\, x} y$. Since $e^{\mathrm{ad}\, x}$ is a linear transformation, it is its own derivative. Hence by Proposition IV.1.19, $e^{\mathrm{ad}\, x}(\mathbf{L}(T \cap U)) \subseteq \mathbf{L}(T)$. By Proposition IV.1.25 and Remark IV.1.17 $e^{\mathrm{ad}\, x}(\mathbf{L}(S)) \subseteq \mathbf{L}(S)$. Similarly $e^{\mathrm{ad}(-x)}(\mathbf{L}(S)) \subseteq \mathbf{L}(S)$. Applying $e^{\mathrm{ad}\, x}$ to both sides, we obtain that $e^{\mathrm{ad}\, x}(\mathbf{L}(S)) = \mathbf{L}(S)$.

Let $z \in H(\mathbf{L}(S))$. Pick $x = (1/n)z \in U$. By Proposition IV.1.26, $x \in \mathbf{L}(H(T))$, and by Corollary IV.1.22, $x \in H(T)$. By the preceding paragraph, $e^{\mathrm{ad}\, x}(\mathbf{L}(S)) = \mathbf{L}(S)$ and hence $e^{\mathrm{ad}\, z}(\mathbf{L}(S)) = (e^{\mathrm{ad}\, x})^n (\mathbf{L}(S)) = \mathbf{L}(S)$. Thus $\mathbf{L}(S)$ is a Lie wedge. ∎

The following is one version of Lie's Fundamental Theorems:

IV.1.28.   Proposition.   *Let M be a closed subalgebra of L. Then $M \cap B$ is a local group. Conversely, if H is a local group with respect to B, then $\mathbf{L}(H)$ is a subalgebra.*

Proof.   Since M is a subspace, a subalgebra, and closed, it follows from the Baker–Campbell–Hausdorff formula that M is closed with respect to the $*$-operation (whenever defined). Thus $M \cap B$ is a local group for any C-H-neighborhood B.

Conversely, suppose that H is a local group with respect to B. Since H is a local semigroup with respect to B, by Proposition IV.1.25, the set $\mathbf{L}(H)$ is a (closed) wedge. Since $H = -H$, $\mathbf{L}(H) = -\mathbf{L}(H)$. Thus $\mathbf{L}(H)$ is a closed subspace.

Suppose that $x, y \in \mathbf{L}(H)$. Then $x = \lim n \cdot x_n$, $y = \lim n \cdot y_n$, where $x_n, y_n \in H$. Then also $-x_n, -y_n \in H$. By Formula (A4) of Appendix 1 we have that

$$[x, y] = \lim n^2 \cdot \big(x_n * y_n * (-x_n) * (-y_n)\big) = \lim n \cdot \Big(n \cdot \big(x_n * y_n * (-x_n) * (-y_n)\big)\Big).$$

Thus it is must be the case that $\lim n \cdot \big(x_n * y_n * (-x_n) * (-y_n)\big) = 0$; hence, for n large enough, the element $k \cdot \big(x_n * y_n * (-x_n) * (-y_n)\big) \in B$ for all k, $1 \le k \le n$, and is therefore a member of H. We conclude that $[x, y] \in \mathbf{L}(H)$.   ∎

IV.1.29.   Corollary.   *Let S be a local semigroup. Then $H\big(\mathbf{L}(S)\big)$ is a subalgebra.*

Proof.   Pick a C-H-neighborhood B such that $S \cap B$ is a local semigroup with respect to B, and let $T = \overline{S \cap B}^B$. Then by Proposition IV.1.25 and Remark IV.1.17, $\mathbf{L}(S) = \mathbf{L}(T)$, and hence $H\big(\mathbf{L}(S)\big) = H\big(\mathbf{L}(T)\big)$. By Proposition IV.1.26, $H\big(\mathbf{L}(T)\big) = \mathbf{L}\big(H(T)\big)$, and by Proposition IV.1.28, $\mathbf{L}\big(H(T)\big)$ is a subalgebra.   ∎

One can use the Lie Wedge Theorem IV.1.27 directly to devise an alternate proof of this corollary (see Corollary II.1.8).

We turn now to the consideration of the local subgroup generated by a local semigroup and its tangent object. For this purpose, we need some convenient notation:

Notation.   For $Y \subseteq L$, let $\langle\!\langle Y \rangle\!\rangle$ denote the closed Lie algebra generated by Y. (Note that in the finite dimensional case all vector subspaces are closed.)

IV.1.30.   Proposition.   *Let S be a local semigroup with respect to B such that $S \subseteq \langle\!\langle \mathbf{L}(S) \rangle\!\rangle$, and let $G_B(S)$ denote the local subgroup with respect to B generated by S. Then $\mathbf{L}\big(G_B(S)\big) = \mathbf{L}(\langle S \cup -S \rangle_B) = \langle\!\langle \mathbf{L}(S) \rangle\!\rangle$.*

Proof.   Since $\mathbf{L}(S) \subseteq \mathbf{L}\big(G_B(S)\big)$, and since by Corollary IV.1.29, the latter is a subalgebra, we have that $\langle\!\langle \mathbf{L}(S) \rangle\!\rangle \subseteq \mathbf{L}\big(G_B(S)\big)$.

By Proposition IV.1.28, $\langle\!\langle \mathbf{L}(S) \rangle\!\rangle \cap B$ is a local group with respect to B. Thus $S \subseteq \langle\!\langle \mathbf{L}(S) \rangle\!\rangle \cap B$ implies $G_B(S) \subseteq \langle\!\langle \mathbf{L}(S) \rangle\!\rangle \cap B$, and hence $\mathbf{L}\big(G_B(S)\big) \subseteq \langle\!\langle \mathbf{L}(S) \rangle\!\rangle$.   ∎

We now consider Lie semialgebras and derive the converse of Remark IV.1.14.

IV.1.31.   Proposition. *Let S be a divisible local semigroup with respect to B. Then $\mathbf{L}(S)$ is a Lie semialgebra, and $\overline{S}^B = \mathbf{L}(S) \cap B$.*

Proof.   By Corollary IV.1.22 we have $\mathbf{L}(S) \cap B \subseteq \overline{S}^B$. Conversely, since S is divisible, so is $\overline{S}^B$. Hence for $x \in \overline{S}^B$, $x = \lim n \cdot (\frac{1}{n} \cdot x)$ implies $x \in \mathbf{L}(\overline{S}^B) = \mathbf{L}(S)$. Thus the reverse containment also holds. It now follows that $\mathbf{L}(S)$ is a Lie semialgebra since its intersection with B is a local semigroup with respect to B. ∎

IV.1.32.   Corollary. *Let S be a locally divisible local semigroup. Then $\mathbf{L}(S)$ is a Lie semialgebra.*

Proof.   By part (ii) of Proposition IV.1.11 a C-H-neighborhood B can be chosen so that $S \cap B$ is both divisible and a local semigroup. By Remark IV.1.17, S and $S \cap B$ have the same tangent wedges; thus by the previous proposition, $\mathbf{L}(S)$ is a Lie semialgebra. ∎

In Proposition II.2.13 we saw that finite dimensional Lie semialgebras span Lie algebras. We shall now show by a direct approach that a suitable version of this result holds even in infinite dimensional Dynkin algebras. We observed that Lie wedges do not have this property in general.

IV.1.33.   Proposition. *Let K be a Lie semialgebra. Then $\langle\!\langle K \rangle\!\rangle = \overline{K - K}$.*

Proof.   Clearly the closure of $K - K$ is a closed subspace of $\langle\!\langle K \rangle\!\rangle$. In order to obtain the reverse inequality, it suffices (by continuity of the Lie bracket) to show that if $x, y \in K$, then $[x, y]$ is in the closure of $K - K$. Since K is a Lie semialgebra, the element $z_n = (\frac{1}{n} \cdot x * \frac{1}{n} \cdot y) - (\frac{1}{n} \cdot y * \frac{1}{n} \cdot x)$ is defined and in $K - K$ for n large enough. It follows that $n^2 \cdot z_n$ is also in $K - K$. Then by Formula (A5) of Appendix 1, $[x, y] = \lim n^2 \cdot z_n$ is in the closure of $K - K$. ∎

Recall that a half-space of $\mathfrak{g}$ has the form $\omega^{-1}\mathbb{R}^+$ where ω is a non-zero continuous linear functional into $\mathbb{R}$. The following result shows through a direct proof that the the characterization of half-space semialgebras given in Corollary II.2.24 remains valid for infinite dimensional Dynkin algebras.

IV.1.34.   Proposition. (The Half-Space Theorem) *Let W be a half-space of $\mathfrak{g}$. The following are equivalent:*

(1) *The edge $H(W)$ of W is a subalgebra.*

(2) *W is a local semigroup.*

(3) *W is a Lie semialgebra.*

(4) *$W \cap B$ is a local semigroup with respect to B for every C-H-neighborhood B.*

Moreover, under these circumstances, W is invariant if and only if $H(W)$ is an ideal.

Proof.   That (2) implies (1) follows from Corollary IV.1.29 since $\mathbf{L}(W) = W$. That (4) implies (3) implies (2) is clear.

Assume (1) and let B be a C-H-neighborhood. Suppose there exist $x, y \in W \cap B$ with $x * y$ not in W. Then $\{tx * y : 0 \le t \le 1\}$ is a connected set meeting both W and its complement. Hence there exists some t with $tx * y \in H(W)$. Let $u = (1-t)x$, $v = tx * y$. Then $u \in W$, $v \in H(W)$, but $u*v = (1-t)x*tx*y = x*y$ is not in W. Again $\{u * sv : 0 \leq s \leq 1\}$ is a connected set meeting both W and its complement. Hence $u * sv \in H(W)$ for some s. By Proposition IV.1.28, $H(W) \cap B$ is a local group with respect to B for every C-H-neighborhood B. Since $u * sv, -sv \in H(W)$, we conclude that $u = u * sv * (-sv) \in H(W)$. But then $u * v \in H(W)$, a contradiction.

Finally, each automorphism which leaves the wedge W invariant also leaves its edge invariant. Conversely, the group $\text{INN}\, L$ of inner automorphisms of L is connected, and thus, if the edge $H(W)$ is invariant under $\text{INN}\, L$, then both half spaces it bounds remain invariant under this group. This remark proves the last assertion.   ∎

It is of interest for later applications that there is a local semigroup variation of the theme of the Half-space Theorem which we now present.

IV.1.35.   Proposition.   *Let B be a C-H-neighborhood in a Dynkin algebra $\mathfrak{g}$, and let S be a local semigroup with respect to B. Then the following statements are equivalent:*

(1) $\mathbf{L}(S)$ *is a half-space.*

(2) $\mathbf{L}(S)$ *is a half-space Lie semialgebra.*

(3) *There is a closed half-space W in $\mathfrak{g}$ such that $\overline{S}^B = B \cap W$.*

Moreover, if these conditions are satisfied, then the local group of units $H(\overline{S}^B)$ is $B \cap H(W)$.

Proof.   Since $\mathbf{L}(S) = \mathbf{L}(\overline{S}^B)$ by Proposition IV.1.25, the implication $(3) \to (1)$ is clear. Further, if (1) is satisfied, then by Corollary IV.1.29, the edge $H(\mathbf{L}(S))$ is a Lie algebra. So by the preceding Proposition IV.1.34, $\mathbf{L}(S)$ is a Lie semialgebra. Thus (1) implies (2). It therefore remains to show that (2) implies (3).

Suppose (2). We set $W = \mathbf{L}(S)$. Then $W \cap B = \mathbf{L}(\overline{S}^B) \cap B \subseteq \overline{S}^B$ by Corollary IV.1.22. We must also show the reverse inclusion. Suppose there exists $x \in S \setminus W$. Then $-x$ is in the interior of W, hence in the interior of $\overline{S}^B$. Thus there is an open subset U of $\overline{S}^B$ such that $x * U \subseteq B$ and $-x \in U$. But then $x * U$ is an neighborhood of 0 entirely contained in $\overline{S}^B$, from which it follows that $\mathbf{L}(S) = \mathfrak{g}$. But $\mathbf{L}(S)$ is a half-space by (2). This contradiction shows $S \setminus W = \varnothing$, and hence $\overline{S}^B \subseteq W \cap B$.

If the equivalent conditions (1), (2), and (3) are satisfied, then $H(\overline{S}^B) = \overline{S}^B \cap -\overline{S}^B = (B \cap W) \cap -(B \cap W) = B \cap (W \cap W) = B \cap H(W)$.   ∎

Further invariance properties of Lie wedges

For the following invariance results on Lie wedges we have to fall back on the special power series f and g which we introduced and discussed in Definition

II.2.1 and Lemma II.2.2. The series for f is given by $f(X) = (1 - e^{-X})/X = 1 + (1/2!)X + (1/3!)X^2 + \dots$, and $g(X)$ is given by the series for $1/f(X)$. For $x \in B$, let λ_x denote the left translation mapping from B into $\mathfrak{g}$ sending y to $x * y$. Then the derivative at 0, $d\lambda_x(0): \mathfrak{g} \to \mathfrak{g}$ is given by $d\lambda_x(0)(y) = g(\operatorname{ad} x)(y)$. (This is discussed in detail in the beginning of the subsection on invariant vector fields in Section IV.5.)

IV.1.36. **Proposition.** *Let S be a local semigroup with respect to B, and let $x \in H(S)$. Then $L_x(S) = g(\operatorname{ad} x)\big(\mathbf{L}(S)\big)$.*

Proof. Pick a C-H-neighborhood $B_1 \subseteq B$ such that $x * B_1 \subseteq B$ and let $T = x * (S \cap B_1)$. We show $T = S \cap x * B_1$. Let $t = x * s \in T$, where $s \in S \cap B_1$. Then $t \in S$ since S is a local semigroup with respect to B and obviously $t \in x * B_1$. Conversely, suppose $x * y \in S$, where $y \in B_1$. Then since S is a local semigroup with respect to B, $y = -x * (x * y) \in S$, so $x * y \in T$. Thus λ_x is a diffeomorphism (with inverse λ_{-x}) from B_1 onto the open neighborhood $x * B_1$ of x which carries $S \cap B_1$ to T. Using Proposition IV.1.19 both ways, we conclude $d\lambda_x(0)\big(\mathbf{L}(S \cap B_1)\big) = L_x(T)$. Using Remark IV.1.17 we thus have that

$$g(\operatorname{ad} x)\mathbf{L}(S) = d\lambda_x(0)(\mathbf{L}(S \cap B_1) = L_x(S \cap x * B_1) = L_x(S).$$

∎

We derive another invariance property of tangent wedges of local semigroups (see Theorem IV.1.27).

IV.1.37. **Proposition.** *Let W be the tangent wedge of a local semigroup S with respect to B. If $x \in H(W)$, then $f(\operatorname{ad} x)(W) \subseteq W$, or, equivalently, $W \subseteq g(\operatorname{ad} x)(W)$.*

Proof. By Proposition IV.1.25, we may assume without loss of generality that $S = \overline{S}^B$. then by Corollary IV.1.22 $W \cap B \subseteq S$. First choose $x \in H(W) \cap B$. Pick an C-H-neighborhood B_1 such that $B_1 + x \subseteq B$. If $y \in W \cap B_1$, then $y + x \in W$ and $y + x \in B$; hence $W \cap B_1 \subseteq (W \cap B) - x$. Thus using Remark IV.1.17, we have

$$W = \mathbf{L}(W) = \mathbf{L}(W \cap B_1) \subseteq \mathbf{L}((W \cap B) - x) = \mathbf{L}_x(W \cap B) \subseteq \mathbf{L}_x(S) = g(\operatorname{ad} x)(W)$$

where the last equality follows from Proposition IV.1.36. Applying $f(\operatorname{ad} x)$ to both sides, we obtain $f(\operatorname{ad} x)(W) \subseteq W$, since $f(\operatorname{ad} x)$ and $g(\operatorname{ad} x)$ are inverses of each other.

Now let $x \in H(W)$ be arbitrary. Suppose that $f(\operatorname{ad} t \cdot x)W \subseteq W$ for all $|t| < s$. We show that this implies $f(\operatorname{ad} t \cdot x)W \subseteq W$ for all $|t| < 2s$. Since by the first part of the proof the supposition holds for $t \cdot x \in B$, it will follow that it holds for all t, and the proof will be complete.

We first observe that the power series $f(X)$ satisfies the equation $f(2X) = \frac{1}{2}(1 + e^{-X})f(X)$, since $1 - e^{-2X} = (1 + e^{-X})(1 - e^{-X})$. If now $|t| < s$, then $(I + e^{\operatorname{ad} t \cdot x})W \subseteq W$, since W is a Lie wedge by the Lie Wedge Theorem IV.1.27, and hence W satisfies $e^{\operatorname{ad} t \cdot x} W \subseteq W$. Also $f(\operatorname{ad} t \cdot x) W \subseteq W$ by assumption. It then follows that

$$f(\operatorname{ad} 2t \cdot x)\, W = \frac{1}{2} \cdot (I + e^{-\operatorname{ad} t \cdot x}) f(\operatorname{ad} t \cdot x)\, W \subseteq W.$$

The equivalent formulation in terms of g follows again from the fact that $f(\operatorname{ad} x)$ and $g(\operatorname{ad} x)$ are inverses of each other.   ∎

The inclusions in Proposition IV.1.37 are not in general equalities as the next example shows.

IV.1.38   Example. Let L be the four dimensional nilpotent Lie algebra spanned by x, y, z, u subject to the multiplication rules $[x, y] = z$, $[u, y] = z$, $[u, x] = y$, and all other brackets zero. Let E be the subspace spanned by x, y, z. Then $\operatorname{ad} u$ restricted to E is the nilpotent operator T given by $Tx = y$, $Ty = z$, and $Tz = 0$. Let C be the Lorentzian cone in E spanned by the conic section $\{x + r \cdot y + \frac{r^2}{2} \cdot z : r \in \mathbb{R}\}$, that is, $C = \{a \cdot x + b \cdot y + c \cdot z : 2ac \geq b^2, \ 0 \leq a, c\}$. Thus C is the cone whose axis is spanned by $x + z$ and which contains x and z in its boundary. A straightforward computation yields that the given generating conic section for C is invariant under $e^{t \cdot T} = I + t \cdot T + \frac{1}{2} t^2 \cdot T^2$, $t \in \mathbb{R}$. Hence C is invariant under $e^{t \cdot T}$. Thus $W = \mathbb{R} \cdot u + C$ is invariant under $e^{t \cdot u}$, $t \in \mathbb{R}$, and hence is a Lie wedge with edge $H(W) = \mathbb{R} \cdot u$. By the end of this chapter we shall have the results which will show that, under these circumstances, W is the tangent wedge of some local semigroup.

The edge is not an ideal. Moreover,

$$g(\operatorname{ad} t \cdot u)(x) = x + \frac{t}{2} \cdot [u, x] + \frac{t^2}{12} \cdot [u, [u, x]] = x + \frac{t}{2} \cdot y + \frac{t^2}{12} \cdot z$$

is not contained in W for all $t > 0$ since $t^2/4 \nleq t^2/6$. Thus the relation $g(\operatorname{ad} x) W \subseteq W$ for sufficiently small $x \in H(W)$ does not hold for Lie wedges.   ∎

2. Tangent wedges and local wedge semigroups

In the last section we showed that the tangent wedge associated with any local semigroup is a Lie wedge. A major objective of the remainder of this chapter is to establish a converse, namely that for every Lie wedge there exists a local semigroup for which it is the tangent wedge. This will require, however, building up a rather significant amount of machinery. In this section we start with wedges which are tangent wedges to some local semigroup and consider some of the more elementary constructions and properties associated with them. We begin with some examples.

IV.2.1. Example. Let $\mathbb{R}^2$ denote the abelian 2-dimensional Lie algebra (with the Baker–Campbell–Hausdorff multiplication given by addition and C-H-neighborhood the whole algebra). Let $S_1 = \{(x,y): -x^2 \le y \le x^2\}$, $S_2 = \{(x,y): 0 \le x,\text{ and } y$ is an integer$\}$, and $S_3 = \{(x,y): y = 0,\ x$ is a non-negative rational$\}$. Then S_j is a semigroup and $\mathbf{L}(S_j) = W = \mathbb{R}^+ \times \{0\}$ for $j = 1,2,3$. Note that the germs of S_1, S_2, and S_3 are quite different, but that all have the same tangent wedge. Note further that W is a subset of S_1 and S_2, but not of S_3 (hence S_1 and S_2 are full, but not S_3). Finally note that $H(W) = \{0\}$, and this is a proper subset of $H(S_2)$. ∎

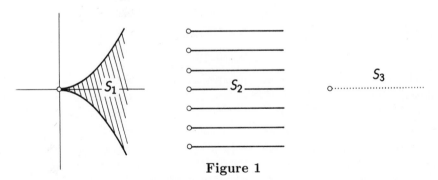

Figure 1

The following example is based on the 3-dimensional Heisenberg algebra and was first introduced at the end of Section 1 of Chapter II and is recorded here for easy reference.

IV.2.2. Example. Let $\mathfrak{g}$ denote the Heisenberg algebra, that is, the 3-dimensional nilpotent Lie algebra with generators $\mathbf{p} = (1,0,0)$, $\mathbf{q} = (0,1,0)$, $\mathbf{e} = (0,0,1)$, $[\mathbf{p},\mathbf{q}] = \mathbf{e}$, and $\mathbf{e}$ in the center. Then the Baker–Campbell–Hausdorff multiplication is defined globally and given by $x * y = x + y + \frac{1}{2}[x, y]$, that is,

$$(x_1, x_2, x_3) * (y_1, y_2, y_3) = \left(x_1 + y_1, x_2 + y_2, x_3 + y_3 + \frac{1}{2}(x_1 y_2 - x_2 y_1)\right).$$

Let $S = \{(x, y, z): 0 \leq x, y; \; -\frac{1}{2}xy \leq z \leq \frac{1}{2}xy\}$. We saw in Section 1, Chapter II that S is a subsemigroup. Since S is closed and hence full (Proposition IV.1.24), it follows that $\mathbf{L}(S)$ consists of all rays in $\mathfrak{g}$ which are completely contained in S. One then sees directly that $\mathbf{L}(S) = \{(x, y, 0): 0 \leq x, y\}$.

We show that the rays through $\mathbf{p}$ and $\mathbf{q}$ generate S. Let $(x, y, z) \in S$. If $z = 0$, then $(\frac{1}{2}x, 0, 0) * (0, y, 0) * (\frac{1}{2}x, 0, 0) = (x, y, z)$. If $z \neq 0$, then one verifies directly that $(a, 0, 0)*(0, y, 0)*(c, 0, 0) = (x, y, z)$, where $a = \frac{1}{2}x + \frac{z}{y}$ and $c = \frac{1}{2}x - \frac{z}{y}$. Note that $z \neq 0$ implies $y > 0$. Thus a and c are well-defined. Furthermore, since $(x, y, z) \in S$, $ay = \frac{1}{2}xy + z \geq 0$; it follows that $a \geq 0$ since $y > 0$. Similarly it is established that $c \geq 0$. ∎

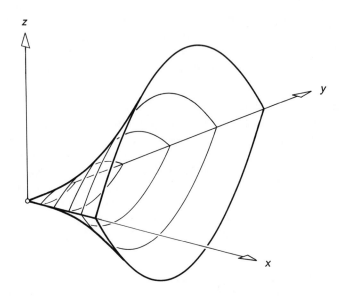

Figure 2

This example exhibits several interesting characteristics. First of all $\mathbf{L}(S)$ generates S, but sits thinly in S (the set $\mathbf{L}(S)$ is 2-dimensional, but S is 3-dimensional). In addition the subspace generated by $\mathbf{L}(S)$ does not contain S and is not a subalgebra.

Example IV.2.1 shows that there are, off-hand, many local semigroups or germs of local semigroups that all have the same tangent wedge. We may still consider, however, a reverse procedure of starting with a tangent wedge and associating with it the local semigroup with respect to some neighborhood that it locally generates.

IV.2.3.   Proposition. *Let $\mathbf{L}(S)$ be the tangent wedge of S, where S is a local semigroup with respect to a C-H-neighborhood B. Let $T = \langle \mathbf{L}(S) \rangle_B$. Then T is the smallest local semigroup with respect to B which contains $\mathbf{L}(S) \cap B$, and*

(i) $\mathbf{L}(T) = \mathbf{L}(\overline{T}^B) = \mathbf{L}(S)$;

(ii) T *is full;*

(iii) $\overline{T}^B \subseteq \overline{S}^B \cap \langle\!\langle \mathbf{L}(S)\rangle\!\rangle$; *hence* $\overline{T}^B$ *is the smallest local semigroup with respect to B closed in B with tangent set $\mathbf{L}(S)$);*

(iv) $T = \langle \mathbf{L}(T)\rangle_B$ *and* $\overline{T}^B$ *is the closure in B of* $\langle \mathbf{L}(\overline{T}^B)\rangle_B$.

Proof. The first assertion follows from Proposition IV.1.6.

(i) By Corollary IV.1.22 we have $\mathbf{L}(S) \cap B \subseteq \overline{S}^B$. Hence $T \subseteq \overline{S}^B$ since the latter is a local subsemigroup with respect to B containing $\mathbf{L}(S) \cap B$ and the former is the smallest such. Thus $\mathbf{L}(T) \subseteq \mathbf{L}(\overline{S}^B) = \mathbf{L}(S)$. Conversely, $\mathbf{L}(S) \cap B \subseteq T$ implies $\mathbf{L}(S) \subseteq \mathbf{L}(T)$. Thus they are equal. By Proposition IV.1.25, we have $\mathbf{L}(T) = \mathbf{L}(\overline{T}^B)$. This completes (i).

(ii) By (i) we have $\mathbf{L}(T) \cap B = \mathbf{L}(S) \cap B \subseteq T$.

(iii) Since $T \subseteq \overline{S}^B$, also $\overline{T}^B \subseteq \overline{S}^B$. Also $\langle\!\langle \mathbf{L}(S)\rangle\!\rangle \cap B$ is a local group with respect to B by Proposition IV.1.28. This implies that it contains T and hence $\overline{T}^B$.

(iv) This part is immediate from (i) and the definition of T. ∎

The subtle distinctions in the next definition reflect the intricacies of local generation of (local) semigroups. We do need all the concepts and have to stick by our conventions.

IV.2.4. **Definition.** A local semigroup S with respect to a C-H-neighborhood B is called a *local wedge semigroup with respect to B* if S is full and is contained in the closure in B of the local semigroup with respect to B generated by $\mathbf{L}(S) \cap B$. We say that S is *strict* in the case that it is locally generated by $\mathbf{L}(S) \cap B$ and *closed* if it is the closure in B of this semigroup. We refer alternately to a (strict) local wedge semigroup with respect to B by saying it is *(strictly) infinitesimally generated*. A local semigroup is a *local wedge semigroup* if its intersection with some C-H-neighborhood B is a local wedge semigroup with respect to B.

IV.2.5. **Remark.** In light of Proposition IV.2.3, a local semigroup S with respect to B is a strict local wedge semigroup with respect to B if and only if it is minimal in the set of all full local semigroups T with respect to B satisfying $\mathbf{L}(S) \subseteq \mathbf{L}(T)$, and S is a closed local wedge semigroup with respect to B if and only if it is minimal in the set of all closed local semigroups T with respect to B satisfying $\mathbf{L}(S) \subseteq \mathbf{L}(T)$.

With these concepts we can now formulate in which way local semigroups with respect to a fixed C-H-neighborhood are determined by their tangent wedges—and vice versa.

IV.2.6. **Theorem.** (The Correspondence Theorem) *Let B be a fixed C-H-neighborhood. The assignment $S \mapsto \mathbf{L}(S)$ from the set of strict (respectively, closed) wedge semigroups with respect to B, to the set of wedges which are tangent wedges for some local semigroup with respect to B, as well as the assignment $W \mapsto \langle W\rangle_B$ (respectively, $W \mapsto \overline{\langle W\rangle_B}^B$) from the set of tangent wedges for some local semigroup* *with respect to B to the set of strict (respectively, closed) local wedge semigroups with respect to B are mutually inverse bijections.*

Proof.   That the composition in one direction is the identity follows immediately from part (iii) of Proposition IV.2.3, and that the composition in the other direction is the identity follows from part (i) of Proposition IV.2.3 (together with Proposition IV.1.25 for the parenthetical case).   ∎

Recall that a Dynkin algebra $\mathfrak{g}$ is called *exponential* if it is the Lie algebra of a Lie group G such that the exponential function $\exp: \mathfrak{g} \to G$ is a diffeomorphism. This means that the Campbell–Hausdorff multiplication has an analytic extension to the whole of $\mathfrak{g} \times \mathfrak{g}$. We saw earlier in Corollary II.2.42 that a wedge W in a finite dimensional exponential Lie algebra is a $*$-subsemigroup if and only if it is a Lie semialgebra.

IV.2.7.   Remark. If $\mathfrak{g}$ is an exponential Dynkin algebra, then the results of Theorem IV.2.6 apply with $B = \mathfrak{g}$, and thus the local semigroups are actually semigroups. For the case of an abelian Dynkin algebra the Baker–Campbell–Hausdorff multiplication is just addition, and the bijections of Theorem IV.2.6 assign to a wedge the same wedge (in one case viewed as a tangent object and in the other as an additive semigroup). In this case there is only one wedge semigroup associated with a given wedge. More generally, a given tangent wedge will have a unique local wedge semigroup with respect to B associated with it if and only if the local semigroup with respect to B that it generates is closed in B (for otherwise, the closure is another).   ∎

In Example IV.2.1, the only wedge semigroup associated with the non-negative real x-axis is the non-negative x-axis itself, although many semigroups have that same tangent object. In Example IV.2.2, the semigroup S was shown to be generated by its tangent wedge $\mathbf{L}(S)$, and hence S is a wedge semigroup. Since S is also closed, it is the only wedge semigroup with tangent object $\mathbf{L}(S)$. In general it is quite difficult to calculate explicitly the (local) semigroup generated by a wedge.

The results of Theorem IV.2.6 are satisfactory for relating local wedge semigroups with respect to B and the tangent wedges of local semigroups with respect to B. But now we consider the case that B is allowed to vary.

IV.2.8.   Proposition. *Let S be a full local semigroup with respect to B and $W = \mathbf{L}(S)$. Let C be C-H-neighborhood in B and set $T = \langle W \cap C \rangle_C$. Then $T \subseteq S$, T is a strict local wedge semigroup with respect to C, $\overline{T}^C$ is a closed local wedge semigroup with respect to C, and $W = \mathbf{L}(T) = \mathbf{L}(\overline{T}^C) = \mathbf{L}(S \cap C)$.*

Proof.   By Remark IV.1.3, the set $S \cap C$ is a local semigroup with respect to C which contains $W \cap B \cap C = W \cap C$. Hence it contains the local semigroup T generated by $W \cap C$. The rest now follows from Theorem IV.2.6.   ∎

This proposition says, in particular, that *the tangent wedge W of any local semigroup with respect to some C-H-neighborhood B has for each C-H-neighborhood $C \subseteq B$ a unique strict (respectively, a unique closed) local wedge semigroup with respect to C for which W is the tangent wedge.* .

IV.2.9.   **Proposition.**   *If S is a strict local wedge semigroup with respect to B, then for any C-H-neighborhood $C \subseteq B$, the set $S \cap C$ is a local subsemigroup with respect to C, and $\mathbf{L}(S \cap C) = \mathbf{L}(S)$. For the strict local wedge semigroup with respect to C, $T = \langle \mathbf{L}(S) \cap C \rangle_C$, we have $\mathbf{L}(T) = \mathbf{L}(S)$, $T \subseteq S \cap C$, and $S = \langle T \rangle_B = \langle \mathbf{L}(S) \cap C \rangle_B$. Analogous statements hold for closed local wedge semigroups with respect to B.*

Proof.   We show $S = \langle T \rangle_B = \langle \mathbf{L}(S) \cap C \rangle_B$. (The preceding assertions follow from the Proposition IV.2.8.) Since $\mathbf{L}(S) \cap C \subseteq S \cap C$, we conclude that $T \subseteq S \cap C$. Hence $\langle T \rangle_B \subseteq S$.

   Let $Q = \langle \mathbf{L}(S) \cap C \rangle_B$. If $x \in \mathbf{L}(S) \cap B$, then $y = \frac{1}{n} \cdot x \in \mathbf{L}(S) \cap C$ for some n. Then $y \in Q$, so $x = n \cdot y \in Q$, since Q is a local semigroup with respect to B. Thus $\mathbf{L}(S) \cap B \subseteq Q$. Since S is the smallest local semigroup with respect to B containing $\mathbf{L}(S) \cap B$, we conclude $S \subseteq Q$. Since $Q \subseteq \langle T \rangle_B$, the equalities follow. The last statements now follow immediately.   ∎

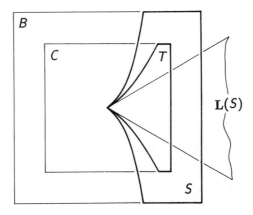

Figure 3

   This proposition tells us that a strict (respectively, closed) local wedge semigroup with respect to B can be reconstructed from the intersection of its tangent wedge with any smaller C-H-neighborhood. This justifies the alternate terminology of calling a local wedge semigroup "infinitesimally generated". (See Definition IV.2.4.)

   We have seen in Section 1 that we can assign to any germ of local semigroups a tangent Lie wedge. Example IV.2.1 shows this assignment to be far from being one-to-one. However, one would like to know whether there is a smallest germ of (closed) local wedge semigroups associated with a given Lie wedge. This question is related to the preceding propositions. Further it is natural to ask whether a strict local wedge semigroup with respect to B can be restricted to a smaller neighborhood in such a way that the restriction is equal to (and not simply contains) a strict local wedge semigroup with respect to the smaller neighborhood. Such questions turn out to be very delicate, and require a deeper knowledge of the structure of local semigroups. We shall return to them as we proceed in the chapter.

   We conclude this section with a very basic fact on the local approximation of a local semigroup by its tangent wedge. Indeed, if S is a local subsemigroup with respect to a suitable C-H-neighborhood B, then we now make precise, in which way, the set $\mathbf{L}(S) \cap C$ approximates the whole of $S \cap C$ for small enough neighborhoods C of 0 in B. To this end we first need a purely geometric idea.

IV.2.10.   Definition. Let W be a wedge in a finite dimensional real vector space V. We say that a wedge W' *surrounds* W if $W \subseteq W'$ and $W \setminus (W \cap -W)$ is contained in the interior of W'.

Except for the edge of the wedge $H(W)$, a surrounding wedge W' is a neighborhood of W. Notice that *a vector subspace or a half-space is surrounded by itself.*

Now we have the following local structure theorem for local semigroups.

IV.2.11.   Theorem. (The Surrounding Wedge Theorem) *Let S be a local semigroup with respect to a C-H-neighborhood B in a finite dimensional Lie algebra $\mathfrak{g}$. If a closed wedge W in $\mathfrak{g}$ surrounds $\mathbf{L}(S)$, then there is an open ball C around 0 in B such that $C \cap S \subseteq W$.*

Proof.   Without loss of generality we may assume that $S = \overline{S}^B$, since $\mathbf{L}(S) = \mathbf{L}(\overline{S}^B)$ and $C \cap S \subseteq C \cap \overline{S}^B$.

Assume that such a C does not exist for some wedge W which surrounds $\mathbf{L}(S)$. Then there is a sequence of elements $x_n \in S \setminus W$ with $0 = \lim x_n$. Pick a vector space complement V for $H(\mathbf{L}(S)) = \mathbf{L}(S) \cap -\mathbf{L}(S)$. We write $x_n = g_n + v_n$ where $g_n \in H(\mathbf{L}(S))$ and $v_n \in V$, $\lim g_n = 0 = \lim v_n$. Since $g_n \in H(\mathbf{L}(S)) \subseteq W$ and $x_n \notin W$, we conclude that $v_n \notin W$. Set $r_n = 1/\|v_n\|$. We may assume (since $\mathfrak{g}$ is finite dimensional) that $v = \lim r_n \cdot v_n$ exists for some suitable subsequence.

Since by Corollary IV.1.22, $\mathbf{L}(S) \cap B \subseteq \overline{S}^B = S$, we conclude that $-g_n, g_n \in S$ for large n. As a consequence of the Baker–Campbell–Hausdorff formula (see (A2) of Corollary A.1.4, Appendix 1), we have that

$$\|(-g_n) * (g_n + v_n) - v_n\| \leq \|[-g_n, g_n + v_n]\| = \|[g_n, v_n]\| \leq M\|g_n\|\|v_n\|$$

for some suitable constant M and large enough n. In view of the preceding and part (i) of the Replacement Lemma IV.1.20, $v = \lim r_n((-g_n) * x_n)$. We conclude that $v \in \mathbf{L}(S)$. Thus $v \in (\mathbf{L}(S) \cap V) \setminus \{0\} \subseteq \mathbf{L}(S) \setminus H(\mathbf{L}(S)) \subseteq \mathrm{int}(W)$ since W surrounds $\mathbf{L}(S)$. But since $v_n \notin W$ and thus $r_n v_n \notin W$, this leads to a contradiction.   ∎

As a special case of this result we retrieve the non-trivial fact: *A local group G in B is locally contained in* $\mathbf{L}(G)$. This is no longer true in the infinite dimensional case.

Notice that, as a consequence, the Surrounding Wedge Theorem tells us in particular that in a finite dimensional Lie algebra $\mathfrak{g}$, the identity 0 is necessarily isolated in any local semigroup S with $\mathbf{L}(S) = \{0\}$.

Let us observe that in a finite dimensional vector space every wedge is indeed surrounded by at least one wedge, and indeed by wedges approximating the given one:

IV.2.12. **Proposition.** *Any wedge W in a finite dimensional vector space L is the intersection of wedges W' surrounding W.*

Proof. Write $L = H(W) \oplus V$ with V a vector space complement for $H(W)$. Then Proposition I.2.12 reduces the claim to pointed cones. The details in this case are left as an exercise since a proof is given in a more general context in Propositions IV.6.8 and IV.6.9 below. ∎

The Surrounding Wedge Theorem is a most useful tool in the Lie theory of semigroups. We give a first application.

IV.2.13. **Corollary.** *Let S be a closed local semigroup with respect to a C-H-neighborhood B in a finite dimensional Lie algebra $\mathfrak{g}$. Then there is a C-H-neighborhood B' in B such that $H(S \cap B') = H(\mathbf{L}(S)) \cap B'$.*

Proof. Since S is closed in B, we have $\mathbf{L}(S) \subseteq B \cap S$ by Corollary IV.1.22. Then $\mathbf{L}(H \cap (S)) \cap B \subseteq S \cap -S = H(S)$. Thus for any C-H-neighborhood B' in B we have $H(\mathbf{L}(S)) \cap B' \subseteq H(S \cap B')$.

Now let us take a wedge W which surrounds $\mathbf{L}(S)$ and satisfies $H(W) = H(\mathbf{L}())$. Such wedges exist by Proposition IV.2.12. By the Surrounding Wedge Theorem IV.2.11, there is a C-H-neighborhood B' such that $B' \cap S \subseteq W$. Now

$$H(S \cap B') = (S \cap B') \cap -(S \cap B') \subseteq W \cap W = H(W) = H(\mathbf{L}(S)).$$

This proves the reverse containment for B'. ∎

3. Locally reachable sets

In the remainder of this chapter we will be concerned primarily with local semigroups which are generated by a family of one-parameter semigroups. Instead of one parameter semigroups we shall also speak of rays. It is the concept of local semigroups generated by rays that is central in the local Lie theory of semigroups. The discourse now attains a distinct control theoretical flavor.

In this section we again let **g** *denote a Dynkin algebra with a complete norm* $\| \cdot \|$ *satisfying* $\| [x, y] \| \leq \|x\| \, \|y\|$ *for all* $x, y \in$ **g** *and let* B *denote a C-H-neighborhood in* **g**.

IV.3.1.   **Definition.**   Let $\Omega \subseteq$ **g** and let B be a C-H-neighborhood. The *local ray semigroup with respect to* B *generated by* Ω is $\langle \mathbb{R}^+ \cdot \Omega \cap B \rangle_B$; this local semigroup is denoted $\mathrm{Sg}(\Omega, B)$.

From this and earlier definitions the following note is clear:

IV.3.2.   **Remark.**   If $\Omega = W$ is a wedge, then $\mathrm{Sg}(\Omega, B) = \langle W \rangle_B$, the local semigroup with respect to B generated by $W \cap B$.   ∎

IV.3.3.   **Proposition.**   *If* W *is the smallest Lie wedge containing* Ω *and* $S = \mathrm{Sg}(\Omega, B)$, *then* $W \subseteq \mathbf{L}(S)$ *and* $\overline{S}^B = \overline{\mathrm{Sg}(W, B)}^B = \overline{\mathrm{Sg}(\mathbf{L}(S), B)}^B$.

Proof.   For $t > 0$ and $x \in \Omega$, if $t \cdot x \in B$, then $t \cdot x \in S$. Hence $\Omega \subseteq \mathbf{L}(S)$. Since by Theorem IV.1.27 $\mathbf{L}(S)$ is a Lie wedge, it follows that $W \subseteq \mathbf{L}(S)$. Now $\Omega \subseteq W \subseteq \mathbf{L}(S)$ implies that $S = \mathrm{Sg}(\Omega, B) \subseteq \mathrm{Sg}(W, B) \subseteq \mathrm{Sg}(\mathbf{L}(S), B)$. Hence the same inclusion holds for the closures. The reverse inclusion for the closures of S and $\mathrm{Sg}(\mathbf{L}(S), B)$ follows from part (iii) of Proposition IV.2.3. Hence the three are equal.   ∎

Reachability and attainability

In order to study the fine structure of local ray semigroups, it is useful to have, for each member of the local semigroup, a measure of the complexity involved in generating it from the given generators; this complexity should be measured in terms of the sum total of the size of the generators needed. By analogy, one might think of the important role played by the length of words in studying free semigroups and groups. One such notion is closely related to the control theoretic notion of attainable sets, as we shall see later in Section 5. Actually a slightly modified notion, which we refer to as "reachability", is better suited to our purposes and will play the more important role in what follows. It may be viewed as a "normalized" version of attainability (see Proposition IV.3.8). The definition given now is presented from a semigroup theoretic viewpoint; an equivalent approach more control theoretic in nature appears in Section 5. Since we consider almost exclusively sets accessible from the origin, we delete the usual reference to the initial point.

IV.3.4. **Definition.** Let $\Omega \subseteq \mathbf{g}$, let $\delta > 0$, and let B be a C-H-neighborhood.
(1) We define $\mathrm{Reach}_\Omega(\delta, B) \subseteq B$ as follows:

$x \in \mathrm{Reach}_\Omega(\delta, B)$ if there exist $u_1, \ldots, u_n \in \Omega$ and $t_1, \ldots, t_n > 0$ satisfying
(i) $t_j \cdot u_j \in B$, for each $j = 1, \ldots, n$,
(ii) $t_1 \cdot u_1 * \cdots * t_{j-1} \cdot u_{j-1} * t \cdot u_j \in B$, for $0 \leq t \leq t_j$, $j = 1, \ldots, n$
(iii) $t_1 \cdot u_1 * \cdots * t_n \cdot u_n = x$,
(iv) $\delta = \sum_{j=1}^n t_j \|u_j\| = \sum_{j=1}^n \|t_j \cdot u_j\|$.

$\mathrm{Reach}_\Omega(\delta, B)$ is called the set of points *reachable by* Ω *in* B *at cost* δ. We further set

$$\mathrm{Reach}_\Omega(0, B) = \{0\},$$
$$\mathrm{Reach}_\Omega(\leq \delta, B) = \bigcup_{0 \leq s \leq \delta} \mathrm{Reach}_\Omega(s, B) \text{ for } 0 \leq \delta \leq \infty,$$
$$\mathrm{Reach}_\Omega(< \delta, B) = \bigcup_{0 \leq s < \delta} \mathrm{Reach}_\Omega(s, B), \text{ for } 0 < \delta \leq \infty,$$
$$\mathrm{Reach}_\Omega(B) = \mathrm{Reach}_\Omega(< \infty, B),$$
$$\mathcal{R}_\Omega(\delta, B) = \overline{\mathrm{Reach}_\Omega(\delta, B)} \cap B,$$
$$\mathcal{R}_\Omega(\leq \delta, B) = \overline{\mathrm{Reach}_\Omega(< \delta, B)} \cap B, \text{ for } 0 < \delta < \infty,$$
$$\mathcal{R}_\Omega(B) = \bigcup_{0 \leq \delta} \mathcal{R}_\Omega(\leq \delta, B).$$

(2) We define $A_\Omega(T, B) \subseteq B$ for $T > 0$ as follows

$x \in A_\Omega(T, B)$ if there exist $u_1, \ldots, u_n \in \Omega$ and $t_1, \ldots, t_n > 0$ satisfying
(i) $t_j \cdot u_j \in B$, for each $j = 1, \ldots, n$,
(ii) $t_1 \cdot u_1 * \cdots * t_{j-1} \cdot u_{j-1} * t \cdot u_j \in B$, for $0 \leq t \leq t_j$, $j = 1, \ldots, n$
(iii) $t_1 \cdot u_1 * \cdots * t_n \cdot u_n = x$,
(iv) $T = \sum_{j=1}^n t_j$.

Then $A_\Omega(T, B)$ is called the set of points *attainable by* Ω *in* B *at time* T. We further set

$$A_\Omega(0, B) = \{0\},$$

$$A_\Omega(\leq T, B) = \bigcup_{0 \leq t \leq T} A_\Omega(t, B), \quad \text{for } 0 \leq T \leq \infty,$$

$$A_\Omega(< T, B) = \bigcup_{0 \leq t < T} A_\Omega(t, B), \quad \text{for } 0 < T \leq \infty,$$

$$A_\Omega(B) = A_\Omega(< \infty, B),$$

$$\mathcal{A}_\Omega(T, B) = \overline{A_\Omega(T, B)} \cap B,$$

$$\mathcal{A}_\Omega(\leq T, B) = \overline{A_\Omega(< T, B)} \cap B, \quad \text{for } 0 < T < \infty,$$

$$\mathcal{A}_\Omega(B) = \bigcup_{0 \leq T} \mathcal{A}_\Omega(\leq T, B).$$

Note that in the preceding definition the notion of attainability is independent of the norm on $\mathfrak{g}$, while the notion of reachability is intimately tied up with the particular choice one makes of a norm.

The next propositions develop some of the basic properties and interrelations of reachable and attainable sets, and their relation with local ray semigroups.

IV.3.5.   Remark.   The following conclusions hold:

$$\mathcal{R}_\Omega(\leq \delta, B) = \overline{\text{Reach}_\Omega(\leq \delta, B)} \cap B,$$

$$\mathcal{A}_\Omega(\leq T, B) = \overline{A_\Omega(\leq T, B)} \cap B.$$

Proof.   Let $x = t_1 \cdot u_1 * \cdots * t_n \cdot u_n \in \text{Reach}_\Omega(\delta, B)$. Then as t approaches t_n from below, $t_1 \cdot u_1 * \cdots * t \cdot u_n \in \text{Reach}_\Omega(< \delta, B)$ approaches x. Hence $\text{Reach}_\Omega(\leq \delta, B) \subseteq \overline{\text{Reach}_\Omega(< \delta, B)}$, and so their closures agree. A similar argument holds for $\mathcal{A}_\Omega(\leq T, B)$.   ∎

IV.3.6.   Proposition.   *Under the current hypotheses the following conclusions hold:*

(i) *For $t > 0$, $x \in \Omega$, and $t \cdot x \in B$, we have $t \cdot x \in A_\Omega(< T, B)$ if $t < T$ and $t \cdot x \in \text{Reach}_\Omega(< \delta, B)$ if $t\|x\| < \delta$.*

(ii) *$\Omega \subseteq \mathbf{L}(A_\Omega(< T, B))$ and $\Omega \subseteq \mathbf{L}(\text{Reach}_\Omega(< \delta, B))$.*

Proof.   (i) This follows immediately by taking products with one factor.

(ii) For $x \in \Omega$ we have $x = \lim n \cdot ((1/n) \cdot x)$. The conclusion now follows from part (i).   ∎

In the global setting of a Lie group it is well known and elementary that the semigroup generated by $\{\exp(t \cdot x) : 0 \leq t, x \in \Omega\}$ and the points attainable from the identity by means of Ω agree. However, the reader is to be warned that the two notions are subtly distinct in the local setting, and there are many difficult problems involved in determining their precise relationship. However, inclusion does always hold in one direction.

IV.3.7.   **Proposition.**   *Let* $\Omega \subseteq \mathbf{g}$. *Then* $A_\Omega(B) = \mathrm{Reach}_\Omega(B) \subseteq \mathrm{Sg}(\Omega, B)$ *and* $\mathcal{A}_\Omega(B) = \mathcal{R}_\Omega(B) \subseteq \overline{\mathrm{Sg}(\Omega, B)}^B$.

Proof.   For an element to get into either $A_\Omega(B)$ or $\mathrm{Reach}_\Omega(B)$, only conditions (i)–(iii) of Definition IV.3.4 need to be met. Hence the two sets are equal. Let $x \in A_\Omega(B)$. Then $x = t_1 \cdot u_1 * \cdots * t_n \cdot u_n$, where $u_j \in \Omega$ are as in Definition IV.3.4. Then each $t_j \cdot u_j \in B \cap \mathbb{R}^+ \cdot \Omega$, hence is in $\mathrm{Sg}(\Omega, B)$. By induction on j and the local semigroup property, $t_1 \cdot u_1 * \cdots * t_j \cdot u_j \in \mathrm{Sg}(\Omega, B)$, and thus $x \in \mathrm{Sg}(\Omega, B)$. The last statement follows by passing to closures in B.   ∎

IV.3.8.   **Proposition.**   *Let* $\Omega \subseteq \mathbf{g}$, *and let* Θ *denote all unit vectors in* $\mathbb{R}^+ \cdot \Omega$. *Then* $\mathrm{Reach}_\Omega(\delta, B) = \mathrm{Reach}_\Theta(\delta, B) = A_\Theta(\delta, B)$ *and* $\mathcal{R}_\Omega(\delta, B) = \mathcal{R}_\Theta(\delta, B) = \mathcal{A}_\Theta(\delta, B)$.

Proof.   Note that the norm of u_j is equal to the norm of $\|u_j\| \cdot (u_j/\|u_j\|)$ and that

$$t_1 \cdot u_1 * \cdots * t_n \cdot u_n = t_1 \|u_1\| \cdot \frac{u_1}{\|u_1\|} * \ldots * t_n \|u_n\| \cdot \frac{u_n}{\|u_n\|}.$$

The first equality then follows from the definitions. (Note that we can assume without loss of generality that $u_j \neq 0$ for every j since 0 contributes nothing to the Baker–Campbell–Hausdorff product nor to the norm sum.) The second equality is immediate since the definitions for attainability and reachability at δ agree for unit vectors. The last string of equalities follows trivially from the first.   ∎

The next proposition gives further comparisons between the notions of attainable and reachable sets.

IV.3.9.   **Proposition.**   *Under the present hypotheses the following statements hold:*

(i) $A_\Omega(< T, B) \subseteq \mathrm{Reach}_\Omega(< NT, B)$ *if* $\|u\| \leq N$ *for all* $u \in \Omega$.

(ii) *If there exists* $\varepsilon > 0$ *such that* $u \in \Omega$ *implies* $u = r \cdot y$ *for some* $r > 0$ *and some* $y \in \Omega$ *with* $\varepsilon < \|y\|$, *then* $\mathrm{Reach}_\Omega(< \varepsilon\delta, B) \subseteq A_\Omega(< \delta, B)$.

(iii) *If* $\| \cdot \|_1 \leq K \| \cdot \|_2$, *then* $\mathrm{Reach}_\Omega(< \delta, B, \| \cdot \|_1) \subseteq \mathrm{Reach}_\Omega(< K\delta, B, \| \cdot \|_2)$.

Analogous statements hold for $\mathcal{A}_\Omega(\cdot, B)$ *and* $\mathcal{R}_\Omega(\cdot, B)$.

Proof.   (i) Suppose $x = t_1 \cdot u_1 * \cdots * t_n \cdot u_n$ where $u_j \in \Omega$. If $T > \sum_{j=1}^n t_j$, then $\sum_{j=1}^n t_j \|u_j\| \leq \sum_{j=1}^n N t_j < NT$. Hence the first inclusion follows.

(ii) Suppose that $x = t_1 \cdot u_1 * \cdots * t_n \cdot u_n$ with $\sum_{j=1}^n t_j \|u_j\| < \varepsilon\delta$. If $u_j = r_j \cdot y_j$ with $\varepsilon < \|y_j\|$, then $x = t_1 r_1 \cdot y_1 * \cdots * t_n r_n \cdot y_n$ and

$$\sum_{j=1}^n t_j r_j \leq \frac{1}{\varepsilon} \sum_{j=1}^n t_j r_j \|y_j\| = \frac{1}{\varepsilon} \sum_{j=1}^n t_j \|u_j\| < \delta.$$

(iii) Immediate.

The analogous statements follow by taking the appropriate closures.   ∎

Campbell–Hausdorff multiplication versus addition

We next derive a fundamental inequality that establishes how closely addition approximates the Baker–Campbell–Hausdorff multiplication near 0. In the remainder of the chapter an important role will be played by those C-H-neighborhoods B which satisfy

$$(\dagger) \qquad\qquad \|x * y - x - y\| \leq \|x\|\,\|y\| \quad \text{for all} \quad x, y \in B.$$

By (A2) of Appendix 1 this happens on sufficiently small C-H-neighborhoods B for a norm satisfying $\|[x,y]\| \leq \|x\|\|y\|$ (which we are assuming).

IV.3.10.   Proposition.   *Let* $0 < \delta < 1$ *and let* B *be a C-H-neighborhood containing the ball of radius* $\delta + \delta^2$ *around* 0 *and satisfying* $(\dagger)$. *Suppose that* $x_1, \ldots, x_n$ *are elements of* B *such that* $\sum_{j=1}^{n} \|x_j\| \leq \delta$. *Then* $x_1 * \cdots * x_j$ *is defined and in* B *for each* $j = 1, \ldots, n$, *and*

$$(1) \qquad \left\| x_1 * \cdots * x_n - \sum_{j=1}^{n} x_j \right\| < \left(\sum_{j=1}^{n} \|x_j\| \right)^2 \leq \delta^2, \ and$$

$$(2) \qquad\qquad \|x_1 * \cdots * x_n\| < \delta + \delta^2.$$

Proof.   The proof of the proposition proceeds by induction on n. The case $n = 1$ is clear. We assume the proposition is true for $n = j$ and establish it for $n = j+1$. Let $y_j = x_1 * \cdots * x_j$, $z_j = x_1 + \ldots + x_j$, and $s_j = \sum_{i=1}^{j} \|x_i\| \leq \delta$. The inductive hypothesis implies that y_j is defined and in B and that

$$\|y_j\| \leq \|z_j\| + \|y_j - z_j\| < s_j + s_j^2 = s_j(1 + s_j) \leq 2s_j.$$

(The preceding argument also shows that (2) is a consequence of (1).) From this we then derive

$$\left\| x_1 * \cdots * x_{j+1} - \sum_{i=1}^{j+1} x_i \right\| = \|y_j * x_{j+1} - z_j - x_{j+1}\|$$

$$\leq \|y_j * x_{j+1} - y_j - x_{j+1}\| + \|y_j - z_j\|$$

$$\leq \|y_j\|\,\|x_{j+1}\| + s_j^2 \ \text{(by hypothesis and induction)}$$

$$< 2s_j\|x_{j+1}\| + s_j^2 \ < \ (s_j + \|x_{j+1}\|)^2$$

$$= \left(\sum_{i=1}^{j+1} \|x_i\| \right)^2 \ \leq \ \delta^2.$$

∎

IV.3.11. **Corollary.** *Let $\Omega \subseteq \mathfrak{g}$ be arbitrary (respectively, let $\Omega \subseteq \mathfrak{g}$ be bounded), and let B be a C-H-neighborhood. Then there exists $\delta > 0$ (resp. $T > 0$) such that if $u_1, \ldots, u_n \in \Omega$ and $\sum_{i=1}^{n} t_i \|u_i\| \leq \delta$ (respectively, $\sum_{i=1}^{n} t_i \leq T$), then $t_1 u_1 * \cdots * t_n u_n$ is defined and in B.*

Proof. We consider first the case that Ω is arbitrary. By (A2) of Appendix 1 we can choose B so that $\|x * y - x - y\| \leq \|x\| \|y\|$ for $x, y \in B$. (If this fails for the given B, we replace it by a smaller one where it holds.) Pick $\delta \leq 1$ so that the ball of radius $\delta + \delta^2$ is contained in B. Let $x_i = t_i u_i$. Then $\|x_i\| = t_i \|u_i\|$, so the desired conclusion follows from Proposition IV.3.10.

The case for Ω bounded follows from the case just established. If Ω is bounded in norm by N and T is chosen less than δ/N, then from $\sum t_i \leq T$ it follows that $\sum t_i \|u_i\| \leq NT \leq \delta$. ∎

The definitions of the sets $\mathrm{Reach}_\Omega(< \delta, B)$ and $A_\Omega(< T, B)$ depended on an initial choice of the C-H-neighborhood B, and, offhand, might change if B is slightly varied, even for very small δ (respectively, T). Corollary IV.3.11 says that this is not the case. In fact let us temporarily say that (δ, B) for a $\delta > 0$ and a C-H-neighborhood is a *compatible pair* if the conditions $u_1, \ldots, u_n \in \Omega$ and $\sum_{i=1}^{n} t_i \|u_i\| \leq \delta$ imply that $t_1 \cdot u_1 * \cdots * t_n \cdot u_n$ is defined and is contained in B. Then for any compatible pair

$$\mathrm{Reach}_\Omega(< \delta, B) = \{t_1 \cdot u_1 * \cdots * t_n \cdot u_n : u_1, \ldots, u_n \in \Omega, \sum_{i=1}^{n} t_i \|u_i\| \leq \delta\}.$$

In particular, if (δ, B_1) and (δ, B_2) are compatible pairs, then $\mathrm{Reach}_\Omega(< \delta, B_1) = \mathrm{Reach}_\Omega(< \delta, B_2)$. Corollary IV.3.11 shows that *there is a $\delta_0 > 0$ so that for every $0 < \delta \leq \delta_0$, there is a C-H-neighborhood B such that (δ, B) is a compatible pair.* Hence the following definition is meaningful:

IV.3.12. **Definition.** The *full reachable set* $\mathrm{Reach}_\Omega(< \delta)$ (respectively *the full attainable set* $A_\Omega(< T)$) is defined to be $\mathrm{Reach}_\Omega(< \delta, B)$ (resp. $A_\Omega(< T, B)$) if there exists a C-H-neighborhood B such that if $u_1, \ldots, u_n \in \Omega$ and if $\sum_{i=1}^{n} t_i \|u_i\| < \delta$ (resp. $\sum_{i=1}^{n} t_i < T$), then $t_1 u_1 * \cdots * t_n u_n \in B$. We also define

$$\mathcal{R}_\Omega(\leq \delta) = \overline{\mathrm{Reach}_\Omega(< \delta)}, \quad \mathcal{A}_\Omega(\leq T) = \overline{A_\Omega(< T)},$$

where the closure is taken in $\mathfrak{g}$.

Corollary IV.3.11 above has the following consequence and thereby shows that the concepts we have just introduced are meaningful:

IV.3.13. **Corollary.** *The set $A_\Omega(< T)$ is defined for all T small enough, provided that Ω is bounded, and $\mathrm{Reach}_\Omega(< \delta)$ is defined for all δ small enough for arbitrary Ω. If $t_1 u_1 * \cdots * t_n u_n$ is in $\mathrm{Reach}_\Omega(< \delta)$ (respectively, $A_\Omega(< T)$), then so is $t_1 u_1 * \cdots * t_{i-1} u_{i-1} * t u_i$ for $0 \leq t \leq t_i$ and $1 \leq i \leq n$.* ∎

We are now ready to piece together the preceding results in the following theorem.

IV.3.14.   **Theorem.**   *Let $\Omega \subseteq \mathfrak{g}$ and let B be an arbitrary C-H-neighborhood. Then for all δ sufficiently small, we have*

$$\text{Reach}_\Omega(< \delta) \subseteq \text{Sg}(\Omega, B) \qquad \text{and} \qquad \mathcal{R}_\Omega(\le \delta) \subseteq \overline{\text{Sg}(\Omega, B)}^B.$$

More specifically, if $\|x * y - x - y\| \le \|x\|\,\|y\|$ for $x, y \in B$ and if the ball of radius $\delta + \delta^2$ is contained in B, then $\text{Reach}_\Omega(< \delta)$ is defined, and $\text{Reach}_\Omega(< \delta) \subseteq \text{Sg}(\Omega, B) \subseteq B$. In this case $x \in \text{Reach}_\Omega(\gamma)$ for some $\gamma < \delta$ implies that $\|x\| < \gamma + \gamma^2$.

Similarly, if Ω is bounded,

$$A_\Omega(< T) \subseteq \text{Sg}(\Omega, B) \text{ and } \mathcal{A}_\Omega(\le T) \subseteq \overline{\text{Sg}(\Omega, B)}^B$$

for all T sufficiently small.

Proof.   It follows immediately from Corollary IV.3.11 that $\text{Reach}_\Omega(< \delta) \subseteq B$ for all δ sufficiently small. For such δ we have $\text{Reach}_\Omega(< \delta) = \text{Reach}_\Omega(< \delta, B)$. Hence by Proposition IV.3.7 this set is contained $\text{Sg}(\Omega, B)$. Similar remarks apply for $A_\Omega(< T)$. Apply these results to a C-H-neighborhood B' whose closure is contained in B. Then the closures of the reachable and attainable sets which are contained in B' will be contained in the closure in B of $\text{Sg}(\Omega, B)$.

The remaining part of the theorem follows directly from Proposition IV.3.10 applied as in the proof of Corollary IV.3.11.   ∎

Local one-parameter semigroups of sets

We consider some further properties of the attainable and reachable sets.

IV.3.15.   **Definition.**   A *local one-parameter semigroup of sets* in a Dynkin algebra $\mathfrak{g}$ is a function $t \mapsto A(t)$ from $[0, T)$ for some $T > 0$ into the non-empty subsets of some C-H-neighborhood B of $\mathfrak{g}$ satisfying $A(t) * A(s) = A(t + s)$ for $t, s \in [0, T)$, $t + s < T$. The local one-parameter semigroup is *continuous* if given any neighborhood U of 0, there exists an $\varepsilon > 0$ such that $A(t) \subseteq U$ for $t < \varepsilon$.

IV.3.16.   **Lemma.**   *Let $t \mapsto A(t)$ be a continuous local one-parameter semigroup of sets in $\mathfrak{g}$, where $\mathfrak{g}$ is finite dimensional. Then $t \mapsto \overline{A(t)}$ is a continuous local one-parameter semigroup of compact sets.*

Proof.   Let $K(t) = \overline{A(t)}$. By continuity the sets $A(t)$ are contained in a compact neighborhood of 0 for small t. Hence their closures $K(t)$ are compact. Now let $s < T$ be arbitrary in the domain of A. By the one-parameter semigroup property $A(s) \subseteq K(s/n)^n$ for all n. For n large enough $K(s/n)$ is compact, and hence by the continuity of $*$, $K(s/n)^n$ is compact. Thus $K(s)$ is compact. By the continuity of $*$, for $s + t < T$ we have $K(s) * K(t) \subseteq K(s + t)$. But since $K(s)$ and $K(t)$ are compact, so is $K(s) * K(t)$. Thus $A(s + t) = A(s) * A(t) \subseteq K(s) * K(t)$ implies $K(s + t) \subseteq K(s) * K(t)$.   ∎

IV.3.17. **Proposition.** *The following conclusions hold:*

(i) *Let Ω be a bounded subset of $\mathfrak{g}$. Choose T so that $A_\Omega(t)$ is defined for all $t < T$. Then $t \mapsto A_\Omega(t)$ and $t \mapsto A_\Omega(< t)$ are continuous local one-parameter semigroups.*

(ii) *Let Ω be an arbitrary subset of $\mathfrak{g}$. Choose δ so that $\mathrm{Reach}_\Omega(t)$ is defined for all $t < \delta$. Then $t \mapsto \mathrm{Reach}_\Omega(t)$ and $t \mapsto \mathrm{Reach}_\Omega(< t)$ are continuous local one-parameter semigroups.*

(iii) *If $\mathfrak{g}$ is finite-dimensional, then $t \mapsto \mathcal{A}_\Omega(t)$, $t \mapsto \mathcal{A}_\Omega(\leq t)$, $t \mapsto \mathcal{R}_\Omega(t)$, and $t \mapsto \mathcal{R}_\Omega(\leq t)$ are continuous local one-parameter semigroups of compact sets.*

Proof.   (i) That $t \mapsto A_\Omega(t)$ is continuous follows from Theorem IV.3.14. Suppose $t + s = r < T$. It is straightforward from the definition that $A_\Omega(t) * A_\Omega(s) \subseteq A_\Omega(t + s)$. Conversely, let $x = t_1 \cdot x_1 * \cdots * t_n \cdot x_n$, where $\sum t_j = r$ and $x_j \in \Omega$ for each j. Pick the least k such that $\theta = \sum_{j=1}^{k} t_j \geq t$. Then $t_k \cdot x_k = (t - \sum_{j=1}^{k-1} t_j) \cdot x_k * (\theta - t) \cdot x_k$; so $x = y * z$ where $y = t_1 \cdot x_1 * \cdots * (t - \sum_{j=1}^{k-1} t_j) \cdot x_k$ and $z = (\theta - t) \cdot x_k * \cdots * t_n \cdot x_n$. One verifies directly that $y \in A_\Omega(t)$ and $z \in A_\Omega(s)$. Thus the reverse containment is established.

That the mapping $t \mapsto A_\Omega(< t)$ is a continuous local one-parameter semigroup of sets follows in a straighforward manner from the previous paragraph.

(ii) The case of the families of reachable sets follows from the one for attainable sets just established since $\mathrm{Reach}_\Omega(t) = A_\Theta(t)$ by Proposition IV.3.8, where Θ is Ω normalized.

(iii) If $\mathfrak{g}$ is finite-dimensional, then the last assertion follows from (i) and (ii) above in view of the previous Lemma IV.3.16. ∎

There is an interesting converse problem associated with Theorem IV.3.14. Given $\Omega \subseteq \mathfrak{g}$ and $\delta > 0$, is there a C-H-neighborhood B such that $\mathrm{Sg}(\Omega, B) \subseteq \mathrm{Reach}_\Omega(< \delta)$? In general this is not the case, and the problem of determining precisely when it is appears quite difficult. However, we are able to provide an affirmative solution for the important special case that there is a germ containing all sufficiently small reachable sets. This observation (in a somewhat varied form) will be crucial to our later construction of a local semigroup for a Lie wedge.

IV.3.18. **Lemma.** *Suppose that $\mathrm{Reach}_\Omega(< \delta)$ and $\mathrm{Reach}_\Omega(< 2\delta)$ are full reachable sets and B is a C-H-neighborhood such that $B \cap \mathrm{Reach}_\Omega(< \delta) = B \cap \mathrm{Reach}_\Omega(< 2\delta)$. Then $S = B \cap \mathrm{Reach}_\Omega(< \delta)$ is a local semigroup with respect to B. An analogous result holds for $\mathcal{R}_\Omega(\leq \delta)$ and $\mathcal{R}_\Omega(\leq 2\delta)$.*

Proof.   Let $x, y \in S$ such that $x * y \in B$. By Proposition IV.3.16(ii), $x * y \in \mathrm{Reach}_\Omega(< 2\delta)$. Hence $x * y \in B \cap \mathrm{Reach}_\Omega(< \delta) = S$.

Since $\mathcal{R}_\Omega(\leq \delta)$ is the closure of $\mathrm{Reach}_\Omega(< \delta)$ and the latter is a continuous local one-parameter semigroup of sets by Proposition IV.3.17, it follows from the continuity of $*$ that $\mathcal{R}_\Omega(\leq \delta) * \mathcal{R}_\Omega(\leq \delta) \subseteq \mathcal{R}_\Omega(\leq 2\delta)$. Hence the proof in the preceding paragraph for Reach applies to the case of $\mathcal{R}$ in a completely analogous fashion. ∎

All of this builds up towards the following result in which we use the concept of a germ as introduced in Definition IV.1.8:

IV.3.19.   Theorem. (The Confluence Theorem) *Let Ω be any subset in the Dynkin algebra $\mathfrak{g}$ and call W the smallest Lie wedge containing Ω. Consider an $\varepsilon > 0$ for which the full reachable set $\mathrm{Reach}_\Omega(< \varepsilon)$ is defined and denote with Γ its germ. Let Γ^* denote the germ of $\mathrm{Reach}_\Omega(\le \varepsilon)$. Then the following conclusions hold:*

(i) *If the full reachable sets $\mathrm{Reach}_\Omega(< \delta)$ all belong to Γ for $\delta \le \varepsilon$, then the local ray semigroups $\mathrm{Sg}(\Omega, B)$ also belong to Γ for all sufficiently small C-H-neighborhoods B.*

(ii) *If the sets $\mathcal{R}_\Omega(\le \delta)$ all belong to Γ^* for $\delta \le \varepsilon$, then the sets $\mathcal{R}_W(\le \delta)$, $\overline{\mathrm{Sg}(\Omega, B)}^B$ and $\overline{\mathrm{Sg}(W, B)}^B$ also belong to Γ^* for all $\delta \le \varepsilon$ and for all sufficiently small C-H-neighborhoods B.*

Proof.   (i) Choose any C-H-neighborhood such that $B \cap \mathrm{Reach}_\Omega(< \varepsilon/2) = B \cap \mathrm{Reach}_\Omega(< \varepsilon)$. We complete the proof by showing that $\mathrm{Sg}(\Omega, B)$ belongs to Γ for any such B. Let $S = B \cap \mathrm{Reach}_\Omega(< \varepsilon)$. By Lemma IV.3.18, S is a local semigroup with respect to B, and $t \cdot y \in \mathrm{Reach}_\Omega(< \varepsilon)$ for all $0 < t < \delta/\|y\|$ and for all $y \in \Omega$ by Proposition IV.3.6. Therefore, $\mathbb{R}^+ \cdot \Omega \cap B \subseteq S$. Thus $\mathrm{Sg}(\Omega, B) \subset S$, in view of the definition of $\mathrm{Sg}(\Omega, B)$.

By Theorem IV.3.14, there is a $0 < \delta_0 < \varepsilon$ such that the full reachable set $\mathrm{Reach}_\Omega(< \delta_0)$ is contained in $\mathrm{Sg}(\Omega, B)$. Since $\mathrm{Reach}_\Omega(< \delta_0)$ and $\mathrm{Reach}_\Omega(< \varepsilon)$ belong to the same germ Γ and $\mathrm{Sg}(\Omega, B)$ is trapped between them, it follows that it also belongs to Γ.

(ii) The proof that the sets $\mathcal{R}_\Omega(\le \delta_0)$ and $\overline{\mathrm{Sg}(\Omega, B)}^B$ lie in Γ^* for suitable δ_0 and B is analogous to that given in the preceding case. The latter is equal to $\overline{\mathrm{Sg}(W, B)}^B$ by Proposition IV.3.3. Since $\mathcal{R}_W(\le \delta)$ is trapped between $\mathcal{R}_\Omega(\le \delta)$ and $\overline{\mathrm{Sg}(W, B)}^B$ for any $\delta \in [0, \varepsilon]$ by Proposition IV.3.7, these sets also belong to Γ for all of these δ.   ■

It is clear that the sets $\mathrm{Reach}_\Omega(< \delta)$ and $\mathrm{Sg}(\Omega, B)$ each form a filter base as δ and B range through their respective domains. A particularly interesting question is whether the filters generated by these bases have a filter base that is a germ. The Confluence Theorem IV.3.19 now allows the following observation:

IV.3.20.   Remark. If the filter generated by the sets $\mathrm{Reach}_\Omega(< \delta)$ is a germ, then the local semigroups $\mathrm{Sg}(\Omega, B)$ generate the same filter and belong to the same germ for B small enough.   ■

If conditions (i) and (ii) of Proposition IV.3.9 are satisfied, then it follows that attainable sets for small times belong to the same germ as the reachable sets (provided either of the two belong to a single germ). Hence in the presence of (i) and (ii), the hypotheses of Theorem IV.3.19 will imply that the sets $A_\Omega(< t)$ will also belong to the given germ for small t. Conditions (i) and (ii) are satisfied if Ω is bounded and bounded away from 0 (e.g., finite and excluding 0).

IV.3.1.   Exercise. Let $\Omega \subset \mathfrak{g}$. Define the *optimal cost function* by

$$C_\Omega^*(x) = \inf\{\delta : x \in \mathrm{Reach}_\Omega(\delta)\}$$

(where the empty set has inf ∞). If Ω is understood, we frequently omit it as a subscript.

Show that

(i) $C^*(x * y) \leq C^*(x) + C^*(y)$, provided $C^*(x * y) \neq \infty$;

(ii) If (†) is satisfied on B, $x \in \text{Reach}_\Omega(\delta)$ where $\delta < 1$, and B contains the ball of radius $\delta + \delta^2$ around 0, then $\|x\| \leq C^*(x) + C^*(x)^2$.

(Proof. Part (i) follows readily from the fact that $\text{Reach}_\Omega(t)$ is a local one-parameter family of sets (Proposition IV.3.17), and part (ii) from Theorem IV.3.10.)

4. Lie's Theorem: Pointed cones — split wedges

We are beginning to approach the main goal of the latter portion of this chapter which is to prove a converse of Theorem IV.1.21 above, namely, Lie's Third Fundamental Theorem for local semigroups in the following form: *If W is a Lie wedge in a Dynkin algebra, then there is a C-H-neighborhood B and a local semigroup S with respect to B such that W = $\mathbf{L}(S)$.* The proof of this theorem is surprisingly involved. The discourse of the preceding section gave a first impression of some of the subtle issues involved in the local theory of semigroups. However, there are important special cases for which a proof of Lie's Fundamental theorem is rather straightforward. Nevertheless even in the simplest case we shall detect some of the features which make local semigroup theory so much different from local group theory. Therefore we discuss these cases independently.

Indeed in this section we first consider the special case that W is a pointed cone. Note that a pointed cone is trivially a Lie wedge. We show how to construct a local semigroup having a given pointed cone as its tangent object. This provides the desired converse of Theorem IV.1.21 for the case of a pointed cone.

Again in the following $\mathfrak{g}$ denotes a Dynkin algebra equipped with a complete norm satisfying $\| [x, y] \| \leq M \|x\| \|y\|$ for some constant M. We recall that the norm is called standard if $M = 1$ and that a scalar multiple of the given norm is standard (see Definition A.1.2 of Appendix 1). As usual, B denotes a C-H-neighborhood.

IV.4.1.   Proposition.   *Let S be a local semigroup with respect to B. Then the first of following statements implies the second, and if $\mathfrak{g}$ is finite dimensional, both are equivalent:*

(1) $\overline{S}^B \cap B'$ *has no units except* 0 *for some C-H-neighborhood B'.*

(2) $\mathbf{L}(S)$ *is a pointed cone.*

Proof.   (1)$\Rightarrow$(2). By Proposition IV.1.25 and Remark IV.1.17 we have

$$\mathbf{L}(S) = \mathbf{L}(\overline{S}^B) = \mathbf{L}(\overline{S}^B \cap B').$$

Proposition IV.1.26 yields

$$H\big(\mathbf{L}(\overline{S}^B \cap B')\big) = \mathbf{L}\big(H(\overline{S}^B \cap B')\big).$$

By Condition (1), however, $H(\overline{S}^B \cap B') = \{0\}$. Thus (2) follows.

$(2) \Rightarrow (1)$. Again by Proposition IV.1.26,

$$\mathbf{L}(H(\overline{S}^B)) = H(\mathbf{L}(\overline{S}^B)) = H(\mathbf{L}(S)) = \{0\}.$$

At this point we assume that $\mathfrak{g}$ is finite dimensional. By the Surrounding Wedge Theorem IV.2.11, as the vector space $\mathbf{L}\big(H(\overline{S}^B)\big)$ surrounds itself, there exists a C-H-neighborhood B' such that $H(\overline{S}^B) \cap B' = \{0\}$. Since $H(\overline{S}^B) \cap B' = H(\overline{S}^B \cap B')$, condition (1) follows. ∎

The proof of this proposition shows that we have to invoke the compactness arguments hidden in the Surrounding Wedge Theorem to conclude that the non-existence of tangent vectors at the origin implies the absence of units near 0. Therefore, in the infinite-dimensional case we need to restrict our attention to a special class of cones.

IV.4.2.   Lemma.   *Let E be a completely normable real vector space and let K be a pointed cone in E. The following statements are equivalent:*

  (1) *There is a continuous linear functional α on E which takes positive values on $K \setminus \{0\}$ and for which $\alpha^{-1}(1) \cap K$ is bounded.*

  (2) *There is a norm $\|\cdot\|$ on E which is compatible with the topology and which is additive on K.*

Furthermore, if these equivalent conditions hold, then the linear functional in part (1) and the norm in part (2) can be chosen so that they agree on K.

Proof. $(1) \Rightarrow (2)$. Let $H = \alpha^{-1}(0)$ and choose $k \in K$ with $\alpha(k) = 1$. We define $f: \mathbb{R} \times H \to E$ by $f(r, x) = r \cdot k + x$. Then f is an isomorphism of topological vector spaces, since its inverse $y \mapsto (\alpha(y), y - \alpha(y) \cdot k)$ is continuous. We set $Q = \alpha^{-1}(1) \cap K$. Then $K = \mathbb{R}^+ \cdot Q$, because $y \in K$ implies $y = 0$ or $\alpha(y) > 0$, hence $1/\alpha(y) \cdot y \in Q$. We set $P = Q - k$ and note that $P \subseteq H$. Then f maps $1 \times P$ isomorphically onto Q, since $f(1, q - k) = 1 \cdot k + (q - k)$, and hence f maps $\mathbb{R}^+ \cdot (\{1\} \times P)$ isomorphically onto K. We may therefore assume without loss of generality that $E = \mathbb{R} \times H$ with $K = \mathbb{R}^+ \cdot (\{1\} \times P)$ with P bounded in H and α the projection onto $\mathbb{R}$. Since E is completely normable, so is H, and since P is bounded, there is a norm $\|\cdot\|_H$ on H for which P is completely contained in the unit ball and which defines the topology on H. We now define $\|(r, x)\| = \max\{|r|, \|x\|_H\}$ on E. This norm certainly defines the topology on E, and if $(r, x) \in K$, then $(1, (1/r) \cdot x) \in \alpha^{-1}(1) \cap K = \{1\} \times P$. Thus $\|(1/r) \cdot x\|_H \le 1$, i.e., $\|x\|_H \le r$. It follows that $\|(r, x)\| = r = \alpha(r, x)$, so $\|\cdot\|$ is additive on K.

$(2) \Rightarrow (1)$. If $x \in K - K$, then $x = c - d$ for $c, d \in K$. If also $x = c' - d'$ with $c', d' \in K$, then $c + d' = c' + d$, and we have $\|c\| + \|d'\| = \|c + d'\| = \|c' + d\| = \|c'\| + \|d\|$ by (2). Hence $\|c\| - \|d\| = \|c'\| - \|d'\|$, and thus there is a well-defined function $\alpha: K - K \to \mathbb{R}$ defined by $\alpha(x) = \|c\| - \|d\|$, where $x = c - d$ and $c, d \in K$. Now α is readily seen to be linear and satisfy $\alpha(x) = \|x\| > 0$ for $x \in K \setminus \{0\}$. If $\alpha(x) > 1$, and $x = c - d$ with $c, d \in K$, then $\|c\| - \|d\| > 1$, and hence $\|x\| > \|c\| - \|d\| > 1$. Thus $\|x\| \le 1$ implies $\alpha(x) \le 1$ and similarly implies $-1 \le \alpha(x)$. By the Hahn–Banach Theorem α extends to a linear functional (again called α) on E satisfying $\alpha(x) \le 1$ whenever $\|x\| \le 1$ (and is hence continuous). Moreover, if $x \in \alpha^{-1} \cap K$, then $1 = \alpha(x) = \|x\|$. This shows both that $\alpha^{-1}(1) \cap K$ is bounded, and that the last statement of the lemma also holds. ∎

This lemma allows the following definition.

IV.4.3.   **Definition.** A pointed cone in a completely normable real vector space is called *strictly positive* if the equivalent conditions of Lemma IV.4.2 are satisfied. Any norm compatible with the topology and additive on K will be called *K-additive*.

IV.4.4.   **Proposition.** *A finite dimensional pointed cone in a completely normable real vector space is strictly positive.*

Proof.   We apply Proposition I.2.28 to the finite dimensional space $L = K - K$ spanned by the cone K and extend the functional $\omega \in \text{algint}\, K^*$ in the dual of K in L to a functional α on the whole space, using the Theorem of Hahn and Banach as in the second part of the proof of Lemma IV.4.2.   ∎

In the following lemmas we assume that we are working in the following situation: K is a strictly positive cone in a Dynkin algebra $\mathfrak{g}$, the set B is a C-H-neighborhood in $\mathfrak{g}$, and $\|\cdot\|$ is a norm on $\mathfrak{g}$ compatible with the topology which is both K-additive and satisfies

$$(\dagger) \qquad\qquad \|x * y - x - y\| \leq \|x\|\,\|y\| \quad \text{for all} \quad x, y \in B.$$

IV.4.5.   **Lemma.** *Suppose the radius of B does not exceed $\sqrt{2} - 1$. Define D_K by*

$$D_K = \{(x, z) \in \mathfrak{g} \times \mathfrak{g} \colon z \in K \;\text{and}\; \|x - z\| \leq \|z\|^2\}.$$

Then $(x_1, z_1), (x_2, z_2) \in D_K$ and $x_1, x_2, z_1, z_2 \in B$ imply $(x_1 * x_2, z_1 + z_2) \in D_K$.

Proof.   We have

$$\|x_1 * x_2 - (z_1 + z_2)\| \leq \|x_1 * x_2 - (x_1 + x_2)\| + \|x_1 - z_1\| + \|x_2 - z_2\|$$
$$\leq \|x_1\|\,\|x_2\| + \|z_1\|^2 + \|z_2\|^2$$

from $(\dagger)$ and the definition of D_K. Because $\|x_j - z_j\| \leq \|z_j\|^2$, we have $\|x_j\| \leq \|z_j\| + \|z_j\|^2 = \|z_j\|(1 + \|z_j\|)$ for $j = 1, 2$, and if $\|z_j\| \leq \sqrt{2} - 1$, then $\|x_1\|\,\|x_2\| \leq 2\|z_1\|\,\|z_2\|$. But then

$$\|x_1 * x_2 - (z_1 + z_2)\| \leq \|z_1\|^2 + 2\|z_1\|\,\|z_2\| + \|z_2\|^2 = (\|z_1\| + \|z_2\|)^2$$

which equals $\|z_1 + z_2\|^2$, since the norm is additive on K.   ∎

We now consider the set

$$S = \{x \in B \colon (\exists z)\, z \in K \cap 2B \;\text{and}\; \|x - z\| \leq \|z\|^2\}.$$

IV.4.6.   **Lemma.** *Suppose the radius of B is less than $1/2$. Then $\mathbf{L}(S) = K$.*

Proof.   Trivially $K \subseteq \mathbf{L}(S)$. Conversely, suppose that $x = \lim r_n x_n$ with $\|x_n - z_n\| \leq \|z_n\|^2$ for suitable elements $z_n \in K \cap 2B$ and with $\lim x_n = 0$. Then applying part (ii) of the Replacement Lemma IV.1.20 with $M = 1$ and r twice the radius of B, we conclude that $x = \lim r_n \cdot z_n \in K$.   ∎

IV.4.7.   Lemma.   *If the radius r of B is less than $1/16$, then $S * S \cap B \subseteq S$.*

Proof.   Suppose that $x, y, x * y \in B$ and $u, v \in 2B$ with $\|x - u\| \leq \|u\|^2$ and $\|y - v\| \leq \|v\|^2$. By Lemma IV.4.5,

$$\|x * y - (u + v)\| \leq \|u + v\|^2.$$

In addition,

$$\|u + v\| \leq \|x * y - (u + v)\| + \|x * y\| \leq$$
$$\|u + v\|^2 + r \leq (2r + 2r)^2 + r = (16r + 1)r < 2r.$$

Thus $u + v \in 2B$.   ∎

We now summarize the results of the preceding discussion in the following theorem. We recall, in particular, that *every pointed cone in a finite dimensional vector space is strictly positive.*

IV.4.8.   Theorem.   *Let $\mathfrak{g}$ be a Dynkin algebra and K a strictly positive cone in $\mathfrak{g}$. Then there exists a norm $\| \cdot \|$ on $\mathfrak{g}$ compatible with the topology which is additive on K and a C-H-neighborhood B satisfying*

(†) $\|x * y - (x + y)\| \leq \|x\| \, \|y\|$ *for* $x, y \in B.$

Furthermore, if B is contained in the open ball of radius $1/16$, then the set

$$S = \{x \in B : \|x - z\| \leq \|z\|^2 \text{ for some } z \in K \cap 2B\}$$

is a local semigroup with respect to B, and $\mathbf{L}(S) = K$.

Proof.   By Lemma IV.4.2 there is a norm $\|\cdot\|_1$ on $\mathfrak{g}$ compatible with the topology which is additive on K (see also Remark IV.4.3). By Formula (A2) of Appendix 1 there exists a C-H-neighborhood B and an $M > 0$ such that $\|x * y - (x + y)\|_1 \leq M\|x\|_1\|y\|_1$ for $x, y \in B$. If we set $\|\cdot\| = M\|\cdot\|_1$, then $\|\cdot\|$ is additive on K and satisfies the desired inequality on B. The remaining assertions now follow from the preceding lemmas.   ∎

Note that in Theorem IV.4.8 the properties satisfied by the norm on B remain true for any closed cone K' contained in K. Hence if K is replaced by K' in the definition of S, we obtain a set S' which is a local semigroup *with respect to B* and which has K' for its tangent cone. Thus local semigroups with respect to B can be found for K and all smaller cones. Thus we have

IV.4.9.   Remark.   *If $\mathfrak{g}$, K, and B are as in Theorem IV.4.8, then for every cone K' contained in K there is a local semigroup S' with respect to B satisfying* $\mathbf{L}(S') = K'$.   ∎

This situation is strongly contrasted with what happens when the cones under consideration have larger and larger opening. This is made precise in the following paragraph:

IV.4.10.   Construction.   Let $\mathfrak{g}$ be a finite-dimensional Lie algebra and let W be a half-space whose edge is not a subalgebra. Let C_n be any sequence of pointed cones in the interior of W such that C_{n+1} surrounds C_n and that $\bigcup_{n=1}^{\infty} C_n = \operatorname{int} W$. Such sequences exist. For each cone C_n, by Theorem IV.4.8, there exists a C-H-neighborhood B_n and a local semigroup S_n with respect to B_n such that $C_n = \mathbf{L}(S_n)$. But $\bigcap_{n=1}^{\infty} B_n$ is never a neighborhood of 0.

Proof.   Let $W = \varphi^{-1}\mathbb{R}^+$, where φ is a linear functional. We may obtain the cones C_n by taking in the affine hyperplane $\varphi^{-1}(1)$ an ascending sequence of compact convex subsets, each one containing the preceding one in its interior in the hyperplane, whose union is the hyperplane, and by letting C_n be the cone generated by the n-th compact convex set.

Assume, by way of contradiction, that a C-H-neighborhood $B \subseteq B_n$ for all n did exist. Then let T_n be the local semigroup with respect to B generated by $C_n \cap B$; by Proposition IV.2.3, we have $\mathbf{L}(T_n) = C_n$. Clearly the sequence T_n is ascending. It is immediate to verify that its union T is a local semigroup with respect to B. We claim that $T \subseteq W$. Assume, on the contrary, that there exists $y \in T_n \cap (\mathfrak{g} \setminus W)$ for some n. Then $y \in B$ and $\varphi(y) < 0$. Since $\varphi(-y) > 0$, there exists some $m > n$ with $-y$ in the interior of C_m. Pick an open set V with $-y \in V \subseteq B \cap \operatorname{int}(C_m)$ and $y * V \subseteq B$. Since $T_n \subseteq T_m$, it follows that $0 = y * (-y) \subseteq y * V \subseteq T_m * T_m \cap B \subseteq T_m$. However, this means that T_m contains a neighborhood of 0, and hence $\mathbf{L}(T_m) = \mathfrak{g}$, a contradiction. We conclude that $T \subseteq W$, as asserted. Hence $\mathbf{L}(T) \subseteq \mathbf{L}(W) = W$.

On the other hand, for each n, $C_n = \mathbf{L}(T_n) \subseteq \mathbf{L}(T)$. Since W is the closure of the union of the C_n, we conclude that $W \subseteq \mathbf{L}(T)$. Hence $W = \mathbf{L}(T)$. But this is impossible, since the edge of a tangent wedge must be a subalgebra by Corollary IV.1.29, and the edge of W is not a subalgebra.   ∎

The proof of Theorem IV.4.8 and the lemmas preceding it relied only on the inequality (†) and the additivity of the norm on the cone. Thus the Theorem would hold for any local binary operation $*$ admitting a norm compatible with the topology which satisfied these two conditions. Indeed, given a finite number of such operations, by rescaling a norm can be found which is additive on the cone K and satisfies (†) on some C-H-neighborhood B for all the operations simultaneously. Thus we have

IV.4.11.   Remark. The set S defined in Theorem IV.4.8 is locally closed with respect to B for *all* locally defined binary operations satisfying (†).   ∎

We consider now the infinitesimally generated local semigroups associated with a pointed cone and the relationships with the notions of reachability developed in the preceding section. In the following results the reader should bear in mind the important special case that the generating set Ω is itself a pointed cone K in a finite dimensional Lie algebra.

IV.4.12.   Theorem. (The Cone Theorem) *Let Ω be a subset of a Dynkin algebra $\mathfrak{g}$ such that the smallest cone K containing Ω is strictly positive. Then a norm $\| \cdot \|$ compatible with the topology, a continuous linear functional α, and a positive number $\delta \leq \frac{1}{16}$ can be chosen so that the following conditions are satisfied:*

(i) $x * y$ *is defined if* $\|x\|, \|y\| < (1 + \delta)\delta$,

(ii) $\|x * y - x - y\| \le \|x\| \|y\|$ *for all* $\|x\|, \|y\| < \delta$,

(iii) $\|x\| = \alpha(x)$ *for* $x \in K$ *and* $\|\alpha\| \le 1$, *and*

(iv) $\mathrm{Reach}_K(< \delta)$ *is defined.*

Let B *be an open ball of radius* $4\varepsilon \le \delta$, *let* $N = B \cap \alpha^{-1}(] - \varepsilon, \varepsilon[)$, *and let* $S = \mathrm{Sg}(\Omega, N)$. *Then the following conclusions hold:*

(a) $S = \mathrm{Reach}_\Omega(< \delta) \cap N = \mathrm{Reach}_\Omega(< 2\varepsilon) \cap N$,

(b) $\mathbf{L}(S) = \mathbf{L}(\langle N \cap K \rangle_N) = K$, *and*

(c) $\overline{S}^N = \overline{\langle K \cap N \rangle_N}^N = \mathcal{R}(\le \delta) \cap N$.

Before we give the proof of the theorem, it is convenient to derive some lemmas. In all the following lemmas, we assume the existence of the norm $\| \cdot \|$, the functional α, and the constant δ satisfying conditions (i)–(iv) of the theorem.

IV.4.13. **Lemma.** *Let* $\|x\|, \|y\| < \delta$, $y \in K$. *Then* $\alpha(x) + (1 - \delta)\alpha(y) \le \alpha(x * y)$.

Proof. We have

$$\alpha(x) + \alpha(y) - \alpha(x * y) = \alpha(x + y - x * y) \le \|x + y - x * y\| \le \|x\| \|y\| < \delta.\alpha(y).$$

Thus $\alpha(x) + (1 - \delta)\alpha(y) < \alpha(x * y)$. ∎

IV.4.14. **Lemma.** *Suppose* $x_1, \ldots, x_n \in K$, $\|x_1 + \cdots + x_n\| < \delta$. *Then*

(i) $\alpha(x_1 * \cdots * x_j) < \alpha(x_1 * \cdots * x_{j+1}) \le \alpha(x_1 * \cdots * x_n)$ *for* $j = 1, \ldots, n - 1$,

(ii) $(1 - \delta)\|x_1 + \cdots + x_n\| < \alpha(x_1 * \cdots * x_n)$, *and*

(iii) $\alpha(x_1 * \cdots * x_n) < \varepsilon$ *implies* $\|x_1\| + \cdots + \|x_n\| < 2\varepsilon$.

Proof. (i) Since $\mathrm{Reach}_K(< \delta)$ is defined, all partial products are defined. Apply Lemma IV.4.13 with $x = x_1 * \cdots * x_j$ and $y = x_{j+1}$.

(ii) Employing (1) of Proposition IV.3.10 and additivity, we obtain

$$\|x_1 * \cdots * x_n - \sum_{j=1}^{n} x_j\| < (\sum_{j=1}^{n} \|x_j\|)^2 = \|\sum_{j=1}^{n} x_j\|^2.$$

Let $c = \|x_1 + \cdots + x_n\| = \alpha(x_1 + \cdots + x_n)$. Then

$$\begin{aligned}
c &= \alpha(x_1 * \cdots * x_n) + \alpha(x_1 + \cdots + x_n - (x_1 * \cdots * x_n)) \\
&\le \alpha(x_1 * \cdots * x_n) + \|x_1 * \cdots * x_n - \sum x_j\| \\
&\le \alpha(x_1 * \cdots * x_n) + \|x_1 + \cdots + x_n\|^2 \\
&= \alpha(x_1 * \cdots * x_n) + c^2.
\end{aligned}$$

Thus $(1 - \delta)c < (1 - c)c = c - c^2 < \alpha(x_1 * \cdots * x_n)$.

(iii) Since $\delta < (1/2)$, by (ii) we have

$$\frac{1}{2}\|x_1 + \cdots + x_n\| \le (1 - \delta)\|x_1 + \cdots + x_n\| \le \alpha(x_1 * \cdots * x_n) < \varepsilon.$$

The desired conclusion follows from this inequality and the K-additivity of the norm. ∎

We are now ready for the proof of Theorem IV.4.12.

Proof.   (Theorem IV.4.12). Since K is strictly positive, there is a norm compatible with the topology which is additive on K. If this norm is rescaled to a standard norm $\|\cdot\|$ on $\mathfrak{g}$, then it remains additive on K, and by Corollary A.1.4 of Appendix 1, condition (ii) is satisfied on some ball of radius r. As in the proof of Lemma IV.4.2, it is possible to choose a linear functional α with $\|\alpha\| \leq 1$ such that $\alpha(x) = \|x\|$ for $x \in K$. Finally we choose δ, $0 < \delta \leq r$, so that conditions (i) and (iv) are also satisfied. We verified that condition (iv) can be satisfied in Corollary IV.3.13.

(a) To verify part (a) of the theorem, we first show that $\mathrm{Reach}_\Omega(< \delta) \cap N = \mathrm{Reach}_\Omega(< 2\varepsilon) \cap N$. Let $x = t_1 x_1 * \ldots * t_n x_n$, where $x_i \in \Omega$ and $\sum t_i \|x_i\| < \delta$ and $\alpha(x) < \varepsilon$. By Lemma IV.4.14.iii, $\sum t_i \|x_i\| < 2\varepsilon$. Hence $x \in \mathrm{Reach}_\Omega(< 2\varepsilon)$. The other containment is immediate. By Lemma IV.3.18, $S' = \mathrm{Reach}_\Omega(< \frac{1}{2}\delta) \cap N$ is a local semigroup with respect to N.

To complete the proof of (a), we show that $S' = S = \mathrm{Sg}(\Omega, N)$ by showing that S' is the smallest local semigroup with respect to N which contains $N \cap \mathbb{R}^+ \cdot \Omega$. Let T be any other such local semigroup, and let $x = r_1 x_1 * \ldots * r_n x_n \in S'$, where $x_1, \ldots, x_n \in \Omega$ and $r_1 \|x_1\| + \cdots + r_n \|x_n\| < (1/2)\delta$. Then for each $j \leq n$, part (i) of Lemma IV.4.14 yields that $\alpha(r_1 \cdot x_1 * \cdots * r_j \cdot x_j) \leq \alpha(r_1 \cdot x_1 * \cdots * r_n \cdot x_n) < \varepsilon$. Applying the left-right dual of part (i) of Lemma IV.4.14 to $r_1 \cdot x_1 * \cdots * r_j \cdot x_j$, we conclude $\alpha(r_j \cdot x_j) < \varepsilon$. Hence $r_1 \cdot x_1 * \cdots * r_j \cdot x_j \in N$ and $r_j \cdot x_j \in N$. It follows by a simple induction argument that $r_1 \cdot x_1 * \cdots * r_n \cdot x_n \in T$ since each $r_j \cdot x_j \in (N \cap \mathbb{R}^+ \cdot \Omega) \subseteq T$. Thus $x \in T$. We conclude that $S' \subseteq T$.

(b) We have $K \subseteq \mathbf{L}(\langle K \cap N \rangle_N)$. Since K is a strictly positive cone and N is contained in the ball B' of radius $\varepsilon \leq 1/16$, we conclude from Theorem IV.4.8 that K is the tangent cone of a local semigroup containing $\langle K \cap N \rangle_N$, so the reverse containment also holds. Hence $K = \mathbf{L}(\langle K \cap N \rangle_N)$. Since $\mathbf{L}(S) = \mathbf{L}(\mathrm{Sg}(\Omega, N))$ is a wedge containing Ω, by minimality $K \subseteq \mathbf{L}(S)$. Conversely, since $\mathrm{Sg}(\Omega, N) \subseteq \langle K \rangle_N$ we have $\mathbf{L}(S) \subseteq K$. Thus these two are also equal.

(c) The first equality follows from Proposition IV.3.3, and the second by taking the closure in N of both sides of the equality in part (a). This completes the proof of the Cone Theorem IV.4.12.   ∎

A first consequence which exhibits the power of the Cone Theorem reads as follows:

IV.4.15.   Corollary.   *Let $\Omega \subseteq K$, where K is a strictly positive cone in $\mathfrak{g}$. Then there exists a basis $\mathcal{B}$ of C-H-neighborhoods with the property that if $B, B' \in \mathcal{B}$, $B' \subseteq B$, then $\mathrm{Sg}(\Omega, B') = \mathrm{Sg}(\Omega, B) \cap B'$. In addition, the complement of $\mathrm{Sg}(\Omega, B')$ in $\mathrm{Sg}(\Omega, B)$ is a local ideal of $\mathrm{Sg}(\Omega, B)$ (see Definition IV.1.2). Similar statements hold for the relative closures of these semigroups in the corresponding neighborhoods.*

Proof.   Let N_ε denote the neighborhood in Theorem IV.4.12, and S_ε the corresponding strictly infinitesimally generated local semigroup with respect to N_ε. For $\gamma < \varepsilon < (1/4)\delta$, we have from IV.4.12

$$S_\gamma = \mathrm{Reach}_\Omega(< \frac{1}{2}\delta) \cap N_\gamma = \mathrm{Reach}_\Omega(< \frac{1}{2}\delta) \cap N_\gamma \cap N_\varepsilon = S_\varepsilon \cap N_\gamma.$$

Suppose that $x, y \in S_\varepsilon$ and $x * y \in S_\gamma$. Then $x, y \in \text{Reach}_\Omega(< \frac{1}{2}\delta)$, so $x = x_1 * \cdots * x_n$ and $y = y_1 * \cdots * y_m$ where $\sum_{j=1}^n \|x_j\|$ and $\sum_{j=1}^m \|y_j\|$ are each less than $\delta/2$. Then $x * y = x_1 * \cdots * x_n * y_1 * \cdots * y_m$ and $\alpha(x * y) < \gamma$ imply by Lemma IV.4.14.iii that

$$\sum_{j=1}^n \|x_j\| + \sum_{j=1}^m \|y_j\| < 2\gamma.$$

Hence $\sum \|x_j\|$ and $\sum \|y_j\|$ are each less than 2γ. Lemma IV.4.14.1 and its dual imply $\alpha(x), \alpha(y) < \gamma$. Thus $x, y \in N_\gamma \cap S_\varepsilon = S_\gamma$.

The corresponding result for the closures follows similarly from Theorem IV.4.12. ∎

IV.4.1.   Exercise.   Use the preceding results to show that the closure in B' of $S = \text{Sg}(\Omega, B')$ can be embedded in a compact topological monoid if the closure of B' is compact. (Hint: Take the one-point compactification (with additional point ∞) of $\overline{S}^B$, define the products to be the usual products if the answer is in $\overline{S}^B$, and ∞ otherwise.)

IV.4.16.   Corollary.   *Let Ω be a subset of a strictly positive cone K in $\mathfrak{g}$, and suppose that K is the smallest cone containing Ω. Then for all sufficiently small C-H-neighborhoods B and all sufficiently small $\delta > 0$, $\text{Sg}(\Omega, B)$ and $\text{Reach}_\Omega(< \delta)$ belong to the same germ. The same is true for $\overline{\text{Sg}(\Omega, B)}^B$, $\overline{\langle K \cap B \rangle}_B$, and $\mathcal{R}_\Omega(\leq \delta)$.*

Proof.   It is immediate from part (a) of Theorem IV.4.12 that the sets $\text{Reach}_\Omega(< \gamma)$ all belong to the same germ for $\gamma \leq \delta$. The corollary now follows from the Confluence Theorem IV.3.19 in light of part (iii) of IV.1.9 for the second case. ∎

In the case that Ω is contained in a strictly positive cone, the reachable sets drift away from 0 as δ increases. The following proposition gives a useful bound concerning the rate at which they move away. (Compare Theorem IV.3.14).

IV.4.17.   Proposition.   *Assume the hypotheses and conditions of Theorem IV.4.12. If $x \in \text{Reach}_\Omega(\gamma) \cap B$, $\gamma < \delta$, then $\gamma - \gamma^2 < \|x\|$. Hence $x \in \text{Reach}_\Omega(< \frac{4}{3}\|x\|)$.*

Proof.   By hypothesis $x = t_1 \cdot x_1 * \cdots * t_n \cdot x_n$, where $x_i \in \Omega$ and $\sum t_i \|x_i\| = \gamma$, $1 \leq i \leq n$. Let $y_i = t_i x_i$. Using additivity of the norm on K and (1) of IV.3.10, we obtain

$$\gamma = \sum \|y_i\| = \|\sum y_i\|$$
$$\leq \|\sum y_i - y_1 * \cdots * y_n\| + \|y_1 * \cdots * y_n\|$$
$$< \left(\sum \|y_i\|\right)^2 + \|x\| = \gamma^2 + \|x\|.$$

Thus $\gamma - \gamma^2 < \|x\|$.

Since $(1 - \frac{1}{4})\gamma \leq (1 - \gamma)\gamma < \|x\|$, we conclude that $\gamma < \frac{4}{3}\|x\|$. ∎

Lie's Fundamental Theorem for split Wedges

At this point we understand the status of Lie's Third Fundamental Theorem for local semigroups to the extent that we are dealing with strictly positive cones in a Dynkin algebra. Now we generalize the discussion of Lie's Fundamental Theorems by considering wedges W that can be decomposed into a direct sum of the edge H and a strictly positive cone which is invariant under the induced action of the edge. Such wedges are Lie wedges, and in this section we show that there exist local semigroups for which they are the tangent wedges. The general idea is quite simple: start with a local group G with tangent Lie algebra the edge H of the wedge and a local ray semigroup S generated by the positive cone. Since $s_1 * g_1 * s_2 * g_2 = s_1 * (g_1 * s_2 * (-g_1)) * (g_1 * g_2)$, and since the cone and hence the semigroup it generates is invariant, $S * G$ should be essentially a semidirect product of S and G, a local subsemigroup, and have tangent object W. However, as is so frequently the case in working locally, a number of technical difficulties must be overcome before some approximation to the preceding program can be carried out.

The results we obtain in the case of split wedges are only slightly more specific and detailed than those we obtain in the general case of Lie wedges, and the casual reader may well be content with the preceding sketch of the general idea and omit the remainder of this section. However, since split wedges are such an important class of Lie wedges, it seems appropriate to have on record some of their basic properties. In addition, they represent the limit of the results that one can get without the introduction of a range of new techniques.

In the sequel we will often refer to the following Definition.

IV.4.18.   Definition.   A wedge W in a completely normable space E is said to be *strongly positive* if the edge $H = H(W) = W \cap -W$ admits a closed complement V such that $W \cap V$ is a strictly positive cone.

The reader should distinguish clearly between a strictly positive cone and a strongly positive wedge.

IV.4.19.   Proposition.   *If W is a strongly positive wedge, then W admits a direct sum decomposition of the form $H + K$, where H is the edge of W and $K = W \cap V$ is a strictly positive cone. If E is finite dimensional, then every wedge is strongly positive.*

Proof.   The first statement is easily verified. Since any subspace of a finite dimensional space has a complement and the intersection of W with that complement must be a proper cone (and hence strictly positive by Remark IV.4.4), the second statement follows.   ∎

Let W be a strongly positive wedge with edge H in a Dynkin algebra $\mathfrak{g}$, let V be a closed complement of H, and let $K = V \cap W$. We choose on V as in Lemma IV.4.2 a norm $\| \cdot \|$ additive on K and compatible with the topology. Pick a complete norm compatible with the topology on H, and define $\|\cdot\|$ on $H + V$ by $\|g + x\| = \|g\| + \|x\|$. Since $(g, x) \mapsto g + x$ is continuous from $H + V$ to $\mathfrak{g}$, and since $\mathfrak{g}$ has a complete norm, the mapping is a homeomorphism (by the Open Mapping Theorem for Banach spaces). Hence the norm just defined may be viewed as a norm on $\mathfrak{g}$ compatible with the topology. Then some scalar multiple will be a standard norm with respect to the Lie algebra structure. Finally by Corollary A.1.4 of Appendix 1, there exists a CH-neighborhood B such that

$$(\dagger) \qquad \|x * y - x - y\| \leq \|x\|\,\|y\| \qquad \text{for all} \qquad x, y \in B.$$

We thus obtain the following

IV.4.20. Lemma. *Let $\mathfrak{g}$ be a Dynkin algebra with a strongly positive wedge W. Let V be a closed complement for the edge H of W such that $K = V \cap W$ is strictly positive. Then $\mathfrak{g}$ admits a norm $\|\cdot\|$ compatible with the topology which is additive on K, satisfies $\|g + x\| = \|g\| + \|x\|$ for $g \in H$ and $x \in V$, and satisfies $(\dagger)$ on some C-H-neighborhood B.* ∎

     In the remainder of this section we assume that we are working in the setting of Lemma IV.4.20 and that the norm has the properties given there. We assume further that B is a C-H-neighborhood of radius $r \leq 1/16$ on which $(\dagger)$ is satisfied.

IV.4.21. Lemma. *Let $S = \langle K \cap B \rangle_B$. If $0 < d \leq r$, $s \in S$, $g \in H \cap B$, and $\|s * g\| \leq d$, then $\|s\| + \|g\| < 2d$.*

Proof. By Theorem IV.4.8 there exists a $z \in K$ with $\|z\| < 2r \leq \frac{1}{8}$ and $\|s - z\| \leq \|z\|^2$. Then

$$(1) \qquad \|s\| \leq \|s - z\| + \|z\| \leq \|z\|^2 + \|z\| = (\|z\| + 1)\|z\| \leq \frac{9}{8}\|z\| < 2\|z\|.$$

Therefore

$$\|s * g - (z + g)\| \leq \|s * g - (s + g)\| + \|s + g - (z + g)\| \leq \|s\|\,\|g\| + \|s - z\| \text{ (by } (\dagger))$$
$$(2) \qquad\qquad < 2\|z\|\,\|g\| + \|z\|^2 \text{ (by (1))} = (2\|g\| + \|z\|)\|z\|$$
$$\qquad\qquad \leq 2(\|g\| + \|z\|)\|z\| = 2\|g + z\|\,\|z\|,$$

the last equality holding by virtue of Lemma IV.4.20. Continuing the inequality, we have $2\|g + z\|\,\|z\| < 4r\|g + z\|$ (since $\|z\| < 2r$) $\leq \frac{1}{4}\|g + z\|$ in view of $r \leq \frac{1}{16}$. From this last inequality we conclude that $\|g + z\| \leq \|s * g\| + \|z + g - s * g\| < d + \frac{1}{4}\|g + z\|$. As a consequence we have

$$(3) \qquad\qquad\qquad \|g + z\| < \frac{4}{3}d.$$

By (1), (3), and Lemma IV.4.20, we have

$$\|s\| + \|g\| < \frac{9}{8}\|z\| + \|g\| \leq \frac{9}{8}\|z + g\| < \frac{9}{8} \cdot \frac{4}{3}d = \frac{3}{2}d < 2d.$$

This concludes the proof. ∎

IV.4.22.   Lemma. *Let W be a wedge in $\mathfrak{g}$, and let V be a closed complement of the edge H of W. The following statements are equivalent:*

(1) $e^{\mathrm{ad}\,x}V = V$ *for all* $x \in H$.

(2) $[x, V] \subseteq V$ *for all* $x \in H$.

Proof.   This is an immediate consequence of Lemma II.1.7.   ∎

Now to the crucial definition!

IV.4.23.   Definition.   A wedge W in a Dynkin algebra $\mathfrak{g}$ is called a *split wedge* if W is a Lie wedge and if there exists a closed complement V to the edge H of W such that $W \cap V$ is strictly positive and V satisfies the equivalent conditions of Lemma IV.4.22.

IV.4.24.   Proposition.   *Let W be a split wedge, and let V be a closed invariant complement for H as in* Definition IV.4.23. *Set $K = W \cap V$. Then the following conclusions hold:*

(i) $e^{\mathrm{ad}\,x}K = K$ *for all* $x \in H$.

(ii) *If B is a C-H-neighborhood, then $g * (K \cap B) * (-g) \subseteq K$ for all $g \in H \cap B$.*

Proof.   (i) Since W is a Lie wedge, $e^{\mathrm{ad}\,x}W = W$ for all $x \in H$ by Definition II.1.3. Since $e^{\mathrm{ad}\,x}$ is an automorphism, it also preserves the intersection $K = W \cap V$.

(ii) This follows immediately from (i) since $y \mapsto g * y * (-g)$ equals $y \mapsto \exp(\mathrm{ad}\,g)(y)$ on B (see (A6) of Appendix 1).   ∎

IV.4.25.   Lemma.   *Let K be a wedge in $\mathfrak{g}$ and let Ψ be a a continuous Lie algebra automorphism carrying K into itself. If B and B' are C-H-neighborhoods such that $\Psi(B') \subseteq B$, $S = \langle K \cap B \rangle_B$, and $S' = \langle K \cap B' \rangle_{B'}$, then $\Psi(S') \subseteq S \cap \Psi(B')$.*

Proof.   In view of the definition of $*$ via the Baker–Campbell–Hausdorff formula, Ψ is a $*$-isomorphism from B' to $\Psi(B')$. Thus $\Psi(S')$ is the smallest local semigroup with respect to $\Psi(B')$ containing $\Psi(K) \cap \Psi(B')$. Since $\Psi(K) \cap \Psi(B') \subseteq K \cap B$, we conclude $\psi(S') \subseteq S \cap \Psi(B')$.   ∎

The next theorem gives the main result of this section, namely, that a split Lie wedge is the tangent wedge for a local semigroup.

IV.4.26.   Theorem.   (The Split Wedge Theorem) *Let $\mathfrak{g}$ be a Dynkin algebra and W a split wedge with edge $H = H(W)$, and let V be an invariant complement for H such that $K = W \cap V$ is strictly positive. For any C-H-neighborhood B we denote $S(B) \overset{\text{def}}{=} \langle K \cap B \rangle_B$ the local semigroup with respect to B which is strictly infinitesimally generated by $K \cap B$. We further set*

$$T(B) \overset{\text{def}}{=} \{x \in B \colon (\exists s, g)s \in S(B), g \in B \cap H \text{ and } x = s * g\}.$$

Then there is a C-H-neighborhood B such that the following conclusions hold:

(i) *There is a 0-neighborhood U_1 such that for all C-H-neighborhoods $B' \subseteq U_1$, the set $T(B) \cap B'$ is a local semigroup with respect to B'.*

(ii) $\mathbf{L}(T(B) \cap B') = W$.

(iii) *Let* $\Omega = H \cup K$. *Given a full reachable set* $\mathrm{Reach}_\Omega(< \delta) \subseteq B$, *then there is a 0-neighborhood* $U_2 = U_2(\delta)$ *such that* $B' \subseteq U_2$ *implies* $T(B) \cap B' \subseteq B' \cap \mathrm{Reach}_\Omega(< \delta)$.

(iv) *If the full reachable set* $\mathrm{Reach}_\Omega(< \gamma)$ *is contained in a C-H-neighborhood* B' *contained in* U_2, *then it belongs to the germ of* $T(B)$.

Proof. (i) First choose a norm compatible with the topology as in Lemma IV.4.20. Then choose a C-H-neighborhood B such that B is contained in the ball of radius $\frac{1}{16}$ around 0, (†) is satisfied on B, and B is in the family $\mathcal{B}$ of Corollary IV.4.15 for $\Omega = K$. Pick a C-H-neighborhood $B_1 \subseteq B$ also in $\mathcal{B}$ such that $B_1 * B_1 * B_1 * B_1 \subseteq B$. Abbreviate $S = S(B)$. Then we have that $S_1 = B_1 \cap S$, where $S_1 = \langle K \cap B_1 \rangle_{B_1}$. Now let U_1 be the ball of some radius d around 0 such that $2U_1 \subseteq B_1$.

Now we consider any C-H-neighborhood $B' \subseteq U_1$ and abbreviate $T = T(B) \cap B'$. Let $x_i = s_i * g_i \in T$, where $s_i \in S$ and $g_i \in H \cap B$ for $i = 1, 2$. By Lemma IV.4.21, we have $\|s_i\| + \|g_i\| < 2d$, and thus $s_i \in S \cap B_1 = S_1$ and $g_i \in B_1$ for $i = 1, 2$. Consider the mapping $y \mapsto g_1 * y * (-g_1) = e^{\mathrm{ad}\, g_1} y$ from B_1 into B. In light of Proposition IV.4.24 and Lemma IV.4.25, we conclude that $g_1 * S_1 * (-g_1) \subseteq S$. Hence $s_1 * g_1 * s_2 * (-g_1) \in S * S \cap B \subseteq S$. In addition, $g_1 * g_2 \in B_1 * B_1 \cap H \subseteq B \cap H$. Multiplying these two products together, we conclude that $x_1 * x_2 \in T$ if $x_1 * x_2 \in B'$. Thus T is a local semigroup with respect to B'.

(ii) Let $x \in \mathbf{L}(T)$. Then $x = \lim n(s_n * g_n)$, where $s_n * g_n \in T$ and $s_n * g_n \to 0$. By Lemma IV.4.21, $s_n \to 0$ and $g_n \to 0$. By Theorem IV.4.8 there exists $z_n \in K$ such that $\|z_n\| \leq \frac{1}{8}$ and $\|s_n - z_n\| \leq \|z_n\|^2$. The inequality (2) in the proof of Lemma IV.4.21 applies here to yield

$$\|s_n * g_n - (z_n + g_n)\| \leq 2\|g_n + z_n\| \|z_n\| \leq 2\|g_n + z_n\|^2.$$

By the Replacement Lemma IV.1.20(ii), we conclude that $x = \lim n(z_n + g_n)$. Thus $x \in W$. We infer that $\mathbf{L}(T) \subseteq W$. Since $S \cap B' \subseteq T$, by Theorem IV.4.8 we have $K = \mathbf{L}(S) \subseteq \mathbf{L}(T)$. Also $H \cap B' \subseteq T$ implies that $H = \mathbf{L}(H) \subseteq \mathbf{L}(T)$. Since $W = H + K$, $W \subseteq \mathbf{L}(T)$. Thus $W = \mathbf{L}(T)$.

(iii) We pick any positive $\varepsilon < \frac{1}{5}\delta$ and let U_2 be the ball around 0 with radius $\min\{d, \varepsilon\}$. Now let B' be in U_2 and abbreviate once more $T = T(B) \cap B'$. Suppose $x = s * g \in T$, where $s \in S$ and $g \in B \cap H$. By Lemma IV.4.21 we have $\|s\|, \|g\| < 2\varepsilon$. Hence $g \in \mathrm{Reach}_\Omega(< 2\varepsilon)$, since $g \in W$. By Theorem IV.4.12 and the choice of B, we know that $s \in \mathrm{Reach}_K(\gamma)$ for some γ. It follows from Proposition IV.4.17 that $\gamma < \frac{3}{2}\|s\| \leq 3\varepsilon$. Thus $s \in \mathrm{Reach}_K(< 3\varepsilon) \subseteq \mathrm{Reach}_\Omega(< 3\varepsilon)$. By Proposition III.3.17 we conclude that $x = s * g \in \mathrm{Reach}_\Omega(< 5\varepsilon) \subseteq \mathrm{Reach}_\Omega(< \delta)$.

(iv) If $\mathrm{Reach}_\Omega(< \gamma) \subseteq B'$, then by Proposition III.3.7, $\mathrm{Reach}_\Omega(< \gamma) \subseteq \mathrm{Sg}(\Omega, B')$. Since $\Omega \cap B' \subseteq T(B) \cap B'$ and the latter is a local semigroup with respect to B', it follows that $\mathrm{Sg}(\Omega, B') \subseteq T(B) \cap B'$. Applying part (iii) to $\mathrm{Reach}_\Omega(< \gamma)$, we conclude that $T(B) \cap B'' \subseteq \mathrm{Reach}_\Omega(< \gamma)$ for B'' small enough. Then $T(B) \cap B'' \subseteq \mathrm{Reach}_\Omega(< \gamma) \subseteq T(B) \cap B'$ in view of what we saw above. Thus $\mathrm{Reach}_\Omega(< \gamma)$ belongs to the same germ of $T(B)$. ∎

IV.4.27.   **Corollary.**   *If W is a split wedge in $\mathfrak{g}$, then the sets $\mathcal{R}_W(\leq \delta)$ and $\overline{\mathrm{Sg}(W, B)}^B$ belong to the same germ for all δ and B sufficiently small.*

Proof.   Define Ω as in Theorem IV.4.26. Since by part (iv), the sets $\mathrm{Reach}_\Omega(< \delta)$ belong to the germ of $T(B)$ for δ small enough, so do their closures by Proposition IV.1.9. Then apply the Confluence Theorem IV.3.19, noting that W is the smallest Lie wedge containing Ω.   ∎

IV.4.28.   **Remark.**   The local semigroups $T(B) \cap B' = S * (H \cap B) \cap B'$ constructed in Theorem IV.4.26 contain $(H \cup K) \cap B'$, but not necessarily $W \cap B$. Hence even though $\mathbf{L}(T(B) \cap B') = W$, the local semigroups $T(B) \cap B'$ need not be full. By contrast, the closures sets in Corollary IV.4.27 are always full by Proposition IV.1.24.   ∎

        We close this section by listing some important cases in which a Lie wedge splits.

IV.4.29.   **Theorem.**   *Let $\mathfrak{g}$ be a finite dimensional Lie algebra and let W be a Lie wedge in $\mathfrak{g}$. Then W splits in each of the following cases:*

   (i) *The edge $H = W \cap -W$ is semisimple modulo the largest ideal of $\mathfrak{g}$ contained in it. In particular, this is the case if the edge is semisimple.*

   (ii) *H is a compactly embedded subalgebra of $\mathfrak{g}$.*

Proof.   These cases both follow from the well-known fact that a finite dimension vector space module over a semisimple or compact Lie algebra is semisimple in the sense that any invariant subspace has an invariant complement.   ∎

5. Geometric control in a local Lie group

In Section 3 we worked with a definition of attainable and reachable sets which was rather unorthodox from a control theory but which was quite natural from a semigroup orientation. In this section we show how these notions admit an alternate, more conventional, characterization as terminal points for solutions to a certain ordinary differential equation with a varying set of steering functions. To this end we first treat explicitly the differential equation satisfied by a curve $x(\cdot): [0, T] \to \mathfrak{g}$ with $x(0) = 0$ which is defined by the property that, up to terms of order two or more, the point $x(t+h)$ is obtained as the Baker–Campbell–Hausdorff product of $x(t)$ with a vector $hu(t)$, where $u(t)$ ranges through Ω in a fashion that allows for sufficiently many discontinuities to accomodate sudden changes in direction. Such sudden changes occur in the context of the algebraic generation of a local semigroup and in applications in geometric control. The differential equation is the one we refer to as the fundamental differential equation.

The fundamental differential equation

We fix some subset Ω (called the *set of controls*) of a Dynkin algebra $\mathfrak{g}$. We consider various classes of functions, called *steering functions* or *control functions*, from some interval I contained in the real numbers into Ω. Suppose that I has endpoints a and b. A function $f: I \to \Omega$ is a *piecewise constant* function if there is an increasing finite sequence $\{x_i : 0 \le i \le n\}$ with $x_0 = a$ and $x_n = b$ such that f is constant on each open interval $]x_{i-1}, x_i[$ for $1 \le i \le n$.

A reasonably general, yet well-behaved, class of control functions with which to operate is the class of *regulated functions* or *fonctions réglées* in the sense of [Bou61], Chap. I. Regulated functions are defined to be those functions having limits from the right and from the left in all points of their domain of definition (wherever such limits make sense). If the domain of definition is a compact interval, then they are characterized as those functions which are the uniform limits of piecewise constant functions. A regulated function exhibits many properties akin to continuity; for instance, it is bounded and is, in fact, continuous on the complement of a countable subset of its domain.

From time to time it is useful to consider subclasses and superclasses of the class of regulated functions. Particular examples are the subclass of piecewise analytic functions, the subclass of piecewise continuous functions, and the superclass of bounded measurable functions.

Throughout this section the functions g and f discussed in Definition II.2.1 and Lemma II.2.2 will play a crucial role. We have used them already extensively in Chapter II, Section 2 and earlier in this chapter in Propositions IV.1.35 and 3.6. In fact one may view the following discourse as a continuation and deepening of arguments we used in Section 2 of Chapter II. Equation (D) in the following theorem gives the basic differential equation arising in the present setting. Equation (M) gives a useful alternate.

IV.5.1.   **Theorem.**   *Let $\mathfrak{g}$ be a Dynkin algebra, let $\Omega \subseteq \mathfrak{g}$, and let B be an open C-H-neighborhood of 0 in $\mathfrak{g}$. Let $u \colon [0, T] \to \Omega$ be a regulated function which is continuous on the complement of some countable set Q. Then for any continuous function $x \colon [0, \varepsilon] \to B$ with $0 < \varepsilon \leq T$ which is differentiable at the points of $[0, \varepsilon] \backslash Q$ and which satisfies $x(0) = x_0 \in B$, the following statements are equivalent:*

(1) *For all $t, t + h \in [0, \varepsilon] \setminus Q$ and all h with $x(t + h), hu(t) \in B$, the equation*

$$\text{(M)} \qquad\qquad x(t + h) = x(t) * h{\cdot}u(t) + o(t, h)$$

holds with some remainder function o satisfying $\lim_{h \to 0} \|o(t, h)\| \, / h = 0$.

(2) *For all $t \in [0, \varepsilon] \setminus Q$, the differential equation*

$$\text{(D)} \qquad\qquad x'(t) = g\big(\operatorname{ad} x(t)\big)u(t)$$

is satisfied.

(3) *For all $t \in [0, \varepsilon] \setminus Q$, the following differential equation holds:*

$$\text{(D')} \qquad\qquad f\big(\operatorname{ad} x(t)\big)x'(t) = u(t).$$

(4) *x satisfies the following Volterra integral equation:*

$$\text{(I)} \qquad\qquad x(t) = x_0 + \int_0^t g\big(\operatorname{ad} x(s)\big)u(s)ds.$$

The function x is uniquely determined on $[0, \varepsilon]$ by (D), (D'), or (I).

Proof.   (1)$\Rightarrow$(2). By Lemma II.2.4 or Formula (A8) of Appendix 1, we have that for $h \neq 0$

$$\text{(*)} \qquad \begin{aligned} x(t) * hu(t) &= x(t) + g\big(\operatorname{ad} x(t)\big)h{\cdot}u(t) + R(t, h) \\ &= x(t) + h{\cdot}\big(g\big(\operatorname{ad} x(t)\big)u(t) + h^{-1}R(t, h)\big) \end{aligned}$$

for some suitable remainder function R satisfying $\lim_{h \to 0} \|R(t, h)\| / h = 0$. It then follows from (M) in condition (1) that

$$\frac{x(t + h) - x(t)}{h} = g\big(\operatorname{ad} x(t)\big)u(t) + \frac{r(t, h) + R(t, h)}{h}.$$

Whenever $t \notin Q$, we can pass to the limit by letting h approach 0 and obtain (D) in condition (2).

$(2) \Rightarrow (1)$. Whenever $x(\cdot)$ is differentiable, in view of (D) we have

$$x(t + h) = x(t) + hx'(t) + \delta(t, h) = x(t) + g\big(\operatorname{ad} x(t)\big) hu(t) + \delta(t, h)$$

with a suitable remainder term δ satisfying $\lim_{h \to 0} \|\delta(t, h)\|/h = 0$. Then condition $(*)$ above yields that

$$x(t + h) = x(t) * hu(t) + r(t, h)$$

with $r(t, h) = \delta(t, h) - R(t, h)$. This proves (1).

Since equation (D') transforms into (D) and vice verse in light of the fact that f and g are inverses of each other, conditions (2) and (3) are equivalent. Equation (I) follows by straightforward integration from equation (D), and (D) follows from (I) by the Fundamental Theorem of Calculus. Hence (2) and (4) are equivalent, too.

The uniqueness follows from the subsequent lemma. ∎

IV.5.2. **Lemma.** (i) *Suppose that $p(X) = a_0 + a_1 X + \cdots$ is a power series with complex coefficients. Then in the ring $\mathbb{C}[[X, Y]]$ of power series in two commuting variables there is a power series $P(X, Y)$ such that*

$$p(X) - p(Y) = (X - Y)P(X, Y).$$

(ii) *If ρ is the radius of convergence of $p(X)$, then the series $P(u, v)$ converges absolutely for $u, v \in \mathbb{C}$, $|u|, |v| < \rho$. If we set $p_+(X) = |a_0| + |a_1| X + \cdots$, then $|p(u) - p(v)| \le |u - v| P_+(|u|, |v|)$.*

(iii) *In any Banach algebra,*

$$\|p(u) - p(v)\| \le \|u - v\| P_+(\|u\|, \|v\|) \text{ for all } \|u\|, \|v\| < \rho.$$

(iv) *The right hand of (D) satisfies a local Lipschitz condition. Thus for every $x_0 \in B$, equation (D) has a unique local solution $x(t)$ with initial value $x(0) = x_0$.*

(v) *There is a C-H-neighborhood B such that $x, y \in B$ implies*

$$\|g(\operatorname{ad} x) - g(\operatorname{ad} y)\| \le \|x - y\|.$$

Proof. (i) If we set $p_n(X, Y) = \sum_{k=0}^{n-1} X^k Y^{n-k-1}$, then

$$p(X) - p(Y) = (X - Y)(a_1 + \cdots + a_n p_n(X, Y) + \cdots.$$

(ii) For complex numbers u and v, the relation $|u|, |v| \le r$ implies

$$|p_n(u, v)| \le p_n(|u|, |v|) \le nr^{n-1}.$$

Thus $P_+(|u|, |v|) \le |a_1| + \cdots + n|a_n| r^{n-1} + \ldots$ is a majorant for $P(u, v)$, and this implies (ii).

(iii) In a Banach algebra, we have $u^n - v^n = u^{n-1}(u-v) + u^{n-2}(u-v)v + \cdots + u(u-v)v^{n-2} + (u-v)v^{n-1}$. Hence $\|u^n - v^n\| \leq \|u\|^{n-1}.\|u-v\| + \cdots + \|u - v\|.\|v\|^{n-1} \leq \|u - v\|. \sum_{k=0}^{n-1} \|u\|^k \|v\|^{n-k-1} = \|u - v\| p_n(\|u\|, \|v\|)$. It follows that $\|p(u) - p(v)\| \leq \|u - v\| P_+(\|u\|, \|v\|)$.

(iv) Now we specialize to $p(X) = g(X)$ and, accordingly, denote $G(X, Y)$ the power series in two variables associated with g according to parts (i),(ii), and (iii). For the purposes of this Lemma there is no harm in assuming that we have a standard norm. Then for $\|x\|, \|y\| < 2\pi$, recalling $\|\operatorname{ad} u\| \leq \|u\|$, we have

(†) $\|g(\operatorname{ad} x) - g(\operatorname{ad} y)\| \leq \|x - y\| G_+(\|x\|, \|y\|)$.

By Lemma II.2.2, the radius of convergence of the power series $g(X)$ is 2π. Hence for x, y in any fixed ball of radius less than 2π, the right hand side of (†) has a norm bound $M_1 \|x - y\|$. Then

$$\|g(\operatorname{ad} x)u - g(\operatorname{ad} y)u\| \leq \|g(\operatorname{ad} x) - g(\operatorname{ad} y)\| \, \|u\| \leq M_1 \|x - y\| \, \|u\|.$$

If M_2 is a norm bound for the range of $u(\cdot)$, then $\|g(\operatorname{ad} x)u(t) - g(\operatorname{ad} y)u(t)\| \leq M_1 M_2 \|x - y\|$. Thus the right-hand side of (D) satisfies a local Lipschitz condition. The existence of a locally unique solution to (D) is now a standard result of differential equations. (See for instance [Bou61], Chap. II.)

(v) Since $G_+(0, 0) = 1/2$, this is a consquence of the preceding information.   ∎

Let us now recall (from the same source, say) the dependence of the solution of an initial value problem on the initial values and parameters and record the following property:

IV.5.3.   Proposition. *If $x_0^{(n)}$ is a sequence of vectors in B converging to x_0 in B and if u_n is a sequence of regulated functions $[0, T] \to \Omega$ converging uniformly to u, and if further the solutions of*

(D$_n$) $x_n'(t) = g(\operatorname{ad} x_n(t))u_n(t), \qquad x_n(0) = x_0^{(n)}$

all stay in B, then they converge uniformly to a solution of (D).   ∎

IV.5.4.   Proposition. *Suppose that $x(t)$ is a solution of* (D) *and that $y \in B$. Assume here that $y * x(t) \in B$ for all $0 \leq t \leq \varepsilon$. Then $t \mapsto y * x(t)$ is a solution of* (D) *with $y * x(0) = y * x_0$.*

Proof.   We have

$$y * x(t + h) = y * \left(x(t) * hu(t) + r(t, h) \right)$$

where $\lim_{h \to 0} \|r(t, h)\|/h = 0$ from condition (M) of Theorem IV.5.1. Fix t and consider curves $\alpha(h) = x(t) * hu(t)$, $\beta(h) = r(t, h)$, and the vector-valued function $A(v) = y * v$. Let

$$\Gamma(h) = y * \left(x(t) * hu(t) + r(t, h) \right) = A\left(\alpha(h) + \beta(h) \right).$$

By the chain rule

$$\Gamma'(0) = A'\big(\alpha(0) + \beta(0)\big)[\alpha'(0) + \beta'(0)] = A'\big(\alpha(0)\big)\big(\alpha'(0)\big)$$

(since $\beta(0) = 0 = \beta'(0)$ from the conditions that r satisfies). Set

$$\Lambda(h) = y * \big(x(t) * hu(t)\big) = A\big(\alpha(h)\big).$$

Then $\Lambda'(0) = A'\big(\alpha(0)\big)[\alpha'(0)]$. Thus Γ and Λ have the same derivative at 0. It follows that the derivative of $\Gamma - \Lambda$ at 0 vanishes, and hence there exists a function $s(t, h)$ satisfying $\lim_{h \to 0} \|s(t, h)\|/h = 0$ defined by

$$\begin{aligned}
s(t, h) &= \Gamma(h) - \Lambda(h) \\
&= y * \big(x(t) * hu(t) + r(t, h)\big) - y * x(t) * hu(t) \\
&= y * x(t + h) - y * x(t) * hu(t).
\end{aligned}$$

By (M) of Theorem IV.5.1, the proof is complete. ∎

We consider now specific solutions to (D) for the class of piecewise constant control functions.

IV.5.5. **Proposition.** *Consider the constant steering function $u(t) = w$ on some interval $[t_0, t_1]$. Then the unique solution to (D) for the initial value $x(t_0) = y$ is given by $x(t) = y * (t - t_0) \cdot w$.*

Proof. We note that the function $t \mapsto t \cdot w$ on $[0, T]$ satisfies (M) since $t \mapsto t \cdot w$ is a local one parameter semigroup with respect to $*$. Hence by Theorem IV.5.1, it is a solution of (D) for the constant function $u(t) = w$. After Propostion IV.5.4, the function $t \mapsto y * t \cdot w$ is also a solution to (D) with $u(t) = w$. The proposition follows by translating to the interval $[t_0, t_1]$. ∎

By a straightforward induction on Proposition IV.5.5 we find

IV.5.6. **Proposition.** *Suppose that $w_1, \ldots, w_n$ are vectors, $t_1, \ldots, t_n$ are positive real numbers, and $t_1 \cdot w_1 * \cdots * t_{k-1} \cdot w_{k-1} * t \cdot w_k \in B$, a C-H-neighborhood, for all k and all t, $0 \le t \le t_k$. We define $s_0 = 0$ and*

$$s_k = t_1 + \cdots + t_k \qquad and \qquad w_k(t) = (t - s_{k-1}) \cdot w_k$$

for $s_{k-1} \le t < s_k$, $k = 1, \ldots, n$, and set $w_n(s_n) = t_n \cdot w_n$. From these functions w_k we construct a continuous, piecewise differentiable function $x : [0, s_n] \to B$ defined by

$$x(t) = t_1 \cdot w_1 * \cdots * t_{k-1} \cdot w_{k-1} * w_k(t) \quad for \quad s_{k-1} \le t < s_k, \quad k = 1, \ldots, n.$$

Furthermore, we let u be the piecewise constant function which is w_k on $[s_{k-1}, s_k[$ and $u(s_n) = w_n$. Then x is a solution of (D) for u with $x(0) = 0$. Moreover this solution is contained in the local semigroup $S \subseteq B$ generated by

$$\bigcup_{k=1,\ldots,n} \mathbb{R}^+ \cdot w_k \cap B.$$

Hence the points reachable by piecewise constant controls are contained in S. ∎

We extend the notions of Definition IV.3.4.

IV.5.7.   Definition. Let $\mathfrak{g}$ be a complete normed Dynkin algebra with fixed norm $\|\cdot\|$ and let $\Omega \subseteq \mathfrak{g}$. Let $\mathcal{U}$ denote some class of bounded measurable functions from $[0, T]$ to Ω, $T > 0$. Let B be a fixed C-H-neighborhood. A point $p \in L$ is said to be $\mathcal{U}$-*attainable (from the origin) in* B *at time* T if there exists $u \in \mathcal{U}$ and an $x \colon [0, T] \to B$ which is continuous and satisfies the boundary conditions $x(0) = 0$ and $x(T) = p$ and the differential equation

$$(\mathrm{D_0}) \qquad\qquad x'(t) = g\big(ad\, x(t)\big) u(t)$$

almost everywhere on $[0, T]$. The set of all such points p is denoted $\mathcal{A}_{\mathcal{U}}(T, B)$. The point p is $\mathcal{U}$-*reachable (from the origin) in* B *at cost* δ if there exists $u \in \mathcal{U}$ and a solution x to $(\mathrm{D_0})$ in B such that $x(T) = p$ and $\int_0^T \|u(t)\|\, dt = \delta$; the set of all such points is denoted $\mathcal{R}_{\mathcal{U}}(\delta, B)$. Other attainable and reachable sets can be defined from these in a fashion strictly analogously to that given in Definitions IV.3.4 and 12. In particular, $\mathcal{R}_{\mathcal{U}}(B) = \mathcal{R}_{\mathcal{U}}(< \infty, B)$, and $\mathcal{R}_{\mathcal{U}}(\delta)$ is the full $\mathcal{U}$-reachable set defined as in Definition IV.3.12. Again we generally omit the words "from the origin".

We shall frequently refer to the initial value problem $(\mathrm{D_0})$ above; also the other concepts introduced here will be used heavily. The next corollary shows that Definition IV.5.7 is indeed an extension of Definition IV.3.4.

IV.5.8.   Corollary. *Suppose that $\Omega \subseteq \mathfrak{g}$, and that B is a C-H-neighborhood, and let $\mathcal{U}$ denote the set of piecewise constant functions into Ω. Then $\mathcal{A}_{\mathcal{U}}(T, B) = A_\Omega(T, B)$ and $\mathcal{R}_{\mathcal{U}}(\delta, B) = Reach_\Omega(\delta, B)$, where the right-hand sets are those of Definition IV.3.4.*

Proof.   Given any product $p = t_1 \cdot w_1 * \cdots * t_n \cdot w_n$ satisfying the conditions of Proposition IV.5.6 and with $w_k \in \Omega$ for each k, then Proposition IV.5.6 shows how to obtain p as the terminal point of a solution of $(\mathrm{D_0})$ for a piecewise constant function with values $w_1, \ldots, w_n$. Conversely, since the solution to $(\mathrm{D_0})$ is unique and independent of the values at the finitely many points of discontinuity of a given piecewise constant function into Ω, the solution must arise as described in Proposition IV.5.6. Hence the first equality follows directly from the definitions. Given a piecewise constant function u as in Proposition IV.5.6 which takes the value w_k on the open interval (s_{k-1}, s_k), then

$$\int_0^{s_n} \|u(t)\|\, dt = \sum_{j=1}^n t_j \|w_j\|.$$

Thus the costs as computed in Definitions IV.5.4 and IV.3.7 agree.   ∎

The preceding corollary and its proof justify the following observation:

IV.5.9.   Remark. For any generating set $\Omega \subseteq \mathfrak{g}$ the notions of attainability and reachability by the $*$–multiplication (as in Definition IV.3.4) or by solutions to $(\mathrm{D_0})$ for the class of piecewise constant functions (as in Definition IV.5.7) are alternate, equivalent approaches to the same constructions.   ∎

We shall henceforth freely pass between these two viewpoints. Also it should be noted that many of the results of Section IV.3 have alternate formulations in the framework of this section for the class of piecewise constant functions.

The proof of the following lemma involves more or less standard function manipulation and is left as an exercise (see Exercise EIV.5.1 at the end of this section).

IV.5.10.   Lemma.   *Suppose that* $u: [0, T] \to \Omega$ *is a regulated function, where* $\Omega \subseteq \mathfrak{g}$. *Then* u *is the uniform limit of of a sequence of step functions* $u_n: [0, T] \to \Omega$. *Furthermore, if* $\delta = \int_0^T \|u(t)\| \, dt$, *then the functions* u_n *may be chosen so that* $\int_0^T \|u_n(t)\| \, dt = \delta$ *for each* n.   ∎

The following proposition gives further relationships between the notions of attainability and reachability introduced in Section IV.3 and this section.

IV.5.11.   Proposition.   *Let* $\Omega \subseteq \mathfrak{g}$, *let* B *be a C-H-neighborhood, and let* $\mathcal{U}$ *be a collection of regulated functions into* Ω *which contains the step functions. Then*

$$A_\Omega(T, B) \subseteq \mathcal{A}_\mathcal{U}(T, B) \subseteq \mathcal{A}_\Omega(T, B),$$
$$\mathrm{Reach}_\Omega(\delta, B) \subseteq \mathcal{R}_\mathcal{U}(\delta, B) \subseteq \mathcal{R}_\Omega(\delta, B).$$

Proof.   The first inclusion in each case follows from Corollary IV.5.8. We prove the second inclusion for the reachability case, the attainability case being similar. Let $y = x(T)$ where x is a solution of (D_0) for some $u \in \mathcal{U}$ with $\int_0^T \|u(t)\| \, dt = \delta$. Pick a sequence of step functions u_n as in Lemma IV.5.10. Let x_n denote the solutions to (D_n) for the control functions u_n. Then $x_n(T) \in \mathrm{Reach}_\Omega(\delta, B)$ by Corollary IV.5.8, and $x_n(T)$ converges to $x(T) = y$ by Proposition IV.5.3.   ∎

IV.5.12.   Proposition.   *The full reachable set* $\mathcal{R}_\mathcal{U}(< \delta)$ *is defined and lies in* $\overline{B}$ *if the full reachable set* $\mathrm{Reach}_\Omega(< \delta)$ *is defined and lies in the C-H-neighborhood* B.

Proof.   For a steering function $u \in \mathcal{U}$ with domain $[0, T]$ and $\int_0^T \|u(t)\| \, dt = \gamma < \delta$, pick a sequence of of piecewise constant functions u_n, each of cost γ, converging to u on $[0, T]$ as in Lemma IV.5.10. By hypothesis, the solutions to (D_n) lies entirely in B. By Proposition IV.5.3 the solutions converse uniformly to the solution for u. Hence that solution must exist and lie in $\overline{B}$.   ∎

We include an alternate version of Proposition IV.3.10 in the current setting.

IV.5.13.   Proposition.   *Let* $\Omega \subseteq \mathfrak{g}$ *and suppose that* $\mathfrak{g}$ *has a complete norm satisfying* $\|x * y - x - y\| \leq \|x\| \|y\|$ *on some C-H-neighborhood* B *containing the closed ball of radius* $\varepsilon < 1$. *If* $u: [0, T] \to \Omega$ *is a regulated function such that* $\int_0^T \|u(t)\| \, dt = \delta < \varepsilon/2$, *then the solution* $x: [0, T] \to B$ *to* (D_0) *exists and satisfies*

(1)
$$\left\| x(t) - \int_0^t u(s) \, ds \right\| \leq \delta^2,$$

(2) $$\|x(t)\| \le \delta + \delta^2.$$

Furthermore, $x(t)$ is a member of $\overline{\mathrm{Sg}(\Omega, B)}^B$.

Proof.   Let u_n be a sequence of step functions from $[0, T]$ into Ω which converge uniformly to u as in Lemma IV.5.10. Using the equivalence of Remark IV.5.9, we deduce as in the proof of Proposition IV.3.10 that the solution x_n of (D_n) lies in B for all n chosen large enough so that $\int_0^T \|u_n(t)\| \, dt < \varepsilon/2$. Reinterpreting (2) of Proposition IV.3.10 in terms of integrals as in the proof of Corollary IV.5.8, we conclude that $\|x_n(t)\| < \delta + \delta^2$. Taking the limit as $n \to \infty$, we conclude that (2) holds. In an analogous fashion we deduce (1) from (1) of Proposition IV.3.10. The last assertion follows directly from Proposition IV.3.7 and IV.5.11.   ∎

Invariant vector fields

In the geometry of Lie groups it is the invariant vector fields that play a crucial role. The translates of one parameter groups give the solution curves to these vector fields. For our purposes in studying the local theory, it is convenient to translate this machinery into the Lie algebra. There it provides a powerful theoretical tool for the investigation of local semigroups and helpful intuitive insights. In achieving our ultimate goal of showing that general Lie wedges are tangent wedges of some local semigroup we shall find invariant vector fields in this setting indispensible.

Let $\mathfrak{g}$ be a Dynkin algebra and let B be a Baker–Campbell–Hausdorff neighborhood of 0. We identify the tangent bundle of B with $B \times \mathfrak{g}$. A *vector field on B* is then a function $X : B \to \mathfrak{g}$.

Each member of $\mathfrak{g}$ gives rise to an analytic vector field on B in the following fashion: For $y \in \mathfrak{g}$, define the $X_y : B \to \mathfrak{g}$ by $X_y(x) = g(\operatorname{ad} x)(y)$ where g is as in the earlier parts of this section. The integral curves of this vector field in B are solutions of $x'(t) = X_y\big(x(t)\big) = g\big(\operatorname{ad} x(t)\big)y = (1 + \frac{1}{2}\operatorname{ad} x(t) + \cdots)y$, that is, they are solutions of (D) of Theorem IV.5.1 for the constant function $u(t) = y$. By Proposition IV.5.5, if the initial value $x(0)$ is z, then the solution is given by $x(t) = z * ty$. Note, in particular, that the tangent vector at z is given by $x'(0) = g(\operatorname{ad} z)(y)$.

Let $\lambda_x : B \to \mathfrak{g}$ be given by $y \mapsto x * y$. Then for x, z, and t chosen small enough so that $*$-products are defined, by Proposition IV.2.6, $d\lambda_x(z)$ maps the tangent vector at z of the curve $t \mapsto z * ty$ to the tangent vector at $x * z$ of the curve $t \mapsto x * z * ty$. By the preceding paragraph, this translates to $d\lambda_x(z)\big(g(\operatorname{ad} z)y\big) = g(\operatorname{ad} x * z)(y)$. Therefore the vector fields X_y of the previous paragraph are left-invariant with respect to the C-H-multiplication $*$. By taking $z = 0$, we retrieve the equality $d\lambda_x(0)(y) = g(\operatorname{ad} x)(y)$ of Lemma II.2.5. Thus any left-invariant vector field X must be of the form X_y, where $y = X(0)$. Hence $y \mapsto X_y$ establishes a one-to-one correspondence between $\mathfrak{g}$ and the left-invariant vector fields on B. It is not difficult to establish that this correspondence is an isomorphism of Lie algebras, where the vector fields are equipped with their usual

Lie product (see Exercise EIV.5.2 at the end of this section). We summarize these results in the following proposition:

IV.5.14.   Proposition. *Let B be a C-H-neighborhood in the Dynkin algebra $\mathfrak{g}$. For $y \in \mathfrak{g}$, define a vector field X_y on B by*

$$X_y(x) = g(\operatorname{ad} x)(y) = d\lambda_x(0)(y).$$

Then $y \mapsto X_y$ is a Lie algebra isomorphism from $\mathfrak{g}$ to the left-invariant vector fields on B. ∎

IV.5.15.   Remark. Let $\Omega \subseteq \mathfrak{g}$ and let $u : [0, T] \to \Omega$ be a regulated function. Let $U_t = X_{u(t)}$ for $0 \le t \le T$. Then

(D″) $$x'(t) = U_t\big(x(t)\big)$$

is an alternate equivalent form of equation (D) of Theorem IV.8.1.

Proof.   Note that $U_t\big(x(t)\big) = X_{u(t)}\big(x(t)\big) = g\big(\operatorname{ad} x(t)\big)\big(u(t)\big).$ ∎

The previous proposition and remark coupled with the results of the preceding section provide a convenient intuitive geometric framework for viewing (and indeed motivating) many of our earlier constructions. For example, to say that $x \in A_\Omega(T)$, that is, that the point $x \in B$ is attainable at time T by means of piecewise constant functions into Ω, is equivalent to saying that one may reach x by starting at 0, following the solution of the left-invariant vector field associated with some member of Ω and arriving at some point x_1 at time t_1, switching to some other vector field and following its solution for some length of time t_2 from x_1 to some point x_2, continuing this process, and finally arriving at x after a total time T has elapsed with only finitely many switches of the vector fields. If there are no more than $n - 1$ switches in this process, we say that x *is attainable in n steps.*

We now apply the machinery of invariant vector fields to the important question of the existence of non-empty interiors of local semigroups. We consider first some elementary properties of the interior.

IV.5.16.   Proposition. *Let S be a local semigroup with respect to a C-H-neighborhood B, and suppose the interior $\operatorname{int}(S)$ of S is non-empty. Then the following conclusions hold:*

   (i) *$\operatorname{int}(S)$ is a local ideal of S.*

   (ii) *If 0 is in the closure of $\operatorname{int}(S)$, then $\operatorname{int}(S) = \operatorname{int}(\overline{S}^B)$ and $\operatorname{int}(S)$ is dense in S.*

Proof.   (i) Let $s \in S$, $y \in \operatorname{int}(S)$. Pick U open such that $y \in U \subseteq \operatorname{int}(S) \cap B$ and $s * U \subseteq B$. Then $s * U$ is open and $s * U \subseteq S * S \cap B \subseteq S$ imply that $s * y \in \operatorname{int}(S)$. Thus $\operatorname{int}(S)$ is a left ideal in B; similarly it is a right ideal and hence an ideal.

(ii) Obviously $\operatorname{int}(S) \subseteq \operatorname{int}(\overline{S}^B)$. Let $s \in U = \operatorname{int}(\overline{S}^B)$. There exists V open containing 0 such that $s * (-V) * V \subseteq U$. Let $W = V \cap \operatorname{int}(S)$; in view of the hypothesis $W \neq \emptyset$. Then $s * (-W) \subseteq U \subseteq \overline{S}^B$. Since $s * (-W)$ is open, there exists $t \in S$, $w \in W$ such that $s * (-w) = t$. Then $s = t * w \in t * W \subseteq \mathrm{int}(S)$ since $t * W \subseteq s * (-V) * V \subseteq B$.

Let s_n be a sequence in $\mathrm{int}(S)$ converging to 0. Then for $s \in S$, $s * s_n$ converges to s and $s * s_n$ is in B and hence in $\mathrm{int}(S)$ for large n by part (i).   ■

We next establish a major result concerning the existence for interior points for local semigroups in the finite dimensional setting.

IV.5.17.   Lemma. *Suppose that* $\dim \mathfrak{g} = n$, *that* B *is a C-H-neighborhood, and that* Ω *is a subset which generates the Lie algebra* $\mathfrak{g}$. *Then for* $T > 0$, *the set of points in* $A_\Omega(< T, B)$ *that are attainable in* n *steps has non-empty interior.*

Proof.   For each $u \in \Omega$, let X_u denote the corresponding left-invariant vector field on B. We consider all C^∞-functions $f : Q \to B$ such that Q is an open subset of $\mathbb{R}^m$ for some $m \leq n$, such that the rank of $df(q)$ is m for all q and that $f(Q)$ is contained in the set of points in $A_\Omega(< T_f, B)$ that are attainable in m steps for some $0 < T_f < T$. Among these functions we find one such that $m \geq 0$ is maximal. If we can show that $m = n$, then we are finished by the Inverse Function Theorem.

First we claim that for all $u \in \Omega$ and all $q \in Q$, the vector $X_u\big(f(q)\big)$ is in the image of $df(q)$. We prove this claim by contradiction and assume that there is a $p \in Q$ such that for $y = f(p)$, there exists $u \in \Omega$ such that $X_u(y)$ is not in the range of $df(p)$. There exists an open set Q_1 with $p \in Q_1 \subseteq Q$ and $0 < \varepsilon < T - T_f$ such that $f(q) * tu$ is defined and in B for $q \in Q_1$ and $|t| < \varepsilon$. Define $F : Q_1 \times (-\varepsilon, \varepsilon) \to B$ by $F(q, t) = f(q) * tu$. Note that $F(q, 0) = f(q)$; hence the range of $dF(x, 0)$ contains the range $df(x)$. The derivative at 0 of the function $t \mapsto F(p, t) = y * tu : \,] - \varepsilon, \varepsilon[\to B$ is $d\lambda_y(u) = g(\mathrm{ad}\, y)(u) = X_u(y)$. Thus, by the chain rule, the range of $dF(p, 0)$ also contains $X_u(y)$. Hence the rank of $dF(p, 0)$ must be at least $m + 1$, and by continuity, the same must be true at all (q, t) in some neighborhood of $(p, 0)$. Pick an open set $Q_2 \subseteq Q_1$ containing p and a $0 < \delta < \varepsilon$ such that $Q_3 \overset{\text{def}}{=} Q_2 \times \,]0, \delta[$ is contained in this neighborhood and the rank of dF is constant $\geq m + 1$ on Q_3. Consider the C^∞ function $F_3 : Q_3 \to B$ obtained by restricting F. For $q \in Q_2$ and $t \in (0, \delta)$,

$$F_3(q, t) = f(q) * tu \in A_\Omega(< T_f, B) * tu \subseteq A_\Omega(< T_f + \varepsilon, B)$$

since $0 < t < \delta < \varepsilon$. Also, since $f(q)$ was attainable in m steps or less by assumption, $f(q) * tu$ is attainable in $m + 1$ steps or less. This contradicts the maximality of m, and our claim is proved.

The remainder of the proof now follows by applying some elementary differential geometry. Since f has full rank m at any point $q \in Q \subseteq \mathbb{R}^m$, by restricting to a small neighborhood the mapping f is an embedding. We assume that Q is chosen small so that $f(Q)$ is an embedded submanifold. Each of the left invariant vector fields $X_u(y)$, $y \in B$, have the property that $X_u\big(f(q)\big)$ is in the image of $df(q)$ by the preceding paragraph, that is, the vector $X_u\big(f(q)\big)$ is tangent to the embedded submanifold $f(Q)$. It is standard and easily verified that taking the Lie algebra of vector fields generated by a family of vector fields and then restricting this family to an embedded submanifold yields the same result as first restricting them to the embedded submanifold and then generating the Lie algebra, as long as each of the original vector fields had the property that evaluated at any point in the embedded submanifold, the vector at that point was tangent to the submanifold. (This is because the vector fields and their restrictions are related with respect to the embedding.) Since we have shown that the restrictions of members of Ω to the embedded submanifold $f(Q)$ are all tangent to the embedded submanifold, the same will be true for all the vector fields in the Lie algebra generated by the restrictions. So the projection of this Lie algebra at a point $p = f(q)$ sending the left invariant vector field X to $X(p)$ will have range a vector space of dimension at most m. However, by hypothesis, the family of left-invariant vector fields corresponding to Ω generates the Lie algebra $\mathfrak{g}$, so that the projection onto p has dimension n. Since these two projections must agree, it follows that $m = n$. This completes the proof. ∎

With this lemma it now possible to establish the following important theorem.

IV.5.18.   Theorem. (The Dense Interior Theorem for Local Semigroups) *Suppose that the finite dimensional Lie algebra $\mathfrak{g}$ is generated by the subset Ω. Then any full reachable set $\mathrm{Reach}_\Omega(< \delta, B)$ has dense interior.*

If B is a C-H-neighborhood, then the local semigroup $\mathrm{Sg}(\Omega, B)$ with respect to B generated by Ω also has dense interior.

Proof.   Let Θ denote all unit vectors in $\mathbb{R}^+ \cdot \Omega$. By Proposition IV.3.8, $\mathrm{Reach}_\Omega(< \gamma) = A_\Theta(< \gamma)$ for $\gamma \leq \delta$. By Lemma IV.5.17, the interior of $A_\Theta(< \gamma)$ is non-empty.

For each $\gamma = 1/n$, pick x_n in the interior of $A_\Theta(< \frac{1}{n}) = \mathrm{Reach}_\Omega(< \frac{1}{n})$. By Theorem IV.3.14 the sequence x_n converges to 0.

Let $x \in \mathrm{Reach}_\Omega(\varepsilon)$, where $\varepsilon < \delta$. Let U be an open set containing x. Choose $\frac{1}{n} < \delta - \varepsilon$ with $x * x_n \in U$. Pick an open set V such that $x_n \in V \subseteq \mathrm{Reach}_\Omega(< \frac{1}{n})$ and $x * V \subseteq U$. Then

$$x * V \subseteq \mathrm{Reach}_\Omega(\varepsilon) * \mathrm{Reach}_\Omega(< \frac{1}{n}) \subseteq \mathrm{Reach}_\Omega(< \delta),$$

the last inclusion coming from Proposition IV.3.17. Since $x * V$ is open, the interior of $\mathrm{Reach}_\Omega(< \delta)$ meets U. Since U was arbitrary, the interior is dense.

We turn now to the semigroup case. From Theorem IV.3.14 we know that $\mathrm{Reach}_\Omega(< \delta) \subseteq \mathrm{Sg}(\Omega, B)$ for some δ small enough. We have just seen that 0 is in the closure of the interior of $\mathrm{Reach}_\Omega(< \delta)$ and hence is in the closure of the interior of $\mathrm{Sg}(\Omega, B)$. The remainder of the theorem now follows from Proposition IV.5.16. ∎

EIV.5.1.   Exercise. Prove Lemma IV.5.10.
(Hint: First choose $u_n: [0, T] \to \mathfrak{g}$ such that u_n is a step function and $\|u_n(t) - u(t)\| < 1/n$ for all $0 \leq t \leq T$. There exist $0 = t_0 < t_1 < \ldots < t_m = T$ such that u_n is constant on each (t_{i-1}, t_i). There exist $x_i, y_i \in u(t_{i-1}, t_i)$ such that

$$\|x_i\|(t_i - t_{i-1}) \leq \int_{t_{i-1}}^{t_i} \|u(t)\| \, dt \leq \|y_i\|(t_i - t_{i-1}).$$

Then there exists s_i such that $t_{i-1} \le s_i \le t_i$ and

$$\int_{t_{i-1}}^{s_i} \|x_i\| \, dt + \int_{s_i}^{t_i} \|y_i\| \, dt = \int_{t_{i-1}}^{t_i} \|u(t)\| \, dt.$$

Define $w_n(\cdot)$ by $w_n(t_i) = u(t_i)$ for $i = 0, 1, \cdots, m$, $w_n(t) = x_i$ on $(t_{i-1}, s_i]$ and $w_n(t) = y_i$ on (s_i, t_i). Then (by the triangular inequality and choice of $u_n(\cdot)$) $\|w_n(t) - u(t)\| \le 2/n$ for all $t \in [0, T]$, $w_n(\cdot)$ is a step function, and $\int_0^T w_n(t) \, dt = \int_0^T u(t) \, dt$ by choice of w_n.)

EIV.5.2.   **Exercise.** Show that the correspondence $y \mapsto X_y$ given in the paragraph before Proposition IV.5.14 is an isomorphism of Lie algebras.

(Hint: Let $x \in B$, where B is a BCH-neighborhood. Pick open sets U containing x and B' containing 0 with $U, B' \subseteq B$ such that $\lambda_x(B') = U$. The restriction of vector field $[X_y, X_z]$ to U and B' respectively agrees with the vector fields obtained by first restricting to U or B' and then taking Lie bracket. By the left invariance of X_y and X_z, we have $d\lambda_x\big(X_y(w)\big) = X_y(x * w)$ for all $w \in B'$. Thus the vector fields $X_y|B'$ and $X_y|U$ are λ_x-related. Similarly $X_z|B'$ and $X_z|U$ are λ_x-related. Thus the corresponding Lie brackets $[X_y|B', X_z|B']$ and $[X_y|U, X_z|U]$ are λ_x-related. In particular, $d\lambda_x(0)\big([X_y|B', X_z|B'](0)\big) = [X_y|U, X_z|U](x)$; it follows that $[X_y, X_z]$ is again left invariant.

Consider now the curves $t \mapsto ty$ and $t \mapsto tz$, which have y and z respectively for tangent vectors at 0 and are integral curves for X_y and X_z respectively. By standard differential geometry $t \mapsto t^{\frac{1}{2}}x * t^{\frac{1}{2}}y * t^{\frac{1}{2}}(-x) * t^{\frac{1}{2}}(-y)$ is a curve with tangent vector at 0 given by $[X_y, X_z](0)$. But a direct application of the Commutator Formula ((A5) of Appendix 1) yields that this tangent vector is $[x, y]$. Since $[X_y, X_z]$ is left-invariant, $[X_y, X_z] = X_{[y,z]}$.)

6. Wedge fields

Let L be a completely normable topological vector space. Recall from Definition I.2.29 that the semiprojective space $\Pi(L)$ is given by collapsing all orbits $\mathbb{P} \cdot w$ in $L \setminus \{0\}$ to points and endowing the set $\Pi(L)$ with the quotient topology for the function π from $L \setminus \{0\}$ to $\Pi(L)$ sending w to $\mathbb{P} \cdot w$. This construction gives a formal way of representing the unit sphere which is independent of the compatible norm chosen.

IV.6.1.   Proposition. *For any norm on L compatible with the topology, the restriction of π to the unit sphere is a homeomorphism onto $\Pi(L)$.*

Proof.   Clearly the restriction to the unit sphere is one-to-one, onto, and continuous. Since $x \mapsto x/\|x\|$ is continuous on $L \setminus \{0\}$, the inverse is also continuous since the topology of $\Pi(L)$ is the quotient topology.   ∎

Given two norms compatible with the topology on L, using the fact that the norms are bounded by scalar multiples of each other, we readily verify that the natural homeomorphism along rays between the respective unit spheres is uniformly continuous in both directions and thus is an isomorphsm of the uniform structures induced by the respective metrics. Hence these uniform structures induce a unique uniform structure on $\Pi(L)$. We record this for easy reference in the following definition:

IV.6.2.   Definition.   For any norm compatible with the topology of L, if the metric induced by this norm on the unit sphere is carried over to $\Pi(L)$ via the restriction of π, then the uniformity associated with this metric is independent from the norm and is called *the induced uniformity.*

The next proposition is a direct consequence of these observations.

IV.6.3.   Proposition.   *Let C and D be pointed cones in L. The following statements are equivalent:*

(1) *$\Pi(D \setminus \{0\})$ is a uniform neighborhood of $\Pi(C \setminus \{0\})$ in $\Pi(L)$.*

(2) *There exists a compatible norm on L and an $\varepsilon > 0$ such that $\|x\| = 1$ and $\|x - y\| < \varepsilon$ for some $y \in C$ with $\|y\| = 1$ imply $x \in D$.*

(3) *For every compatible norm there exists an $\varepsilon > 0$ dependent on the norm such that $\|x\| = 1$ and $\|x - y\| < \varepsilon$ for some $y \in C$ with $\|y\| = 1$ imply $x \in D$.*   ∎

In Definition IV.2.10 we explained what we mean when we say that one wedge surrounds another *in a finite dimensional vector space*. We give now a definition of a surrounding cone appropriate to both the finite *and infinite dimensional* setting. After Proposition IV.6.9 below we shall know that the two definitions are equivalent in the finite dimensional setting. We shall know after these discussions that, in particular, in a finite dimensional vector space *every wedge is surrounded by some wedge*.

IV.6.4.   Definition.   Let C and D be pointed cones in a completely normable vector space L. If C and D satisfy the equivalent conditions of Proposition IV.6.3, then we say that D *surrounds* C. We say that a wedge W' *surrounds* a wedge W, written

$$W \subset\subset W',$$

if $H(W) = H(W')$ and $W'/H(W)$ surrounds $W/H(W)$ in $L/H(W)$.

We consider a standard construction for obtaining a basis of surrounding cones of a strictly positive cone. Let C be a strictly positive cone. We use Lemma IV.4.2 to find some continuous linear functional ω and norm compatible with the topology such that $\|\omega\| = 1$ and ω agrees with the norm on C. Let $K = \omega^{-1}(1) \cap C$. Let $0 < \varepsilon < \frac{1}{3}$. Define

$$K(\varepsilon) = \overline{\left(K + N(0, \varepsilon)\right) \cap \omega^{-1}(1)},$$

where $N(0, \varepsilon)$ is the open ball of radius ε around 0. Then $K(\varepsilon)$ is a closed convex subset of $\omega^{-1}(1)$ and every point of $K(\varepsilon)$ is of distance less than or equal to ε from K.

IV.6.5.   Definition.   Set $C(\varepsilon) = \mathbb{R}^+ \cdot K(\varepsilon)$.

IV.6.6.   Lemma. $C(\varepsilon)$ *is a (closed) strictly positive cone.*

Proof.   Since $K(\varepsilon)$ is convex, so is $C(\varepsilon)$. The latter is clearly closed under multiplication by non-negative scalars. We show $C(\varepsilon)$ is closed. Let $r_n \cdot x_n \to y$, $x_n \in K(\varepsilon)$, $r_n \geq 0$. Then for some subsequence, $r_n \to r$, for otherwise $\omega(r_n \cdot x_n) = r_n \omega(x_n) = r_n \to \infty$, which contradicts $\omega(r_n \cdot x_n) \to \omega(y)$. Then $x_n \to \frac{1}{r} \cdot y \in K(\varepsilon)$, so $y \in C(\varepsilon)$ if $r \neq 0$. If $r = 0$, then $\|r_n \cdot x_n\| = r_n \to 0$, so $y = 0 \in C(\varepsilon)$. Thus $C(\varepsilon)$ is a closed cone. Now ω remains strictly positive on $C(\varepsilon)$. Also $\omega^{-1}(1) \cap C(\varepsilon) = K(\varepsilon)$, and hence is bounded in norm by $1 + \varepsilon$. Thus $C(\varepsilon)$ is strictly positive.   ■

IV.6.7.   Lemma.   *Under the present circumstance, the following conclusions hold:*

   (i)   *If $\omega(x) = 1$ and $\|x - y\| < \varepsilon < 1/6$ for some $y \in K$, then $\|\frac{1}{\|x\|} \cdot x - y\| < 5\varepsilon$.*

   (ii)   *If $x \in K$, then $N(x, \frac{1}{3}\varepsilon) \subseteq C(\varepsilon)$ for $\varepsilon < \frac{1}{3}$.*

Proof.   (i) We have

$$|1 - \|x\|\,| = |\omega(x) - \|x\|\,| \leq |\omega(x) - \omega(y)| + |\omega(y) - \|y\|\,| + |\,\|y\| - \|x\|\,|$$

$$\leq \|\omega\|\,\|x - y\| + 0 + \|x - y\| < 2\varepsilon < \frac{1}{3}.$$

Hence $\|x\| - 1 < 2\varepsilon$, that is, $\|x\| < 1 + 2\varepsilon$. *Note:* For any real number r, the estimate $|1 - r| < \varepsilon < \frac{1}{3}$ implies $|1 - 1/r\| < \frac{3}{2}\varepsilon$. Hence $|\frac{1}{\|x\|} - 1| < (3/2)(2\varepsilon) = 3\varepsilon$. Thus

$$\|\frac{1}{\|x\|}\cdot x - y\| \leq \|\frac{1}{\|x\|}\cdot x - x\| + \|x - y\| \leq |\frac{1}{\|x\|} - 1|\,\|x\| + \varepsilon$$

$$\leq 3\varepsilon(1 + 2\varepsilon) + \varepsilon < 4\varepsilon + (6\varepsilon)\varepsilon < 5\varepsilon.$$

(ii) Suppose $\|y - x\| < \frac{1}{3}\varepsilon$. Then

$$|1 - \omega(y)| = |\omega(x) - \omega(y)| = |\omega(x - y)| \leq \|\omega\|\,\|x - y\| < \frac{1}{3}\varepsilon.$$

For $r = 1/\omega(y)$, again by the *Note* above, $|1 - r| < (3/2)(\varepsilon/3) = \varepsilon/2$. Then, using $\varepsilon < 1/3$, we obtain

$$|1 - r|.\|y\| \leq |1 - r|\cdot(\|x\| + \|y - x\|) < \frac{\varepsilon}{2}\cdot\frac{3 + \varepsilon}{3} = \frac{3\varepsilon + \varepsilon^2}{6} < \frac{4\varepsilon}{6} = \frac{2\varepsilon}{3}.$$

Thus

$$\|x - r\cdot y\| \leq \|x - y\| + \|y - r\cdot y\| \leq \frac{1}{3}\varepsilon + |1 - r|.\|y\| < \varepsilon.$$

Since $\omega(ry) = r\omega(y) = 1$, it follows that $r\cdot y \in K(\varepsilon)$, and thus $y \in C(\varepsilon)$. ∎

IV.6.8. **Proposition.** *The family* $\mathcal{S} = \{C(\varepsilon): \varepsilon > 0\}$ *is a basis of the filterbasis of all cones surrounding* C. *All members of* $\mathcal{S}$ *are strictly positive, and* C *is in the interior of* $C(\varepsilon) \setminus \{0\}$ *for all* $\varepsilon > 0$.

Proof. First we note that by Lemma IV.6.6, all $C(\varepsilon)$ are strictly positive. Now we shall show that each $C(\varepsilon)$ surrounds C. Let S denote the unit sphere for the fixed norm used for Definition IV.6.5. If $x \in C \cap S$, then $1 = \|x\| = \omega(x)$, that is, $x \in K$. Thus Lemma IV.6.7 tells us $N(x, \frac{1}{3}\varepsilon) \subseteq C(\varepsilon)$. In view of Proposition IV.6.3 and Definition IV.6.4, this shows that $C(\varepsilon)$ surrounds C. Since K is contained in the interior of $C(\varepsilon)$, so is $C \setminus \{0\}$.

Next suppose that D is an arbitrary cone surrounding C. In order to complete the proof we must show that $C(\gamma) \subseteq D$ for some positive γ. Again by Proposition IV.6.3, there exists a positive $\varepsilon < (1/6)$ such that $x \in D$ whenever $\|x\| = 1$ and $\|x - y\| < \varepsilon$ for some $y \in K$. Let $z \in C(\frac{1}{6}\varepsilon)$, $\|z\| = 1$. Then for $u = (1/\omega(z))z$, we have $\|u - v\| < \frac{1}{5}\varepsilon$ for some $v \in K$. Note that $u/\|u\| = z$. By Lemma IV.6.7, $\|z - v\| = \|(1/\|u\|)u - v\| < \varepsilon$. Thus $z \in D$. Since D contains all unit vectors of the cone $C(\frac{1}{6}\varepsilon)$, it contains the cone itself. ∎

The following Proposition will prove the link between the present more general setting and the old Definition IV.2.10. It also gives a proof of Proposition IV.2.12.

IV.6.9. **Proposition.** *Let* W *and* W' *be wedges in* L *such that* $W/H(W)$ *is strictly positive. If* $W \subset\subset W'$, *then* $W \setminus H(W) \subseteq \text{int}(W' \setminus H(W))$. *If* L *is finite dimensional, then the converse also holds.*

Proof. By first passing to $L/H(W)$, it suffices to consider the case that W is a pointed cone. Let S denote the unit sphere. By Proposition IV.6.3, $S \cap W$ is contained in the interior of $S \cap W'$. Hence $\pi(S \cap W) \subseteq \operatorname{int} \pi(S \cap W')$, since π is a homeomorphism on S by Proposition IV.6.1. Finally,

$$W = (\pi)^{-1}\pi(S \cap W) \subseteq \operatorname{int}(\pi)^{-1}\pi(S \cap W') = W'.$$

Now suppose that L is finite dimensional. Again it suffices to consider the case that W is a pointed cone. Then W is strictly positive by Remark IV.4.4. Define K and $K(\varepsilon)$ for W as for Definition IV.6.4. Now that we are in a finite dimensional vector space, the subsets $K(\varepsilon)$ are compact. Since their intersection is K, one of them must be in W'. Hence W' contains a $C(\varepsilon)$ and thus must surround W. ∎

We shall need a strengthened version of Lemma IV.5.3 which we shall present now.

IV.6.10.   Proposition. *If W is a strongly positive wedge in a Dynkin algebra $\mathfrak{g}$ with closed complement L_1 for its edge $H = H(W)$, then there exist continuous projections p_H and $p_1 = I - p_H$ onto H and L_1 respectively. The mapping $x \mapsto (p_1(x), p_H(x))$ is an isomorphism of topological vector spaces which carries W to $(L_1 \cap W) \times H$. Given norms on L_1 and H, we can take the sup norm on $L_1 \times H$ and obtain an equivalent norm on L such that the following conditions hold:*

(i) *If $x = x_1 + h$, $x_1 = p_1(x)$, $h = p_H(x)$, then $\|x\| = \max\{\|x_1\|, \|h\|\}$.*

(ii) *On some C-H-neighborhood B, the norm satisfies the following inequalities for $x, y \in B$ and $u \in \mathfrak{g}$:*

(†) $\|x * y - x - y\| \leq \|x\|\,\|y\|,$

(‡) $\|g(\operatorname{ad} x)(u) - g(\operatorname{ad} y)(u)\| \leq \|x - y\|\,\|u\|.$

Furthermore, there exists a continuous linear functional ω on $\mathfrak{g}$ such that $\|\omega\| = 1$ and $\|x\| = \omega(x)$ for all $x \in C(\varepsilon)$, where $C(\varepsilon)$ is some strictly positive cone in L_1 which surrounds $C = L_1 \cap W$ in L_1.

Proof.   We assume that we are working with a standard norm. For the strictly positive cone $C = L_1 \cap W$ and for some $0 < \varepsilon < 1/3$ as in Definition IV.6.5, we construct $C(\varepsilon)$ in L_1. Now apply Lemma IV.4.20 to the wedge $W(\varepsilon) = H + C(\varepsilon)$. If we replace the sum norm on $L_1 \times H$ by the sup norm, we obtain everything except (‡). But this follows from Lemma IV.5.2(v), since we may choose B so small that $x, y \in B$ implies

$$\|g(\operatorname{ad} x) - g(\operatorname{ad} y)\| \leq \|x - y\|.$$

In the presence of a standard norm, (‡) is now an immediate consequence. ∎

IV.6.11.   Proposition. *Let W be a strongly positive wedge in a Dynkin algebra $\mathfrak{g}$ with edge H and closed complement L_1. Suppose the norm on L is chosen as in Proposition IV.6.10. Let T be a continuous linear operator on $\mathfrak{g}$ satisfying $\|I_L - T\| \leq \gamma < \frac{1}{9}\varepsilon$ where $\varepsilon < 1$. Then $T(W) \cap L_1 \subseteq C(\varepsilon)$.*

Proof.   Let $x \in W$ such that $T(x) \in L_1$. Now $x = x_1 + h$ for some $x_1 \in L_1 \cap W$ and $h \in H$. Let $k = h - T(h)$. Then

$$\|k\| = \|I(h) - T(h)\| \leq \|I - T\|\,\|h\| \leq \gamma \|h\|.$$

Similarly for $y = x_1 - T(x_1)$ we obtain $\|y\| \leq \gamma\|x_1\|$.

Now $T(x) \in C(\varepsilon)$ if and only if $T(r \cdot x) \in C(\varepsilon)$. So we assume without loss of generality that $\|x_1\| = 1$, by first scaling by r if necessary (we consider the case $x_1 = 0$ later).

For $b \in L_1$, $\|h - b\| = \max\{\|h\|, \|b\|\} \geq \|h\|$. We have $T(x), x_1 \in L_1$, so $h - k - y = T(h) + T(x_1) - x_1 = T(x) - x_1 \in L_1$. Thus $\|h - (h - k - y)\| \geq \|h\|$, i.e.,

$$\|h\| \leq \|k + y\| \leq \|k\| + \|y\| \leq \gamma\|h\| + \gamma\|x_1\| = \gamma\|h\| + \gamma.$$

We conclude $(1 - \gamma)\|h\| \leq \gamma$, i.e., $\|h\| \leq \gamma/(1 - \gamma)$. Hence

$$\|k\| \leq \gamma\|h\| \leq \frac{\gamma^2}{1 - \gamma}.$$

We now have

$$\|T(x) - x_1\| = \|T(x_1) - x_1 + T(h)\| \leq \|T(x_1) - x_1\| + \|h - k\|$$

$$\leq \|y\| + \|h\| + \|k\| \leq \gamma + \frac{\gamma}{1 - \gamma} + \frac{\gamma^2}{1 - \gamma} = \frac{2\gamma}{1 - \gamma}.$$

But since $\gamma < 1/3$, we conclude $2\gamma/(1 - \gamma) < 2\gamma/(2/3) = 3\gamma < (1/3)\varepsilon$. By Lemma IV.6.7, $T(x) \in C(\varepsilon)$.

If $x_1 = 0$, then $T(h) \in L_1$. Thus, as before, $\|h - T(h)\| \geq \|h\|$. But

$$\|h - T(h)\| \leq \|I - T\| \, \|h\| \leq \gamma\|h\| < \|h\|,$$

a contradiction. So this case cannot happen. ∎

We return to the case that L is a completely normable vector space. In Definition I.5.19 we introduced the idea of a *vector field distribution on a manifold*. We specialize this notion for the present purposes.

IV.6.12. **Definition.** A *wedge field* on an open subset B of L is a function $V: B \to \mathcal{W}(L)$, where $\mathcal{W}(L)$ denotes the collection of wedges in L. A continuous function $x: [0, T] \to B$ which is differentiable off of a countable set and satisfies $x'(t) \in V(x(t))$ whenever $x'(t)$ is defined is called a *solution of the wedge field V*.

If a wedge field satisfies some type of continuity condition, then we can frequently obtain local information about solutions of the wedge field by passing to a slightly larger constant surrounding wedge field. Hence it is useful to have on record the following basic proposition concerning constant wedge fields.

IV.6.13. **Proposition.** *Let B be an open subset of L, and let W be a closed wedge. If $x: [0, T] \to B$ is a solution of the constant wedge field W (that is, $V(x) = W$ for all $x \in B$), then*

(i) *$t_1 \leq t_2$ implies $x(t_2) \in x(t_1) + W$ for all $0 \leq t_1 \leq t_2 \leq T$.*

(ii) *If $\gamma \in W^*$, then $t \mapsto \gamma(x(t))$ is non-decreasing on $[0, T]$.*

Proof. Let us first establish (ii). Since γ is linear, the derivative at t of $t \mapsto \gamma(x(t))$ is $\gamma(x'(t)) \in \gamma(W)$. Since $\gamma \in W^*$, $\gamma(x'(t)) \geq 0$. Thus the function $t \mapsto \gamma(x(t))$ is non-decreasing. By Proposition I.1.4, we have the duality $W^{**} = W$. Hence $y \in x(t_1) + W$ if and only if $\gamma(y - x_1) \geq 0$ for all $\gamma \in W^*$. Thus

$$x(t_1) + W = \{y \in \mathfrak{g} : \gamma(y) \geq \gamma(x(t_1)) \text{ for all } \gamma \in W^*\}.$$

Therefore, the first assertion follows from the second. ∎

7. The rerouting technique

Let Ω be a subset of a Dynkin algebra $\mathfrak{g}$, and let $\mathcal{U}$ be a class of steering functions with codomain Ω. If some point very near 0 is reachable at a high cost, then by adjusting the steering function, it may be possible to reach the point at a much lower cost. Such a replacement procedure we call rerouting. We first develop a local rerouting theory.

Local rerouting

IV.7.1.   Definition. A class $\mathcal{U}$ of steering functions *admits rerouting locally* if there exists $\xi > 0$ such that the full reachable set $\mathcal{R}_{\mathcal{U}}(< \xi)$ is defined and given any $0 < \delta \leq \xi$, there exists a C-H-neighborhood B such that any point in B which is $\mathcal{U}$-reachable at some cost less than ξ is also $\mathcal{U}$-reachable at some cost less than δ, that is, $\mathcal{R}_{\mathcal{U}}(< \xi) \cap B \subseteq \mathcal{R}_{\mathcal{U}}(< \delta)$.

The hypothesis of the Confluence Theorem IV.5.19 that the reachable sets belong to one and the same germ is an alternate formulation of this idea.

IV.7.2.   Proposition. *Let $\mathcal{U}$ be a class of steering functions. The following statements are equivalent:*

(1) *The class $\mathcal{U}$ admits rerouting locally.*

(2) *There exists $\xi > 0$ such that if x_n is a sequence in $\mathcal{R}_{\mathcal{U}}(< \xi)$ converging to 0, then x_n is $\mathcal{U}$-reachable at cost δ_n, where $\lim \delta_n = 0$.*

(3) *There exists $\xi > 0$ such that the sets $\mathcal{R}_{\mathcal{U}}(< \delta)$ all belong to the same germ for $0 < \delta \leq \xi$.*

Proof. $(1) \Rightarrow (3)$. Given $0 < \delta \leq \xi$, there exists a C-H-neighborhood B such that $\mathcal{R}_{\mathcal{U}}(< \varepsilon) \cap B \subseteq \mathcal{R}_{\mathcal{U}}(< \delta)$, and hence $\mathcal{R}_{\mathcal{U}}(< \varepsilon)$ and $\mathcal{R}_{\mathcal{U}}(< \delta)$ are in the same germ. Since δ was arbitrary, conclusion (3) follows.

$(3) \Rightarrow (2)$. Let $\lim x_n = 0$, where each $x_n \in \mathcal{R}_{\mathcal{U}}(< \xi)$. If $x_n = 0$, let $\delta_n = 0$. Otherwise pick δ_n such that $x_n \in \mathcal{R}_{\mathcal{U}}(< \delta_n) \setminus \mathcal{R}_{\mathcal{U}}(< \delta_n/2)$. Given any $0 < \varepsilon < \xi$, there exists a C-H-neighborhood B such that $\mathcal{R}_{\mathcal{U}}(< \varepsilon) \cap B = \mathcal{R}_{\mathcal{U}}(< \xi) \cap B$. For all $x_n \in \mathcal{R}_{\mathcal{U}}(< \xi) \cap B$, it follows that $\delta_n < 2\varepsilon$. Since ε was arbitrary, $\lim \delta_n = 0$.

$(2) \Rightarrow (1)$. Assume $\mathcal{U}$ does not admit rerouting locally. Then there exists a positive $\delta < \xi$ such that $\mathcal{R}_{\mathcal{U}}(< \delta) \cap B$ is a strict subset of $\mathcal{R}_{\mathcal{U}}(< \xi) \cap B$ for every C-H-neighborhood B. Let B_n denote the open ball of radius $1/n$ and pick x_n in $(\mathcal{R}_{\mathcal{U}}(< \xi) \cap B_n) \setminus (\mathcal{R}_{\mathcal{U}}(< \delta) \cap B_n)$. Then $\lim x_n = 0$, but $x_n \notin \mathcal{R}_{\mathcal{U}}(< \delta)$ for every n, a contradiction. ∎

This is the place to generalize some of the ideas and results of Section 3. We introduce certain classes of steering functions that are appropriate for our present purposes.

IV.7.3. **Definition.** An *admissible class of steering functions for* Ω is a class of regulated functions which satisfy the following conditions:

(i) The domain of each function is some finite interval of real numbers and the codomain is Ω.

(ii) The class contains any piecewise constant function satisfying the conditions of (i).

(iii) The restriction of any member of the class to any smaller interval is a member of the class.

(iv) The composition of a translation of $\mathbb{R}$ with any member of the class is a member of the class.

(v) If the domain of a function is an interval which is the union of two subintervals and if the restriction of the function to each of the subintervals is in the class, then the function itself is in the class.

Given two members $u: [0, T_u] \to \Omega$ and $v: [0, T_v] \to \Omega$ from an admissible class, their *concatenation* $ubv: [0, T_u + T_v] \to \Omega$ defined by $ubv(t) = u(t)$ for $0 \le t \le T_u$ and $ubv(t) = u(T_n) * v(t - T_u)$ for $T_u < t \le T_u + T_v$ is again in the class in view of conditions (iv) and (v). The subclass $\mathcal{U}$ of functions u in the class with domain of the form $[0, T_u]$ is called a *semigroup class of steering functions*, as it is a semigroup under concatenation.

Note that the classes of piecewise constant functions, piecewise analytic functions, piecewise continuous functions, and regulated functions with codomain Ω are all admissible classes.

IV.7.4. **Proposition.** *Let x and y be solutions of the differential equation* (D_0) *of Definition IV.5.7 on some C-H-neighborhood B for the steering functions* $u(\cdot)$ *and* $v(\cdot)$, *respectively. Then the solution of* (D_0) *for the steering function* ubv *is given by* $z(t) = x(t)$ *for* $0 \le t \le T_u$ *and* $z(t) = x(T_u) * y(t - T_u)$ *for* $T_u < t \le T_u + T_v$.

Proof. The proof follows immediately from the definition of concatenation and Proposition IV.5.4. ∎

In Section IV.5 we saw that many of the results of Section IV.3 could be reinterpreted in the language of piecewise constant steering functions. In the latter framework many of them extend to arbitrary semigroup classes of steering functions. We have, for example, the following analog of Proposition IV.3.17:

IV.7.5.   Proposition. *Let Ω be an arbitrary subset of a Dynkin algebra $\mathfrak{g}$, and let $\mathcal{U}$ be a semigroup class of steering functions into Ω. Choose δ so that $\mathcal{R}_\mathcal{U}(< t)$ is defined for all $t < T$. Then $t \mapsto \mathcal{R}_\mathcal{U}(t)$ and $t \mapsto \mathcal{R}_\mathcal{U}(< t)$ are continuous local one parameter semigroups.*

If Ω is bounded, then the same conclusions are true for $t \mapsto \mathcal{A}_\mathcal{U}(t)$ and $t \mapsto \mathcal{A}_\mathcal{U}(< t)$.

Proof.   That $t \mapsto \mathcal{R}_\mathcal{U}(t)$ is continuous follows from Proposition IV.5.13 or from Propositions IV.3.17 and IV.5.11. One uses Proposition IV.7.4 in a straightforward manner to establish that $\mathcal{R}_\mathcal{U}(t) * \mathcal{R}_\mathcal{U}(s) = \mathcal{R}_\mathcal{U}(t + s)$. Indeed, to show inclusion in one direction, one concatenates, and in the other direction, one breaks the functions up into component parts. The conditions (i)-(v) of Definition IV.7.3 guarantee that all the needed functions are again back in $\mathcal{U}$, and the costs add up correctly since the integral is translation invariant. The proofs for the other cases are analogous.∎

We shall see now that we also have generalizations of Lemma IV.3.18 and of the Confluence Theorem IV.3.19.

IV.7.6.   Lemma. *Suppose that $\mathcal{R}_\mathcal{U}(< \delta)$ and $\mathcal{R}_\mathcal{U}(< 2\delta)$ are full reachable sets and B is a C-H-neighborhood such that $B \cap \mathcal{R}_\mathcal{U}(< \delta) = B \cap \mathcal{R}_\mathcal{U}(< 2\delta)$. Then $S = B \cap \mathcal{R}_\mathcal{U}(< \delta)$ is a local semigroup with respect to B.*

Proof.   Let $x, y \in S$ such that $x * y \in B$. By Proposition IV.7.5, $x * y \in \mathcal{R}_\mathcal{U}(< 2\delta)$, Hence $x * y \in \mathcal{R}_\mathcal{U}(< \delta) \cap B = S$.   ∎

IV.7.7.   Theorem. (The General Confluence Theorem) *Let $\Omega \subseteq \mathfrak{g}$ and $\mathcal{U}$ be a semigroup of control functions with codomain Ω. If there exists $\xi > 0$ such that the full reachable sets $\mathcal{R}_\mathcal{U}(< \delta)$ all belong to a germ Γ for $0 < \delta \leq \xi$ (equivalently, if $\mathcal{U}$ admits rerouting locally), then Γ contains a local semigroup S.*

Furthermore, the following statements hold:

(i) *There exists a C-H-neighborhood B such that $\mathrm{Sg}(\Omega, B) \subseteq S$.*

(ii) *Given a C-H-neighborhood B', there exists a smaller C-H-neighborhood B such that $S \cap B \subseteq \overline{\mathrm{Sg}(\Omega, B')}$.*

(iii) *If W is the smallest Lie wedge containing Ω, and $\overline{\Gamma}$ is the germ of $\mathcal{R}_\Omega(< \delta)$, then $\mathcal{R}_W(\leq \delta)$, $\overline{\mathrm{Sg}(\Omega, B)}^B$, and $\overline{\mathrm{Sg}(W, B)}^B$ all belong to $\overline{\Gamma}$ for all $\delta \leq \gamma$ and all B sufficiently small.*

Proof.   Recall first from Proposition IV.7.2, that $\mathcal{U}$ admits rerouting locally if and only if the sets $\mathcal{R}_\mathcal{U}(< \delta)$ belong to one and the same germ Γ for all small δ.

Pick a C-H-neighborhood B such that $B \cap \mathcal{R}_\mathcal{U}(< \xi/2) = B \cap \mathcal{R}_\mathcal{U}(< \xi)$ and call this set S. Then by Lemma IV.7.4, S is a local semigroup with respect to B. Clearly S belongs to Γ.

(i) By Proposition IV.3.6 we have $t \cdot x \in B \cap \mathrm{Reach}_\Omega(< \xi, B)$ for small enough $t \in \mathbb{R}^+$ and $x \in \Omega$. By Proposition IV.5.11,

$$\mathrm{Reach}_\Omega(< \xi, B) \subseteq \mathcal{R}_\mathcal{U}(< \varepsilon) \cap B = S.$$

Since S is a local semigroup with respect to B, therefore, $\mathbb{R}^+ \cdot \Omega \cap B \subseteq S$, and hence $\mathrm{Sg}(\Omega, B) \subseteq S$.

(ii) By the continuity of $t \mapsto \mathcal{R}_\mathcal{U}(, t)$ in Proposition IV.7.5, we find a $\delta > 0$ such that $\delta \leq \xi$ and $\mathcal{R}_\mathcal{U}(< \delta) \subseteq B'$. Since $\mathcal{R}_\mathcal{U}(< \delta)$ and S both belong to Γ, we find a C-H-neighborhood B in B' such that $\mathcal{R}_\mathcal{U}(< \delta) \cap B = S \cap B$. Then

$$S \cap B = \mathcal{R}_\mathcal{U}(< \delta) \cap B = \mathcal{R}_\mathcal{U}(< \delta, B') \cap B$$
$$\subseteq \mathcal{R}_\Omega(\leq \delta, B') \subseteq \overline{\mathrm{Sg}(\Omega, B')}^{B'},$$

where the second equality holds from the choice of δ, the first inclusion holds from Proposition IV.5.11, and the last inclusion holds from Proposition IV.3.7.

(iii) We may assume that ξ is such that for a C-H-neighborhood B, the closure of $\mathcal{R}_\mathcal{U}(< \xi)$ is still contained in B. Then $\mathcal{R}_\mathcal{U}(< \delta) = \mathcal{R}_\mathcal{U}(< \delta, B)$ for $\delta \leq \xi$. It follows from Proposition IV.5.11 that $\mathcal{R}_\Omega(\leq \delta, B)$, the closure in B of $\mathrm{Reach}_\Omega(< \delta, B)$, is also the closure in B of $\mathcal{R}_\mathcal{U}(< \delta, B)$. Since B is large enough, this translates to the statement that $\mathcal{R}_\Omega(\leq \delta)$ is equal to the closure of $\mathcal{R}_\mathcal{U}(< \delta)$. By (iii) of Proposition IV.1.9 the sets $\mathcal{R}_\Omega(\leq \delta)$ all belong to the same germ $\overline{\Gamma}$. The desired conclusion now follows from the Confluence Theorem IV.3.19(ii). ∎

IV.7.8. **Corollary.** *Let $\Omega \subseteq \mathfrak{g}$ and let $\mathcal{U}$ be a semigroup class of steering functions with codomain Ω. Then $\mathcal{U}$ admits local rerouting if and only if the sets $\mathcal{R}_\mathcal{U}(< \delta)$ are local semigroups for all δ small enough. In this case, the sets $\mathcal{R}_\Omega(\leq \delta)$ are also local semigroups.*

Proof. If $\mathcal{U}$ admits local rerouting, then by the preceding General Confluence Theorem there exists a local semigroup $S \in \mathrm{Germ}(\mathcal{R}_\mathcal{U}(< \delta))$ for a sufficiently small δ. Then also the local semigroup $\overline{S}$ is in $\mathrm{Germ}(\mathcal{R}_\Omega(\leq \delta))$. It follows that both $\mathcal{R}_\mathcal{U}(< \delta)$ and $\mathcal{R}_\mathcal{U}(\leq \delta)$ also are local semigroups.

Conversely, suppose that $\mathcal{R}_\mathcal{U}(< \delta)$ is a local semigroup for each $\delta \leq \varepsilon$, where ε is fixed. Let $\delta = \varepsilon/2^n$ and let $T = \mathcal{R}_\mathcal{U}(< \delta)$. By Proposition IV.1.12, there exists a C-H-neighborhood B such that $B \cap T$ is a local semigroup with respect to B. Then $T^2 \cap B \subseteq T$. By Proposition IV.7.5 we have $T^2 = \mathcal{R}_\mathcal{U}(< 2\delta)$. Thus $\mathcal{R}_\mathcal{U}(< 2\delta) \cap B = T^2 \cap B \subseteq T \cap B$. Since the reverse inclusion is immediate, we conclude that $\mathcal{R}_\mathcal{U}(< 2\delta)$ and $\mathcal{R}_\mathcal{U}(< \delta)$ are in the same germ Γ, and hence every $\mathcal{R}_\mathcal{U}(< \gamma)$ is in Γ for $\delta \leq \gamma \leq 2\delta$. An inductive argument now completes the proof. ∎

Achieving rerouting

We shall now work in the the following context:

Let $\mathfrak{g}$ be a Dynkin algebra, let W be a strongly positive Lie wedge in $\mathfrak{g}$, let $H = H(W)$ be the edge of W, and let L_1 be a closed complement for H. See Definition IV.4.18 and Section 6 for many of the basic concepts and background results.

Our ultimate goal is to show that if we take our control set $\Omega = W$ and a suitable class $\mathcal{U}$ of steering functions into W, then W admits rerouting locally. To accomplish this, we give explicitly an efficient method of rerouting any trajectory satisfying (D_0) and lying in some fixed C-H-neighborhood B. The idea of the rerouting is elementary. We use the C-H-multiplication $*$ to write each member of B uniquely as a product $x * h$, where $x \in L_1$ and $h \in H$. We use this factorization to project the trajectory onto L_1 to form the first leg of the rerouting. The second leg follows the appropriate left invariant vector field (which corresponds to some member of H) from the terminal point of the first leg to the terminal point of the original trajectory. These two legs together make up the new trajectory.

There are two things that must be checked in the preceding procedure. First of all, it must be established that the new trajectory arises from a steering function which is again in $\mathcal{U}$. This will be the essential content of the Rerouting Theorem IV.7.12. It is at this point (and at this point alone) that the fact that W is a Lie wedge (that is, that it is invariant under the action of the group generated by the edge) comes into play. Secondly, it must be checked that the rerouting is efficient in the sense that the cost of the new trajectory is low enough to insure that $\mathcal{U}$ admits local rerouting.

One has a rather wide choice for the admissible class of functions to use. Since the new trajectory constructed includes a composition with the projection from B into L_1 arising from the factorization of the C-H-multiplication $*$, the class of piecewise constant functions no longer suffices, but one can take the class of piecewise analytic functions (since the C-H-multiplication is analytic) or the increasingly larger classes of piecewise C^∞-functions, piecewise continuous functions, or regulated functions. The semigroup class $\mathcal{U}$ then consists of all such functions with domain $[0, T]$ for some $T > 0$ and codomain W. In the remainder we consider the class $\mathcal{U}$ of piecewise continuous functions, and leave the minor adjustments necessary for the other cases to the reader.

Let B be a C-H-neighborhood in $\mathfrak{g}$ and let $x \in B$. Then $\lambda_x : B \to \mathfrak{g}$ defined by $\lambda_x(y) = x * y$ has derivative $\lambda'_x(0) = d\lambda_x(0) = g(\operatorname{ad} x)$ (see for instance, Lemma II.2.5, or Section 5 above).

We note some basic properties of the derivatives.

IV.7.9.   **Lemma.**   *For $y \in B * B$, let $\rho_y : B * B \to \mathfrak{g}$ be defined by $\rho_y(x) = x * y$. Then for all $x, y \in B$ we have the following conclusions:*

    (i) $\left(d\lambda_y(0)\right)^{-1} = d\lambda_{-y}(y); \; \left(d\rho_y(0)\right)^{-1} = d\rho_{-y}(y),$

    (ii) $d\lambda_x(y) = d\lambda_{x*y}(0)\left(d\lambda_y(0)\right)^{-1}, \; d\rho_y(x) = d\rho_{x*y}(0)\left(d\rho_x(0)\right)^{-1},$

    (iii) $d\lambda_x(0) = d\rho_x(0) \, e^{\operatorname{ad} x}$, *and*

    (iv)   *if $y = x * h$, $d\lambda_y(0) = d\rho_h(x)d\lambda_x(0)e^{\operatorname{ad} h}$.*

Proof.   (i) We have by the chain rule

(1) $$d\lambda_{x*y}(0) = d(\lambda_x \lambda_y)(0) = d\lambda_x(y)d\lambda_y(0).$$

In particular, $I = d\lambda_0(0) = d\lambda_{-y*y}(0) = d\lambda_{-y}(y)d\lambda_y(0)$. Since λ_x is a local diffeomorphism (see the proof of Proposition IV.1.35), this establishes the first part of (i). The second follows similarly.

(ii) The first half of (ii) follows by applying $(d\lambda_y(0))^{-1}$ on the right to both sides of equation (1) above. The second half follows analogously.

(iii) For all $y \in B$,

$$\lambda_x(y) = x * y = (x * y * (-x)) * x = (e^{\operatorname{ad} x} y) * x = \rho_x(e^{\operatorname{ad} x} y).$$

Hence $d\lambda_x(0) = d(\rho_x e^{\operatorname{ad} x})(0) = d\rho_x(0) de^{\operatorname{ad} x}(0) = d\rho_x(0) e^{\operatorname{ad} x}$, since $e^{\operatorname{ad} x}$ is linear.

(iv) Keeping (iii) in mind, we compute

$$\begin{aligned}
d\lambda_y(0) &= d(\lambda_x \lambda_h)(0) = d\lambda_x(h) d\lambda_h(0) \\
&= d\lambda_x(h) d\rho_h(0) e^{\operatorname{ad} h} = d(\lambda_x \rho_h)(0) e^{\operatorname{ad} h} \\
&= d(\rho_h \lambda_x)(0) e^{\operatorname{ad} h} = d\rho_h(x) d\lambda_x(0) e^{\operatorname{ad} h}.
\end{aligned}$$

∎

IV.7.10. **Lemma.** *Let* $\pi_1 \colon L_1 \times H \to L_1$, $\pi_H \colon L_1 \times H \to H$ *denote the projections to* L_1 *and* H *respectively, and let* $B_1 = L_1 \cap B$ *and* $B_H = H \cap B$. *Let* $m \colon B_1 \times B_H \to \mathfrak{g}$ *be defined by* $m(x, h) = x * h$, *where* $*$ *is the Baker–Campbell–Hausdorff multiplication. Then* m *is an analytic diffeomorphism for* B *small enough and* $dm \colon L_1 \times H \to \mathfrak{g}$ *is given by*

$$dm(x, h) = d\rho_h(x) \pi_1 + d\lambda_{x*h}(0) f(\operatorname{ad} h) \pi_H.$$

Moreover,

$$dm(x, h)(u, v) \in d\rho_h(x) u + d\lambda_{x*h}(0) H.$$

Proof. Let $\eta \colon B \times B \to \mathfrak{g}$ be defined by $\eta(x, y) = x * y$. It follows directly from the Baker–Campbell–Hausdorff formula (see Appendix 1) that $d\eta(0, 0) \colon \mathfrak{g} \times \mathfrak{g} \to \mathfrak{g}$ is given by $d\eta(0, 0)(u, v) = u + v$. Then $m = \eta \circ (\pi_1 \times \pi_H)$ implies $dm(0, 0) = d\eta(0, 0) \circ (\pi_1 \times \pi_H) = \pi_1 + \pi_H$. Thus dm is an isomorphism, so by the Inverse Function Theorem m is locally a diffeomorphism at $(0, 0)$. It is analytic since the C-H-multiplication is.

Let $(x, h) \in B_1 \times B_H$. Then

$$dm(x, h)(u, v) = \left(\frac{\partial}{\partial x} m\right)(x, h)(u) + \left(\frac{\partial}{\partial h} m\right)(x, h)(v),$$

and we observe that the partial derivatives are computed as follows:

$$\left(\frac{\partial}{\partial x} m\right)(x, h)(u) = d(\xi \mapsto \xi * h)(x, h)(u) = d\rho_h(x)(u),$$

and

$$\left(\frac{\partial}{\partial h} m\right)(x, h)(v) = d(\xi \mapsto x * \xi)(x, h)(v) = d\lambda_x(h)(v).$$

So

$$dm(x, h)(u, v) = d\rho_h(x)(u) + d\lambda_x(h)(v) = d\rho_h(x) u + d\lambda_{x*h}(0) \big(d\lambda_h(0)\big)^{-1} v$$

by Lemma IV.7.9(ii).

Finally $\big(d\lambda_h(0)\big)^{-1}(v) = \big(g(\operatorname{ad} h)\big)^{-1}(v) = f(\operatorname{ad} h)(v) \in H$ since $h, v \in H$ and H is a subalgebra. The last assertion then follows from the preceding equality. ∎

In light of the preceding lemma we can replace B if necessary by a smaller open neighborhood of 0 (again called B) on which $m^{-1}: B \to L_1 \times H$ is defined and is an analytic diffeomorphism and such that $B = B_1 * B_H$, where B_1 and B_H are balls in the respective norms of L_1 and H.

IV.7.11.   Remark. Using the notation just introduce, let $p_1 = \pi_1 m^{-1}$ and $p_H = \pi_H m^{-1}$. Then p_1 and p_H are analytic mappings, and $m^{-1} = p_1 \times p_H$; hence $m: B_1 \times B_H \to B$ and $p_1 \times p_H: B \to B_1 \times B_H$ are mutually inverse analytic diffeomorphisms. Thus for $x \in B$, $x = p_1(x) * p_H(x)$, where $p_1(x) \in L_1$ and $p_H(x) \in H$, and this factorization is unique.   ∎

We come now to a crucial ingredient in the proof of the characterization of Lie wedges. Recall from Definition II.1.3 that W is a Lie wedge if and only if $e^{\operatorname{ad} z}W = W$ for all z in the edge H of W. The initial value problem (D_0) is to be found in Definition IV.5.7.

IV.7.12.   Theorem. (The Rerouting Theorem) *Let W be a Lie wedge in the Dynkin algebra $\mathfrak{g}$, and let $\mathcal{U}$ be the semigroup class of piecewise continuous steering functions into W. Let $x \in \mathcal{R}_{\mathcal{U}}(< \delta, B)$, and let $u: [0,T] \to W$ be a steering function in $\mathcal{U}$ such that the solution x of (D_0) for u satisfies $x(T) = x$. We set $h(t) = p_H\big(x(t)\big)$, $x_1(t) = p_1\big(x(t)\big)$, and*

$$\overline{x}(t) = \begin{cases} x_1(t) & \text{for } 0 \leq t \leq T, \\ x_1(T) * (t - T) \cdot h(T) & \text{for } T \leq t \leq T + 1. \end{cases}$$

Then $\overline{x}$ is a solution of (D_0) for some $\overline{u} \in \mathcal{U}$ and $\overline{x}(T+1) = x(T) = x$.

Moreover, $\overline{x}(t) \in B_1$ for all $t \in [0,T]$ and $\overline{x}(t) \in B_1 * [0,1] \cdot h(T)$ otherwise.

Proof.   By (D_0), the function x has the piecewise continuous derivative

$$x'(t) = g\big(\operatorname{ad} x(t)\big)u(t).$$

Then the function $x_1 = p_1 \circ x$ is continuous and has a piecewise continuous derivative. By (D') of Theorem IV.5.1, the steering function $u_1(t)$ for $x_1(t)$ is given by $f(\operatorname{ad} x_1(t))x_1'(t)$, and hence is piecewise continuous. Thus to show that $u_1 \in \mathcal{U}$, it suffices to show that $x_1'(t) = g\big(\operatorname{ad} x(t)\big)u_1(t) \in g(\operatorname{ad} x(t)W$ whenever $x_1'(t)$ exists.

We have $x(t) = x_1(t) * h(t) = m(x_1(t), h(t))$, where $x_1(t) \in B_1$ and $h(t) \in B_H$. By the Chain Rule and Lemma IV.7.10,

$$x'(t) = dm\big(x_1(t), h(t)\big)\big(x_1'(t), h'(t)\big) \in d\rho_{h(t)}\big(x_1(t)\big)(x_1'(t)) + d\lambda_{x(t)}(0)H.$$

Thus

$$d\rho_{h(t)}\big(x_1(t)\big)\big((x_1'(t)\big) \in x'(t) + d\lambda_{x(t)}(0)H$$
$$= g\big(\operatorname{ad} x(t)\big)u(t) + g\big(\operatorname{ad} x(t)\big)W \subseteq g\big(\operatorname{ad} x(t)\big)W.$$

We conclude that

$$x_1'(t) \in \left(d\rho_{h(t)}(x_1(t))\right)^{-1} g(\operatorname{ad} x(t))W = \left(d\rho_{h(t)}x_1(t)\right)^{-1} d\lambda_{x(t)}(0)W$$
$$= \left(d\rho_{h(t)}x_1(t)\right)^{-1} d\rho_{h(t)}\left(x_1(t)\right) d\lambda_{x_1(t)}(0) e^{\operatorname{ad} h(t)} W \quad \text{(by Lemma IV.7.9(iv))}$$
$$= d\lambda_{x_1(t)} W \quad (\text{since } e^{\operatorname{ad} h(t)} W = W \text{ by hypothesis})$$
$$= g(\operatorname{ad} x_1(t))W.$$

By Proposition IV.5.5, the function $\overline{x}$ has constant steering function $h(T) \in H \subseteq W$ on the interval $[T, T{+}1]$. Thus $\overline{u}$, the concatenation of u_1 and the constant function $h(T)$ is a piecewise continuous function into W and a steering function for $\overline{x}$.

The last assertion of the theorem follows immediately from the construction of $\overline{x}$. ∎

By Proposition IV.6.10, there exists a norm $\|\cdot\|$ on L, a linear functional ω on $\mathfrak{g}$ with $\|\omega\| = 1$ and $\omega(H) = \{0\}$, a C-H-neighborhood B, and a strictly positive cone $C(\varepsilon)$ surrounding $C = W \cap L_1$ in L_1 satisfying

(i) if $x = x_1 + h$, $x_1 \in L_1$, $h \in H$, then $\|x\| = \max\{\|x_1\|, \|h\|\}$;

(ii) if $x \in C(\varepsilon)$, then $\|x\| = \omega(x)$;

(iii) if $x \in B$ and $u \in \mathfrak{g}$, then

$$(\ddagger) \qquad\qquad \|g(\operatorname{ad} x)(u) - u\| \le \|x\| \, \|u\|.$$

Observe that the inequality $(\ddagger)$ above is a special case of the inequality $(\ddagger)$ in Proposition IV.6.10 from which we obtain it by taking $y = 0$ and $g(\operatorname{ad} y)$ the identity of $\mathfrak{g}$. We further assume that B is chosen small enough so that there exist unique continuous projections $p_H : B \to H$ and $p_1 : B \to L_1$ satisfying $x = p_1(x) * p_H(x)$ for each $x \in B$ as in Remark IV.7.11.

IV.7.13. **Theorem.** (The Local Rerouting Theorem) *Any Lie wedge W in a Dynkin algebra admits rerouting locally for the semigroup class of piecewise continuous functions.*

Proof. Assume that everything is chosen and fixed to satisfy the preceding conditions. Pick an open ball B_γ of radius γ around 0 contained in the open set B such that $p_1(B_\gamma)$ and $p_H(B_\gamma)$ are contained in B and in the ball around 0 of radius $\varepsilon/9$. By Proposition IV.7.5, the family $\mathcal{R}_\mathcal{U}(< t)$ is continuous. Hence we find a $0 < \xi < 1/3$ such that $\mathcal{R}_\mathcal{U}(< \xi)$ is defined and contained in B_γ.

Fix some δ with $0 < \delta < \xi$. Pick a C-H-neighborhood $B' \subseteq B_\gamma$ such that $p_1(B')$ and $p_H(B')$ are both contained in the open ball around 0 of radius $\delta/3$. Suppose $x \colon [0, T] \to B_\gamma$ is a solution to (D_0) with $x(T) \in B'$ for some piecewise continuous steering function $u \colon [0, T] \to W$. Let $x_1(t) = p_1\big(x(t)\big) \in p_1(B_\gamma)$, a subset of the ball around 0 of radius $\varepsilon/9$. Since also $x_1(t) \in B$, it follows from $(\ddagger)$ that

$$\|g(\operatorname{ad} x_1(t)) - I\| \le \|x_1(t)\| < \frac{\varepsilon}{9}.$$

By the Rerouting Theorem IV.7.12, we know that x_1 is a solution of (D_0) for some piecewise continuous function $u_1 \colon [0, T] \to W$. Since the solution x_1 lies entirely in the subspace L_1, we have $x_1'(t) \in L_1$; thus $x_1'(t) = g(\operatorname{ad} x_1(t))u_1(t) \in L_1 \cap g(\operatorname{ad} x_1(t))W$. From the preceding paragraph and Proposition IV.6.11, we conclude that $x_1'(t) \in C(\varepsilon)$ for $0 \le t \le T$. Applying Proposition IV.6.13 to the constant wedge field $C(\varepsilon)$, we conclude that $x_1(t) \in x_1(0) + C(\varepsilon) = C(\varepsilon)$ and

$$(\nearrow) \qquad\qquad t \mapsto \omega\big(x_1(t)\big) = \|x_1(t)\| \text{ is non-decreasing on } [0, T].$$

Since the norm is additive on $C(\varepsilon)$, we have

$$\|x_1(T)\| = \|\int_0^T x_1'(t)\, dt\| = \int_0^T \|x_1'(t)\|\, dt.$$

Thus

$$\int_0^T \|u_1(t)\|\, dt - \|x_1(T)\| \le \left| \int_0^T \|u_1(t)\|\, dt - \int_0^T \|x_1'(t)\|\, dt \right|$$

$$\le \int_0^T \big| \|x_1'(t)\| - \|u_1(t)\| \big|\, dt$$

$$\le \int_0^T \|x_1'(t) - u_1(t)\|\, dt$$

$$= \int_0^T \|g(\operatorname{ad} x_1(t))u_1(t) - u_1(t)\|\, dt$$

$$\le \int_0^T \|x_1(t)\|\, \|u_1(t)\|\, dt \quad (\text{by } (\ddagger))$$

$$\le \|x_1(T)\| \int_0^T \|u_1(t)\|\, dt \quad (\text{by } (\nearrow)).$$

Solving for $\int \|u_1(t)\|\, dt$, we obtain

$$\int_0^T \|u_1(t)\|\, dt \le \|x_1(T)\| / (1 - \|x_1(T)\|) < \frac{\delta/3}{2/3} = \frac{1}{2}\delta.$$

Since $x(T) \in B'$ we know $\|p_H\big(x(T)\big)\| < \frac{1}{3}\delta$. Thus using the constant steering function with value $p_H\big(x(T)\big)$ on the interval $[0, \|p_H\big(x(T)\big)\|]$, we see that $p_H\big(x(T)\big) \in \operatorname{Reach}_W(< \frac{1}{3}\delta)$. Concatenating this control with u_1 as, for example, in Proposition IV.7.4, we see that

$$x(T) = p_1\big(x(T)\big) * p_H\big(x(T)\big) \in \mathcal{R}_\mathcal{U}(< \frac{1}{2}\delta) * \operatorname{Reach}_W(< \frac{1}{3}\delta) \subseteq \mathcal{R}_\mathcal{U}(< \delta).$$

Thus $\mathcal{R}_\mathcal{U}(< \tau, B_\gamma) \cap B' \subseteq \mathcal{R}_\mathcal{U}(< \delta)$ for any $\tau > 0$; in particular, $\mathcal{R}_\mathcal{U}(< \xi) \cap B' \subseteq \mathcal{R}_\mathcal{U}(< \delta)$. Therefore W admits local rerouting. ∎

We note that the proof of Theorem IV.7.13 actually yields a slightly stronger result than the statement of the Theorem, namely, the following:

IV.7.14. **Remark.** For a Lie wedge W in a Dynkin algebra, there exists a C-H-neighborhood $B(= B_\gamma)$ such that for $\delta > 0$, there exists a C-H-neighborhood B' such that $\mathcal{R}_\mathcal{U}(B) \cap B' \subseteq \mathcal{R}_\mathcal{U}(< \delta, B)$, where

$$\mathcal{R}_\mathcal{U}(B) = \bigcup_{0 < t} \mathcal{R}_\mathcal{U}(< t, B).$$

∎

Another useful bit of information was uncovered in the proof of Theorem IV.7.13 and deserves to be recorded separately:

IV.7.15. **Remark.** Given $\|\cdot\|$, ω, L_1, and $C(\varepsilon)$ as above, there is an open ball B_γ of radius γ and a C-H-neighborhood $N \subseteq B_\gamma$ such that for a solution $x \colon [0, T] \to B_\gamma$ of (D_0) for a piecewise continuous steering function, the relation $x(T) \in N$ implies $p_1\big(x(t)\big) \in C(\varepsilon)$ for all $0 \leq t \leq T$. ∎

8. The Edge of the Wedge Theorem

In Theorem IV.1.27 we showed that the tangent set of a local semigroup is a Lie wedge. In Theorem IV.4.8 (with a more detailed version in the Cone Theorem IV.4.12) and the Split Wedge Theorem IV.4.26, we proved converses for special Lie wedges. The purpose of this section is to derive finally in full generality the converse statement, the Edge of the Wedge Theorem, which asserts that every strongly positive Lie wedge W arises as the tangent object of some local semigroup. Indeed, we saw in Section 7 that, for the class $\mathcal{U}$ of piecewise continuous functions, the sets $\mathcal{R}_{\mathcal{U}}(\delta)$ are local semigroups and belong to the same germ for all δ sufficiently small. It will result that their tangent set is W. Our method of proof will be to construct one specific local semigroup S, show that it is in the same germ as the preceding sets, and establish that its tangent wedge is W. Along the way we develop some of the special properties of the local semigroup S.

We work in essentially the same context as the preceding section. Specifically, we let $\mathfrak{g}$ be a *Dynkin algebra*, W be a *strongly positive Lie wedge* in $\mathfrak{g}$, further let $H = H(W)$ be the *edge of* W, and L_1 be a *closed complement for* H such that $L_1 \cap W$ is *strictly positive in* L_1. By Proposition IV.6.10, there exists a norm $\|\cdot\|$ on $\mathfrak{g}$, a linear functional ω on $\mathfrak{g}$ with $\|\omega\| = 1$ and $\omega(H) = \{0\}$, a C-H-neighborhood $\widetilde{B}$, and a strictly positive cone $C(\varepsilon)$ surrounding $C = W \cap L_1$ in L_1 satisfying the following conditions:

(i) If $x = x_1 + h$, $x_1 \in L_1$, $h \in H$, then $\|x\| = \max\{\|x_1\|, \|h\|\}$;

(ii) if $x \in C(\varepsilon)$, then $\|x\| = \omega(x)$;

(iii) if $x \in \widetilde{B}$ and $y \in \mathfrak{g}$, then

$$(\ddagger) \qquad\qquad \|g(\operatorname{ad} x)(y) - y\| \le \|x\|\,\|y\|.$$

We further assume that $\widetilde{B}$ is chosen so that there exist unique continuous projections $p_H : \widetilde{B} \to H$ and $p_1 : \widetilde{B} \to L_1$ satisfying $x = p_1(x) * p_H(x)$ for each $x \in \widetilde{B}$. Now choose an open ball $B_\gamma \subseteq \widetilde{B}$ of radius γ around 0 for $\widetilde{B}$ as in the proof of Theorem IV.7.13, that is, $p_1(B_\gamma)$ and $p_H(B_\gamma)$ are both contained in $\widetilde{B}$ and in the ball of radius $\varepsilon/9$ around 0. Finally pick an open neighborhood B of 0 so that $B * B \subseteq B_\gamma$ and so that the conditions of Remark IV.7.3 are satisfied, that is, $B_1 = p_1(B)$ and $B_H = p_H(B)$ are open balls in the respective norms of L_1 and H, and $m : B_1 \times B_H \to B$ is a diffeomorphism. We then recall the definition of $\mathcal{R}_{\mathcal{U}}(B)$ from Definition IV.5.7 and set

$$S = \mathcal{R}_{\mathcal{U}}(B) = \bigcup_{0 < t} \mathcal{R}_{\mathcal{U}}(< t, B),$$

where $\mathcal{U}$ is the semigroup class of piecewise continuous functions into W. We say that *S is the set of points reachable in B by means of piecewise continuous steering functions into W*.

Recall that the initial value problem (D_0) was defined in Definition IV.5.7.

IV.8.1.   Lemma.   *The set S is a local semigroup with respect to B.*

Proof.   Let $x, y \in S$ such that $x * y \in B$. Let $u : [0, T_x] \to W$ be a steering function for x, and $v : [0, T_y] \to W$ a steering function for y. Then the concatenation ubv is a steering function for $x * y$, that is, the solution z of (D_0) for ubv satisfies $z(0) = 0$, $z(T_x + T_y) = x * y$, and $z(t) \in B * B \subseteq B_\gamma$ for all t in view of Proposition IV.7.4. Let $z_1 = p_1 \circ z$. We saw in the proof of Theorem IV.7.13 in statement $(\nearrow)$ that the function $t \mapsto \|z_1(t)\|$ is non-decreasing on $[0, T_x + T_y]$. Since we are assuming $x * y \in B$, if follows that $z_1(T_x + T_y) = p_1(x * y) \in B_1$. Since B_1 is an open ball in L_1 and $\|z_1(t)\| \le \|z_1(T_x + T_y)\|$ for $0 \le t \le T_x + T_y$, we conclude $z_1(t) \in B_1$ for $t \in [0, T_x + T_y]$.

Also $x * y \in B$ implies $p_H(x * y) \in B_H$, and hence $r \cdot p_H(x * y) \in B_H$ for $0 \le r \le 1$. Thus for $T_x + T_y \le t \le T_x + T_y + 1$, we have

$$z_1(T_x + T_y) * (t - T_x - T_y) \cdot p_H(x * y) \in B_1 * B_H = B.$$

Hence the whole rerouting for z given in the Rerouting Theorem IV.7.12 lies in B, that is, $x * y \in \mathcal{R}_{\mathcal{U}}(B)$. This shows that S is a local semigroup with respect to B. ∎

IV.8.2.   Lemma.   *For all δ sufficiently small we have $\mathcal{R}_{\mathcal{U}}(< \delta) \in \mathrm{Germ}(S)$.*

Proof.   By Remark IV.7.6, there exists a C-H-neighborhood C such that

$$\mathcal{R}_{\mathcal{U}}(B) \cap C \subseteq \mathcal{R}_{\mathcal{U}}(B_\gamma) \subseteq \mathcal{R}_{\mathcal{U}}(< \delta, B_\gamma) \cap C.$$

The inclusion $\mathcal{R}_{\mathcal{U}}(< \delta, B) \cap C \subseteq \mathcal{R}_{\mathcal{U}}(B) \cap C$ always holds. Since $\mathcal{R}_{\mathcal{U}}(< \delta, B) = \mathcal{R}_{\mathcal{U}}(< \delta, B_\gamma) = \mathcal{R}_{\mathcal{U}}(< \delta)$ for all δ sufficiently small, $S \cap C = \mathcal{R}_{\mathcal{U}}(< \delta) \cap C$. ∎

IV.8.3.   Lemma.   *Let $S_1 = p_1(S)$ and $S_H = p_H(S)$. Then $(x_1, h) \mapsto x_1 * h : S_1 * S_H \to S$ is a diffeomorphism. Furthermore, $S_1 = S \cap L_1$ and $S_H = B_H = S \cap H$.*

Proof.   Since $x = p_1(x) * p_H(x)$ for $x \in S$, we have $S \subseteq S_1 * S_H$. Conversely, suppose $x_1 \in S_1$ and $h \in S_H$. Then $x_1 = p_1(x)$ for some $x \in S$. If $x : [0, T] \to B$ is a solution of (D_0) with $x(T) = x$ for some piecewise continuous steering function, then by the Original Rerouting Theorem IV.7.12, the function $\xi_1 = p_1 \circ x$ is also a solution of (D_0) in $B_1 \subseteq B$, and $\xi_1(T) = x_1$. Thus $x_1 \in S$, and the function $\overline{x}$ defined by $\overline{x}(t) = \xi_1(t)$ for $0 \le t \le T$ and $\overline{x}(t) = x_1 * (t - T) \cdot h$ for $T \le t \le T + 1$ satisfies (D_0) (again by the Original Rerouting Theorem) with $\overline{x}(T + 1) = x_1 * h$. Hence $x_1 * h \in S$. We conclude that $S_1 * S_H \subseteq S$, and so sets S and $S_1 \times S_H$ are carried to each other by the mutually inverse diffeomorphisms $p_1 \times p_H$ and m of Remark IV.7.3.

We saw in the preceding paragraph that if $x \in S$, then also $x_1 = p_1(x) \in S$. Hence $p_1(S) \subseteq S \cap L_1$. Conversely, for $x \in S \cap L_1$, we have $x = p_1(x) \in p_1(S)$. Thus the two sets are equal.

Let $h \in B_H$. Then $t \mapsto t \cdot h$ for $0 \leq t \leq 1$ is a solution of (D_0) for the constant steering function $h \in W$ by Proposition IV.5.5. Thus $h \in S \cap B_H$, so $h = p_H(h) \in p_H(S)$. Therefore

$$B_H \subseteq S \cap H \subseteq p_H(S) \subseteq B_H,$$

and so all three sets are equal.   ∎

IV.8.4.   Lemma. $\mathbf{L}(S_1) = W \cap L_1$.

Proof.   Let $x \in W \cap L_1$. As in the last paragraph of the preceding lemma, the mapping $t \mapsto t \cdot x$ for $0 \leq t \leq 1$ satisfies (D_0) for the constant steering function x, so $t \cdot x \in S$ for all $0 \leq t \leq 1$. Hence $x \in \mathbf{L}(S)$. We conclude $W \cap L_1 \subseteq \mathbf{L}(S_1)$.

The more difficult portion of the proof is the reverse containment. We easily note that $S_1 \subseteq L_1$ implies $\mathbf{L}(S_1) \subseteq \mathbf{L}(L_1) = L_1$. But for the remainder we shall show that $\mathbf{L}(S_1)$ is contained in the cone $C(\varepsilon')$ surrounding $W \cap L_1$ for arbitrarily small ε'. Since $W \cap L_1$ is the intersection of the $C(\varepsilon')$, $\varepsilon' > 0$ by Proposition IV.10.8, this will show that $\mathbf{L}(S_1) \subseteq W \cap L_1$ and thereby complete the proof.

Now let $\varepsilon \geq \varepsilon' > 0$ be given and consider an arbitrary element $x \in \mathbf{L}(S_1)$. We must show $x \in C(\varepsilon')$. We now repeat the construction of the earlier parts of this section up to this point with ε' in place of ε. This will result in C-H-neighborhoods C', $B' \subseteq B'_\gamma \subseteq \widetilde{B}$ and a local semigroup S' with respect to B'. However, by Lemma IV.7.2 above, S' will lie in the same germ as the sets $\mathcal{R}_\mathcal{U}(< \delta)$ for small δ, and hence in Germ(S) Thus the local semigroups S and S' will have the same tangent object, and thus $x \in \mathbf{L}(S'_1)$, too. Hence $x = \lim r_n \cdot s_n$, where $s_n \in S'_1$ and $\lim s_n = 0$. By Lemma IV.8.3, we have $s_n \in S'$; hence there exists $x_n \colon [0, T_n] \to B' \subseteq B'_\gamma$, a solution of (D_0) for some piecewise continuous steering function into W, such that $x_n(T_n) = s_n$. By Remark IV.7.7, there is a C-H-neighborhood N' such that for all n for which $x_n(T_n) = s_n \in N'$ we have $p_1 x_n(t) \in C(\varepsilon')$ for all $0 \leq t \leq T_n$. Hence $s_n = p_1(s_n) = p_1 x_n(T_n) \in C(\varepsilon')$. It follows that $x = \lim r_n \cdot s_n \in C(\varepsilon')$, and this is what we had to show.   ∎

IV.8.5.   Lemma. $\mathbf{L}(S) = W$.

Proof.   Consider the diffeomorphism $m \colon B_1 \times B_H \to B$. By Lemma IV.8.3, we know $m(S_1 \times S_H) = S$, so $dm(0,0)\bigl(\mathbf{L}(S_1 \times S_H)\bigr) = \mathbf{L}(S)$ by Proposition IV.1.19 applied to m and its inverse. Since $S_H = B_H$ by Lemma IV.8.3 and $\mathbf{L}(S_1) = W \cap L_1$ by Lemma IV.8.4, it now follows readily that

$$\mathbf{L}(S_1 \times S_H) = \mathbf{L}(S_1 \times B_H) = \mathbf{L}(S_1) \times \mathbf{L}(B_H) = (W \cap L_1) \times H.$$

As a special case of Lemma IV.7.10, we have $dm(0,0) = \pi_1 + \pi_H$, so

$$\mathbf{L}(S) = dm(0,0)\bigl(\mathbf{L}(S_1 \times S_H)\bigr) = (W \cap L_1) + H = W.$$

                                                                            ∎

We are now ready for the fundamental theorem of Lie wedges and the major objective of this chapter. Notice that to achieve the desirable factorization, the neighborhood B we find ourselves working in need not be a C-H-neighborhood (that is, it need not be symmetric nor closed under multiplication by scalars between 0 and 1). Also recall the definition of a strongly positive wedge from Definition IV.4.18.

IV.8.6. **Theorem.** (The Edge of the Wedge Theorem) *Let W be a strongly positive Lie wedge in a Dynkin algebra $\mathfrak{g}$ (or a Lie wedge in a finite dimensional real Lie algebra $\mathfrak{g}$). Let $\mathcal{U}$ denote the semigroup class of piecewise continuous steering functions. Then there exists an open neighborhood B of 0 such that if $S = \mathcal{R}_{\mathcal{U}}(B)$, the set of points $\mathcal{U}$-reachable in B, then*

   (i) *S is a local semigroup with respect to B and $\mathbf{L}(S) = W$.*

   (ii) *If $S_1 = S \cap L_1$ and $S_H = S \cap H(W)$, then S_H is a ball around 0 in $H(W)$ and a local group with respect to B and the mapping $(x_1, h) \mapsto x_1 * h \colon S_1 \times S_H \to S$ is the restriction of a diffeomorphism.*

   (iii) *The family $\mathcal{U}$ admits rerouting locally and S belongs to the same germ as the sets $\mathcal{R}_{\mathcal{U}}(< \delta)$ for all δ sufficiently small.*

   (iv) *$\overline{S}$ lies in the same germ as $\mathcal{R}_W(\leq \delta)$ for all δ sufficiently small and in the same germ as $\overline{\mathrm{Sg}(W, B')} \cap B'$ for all neighborhoods $B' \subseteq B$ of 0.*

Proof. The theorem is essentially a collection of earlier results. We first observe that by Remark IV.7.2, every wedge is strongly positive if $\mathfrak{g}$ is finite dimensional. Part (i) follows from Lemmas IV.8.1 and IV.8.5. Part (ii) follows from Lemma IV.8.3. Part (iii) follows from Theorem IV.7.5 and Lemma IV.8.2. Part (iii) of the General Confluence Theorem IV.7.7 implies that the sets $\mathcal{R}_W(\leq \delta)$ and $\overline{\mathrm{Sg}(W, B')} \cap B'$ all belong to the same germ for all δ and B' sufficiently small. In the proof of part (iii) of Theorem IV.7.7 it was shown that $\mathcal{R}_W(\leq \delta)$ is the closure of $\mathcal{R}_{\mathcal{U}}(< \delta)$; since the latter is in the same germ as S, it follows by IV.1.9(iii) that the former is in the same germ as $\overline{S}$. Let $B' \subseteq B$ be a neighborhood of 0. We have just seen that there exists a neighborhood $B'' \subseteq B'$ such that $\overline{\mathrm{Sg}(W, B'')} \cap B''$ is in the same germ as $\overline{S}$. Since by Proposition IV.2.3,

$$\overline{\mathrm{Sg}(W, B'')} \cap B'' \subseteq \overline{\mathrm{Sg}(W, B')} \cap B'' \subseteq \overline{S} \cap B'',$$

we conclude that $\overline{\mathrm{Sg}(W, B')} \cap B'$ must also be in this same germ. $\blacksquare$

The Edge of the Wedge Theorem is frequently stated in the following more compact form. This version is an immediate corollary of Theorems IV.8.6 and IV.1.27.

IV.8.7. **Corollary.** (Sophus Lie's Fundamental Theorem for Semigroups) *Let $\mathfrak{g}$ be a finite dimensional real Lie algebra. A subset W of $\mathfrak{g}$ is a Lie wedge if and only if it is the tangent set of some local semigroup.* $\blacksquare$

Now we can also sharpen Proposition IV.3.3 in finite dimensional Lie algebras.

IV.8.8. **Corollary.** *Let Ω be any subset of a finite dimensional real Lie algebra $\mathfrak{g}$. If W is the smallest Lie wedge containing Ω, then $\mathbf{L}\big(\mathrm{Sg}(\Omega, B')\big) = W$ for all C-H-neighborhoods B' sufficiently small.*

Proof. By the Edge of the Wedge Theorem IV.8.6, there exists an open neighborhood B of 0 and a local semigroup S with respect to B such that $\mathbf{L}(S) = W$. Let B' be a C-H-neighborhood contained in B. Then $S \cap B'$ is a local semigroup with respect to B' and $\mathbf{L}(S \cap B') = \mathbf{L}(S) = W$. By Theorem IV.4.6, $\mathbf{L}\big(\mathrm{Sg}(W, B')\big) = W$. Since $\Omega \subseteq W$, we have $\mathbf{L}\big(\mathrm{Sg}(\Omega, B')\big) \subseteq \mathbf{L}\big(\mathrm{Sg}(W, B')\big) = W$. The reverse inclusion was shown in Proposition IV.3.3. $\blacksquare$

Under the hypotheses of the Edge of the Wegde Theorem IV.8.6, for small enough C-H-neighborhoods B, any $x \in B$ factors uniquely as $x = x_1 * h$, $x_1 \in L_1$, $h \in H(W)$. The next exercise gives sufficient conditions for the semigroup S to have an analogous factorization in terms of the cone $C = L_1 \cap W$ and H.

EIV.8.1.   Exercise. Let $\mathfrak{g}$ be a finite dimensional Lie algebra which is the direct sum of a subalgebra $\mathfrak{h}$ and a vector subspace $\mathfrak{p}$ containing a pointed cone C satisfying

(i) $[\mathfrak{p}, \mathfrak{p}] \subseteq \mathfrak{h}$,

(ii) $e^{\operatorname{ad} \mathfrak{h}}(C) \subseteq C$.

For sufficiently small open balls B_1 and $B_\mathfrak{h}$ around 0 in $\mathfrak{p}$ and $\mathfrak{h}$, respectively, $S = (B_1 \cap C) * B_\mathfrak{h}$ is a closed strictly infinitesimally generated local semigroups with respect to $B = B_1 * B_\mathfrak{h}$ with tangent wedge $W = \mathfrak{h} \oplus C$, $H(W) = \mathfrak{h}$.

(Outline of solution.   By the unique product representation locally, $\overline{(B_1 \cap C} * \overline{B_\mathfrak{h}}) \cap B = (B_1 \cap C) * B_\mathfrak{h}$. Hence S is closed in B. To show that S is a local semigroup, it suffices to show locally that $C * C \subseteq C * H$, since then for $s_1 * s_2 \in B$, we have $s_1 * s_2 = s_1 * c_2 * h_2 = c_3 * h_3 * h_2 \in C * H$, and by unique factorization $c_3 \in C \cap B_1$ and $h_3 * h_2 \in B_\mathfrak{h}$.

For any choice of B_1 and $B_\mathfrak{h}$ with unique factorization, let $m: B_1 \times B_\mathfrak{h} \to B$ have inverse $\psi: B \to B_1 \times B_\mathfrak{h}$. Let $s \in S$, $a \in C \cap B_1$, $s * a \in B$, and consider the curve $\gamma(t) = s * t \cdot a$, $0 \le t \le 1$. To show that γ lies in B it suffices to show its image under ψ lies in $C \times \mathfrak{h}$, which we show with the help of Theorem I.5.17. We consider the vector field $X(y) = d\lambda_y(0)(a)$. To apply Theorem I.5.17, we must show $d\psi(y)X(y) \in L_{\psi(y)}(C \times \mathfrak{h}) = L_c(C) \times \mathfrak{h}$ for $y = c * h \in C * \mathfrak{h}$. Let $d\psi(y)\big(d\lambda_y(0)(a)\big) = (u, v) \in \mathfrak{p} \times \mathfrak{h}$. By Lemmas IV.7.9(iv) and 7.10 and by manipulations as in IV.7.12, we obtain $u \in g(\operatorname{ad} c)(e^{\operatorname{ad} h}a + \mathfrak{h}) \subseteq g(\operatorname{ad} c)(C + \mathfrak{h})$; as in Lemma V.4.56, the latter is in $L_c(W)$. Thus $u \in L_c(W) \cap \mathfrak{p} = L_c(C)$, as desired. The end of the proof of Theorem V.4.57 yields $\mathbf{L}(S) = W$.)

Problems for Chapter IV

PIV.1.   Problem. *Formulate the notion of a free (topological) semigroup over a local semigroup. Does this free object admit a differentiable semigroup structure? (Cf. Chapter VII.)*   ■

PIV.2.   Problem. *Characterize those subsets Ω of a finite dimensional Lie algebra L for which the set $\mathcal{R}_\mathcal{U}(B)$ of locally $\mathcal{U}$- reachable points for a class of steering functions $\mathcal{U}$ is a local semigroup with respect to B. (In Theorem IV8.6 this question is discussed for piecewise continuous steering functions and Lie wedges.)*■

PIV.3.   Problem. *Consider vectors X and Y in a Lie algebra L and denote with W the smallest Lie wedge containing X, Y, and $-Y$. Note that $[Y, X] \in L_X(W) = \mathbf{L}(W - X)$. Determine the set of all monomials p in the free Lie algebra $\operatorname{Lie}[X, Y]$ generated by X and Y such that $p(X, Y) \in L_X(W)$.*   ■

Notes for Chapter IV

Section 1.   Local semigroups have been considered in various contexts, see for instance [HM66], [Ol81]. Our treatment here emphasizes local semigroups in a suitable Campbell–Hausdorff neighborhood of a Dynkin algebra. The view point of this chapter and many results in this section were presented by Hofmann and Lawson in [HL83a]. Proposition IV.1.36 is in [HL88].

Section 2.   Much of this material, notably the Surrounding Wedge Theorem IV.2.11 comes from [HL83a].

Section 3.   In this section we adapt the notions of geometric control theory to our setting of local semigroups. The notions appear somewhat unorthodox since we give them a semigroup theoretic twist. Alternate formulations more consistent with usual formulations in control theory appear in Section 5. Variants of Proposition IV.3.10 can be found in many papers where differentiable multiplications are studied (see e.g. the paper of Birkhoff [Bi38] for a very early example). It first appeared in the current context in [HL88]. One-parameter semigroups of sets in Lie groups have been considered in detail by Rådström [Rå52], who shows that each one has an infinitesimal generator consisting of a compact convex subset of the Lie algebra. His results nicely complement our current considerations. One-parameter semigroups of sets were also studied by Gleason in conjunction with Hilbert's fifth problem [Gl52], and have been studied in more general semigroups [HM66]. The study of the relationship between local sets of reachability and local semigroups originated in [HL88]. The Confluence Theorem IV.3.19 appears here for the first time.

Section 4.   The Cone Theorem IV.4.12 was published in [HL83c], Part 1, and the Split Wedge Theorem IV.4.26 in [HL83c], Part 2. The connections made with local sets of reachability, however, have their roots in [HL88] and have their sharpest formulation as presented here.

Section 5.   The first subsection on the fundamental differential equation appears in [HL88]. The basic theory of left-invariant vector fields in the Lie algebra can be found, for example, in [Bou75]. Some of the results are essentially translations to the context of local semigroups of more general results of control theory. For example, a control theoretic version of the Dense Interior Theorem IV.5.18 may be found in [SJ72]. The proof we give is essentially due to Krener [Kr74].

Section 6.   Both the concept and importance of wedge fields in the context under consideration appeared in [HH86b], and was suggested by the work of Ol'shanskiĭ [Ol81]. General ideas of wedge fields occur widely in the area of control theory. The treatment here runs along slightly different lines from that of [HH86b].

Section 7.   A first version of the crucial rerouting technique was introduced in [HH86b]. The presentation given here explores additionally the connections with locally reachable sets.

Section 8.   After partial solutions for pointed cones and split wedges [HL83c], the characterization in complete generality of the tangent sets of local semigroups as Lie wedges was given in by Hilgert and Hofmann in [HH86b]. An alternate approach for the finite dimensional case in which the edge is the Lie algebra for a closed subgroup upstairs was outlined previously in [Ol81]. Again the connections with germs of locally reachable sets is new.

Chapter V

Subsemigroups of Lie groups

The first three chapters were devoted to what we might call *the infinitesimal Lie theory of semigroups*; its methods came out of a new blend of traditional Lie algebra with convex geometry. The last chapter was devoted to *the local Lie theory of semigroups*; the methods used there resulted from amalgamating traditional local Lie group theory with the methods of geometric control theory on an open domain of $\mathbb{R}^n$ and the ingredient of convex cones. Now we turn for the first time to an attempt at a *global Lie theory of semigroups*, that is, we shall deal with subsemigroups of Lie groups. In order that a subsemigroup S of a Lie group G be amenable to analysis through any sort of Lie theory it will certainly have to be "infinitesimally generated" in a suitable sense. Even after we have developed a good deal of the theory, it remains still somewhat vague what we should actually mean by this expression. Indeed, in contrast with the infinitesimal and local Lie theories of semigroups we have not yet reached a completely satisfactory state of the global theory. In this chapter we will make a concerted effort to illustrate the typical difficulties by discussing quite a number of concrete examples. We propose a variety of concepts which may serve as measuring the various degrees by which a semigroup may be determined infinitesimally. The situation is already subtle in the case of analytic *groups* as is well-known from traditional Lie theory; it is substantially more delicate in the case of semigroups. As these semigroups nevertheless exist and arise in applications one must finally face them no matter what obstacles they may provide. Previous chapters often gave rise to open questions and research problems. To a much higher degree, this is the case with the present chapter, which shall remain more open ended than the previous ones.

The developments of the preceding chapter showed us that *the infinitesimal generating sets of local semigroups in Lie groups are exactly the Lie wedges*. The central problem of the global theory is to determine which Lie wedges are the tangent object of a global subsemigroup of a Lie group. We know that this is not always the case; this needs to be carefully documented. It is probably useful for the reader to keep this problem in mind as a guiding idea for a reading of the chapter. The answer to the problem of characterizing those Lie wedges which arise as tangent wedges of subsemigroups of a Lie group will be given in the next chapter.

In the previous chapter, we developed the local theory for the infinite dimensional case. Some of the concepts introduced in the present chapter are quite suitable for not necessarily finite dimensional Lie groups. Yet, at the present state of affairs there are numerous good reasons for us to restrict our attention to *finite dimensional* Lie groups.

However, before we return to Lie theory we provide, in a preliminary section, background material on the general relation of preorders and semigroups. After that, in Section 1, we present the general definition and the basic theory of infinitesimally generated semigroups in a Lie group. In Section 2 we study the subgroups naturally associated with an infinitesimally generated subsemigroup S of a Lie group G, namely, the analytic subgroup generated by S in G (no problem!) and the largest subgroup contained in S (much harder!). Section 3 contains a dicourse of semidirect products in the context of infinitesimally generated subsemigroups. Section 4 is a fairly extensive catalog of examples, many of which we dissect meticulously. For instance, a full description of the subsemigroups of the group $\mathrm{Sl}(2, \mathbb{R})$ and its universal covering group and the general development of contraction semigroups are given in great detail. Section 5 contains a general discourse on maximal subsemigroups in Lie groups. A good portion if this material pertains to abstract groups and to topological groups, but the most relevant conclusions apply to the Lie group case and thus belong to the topic of subsemigroups of Lie groups. In Section 6 we discuss the present state of knowledge of closed divisible semigroups in Lie groups. In Section 7 we develop a theory of congruence relations on open subsemigroups of Lie groups whose closure contains the identity; if the congruence classes are closed, every such congruence relation defines local foliations, except at a small set of singular points.

0. Background on semigroups in groups

In this section we present some basic facts on the algebraic and topological theory of subsemigroups of groups and their relationship to certain natural order relations. *Throughout this section G denotes a group with identity $\mathbf{1}$.*

A subset S of G is a *subsemigroup* if $SS \subseteq S$ and a *submonoid* if it is a subsemigroup containing $\mathbf{1}$. A non-empty subset I of S is a *left ideal* of S of $SI \subseteq I$, a *right ideal* if $IS \subseteq I$, and an *ideal* if it is both a left and a right ideal.

If X is any subset of G, we shall write

$$\langle X \rangle = X \cup X^2 \cup X^3 \cup \cdots$$

for the semigroup generated in G by X. We shall also write $G(X) = \langle X \cup X^{-1} \rangle$; $G(X)$ is the subgroup generated by X.

V.0.1.   **Proposition.**   *For a submonoid S of G, the set*

$$H(S) = S \cap S^{-1} = \{ g \in S : g^{-1} \in S \}$$

is a subgroup, and is the largest subgroup contained in S in the sense that it contains all other subgroups which are subsets of S. If $S \neq H(S)$, then $I = S \setminus H(S)$ is an ideal of S which contains all proper left, respectively right, respectively two-sided ideals. If G is a topological group and S is closed in G, then $H(S)$ is closed in G. ∎

The set $H(S)$ is called the *group of units* or *maximal group* of S. The ideal $I = S \setminus H(S)$ is called the *maximal ideal*.

A subset $A \subseteq G$ is *invariant* or *normal* if $gAg^{-1} = A$ for all $g \in G$. Given any $B \subseteq G$, the largest normal subset of G contained in B is given by

$$B_N = \bigcap \{gBg^{-1} : g \in G\}.$$

The largest normal subgroup contained in a semigroup S, called the *core* of S, is given by

$$\mathrm{Core}(S) = H(S_N) = \big(H(S)\big)_N = \bigcap \{gH(S)g^{-1} : g \in G\}.$$

We say that a submonoid S is *reduced* in G if $\mathrm{Core}(S) = \{\mathbf{1}\}$.

V.0.2. **Proposition.** *If $S \subseteq G$ is a submonoid, then $S/\mathrm{Core}(S)$ is reduced in $G/\mathrm{Core}(S)$. If $H(S)$ is closed in G, then $\mathrm{Core}(S)$ is also closed.* ∎

The *reduction* of the pair (G, S) is the pair (G_R, S_R), where $G_R = G/\mathrm{Core}(S)$ and $S_R = S/\mathrm{Core}(S)$.

All the preceding assertions are straightforward and elementary in nature and are left as an exercise.

EV.0.1. **Exercise.** Derive the preceding assertions.

Preorders on groups and semigroups of positivity

There is a close relationship between orders on a group and subsemigroups, where we think of a subsemigroup as the semigroup of positive elements. This equivalence is of an elementary nature, but provides a useful alternative viewpoint in several contexts.

We recall that a *preorder* $\preceq$ on a set G is a transitive and reflexive binary relation. The relation $\approx$ defined on G by setting

$$x \approx y \qquad \Longleftrightarrow x \preceq y \text{ and } y \preceq x$$

is an equivalence relation on G called the *equivalence relation associated with* $\preceq$. A *partial order* $\leq$ on G is a preorder which is also antisymmetric, that is, satisfies

$$x \leq y \text{ and } y \leq x \qquad \Longrightarrow \qquad x = y.$$

If a preorder $\preceq$ is given on G, then on the quotient space $G/\approx$ of G modulo the associated equivalence relation we have a natural partial order defined unambiguously by

$$\mathrm{coset}(x) \leq \mathrm{coset}(y) \Longleftrightarrow x \preceq y,$$

called the *associated partial order*.

V.0.3.   Definition.   Let G be a group. A *left preorder* (sometimes also called a *left compatible preorder*) on G is a preorder $\preceq$ such that the following condition is satisfied:

$$x \preceq y \Longrightarrow gx \preceq gy \quad \text{for all} \quad g, x, y \in G.$$

We also call the pair $(G, \preceq)$ a *left preordered group*. Analogously we define a *right preorder*. A preorder on G is called a *group preorder* if it is simultaneously a left and a right preorder, and in this case $(G, \preceq)$ is called a *preordered group*. Similar terminology applies to partial orders.

The following well known facts have straighforward and elementary proofs which again can be safely left to the reader. We state the results for left preorders. There are of course right analogues.

V.0.4.   Proposition.   (i) *If $\preceq$ is a left preorder on a group G, then the set $S_\preceq = \{x \in G : 1 \preceq x\}$ is a subsemigroup and $H_\preceq = \{x \in G : 1 \approx x\}$ is the subgroup $H(S)$, the group of units of S. The relation $x \preceq y$ is equivalent to $y \in xS_\preceq$, and the relation $x \approx y$ means $y \in xH_\preceq$. The quotient space $G/\approx$ is the homogeneous space $G/H_\preceq$ of all left cosets $xH_\preceq$, and the associated partial order is given by $xH_\preceq \leq yH_\preceq \Longleftrightarrow x \preceq y$. The associated partial order on $G/H_\preceq$ satisfies*

$$(1) \qquad\qquad \xi \leq \eta \Longrightarrow g\xi \leq g\eta \quad \text{for all} \quad \xi, \eta \in G/H_\preceq, \quad g \in G.$$

(ii) *Let S be a submonoid of a group G and let $H = H(S)$ be the group of units. Then the prescription*

$$x \preceq_S y \Leftrightarrow y \in xS \Leftrightarrow x^{-1}y \in S$$

defines a preorder on G such that $x \approx_S y$ if an only if $x \in yH$. The quotient space $G/\approx_S$ is G/H and the associated partial order on G/H is unambiguously given by

$$xH \leq_S yH \Longleftrightarrow y \in xS.$$

(iii) *For all preorders $\preceq$ on a group G one has $\preceq_{(S_\preceq)} = \preceq$ and for all submonoids S one has $S_{(\preceq_S)} = S$.*

(iv) *Let $S \subseteq G$ be a submonoid. The preorder $\preceq_S$ is antisymmetric (and hence a partial order) if and only if $H(S) = \{1\}$.*   ∎

We shall call the monoid $S_\preceq$ the *positivity semigroup* of the preorder.

V.0.5.   Definition.   If G is a group and H a subgroup, then a *left partial order* on G/H is a partial order satisfying $gx \leq gy$ whenever $x \leq y$ for all $g \in G$, $x, y \in G/H$.

V.0.6.   Proposition.   *Given a left partial order on a homogeneous space G/H of left cosets of a group modulo some subgroup, the relation $x \preceq y$ given by $xH \leq yH$ is a left preorder on G for which the given partial order is the associated one. Conversely, given a submonoid S of G, the group G acts transitively on $G/H(S)$ on the left as a group of order preserving bijections (where $G/H(S)$ is given the induced order $\leq_S$), and hence the induced order on $G/H(S)$ is a left partial order.*∎

These statements express the fact that on a group G there are natural bijections between the set of left preorders, the set of submonoids, and the set of left partial orders on all possible left coset spaces G/H.

V.0.7. **Proposition.** *Let $\preceq$ be a preorder on a group G and S its positivity semigroup. Let H denote the maximal subgroup of S. Then the following statements are equivalent:*

(1) *$\preceq$ is a group preorder.*

(2) *$xSx^{-1} = S$ for all $x \in S$, that is, S is invariant under all inner automorphisms of G.*

(3) *The subgroup H is normal, and the associated partial order on the factor group G/H is a group partial order.* ∎

V.0.8. **Definition.** We say that a submonoid $S \subseteq G$ is *total* if $G = S \cup S^{-1}$.

V.0.9. **Proposition.** *Let $S \subset G$ be a submonoid. Then S is total if and only if either $g \preceq_S h$ or $h \preceq_S g$ for all $g, h \in G$.* ∎

Preorders satisfying the law of Proposition V.0.9 are also called *total*.

In general, a topological property of a submonoid S of a topological group G is reflected in some corresponding order-theoretical topological property of the preorder S induces on G or of the partial order that S induces on the space $G/H(S)$. Some of these correspondences are worked out in following propositions.

Let G be a topological group. A preorder $\preceq$ on G is said to be *closed* if the graph of $\preceq$ is closed in $G \times G$.

V.0.10. **Proposition.** *Let S be a submonoid of the topological group G. The left preorder $\preceq_S$ on G is closed if and only if the semigroup S is closed. The left preorder $\preceq_S$ on G has dense interior if and only if the semigroup S has dense interior.*

Proof. If S is closed and if $x^{-1}y \notin S$, then there exist an open set U containing x and an open set V containing y such that $U^{-1}V$ misses S. Thus the open set $U \times V$ misses the graph of $\preceq_S$, so it is closed. Conversely, if the graph $\mathrm{Gr}(\preceq_S)$ is closed, then $\{1\} \times S = \mathrm{Gr}(\preceq_S) \cap \{1\} \times G$ is closed in $\{1\} \times G$. Thus S is closed in G.

Suppose that $\mathrm{int}(S)$ is dense in S and $x \preceq_S y$. Then $x^{-1}y \in S$. Let U be an open set containing 1. Then there exists $s \in V = x^{-1}yU \cap \mathrm{int}(S)$. Pick a symmetric open set W containing 1 and an open set U_s containing s such tht $WU_s \subseteq V \subseteq S$. Then $W \times U_s$ is contained in the interior of the graph of $\preceq_S$, so $xW \times xU_s$ is also contained in the interior of the graph, since $\preceq_S$ is a left preorder. Then $(x, y) \in xU \times yU$ and $(x, xs) \in (xW \times xU_s) \cap (xU \times yU)$. Since the sets $xU \times yU$ form a basis of open sets around (x, y), it follows that the interior of the graph is dense in the graph.

Conversely assume the interior of the graph of $\preceq_S$ is dense in the graph. Let $s \in S$ and let W be an open set containing s. Pick a symmetric open set U containing 1 such that $U(sU) \subseteq W$. Then by hypothesis there exist open sets $A \subseteq U$ and $B \subseteq sU$ such that $A \times B$ is contained in the graph of $\preceq_S$. Then $A^{-1}B \subseteq S \cap W$; since also $A^{-1}B$ is open, it follows that $\mathrm{int}(S)$ is dense. ∎

If G is a locally compact topological group and H any closed subgroup, then the quotient space G/H of cosets gH, $g \in G$ is a locally compact homogeneous space and the quotient map $g \mapsto gH\colon G \to G/H$ is open. As a consequence, if S

is any closed subset of G satisfying $SH = S$, then the complement is open and H-stable and thus $S/H = (G/H) \setminus ((G \setminus S)/H)$ is closed.

In particular, *if S is a closed subsemigroup of a locally compact topological group and $H = H(S) = S \cap S^{-1}$, then the space $S = S/H(S)$ is a closed subset of the coset space G/H and is, in particular, locally compact.*

If $(X, \leq)$ is a topological space together with a closed partial order we call it a *partially ordered space*, or sometimes briefly a *pospace*.

V.0.11.   Proposition.   *If S is a closed submonoid of a locally compact group, then $G/H(S)$ is a locally compact pospace with the induced order $xH \leq yH$ if and only if $x \preceq_S y$. The set $S/H(S)$ is a closed, hence locally compact, subspace with unique minimal element $H(S)$.*

Proof.   Let $H = H(S)$. Let $\Gamma = \{(\xi, \eta) \in G/H \times G/H : \xi \leq_S \eta\}$ denote the graph of $\leq_S$. If $p : G \to G/H$ is the quotient map, then p is open and hence so is $p \times p : G \times G \to G/H \times G/H$. Thus Γ is closed if and only if $\Gamma^* \overset{\text{def}}{=} (p \times p)^{-1}(\Gamma)$ is closed in $G \times G$. But $(x, y) \in \Gamma^*$ means $p(x) \leq_S p(y)$ and this means $x \preceq y$ in G by Proposition V.0.4. Thus Γ^* is the graph of $\preceq_S$ and hence closed by Proposition V.0.10.

Since $S = SH$ and p is a quotient mapping, it follows that $p(S)$ is closed in G/H. The fact that H is the unique minimal element in $p(S)$ follows directly from the definition of the order in G/H.   ∎

One says that a subset A of a preordered set X is *increasing* or is *an upper set* if

$$A = \uparrow A = \{y \in X : y \geq x \quad \text{for some} \quad x \in A\}$$

and is *decreasing* or is *a lower set* if

$$A = \downarrow A = \{z \in X : z \leq x \quad \text{for some} \quad x \in A\}.$$

V.0.12.   Proposition.   *Let G be a topological group, and let S be a submonoid. If U is an open set in G or $G/H(S)$, then $\uparrow U$ and $\downarrow U$ are again open.*

Proof.   We have in G that $\uparrow U = US$ and $\downarrow U = US^{-1}$, which are both open in G if U is open in G.

Let $H = H(S)$ and let $p : G \to G/H$. If V is an open subset of G/H, then since the order on G/H is that induced from G, one readily verifies that $p^{-1}(\uparrow V) = \uparrow (p^{-1}(V))$. The latter is open by the preceding paragraph, so the former is open since p is open.   ∎

Green's preorders and relations

There are also natural standard orders and relations, called Green's orders and relations, that a semigroup induces on itself. We consider these briefly and their relationship with the preceding notions of order. If S is a monoid, we have a reflexive transitive relation on S given by $x \preceq_{\mathcal{R}} y$ if and only if $xS \subseteq yS$. The largest equivalence relation contained in it is called *Green's $\mathcal{R}$-relation* and we write $x\,\mathcal{R}\,y$ if and only if $xS = yS$. The quotient space $S/\mathcal{R}$ is equipped with a partial order $\leq_{\mathcal{R}}$ induced by $\preceq_{\mathcal{R}}$. In other words we have $\mathcal{R}(x) \leq_{\mathcal{R}} \mathcal{R}(y)$ if and only if $xS \subseteq yS$.

There is no preference for the right side and completely analogous concepts pertain to what is known as *Green's $\mathcal{L}$-relation.*

A slight variation of the same theme leads to a preorder on S defined by $x \preceq_{\mathcal{J}} y \Leftrightarrow SxS \subseteq SyS$. The equivalence relation derived from this preorder is called *Green's $\mathcal{J}$-relation* given by $x\mathcal{J}y \leftrightarrow SxS = SyS$. The partial order derived on $S/\mathcal{J}$ from the $\mathcal{J}$-preorder is given by $\mathcal{J}(x) \leq_{\mathcal{J}} \mathcal{J}(y) \Leftrightarrow x \in SyS$.

For these and related matters we refer to any standard text on semigroup theory (for instance [CP61], notably Chapter 2). For a submonoid S of a group, Green's $\mathcal{R}$-relation is particularly simple:

V.0.13. **Proposition.** *Let S be a submonoid of a group G. Then*

$$x \preceq_{\mathcal{R}} y \Leftrightarrow y^{-1}x \in S \qquad and \qquad x\,\mathcal{R}\,y \Leftrightarrow y^{-1}x \in H(S).$$

Hence $x \preceq_{\mathcal{R}} y \Leftrightarrow y \preceq_S x$ in $(G, \preceq_S)$. Analogous results hold for $\mathcal{L}$. Similarly,

$$x\mathcal{J}y \Leftrightarrow x \in H(S)yH(S).$$

In other words, $\mathcal{R}(x) = xH(S)$ and $\mathcal{J}(x) = H(S)xH(S)$ for all $x \in S$. The quotient space $S/\mathcal{R}$ and $S/\mathcal{J}$ are therefore a subspace of the homogeneous spaces $G/H(S)$ of right cosets $gH(S)$, $g \in G$, respectively, $G/\bigl(H(S) \times H(S)\bigr)$ of double cosets $H(S)gH(S)$, $g \in G$. The maximal element of the poset $S/\mathcal{R}$ is $\mathcal{R}(1) = H(S)$, and the maximal element of the poset $S/\mathcal{J}$ is $\mathcal{J}(1) = H(S)$.

Proof. The relation $x \preceq_{\mathcal{R}} y$ is defined by $x \in yS$, which means $y^{-1}x \in S$. But this is just the condition for $y \preceq_S x$. Further, the relation $x\,\mathcal{R}\,y$ is equivalent to $xS = yS$, that is to the existence of elements $s, t \in S$ such that $x = ys$ and $xt = y$. In particular, $xts = ys = x$ and, similarly, $yst = y$. Since we are inside a group, this entails $st = 1$, that is, $t = s^{-1}$, whence $t \in H(S)$ and thus $y \in xH(S)$. But $y \in xH(S)$ implies $y \in xS$ and $x \in yH(S)^{-1} \subseteq yS$ and thus $x\,\mathcal{R}\,y$. Thus $x\,\mathcal{R}\,y$ is equivalent to $y^{-1}x \in H(S)$. The fact that $H(S)$ is the maximal $\mathcal{R}$-class is clear.

The $\mathcal{J}$-relation is treated analogously. The details can be safely left as an exercise. ∎

If $J \subseteq S$ is a left ideal, then it is $\mathcal{L}$-saturated and $J/\mathcal{L}$ is a well-defined subset of the quotient space $S/\mathcal{L}$. It is straightforward that any non-empty subset J of S is a left ideal if and only if it is $\mathcal{L}$-saturated and $J/\mathcal{L}$ is a lower set.

V.0.14. **Remark.** By the preceding proposition the preorder $\preceq_{\mathcal{R}}$ is the restriction to S of the dual order of the left preorder $\preceq_S$ on G. An analogous statement holds for $\preceq_{\mathcal{L}}$.

Subsemigroups of topological groups

In this subsection we present some of the elementary facts about subsemi-groups of topological groups. We refer to [HM66] and [CHK83] for the basic theory of topological semigroups. We begin with a global variant of Proposition IV.5.16.

V.0.15.   Proposition. *Let S is a subsemigroup of a topological group G, and suppose that $\operatorname{int}(S) \neq \emptyset$.*

(i) *$(\operatorname{int} S)S \cup S(\operatorname{int} S) \subseteq \operatorname{int} S$. In other words, the interior of S is an ideal of S.*

(ii) *If $\mathbf{1}$ in in the closure of $\operatorname{int}(S)$, then $\operatorname{int}(S) = \operatorname{int}(\overline{S})$ and $\operatorname{int}(S)$ is dense in S.*

Proof.   (i) Let $g \in \operatorname{int} S$ and let U be an open neighborhood of g which is contained in S. Then for any $s \in S$ we have $gs \in Us \subseteq S$. Since G is a topological group, the set Us is an open neighborhood of gs, whence $gs \in \operatorname{int} S$. Similarly, $sg \in \operatorname{int} S$.

(ii) Obviously $\operatorname{int}(S) \subseteq \operatorname{int}(\overline{S})$. Let $s \in U = \operatorname{int}(\overline{S})$. There exists V open containing $\mathbf{1}$ with $sV^{-1} \subseteq U$. Let $W = V \cap \operatorname{int}(S)$; by hypothesis $W \neq \emptyset$. Then $sW^{-1} \subseteq U \subseteq \overline{S}$; since sW^{-1} is open, there exists $t \in S$, $w \in W$ such that $sw^{-1} = t$. Then $s = tw \in tW \subseteq \operatorname{int}(S)$.   ∎

The following is one of the most basic facts about compact topological semigroups. We refer the reader to the previously mentioned references.

V.0.16.   Proposition. *Let S be a compact semigroup. Then S contains an idempotent.*   ∎

V.0.17.   Proposition. *Let S be a non-empty compact subsemigroup of a topological group G. Then S is a compact group.*

Proof.   By Propostion V.0.16 S contains $\mathbf{1}$. If $g \in S$, then gS is also a compact semigroup and hence must contain $\mathbf{1}$. Hence $g^{-1} \in S$.   ∎

V.0.18.   Proposition. *Let S be a subsemigroup of a compact group G and suppose that $\operatorname{int}(S) \neq \emptyset$. Then S is an open and compact subgroup of G. Hence $S = G$ if G is connected.*

Proof. $\overline{S}$ is a compact subsemigroup of G, hence a group by Proposition V.0.17. Let U be a non-empty open set contained in S. Then $U^{-1} \subseteq \overline{S}$, so there exists $s^{-1} \in U^{-1} \cap S$. Then $\mathbf{1} \in s^{-1}U \subseteq S$, so $\mathbf{1} \in \operatorname{int}(S)$.

As $\overline{S}$ is a subgroup with interior, it is a standard result of topological group theory that it is open and closed. Then $\overline{S} = \operatorname{int}(\overline{S}) = \operatorname{int}(S) \subseteq S$ by Proposition V.0.15.ii. Therefore S is also closed, hence compact. If G is connected, it follows that $S = G$ ∎

V.0.19.   Proposition. i) *Let S be a topological monoid and suppose that $H(S)$ is compact. Then $H(S)$ has a basis of open neighborhoods U with $H(S)UH(S) = U$.*

ii) *In a compact topological monoid, $H(S)$ is automatically compact and $H(S)$ has a basis of open neighborhoods U such that $S \setminus U$ is a 2-sided ideal.*

Proof. i) For any subset X in S we set $X^* = H(S)XH(S)$. We claim, that A^* is closed, if A is closed: Indeed if $s \in \overline{A^*}$, then there is a net $(h_j, a_j, k_j)_{j \in J}$ on $H(S) \times A \times H(S)$ with $s = \lim h_j a_j k_j$. Since $H(S)$ is compact, we may, after passing to a convenient subnet and renaming, assume that $(h, k) = \lim(h_j, k_j)$ exists. Then $h^{-1}sk^{-1} = \lim h_j^{-1} h_j a_j k_j k_j^{-1} = \lim a_j \in A$ since A is closed. Hence $s \in H(S)AH(S)$ and this proves the claim. For any open set V containing H(S) we set $A = S \setminus V$ and $U = S \setminus \big(H(S)AH(S)\big)$. Then U is an open set satisfying $H(S)UH(S) = U \subseteq V$. Since $H(S) \cap A = \emptyset$ implies $H(S) \cap H(S)AH(S) = \emptyset$, the set U is a neighborhood of $H(S)$ and part i) of the proposition is proved.

If S is compact, V is an open neighborhood of $H(S)$, and $A = S \setminus V$, we set $J = SAS$, then J is a compact 2-sided ideal and thus $U = S \setminus J$ is open. Since $S \setminus H(S)$ is a 2-sided ideal of S and A is contained in this ideal, then so is J and thus $H(S) \subseteq S \setminus J \subseteq V$. The fact that $H(S)$ is closed in a compact topological monoid we leave as an exercise to the reader. ∎

Closed partial orders and order convexity

In this subsection we present some of the basic results on closed partial orders and link it with the theory of subsemigroups of topological groups. See [Nach65] or [GHKLMS80], Chapter VI.1 for the first two of the following propositions and for further results on closed partial orders.

V.0.20. **Proposition.** *Let K be a compact subset of a space $(X, \leq)$ with a closed partial order. Then $\downarrow K$ and $\uparrow K$ are closed in X.* ∎

A subset A of a partially ordered space X is *order convex* if $x, z \in A$, $y \in X$, and $z \leq y \leq x$ imply $y \in A$; equivalently $A = \downarrow A \cap \uparrow A$. The space X is *locally order convex* if there exists a basis of open sets for the topology consisting of order-convex sets.

V.0.21. **Proposition.** *Let X be a compact space equipped with a closed partial order. Then X is locally order convex.* ∎

A locally compact space equipped with a closed partial order need not be locally order convex. The next proposition gives equivalent formulations for order convexity in coset spaces.

V.0.22. **Lemma.** *Let S be a submonoid of a topological group and suppose that the group $H(S)$ of units is closed in G. If $S/H(S)$ is endowed with the order induced by S and if U is an open subset in $S/H(S)$, then the order convex hull of U, given as $\uparrow U \cap \downarrow U$, is again open.*

Proof. The proof follows immediately from Proposition V.0.12.

V.0.23. **Proposition.** *Let S be a closed submonoid of a topological group G such that its group $H(S)$ of units is closed in G. Then the following are equivalent:*

(1) $G/H(S)$ *endowed with the induced partial order is locally order convex.*

(2) $S/\mathcal{R}$ *with the partial order induced by $\preceq_{\mathcal{R}}$ is locally order convex.*

(3) *Given U open containing $\mathbf{1}$, there exists a closed proper right ideal I of S such that $S \subseteq UH(S) \cup I$.*

Proof.   (1)$\Rightarrow$(2). Let $H = H(S)$ and let $p\colon G \to G/H$ be the quotient mapping onto the coset space equipped with the parital order induced by S. For $s \in S$, let R_s denote the $\mathcal{R}$-class R_s of s. It follows from Proposition V.0.13 and Remark V.0.14 that mapping $R_s \mapsto p(s) = sH\colon S/\mathcal{R} \to p(S)$ is an order anti-isomorphism onto $p(S)$. Since $H \subseteq S$ and S is closed, it follows that $p|S\colon S \to p(S)$ is a quotient mapping. Thus the correspondence $R_s \mapsto p(s)$ is actually a homeomorphism. Since the order on G/H is locally order convex, its restriction to $p(S) = S/H$ is also locally order convex. The implication follows.

     (2)$\Rightarrow$(3). Let U be an open neighborhood of $\mathbf{1}$. Let Q be an order convex open set such that $p(\mathbf{1}) \in Q \subseteq p(U)$. Let $P = p^{-1}(Q)$. Then $P \subseteq p^{-1}p(U) = UH$. Let $I = \{s \in S\colon s \notin P\}$. Then $S \subseteq I \cup P \subseteq I \cup UH$, I is closed, and $\mathbf{1} \notin I$. Let $s \in I$, $t \in S$. If $st \notin I$, then $st \in P$, so $p(st) \in Q$. By the definition of the induced order on G/H, we have $p(\mathbf{1}) \leq p(s) \leq p(st)$. Since Q is order convex, we conclude that $p(s) \in Q$, i.e., $s \in P$, a contradiction. Hence I is also a right ideal.

     (3)$\Rightarrow$(1). It suffices to show that G/H is locally convex at $p(\mathbf{1})$ since G acts transitively on G/H on the left by order preserving homeomorphisms (see Proposition V.0.6).

     Let $p(U)$ be a basic open set around $p(\mathbf{1})$, where U is an open neighborhood of $\mathbf{1}$. Pick an open set V containing $\mathbf{1}$ such that $V^2 \subseteq U$. Pick a proper closed right ideal I of S such that $S \subseteq VH \cup I$. Choose a symmetric open set W containing $\mathbf{1}$ such that $W^2 \subseteq V \setminus I$. By Lemma V.0.22 the order convex hull of $p(W)$ is an open subset of G/H. To complete the proof, we show the order convex hull is a subset of $p(U)$.

     Let $x, z \in W$ and suppose that $p(x) \leq p(y) \leq p(z)$. We need to show that $p(y) \in U$. By the characterization of the order in G/H, there exist $s, t \in S$ such that $xs = y$ and $yt = z$. Then $y = xs = zt^{-1}$. It follows that

$$x^{-1}z = st \in W^2 \cap S \subseteq V \setminus I.$$

If it were the case that $s \in I$, then $st \in It \subseteq I$, a contradiction. So $s \in VH$, and thus $y = xs \in WVH \subseteq UH$. We conclude that $p(y) \in p(UH) = p(U)$. $\blacksquare$

     We extract a useful bit of information from the proof that (1) implies (2).

V.0.24.   **Remark.** For S a closed submonoid of G, the space $S/\mathcal{R}$ may be identified via the correspondence $R_s \leftrightarrow sH$ with the image of S in $G/H(S)$. This correspondence is a homeomorphism and an order anti-isomorphism. $\blacksquare$

1. Infinitesimally generated semigroups

We begin by reporting some basic facts from Lie group theory. In particular we recall that a a subgroup A of a Lie group G is called *analytic* if and only if there is a connected Lie group $G(A)$ and an injective morphism $f: G(A) \to G$ of Lie groups with $A = f(G(A))$. We write $\mathbf{L}(A) = \mathbf{L}(f)\big(\mathbf{L}(G(A))\big)$, where $\mathbf{L}$ is the functor which associates with a Lie group its Lie algebra. With the preceding convention this functor is extended so that it also assigns a Lie algebra to an analytic subgroup of a Lie group. It remains good pedagogical advice to keep in mind the example of the Lie group $G = \mathbb{R}^2/\mathbb{Z}^2$, the 2-torus, its Lie algebra $\mathbf{L}(G) = \mathbb{R}^2$ with the exponential function given by $\exp X = X + \mathbb{Z}^2$ and the analytic 1-dimensional subgroup $A = (\mathbb{R}{\cdot}(1, \sqrt{2}) + \mathbb{Z}^2)/\mathbb{Z}^2$. In this case we have $G(A) = \mathbb{R}$ and $f(t) = (t, t\sqrt{2})$, so that $\mathbf{L}(A) = \mathbb{R}{\cdot}(1, \sqrt{2})$. The analytic subgroup A is dense in G. Many names circulate for this 1-one parameter subgroup of the torus; we shall adopt for occasional use the short and friendly name of *"the dense wind"*. Frequently one identifies the underlying groups of A and $G(A)$ and says that $G(A)$ *is the group A with its inherent Lie group structure.* Under these circumstances, f is just an inclusion map and it is reasonable to identify $\mathbf{L}(G(A))$ and $\mathbf{L}(A)$, too.

Since we do accept in this book the attitude that the exponential function is the primary structural characteristic of a Lie group, our definition of an analytic subgroup was chosen accordingly. Its name derives from the fact that with respect to the analytic structure of the Lie group G, an analytic subgroup A is an immersed submanifold of G and indeed the integral manifold of the distribution given by transporting the Lie algebra $\mathbf{L}(A)$ by left translations, say, throughout the entire tangent bundle of G.

It is a fundamental fact, however, that analytic subgroups of finite dimensional Lie groups may be characterized in purely topological terms.

V.1.1.   Theorem. (Yamabe's Theorem on Path Connectivity) *A subgroup A of a finite dimensional Lie group G is analytic if and only if it is path connected.*

Proof.   For the non-trivial proof we refer to [Bou75], p.275 or [Go69]. Yamabe's original proof is too condensed for easy absorption.   ∎

This theorem is demonstrably restricted to finite dimensional Lie groups, and indeed Brouwer's Fixed Point Theorem enters the proof ineluctably. This theorem is one factual reason why we restrict ourselves to finite dimensional Lie groups.

Preanalytic semigroups and their tangent objects

If all of these complications arise on the level of group theory, we certainly have to take precautions that, in any Lie theory of subsemigroups of Lie groups, these difficulties are adequately taken into account. Yamabe's Theorem on Path Connectivity motivates the following definition.

V.1.2.   Definition.   A subsemigroup S of a Lie group G is called *preanalytic* if and only if the subgroup $\langle S \cup S^{-1} \rangle$ generated by S in G is arcwise connected. This group is an analytic subgroup of G by Yamabe's Theorem V.1.1 and it has, therefore, an inherent Lie group structure; equipped with this structure we shall denote this group with $G(S)$. In particular, $\mathbf{L}\big(G(S)\big)$ is a well-defined Lie subalgebra of $\mathbf{L}(G)$.

If S is an analytic subgroup, then the notation $G(S)$ agrees with the one we used in the introduction.

We observe right away that the closure of $\langle S \cup S^{-1} \rangle$ in G is a closed connected subgroup of a *finite dimensional* Lie group and is, therefore, itself a connected Lie group. As far as S is concerned, for most purposes we can restrict our attention to this group and thus may assume that it is G itself. In this case, $\mathbf{L}\big(G(S)\big)$ is an ideal of $\mathbf{L}(G)$ and, as a subgroup of G, the group $G(S)$ is normal in G and contains the commutator group of G. (See [Bou75] Chap.III, §9,n° 3, Proposition; p.232.) For many aspects of our problem we can even assume that we are working inside $G(S)$. If we proceed in this fashion we should not forget, however, that a refinement of the topology may have intervened.

The following remark is an immediate consequence of Yamabe's Theorem V.1.1:

V.1.3.   Remark.   Every arcwise connected semigroup containing the identity in a Lie group is preanalytic. A subsemigroup of a *connected* Lie group is preanalytic if it has non-empty interior.   ■

We illustrate the concept of a preanalytic semigroup with a few examples.

V.1.4.   Example.   (i) Let $G = \mathbb{R}^2$ and $S = \{(x,y) \in G : x = y = 0 \text{ or } x, y > 0\}$. Then S is a preanalytic semigroup containing the identity which is arcwise connected but neither closed nor open. We have $\langle S \cup S^{-1} \rangle = G(S) = G$.

(ii) Let $G = \mathbb{R}^2/\mathbb{Z}^2$ be the 2-torus and $S = \big(\mathbb{R}^+\cdot(1, \sqrt{2}) + \mathbb{Z}^2\big)/\mathbb{Z}^2$. Then S is a preanalytic semigroup which is pathwise connected and contains the identity. Further $\langle S \cup S^{-1} \rangle \neq G(S) \cong \mathbb{R}$.

(iii) Let G and S be as in the preceding Example (ii). Set

$$T = \big(\mathbb{R}^+\cdot(1, \sqrt{3}) + \mathbb{Z}^2\big)/\mathbb{Z}^2.$$

Then T has the same general properties as S while $S \cap T$ is a dense subsemigroup of the torus containing the identity which is not preanalytic. Among the preanalytic subsemigroups containing $S \cap T$ there is no unique smallest one, not even among the arcwise connected preanalytic subsemigroups. ∎

We shall now associate with any preanalytic subsemigroup S of a Lie group G a tangent object at the origin. For any C-H-neighborhood B of $\mathbf{L}(G)$ on which the exponential function $\exp: \mathbf{L}(G) \to G$ is injective, the set $S_B \overset{\text{def}}{=} \exp^{-1}(S \cap \exp B)$ is a local semigroup with respect to B in the sense of Definition IV.1.1. The tangent object of S at the origin should be determined by the local semigroup S_B. However, the example of the dense wind should warn us that we should not take the set $L(S_B) = L_0(S)$ of Definition IV.1.15 as the tangent object of S. Indeed, if S happens to be an analytic subgroup of G, then the tangent object of S at the origin should be precisely $\mathbf{L}(S)$ in the Lie group sense about which we spoke at the beginning of this section. We therefore choose the following definition:

V.1.5.   Definition.   Let S be a preanalytic subsemigroup of a Lie group G and let $\exp_{G(S)}: \mathbf{L}\big(G(S)\big) \to G(S)$ be the exponential function of the Lie group G(S) (that is, the analytic group $\langle S \cup S^{-1} \rangle$ with its inherent Lie group structure). Then the *tangent wedge* $\mathbf{L}(S)$ of S is the set $L_0\big((\exp_{G(S)}^{-1}(S)\big)$ of subtangent vectors at 0 of the pull-back of S under $\exp_{G(S)}$ in the sense of Definition IV.1.15.

We note right away that *if $\langle S \cup S^{-1} \rangle$ happens to be closed and therefore equal to $G(S)$, then $\mathbf{L}(S)$ and $L_0(S_B)$ agree*; if not, then the second set may be substantially larger as the dense wind immediately illustrates. At any rate, the definition immediately implies the following observations:

V.1.6.   Proposition.   *Let S be a preanalytic semigroup containing the identity. For each C-H-neighborhood B in $\mathbf{L}\big(G(S)\big)$, the following conclusions hold:*

(i) *$B \cap (\exp_{G(S)})^{-1}(S)$ is a local semigroup with respect to B.*

(ii) *$\mathbf{L}(S) = L_0\big(B \cap (\exp_{G(S)})^{-1}(S)\big)$.*

(iii) *$\mathbf{L}(S)$ is a Lie wedge.*

Proof.   (i) Set $T = B \cap (\exp_{G(S)})^{-1}(S)$. Clearly $0 \in T$. Suppose that $x, y \in T$ and $x * y \in B$. Then $\exp x, \exp y \in S$ and $\exp(x * y) = \exp(x) \exp(y) \in S$. By the definition of T it follows that $x * y \in T$ which we had to verify.

(ii) is an immediate consequence of Remark IV.1.17.

(iii) This follows from the Lie Wedge Theorem IV.1.27 in view of (i) and (ii) above. ∎

Next we aim for an alternative description of $\mathbf{L}(S)$ in the spirit of Proposition IV.1.21. If A is an analytic subgroup of a Lie group G and H this same group *equipped with its intrinsic Lie group structure, then for $X \subseteq A$ we shall denote with $\mathrm{cl}_H X$ the closure of X in H.*

V.1.7.   Proposition.   *For a preanalytic subsemigroup S of a Lie group G and and element $x \in \mathbf{L}(G)$, the following statements are equivalent:*

(1) $x \in \mathbf{L}(S)$.

(2) $\exp \mathbb{R}^+ \cdot x \subseteq \mathrm{cl}_{G(S)} S$.

In particular, $\mathbf{L}(S) = \mathbf{L}(\mathrm{cl}_{G(S)} S)$. If S is closed in G, then conditions (1) and (2) are also equivalent to

(0) $\exp \mathbb{R}^+ \cdot x \subseteq S$.

Proof.   Let B be a C-H-neighborhood of $\mathbf{L}(S)$. Then Proposition V.1.6(ii) and Proposition IV.1.21 show that (1) above is equivalent to

(3) $\mathbb{R}^+ \cdot x \cap B \subseteq \overline{(\exp_{G(S)})^{-1}(S)} \cap B$.

Now suppose that B is chosen so that $\exp_{G(S)}$ induces a diffeomeorphism from B onto an open identity neighborhood U of the Lie group $G(S)$. Then

$$\overline{(\exp_{G(S)})^{-1}(S)} \cap B = (\exp_{G(S)})^{-1}(\mathrm{cl}_{G(S)}(S) \cap U).$$

Hence, for this B, condition (3) is equivalent to

(4) $\exp_{G(S)}(B \cap \mathbb{R}^+ \cdot x) \subseteq U \cap \mathrm{cl}_{G(S)}(S)$.

Now $\exp_{G(S)} \mathbb{R}^+ \cdot x$ is a semigroup generated by $\exp(B \cap \mathbb{R}^+ \cdot x)$. It follows that (4) implies (2). And since $\exp_{G(S)}(B \cap \mathbb{R}^+ \cdot x) \subseteq U \cap \exp(\mathbb{R}^+ \cdot x)$ also (2) implies (4). Thus the equivalence of (1) and (2) is established. The final observations follow at once from this equivalence.   ∎

We must again draw attention to the fact that in Condition (2) of Proposition V.1.7, we use the closure of S with respect to the Lie group topology of $G(S)$ and not the topology induced from G. The example of the dense wind shows that the latter will be larger in general.

A contemplation of Proposition V.1.7 might suggest that the set $\Omega \overset{\mathrm{def}}{=} \{x \in \mathbf{L}(G) : \exp \mathbb{R}^+ \cdot x \subseteq S\}$ could conceivably serve as a candidate for a tangent object of the preanalytic semigroup S. However, Example V.1.4(i) above shows that this set, which is equal to S in this case, may even fail to be closed, let alone be a Lie wedge. Hence this candidate is useful at best in those situations for which it is stipulated that this set be a Lie wedge. We shall have occasion, nevertheless, to consider this set, too, as we shall see in the following subsection.

Ray semigroups and infinitesimally generated semigroups

In Example V.1.4(i) we have elements $x \in \mathbf{L}(S)$ for which $\exp \mathbb{R}^+ \cdot x$ is not contained in S; one such element is $x = (1, 0)$. On the other hand if, in the same group $G = \mathbb{R}^2$ we consider the subsemigroup $T = \{(0, 0)\} \cup {]}1, \infty{[}^2$, then $\langle T \cup T^{-1} \rangle = G$. In particular, T is preanalytic and $G(T) = G$. But ostensibly, $\mathbf{L}(T) = \{0\}$. This situation reminds us that we expect any Lie theory of semigroups to work reasonably only for semigroups which are, in one way or another determined by their one-parameter subsemigroups. We have to make precise what we mean by that.

Sometimes, one-parameter subsemigroups, being homomorphic images of rays in the Lie algebra, are themselves called rays. This explains the terminology introduced in the next definition.

V.1.8.   Definition. A subsemigroup S of a Lie group G is called a *ray semigroup* if it is algebraically generated by a set of one-parameter semigroups, that is, if there is a set $\Omega \subseteq \mathbf{L}(G)$ such that $S = \langle \exp \mathbb{R}^+ \cdot \Omega \rangle$. We shall call Ω a *set of infinitesimal generators of S*.

Of course, any ray semigroup contains the identity and is pathwise connected hence is preanalytic by Remark V.1.3(i). The tangent wedge $\mathbf{L}(S)$ of a ray semigroup is therefore well-defined and contains the generating set Ω of infinitesimal generators. It is clear that every semigroup S in a Lie group G contains a unique largest ray semigroup, namely the one with the set $\Omega = \{x \in \mathbf{L}(G) : \exp \mathbb{R}^+ \cdot x \subseteq S\}$ as a set of infinitesimal generators. For a ray semigroup S there is an easy way to get at $\mathbf{L}(G(S))$ and thus at $G(S)$ itself. For a subset X in a Lie algebra L as in Chapter IV, we shall denote with $\langle\langle X \rangle\rangle$ the Lie algebra generated by X in L. (See Notation introduced in the paragraph preceding Proposition IV.1.30.)

V.1.9.   Proposition. *If S is a ray semigroup in a Lie group and Ω a set of infinitesimal generators, then $\mathbf{L}(G(S)) = \langle\langle \mathbf{L}(S) \rangle\rangle = \langle\langle \Omega \rangle\rangle$.*

Proof.   Trivially, the third set is contained in the second and that one in the first. However, since $S = \langle \exp \mathbb{R}^+ \cdot \Omega \rangle$ we may conclude that the underlying group of $G(S)$ is generated by $\exp \mathbb{R}^+ \cdot \Omega \cup \exp -\mathbb{R}^+ \cdot \Omega = \exp \mathbb{R} \cdot \Omega \subseteq \exp\langle\langle \Omega \rangle\rangle$. Hence it is necessarily contained in the analytic subgroup A with $\mathbf{L}(A) = \langle\langle \Omega \rangle\rangle$. But this shows $\mathbf{L}(G(S)) \subseteq \langle\langle \Omega \rangle\rangle$ which proves the proposition.   ∎

Ray semigroups have one important feature: They have a large interior as we shall see in the following theorem.

V.1.10.   Theorem.   (The Dense Interior Theorem for Ray Semigroups) *Let G be a finite dimensional Lie group and S a ray semigroup with $G(S) = G$. Then the following conclusions hold:*

(i) $S \subseteq \overline{\operatorname{int} S}$.

(ii) $\operatorname{int} \overline{S} = \operatorname{int} S$.

Proof.   By Proposition V.1.9 we know that $\mathbf{L}(G) = \langle\langle \mathbf{L}(S) \rangle\rangle$. Let B be a C-H-neighborhood of $\mathbf{L}(G)$ and let $T = \operatorname{Sg}(\Omega, B)$ be the local ray semigroup with respect to B generated by Ω according to Definition IV.3.1. Then $\mathbf{L}(T) \subseteq \mathbf{L}(S)$ and

(1) $$\exp T \subseteq S.$$

By the Dense Interior Theorem for Local Semigroups IV.5.18,

(2) $$T \subseteq \overline{\operatorname{int} T}.$$

We may assume that B was chosen so small that $\exp$ induces a diffeomorphism from B onto an open neighborhood U of 1 in G. Since $\exp(\operatorname{int} T) \subseteq \operatorname{int} S$ we conclude from (1) and (2) that

(3) $$U \cap \operatorname{int} S \neq \emptyset.$$

Lemma V.0.15(i) now shows that

(4) $$(\forall s \in S) \, sU \cap \operatorname{int} S \neq \emptyset.$$

Hence for each $s \in S$ we have $s \in \overline{\text{int } S}$ and (i) is proved. In order to show (ii) consider $s \in V \subseteq \overline{S}$ for some open set V in G. Then there is an open neighborhood V_1 of 1 so that $sV_1^{-1} \subseteq V$. Then the set $W = V_1 \cap \text{int } S$ is not empty by Part (i) above. Moreover, $sW^{-1} \subseteq V \subseteq \overline{S}$. But since sW^{-1} is open, there exist elements $t \in S$ and $w \in W$ such that $sw^{-1} = t$. Then $s = tw \in tW \subseteq \text{int } S$. We have shown $\text{int } \overline{S} \subseteq \text{int } S$, and since the reverse containment is trivial, we have completed the proof of the theorem. ∎

Not every semigroup of interest to us from a Lie semigroup theoretical view point is a ray semigroup. From Example V.1.4(i) we recall that $\exp \mathbf{L}(S)$ need not be contained in S, but under all circumstances, $\exp \mathbf{L}(S)$ *is* contained in the closure of S in $G(S)$ by Proposition V.1.7. These remarks motivate the following definition.

V.1.11.   Definition.   A subsemigroup S of a Lie group G is called *infinitesimally generated* if all of the following conditions are satisfied:

(i) S is preanalytic.

(ii) $\exp \mathbf{L}(S) \subseteq S \subseteq \text{cl}_{G(S)} \langle \exp \mathbf{L}(S) \rangle$.

(iii) $G(S) = \langle \exp \mathbf{L}(S) \cup \exp - \mathbf{L}(S) \rangle$.

The semigroup S will be called *strictly infinitesimally generated* if

$$S = \langle \exp \mathbf{L}(S) \rangle.$$

Obviously, every strictly infinitesimally generated semigroup is infinitesimally generated. Condition (iii) is introduced into the definition to make sure that the group $G(S)$ generated by S can be recovered from the Lie algebra generated by $\mathbf{L}(S)$ via Proposition V.1.9. We do not know whether this condition is a consequence of the other two.

Problem. Clarify the independence or dependence of condition (iii) from conditions (ii) and (iii) of Definition V.1.11 above.

The concept of a strictly infinitesimally generated semigroup is straightforward by comparison with the somewhat more elusive concept of an infinitesimally generated semigroup; but we need this one, too. The following remark is an immediate consequence of the definitions.

V.1.12.   Remark.   Every infinitesimally generated subsemigroup S of a Lie group G contains a unique smallest strictly infinitesimally generated subsemigroup with the same tangent wedge $\mathbf{L}(S)$, namely the ray semigroup $\langle \exp \mathbf{L}(S) \rangle$. Every strictly infinitesimally generated semigroup is a ray semigroup. ∎

The close relationship between ray semigroups and infinitesimally generated semigroups is described in the following theorem. For an easy formulation we shall write $T^* = \text{cl}_{G(S)} T$ and $\overline{T} = \text{cl}_G T$.

V.1.13.   Theorem.   (The Ray Semigroup Theorem) *Let T be a ray semigroup in a finite dimensional Lie group G. Then $S = \langle \exp \mathbf{L}(T) \rangle$ is a strictly infinitesimally generated subsemigroup satisfying the following conditions:*

(i) $T \subseteq S \subseteq T^* \subseteq \overline{T}$.

(ii) $\mathbf{L}(T) = \mathbf{L}(S) = \mathbf{L}(T^*)$.

(iii) $G(T) = G(S)$.

(iv) $\operatorname{int}_{G(S)} T = \operatorname{int}_{G(S)} S = \operatorname{int}_{G(S)} T^*$.

Condition (ii) *determines the strictly infinitesimally generated semigroup* S *in a unique way.*

Proof. Since T is a ray semigroup, clearly $T \subseteq S$. Since $\exp \mathbf{L}(T)$ is contained in $G(T)$ by the definition of $\mathbf{L}(T)$ we know $G(T) = G(S)$. Now Proposition V.1.7 shows that $\exp \mathbf{L}(T) \subseteq T^*$ which implies $S \subseteq T^*$. Likewise, $\mathbf{L}(T) = \mathbf{L}(T^*)$ by the same proposition. Hence we have $\mathbf{L}(T) = \mathbf{L}(S) = \mathbf{L}(T^*)$ and thus condition (iv) is the only one left to show. But since T is a ray semigroup this conclusion follows from the Dense Interior Theorem for Ray Semigroups V.1.10(ii).

The uniqueness of S: Suppose that S is a strictly infinitesimally generated semigroup in G satisfying (ii). Then $S = \langle \exp \mathbf{L}(S) \rangle = \langle \exp \mathbf{L}(T) \rangle$, and this proves the asserted uniqueness. ∎

The preceding theorem tells us that, inasmuch as infinitesimal generators, the canonically generated analytic subgroups, and relative interiors are concerned, a ray semigroup is virtually indistinguishable from a canonically associated strictly infinitesimally generated semigroup.

In order to pursue this line a bit further, consider an arbitrary subsemigroup Σ of a Lie group G containing $\mathbf{1}$. We set $\Omega = \{x \in L(G) : \exp \mathbb{R}^+ \cdot x \subseteq \Sigma\}$. Now we define the ray semigroup $T = \langle \exp \Omega \rangle$, which is clearly the largest ray semigroup contained in Σ. The question is: When is T infinitesimally generated? At any rate, T is strictly infinitesimally generated if and only if $T = \langle \exp \mathbf{L}(T) \rangle$ by Definition V.1.11. Since $\Omega \subseteq \mathbf{L}(T)$, this is the case if $\mathbf{L}(T) \subseteq \Omega$. By the definition of Ω, this condition is also necessary. The following sufficient conditions on Σ for this to happen give a first indication of the effectiveness of the concepts.

V.1.14. **Proposition.** *Let Σ be a subsemigroup of a Lie group G. Set $\Omega = \{x \in \mathbf{L}(G) : \exp \mathbb{R}^+ \cdot x \subseteq \Sigma\}$ and let $T = \langle \exp \Omega \rangle$ be the maximal ray semigroup contained in Σ. Then T is strictly infinitesimally generated if and only if*

$$(*) \qquad\qquad \mathbf{L}(T) \subseteq \Omega,$$

and this condition is satisfied if at least one of the following conditions hold:

(i) Σ *is closed in* G.

(ii) Σ *is infinitesimally generated.*

(iii) $\exp \mathbf{L}(\Sigma) \subseteq \Sigma$.

Proof. We first notice that (i) implies (iii) by Proposition V.1.7, and that (ii) implies (iii) trivially. Then the remarks preceding the proposition reduce the proof to the verification that (iii) implies $(*)$. But if $\exp \mathbf{L}(\Sigma) \subseteq \Sigma$, then, by the definition of Ω, we have $\mathbf{L}(\Sigma) \subseteq \Omega$. Now $T \subseteq \Sigma$ implies $\mathbf{L}(T) \subseteq \mathbf{L}(\Sigma)$, and the assertion $(*)$ follows. ∎

We need a characterization of invariant subsemigroups in terms of their infinitesimal generators.

V.1.15.   Proposition. *Let S be a preanalytic subsemigroup of a finite dimensional connected Lie group G and set $S_0 = \langle \exp \mathbf{L}(S) \rangle$. Consider the following statements:*

(1) *$gSg^{-1} = S$ for all $g \in G$.*

(2) *$L(S)$ is an invariant wedge.*

(3) *$gS_0g^{-1} = S_0$ for all $g \in G$.*

Then (1) $\Rightarrow$ (2) $\Rightarrow$ (3), *and if S is strictly infinitesimally generated or closed in $G(S)$ and infinitesimally generated, then all three conditions are equivalent.*

Proof.   (1)$\Rightarrow$(2). Let $X \in \mathbf{L}(S)$ and $g \in G$. From (1) we conclude that $\langle S \cup S^{-1} \rangle$ is invariant under inner automorphisms. Hence each inner automorphsm $x \mapsto gxg^{-1}$ of G induces an automorphism I_g of $G(S)$, the Lie group whose underlying analytic subgroup is $\langle S \cup S^{-1} \rangle$. Hence $\mathrm{cl}_{G(S)} S$ is invariant under I_g. Thus $\exp_G t \cdot \mathrm{Ad}(g)(X) = g(\exp_G t \cdot X)g^{-1} = I_g(\exp_{G(S)} t \cdot X) \in \mathrm{cl}_{G(S)} S$ for all $t \in \mathbb{R}^+$. Hence $\mathrm{Ad}(g)(X) \in \mathbf{L}(S)$ by Proposition V.1.7. Thus for any $Y \in \mathbf{L}(S)$ we have $e^{\mathrm{ad}\, Y} X = \mathrm{Ad}(\exp Y)(X) \in \mathbf{L}(S)$. Thus $L(S)$ is invariant.

(2)$\Rightarrow$(3). If $g \in G$, then (2) implies $\mathrm{Ad}(g)\mathbf{L}(S) = \mathbf{L}(S)$ because of $\mathrm{Ad}(\exp Y) = e^{\mathrm{ad}\, Y}$ and because G is generated by $\exp \mathfrak{g}$. Then $gS_0g^{-1} = \langle g(\exp \mathbf{L}(S))g^{-1} \rangle = \langle \exp \mathrm{Ad}(g)(\mathbf{L}(S)) \rangle = \langle \exp \mathbf{L}(S) \rangle = S_0$.

(3)$\Rightarrow$(1) If S is strictly infinitesimally generated, then $S = S_0$ and the implication is trivial. If $S = \mathrm{cl}_{G(S)} S$ and S is infinitesimally generated, then the implications (3)$\Rightarrow$(1) follows by the continuity of I_g on $G(S)$. ■

It will be useful to conclude the section with a summary on basic properties of infinitesimally generated semigroups in Lie groups as follows

V.1.16.   Theorem. (The Infinitesimal Generation Theorem) *Let S be an infinitesimally generated subsemigroup of a finite dimensional Lie group G. Then $S_0 \overset{\text{def}}{=} \langle \exp \mathbf{L}(S) \rangle$ is the unique largest strictly infinitesimally generated subsemigroup of S, and the following conclusions hold:*

(i) *$G(S_0) = G(S)$ and $S \subseteq \mathrm{cl}_{G(S)}(S_0)$.*

(ii) *$\mathbf{L}(S_0) = \mathbf{L}(S)$.*

(iii) *$\mathrm{int}_{G(S)} S_0 = \mathrm{int}_{G(S)} S$, and this set is a dense ideal of S.*

(iv) *If $S = S_0$ or if $S = \mathrm{cl}_{G(S)} S$ then S is invariant in G if and only if $\mathbf{L}(S)$ is invariant in $\mathbf{L}(G)$.*

Proof.   By Proposition V.1.14, the semigroup S_0 is strictly infinitesimally generated, and its definition clearly makes it the unique largest such semigroup contained in S. Condition (i) is a consequence of Definition V.1.11(ii) and (iii). Proposition V.1.7 together with Definition V.1.11(ii) establish (ii). In order to prove (iii) we first invoke the Dense Interior Theorem for Ray Semigroups V.1.10 and derive that $\mathrm{int}_{G(S_0)} S_0 = \mathrm{int}_{G(S)} S_0$ is dense in $\mathrm{cl}_{G(S_0)} S_0 = \mathrm{cl}_{G(S)} S_0$ which equals $\mathrm{cl}_{G(S)} S$ by (i) above. Trivially, $\mathrm{int}_{G(S)} S_0 \subseteq \mathrm{int}_{G(S)} S$. But by (i) above and Conclusion (ii) of the Dense Interior Theorem V.1.10 we know $\mathrm{int}_{G(S)} S \subseteq \mathrm{int}_{G(S)}(\mathrm{cl}_{G(S)} S_0) = \mathrm{int}_{G(S_0)}(\mathrm{cl}_{G(S_0)} S_0) = \mathrm{int}_{G(S_0)} S_0 = \mathrm{int}_{G(S)} S_0$. This allows us to conclude $\mathrm{int}_{G(S)} S_0 = \mathrm{int}_{G(S)} S$. By Lemma V.0.15(i), the set $\mathrm{int}_{G(S)} S$ is an ideal of S. Thus (iii) is proved. Finally, (iv) is an immediate consequence of Proposition V.1.15. ■

The Infinitesimal Generation Theorem shows that, as far as the tangent wedges, canonically generated analytical subgroups, and relative interiors are concerned, an infinitesimally generated semigroup is indistinguishable from a canonically attached strictly infinitesimally generated subsemigroup.

2. Groups associated with semigroups

Every subsemigroup S of a group G with $1 \in S$ determines two groups canonically: Firstly the smallest subgroup $\langle S \cup S^{-1} \rangle$ of G containing S and the largest group $H(S) = S \cap S^{-1}$ contained in S. We are dealing here with a Lie group G and an infinitesimally generated subsemigroup S. With the information we have it is easy to formulate what there is to know about the former:

V.2.1.   Proposition. *If S is an infinitesimally generated semigroup in a Lie group G, then $\mathbf{L}\big(G(S)\big) = \langle\!\langle \mathbf{L}(S) \rangle\!\rangle$.*

Proof.   There is no loss in assuming $G = G(S)$. By Definition V.1.12(iii), the largest ray semigroup $S_0 = \langle \exp \mathbf{L}(S) \rangle$ also generates $G(S)$ as a group. Thus $\mathbf{L}\big(G(S)\big) = \langle\!\langle \mathbf{L}(S_0) \rangle\!\rangle$ by Proposition V.1.1.9. Since $\mathbf{L}(S) = \mathbf{L}(S_0)$ in view of Theorem V.1.16, the assertion follows.   ∎

Thus the definition of an infinitesimally generated semigroup S of a Lie group G yields at once a characterization of the group $G(S)$ generated by S in terms of its tangent wedge $\mathbf{L}(S)$. It is much harder to describe the group $H(S)$ of units of S in terms of tangent objects. The difficulty is due to the generality we allowed in Definition V.1.11 for infinitesimally generated semigroups. The objective of the remainder of this section, nevertheless, is to show that $H(S)$ is an analytic subgroup of G with $\mathbf{L}\big(H(S)\big) = H\big(\mathbf{L}(S)\big)$ (in the terminology of Definition I.1.1) and that it is in fact closed in the Lie group $G(S)$.

We begin with the much more straightforward case that S is *strictly infinitesimally generated*. Even this case shows some of the obstacles one has to overcome.

V.2.2.   Lemma. *Let S be a strictly infinitesimally generated subsemigroup of a Lie group G. Then its group of units $H(S) = S \cap S^{-1}$ is an analytic subgroup of G and $\mathbf{L}\big(H(S)\big)$ is exactly $H\big(\mathbf{L}(S)\big)$.*

Moreover, $H(S)$ is closed in $G(S)$.

Proof.   Let A denote the arc component of 1 in $H(S)$. By Yamabe's Path Connectivity Theorem V.1.1, A is the unique largest analytic subgroup contained in $H(S)$. From $A \subseteq H(S) \subseteq S$ we have $\mathbf{L}(A) \subseteq \mathbf{L}(S)$ and since $\mathbf{L}(A)$ is a vector space, $\mathbf{L}(A) \subseteq H\big(\mathbf{L}(S)\big)$ follows.

Conversely, let $x \in H\big(\mathbf{L}(S)\big)$. Then x and $-x$ are in $\mathbf{L}(S)$. By hypothesis we have $S = \langle \exp \mathbf{L}(S) \rangle$, whence $\exp \mathbb{R} \cdot x \subseteq S$ and thus $\exp \mathbb{R} \cdot x \subseteq S \cap S^{-1} = H(S)$. Thus $\exp \mathbb{R} \cdot x \subseteq A$ and hence $x \in \mathbf{L}(A)$. We have shown $\mathbf{L}(A) = H\big(\mathbf{L}(S)\big)$.

In order to show that $H(S)$ is an analytic subgroup with $H(\mathbf{L}(S)) = \mathbf{L}(H(S))$, it now suffices to prove that $H(S) \subseteq A$. Let $h \in H(S)$. Since S is strictly infinitesimally generated, there are elements $x_1, \ldots, x_n \in \mathbf{L}(S)$ such that $h = \exp x_1 \cdots \exp x_n$. We shall now show that $\exp x_k \in A$ for all $k = 1, \ldots, n$, which will imply $h \in A$. From Proposition V.0.1 we know that $S \setminus H(S)$ is an ideal and thus conclude that $\exp x_k \in H(S)$ for all $k = 1, \ldots, n$. For any such k and any $t \in [0, 1]$ we have $(\exp t \cdot x_k)(\exp(1 - t) \cdot x_k) = \exp x_k \in H(S)$. Hence, by Proposition V.0.1 once again, we conclude $\exp t \cdot x_k \in H(S)$. Therefore, $\exp x_k$ is in the arc component of 1 in $H(S)$, that is, in A, as asserted.

It remains to verify that $H(S)$ is closed in $G(S)$. We may assume that $G(S) = G$ and have to show that $H(S)$ is closed. Now $\overline{H(S)}$ is a closed connected Lie subgroup of G which is contained in $\overline{S} \cap (\overline{S})^{-1} = H(\overline{S})$. By Proposition V.1.7 we have $\mathbf{L}(S) = \mathbf{L}(\overline{S})$. Hence we have $\mathbf{L}(\overline{H(S)}) \subseteq \mathbf{L}(H(\overline{S})) \subseteq \mathbf{L}(\overline{S}) \cap \mathbf{L}(\overline{S}^{-1}) = \mathbf{L}(S) \cap \mathbf{L}(S^{-1}) = \mathbf{L}(H(S))$ in view of the preceding results. The last Lie algebra is obviously contained in $\mathbf{L}(\overline{H(S)})$. But the relation $\mathbf{L}(H(S)) = \mathbf{L}(\overline{H(S)})$ for the analytic group $H(S)$ implies $H(S) = \overline{H(S)}$. ∎

At this point we shall improve Lemma V.2.2 by replacing the hypothesis that S be strictly infinitesimally generated by the weaker hypothesis that it be merely infinitesimally generated. For the moment we pay a price in the form of a weaker conclusion since we shall not assert the connectedness of $H(S)$. However, we add some information needed on automorphisms of G which is of independent interest.

V.2.3.   Proposition.   i) *Let S be an infinitesimally generated subsemigroup of a Lie group G. Then its group of units $H(S)$ is a Lie subgroup of $G(S)$ with $\mathbf{L}(H(S)) = H(\mathbf{L}(S))$. Moreover, $H(S)$ is a $G(S)$-open subgroup of $H(\mathrm{cl}_G(S))$.*

   ii) *If $\mathbf{L}(\alpha): \mathbf{L}(G) \to \mathbf{L}(G)$ is the morphism induced by an automorphism α of G, then $\alpha(S) \subseteq S$ implies $\mathbf{L}(\alpha)\,\mathbf{L}(S) \subseteq \mathbf{L}(S)$.*

   iii) *If $S_0 = \langle \exp \mathbf{L}(S) \rangle$ is the largest strictly infinitesimally generated subsemigroup of S, then $\alpha(S_0) \subseteq S_0$ for every automorphism α of G with $\alpha(S) \subseteq S$.*

   iv) *In particular, if $\mathrm{Ad}(h): \mathbf{L}(G) \to \mathbf{L}(G)$ denotes $\mathbf{L}(g \mapsto hgh^{-1})$, then*

$$\mathrm{Ad}(h)\,\mathbf{L}(S) = \mathbf{L}(S)$$

   and $hS_0h^{-1} \subseteq S_0$ for all $h \in H(S)$.

Proof.   i) Once more we shall assume $G(S) = G$ without losing generality. Let $S_0 = \langle \exp \mathbf{L}(S) \rangle$ by the largest strictly infinitesimally generated subsemigroup of the Infinitesimal Generation Theorem V.1.16. In particular we recall from this theorem that $G(S_0) = G(S) = G$. Then Lemma V.2.2 above applies to S_0 and shows that $H(S_0)$ is a closed connected Lie subgroup of G with $\mathbf{L}(H(S_0)) = H(\mathbf{L}(S_0)) = H(\mathbf{L}(S))$. Now $H(\overline{S})$ is a closed subgroup of G, hence is a Lie group and $\mathbf{L}(H(\overline{S})) \subseteq \mathbf{L}(\overline{S}) \cap - \mathbf{L}(\overline{S}) = \mathbf{L}(S) \cap - \mathbf{L}(S)$ since S is infinitesimally generated, hence preanalytic. It follows that $\mathbf{L}(H(\overline{S})) = \mathbf{L}(S) \cap - \mathbf{L}(S) = L(H(S_0))$. This proves that $H(S_0)$ is the identity component $H(\overline{S})_0$ of the Lie group $H(\overline{S})$. Now

$H(\overline{S})_0 \subseteq H(S_0) \subseteq H(S) \subseteq H(\overline{S})$. Thus $H(S)$ is an open subgroup of the Lie group $H(\overline{S})$ and is, therefore, a Lie group with the same Lie algebra.

ii) If α is an automorphism of G, and if $\mathbf{L}(\alpha)$ is the induced automorphism of $\mathbf{L}(G)$ then we have $\alpha(\exp x) = \exp \mathbf{L}(\alpha)(x)$ for all $x \in \mathbf{L}(G)$. Then $\alpha(S) \subseteq S$ implies $\mathbf{L}(\alpha)(\mathbf{L}(S)) \subseteq \mathbf{L}(S)$ by the Definition V.1.5 of $\mathbf{L}(S)$ (or via Proposition V.1.7).

iii) Under these circumstances the following holds:

$$\alpha(S_0) = \langle \alpha(\exp \mathbf{L}(S)) \rangle \subseteq \langle \exp\left(\mathbf{L}(\alpha)(\mathbf{L}(S))\right) \rangle \subseteq \langle \exp \mathbf{L}(S) \rangle = S_0.$$

iv) If $h \in H(S)$, then the inner automorphism of G given by $\alpha(x) = hxh^{-1}$ satisfies $\alpha(S) \subseteq S$. Likewise, the assumption $h^{-1}Sh \subseteq S$ yields $\alpha^{-1}(S_0) \subseteq S_0$, and thus Conclusion iv) follows from the preceding. ∎

It remains to show that $H(S)$ is in fact connected. This requires some machinery of a transformation group and semigroup theory flavor. Thus our next step uses a version of the local cross-section theorem to describe a product decomposition of small neighborhoods of $H(S)$ in S. This result is of independent interest. It applies to all submonoids S of a Lie group G as soon as their group $H(S)$ of units is closed in G.

V.2.4.   Proposition. (The Units Neighborhood Theorem) *Let S be a subsemigroup of a finite dimensional Lie group G containing $\mathbf{1}$ and assume that $H(S)$ is closed in G. Let E be a vector space complement for the subalgebra $H(\mathbf{L}(S))$ in $\mathbf{L}(G)$. Then we find an arbitrarily small open cell neighborhood C of 0 in E such that the following conclusions hold:*

> (i) *If we let $H(S)$ act on the cartesian product $C \times H(S)$ on the right by right multiplication on the right factor, and on $U \stackrel{\text{def}}{=} (\exp C)H(S)$ by right multiplication, then the map $(x,h) \mapsto (\exp x)h\colon C \times H(S) \to U$ is an $H(S)$-equivariant diffeomorphism onto an open neighborhood of $H(S)$ in G. In particular, this statement applies to the case that $S = H(S)$.*
>
> (ii) *The map $(c,h) \mapsto (\exp c)h\colon (C \cap \exp^{-1} S) \times H(S) \to S \cap U$ is a diffeomorphism.*

Proof.   We choose a C-H-neighborhood C_1 of $\mathbf{L}(G)$ so small that exp induces a diffeomorphism from C_1 onto an open identity neighborhood in G. Let us abbreviate the subalgebra $\mathbf{L}(H(S))$ with F. Since the differential of the function $(x,y) \mapsto x * y\colon (B \cap E) \times (B \cap F) \to \mathbf{L}(G)$ at $(0,0)$ is the isomorphism $(x,y) \mapsto x + y\colon E \times F \to \mathbf{L}(G)$, we may assume that there are open cell neighborhoods C_E and C_F of 0 in E and F, respectively, such that $C_1 = C_E * C_F$. By hypothesis, $H(S)$ is closed in G; hence it is possible to choose C_1 so small that

$$(1) \qquad\qquad\qquad\qquad H(S) \cap \exp C_1 = \exp C_F.$$

Finally, we choose a symmetric open cell neighborhood $C = -C$ of 0 in E inside C_E so that

$$(2) \qquad\qquad\qquad C * C \subset C_1 \quad \text{and} \quad C * C \cap C_F = \{0\}.$$

Now we note, firstly, that $\big(\exp(C * C_F)\big)H(S)$ is an open neighborhood of $H(S)$ in G since $\exp(C * C_F)$ is an open identity neighborhood in G. Next we observe

(3) $\big(\exp(C * C_F)\big)H(S) = (\exp C)(\exp C_F)H(S) = (\exp C)H(S) = U.$

We claim

(4) $H(S) \cap (\exp C)^2 = \{1\}.$

Indeed if $h = (\exp c_1)(\exp c_2)$ with $h \in H(S)$, and c_1, $c_2 \in C$, then $h = \exp(c_1 * c_2) \in H(S) \cap \exp C_1 = \exp C_F$ by (2) and (1) above. Hence there is an $x \in C_F$ such that $\exp x = \exp(c_1 * c_2)$, which implies $x = c_1 * c_2 \in C * C \cap C_F = \{0\}$ by (2). Hence $h = \exp 0 = 1$ and this proves our claim.

Now we prove (i): The function in (i) is surjective by definition and its range U is open by (3). The function is a local diffeomorphism at all points of $C \times \{1\}$ since $(c_E, c_F) \mapsto c_E * c_F \mapsto \exp(c_E * c_F) = (\exp c_E)(\exp c_F): C_E \times C_F \to \exp C_1$ is a diffeomorphism onto an open subset of G. Since right translations with any element $h \in H(S)$ is a diffeomorphism of U, we readily observe that the function $(c, h) \mapsto (\exp c)h : C \times H(S) \to U$ is a surjective local diffeomorphism at all points of its domain. It remains to show that it is injective. Thus suppose that $(\exp c_1)h_1 = (\exp c_2)h_2$ with h_1, $h_2 \in H(S)$ and $c_1, c_2 \in C$. Then $h_1 h_2^{-1} = (\exp c_1)^{-1}(\exp c_2) = \exp\big(-c_1 * c_2\big)) \in H(S) \cap \exp(C * C) = \{1\}$ by (4). We conclude $h_1 = h_2$ and $c_1 = c_2$ and thereby finish the proof of (i).

Finally we prove (ii). By (i) above it now suffices to verify that the open neighborhood $S \cap U$ of $H(S)$ in S is of the form $(S \cap \exp C)H(S)$. Since the last set is clearly contained in U as well as in S, it remains to show that

(5) $S \cap U \subseteq (S \cap \exp C)H(S).$

Thus let $s \in S \cap U$. Then $s = (\exp c)h$ with some $h \in H(S)$ and $c \in C$ by the definition of U. Hence $\exp c = sh^{-1} \in S \cap \exp C$, and this proves (5).   ∎

The Units Neighborhood Theorem gives us rather accurate information on the geometric nature of each of the members U of an open neighborhood base of the group $H(S)$ of units of any subsemigroup S of a Lie group inasmuch as they are equivariantly diffeomorphic to a product of a subspace $C \cap \exp^{-1} S$ of an open cell C in a real vector subspace and the manifold $H(S)$.

We shall now show the rather non-trivial result that for these U the complements $J = S \setminus U$ are right ideals, that is, satisfy $JS \subseteq J$ —provided S is infinitesimally generated and the U are sufficiently small.

It is instructive to visualize the limitations of this result by contemplating some examples.

V.2.5.   Example.   (i) Let $G = \mathbb{R}^2$ and let $S_1 = \{(x, y) \in G : y \geq 0, |x| \leq y\}$ and $S_2 = \big(\mathbb{Z} \times \{0\}\big) + S_1$. Notice that S_1 is closed and strictly infinitesimally generated, while S_2 is closed, but not infinitesimally generated. We have $H(S_2) = \mathbb{Z} \times \{0\}$, its identity component being $\{0\}$. Finally let $S = S_2 \setminus (-\mathbb{N} \times \{0\})$ with $\mathbb{N} = \{1, 2, \ldots\}$.

Then S is a locally compact subsemigroup of G with $\mathbf{L}(S) = S_1$, $\overline{S} = S_2$, $H(S) = \{0\} \neq H(\overline{S})$; in fact $H(\overline{S}) \cap S$ is a proper submonoid of $H(\overline{S})$.

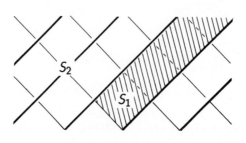

Figure 1

(ii) Let $G =]0, \infty[\times \mathbb{R}$ be the group whose elements multiply in such a way that the function $(x, y) \mapsto A(x, y)$ with

$$A(x, y) = \begin{pmatrix} x & y \\ 0 & 1 \end{pmatrix}, x > 0, y \in \mathbb{R},$$

is a morphism into $\mathrm{Gl}(2, \mathbb{R})$, that is, the multiplication is defined by $(u, v)(x, y) = (ux, v + uy)$. We set $S =]0, \infty[\times \mathbb{R}^+$. Then S is a half-space semigroup with $H(S) =]0, \infty[\times \{0\}$.

If we consider $s = (1, t)$, then

(i) $sH(S) =]0, \infty[\times \{t\}$,

(ii) $H(S)s = \{(x, tx): x > 0\}$, and

(iii) $H(S)sH(S) =]0, \infty[\times]0, \infty[$.

In particular, if we set $U = (\{1\} \times]-\varepsilon, \varepsilon[)H(S) =]0, \infty[\times]-\varepsilon, \varepsilon[$, then $S \setminus U =]0, \infty[\times [\varepsilon, \infty[$ is a *right* ideal, but not a *left* ideal. In fact *any subset J which is right-invariant under multiplication by $H(S)$, which is a left ideal, and which is not contained in $H(S)$ contains all of* $]0, \infty[\times]0, \infty[$. A similar statement holds if "right" and "left" are interchanged.

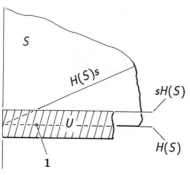

Figure 2

This example shows that what we can expect is that the sets $U \cap S = (\exp(C \cap \exp^{-1} S)H(S)$ are complements of *right*-ideals at best.

In order to prove the best possible result on the neighborhoods of the group of units we first prove a lemma on the partial order induced on a Lie group by a ray semigroup according to Proposition V.0.4. Since we do not make reference to an analytic structure on the underlying topological group, we agree that a *ray semigroup* S in an arbitrary topological group G is a subsemigroup for which there is a family $\{\sigma_j : j \in \Omega\}$ of one-parameter semigroups $\sigma_j : \mathbb{R}^+ \to S$ such that $S = \langle \bigcup_{j \in \Omega} \sigma_j(\mathbb{R}^+) \rangle$. This Definition reduces to Definition V.1.8 in case G is a Lie group.

V.2.6. **Lemma.** *Let S be a ray semigroup in a topological group G and let $\preceq = \preceq_S$ be the left preorder associated with S as in Proposition V.0.4(ii). Then for any pair of elements $x, y \in G$ such that $y \in xS$ the interval $[x, y] \overset{\text{def}}{=} \{w \in G : x \preceq w \preceq y\}$ is connected.*

Proof. It suffices to show that for every $g \in [x, y]$ there is a connected set A with $g, y \in A \subseteq [x, y]$. Since $g \in [x, y]$, there exist elements $p, q \in S$ such that $g = xp$ and $y = gq$. Since S is a ray semigroup we have $q = u_1 \cdots u_n$ with $u_k = \sigma_k(1)$ for suitable one-parameter subsemigroups σ_k of S, $k = 1, \ldots, n$. We set $u_0 = \mathbf{1}$ and define a path $\beta : [0, n] \to G$ by concatenation $\beta(t) = u_1 \cdots u_{j-1} \sigma_j(t - j + 1)$ for $j - 1 \le t < j$, $j = 1, \ldots, n$, and $\beta(n) = q$. Then we have a continuous path $\alpha : [0, n] \to G$ given by $\alpha(t) = g\beta(t) = xp\beta(t)$, and $\alpha(0) = g$, $\alpha(n) = gq = y$. For any $t \in [j - 1, j]$ we have $y = gq = g\beta(t)\sigma_j(j - 1)u_{j+1} \cdots u_n$. Since all $\sigma_j(t)$ are in S we have $x \le \alpha(t_1) \preceq \alpha(t_2) \le y$ whenever $t_1 \le t_2$. Now we set $A = \alpha([0, n])$. This is the set we need. ∎

In a somewhat loose manner of speaking, we call a neighborhood of $H(S)$ a *tube* if it is of the form $UH(S)$. The following theorem is crucial.

V.2.7. **Theorem.** (The Tube Theorem) *Let S be a submonoid of a locally compact group G containing a dense ray semigroup, and suppose that there is an identity neighborhood N such that*

$$(6) \qquad\qquad H(\overline{S}) \cap N = H(S) \cap N.$$

Then for any open identity neighborhood U there exists a closed proper right ideal I in S such that $S \subseteq UH(S) \cup I$.

Proof. Since $N \cap H(\overline{S})$ is a neighborhood of $\mathbf{1}$ in $H(\overline{S})$, it follows from (6) that $H(S)$ contains a non-empty open subset of $H(\overline{S})$ and thus is an open, hence closed subgroup of $H(\overline{S})$ which is itself a closed subgroup of G. In particular, $H(S)$ is closed in G. For abbreviation we write $H = H(S)$.

Now we pick an open neighborhood W of $\mathbf{1}$ such that $\overline{W}$ is compact and contained in $N \cap U$. We set $A = S \setminus \overline{W}H \supseteq S \setminus UH$. Since $\overline{W}$ is compact and H is closed, the product $\overline{W}H$ is closed, whence A is open. If $\operatorname{cls}_S(AS) \ne S$, then we set $I = \operatorname{cls}_S(AS)$ and the proof is completed. We finish the proof by showing that the assumption $\operatorname{cls}_S(AS) = S$ leads to a contradiction. Assuming $S = \operatorname{cls}_S(AS)$ we have $\mathbf{1} \in \operatorname{cls}_S(AS)$, and so for each open identity neighborhood V there exist elements $a_V \in A$ and $s_V \in S$ such that $a_V s_V \in V$. By continuity of multiplication there is an open set $U_V \subseteq A$ containing a_V such that $U_V s_V \subseteq V$. Now let T denote a dense ray semigroup contained in S. There is an element $u_V \in U_V \cap T$. Then $u_V s_V \in V$, and $u_V \in A = S \setminus \overline{W}H$. We consider on G the left preorder defined by T. Then by Lemma V.2.6 the interval $[\mathbf{1}, u_V]$ is connected. It contains a point inside WH, namely, $\mathbf{1}$, and a point outside $\overline{W}H$, namely, u_V.

Hence there exists an element $b_V \in \partial WH = \overline{W}H \setminus WH$, the boundary of WH such that $1 \preceq b_V \preceq u_V$. In particular, $b_V \in T$. Thus $d_V = b_V^{-1} u_V \in T$. We write $b_V = w_V h_V$ with $w_V \in \overline{W}$ and $h_V \in H$. Since $b_v \notin WH$ we know that $w_V \notin W$, and since $h_V^{-1} \in S$ we know $w_V = b_V h_V^{-1} \in S$. As $\overline{W}$ is compact, there is a net $(V(j))_{j \in J}$ of open identity neighborhoods in G which is cofinal in the filterbasis of open identity neighborhoods and is such that $w = \lim_{j \in J} w_{V(j)}$ exists and is contained in ∂W. But $w_{v(j)} h_{V(j)} d_{V(j)} s_{V(j)} = u_{V(j)} s_{V(j)} \in V(j)$; hence the net $h_{V(j)} d_{V(j)} s_{V(j)}$, which is contained in S converges to w^{-1}. Therefore $w \in \overline{S} \cap \overline{S}^{-1} = H(\overline{S})$. Since $\overline{W} \subseteq N$ we also have $w \in H(\overline{S}) \cap N = H(S) \cap N$ in view of (6). But now WH is an open set containing w. Hence for all sufficiently large indices $j \in J$ we have $b_{V(j)} = w_{V(j)} h_{V(j)} \in WHH = WH$, contradicting the very choice of the b_V. This contradiction completes the proof. ∎

A first important consequence of the last theorem is the result on the group of units of an infinitesimally generated semigroup in a Lie group which we announced at the beginning of the section.

V.2.8.   Theorem.   (The Unit Group Theorem) *Let S be an infinitesimally generated subsemigroup of a Lie group G. Then its group $H(S)$ of units is an analytic subgroup of G and in fact a connected Lie subgroup of $G(S)$.*

Proof.   We may assume $G = G(S)$. Then by Proposition V.2.3(i), the group $H(S)$ of units is closed and thus hypothesis (6) of the Tube Theorem V.2.7 is satisfied in view of Proposition IV.1.26 and Corollary IV.2.13. It remains to show that $H(S)$ is connected. We prove this assertion by contradiction and assume that we could find an element $h \in H(S) \setminus H(S)_0$ where $H(S)_0$ denotes the identity component of the Lie group $H(S)$. We find an open connected identity neighborhood U such that $UH(S)_0 \cap U(H(S) \setminus H(S)_0) = \emptyset$. Note that $UH(S)_0$ and $UH(S)_0 h$ are components of $UH(S)$. By the Tube Theorem V.2.7, there is a closed proper right ideal I with $S \subseteq UH(S) \cup I$. Since $S_0 = \langle \exp \mathbf{L}(S) \rangle$ is dense in S, by the Infinitesimal Generation Theorem V.1.16, we find $s', s'' \in \operatorname{int} S_0 \cap UH(S)$ so close to h and h^{-1}, respectively, that $s's'' \in S \setminus I$ and $s' \in UH(S)_0 h$. Further we find elements $x_1, \ldots, x_k, x_{k+1}, \ldots, x_m \in \mathbf{L}(S)$ such that $s' = \exp x_1 \cdots \exp x_k$ and $s'' = \exp x_{k+1} \cdots \exp x_m$. We define a continuous function $f : [0, m] \to S_0$ by $f(m) = s's''$ and $f(t) = \exp x_1 \cdots \exp x_{p-1} \exp(t - p + 1) \cdot x_p$ for $t \in [p - 1, p[$, $p = 1, \ldots, m$. Since $f(k) = s'$ and $f(m) = s's''$ are in two disjoint components of $UH(S)$, there must be a number $\tau \in]k, m[$ such that $f(\tau) \notin UH(S)$. Suppose that p is that natural number in $\{k, \ldots, m - 1\}$ for which $\tau \in [p, p + 1[$. We set $j = \exp x_1 \ldots \exp x_{p-1} \exp(\tau - p) \cdot x_p$ and $s = \exp(p + 1 - \tau) \cdot x_p \exp x_{p+1} \ldots \exp x_m$. Then $j = f(\tau) \in S \setminus U \subseteq I$ and $s's'' = js \in I$ since I is a right ideal. But this is a contradiction to $s's'' \in S \setminus I$. This contradiction proves that $H(S)_0 = H(S)$. ∎

EV.2.1.   Exercise.   Let S be a submonoid of a locally compact group with a dense ray semigroup. Suppose that $H(S)$ is closed. Then $H(S)$ is connected. ∎

We recall from Definition IV.4.23 the concept of a split wedge W in a Lie algebra. If $\mathfrak{g}$ is a *finite dimensional* Lie algebra, then a split wedge is simply a Lie wedge for which the edge $H(W)$ of the wedge has a vector space complement E

in $\mathfrak{g}$ such that $[H(W), E] \subseteq E$. Let us consider what this means if G is a Lie group with an infinitesimally generated subsemigroup S of G, and $\mathfrak{g}$ is the Lie algebra $\langle\!\langle \mathbf{L}(S) \rangle\!\rangle$ and W is the tangent wedge $\mathbf{L}(S)$ of S. Then $\langle S \cup S^{-1} \rangle$ and $H(S)$ are analytic subgroups by Proposition V.2.1 and the Unit Group Theorem V.2.8 with associated Lie algebras $\langle\!\langle \mathbf{L}(S) \rangle\!\rangle$ and $H(W)$, respectively, as we know from Propositions V.2.1 and 2.

V.2.9.   Lemma.   *Let S be an infinitesimally generated subsemigroup of a Lie group G_1. Then the following statements are equivalent:*

(1) $\mathbf{L}(S)$ *is a split wedge in* $\mathfrak{g} \overset{\text{def}}{=} \langle\!\langle \mathbf{L}(S) \rangle\!\rangle$, *the Lie algebra generated by* $\mathbf{L}(S)$ *in the Lie algebra $\mathfrak{g}_1$ of G.*

(2) *There is a vector space complement E of* $\mathfrak{h} \overset{\text{def}}{=} \mathbf{L}\big(H(S)\big)$ *in $\mathfrak{g}$ such that* $\big(\operatorname{Ad} H(S)\big) E \subseteq E$, *where* $\operatorname{Ad} g = \mathbf{L}(g' \mapsto g g' g^{-1})$.

Proof. $(1) \Rightarrow (2)$. Let E be a vector space complement of $\mathfrak{h}$ in $\mathfrak{g}$ such that $[\mathfrak{h}, E] \subseteq E$. Then $e^{\operatorname{ad} x} E \subseteq E$ for all $x \in \mathfrak{h}$ by Lemma IV.4.22. Now suppose that $h = \exp x$ with $x \in \mathfrak{h}$. Then $\operatorname{Ad} h = e^{\operatorname{ad} x}$ in view of the general formula $\operatorname{Ad} \circ \exp = \exp \circ \operatorname{ad}$ (see Appendix). Hence $(\operatorname{Ad} h) E \subseteq E$ for all $h \in \exp \mathfrak{h}$. Since $H(S) = \langle \exp \mathfrak{h} \rangle$ by the Unit Group Theorem V.2.7, assertion (2) follows.

$(2) \Rightarrow (1)$. Let E be the vector space guaranteed by (2) and let $x \in \mathfrak{h}$. Then $e^{\operatorname{ad} x} E = \big(\operatorname{Ad}(\exp x)\big)(E) \subseteq E$ by (2). Hence $[x, E] \subseteq E$ by Lemma IV.4.22 again. This proves (1).   ∎

This allows the following definition:

V.2.10.   Definition.   A subsemigroup S of a Lie group G will be called *split* if it is infinitesimally generated and satisfies the equivalent conditions of Lemma V.2.7 above.

V.2.11.   Proposition.   *Any of the following conditions is sufficient for an infinitesimally generated subsemigroup S of a Lie group G to be split:*

(i) $H(S)$ *is compact.*

(ii) $H(S)$ *is semisimple.*

Proof.   If $H(S)$ is compact, then certainly $\mathbf{L}\big(H(S)\big)$ is compactly embedded in $\mathbf{L}(G)$, and $H(S)$ is semisimple if and only if $\mathbf{L}\big(H(S)\big)$ is semisimple. The proposition then follows from Theorem IV.4.29.   ∎

For split subsemigroups *with compact group of units* the Unit Separation Theorem can be improved. We saw in Proposition V.0.19 that in a *compact* monoid the group $H(S)$ has a basis of open neighborhoods U such that $S \backslash U$ is a two-sided ideal. In an infinitesimally generated semigroup in a Lie group this continues to hold if one assumes only the compactness of $H(S)$.

V.2.12.   Corollary.   (The Tube Theorem for Compact Unit Groups) *Let S be an infinitesimally generated subsemigroup of a Lie group G. If the group $H(S)$ is compact, then for any open neighborhood U of $H(S)$ there is a two-sided proper ideal J of S such that $S \subseteq U \cup J$.*

Proof.   Let U be an open neighborhood of $H(S)$. Since $H(S)$ is compact, there is a compact neighborhood C of $\mathbf{1}$ such that $CH(S) \subseteq U$. By the Tube Theorem V.2.7 there is a closed proper right ideal I such that $S \subseteq CH(S) \cup I$. Then $J \overset{\text{def}}{=} SI$ is a two-sided ideal satisfying $S \subseteq U \cup J$. Further, $J = SI = CH(S)I \cup I^2 = CH(S)I$, since $I^2 \subseteq I \subseteq CH(S)I$. As $CH(S)$ is compact and I is closed, J is closed. Finally, we claim that J is proper, for if not, then $\mathbf{1} \in J$ and there are elements $c \in C$, $h \in H(S)$, and $i \in I$ such that $\mathbf{1} = chi$. But then $\mathbf{1} = (ch)^{-1}\mathbf{1}ch = ich \in I$ since I is a right ideal. Yet this is impossible since $I \neq S$. This contradiction proves the claim and finishes the proof of the corollary.∎

The 2-dimensional Example V.2.5(ii) shows that without the hypothesis of compactness it can happen that an open neighborhood U of $H(S)$ satisfying $H(S)UH(S) = U$ is necessarily all of S. Corollary V.2.12 shows among other things that this phenomenon occurs only if $H(S)$ is not compact.

It is interesting to note a result on ray semigroups which is related to Exercise EV.2.1.

EV.2.2.   Exercise.   If a subsemigroup S of a topological group G is generated by a set of one parameter subsemigroups, then its group of units $H(S)$ is generated by the one parameter subsemigroups it contains (hence is, in particular, arcwise connected). As a consequence, the group of units of a ray semigroup in a Lie group is analytic. (Hint. The set $S \setminus H(S)$ is an ideal of S. The ideas of the proof of Theorem V.2.8 can be applied to prove the assertion.)   ∎

3. Homomorphisms and semidirect products

This section deals with the effect of morphisms between Lie groups on preanalytic subsemigroups and their tangent objects. Particular attention will be given to semidirect products.

V.3.1. **Proposition.** *Let $\varphi: G \to H$ be a morphism of Lie groups and $\mathbf{L}(\varphi): \mathbf{L}(G) \to \mathbf{L}(H)$ the associated morphism of Lie algebras. The the following conclusions hold:*

(i) *If S is a subsemigroup of G with $G(S) = G$ and G is connected, then $\varphi(S)$ is preanalytic and $\mathbf{L}(\varphi)\big(\mathbf{L}(S)\big) \subseteq \mathbf{L}\big(\varphi(S)\big)$.*

(ii) *If T is a submonoid of H with $G(T) = H$ and φ is surjective, then $G\big(\varphi^{-1}(T)\big) = G$ and $\mathbf{L}\big(\varphi^{-1}(T)\big) = \mathbf{L}(\varphi)^{-1}\big(\mathbf{L}(T)\big)$.*

Proof. (i) The group which is algebraically generated by $\varphi(S)$ in H is $\varphi(G)$. Hence it is an analytic group since G is connected and thus $\varphi(S)$ is preanalytic according to Definition V.1.2. Since $\mathbf{L}\big(\varphi(S)\big)$ is defined in reference to $G\big(\varphi(S)\big)$, it is no loss of generality if we assume that φ is surjective, thus open by the Open Mapping Theorem for Locally Compact Groups, since G is connected, hence countable at infinity. Thus we assume that φ is a quotient map. Finally, since $\varphi(\overline{S}) \subseteq \overline{\varphi(S)}$, we may assume that S is closed. Now $x \in \mathbf{L}(S)$ means $\exp_G \mathbb{R}^+ \cdot x \subseteq S$ by Proposition V.1.7. This entails $\exp_H \mathbb{R}^+ \cdot \mathbf{L}(\varphi)(x) = \varphi(\exp_G \mathbb{R}^+ \cdot x) \subseteq \varphi(S)$. Hence $\mathbf{L}(\varphi)(x) \in \mathbf{L}\big(\varphi(S)\big)$ by Proposition V.1.7. Conclusion (i) is proved.

(ii) Let us set $S = \varphi^{-1}(T)$. Since φ is surjective, $\varphi(S) = T$, whence $\varphi\big(G(S)\big) = G(T) = H$. Since $\ker \varphi = \varphi^{-1}(1) \subseteq S$ on account of $1 \in T$, certainly $\ker \varphi \subseteq G(S)$, whence $G(S) = \varphi^{-1}\big(\varphi(G(S))\big) = \varphi^{-1}(H) = G$. Since $\varphi\big(\varphi^{-1}(T)\big) = T$, conclusion (i) above implies $\mathbf{L}(\varphi)\Big(\mathbf{L}\big(\varphi^{-1}(T)\big)\Big) \subseteq \mathbf{L}(T)$, that is,

$$\mathbf{L}\big(\varphi^{-1}(T)\big) \subseteq \mathbf{L}(\varphi)^{-1}\,\mathbf{L}(T).$$

We have to show the reverse containment. So let $x \in \mathbf{L}(\varphi)^{-1}\big(\mathbf{L}(T)\big)$. Then $\mathbf{L}(\varphi)(x) \in \mathbf{L}(T)$, that is,

$$\varphi(\exp_G \mathbb{R}^+ \cdot x) = \exp_H \mathbb{R}^+ \cdot \mathbf{L}(\varphi)(x) \subseteq \overline{T}.$$

Thus $\exp_G \mathbb{R}^+ \cdot x \subseteq \varphi^{-1}(\overline{T})$, whence $x \in \mathbf{L}\big(\varphi^{-1}(\overline{T})\big)$. Since every Cauchy sequence in H can be lifted to a Cauchy sequence in G mapping onto it under φ (see e.g. [Bou58], Chap. IX, §2, n° 10, Proposition 18), we know $\varphi^{-1}(\overline{T}) \subseteq \overline{\varphi^{-1}(T)}$. Now we can conclude $x \in \mathbf{L}\big(\varphi^{-1}(T)\big)$ by Proposition V.1.7. This is what we had to show. ∎

Proposition V.3.1(i) cannot be improved. It is easy to construct examples of a strictly infinitesimally closed semigroup S in a Lie group G and quotient homomorphism $\varphi: G \to G/N$ such that the tangent wedge of $\varphi(S)$ is much bigger than $\mathbf{L}(\varphi)\big(\mathbf{L}(S)\big)$.

V.3.2.   Example.   Let S be any Lorentzian cone in $G = \mathbb{R}^3$ and N the discrete cyclic subgroup generated by a nonzero point in the boundary of S. Then the quotient map $\varphi: G \to G/N$ is a covering homomorphism inducing an isomorphism $\mathbf{L}(\varphi): \mathbf{L}(G) \to \mathbf{L}(G)$ (which we may visualize as the identity if we take $\mathbf{L}(G) = \mathbf{L}(G/N) = \mathbb{R}^3$, $\exp_G = 1_G$ and $\exp_{G/N} = \varphi$!). But $\varphi(S)$ is an infinitesimally generated ray semigroup which is dense in a half-space semigroup such that $\mathbf{L}\big(\varphi(S)\big)$ is a half-space.

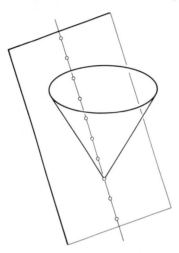

Figure 1

Suppose that G_1 and G_2 are two Lie groups with infinitesimally generated subsemigroups $S_j \subseteq G_j$, $j = 1, 2$. It is a straightforward matter to verify that $S = S_1 \times S_2$ *is infinitesimally generated in* $G = G_1 \times G_2$ *and* $\mathbf{L}(S) = \mathbf{L}(S_1) \times \mathbf{L}(S_2)$ *if we identify in the obvious fashion* $\mathbf{L}(G)$ *with* $\mathbf{L}(G_1) \times \mathbf{L}(G_2)$. It is important for us to generalize this simple fact to the situation of semidirct products. In order to fix notation we give a self-contained run-down of the Lie theory of semidirect products:

V.3.3.   Lemma.   *Let* $\mathbf{n}$ *and* $\mathbf{h}$ *be two Dynkin algebras and let* $\delta: \mathbf{h} \to \operatorname{Der} \mathbf{n}$ *be a morphism of Dynkin algebras into the Dynkin algebra of derivations of* $\mathbf{n}$. *Then the topological product vector space* $\mathbf{n} \times \mathbf{h}$ *becomes a Dynkin algebra* $\mathfrak{g} = \mathbf{n} \oplus_\delta \mathbf{h}$ *when equipped with the bracket*

$$[(x, y), (x', y')] = \big(\delta(y)x' - \delta(y')x + [x, x'], [y, y']\big).$$

There exist C-H-neighborhoods $B_\mathbf{n}$, $B_\mathbf{h}$ *and* B *with C-H-multiplications* $*_\mathbf{n}$, $*_\mathbf{h}$, *and* $*$ *in* $\mathbf{h}$, $\mathbf{n}$ *and* $\mathfrak{g}$, *respectively, such that the following conclusions hold:*

(i)   *The function* $\nu: B_\mathbf{n} \times B_\mathbf{h} \to B$ *given by* $\nu(x, y) = (x, 0) * (0, y)$ *is an analytic diffeomorphism.*

(ii)   *The function*

$$\sharp: (B_\mathbf{n} \times B_\mathbf{h})^2 \to \mathbf{n} \times \mathbf{h}, \qquad (x, y) \sharp (x', y') = (x *_\mathbf{n} e^{\delta(y)} x', y *_\mathbf{h} y'),$$

makes $B_\mathbf{n} \times B_\mathbf{h}$ *into a local Lie group.*

(iii) *The map* $\nu\colon (B_{\mathbf{n}} \times B_{\mathbf{h}}, \natural) \to (B, *)$ *is an isomorphism of local Lie groups.*

(iv) $\nu(x, 0) = (x, 0)$ *for all* $x \in B_{\mathbf{n}}$ *and* $\nu(0, y) = (0, y)$ *for all* $y \in B_{\mathbf{h}}$. *In other words*, ν *agrees on the first and second direct factor with the isomorphism of vector spaces* $(x, y) \mapsto x + y$.

Proof.   It is straightforward to check that $\mathfrak{g}$ is a Dynkin algebra, and the Theorem of the Local Inverse shows that ν is locally invertible. Since (iii) will imply (ii), it now suffices to verify that ν transforms $\natural$ into $*$.

We first show that

$$(0, y) * (x, 0) = (e^{\delta(y)} x, 0) * (0, y).$$

In fact we have $(0, y) * (x, 0) * (0, -y) = e^{\mathrm{ad}_{\mathfrak{g}}(0, y)}(x, 0)$. But $\mathrm{ad}_{\mathfrak{g}}(0, y)(x', y') = \bigl(\delta(y)x', (\mathrm{ad}_{\mathbf{h}} y)(y')\bigr)$ so that $e^{\mathrm{ad}_{\mathfrak{g}}(0, y)} = e^{\delta(y)} \times e^{\mathrm{ad}_{\mathbf{n}} y}$, and hence

$$e^{\mathrm{ad}_{\mathfrak{g}}(0, y)}(x, 0) = (e^{\delta(y)} x, 0).$$

Now we calculate

$$
\begin{aligned}
\nu(x, y) * \nu(x', y') &= (x, 0) * (0, y) * (x', 0) * (0, y')\\
&= (x, 0) * (e^{\delta(y)} x', 0) * (0, y) * (0, y')\\
&= \nu(x *_{\mathbf{n}} e^{\delta(y)} x', y *_{\mathbf{h}} y')\\
&= \nu\bigl((x, y)\natural(x', y')\bigr),
\end{aligned}
$$

and the proof is finished. Claim (iv) is an immediate consequence of the definition of ν in (i) above.   ∎

We shall call $\mathfrak{g} = \mathbf{n} \oplus_{\delta} \mathbf{h}$ *the semidirect sum of* $\mathbf{n}$ *and* $\mathbf{h}$. From the preceding lemma we can derive the interaction of semidirect products of Lie groups with semidirect sums of their Lie algebras via the exponential function.

V.3.4.   Proposition. *Let* H *and* N *be Lie groups and* $\alpha\colon H \to \mathrm{Aut}\, H$ *a morphism of Lie groups. Write* $\lambda\colon \mathrm{Aut}\, N \to \mathrm{Aut}\, \mathbf{L}(N)$ *for the morphism of Lie groups given by* $\lambda(\varphi) = \mathbf{L}(\varphi)$ *for any automorphism* $\varphi\colon N \to N$. *Furthermore, let* $\delta\colon \mathbf{L}(H) \to \mathrm{Der}\, \mathbf{L}(N)$ *be the morphism of Dynkin algebras given by* $\delta = \mathbf{L}(\lambda \circ \alpha) = \mathbf{L}(\lambda)\, \mathbf{L}(\alpha)$. *Then we have the following conclusions:*

(i) $e^{\delta(y)} = \lambda(\alpha \exp_H y)$.

(ii) *The product space* $N \times H$ *is a topological group* $N \rtimes_{\alpha} H$ *with respect to the multiplication* $(n, h)(n', h') = (n\alpha(h)(n'), hh')$.

(iii) $G = N \rtimes_{\alpha} H$ *is a Lie group with Lie algebra* $\mathbf{L}(G) = \mathbf{L}(N) \oplus_{\delta} \mathbf{L}(H)$ *whose exponential function* $\exp_G$ *is defined locally on a suitable C-H-neighborhood* B *of* $\mathbf{L}(G)$ *by* $\exp_G |B = (\exp_N |B_{\mathbf{n}} \times \exp_H |B_{\mathbf{h}}) \circ \nu^{-1}$, *where* $\nu\colon B_{\mathbf{h}} \times B_{\mathbf{n}} \to B$ *is the diffeomorphism of Lemma V.3.3.*

(iv) *If the Lie algebras* $\mathbf{L}(N)$ *and* $\mathbf{L}(H)$ *are identified with the first and second summand of* $\mathbf{L}(N) \oplus_{\delta} \mathbf{L}(H)$, *and* N *and* H *with the first and second factor of* $N \rtimes_{\alpha} H$, *respectively, then the restriction of* $\exp : \mathbf{L}(N) \oplus_{\delta} \mathbf{L}(H) \longrightarrow N \rtimes_{\alpha} H$ *to* $\mathbf{L}(N)$ *and to* $\mathbf{L}(H)$ *is* $\exp_N$ *and* $\exp_H$, *respectively.*

Proof.   Conclusion (i) is simply a reformulation of the commutativity of the following diagram:

$$
\begin{array}{ccccc}
\mathbf{L}(H) & \xrightarrow{\ \mathbf{L}(\alpha)\ } & \mathbf{L}(\operatorname{Aut} N) & \xrightarrow{\ \mathbf{L}(\lambda)\ } & \operatorname{Der} \mathbf{L}(N) \\
{\scriptstyle \exp_H}\big\downarrow & & {\scriptstyle \exp_{\operatorname{Aut} N}}\big\downarrow & & \big\downarrow{\scriptstyle D \mapsto e^D} \\
H & \xrightarrow[\ \alpha\ ]{} & \operatorname{Aut} N & \xrightarrow[\ \lambda\ ]{} & \operatorname{Aut} \mathbf{L}(N).
\end{array}
$$

Conclusion (ii) is easy to check.

In order to verify (iii), it suffices to show that the function $\exp \colon B \to G$ given by $\exp = (\exp_N \times \exp_H) \circ \nu^{-1} \colon B_{\mathbf{L}(N)} \times B_{\mathbf{L}(H)} \to G$ with sufficiently small C-H-neighborhoods $B_{\mathbf{L}(N)}$ and $B_{\mathbf{L}(H)}$ as in Lemma V.3.3 is a local homeomorphism satisfying $\exp(z * z') = \exp z \exp z'$ for $z, z' \in B$.

Now for small $x, x' \in \mathbf{h}$ and $y, y' \in \mathbf{n}$ we calculate

$$
\begin{aligned}
& (\exp_N x, \exp_H y)(\exp_N x', \exp_H y') \\
&= \big((\exp_N x)\alpha(\exp_H y)(\exp_N x'), \exp_H y \exp_H y'\big) \\
&= \big(\exp_N x \exp_N \big(\mathbf{L}\big(\alpha(\exp_H y)\big)x'\big), \exp_H y \exp_H y'\big) \\
&= \big(\exp_N x \exp_N \big(e^{\delta(y)} x'\big), \exp_H y \exp_H y'\big) \\
&= \big(\exp_N \big(x *_{\mathbf{n}} e^{\delta(y)} x'\big), \exp_H (y *_{\mathbf{h}} y')\big) \\
&= (\exp_N \times \exp_H)\big((x,y)\sharp(x',y')\big).
\end{aligned}
$$

If we now set $(\xi, \eta) = \nu(x, y)$ and $(\xi', \eta') = \nu(x', y')$, then we have

$$
\nu\big((x,y)\sharp(x',y')\big) = (\xi, \eta) * (\xi', \eta')
$$

by Lemma V.3.3(iii) above, and thus the preceding calculation shows in view of $\exp|B \circ \nu = \exp_N |B_{\mathbf{n}} \times \exp_H |B_{\mathbf{h}}$ that

$$
\begin{aligned}
\exp(\xi, \eta) \exp(\xi', \eta') &= (\exp \circ \nu)(x, y)(\exp \circ \nu)(x', y') \\
&= (\exp_N \times \exp_H)(x, y)(\exp_N \times \exp_H)(x', y') \\
&= (\exp_N x, \exp_H y)(\exp_N x', \exp_H y') \\
&= (\exp_N \times \exp_H)\big((x,y)\sharp(x',y')\big) \\
&= (\exp_N \times \exp_H)\Big(\nu^{-1}\big((\xi, \eta) * (\xi', \eta')\big)\Big) \\
&= \exp\big((\xi, \eta) * (\xi', \eta')\big),
\end{aligned}
$$

which we had to show.

(iv) The exponential function of a Lie group is uniquely determined by its restriction to any connected open neighborhood of 0 by analytic extension (or by extension along one parameter groups). The assertion is then straightforward from Lemma 3.3(iv). ∎

V.3.5. **Proposition.** *Under the circumstances of* Proposition V.3.4, *suppose that A_N and A_H are analytic subgroups of N and H, respectively, such that $\alpha(h)A_N \subseteq A_N$ for all $h \in A_H$. Then we have the following conclusions:*

(i) *For each $h \in A_H$ the morphism $\mathbf{L}(\alpha(h)) = \lambda(\alpha(h))$ (in the notation of Proposition V.3.4) maps $\mathbf{L}(A_N)$ into itself, and $\alpha(h)$ induces an automorphism $\alpha'(h)$ of $G(A_N)$ with $\mathbf{L}(\alpha'(h)) = \mathbf{L}(\alpha(h))|\,\mathbf{L}(A_H)$.*

(ii) *$\alpha': G(A_H) \to \mathrm{Aut}(G(A_N))$ is a morphism of Lie groups and if in the usual way, $\mathrm{Aut}(G(A_N))$ is identified with a subgroup of $\mathrm{Aut}\,\mathbf{L}(A_H)$ via a morphism λ' given by $\lambda'(\varphi) = \mathbf{L}(\varphi)$, then $\lambda'(\alpha'(\exp_{G(A_H)} x)) = e^{\delta(x)}$ for all $x \in \mathbf{L}(A_H)$, where $\delta = \mathbf{L}(\lambda \circ \alpha)$.*

(iii) *The morphism $\delta': \mathbf{L}(A_H) \to \mathrm{Der}(\mathbf{L}(A_N))$ given by $\delta' = \mathbf{L}(\lambda')\,\mathbf{L}(\alpha')$ is induced by δ.*

(iv) *$A_H \times A_N$ is an analytic subgroup A of $N \rtimes_\alpha H$ and $G(A)$ may be identified with $G(A_N) \rtimes_{\alpha'} G(A_H)$, and $\mathbf{L}(G(A)) = \mathbf{L}(A)$ with $\mathbf{L}(A_N) \oplus_{\delta'} \mathbf{L}(A_H)$.*

Proof. (i) Let $h \in A_H$. Then $\alpha(h)$ maps A_N into itself by hypothesis, hence preserves the one parameter subgroups of A_N. Thus the automorphism $\mathbf{L}(\alpha(h))$ of $\mathbf{L}(N)$ maps $\mathbf{L}(A_N)$ into itself. The same holds for $\mathbf{L}(\alpha(h^{-1}))$. Hence it induces an automorphism ψ of $\mathbf{L}(A_N)$ which satisfies for all $x \in \mathbf{L}(G(A_N)) = \mathbf{L}(A_N)$ the equation $\exp_{G(A_N)} \psi(x) = \exp_N \psi(x) = \alpha(h)(\exp_N x) = \alpha'(h)(\exp_{G(A_N)} x)$. It follows that $\alpha'(h)$ is an automorphism of $G(A_H)$ with $\mathbf{L}(\alpha'(h)) = \psi$.

(ii) The function $\alpha': G(A_H) \to \mathrm{Aut}\,G(A_N)$ is readily seen to be a group homomorphism. Since λ' is an embedding, the continuity of α' follows from the remainder of the assertion which we prove now. For all sufficiently small $x \in \mathbf{L}(A_H)$ we have in view of the formalism developed in Proposition V.3.4 and its proof $e^{\mathbf{L}(\lambda')\,\mathbf{L}(\alpha')(x)} = \lambda'(\alpha'(\exp_{G(A_H)} x)) = \mathbf{L}(\alpha(\exp_H x)) = \lambda(\alpha(\exp_H x)) = e^{\mathbf{L}(\lambda)\,\mathbf{L}(\alpha)(x)} = e^{\delta(x)}$.

(iii) is a direct consequence of the preceding calculation.

(iv) Since $A = A_N \times A_H$ is a clearly a subgroup and since this subgroup is arcwise connected, it is analytic by Yamabe's Theorem V.1.1. The Lie group $G(A_H) \rtimes_{\alpha'} G(A_N)$ has the Lie algebra $\mathbf{L}(A_N) \oplus_{\delta'} \mathbf{L}(A_H)$, and the inclusion map induces an injection $j: G(A_N) \rtimes_{\alpha'} G(A_H) \to N \rtimes_\alpha H$ with $\mathbf{L}(j)$ the inclusion $\mathbf{L}(A_N) \oplus_{\delta'} \mathbf{L}(A_N) \to \mathbf{L}(N) \oplus_\delta L(H)$. This proves the assertion. ∎

With this background we can now prove a semidirect product theorem for semigroups.

V.3.6. **Theorem.** (The Semidirect Product Theorem) *Let N and H be finite dimensional Lie groups and $\alpha: H \to \mathrm{Aut}\,N$ be a morphism of Lie groups. Suppose that S_N and S_H are infinitesimally generated subsemigroups of N and H, respectively, such that $\alpha(h)S_N \subseteq S_N$ for all $h \in S_H$.*

Then $S \overset{\mathrm{def}}{=} S_N \times S_H$ is a preanalytic subsemigroup of $G \overset{\mathrm{def}}{=} N \rtimes_\alpha H$ such that

(*) $$G(S) = G(S_N) \rtimes_{\alpha'} G(S_H) = \langle \exp \mathbf{L}(S) \cup \exp - \mathbf{L}(S) \rangle,$$

and

(**) $$\mathbf{L}(S) = \mathbf{L}(S_N) \oplus \mathbf{L}(S_H) \subseteq \mathbf{L}(A_N) \oplus_{\delta'} \mathbf{L}(A_H),$$

where $A_N = G(S_N)$, $A_H = G(S_H)$, and $\alpha': G(S_H) \to \operatorname{Aut} G(S_N)$ *is induced by* α *and* $\delta': \mathbf{L}(A_H) \to \operatorname{Der} \mathbf{L}(A_N)$ *is induced by* δ *as described in* Proposition V.3.5.
  Moreover,

$$(\dagger) \qquad\qquad S \subseteq \operatorname{cl}_{G(S)}\langle \exp(\mathbf{L}(S))\rangle.$$

Proof. We observe, firstly, that $S \stackrel{\mathrm{def}}{=} S_N \rtimes_\alpha S_H$ is closed under the multiplication of $G \stackrel{\mathrm{def}}{=} N \rtimes_\alpha H$, since

$$(1) \qquad\qquad \alpha(S_H)S_N \subseteq S_N.$$

Next we note that $A_N = \langle S_N \cup S_N^{-1}\rangle$ and $A_H = \langle S_H \cup S_H^{-1}\rangle$ are analytic groups since S_N and S_H are preanalytic. Hence $A = \langle S \cup S^{-1}\rangle = A_H \times A_N$ by Proposition V.3.5(iv). Hence S is preanalytic; thus $\mathbf{L}(S)$ is defined. Relation (1) implies that $\alpha(h)$ maps the analytic group A_N into itself for each $h \in S_H$. Therefore, the automorphism $\mathbf{L}(\alpha(h))$ of $\mathbf{L}(G)$ maps $\mathbf{L}(A_N) = \langle\!\langle\mathbf{L}(S_N)\rangle\!\rangle$ into itself, hence induces on this subalgebra an automorphism since all vector spaces in sight are finite dimensional. Hence $\alpha(h)$ induces an automorphism $\alpha'(h)$ of the analytic group A_N, and since α is a morphism, we have $\alpha(h)^{-1} = \alpha(h^{-1})$, which induces the inverse of $\alpha'(h)$ on A_N. Thus we have $\alpha(A_H)(A_N) \subseteq A_N$. Now in view of Proposition V.3.5 it is no loss of generality to assume that $N = G(A_N) = G(S_N)$, that $H = G(A_H) = G(S_H)$, and that the first equation of $(*)$ is fulfilled.

We shall now compute $\mathbf{L}(S)$ in terms of $\mathbf{L}(S_N)$ and $\mathbf{L}(S_H)$. From the definition of exp in Proposition V.3.4(iii) we know $\exp|B = (\exp_N \times \exp_H) \circ \nu^{-1}$. Since $d\nu(0) = \mathbf{1}_{\mathbf{L}(G)}$ in view of Lemma V.3.3(i), the wedge of subtangents $\mathbf{L}(S) = L_0(\exp^{-1}\overline{S})$, by Proposition IV.1.19, agrees with the wedge of subtangents of $(\exp_N \times \exp_H)^{-1}(\overline{S_N} \times \overline{S_H}) = (\exp_N^{-1}\overline{S_N}) \oplus (\exp_H^{-1}\overline{S_H})$ which is $\mathbf{L}(S_N) \oplus \mathbf{L}(S_H)$. Thus

$$(2) \qquad\qquad \mathbf{L}(S) = \mathbf{L}(S_N) \oplus \mathbf{L}(S_H).$$

This proves the assertion $(**)$.

We still have to verify $(\dagger)$ and the second equality in $(*)$. If we set $(S_N)_0 = \langle\exp\mathbf{L}(S_N)\rangle$ and define $(S_H)_0$ similarly, then we have $(S_N)_0 \subseteq S_N \subseteq \overline{(S_N)_0}$, and S_H is sandwiched in an analogous chain of containments by Definition V.1.12. Since the identity is fixed under all endomorphisms, every endomorphism of S_N leaves $(S_N)_0$ invariant. Hence

$$(3) \qquad\qquad \alpha\big((S_H)_0\big)(S_N)_0 \subseteq (S_N)_0.$$

Hence $S_1 \stackrel{\mathrm{def}}{=} (S_N)_0 \times (S_H)_0$ is an arcwise connected submonoid of S. Since $(S_N)_0 \times \{1\} \subseteq S_0$ and $\{1\} \times (S_H)_0 \subseteq S_0$ we have $S_1 \subseteq S_0$. It follows that $S \subseteq \overline{S_N} \times \overline{S_H} = \overline{(S_N)_0} \times \overline{(S_H)_0} = \overline{S_1} \subseteq \overline{S_0}$ which gives $(\dagger)$.

Since $(S_N)_0 \cup (S_N)_0^{-1}$ generates N and $(S_H)_0 \cup (S_H)_0^{-1}$ generates H, the set $S_1 \cup S_1^{-1}$ generates $N \times H = G$, and thus, a fortiori, $S_0 \cup S_0^{-1}$ generates G. Hence the second equality of Condition $(*)$ is also satisfied. ∎

We notice that the semigroup S is almost an infinitesimally generated sub-semigroup. What is missing is the first containment of Condition (ii) in Definition V.1.11:

$$\exp \mathbf{L}(S) \subseteq S.$$

We deal with this question, which appears to be of a delicate nature, in the following complement to the preceding theorem.

V.3.7.   Corollary.   *Under the circumstances of* Theorem V.3.6, *the following conditions are equivalent:*

(1) *S is infinitesimally generated.*

(2) *For all $x_N \in \mathbf{L}(S_N)$ and $x_H \in \mathbf{L}(S_H)$, the local solutions of the initial value problem*

$$x'(t) = g(\operatorname{ad} x(t))e^{t \cdot \delta(x_H)} x_N, \qquad 0 \le t \le T, \qquad x(0) = 0$$

satisfy

$$x(t) \in \exp^{-1} S_N.$$

If N is abelian, then (1) and (2) are equivalent to

(3) *For all $x_N \in \mathbf{L}(S_N)$ and $x_H \in \mathbf{L}(S_H)$, the following condition holds for some $T > 0$:*

$$\frac{e^{t \cdot \delta(x_H)} - 1}{\delta(x_H)} x_N \in \exp^{-1} S_N \qquad for\ 0 \le t \le T.$$

Proof.   By Theorem V.3.6, the semigroup S is infinitesimally generated if and only if $\exp \mathbf{L}(S) \subseteq S$. For $x \in \mathbf{L}(S) = \mathbf{L}(S_N) \oplus \mathbf{L}(S_H)$, we write $x = x_N + x_H$ with $x_N \in S_N$ and $x_H \in S_H$. Then we can write $\exp t \cdot x = (F(t), \exp_H t \cdot x_H) \in G$ since the second projection is a homomorphism on all levels. Then $\exp \mathbb{R}^+ \cdot x \subseteq S$ if and only if $F(t) \in S_N$ for all $t \ge 0$, and this holds if and only if there is a $T > 0$ such that $F(t) \in S_N$ for all $t \in [0, T]$. If t is small enough we can write $F(t) = \exp_N x(t)$, and upon transporting $\exp t \cdot x$ into $B_{\mathbf{n}} \times B_{\mathbf{h}}$ via $(\exp_N \times \exp_H)^{-1}$ we have locally the equality

$$(x(t+h), (t+h) \cdot x_H) = (x(t), t \cdot x_H)(x(h), h \cdot x_H) = (x(t) *_{\mathbf{n}} e^{\delta(t \cdot x_H)} x(h), t \cdot x_H + h \cdot x_H),$$

so that $x(\bullet)$ satisfies the functional equation

(5) $x(t + h) = x(t) *_{\mathbf{n}} e^{t \cdot \delta(x_H)} x(h)$ for all sufficiently small $t, h \in \mathbb{R}.$

We notice $x'(0) = x_N$. Then $x(h) = h \cdot x_N + o(h)$. From (5) we have $x(t + h) = x(t) *_{\mathbf{n}} h \cdot e^{t \cdot \delta(x_H)} x_N + o(t, h)$ and thus, from Theorem IV.5.1, we see that $x(\bullet)$ is the local solution of the initial value problem

(6) $x'(t) = g(\operatorname{ad} x(t))e^{t \cdot \delta(x_H)} x_N, \qquad x(0) = 0.$

This establishes the equivalence of (1) and (2).

Now suppose that N is abelian. Then $\operatorname{ad} x(t) = 0$ and thus $g(\operatorname{ad} x(t)) = 1_{\mathbf{n}}$. Thus the initial value problem (6) has exactly the solution

$$x(t) = \sum_{n=1}^{\infty} \frac{t^n}{n!} \cdot \delta(x_H)^{n-1} x_N = \frac{e^{t \cdot \delta(x_H)} - 1}{\delta(x_H)} x_N.$$

Then the equivalence of (3) with (2) is now clear.   ∎

V.3.8.   Corollary.   *If, under the circumstances of* Theorem V.3.6, *in addition* S_N *is closed, then* S *is infinitesimally generated.*

Proof.   If this hypothesis is satisfied, then $\overline{S} = S_N \times \overline{S_H}$ and Proposition V.1.7 shows that $\exp \mathbf{L}(S) \subseteq \overline{S} = S_N \times \overline{S_H}$. Then, if $x \in \mathbf{L}(S)$ as in the proof of Corollary V.3.7, we have $\exp t \cdot x = (\exp_N x(t), \exp_H t \cdot x_H)$ for small enough t, whence $\exp x(t) \in S_N$, that is, $x(t) \in \exp_N^{-1} S_N$ for all small enough t. The claim now is proved in view of Corollary V.3.7 above.   ∎

For the following remark, we retain the notation of the proof of Theorem V.3.6.

V.3.9.   Remark.   The subsemigroup $(S_N)_0 \times (S_H)_0$ is a ray semigroup.

Proof. $\exp\Big(\big(\mathbf{L}(N) \oplus \{0\} \big) \cup \big(\{0\} \oplus \mathbf{L}(H) \big) \Big) = \big(\exp \mathbf{L}(N) \times \{1\} \big) \cup \big(\{1\} \times \exp \mathbf{L}(H) \big)$ generates $(S_N)_0 \times (S_H)_0$.   ∎

We observe from our discussion of the semidirect product formalism that in the case of groups, it has its technical complications, but is reasonably straightforward, yet that, in the case of semigroups, some very subtle questions arise which, so it seems, will have to be treated individually in each particular example. The caliber of these questions is reminiscent of the questions we encountered in Chapter IV.

The following results are a source of important examples.

V.3.10.   Corollary.   *Let N be a finite dimensional vector space and $\alpha : H \to$ Aut N a representation of a finite dimensional Lie group in N. Then $G = N \rtimes_\alpha H$ is a well defined Lie group. Suppose that W is a wedge in N which is invariant under $\alpha(H)$. If S_H is any infinitesimally generated subsemigroup of H, then $S = W \times S_H$ is an infinitesimally generated subsemigroup of G with $\mathbf{L}(S) = W \oplus \mathbf{L}(S_H) \subseteq W \oplus_\delta \mathbf{L}(H)$, where δ is associated with α as in Proposition V.3.4.*

Proof.   This is now straightforward from Theorem V.3.6 and Corollaries V.3.7 and 8.   ∎

One standard example of a semidirect product of Lie groups is the tangent bundle of any Lie group. If H is a connected Lie group, we let N be the underlying vector group of $\mathbf{L}(H)$. The adjoint representation $\mathrm{Ad} : H \to$ Aut N allows the construction of $T(H) \overset{\text{def}}{=} N \rtimes_{\mathrm{Ad}} H$, and this group is isomorphic to the tangent bundle of H (see [Bou75], Chap.3, §2, n°2).

V.3.11.   Remark.   If S_H is an infinitesimally generated subsemigroup of H and W an invariant wedge in $\mathbf{L}(H)$, then $S = W \times S_H$ is an infinitesimally generated subsemigroup of the tangent bundle Lie group $T(H)$ with $\mathbf{L}(S) = W \oplus \mathbf{L}(S_H)$.   ∎

4. Examples

In this section we discuss examples. These help to develop our intuition and illustrate the typical problems arising around the investigation of infinitesimally generated subsemigroups of Lie groups.

Abelian examples show how crucial the global topology of the Lie group influences the structure of global subsemigroups. Many typical phenomena of semigroup theory in Lie groups can be detected inside abelian Lie groups. For instance, we can recognize the great variety of congruence relations on semigroups in Lie groups already by considering subsemigroups of abelian Lie groups. Thus abelian Lie groups furnish a rich supply of examples and counterexamples.

In order to see that the existence of global subsemigroups with a given Lie wedge is not a question of global topology alone we have to turn to non-abelian groups. The Heisenberg group is a Lie group homeomorphic to $\mathbb{R}^3$, yet it contains local infinitesimally generated semigroups which cannot be extended to global ones with the same tangent wedge; also it contains strictly infinitesimally generated semigroups in which the exponential image of its tangent wedge is quite thin in the semigroup.

We turn to solvable but not nilpotent groups. An example for a class of groups in which there are almost no obstructions for global semigroups is the class of almost abelian groups (cf. II.2.13). This is due to the fact that there are abundantly many subgroups of codimension one. Moreover, we will study the universal covering group of the three dimensional group of motions of the euclidean plane. In it we shall find an infinitesimally generated semigroup which is not strictly infinitesimally generated. We shall also investigate the harmonic oscillator group, whose Lie algebra plays an important role in the classification of the Lorentzian semialgebras.

Subsequently, we present an extensive study of the special linear group in two dimensions, which may be viewed as a prototype of a non-compact semisimple Lie group. Its infinitesimal theory was prepared in Section 3 of Chapter II.

We find an important class of subsemigroups of Lie groups by considering contractions; a great variety of semigroups is discovered if we approach the idea of contractions in sufficient generality. The remainder of the section is devoted a general theory of contraction semigroups.

Semigroups in abelian Lie groups

V.4.1.   Example. The simplest, but nevertheless instructive, type of examples is the case where the Lie group G under consideration is abelian. If G is a vector group then the infinitesimally generated semigroups are identical with their tangent wedges under the identification $\mathbf{L}(G) = G$. If G is a torus any closed infinitesimally generated semigroup is a group (cf. Example V.1.4). If G is a cylinder then there are wedges which are tangent wedges of infinitesimally generated semigroups as well as wedges which are not (cf. Figure 1).

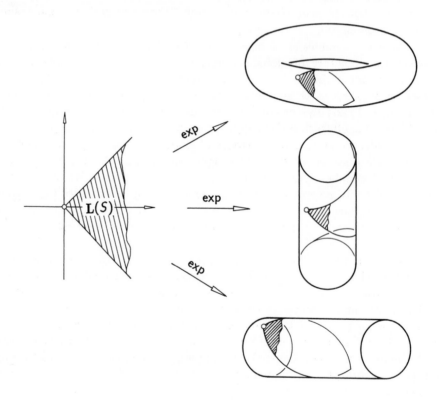

Figure 1

It is important to know for which cones and wedges W in $\mathbf{L}(G)$ we can actually find preanalytic semigroups S in G with $W = \mathbf{L}(S)$. These examples show that the global topology of the group and the position of the wedge in the algebra will play an important role.

V.4.2.   Example. Of course there are many subsemigroups S in G which are not infinitesimally generated. This may even happen if the quotient of S by a discrete subgroup is infinitesimally generated by the same tangent wedge (cf. Figure 2).

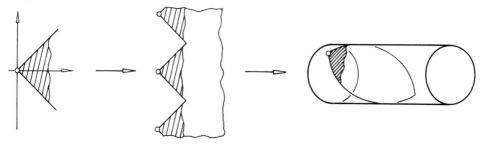

Figure 2

V.4.3.   Example. Another question which exhibits some of its difficulties already
in the abelian case is that of global and
local divisibility. Given a divisible subsemigroup S of G it may be necessary to
choose a very small neighborhood U of
1 in order to have that $U \cap S$ is locally
divisible (cf. chap. IV and the accompany-
ing Figure 3 which shows a one-parameter
semigroup "winding forward at a leisure
pace").

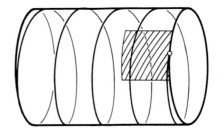

Figure 3

In Section 7 of this Chapter we shall investigate *congruences on open
subsemigroups of Lie groups*. In this context we are only interested in semigroups
whose closure contains the identity. Such semigroups occur as the interior of
infitesimally generated semigroups which generate the group as we saw in the
Infintesimal Generation Theorem V.1.16.

Even in abelian groups congruences on such semigroups can be complicated
in spite of certain restrictions which we shall describe in Section 7 below. The
following examples illustrate this complexity and the natural limitations to the
results of Section 7.

V.4.4.   Example. The simplest case: $S = \mathbb{R}^+ \setminus \{0\}$. Even in this well-known
semigroup there are examples of congruences with interesting features. We recall
the list of all types of congruences on S which have closed congruence classes.

(a) The trivial case: $\kappa = S \times S$ or $\kappa = \Delta$, where Δ denotes the diagonal
in $S \times S$.

(b) There exists an element x in S and a positive "period" $p \in \mathbb{R}$ such
that $x \sim_\kappa x + p$ and such that for any positive q the relation $x \sim_\kappa x + q$ always
implies $q \geq p$. (It is easy to see that if such a pair (x,p) exists at all then p is
uniquely determined.) Write x_0 for the infimum of all such x's. We distinguish
three subcases:

(bi) $x_0 = 0$. Then $x \sim_\kappa x + p$ for a fixed p and all $x \in S$, so $S/\kappa \cong \mathbb{R}/p\mathbb{Z}$. In
    this case κ is closed in $S \times S$.

(bii) $x_0 \in S$, but x_0 is not κ-congruent with $x_0 + p$. Then κ is not closed in $S \times S$; it is given by

$$x \sim_\kappa y \text{ if and only if } x = y \text{ or } x > x_0 \text{ and } x + \mathbb{N}p \cap y + \mathbb{N}p \neq \emptyset.$$

The quotient semigroup S/κ is a T_1-space, but $\kappa(x_0)$ cannot be separated from $\kappa(x_0 + p)$.

(biii) $x_0 \in S$ and $x_0 \sim_\kappa x_0 + p$. Then κ is closed in $S \times S$ and is given by

$$x \sim_\kappa y \qquad \Leftrightarrow \qquad x = y \text{ or } x \geq x_0 \text{ and } x + \mathbb{N}p \cap y + \mathbb{N}p \neq \emptyset.$$

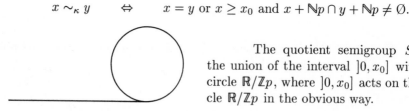

The quotient semigroup S/κ is the union of the interval $]0, x_0]$ with the circle $\mathbb{R}/\mathbb{Z}p$, where $]0, x_0]$ acts on the circle $\mathbb{R}/\mathbb{Z}p$ in the obvious way.

Figure 4

(c) The only remaining type is the Rees congruence associated with an ideal $x_0 + \mathbb{R}^+$. (Note that we suppose the κ-classes to be closed.) In this case the κ-class of x_0 is the interval $[x_0, \infty[$ (hence is not a manifold); all other congruence classes are singletons. Obviously, κ is closed in $S \times S$.

V.4.5.   Example.   (Congruences on subsemigroups of the plane $G = \mathbb{R}^2$) In the following set of examples all semigroups S occuring will be subsemigroups of the additive group $\mathbb{R}^2$ and all congruences will be subordinated to the subgroup $N = \{0\} \times \mathbb{R}$; that is, are contained in the set $\{(u, v) \in G \times G \mid u - v \in N\}$.

Definition. We define congruences κ_N, κ_1, κ_2, and κ_ε, for $\varepsilon > 0$, on each of the semigroups S defined in the sequel. Let (a, b), (a', b') be two elements in S, say $a \leq a'$. Then (a, b), (a', b') in S are said to be congruent with respect to

(i) κ_N if $b = b'$;

(ii) κ_1 if $b = b'$ and $[a, a'] \times \{b\} \subset S$;

(iii) κ_2 if $b = b'$ and $[a, a'] \times \{b\} \subset \overline{S}$;

(iv) κ_ε if $b = b'$ and $[a, a'] \times \{b\} \subset S + ([-\varepsilon, \varepsilon] \times \{0\})$.

Note that κ_2 is the intersection of the congruences κ_ε with $\varepsilon > 0$. The congruence classes with respect to the congruences defined in 4.4 are always closed but the graphs of the congruence relations are not closed.

(ai) S is the interior of the set $\mathbb{R}^+ \times \mathbb{R}^+ \setminus (\omega \times \{0\}) + D$, where D is the closed disc $D = \{(x, y) \in \mathbb{R}^2 \mid (x - \frac{1}{2})^2 + y^2 \leq \frac{1}{4}\}$ and ω denotes the set $\{0, 1, 2, ...\}$.   In this case, the congruence κ_2 is closed in $S \times S$ and S/κ_2 is a Hausdorff k-space, but not locally compact. (In fact, S/κ_2 is locally compact at all points except the κ-class corresponding to $(0, 1)$.)   The space S/κ_2 can be thought of as the union of $\mathbb{R}^+ \setminus \{0\}$ with countably many copies of the interval $]0, 1]$ which meet at 1. Note that κ_1 is not closed in $S \times S$.

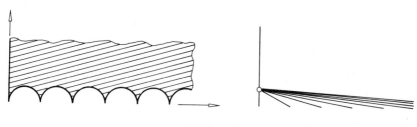

Figure 5

(aii)Replacing in (ai) the set ω by some finite cardinal $\{0, 1, 2, ..., k\}$ we get an open subsemigroup S of $\mathbb{R} \times \mathbb{R}$ such that S/κ_2 is locally compact; the quotient space S/κ_2 is the union of finitely many copies of $]0, 1]$, meeting at a point.

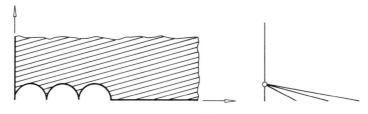

Figure 6

(aiii) Still more sophisticated examples can be obtained by subtracting discs of variable diameter (or isoscele triangles, or other fancy shapes — as long as the difference set is a semigroup). For instance, we may define S as the interior of the set $\mathbb{R}^+ \times \mathbb{R}^+ \setminus \bigcup\{D_n \mid n \in \omega\}$; where $D_n = \{(x, y) \in \mathbb{R}^2 \mid \left(x - 2 + 3(\frac{1}{2})^n\right)^2 + y^2 < (\frac{1}{2})^{2n}\}$.

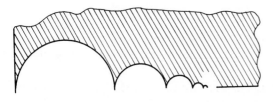

Figure 7

(b) The next set of examples provides congruences where in every neighborhood U of $(0, 0)$ there is an element s of S such that $\kappa(s) \cap U$ is disconnected.

(bi) For every $n \in \mathbb{N}$ define $S_n = \{(x, y) \in \mathbb{R}_+ \times \mathbb{R}_+ \mid \frac{1}{n} < nx < y\}$. Then each S_n is a semigroup and if $m < n$ then $S_m + S_n \subset S_m$. (Indeed, let $(x, y) \in S_m$, $(x', y') \in S_n$. Then $\frac{1}{m} < m(x + x') < mx + nx' < y + y'$, hence $(x + x', y + y') \in S_m$.) Thus the union $S \stackrel{\text{def}}{=} \bigcup\{S_n \mid n \in \mathbb{N}\}$ is an open semigroup; obviously, $(0, 0) \in \overline{S}$ (for instance, $(0, 0) = \lim\left(\frac{(n+1)}{n^3}, \frac{1}{n}\right)$).

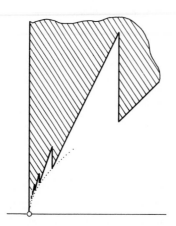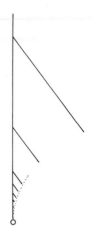

Figure 8

(bii) The example in (bi) can be modified so that the resulting semigroup contains arbitrarily small elements whose κ_N-classes have infinitely many connected components.

For every $m \in \mathbb{N}$ define $C_m = \{(x,y) \in \mathbb{R}_+ \times \mathbb{R}_+ \mid 0 < mx < y\}$ and let $T_m = \bigcup\{(-\frac{1}{2^m} + \frac{1}{(2m)^n}, \frac{1}{2^m} - \frac{1}{(2m)^n}) + C_m \mid n \in \mathbb{N}\}$. Then we define a new open semigroup S_* as the union of the semigroup S in (bi) with translates of the sets T_m:

$$S_* \stackrel{\mathrm{def}}{=} S \cup \bigcup\{(\frac{1}{m^2}, \frac{1}{m}) + T_m\}.$$

(c) (An example with discrete κ-classes) Let S be the half-plane $\mathbb{R}_+ \times \mathbb{R}$. Then we define

$$(a,b) \sim_\kappa (a',b') \text{ if for some } n \in \mathbb{N} \ a = a' \geq \frac{1}{2^n} \text{ and } b \in b' + \mathbb{Z}2^n.$$

V.4.6.   Example. The next example shows that a *non-closed* analytic normal subgroup N in a Lie group G may induce on an open subsemigroup S a congruence *with closed congruence classes*.

Let us start with simply connected objects. In the additive group $\mathbb{R}^3$ write S_0 for the pointed cone $\{(x,y,z) \in \mathbb{R}^3 \mid 0 < y < y + z < x\}$, and N_0 for the closed one-parameter group $\mathbb{R}.(0,1,\sqrt{2})$.

We let H be the discrete subgroup $\{(0,y,z) \mid \{y,z\} \subset \mathbb{Z}\}$, and write G for the quotient group G_0/H. Then the image S of S_0 under the quotient map $q: G_0 \to G_0/H = G$ is an open subsemigroup of G and $1 \in \overline{S}$. Now we define κ to be the congruence whose congruence class at a point $v = (x,y,z)$ is given by:

$$\kappa(q(v)) = \begin{cases} q\big((v + N_0) \cap S_0\big), & \text{if } 0 < x < 1, \\ q(\{x\} \times \mathbb{R} \times \mathbb{R}) \cap S, & \text{if } 1 \leq x. \end{cases}$$

It is not difficult to see that κ is a closed congruence on S and that $N = q(N_0)$; obviously, N is not closed in G.

V.4.7. **Example.** Let S be the semigroup $\{(x, y) \in \mathbb{R} \times \mathbb{R} : x > 0, y > 0\}$. For every natural number $k \in \mathbb{N}$ we set

$$p_k = \left(1 - \frac{1}{2^{k-1}}, \frac{1}{2^k}\right) \in \overline{S},$$

$$q_k = \left(1 - \frac{1}{2^{k-1}}, \frac{1}{2^{k-1}}\right) \in \overline{S},$$

and define

$$N_k = \mathbb{R} \cdot (1, -k)$$

$$J_k = (p_k + \overline{S}) \cap S,$$

$$Z_k = (q_k + \overline{S} + N_k) \cap J_k.$$

(Thus Z_k consists of the points $s \in J_k$ which lie on or above the line $q_k + N_k$.) We now define κ to be the congruence on S whose congruence classes are given by

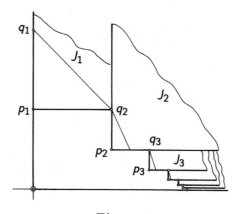

Figure 9

$$\kappa(x) = \begin{cases} \{x\}, & \text{if } x \notin \bigcup_{k \in \mathbb{N}} J_k; \\ (x + N_k) \cap J_k, & \text{if } x \in J_k \backslash Z_k \text{ for some } k \in \mathbb{N}; \\ \bigcup_{k \in \mathbb{N}} Z_k, & \text{otherwise.} \end{cases}$$

The congruence κ is closed; the quotient space S/κ is Hausdorff but not locally compact. Extending κ in the natural way to $\overline{S}$ we get a two-cell as quotient space. Note that the congruence classes of points in the interior of some set $\bigcup_{k \in \mathbb{N}} J_k \backslash Z_k$ are line segments parallel to N_k. Thus there are countably many "different slopes" of such κ-classes. On the other hand, every congruence class is homeomorphic to one of the spaces $\{0\}$, $]0, 1]$, $[0, 1]$, $]0, 1] \times]0, 1[$.

V.4.8. **Example.** It is not difficult to modify the above example so that congruence classes homeomorphic to any given closed subset of the interval $[0, 1]$ occur. For instance, let C be Cantor's middle-third set and define κ_1 by

$$\kappa_1(x) = \begin{cases} \{x\}, & \text{if } x \notin \bigcup_{k \in \mathbb{N}} J_k; \\ \{\frac{1}{8} \cdot ((1, 5) + C \cdot (1, -1)), & \text{if } x \in \{\frac{1}{8} \cdot ((1, 5) + C \cdot (1, -1)); \\ (x + N_1) \cap J_1, & \text{if } x = (a, b) \in J_1 \text{ with } b > \frac{5-a}{8}; \\ (x + N_k) \cap J_k, & \text{if } x \in J_k \backslash Z_k \text{ for some } k > 1; \\ \bigcup_{k \in \mathbb{N}} Z_k, & \text{otherwise.} \end{cases}$$

Semigroups in nilpotent Lie groups

V.4.9. **Example.** Let G be the Heisenberg group, i.e. the group of all real 3×3-matrices of the form

$$(a, b, c) = \begin{pmatrix} 1 & a & c \\ 0 & 1 & b \\ 0 & 0 & 1 \end{pmatrix}$$

This group is a good example for the fact that not only the topology but also the algebraic structure of a Lie group can make it impossible to find a semigroup with some prescribed tangent wedge. In fact, if the interior of a wedge W in $\mathbf{L}(G)$ meets the center of $\mathbf{L}(G)$ then the semigroup generated by $\exp W$ is all of G. We prove this by showing the following slightly more general lemma:

V.4.10.   Lemma.   *Let S be a subsemigroup of Heisenberg group G containing central elements in its interior. Then $S = G$.*

Proof.   Note first that we may identify G with $\mathbf{L}(G)$ if we endow $\mathbf{L}(G)$ with the Campbell–Hausdorff-multiplication

$$(1) \qquad\qquad x * y = x + y + \frac{1}{2}[x, y] \quad \text{for all} \quad x, y \in \mathbf{L}(G).$$

The Lie algebra $\mathbf{L}(G)$ can be represented as the real 3×3-matrices of the form

$$(\alpha, \beta, \gamma) = \begin{pmatrix} 0 & \alpha & \gamma \\ 0 & 0 & \beta \\ 0 & 0 & 0 \end{pmatrix}$$

If we set $x = (1,0,0)$, $y = (0,1,0)$ and $z = (0,0,1)$ then $\mathbb{R}\cdot z$ is the center of G. For $w = \xi\cdot x + \eta\cdot y$ and $w' = x - \xi\cdot y$ we calculate for $z_0 = \lambda\cdot z$ and $n \in \mathbb{N}$

$$u \overset{\text{def}}{=} (n\cdot(z_0 + w)) * (n\cdot(z_0 + w')) = n\cdot(\xi + 1)\cdot x + n(\eta - \xi)\cdot y + n\big(2\lambda - \frac{1}{2}n\cdot(\xi^2 + \eta)\big)\cdot z.$$

If S is a semigroup containing a whole neighborhood of z_0 this calculation shows that for a suitable choice of n, ξ, and η, the element u lies both in $\operatorname{int} S$ and in the x-y-plane. Rotating w and w' around the z-axis we also find $-u \in \operatorname{int} S$. But then $\mathbf{1} = u * -u \in \operatorname{int} S$, so that $S = G$.   ∎

Note that Example V.4.9 contrasts with the assertion of the local theory that for any cone K in $\mathbf{L}(G)$ there exists a *local* semigroup (S, B) having K as tangent object. The crux here is that the neighbourhood B must be chosen small enough (cf. Figure 10).

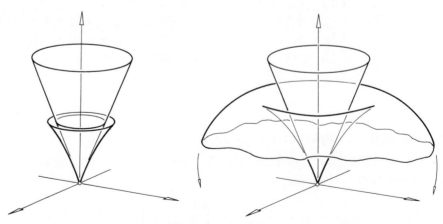

Figure 10

V.4.11. **Example.** In spite of Lemma V.4.5 the Heisenberg group is full of infinitesimally generated semigroups. If we still identify G with $\mathbf{L}(G)$ under the Campbell Hausdorff multiplication then we see that one particularly simple class consists of all wedges containing the center (cf. Figure 11).

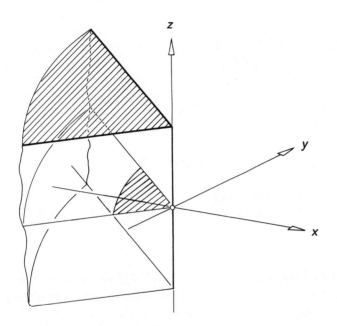

Figure 11

There are also strictly infinitesimally generated subsemigroups in the Heisenberg group which do not contain a non-trivial group:

V.4.12. **Example.** Let G be the Heisenberg group and $S = \{(a, b, c) \in G : 0 \leq a, b;\ 0 \leq c \leq ab\}$. Then S is a strictly infinitesimally generated semigroup with

$$\mathbf{L}(S) = \{(\alpha, \beta, \gamma) \in \mathbf{L}(G) : \gamma = 0,\ 0 \leq \alpha, \beta\}.$$

Proof. It is straightforward to check that S is a closed semigroup. Identifying G with a three dimensional vector space we may visualize S as the region in the first octant bounded by the surface $x_3 = x_1 x_2$ and the x_1-x_2-plane (cf. Figure 12).

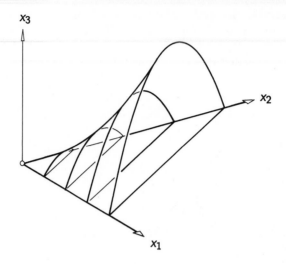

Figure 12

The one-parameter semigroup $\sigma(t) = \exp t{\cdot}(1,0,0) = (t,0,0)$ and $\tau(t) = \exp t{\cdot}(0,1,0) = (0,t,0)$ generate S. In fact, if $(a,b,c) \in S$ and $b > 0$ then

$$(2) \qquad\qquad (a,b,c) = \sigma\!\left(\frac{c}{b}\right)\tau(b)\sigma\!\left(a - \frac{c}{b}\right).$$

If $b = 0$ then $c = 0$, hence $(a,b,c) = \sigma(a)$. It remains to show that $\mathbf{L}(S) \subset \{(\alpha,\beta,\gamma) \in \mathbf{L}(G) : \gamma = 0,\ 0 \le \alpha,\beta\}$. Since S is a closed ray semigroup by what we have just seen, Proposition V.1.7 shows that $\mathbf{L}(S) = \{x \in \mathbf{L}(G) : \exp \mathbb{R}^+{\cdot}x \subset S\}$. But $\exp t{\cdot}(\alpha,\beta,\gamma) = (t\alpha, t\beta, t\gamma + t^2\alpha\beta)$ so that $\mathbf{L}(S) \subset \{(\alpha,\beta,\gamma) \in \mathbf{L}(G) : \gamma = 0,\ \alpha,\beta \ge 0\}$.   ∎

Semigroups in solvable non-nilpotent Lie groups

We have seen in Chapter II that almost abelian algebras were full of semi-algebras because of the abundance of hyperplane subalgebras. Similarly we find a lot of infinitesimally generated subsemigroups of the corresponding groups:

V.4.13.   Example. Let G be the group of real $(n+1) \times (n+1)$-matrices of the form

$$\begin{pmatrix} r{\cdot}E_n & v \\ 0 & 1 \end{pmatrix}$$

where $r \in \mathbb{R}$, $v \in \mathbb{R}^n$ and $E_n : \mathbb{R}^n \to \mathbb{R}^n$ is the identity. Such a group we call an *almost abelian* group. The Lie algebra $\mathbf{L}(G)$ is an almost abelian algebra and can be represented as the matrices of the form

$$\begin{pmatrix} r{\cdot}E_n & v \\ 0 & 0 \end{pmatrix} \qquad r \in \mathbb{R},\ v \in \mathbb{R}^n.$$

In view of the calculations following Proposition II.2.31, the exponential map exp : $\mathbf{L}(G) \to G$ is a diffeomorphism. Any wedge in $\mathbf{L}(G)$ is a semialgebra by Theorem II.2.30. Hence we can apply Corollary II.2.42 to see that exp maps every wedge W in $\mathbf{L}(G)$ homeomorphically onto a subsemigroup of G. In particular there exists a subsemigroup S of G with $\mathbf{L}(S) = W$ for any wedge W in $\mathbf{L}(G)$. ■

V.4.14. **Example.** Let G be the semidirect product of $\mathbb{C}$ by $\mathbb{R}$ where $\mathbb{R}$ acts on $\mathbb{C}$ by rotation. Then the set $S = \{(c, r) \in G \ : \ |c| \leq r\}$ is a closed infinitesimally, but not strictly infinitesimally generated semigroup with $\mathbf{L}(S) = \{(\gamma, \rho) \in \mathbf{L}(G) \ : \ |\gamma| \leq \rho, \ \gamma \in \mathbb{C} \, ; \ \rho \in \mathbb{R}\}$ where $\mathbf{L}(G)$ is the corresponding Lie algebra semidirect sum of $\mathbb{C}$ and $\mathbb{R}$.

Proof. Note first that we may represent G as the set of 3×3-matrices of the form:

$$(c, r) = \begin{pmatrix} e^{ir} & c & 0 \\ 0 & 1 & 0 \\ 0 & 0 & e^r \end{pmatrix}$$

Note that the resulting multiplication is given by $c, r)(c', r') = (c + e^{ir}c', r + r')$. Then the Lie algebra $\mathbf{L}(G)$ is given by

$$(\gamma, \rho) = \begin{pmatrix} i\rho & \gamma & 0 \\ 0 & 0 & 0 \\ 0 & 0 & \rho \end{pmatrix}$$

and the exponential function is the usual matrix exponential function. It is not surjective. It is easy to check that S is a closed subsemigroup of G. Since S also contains inner points it is preanalytic and $\mathbf{L}(G) = \{(\gamma, \rho) \in \mathbf{L}(G) \ : \ \exp \mathbb{R}^+ \cdot ((\gamma, \rho)) \subset S\}$. A one parameter group in G is given by

$$(3) \qquad \exp t(\gamma, \rho) = \left(\frac{\gamma(e^{it\rho} - 1)}{i\rho}, t\rho \right) \quad \text{for} \quad \rho \neq 0.$$

Hence $\exp \mathbb{R}^+ \cdot (\gamma, 1) \subset S$ if and only if

$$|\gamma(e^{it} - 1)| \leq t \quad \text{for all} \quad t \in \mathbb{R}^+.$$

But this is equivalent to each of the following statements:

$$2|\gamma|^2(1 - \cos t) \leq t^2 \qquad \text{for all} \quad t \in \mathbb{R}^+,$$
$$4|\gamma|^2 \sin^2 \frac{t}{2} \leq t^2 \qquad \text{for all} \quad t \in \mathbb{R}^+,$$
$$|\gamma|| \sin \frac{t}{2}| \leq \frac{t}{2} \qquad \text{for all} \quad t \in \mathbb{R}^+,$$
$$|\gamma| \frac{|\sin t|}{t} \leq 1 \qquad \text{for all} \quad t \in \mathbb{R}^+,$$
$$|\gamma| \leq 1.$$

Since $\exp(\gamma, 0) = (\gamma, 0)$ is in S if and only if $\gamma = 0$ this shows that $\mathbf{L}(S) = \{(\gamma, \rho) \in \mathbf{L}(G) : |\gamma| \le \rho\}$ (cf. Figure 13).

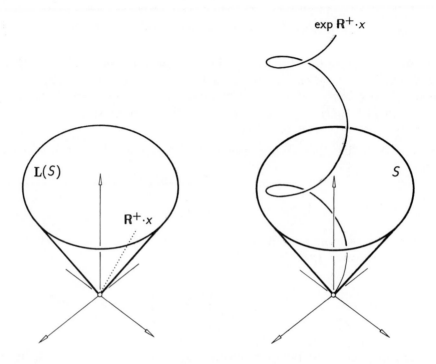

Figure 13

In order to show that S is infinitesimally but not strictly infinitesimally generated we consider the semigroup S_R generated by $\exp \mathbf{L}(S)$. Recall that

$$\exp \mathbf{L}(S) = \{(\frac{\gamma(e^{i\rho t} - 1)}{i\rho}, \rho t) : \rho > 0, |\gamma| \le \rho, t \ge 0\}.$$

We have $|\gamma| \frac{|\sin t|}{t} < 1$ for $|\gamma| \le 1$ and $t > 0$. In view of the earlier calculation this is equivalent to $|\gamma(e^{it} - 1)| < 1$. Thus $\exp \mathbf{L}(S) \subset T = \{(c, r) \in G : |c| < r\} \cup \{(0, 0)\}$. But it is easy to check that T is a semigroup so that $S_R \subset T$.

Conversely, let $(c, r) \in T$. If we set

$$c_k = de^{ia_k}(e^{ir/n} - 1) \quad \text{where} \quad |d| \le 1, \ a_k \in \mathbb{R}$$

we obtain

$$(c_1, \frac{r}{n}) \cdots (c_n, \frac{r}{n}) = (d(e^{ir/n} - 1) \cdot \sum_{k=1}^{n} e^{i(a_k + \frac{(k-1)}{n})}, r)$$

Thus if we set $a_k = \frac{-(k-1)}{n}$ we get

$$(c_1, \frac{r}{n}) \cdots (c_n, \frac{r}{n}) = (dn(e^{ir/n} - 1), r)$$

Now we choose n so that

$$\frac{|c|}{r} < |e^{ir/n} - 1| \cdot \frac{n}{r} \leq 1.$$

Then we can find $d \in \mathbb{C}$ with $|d| < 1$ such that $c = d(e^{ir/n} - 1)n$. Since $(c_k, \frac{r}{n}) = \exp(\frac{ir}{n} d e^{i\alpha_k}, \frac{r}{n})$ this shows that $T = S_R$. Thus the claim follows since T is dense in S. ∎

V.4.15. **Example.** Let H be the Heisenberg group, represented as pairs $(v, z) \in \mathbb{R}^2 \times \mathbb{R}$ endowed with the multiplication

$$(4) \qquad (v, z) \cdot (v', z') = (v + v', z + z' + \frac{1}{2} \langle dv \mid v' \rangle),$$

where $\langle \bullet \mid \bullet \rangle \colon \mathbb{R}^2 \times \mathbb{R}^2 \to \mathbb{R}$ is the scalar product and $d \colon \mathbb{R}^2 \to \mathbb{R}^2$ is given by the matrix $\begin{pmatrix} 0 & -1 \\ +1 & 0 \end{pmatrix}$. Note that this is just the Campbell–Hausdorff multiplication on the Heisenberg algebra $\mathbf{L}(H)$ represented as pairs $(\zeta, \xi) \in \mathbb{R}^2 \times \mathbb{R}$ with bracket

$$[(\zeta, \xi), (\zeta', \xi')] = (0, \langle d\zeta \mid \zeta' \rangle).$$

For $t \in \mathbb{R}$ set

$$R(t) = \begin{pmatrix} \cos t & -\sin t \\ \sin t & \cos t \end{pmatrix}$$

and let $\mathbb{R}$ act on H by $\alpha(r)(v, z) = (R(r)v, z)$. The semidirect product $H \rtimes_\alpha \mathbb{R}$ with respect to this action is called the *harmonic oscillator group* $\mathcal{O}$. The product on $\mathcal{O}$ is given by

$$(5) \qquad (v, z, r)(v', z', r') = (v + R(r)v', z + z' + \frac{1}{2} \langle dv \mid R(r)v' \rangle, r + r').$$

Then we can show that for any invariant generating cone W in $\mathbf{L}(\mathcal{O})$ the closed subsemigroup S of $\mathcal{O}$, generated by $\exp W$, has tangent cone W and there is a neighborhood U of $\mathbf{1}$ in $\mathcal{O}$ such that $S \cap U = \exp W \cap U$.

In order to prove this statement we need to have a good description of the exponential function. We start by calculating the one-parameter subgroups of $\mathcal{O}$:

Note first that Theorem V.3.6 shows that we may identify the underlying spaces of $\mathcal{O}$ and $\mathbf{L}(\mathcal{O})$ so that the generator x of a one-parameter subgroup $\gamma(t)$ is just $x = \gamma'(0)$, since in our representation of $\mathcal{O}$ the multiplication is globally given by the multiplication $\sharp$ and the differential of ν (see Lemma V.3.3(i)!) at zero is the identity. Now let $\Phi(t) = (v(t), z(t), r(t))$ be a one parameter subgroup of $\mathcal{O}$. Then $r(t) = tr_0$ and for $t, s \in \mathbb{R}$

$$(v(s) + R(sr_0)v(t), z(s) + z(t) + \frac{1}{2} \langle dv(x) \mid R(sr_0)(t) \rangle) = (v(s + t), z(s + t)).$$

Fixing s and letting t tend to zero we obtain

$$(\dot{v}(s), \dot{z}(s)) = (R(sr_0)\dot{v}(0), \dot{z}(0) + \frac{1}{2} \langle dv(s) \mid R(sr_0)\dot{v}(0) \rangle)$$

since $v(0) = 0$ and $z(0) = 0$. But since $R(sr_0) = e^{sr_0 d}$ we conclude $v(s) = (e^{sr_0 d} - 1)v_0$, where $r_0 dv_0 = \dot{v}(0)$ for $r_0 \neq 0$. Thus we have in this case

$$\dot{z}(s) = \dot{z}(0) + \frac{1}{2}\langle d(e^{sr_0 d} - 1)v_0 \mid r_0 e^{sr_0 d} d(v_0)\rangle =$$

$$= \dot{z}(0) + \frac{r_0}{2}\langle e^{sr_0 d} dv_0 \mid e^{sr_0 d} dv_0\rangle - \frac{r_0}{2}\langle dv_0 \mid e^{sr_0 d} dv_0\rangle =$$

$$= \dot{z}(0) + \frac{r_0}{2}\langle dv_0 \mid dv_0\rangle - \frac{r_0}{2}\langle dv_0 \mid de^{sr_0 d} v_0\rangle =$$

$$= \dot{z}(0) + \frac{r_0}{2}\|v_0\|^2 - \frac{r_0}{2}\langle v_0 \mid e^{sr_0 d} v_0\rangle$$

since d and $e^{sr_0 d}$ are orthogonal. Integration now yields

$$z(s) = s \cdot \left(\dot{z}(0) + \frac{r_0}{2} \cdot \|v_0\|^2\right) - \frac{1}{2} \cdot \langle dv_0 \mid e^{sr_0 d} \cdot v_0\rangle$$

since d is skew symmetric. Thus the exponential function $\exp: \mathbf{L}(\mathcal{O}) \mapsto \mathcal{O}$ is given by $\exp(v, z, 0) = (v, z, 0)$, and for $r \neq 0$ by

$$(6)\ \exp(v, z, r) = \left(\tfrac{1}{r}(1 - e^{rd}) \cdot dv,\ z + \tfrac{1}{2r}\|v\|^2 - \tfrac{1}{2r^2} \cdot \langle dv \mid e^{rd} \cdot v\rangle,\ r\right).$$

In fact, we only need to note that $\frac{d}{dt}\exp t \cdot (v, z, r)\big|_{t=0} = (v, z, r)$ and use the fact that $d^{-1} = -d$ is orthogonal in the above calculations. From this we calculate easily that $\exp\big|_B: B \mapsto B$ is a diffeomorphism, where $B = \mathbb{R}^2 \times \mathbb{R} \times]-2\pi, 2\pi[$. Therefore the set $C = \{((v, z, r), (v', z', r')) \in \mathbf{L}(\mathcal{O}) \times \mathbf{L}(\mathcal{O}): -2\pi < r + r' < 2\pi\}$ is contained in the set $\{((v, z, r), (v', z', r')) \in \mathbf{L}(\mathcal{O}) \times \mathbf{L}(\mathcal{O}): \exp(v, z, r) \cdot \exp t \cdot (v', z', r')) \in \exp B$ for all $t \in [0, 1]\}$ and hence we can apply Corollary II.2.42 to obtain that

$$\exp((v, z, r)) \exp((v', z', r')) \in \exp W \text{ for all } ((v, z, r), (v', z', r')) \in C \cap (W \times W)$$

where W is any generating semialgebra in $\mathbf{L}(\mathcal{O})$.

Note that $\mathbb{R}^2 \times \mathbb{R} \times [2\pi, \infty[$ is a semigroup ideal in $\mathbb{R}^2 \times \mathbb{R} \times \mathbb{R}^+$. Therefore for any semialgebra W in $\mathbf{L}(\mathcal{O})$ which is contained in $\mathbb{R}^2 \times \mathbb{R} \times \mathbb{R}^+$ the set $S = \exp W \cup (\mathbb{R}^2 \times \mathbb{R} \times [2\pi, \infty[)$ is a subsemigroup of $\mathcal{O}$. But clearly $\mathbf{L}(S) = W$. Finally we recall from Corollary II.3.15 and from the Uniqueness Theorem for Invariant Cones III.2.15 that any invariant cone in $\mathbf{L}(\mathcal{O})$ is isomorphic to the one given by the invariant form

$$q((v, z, r), (v', z', r')) = rz' + r'z + \langle v \mid v'\rangle$$

and the restriction $r \geq 0$. Thus the argument above applies and proves the statement following (5).

We want to give a geometric description of the semigroup generated by $\exp W$ where W is a generating invariant cone in $\mathbf{L}(\mathcal{O})$. By what we have just seen we may assume that $W = \{(v, z, r) \in \mathbf{L}(\mathcal{O}) : 2rz + \|v\|^2 \leq 0,\ r \geq 0\}$.

Let $r \in]0, 2\pi[$ and consider $(\exp W) \cap (\mathbb{R}^2 \times \mathbb{R} \times \{r\})$. Note first that for $\exp(v, z, r) = (v', z', r')$ we have $\exp(e^{td}v, z, r) = (e^{td}v', z', r')$, i.e., the set $(\exp W) \cap (\mathbb{R}^2 \times \mathbb{R} \times \{r\})$ is invariant under rotations in the v-plane. If now $v = (x, 0)$

then $dv = (0, x)$ and $e^{rd}v = (x \cos r, x \sin r)$. Therefore $\langle dv \mid e^{rd}v \rangle = x^2 \sin r$. Moreover

$$\|(1 - e^{rd})dv\|^2 = \|(1 - e^{rd})v\|^2 = 2\|v\|^2 - 2\langle v \mid e^{rd}v \rangle = 2\|v\|^2(1 - \cos r)$$

since $\langle v \mid e^{rd}v \rangle = x^2 \cos r$. As $2rz + \|v\|^2 \leq 0$ just means $z + \frac{1}{2r}\|v\|^2 \leq 0$, this shows that $(\exp W) \cap (\mathbb{R}^2, \mathbb{R}, r)$ is the region below the paraboloid given by

(7) $$\left(v, \|v\|^2 \cdot \frac{-\sin r}{4(1 - \cos r)}, r\right), \quad v \in \mathbb{R}^2 \quad \text{(cf. Figure 14)}$$

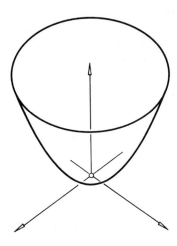

Figure 14

Note that $\lim_{r \to 0} \frac{1 - \cos r}{\sin r} = \lim_{r \to 0} \frac{\sin r}{\cos r} = 0$ so that $\frac{\sin r}{2(1 - \cos r)}$ approaches $\pm \infty$ as r approaches $2\pi n$ with $n \in \mathbb{N}$ depending on whether one approaches from the left or from the right (cf. Figure 15).

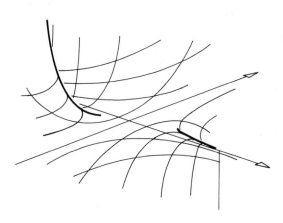

Figure 15

V.4.16.   Example. It is not possible in Example V.4.15 to replace the oscillator group by another group with the same Lie algebra:

Let $\mathcal{O}$ be the oscillator group and $0 \neq z \in Z(\mathbf{L}(\mathcal{O}))$ where $Z(\mathbf{L}(\mathcal{O}))$ is the center of the oscillator algebra. Let S be the closed subsemigroup of $\mathcal{O}$ generated by $\exp W$ where W is a generating invariant cone in $\mathbf{L}(\mathcal{O})$. Then $N = \exp(\mathbb{Z} \cdot z)$ is a discrete Lie subgroup of $\mathcal{O}$ and the subsemigroup SN/N of $\mathcal{O}/N$ has a half-space bounded by the hyperplane ideal of $\mathbf{L}(\mathcal{O})$ as tangent wedge.

Proof.   Set $S_0 = \mathbb{R}^2 \times \mathbb{R} \times \mathbb{R}^+$ with $\mathbb{R}^+ = \{r \in \mathbb{R} : 0 \leq r\}$. This is a half space semigroup in $\mathcal{O}$. Now note that SN/N clearly is contained in $S_0 N/N = S_0/N$, hence $\mathbf{L}(SN/N) \subset \mathbf{L}(S_0/N) = \mathbf{L}(S_0)$. This is a half-space semialgebra bounded by the Heisenberg algebra. Conversely, we know that $\exp$ is a diffeomorphism from a tube around the center of $\mathbf{L}(\mathcal{O})$ onto the image of this tube. But the projection of W along $Z(\mathbf{L}(\mathcal{O}))$ onto $\mathbb{R}^2 \times \{0\} \times \mathbb{R}$ is $(\mathbb{R}^2 \times \{0\} \times (\mathbb{R}^+ \setminus \{0\})) \cup \{(0,0,0)\}$. Hence for any $x \in \mathbb{R}^2 \times \{0\} \times (\mathbb{R}^+ \setminus \{(0,0,0)\})$ we can find a $y \in W \cap \{\mathbb{R}^2 \times \mathbb{Z} \cdot z \times (\mathbb{R}^+ \setminus \{0\})\}$ projecting down to x. Hence $(\exp x)N = (\exp y)N \in SN/N$. Thus, choosing x close to the origin, we derive $\mathbf{L}(SN/N) = \mathbb{R}^2 \times \mathbb{R} \times \mathbb{R}^+$.   ∎

Semigroups in semisimple Lie groups

This part of the section is devoted to the study of the special linear group $\mathrm{Sl}(2, \mathbb{R})$. In a sense this group is the epitome of a (semi-)simple non-compact Lie group and its Lie algebra $\mathrm{sl}(2, \mathbb{R})$ is one of the basic building blocks in the theory and classification of semisimple Lie algebras. Thus $\mathrm{Sl}(2, \mathbb{R})$ is the canonical starting point for the study of infinitesimally generated semigroups in semisimple Lie groups. Since the information available on semigroups in general semisimple Lie groups is rather sparse at the moment, we present a fairly extensive study of the situation in $\mathrm{Sl}(2, \mathbb{R})$ and its simply connected covering group $\widetilde{\mathrm{Sl}(2, \mathbb{R})}$.

Recall that $\mathrm{Sl}(2, \mathbb{R})$ is the set of real 2×2-matrices with determinant 1. Its Lie algebra $\mathrm{sl}(2, \mathbb{R})$ is the set of real 2×2-matrices of zero trace. The exponential function $\exp \colon \mathrm{sl}(2, \mathbb{R}) \to \mathrm{Sl}(2, \mathbb{R})$ is given by the usual matrix exponential function. We will simply write $\mathrm{sl}(2)$ and $\mathrm{Sl}(2)$ instead of $\mathrm{sl}(2, \mathbb{R})$ and $\mathrm{Sl}(2, \mathbb{R})$.

For computational convenience we introduce the normalized trace $\tau \colon \mathrm{sl}(2) \to \mathbb{R}$ defining $\tau(\begin{pmatrix} a & b \\ c & d \end{pmatrix}) = \frac{1}{2}(a + d)$ and a normalized Cartan–Killing form $k \colon \mathrm{sl}(2) \times \mathrm{sl}(2) \to \mathbb{R}$ defined by $k(X, Y) = \tau(XY)$. Then $k(X, Y) = \frac{1}{8}\chi(X, Y)$ with $\chi \colon \mathrm{sl}(2) \times \mathrm{sl}(2) \to \mathbb{R}$, defined by $\chi(X, Y) = \mathrm{tr}(\mathrm{ad}\, X\, \mathrm{ad}\, Y)$, where the trace is taken on the space of endomorphism of the vectorspace $\mathrm{sl}(2)$, i.e. χ is the Cartan–Killing form. By abuse of notation we will write $k(X)$ for $k(X, X)$.

For g in $\mathrm{Sl}(2)$ we obtain a Lie algebra automorphism I_g of $\mathrm{sl}(2)$ via $I_g(X) = gXg^{-1}$. The map $g \mapsto I_g$ from $\mathrm{Sl}(2)$ to $\mathrm{Aut}(\mathrm{sl}(2))$ is a Lie group morphism whose kernel is the two element group $(\mathbf{1}, -\mathbf{1})$. Its image is the connected component $\mathrm{Aut}_0(\mathrm{sl}(2))$ at the identity, and this group is generated by all automorphisms of the form $e^{\mathrm{ad}\, X}$ with X in $\mathrm{sl}(2)$. This means that $\mathrm{Aut}_0(\mathrm{sl}(2))$ is the adjoint group of $\mathrm{Sl}(2)$.

We will write $g \cdot X = gXg^{-1}$ for g in sl(2) and $X \in A$ where A is the Banach algebra of linear operators on $\mathbb{R}^2$. In this fashion sl(2) acts linearly and automorphically on A and sl(2), and automorphically on Sl(2). The exponential function is equivariant relative to these actions, i.e. exp: sl(2) → Sl(2) satisfies $g \cdot \exp X = \exp(g \cdot X)$. By a slight abuse of language we call the action of Sl(2) on sl(2) the adjoint action. This action needs to be understood very well in the following.

The form k is bilinear, symmetric, non-degenerate and *invariant* in the sense that $k(g \cdot X, g \cdot Y) = k(X, Y)$ for g in Sl(2) and that $k([X,Y], z) = k(X, [Y, Z])$. In fact if Φ is any automorphism of sl(2), then $k(\Phi X, \Phi Y) = \frac{1}{8} \operatorname{tr}(\operatorname{ad} \Phi X \operatorname{ad} \Phi Y) = \frac{1}{8} \operatorname{tr}(\Phi \operatorname{ad} X \operatorname{ad} Y \Phi^{-1}) = \frac{1}{8} \operatorname{tr}(\operatorname{ad} X \operatorname{ad} Y) = k(X, Y)$. There is, up to scalar multiplication, only one invariant form.

The points X in sl(2) for which $k(X) \le 0$ will play an important role in our further discussion. They form a double cone, called *the standard double cone*. In obvious ways this cone is reminiscent of the light cone in special relativity. We want to distinguish one of the two cones. For this purpose, we need to introduce a basis for sl(2); there will be involutive automorphisms of sl(2) that interchange the two cones.

We identify A with the algebra of real 2 by 2 matrices and set

(8)
$$H = \begin{pmatrix} 1 & 0 \\ 0 & -1 \end{pmatrix}, \quad P = \begin{pmatrix} 0 & 1 \\ 0 & 0 \end{pmatrix}, \quad Q = \begin{pmatrix} 0 & 0 \\ 1 & 0 \end{pmatrix},$$

$$T = P + Q = \begin{pmatrix} 0 & 1 \\ 1 & 0 \end{pmatrix}, \quad U = P - Q = \begin{pmatrix} 0 & 1 \\ -1 & 0 \end{pmatrix}.$$

Call $\{H, P, Q\}$ the *first basis* and $\{H, T, U\}$ the *second basis* for sl(2). The multiplication is given by

(9a) $\quad [H, P] = 2 \cdot P, \quad [H, Q] = -2 \cdot Q, \quad [P, Q] = H, \quad$ and

(9b) $\quad [H, T] = 2 \cdot U, \quad [H, U] = 2 \cdot T, \quad [U, T] = 2 \cdot H.$

We observe that

(9c) $\quad k(h \cdot H + p \cdot P + q \cdot Q) = h^2 + pq; \quad k(h \cdot H + t \cdot T + u \cdot U) = h^2 + t^2 - u^2.$

In particular, $k(H) = k(T) = 1$, $k(P) = k(Q) = 0$, and $k(U) = -1$. The first basis is adapted to the general theory of semisimple algebras, but for the purpose of geometric representation, we prefer the second basis and denote the plane $\mathbb{R} \cdot H + \mathbb{R} \cdot T$ with the letter $\mathcal{E}$, and call it *the horizontal plane*, while the line $\mathbb{R} \cdot U$ will be called *the vertical line*. Once and for all, we will write $X = X' + x \cdot U$ with X' horizontal. Moreover we introduce a non-canonical Hilbert space structure on sl(2) through the inner product $\langle X_1 \mid X_2 \rangle = h_1 h_2 + t_1 t_2 + x_1 x_2$ for $X_j = h_j \cdot H + t_j \cdot T + x_j \cdot U$ and $j = 1, 2$. We will write $|X| = \langle X \mid X \rangle^{1/2}$, and observe that $k(X) = |X'|^2 - x^2$.

With respect to these definitions, the standard double cone is given by the set $\{X : |X'| \le |x|\}$. We distinguish one part of the double cone by setting

$\mathcal{K} = \{X : |X'| \le x\}$. The boundary of $\mathcal{K}$ given by $\{X : k(X) = 0, \ x \ge 0\}$ is denoted by $\mathcal{N}$. We observe that $\mathcal{K}$ as well as $\mathcal{N}$ is invariant as a consequence of the connectivity of the adjoint group. Note here that $\mathrm{Aut}_0(\mathrm{sl}(2))$ has index two in the full automorphism group and that one convenient representative $\alpha: \mathrm{sl}(2) \to \mathrm{sl}(2)$ of the second coset is defined by $\alpha \cdot X = TXT$. i.e., by

(10) $\alpha(H) = -H, \quad \alpha(P) = Q, \quad \alpha(Q) = P, \quad \alpha(U) = -U, \quad \alpha(T) = T.$

Note that α exchanges $\mathcal{K}$ and $-\mathcal{K}$ and $\mathcal{N}$ and $-\mathcal{N}$.

Recall from Proposition II.3.5 that we obtain the two dimensional subalgebras of $\mathrm{sl}(2)$ as follows.

V.4.17.   Proposition. *For a plane B in $\mathrm{sl}(2)$ the following statements are equivalent:*

   a) *B is a subalgebra.*
   b) *$B = X^\perp$ for some $X \ne 0$ with $k(X) = 0$.*
   c) *$B = X^\perp$ for some $X \in B$.*
   d) *$B^\perp \subset B$.*
   e) *B is tangent to $\mathcal{N} \cup -\mathcal{N}$. All such B are conjugate under $\exp \mathbb{R} \cdot U$.*

Note also that we can completely describe the orbits in $\mathrm{sl}(2)$ under the adjoint action (cf. Proposition II.3.6).

V.4.18.   Proposition. (cf. Figure 3 in Section 3 of Chapter II) *The orbits in $\mathrm{sl}(2)$ under the adjoint action are:*

   a) *In the interior of the standard double cone*

$$\mathrm{Sl}(2) \cdot (uU) = \{X = X' + xU : x^2 - |X'|^2 = u^2, \ xu > 0\}.$$

   b) *In the exterior of the standard double cone*

$$\mathrm{Sl}(2) \cdot (tT) = \{X = X' + xU : |X'|^2 - x^2 = t^2\}.$$

   c) *On the boundary of the standard double cone*

$$\mathrm{Sl}(2) \cdot 0 = \{0\}; \quad \mathrm{Sl}(2) \cdot P = \mathcal{N} \backslash \{0\}; \quad \mathrm{Sl}(2) \cdot Q = -\mathcal{N} \backslash \{0\}.$$

The hyperboloids of a) *and* b) *constitute the level sets of k for non-zerovalues.*

Next we will develop an explicit description of the exponential function $\exp: \mathrm{sl}(2) \to \mathrm{Sl}(2)$. For any X in $\mathrm{sl}(2)$, we have $X^2 = k(X) \cdot \mathbf{1}$, whence all even powers of X are scalar multiples of X. We define the power series

(11) $C(z) = 1 + \dfrac{z}{2!} + \dfrac{z^2}{4!} + \dots \quad \text{and} \quad S(z) = 1 + \dfrac{z}{3!} + \dfrac{z^2}{5!} + \dots$

and note the formulae

(12a) $C(z^2) + zS(z^2) = e^z.$

(12b) $\qquad C(x) = \begin{cases} \cosh\sqrt{x} & \text{for } 0 \le x, \\ \cos\sqrt{-x} & \text{for } 0 > x, \end{cases} \quad \sqrt{|x|}S(x) = \begin{cases} \sinh\sqrt{x} & \text{for } 0 \le x, \\ \sin\sqrt{-x} & \text{for } 0 > x. \end{cases}$

(12c) $\qquad\qquad\qquad\qquad C(z)^2 - zS(z)^2 = 1$

(12d) $\qquad\qquad\qquad\qquad C'(z) = \dfrac{1}{2}S(z).$

Formulae (12a) and (12b) are obvious, the last two identities can easily be shown by considering just positive z and using the analyticity of S and C.

We have *the fundamental formula for the exponential function*

(13) $\qquad\qquad\qquad \exp X = C\big(k(X)\big)\cdot\mathbf{1} + S\big(k(X)\big)\cdot X.$

In particular, we find the element $\exp X$ inside A always in the plane spanned by $\mathbf{1}$ and X, and, since $\operatorname{tr} X = 0$, we have $\tau(\exp X) = C\big(k(X)\big)$. Moreover we note that $k(X) = 0$ implies $\exp X = \mathbf{1} + X$ so that on $\mathcal{N}\cup-\mathcal{N}$ the exponential function is affine.

To discuss the singularities of the exponential function, consider the function C which is holomorphic on the whole complex plane. From (12) it follows that for $x > -\pi^2$ we have $C'(x) > 0$. Hence $C: [-\pi^2, \infty[\to [-1, \infty[$ is a homeomorphism, and thus has an inverse $c: [-1, \infty[\to [-\pi^2, \infty[$ which must be real analytic on $]-1, \infty[$.

Now let us introduce a half space in A given by $A^* = \tau^{-1}]-1, \infty[$. We define a function which we provisionally call logarithm $\mathrm{Log}: A^* \to \mathrm{sl}(2)$ by

(14) $\qquad\qquad \mathrm{Log}\, g = \dfrac{1}{S\big(c(\tau(g))\big)}\cdot\big(g - \tau(g)\cdot\mathbf{1}\big) \quad \text{for } g \in A^*.$

By the definition of k we have $k(g - \tau(g)\cdot\mathbf{1}) = \tau\big((g - \tau(g)\cdot\mathbf{1})^2\big) = \tau(g^2) - 2\big(\tau(g)\big)^2 + \big(\tau(g)\big)^2 = \tau(g^2) - \big(\tau(g)\big)^2 = -\det g + \big(\tau(g)\big)^2$. Thus $k(\mathrm{Log}\,g) = S\big(c(\tau(g))\big)^{-2}\cdot\big(\tau(g)^2 - \det g\big)$.

Now we specialize to $g \in \mathrm{Sl}(2) \cap A^*$, i.e. to $\det g = 1$, and find $k(\mathrm{Log}\,g) = S\big(c(\tau(g))\big)^{-2}\cdot\big(\tau(g)^2 - 1\big)$. If for the moment we set $y = \tau(g)$ and $x = c(y)$, then $y = C(x)$ and from (12c) we find $c(y) = x = S(x)^{-2}\cdot(1 - y^2) = S\big(c(y)\big)^{-2}\cdot(1 - y^2)$. Thus for $g \in \mathrm{Sl}(2) \cap A^*$ we have $k(\mathrm{Log}\,g)c(\tau(g)) > \pi^2$ and then

$\exp\mathrm{Log}\,g = C\big(k(\mathrm{Log}\,g)\big)\cdot\mathbf{1} + S\big(k(\mathrm{Log}\,g)\big)\,\mathrm{Log}\,g = y\cdot\mathbf{1} + S(x)\cdot\dfrac{1}{S(x)}\cdot(g - y\cdot\mathbf{1}) = g.$

So we have

$\qquad\qquad\qquad \exp\mathrm{Log}\,g = g \quad \text{for } g \in \mathrm{Sl}(2) \cap A^*.$

Now $\tau(\exp X) > -1$ iff $C\big(k(X)\big) > -1$ by (13), hence $\exp X \in A^*$ iff $k(X) > -\pi^2$ iff $|X'|^2 - x^2 > -\pi^2$. This gives an invariant open domain $\mathcal{D} = \{X \in \mathrm{sl}(2) : k(X) > -\pi^2\} = \{X = X' + xU : x^2 < |X'|^2 + \pi^2\}$. For $x \in \mathcal{D}$ we have $\exp X \in A^*$ so that we may consider the analytical function $X \mapsto \mathrm{Log}\exp X: \mathcal{D} \mapsto \mathrm{sl}(2)$. Since $\exp$ is a local diffeomorphism around zero, every X near zero may be represented in the form $X = \mathrm{Log}\,g$ for some g near $\mathbf{1}$. For these X we have $\mathrm{Log}\exp X = \mathrm{Log}\exp\mathrm{Log}\,g = \mathrm{Log}\,g = X$. The analytical function $X \mapsto \mathrm{Log}\exp X$ thus agrees on a neighborhood of zero with the identity function. We thus have the first part of

V.4.19.   Theorem.   a) *The exponential function* $\exp\colon \mathrm{sl}(2) \to \mathrm{Sl}(2)$ *induces an isomorphism of real analytical manifolds from* $\mathcal{D}$ *onto* $\mathrm{Sl}(2) \cap A^*$ *whose inverse is given by* $\mathrm{Log}\colon \mathrm{Sl}(2) \cap A^* \mapsto \mathcal{D}$, $\mathrm{Log}\, g = S\Big(C^{-1}\big(\tau(g)\big)\Big)^{-1}\cdot\big(g - \tau(g)\cdot\mathbf{1}\big)$.

b) $\mathrm{Sl}(2) \cap \tau^{-1}\big(]-\infty, -1]\big) \cap \exp\big(\mathrm{sl}(2)\big) = \{-\mathbf{1}\}$.

c) *The set of singular points of* $\exp$ *is* $\exp^{-1}(-\mathbf{1}) = \{\mathrm{Sl}(2)\cdot(\pi + 2\pi\mathbb{Z})\cdot U\}$.

Proof.   To show b) and c) simply note that from (12.b) and (13) it follows that $\exp X \in \exp\big(\mathrm{sl}(2)\big) \cap \tau^{-1}\big(]-\infty, -1]\big)$ is equivalent to $\sqrt{-k(X)} \in \pi + 2\pi\mathbb{Z}$.   ∎

Let us finally remark that, as a consequence of Theorem V.4.19(a) any 2-dimensional subalgebra such as $\mathcal{B} = \mathbb{R}\cdot H + \mathbb{R}\cdot P$ is mapped diffeomorphically under $\exp$ since $\mathcal{D}$ is an open neighborhood of $\mathcal{B}$ which is invariant under inner automorphisms.

We need to recall some facts from the local theory of semigroups in $\mathrm{Sl}(2)$. We start by fixing the notation for the relevant sets.

V.4.20.   Definition.   For $X \neq 0$ and $k(X) = 0$ we set
$X^+ = \{Y \in \mathrm{sl}(2) : k(X, Y) \leq 0\}$, $X^- = -X^+$;
$\mathcal{P}^+ = P^- \cap Q^- = \{hH + pP + qQ : (h, p, q) \in \mathbb{R} \times \mathbb{R}^+ \times \mathbb{R}^+\}$;
$\mathcal{P}^- = P^+ \cap (-Q)^+ = \{hH + pP + qQ : (h, p, q) \in \mathbb{R} \times \mathbb{R}^+ \times -\mathbb{R}^+\}$.
(The reader should keep the distinction between $\mathcal{P}^+$ and P^+ etc. in mind! Cf. Figure 2 of Section 3 in Chapter II.) With these definitions we have (cf. Theorem II.3.7)

V.4.21.   Proposition.   *Any generating semialgebra on* $\mathrm{sl}(2)$ *is the intersection of half space semialgebras each of which is of the form* X^+ *for some* $X \neq 0$ *with* $k(X) = 0$.   ∎

V.4.22.   Theorem.   *If* W *is a wedge in* $\mathrm{sl}(2)$, *and* $\dim H(W) = 1$, *then* W *is a Lie semialgebra iff it is conjugate to one of the following.*

   a) $\mathcal{P}^+, \mathcal{P}^-$ *or* $-\mathcal{P}^-$, *if* $\dim(W - W) = 3$,

   b) *the half-planes in* $P^\perp$ *bounded by* $\mathbb{R}\cdot H$ *if* $\dim(W - W) = 2$ *and* $k(X) > 0$ *for some* $X \in H(W)$,

   c) *the half-planes in* $P^\perp$ *bounded by* $\mathbb{R}\cdot P$ *if* $\dim(W - W) = 2$ *and* $k(X) = 0$ *for all* $X \in H(W)$,

   d) $\mathbb{R}\cdot H, \mathbb{R}\cdot U$ *or* $\mathbb{R}\cdot P$ *if* $\dim(W - W) = 1$.   ∎

V.4.23.   Theorem.   *A wedge* W *in* $\mathrm{sl}(2)$ *is a Lie wedge iff it is either a cone or else a semialgebra.*

Proof.   If $\dim H(W) = 0, 3$ then the assertion is clear. If $\dim H(W) = 2$ then W is a half-space and the assertion follows from Corollary II.2.24. Now suppose that $\dim H(W) = 1$ and that W is a Lie wedge. Now W is the intersection of two different half-spaces. By the Characterization Theorem for Lie Wedges II.1.12, we have $[x, H(W)] \subseteq T_x$ for all $x \in C^1(W)$. Thus if T is the boundary of one of the two half-spaces, then $[x, H(W)] \subseteq T$ for all $x \in T$. Now for all $t \in T \setminus H(W)$ we have $T = H(W) + \mathbb{R} \cdot t$ and thus $[t, T] = [t, H(W)] \subseteq T$. By continuity, this implies $[T, T] \subseteq T$. Thus T is a subalgebra. Thus the two half-spaces are semialgebras by Corollary II.2.24 again, and hence so is their intersection W. ∎

We have now laid the ground to study infinitesimally generated semigroups in $\mathrm{Sl}(2)$. There is one which will turn out to play a special role. If we identify A with the set of two by two matrices as we did before, this semigroup can be described as the set of matrices in $\mathrm{Sl}(2)$ with non-negative entries. *We denote this semigroup by* $\mathrm{Sl}(2)^+$.

V.4.24.   Proposition.   *The exponential function induces on isomorphism*

$$\exp \colon \mathcal{P}^+ \to \mathrm{Sl}(2)^+$$

of analytic manifolds with boundary (cf. Definition V.4.20).

Proof.   After (14), it suffices to show (i) $\exp \mathcal{P}^+ \subseteq \mathrm{Sl}(2)^+$ and (ii) $\mathrm{Log}\,\mathrm{Sl}(2)^+ \subseteq \mathcal{P}^+$ with Log given in Theorem V.4.19. To show (i), let $X = hH + pP + qQ$ with $p, q \geq 0$ and set $\exp(X) = \begin{pmatrix} a & b \\ c & d \end{pmatrix}$. We have $k(X) = h^2 + pq$ and set $t = k(X)$. If $k(X) = 0$, then $X = P$ or $X = Q$ and $\exp X = \mathbf{1} + P$ (resp. $\mathbf{1} + Q$) which is contained in $\mathrm{Sl}(2)^+$. So assume $t > 0$, and conclude from (13) that $a = (\cosh t) + \frac{(\sinh t)}{t} h$, $b = \frac{\sinh t}{t} p$, $c = \frac{\sinh t}{t} q$, and $d = \cosh t - \frac{\sinh t}{t} h$. Since $t > 0$ we have $\sinh t > 0$, so that $a, b, c \geq 0$. But from $\frac{h}{t} = \frac{h}{\sqrt{h^2 + pq}} \leq 1$ it follows that $d \geq 0$, too.

To prove (ii), let $g = \begin{pmatrix} a & b \\ c & d \end{pmatrix} \in \mathrm{Sl}(2)^+$. Then $\det g = 1$ implies $ad = 1 + cb \geq 1$, whence $1 \leq \sqrt{ad} \leq \frac{a+d}{2}$ which means $\tau(g) \geq 1$. Thus $C^{-1}\big(\tau(g)\big) = (\mathrm{arcosh}\,\tau(g))^2 \geq 0$ and $S\big(C^{-1}\big(\tau(g)\big)\big) \geq 0$ (with equality precisely for $g = \mathbf{1}$). By Theorem V.4.19 we have $\mathrm{Log}\,g \in \mathcal{P}^+$ if $b \geq 0$ and $c \geq 0$, which is then the case. ∎

As a consequence of this proposition, we know that $\mathrm{Sl}(2)^+$ is a uniquely divisible semigroup whose tangent object $\mathbf{L}(\mathrm{Sl}(2)^+)$ is $\mathcal{P}^+$. Here we mean by uniquely divisible that for any $s \in \mathrm{Sl}(2)^+$ and any $n \in \mathbb{N}$ there exists a unique $s_1 \in \mathrm{Sl}(2)^+$ with $(s_1)^n = s$. Moreover, the Campbell–Hausdorff multiplication $(X, Y) \mapsto X * Y$ allows an analytic extension to a semigroup multiplication $* \colon \mathcal{P}^+ \times \mathcal{P}^+ \mapsto \mathcal{P}^+$.

The preceding calculations permit us to demonstrate that the Lie wedges of quite reasonable semigroups in $\mathrm{Sl}(2)^+$ are not semialgebras. Indeed, let S be the set of all matrices $g = \begin{pmatrix} a & b \\ c & d \end{pmatrix}$ in $\mathrm{Sl}(2)^+$ with $a \geq 1$. Except for $g = \mathbf{1}$ we have $\tau(g) > 1$ and then $t = S\big(\mathrm{arcosh}\,\tau(g)\big) > 0$. Thus $\mathrm{Log}\,g = (2t)^{-1} \begin{pmatrix} a - d & b \\ c & d - a \end{pmatrix}$ with $d = (1 + bc)a^{-1}$ whence $\mathrm{Log}\,g = hH + pP + qQ$ with $h = \frac{1}{2ta}(a^2 - 1 - bc) \geq -\frac{bc}{2ta}$. One checks that $L(S) = \{hH + pP + aQ : (h, p, q) \in (\mathbb{R}^+)^3\}$: Indeed , if $\begin{pmatrix} a & b \\ c & d \end{pmatrix} = \exp r(hH + pP + qQ) \in S$ for $r > 0$, then $a = a(r) = C(r^2 k) + S(r^2 k) rh \geq 0$, whence $h = \lim_{r \to 0} \frac{a(r)}{r} \geq 0$.

Whether S is in fact the smallest closed subsemigroup $\bar{S}_R$ (cf. Theorem V.1.13) containing $\exp L(S)$ we do not know, but our calculation shows at any rate that $\exp\big(L(S)\big) \subseteq \bar{S}_R \subseteq S$, and since $L(S)$ is no semialgebra, $\exp L(S)$ is not a neighborhood of $\mathbf{1}$ in $\bar{S}_R$, let alone S.

V.4.25.   Proposition. *Let S be a preanalytic semigroup in $\mathrm{Sl}(2)$ and $\bar{S}_R$ the smallest closed semigroup in $\mathrm{Sl}(2)$ containing $\exp\big(\mathbf{L}(S)\big)$. Then we have the following possibilities:*

  a) $\bar{S}_R$ *is a circle group* $\exp \mathbb{R}{\cdot}X$ *with* $k(X) < 0$,
  b) $\bar{S}_R = G$,
  c) *a conjugate of $\bar{S}_R$ is contained in* $\mathrm{Sl}(2)^+$.

Proof.   Without loss of generality we set $S = \bar{S}_R$. Then S is infinitesimally generated, and is therefore completely characterized by its tangent Lie wedge $L(S)$. If $L(S)$ contains an element X with $k(X) < 0$, then S contains $\exp \mathbb{R}{\cdot}X$, a circle group, whence $\mathbb{R}{\cdot}X \subseteq L(S)$. Thus $L(S)$ is not a cone and hence, by Theorem V.4.23 a semialgebra. This means that $L(S) = \mathrm{sl}(2)$ or $\mathbb{R}{\cdot}X$ since no other semialgebras contain elements of negative k-length and have $\mathbb{R}{\cdot}X$ in the edge. Now assume that $L(S) \cap \mathrm{int}(\mathcal{K} \cup -\mathcal{K}) = \varnothing$.

   Find a half-space X^+ with $k(X) = 0$ and $\mathcal{K} \subseteq X^+$ as well as $L(S) \subseteq -X^+$, and a half space Y^+ with $k(Y) = 0$ and $-\mathcal{K} \subseteq Y^+$ as well as $L(S) \subseteq -Y^+$. Then $-X^+ \cap -Y^+$ is a semialgebra containing $L(S)$, and by Theorem V.4.2 it is conjugate to $\mathcal{P}^+$. Thus a conjugate of S is contained in $\mathrm{Sl}(2)^+$.   ∎

   In order to sharpen this result we will show that $\mathrm{Sl}(2)^+$ is a maximal proper connected subsemigroup of $\mathrm{Sl}(2)$, i.e. that any connected subsemigroup T of $\mathrm{Sl}(2)$ containing $\mathrm{Sl}(2)^+$ is either $\mathrm{Sl}(2)^+$ of $\mathrm{Sl}(2)$. For this purpose we need a lemma.

V.4.26.   Lemma. *For any element $X \in \mathrm{sl}(2)$ there are the following mutually exclusive possibilities for the element $\exp X = g$:*

  (i) $X \in \mathcal{P}^+ \cup -\mathcal{P}^+$ *(i.e. $g \in \mathrm{Sl}(2)^+ \cup \big(\mathrm{Sl}(2)^+\big)^{-1}$).*

  (ii) *There is a positive number s such that $k(sH * X) < 0$ $\big($i.e., $(\exp sH)g$ lies on a circle group in $\mathrm{Sl}(2)\big)$.*

Remark.   *For X, Y and $Z \in D$ with $\exp X \exp Y = \exp Z$ we write $Z = X * Y$.*

Proof.   We set $g = \begin{pmatrix} x & y \\ z & w \end{pmatrix}$; then (14) says that for a suitable number $a \in \mathbb{R}$ we

have $X = \mathrm{Log}\, g = \alpha\big(g - \tau(g){\cdot}\mathbf{1}\big) = \tfrac{\alpha}{2}\begin{pmatrix} x - w & 2y \\ 2z & w - y \end{pmatrix} = \alpha\big(\tfrac{1}{2}(x-w)H + yP + zQ\big)$.

Thus case (i) occurs precisely when $yz \geq 0$. We now assume $yz < 0$, i.e. $xw - 1 = yz < 0$, i.e., $xw < 1$.

Now $\exp(sH * X) = (\exp sH)g = \begin{pmatrix} tx & ty \\ \frac{z}{t} & \frac{w}{t} \end{pmatrix}$ with $t = e^s$.

Thus with a suitable scalar β we then have

$$sH * X = \beta\big(\tfrac{1}{2}(tx - \tfrac{w}{t})H + ty + \tfrac{z}{t}Q\big).$$

Then $k(sH * X) < 0$ iff $\frac{1}{4}(tx - \frac{w}{t})^2 + xw - 1 = \frac{1}{4}(tx - \frac{w}{t})^2 + yz < 0$ iff $(tx + \frac{w}{t})^2 < 4$. There are two cases to consider: If $xw \leq 0$, then it is easy to find a $t > 0$ with $(tx + \frac{w}{t})^2 < 4$. If $xw > 0$, then the function $u \mapsto (ux + \frac{w}{u})^2$ attains a minimum for $t = (w/x)^{1/2}$, and this minimum is equal to $xw < 4$. In either case if we take $s = \log t$ we have $k(sH * X) < 0$. ∎

V.4.27. **Lemma.** *Let S be a subsemigroup of* $\mathrm{Sl}(2)$ *containing* $\mathrm{Sl}(2)^+$. *If S meets the interior of* $\exp(\mathcal{K} \cup -\mathcal{K}) \cup (\mathrm{Sl}(2)^+)^{-1}$, *then* $S = \mathrm{Sl}(2)$.

Proof. Let $s \in S$, then each neighborhood U of x contains inner points of S: Indeed the identity neighborhood $s^{-1}U$ contains inner points of $\mathrm{Sl}(2)^+$ hence of S, and so $U = s(s^{-1}U)$ contains inner points of S. Therefore, if S meets the interior of $\exp(\mathcal{K} \cup -\mathcal{K})$, then an open subset of some circle group is in S, and it then follows that this whole circle group and $\exp \mathbb{R} \cdot H$. Thus $S = \mathrm{Sl}(2)$ in this case. Now assume that S contains a point s in the interior of $(\mathrm{Sl}(2)^+)^{-1}$. Then $(\mathrm{Sl}(2)^+)^{-1}$ contains an open subset V of S. But then $V^{-1} \subset \mathrm{Sl}(2)^+$, whence the identity neighborhood VV^{-1} is contained in S. But since $\mathrm{Sl}(2)$ is connected, the semigroup generated by any symmetric identity neighborhood is $\mathrm{Sl}(2)$. ∎

V.4.28. **Lemma.** *Let S be a subsemigroup of* $\mathrm{Sl}(2)$ *containing* $\mathrm{Sl}(2)^+$. *Suppose that S contains a boundary point s of* $(\mathrm{Sl}(2)^+)^{-1}$ *which is not contained in* $\mathrm{Sl}(2)^+$. *Then* $S = \mathrm{Sl}(2)$.

Proof. We may assume that $s \in \exp(\mathbb{R} \cdot H + \mathbb{R} \cdot P)$; the case $s \in \exp(\mathbb{R} \cdot H + \mathbb{R} \cdot Q)$ is treated analogously. If $B = \exp(\mathbb{R} \cdot H + \mathbb{R} \cdot P)$, then the semigroup $S \cap B$ contains the half space semigroup $S' = \exp(\mathbb{R} \cdot H + \mathbb{R}^+ \cdot P)$ in B, and the element s outside S'. But then sS' is a neighborhood of the identity in B and since B is generated as a semigroup by any neighborhood of the identity, we have $B \subseteq S$. Thus S contains the semigroup generated by $B \cup (\mathrm{Sl}(2)^+)$ which, by Proposition V.4.25 is dense in $\mathrm{Sl}(2)$. But it also contains inner points, namely the ones of $\mathrm{Sl}(2)^+$. The assertion then is a consequence of the following Lemma:

V.4.29. **Lemma.** *If S is a dense subsemigroup of a topological group G and if the interior of S is not empty, then $S = G$.*

Proof. Let U be the interior of S. Since $U \neq \emptyset$, there is an $s \in S \cap U^{-1}$. Then sU is an open identity neighborhood which is contained in S. Thus the subgroup $H = S \cap S^{-1}$ is open in G and contained in S. If $g \in G$, then the neighborhood gH of g contains a semigroup element $t \in S$, whence $g \in sH^{-1} = sH \subseteq sS \subseteq S$. ∎

Now we have the following Proposition:

V.4.30. **Proposition.** *Let S be any proper subsemigroup of* $\mathrm{Sl}(2)$ *containing* $\mathrm{Sl}(2)^+$. *Then* $S \cap \mathrm{im} \exp = \mathrm{Sl}(2)^+$ *and* $S \subseteq \mathrm{Sl}(2)^+ \cup - \mathrm{Sl}(2)^+ = \{1, -1\} \cdot \mathrm{Sl}(2)^+$.

Proof. Since $S \cup -S = \{1, -1\} \cdot S$ is a semigroup containing S, we may assume without losing generality that $S = -S$. By Lemmas V.4.26, 27 and V.4.28 we know $S \cap \mathrm{im} \exp \subseteq \mathrm{Sl}(2)^+$, which proves the first assertion. Now suppose $s \in S$. Since $\mathrm{im} \exp \cup - \mathrm{im} \exp = \mathrm{Sl}(2)$ (see Theorem V.4.19), we know that s or $-s$ is in $\mathrm{im} \exp$, hence s or $-s$ is in $\mathrm{Sl}(2)^+$. Thus $s \in \mathrm{Sl}(2)^+ \cup - \mathrm{Sl}(2)^+$. ∎

The sets $\text{Sl}(2)^+$ and $-\,\text{Sl}(2)^+$ are obviously disjoint, hence:

V.4.31.   Example. a) The subsemigroup $\text{Sl}(2)^+$ is a maximal connected proper subsemigroup of $\text{Sl}(2)$.
b) The image $\text{PSl}(2)^+$ of $\text{Sl}(2)^+$ in $\text{PSl}(2)^+$ is a maximal proper subsemigroup of $\text{PSl}(2)^+$. ∎

The study of the universal covering group $\widetilde{\text{Sl}(2)}$ of $\text{Sl}(2)$ will also allow us to show that any proper subsemigroup of $\text{Sl}(2)$ containing a circle group coincides with that circle group. Thus we can sharpen Proposition V.4.25 as follows.

V.4.32.   Example. Let S be closed preanalytic subsemigroup of $\text{Sl}(2)$. Then we have the following possibilities:

    a) S is a circle group.

    b) $S = \text{Sl}(2)$.

    c) There exists a $g \in \text{Sl}(2)$ such that $g\overline{S}_R g^{-1} \subset \text{Sl}(2)^+$. If $g\overline{S}_R g^{-1} = \text{Sl}(2)^+$ then S is either $\text{Sl}(2)^+$ or $\{\mathbf{1}, -\mathbf{1}\}\,\text{Sl}(2)^+$. ∎

We do not know whether there are closed connected proper semigroups S with $\overline{S}_R \subset \text{Sl}(2)^+$ which are not contained in $\text{Sl}(2)^+$.

Proposition V.4.20 shows that no semialgebra intersecting the interior of the standard double cone manifests itself as a global semigroup in $\text{Sl}(2)$ - unless it is a subalgebra; on the other hand, according to the basic theorem of the local Lie theory of semigroups, all of them define local semigroups.

This situation becomes radically different if we ascend to the universal covering group $\widetilde{\text{Sl}(2)}$ of $\text{Sl}(2)$. The polar decomposition of each element of $\text{Sl}(2)$ into a product of an element of $\text{so}(2)$ and a triangular matrix shows quickly that $\text{Sl}(2)$ is topologically the product of a one sphere and a plane. Thus the universal covering space is $\mathbb{R}^3$. Therefore, in order to present the universal covering group all that is required is the fixing of a covering map $f\colon \mathbb{R}^3 \to \text{Sl}(2)$, presumably one which respects the polar decomposition. The general theory of simple connectivity and universal covering spaces then gives a unique Lie group structure on $\mathbb{R}^3$ for any fixed identity element such that f becomes a covering morphism, and the lifting of the exponential function $\exp\colon \text{sl}(2) \to \text{Sl}(2)$ to a function $\text{Exp}\colon \text{sl}(2) \to \mathbb{R}^3$ gives the exponential function of the universal covering. Thus, theoretically, there is nothing left to do. Except, that for calculations and even for the formation of a geometric intuition of the structure of the covering group, a lot depends on a explicit choice of the "parametrization" f. We propose here a particular one which we find to have many good features. Notably, as convenient domain for f we will take $\text{sl}(2)$ itself and respect as much as we can the symmetries defined by the adjoint action of the circle group $\exp \mathbb{R}\cdot U$.

Of course, we retain the notation and concepts introduced above.

V.4.33.   Lemma. *For $X = X' + xU$ we have the following identities:*

$$(\exp xU)\exp(-xU + e^{-(x/2)\,\text{ad}\,U}X) = \exp xU \exp(e^{-(x/2)\,\text{ad}\,U}X') =$$
$$= C(|X'|^2)\exp(xU) + S(|X'|^2)X' = C(k(X))\exp(xU) + S(k(X))X' =$$
$$= \cosh(|X'|)\exp(xU) + \sinh(|X'|)(|X'|^{-1}X').$$

Proof. The first equality is immediate from $X = X' + xU$, whence $e^{t \operatorname{ad} U} X = e^{t \operatorname{ad} U} X' + xU$. By the invariance of k we have $k(e^{-(x/2) \operatorname{ad} U} X') = k(X') = |X'|^2$. From (13) it then follows that

$$\exp(e^{-(x/2) \operatorname{ad} U} X') = C(|X'|^2) \cdot \mathbf{1} + S(|X'|^2)(e^{-(x/2) \operatorname{ad} U} X').$$

Now we multiply through with $\exp tU$ and note from

$$(15) \qquad\qquad e^{r \operatorname{ad} U} = \begin{pmatrix} \cos 2r & \sin 2r & 0 \\ -\sin 2r & \cos 2r & 0 \\ 0 & 0 & 1 \end{pmatrix}$$

that $e^{t \operatorname{ad} U}(X') = \exp(2tU) \cdot X'$ since

$$\exp tU = \begin{pmatrix} \cos t & \sin t \\ -\sin t & \cos t \end{pmatrix} \in A.$$

Thus in view of (12) all identities are proved. $\blacksquare$

We are now ready to give the core definition:

V.4.34. **Definition.** We define a function $f : \mathrm{sl}(2) \to \mathrm{Sl}(2)$ by

$$f(X) = C\big(k(X')\big) \exp xU + S\big(k(X')\big)X' = (\exp xU) \exp(e^{-(x/2) \operatorname{ad} U} X').$$

Note that $f(X) = \cosh |X'| \exp xU + \sinh |X'| \frac{X'}{|X'|}$ for $x' \neq 0$ and $f(xU) = \exp xU$. We will show that this function is a covering map, thus use it to introduce on $\mathrm{sl}(2)$ the structure of the universal covering group of $\mathrm{Sl}(2)$. First we observe that f is analytic and that $f(X)$ is contained in the plane spanned in A by $\exp xU$ and X'. Recall that $\exp xU = (\cos x) \cdot \mathbf{1} + (\sin x)U$, and hence

$$(16) \qquad f(X) = (\cosh |X'| \sin x)U + (\cosh |X'| \cos x) \cdot \mathbf{1} + \frac{\sinh |X'|}{|X'|} X'.$$

From $\tau(U) = \tau(X') = 0$, we immediately obtain

$$(*) \qquad\qquad \tau\big(f(X)\big) = \cosh |X'| \cos x.$$

If we denote by p_u the projection of A onto $\mathbb{R} \cdot U$ with kernel spanned by $\mathbf{1}$, H and T, then

$$(**) \qquad\qquad p_u\big(f(X)\big) = \cosh |X'| \sin x.$$

For any element $a \in A$, we define a complex number $z(a) = \tau(a) + i p_u(a)$ and call it *the characteristic number* of $a \in A$. In this way we can write $(*)$ and $(**)$ as

$$(17) \qquad\qquad z\big(f(X)\big) = \cosh |X'| e^{ix}.$$

We extend $X \mapsto X'$ to a projection $a \mapsto a' : A \mapsto \mathbb{R} \cdot H + \mathbb{R} \cdot T$ with kernel $\mathbb{R} \cdot \mathbf{1} + \mathbb{R} \cdot U$.

V.4.35.   Lemma. *If* $a \in A$, *then* $a = \tau(a) \cdot 1 + p_u(a)U + a'$ *and* $\det a = |z(a)|^2 - |a'|^2$. *In particular,* $g \in \mathrm{Sl}(2)$ *implies* $|z(g)|^2 = 1 + |g'|^2 \geq 1$.

Proof.   The claim follows immediately from the definition of $z(a)$ and the fact that with $a' = hH + tT$ we have

$$a = \begin{pmatrix} \tau(a) + h & p_u(a) + t \\ -(p_u(a) - t) & \tau(a) - h \end{pmatrix}. \qquad\blacksquare$$

We apply the lemma to $f(X)$ and obtain for $X' \neq 0$

$$(18) \qquad \frac{X'}{|X'|} = \left(|z(f(X))|^2 - 1 \right)^{-1/2} f(X)'.$$

Formulae (17) and (18) show that $f(X)$ determines X' completely and we see from (17) that $f(X_1) = f(X_2)$ iff $x_2 - x_1 \in 2\pi \cdot \mathbb{Z}$. Moreover if u is a complex number of modulus greater than one and E a horizontal unit vector in $\mathrm{sl}(2)$, then there is an X in $\mathrm{sl}(2)$ such that $u = z(f(X))$ and $X' = |X'|E$. Note that f is *surjective*. In fact, let $g \in \mathrm{Sl}(2)$, then $|z(g)| \geq 1$, and by the preceding remarks, we find an X in $\mathrm{sl}(2)$ such that $z(f(X)) = z(g)$ and $f(X)' = g'$, since $|g'| = (|z(g)|^2 - 1)^{1/2}$ by Lemma V.4.30. From this we conclude $f(X) = g$ by (17).

We decompose f canonically into the quotient map $\mathrm{sl}(2) \to \mathrm{sl}(2)/2\pi \cdot \mathbb{Z}U$ and the induced continuous bijection $f^* \colon \mathrm{sl}(2)/2\pi \cdot \mathbb{Z} \to \mathrm{Sl}(2)$. Since f has no singular points, as is readily verified from the definition, f^* is also open. Hence f^* is a homeomorphism and thus f is a covering map. Since f is analytic, we know that there is a Lie group multiplication $(X, Y) \mapsto X \circ Y \colon \mathrm{sl}(2) \times \mathrm{sl}(2) \to \mathrm{sl}(2)$ satisfying

$$(19) \qquad\qquad f(X \circ Y) = f(X)f(Y)$$

We denote $(\mathrm{sl}(2), \circ)$ by G. In order to establish some basic properties of the multiplication, we observe a number of equivariance properties of f.

First we define an action of the additive group $\mathbb{R}$ on $\mathrm{sl}(2)$ as a combination of a rotation around the vertical and a vertical translation: For $r \in \mathbb{R}$ and $X \in \mathrm{sl}(2)$ we set

$$(20) \qquad r \cdot X = e^{r/2 \operatorname{ad} U} X' + (r + x)U = (\exp rU)X' + (r + x)U.$$

Now we can establish the following lemma.

V.4.36.   Lemma.   a) $f(r \cdot X) = (\exp rU)f(X)$,
    b) $f(e^{r \operatorname{ad} U} X) = (\exp rU)f(X)(\exp -rU) = e^{r \operatorname{ad} U} f(X)$,
    c) $f(\alpha X) = Tf(X)T$. $\left(Cf. (10)\right)$.

Proof.   For a proof of a) we compute

$$\begin{aligned}
f(r \cdot X) &= f(e^{r/2 \operatorname{ad} U} X' + (r + x)U) \\
&= C\big(k(X')\big)\exp(r + x)U + S\big(k(X')\big)e^{r/2 \operatorname{ad} U} X' \\
&= \exp rU\left(C\big(k(X')\big)\exp xU + S\big(k(X')\big)X' \right) = (\exp rU)f(X).
\end{aligned}$$

Next we verify b):

$$f(e^{r\,\mathrm{ad}\,U}X) = f(e^{r\,\mathrm{ad}\,U}X' + xU) = C\big(k(X')\big)\exp xU + S\big(k(X')\big)e^{r\,\mathrm{ad}\,U}X'$$

$$= e^{r\,\mathrm{ad}\,U}\Big(C\big(k(X')\big)\exp xU + S\big(k(X')\big)X'\Big) = e^{r\,\mathrm{ad}\,U}f(X)$$

$$= (\exp rU)f(X)\exp(-rU),$$

since $e^{r\,\mathrm{ad}\,U}Y = \exp(rU)Y\exp(-rU)$. Finally, c) is left to the reader as an easy exercise. ∎

From the discussion above we derive the following

V.4.37. **Theorem.** *There is a Lie group multiplication* $(X, Y) \mapsto X \circ Y$ *on* sl(2) *such that* $f\colon G \to \mathrm{Sl}(2)$ *is the universal covering morphism, where* G *denotes the group* $(\mathrm{sl}(2), \circ)$, *and that the following properties are satisfied:*
  (a) $(rU) \circ (sU) = (r+s)U$.
  (b) *If* $E \in \mathcal{E}$, *then* $rE \circ sE = (r+s)E$.
  (c) $rU \circ X \circ (-rU) = e^{r\,\mathrm{ad}\,U}X$. *Thus the decomposition* $\mathcal{E} \oplus \mathbb{R}{\cdot}U$ *is invariant under inner automorphism induced by* rU.
  (d) $rU \circ X = e^{r/2\,\mathrm{ad}\,U}X' + (r+x)U$ *and* $rU \circ (\mathcal{E} + uU) = \mathcal{E} + (r+u)U$.
  e) $X \circ (rU) = e^{-r/2\,\mathrm{ad}\,U}X' + (r+x)U$ *and* $(\mathcal{E} + uU) \circ rU = \mathcal{E} + (r+u)U$.
  (f) $X \circ (-X) = (-X) \circ X = 0$.
  (g) $\pi\mathbb{Z}{\cdot}U$ *is the center of* G, *and we have* $X \circ 2n\pi U = X + 2n\pi U$,
    $X \circ (2n+1)\pi U = -X + (2n+1)\pi U$ *for* $n \in \mathbb{Z}$.

Proof. (a) We have

$$f(rU \circ sU) = f(rU)f(sU) = \exp rU \exp sU = \exp(r+s)U = f\big((r+s)U\big).$$

For $r = s = 0$ we note $rU \circ sU = 0 = (r+s)U$. Since liftings are unique, we conclude $rU \circ sU = (r+s)U$.
    (b) Let $E \in \mathcal{E}$ be of norm one, i.e., $E = hH + tT$ with $h^2 + t^2 = 1$. Then $E^2 = k(E){\cdot}1 = 1$ and $f(rE \circ sE) = \big((\cosh r){\cdot}1 + (\sinh r)E\big)\big((\cosh s){\cdot}1 + (\sinh s)E\big) = \cosh(r+s){\cdot}1 + \sinh(r+s)E = f\big((r+s)E\big)$. As before, we conclude $rE \circ sE = (r+s)E$.
    (c), (d) and (e): These assertions follow from calculations of the following type:

$$f\big(rU \circ X \circ (-rU)\big) = (\exp rU)f(X)\big(\exp(-rU)\big) = e^{r\,\mathrm{ad}\,U}f(X) =$$

$$= e^{r\,\mathrm{ad}\,U}\Big(C\big(k(X')\big)\exp xU + S\big(k(X')\big)\Big) =$$

$$= C\big(k(e^{r\,\mathrm{ad}\,U}X')\big)e^{r\,\mathrm{ad}\,U}\exp xU + S\big(k(e^{r\,\mathrm{ad}\,U}X')\big)e^{r\,\mathrm{ad}\,U}X' =$$

$$= f(e^{r\,\mathrm{ad}\,U}X)$$

since $e^{r\,\mathrm{ad}\,U}X' = (e^{r\,\mathrm{ad}\,U}X)'$.

(f) We have

$$f(X \circ -X) = (\exp xU)\exp(e^{-(x/2)\,\mathrm{ad}\,U}X')\exp(-xU)\cdot\exp(-e^{(x/2)\,\mathrm{ad}\,U}X')$$
$$= \exp(e^{x\,\mathrm{ad}\,U}e^{-(x/2)\,\mathrm{ad}\,U}X')\exp(-e^{(x/2)\,\mathrm{ad}\,U}X') = 1 \quad.$$

Thus $f(X \circ (-X)) = f(0)$ and as before $X \circ (-X) = 0$.

(g) If X is central in G, then $f(X) = \pm 1$, but this implies $X \in \pi\mathbb{Z}\cdot U$. Moreover $X \circ 2n\pi U = e^{-n\pi\,\mathrm{ad}\,U}X' + (x + 2\pi n)U = X' + xU + 2\pi nU$ by (15). The last inequality is shown similarly. ∎

Theorem V.4.37 implies that all horizontal and vertical lines through 0 are one-parameter groups in G, and that the inverse agrees with the additive inverse. We will show presently, that *each* one parameter group of G lies in a plane containing the vertical line $\mathbb{R}\cdot U$. The group of rotations around the vertical is an automorphism group, and thus we can reduce our structural description to one plane containing the vertical, say the plane spanned by H and U, and derive the general information by rotation. This is an important advantage of our parametrization.

We apply this strategy to determine the exponential function $\mathrm{Exp}\colon \mathrm{sl}(2) \to G$. From Theorem V.4.37(a,b) it follows that Exp agrees with the identity function on $\mathbb{R}\cdot U$ and $\mathcal{E}$. Moreover Exp is uniquely determined through $\mathrm{Exp}\,0 = 0$ and $f(\mathrm{Exp}\,X) = \exp X$. We write $\mathrm{Exp}\,X = \bar{X} + \bar{x}U$ with $\bar{X} = (\mathrm{Exp}\,X)'$. Then the defining equation $f(\mathrm{Exp}\,X) = \exp X$ reads

$$(21) \qquad C\big(k(\bar{X})\big)\exp\bar{x}I + S\big(k(\bar{X})\big)\bar{X} = C\big(k(X)\big)\cdot 1 + S\big(k(X)\big)X$$

From this we derive the following

V.4.38. **Lemma.** *For X in $\mathrm{sl}(2)$ we have*
(a) $S(|\bar{X}|^2)\bar{X} = S(k(X))X'$.
(b) $C(|\bar{X}|^2)\cos\bar{x} = C\big(k(X)\big)$; $C(|\bar{X}|^2)\sin\bar{x} = S\big(k(X)\big)x$
(c) $z(\exp x) = \cosh(|\bar{X}|)e^{i\bar{x}} = C\big(k(X)\big) + ixS\big(k(X)\big)$.

Proof. The first three equations follow directly from (21) and the fact that $|\bar{X}|^2 = k(\bar{X})$. The last identity is just a reformulation of (17) in view of (21). ∎

The equations of Lemma V.4.38 allow to us to compute the exponential function. In fact they tell us that $\bar{X}$ is a scalar multiple of X'. Thus if we know the functions $\rho = \rho(X) = |\bar{X}|$ and $\mathrm{sgn}\,S\big(k(X)\big)$ then we know $\bar{X}$. In particular, as we already announced, $\mathrm{Exp}\,X \in \mathrm{span}\{X', U\}$. But the complete information on ρ and $\bar{x}$ and thus on Exp is contained in the last equation of Lemma V.4.38 which we call *the characteristic equation*. By a slight abuse of language we call the complex number $z(\mathrm{Exp}\,x)$ *the characteristic number* of X.

We want to determine the shape of the one-parameter groups of G from the characteristic equation. Since $\mathrm{Exp}\,X$ is contained in the plane spanned by X' and U, it suffices up to sign to present ρ as a function of $\bar{X}$. or vice versa. The special form of the functions C and S forces us to treat vectors with positive, negative and zero k-length separately.

The characteristic equation reads

$$(22) \quad \cosh \rho \cdot e^{i\bar{x}} = \begin{cases} 1 + ir, & \text{for } X = 2rP; \\ \cos r + is \sin r, & \text{for } X = r\left(\sqrt{s^2 - 1}\cdot H + sU\right), \quad s \geq 1; \\ \cosh r + is \sinh r, & \text{for } X = r\left(\sqrt{s^2 + 1}\cdot H + sU\right), \quad s \geq 0. \end{cases}$$

Note that it suffices to consider the elements X listed in (22). Indeed, any element of sl(2) is conjugate to such an element X under suitable rotation around $\mathbb{R}\cdot U$. Thus any one-parameter group of G is the result of rotating one those described in (22).

To get a rough intuition of what these one-parameter groups look like we consider the following figure that depicts graphically how (22) determines the pair $(\rho, \bar{x})$ from the given data r, s (cf. Figure 16).

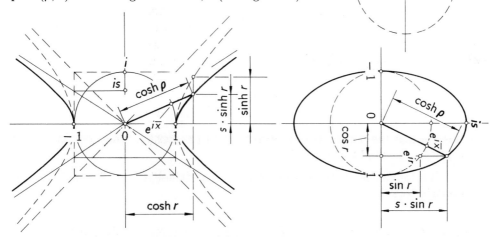

Figure 16

An analytic description of the point sets $\operatorname{Exp}\mathbb{R}\cdot X$ in span$\{H, U\}$ is given by

V.4.39. **Proposition.** *Let X be in $\mathbb{R}^+\cdot H + \mathbb{R}^+\cdot U$. Then*

(a) $\operatorname{Exp}(\mathbb{R}^+\cdot X) = \{\rho H + \xi U : \cosh\rho\cos\xi = 1, \quad \rho \geq 0, \text{ where } 0 \leq \xi \leq \frac{\pi}{2}\}$ *for* $k(X) = 0$,

(b) $\operatorname{Exp}(\mathbb{R}^+\cdot X) = \{\rho\operatorname{sgn}(\sin\xi)H + \xi U : \cosh^2\rho\big((a-1)\cos^2\xi + 1\big) = a,$ *where* $\rho, \xi \geq 0$ *and* $a = x^2 k(X)^{-1}\}$ *for $k(X) < 0$,*

(c) $\operatorname{Exp}(\mathbb{R}^+\cdot X) = \{\rho H + \xi U : \cosh^2\rho\big((a+1)\cos^2\xi - 1\big) = a,$ *where* $\rho \geq 0, 0 \leq \xi \leq \frac{\pi}{2}$ *and* $a = x^2 k(X)^{-1}\}$ *for $k(X) > 0$,*

(d) $\operatorname{Exp}(\mathbb{R}^-\cdot X) = -\operatorname{Exp}(\mathbb{R}^+\cdot X)$.

We recall once more that we get all one-parameter groups in G by rotating those described by Proposition V.4.39.

Proof. (a) It suffices to consider $X = 2rP$ so that by (22 a)) we get $\cosh\rho|\cos\bar{x}| = \sqrt{1 + r^2}(1 + \tan^2\bar{x})^{-1/2} = 1$ and $\cos\bar{x} > 0$.

(b) Without losing generality, we consider $X = r\left(\sqrt{s^2 - 1}\cdot H + sU\right)$, where $s \geq 1$ and obtain $a = s^2$. Thus by (22 b)), we calculate

$$\cosh^2 \rho = \cos^2 r + a \sin^2 r = (\cot^2 r + a)(1 + \cot^2 r)^{-1}$$
$$= a(1 + \cot^2 \bar{x})(1 + a \cot^2 \bar{x})^{-1} = a(\sin^2 \bar{x} + a \cos^2 \bar{x})^{-1} = a(1 + (a-1)\cos^2 \bar{x})^{-1}.$$

(c) We consider $X = r\left(\sqrt{s^2 + 1} \cdot H + sU\right)$, where $s \geq 0$ and obtain $a = s^2$ so that by (22 c)), $\cos \bar{x} = \cosh r(\cosh \rho)^{-1}$ and $\cosh^2 \rho = a \sinh^2 r + \cosh^2 r = (a+1)\cos^2 r - a$. Thus $\cos^2 x \cosh^2 \rho = \frac{\cosh^2 \rho + a}{a+1}$, and the claim follows.

(d) Clear with Theorem V.4.37.   ∎

Using the identity $\operatorname{arccosh} s = \log\left(s - \sqrt{s^2 - 1}\cdot\right)$ it is now a matter of elementary calculation to derive

(23a) $$|\bar{x}| = \arccos\left(\frac{1}{\cosh \rho}\right) \quad \text{for } k(X) = 0,$$

(23b) $$\rho = \log(s + \sqrt{s^2 - 1}|\sin \bar{x}|) - \frac{1}{2}\log\left(1 + (s^2 - 1)\cos^2 \bar{x}\right)$$

$$\text{for } k(X) < 0 \text{ and } s = \frac{|x|}{\sqrt{-k(X)}},$$

(23c) $$|\bar{x}| = \arccos\left(\left(\frac{s^2}{s^2 + 1}\frac{1}{\cosh^2 \rho} + \frac{1}{s^2 + 1}\right)^{\frac{1}{2}}\right)$$

$$\text{for } k(X) > 0 \text{ and } s = \frac{|x|}{\sqrt{k(X)}}.$$

We observe that the point sets described by (23 b) and (23c), "converge" to the set described by (23 a) if s tends to infinity.

It is important to develop an intuitive idea of these results. Figure 17 should help in this regard.

The dark area in Figure 17 is the complement of the image of the exponential function. Reading Figure 17 "modulo 2π", i.e., considering the plank between level $-\pi$ and π and identifying opposite boundary points we obtain a picture of $Sl(2)$. Proceeding in the same way with levels $-\frac{\pi}{2}$ and $\frac{\pi}{2}$ we get a representation of $PSl(2)$. In particular, $PSl(2)$ is exponential, that is, has a surjective exponential function.

V.4.40.   Theorem. a) Exp *induces an isomorphism of real analytical manifolds from* $\mathcal{D} = \{X : k(X) > -\pi^2\}$ *onto the open area between the surfaces* $\operatorname{Exp}(-\mathcal{N}) + \pi u$ *and* $\operatorname{Exp} \mathcal{N} - \pi u$, *i.e. the open set*

$$\{X = \rho E + \xi U : E \in \mathcal{E}, |E| = 1, \xi \in]-\pi, \pi[\text{ and } \cosh \rho \cos \xi > -1\}.$$

b) *The exterior of the standard double cone gets mapped onto*

$$\{X = \rho E + \xi U : E \in \mathcal{E}, |E| = 1, \xi \in]-\frac{\pi}{2}, \frac{\pi}{2}[\text{ and } \cosh \rho \cos \xi > 1\}.$$

c) *The singular points of* Exp *are* $\operatorname{Exp}^{-1}\{n\pi U : n \in \mathbb{Z}\backslash\{0\}\}$. *This set arises from the following upon rotation about the* U-*axis:*

$$\{X = \pi n\left(\sqrt{s^2 - 1}\cdot H + sU\right) : s \geq 1, n \neq 0\}.$$

Proof. a) By Theorem V.4.19, $\exp = f \circ \mathrm{Exp}$ induces an analytic isomorphism from $\mathcal{D}$ onto $\mathrm{Sl}(2) \cap A^*$. Thus $\exp$ induces an analytic isomorphism from $\mathcal{D}$ onto the component of zero in $f^{-1}(\mathrm{Sl}(2) \cap A^*) = \{X \in \mathbf{L} : \tau(f(X)) > -1\}$. Since $\tau(f(X)) = \cosh|X'| \cos x$ by (16), we are looking for the zero component of the set of all $X = \rho E + \xi U$ with $E \in \mathcal{E}$ and $\cosh \rho \cos \xi > -1$. But the horizontal plane through the ξU with $\cos \xi = -1$ separate the set $f^{-1}(\mathrm{Sl}(2) \cap A^*)$ into the components, so the claim follows.

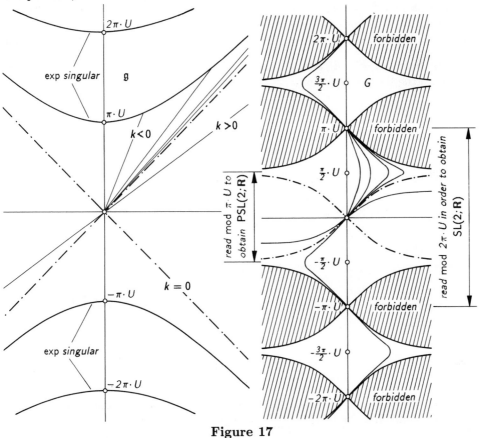

Figure 17

b) Note first that by Proposition V.4.39, the inequality $k(X) < 0$ implies $\cosh^2 \rho(1 + (a-1)\cos^2 \bar{x}) = a$ for some $a \in]1, \infty[$, hence $\cosh^2 \rho \cos^2 \bar{x} < 1$. Conversely, $k(X) > 0$ implies $\cosh^2 \rho((a+1)\cos^2 \bar{x} - 1) = a$ for some $a \in]0, \infty[$, hence $\cosh^2 \rho \cos^2 \bar{x} > 1$ and the claim follows.

c) The set of singular points of Exp is invariant under the adjoint action, hence, after a), is the union of the orbits of the singular points of the form $X = tU$ (cf. Proposition V.4.18). The derivative of Exp in the point X has the kernel $\oplus\{\ker(\mathrm{ad}\, X - 2in\cdot\mathbf{1}) : n = \pm 1, \pm 2, \ldots\}$ (after extension of the scalars to $\mathbf{C}$) (see [Bou72]). Now

$$\mathrm{ad}\,\pi t U - 2\pi i n\cdot\mathbf{1} = 2\pi \begin{pmatrix} -in & t & 0 \\ -t & -in & 0 \\ 0 & 0 & -in \end{pmatrix}$$

relative to the second basis, and the determinant of this vector space endomorphism of sl(2) is $2\pi ni(n^2 - t^2)$. Thus the singular points on the vertical axis are precisely $n\pi U$, $n = \pm 1, \pm 2, \dots$ ∎

To conclude the general description of $\widetilde{Sl(2)}$ we describe the 2-dimensional subgroup $\operatorname{Exp}\mathcal{B}$, $\mathcal{B} = \mathbb{R}{\cdot}H + \mathbb{R}{\cdot}P$ by giving the level lines in sl(2). This means that for a given fixed $\bar{x} \in [0, \frac{\pi}{2}]$ we wish to determine the set $\{\bar{X} : \operatorname{Exp} X = \bar{X} + \bar{x}U; X \in \mathcal{B}\}$. But $X = hH + 2pP = (hH + pT) + pU$ gives rise to the characteristic equation $\cosh \rho e^{i\bar{x}} = \cosh h + i\frac{p}{h}\sinh h$. If we set $s = \tan\bar{x} = \frac{p}{h}\tanh h$ we get $p = sh\coth h$ and the characteristic equation becomes $\cosh e^{i\bar{x}} = \cosh h(1 + is)$. Now $\bar{X} = \frac{\rho}{\sqrt{h^2 + p^2}}(h{\cdot}H + p{\cdot}T) = h^*H + p^*P$ and $\frac{h^*}{p^*} = \frac{h}{p} = s\coth h$ tends to s for large h. Since for $h = h^* = 0$ we find $p^* = \rho = \operatorname{arsinh}s = \log(s + \sqrt{1 + s^2})$, and we obtain the following picture

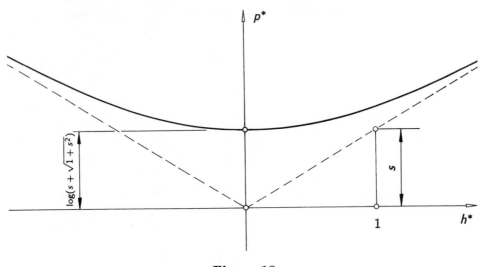

Figure 18

Note in particular that the level lines are invariant under reflection at the plane $\mathbb{R}{\cdot}U + \mathbb{R}{\cdot}T$ and that the level line of $-\bar{x}$ is just the negative of the level line of $\bar{x}$. Thus, if
$$\operatorname{Exp}\mathcal{B} = \{(X', \beta(X')) \in sl(2); X' \in \mathcal{E}\}, \text{ where } \beta \text{ is the appropriate analytic}$$
function, we have $\beta(hH + tT) = \beta(-hH + tT) = -\beta(-hH - tT) = \beta(hH - tT)$.

At this point we are ready to complete the proof of Example V.4.32, i.e. to show that any proper subsemigroup of $Sl(2)$ containing a circle group coincides with that circle group. This will be a corollary of the following

V.4.41.   **Theorem.** *Let S be a subsemigroup of $G = sl(2)$ which contains a conjugate K of $\mathbb{R}{\cdot}U$ $\bigl($i.e. the lifting of a circle group in* sl(2)$\bigr)$. *Then $S = K$ or $S = G$.*

Proof.   Without losing generality we take $K = \mathbb{R}{\cdot}U$ and suppose that there is an $X \in S\backslash\mathbb{R}{\cdot}U$.
We recall $X = X' + x{\cdot}U$ according to (16) and note that by Theorem V.4.37 d) we have $(-x)U \circ X \in \mathcal{E}$. But then $0 \neq (-x)U \circ X \in S$. We may therefore assume

that $x \in E$. By Theorem V.4.37 c) and (15), we conclude $-X = e^{(\pi/2)U} X = (\pi/2)U \circ X \circ (-\pi/2)U \in S$. Hence by Theorem V.4.37(f), we note that $r e^{\operatorname{ad} X} U = X \circ (rU) \circ (-X) \in S$ for all $r \in \mathbb{R}$. Thus S contains the analytical subgroup generated by $\mathbb{R} \cdot U$ and $\mathbb{R} \cdot (e^{\operatorname{ad} X} U)$, whose Lie algebra is generated by U and $e^{\operatorname{ad} X} U$, and hence agrees with sl(2). Thus $S = G$. ∎

V.4.42. **Corollary.** *Let $G^\sharp$ be a quotient group of G modulo a non-degenerate central subgroup Z and let $S^\sharp$ be a subsemigroup of $G^\sharp$ containing a circle group K. Then $S^\sharp = K^\sharp$ or $S^\sharp = G^\sharp$.*

Proof. Let $p: G \mapsto G/Z = G^\sharp$ by the quotient morphism and consider $S = p^{-1}(S^\sharp)$. Then S is a subsemigroup of G containing $K = p^{-1}(K^\sharp)$. Since all one-parameter groups of G whose image in $G^\sharp$ is a circle group contain the center of G, we conclude that K is connected and thus is a one-parameter group. Now K is of the form $\operatorname{Exp} \mathbb{R} \cdot W$ with a conjugate W of U by Proposition V.4.13. Hence Theorem V.4.36 applies and shows $S = K$ or $S = G$. But this implies that either $S^\sharp = p(S) = p(K) = K^\sharp$ or $p(S^\sharp) = p(G) = G^\sharp$. ∎

Note that via Proposition V.3.1 we find a lot of examples of infinitesimally generated subsemigroups of G by pulling back semigroups from Sl(2) and then considering infinitesimally generated subsemigroups of these pull backs (cf. Theorem V.1.14). Here we will concentrate on semigroups which do not arise in this way.

Recall from Theorem V.4.40 that $\mathcal{B} = P^\perp$ is mapped diffeomorphically under Exp onto a surface which we may describe by $\operatorname{Exp} \mathcal{B} = \{X \in \operatorname{sl}(2): x = \beta(X')\}$ with a suitable analytical function β from the horizontal plane $\mathcal{E}$ in $\mathbb{R}$. We set $\Omega^+ = \{X \in \operatorname{sl}(2): x \geq \beta(X')\}$ and $\Omega^- = \{X \in \operatorname{sl}(2): x \leq \beta(X')\}$. Note that $\Omega^- = \alpha(e^{\pi/2 \operatorname{ad} U} \Omega^+)$, since $\beta(hH + tT) = -\beta(hH - tT)$. In contrast to the situation for Sl(2) we find that Ω^+ is a semigroup. In order to show this we need the following lemma, which is of separate interest.

V.4.43. **Lemma.** *Let G be a connected locally compact group and let H be a closed subgroup of G. If A is a closed subgroup of G, isomorphic to $\mathbb{R}$ such that the multiplication $A \times H \mapsto G$ is a homeomorphism, then $G \backslash H$ has two connected components which are both subsemigroups of G and whose boundary is H.*

Proof. Using the inversion $g \mapsto g^{-1}: G \mapsto G$ we see that the multiplication $H \times A \mapsto G$ is also a homeomorphism. Let C be one of the connected components of $A \backslash \{1\}$. Define $S = CH$. Then S is one of the connected components of $G \backslash H$. But HC is also a component of $G \backslash H$. Since HC and S intersect at least in C they are equal. From $HC = CH$ and the fact that C and H are semigroups it follows that S is a semigroup. Moreover it follows that $S^{-1} \cap S = \emptyset$. Thus $G = S \dot\cup H \dot\cup S^{-1}$ and the other claims follow. (Here $\dot\cup$ means disjoint union). ∎

V.4.44. **Example.** We have $(\operatorname{int} \Omega^+) = (\mathbb{R}^+ \backslash \{0\}) U \circ \operatorname{Exp} \mathcal{B}$ and Ω^+ is a closed semigroup bounded by a 2-dimensional subgroup. Analogous statements hold for Ω^-.

Proof. Consider the function $F: \operatorname{sl}(2) \mapsto \operatorname{Sl}(2)$ defined by $F(hH + pP + uU) = (\exp uU)(\exp(hH + pP))$. From (13) we see that

$$F(hH + pP + uU) = \begin{pmatrix} \cos u & \sin u \\ \sin -u & \cos u \end{pmatrix} \begin{pmatrix} e^h & \frac{p}{h} \sin h \\ 0 & e^{-h} \end{pmatrix}$$

Note that F maps $\mathcal{B}$ diffeomorphically onto the subgroup of upper triangular matrices in $\mathrm{Sl}(2)$. Moreover the restriction of F to $\mathbb{R}\cdot U$ is a covering map of the subgroup $\mathrm{SO}(2)$ sitting in $\mathrm{Sl}(2)$. Thus F is a covering map. If we now consider $\mathrm{sl}(2)$ together with the group multiplication $\diamond$ provided by F, we see that $\mathbb{R}\cdot U$ is a closed one-parameter subgroup and $\mathcal{B}$ a connected Lie subgroup of codimension one such that the multiplication $\mathbb{R}\cdot U\times\mathcal{B}\to\widetilde{\mathrm{Sl}(2)}$ given by $(uU, hH+pP)\mapsto (uU)\diamond(hH+pP)$ is a homeomorphism. Now Lemma V.4.43 applies and shows that $(\mathbb{R}^+\backslash\{0\})U\diamond B$ is an open subsemigroup with boundary $\mathcal{B}$. The uniqueness of the simply connected covering group shows that we find an isomorphism $\Phi\colon(\mathrm{sl}(2),\diamond)\to\mathrm{Sl}(2)$ of Lie groups such that the following diagram commutes

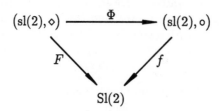

Thus $\Phi(\mathcal{B}) = \mathrm{Exp}\,\mathcal{B}$ and $\Phi(uU) = \mathrm{Exp}(uU)$ and the claim follows.   ∎

Example V.4.44 shows that $e^{r\,\mathrm{ad}\,U}\Omega^+$ and $e^{r\,\mathrm{ad}\,U}\Omega^-$ are half-space semigroups in G. In fact, since every two dimensional connected subgroup of G is conjugate to $\mathrm{Exp}\,\mathcal{B}$ und a rotation, these are only half-space semigroups in G. This allows us to prove the following theorem:

V.4.45.   Theorem. a) *Let S be an infinitesimally generated closed subsemigroup in G. If $L(S)$ is a semialgebra, then S is contained in the intersection $\widetilde{S}$ of a family of half-space semigroups in G, each of which is conjugate either to Ω^+ or to Ω^-, such that $L(\widetilde{S}_R) = L(S)$, i.e. $\widetilde{S}_R = S$, where $\widetilde{S}_R$ is the semigroup generated by all one-parameter semigroups in $\widetilde{S}$.*

b) *For each semialgebra W in $\mathrm{sl}(2)$ there exists exactly one infinitesimally generated closed subsemigroup S with $L(S) = W$.*

Proof.   By Proposition V.4.21, every semialgebra W is the intersection of a family of half-space semialgebras W_j, $j\in I$, each of which determines a unique half-space semigroup S_j in G with $L(S_j) = W_j$. In view of the definition of $L(S_j)$ we have $L(\bigcap S_j) = \bigcap W_j = W$, thus we set $\widetilde{S} = \bigcap S_j$. If S is infinitesimally generated by $W = L(S)$ and since $\exp W = \exp(\bigcap W_j) \leq \bigcap \exp W_j \leq \bigcap S_j = \widetilde{S}$, then S is contained in $\widetilde{S}$, as $\widetilde{S}$ is closed. Now the rest follows.   ∎

Note that Theorem V.4.45 leaves open the problem whether $\widetilde{S}$ is equal to S or not.

We now generalize the notion of invariance to an arbitrary group $\mathcal{G}$. We say a subsemigroup S in $\mathcal{G}$ is *invariant* iff $gSg^{-1} = S$ for all $g\in\mathcal{G}$. A subgroup is invariant exactly if it is normal. It is also clear that for an invariant subsemigroup S containing the identity in a Lie group $\mathcal{G}$ the Lie wedge $L(S)$ is an invariant wedge (see Proposition V.1.15). If $\mathcal{G} = G$ we have only two invariant wedges and we will show that they give rise to invariant subsemigroups. This is remarkable, since G is a simple group. We set

(24) $\qquad \Sigma^+ = \{X \in G = \mathrm{sl}(2): X \text{ is on or above the surface } \exp \mathcal{N}\}$

(see Figure 17). Note that

(25)
$$\bigcap_{r \in \mathbf{R}} \mathrm{Exp}\, rU \circ \Omega^+ \, \mathrm{Exp}\, -rU = \bigcap_{r \in \mathbf{R}} e^{r \, \mathrm{ad}\, U} \Omega^+$$
$$= \{X: \cos x \cosh |X'| \le 1, x \ge 0\} = \Sigma^+.$$

In fact, the first equality follows from Theorem V.4.37(c), the last equality $\Sigma^+ = \bigcap_{r \in \mathbf{R}} e^{r \, \mathrm{ad}\, U} \Omega^+$ is straightforward, since $e^{r \, \mathrm{ad}\, U}$ is just a rotation under which the ray $\mathbf{R} \cdot P$ sweeps out the surface $\mathcal{N}$ while its image $\exp \mathbf{R} \cdot P$ sweeps out the surface $\exp \mathcal{N}$. ∎

V.4.46. **Theorem.** (a) Σ^+ *is a closed invariant non-divisible semigroup with* $L(\Sigma^+) = \mathcal{K}$.

(b) $\Sigma^+ = [0, \pi]U \circ \mathrm{Exp}\, \mathcal{K}$. *In particular every element in* Σ^+ *is the product of two exponentials.*

(c) Σ^+ *is generated by each of its identity neighborhoods.*

(d) $(-NU) \circ \Sigma^+ = G$.

(e) *If we set* $\Sigma^- = -\Sigma^+$, *then the analogous statements hold for* Σ^-.

Proof. Since $\Sigma^- = \alpha(\Sigma^+)$ it suffices to treat Σ^+.

a) Σ^+ is the intersection of closed halfspace semigroups, hence is a closed semigroup. Obviously, $L(\Sigma^+) = \mathcal{K}$ since all conjugates of B are tangent to $\mathcal{K}$. The semigroup Σ^+ cannot be divisible, since there are open sets in $\Sigma^+ \setminus \mathrm{Exp}(\mathrm{sl}(2))$ (see Figure 17).

By b) below, $\Sigma^+ \subset (\mathrm{Exp}\, \mathcal{K}) \subset (\mathrm{Exp}\, \mathcal{K})\Sigma^+$. Since $\mathcal{K}$ is invariant, so is $\mathrm{Exp}\, \mathcal{K}$ and hence Σ^+.

b) It is clear from Theorem V.4.35 and the invariance of $\mathrm{Exp}\, \mathcal{K}$ that $\Sigma^+ \supset [0, \pi]U \circ \mathrm{Exp}\, \mathcal{K}$. Conversely, $\Sigma^+ \subset [0, \pi]U \circ \mathrm{Exp}\, \mathcal{K}$, as follows from Theorem V.4.37(d) and Figure 17.

c) Since $\mathrm{Exp}(\mathcal{K})$ is contained in the semigroup generated by each of the sets $\exp(\mathcal{K} \cap V)$ where V is a zero-neighborhood in $\mathrm{sl}(2)$, and since $rU \circ sU = (r+s)U$, claim c) follows from b) above.

d) follows from $(-nU) \circ \Sigma^+ = \Sigma^+ - \{nU\}$ according to Theorem V.4.37(d). ∎

V.4.47. **Theorem.** *If S is a non-zero invariant closed semigroup in G, then* $\overline{S}_R = \Sigma^+$ *or* Σ^-, *where S_R is the subsemigroup generated by all one-parameter semigroups in S and $\overline{S}_R$ is the closure of S_R. In particular $S = \Sigma^+$ or Σ^-, if S is infinitesimally generated.*

Proof. $L(S)$ is invariant, hence equal to $\mathcal{K}$ or $-\mathcal{K}$. The closed infinitesimally generated semigroup $\overline{S}_R$ is therefore equal to Σ^+ or Σ^-. The rest is clear. ∎

The existence of the semigroup Σ^+ secures on the Lie group G a partial order compatible with the group structure. Let us summarize what this means for G:

V.4.48. Example. (a) The group G allows a partial order which is defined by $g \leq h$ iff $(-g) \circ h \in \Sigma^+$ iff $h \in g \circ \Sigma^+ = \Sigma^+ \circ g$, and which is compatible with the group structure (i.e. satisfies $f \circ g \leq f \circ h$ and $g \circ f \leq h \circ f$ for all f, g, and h with $g \leq h$).

(b) For each $g \in G$ there is a natural number n such that $g \leq nU$ $(= U \circ \cdots \circ U$ (n times)$)$. In particular, $(G, \leq)$ is directed, i.e. for any pair $g, h \in G$ there is an $f \in G$ with $g \leq f$ and $h \leq f$.

(c) The partial order is compatible with the topology in the sense that the graph of $\leq$ is closed in $G \times G$.

Proof.   Reflexivity, transitivity, and antisymmetry are shown as usual, and the monotonicity laws follow in the standard fashion from the invariance of Σ^+ under inner automorphism. By Theorem V.4.46(d) we have $G = (-\mathbb{N} \cdot U) \circ \Sigma^+ = \Sigma^+ \circ (-\mathbb{N} \cdot U)$. Since $g \leq n \cdot U$ is equivalent to $g \in \Sigma^+ \circ (-nU)$, this proves (b) since the remainder is clear. The graph $\{(g, h) \in G \times G : g \leq h\} = \{(g, h) \in G \times G : (-g) \circ h \in \Sigma^+\}$ is closed since Σ^+ is closed. This shows (c).   ∎

A partially ordered group is called *archimedean* iff $a^n \leq b$ for all $n \in \mathbb{Z}$ implies $a = 1$. In the partially ordered group $(G, \leq)$ we have

$$\mathcal{E} \circ tU \leq \Sigma^+ \qquad \text{for} \quad t \geq \pi/2$$

by the definition of Σ^+. Hence whenever $E \in \mathcal{E}$, then by Theorem V.4.37(b) we have $tU \in nE \circ \Sigma^+$ for $n \in \mathbb{Z}$: Thus $nE \leq tU$ for all horizontal vectors E and all $t \geq \pi/2$. In view of the fact that every vector X with $k(X) > 0$ is conjugate to a scalar multiple of H by V.4.13, we can say that for any X with $k(X) > 0$ there is a vector Y with $k(Y) < 0$ such that all powers of X are dominated by Y. Thus $(G, \leq)$ *is not archimedean*, even though Example V.4.48(b) gives a weak archimedean type property. We could have indirectly verified this observation by recalling a theorem on partially ordered groups which says that a directed, partially ordered archimedean group must be commutative. (See [Bi73] p. 317, or [Fu63] p. 95.) This theorem is based on a theorem of Iwasawa (see loc.cit.) according to which every complete lattice-ordered group is commutative, and on the theorem, that every directed archimedean group can be embedded in a complete lattice-ordered group (see [Fu], p. 95). Thus the group $(G, \leq)$ is pretty far from being lattice-ordered, and we have seen the reason clearly: The set of non-negative elements Σ^+ contains the translates of many subgroups. In fact it is easy to verify that it contains translates of every two dimensional subgroup of G.

Contraction semigroups in Lie groups

In the last part of this section we will discuss a class of semigroups which can be described as contractions with respect to some function. Let V be a set and

G a group acting on V. Furthermore let $f: V \to A$ be an arbitrary map where $(A, \leq)$ is a partially ordered set. Under these circumstances it is clear that

$$(26a) \qquad S_f = \{g \in G: f(g \cdot v) \leq f(v) \ \forall v \in V\}$$

is a subsemigroup. Its inverse is given by

$$(26b) \qquad S_f^{-1} = \{g \in G: f(g \cdot v) \geq f(v) \ \forall v \in V\}.$$

     Thus it is clear that $H(S_f) = S_f \cap S_f^{-1}$ is the group of f-isometries. On the other hand we don't have such an easy description for the group $G(S_f)$ generated by S_f. In order to avoid technical complications (cf. Section V.1) we only want to consider the following situation: Suppose that G is a Lie group acting linearly and continuously on a finite dimensional vector space V and that that the set $f^{-1} \downarrow f(v) = \{w \in V: f(w) \leq f(v)\}$ is closed for any $v \in V$. This is a weak form of semicontinuity, and we shall say that $f: V \to A$ is *hemicontinuous*. Now the continuity of the action shows that S_f is closed. Therefore we have

$$(27) \qquad \mathbf{L}(S_f) = \{x \in \mathbf{L}(G) : \exp \mathbb{R}^+ x \subseteq S\}.$$

V.4.49.   **Example.**   (*Invariant Sets*) Let G be a Lie group acting linearly and continuously on a finite dimensional vector space V. Moreover let $M \subseteq V$ be closed subset. If $f = 1 - \chi_M$ where χ_M is the characteristic function of M then

$$(28a) \qquad S_f = \{g \in G: g \cdot M \subseteq M\},$$

$$(28b) \quad \mathbf{L}(S_f) = \{x \in \mathbf{L}(G): \forall (v \in M) \, x \cdot v \in L_v(M)\}, \ x \cdot v = \left.\frac{d}{dt}\right|_{t=0} \left((\exp tx) \cdot v\right).$$

Proof.   It is clear that we have $S_f = \{g \in G: g \cdot M \subseteq M\}$ and since M is closed we know that f is semicontinuous. All we have to note is that we may apply Theorem I.5.17 with the vector field $X(v) = x \cdot v$ whose flow is given by $(\exp tx) \cdot v$.∎

V.4.50.   **Example.**   (*non-negative Matrices*) Let S be the semigroup of all invertible $n \times n$-matrices with non-negative entries. Then for $v = (v_1, ..., v_n) \in (\mathbb{R}^+)^n$ the set of subtangent vectors $L_v((\mathbb{R}^+)^n)$ is equal to $\mathbb{R}^{o_1} \times ... \times \mathbb{R}^{o_n}$ where o_j stands for $+$ if $v_j = 0$ and for 1 if $v_j \neq 0$. Testing (28b) against the standard basis for $\mathbb{R}^n$ we find that $\mathbf{L}(S_f) = \{x = (x_{ij})_{i,j=1,...,n} \in gl(n): x_{ij} \geq 0 \text{ for } i \neq j\}$.
     We can use Example V.4.49 to give a general characterization of $\mathbf{L}(S_f)$.

V.4.51.   **Theorem.**   *Let G be a Lie group acting linearly and continuously on a finite dimensional vector space V. Moreover let $f: V \to A$ be a hemicontinuous map into the ordered set $(A, \leq)$. Then*

$$(29) \qquad \mathbf{L}(S_f) = \{x \in \mathbf{L}(G): x \cdot v \in \mathbf{L}_v\left(f^{-1} \downarrow f(v)\right) \text{ for all } v \in V\}.$$

Proof.   Note first that $f(g \cdot v) \leq f(v)$ for all $v \in V$ if and only if for all $v \in V$ the set $f^{-1} \downarrow f(v)$ is invariant under g. In fact, if $f(g \cdot v) \leq f(v)$ for all $v \in V$

and $v \in f^{-1} \downarrow f(w)$ then $f(g \cdot v) \leq f(v) \leq f(w)$ which means that $g \cdot f^{-1} \downarrow f(w) \subseteq f^{-1} \downarrow f(w)$. Conversely, since $v \in f^{-1} \downarrow f(v)$, the invariance of $f^{-1} \downarrow f(v)$ implies $f(g \cdot v) \leq f(v)$ for all $v \in V$. Thus the semigroup S_f is equal to the intersection of all semigroups $S_{f,v}$ of the form $\{g \in G : g \cdot (f^{-1} \downarrow f(v)) \subseteq f^{-1} \downarrow f(v)\}$, with $v \in V$. But the we know from Example V.4.49 that

$$(30) \quad \mathbf{L}(S_f) = \bigcap_{v \in V} \mathbf{L}(S_{f,v}) = \bigcap_{v \in V} \{x : x \cdot w \in \mathbf{L}_w(f^{-1} \downarrow f(v)) \text{ for all } w \in f^{-1} \downarrow f(v)\}.$$

It is clear that the right hand side of (30) is contained in $\{x \in \mathbf{L}(G) : x \cdot v \in \mathbf{L}_v(f^{-1} \downarrow f(v)) \text{ for all } v \in V\}$. On the other hand, if $x \cdot v \in L_v(f^{-1} \downarrow f(v))$ for *all* $v \in V$ then we get $x \cdot w \in \mathbf{L}_w(f^{-1} \downarrow f(w)) \subseteq \mathbf{L}_w(f^{-1} \downarrow f(v))$ for all $v \in V$ and all $w \in f^{-1} \downarrow f(v)$ since for these w one has $f^{-1} \downarrow f(w) \subseteq f^{-1} \downarrow f(v)$. This proves the theorem. ∎

If the ordered set $(A, \leq)$ is $\mathbb{R}$ with the usual order and the function $f : V \to \mathbb{R}$ happens to be differentiable then the set of subtangent vectors $L_v(f^{-1} \downarrow f(v))$ is closely related to the derivative of f. If v is a regular point of f, then $\mathbf{L}_v(f^{-1} \downarrow f(v))$ is just a half-space bounded by the kernel of $df(v)$ and the relation $x \cdot v \in \mathbf{L}_v(f^{-1} \downarrow f(v))$ reads $df(v)(x \cdot v) \leq 0$. If v is singular these two expressions are no longer equivalent since $\mathbf{L}_v(f^{-1} \downarrow f(v))$ may still be non-trivial, but it turns out that it suffices to consider only regular points:

V.4.52.   Proposition.   *Let A be a Banach space and let $f : V \to A$ be continuously differentiable, then we have*

$$(31) \qquad\qquad \mathbf{L}(S_f) = \{x \in \mathbf{L}(G) : (\forall v \in V)\, df(v)(x \cdot v) \leq 0\}.$$

Proof.   For any $x \in \mathbf{L}(G)$ and $v \in V$ we consider the function $\gamma_{x,v} : \mathbb{R} \to A$ defined by $\gamma_{x,v}(t) = f((\exp tx) \cdot v)$. This function is differentiable, and since $f(v) = \gamma_{x,v}(0)$, it suffices to show that $\gamma_{x,v}$ is (not necessarily strictly) decreasing on $[0, \infty[$ for all $v \in V$ if and only if $df(v)(x \cdot v) \leq 0$ for all $v \in V$. To this end we note that $\gamma'_{x,v}(t) = df((\exp tx) \cdot v)(x \cdot ((\exp tx) \cdot v))$. Since we deal with arbitrary $v \in V$ we see that $\gamma'_{x,v}(t) \leq 0$ for all $v \in V$ and all $t \in \mathbb{R}^+$ if and only if $df(v)(x \cdot v) \leq 0$ for all $v \in V$, which proves our claim. ∎

V.4.53.   Example.   (*Quadratic forms*) Let V be a finite-dimensional real vector space and $G = \mathrm{Gl}(n)$ with the identity representation. Further let $B : V \times V \to \mathbb{R}$ be a symmetric bilinear map and suppose that $q : V \to \mathbb{R}$ is defined by $q(v) = B(v, v)$, the associated quadratic form. Then we have

$$(32) \qquad\qquad \mathbf{L}(S_q) = \{x \in \mathbf{L}(G) : B(v, x \cdot v) \leq 0 \text{ for all } v \in V\}.$$

V.4.54.   Example.   (*Quadratic forms on $\mathbb{R}^2$*). Let $V = \mathbb{R}^2$ and $G = \mathrm{Gl}(n)$ with the identity representation. Moreover let $B : V \times V \to \mathbb{R}$ be an arbitrary bilinear map and $q(v) = B(v, v)$. We have $g \in S_q$ if and only if $B(g \cdot v, g \cdot v) \leq B(v, v)$ for all $v \in V$. This is equivalent to

$$(*) \qquad\qquad \alpha r^2 + \beta rs + \gamma s^2 \leq 0 \qquad \text{for all } r, s \in \mathbb{R},$$

where α, β and γ depend on g. Now $(*)$ is equivalent to

$$(**) \qquad\qquad \alpha, \gamma \leq 0 \quad \text{and} \quad 4\alpha\gamma \leq \beta^2.$$

In view of (32), similar inequalities characterize the elements of $\mathbf{L}(S_q)$. We give two examples for $V = \mathbb{R}^2$. Then $g \in \mathrm{Gl}(V)$ and $x \in \mathrm{gl}(V)$ are represented as 2×2-matrices

$$\begin{pmatrix} a & b \\ c & d \end{pmatrix},$$

so that $\alpha, \ldots, \gamma$ are now functions of $a, \ldots, d$.

(a) $B\big((r, s), (r', s')\big) = rr' + ss'$.

Then $q\big(g{\cdot}(r, s)\big) \leq q(r, s)$ reads $(ar + bs)^2 + (cr + ds)^2 \leq r^2 + s^2$ and a simple calculation shows that $\alpha = a^2 + c^2 - 1$, $\beta = 2ab + 2cd$ and $\gamma = b^2 + d^2 - 1$. Therefore $(**)$ yields $4(a^2 + c^2 - 1)(b^2 + d^2 - 1) \leq 4(ab + cd)^2$ which can be rewritten as $(\det g)^2 \leq a^2 + b^2 + c^2 + d^2 - 1$. Thus $g \in S_q$ if and only if

$$a^2 + c^2 - 1, \quad b^2 + d^2 - 1 \leq 0, \qquad 1 + (\det g)^2 \leq a^2 + b^2 + c^2 + d^2.$$

In order to determine the tangent cone $\mathbf{L}(S_q)$ we note that it consists of all x which satisfy $(ar + bs)r + s(cr + ds) \leq 0$, that is, we have $\alpha = a$, $\beta = b + c$ and $\gamma = d$. Thus $x \in \mathbf{L}(S_q)$ if and only if $a, d \leq 0$, $(a + d)^2 \leq (b + c)^2 + (a - d)^2$.

(b) $B\big((r, s), (r', s')\big) = rr' - ss'$.

In this case $q\big(g{\cdot}(r, s)\big) \leq q(r, s)$ reads $(ar + bs)^2 - (cr + ds)^2 \leq r^2 - s^2$ and again a simple calculation shows that $\alpha = a^2 - c^2 - 1$, $\beta = 2ab - 2cd$ and $\gamma = b^2 - d^2 - 1$. Now $(**)$ yields $4(a^2 - c^2 - 1)(b^2 - d^2 - 1) \leq 4(ab - cd)^2$ which can be rewritten as $-(\det g)^2 \leq a^2 + b^2 - c^2 - d^2 - 1$. Thus $g \in S_q$ if and only if

$$a^2 - c^2 - 1, \quad b^2 - d^2 - 1 \leq 0, \qquad 1 - (\det g)^2 \leq a^2 + b^2 - c^2 - d^2.$$

This time the tangent cone $\mathbf{L}(S_q)$ consists of all x which satisfy $(ar + bs)r - s(cr + ds) \leq 0$, that is, we have $\alpha = a$, $\beta = b - c$ and $\gamma = -d$. Thus $x \in \mathbf{L}(S_q)$ if and only if $a, -d \leq 0$, $(a - d)^2 \leq (b - c)^2 + (a + d)^2$.

In these examples we started with a global semigroup S, which determines its tangent wedge $W = \mathbf{L}(S)$. We don't have to worry about the existence of a semigroup with tangent wedge W of the type given in (29). However, we still have to find *the smallest semigroup S_0 with tangent wedge W*.

V.4.55. **Example.** Let $B : V \times V \to \mathbb{R}$ be a positive definite inner product and $q : V \to \mathbb{R}$ the corresponding quadratic form. Then the smallest semigroup containing the exponential image of $\mathbf{L}(S_q)$ is $S = \{g \in S_q : \det g > 0\}$.

Proof. We can write any $g \in \mathrm{Gl}(V)$ in a unique fashion (polar decomposition) as $(\exp x)k$ where $k \in \{g \in \mathrm{Gl}(V) : q(g{\cdot}v) = q(v) \text{ for all } v \in V\} = H(S_q) = \mathrm{O}(q, V)$ is orthogonal and x is a B-symmetric matrix in $\mathrm{gl}(V)$. If $g \in S_q$ then so is $\exp x$. Since x is symmetric we can diagonalize it by conjugation with elements from $\mathrm{O}(q, V)$. Thus we find a diagonal matrix $y = hxh^{-1}$ with $h \in \mathrm{O}(q, V)$, so that $\exp y = h(\exp x)h^{-1}$ is a diagonal matrix in S_q. This shows that all entries of $\exp y$ are less than or equal to 1, so that all entries of y are non-positive. It follows from this argument that $g = (\exp x)k$ is in S_q if and only if the spectrum of x is in the negative (closed) half-line. In particular we see that a symmetric matrix is in $\mathbf{L}(S_q)$ if and only if its spectrum is in the negative (closed) half-line. Thus any element $g = (\exp x)k$ with $\det k > 0$ is in the semigroup generated by $\exp \mathbf{L}(S_q)$. Conversely the semigroup generated by $\exp \mathbf{L}(S_q)$ is path-connected and therefore consists of elements with positive determinant. This proves the claim since $\det(\exp x) > 0$ for any x. ∎

The proof of Example V.4.55 shows that the map $\varphi: C_B \times \mathrm{O}(q, V) \to S_q$ defined by $\varphi(x, k) = (\exp x)k$ is a homeomorphism, where C_B denotes the cone of all B-symmetric matrices in $\mathbf{L}(S_q)$. The same result is true for arbitrary non-degenerate quadratic forms (cf. [BK79]). However, the existence of a polar decomposition for non-definite quadratic forms is harder to prove. Instead of this proof we give another which shows directly that $\bigl(\exp(C_B)\bigr)\mathrm{O}(q, V)_0$ is the smallest semigroup containing $\exp \mathbf{L}(S_q)$ where $\mathrm{O}(q, V)_0$ is the connected component of the identity of $\mathrm{O}(q, V)$. The argument is valid in a more general context (cf. [Ol81]). We proceed through a few lemmas. From Definition II.2.1, we recall $g(\xi) = f(\xi)^{-1}$ with $f(\xi) = (1 - e^{-\xi})\xi^{-1}$.

V.4.56.   Lemma.   *Let $\mathfrak{g}$ be a Lie algebra which is the direct sum of a subalgebra $\mathfrak{k}$ and a vector subspace $\mathfrak{p}$ containing a pointed cone C. Suppose that*

(i) $[\mathfrak{p}, \mathfrak{p}] \subseteq \mathfrak{k}$.

(ii) $e^{\mathrm{ad}\,\mathfrak{k}} C \subseteq C$.

(iii) *For each $c \in C$ the spectrum of $\mathrm{ad}\, c$ is real.*

Then $W \overset{\mathrm{def}}{=} \mathfrak{k} \oplus C$ is a Lie wedge with edge $\mathfrak{k}$ satisfying

$$g(\mathrm{ad}\, c)W \subseteq g(\mathrm{ad}\, c)L_c(W) = L_c(W) \qquad \text{for all } c \in C.$$

Proof.   The Lie wedge property of W (see Definition II.1.3) is immediate from (ii). In particular, if $k \in \mathfrak{k}$ and $x \in W$, then $[x, k] \in T_x(W)$. If $w = k + c' \in W$ with $k \in \mathfrak{k}$ and $c' \in C$, then for any $c \in C$ we have $[c, w] = [c, k] + [c, c'] \in T_c(W) + [\mathfrak{p}, \mathfrak{p}]$. From Definition I.1.8 and Corollary I.5.4 we know $T_c(W) = (c^\perp \cap W^*)^\perp$, and (i) implies $[\mathfrak{p}, \mathfrak{p}] \subseteq \mathfrak{k} = H(W) \subseteq (c^\perp \cap W^*)^\perp$. Hence $(\mathrm{ad}\, c)(W - W) \subseteq T_c(W)$. Since $\mathrm{ad}\, c$ has real spectrum for $c \in C$ the element $f(\mathrm{ad}\, c)$ has an inverse $g(\mathrm{ad}\, c)$. Hence $f(\mathrm{ad}\, c)T_c(W) = T_c(W)$.

We claim that $f\bigl(\mathrm{ad}(c)\bigr)L_c(W) = L_c(W)$. To see this let $y \in L_c(W)$. Then $z \overset{\mathrm{def}}{=} f(\mathrm{ad}\, c)(y) - y = -\frac{1}{2}\cdot(\mathrm{ad}\, c)y + \frac{1}{3!}\cdot(\mathrm{ad}\, c)^2 y \pm \cdots \in T_c(W)$. Pick $w \in T_c(W)$ so that $z = f(\mathrm{ad}\, c)w$. Then $f(\mathrm{ad}\, c)(y - w) = (z + y) - z = y$ and $y - w \in L_c(W)$. This proves the claim. Then applying $g(\mathrm{ad}\, c)$ we find $g(\mathrm{ad}\, c)(W) \subseteq g(\mathrm{ad}\, c)L_c(W) = L_c(W)$. ∎

V.4.57.   Theorem.   *Let G be a Lie group with a closed connected subgroup H and suppose that its Lie algebra $\mathfrak{g} = \mathbf{L}(G)$ is the direct sum of $\mathfrak{k} = \mathbf{L}(H)$ and a vector subspace $\mathfrak{p}$ containing a pointed cone C. Set $W = C \oplus \mathfrak{k}$ and $S = (\exp C)H$. Suppose that*

(i) $[\mathfrak{p}, \mathfrak{p}] \subseteq \mathfrak{k}$ *and* $[\mathfrak{p}, \mathfrak{k}] \subseteq \mathfrak{p}$.

(ii) $\operatorname{Ad} H(C) \subseteq C$.

(iii) *For each $c \in C$ the spectrum of $\operatorname{ad} c$ is real.*

Then S is a semigroup if it is closed. Moreover, suppose that the following conditions are also satisfied:

(iv) $(\exp -C \exp C) \cap H = \{1\}$.

(v) $z \in \mathfrak{z} \cap (C - C)$ *with $\mathfrak{z}$ the center of $\mathfrak{g}$ and $\exp z = 1$ implies $z = 0$.*

Then the function $\varphi \colon \mathfrak{p} \times H \to G$ defined by $\varphi(y, h) = (\exp y)h$ maps $C \times H$ homeomorphically onto S and $\mathbf{L}(S) = W$. In this case, if S is closed, then S is strictly infinitesimally generated.

Proof. We have $SH \subseteq S$. Thus in order to show $SS \subseteq S$ it suffices to show $S \exp C \subseteq S$. Suppose $s \in S$ and $x \in C$; we shall show that there is an $\varepsilon > 0$ such that $s \exp t{\cdot}x \in S$ for all $0 \le t \le \varepsilon$. Then the set $\{t \in \mathbb{R}^+ \colon s \exp \tau{\cdot}x$ for all $0 \le \tau \le t\}$ is open in $\mathbb{R}^+$. If S is closed, then it is also closed and hence all of $\mathbb{R}^+$. Now $s = \varphi(p, k)$ for some $p \in \mathfrak{p}$ and $k \in H$. We claim that φ is regular at (p, k). Since right translation by k is a diffeomorphism of G the claim holds if φ is regular at $(p, 1)$. But if $(u, v) \in \mathfrak{p} \times \mathfrak{k}$, then in view of [He78], Chap. II we have

(33)
$$\begin{aligned}
d\varphi(y, h)(u, v) &= d\lambda_{\exp y}(h)v + d\rho_h(\exp y)\big(d\exp(y)u\big) \\
&= d\lambda_{\exp y}(h)v + d\rho_h(\exp y)d\lambda_{\exp y}(1)\big(f(\operatorname{ad} y)u\big) \\
&= d\lambda_{\exp y}(h)v + d\lambda_{\exp y}(h)d\rho_h(1)\big(f(\operatorname{ad} y)u\big) \\
&= d\lambda_{\exp y}(h)\big(v + d\rho_h(1)f(\operatorname{ad} y)u\big).
\end{aligned}$$

where ρ_g is *right* translation by g. In particular, $0 = d\varphi(p, 1)(u, v) = d\lambda_{\exp p}(1)(v + f(\operatorname{ad} p)u)$ implies $f(\operatorname{ad} p)u = -v \in \mathfrak{k}$. Now $f(\xi) = (1 - \cosh -\xi)/\xi - (\sinh -\xi)/\xi$. The first summand is a power series $f_1(\xi)$ of odd powers of ξ. Thus $f_1(\operatorname{ad} p)u \in \mathfrak{k}$ by (i). The second summand is a power series $f_2(\xi)$ of even powers of ξ, and $z \mapsto f_2(z)$ has no real zeros. Hence $f_2(\operatorname{ad} p)$ is invertible by (iii). The relation $f(\operatorname{ad} p)u \in \mathfrak{k}$ then implies $f_2(\operatorname{ad} p)u = 0$ and thus $u = 0$. Then also $v = 0$. Thus $(u, v) = (0, 0)$. This proves the claim that φ is regular at (p, k). So we find a neighborhood U of s and a smooth map $\psi \colon U \to \mathfrak{p} \times H$ such that $\varphi \circ \psi = \operatorname{id}_U$. We shall show that $\psi(s \exp t{\cdot}x) \in C \times H$ for all $0 \le t \le \varepsilon$ by applying Theorem I.5.17. So we consider the vector field $X(g) = d\lambda_g(1)(x)$, where λ_g is the left translation by g. The integral curve of X starting at s is $s \exp tx$. We have to show that $d\psi(g)X(g) \in L_{\psi(g)}(C \times H)$ for $g \in U$. If we set $g = (\exp y)h$, that is, $(y, h) = \psi(g)$, and define $u \in \mathfrak{p}$ and $v \in \mathfrak{k}$ by the equation

(*)
$$(u, v) = d\psi(g)(d\lambda_g(1)x),$$

then, in view of $W = C \oplus \mathfrak{k}$, we have to show $(u, v) \in L_{(y, h)}(C \times H) = L_y(C) \times \mathbf{L}(H)$, that is, $u \in L_y(C)$. Relation (*) transforms into $d\lambda_{\exp y}(h)d\lambda_h(1)x = d\lambda_g(1)x = \big(d\psi(g)\big)^{-1}(u, v) = d\varphi(y, h)(u, v)$. Then (33) implies

$$d\lambda_h(1)x = v + d\rho_h(1)f(\operatorname{ad} y)u.$$

In view of $d\rho_h(1)^{-1}d\lambda_h(1) = \operatorname{Ad}(h)$, this transforms into

$$\operatorname{Ad}(h)x = d\rho_h(1)^{-1}v + f(\operatorname{ad} y)u,$$

which implies

$$u = g(\operatorname{ad} y)(\operatorname{Ad}(h)x - d\rho_h(\mathbf{1})^{-1}v).$$

By (ii) we know $\operatorname{Ad}(h)x \in C$, and $v \in \mathfrak{k}$ implies $d\rho_h(\mathbf{1})^{-1}v \in \mathfrak{k}$. Thus $\operatorname{Ad}(h)x - d\rho_h(\mathbf{1})^{-1}v \in C + \mathfrak{k} = W$. Hence $u \in g(\operatorname{ad} y)W \cap \mathfrak{p}$. Finally Lemma V.4.56 applies and gives us $g(\operatorname{ad} y)W \subseteq L_y(W)$, that is $u \in L_y(W) \cap \mathfrak{p} = L_y(C)$ as we had to show. This completes the proof that S is a semigroup.

Now we assume (iv) and (v), too. We have seen that φ is regular in all points of $C \times H$; hence it is continuous and open at all these points. In order to show that φ maps $C \times H$ homeomorphically onto S it therefore suffices to show that φ is injective on $C \times H$. So let $(\exp x)k = (\exp x')k'$, with $x, x' \in C$ and $k, k' \in H$; we shall show $x = x$ and $k = k'$. Now $(\exp x')^{-1}\exp x = k'k^{-1} \in (\exp -C \exp C) \cap H = \{\mathbf{1}\}$ by (iv). Hence $k = k'$ and $\exp x = \exp x'$. By (iii), $\exp$ is regular at x. Then it follows from Lemma V.6.7 that $[x, x'] = 0$ and $\exp(x - x') = \mathbf{1}$. Now $e^{\operatorname{ad}(x - x')} = \operatorname{Ad}(\exp(x - x')) = \operatorname{Id}$. By (iii), since $\operatorname{ad} x$ and $\operatorname{ad} x'$ commute, $\operatorname{ad}(x - x')$ has real spectrum. Hence $\operatorname{ad}(x - x') = 0$ and $x - x' \in \mathfrak{z} \cap \mathfrak{p}$. Condition (v) then implies $x = x'$ as claimed.

It remains to show that $\mathbf{L}(S) = W$. If we identify the tangent space of $\mathfrak{p} \times H$ at $(0, \mathbf{1})$ with $\mathfrak{p} \times \mathfrak{k}$, then $d\varphi(0, \mathbf{1}): \mathfrak{p} \times \mathfrak{k} \to G$ is given by $d\varphi(0, \mathbf{1})(u, v) = u + v$. Hence $d\varphi(0, \mathbf{1})$ maps $C \times \mathfrak{k}$ isomorphically onto $C \oplus \mathfrak{k} = W$. Since φ is regular at $(0, \mathbf{1})$ and maps $C \times H$ homeomorphically onto S, the derivative $d\varphi(0, \mathbf{1})$ maps $L_{(0,1)}(C \times H)$ onto $L_1(S) = \mathbf{L}(S)$. Hence $\mathbf{L}(S) = W$. Since $(\exp C) \cup H \subseteq \langle \exp \mathbf{L}(S) \rangle$ we have $S = (\exp C)H \subseteq \langle \exp \mathbf{L}(S) \rangle$. The reverse inclusion follows from Proposition V.1.7 if S is closed.   ∎

The middle example in Figure 1 of Example V.4.1 illustrates how we can have that S is a semigroup not satisfying (v) even when $H = \{\mathbf{1}\}$.

We now show how the hypotheses (i)–(v) in Theorem V.4.57 may arise.

V.4.58.   Corollary.   *Suppose that G is a Lie group and σ an involutive automorphism of G with H as the group of fixed elements. Suppose that C is a pointed cone in $\mathbf{L}(G)$ such that the following conditions are satisfied:*

(a) *C is invariant under $\operatorname{Ad} H_0$.*

(b) *$\mathbf{L}(\sigma)(c) = -c$ for all $c \in C$.*

(c) *$\operatorname{ad} c$ has real spectrum for $c \in C$.*

(d) *If $z \in C - C$ is a central element of $\mathbf{L}(G)$ with $\exp z = \mathbf{1}$ then $z = 0$.*

Then $S = (\exp C)H_0$ is a strictly infinitesimally generated semigroup with sub-tangent wedge $\mathbf{L}(S) = C \oplus \mathbf{L}(H)$ at the origin, provided S is closed.

Proof.   We shall apply Theorem V.4.57 with H_0 in place of H. We let $\mathfrak{p}$ be the eigenspace of $\mathbf{L}(\sigma)$ for the eigenvalue -1. Since $\mathfrak{k} = \mathbf{L}(H)$ is the space of $\mathbf{L}(\sigma)$-fixed points, we have $\mathbf{L}(G) = \mathfrak{p} \oplus \mathfrak{k}$, and $[\mathfrak{p}, \mathfrak{p}] \subseteq \mathfrak{k}$, and $[\mathfrak{k}, \mathfrak{p}] \subseteq \mathfrak{p}$. Thus Theorem V.4.57(i) holds. Conditions (ii), (iii) and (v) of that theorem are conditions (a),(c) and (d). It remains to show that (iv) holds.

Suppose that $p^{-1}p' = k \in H$ with $p = \exp x$, $p' = \exp x'$, $x, x' \in \mathfrak{p}$. Because of (b) and the definition of H, this implies $\sigma(p') = \sigma(\exp x') = \exp(\mathbf{L}(\sigma)(x')) = \exp(-x') = p'^{-1}$, likewise $\sigma(p) = p^{-1}$. But also $\sigma(k) = k$.

Thus $p^{-1}p' = k = \sigma(k) = \sigma(p^{-1})\sigma(p') = pp'^{-1}$. Hence $p^2 = p'^2$, that is, $\exp 2x = \exp 2x'$. Condition (d), as we have seen in the proof of Theorem V.4.57, implies $2x = 2x'$ and thus $p = p'$ and therefore $k = 1$. ∎

V.4.59. **Example.** In the preceding corollary let $G = \text{Sl}(2, \mathbb{C})$ (as a real Lie group) with complex conjugation as σ. Then $H = H_0 = \text{Sl}(2, \mathbb{R})$, $\mathfrak{p} = i\,\text{sl}(2, \mathbb{R})$. We take $C = i \cdot \mathcal{K}$ with $\mathcal{K}$ the upper half of the standard double cone in $\text{sl}(2, \mathbb{R})$, and set $W = i \cdot \mathcal{K} \oplus \text{sl}(2, \mathbb{R})$. Then $S = (\exp i \cdot \mathcal{K})\,\text{Sl}(2, \mathbb{R})$ is an infinitesimally generated closed subsemigroup of G, homeomorphic to $\mathcal{K} \times \mathbb{R}^2 \times S^1$ and with $\mathbf{L}(S) = W$. The Lie wedge W is not a Lie semialgebra and S is not locally divisible. The universal covering semigroup $\widetilde{S}$ of S is a semigroup locally isomorphic to S and homeomorphic to $\mathcal{K} \times \mathbb{R}^3$ which is not isomorphic to any subsemigroup of a Lie group. ∎

EV.4.1. **Exercise.** Prove (i) that S is closed, (ii) that W is not a Lie semialgebra, and (iii) that $\widetilde{S}$ is not embeddable into a Lie group. (Hint: (i) Define a sesquilinear form on $\text{sl}(2, \mathbb{C})$ by $\langle x, y \rangle = B(x, \overline{y})$ with the Cartan–Killing form B. Show that $S = \{g \in \text{Sl}(2, \mathbb{C}) : \text{Spec}\, gg^* \text{ is positive }\}$, where g^* is the adjoint of g with respect to $\langle \bullet, \bullet \rangle$ (polar decomposition with respect to $\langle \bullet, \bullet \rangle$!). (ii) Use the Second Triviality Theorem II.7.6. (iii) Suppose $\widetilde{S} \subseteq G^*$ with a Lie group G^*; without loss of generality $G^* = \langle \widetilde{S} \rangle$. Show that $D = \ker(\widetilde{S} \to S)$ is a central discrete subgroup of G^* and that G^*/D is a Lie group containing and generated by $\widetilde{S}/D \cong S$. Conclude that G^*/D is a homomorphic image of G by Proposition VII.3.28 below. Hence $\mathbf{L}(G^*) \cong \text{sl}(2, \mathbb{C})$ and G^* contains a copy of the simply connected covering group of $\text{Sl}(2, \mathbb{R})$. Impossible.) ∎

Next we show that we can apply Corollary V.4.58 to the tangent wedge $\mathbf{L}(S_q)$ for any non-degenerate quadratic form $q: V \to \mathbb{R}$.

V.4.60. **Remark.** Let V be a finite dimensional real vector space and $B: V \times V \to \mathbb{R}$ a non-degenerate symmetric bilinear form with associated quadratic form q. Moreover let $\top: \text{gl}(V) \to \text{gl}(V)$ be the adjoint operation with respect to B. Then

(i) $\sigma: \text{Gl}(V) \to \text{Gl}(V)$, defined by $\sigma(g) = (g^{-1})^\top$, is an automorphism.

(ii) The derivative $d\sigma: \text{gl}(V) \to \text{gl}(V)$ of σ at $\mathbf{1}$ is given by $d\sigma(x) = -x^\top$.

(iii) The fixed point set of σ is $O(q) = \{g \in \text{Gl}(V) : g^{-1} = g^\top\}$. Its Lie algebra is $\mathfrak{k}_B = \{x \in \text{gl}(V) : x^\top = -x\}$.

(iv) Let $W = \{x \in \text{gl}(V) : (\forall v \in V)\, B(x \cdot v, v) \leq 0\}$, then the set $C = W \cap \mathfrak{p}_B$ with $\mathfrak{p}_B = \{x \in \text{gl}(V) : x^\top = x\} = \{x \in \text{gl}(V) : d\sigma(x) = -x\}$ is a pointed cone in $\mathfrak{p}_B$ which is invariant under inner automorphisms coming from $O(q)$, i.e. under $\text{Ad}(O(q))$.

Proof. The proof is left to the reader. ∎

In order to be able to apply Corollary V.4.58 to the situation of Remark V.4.60 it remains to show that the spectrum of $\text{ad}\, x$ is real for any $x \in C$. The tool for dealing with this issue is complexification.

V.4.61.   Remark. Let V be a finite-dimensional real vector space and B a non-degenerate symmetric bilinear form. If $V_{\mathbb{C}} = V + iV$ is the complexification of V we define $B_{\mathbb{C}} \colon V_{\mathbb{C}} \times V_{\mathbb{C}} \to \mathbb{C}$ by $B_{\mathbb{C}}(v + iw, v' + iw') = B(v, v') + B(w, w') + iB(w, v') - iB(v, w')$. Then $B_{\mathbb{C}}$ is a non-degenerate sesquilinear form. Let $* \colon \mathrm{gl}(V_{\mathbb{C}}) \to \mathrm{gl}(V_{\mathbb{C}})$ be the adjoint operation with respect to $B_{\mathbb{C}}$. We set $\mathfrak{k}_{B_{\mathbb{C}}} = \{x \in \mathrm{gl}(V_{\mathbb{C}}) \colon x^* = -x\}$ and $\mathfrak{p}_{B_{\mathbb{C}}} = \{x \in \mathrm{gl}(V_{\mathbb{C}}) \colon x^* = x\}$. For $S = \{g \in \mathrm{Gl}(V_{\mathbb{C}}) \colon (\forall v_c \in V_{\mathbb{C}})\, B_{\mathbb{C}}(g \cdot v_c, g \cdot v_c) \leq B_{\mathbb{C}}(v_c, v_c)\}$, the tangent wedge $\mathbf{L}(S)$ is $\mathfrak{k}_{B_{\mathbb{C}}} + C_{\mathbb{C}}$, where $C_{\mathbb{C}} = \mathbf{L}(S) \cap \mathfrak{p}_{B_{\mathbb{C}}}$. Moreover we have

(i) $i\mathfrak{p}_{B_{\mathbb{C}}} = \mathfrak{k}_{B_{\mathbb{C}}}$, and $\mathfrak{k}_{B_{\mathbb{C}}}$ is a real form of $\mathrm{gl}(V_{\mathbb{C}})$.

(ii) $iC_{\mathbb{C}}$ is a generating invariant cone in $\mathfrak{k}_{B_{\mathbb{C}}}$.

Proof.   Let $\{e_1, ..., e_n\}$ be a B-orthogonal basis for V such that $q(e_j) = 1$ for $j = 1, ..., p$ and $q(e_j) = -1$ for $j = p + 1, ..., n$. Then we can express the adjoint operation $*$ in terms of the usual complex conjugate transpose $*$. In fact, if $x \in \mathrm{gl}(V_{\mathbb{C}})$ is given as a block matrix

$$x = \begin{pmatrix} A & B \\ C & D \end{pmatrix},$$

then x^* is given as the block matrix

$$x^* = \begin{pmatrix} A^* & -C^* \\ -B^* & D^* \end{pmatrix}.$$

Using this matrix representation of the $B_{\mathbb{C}}$-adjoint we can directly verify all assertions except (ii). In order to prove (ii) we note first that $\mathfrak{p}_{B_{\mathbb{C}}}$ is a $\mathfrak{k}_{B_{\mathbb{C}}}$-module, so that $C_{\mathbb{C}}$ is invariant under all $e^{\mathrm{ad}\, x}$ with $x \in \mathfrak{k}_{B_{\mathbb{C}}}$ since $\mathbf{L}(S)$ is a Lie wedge. Thus the invariance of $C_{\mathbb{C}}$ in $\mathfrak{k}_{B_{\mathbb{C}}}$ follows as $\mathrm{ad}\, x$ is a *complex* linear map on $\mathrm{gl}(V_{\mathbb{C}})$. Now it only remains to show that $iC_{\mathbb{C}}$ is generating in $\mathfrak{k}_{B_{\mathbb{C}}}$. To this end we note that the span of $iC_{\mathbb{C}}$ in $\mathfrak{k}_{B_{\mathbb{C}}}$ has to be an ideal. If $p = n$ there is nothing to prove; otherwise, $x \in \mathrm{gl}(V_{\mathbb{C}})$ given by $x \cdot e_j = 0$ for $j = 1, ..., p - 1, p + 2, ..., n$, $x \cdot e_p = iae_p$ and $x \cdot e_{p+1} = -ide_{p+1}$ with $a, d \geq 0$ is contained in $iC_{\mathbb{C}}$. Thus $iC_{\mathbb{C}}$ is neither contained in $i\mathbb{R} \cdot \mathbf{1}$ nor in $[\mathfrak{k}_{B_{\mathbb{C}}}, \mathfrak{k}_{B_{\mathbb{C}}}]$, hence in no non-trivial ideal of the real form $\mathfrak{k}_{B_{\mathbb{C}}}$ of $\mathrm{gl}(V_{\mathbb{C}})$.   ∎

V.4.62.   Example. Let V be a finite-dimensional real vector space and $B \colon V \times V \to \mathbb{R}$ a non-degenerate symmetric bilinear form with associated quadratic form q. Further, let $C = \{x \in \mathrm{gl}(V) \colon (\forall v \in V)\, x = x^\top, B(x \cdot v, v) \leq 0\}$. Then the map $\varphi \colon C \times O(q)_0 \to \mathrm{Gl}(V)$ defined by $\varphi(x, k) = (\exp x)k$ is a homeomorphism onto a closed subset of $\mathrm{Gl}(V)$. Moreover the set $S = (\exp C)O(q)_0$ is a semigroup.

Proof.   Since $C \subseteq C_{\mathbb{C}}$ we conclude from Remark V.4.61 and Theorem III.2.12 that the spectrum of $(\mathrm{ad}\, ix)|_{\mathfrak{k}_{\mathbb{C}}}$ is purely imaginary for all $x \in C$. But then $\mathfrak{p}_{B_{\mathbb{C}}} = i\mathfrak{k}_{B_{\mathbb{C}}}$ and $\mathrm{gl}(V_{\mathbb{C}}) = \mathfrak{p}_{B_{\mathbb{C}}} + \mathfrak{k}_{B_{\mathbb{C}}}$ imply that the spectrum of $\mathrm{ad}\, ix$ is purely imaginary. This just means that $\mathrm{ad}\, x$ has real spectrum for all $x \in C$, which is what we had to show. As in Exercise EV.4.1, show that S is closed.   ∎

This example has to be seen in the light of the Split Wedge Theorem IV. 4.26. The split wedge in Example V. 4.62 is $W = C \oplus \mathrm{so}(q)$. In Chapter IV we constructed a (local) semigroup by forming the product of a (local) semigroup generated by $\exp C$ and a (local) group generated by the edge of W —here $O(q)_0$. However, in Example V.4.61, it suffices to take the *set* $\exp C$ *itself* instead of the semigroup generated by it.

5. Maximal Semigroups

Suppose that G is a connected Lie group with Lie algebra $\mathbf{L}(G)$ and let $\Omega \subseteq \mathbf{L}(G)$. We could say that the set Ω gives rise to a controllable system on G if the semigroup generated by $\exp(\mathbb{R}^+\Omega)$ is all of G. We would like to be able to test a subset Ω to determine whether or not it gives rise to a controllable system. The approach adopted in this section is to try to classify all the maximal subsemigroups of G and their tangent objects in $\mathbf{L}(G)$ in a reasonably concrete fashion. Then Ω will give rise to a controllable system if and only if Ω is not contained in the tangent set of any maximal semigroup.

This machinery is also sometimes helpful in the problem of determining whether a Lie wedge W is global, that is, whether it is the tangent wedge of a subsemigroup of G. If W is not contained in the tangent set of a maximal semigroup, then the exponential image of W must generate all of G and is hence not global. On the other hand, if it is known that W is contained in a wedge that is global, this information can be useful in determining that W itself is global.

Traditionally in the study of topological groups it is the open maximal semigroups that have been considered ([Wr57] and [Hi87b]). Our development proceeds along the lines of [Law 86] by considering semigroups that are maximal as subsemigroups. This allows the application of the algebraic machinery of maximal semigroups.

Algebraic preliminaries

In this subsection we develop some of the basic algebraic machinery of maximal semigroups. The results are rather straightforward, but will be useful to have on record in the later developments.

In this subsection G denotes a group (with no topological structure).

V.5.1.   **Definition.**   A subsemigroup M of G is a *maximal subsemigroup* of G if

   (i) the only subsemigroups containing M are M and G; and

   (ii) M is not a group.

V.5.2.   Remark. Condition (ii) of Definition V.5.1. is a technical convenience which insures the existence of a non-empty $M^\sharp = M \setminus H(M)$, which is the maximal ideal in M. Note also that $1 \in M$ if M is maximal (otherwise consider $\{1\} \cup M$).

V.5.3.   Lemma. *Let M be a maximal subsemigroup of G, and T a submonoid with $T^{-1} \not\subseteq M$. If $MT^{-1} \subseteq T^{-1}M$, then $T^{-1}M = G$.*

Proof.   We have $T^{-1}MT^{-1}M \subseteq T^{-1}T^{-1}MM \subseteq T^{-1}M$, so $T^{-1}M$ is a subsemigroup containing T^{-1} and M (since $1 \in M$ and $1 \in T^{-1}$). Then $G = T^{-1}M$ by maximality of M.   ∎

   Although elementary in nature, the next proposition is constantly applied in the theory of maximal semigroups. It is often applied by showing that S cannot meet M "deeply" (condition (i)), and hence M "swallows" S^{-1}.

V.5.4.   Proposition.   (The Swallowing Lemma) *Let M be a maximal subsemigroup of G and S a subsemigroup satisfying $MS^{-1} \subseteq S^{-1}M$ (which is the case if S or M is normal). Then either*

   (i) *$S \cap I \neq \emptyset$ for every left ideal I of M, or*

   (ii) *$S^{-1} \subseteq M$.*

Proof.   Suppose $S^{-1} \not\subseteq M$. Then we show (i) holds. Let $T = S \cup \{1\}$. Then $MT^{-1} \subseteq T^{-1}M$. Then $T^{-1}M = G$ by Lemma V.5.3. Let I be a left ideal of M. Pick $x \in M^\sharp$ and $y \in I$. Then $z = xy \in M^\sharp \cap I$. Since $T^{-1}M = G$, there exist $s \in T$, $m \in M$ such that $z^{-1} = s^{-1}m$. Hence $s = mz \in MI \subseteq I$. Also $s \neq 1$ since $z \in M^\sharp$ implies $z^{-1} \notin M$. Thus $s \in S$, and $S \cap I \neq \emptyset$.   ∎

V.5.5.   Lemma. *Let S be a subsemigroup of the integers $(\mathbb{Z}, +)$ containing both a positive and a negative number. Then S is a subgroup of $\mathbb{Z}$.*

Proof.   Suppose the maximal ideal $S^\sharp = S \setminus H(S)$ is non-empty. Let m be the number in $S^\sharp$ closest to 0, and let $n \in S$ be the number closest to 0 of the opposite sign. Then $m + n \in S^\sharp$ and is closer to 0 than one of m or n, a contradiction.   ∎

V.5.6.   Proposition. *Let M be a maximal subsemigroup of G and let $x \in G$ satisfy $xM \subseteq Mx$. Then $x \in M \cup M^{-1}$.*

Proof.   Let $T = \{x^n : n \geq 1\}$. Since $Q = \{y : yM \subseteq My\}$ is a subsemigroup, $T \subseteq Q$. Hence $TM \subseteq MT$. Similarly $Mx^{-1} \subseteq x^{-1}M$ implies $MT^{-1} \subseteq T^{-1}M$.

   If $T \subseteq M$ or $T^{-1} \subseteq M$, then the proof is complete. If neither of these were to hold, then by the Swallowing Lemma V.5.4 and its dual, $T \cap M^\sharp \neq \emptyset$, and $T^{-1} \cap M^\sharp \neq \emptyset$. Let $S = \{m \in \mathbb{Z} : x^m \in M^\sharp\}$. It is immediate that S is a semigroup and S contains both positive and negative numbers. Then $0 \in S$ by Lemma V.5.5. Hence $1 = x^0 \in M^\sharp$, a contradiction. Thus this final case cannot occur.   ∎

V.5.7.   Corollary. *Let M be a maximal subsemigroup of G. If M is invariant, then M is total (that is, $G = M \cup M^{-1}$).*   ∎

   Recall from Proposition V.0.9 that a semigroup is total if and only if the left preorder it induces on G is a total preorder.

V.5.8. **Corollary.** *Let M be a maximal subsemigroup of G. Let $Z(G)$ denote the center of G. Then $M \cap Z(G)$ is total in $Z(G)$.*

Proof. $M \cap Z(G)$ is total in $Z(G)$ if and only if $Z(G) \subseteq M \cup M^{-1}$. The latter follows from Proposition V.5.6. ∎

V.5.9. **Corollary.** *A maximal subsemigroup of an abelian group is total.* ∎

The preceding results indicate how maximal semigroups interact with central subgroups. We consider now their interaction with normal abelian subgroups.

V.5.10. **Lemma.** (The Purity Lemma) *Let M be a maximal subsemigroup of G, let H be a normal abelian subgroup, and let $y \in H$. If $y^n \in M$ for some $n > 1$, then $y \in M$.*

Proof. Suppose $y \notin M$. Then the subsemigroup generated by $M \cup \{y\}$ is all of G (by maximality of M). Note that since $1 \in M$, any member of G is either in M or has a representation $m_1 y m_2 y \cdots m_{k-1} y m_k$ (since such products together with M form a subsemigroup containing M and y). There exists $g \in M^\sharp \cap H$ by the Swallowing Lemma V.5.4. Then for some $m_1, \ldots, m_k \in M$,

$$g^{-1} = m_1 y \cdots y m_k = (m_1 \cdots m_k) y^{m_2 \cdots m_k} \cdots y^{m_k} = mb,$$

where $z^w = w^{-1} z w$, $m = m_1 \cdots m_k$, and b is the product of the remaining factors. Since $y \in H$ and H is normal, $b \in H$. Thus $m = g^{-1} b^{-1} \in H$. So m commutes with b. Hence

$$g^{-n} = m^n b^n = m^{n-1} m (y^n)^{m_2 \cdots m_k} \cdots (y^n)^{m_k} = m^{n-1} m_1 y^n m_2 y^n \cdots y^n m_k \in M$$

since $y^n \in M$. But $g \in M^\sharp$ implies $g^n \in M^\sharp$. Therefore $1 = g^n g^{-n} \in M^\sharp M \subseteq M^\sharp$, a contradiction. ∎

We close this section with an elementary, but useful, lemma.

V.5.11. **Lemma.** (The Reduction Lemma) *Let $\varphi : G \to H$ be a homomorphism onto H, and let S be a submonoid of H. Then S is maximal (respectively total, respectively invariant) if and only if $\varphi^{-1}(S)$ is maximal (respectively total, respectively invariant) in G.*

Proof. Suppose that S is maximal in H. Then $\varphi^{-1}(S)$ is not a group since S is not a group. If T is a subsemigroup containing $\varphi^{-1}(S)$, then T contains the kernel of φ, so $T = \varphi^{-1}(\varphi(T))$. Since $S \subseteq \varphi(T)$, either $\varphi(T) = S$ or $\varphi(T) = H$. Thus $T = \varphi^{-1}\varphi(T) = \varphi^{-1}(S)$ or $T = \varphi^{-1}\varphi(T) = \varphi^{-1}(H) = G$. Hence $\varphi^{-1}(S)$ is maximal.

The remaining arguments are all similarly straightforward. ∎

Recall from Proposition V.0.2 that $\mathrm{Core}(S)$ is the largest normal subgroup contained in S, that the semigroup S is said to be reduced in G if $\mathrm{Core}(S) = \{1\}$, and that given any group G and any subsemigroup S we can form the reduction (G_R, S_R) of (G, S) by dividing out $\mathrm{Core}(S)$.

Note that since $\mathrm{Core}(S) \subseteq S$, we have $S = \varphi^{-1}(\varphi(S)) = \varphi^{-1}(S_R)$. This observation allows us to apply the Reduction Lemma V.5.11 to obtain

V.5.12.   Corollary.   *Let S be a submonoid of G. Then S is maximal in G if and only if S_R is maximal in $G_R = G/\mathrm{Core}(S)$.*   ∎

Corollary V.5.12 shows that all maximal subsemigroups arise as inverse images of reduced ones. It is then the latter that we seek to characterize.

Topological generalities

Throughout this subsection G denotes a connected topological group. We recall the elementary fact that every connected topological group is locally generated in the sense that given any neighborhood of the identity, every member of G can be written as a finite product of elements coming out of this neighborhood.

V.5.13.   Lemma.   *Let $U \subseteq S \subset G$ where U is open and S is a subsemigroup. Then $S \cap U^{-1} = \varnothing$.*

Proof.   Suppose $s \in S \cap U^{-1}$. Then $s^{-1} \in U \subseteq S$, and thus $\mathbf{1} = ss^{-1} \in sU \subseteq sS \subseteq S$. Thus S contains the open neighborhood sU of $\mathbf{1}$, and hence $S = G$ by local generation, a contradiction.   ∎

V.5.14.   Proposition.   *Let S be a proper (open) subsemigroup of G with $\mathrm{int}(S) \neq \varnothing$. Then S is contained in a maximal (open) subsemigroup.*

Proof.   Let $\mathcal{M}$ be a maximal tower of proper (open) subsemigroups of G containing S and let M be their union. If $U = \mathrm{int}(S)$, then by Lemma V.5.13, $T \cap U^{-1} = \varnothing$ for all $T \in \mathcal{M}$. Hence $M \cap U^{-1} = \varnothing$, so M is proper and also not a group. Clearly M is a maximal (open) subsemigroup.   ∎

V.5.15.   Corollary.   *Suppose $B \subseteq G$ has the property that the semigroup it generates has non-empty interior in G. If B is not contained in a maximal subsemigroup with non-empty interior, then the semigroup generated by B is all of G.*   ∎

Corollary V.5.15 is essentially a restatement of Proposition V.5.14. It motivates our approach in what follows. If we can classify the maximal subsemigroups with non-empty interior of G, then we need only check whether B is a subset of one of these maximal subsemigroups to determine whether it generates G. It is possible to classify the maximal subsemigroups in several cases, while the problem of determining the semigroup generated by a given set can be quite difficult.

V.5.16.   Proposition.   *If M is a maximal semigroup with non-empty interior, then M is closed.*

Proof.   Let $U = \mathrm{int}\, M$. By Lemma V.5.13, $M \cap U^{-1} = \varnothing$, so $\overline{M} \cap U^{-1} = \varnothing$. Thus $\overline{M}$ is a proper subsemigroup containing M, and so is equal to M by maximality of M.   ∎

V.5.17. **Proposition.** *Let M be a maximal subsemigroup of G with $\mathrm{int}(M) \neq \emptyset$. If H is a compact subgroup, then $H \cap \mathrm{int}(M) = \emptyset$. If H is compact and normal, then $H \subseteq \mathrm{Core}(M)$.*

Proof. Suppose that $H \cap \mathrm{int}(M) \neq \emptyset$. Since $H \cap \mathrm{int}(M)$ is an open subsemigroup of the compact group H, it is a compact open subgroup by Proposition V.0.18. Hence $1 \in \mathrm{int}(M)$, so $M = G$, a contradiction. Then by the Swallowing Lemma V.5.4 $H \subseteq M$ if H is normal. It follows that $H \subseteq \mathrm{Core}(M)$ in this case, since $\mathrm{Core}(M)$ is the largest normal subgroup of M. ∎

V.5.18. **Proposition.** *If M is a maximal subsemigroup of G and if $H(M) \cap \overline{\mathrm{int}(M)} \neq \emptyset$, (in particular, if $1 \in \overline{\mathrm{int}(M)}$), then $\mathrm{int}(M)$ is a maximal open subsemigroup and $M = \overline{\mathrm{int}(M)}$.*

Proof. Let $I = \mathrm{int}\, M$. By Proposition V.0.15, I is an ideal of M, so I is an open subsemigroup. Then $\overline{I} \subseteq \overline{M} = M$ by Proposition V.5.16. Conversely if $g \in H(M) \cap \overline{\mathrm{int}(M)}$, then $M = Mg \subseteq M\overline{I} \subseteq \overline{MI} \subseteq \overline{I}$.

If $T \neq G$ is an open subsemigroup containing I, then $M = \overline{I} \subseteq \overline{T}$ and $\overline{T} \neq G$ (Lemma V.5.13) imply $M = \overline{T}$. Thus $T = \mathrm{int}(T) \subseteq \mathrm{int}(M)$. Thus $\mathrm{int}(M)$ is a maximal open subsemigroup. ∎

Total semigroups

Corollaries V.5.7 and V.5.9 gave sufficient conditions for a maximal subsemigroup to be total. In this subsection we consider this situation in more detail. *Throughout G denotes a connected topological group.*

V.5.19. **Proposition.** *Let M be a closed total subsemigroup of G. Then M is maximal.*

Proof. Suppose that $x \notin M$. Let $U = G \setminus M$. Then U is open and $U \subseteq M^{-1}$ since $G = M \cup M^{-1}$. Thus $U^{-1} \subseteq M$. The subsemigroup T generated by $\{x\} \cup M$ contains xU^{-1}, an open set containing 1. Thus $T = G$ by connectedness of G. ∎

The next proposition gives basic properties of closed total subsemigroups. Recall $M^{\sharp} = M \setminus H(M)$.

V.5.20. **Proposition.** *Let M be a closed subsemigroup of G. Then M is total if and only if $G = M^{\sharp} \cup H(M) \cup (M^{\sharp})^{-1}$. In this case $M^{\sharp} = \mathrm{int}(M)$, $M = \overline{M^{\sharp}}$, and $H(M) = \partial(M^{\sharp})$ $(= \overline{M^{\sharp}} \setminus \mathrm{int}(M))$.*

Proof. The equivalence follows from the equality

$$M \cup M^{-1} = M^{\sharp} \cup H(M) \cup H(M)^{-1} \cup (M^{\sharp})^{-1} = M^{\sharp} \cup H(M) \cup (M^{\sharp})^{-1}.$$

If M is total, then since $M = M^{\sharp} \cup H(M)$ and $M \cap (M^{\sharp})^{-1} = \emptyset$, we have $(M^{\sharp})^{-1} = G \setminus M$ is open, and hence $M^{\sharp}$ is open. Thus $M^{\sharp} \subseteq \mathrm{int}(M)$. Since $\mathrm{int}(M)$ is an ideal (Proposition V.0.15(i)) and $M^{\sharp}$ is the largest ideal (Proposition V.0.1), the other inclusion always obtains.

If $M^\sharp = \overline{M^\sharp}$, then $M^\sharp$ would be open and closed, contradicting the connectedness of G. So there exists $g \in H(M) \cap \overline{M^\sharp}$. Then by Proposition V.5.18, $M = \overline{M^\sharp}$. Thus $\partial(M^\sharp) = M \setminus M^\sharp = H(M)$.   ∎

Closed total subsemigroups are maximal, but the converse does not hold in general. We are interested in determining sufficient conditions to insure the converse. We begin with two important examples where we can calculate all closed maximal subsemigroups. We have seen in Chapter II in the theory of Lie semialgebras that the lower-dimensional examples played an important role in the general theory and that is also the case in the theory of maximal semigroups.

V.5.21.   Example.   Let M be a maximal closed subsemigroup of $(\mathbb{R}, +)$. Then either $M = \mathbb{R}^+$ or $M = -\mathbb{R}^+$.

Proof.   Let $y \in M$, $y > 0$. By the Purity Lemma V.5.10, $\frac{1}{n}y \in M$ for all positive integers n. Thus $\frac{m}{n}y \in M$ for all positive rationals. Since M is closed, $\mathbb{R}^+ \subseteq M$. Since $\mathbb{R}^+$ is closed and total, by Proposition V.5.19 it is maximal, and hence must be equal to M. Similarly if M contains a negative number, $M = -\mathbb{R}^+$.   ∎

We show how the result of Example V.5.21 can be used for generalizations. First we record a fact we need at this point.

V.5.22.   Lemma.   *Let G be a connected locally compact group and H a closed normal subgroup. If G/H is simply connected, then H is connected.*

Proof.   Let H_0 be the identity component of H, and suppose that $H_0 \neq H$. It is a closed normal subgroup, and the image of H in G/H_0 is a closed normal and totally disconnected subgroup of a connected group, hence is central. Since G/H_0 is locally compact and connected, there exists a compact subgroup N strictly contained and open in H/H_0. Let N_1 be its inverse image in G. Then H/N_1 is a discrete subgroup of G/N_1 and thus G/N_1 a covering group of G/H, which contradicts the hypothesis.   ∎

V.5.23.   Theorem.   *Let G be a connected topological group which is either locally compact or locally connected, and let M be a closed subsemigroup. The following are equivalent:*

(1) *M is maximal and invariant;*

(2) *M is total and invariant;*

(3) *M is total and $H(M) = \operatorname{Core}(M)$, i.e. $H(M)$ is normal;*

(4) *there exists a continuous, open homomorphism of G onto $(\mathbb{R}, +)$ such that the image of M is $\mathbb{R}^+$;*

(5) *M is maximal and G_R is topologically isomorphic to the additive reals. (In this case the topological isomorphism must carry M_R to $\mathbb{R}^+$ or $-\mathbb{R}^+$.)*

Furthermore, in the locally compact case, these conditions imply that $H(M)$ is connected.

Proof.   The equivalence of (1) and (2) follows from Corollary V.5.7 and Proposition V.5.19. If M is invariant, then $H(M) = M \cap M^{-1}$ is normal, and thus $H(M) = \operatorname{Core}(M)$. Thus (2) implies (3).

We show (3) implies (5). In this case $G_R = G/H(M)$ is a totally ordered (see the Reduction Lemma V.5.11) connected topological group which is locally compact or locally connected (since the quotient mapping is open). The set M_R is a closed set of positive elements. Note that $G_R \setminus \{1_R\}$ is the disjoint union of the relatively closed sets $M_R \setminus \{1_R\}$ and $(M_R)^{-1} \setminus \{1_R\}$, and is hence not connected. By standard characterizations of the the reals as a topological group, G_R is abelian and topologically isomorphic to $(\mathbb{R}, +)$ (see [Bou55], Chapter V, §3, Exercise 4). The image of M_R in $\mathbb{R}$ is a closed maximal subsemigroup, hence equal to $\mathbb{R}^+$ or $-\mathbb{R}^+$ (Example V.5.21). This completes the proof of (5) and the parenthetical remark.

The implication (5) implies (4) quickly follows (by composing with inversion if necessary). That (4) implies (1) follows from the Reduction Lemma (V.5.11). Also (4) together with Lemma V.5.22 implies that $H(M)$ is connected in the locally compact case. ∎

In the abelian case all subsemigroups are invariant, so Proposition V.5.23 yields

V.5.24. **Corollary.** *The maximal closed subsemigroups of topological vector spaces are half-spaces.* ∎

We next determine the closed maximal subsemigroups of the unique 2-dimensional non-abelian connected Lie group. The corresponding Lie algebra is the almost abelian Lie algebra of dimension two (see Definition II.2.29). This group may be thought of as the identity component in the group of affine motions on $\mathbb{R}$. Hence we denote it $\mathrm{Aff}(\mathbb{R})$. We show that the closed maximal subsemigroups are the half-spaces with boundary some 1-dimensional subgroup. With one exception these are all isomorphic to $\mathrm{Aff}(\mathbb{R})^+$, the semigroup of affine motions with translation term non-negative. The exception is when the boundary is the unique one dimensional normal subgroup of $\mathrm{Aff}(\mathbb{R})$. This is the one case that the semigroup is not reduced in $\mathrm{Aff}(\mathbb{R})$.

Alternately $\mathrm{Aff}(\mathbb{R})$ may be identified with the multiplicative matrix group $\{ \begin{bmatrix} x & y \\ 0 & 1 \end{bmatrix} : 0 < x \}$, or with the set of ordered pairs $\{(x, y) : 0 < x\}$ with multiplication $(a, b)(x, y) = (ax, ay + b)$. The semigroup $\mathrm{Aff}(\mathbb{R})^+$ then consists of the upper right hand quadrant of the plane determined by $y \geq 0$. (Cf. Figure 2 of Section 2 in Chapter V.)

V.5.25. **Lemma.** *Let G be the group of positive reals under multiplication. Given $s, t \in G$ with $0 < s < 1 < t$, a positive integer N, and $\varepsilon > 0$, there exist positive integers j, k with $j \geq N$ such that $|s^j t^k - 1| < \varepsilon$.*

Proof. Consider first the additive group $\mathbb{R}$ and positive real numbers x and y. The set $\{nx : n \geq 1\}$ is a cyclic semigroup in the compact group $\mathbb{R}/\mathbb{Z} \cdot y$. Hence its closure is a compact semigroup and thus is a compact group by Proposition V.0.17. So the elements nx cluster to the identity of $\mathbb{R}/\mathbb{Z} \cdot y$. Hence there exists $j \geq N$ such that $|jx - ky| < \varepsilon$ for some $k > 0$, that is, $|j(-x) + ky| < \varepsilon$. The lemma now follows from this derivation by applying the exponential function from the additive reals to the multiplicative positive reals. ∎

V.5.26. **Example.** Let $G = \mathrm{Aff}(\mathbb{R})$ be the (unique) Lie group with Lie algebra the 2-dimensional non-abelian Lie algebra. If a closed semigroup M is maximal, then there exists a 1-dimensional group such that M is the union of this group and one of the two components of its complement. In particular, M is total.

Proof.   We identify G as $\{(x, y): x > 0, \ y \in \mathbb{R}\}$ with multiplication $(u, v)(x, y) = (ux, uy + v)$. Then G has identity $\mathbf{1} = (1, 0)$ and the one dimensional groups are the straight lines through $\mathbf{1}$, i.e., the set of all (x, y) such that $y = mx - m$ for some fixed slope m. The vertical line $H = \{(1, y): y \in \mathbb{R}\}$ is the only non-trivial normal subgroup. The sets $H^+ = \{(1, y): y \geq 0\}$ and $H^- = -H^+$ are invariant subsemigroups.

By the Swallowing Lemma V.5.4 either $H \subseteq M$ or $M^\sharp \cap H \neq \emptyset$. Suppose that $H \subseteq M$. Then G/H is isomorphic to the group of positive multiplicative reals, which in turn is isomorphic to $(\mathbb{R}, +)$, and M/H is a closed maximal subsemigroup by the Reduction Lemma V.5.11. By Example V.5.21, M/H corresponds either to $(0, 1]$ or $[1, \infty)$. Then $M = \{(x, y): 0 < x \leq 1\}$ or $M = \{(x, y): 1 \leq x\}$. In either case M is a "half-space" of the desired type with boundary H.

The more complicated case is $M^\sharp \cap H \neq \emptyset$. Let us assume $(1, y) \in M^\sharp$ for some $y > 0$. By the Purity Lemma V.5.10, $(1, \frac{1}{n}y) \in M$ for all positive n, $(1, \frac{m}{n}y) \in M$ by the semigroup property for $\frac{m}{n} > 0$, and so $H^+ \subseteq M$ since M is closed. Since H^+ is total in H and hence maximal, and since $H \not\subseteq M$ in this case, then $M \cap H = H^+$.

Since H^- is normal, $H^- M = G$. Since multiplying any element of G on the left by $(1, y)$ shifts the element y units vertically, it must be the case that $(\{x\} \times \mathbb{R}) \cap M \neq \emptyset$ for all $x > 0$.

Let $(s, ms - m)$, $(t, \mu t - \mu) \in M$ with $0 < s < 1 < t$. Using the fact that the straight lines through $\mathbf{1} = (1, 0)$ are subgroups or by direct computation, one obtains that the powers of these elements are given by $(s^j, ms^j - m)$ and $(t^k, \mu t^k - \mu)$. The product is again in M and is given by

$$(s^j t^k, \mu s^j t^k - \mu s^j + ms^j - m) = \left(s^j t^k, \mu(s^j t^k - 1) + (m - \mu)(s^j - 1)\right).$$

If j, k are chosen as in Lemma V.5.25, we see that this product can be made arbitrarily close to $(1, \mu - m)$, which must thus be a member of M. Hence $\mu - m \geq 0$, i.e., $m \leq \mu$.

From the preceding paragraph it follows that

$$a = \sup\{m: (s, ms - m) \in M \quad \text{for some} \quad s < 1\}$$
$$\leq \inf\{\mu: (t, \mu t - \mu) \in M \quad \text{for some} \quad t > 1\} = b.$$

If d is chosen so that $a \leq d \leq b$, then if follows that the region above the straight line of slope d through $(1, 0)$ contains M.   ∎

Nilpotent groups

In this subsection we let denote by $[g, h]$ the commutator $g^{-1}h^{-1}gh$ in a group G. If H is a subgroup, let $[G, H]$ denote the subgroup generated by the set $\{[g, h]: g \in G, h \in H\}$. Define recursively G_n by $G_0 = G$ and $G_{n+1} = [G, G_n]$. The group G is nilpotent if $G_{n+1} = \{\mathbf{1}\}$ for some n. Note that if $G_{n+1} = \{\mathbf{1}\}$, then G_n is contained in the center of G.

V.5.27. **Lemma.** *Suppose that G is a group, M is a maximal subsemigroup of G, and $g, h \in G$ are such that $[g, h]$ is in the center of G. Then $[g, h], [g, h]^{-1} \in M$.*

Proof. Let $w = [g, h]$. If $w = 1$, then the lemma is trivial since $1 \in M$. By Corollary V.5.8, $w \in M$ or $w^{-1} \in M$. We suppose without loss of generality that $w \in M$ and show also $w^{-1} \in M$.

Suppose on the contrary that $w^{-1} \notin M$. Let $S = \{w^{-n} : n \geq 0\}$. Then SM is a semigroup containing w^{-1} and M; by maximality of M, $SM = G$. Thus $g = w^{-r}u$, $h = w^{-m}v$, $g^{-1} = w^{-k}x$, $h^{-1} = w^{-p}y$ for some $r, m, k, p \geq 0$ and $u, v, x, y \in M$. Thus $w^r g = u \in M$ and similarly $w^m h, w^k g^{-1}, w^p h^{-1} \in M$. Hence $w^q g, w^q h, w^q g^{-1}, w^q h^{-1} \in M$ where $q = r + m + k + p$ since $w \in M$ by assumption.

Now $gh = hg[g, h]$. Since $[g, h]$ is central, an easy induction yields that $g^n h^n = h^n g^n [g, h]^{n^2}$, i.e., $g^n h^n [g, h]^{-n^2} = h^n g^n$. Then for $z = w^q$, we have $(zg^{-1})^n (zh)^n (zg)^n (zh^{-1})^n \in M$ for all $n \geq 1$ and

$$(zg^{-1})^n (zh)^n (zg)^n (zh^{-1})^n = z^{4n} g^{-n} (h^n g^n) h^{-n} = z^{4n} g^{-n} (g^n h^n [g, h]^{-n^2}) h^{-n}$$
$$= z^{4n} w^{-n^2} = w^{4nq} w^{-n^2} = w^{4nq - n^2}.$$

Since for large n, $4nq - n^2 < 0$ and since $w \in M$ is central from Lemma V.5.10, we conclude that $w^{-1} \in M$, a contradiction. This completes the proof. ∎

V.5.28. **Proposition.** *Let M be a maximal subsemigroup of G which is reduced in G. Then $G/Z(G)$ has trivial center.*

Proof. Suppose not. Let $\varphi : G \to G/Z(G)$. Then there exists $g \in G$ such that $\varphi(g)$ is in the center of $G/Z(G)$, but $\varphi(g)$ is not the identity. Since $g \notin Z(G)$, there exists $h \in G$ such that $gh \neq hg$. Then $[g, h] = g^{-1} h^{-1} gh \neq 1$. But

$$\varphi([g, h]) = \varphi(g^{-1})\varphi(h^{-1})\varphi(g)\varphi(h) = \varphi(g^{-1})\varphi(g)\varphi(h^{-1})\varphi(h) = \varphi(1).$$

Thus $[g, h] \in Z(G)$. By Lemma V.5.27, $[g, h], [g, h]^{-1} \in M$. Since $[g, h] \in Z(G)$, the subgroup it generates is normal. But this contradicts the assumption that M is reduced in G. ∎

V.5.29. **Theorem.** *Let M be a maximal subsemigroup of a nilpotent group G. Then M is total and invariant in G and $[G, G] \subseteq H(M)$. Hence $G_R = G/H(M)$ is abelian and totally ordered.*

Proof. Let (G_R, M_R) be the reduction of (G, M). By the Reduction Lemma V.5.11, M_R is maximal in G_R. Since the quotient of a nilpotent group is nilpotent, any non-trivial quotient of G_R is nilpotent, hence has non-trivial center. By Proposition V.5.28 it follows that G_R must be abelian. Thus $[G, G] \subseteq \text{Core}(M)$. By Corollary V.5.9, M_R is total, and by the Reduction Lemma V.5.11, M is total in G. For $g \in G$, $m \in M$ we have $(g^{-1} m g m^{-1}) m = g^{-1} m g \in M$ since $[G, G] \subseteq H(M) \subseteq M$. Thus M and hence $H(M)$ are invariant. Then $\text{Core}(M) = H(M)$. ∎

V.5.30. **Corollary.** *Let G be a connected nilpotent topological group which is locally compact or locally connected. Let M be a maximal subsemigroup with $\text{int}(M) \neq \varnothing$. Then (G_R, M_R) is topologically isomorphic to $(\mathbb{R}, \mathbb{R}^+)$. Hence*

$\text{Core}(M) = H(M)$ *is a closed normal subgroup, which is also connected in the locally compact case.*

Proof. By Theorem V.5.29, M is total and invariant, and by Proposition V.5.16 M is closed. The rest now follows from Theorem V.5.23. ∎

V.5.31.   **Proposition.** *Let H be a connected normal nilpotent subgroup of a connected topological group G and let M be a maximal subsemigroup with $\text{int}(M) \neq \emptyset$. Then $[H, H] \subseteq \text{Core}(M)$.*

Proof. We first show $[H, H] \cap \text{int}(M) = \emptyset$. Suppose not. Since $\mathbf{1} \notin \text{int}(M)$ (otherwise $M = G$), $\text{int}(M) \cap H$ is a proper subsemigroup of H with interior in H. By Proposition V.5.14 there exists a maximal subsemigroup S of H containing $\text{int}(M) \cap H$. By Theorem V.5.29, $[H, H] \subseteq S$. If $g \in [H, H] \cap \text{int}(M)$, then $g^{-1} \in [H, H] \cap (\text{int} \, M)^{-1} \subseteq S \cap (\text{int} \, M)^{-1}$; this contradicts Lemma V.5.13 applied to H (with $U = \text{int}(M) \cap H$). Thus $[H, H] \cap \text{int}(M) = \emptyset$.

Since $\text{int}(M)$ is an ideal in M, the Swallowing Lemma V.5.4 implies $[H, H] \subseteq M$ (since H is normal implies $[H, H]$ is normal). Thus $[H, H] \subseteq \text{Core(M)}$. ∎

V.5.32.   **Corollary.** *Let G be a connected nilpotent group. If $A \subseteq G$ and $\text{int}(A) \cap [G, G] \neq \emptyset$, then the subsemigroup generated by A is all of G.*

Proof. Let S be the subsemigroup generated by A. If $S \neq G$, then S is contained in a maximal semigroup M by Proposition V.5.14. By Proposition V.5.31, $[G, G] \subseteq \text{Core}(M)$. But by hypothesis $\emptyset \neq \text{int}(A) \cap [G, G] \subseteq \text{int}(M) \cap [G, G]$, a contradiction since $H(M) \cap \text{int}(M) = \emptyset$. Thus $S = G$. ∎

     The preceding results on maximal semigroups apply in a very general fashion to nilpotent groups. To extend these results to the solvable setting however, we will need to restrict our attention to the Lie group case.

Frobenius–Perron Groups

     Throughout this section G denotes a connected Lie group with Lie algebra $\mathbf{L}(G)$. *For $g \in G$, the inner automorphism $I_g \colon G \to G$ defined by $h \mapsto ghg^{-1}$ induces an automorphism $\text{Ad} \, g \colon \mathbf{L}(G) \to \mathbf{L}(G)$ such that the following diagram commutes:*

$$\begin{array}{ccc} L(G) & \xrightarrow{\ \text{Ad} \, g\ } & L(G) \\ {\scriptstyle \exp} \downarrow & & \downarrow {\scriptstyle \exp} \\ G & \xrightarrow[\ I_g\ ]{} & G. \end{array}$$

Thus $\exp\big(\text{Ad} \, g(x)\big) = g \exp(x) g^{-1}$. On the other hand, for $x \in \mathbf{L}(G)$,

$$\text{Ad} \exp(x) = e^{\text{ad} \, x} = \text{id}_L + \text{ad} \, x + \frac{1}{2}(\text{ad} \, x)^2 + \cdots.$$

We come now to a crucial lemma.

V.5.33. **Lemma.** (The Invariant Wedge Lemma) *Let G be a connected Lie group, let* $\exp \colon \mathbf{L}(G) \to G$ *be the exponential mapping, and let I be an abelian ideal of* $\mathbf{L}(G)$. *If M is a maximal subsemigroup of G with* $\operatorname{int}(M) \neq \emptyset$, *then* $W = \{x \in I \colon \exp(x) \in M\}$ *is a closed wedge in* $\mathbf{L}(G)$ *which is invariant under the adjoint action of G and satisfies* $W - W = I$.

Proof. Since I is an abelian ideal, exp restricted to I is a homomorphism and $\exp(I)$ is a normal subgroup. If $\exp(I) \subseteq M$, then we are finished since I is an ideal, and hence invariant under the adjoint action.

If $\exp(I) \not\subseteq M$, then by the Swallowing Lemma V.5.4, $\exp(I) \cap \operatorname{int}(M) \neq \emptyset$. Pick $h = \exp(y)$, $h \in \operatorname{int}(M)$, $y \in I$.

Since M is closed (Proposition V.5.16) and exp is continuous, $W = \exp^{-1}(M) \cap I$ is closed. Since M is a subsemigroup and exp restricted to I is a homomorphism, W is closed under addition. By the Purity Lemma V.5.10 applied to $H = \exp(I)$, we conclude that W is closed under scalar multiplication by positive rationals, and then by all positive reals by continuity. It follows that W is a wedge. Since $\exp(y) \in \operatorname{int}(M)$, it follows that W has y as an interior point, and hence $I = W - W$.

Let x be any other interior point of W. Then $n \cdot x = y + z_n$ for some $z_n \in W$ for all sufficiently large n. (To see this, let U be open in I, $x \in U \subseteq W$. Then $\frac{1}{n} \cdot y \in x - U$ for large n. Thus $\frac{1}{n} \cdot y + u_n = x$, i.e., $y + z_n = n \cdot x$ where $z_n = n \cdot u_n \in W$.)

Since $h \in \operatorname{int}(M)$, there exists an open set $N = N^{-1}$, $\mathbf{1} \in N$ with $Nh \subseteq \operatorname{int}(M)$. Let $g \in N$ and let x be an interior point of W. For large n, pick $z_n \in W$ such that $n \cdot x = y + z_n$. Let $a = \exp(x)$, $b_n = \exp(z_n)$. We have

$$\exp\big((\operatorname{Ad} g)(n \cdot x) + y\big) = \exp\big((\operatorname{Ad} g)(n \cdot x)\big) \exp(y) = g\big(\exp(n \cdot x)\big) g^{-1} h$$
$$= g\big(\exp(y + z_n)\big) g^{-1} h = ghb_n g^{-1} h \in Nh \cdot M \cdot Nh \subseteq M.$$

Thus $(\operatorname{Ad} g)(n \cdot x) + y \in W$ for large n. Since W is a wedge, $(\operatorname{Ad} g)(x) + \frac{1}{n} \cdot y \in W$ for large n. Since W is closed, $(\operatorname{Ad} g)(x) \in W$. Thus $\operatorname{Ad} g$ carries the interior of W into W and hence preserves W. This is true for all $g \in N$ and by connectivity for all of G (since $\operatorname{Ad}(g_1 \cdots g_n) = \operatorname{Ad} g_1 \circ \cdots \circ \operatorname{Ad} g_n$). ∎

V.5.34. **Definition.** A Lie group G is called a *Frobenius–Perron group* if for every continuous linear action of G on a finite-dimensional real vector space V that leaves a pointed cone C invariant, there exists $v \in C$ with $G \cdot v \subseteq \mathbb{R}^+ \cdot v$.

V.5.35. **Remark.** The dual $C^* \subseteq V^*$ of a pointed cone $C \subseteq V$ is a generating wedge in V^*, and the dual of a ray is a half-space. By duality one obtains the following equivalent formulation of a Frobenius–Perron group: A Lie group G is a Frobenius–Perron group if and only if for every continuous linear action of G on a finite-dimensional real vector space V that leaves a generating wedge invariant, there exists an invariant half-space containing the wedge. (See Lemma III.2.3 for a similar result.)

We have presented a Frobenius–Perron theory of groups in Chapter III.3. By Corollary III.3.7, a connected Lie group is a Frobenius–Perron group if it is compact modulo its radical (where the radical $\operatorname{Rad}(G)$ is the unique maximal normal solvable subgroup). We combine the theory developed previously in this section with the Frobenius–Perron theory to derive our major results.

V.5.36.   **Theorem.**   *Let G be a finite-dimensional connected Lie group which is a Frobenius–Perron group, and let M be a maximal subsemigroup which is reduced in G and satisfies $\operatorname{int} M \neq \emptyset$. Then one of the following holds:*

(i) $\operatorname{Rad}(G) = \{1\}$, *i.e., G is semisimple;*

(ii) $(\operatorname{Rad}(G), \operatorname{Rad}(G) \cap M)$ *is topologically isomorphic to $(\mathbb{R}, \mathbb{R}^+)$;*

(iii) $\operatorname{Rad}(G)$ *is topologically isomorphic to $\operatorname{Aff}(\mathbb{R})$.*

Proof.   If the radical $R = \operatorname{Rad}(G)$ is trivial, then G is semisimple. We consider the case $R \neq \{1\}$, and show that either (ii) or (iii) obtains. The proof consists of a series of reductions.

1. *The nil radical N is abelian.*

It is a standard fact of Lie theory that the nil radical is the largest connected normal nilpotent subgroup. The assertation then follows immediately from Proposition V.5.31.

2. *The radical R is metabelian (i.e., $[R, R]$ is abelian).*

Again it is standard that $[R, R] \subseteq N$, which is abelian.

3. $\dim[\mathbf{L}(R), \mathbf{L}(R)] = 1$ or $[R, R] = \{1\}$.

Let I be the Lie algebra for $[R, R]$. Since $[R, R]$ is normal and abelian, I is an abelian ideal in $\mathbf{L}(G)$. Let $W = \{x \in I : \exp(x) \in M\}$. By the Invariant Wedge Lemma V.5.33, W is an invariant wedge which generates I. If $W = I$, then $\exp(I) = [R, R]$ (since I is abelian) is a normal subgroup contained in M. Since M is reduced, $[R, R] = \{1\}$ in this case.

If $W \neq I$, then since G is a Frobenius–Perron group and since W is generating, there exists an invariant half-space Q with $W \subseteq Q$ by Remark V.5.35. Then $Q \cap -Q$ is an invariant hyperplane in I, so $F = \exp(Q \cap -Q)$ is a normal subgroup. Now

$$\exp^{-1}(\operatorname{int} M) \cap I \subseteq \operatorname{int}(W) \subseteq \operatorname{int}(Q) \subseteq Q \setminus (Q \cap -Q).$$

Thus $\operatorname{int}(M) \cap F = \emptyset$. By the Swallowing Lemma V.5.4, $F \subseteq M$. Since F is normal, $F \subseteq \operatorname{Core}(M) = \{1\}$. Thus $Q \cap -Q = \{0\}$, so I is 1-dimensional, as is $\exp(I) = [R, R]$.

4. *If $[R, R] = \{1\}$, then case (ii) obtains.*

In this case, R is abelian. Let $\mathbf{L}(R)$ be the Lie algebra for R. We again obtain an invariant wedge $W = \{x \in \mathbf{L}(R) : \exp(x) \in M\}$. One repeats the arguments of step 3 to conclude that W is an invariant generating wedge, that $W \neq \mathbf{L}(R)$ since M is reduced and we are not in case (i), and finally that $\mathbf{L}(R)$ is 1-dimensional. By Proposition V.5.17 we have that R cannot be the circle group. It follows that the exponential mapping must carry $\mathbf{L}(R)$ onto a copy of $\mathbb{R}$ and W onto a ray, which we take to be $\mathbb{R}^+$.

5. *If $I = [\mathbf{L}(R), \mathbf{L}(R)]$ is 1-dimensional, then the centralizer of I in $\mathbf{L}(R)$ is I.*

Since I is an abelian ideal, its centralizer is easily verified to be an ideal which contains I. Hence A, the intersection of the centralizer with $\mathbf{L}(R)$, is an ideal containing I. Now $[A, A] \subseteq [\mathbf{L}(R), \mathbf{L}(R)] = I$; thus $[A, [A, A]] \subseteq [A, I] = \{0\}$. Therefore A is a nilpotent ideal, so $\exp(A)$ is a normal nilpotent group. It follows from Proposition V.5.31 and the fact that M is reduced that $\exp(A)$ is abelian. Thus A is abelian. One now applies an argument to A completely analogous to that given in Step 3 to conclude that A is 1-dimensional. Since $I \subseteq A$, we have $I = A$.

  6. *If $I = [\mathbf{L}(R), \mathbf{L}(R)]$ is 1-dimensional, then case (iii) obtains.*
Let $z \in I$, $z \neq 0$. Consider $\operatorname{ad} z \colon \mathbf{L}(R) \to I = \mathbb{R} \cdot z$. By Reduction 5 the kernel of this mapping is I, which is 1-dimensional. Thus $\mathbf{L}(R)$ is 2-dimensional. Since $[\mathbf{L}(R), \mathbf{L}(R)] = I \neq 0$, it follows that $\mathbf{L}(R)$ must be a non-abelian 2-dimensional Lie algebra, and hence isomorphic to $\mathrm{Aff}(\mathbb{R})$.   ∎

V.5.37.   **Corollary.**   *Let G be a finite-dimensional connected solvable Lie group, and let M be a maximal subsemigroup with $\operatorname{int}(M) \neq \varnothing$. Then M is total and one of the following holds:*

  (i) *(G_R, M_R) is topologically isomorphic to $(\mathbb{R}, \mathbb{R}^+)$;*

  (ii) *(G_R, M_R) is topologically isomorphic to $(\mathrm{Aff}(\mathbb{R}), \mathrm{Aff}(\mathbb{R})^+)$.*

Proof.   By Theorem III.3.7, G is a Frobenius–Perron group. Now G_R is solvable, so case (ii) or (iii) of Theorem V.5.36 must hold.

  If $\mathbf{L}(G_R)$ is the non-abelian 2-dimensional Lie algebra, then M_R does not contain the 1-dimensional normal subgroup, since it is reduced in G_R. By Example V.5.26, M is total and must be a half-space semigroup with boundary a non-normal 1-dimensional group, and an inner automorphism then carries M to the upper half space $\mathrm{Aff}(\mathbb{R})^+$.   ∎

  The preceding corollary remains valid for extensions of solvable groups by compact groups.

V.5.38.   **Lemma.**   *The following are equivalent:*

  (1) *$G/\mathrm{Rad}(G)$ is compact.*

  (2) *The Levi subalgebras of $\mathbf{L}(G)$ are compact (i.e., are the Lie algebras of some compact group).*

  (3) *The analytic subgroups corresponding to the Levi subalgebras are compact.*

  (4) *G contains a connected solvable normal subgroup H and a compact group K such that $G = \overline{H}K$.*

Proof. $(1) \Rightarrow (2)$: The Levi subalgebras arise as cross-sections of the Lie algebra homomorphism from $\mathbf{L}(G)$ to $\mathbf{L}(G/\mathrm{Rad}\,G)$, and hence are isomorphic to $\mathbf{L}(G/\mathrm{Rad}\,G)$.

  $(2) \Rightarrow (3)$: If one of the Lie groups associated with a semisimple Lie algebra is compact, then all of them are compact (see, e.g., [Hoch65], p.144).

  $(3) \Rightarrow (4)$: $G = \mathrm{Rad}(G) \cdot K$ where K is the analytic group associated with some Levi factor.

  $(4) \Rightarrow (1)$: Since $\mathrm{Rad}\,G$ is closed and is the largest connected solvable normal subgroup, $H \subseteq \mathrm{Rad}\,G$. Then the image of K in $G/\mathrm{Rad}\,G$ must be all of $G/\mathrm{Rad}\,G$.   ∎

V.5.39.   **Theorem.**   *Suppose $G/\mathrm{Rad}\,G$ is compact. If M is a maximal subsemigroup of G with $\operatorname{int}(M) \neq \varnothing$, then M is total, $\mathrm{Core}\,(M)$ is a normal, connected subgroup containing every semisimple analytic subgroup, and one of the following holds:*

(i) (G_R, M_R) *is topologically isomorphic to* $(\mathbb{R}, \mathbb{R}^+)$,

(ii) (G_R, M_R) *is topologically isomorphic to* $(\mathrm{Aff}(\mathbb{R}), \mathrm{Aff}(\mathbb{R})^+)$.

Proof.   If $\varphi: G \to G_R$, then $\varphi(\mathrm{Rad}\, G) = \mathrm{Rad}\, G_R$ ([Bou65], Chapter III, Section 9, §7, Proposition 24). Thus $G_R/\mathrm{Rad}\, G_R$ is a continuous image of $G/\mathrm{Rad}\, G$, hence compact. By Lemma V.5.38 if K is an analytic semisimple subgroup corresponding to some Levi factor of $\mathbf{L}(G_R)$, then K is compact and normalizes $\mathrm{Rad}\, G_R$. By Theorem III.3.7, G_R is a Frobenius–Perron group. Hence we may apply Theorem V.5.36 to G_R.

Case (i) of Theorem V.5.36 is impossible, for then $G_R = K$ would be compact, an impossibility (see V.5.17). Suppose that case (ii) obtains, i.e., $\mathrm{Rad}\, G_R$ is topologically isomorphic to $\mathbb{R}$. Then K acts on $\mathbb{R}$ by inner automorphisms, and hence must act trivially (since K is compact and connected). Thus elements of $\mathbb{R}$ and K commute, so K is normal (since $G_R = \mathbb{R}K$). Then Proposition V.5.17 implies $K = \{1\}$ (since M_R is reduced). So $G_R = \mathrm{Rad}\, G_R$. Since the image of any analytic semisimple subgroup of G must be a semisimple subgroup of $\mathbb{R}$, the image must be trivial. Theorem V.5.36 implies the rest.

Finally consider the case of Theorem V.5.36 that $\mathrm{Rad}\, G_R$ is topologically isomorphic to $\mathrm{Aff}(\mathbb{R})$. Again K must act on $\mathrm{Aff}(\mathbb{R})$ by inner automorphisms. But the identity component of the automorphism group of $\mathrm{Aff}(\mathbb{R})$ is again topologically isomorphic to $\mathrm{Aff}(\mathbb{R})$ (see [Jac57], p.10), and thus the automorphism group contains no non-trivial compact connected subgroups. Thus again K acts trivially, is thus normal, and hence $K = \{1\}$. So $G_R = \mathrm{Rad}\, G_R$. Since M_R is closed and maximal in G_R, it must be topologically isomorphic to $\mathrm{Aff}(\mathbb{R})^+$ by Corollary V.5.37.

Finally Lemma V.5.22 implies that $\mathrm{Core}(M)$, the kernel of φ, is connected.   ∎

EV.5.1.   **Exercise.** Show Theorem V.5.38 extends to locally compact connected groups G which contain a compact normal subgroup K such that $H = G/K$ is a Lie group with $H/\mathrm{Rad}\, H$ compact.

V.5.40.   **Corollary.**   *Suppose that* $G/\mathrm{Rad}\, G$ *is compact. If* G *contains a subsemigroup* $S \neq G$ *with* $\mathrm{int}(S) \neq \varnothing$, *then one of the following holds:*

(i) *There exists a continuous homomorphism* $\varphi: G \to \mathbb{R}$ *such that* $\varphi(S) \subseteq \mathbb{R}^+$.

(ii) *There exists a continuous homomorphism* $\varphi: G \to \mathrm{Aff}(\mathbb{R})$ *such that* $\varphi(S) \subseteq \mathrm{Aff}(\mathbb{R})^+$.

Proof.   Extend S to a maximal semigroup M and apply Theorem V.5.39.   ∎

The next corollary gives a classification of maximal semigroups in terms of the Lie algebra for the simply connected case.

V.5.41.   **Corollary.**   *The maximal subsemigroups* M *with non-empty interior of a simply connected Lie group with* $G/\mathrm{Rad}\, G$ *compact are in one-to-one correspondence with their tangent objects* $\mathbf{L}(M)$ *and the latter are precisely the closed half-spaces whose boundary is a subalgebra, i.e., the half-space Lie semialgebras. Furthermore,* M *is the semigroup generated by* $\exp(\mathbf{L}(M))$.

Proof.   Let M be maximal with $\mathrm{int}(M) \neq \varnothing$. By Theorem V.5.39, G_R is either $\mathbb{R}$ or $\mathrm{Aff}(\mathbb{R})$ and $\mathrm{Core}(M)$ is connected. Hence $\mathrm{Core}(M)$ is generated by the exponential image of its tangent subalgebra. One verifies directly in each case that M_R is (and hence is generated by) the exponential of its tangent set. Let $\varphi: G \to G_R$. It follows from chasing the diagram

$$
\begin{array}{ccccc}
\mathbf{L}\big(\mathrm{Core}(M)\big) & \xrightarrow{\ \mathbf{L}(\mathrm{incl})\ } & \mathbf{L}(G) & \xrightarrow{\ \mathbf{L}(\varphi)\ } & \mathbf{L}(G_R) \\
{\scriptstyle \exp}\downarrow & & {\scriptstyle \exp}\downarrow & & {\scriptstyle \exp}\downarrow \\
\mathrm{Core}(M) & \xrightarrow[\ \mathrm{incl}\ ]{} & G & \xrightarrow[\ \varphi\ ]{} & G_R
\end{array}
$$

that $\mathbf{L}(M) = \mathbf{L}(\varphi)^{-1}\big(\mathbf{L}(M_R)\big)$ and that M is generated by the exponential image of $\mathbf{L}(M)$. Also since M_R has tangent set a half-space of $\mathbf{L}(G_R)$, this property pulls back so that $\mathbf{L}(M)$ is a half-space of $\mathbf{L}(G)$. Thus associated with each maximal subsemigroup with non-empty interior is a half-space whose exponential image generates it. This guarantees that the assignment is one-to-one.

Conversely suppose that G is simply connected and A is a half-space bounded by a subalgebra Q of codimension 1. Then Q is an ideal or contains an ideal of codimension 2 or 3 (see [Hof65] or [Ti59]), and furthermore there exists a Lie algebra homomorphism $\mathbf{L}(\varphi)$ either onto $\mathbb{R}$ with kernel Q or onto the non-abelian 2-dimensional Lie algebra or onto $sl(2, \mathbb{R})$. But the last case is impossible since then $\mathbf{L}(\mathrm{Rad}\,G)$ must necessarily map to $\{0\}$, and then $sl(2, \mathbb{R})$ would be the image of a compact Lie algebra (Lemma V.5.38), hence itself compact, a contradiction. Also $\mathbf{L}(\varphi)(A)$ will be a half-space in $\mathbb{R}$ or $\mathrm{Aff}(\mathbb{R})$. Since G is simply connected, there is a corresponding $\varphi: G \to \mathbb{R}$ or $\varphi: G \to \mathrm{Aff}(\mathbb{R})$. Pulling back the subsemigroups of $\mathbb{R}$ or $\mathrm{Aff}(\mathbb{R})$ corresponding to the half-space $\mathbf{L}(\varphi)(A)$, one obtains a maximal subsemigroup M of G containing $\exp(A)$. Since $A \subseteq \mathbf{L}(M) \neq \mathbf{L}(G)$ and A is a half-space, $A = \mathbf{L}(M)$. ∎

V.5.42. **Corollary.** *Let Ω generate (as a Lie algebra) $\mathbf{L}(G)$, and assume that $G/\mathrm{Rad}\,G$ is compact. If Ω is not contained in any half-space with boundary a subalgebra, then the semigroup generated by $\exp(\mathbb{R}^+\cdot\Omega)$ is all of G. The converse holds if G is simply connected.*

Proof. Let S be the semigroup generated by $\exp(\mathbb{R}^+\cdot\Omega)$. Since Ω generates $\mathbf{L}(G)$, we have by the Density Theorem for Ray Semigroups V.1.11 in view of Proposition V.1.9 that $\mathrm{int}(S) \neq \varnothing$. If $S \neq G$, extend S to a maximal subsemigroup M. Then $\mathbb{R}^+\Omega \subseteq \mathbf{L}(S) \subseteq \mathbf{L}(M)$, and by Theorem V.5.41 the latter is a half-space with boundary a subalgebra (simple connectivity was not needed for this direction of the proof). But this is a contradiction, so $S = G$. Similarly we deduce the converse using Corollary V.5.40. ∎

We remark that the finite-dimensionality assumption is necessary in order to have a Frobenius–Perron theory for solvable groups. One obtains in general such a theory for compact groups, and hence a version of Theorem V.5.40 (and the other theorems) could be carried out for infinite-dimensional Lie groups for the case G was the product of a compact group and a normal connected nilpotent group.

The classification of maximal subsemigroups in the case of semisimple Lie groups and of general Lie groups remains open. We give a construction for maximal subsemigroups for which the tangent wedges are not half-spaces.

V.5.43. **Example.** Let W be a generating wedge in $\mathbb{R}^n$ with $W \neq \mathbb{R}^n$. If W is a maximal invariant wedge in $\mathbb{R}^n$ under the linear action of a connected Lie group G, then $S = W \times G$ is a maximal subsemigroup of $\mathbb{R}^n \times G$, where the latter is given the semidirect product group structure.

Proof.   It is directly verified that S is a closed subsemigroup with non-empty interior. Hence it is contained in a maximal subsemigroup by Proposition V.5.14. We may identify $\mathbf{L}(\mathbb{R}^n \times G)$ with $\mathbb{R}^n \times \mathbf{L}(G)$; then by the Invariant Wedge Lemma V.5.33, $M \cap \mathbb{R}^n$ is an invariant wedge. Clearly $M \cap \mathbb{R}^n \neq \mathbb{R}^n$ (otherwise $\mathbb{R}^n \times G = \mathbb{R}^n \cdot S \subseteq M$). Thus $M \cap \mathbb{R}^n = W$ since $W \subseteq M \cap \mathbb{R}^n$ and W is a maximal invariant wedge. If $(v, g) \in M$, then $(v, g)(0, g^{-1}) = (v, \mathbf{1}_G) \in M$. Thus $v \in W$ and hence $(v, g) \in W \times G = S$. We conclude $M = S$, i.e., S is maximal.   ∎

It is conjectural that all maximal subsemigroups of reductive Lie groups for which the interior meets $\operatorname{Rad} G$ arise in such a fashion.

6. Divisible Semigroups

A semigroup S is called *divisible* provided it satisfies the following condition:

(D) For each $s \in S$ and each natural number $n \geq 2$, there is an $x \in S$ such that $x^n = s$.

V.6.1.   Remark.   We know from the local theory that for a local semigroup S with respect to a C-H-neighborhood B an analogous statement holds only if $\mathbf{L}(S)$ is a Lie semialgebra (see Corollary IV.1.32). Using the local isomorphism of the exponential mapping we conclude that if S is a locally divisible local semigroup in a Lie group G, then $\mathbf{L}(S)$ is a Lie semialgebra. Conversely, if $\mathbf{L}(S)$ is a Lie semialgebra for a local semigroup S and $S \cap U$ is closed in U for some neighborhood U of $\mathbf{1}$, then S is locally divisible (this follows from Proposition IV.1.31).

The example of the semigroup S generated in $\widetilde{\mathrm{Sl}}(2, \mathbb{R})$ by the exponential image of the invariant cone shows that not every semigroup generated by a Lie semialgebra needs to be divisible (see Theorem V.4.42). On the other hand it remains an open question whether $\mathbf{L}(S)$ is a Lie semialgebra for every closed divisible subsemigroup of a Lie group. In this section we derive this result for certain special cases.

V.6.2.   Lemma.   *Let G be an abelian Lie group and $D \subseteq G$ a divisible subgroup. Then $\overline{D}$ is divisible.*

Proof.   We may assume that D is dense in G, i.e., we restrict our attention to $G = \overline{D}$. Fix $n > 1$ and let $\alpha: G \to G$ be the homomorphism defined by $\alpha(g) = g^n$. By the structure theorem for abelian Lie groups, G is of the form $K \times \mathbb{R}^n \times \Delta$, where K is compact and connected and Δ is discrete. Clearly $\alpha(\mathbb{R}^n) = \mathbb{R}^n$. Since K is connected and abelian, the exponential mapping from its Lie algebra is a surjective homomorphism. Since the image of a divisible group is divisible, K is divisible. Let $\delta \in \Delta$. Then since $K \times \mathbb{R}^n \times \{\delta\}$ is open, there exists $d \in D$ which is also in this open set. Pick $b \in D$ such that $b^n = d$. Then the coordinate of b in Δ is an n-th root for δ. Thus $\alpha(G) = G$ and G is divisible.   ∎

V.6.3.   Remark.   Fix $s \in S$ and suppose that S is divisible. Set $s_1 = s$ and define s_n recursively to be any n-th root of s_{n-1} in S for $n = 2, \ldots$ Then the function $f_s: \mathbb{Q}^+ \to S$ which sends 0 to $\mathbf{1}$ and $m/n!$ to s_n^m is unambiguously defined and is a morphism of semigroups.

V.6.4. **Proposition.** *Any closed divisible subgroup H in a connected Lie group G is connected.*

Proof. Let $g \in H$. Then by Remark V.6.3 there exists an algebraic homomorphism $f: \mathbb{Q}^+ \to H$ such that $f(1) = g$. We extend f in the obvious way to a homomorphism from all of $\mathbb{Q}$ into H and again call the extension f. We set $A = \overline{f(\mathbb{Q})}$; then A is a closed abelian Lie subgroup of G with a dense divisible subgroup. Hence by Lemma V.6.2, A is divisible. If A_0 is the identity component in A, then by a result of Mostow [Mo 57] A/A_0 has finite rank. But A/A_0 is also divisible. It follows from the fundamental theorem of finitely generated abelian groups that $A = A_0$. Since H was closed, $A \subseteq H$; thus g is in the identity component of H. Since g was arbitrary in H, it follows that H is connected. ∎

V.6.5. **Theorem.** *A closed subsemigroup S of a connected Lie group G is divisible if and only if $\exp \mathbf{L}(S) = S$.*

Proof. If $S = \exp \mathbf{L}(S)$, then S is clearly divisible, since $X \in \mathbf{L}(S)$ if and only if $\exp(tX) \in S$ for all $t \geq 0$.

Assume conversely that S is divisible, and let $s \in S$. By Remark V.6.3 there exists a homomorphism from $\mathbb{Q}^+$ into G carrying 1 to s, and this homomorphism extends to a homomorphism $f: \mathbb{Q} \to G$ as in the proof of Proposition V.6.4. The group $A = \overline{f(\mathbb{Q})}$ is divisible and hence is a closed connected Lie subgroup of G by Proposition V.6.4. Since it is abelian, it is of the form $K \times \mathbb{R}^n$, where K is a torus group. If $p: A \to \mathbb{R}^n$ is the projection, then $p \circ f: \mathbb{Q} \to \mathbb{R}^n$ is a group homomorphism with dense image. Since it is clearly a $\mathbb{Q}$-vector space homomorphism (or, alternately, since $\mathbb{R}^n$ is uniquely divisible with roots converging to 0), we conclude that $n = 0$ or $n = 1$. If $n = 0$, then A is a torus. Then $\overline{f(\mathbb{Q}^+)}$ is a compact subsemigroup of the compact group A and is therefore itself a group (Proposition V.0.17). The definition of A implies that $A = \overline{f(\mathbb{Q}^+)}$. Since A is a torus, the exponential mapping is onto, and hence s lies on a one-parameter subgroup of A, a subset of S. Hence $s = \exp X$ for some $X \in \mathbf{L}(S)$. This completes the case $n = 0$.

We consider now the case that $A = K \times \mathbb{R}$ and that $p(f(\mathbb{Q}^+))$ is dense in $\mathbb{R}^+$. We may assume without loss of generality that $p \circ f(1) = 1$, whence in fact $p \circ f(q) = q$ for all $q \in \mathbb{Q}^+$. We now apply the techniques of the one-parameter semigroup theorem (cf. [Hof60], [HM66], [Hey77]). Define $C = \bigcap \{ \overline{f(\{q \in \mathbb{Q}^+ : q < t\})} : 0 < t \}$. Then $C \subseteq S$, and the sets $\overline{f(\{q \in \mathbb{Q}^+ : q < t\})}$ are contained in the compact set $p^{-1}([0,t])$ and are therefore compact. Hence C is a compact divisible abelian group (see, for example, [Hey 77]). (In fact $C = K$, but we do not need that here.) Thus C is connected and hence a torus. The theory we are quoting guarantees the existence of a continuous one-parameter semigroup $F: [0, \infty) \to \overline{f(\mathbb{Q}^+)}$ with $p \circ F(r) = r$ for all $r > 0$ and $F(1) = cf(1) = cs$ for some $c \in C$. Let $g: \mathbb{R} \to C$ be a one parameter group in C with $g(1) = c$. Then $t \mapsto g(t)^{-1}F(t): [0, \infty) \to \overline{f(\mathbb{Q}^+)} \subseteq S$ is a continuous one-parameter semigroup mapping 1 to $g(1)^{-1}F(1) = c^{-1}cs = s$. Once again we have found an $X \in \mathbf{L}(S)$ with $\exp X = s$. This completes the proof. ∎

As a byproduct of Theorem V.6.5 we have the following observation.

V.6.6. **Corollary.** *The exponential map of a divisible Lie group is surjective.* ∎

Better results than this are known. McCrudden [McC 81] has shown that if an element g in a connected Lie group G has roots of all orders, then it lies in the image of the exponential function (i.e., on a one-parameter group).

We now derive some partial results on the problem of showing that $\mathbf{L}(S)$ is a Lie semialgebra if S is divisible. At first glance this might seem to be almost trivially true in light of Theorem V.6.5, but delicate and difficult problems arise from the fact the one-parameter semigroups can bend back near the identity. The semigroup $S = \mathbb{R}^+ \times S^1$ in the group $G = \mathbb{R} \times S^1$, the product of the reals and the circle group, is a good example to illustrate this phenomenon.

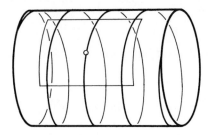

Figure 1

V.6.7.   **Lemma.**   *Suppose that two elements* X, Y *in the Lie algebra* $\mathbf{L}(G)$ *of a Lie group* G *satisfy* $\exp X = \exp Y$ *and that* $\exp$ *is non-singular at* X. *Then* $[X, Y] = 0$, $\exp(Y - X) = 1$, *and* $\exp([0, 1] \cdot (Y - X))$ *is a circle subgroup of* G.

Proof.   All elements $\exp tY$ commute with $\exp X = \exp Y$. Thus

$$\exp X = \exp(tY) \exp(X) \exp(-tY) = \exp(e^{t \operatorname{ad} Y} X)$$

for all t. Since $\exp$ is non-singular at X, for all sufficiently small t we conclude that $(e^{t \operatorname{ad} Y} X) - X = 0$. If we divide by $t \neq 0$ and evaluate the limit at $t \to 0$, we obtain $[Y, X] = (\operatorname{ad} Y) X = 0$. Thus $\operatorname{span}\{X, Y\}$ is an abelian subalgebra of $\mathbf{L}(G)$ and $\exp(Y - X) = \exp(Y) \exp(-X) = \exp(X) \exp(-X) = 1$. Since the mapping $t \mapsto \exp t(Y - X) : \mathbb{R} \to G$ is a one parameter group with kernel containing $1 \in \mathbb{Z}$, it follows that the image is a circle subgroup and that the interval $[0, 1]$ maps onto the image.   ∎

V.6.8.   **Definition.**   Let G be a Lie group with Lie algebra $\mathbf{L}(G)$. A wedge $W \subseteq \mathbf{L}(G)$ *disperses* in G if there exists an open set B containing 0 in $\mathbf{L}(G)$ such that $B \cap W = B \cap \exp^{-1}(\exp(W))$.

V.6.9.   **Lemma.**   *Let* S *be a closed divisible subsemigroup of the Lie group* G. *If* $\mathbf{L}(S)$ *disperses in* G, *then* $\mathbf{L}(S)$ *is a Lie semialgebra.*

Proof.   By Theorem V.6.5, $S = \exp W$, where $W = \mathbf{L}(S)$. Pick B open containing 0 as in Definition V.6.8. Note that any smaller B also works so that we may assume that B is a C-H-neighborhood on which $\exp$ is one-to-one. Since $\exp W = S$, $\exp W$ is a semigroup, and hence $W \cap B = \exp^{-1}(\exp(W)) \cap B$ is a local semigroup with respect to B, since $\exp$ is a local isomorphism. Thus W is a Lie semialgebra.   ∎

V.6.10.   Theorem. *Let S be a closed divisible subsemigroup of a Lie group G with group of units $H = H(S)$. If there exists an open set U containing 1 such that every compact connected subgroup of G contained in the tube UH is contained in H, then $\mathbf{L}(S)$ is a Lie semialgebra which disperses in G.*

Proof.   By Theorem V.6.5, $S = \exp(\mathbf{L}(S))$, so S is infinitesimally generated. Hence $G(S)$ is a Lie group, and without loss of generality we can assume that $G = G(S)$. Let U be the open set containing 1 guaranteed by the hypotheses. Pick an open set U' containing 1 such that $U'U' \subseteq U$. By the Tube Theorem V.2.7, pick a closed right ideal I such that $S \subseteq I \cup U'H$. Pick an open set V containing 1 such that $V \subseteq U' \setminus I$ and V is the homeomorphic image of a C-H-neighborhood B in $\mathbf{L}(G)$.

Let $W = \mathbf{L}(S)$ and $W' = \exp^{-1}(\exp(W))$. We show that W disperses in G. If not, then $W \cap B \neq W' \cap B$, so there exists $X \in (B \cap \exp^{-1}(\exp W)) \setminus B \cap W$ and $Y \in W$ such that $\exp X = \exp Y$. If $\exp t \cdot Y \in I$ for some $t \in [0,1]$, then

$$\exp X = \exp Y = \exp(t \cdot Y)\exp((1-t)\cdot Y) \in I$$

since I is a right ideal. But $\exp X \in \exp(B) = V$, and the latter set misses I. Hence $\exp([0,1]\cdot Y) \subseteq U'H$. Since $X \in B$ and B is symmetric, we conclude that $[-1,0]\cdot X \subseteq B$. Thus for $0 \leq t \leq 1$, we conclude from Lemma V.6.7 that

$$\exp(t\cdot(Y - X)) = \exp(t\cdot Y)\exp(-t\cdot X) = \exp(-t\cdot X)\exp(t\cdot Y) \in V(U'H) \subseteq UH.$$

Again by Lemma V.6.7, the set $\exp([0,1]\cdot(Y - X))$ is a circle subgroup. Therefore $X - Y \in \mathfrak{h}$ so that $X = (X - Y) + Y \in W$, contradicting our assumption. Hence W disperses in G. By Lemma V.6.9, W is a Lie semialgebra.   ∎

V.6.11.   Corollary. *If S is a closed divisible subsemigroup of a Lie group G such that $H(S) = 1$, then $\mathbf{L}(S)$ is a Lie semialgebra.*

Proof.   Since 1 possesses neighborhoods which contain no non-trivial subgroups, the corollary follows immediately from Theorem V.6.10.   ∎

Problem. Is $\mathbf{L}(S)$ a Lie semialgebra whenever S is a closed divisible subsemigroup of a Lie group? Does it perhaps even suffice that S is an infinitesimally generated divisible subsemigroup?

7. Congruences on open subsemigroups

Let S be a semigroup and write Δ for the diagonal $\{(s,s): s \in S\}$ of $S \times S$. A *right congruence* κ on S is a binary relation $\kappa \subseteq S \times S$ which is an equivalence relation and satisfies $\kappa\Delta \subseteq \kappa$. *Left congruences* are defined similarily. A *congruence* κ is an equivalence on S satisfying $\Delta\kappa \cup \kappa\Delta \subseteq \kappa$. (It follows that κ is a congruence on S if it is an equivalence relation on S and a subsemigroup of $S \times S$.)

If S is a topological semigroup then a binary relation is called *closed* if it is closed as a subspace of $S \times S$.

A right congruence κ on a group S is always determined by a subgroup H of S such that an equivalence class $\kappa(s)$ is exactly the coset Hs. If S is a topological group, then closed right congruences correspond, in this fashion, to closed subgroups. If S is a subsemigroup of a topological group G, then every closed subgroup H induces a closed right congruence κ_H on S via $\kappa_H(s) = Hs \cap S$. In general, however, right congruences on subsemigroups S of groups are vastly more complicated. Even for the half-line $S = \mathbb{R}^+$ in $G = \mathbb{R}$ there are numerous non-trivial congruences which are not obtained in this fashion. Typically, the so-called *Rees congruence* whose cosets $\kappa(x)$ are singleton for $x \in [0, a[$ and equal $[a, \infty[$ for $x \geq a$ is one of them. (See [CHK83] or [HM66] for more details.)

In the present Section we shall study right congruences (and congruences) κ on an open subsemigroup S of a Lie group G such that $1 \in \overline{S}$. Rather than supposing that κ is closed we make the weaker assumption that all equivalence classes $\kappa(s)$ are closed in S.

By the Infinitesimal Generation Theorem V.1.16 we know that for any infinitesimally generated subsemigroup S of a Lie group G such that $G = G(S)$, the interior $\operatorname{int} S$ is an open subsemigroup of G with 1 in its closure. Hence the results of this section apply to $\operatorname{int} S$ and the restriction $\kappa | \operatorname{int} S$ of any congruence κ on S with closed congruence classes.

The aim of our investigations is to describe right congruences on S under the "infinitesimal point of view" and to unveil the induced foliation structure:

At every point $p \in \overline{S}$ a right congruence κ on S with closed equivalence classes determines a subalgebra I_p of $L(G)$ and an open neighborhood U of p in G with the following property:

To any vector space complement E_p of I_p in $L(G)$ and any point $u \in U \cap pS$ we can find open cells A_u and B_u around 0 in E_p and I_p, respectively, such that the map

$$\Phi_u: A_u \times B_u \to V = u(\exp A_u)(\exp B_u), \quad (X, Y) \mapsto u(\exp X)(\exp Y)$$

is a diffeomorphism and satisfies

$$\kappa(\Phi_u(X,0) \cap U \cap pS = \Phi_u(\{X\} \times B_u),$$

for all $X \in A_u$. Moreover, if κ is a congruence then I_p is an ideal.

In particular, this result shows that the subsets $\kappa(u) \cap pS \cap U$, with $u \in U$, are analytic submanifolds of $pS \cap U$.

For the purposes of this section we need a generalization of the pre-order $\preceq_S$ introduced in Proposition V.0.4(ii).

V.7.0.   Definition. On any group G containing a semigroup S we define a transitive relation $\prec$ by

$$x \prec y \quad \text{if and only if} \quad y \in xS.$$

The relation $\prec$ is a preorder if and only if it is reflexive. This is the case if and only if S contains the identity. If $\mathbf{1} \in S$ then it agrees with the preorder $\prec_S$ introduced in Proposition V.0.4. We are, however, interested in the case that $\mathbf{1} \notin S$ (because if $\mathbf{1} \in S$ then S contains the identity component of G and the issue becomes trivial from the present point of view).

The open set $S \cap pU$ is a region of points $\prec$-above and near p. The main result of this section then says that to every point $p \in \overline{S}$ there is a region of points $\prec$-above and near p which is foliated by the equivalence classes modulo κ.

The Foliation Lemma

We start our investigations by introducing a notation for the set obtained by transporting κ-classes to the identity via right translations.

V.7.1.   Definition. For $s \in S$ and $x \in \overline{S}$ we write $K_s = \kappa(s)s^{-1}$. The set $L_0(\exp^{-1} K_s)$ of subtangent vectors of K_s at the origin will be denoted with $\mathbf{L}[s]$. We define

(1)
$$I \stackrel{\text{def}}{=} \bigcap_{s \in S} \mathbf{L}[s].$$

Clearly, $\mathbf{L}[s]$ is a closed subset of $\mathbf{L}(G)$ which is closed under non-negative scalar multiplications. Hence I is a closed subset satisfying $\mathbb{R}^+ \cdot I = I$. Most of our efforts will focus on this set.

V.7.2.   Lemma. (i) *If $s \prec t$ then $K_s \subseteq K_t$ and thus $\mathbf{L}[s] \subseteq \mathbf{L}[t]$.*

(ii) *The set $(S, \prec)$ is filtered. That is, for each pair $t, t' \in S$, there is an $s \in S$ with $s \prec t$ and $s \prec t'$.*

(iii) *The families $\{K_s \colon s \in S\}$ and $\{\mathbf{L}[s] \colon s \in S\}$ are filter bases.*

Proof.   (i) If $s \prec t$, then there is an $s' \in S$ with $t = ss'$. Since κ is a congruence, $\kappa(s)s' \subseteq \kappa(ss')$ and thus $\kappa(s) \subseteq \kappa(ss')s'^{-1}$. Therefore $K_s = \kappa(s)s^{-1} \subseteq \kappa(ss')s'^{-1}s^{-1} = K_{ss'} = K_t$. The remainder then follows.

(ii) If $t, t' \in S$, then tS^{-1} and $t'S^{-1}$ are open neighborhoods of 1 in G. Since $1 \in \overline{S}$ we find an element $s \in tS^{-1} \cap t'S^{-1} \cap S$. Then $t \in sS$ and thus $s \prec t$. Similarly, $s \prec t'$.

Assertion (iii) is a consequence of (i) and (ii).   ∎

EV.7.1.   **Exercise.**   Prove the following complements to the propositions of the preceding lemma:

(i) Every sequence $(s_n)_{n \in \mathbb{N}}$ in S converging to 1 is $\prec$-cofinal in S. Consequently, $\{K_{s_n} : n \in \mathbb{N}\}$ is cofinal in $\{K_s : n \in \mathbb{N}\}$, and $\{\mathbf{L}[s_n] : n \in \mathbb{N}\}$ is cofinal in $\{\mathbf{L}[s] : n \in \mathbb{N}\}$.

(ii) If $(s_n)_{n \in \mathbb{N}}$ is any sequence in S converging to $\mathbf{1}$, then there is a an increasing sequence $\big(n(k)\big)_{k \in \mathbb{N}}$ of natural numbers such that $k \mapsto s_{n(k)}$ is decreasing with respect to $\prec$.

(Hints: (i) Let $s \in S$. Then sS^{-1} is an open neighborhood of 1 in G, and since s_n converges to 1, there is an n so that $s_n \in sS^{-1}$. Then $s \in s_n S$, whence $s_n \prec s$. The remainder of the assertion follows from V.7.2(i).

(ii) Set $n(1) = 1$ and suppose that $n(j)$ is constructed for $j = 1, \ldots, k$ so that $n(j) < n(j+1)$ and $s_{n(j+1)} \prec s_{n(j)}$ for $j = 1, \ldots, k-1$. Since $\lim s_n = \mathbf{1}$, the sequence $(s_n)_{n \in \mathbb{N}}$ is cofinal in S by (iv). Hence we find a natural number $n(k+1) > n(k)$ such that $s_{n(k+1)} \prec s_{n(k)}$. By recursion, we construct the sequence $n(k)$, $k = 1, \ldots$ such that $k \mapsto n(k)$ is strictly increasing and $k \mapsto s_{n(k)}$ is $\prec$-decreasing.)

Our next lemma is purely algebraic.

V.7.3.   **Lemma.**   (i) *Suppose $s \prec t$ in S. If $g_1, \ldots, g_n \in K_s$ are such that $s \prec g_k \cdots g_1 t$ for $k = 1, \ldots, n$, then $g_n \cdots g_1 \in K_t$.*

(ii) *If $g \in K_s$ and $gs \prec t$, then $g^{-1} \in K_t$.*

Proof.   (i) We prove the inclusion $g_k \cdots g_1 \in K_t$, that is, $g_k \cdots g_1 t \sim_\kappa t$ for $k = 1, \ldots, n$ by induction. For $k = 1$ the assertion follows from Lemma V.7.2(i). Suppose that $g_k \ldots g_1 t \sim_\kappa t$ for $k < n$. By hypothesis, there is an $s_k \in S$ such that $ss_k = g_k \ldots g_1 t$. Then $ss_k \sim_\kappa t$. Also $g_{k+1} \in K_s$, that is, $g_{k+1} s \sim_\kappa s$. Hence $g_{k+1} ss_k \sim_\kappa ss_k$, since κ is a congruence. Transitivity now implies $g_{k+1} ss_k \sim_\kappa t$ and thus $g_{k+1} g_k \ldots g_1 t \sim_\kappa t$ which we had to show.

(ii) Since $gs \prec t$ there is an $s' \in S$ with $gss' = t$, that is $s^{-1} g^{-1} t = s'$. Also, $g \in K_s$ means $gs \sim_\kappa s$. Now $g^{-1} t = ss' \sim_\kappa gss' = gss^{-1} g^{-1} t = t$, which we had to show.   ∎

In the sequel we need a lemma on tangent sets which belongs to the general context of subtangent vectors (cf. Definition I.5.1 and Proposition I.5.2; also Definition IV.1.15 through Lemma IV.1.20).

V.7.4.   **Lemma.**   (i) *Let A be a subset of a Banach space with $0 \in \overline{A}$. If X is a subtangent vector of A at 0, that is, $X \in L_0(A)$, then there exist sequences*

$(k_n)_{n\in\mathbb{N}}$ of natural numbers and $(X_n)_{n\in\mathbb{N}}$ of elements in A such that

(2) $$X = \lim nk_n{\cdot}X_n.$$

In particular,

(3) $$\lim k_n{\cdot}X_n = 0.$$

(ii) *If L is a Dynkin algebra and A, A' are subsets of L containing 0 in their closures, and if $X \in L_0(A)$, $X' \in L_0(B)$, then there exist sequences $(k_n)_{n\in\mathbb{N}}$, $(k'_n)_{n\in\mathbb{N}}$ of natural numbers and $(X_n)_{n\in\mathbb{N}}$, $(X'_n)_{n\in\mathbb{N}}$ of elements in A and A', respectively, such that*

(4) $$X + X' = \lim n(k_n{\cdot}X_n * k'_n{\cdot}X'_n).$$

Proof. (i) Suppose that $Y \in L_0(A)$. Then by Definition IV.1.15 there are real numbers $r_n > 0$ such that $Y = \lim r_n{\cdot}Y_n$ with a sequence $Y_n \in A$ tending to 0. Now r_n grows beyond bounds and $r_n = k_n + d_n$ with the largest natural number k_n not exceeding r_n and $0 \le d_n < 1$. Then $\lim d_n{\cdot}Y_n = 0$ and thus $Y = \lim r_n{\cdot}Y_n = \lim k_n{\cdot}Y_n$.

Let $n \in \mathbb{N}$. By Remark IV.1.16, $\frac{1}{n}{\cdot}X \in L_0(A)$, and thus by what we just observed, there is a natural number k_n and an element X_n such that $\|\frac{1}{n}{\cdot}X - k_n{\cdot}X_n\| < 1/n^2$. Then $\|X - nk_n X_n\| < 1/n$, whence the claim.

(ii) We know that whenever $X = \lim n{\cdot}Z_n$ and $X' = \lim n{\cdot}Z'_n$, then $X + X' = \lim n(Z_n * Z'_n)$. By Part (i) of the Lemma we know $X = \lim nk_n{\cdot}X_n$ and $X' = \lim nk'_n{\cdot}X'_n$ with natural numbers k_n, k'_n and $X_n \in A$ and $X'_n \in A'$. Now we take $Z_n = k_n{\cdot}X_n$, $Z'_n = k'_n{\cdot}X'_n$ and immediately obtain the assertion. ∎

V.7.5.   Lemma. *For each element $s \in S$ there exists an open neighborhood B of 0 in $\mathbf{L}(G)$ such that*

(5) $$((I \cap B) + (I \cap B)) \cap B \subseteq L[s] \cap \exp^{-1} K_s.$$

Proof. We fix a Campbell–Hausdorff neighborhood B_0 of $\mathbf{L}(G)$. Next we find an element $s' \in S$ with $s' \prec s$ by Lemma V.7.2.(ii). Thus $s = s's''$ for some $s'' \in S$ and therefore $s'Ss^{-1} = s'Ss''^{-1}s'^{-1}$ is an open neighborhood of $\mathbf{1}$. Then we can take for B an open convex symmetric neighborhood such that $B * B * B \subseteq B_0$ and that

(∗) $$B * B * B \subseteq \exp^{-1}(s'Ss^{-1}).$$

Furthermore, since K_s is closed in S and S is open in G, we may choose B so small that also

(∗∗) $$\overline{B \cap \exp^{-1} K_s} \subseteq exp^{-1}K_s.$$

Now we let $X, X', X + X' \in B$ with $X, X' \in I$. Then, by the definition of I we know $X, X' \in L[s']$. This means, in view of Lemma V.7.4, there are sequences of natural numbers k_n, k'_n and elements X_n, $X'_n \in \mathbf{L}(G)$ such that

(6)
$$X = \lim nk_n \cdot X_n, \qquad \exp X_n \in K_s,$$
$$X' = \lim nk'_n \cdot X'_n, \qquad \exp X'_n \in K_{s'},$$

and

(7)
$$X + X' = \lim n \cdot (k_n \cdot X_n * k'_n \cdot X'_n).$$

Because of (6), the sequences $k_n \cdot X_n$ and $k'_n \cdot X'_n$ converge to 0, and in conjunction with (7) we can now pick a natural number N so large that

(8)
$$k_n \cdot X_n, \ k'_n \cdot X'_n, \ n \cdot (k_n \cdot X_n * k'_n \cdot X'_n) \in B \text{ for all } n \geq N.$$

Let us fix $n \geq N$. Since B is convex, we conclude

(9)
$$j \cdot X_n, \ j' \cdot X'_n, \ k \cdot (k_n \cdot X_n * k'_n \cdot X'_n) \in B \text{ for all } j \leq k_n, \ j' \leq k'_n, \ k \leq n.$$

Now we consider the product

$$n \cdot (k_n \cdot X_n * k'_n \cdot X'_n) = \underbrace{X_n * \cdots * X_n}_{k_n \text{ times}} * \underbrace{X'_n * \cdots * X'_n}_{k'_n \text{ times}} * \cdots * X'_n,$$

which is a product of $m = n(k_n + k'_n)$ factors $F_m * F_{m-1} * \cdots * F_2 * F_1$ with $F_p \in \{X_n, X'_n\}$, $p = 1, \ldots, m$. The partial products $F_p * \cdots * F_1$ have the form

$$F_p * \cdots * F_1 = \begin{cases} j' \cdot X'_n * k(k_n \cdot X_n * k'_n \cdot X'_n), & \text{with } j' \leq k'_n, \ k \leq n, \text{ or} \\ j \cdot X_n * k'_n \cdot X'_n * k(k_n \cdot X_n * k'_n \cdot X'_n) & \text{with } j \leq k_n, \ k \leq n. \end{cases}$$

In the first case, $F_p * \cdots * F_1 \in B * B$ by (9), and in the second case $F_p * \cdots * F_1 \in B * B * B$ by (9). Thus

(10)
$$\exp F_p \cdots \exp F_1 = \exp(F_p * \cdots * F_1) \in \exp(B * B * B) \subseteq s'Ss^{-1}$$

for all $p = 1, \ldots, m$ by $(*)$. But also

(11)
$$\exp F_p \in \{\exp X_n, \exp X'_n\} \subseteq K_{s'}$$

by (6). Since $p \in s'Ss^{-1}$ says $s' \prec ps$, conditions (10) and (11) secure the hypotheses of Lemma V.7.3(i); the Lemma applies and yields

$$\exp(k_n \cdot X_n * k'_n \cdot X'_n) = F_{k_n + k'_n} * \cdots * F_1 \in K_s$$

and

$$\exp\big(n \cdot (k_n \cdot X_n * k'_n \cdot X'_n)\big) = F_m * \cdots * F_1 \in K_s.$$

Thus we conclude

(12)
$$X + X' = \lim n \cdot (k_n \cdot X_n * k'_n \cdot X'_n) \in L_0(\exp^{-1} K_s) \cap \exp^{-1} K_s$$

since $B \cap \exp^{-1} K_s$ is closed in B by $(**)$. This completes the proof of the lemma.∎

V.7.6.   Lemma.   *For each $s \in S$ there is a neighorhood B of 0 in $\mathbf{L}(G)$ such that $-B \cap I \subseteq \mathbf{L}[s] \cap \exp^{-1} K_s$.*

Proof.   We choose $s' \prec s$ and B as in the proof of the preceding Lemma V.7.5. but with an additional requirement. Recall that $s = s's''$, whence $sS^{-1}s'^{-1} = s's''S^{-1}s'^{-1}$ is an open neighborhood of $\mathbf{1}$. Hence we may choose B in fact so small that also

(13) $$B \subseteq \exp^{-1}(sS^{-1}s'^{-1}).$$

Now let $X \in B \cap I$. Then $X \in \mathbf{L}[s']$ and thus, as in the proof of the preceding Lemma, $X = \lim nk_n \cdot X_n$ with integers k_n and $\exp X_n \in K_s$. If we choose N so large that $n \geq N$ implies $nk_n \cdot X_n \in B$, then, by the convexity of B again, $X_n \in B$. Since B is symmetric,

$$-X_n \in B.$$

By (13), $g = \exp X_n$ satisfies $g^{-1} = \exp(-X_n) \in sS^{-1}s'^{-1}$, that is, we have $gs \in s'S$. Consequently,

$$(\exp X_n)s \prec s' \text{ for all } n \geq N.$$

By Lemma V.7.3(ii), this implies

(14) $$\exp(-X_n) \in K_s \text{ for all } n \geq N.$$

Thus $-X = \lim nk_n \cdot (-X_n) \in L[s]$ by definition of $\mathbf{L}[s]$. Likewise, since $B \cap \exp^{-1} K_s$ is closed in B we also have $-X \in \exp^{-1} K_s$.   ∎

Now we are ready for the crucial conclusion.

V.7.7.   Lemma.   (A Key Lemma) *The set I is a vector subspace of $\mathbf{L}(G)$, and for each $s \in S$ there is an open neighborhood B of 0 in $\mathbf{L}(G)$ such that*

(15) $$\exp(I \cap B) \subseteq K_s.$$

Proof.   Using Lemma V.7.5, for each $s \in S$ we find an open neighborhood B such that

$$\big((I \cap B) + (I \cap B)\big) \cap B \subseteq \mathbf{L}[s].$$

Since $\mathbb{R}^+ \cdot \mathbf{L}[s] = \mathbf{L}[s]$ we conclude

$$I + I \subseteq \mathbf{L}[s] \text{ for all } s \in S.$$

By the definition of I as intersection of all $\mathbf{L}[s]$ we thus have

(16) $$I + I \subseteq I.$$

By Lemma V.7.6,

$$-I \cap B \subseteq \mathbf{L}[s].$$

Once again by $\mathbb{R}^+ \cdot I = I$ it follows that

$$\mathbb{R} \cdot I \subseteq \mathbf{L}[s] \text{ for all } s \in S.$$

Thus

(17) $$\mathbb{R} \cdot I \subseteq I.$$

In Lemma V.7.5 it was also shown that

(18) $$I \cap B \subseteq \exp^{-1} K_s.$$

The Key Lemma now follows from (16),(17), and (18).   ∎

There is an alternative and more direct way of characterizing the elements of I. We leave the details as an exercise.

EV.7.2. **Exercise.** An element $X \in L(G)$ is in I if and only if there are sequences k_n of natural numbers and X_n, Y_n of elements in $\mathbf{L}(G)$ such that $X = \lim k_n(X_n * -Y_n)$, $\lim X_n = \lim Y_n = 0$, and $\exp X_n \sim_\kappa \exp Y_n$.

This characterization allows the direct conclusion that $-I = I$ and thus that $\mathbb{R} \cdot I = I$ without the preceding two lemmas. ∎

We now formulate and prove a simple lemma on vector spaces.

V.7.8. **Lemma.** *Let L denote a finite-dimensional vector space und $\mathcal{F}$ a filter basis of closed subsets C satisfying $\mathbb{R}^+ \cdot C = C$. If $D = \bigcap \mathcal{F}$ is a vector space, then there is a $C \in \mathcal{F}$ with $D = C$.*

Proof. Passing to the factor space $\mathbf{L}/D$ we may assume that $D = \{0\}$ and then prove that there is a $C \in \mathcal{F}$ such that $C = \{0\}$. Assume that this is not the case. Let Σ be the boundary sphere of some compact ball neighborhood of 0 with respect to any norm on L. Now $\{C \cap \Sigma : C \in \mathcal{F}\}$ is a filter basis of compact sets due to the fact that $C \cap \Sigma \neq \emptyset$ as $\mathbb{R}^+ C = C \neq \{0\}$. Hence it has an element $s \in S$ in its intersection, and this is a contradiction to $D = \{0\}$. ∎

An immediate application of this lemma to the filterbasis $\mathcal{F} = \{\mathbf{L}[s] : s \in S\}$ (see Lemma V.7.2(iii)) shows that there is an $s \in S$ such that $I = \mathbf{L}[s]$. If $s' \prec s$, then this implies $I = \mathbf{L}[s']$. Since $\{s' : s' \prec s\} = sS^{-1}$, this set is an open neighborhood of 1 in G. We therefore have

V.7.9. **Lemma.** *There is an open neighborhood U of 1 in G such that*

(19) $$\mathbf{L}[u] = I \quad \text{for all} \quad u \in S \cap U.$$

∎

V.7.10. **Lemma.** *For each $u \in U$ with U as in Lemma V.7.9 there is a neighborhood B of 0 in $\mathbf{L}(G)$ such that $\exp(I \cap B)$ is a neighborhood of 1 in K_u.*

Proof. Assume the contrary. Then we find a $u \in U$ and a sequence $X_n \to 0$ such that $\exp X_n \in K_u$ but $X_n \notin I$. Now we let E denote a vector space complement for I in $\mathbf{L}(G)$, and we write $X_n = X'_n * X''_n$ with $X'_n \in E$, $X''_n \in I$. (This decomposition is possible and unique since for sufficiently small 0-neighborhoods $B' \subseteq E$ and $B'' \subseteq I$ the map

(20) $$(X', X'') \mapsto X' * X'' : B' \times B'' \to B' * B''$$

is a diffeomorphism onto a neighborhood of 0 in $\mathbf{L}(G)$.) Since $X_n \notin I$ we have $X'_n \neq 0$. Now $\exp X'_n = \exp(X_n * -X''_n) = (\exp X_n)(\exp -X''_n)$. Further $\exp(-X''_n) \in K_u$ for all sufficiently large n by Lemma V.7.7(15). Hence $\exp(-X''_n)u \sim_\kappa u$ and thus $\exp X'_n = (\exp X_n)(\exp -X''_n) \sim_\kappa (\exp X_n)u \sim_\kappa u$ since $\exp X_n \in K_u$. Therefore

(21) $$\exp X'_n \in K_u \quad \text{and} \quad 0 \neq X'_n \in E.$$

If we fix any norm in $\mathbf{L}(G)$, then the sequence $\|X'_n\|^{-1} \cdot X'_n$ has a cluster point Y with $\|Y\| = 1$, hence $Y \neq 0$, and with $Y \in L_0(\exp^{-1} K_u) = \mathbf{L}[u] = I$. But also $Y \in E$. Hence $0 \neq Y \in I \cap E = \{0\}$, a contradiction. This contradiction proves the claim. ∎

V.7.11. **Lemma.** *The vector space I is a subalgebra of $\mathbf{L}(G)$.*

Proof. We take U, u, and B as in Lemma V.7.10. Let V be a neighborhood of 1 in G such that $V \cap K_u \subseteq \exp(I \cap B)$; such a V exists by Lemma V.7.10.

Fix $Y \in B \cap I \cap \exp^{-1} U u^{-1} \cap \exp^{-1} V$. Then $(\exp Y)u \sim_\kappa u$. Next let X be an arbitrary element of I. By Lemma V.7.10, there exists a $T > 0$ such that

$$\exp t \cdot X \in K_{(\exp Y)u} \cap V \exp -Y \qquad \text{and} \qquad t \cdot X * Y \in B$$

for all $|t| < T$. Then, since $(\exp t \cdot X \exp Y)u \sim_\kappa (\exp Y)u \sim_\kappa u$, we have

$$\exp(tX * Y) \in K_u \cap V \subseteq \exp(B \cap I) \qquad \text{for all } |t| < T.$$

Thus

$$\varphi(t) \stackrel{\text{def}}{=} t \cdot X * Y \in I \qquad \text{for all } |t| < T.$$

We compute $\dot{\varphi}(0)$: Notice first that $\varphi(t) = e^{-\operatorname{ad} Y}(Y * t \cdot X)$. By Lemma II.2.4 we then have $\dot{\varphi}(0) = e^{-\operatorname{ad} Y} g(\operatorname{ad} Y)(X) = X - \frac{1}{2} \cdot [Y, X] + \mathrm{o}(Y)(X)$. Since $\dot{\varphi}(0) \in I$ and $X \in I$ we conclude that $[Y, X] + \mathrm{o}(Y)(X) \in I$.

We have shown that there is a 0-neighborhood B' in I such that for all $Y \in B'$ and all $X \in I$

$$[Y, X] + \mathrm{o}(Y)(X) \in I.$$

Now let $Z \in I$ be arbitrary. Then there is a $\tau_0 > 0$ such that for all $|\tau| < \tau_0$ we have $\tau \cdot Z \in B'$. Thus for $0 < |\tau| < \tau_0$ it follows that

$$[Z, X] + \tau^{-1} \mathrm{o}(\tau \cdot Z)(X) \in I.$$

Since $\lim_{\tau \to 0} \tau^{-1} \mathrm{o}(\tau \cdot Z) = 0$, we obtain $[Z, X] \in I$. Since X and Z were arbitrary in I, this completes the proof. ∎

In view of $K_u = \kappa(u)u^{-1}$ the results so far obtained can be summarized as follows:

V.7.12. **Lemma.** (The Foliation Lemma) *Suppose that G is a finite-dimensional Lie group and S an open subsemigroup of G with $1 \in \overline{S}$. Suppose that κ is a right congruence with closed classes on S. Then there exists a subalgebra I of $\mathbf{L}(G)$ and an open neighborhood U of 1 in G such that for each $u \in U \cap S$ there is an open cell neighborhood B_u of 0 in I for which $(\exp B_u)u$ is a neighborhood of u in $\kappa(u)$.* ∎

One could express this result by saying that under the assumptions of the lemma, the manifold $U \cap S$ *is foliated by the equivalence classes of κ and that this foliation is induced by the foliation of G determined by the cosets Hg in G where H is an analytic subgroup of G, namely, the analytic subgroup H with $I = \mathbf{L}(H)$.* We shall make this statement more precise in the *Foliation Theorem* V.7.21 below.

Consequences of the Foliation Lemma

Let us now gather some important consequences of the Foliation Lemma. To this end we first recall the general fact that the connected components of the classes of an equivalence relation μ on a topological space T are themselves the classes of an equivalence relation μ_0. If μ has closed classes, then the same holds for μ_0 since the components of a topological space are closed. If T_1 is a subspace of T, then we denote with $\mu|T_1$ the induced equivalence relation $\mu \cap (T_1 \times T_1)$. In the following, given an open set U of G, *we shall abbreviate the equivalence relation* $\big(\kappa|(U \cap S)\big)_0$ *with* κ'.

V.7.13.   Lemma.   *With the notation of the* Foliation Lemma V.7.12, *for each* $u \in U \cap S$, *the class* $\kappa'(u)$ *is a closed connected submanifold of* $\kappa(u) \cap U$ *and is the connected maximal integral submanifold on* $U \cap S$ *for the distribution on* G *which assigns to* $g \in G$ *the vector subspace* $d\rho_g(\mathbf{1})(I)$ *with the right translation* $\rho_g = (x \mapsto xg)$.

Proof.   If H is the analytic subgroup generated by $\exp I$, then H is a maximal integral submanifold through $\mathbf{1}$ for the given distribution. Accordingly, Hu is a maximal connected integral submanifold for this distribution through u. If $u \in U \cap S$ then, with the notation of Lemma V.7.12, $(\exp B_u)u$ is an open connected submanifold of Hu *in its intrinsic manifold topology* and it also belongs to $\kappa(u)$ by the Foliation Lemma V.7.12. Hence for each $u \in U \cap S$, the closed subset $\kappa(u) \cap U$ is an integral manifold on $U \cap S$ of the distribution. Then, by definition of κ', the classes $\kappa'(u)$ with $u \in U \cap S$ are the maximal connected integral submanifolds of the distribution $g \mapsto d\rho_g(\mathbf{1})(I)$ on $U \cap S$.   ∎

In other words, for all elements $u \in U \cap S$, the connected components of $\kappa(u) \cap U$ are uniquely determined by the subalgebra I and the distribution determined by I on G. The decomposition of $\kappa(u) \cap U$ itself into connected components depends also on the geometric shape of $U \cap S$.

The preceding lemma has a very useful consequence. In dealing with a congruence on S, supposing that translates gx and gy of κ-congruent elements $x, y \in S$ are still in S we cannot in general conclude that gx, gy are congruent, unless $g \in S$. In the present situation, however, we have the following fact:

V.7.14.   Lemma.   *Let everything be as in the* Foliation Lemma V.7.12. *If* $x \sim_{\kappa'} y$ *and* C *is a smooth path joining* x *and* y *in* $\kappa'(x)$, *then for all* $g \in G$ *with* $Cg \subseteq U \cap S$ *we have* $xg \sim_{\kappa'} yg$, *in particular,* $xg \sim_\kappa yg$.

Proof.   In the following, we write λ_g and ρ_g for the left and the right translations by elements $g \in G$. Let $c: [0, 1] \to \kappa'(x)$ with $c(0) = x$ and $c(1) = y$ be a smooth curve such that $\dot{c}(t) \in d\rho_{c(t)}(\mathbf{1})(I)$. If we consider an arbitrary $g \in G$ and define

$c_g(t) = c(t)g$, then

$$\frac{d}{ds}c_g(s)|_{s=t} = d\rho_g\big(c(t)\big)\dot{c}(t) \in d\rho_g\big(c(t)\big)d\rho_{c(t)}(\mathbf{1})(I)$$

$$= d\rho_{c(t)g}(\mathbf{1})(I) = d\rho_{c_g(t)}(\mathbf{1})(I).$$

Thus $c_g([0,1])$ is contained in the integral manifold through $c_g(0) = xg$ given by the distribution $g \mapsto d\rho_g(\mathbf{1})(I)$. If g is such that $c_g([0,1]) \subseteq U \cap S$, then $c_g([0,1])$ is contained in the integral manifold through xg for this distribution *on the open submanifold* $U \cap S$, which is $\kappa'(xg)$ by Lemma V.7.13 above. Hence $c_g([0,1]) \subseteq \kappa'(xg)$ and thus $xg \sim_{\kappa'} yg$ as asserted.   ∎

Since the component $\kappa'(x)$ is a manifold of the kind described by Lemma V.7.13, a curve such as C in the preceding lemma exists for any pair of κ'-related elements x and y in $U \cap S$.

As a consequence of Lemma V.7.14, the κ'-saturations of open subsets of $U \cap S$ are open:

V.7.15.   Lemma. *Let V be an open subset of $U \cap S$, and let $\kappa'(V)$ denote the saturation $\bigcup_{v \in V} \kappa'(v)$ of V with respect to κ' on $U(S)$. Then $\kappa'(V)$ is open. In particular, the quotient map $p: U \cap S \to (U \cap S)/\kappa'$ is open.*

Proof.   We consider $s \in \kappa'(V)$; we must show that there is a whole open neighborhood W of s in S with $W \subseteq \kappa'(V)$. By assumption there is a $v \in V$ with $s \sim_{\kappa'} v$. Since the component $\kappa'(v)$ is an integral manifold of through v for the distribution determined by I on $U \cap S$ there is is a smooth curve $c: [0,1] \to \kappa'(v)$ with $c(0) = v$ and $c(1) = s$ such that $\dot{c}(t) \in d\rho_{c(t)}(1)(I)$ where $\rho_g(x) = xg$ in G.

Since $c([0,1])$ is compact there is an open neighborhood V_0 of $\mathbf{1}$ in G such that $V_0 c([0,1]) \in U(S)$ and that $V_0 v \subseteq V$. By the preceding, every element in the open set $W = V_1 c([0,1])$ is κ'-related to some element in $V_0 v \subseteq V$. However, W contains $s = c(1)$ and thus is the required open neighborhood of s.   ∎

The preceding results have global consequences for the relations between κ-classes and cosets $Hs \cap S$, where H is the analytic subgroup generated by $\exp I$.

V.7.16.   Proposition. *Let G be a Lie group with an open subsemigroup S satisfying $1 \in \overline{S}$. Suppose that κ is a right-congruence on S with closed classes; let I be the subalgebra specified in the Foliation Lemma V.7.12 and H the analytic subgroup generated by $\exp I$. Denote with H_L the group H with its intrinsic Lie group topology. Then for any element $s \in S$ the set*

$$C_s = \{h \in H: \quad hs \in \kappa(s)\}$$

is an open closed neighborhood of 1 in the subspace

$$D_s = \{h \in H_L: hs \in S\}$$

of H_L. In particular, the identity component $(D_s)_0$ of D_s is contained in C_s.

Proof.   As a subset of G, the set D_s equals $H \cap Ss^{-1}$. Since $H \cap Ss^{-1} \cap \big(\kappa(s)\big)s^{-1}$ is closed and because the congruence class $\kappa(s)$ is closed, C_s is closed in D_s since the topology of H_L is finer than or equal to that of H. Now we show that C_s is open in D_s. Suppose that $h \in C_s$. Then $hs = \kappa(s)$. Let U be an open set in G such as it was constructed in the Foliation Lemma V.7.12. Now hsS^{-1} is an open neighborhood of 1 in G, whence $U \cap S \cap hsS^{-1} \neq \emptyset$. Hence we can write $hs = us'$ with suitable elements $u \in U$, $s' \in S$. By The Foliation Lemma V.7.12, there is an open neighborhood B_u of 0 in I such that $(\exp B_u)u$ is a neighborhood of u in $\kappa(u)$. It follows that $\big((\exp B_u)h\big)s = (\exp B_u)us' \subseteq \kappa(u)s' \subseteq \kappa(us') = \kappa(s)$. This implies that the open neighborhood $(\exp B_u)h$ of h in H_L is contained in C_s. Thus C_s is open as asserted. ∎

If H is the analytic subgroup generated by I, then the maximal integral manifolds for the distribution determined by I are the cosets Hg. Notice that one and the same leaf Hg of the foliation of G may—and will in general—intersect $U \cap S$ in countably many leaves. This pathology does not occur when H is closed. If H happens to be normal and $\pi_1(G)$ is finite, this is automatically the case. We shall return to this situation later when dealing with two-sided congruences.

V.7.17. **Corollary.** *Let G, S, κ and H be as in* Proposition V.7.16 *but assume H to be closed. Then for each $s \in S$ the set $Hs \cap \kappa(s)$ is open and closed in $Hs \cap S$. In particular, the connected component of s in the submanifold $Hs \cap S$ of Hs is contained in $\kappa(s)$.*

Proof. If H is closed then $H_L = H$, and Proposition V.7.16 says that $Hs \cap \kappa(s)$ is open and closed in Hs. ∎

V.7.18. **Corollary.** *If, in* Corollary V.7.17, *the set $Hs \cap S$ is connected, then $Hs \cap S \subseteq \kappa(s)$.* ∎

The Foliation Theorem

Our next goal is to strengthen the Foliation Lemma V.7.12 and to obtain the more precise informations of the Foliation Theorem. We retain the notation of the preceding section and start with a lemma on local cross sections.

V.7.19. **Lemma.** *Let $u \in (U \cap S)S^{-1}$ and E a vector space complement of I in $\mathbf{L}(G)$. Then there is an open neighborhood B' of 0 in E such that $C \overset{\text{def}}{=} (\exp B')u$ has the property that $c_1, c_2 \in C \cap U \cap S$ and $c_1 \sim_\kappa c_2$ imply $c_1 = c_2$. In other words, $C \cap U \cap S$ is a local cross section for the κ-classes. ($C \cap U \cap S$ may be empty if $u \notin U \cap S$!)*

Proof. Let $A_u \subseteq E$ and $B_u \subseteq I$ be neighborhoods of 0 such that the map $A_u \times B_u \to A_u * B_u$, $(X', X'') \mapsto X' * X''$, is a diffeomorphism onto a neighborhood of 0 in $\mathbf{L}(G)$, just as in the proof of Lemma V.7.10. We claim that if A_u is chosen sufficiently small, then for two elements $X, Y \in A_u$ the relations $\{(\exp X)u, (\exp Y)u\} \subset U \cap S$ and $(\exp X)u \sim_\kappa (\exp Y)u$ always imply $X = Y$.

Suppose that this claim is false. Then for any natural number n we find a neighborhood $A_u^{(n)}$ of 0 in A_u whose diameter is at most $1/n$ such that $A_u^{(n)}$

contains two different elements X_n and Y_n with $(\exp X_n)u, (\exp Y_n)u \in U \cap S$ and $(\exp X_n)u \sim_\kappa (\exp Y_n)u$. Since $0 \neq P_n = Y_n - X_n \in E$, we may, after passing to a subsequence, assume that the sequence $Z_n \overset{\text{def}}{=} \|P_n\|^{-1} \cdot P_n$ converges to a unit vector Z in E:

$$(22) \qquad\qquad Z = \lim Z_n \in E.$$

Since $u \in (U \cap S)S^{-1}$ there exist elements $v \in U \cap S$ and $s \in S$ with $u = vs^{-1}$. Then u is in the open set vS^{-1}, hence so is $(\exp X_n)u$ for all large enough n. We may and shall assume that this is the case for all $n \in \mathbb{N}$. Thus there are elements $s_n \in S$ with $(\exp X_n)us_n = v$; note that

$$(23) \qquad \lim(\exp X_n)us_n = v = \lim us_n = \lim(\exp Y_n)us_n \text{ and } \lim s_n = s,$$

since $\lim \exp X_n = \lim \exp Y_n = \mathbf{1}$. Moreover, $(\exp X_n)u \sim_\kappa (\exp Y_n)u$, hence $(\exp Y_n)us_n \sim_\kappa v$ for all $n \in \mathbb{N}$. Now the Foliation Lemma applies and shows that we can find elements $X_n'' \in B_u \subseteq I$ such that $(\exp Y_n)us_n = (\exp X_n'')v = (\exp X_n'')(\exp X_n)us_n$ for all sufficiently large n. Thus

$$\exp\bigl(Y_n * (-X_n)\bigr) = (\exp Y_n)(\exp X_n)^{-1} = \exp X_n''$$

and hence

$$(24) \qquad\qquad Y_n * (-X_n) = X_n'' \in I \text{ for all large enough } n.$$

Now we recall from general Lie theory that for all sufficiently small X, P in $\mathbf{L}(G)$ we have $(X + P) * (-X) = f(\operatorname{ad} -X)(P) + \mathrm{o}(P)$ with $f(-z) = (e^z - 1)/z = \sum_{n=0}^\infty \dfrac{z^n}{(n+1)!}$ (cf., e.g., [Bou75], Chap.II, §6, n° 5, Proposition 5.) Thus $\|P_n\|^{-1} \cdot \bigl(Y_n * (-X_n)\bigr) = \|P_n\|^{-1} \cdot (X_n + P_n) * X_n = f(\operatorname{ad} -X_n)\bigl(\|P_n\|^{-1} \cdot P_n\bigr) + \mathrm{O}(P_n)$ with $\lim \mathrm{O}(P_n) = 0$, that is,

$$(25) \qquad\qquad \|P_n\|^{-1} \cdot \bigl(Y_n * (-X_n)\bigr) = f(\operatorname{ad} -X_n)(Z_n) + \mathrm{O}(P_n)$$

for all sufficiently large n. Since $\lim X_n = 0$ we have $\lim f(\operatorname{ad} -X_n) = \mathbf{1}$ and thus (22) and (25) imply

$$(26) \qquad\qquad \lim \|P_n\|^{-1} \cdot \bigl(Y_n * (-X_n)\bigr) = Z \in E.$$

But (24) implies

$$Z = \lim \|P_n\|^{-1} \cdot \bigl(Y_n * (-X_n)\bigr) \in I.$$

Since $E \cap I = \{0\}$ we conclude $Z = 0$, a contradiction to the fact that Z is a unit vector. $\blacksquare$

V.7.20.   **Remark.** Suppose that U is an open neighborhood of $\mathbf{1}$ in G. If $u \in U \cap S$, then $u^{-1}(U \cap S)$ is an open neighborhood of $\mathbf{1}$ in G, hence contains an element $s \in S$. This means that $s = u^{-1}v$ for some $v \in U \cap S$ and thus $u = vs^{-1} \in (U \cap S)S^{-1} = \{g \in G : (\exists u)u \in U \cap S \text{ and } g \prec u\}$. Also, $\mathbf{1} = uu^{-1} \in (U \cap S)S^{-1}$. It is not clear that $\overline{U \cap S} \subseteq (U \cap S)S^{-1}$. But in this direction we have at least the inclusion

$$(U \cap S) \cup \overline{V \cap S} \subseteq (U \cap S)S^{-1},$$

where V is any compact subset of U. (If V is a compact subset of U then there is an open neighborhood W of $\mathbf{1}$ in G such that $VW \subset U$. Then for any $v \in \overline{V \cap S} = V \cap \overline{S}$ and any $w \in W \cap S$ we have $vw \in S$ (since S is an ideal of the semigroup $\overline{S}$), hence $v \in (S \cap U)(W \cap S)^{-1} \subseteq (S \cap U)S^{-1}$.)

Now we have the following main result:

V.7.21.   Theorem.   (The Foliation Theorem). *Suppose that G is a finite-dimensional Lie group and S an open subsemigroup of G with $1 \in \overline{S}$. Let κ be a right congruence with closed classes on S. Then there exists a subalgebra I of $\mathbf{L}(G)$ and an open neighborhood U of 1 in G such that to each $u \in U \cap S$ and for each vector space complement E for I in $\mathbf{L}(G)$ there are open neighborhoods A_u of 0 in I and B_u of 0 in E such that the function*

$$(27) \qquad \Phi_u : A_u \times B_u \to U \ given \ by \ \Phi_u(X,Y) = (\exp X)(\exp Y)u$$

is a diffeomorphism onto an open neighborhood V of u in U in such a fashion that $V \subseteq U \cap S$ and

$$(28) \qquad \Phi_u(A_u \times \{Y\}) = \kappa\big(\Phi_u(0,Y)\big) \cap V \ for \ all \ Y \in B_u.$$

EV.7.2.   Exercise.   Under the hypotheses of the Foliation Theorem V.7.21, one also obtains information for points u which are not necessarily in $S \cap U$ but in the boundary of $S \cap U$. Indeed, I, U, and E may be chosen so that, in addition to the conclusion of V.7.21, for all $u \in (U \cap S)S^{-1} \setminus U \cap S$ we still have the following conclusion:

For all $X \in A_u$ and $Y \in B_u$ with $\Phi_u(X,Y) \in S$, the diffeomorphism Φ_u maps the connected component of (X,Y) in $(A_u \times \{Y\}) \cap \Phi_u^{-1}(S)$ diffeomorphically onto the component of $\Phi_u(X,Y)$ in $\kappa\big(\Phi_u(X,Y)\big) \cap V \cap S$.

Proof of the Foliation Theorem.   (The proof is organized in such a manner that it allows a proof of Exercise EV.7.2 at the same time.)

We choose A_u and B_u so small that B_u satisfies the hypotheses of Lemma V.7.20, that $\big(\exp(A_u * B_u)\big)u \subseteq U$, and that the function Φ_u defined in (27) is a diffeomorphism onto its image V. Furthermore we require that $V \subseteq U \cap S$ if $u \in S \cap S$, and that A_u is connected (for instance is an open cell). Let $Y \in B_u$. We consider the set

$$M_Y \overset{\text{def}}{=} \{X \in A_u : \quad \Phi_u(X,Y) \in \kappa\big(\Phi_u(0,Y)\big)\}.$$

First we observe that $\Phi_u(M_Y \times \{Y\}) = \Phi_u(A_u \times \{Y\}) \cap \kappa\big(\Phi_u(0,Y)\big)$. Since κ-classes are closed, this set is closed in $\Phi_u\big((A_u \times \{Y\}) \cap \Phi_u^{-1}(S)\big)$. Since Φ_u is a diffeomorphism, this shows that M_Y is closed in the open submanifold $\{X'' \in A_u : \Phi_u(X'',Y) \in S\}$ of the open cell A_u. Next we claim that M_Y is open in A_u. For a proof of this claim pick $X \in M_Y$. Then $\Phi_u(0,Y) \sim_\kappa \Phi_u(X,Y)$ by our definition of M_Y. By the Foliation Lemma V.7.12, there is an open zero neighborhood A'' in I such that $(\exp A'')\Phi_u(X,Y)$ is a neighborhood of $\Phi_u(X,Y)$ in $\kappa\big(\Phi_u(X,Y)\big)$. Let us consider an element $Z \in A''$. Then, on one hand $(\exp Z)\Phi_u(X,Y) \sim_\kappa \Phi_u(X,Y)$. Since $\Phi_u(X,Y) \sim_\kappa \Phi_u(0,Y)$ this implies

$$(29) \qquad (\exp Z)\Phi_u(X,Y) \sim_\kappa \Phi_u(0,Y).$$

On the other hand, if $Z \in \mathbf{L}(G)$ is not too large then

$$(30) \qquad \begin{aligned} (\exp Z)\Phi_u(X,Y) &= (\exp Z)(\exp X)(\exp Y)u \\ &= \big((\exp(Z * Y))\big)(\exp Y)u. \end{aligned}$$

If, in addition, $Z \in I$ then, since I is a subalgebra, $Z * X \in I$. Thus, for all sufficiently small $Z \in A''$, we find

$$(31) \qquad (\exp Z)\Phi_u(X, Y) = \Phi_u(Z * X, Y) \in \Phi_u(A_u \times \{Y\}).$$

In view of (29) and the definition of M_Y we find $Z * X \in M_Y$. Hence an entire neighborhood of X in I is contained in M_Y. This shows that M_Y is also open in A_u. Thus $M_Y \times \{Y\}$ is open and closed in the open submanifold $\{(X'', Y) : \Phi_u(X'', Y) \in S\}$ of the open cell $A_u \times \{Y\}$, and therefore contains the connected component of (X, Y) in $(A_u \times \{Y\}) \cap \Phi_u^{-1}(S)$. If $u \in U \cap S$, since $\Phi_u(A_u \times \{Y\}) \subseteq S$ and since A_u was assumed to be connected, we have $M_Y = A_u$. Thus the left side of (28) is contained in the right side. In order to prove the converse, assume that $v \in \kappa(\Phi_u(0, Y)) \cap V$. We write v in the form $v = \Phi_u(X', Y')$ with $X' \in A_u$ and $Y' \in B_u$. By what we have just seen, $v \sim_\kappa \Phi(0, Y')$. By the transitivity of κ we now know $\Phi_u(0, Y) \sim_\kappa \Phi_u(0, Y')$. By Lemma V.7.13, in view of the choice of B_u, this relation implies $\Phi_u(0, Y) = \Phi_u(0, Y')$, and since Φ_u is a diffeomorphism, $Y = Y'$. Thus $v = \Phi_u(X', Y) \in \Phi_u(A_u \times \{Y\})$. This completes the proof of Theorem V.7.21.

(For a proof of the exercise, observe that the component of $\Phi_u(X, Y)$ in $\kappa(\Phi_u(X, Y)) \cap V$ is the maximal integral manifold through $\Phi_u(X, Y)$ on the manifold $V \cap S$ of the distribution given by I. But in V, the maximal integral manifold through $\Phi_u(X, Y)$ in V is $\Phi_u(A_u \times \{Y\})$, as (31) shows. Hence this component is contained in $\Phi_u(A_u \times M_Y)$, and thus in the image under Φ_u of the component of $(A_u \times \{Y\}) \cap \Phi_u^{-1}(S)$.) ∎

Note that the stronger assumption $u \in U \cap S$ of the theorem yields the much stronger conclusion (28). The conclusion of the exercise is, in reality, only an assertion about κ'.

We now assume the situation of the Foliation Theorem V.7.21 and suppose, in addition, that the analytic subgroup H generated by $\exp I$ is closed. Let κ_H denote the right-congruence on S whose cosets are $\kappa_H(s) = Hs \cap S$. From Lemma V.7.14 we know that for *the open set U of the Foliation Theorem we have $\kappa' = \kappa'_K$* with $\kappa' = (\kappa|U \cap S)_0$ and $\kappa'_H = (\kappa_H|U \cap S)_0$.

V.7.22.   **Proposition.** *Let G be a Lie group, H a closed connected subgroup and S an open subsemigroup with $1 \in \overline{S}$. Let U be an open neighborhood of 1. Then the map sending the component of u in $Hu \cap U \cap S$ to Hu is a local homeomorphism $\pi : (U \cap S)/\kappa'_H \to H(U \cap S)/H$.*

Proof.   Let $u \in U \cap S$ and let E be a vector space complement for I in $\mathbf{L}(G)$. Choose an open neighborhood A_u of 0 in I and an open neighborhood B_u of 0 in E such that $\Phi_u : A_u \times B_u \to V_u$ is a diffeomeorphism with the properties given in the Foliation Theorem V.7.21 and assume, in addition, that $\exp B_u$ is a cross section for the orbits $H(\exp Y)$ with $Y \in B_u$. This additional property of B_u can be secured by the fact that H is closed. Now we apply Lemma V.7.15 to κ_H and conclude that the saturation $\kappa'_H(V_u)$ is open, whence $W_u \stackrel{\text{def}}{=} \kappa'_H(V_u)/\kappa'_H$ is an open neighborhood of $\kappa'_H(u) \in (U \cap S)/\kappa'_H$. The map $\alpha = \{Y \mapsto (H(\exp Y)u)\} : B_u \to HV_u/H$ is bijective, continuous and open since $(\exp B_u)u$ is a cross section for the H-orbits and the map $\beta = \{Y \mapsto \kappa'_H((\exp Y)u)\} : B_u \to W_u$ is bijective and continuous.

Furthermore, $\alpha = (\pi|W_u) \circ \beta$. Hence $(\pi|W_u)$ is bijective and open. But it is also continuous since it is the canonical map (arising from the factorization of the continuous function $v \mapsto Hv: \kappa'_H(V_u) \to HV_u/H$ through the quotient modulo its kernel relation. Hence $\pi|W_u: W_u \to HV_u/H$ is a homeomorphism. Thus we have shown that some open neighborhood of every point in $(U \cap S)/\kappa'_H$ is mapped homeomorphically onto some open neighborhood of its image, and this is the assertion of the proposition. ∎

The previous proposition does not refer to a previously given right congruence κ on S, but to a closed connected Lie subgroup H of G and the geometric structure of $U \cap S$ via κ'_H. The proposition shows, in particular, that $(U \cap S)/\kappa_H$ is a manifold, although in general not a Hausdorff manifold. The remark preceding the proposition shows that this gives information on the space $(U \cap S)/\kappa' = (U \cap S)/\kappa_N$ when N and U arise as in the Foliation Theorem.

Let us inspect the proof of the previous Proposition V.7.22 for the case that H and U do arise as in the Foliation Theorem, V.7.21. Then we also have a function $\gamma = \{Y \mapsto \kappa((\exp Y)u)\} : B_u \to \kappa(W_u)/\kappa$. By the Foliation Theorem V.7.21, γ is bijective and continuous. Unlike β, the map γ will not be open in general. If we define $\pi': (U \cap S)/\kappa' \to \kappa(U \cap S)/\kappa$ by $\pi'(\kappa'(u)) = \kappa(u)$ (unambiguously!), then $\gamma = (\pi'|W_u) \circ \beta$. This shows that $\pi'|W_u$ is bijective, and it is continuous for the same reason for which $\pi|W_u$ was continuous. But it is not open in general and $\kappa(W_u)/\kappa$ in general fails to be a neighborhood of $\kappa(u)$ in S/κ.

However, the function $(\pi'|W_u) \circ (\pi|W_u)^{-1}: HV_u/H \to \kappa(W_u)/\kappa$ is a continuous bijection from HV_u/H onto the $\kappa(V_u)/\kappa$.

Thus we have proved the following theorem:

V.7.23. **Theorem.** (*The Local Factorization Theorem*) *Let G be a Lie group with an open subsemigroup S containing $\mathbf{1}$ in its closure. Let I be the subalgebra associated with κ by the* Foliation Lemma *and assume that the analytic subgroup H generated by I is closed. Then there is an open neighborhood U of $\mathbf{1}$ such that every $u \in U \cap S$ has an open neighborhood V_u such that the restriction of the quotient map $s \mapsto \kappa(s): S \to S/\kappa$ to V_u factors through the continuous open map $s \mapsto Hs: V_s \to HV_s/H$ and a well-defined continuous bijection $Hv \mapsto \kappa(v): HV_u/H \to \kappa(V_s)/\kappa$ where $\kappa(V_u) = \bigcup_{v \in V_u} \kappa(v)$ is the κ-saturation of V_u. If all κ-classes $\kappa(v)$ with $v \in V_u$ are connected, then $\kappa(V_u)$ is open in S and $\kappa(V_u)/\kappa$ is an open neighborhood of $\kappa(u)$ in S/κ.* ∎

The saturation $\kappa(V_u)$ in general fails to be open in S.

Transporting right congruences

The assertions of the Foliation Theorem and the Local Factorization Theorem were established only for points u in a neighborhood U of the identity. We now extend these results to points in a neighborhood of an arbitrary point $p \in \overline{S}$ by "transporting" the given right congruence κ "backwards" from p to $\mathbf{1}$.

Let us first prepare the algebraic background.

V.7.24. **Definition.** Suppose that S is a subsemigroup of a group G.

(i) Then we define a new semigroup S_l by

$$(32) \qquad\qquad S_l = \{g \in G : gS \subseteq S\}.$$

(ii) On the set $\mathcal{K}_r$ of all right congruences on S we define a right action $\mathcal{K}_r \times S_l \to \mathcal{K}_r$ of S_l by the rule

$$(33) \quad \kappa g = \{(x,y) \in S \times S : gx \sim_\kappa gy\} \quad \text{for all } \kappa \in \mathcal{K}_r \text{ and all } g \in S_l.$$

V.7.26. **Remark.** (i) It is straightforward to show that the above action is well-defined and that for topological groups G the right congruence κg has closed classes if κ has closed classes. (Similarily, if κ is a closed right congruence then so is κg.)

(ii) If we consider κ as a subset of the semigroup $S \times S$ then (33) reads as

$$\kappa g = (g,g)^{-1} \kappa \cap (S \times S).$$

(iii) If G is a topological group and S is open in G then Proposition V.0.15(i) shows that $\overline{S} \subseteq S_l$.

(iv) If κ is a right congruence on S and T is a semigroup with $S \subseteq T \subseteq S_l$ then there is a unique smallest extension κ_T of κ to T; this smallest extension is given by

$$x \sim_{\kappa_T} y \quad \text{if and only if} \quad xs \sim_\kappa ys \text{ for all } s \in S.$$

Moreover, the so defined right congruence on T has closed classes if and only if κ has closed classes; and κ_T is closed in $T \times T$ if and only if κ is closed in $S \times S$. (To prove this claim, assume that κ has closed congruence classes and let $(y_n)_{n \in D}$ be a convergent net of κ_T-congruent elements in T. Then for every $s \in S$ the elements $y_n s$ belong to a fixed κ-class in S, hence $(\lim y_n)s$ also belongs to this κ-class. By the definition of κ_T this means that $\lim y_n$ belongs to the κ_T-class of the y_n's. Thus κ_T has closed classes if κ has closed classes; the converse is trivial. In the same way we see that κ_T is a closed relation if and only if κ is a closed relation.)

Note that even if $T = \overline{S}$ the extension of κ to a (closed) right congruence on T need not be unique.

V.7.27. **Lemma.** *For any* $x \in S$, $\kappa \in \mathcal{K}_r$ *and* $g \in S_l$ *the equivalence class* $\kappa g(x)$ *of* κg *at* x *is given by the formula*

$$(\kappa g)(x) = g^{-1} \kappa(gx) \cap S.$$

Equivalently,

$$(34) \qquad\qquad g(\kappa g)(x) = \kappa(gx) \cap gS.$$

Proof. The proof is immediate.   ∎

V.7.28. **Theorem.** (The General Foliation Theorem). *Let S be an open subsemigroup of a finite-dimensional Lie group G such that $1 \in \overline{S}$, and let κ be a right congruence with closed classes on S. Then to every point $p \in \overline{S}$ we can find a Lie subalgebra I_p of $\mathbf{L}(G)$ and an open neighborhood U of p in G with the following property:*

For every point $u \in U \cap pS$ and any vector space complement E_p to I_p in $\mathbf{L}(G)$ there are open neighborhoods A_u of 0 in I_p and B_u of 0 in E_p such that the function

$$(35) \qquad \Phi_u \colon A_u \times B_u \to U \text{ given by } \Phi_u(X, Y) = p(\exp X)(\exp Y)p^{-1}u$$

maps $A_u \times B_u$ diffeomorphically onto an open neighborhood V of u in $U \cap pS$ and

$$(36) \qquad \Phi_u(A_u \times \{Y\}) = \kappa\big(\Phi_u(0, Y)\big) \cap V \text{ for all } Y \in B_u.$$

Proof. We apply the Foliation Theorem V.7.21 to the right congruence κp and conclude: There is an open neighborhood U_p of 1 in G and a subalgebra I_p of $\mathbf{L}(G)$ such that for each $u_p \in U_p \cap S$ and for each vector space complement E_p of I_p in $\mathbf{L}(G)$ there are open neighborhoods $A_{u_p}^{(p)}$ of 0 in I_p and $B_{u_p}^{(p)}$ of 0 in E_p such that the map

$$(X, Y) \mapsto (\exp X)(\exp Y)u_p \colon A_{u_p}^{(p)} \times B_{u_p}^{(p)} \to V^{(p)}$$

is a diffeomorphism onto an open neighborhood of u_p in $U_p \cap S$ and that

$$\big(\exp A_{u_p}^{(p)}\big)(\exp Y)u_p = (\kappa p)\big((\exp Y)u_p\big) \cap V^{(p)} \text{ for all } Y \in B_{u_p}^{(p)}.$$

Now we set $U = pU_p$, $u = pu_p$, $A_u = A_{u_p}^{(p)}$, $B_u = B_{u_p}^{(p)}$, $V = pV^{(p)}$ and define

$$\Phi_u \colon A_u \times B_u \to V, \qquad \Phi_u(X, Y) = p(\exp X)(\exp Y)u_p.$$

Then Φ_u is a diffeomorphism and V is an open neighborhood of $u = pu_p$ in $U \cap pS$. Furthermore,

$$\begin{aligned} \Phi_u(A_u \times \{Y\}) &= p(\exp A_u)(\exp Y)u_p = p\big((\kappa p)\{(\exp Y)u_p\} \cap V^{(p)}\big) \\ &= \kappa(p(\exp Y)u_p) \cap pS \cap V \\ &= \kappa\big(\Phi_u(0, Y)\big) \cap V, \end{aligned}$$

since $V \subseteq pS$. This completes the proof. ■

V.7.29. **Proposition.** *The subalgebra I_p of the General Foliation Theorem V.7.28 is given by the following formula:*

$$\mathrm{Ad}(p)(I_p) = \bigcap_{s \in pS} \mathbf{L}[s]$$

with $\mathbf{L}[s] = L_0(\exp^{-1} K_s)$, as in Definition V.7.1.

Proof. By Definition V.7.1 we have $I_p = \bigcap_{s \in S} L_0\big(\exp^{-1}(\kappa p)(s)s^{-1}\big)$. From Lemma V.7.27 we obtain $(\kappa p)(s)s^{-1} = p^{-1}\kappa(ps)s^{-1} \cap Ss^{-1} = p^{-1}\big(\kappa(ps)(ps)^{-1}\big)p \cap Ss^{-1}$. Since Ss^{-1} is a neighborhood of the identity, $\exp^{-1} Ss^{-1}$ is a neighborhood of 0 in $\mathbf{L}(G)$. Therefore $L_0\big(\exp^{-1}(\kappa p)(s)s^{-1}\big) = L_0\big(\exp^{-1} p^{-1}\big(\kappa(ps)(ps)^{-1}\big)p\big)$. But $\exp^{-1} p^{-1}gp = \mathrm{Ad}(p)^{-1}\exp^{-1} g$ whence

$$L_0\big(\exp^{-1}(\kappa p)(s)s^{-1}\big) = \mathrm{Ad}(p)^{-1}L_0\big(\kappa(ps)(ps)^{-1}\big)$$
$$= \mathrm{Ad}(p)^{-1}L_0(K_{ps}) = \mathrm{Ad}(p)^{-1}\mathbf{L}[ps]$$

in the notation of Definition V.7.1. It follows that $I_p = \mathrm{Ad}(p)^{-1}(\bigcap_{s \in S} \mathbf{L}[ps])$ and this establishes the assertion. ∎

Since by the General Foliation Theorem all congruence classes in the vicinity of and $\prec$-above p coincide locally with the cosets of the analytic subgroup of G generated by I_p, the above observation implies the following Corollary.

V.7.30.   Corollary.   *Let p be a point in $\overline{S}$. Then the following assertions hold:*

(i) *p has a neighborhood U in G such that, for all $u \in U \cap pS$,*

(37) $\mathrm{Ad}(p)(I_p) = \mathbf{L}[u] = \mathrm{Ad}(u)(I_u).$

(ii) *The interior O_p of the set $\{x \in \overline{S} : \mathrm{Ad}(x)(I_x) = \mathrm{Ad}(p)(I_p)\}$ in G is non-void.*

(iii) *The set*

$$J_p = \{x \in \overline{S} : \mathrm{Ad}(p)(I_p) \subseteq \mathrm{Ad}(x)(I_x)\}$$

is a closed right ideal of $\overline{S}$.

Note that, since S is second countable, assertion (ii) implies that the set $\{\mathrm{Ad}(p)(I_p) : p \in \overline{S}\}$ is countable. (Obviously, O_p intersects O_q if and only if $\mathrm{Ad}(p)(I_p) = \mathrm{Ad}(q)(I_q)$.)

Proof.   The assertions (i) and (ii) are immediate consequences of Proposition V.7.29. To show (iii) we first observe that for $x \in J_p$ and $y \in \overline{S}$ Proposition V.7.29 shows

$$\mathrm{Ad}(p)(I_p) \subseteq \mathrm{Ad}(x)(I_x) = \bigcap_{s \in xS} \mathbf{L}[s] \subseteq \bigcap_{s \in xyS} \mathbf{L}[s] = \mathrm{Ad}(xy)(I_{xy}),$$

hence J_p is a right ideal of $\overline{S}$. Now let $t \in \overline{J}_p$. Then there exists an element $u \in tS$ with $\mathrm{Ad}(t)(I_t) = \mathrm{Ad}(u)(I_u)$; since $tS \subseteq J_p$ we see that $\mathrm{Ad}(t)(I_t) = \mathrm{Ad}(u)(I_u)$, that is, $t \in J_p$. Thus J_p is closed. ∎

EV.7.3.   Exercise. Use the General Foliation Theorem V.7.28 to formulate and prove a General Factorization Theorem along the lines of Theorem V.7.23.

Two-sided congruences

We now show that the subalgebras I_p of the preceding subsection are ideals of $\mathbf{L}(G)$ if κ is a two-sided congruence.

V.7.31. **Proposition.** *In addition to the assumptions of the* General Foliation Theorem V.7.28, *suppose that κ is a two-sided congruence. Then the following assertions hold:*

(i) *$sK_t s^{-1} \subseteq K_{st}$ and therefore $\mathrm{Ad}(s)(\mathbf{L}[t]) \subseteq \mathbf{L}[st]$, for all $s, t \in S$.*

(ii) *For every $p \in \overline{S}$ the subalgebra I_p of the General Foliation Theorem is an ideal. (Hence $\mathrm{Ad}(p)(I_p) = I_p$.)*

(iii) *The set J_p defined in Corollary V.7.30 is a closed ideal of $\overline{S}$.*

Proof. (i) If $s, t \in S$ then, since κ is also a left congruence, $sK_t = s\kappa(t)t^{-1} \subseteq \kappa(st)t^{-1} = \kappa(st)(st)^{-1}s = K_{st}s$, hence $sK_t s^{-1} \subseteq K_{st}$.

(ii) Fix $p \in S$. By Corollary V.7.30 there is a neighborhood U of p such that $\mathrm{Ad}(p)(I_p) = \mathbf{L}(u)$, for every $u \in U \cap pS$. Pick a point $t \in U \cap pS$ and an open $\mathbf{1}$-neighborhood V in G with $Vt \subset U \cap pS$. Then for all $v \in V \cap S$ we have $\mathbf{L}[vt] = \mathbf{L}[t]$, hence, by (i), $\mathrm{Ad}(v)(\mathbf{L}[t]) \subseteq \mathbf{L}[vt] = \mathbf{L}[t]$, and thus

$$\mathrm{Ad}(v)(\mathbf{L}[t]) = \mathbf{L}[t],$$

since $\dim \mathbf{L}[t] = \dim \mathrm{Ad}(v)(\mathbf{L}[t])$. Now the subgroup $\{g \in G : \mathrm{Ad}(g)(I_p) = I_p\}$ of G contains the non-empty open set $V \cap S$, hence is both closed and open, thus coincides with G, since G is connected. It follows that for any $Y \in \mathbf{L}(G)$ we have $e^{\mathrm{ad}\,Y} I_p = \mathrm{Ad}(\exp Y) I_p = I_p$ and therefore I_p is an ideal by Lemma II.1.7.

(iii) Let $x \in J_p$ and $y \in \overline{S}$. By V.7.30 it is possible to choose $u, v \in S$ with $I_p = \mathbf{L}[u]$ and $I_{yx} = \mathbf{L}[vu]$. Then, by the above assertion (i), $I_p = \mathbf{L}[u] \subseteq \mathbf{L}[vu] = I_{yx}$. Thus J_p is also a left ideal and the proof is complete. ∎

In view of this result, the analytic subgroups N_p generated by I_p of $\mathbf{L}(G)$ are normal. An analytic normal subgroup of a Lie group G is automatically closed if G is simply connected. It is therefore useful to observe that when dealing with two-sided congruences, we may just as well pass to the universal covering group.

V.7.32. **Proposition.** *Let G be a connected Lie group, S an open subsemi-group with $\mathbf{1} \in \overline{S}$ and κ a congruence relation with closed classes on S. Let $\pi: \widetilde{G} \to G$ the universal covering and $S^* = \pi^{-1}(S)$. Then the following conclusions hold:*

(i) *S^* is an open subsemigroup in the simply connected Lie group $\widetilde{G}$ with $\mathbf{1} \in \overline{S^*}$ and $\pi | S^*: S^* \to S$ is a covering morphism.*

(ii) *If we set $\kappa^* = \{(x, y) \in \widetilde{G} \times \widetilde{G} : \pi(x) \sim_\kappa \pi(x)\}$, then κ^* is a congruence relation on S^* with closed classes mapping onto the κ-classes of S under $\pi | S^*$.*

(iii) *If we identify $\mathbf{L}(\widetilde{G})$ with $\mathbf{L}(G)$ in such a fashion that $\pi \circ \exp_{\widetilde{G}} = \exp_G$. then I is the ideal constructed for κ^* as in the Foliation Lemma V.7.12. and if U is an open neighborhood of $p \in \overline{S}$ in G, satisfying the conclusions of the Foliation Lemma, then $U^* = \pi^{-1}(U)$ satisfies the conclusions of this lemma in $\widetilde{G}$.*

(iv) *If $p \in \overline{S}$ and N_p is the analytic subgroup generated by $\exp_{\widetilde{G}} I_p$ then N_p is closed, and the set of connected components of the manifolds $uN_p \cap U^*$ as u ranges through $U^* \cap S^*$ agrees with the set of connected components of the submanifolds $\kappa^*(u)$.*

Proof.   (i) Clearly S^* is an open subsemigroup of $\widetilde{G}$. Let W be an evenly covered cell neighborhood of 1 in G and W^* the component of 1 in $\pi^{-1}(W)$. Then $\pi|W^*\colon W^* \to W$ is a homeomorphism, and if $s_n \in W \cup S$ is a sequence converging to 1 in G, then $(\pi|W^*)^{-1}(s_n)$ is a sequence in $W^* \cup S^*$ converging to 1 in $\widetilde{G}$. Hence $1 \in \overline{S^*}$. It is clear that $\pi|S^*\colon S^* \to S$ is a covering morphism.

(ii) is straightforward.

(iii) It suffices to prove (iii) with $p = 1$. For $s \in S^*$ we write $K_s^* = \kappa^*(s)s^{-1}$ and note

$$\pi(K_s) = \kappa\big(\pi(s)\big)\pi(s)^{-1} = K_{\pi(s)}.$$

With our identification of $\mathbf{L}(G)$ with the Lie algebra of $\widetilde{G}$ we now observe that $\mathbf{L}[\pi(s)] = L_0(\exp_G^{-1} K_{\pi(s)})$ of subtangent vectors of $K_{\pi(s)}$ at the origin agrees with the set $\mathbf{L}[s] = L_0(\exp_{\widetilde{G}}^{-1} K_s)$ of subtangent vectors of K_s at the origin. Hence, since $S = \pi(S^*)$, we have $\bigcap_{s \in S^*} \mathbf{L}[s] = \bigcap_{t \in S} \mathbf{L}[t] = I$. This shows that I is the ideal constructed in the Foliation Lemma for $\widetilde{G}$ and S^*. Moreover, if $u \in U^*$ then $\pi(u) \in U$ and thus $I = \mathbf{L}[\pi(u)] = \mathbf{L}[u]$ and if B_u is a an open neighborhood of 0 in I such that $(\exp_G B_u)\pi(u)$ is a neighborhood of $\pi(u)$ in $\kappa\big(\pi(u)\big)$, then $(\exp_{\widetilde{G}} B_u)u$ is also a neighborhood of u in $\kappa^*(u) = \pi^{-1}\kappa\big((\pi(u)\big)$ since π is a local isomorphism.

(iv) Since $\widetilde{G}$ is simply connected, every normal analytic subgroup is closed. The remainder then follows from an application of the General Foliation Theorem V.7.28 to $\widetilde{G}$ and S^*.   ∎

The stratified domain

The Foliation Theorem V.7.21. enables us to find a canonical form for the open set $S \cap U$ which has played such a crucial role from the Foliation Lemma onward.

V.7.33.   Definition. Let G be a Lie group, S an open subsemigroup with $1 \in \overline{S}$, and let I be the subalgebra constructed in the Foliation Lemma V.7.12. We fix once and for all a vector space complement E for I in $\mathbf{L}(G)$. Then we define *the stratified domain* $U(S)$ to be the set of all points $s \in S$ such that for suitable open cell neighborhoods A_s of 0 in I and B_s in E the function $\Phi_s = \big((X,Y) \mapsto (\exp X)(\exp Y)s\big)$ maps $A_s \times B_s$ diffeomorphically onto an open neighborhood V of s such that for each $v = \Phi(X,Y) \in V$ the intersection of the congruence class $\kappa(v)$ with V is exactly $\Phi_s(A_s \times \{Y\})$. The function Φ_u is called *a local canonical coordinate system at u.*

V.7.34.   Lemma. *$U(S)$ is an open subset of S and there are open neighborhoods U of 1 in G such that $U \cap S = U(S)$.*

Proof.   Let $s \in U(S)$. Then by definition there is a local canonical coordinate system at s in the form $\Phi_s\colon A_s \times B_s \to V$ as described in Definition V.7.24. We choose A_s and B_s so small that for all $X \in A_s$ and $Y \in B_s$ we have

$$(38) \qquad\qquad d\rho_{X*Y}^*(0)(E) \cap I = \{0\},$$

where $\rho_Y^*(X) = X * Y$. Since $d\rho_0(0) = \mathbf{1}$, this choice is possible. Now let $v = \Phi_s(X_v, Y_v)$ be an arbitrary point in V. We shall show that $v \in U(S)$. First we choose zero neighborhoods A_v in I and B_v in E so that $\Phi_v: A_v \times B_v \to V_v \subseteq V$ given by $\Phi_v(X, Y) = (\exp X)(\exp Y)v$ is a diffeomorphism onto an open subset V_s of V, and, moreover, that the following condition is satisfied: In the zero neighborhood $A_s * B_s$ every element Z is uniquely of the form $Z = Z_I * Z_E$ with $Z_I \in A_s$ and $Z_E \in B_s$.

We require that

$$(39) \qquad (A_v * (A_v * X_v * Y_v))_I \subseteq A_s,$$

$$(40) \qquad (Y_v * B_v)_E \subseteq B_s,$$

and that the function

$$(41) \qquad Y \mapsto (Y * X_v * Y_v)_E : A_v \to A_s \text{ is injective.}$$

The last choice is possible since the derivative of $Y \mapsto (Y * X_v * Y_v)_E : A_s \to E$ at 0 is $\mathrm{pr}_E \circ d\rho_{X_v * Y_v}^*(0)$ and this linear map is injective as $d\rho_{X_v * Y_v}^*(E) \cap I = \{0\}$ by (38).

Let $X \in A_v$ and $Y \in B_v$. Then we set $Z = Y * X_v * Y_v$ and observe

$$\begin{aligned}
\Phi_v(X, Y) &= (\exp X)(\exp Y)v = (\exp X)(\exp Y)(\exp X_v)(\exp Y_v)s \\
&= (\exp X)(\exp Z)s = \exp(X * Z_I)(\exp Z_E)s \\
&= \Phi_s(X * Z_I, Z_E) \in \kappa\big(\Phi_s(0, Z_E)\big),
\end{aligned}$$

by the choice of Φ_s. But also

$$\begin{aligned}
\Phi_s(0, Z_E) &\sim_\kappa \Phi_s(Z_I, Z_E) = (\exp Y)(\exp X_v)(\exp Y_v)s \\
&= (\exp Y)v = \Phi_v(0, Y).
\end{aligned}$$

Hence we conclude $\Phi_v(A_v \times \{Y\}) \subseteq \kappa\big(\Phi_v(0, Y)\big)$ for all $Y \in B_v$. Now let $\Phi_v(X, Y) \in \kappa\big(\Phi_v(0, Y')\big)$ with $(X, Y) \in A_v \times B_v$, $Y' \in B_v$. We write $Z' = Y' * X_v * Y_v$. But $\Phi_v(X, Y) = \Phi_s(X * Z_I, Z_E)$ and $\Phi_v(0, Y') = \Phi_s(Z_I', Z_E')$ as we saw above. In view of the properties of Φ_s we may conclude that $Z_E' = Z_E$, that is $Y' * X_v * Y_v = Y * X_v * Y_v$. From (41) we infer $Y = Y'$. This shows $\Phi_v(A_v \times \{Y\}) = \kappa\big(\Phi_v(0, Y)\big) \cap V_v$ and this completes the proof that $v \in U(S)$. In turn this shows that $U(S)$ is indeed open. Finally, by the Foliation Theorem V.7.21, there is an open set U' in G containing $\mathbf{1}$ such that $U' \cap S \subseteq U(S)$. If we set $U = U' \cup U(S)$, then the set U is open in G, contains $\mathbf{1}$ and satisfies $U \cap S = U(S)$. ∎

V.7.35. **Remark.** In all the results such as the Foliation Lemma, the Foliation Theorem and the Local Factorization Theorem we may replace the open set $U \cap S$ of reference by the stratified domain $U(S)$. ∎

Problems for Chapter V

PV.1.   Problem. *Find all maximal subsemigroups in simple Lie groups. Cf.* [Ol82a], [GKS84]. ∎

PV.2.   Problem. *Suppose that S is a closed divisible semigroup in a Lie group G. Is $\mathbf{L}(S)$ a Lie semialgebra? (Cf. Theorem V.8.6.10 and Corollary V.6.11. Does it perhaps suffice to assume that S is divisible and infinitesimally generated without being closed?* ∎

PV.3.   Problem. *Consider Definition V.1.11. Do conditions* (i) *and* (ii) *already imply* (iii)? ∎

PV.4.   Problem. *Consider carefully the remark following V.4.5 and Figure 4. (See also* [HH86b] *and the computations made there.) Compute explicitly the local semigroups generated by the cone W in the Heisenberg algebra for various neighborhoods B of reference.* ∎

PV.5.   Problem. *Consider a subsemigroup S of the semigroup $M_n(\mathbb{R})$ of all $n \times n$ real matrices with tangent wedge $W = \mathbf{L}(S)$. Characterize the strictly infinitesimally generated semigroup $S_0 = \langle \exp W \rangle$ generated by W. The following examples are of particular interest:*

(a) $S = \{(p_{jk})_{j,k=1,\ldots,n} : p_{jk} \geq 0\}$,

(b) $S = \{(p_{jk})_{j,k=1,\ldots,n} : p_{jk} \geq 0, \quad \sum_{k=1}^{n} p_{jk} = 1\}$.

For $n = 3$, example (b) *was settled in* [Joh73]. ∎

PV.6.   Problem. *Suppose that $W = H(W) \oplus C$ is a split Lie wedge in $\mathbf{L}(G)$ and that*

(1) *$H(W)$ is the Lie algebra of a closed analytic subgroup T of G,*

(2) *C is global giving rise to a semigroup S in G.*

Is W global with TS being a subsemigroup such that $L(TS) = W$? ∎

PV.7.   Problem. *Find conditions under which the one-point compactification of a closed infinitesimally generated semigroup is a compact topological semigroup.* ∎

PV.8.   Problem. *Suppose that S is a closed infinitesimally generated semigroup with group H of units. Under which circumstances are the intervals of the partially ordered space G/H compact (global hyperbolicity)?* ∎

Notes for Chapter V

Section 0. This is standard material on groups and their subsemigroups in the algebraic and topological setting. Some standard references are given in the text. Some results, however, are of independent interest, such as Proposition V.0.23.

Section 1. The definitions of preanalytic semigroups and infinitesimally generated semigroups in a Lie group (see Definitions V.1.2 and 11) are used here for the first time and replace earlier similar definition in [HL83a]. One parameter semigroups are sometimes called rays [HM66], whence semigroups generated by rays can be referred to as ray semigroups (see [Hir73], where they are called ray mobs, or [HL83a]). The results of this section culminate in the Infinitesimal Generation Theorem V.1.16 which is new. It rests on the new definitions. Some results in [HL83a] forshadow this theorem.

Section 2. Proposition V.2.1 and Lemma V.2.2 and the Units Neighborhood Theorem V.2.4 appear in [HL83a]. However, the Tube Theorem V.2.7 and the Unit Group Theorem V.2.8 are new. (For the latter see [Hi87].)

Section 3. This section reviews the theory of semidirect products of Lie groups and extends it to the context of semigroups.

Section 4. The examples described here are not always easy to track. The examples of open subsemigroups of euclidean space and their congruences in Example V.4.4 ff. stem from Hofmann and Ruppert [HR88]. The subsemigroups of the Heisenberg group were discussed in [HL81] and [HL83a]. Lemma 4.10 is new in the present form but was discussed in a somewhat more specialized form in [HH86b]. Example V.4.14 is due to Hofmann and Lawson in [HL83a]. Example V.4.15 is due to Hilgert [Hi86c]. The study of examples in Sl(2) and its universal covering group is taken from [HH85b]. The examples of contraction semigroups are standard for positive definite sesquilinear forms (see e.g. [Law 86]); for arbitrary sesquilinear forms see [BK79], [Ol81], and [Hi88a].

Section 5. The discussion of this section follows a paper by Lawson [Law87a]. The forerunners and applications of this material to control theory see [BJKS82], [HHL85], and [Hi86b]. **Section 6.** Closed divisible subsemigroups of Lie groups were studied by Hofmann and Lawson in [HL83b]. **Section 7.** The results on congruences and foliations are due to Hofmann and Ruppert [HR88]. Some parts of the material are published here fore the first time.

Chapter VI

Positivity

There is a natural bijection between the set of subalgebras A of the Lie algebra $\mathbf{L}(G)$ of a Lie group G and the analytic subgroups H of G. This bijection is implemented by the assignment $A \mapsto \langle \exp A \rangle$ and its inverse function $H \mapsto \mathbf{L}(H)$. The situation is much more complicated for the relation between the set of all Lie wedges W in $\mathbf{L}(G)$ and the set of all infinitesimally generated subsemigroups S of G. The assignments $W \mapsto \langle \exp W \rangle$ and $S \mapsto \mathbf{L}(S)$ are still well defined, but fail to be inverses of each other. In one direction the situation is nearly as good as that in the group case: The strictly infinitesimally generated subsemigroup $S_0 = \langle \exp \mathbf{L}(S) \rangle$ differs very little from S as we know from the Infinitesimal Generation Theorem V.1.16. But in stark contrast with this situation, $\mathbf{L}(\langle \exp W \rangle)$ is much larger than W in general as we have seen in many examples. (See Examples V.4.1, V.4.5, V.4.16,17, and 20.) In Chapter IV we saw that every Lie wedge W in $\mathbf{L}(G)$ generates a *local* semigroup S with respect to some open neighborhood U of $\mathbf{1}$ in G for which $SS \cap U \subseteq S$ and $W = \mathbf{L}(S)$ holds (see Sophus Lie's Fundamental Theorem for Semigroups IV.8.7). But the question whether W is of the form $\mathbf{L}(S)$ for some *global* subsemigroup of G is still a hard question even if G is simply connected (see Example V.4.5). We shall call a Lie wedge W in the Lie algebra $\mathbf{L}(G)$ global if G is simply connected and there is a subsemigroup S in G such that $W = \mathbf{L}(S)$.

In this chapter we shall provide tools which allow us to forge necessary and sufficient conditions for a given Lie wedge to be global. Our tools will pertain to Lie wedges W in the Lie algebra $\mathbf{L}(G)$ of a Lie group G for which the edge $W \cap -W$ is is the Lie algebra of a *closed* Lie subgroup H of G; they will involve suitable exact 1-forms on the manifold G. The crucial idea is that of positivity of forms with respect to wedge fields and monotonicity of measures and smooth functions with respect to the natural preorder induced on G by the given infinitesimally generated semigroup.

1. Cone fields on homogeneous spaces

Our standard assumption in this section is that G is a Lie group and H a closed subgroup.

We recall the adjoint representation $\mathrm{Ad}\colon G \to \mathrm{Aut}\big(\mathbf{L}(G)\big)$, which is characterized by the relation

$$(*) \qquad \exp \mathrm{Ad}(g)(X) = g(\exp X)g^{-1} \text{ for all } g \in G, X \in \mathbf{L}(G).$$

It is useful to remember that the induced representation $\mathbf{L}(Ad)\colon \mathbf{L}(G) \to \mathrm{Der}\big(\mathbf{L}(G)\big)$ is none other than the adjoint representation ad. This means that we have the commutative diagram

$$
\begin{array}{ccc}
\mathbf{L}(G) & \xrightarrow{\ \mathrm{ad}\ } & \mathrm{Der}\,\mathbf{L}(G) \\
{\scriptstyle \exp_G}\downarrow & & \downarrow{\scriptstyle D \mapsto e^D} \\
G & \xrightarrow[\ \ \mathrm{Ad}\ \ ]{} & \mathrm{Aut}\,\mathbf{L}(G).
\end{array}
$$

In other words,

$$(**) \qquad \mathrm{Ad}(\exp_G X) = e^{\mathrm{ad}\,X} \text{ for all } X \in \mathbf{L}(G).$$

The homogeneous space G/H

Now we bring H into the picture. The homogeneous space $M \stackrel{\mathrm{def}}{=} G/H = \{gH\colon g \in G\}$ is an analytic manifold and a left G-space under the action $(g, \xi) \mapsto g\xi\colon G \times M \to M$ with $\xi = g'H$ and $g\xi = gg'H$. We shall write

$$(1) \qquad \mu_g\colon M \to M, \qquad \mu_g(\xi) = g\xi \ \text{ for } \ g \in G, \xi \in M.$$

We shall consider the tangent bundles $T(G)$, $T(H)$, and $T(M)$ of G, H and M. The tangent space of G at g will be denoted $T(G)_g$, and similarly for the other manifolds.

If we set $\mathfrak{g} = \mathbf{L}(G) = T(G)_1$ and $\mathfrak{h} = \mathbf{L}(H) = T(H)_1$ and write $\varepsilon = H \in M$, then the orbit projection

$$(2) \qquad \pi\colon G \to M, \qquad \pi(g) = gH$$

induces a linear surjection

(3) $$d\pi(\mathbf{1})\colon \mathfrak{g} = T(G)_{\mathbf{1}} \to T(M)_\varepsilon$$

with kernel $\mathfrak{h}$, so that $d\pi(\mathbf{1})$ induces an isomorphism

(4) $$X + \mathfrak{h} \mapsto d\pi(\mathbf{1})(X)\colon \mathfrak{g}/\mathfrak{h} \overset{\cong}{\to} T(M)_\varepsilon$$

under which we shall henceforth identify $T(M)_\varepsilon$ with $\mathfrak{g}/\mathfrak{h}$. Given this identification, we identify the surjection $d\pi(\mathbf{1})$ with the quotient map $p\colon \mathfrak{g} \to \mathfrak{g}/\mathfrak{h}$, where $p(X) = X + \mathfrak{h}$.

The stability subgroup of G acting on M at ε is H. Hence each μ_h with $h \in H$ fixes ε, thus induces a vector space isomorphism $d\mu_h(\varepsilon)\colon T(M)_\varepsilon \to T(M)_\varepsilon$ so that

(5) $$\eta\colon H \to \mathrm{Aut}(T(M)_\varepsilon), \qquad \eta(h) = d\mu_h(\varepsilon)$$

is a representation of H which we now describe in terms of our identification of of $T(M)_\varepsilon$ with $\mathfrak{g}/\mathfrak{h}$:

VI.1.1. **Lemma.** *Under the natural identification of $T(M)_\varepsilon$ with $\mathfrak{g}/\mathfrak{h}$ we have*

(6) $$\eta(h)(X + \mathfrak{h}) = \mathrm{Ad}(h)(X) + \mathfrak{h} \quad \text{for all } h \in H, \quad X \in \mathfrak{g}.$$

Proof. We fix $h \in H$ and let $I_h\colon G \to G$ denote the inner automorphism given by $I_h(x) = hxh^{-1}$. Then $\pi I_h(g) = I_h(gH) = hgh^{-1}H = hgH = \mu_h(gH)$. We thus have one commutative diagram on the group level and one induced on the level of the tangent spaces:

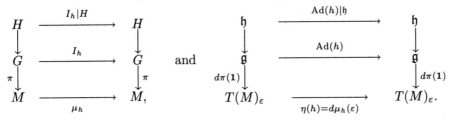

In view of the identification via (4), this diagram proves the asserted relation (6).∎

Next let W be a wedge in $\mathfrak{g}$.

VI.1.2. **Lemma.** *If H is a closed subgroup of the finite dimensional Lie group G and W is a wedge in $\mathbf{L}(G)$ with $W \cap -W = \mathbf{L}(H)$, then the following statements are equivalent:*

(1) *W is a Lie wedge.*

(2) *$\mathrm{Ad}(h)(W) = W$ for all $h \in H_0$, where H_0 denotes the identity component of H.*

Proof. $(1) \Rightarrow (2)$. Let $X \in \mathbf{L}(H)$. If we set $h = \exp_G X$, then $\mathrm{Ad}(h) = \mathrm{Ad}(\exp_G X) = e^{\mathrm{ad}\,X}$ by $(**)$. Since $\mathbf{L}(H) = W \cap -W$ we have $e^{\mathrm{ad}\,X}W = W$ by (1) and Definition II.1.3. Hence $\mathrm{Ad}(h)W = W$. Since $\exp \mathbf{L}(H)$ generates H_0, we may conclude $\mathrm{Ad}(h)W = W$ for all $h \in H$. This proves (2).

$(2) \Rightarrow (1)$. Let $X \in W \cap -W$. Then $W \cap -W = \mathbf{L}(H)$ implies that $h = \exp_G X \in H_0$, whence $\mathrm{Ad}(h)W = W$ by (2). Since $e^{\mathrm{ad}\,X} = \mathrm{Ad}(h)$ again, assertion (1) follows. ∎

Henceforth, in this section, we shall denote with W a Lie wedge whose edge $W \cap -W$ is exactly $\mathfrak{h}$. The image $p(W)$ will be denoted V. From Section I.2 we recall that $V \subseteq \mathfrak{g}/\mathfrak{h}$ is the associated pointed cone.

VI.1.3.   Lemma.   *If H is a closed subgroup of G and W a wedge in $\mathbf{L}(G)$ with edge $\mathbf{L}(H)$ such that $\mathrm{Ad}(h)(W) = W$ for all $h \in H$, then we have the following conclusions:*

(i) *The associated pointed cone V in $T(M)_\varepsilon = \mathfrak{g}/\mathfrak{h}$ is invariant under the group $\eta(H)$.*

(ii) *For $g, g' \in G$ with $\pi(g) = \pi(g')$ we have*

$$d\mu_g(\varepsilon)(V) = d\mu_{g'}(\varepsilon)(V).$$

Proof.   (i) This is an immediate consequence of the preceding Lemmas VI.1.1 and 2.

(ii) The relation $\pi(g) = \pi(g')$ is tantamount to the existence of an $h \in H$ such that $g' = gh$. Thus $d\mu_{g'}(\varepsilon) = d\mu_g(\varepsilon) \circ d\mu_h(\varepsilon) = d\mu_g \circ \eta(h)$ in view of (5). The assertion now follows from (i) above.   ∎

Invariant wedge fields on G and G/H

Now we generalize the ideas of a wedge field on an open subset B of a vector space given in Definition IV.6.12 to an arbitrary differentiable manifold M. For this purpose we need some notation. Recall that for a (finite dimensional) vector space L the set of wedges of L is denoted $\mathcal{W}(L)$. For a differentiable manifold M we set

$$(7) \qquad\qquad \mathcal{W}(M) = \bigcup_{\xi \in M} \mathcal{W}(T(M)_\xi).$$

VI.1.4.   Definition.   (i) A *wedge field* or *cone field* on a C^1-manifold M is a function

$$(8) \qquad\qquad \Xi: M \to \mathcal{W}(M) \text{ with } \Xi(\xi) \subseteq T(M)_\xi.$$

(ii) If G is any group acting on the left of M as a group of diffeomorphisms $\mu_g = (\xi \mapsto g\xi)$, then a wedge field Ξ is called G-*invariant* if

$$(9) \qquad\qquad d\mu_g(\xi)\big(\Xi(\xi)\big) \subseteq \Xi(g\xi) \text{ for all } g \in G, \xi \in M.$$

We notice that a wedge field Ξ is G-invariant if and only if equality holds in (9), as $d\mu_{g^{-1}}(g\xi)\big(\Xi(g\xi)\big) \subseteq \Xi(\xi)$ and $d\mu_{g^{-1}}(g\xi) = \big(d\mu_g(\xi)\big)^{-1}$, whence the assertion.

From Lemma VI.1.3 we obtain directly the following example of an invariant wedge field:

VI.1.5. **Lemma.** *Let G be a Lie group with a closed subgroup H and a Lie wedge W with edge $\mathfrak{h}$ such that $\mathrm{Ad}(H)(W) = W$. As usual in this section set $M = G/H$. Define*

(10) $$\Xi \colon M \to \mathcal{W}(M) \text{ by } \Xi(\xi) = d\mu_g(\varepsilon)(V), \text{ where } \xi = gH.$$

and

(11) $$\Theta \colon G \to \mathcal{W}(G) \text{ by } \Theta(g) = d\lambda_g(\mathbf{1})(W) \text{ for } g \in G.$$

Then we have the following conclusions

(i) *Ξ is a well-defined invariant cone field on M.*

(ii) *Θ is an invariant wedge field on G.*

(iii) *$d\pi(g)\big(\Theta(g)\big) = \Xi\big(\pi(g)\big)$ for all $g \in G$.*

Proof. (i) By Lemma VI.1.3(ii), the definition does not depend on the representation of ξ in the form gH. Hence Ξ is well defined. Now let $g \in G$ and $\xi = g'H \in M$. Then $d\mu_g(\xi)\Xi(\xi) = d\mu_g(\xi)d\mu_{g'}(\varepsilon)(V) = d(\mu_g \circ \mu_{g'})(\varepsilon)(V) = d\mu_{gg'}(\varepsilon)(V) = \Xi(gg'H) = \Xi(g\xi)$ which is what we had to show.

(ii) The verification is similar to the preceding one (and is simpler).

(iii) $d\pi(g)\big(\Theta(g)\big) = d\pi(g)\big(d\lambda_g(\mathbf{1})(W)\big) = d(\pi \circ \lambda_g)(\mathbf{1})(W)$, and since $\pi \circ \lambda_g = \mu_g \circ \pi$ by (1), this is equal to $d\mu_g(\varepsilon)\big(d\pi(\mathbf{1})(W)\big) = d\mu_g(\varepsilon)(V) = \Xi(gH) = \Xi\big(\pi(g)\big)$ in view of the definition of Ξ in (10). ∎

It is instructive to visualize a wedge field in a particular example. An example which presents no topological difficulty is that of the Heisenberg group (See Example V.4.4 and Lemma V.4.5.) We realize the Heisenberg algebra $\mathfrak{g}$ on $\mathbb{R}^3$ in such a fashion that the three standard basis vectors e_j, $j = 1, 2, 3$ satisfy the relation $[e_1, e_2] = e_3$ so that the third axis is the center. We realize G on the underlying vector space $\mathfrak{g}$ by writing the multiplication as $XY = X + Y + \frac{1}{2} \cdot [X, Y]$ for $X, Y \in G$. The exponential function $\exp \colon \mathfrak{g} \to G$ is simply the identity function of $\mathbb{R}^3$. We have $T(G) = \mathbb{R}^3 \times \mathbb{R}^3$ and $T(G)_X = \mathbb{R}^3$. One calculates readily that

$$\mathrm{ad}(x, y, z) = \begin{pmatrix} 0 & 0 & 0 \\ 0 & 0 & 0 \\ -y & x & 0 \end{pmatrix},$$

and, consequently

$$g\big(\mathrm{ad}(x, y, z)\big) = d\lambda_{(x,y,z)}(0) = \begin{pmatrix} 1 & 0 & 0 \\ 0 & 1 & 0 \\ -\frac{y}{2} & \frac{x}{2} & 1 \end{pmatrix}.$$

We shall need the inverse of the adjoint of this map relative to the standard inner product of $\mathbb{R}^3$:

$$\big(g\big(\mathrm{ad}(x, y, z)\big)^*\big)^{-1} = \begin{pmatrix} 1 & 0 & \frac{y}{2} \\ 0 & 1 & -\frac{x}{2} \\ 0 & 0 & 1 \end{pmatrix}.$$

It is useful to recall that the group of rotations around the z-axis is an automorphism group of $\mathfrak{g}$ and G. It is thefore no loss of generality to perform calculations for $y = 0$ and rotate.

VI.1.6.   **Example.**   (i) In the Heisenberg algebra $\mathfrak{g}$ above we let W be the pointed cone (and therefore Lie wedge) $\{(x, y, z): 0 \leq z \text{ and } x^2 + y^2 \leq z^2\}$ This cone is the $\mathbb{R}^+$-span of the circle $\{(\cos t, \sin t, 1): t \in \mathbb{R}\}$. Then $\Theta(x, y, z) = g\big(\mathrm{ad}(x, y, z)\big)(W)$ and this is the $\mathbb{R}^+$-span of the ellipse

$$\{(\cos t, \sin t, 1 + \frac{x}{2} \sin t - \frac{y}{2} \cos t): t \in \mathbb{R}\}$$

which is in the cylinder around the z-axis with radius 1, and whose short axis, in the case $y = 0$, is spanned by $(\pm 1, 0, 1)$ and whose long axis is spanned by $(0, 1, 1 \pm \frac{x}{2})$. Notice that $\Theta(x, y, z)$ does not depend on z so that it is no loss of generality to perform calulations for $z = 0$ and to translate vertically.

     (ii) We compute the dual wedge $\Theta(x, y, z)^*$: For vectors X and ω of $\mathbb{R}^3$ we write $\langle \omega, X \rangle$ for the standard scalar product on $\mathbb{R}^3$. Then $\omega \in \Theta(Y)^*$ if and only if $\langle \omega, g(\mathrm{ad}\, Y)(X) \rangle \geq 0$ for all $X \in W$, if and only if $\langle g(\mathrm{ad}\, Y)^*(\omega), X \rangle \geq 0$ for all $X \in W$ if and only if $g(\mathrm{ad}\, Y)^*(\omega) \in W^* = W$ if and only if $\omega \in \big(g(\mathrm{ad}\, Y)^*\big)^{-1}(W)$. Thus

$$\Theta(x, y, z)^* = \Big(g\big(\mathrm{ad}(x, y, z)\big)^*\Big)^{-1}(W),$$

and this cone is the $\mathbb{R}^+$-span of the circle of radius 1

$$\{(\frac{y}{2} + \cos t, -\frac{x}{2} + \sin t, 1): t \in \mathbb{R}\}.$$

This circle lies in the horizontal plane at height 1 with center $(\frac{y}{2}, \frac{-x}{2}, 1)$.

     (iii) The vector field $X(x, y, z) = (2, 2, 4 + x - y)$ satisfies $\langle (\frac{y}{2} + \cos t, -\frac{x}{2} + \sin t, 1), (2, 2, 4 + x - y) \rangle = 2(2 + \cos t + \sin t) > 0$ for all t and hence is in the interior of $\Theta(x, y, z)$ for all (x, y, z). The curve $u: \mathbb{R} \to G$ given by $u(t) = (2t, 2t + 5, 6 - t)$ is a solution of the initial value problem $\dot{u}(t) = X\big(u(t)\big)$ and $u(0) = (0, 5, 6) \in W$.

     (iv) The vector field $\omega: \mathfrak{g} \to \mathfrak{g}$ with $\omega(x, y, z) = (y, -x, 2)$ satisfies $\omega(g) \in \mathrm{int}\big(\Theta(g)^*\big)$ for all $g \in G$. In particular, $\langle \omega(g), X \rangle > 0$ for all $0 \neq X \in \Theta(g)$.

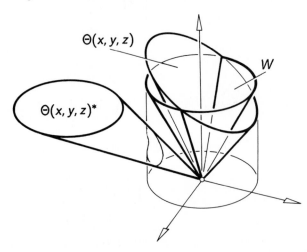

W-admissible piecewise differentiable curves

The reader should notice that a cone field is a special vector field distribution in the sense of Definition I.5.19. We recall the ideas of Definition I.5.20 and consider a piecewise smooth curve $x: I \to M$ for some interval $I \subseteq \mathbb{R}$. The tangent space of the manifold $\mathbb{R}$ at each $t \in \mathbb{R}$ is $\mathbb{R}$, and we write $\dot{x}(t) = dx(t)(1) \in T(M)_{x(t)}$ for each $t \in I$ at which x is differentiable.

The following definition pursues the lines of Definitions I.5.20 and IV.6.12.

VI.1.7.   Definition. i) A continuous curve $x: I \to M$ is called a *piecewise smooth chain* or simply a *chain* if it is piecewise smooth, that is if there is a finite sequence of numbers $r_1 < \ldots < r_n$ in I such that x is infinitely often differentiable on the complements of these points and that it has one-sided derivatives of any order in *all* points of I. *We shall write $\dot{x}(t)$ or $x'(t)$ for the right derivative in all points except the largest point of I, if it exists, where these vectors shall mean the left derivative.* In all points where the chain is differentiable, this agrees with the usual notation.

ii) A chain x is said to be *subordinate to, or a solution of a wedge field* Λ if

$$(12) \qquad\qquad \dot{x}(t) \in \Lambda\big(x(t)\big) \text{ for all } t \in I.$$

iii) A chain in G is called W-*admissible* if it is subordinate to Θ, and a chain in G/H is called W-*admissible* if it is subordinate to Λ.

Recall that
$$\Theta(g) = d\lambda_g(\mathbf{1})(W)$$
and
$$\Xi(gH) = d\mu_g(\varepsilon)\big(d\pi(\mathbf{1})(W)\big) = d\pi(g)\big(\Theta(g)\big).$$

Notice that the derivative of a chain has only jump discontinuities due to the existence of one-sided derivatives of all orders.

Let us clarify the relationship between W-admissible chains on G and those on G/H. We continue our standard notation and assumption.

VI.1.8.   Proposition.   (i) $\Theta(g) = d\pi(g)^{-1}\big(\Xi(gH)\big)$ *for all $g \in G$.*

(ii) *For a chain $x: I \to G$ the following statements are equivalent:*

(1) x *is W-admissible in G.*

(2) $\pi \circ x$ *is W-admissible in G/H.*

(iii) *If $x: I \to G$ is a W-admissible chain and $h: I \to H$ is a chain, then $xh = \big(t \mapsto x(t)h(t)\big): I \to G$ is W-admissible.*

Proof.   (i) We consider an element $u \in T(M)_g$. Then there is a unique $v \in \mathfrak{g}$ such that $u = d\lambda_g(\mathbf{1})(v)$. Then $d\pi(g)(u) \in \Xi(gH)$ if and only if $d\mu_g(\varepsilon)d\pi(\mathbf{1})(v) =$

$d\pi(g)d\lambda_g(\mathbf{1})(v) = d\pi(g)(u) \in \Xi(gH) = d\mu_g(\varepsilon)\big(d\pi(\mathbf{1})(W)\big)$ which is equivalent to $d\pi(\mathbf{1})(v) \in d\pi(\mathbf{1})(W)$ and this is tantamount to $v \in W$. This relation, however, is equivalent to $u = d\lambda_g(\mathbf{1})(v) \in d\lambda_g(\mathbf{1})(W) = \Theta(g)$. This proves the claim.

(ii) Condition (1) means that for all $t \in I$ in which x is differentiable we have $\dot{x}(t) \in \Theta\big(x(t)\big)$, and (2) is equivalent to $d\pi\big(x(t)\big)\big(\dot{x}(t)\big) = (\pi \circ x)'(t) \in \Xi\big(x(t)H\big)$ for all t in the domain of definition of $\dot{x}$. The equivalence of (1) and (2) then follows from (i) above.

(iii) The curve $\pi \circ xh = \big(t \mapsto x(t)h(t)H\big)$ agrees with $\pi \circ x = \big(t \mapsto x(t)H\big)$ and is, therefore, W-admissible in M by (ii) above. But then once more by (ii), the curve xh itself is W-admissible in G as was asserted. ∎

Let us recall from our discussion of Proposition V.2.4, that the bundle map $\pi\colon G \to G/H$ is locally trivial. Specifically, if E is a vector space complement of $\mathfrak{h}$ in $\mathfrak{g}$, then there is an open cell C around 0 in E such that $(c, h) \mapsto (\exp c)h\colon C \times H \to U'$ is a diffeomorphism onto an open neighborhood of H in G which is stable under the multiplication by H on the right. Hence $U = \pi(U') = U'/H$ is an open neighborhood of ε in $M = G/H$ on which we may define $\sigma\colon U \to G$ by $\sigma(\xi) = \exp c$ with the unique $c \in C$ for which we have $\xi = (\exp c)H$ for $\xi \in U$. This σ is a local cross section satisfying

$$(13) \qquad\qquad \pi\big(\sigma(\xi)\big) = \xi \quad \text{for all} \ \ \xi \in U.$$

If $g \in G$, we set $U_g = \mu_g(U)$ and define $\sigma_g\colon U_g \to G$ by $\sigma_g(g\xi) = g\sigma(\xi)$ for $\xi \in U$. Then σ_g satisfies

$$(14) \qquad\qquad \pi\big(\sigma_g(\eta)\big) = \eta \quad \text{for all} \ \ \eta \in U_g.$$

and is, therefore, a local cross section satisfying $\sigma_g(gH) = g$. If $\xi_0 \in U_g$ and g_0 is an arbitrary element in ξ_0, we set $h = \sigma_g(\xi_0)^{-1}g_0$ and observe $h \in H$. Then the function $\xi \mapsto \sigma_g(\xi)h\colon U_g \to G$ is a local cross section σ_{g,ξ_0,g_0} mapping ξ_0 to g_0.

VI.1.9.   Proposition. *Let* $\xi\colon [0,1] \to G/H$ *be a chain. Then for each* g_0 *with* $\pi(g_0) = \xi(0)$ *there exists a chain* $x\colon [0,1] \to G$ *with* $\pi \circ x = \xi$ *and* $x(0) = g_0$. *If* ξ *is* W-*admissible, then* x *is* W-*admissible, too.*

Proof.   We consider the set $P \subseteq [0,1]$ of all $p \in [0,1]$ such that there is a chain $y\colon [0,p] \to G$ with $\pi \circ y = \xi|[0,p]$ and $y(0) = g_0$. We claim that $1 \in P$. Clearly $0 \in P$ since the function $y\colon \{0\} \to G$ with $y(0) = g_0$ satisfies the requirements. Furthermore, P is an interval as one observes readily from the definition of P. We let $s = \sup P$. It now suffices to show that $s \in P$ and $s = 1$. In order to prove these claims, take any $g \in \xi(s)$. Then, by the continuity of ξ, there is an interval neighborhood I of s in $[0,1]$ with $\xi(I) \subseteq U_g$. Hence, by the definition of s there is a $p \in P \cap I$. Then there is a chain $y\colon [0,p] \to G$ with $\pi \circ y = \xi|[0,p]$. Now define the function $z\colon [0,p] \cup I \to G$ by

$$(16) \qquad\qquad z(t) = \begin{cases} y(t), & \text{if } t \in [0,p]; \\ \sigma_{g,\xi(p),y(p)}\big(\xi(t)\big), & \text{if } p < t \in I. \end{cases}$$

Then z is a chain since $\sigma_{g,\xi(p),y(p)}$ is analytic, hence smooth. Moreover, $\pi \circ z = \xi|\big([0,p] \cup I\big)$. It follows that $[0,p] \cup I \subseteq P$. Thus $s \in P$, and since I is a neighborhood of $s = \sup P$ in $[0, 1]$ we simultaneously conclude $s = 1$, as asserted. It follows that there is a chain differentiable continuous lifting $x: [0, 1] \to G$ with $\pi \circ x = \xi$. It remains to observe that x is W-admissible if ξ has this property. But this follows from Lemma VI.1.8(ii). $\blacksquare$

The preceding proposition implies at once that every W-admissible chain $\xi: I \to G/H$ has a lifting $x: I \to G$ with the same properties and with one prescribed value $x(a) = g_0$ for a given $g_0 \in \xi(a)$. Let us now summarize the previous results. We need some notation:

Let $\mathcal{D}(G)$ denote the set of all chains $x: [0, T] \to G$ for some $T \in \mathbb{R}^+ = [0, \infty[$ with $x(0) = \mathbf{1}$. Denote with $\mathcal{A}(G)$ the subset of all W-admissible chains in $\mathcal{D}(G)$. Analogously, define $\mathcal{D}(G/H)$ to be the set of all chains $\xi: [0, T] \to G/H$ for some $T \in \mathbb{R}^+$ with $\xi(0) = \varepsilon$. Let $\mathcal{A}(G/H)$ denote the subset of all W-admissible members of $\mathcal{D}(G/H)$. Define

$$\Pi: \mathcal{D}(G) \to \mathcal{D}(G/H) \quad \text{by} \quad \Pi(x) = \pi \circ x.$$

VI.1.10.   Corollary.   *Let G be a connected Lie group with a closed subgroup H and let W be an $\mathrm{Ad}(H)$-invariant wedge in $\mathfrak{g}$ whose edge is $\mathfrak{h}$. Then the following conclusions hold:*

(i) Π *is surjective.*

(ii) $\Pi\big(\mathcal{A}(G)\big) = \mathcal{A}(G/H)$.

(iii) $\Pi^{-1}\big(\mathcal{A}(G/H)\big) = \mathcal{A}(G)$. $\blacksquare$

In the following paragraphs we shall attempt to elucidate the significance of the set of W-admissible chains. On the set $\mathcal{D}(G)$ we define the operation of *concatenation* which, in a slightly different context, we have seen in Definition IV.7.5. For $x_j \in \mathcal{D}(G)$, $j = 1, 2$ with $x_j: [0, T_j] \to G$ we define $x_1 \flat x_2: [0, T_1 + T_2] \to G$ by

$$(17) \qquad (x_1 \flat x_2)(t) = \begin{cases} x_1(t), & \text{if } t \in [0, T_1]; \\ x_1(T_1)x_2(t - T_1), & \text{if } t \in]T_1, T_1 + T_2]. \end{cases}$$

VI.1.11.   Lemma.   *The set $\mathcal{D}(G)$ is a monoid with respect to concatenation $\flat$, the identity being the constant function with domain 0 and value $\mathbf{1}$. The subset $\mathcal{A}(G)$ is a submonoid.*

Proof.   The associativity of $\flat$ is straightforward from the definition and the claim about the identity is clear, too. Let us briefly observe that the concatenation of W-admissible chain is W-admissible: Let $T_1 < t \le T_1 + T_2$ be such that $\dot{x}_2(t - T_1)$ exists. Then

$$(x_1 \flat x_2)'(t) = d\lambda_{x_1(T_1)}\big(x_2(t - T_1)\big)\big(\dot{x}_2(t - T_1)\big)$$

$$\in d\lambda_{x_1(T_1)}\big(x_2(t - T_1)\big)\Big(\Theta\big(x_2(t - T_1)\big)\Big)$$

$$= \Theta\big(x_1(T_1)x_2(t - T_1)\big) = \Theta\big((x_1 \flat x_2)(t)\big)$$

in view of the invariance of Θ, and this proves the claim. $\blacksquare$

The concatenation operation is the reason which necessitates the restriction that smoothness conditions can be postulated only piecewise for the class of trajectories which is of interest to us.

In the Example VI.1.6 of the Heisenberg group we may form the concatenation

$$x(t) = \begin{cases} (0, 5t, 6t), & \text{if } 0 \leq t \leq 1; \\ (2(t-1), 2t+3, 7-t), & \text{if } t > 1 \end{cases}.$$

In view of Example VI.1.6(iii) $x: \mathbb{R}^+ \to G$ is a W-admissible chain for which $\dot{x}(t) \in \operatorname{int} \Theta(x(t))$. One notices that this chain ascends on a straight segment and then descends on a straight line indefinitely.

The smallest class of trajectories inside $\mathcal{A}(G)$ which is of relevance here is introduced in the following definition:

VI.1.12.   Definition. If $X_1, \ldots, X_n \in W$ we define $e_k: [0, T_k] \to \langle \exp W \rangle \subseteq G$ by $e_k(t) = \exp t \cdot X_k$ for $k = 1, \ldots, n$. Then $f = e_1 \flat \cdots \flat e_n \in \mathcal{A}(G)$ and we shall occasionally call f a *pc-chain* indicating a *piecewise constant* steering function. The set of all pc-chains will be denoted $\mathcal{E}(G)$.

Clearly the set $\mathcal{E}(G)$ of pc-chains is a subsemigroup of $\mathcal{A}(G)$. Notice that the concept of a pc-chain depends heavily on the choice of the wedge W.

The *evaluation function*

$$\operatorname{ev}: \mathcal{D}(G) \to G \quad \text{given by} \quad \operatorname{ev}(x) = x(T) \quad \text{for} \quad x: [0, T] \to G$$

obviously satisfies

(18) $$\operatorname{ev}(x \flat y) = \operatorname{ev}(x)\operatorname{ev}(y),$$

and maps the identity to $\mathbf{1}$. It is, therefore, a homomorphism of monoids, and the image of $\mathcal{D}(G)$ under ev is G, the image of $\mathcal{A}(G)$ is the subsemigroup

(19) $$S(W) = \{g \in G: (\exists x \in \mathcal{A}(G), T \in \mathbb{R}^+) \quad g = x(T)\},$$

and the image of $\mathcal{E}(G)$ under ev is exactly $\langle \exp W \rangle$. Thus we have clearly

(20) $$\langle \exp W \rangle \subseteq S(W).$$

These remarks are global parallels to the material in Proposition IV.5.6. We are aiming for a result which shows that $\langle \exp W \rangle$ is in fact dense in $S(W)$. This is not as easy as it may seem at first. The insights of Section 5 of Chapter IV are also utilized in the proof of the following proposition:

VI.1.13.   Proposition. *Let N be a neighborhood of $\mathbf{1}$ in G and $x: [0, 1] \to G$ a member of $\mathcal{A}(G)$. Then there is a pc-chain f such that*

(21) $$f(t) \in x(t)N \quad \text{for all} \quad t \in [0, 1].$$

Proof.   We prove this lemma in two steps.

Step 1. The first step deals with the situation in a sufficiently small neighborhood of **1** and belongs to the context of Section 5 of Chapter IV. We let B denote a C-H-neighborhood in $\mathfrak{g}$ such that exp maps B homeomorphically onto an open neighborhood $\exp B$ of **1**, and equip B with a metric d compatible with the uniform structure of B induced by the unique invariant uniform structure given by any norm on $\mathfrak{g}$. We let $X: [0, T] \to B$ with $X(0) = X_0$ denote a piecewise smooth curve with $X'(t) \in g\big(\operatorname{ad} X(t)\big)(W)$ and set $u: [0, T] \to W, u(t) = f\big(\operatorname{ad} X(t)\big)\big(X'(t)\big)$. Then $X'(t) = g\big(\operatorname{ad} X(t)\big)\big(u(t)\big)$ as in Theorem IV.5.1.

Now we claim that there is a function $\delta:]0, 1] \to]0, 1]$, with $\delta(t) \le t$ such that for each r, whenever $d(X_0, 0) < \delta(r)$ there are elements $X_k, k = 1, \cdots, n$ in W such that the function $F: [0, 1] \to B$ given by

$$(22) \qquad F(t) = X_0 * \frac{T}{n} \cdot X_1 \cdots * \frac{(k-1)T}{n} \cdot X_{k-1} * (t - \frac{(k-1)T}{n}) \cdot X_k$$

for $\frac{(k-1)T}{n} \le t < \frac{kT}{n}, k = 1, \ldots, n$, and by $F(1) = X_0 * \cdots * X_n$ satisfies

$$(23) \qquad d\big(X(t), F(t)\big) \le r \quad \text{for all} \ \ t \in [0, T].$$

We propose to take $X_k = u(\frac{kT}{n})$, and define the piecewise constant function $v: [0, 1] \to [0, 1]$ by

$$v(t) = u(\frac{kT}{n}) \quad \text{for} \quad \frac{(k-1)T}{n} \le t < \frac{kT}{n}, k = 1, \ldots, n,$$

and $v(1) = u(1)$. Now given $r \in]0, 1]$, by Proposition IV.5.3, we find a $\delta(r) \in]0, r]$ so that for every regulated function $w: [0, 1] \to [0, 1]$ such that $d\big(w(t), u(t)\big) \le \delta(r)$ for all $t \in [0, 1]$ and every initial value X_0 with $d(X_0, 0) < \delta(r)$ the unique solution X of

$$(D) \qquad X'(t) = g\big(\operatorname{ad} X(t)\big)\big(w(t)\big) \ \ \text{with} \ \ X(0) = X_0$$

satisfies $d\big(X(t), F(t)\big) < r$. In particular, for $t \in [\frac{(k-1)T}{n}, \frac{kT}{n}[$ we have

$$(24) \qquad X'(t) = g\Big(\big(\operatorname{ad} X(\frac{(k-1)T}{n})\big)\Big)\big(u(\frac{(k-1)T}{n})\big) \in g\Big(\operatorname{ad}\big(X(t)\big)\Big)(W).$$

In order to establish our claim it now suffices to invoke the piecewise uniform continuity of u on the compact interval $[0, 1]$ in order to find a natural number so large that the function v constructed above satisfies $d\big(v(t), u(t)\big) \le \delta(r)$ for all $t \in [0, 1]$. This completes the proof of Step 1. We shall call a function of the type of F a *concatenation of W-rays*.

Step 2. Now we consider the global situation. Let $x: [0, 1] \to G$ be a member of $\mathcal{A}(G)$. Let $D: G \times G \to \mathbb{R}$ be a metric on G such that all left translations are isometries. It is no loss of generality to assume that N is the open D-unit ball around **1**. We define a metric d on B by $d(X, Y) = D(\exp X, \exp Y)$; then d is compatible with the uniform structure on B induced by any norm of $\mathfrak{g}$. We may also assume that $N \subseteq \exp B$, so that the d-unit ball around 0 is mapped homeomorphically onto N under exp. Choose a natural number m so large that $x_k(t) \overset{\text{def}}{=} x(\frac{k-1}{m})^{-1}x(t - \frac{k-1}{m}) \in \exp B$ for all $t \in [\frac{k-1}{m}, \frac{k}{m}]$ and for $k = 1, \ldots, m$. Then we define a piecewise smooth function $X_k : [0, \frac{1}{m}] \to B$ by $\exp X_k(t) = x_k(t)$. We are posed to apply the result of Step 1, but we have to proceed judiciously in order to assemble the pieces correctly. We now define a sequence $0 < r_1 < \ldots < r_m$ of positive numbers by setting $r_m = 1$ and defining inductively, coming down from above, $r_{m-k} = \delta(r_{m-k+1})$ for $k = 1, \ldots, m-1$ with the function δ of Step 1. (In other words, $r_k = \delta^{m-k}(1)$.) Now we apply Step 1 and find a concatenation of W-rays $F_1 : [0, \frac{1}{m}] \to B$ such that

$$F_1(0) = 0 \text{ and } d\big(F_1(t), X_1(t)\big) < r_1 \text{ for all } \quad t \in [0, \frac{1}{m}].$$

We set $e_1 = \exp \circ F_1$. Then we have the following facts:

(i_1) $D\big(x(t), e_1(t)\big) < r_1$ for all $t \in [0, \frac{1}{m}]$.

(ii_1) e_1 is a concatenation of local one parameter semigroups $t \mapsto \exp tX : [0, \tau] \to G$ with $X \in W$.

Now we proceed by induction and suppose that we found for $1 < p < k$ with $k < m$ functions $e_p : [0, \frac{p}{m}] \to G$ such that $e_p |[0, \frac{p-1}{m}] = e_{p-1}$ and that

(i_p) $D\big(x(t), e_p(t)\big) < r_p$ for all $t \in [\frac{p-1}{m}, \frac{p}{m}]$

(ii_p) e_p is a pc-chain.

Then the element $y = x(\frac{k}{m})^{-1}e_k(\frac{k}{m})$ satisfies $d(y, \mathbf{1}) = d\big(e_k(\frac{k}{m}), x_k(\frac{k}{m})\big) < r_k$ by the left invariance of d and by (i_k). If we let X_0 be the unique element in B with $\exp X_0 = y$, then $d(X_0, 0) < r_k$ by the definition of d and we apply Step 1 with the initial value X_0 construct a concatenation of W-rays $F_{k+1} : [0, \frac{1}{m}] \to B$ satisfying $F_{k+1}(0) = X_0$ and

$$(25) \qquad d\big(F_{k+1}(t), X_{k+1}(t)\big) < r_{k+1} \text{ for all } t \in [0, \frac{1}{m}].$$

Then $F_{k+1} = X_0 * E$ with a unique E such that $\exp E$ is a pc-chain. We define $e_{k+1} : [0, \frac{k+1}{m}] \to G$ by $e_{k+1} = e_k \flat \exp E$. Then e_{k+1} satisfies (ii_{k+1}). Furthermore, $\exp F_{k+1}(t) = \exp X_0 \exp E(t) = y \exp E(t) = x(\frac{k}{m})^{-1}e_p(\frac{k}{m})\exp E(t)$ for $t \in [0, \frac{1}{m}]$. Hence $x(\frac{k}{m})\exp F_{k+1}(t - \frac{k}{m}) = e_{k+1}(t)$ for $t \in [\frac{k}{m}, \frac{k+1}{m}]$. Hence for these same t we compute

$$D\big(x(t), e_{k+1}(t)\big) = D\big(x(\frac{k}{m})x_k(t - \frac{k}{m}), x(\frac{k}{m})\exp F_{k+1}(t - \frac{k}{m})\big)$$

$$= D\big(x_k(t - \frac{k}{m}), \exp F_{k+1}(t - \frac{k}{m})\big)$$

by the left invariance of d. However, by the definition of X_k, this last number is $d\big(X_k(t - \frac{k}{m}), F_{k+1}(t - \frac{k}{m})\big) < r_{k+1}$ by (25). This shows that (i_{k+1}) holds, too. Thus our recursive construction is complete. Then the function $f = e_m$ is a pc-chain and satisfies $D\big(x(t), f(t)\big) < r_m = 1$, which means $D(x(t)^{-1}f(t), \mathbf{1}) < 1$ and thus $x(t)^{-1}f(t) \in D^{-1}([0, 1[) = N$. This completes the proof. ∎

This shows that every chain can be uniformly approximated by pc-chains. Clearly, the choice $T = 1$, that is the choice of the unit interval as parameter interval is no restriction of generality. We now observe that every chain can also be uniformly approximated by W-admissible smooth paths. For this purpose we let $\mathcal{C}^\infty(G)$ denote the set of all smooth curves $x : [0, T] \to G$ with $\dot{x}(t) \in \Xi\big(x(t)\big)$.

VI.1.14. **Proposition.** *Let N be a neighborhood of the identity in G and let $x:[0,1] \to G$ be a W-admissible chain in G. Then there is a W-admissible smooth curve $y:[0,1] \to G$ such that $y(t) \in x(t)N$ for all $t \in [0,1]$.*

Proof. By Proposition VI.1.13 it suffices to approximate uniformly every pc-chain.

Let $x:[0,T] \to G$ be a concatenation of W-rays given by

$$(26) \qquad x(t) = (\exp r_1 \cdot X_1) \cdots (\exp r_{k-1} \cdot X_{k-1})(\exp(t - r_k) \cdot X_k) \text{ for } t \in [r_{k-1}, r_k[,$$

and $k = 1, \ldots, n$. We define n chains $\varphi_k : \mathbb{R} \to \mathbb{R}$ with $k = 1, \ldots, n$ in R by

$$(27) \qquad \varphi_k(t) = \begin{cases} 0, & \text{for } t < r_{k-1}; \\ t - r_{k-1}, & \text{for } r_{k-1} \le t < r_k; \\ r_k, & \text{for } r_k \le t. \end{cases}$$

With these chains we can write

$$(28) \qquad x(t) = (\exp \varphi_1(t) \cdot X_1) \cdots (\exp \varphi_n(t) \cdot X_n) \text{ for all } t \in [0, T].$$

Now we can approximate uniformly on all of $\mathbb{R}$ the functions φ_k by monotone C^∞-functions ψ_k which vanish on $]-\infty, r_{k-1}]$ and take the value r_k on $[r_k, \infty[$. We set

$$(29) \qquad y(t) = (\exp \psi_1(t) \cdot X_1) \cdots (\exp \psi_n(t) \cdot X_n) \text{ for all } t \in [0, T].$$

Then y uniformly approximates x. For $t \notin \{r_1, \ldots, r_n\}$, say, $r_{k-1} < t < r_k$, the functions φ_p are constant near t for $p \ne k$ and thus we have $y(s) = (\exp r_1 \cdot X_1) \cdots (\exp r_{k-1} \cdot X_{k-1})(\exp \psi_k(s) \cdot X_k)$ for all s near t, whence

$$(30) \qquad \dot{y}(t) = d\lambda_{y(t)}(\mathbf{1})(\dot{\psi}_k(t) \cdot X_k) \in d\lambda_{y(t)}(\mathbf{1})(W).$$

On the other hand $\dot{\psi}_p(t_k) = 0$ for all p and k. Hence, for any smooth function Ω on an open neighborhood of $(r_1, \ldots, r_k, 0, \ldots, 0)$ in $\mathbb{R}^n$ with values in G we find $\frac{d}{dt}|_{t=t_k} \Omega(\psi(t)) = d\Omega(\psi(t_k))(\dot{\psi}(t_k)) = 0$ with $\psi(t) = (\psi_1(t), \ldots, \psi_n)$. In particular we have

$$(31) \qquad \dot{y}(t_k) = 0 \text{ for } k = 1, \ldots, n.$$

In any case, $\dot{y}(t) \in d\lambda_{y(t)}(\mathbf{1})(W) = \Theta(y(t))$ and thus is a W-admissible smooth curve approximating the pc-chain x uniformly as well as we like. ∎

Let us write S for the ray semigroup $\langle \exp W \rangle$. The Lie wedge W generates a Lie subalgebra $\langle\langle W \rangle\rangle$ which is equal to $\mathbf{L}(G(S))$ by Proposition V.1.9. Let $A = \langle S \cup S^{-1} \rangle$ be the subgroup generated by $\exp W$. Then the underlying groups of A and $G(S)$ agree; the latter possibly carries a finer topology. The wedge field $g \mapsto d\lambda_g(0)(\langle\langle W \rangle\rangle)$ is none other than the distribution whose integral manifolds are the cosets gA. In particular, let $x \in \mathcal{A}(G)$ with $x:[0,T] \to G$ be a W-admissible chain. Then it is certainly a $\langle\langle W \rangle\rangle$-admissible chain and thus $x([0,T])$ is contained in the integral manifold of this distribution through $\mathbf{1}$, namely, A. Therefore we have the following remark:

VI.1.15.   **Remark.**   If $x \in \mathcal{A}(G)$, then $\operatorname{im} x \subseteq \langle \exp W \cup \exp -W \rangle$.   ∎

The following summary shows the significance of the W-admissible chains for the investigation of subsemigroups in a Lie group.

VI.1.16.   **Theorem.**   (The Approximation Theorem for Chains) *Let G be a Lie group, H a closed connected subgroup, W a Lie wedge in $\mathfrak{g}$ with $\mathfrak{h}$ as edge. The semigroup $S \overset{\text{def}}{=} \langle \exp W \rangle$ generated by W and the semigroup $S(W)$ of all elements $g \in G$ which are the endpoints of some W-admissible smooth chain are related by*

$$S \subseteq S(W) \subseteq \operatorname{cl}_{G(S)} S.$$

In particular, $\operatorname{cl}_{G(S)} S(W)$ is the smallest $G(S)$-closed subsemigroup containing $\exp W$. Moreover, this semigroup is the $G(S)$-closure of the set of endpoints of all W-admissible smooth curves.

Proof.   The first containment follows from (20) above. By Remark VI.1.15 it is no loss of generality to assume that $G(S) = G$. The second containment then follows from Proposition VI.1.13. Finally, Proposition VI.1.14 proves the last assertion.   ∎

We recall from Proposition VI.1.8 that a chain x in G is W admissible in G if and only if its image $\Pi(x) = \pi \circ x$ is W admissible in G/H and from Proposition VI.1.9, that every W-admissible chain (or smooth curve, for that matter) lifts to a W-admissible chain (or smooth curve) in G. Of course, all semigroups in sight are preanalytic (see Definition V.1.2), and $\langle \exp W \rangle$ is a ray semigroup (see Definition V.1.8).

For the following corollary, recall the definition of the tangent wedge $\mathbf{L}(S)$ of a preanalytic semigroup S given in Definition V.1.5.

VI.1.17.   **Corollary.**   *Under the circumstances of* Theorem VI.1.16,

$$(32) \qquad\qquad W \subseteq \mathbf{L}(S) = \mathbf{L}\big(S(W)\big),$$

and

$$(33) \qquad\qquad \operatorname{int}_{G(S)} S = \operatorname{int}_{G(S)} S(W).$$

Proof.   In view of Proposition V.1.7, the first containment of (32) is trivial. The remainder of the proposition then follows from the Ray Semigroup Theorem V.1.13 and from Theorem VI.1.16 above.   ∎

Example V.4.5 shows that the first containment may be proper even though there exist in G local semigroups whose tangent wedge is precisely W (see Sophus Lie's Fundamental Theorem for Semigroups IV.8.7). The dense wind S on the torus $G = \mathbb{R}^2/\mathbb{Z}^2$ (see Example V.1.4(i)) generated by a half ray $W = \mathbb{R}^+ \cdot (1, \sqrt{2})$ is an example with $S = \langle \exp W \rangle = S(W)$ and with $W = \mathbf{L}\big(S(W)\big)$. But $\overline{S} = G$; hence $\mathbf{L}\big(S(W)\big) \neq \mathbf{L}(\overline{\langle \exp W \rangle})$.

Theorem VI.1.16 and its Corollary VI.1.17 show precisely why the W-admissible chains and W-admissible smooth curves can contribute to an answer of the very crucial question when a Lie wedge W in the Lie algebra $\mathfrak{g}$ of a Lie group G is the tangent wedge $\mathbf{L}(S)$ of a subsemigroup S of G.

2. Positive forms

We continue the standard assumptions and notations of Section 1: We deal with a Lie group G, a closed subgroup H, and their respective Lie algebras $\mathfrak{g}$ and $\mathfrak{h}$. The coset projection $g \mapsto gH: G \to G/H$ is denoted π. The coset $\pi(\mathbf{1}) = H \in G/H$ is written ε. Furthermore, W shall be an $\mathrm{Ad}(H)$-invariant Lie wedge in $\mathfrak{g}$ with edge $\mathfrak{h}$. Also, we deal with the left invariant wedge fields Ξ on G/H and Θ on G defined in (10) and (11) of Lemma VI.1.5.

1-Forms

Let M be an arbitrary C^∞-manifold and $T(M)$ its tangent bundle. The tangent space at x is $T(M)_x$. The vector space dual $T(M)_x{}^{\widehat{}}$ is denoted $\widehat{T}(M)_x$. The *cotangent bundle* is therefore denoted $\widehat{T}(M)$. A *1-form* or briefly a *form* is a smooth cross section $\omega: M \to \widehat{T}(M)$ with $\omega(x) \in T(M)_x{}^{\widehat{}}$.

If $f: M \to N$ is a smooth map and ω a 1-form on N, then there is a *pull-back form* $f^*(\omega)$ on M given for $x \in M$ and $X \in T(M)_x$ by

$$(*) \qquad \langle f^*(\omega)(x), X \rangle = \langle \omega(f(x)), df(x)(X) \rangle.$$

Suppose that $\Lambda: M \to \mathcal{W}(M)$ is a wedge field. We write $\Lambda^*(x)$ for the dual $(\Lambda(x))^*$ in $T(M)_x{}^{\widehat{}}$. Thus Λ^* is a function which associates with a point x of M a subset of the cotangent bundle $\widehat{T}(M)$ of M, namely, a wedge in the fiber $\widehat{T}(M)_x$.

VI.2.1. **Definition.** i) A form ω is called *positive* if $\omega(x) \in \Lambda^*(x)$ for all $x \in M$. It is called *strictly positive at a point* x if $\langle \omega(x), X \rangle > 0$ for all $X \in \Lambda(x) \setminus -\Lambda(x)$. It is said to be *strictly positive* if it strictly positive in all points.

ii) If, in particular, M is an open subset of G/H and $\Lambda = \Xi|M$, then a (strictly) positive form ω is called *(strictly) W-positive*. Similarly, a form on an open subset M of G is called *(strictly) W-positive* if it is (strictly) positive for the wedge field $\Theta|M$.

One notices readily that ω is strictly positive at x if and only if $\omega(x) \in$ algint $\Lambda^*(x)$, the algebraic interior of $\Lambda^*(x)$ (see Definition I.2.20 and the following material). We recall from Proposition I.1.7 that $\Lambda(x)$ must be pointed in order for $\Lambda^*(x)$ to have a non-void interior. The aspect of strictly positive forms, therefore, is topologically most relevant if the cone field Λ consists of a *pointed* cones $\Lambda(x)$. The passage to the homogeneous space becomes advisable here.

VI.2.2.   Proposition. *For a form* $\omega:U \to \widehat{T}(U)$, *where* U *is open in* G/H *the following statements are equivalent:*

(1) ω *is* W-*positive (respectively, strictly positive) on* U.

(2) $\pi^*(\omega)$ *is* W-*positive (respectivley, strictly positive) on* $\pi^{-1}(U)$.

Proof.   By Lemma VI.1.5(iii) we have $d\pi(g)\big(\Theta(g)\big) = \Xi\big(\pi(g)\big)$ for all $g \in G$. Thus an $\alpha \in \widehat{T}(G/H)_{\pi(g)}$ belongs to $\Xi\big(\pi(g)\big)^*$ if and only if $\alpha \circ d\pi(g) \in \widehat{T}(W)_g$ belongs to $\Theta^*(g)$. If we apply this to $\alpha = \omega\big(\pi(g)\big)$ then we obtain the desired conclusion for positivity. Strict positivity is defined in such a fashion that the asserted equivalence holds likewise in the case of strict positivity. ∎

VI.2.3.   Lemma. *Let* M *be a paracompact* C^∞-*manifold with a wedge field* Λ. *Let* $\{U_j: j \in J\}$ *be a locally finite open cover such that for each* $j \in J$ *there is a (strictly) positive form* ω_j *on the manifold* U_j. *Then there exists a (strictly) positive form* ω *on* M. *Moreover, given a fixed point* x *and an index* k *with* $x \in U_k$, *then* ω *may be chosen so that it agrees with* ω_k *on a neighborhood of* x.

Proof.   We find a smooth partition of unity $\{f_j: j \in J\}$ such that the support of f_j is contained in U_j. (See for instance [He78], I, Theorem 1.3.) We understand $f_j \cdot \omega_j: M \to \widehat{T}(M)$ to be that smooth section which on U_j takes the value $f_j(x)\omega_j(x) \in \widehat{T}(M)_x$ and the value $0_x \in \widehat{T}(M)_x$ outside U_j. Now we set

$$(1) \qquad\qquad \omega = \sum_{j \in J} f_j \cdot \omega_j.$$

Let $X \in \Xi(x)$, then $\langle \omega(x), X \rangle = \sum_{j \in J} f_j(x)\langle \omega_j(x), X \rangle \geq 0$ since $f_j(x) \geq 0$ and $\langle \omega_j(x), X \rangle \geq 0$ as ω_j is positive. If $X \in \Xi(x) \setminus -\Xi(x)$ and all ω_j are strictly positive, then we note that because of $\sum_{j \in J} f_j(x) = 1$ there is at least one index $j_x \in J$ such that $f_{j_x}(x) > 0$ and thus $f_{j_x}(x_j)\langle \omega_{j_x}(x), X \rangle > 0$ and thus $\langle \omega(x), X \rangle > 0$. Finally suppose that $x \in U_k$ is given. Since $\{U_j: j \in J\}$ is locally finite, there is a *closed* neighborhood N of x such that N is contained in U_k and $\{j \in J: U_j \cap N \neq \emptyset$ and $j \neq k\}$ is a finite set F. Now let

$$U_j' = \begin{cases} U_j, & \text{if } j = k \text{ or } j \notin F; \\ U_j \setminus N, & \text{if } j \in F. \end{cases}$$

Then $\{U_j': j \in J\}$ is a cover of M which refines $\{U_j: j \in J\}$ such that N is contained in exactly one set of the cover and does not meet any other covering set. Now suppose $y \in N$. Since the support of f_j is contained in U_j we have $f_j(y) = 0$ for all $j \in J \setminus \{k\}$. Since $\sum_{j \in J} f_j(y) = 1$, we conclude $f_k(y) = 1$. Hence $\langle \omega(y), y \rangle = \langle \omega_j(x), X \rangle$. Thus ω is the desired form. ∎

This lemma will allow us to show the existence of strictly positive W-admissible forms on G and G/H. Obviously we have to deal with the local situation first.

VI.2.4.   Lemma. *There is an open neighborhood U of ε in G/H and a strictly W-positive form ω on U.*

Proof.   We fix a vector space complement E of $\mathfrak{h}$ in $\mathfrak{g}$. By Lemma IV.4.2 we find a linear functional $\alpha:\mathfrak{g} \to \mathbb{R}$ such that

(2) $$\alpha(\mathfrak{h}) = \{0\} \text{ and } \alpha(X) > 0 \text{ for all } X \in W \setminus -W.$$

Now we choose a C-H-neighborhood B in $\mathfrak{g}$ so small that

(3) $$(X, h) \mapsto (\exp X)h : (E \cap B) \times H \to U'$$

is a diffeomorphism onto an open neighborhood of H in G. (Compare the Units Neighborhood Theorem V.2.4!) Now we define an open neighborhood U of ε in G/H by $U = \pi\big(\exp(E \cap B)\big)$ and a local cross section $\sigma:U \to U' \subseteq G$ by $\sigma\big((\exp X)H\big) = \exp X$ for all $X \in E \cap B$. Notice $U' = \pi^{-1}(U)$.

Next we define ω on U as follows: Every $\xi \in U$ is of the form $\xi = xH$ with $x = \exp X$ with a unique $X \in E \cap B$. Let $Y \in T(G/H)_\xi$. Then $Y = d\pi(x)(Y')$ for some $Y' \in T(G)_x$ which we can write as $Y' = d\lambda_x(\mathbf{1})(Z)$ with a $Z \in T(G)_{\mathbf{1}} = \mathfrak{g}$. Then $Y = d\pi(x)d\lambda_x(\mathbf{1})(Z) = d\mu_x(\varepsilon)d\pi(\mathbf{1})(Z) = d\mu_x(\varepsilon)(Z + \mathfrak{h})$ and we may choose a unique representative Z in E for the coset $Z + \mathfrak{h}$. We now define

(4) $$\langle \omega(\xi), Y \rangle = \alpha(Z).$$

From the definition of $\Xi(\xi) = d\mu_x(\varepsilon)d\pi(\mathbf{1})(W)$ in Lemma VI.1.5 we know that Y is in $\Xi(\xi)$ if and only if Z is in $E \cap W$.

Now α is strictly positive on $(E \cap W) \setminus \{0\}$ by (2). It now follows that

(5) $$\alpha(Z) > 0 \text{ for all } 0 \neq Z \in E \cap W.$$

By (4) this means

(6) $$\langle \omega(\xi), Y \rangle > 0 \text{ for all } 0 \neq Y \in \Xi(\xi).$$

This shows that ω is strictly positive on U.   ∎

VI.2.5.   Lemma. *Let U be an open subset of G/H and $g \in G$. Let $U_g = \mu_g(U)$ and define the diffeomorphism $f:U \to U_g$ by the restriction $\mu_g|U$. If ω is a (strictly) W-positive form on U then $(f^{-1})^*(\omega)$ is a (strictly) W-positive form ω_g on U_g.*

Proof.   Let $\xi \in U_g$. Then $\eta = \mu_{g^{-1}}(\xi) = f^{-1}(\xi) \in U$. Let $X \in \Xi(\xi)$. Then

(7) $$Y \overset{\text{def}}{=} d\mu_{g^{-1}}(\eta)(X) \in \Xi(\eta)$$

since Ξ is an invariant wedge field. Now $\langle \omega_g(\xi), X \rangle = \langle \omega(\eta), Y \rangle \geq 0$ since ω is a W-positive form. This shows that ω_g is a W-positive form. The proof for strict positivity is completely analogous.   ∎

VI.2.6.   Theorem.   (The Existence Theorem for Strictly W-Positive Forms)
*If G is a Lie group, H a closed subgroup, and W any Lie wedge in $\mathfrak{g}$ whose edge
is $\mathfrak{h}$, then there are strictly W-positive forms on G and G/H.*

Proof.   The assertion for G follows at once from the assertion for G/H via
Proposition VI.2.2. Thus we have to prove the assertion for G/H. By Lemma
VI.2.4, there is an open neighborhood $U = U_1$ of ε in G/H and a strictly W-
positive form $\omega = \omega_1$ on U_1. We apply Lemma VI.2.5 and obtain W-positive
forms ω_g on $U_g = \mu_g(U)$. Now we apply Lemma VI.2.3 to the paracompact
manifold G/H and a locally finite refinement of the open cover $\{U_g : g \in G\}$ with
the family $\{\omega_g : g \in G\}$ of strictly positive forms and conclude that there is a strictly
W-positive form ω on G/H. This completes the proof.   ∎

In view of the precise statement of Lemma VI.2.4, the proof of the preceding
theorem yields the following additional information:

VI.2.7.   Corollary.   *Under the circumstances of the preceding theorem and
with the notation of Lemma VI.2.4, there is an open neighborhood U_0 of ε and there
is a strictly W-positive form on G/H which agrees with the form ω of Lemma
VI.2.4 on U_0.*   ∎

For the following we recall that a differentiable form ω is called *closed*
if its exterior derivative $d\omega$ vanishes; in the case of a 1-form this means simply
$0 = 2d\omega(X,Y) = X\big(\omega(Y)\big) - Y\big(\omega(X)\big) - \omega([X,Y])$. A 1-form ω is *exact* if there is
a smooth function $f: M \to \mathbb{R}$ with $\omega = df$. The Poincaré Lemma says that a closed
differential 1-form on any manifold M which is diffeomorphic to a star shaped open
set of $\mathbb{R}^n$ is exact. However, we do not have to invoke the Poincaré Lemma in order
to sharpen Lemma VI.2.4 as follows:

VI.2.8.   Proposition.   *There is an open neighborhood U' of ε in G/H and
an exact strictly positive W-form ω on U.*

Proof.   Let U and ω be as in Lemma VI.2.4. For $\xi \in U$ we set $f(\xi) =
\alpha\big((\exp|B)^{-1}\sigma(\xi)\big)$. Then $f: U \to \mathbb{R}$ is differentiable and

$$df(\xi) = \alpha\big(d(\exp|B)^{-1}\big(\sigma(\xi)\big)d\sigma(\xi)\big).$$

For $\xi = \varepsilon$ and for Y and Z as in the proof of Lemma VI.2.4 we find $\langle df(\varepsilon), Y \rangle =
\alpha(Z) = \langle \omega, Y \rangle$. Thus df agrees with ω at ε, hence is strictly positive at ε and
then also in a whole neighborhood U' of ε in U.   ∎

Let us call a form ω on a manifold M *exact on the open set U* if $\omega|U$ is
exact. Such a form is closed on U; for if $\omega|U = df$ with a smooth function f on
U, then $d\omega|U = ddf = 0$.

If $\psi: M \to N$ is a smooth map and ω is exact on N, then $\psi^*(\omega)$ is exact
on M, for if $\omega = df$, then $\varphi^*(\omega) = d(f \circ \psi)$.

From the preceding results we now have at once the following observation:

VI.2.9.   Corollary.   *On G and G/H there are strictly W-positive 1-forms
which are exact on open neighborhoods of 1 and ε, respectively.*   ∎

We mention in passing that the existence of globally exact forms on a manifold M is a cohomological question once closed forms are known. If we write $H^1(M)$ for the quotient vector space of the vector space of all closed forms modulo the vector space of all exact forms, then the vanishing of $H^1(M)$ is equivalent to the statement that closed forms are exact. However in our situation, the Example VI.1.6 of the Heisenberg group shows that the problem has additional complications: The form ω produced there in Example VI.1.6(iv) is strictly W-positive and but nowhere locally closed. Our results show, however, that there are strictly W-positive forms on the Heisenberg group for the given W which are locally exact at 0 (in fact agree with the exact form ω' given by $\langle \omega'(g), (x, y, z) \rangle = z$ for all g on a neighborhood of 0.) Yet for cohomological reasons, every closed form on $\mathbb{R}^3$ is exact. Our later results will show that in the present case there are no strictly W-positive exact forms. Hence there cannot be any strictly W-positive closed forms.

3. W-admissible chains revisited

We continue the set-up of the preceding sections: G is a Lie group with a closed subgroup H, and $\mathfrak{g}$ and $\mathfrak{h}$ are their respective Lie algebras. We assume that we are given an $\mathrm{Ad}(H)$-invariant Lie wedge W in $\mathfrak{g}$ whose edge is exactly $\mathfrak{h}$. In Definition VI.1.7, we introduced the concepts of smooth chains and of W-admissible chains. In this section we investigate how the idea of W-admissible chains which is defined in terms of the tangent bundles of G and G/H relates to the concept of W-positive forms which we introduced in the preceding section and which are defined in terms of the cotangent bundle of G and G/H.

VI.3.1.   Proposition. i) *If $\xi:[0,T] \to U$ with U open in G/H is a W-admissible chain, and ω is a W-positive form on G/H, then $\langle \omega(\xi(t)), \dot{\xi}(t) \rangle \geq 0$ for all $t \in [0,T]$*

ii) *If ω is strictly W-positive at $\xi(t)$, then $\langle \omega(\xi(t)), \dot{\xi}(t) \rangle > 0$ provided that $\dot{\xi}(t) \neq 0$.*

Proof.   i) $\langle \omega(\xi(t)), \dot{\xi}(t) \rangle \in \omega(\xi(t))\big(\Xi(\xi(t))\big) \subseteq R^+$, because ξ is W-admissible and ω is W-positive.

ii) If ω is strictly W-positive at $\xi(t)$, then $\omega(\xi(t))$ takes positive values on the non-zero elements of $\Xi(\xi(t))$.   ∎

Once a W-positive form ω is given, then for any W-admissible chain $\xi:[0,T] \to G/H$, the integral

$$(1) \qquad \rho_\xi:[0,T] \to \mathbb{R}^+, \qquad \rho_\xi(t) = \int_0^t \langle \omega(\xi(\tau)), \dot{\xi}(\tau) \rangle d\tau$$

is a kind of arc-length measured in the direction of ω. It is a monotone piecewise smooth function. We might call it ω-*time*. In the following proposition we assume that ω is exact on a neighborhood U of ε in G/H.

VI.3.2.   Proposition. *Suppose that that $\xi(0) = \varepsilon$ and $\mathrm{im}\,\xi \subseteq U$ for an open neighborhood of ε on which ω is exact, that is, there is a smooth function $f:U \to \mathbb{R}$ with $df = \omega|U$ with $f(\varepsilon) = 0$. Then*

$$(2) \qquad \rho_\xi(t) = f(\xi(t)) \quad \text{for all } t \in [0,T].$$

Proof. We have $\rho_\xi(0) = 0 = f\big(\xi(0)\big)$. Then it suffices to observe that the two functions $t \mapsto \rho_\xi(t), f\big(\xi(t)\big)$ have the same derivative. However, the derivative of the first one is $\langle \omega\big(\xi(t)\big), \dot\xi(t)\rangle$ while that of the second is $df\big(\xi(t)\big)\big(\dot\xi(t)\big)$, and these are ostensibly the same. ∎

If an open neighborhood U of ε in G/H and a chain $\xi:[0,T] \to G/H$ with $\xi(0) = \varepsilon$ and with $\xi([0,T]) \not\subseteq U$ are given, we set

$$(3) \qquad \tau(\xi, U) = \sup\{t \in [0,T]: \xi([0,t[) \subseteq U\} = \min\{t \in [0,T]: \xi(t) \notin U\}.$$

VI.3.3. Theorem. (The Escape Theorem) *Let G be a Lie group with a closed subgroup H and W an $\mathrm{Ad}(H)$-invariant Lie wedge in $\mathfrak{g}$ with edge $\mathfrak{h}$. Suppose that ω is a W-positive form on G/H (see Definition VI.2.11). Suppose further that ω is positive on an open neighborhood U and strictly positive on open neighborhood U_+ of ε. Then for each open neighborhood U_0 of ε inside $U \cap U_+$ there is a positive number m such that for every r with $0 < r \le m$ there is an open neighborhood $N \subseteq U_0$ of ε such that for every W-admissible chain $\xi:[0,T] \to G/H$ with $\xi(0) = \varepsilon$ and with $\xi([0,T]) \not\subseteq N$ one has*

$$(4) \qquad \rho_\xi\big(\tau(\xi, N)\big) = r.$$

Moreover, if the smooth function $f:U \to \mathbb{R}$ is such that $df = \omega$ and $f(\varepsilon) = 0$, then

$$(5) \qquad \rho_\xi(t) = f\big(\xi(t)\big)$$

for each $\xi:[0,T] \to U$ in $\mathcal{A}(G/H)$.

Proof. Suppose that $\xi:[0,T] \to U$, $\xi(0) = \varepsilon$ is W-admissible. In view of Proposition VI.3.2 for all $t \in [0,T]$ we have

$$\rho_\xi(t) = f\big(\xi(t)\big),$$

as asserted in the last part of the theorem.

Now we go into $\mathfrak{g}$ and choose $\alpha \in W^*$ in such a fashion that

$$\alpha(X) > 0 \quad \text{for all} \quad X \in W \setminus -W.$$

As in the proof of Lemma VI.2.4 we introduce a vector space complement E for $\mathfrak{h}$ in $\mathfrak{g}$ and a C-H-neighborhood B so small that the function $\varphi_0: B \to U_0$, $\varphi_0 = \pi \circ \exp$ induces a diffeomorphism φ from $E \cap B$ onto a neighborhood U_{00} of ε in U_0. This allows us to define $F_0: B \to \mathbb{R}$ such that $F_0(X) = f\big(\varphi(X)\big)$ and $F = F_0|(E \cap B)$. We note that $(\varphi^{-1})^*(dF) = df|U_{00} = \omega|U_{00}$. If we now assume $\xi([0,T]) \subseteq U_{00}$ and define $X:[0,T] \to B$ by $\varphi\big(X(t)\big) = \xi(t)$, then (5) implies

$$(6) \qquad \rho_\xi(t) = F\big(X(t)\big).$$

Next we choose a positive number r' so small that the compact base $C = E \cap W \cap \alpha^{-1}(r')$ is contained in B. Now we use the fact that ω is strictly *positive on* $U_{00} \subseteq U_+$ and conclude that $dF_0(0)(Y) > 0$ for all Y in the compact set C. Hence $dF_0(0)$ remains positive on a compact convex neighborhood C' of C in $E \cap \alpha^{-1}(r')$. Then $W' = \mathfrak{h} + \mathbb{R}^+ \cdot C'$ is a wedge with $W \subset\subset W'$ and $C' = E \cap W' \cap \alpha^{-1}(r')$. Now we find an open C-H-neighborhood B' with $\overline{B'} \subseteq B$ and a positive number δ satisfying the following conditions:

(i) $\langle dF_0(X), Y \rangle \geq \delta$ for $X \in \overline{B'}$ and $Y \in C'$,

(ii) $E \cap g(\operatorname{ad} X)(W) \subseteq W'$ for $X \in \overline{B'}$.

(See Proposition IV.6.11.) Set $U_1 = \varphi(E \cap B')$.

We consider $Y \in C'$, and define $\eta: \mathbb{R}^+ \to G/H$ by $\eta(t) = \pi(\exp t \cdot Y)$. For all $t \in [0, \tau(\eta, U_1)]$ notice $F(t \cdot Y) = \int_0^t \langle dF(\tau \cdot Y), Y \rangle d\tau \geq t\delta$ in view of (i) above. Since every $X \in E \cap W' \cap \partial B'$ where $\partial B'$ denotes the boundary of B' in $\mathfrak{g}$ is of the form $\tau(\eta, U_1) \cdot Y$ with some $Y \in C'$, we conclude that

(7)
$$m \overset{\text{def}}{=} \min\{F(X): X \in E \cap W' \cap \partial B'\} > 0.$$

Let $0 < r \leq m$ and define $B'' = B' \cap F_0^{-1}(] - \infty, r[)$ and set $N = \varphi(E \cap B'') \subseteq U_1 \subseteq \overline{U_1} \subseteq U_0 \subseteq U$.

If $\xi: [0, T] \to G/H$ is a W-admissible chain with $\xi([0, T]) \not\subseteq N$, it is no loss of generality to assume that $\xi([0, T]) \subseteq \overline{U_1}$ and $T = \tau(\xi, U_1)$. Again we define $X: [0, T] \to E \cap \overline{B'}$ by $\varphi(X(t)) = \xi(t)$. Since ξ is W-admissible, we know that $\dot{X}(t) \in E \cap g(\operatorname{ad} X(t))(W)$ for all $t \in [0, T]$ as $g(\operatorname{ad} Z) = \lambda_Z^*(0)$ with $\lambda_Z^*(Z') = Z * Z'$ (compare the proof of Lemma VI.2.4). For these t, however, we have $X(t) \in \overline{B'}$, hence $\dot{X}(t) \in W'$ in view of relation (ii) above. We apply the Invariance Theorem for Vector Fields I.5.17 and the simple observation that $W' \subseteq L_{X(t)}(W')$ (see Remark I.1.10) in order to conclude that $X(t) \in W'$ for all $t \in [0, T]$. From (6) and (7) above we now infer that

(8)
$$\rho_\xi(\tau(\xi, U_1)) \geq m \geq r,$$

since $\xi(t) \in U_1$ if and only if $X(t) \in B'$. Now we claim that $\rho_\xi(\tau(\xi, N)) = r$, and this will finish the proof. Let us abbreviate $\sigma = \tau(\xi, N)$. By the definition of B'' and $\tau(\xi, N)$ we have $F(X([0, \sigma])) \subseteq [0, r]$, whence

(9)
$$\rho_\xi(\sigma) \leq r$$

in view of (6). If we assume $\rho_\xi(\sigma) < r$, then from (8) and (9), by the Intermediate Value Theorem we conclude that here must be a t with

$$\sigma < t \leq \tau(\xi, U_1) \quad \text{and} \quad \rho_\xi(t) = r,$$

and that t is minimal with respect to these properties. Now let $\sigma < s < t$. Then $F(X([0, s])) = \rho_\xi([0, s]) \subseteq \rho_\xi([0, t[) \subseteq [0, r]$. Hence

$$X([0, s]) \subseteq B' \cap F^{-1}(] - \infty, r[) = B'',$$

which is tantamount to $\xi([0, s]) \subseteq N$. Hence $\sigma = \tau(\xi, N) \geq s$ by the definition of $\tau(\xi, N)$. But this contradicts $\sigma < s$. Thus our assumption must be false and the asserted equality $\rho_\xi(\sigma) = r$ follows. The proof is complete. ∎

This theorem says that for $0 < r \leq m$ we find $N = N(r)$ so that every W-admissible chain $\xi \in \mathcal{A}(G/H)$ which escapes from N escapes *exactly* at ω-time r. Corollary VI.2.9 shows that forms ω such as are required in the Escape Theorem do exist.

This illustrates how we may use W-positive forms in order to show how W-admissible chains in G/H starting at ε move away from ε. This allows us some control over the set of all W-admissible chains in G which stay inside $\pi^{-1}(U)$. For this purpose, for any open neighborhood U of ε in G/H, we shall write

$$\mathcal{A}_U(G) = \{x \in \mathcal{A}(G): \pi(\operatorname{im} x) \subseteq U\},$$

and $\mathcal{A}_U(G/H) = \Pi\big(\mathcal{A}_U(G)\big)$. Accordingly, we shall set

$$\begin{aligned} S_U(W) &= \operatorname{ev}\big(\mathcal{A}_U(G)\big) \\ &= \{g \in G: (\exists x \in \mathcal{A}_U(G)) \quad x: [0, T] \to G, \quad x(T) = g\}. \end{aligned}$$

The set $S_U(W)$ is an excellent candidate for a local subsemigroup with respect to the open set $U' = \pi^{-1}(U)$ with tangent wedge W, that is, a subset S of U' satisfying

(i) $H \subseteq S$ and $SS \cap U' \subseteq S$,

(ii) $\mathbf{L}(S)$ (defined as $L_0(\exp^{-1} S)$) equals W.

However, (i) appears to be very delicate; we do not know whether a pair (S, U') satisfying (i) and (ii) exists. This is somewhat frustrating in view of Lie's Fundamental Theorem IV.8.7; the local semigroups there were in fact constructed locally according to the scheme which gives us $S_U(W)$. When contemplating the problem of the existence of (S, U'), we find it particularly helpful to visualize Example V.2.5 and its implications. It illustrates the additional complications beyond the ones of the purely local situation which we encounter when we ascend to a situation which is intermediate between the purely local and the global one. At any rate, we shall now observe that at least condition (ii) above is satisfied for $S = S_U(W)$. For this purpose we generalize the concept of the set of subtangents at a point x of a C^∞-manifold M (see Definitions I.5.1 and IV.1.15). Let S be a subset of a C^∞-manifold and $x \in M$ and suppose that there is a function $\alpha: D \to \overline{S}$, for which $0 \in D \subseteq \mathbb{R}^+$, $\alpha(t) \in S$ for $0 < t \in D$, 0 is a cluster point of $D \setminus \{0\}$ and that, for all smooth functions $f: M \to \mathbb{R}$, the following limit exists

$$\lim_{\substack{h \searrow 0 \\ h \in D}} \frac{1}{h}\Big(f\big(\alpha(h)\big) - f\big(\alpha(0)\big)\Big)$$

Let us call this limit $X(f)$. It is an easy exercise to verify that the function $f \mapsto X(f)$ is a derivation on $C^\infty(M)$, hence is a member $X \in T(M)_x$.

VI.3.4. **Definition.** The element X in the tangent space $T(M)_x$ of M at x is called a *subtangent vector of S at x*. We let $L_x(S) \subseteq T(M)_x$ denote the set of all subtangent vectors of S at x.

We note right away that if $S \subseteq L$ for a vector space L, then this concept agrees with that defined in Definitions I.5.1 and IV.1.15 if we identify L with the tangent space at each point of L. Moreover, if $S \subseteq G$, then the equation $L_1(S) = L_0(\exp^{-1} S)$ links the new concept in a consistent fashion with the old ones.

We may leave it as an exercise to the reader to verify the following version of Proposition IV.1.19: *If* $f: M_1 \to M_2$ *is a smooth map between smooth manifolds and* $S \subseteq M_1$, $x \in M_1$, *then*

$$df(x)\big(L_x(S)\big) \subseteq L_{f(x)}\big(f(S)\big).$$

We set $\mathbf{L}\big(S_U(W)\big) = L_0\big(\exp^{-1} S_U(W)\big) = L_1\big(S_U(W)\big)$.

VI.3.5.   Lemma.   *Under the hypotheses of the* Escape Theorem VI.3.3 *we have* $\mathbf{L}\big(S_U(W)\big) = W$.

Proof.   Clearly, $W \subseteq \mathbf{L}\big(S_U(W)\big)$. We must show the reverse inequality. Now we observe

$$d\pi(0)\Big(L_1\big(S_U(W)\big)\Big) \subseteq L_\varepsilon\Big(\pi\big(S_U(W)\big)\Big).$$

Let us set $\Sigma_U(W) = \pi\big(S_U(W)\big)$. In view of Propositions VI.1.8 and 9, this set is equal to

$$\mathrm{ev}\big(\mathcal{A}_U(G/H)\big) = \{\gamma \in G/H: \quad (\exists \xi)\, \xi: [0,T] \to G/H, \quad \xi \in \mathcal{A}_U(G/H), \quad \xi(T) = \gamma\}.$$

Since $W = \big(d\pi(0)\big)^{-1}(V)$ by Proposition VI.1.8(i), it now suffices to verify

$$(10) \qquad\qquad\qquad L_\varepsilon\big(\Sigma_U(W)\big) \subseteq V.$$

We prove this by showing that for each pointed cone V_1 in $T(G/H)_\varepsilon$ with $V \subset\subset V_1$ we have

$$(11) \qquad\qquad\qquad L_\varepsilon\big(\Sigma_U(W)\big) \subseteq V_1.$$

Now $W_1 = \big(d\pi(1)\big)^{-1}(V_1)$ is a wedge with edge $\mathfrak{h}$ and with $W \subset\subset W_1$. We select the set-up of the proof of Theorem VI.3.3. We may assume that $W_1 \subseteq W'$. By Proposition IV.6.11 we choose the C-H-neighborhood B' in B so small that condition (i) of the proof of Theorem VI.3.3 and, in place of condition (ii) of the proof of Theorem VI.3.3, the stronger condition

(ii′) $E \cap g(\mathrm{ad}\, X)(W) \subseteq W_1$ for all $X \in \overline{B'}$

is satisfied. We then proceed to construct m and B'' with $N = \varphi(E \cap B'')$ as in Theorem VI.3.3. We choose $r = m$. If $\xi \in \mathcal{A}_N(G/H)$ and $X: [0,T] \to E \cap B''$ is again defined by $\xi(t) = \varphi\big(X(t)\big)$, then $\dot{X}(t) \in E \cap g\big(\mathrm{ad}\, X(t)\big)(W) \subseteq W_1$, whence $X(t) \in W_1$ for all $t \in [0,T]$. Using $d\varphi(0)(L_0(\varphi^{-1}\Sigma_N W)) = L_\varepsilon(\Sigma_N W)$ we find

$$L_\varepsilon\big(\Sigma_N(W)\big) \subseteq V_1.$$

Now we claim that

$$(12) \qquad\qquad\qquad \Sigma_N(W) = N \cap \Sigma_U(W).$$

Once this claim is established, (11) is an immediate consequence and the proof is complete. The left hand side of (12) is trivially contained in the right hand side. We have to show the converse. Thus let $\xi_0 \in N \cap \Sigma_U(W)$. Then there is a $\xi\colon [0,T] \to U$ in $\mathcal{A}_U(G/H)$ with $\xi(T) = \xi_0$. We claim that $\xi([0,T]) \subseteq N$; if this is shown, the proof is complete since then $\xi \in \mathcal{A}_N(G/H)$ and thus $\xi_0 = \xi(T) \in \Sigma_N(W)$. But by the Escape Theorem VI.3.3, we have a real valued smooth function f on U such that $\rho_\xi(t) = f\big(\xi(t)\big)$ and that $f(N) \subseteq]-\infty, m[$. By way of contradiction, assume that $\xi([0,T]) \not\subseteq N$. As $\xi(T) \in N$ we have $\tau(\xi, N) < T$ by the definition of $\tau(\xi, N)$. But since ρ_ξ is monotone, $m = \rho_\xi\big(\tau(\xi, N)\big) \le \rho_\xi(T) = f\big(\xi(T)\big)$. Since $\xi(T) = \xi_0 \in N$, we conclude $f\big(\xi(T)\big) < m$. This leads to the contradiction $m < m$. The proof is complete. ∎

We summarize our results in the following theorem:

VI.3.6.   Theorem. *Let G be a Lie group with a closed subgroup H and W an $\mathrm{Ad}(H)$-invariant Lie wedge in $\mathfrak{g}$ with edge $\mathfrak{h}$. Let ω be a 1-form on G/H. Assume the following hypotheses:*

(i) *ω is W-positive.*

(ii) *ω is strictly W-positive at ε*

(iii) *ω is exact on an open neighborhood U of ε in G/H.*

Then $\mathbf{L}\big(S_U(W)\big) = W = L_0\big(\exp^{-1} S_U(W)\big)$. ∎

This theorem shows that the construction of elements reachable from the identity by W-admissible chains can be controlled in the sense that the subtangent set at $\mathbf{1}$ remains W—but only as far as exactness reaches. Taking for U the entire manifold G/H we recall that $S_{G/H}(W) = S(W)$ is a semigroup in which the ray semigroup $S = \langle \exp W \rangle$ is dense by the Approximation Theorem for Chains VI.1.16. We then have $W \subseteq \mathbf{L}(S) \subseteq \mathbf{L}\big(S(W)\big) = W$ by Theorem VI.3.6. Thus $\mathbf{L}(S) = W$, and we have obtained a fundamental result giving sufficient conditions for the global validity of Lie's Fundamental Theorem. We shall use the following notation:

VI.3.7.   Definition. Let G be a Lie group and W a Lie wedge in $\mathfrak{g}$. We say that W is *global in G* if $\mathbf{L}(\langle \exp W \rangle) = W$.

Notice that in this case, $S = \langle \exp W \rangle$ is strictly infinitesimally generated by Definition V.1.12. The half-line $\mathbb{R}^+$ in $\mathbb{R} = \mathbf{L}(\mathbb{R}) = \mathbf{L}(\mathbb{R}/\mathbb{Z})$ is global in $\mathbb{R}$ but not in $\mathbb{R}/\mathbb{Z}$.

VI.3.8.   Theorem. (The Globality Theorem) *For a Lie wedge W in the Lie algebra $\mathfrak{g}$ of a finite dimensional Lie group G, the following conditions are sufficient for W to be global:*

(I) *There is a closed connected subgroup H such that $\mathfrak{h}$ is the edge of W.*

(II) *There is an exact 1-form ω on G/H which satisfies the following conditions:*

(i) *ω is W-positive.*

(ii) *ω is strictly W-positive at $\varepsilon = H$ in G/H.*

Proof.   By Lemma VI.1.2, the connectivity of H implies that W is $\mathrm{Ad}(H)$-invariant. Then the theorem is a consequence of Theorem VI.3.6. ∎

This theorem contains much more information than meets the eye:

VI.3.9.   Corollary.   *Under the circumstances of the* Globality Theorem VI.3.8, *let* $S = \langle \exp W \rangle$ *and let* $S(W)$ *be the semigroup of all points reachable from* $\mathbf{1}$ *by* W*-admissible chains. Then* $S(W)$ *is an infinitesimally generated semigroup and* S *is the largest strictly infinitesimally generated subsemigroup of* $S(W)$. *Moreover,* $\mathbf{L}(\overline{S}) = W$.

Proof.   By the Approximation Theorem for Chains VI.1.16, we have

$$(13) \hspace{3cm} S \subseteq S(W) \subseteq \mathrm{cl}_{G(S)} S \subseteq G(S).$$

If A is the subgroup generated by $\exp W$ and $A(W)$ the subgroup generated by $S(W)$, then $S \subseteq S(W) \subseteq A$ by (13) since the underlying groups of A and $G(S)$ agree. But then $A = A(W)$, whence $G(S) = G\big(S(W)\big)$. Hence $G\big(S(W)\big) = G(S) = \langle \exp W \cup \exp -W \rangle = \langle \exp \mathbf{L}\big(S(W)\big) \cup \exp -L\big(S(W)\big) \rangle$. Thus condition (iii) of Definition V.1.12 is satisfied. Also, $S(W) \subseteq \mathrm{cl}_{G(S)} S = \mathrm{cl}_{G\big(S(W)\big)} S \subseteq \mathrm{cl}_{G\big(S(W)\big)} S(W)$ verifies condition (ii) of Definition V.1.12, and condition (i) is trivial.

Finally, we recall now that $L_0\big(\exp^{-1} S(W)\big) = L_0\big(\exp^{-1} \overline{S(W)}\big)$ since Proposition I.5.6 applies to subtangent vectors of sets on manifolds. Because of $\overline{S} = \overline{S(W)}$, the last assertion follows.   ∎

Thus the sufficient conditions of the Globality Theorem in effect guarantee the existence of a *closed* subsemigroup $T = \overline{S}$ of G such that $\mathbf{L}(T) = W$.

Occasionally, the Globality Theorem VI.3.8 yields negative results. If G is the Heisenberg group of Example VI.1.6 and W is any Lie wedge in $\mathfrak{g} = \mathbb{R}^3$ not containing the z-axis but having at least one point of the z-axis in its interior, then W is not global in G by Lemma V.4.5. Moreover, every closed form is exact on $\mathbb{R}^3$. Hence by the Globality Theorem, no closed form can be at the same time W-positive and strictly W-positive at 0, while strictly W-positive forms exist which are exact on some neighborhood of 0 by Corollary VI.2.9.

In Section 5 below we shall see that the conditions of the Globality Theorem are also necessary when $G = G(S)$.

4. Ordered groups and homogeneous spaces

In the previous section we gave sufficient conditions for a Lie wedge to be global. We shall investigate now to which extent these conditions are also necessary. For this purpose we use the order theoretical frame work whose background was prepared in the preliminary Section 0 of the preceding Chapter V.

Monotone functions and measures

On a left preordered group G one can talk about monotonicity of real valued functions and Radon measures. *In this subsection, $C_{00} = C_{00}(G)$ shall denote the algebra of continuous functions $f: G \to \mathbb{R}$ with compact support and $C_{00}^+ = C_{00}^+(G)$ the cone $\{f \in C_{00}(G): f \geq 0\}$.*

VI.4.1. **Definition.** i) A real valued function $f: G \to \mathbb{R}$ on a left preordered group is called *monotone* if

$$x \preceq y \Longrightarrow f(x) \leq f(y) \text{ for all } x, y \in G.$$

If S is a monoid in a group G and f is a monotone function with respect to the preorder defined by S, then we shall briefly say that *f is monotone with respect to S.*

ii) A Radon measure μ on a locally compact groups G is called *monotone* if it is positive and for all $g, g' \in G$ with $g \preceq g'$ and all $f \in C_{00}^+$ one has

(1) $$\int f(xg^{-1})d\mu(x) \leq \int f(xg'^{-1})d\mu(x).$$

If S is the positivity semigroup then *f is monotone if and only if $f(x) = \min f(xS)$ for all $x \in G$.* We also notice right away that *every right invariant Haar measure is monotone.* The nomenclature of monotonicity for measures is justified by the following observation:

VI.4.2.   Remark. A Radon measure μ on a locally compact group G is monotone if an only if for all $g, g' \in G$ with $g \preceq g'$ and for all compact subsets $A \subseteq G$ one has

$$(2) \qquad\qquad\qquad\qquad \mu(Ag) \leq \mu(Ag').$$

Proof.   Let χ_A denote the characteristic function of A. Note that $\mu(Ag) = \int \chi_{Ag}(x)d\mu(x) = \int \chi_A(xg^{-1})d\mu(x)$ and that χ_A can be approximated by functions $f_j \in C_{00}^+$ with $f_j \searrow \chi_A$ in the sense that $\int f_j d\mu \searrow \int \chi_A d\mu$. If (1) holds then $\int f_j(xg^{-1}d\mu(x) \leq \int f_j(xg'^{-1})d\mu(x)$, whence (2). The proof of the converse is left to the reader as an exercise in measure theory.   ∎

Monotone functions exist on every left preordered group:

VI.4.3.   Remark. Suppose that S is a monoid in a group G and χ is the characteristic function of S in G. Let $\preceq$ be a left preorder whose positivity semigroup $S_{\preceq}$ is contained in S. Then χ is monotone.

Proof.   Let $x \preceq y$. Then $y \in xS_{\preceq}$, and if $x \in S$, then $y \in SS_{\preceq} \subseteq SS = S$. Thus, if x and y are neither both inside nor both outside S and thus have both χ-value 1 or 0, repectively, then $x \notin S$ and $y \in S$. Hence $\chi(x) = 0 < 1 \leq \chi(y)$.   ∎

VI.4.4.   Remark. If f is a monotone function on G and if H is the maximal subgroup of the positivity semigroup S, then f is constant on the cosets xH, hence defines a monotone function $\underline{f}$ on G/H such that $\underline{f}(xH) = f(x)$.

Proof.   If $y \in xH$ then $x \approx y$ and thus $x \preceq y$ and $y \preceq x$, whence $f(x) \leq f(y)$ and $f(y) \leq f(x)$. Hence $f(x) = f(y)$. The remainder is clear.   ∎

We shall observe now that once we have monotone functions on a locally compact group, we also have monotone measures, as monotone functions and measures are related. Here and in the remainder of this section *we shall fix on a locally compact group G a right invariant Haar measure m.* Instead of $\int f(x)dm(x)$ we shall also write $\int f(x)dx$ or even $\int f$.

VI.4.5.   Lemma. (i) *Let G be a left preordered locally compact group and S the positivity semigroup. If $f \in L^{\infty}(G, m)$ is monotone, then $f \cdot m$ is a monotone measure. In particular, $\chi_{\overline{S}} \cdot m$ is a monotone measure.*

(ii) *If μ is a monotone measure on G and $0 \leq f \in L^1(G, \mu)$, then the function $f \natural \mu$ defined by*

$$(3) \qquad\qquad\qquad (f \natural \mu)(g) = \int f(xg^{-1})d\mu(x)$$

is a monotone function.

(iii) *If G is a Lie group and $f \in C_{00}^+ \cap C^{\infty}$, and if μ is a Radon measure on G, then $f \natural \mu \in C^{\infty}$.*

Proof.   (i) Let $p \in C_{00}^+$ and $g \preceq g'$. Then $\int p(xg^{-1})f(x)dx = \int p(x)f(xg)dx$ by the right invariance of Haar measure. Now $xg \preceq xg'$ and thus $\int p(x)f(xg)dx \leq \int p(x)f(xg')dx = \int p(xg'^{-1})f(x)dx$ as f is monotone. This proves the first claim.

By Remark VI.4.3 above, $\chi_{\overline{S}}$ is a monotone function. Hence $\chi_{\overline{S}} \cdot m$ is a monotone measure.

(ii) We suppose $g \preceq g'$. Then

$$(f \natural \mu)(g) = \int f(xg^{-1}) d\mu(x) \le \int f(xg'^{-1}) d\mu(x) = (f \natural \mu)(g')$$

since μ is monotone and any positive L^1-function may be approximated by C_{00}^+-functions in measure.

(iii) Let $g \in G$ and let X be a smooth vector field on G. Set $\Phi_1(x)(g) = f(xg^{-1})$ and $\varphi_1(g) = \int \Phi_1(x)(g) d\mu(x)$. Write $\varphi_2(g) = \langle d\varphi_1(g), X_g \rangle = X_g(\varphi_1)$. Since differentiation and integration commute, this expression is well defined and equals $X_g(\int \Phi_1(x)(\bullet) d\mu(x)) = \int X_g(\Phi_1(x)(\bullet)) d\mu(x)$. Now we set

$$\Phi_2(x)(g) = X_g(\Phi_1(x)(\bullet)).$$

Then $\varphi_2(g) = \int \Phi_2(x)(g) d\mu(x)$. But then the path to induction is well paved, and along this path we obtain indeed that $f \natural \mu$ is a smooth function. ∎

It is helpful to our geometric intuition if we realize that for any compact subset A of G and the measure $\mu = \chi_{\overline{S}} \cdot m$ we have $\mu(A) = \int \chi_A \chi_{\overline{S}} dm = m(A \cap \overline{S})$. If $g \in G$ and $s \in S$, then $\mu(Ags) = m(Ags \cap \overline{S}) = m(Ag \cap \overline{S}s^{-1})$ by the right invariance of m. Thus, since $\overline{S} \subseteq \overline{S}s^{-1}$ and as $\mu(Ag) = m(Ag \cap \overline{S})$, we observe once more that $\mu(Ag) \le \mu(Ags)$.

VI.4.6. Lemma. *Let μ be a monotone measure on a left preordered locally compact group G and let H be the group of invertible elements of the positivity semigroup S. Then μ is H-right invariant.*

Proof. Let $f \in C_{00}$ and $h \in H$. Then $1 \approx h^{-1}$, since both h and h^{-1} are in S. Thus $\int f(xh) d\mu(x) = \int f(x(h^{-1})^{-1}) d\mu(x) = \int f(x1^{-1}) d\mu(x) = \int f(x) d\mu(x)$.

For each homeomorphism T of G and each Radon measure μ, a Radon measure $T(\mu)$ is defined by the equation

$$\int f dT(\mu) = \int (f \circ T) d\mu \text{ for } f \in C_{00}.$$

The following simple remark is clear:

VI.4.7. Lemma. *The group G acts on the vector space of Radon measures on the left by*

$$g \bullet \mu = \rho_g^{-1}(\mu),$$

and

$$(f \natural \mu)(g) = \int f d(g \bullet \mu).$$

∎

Note that $(g \bullet \mu)(A) = \mu(Ag)$ for all compact sets A. The following technical lemma begins to point into the direction of the construction of 1-forms which are strictly W-positive at the origin.

VI.4.8.   **Lemma.**   *Assume that G is a Lie group and that $X \in \mathfrak{g}$. Take a function $f \in C^\infty \cap C_{00}$ and set*

$$\psi(g) = \int f(xg^{-1})d\mu(x) = \int fd(g \bullet \mu).$$

Then

(i) $\langle d\psi(1), X \rangle = \lim_{t \to \infty} \int \frac{1}{t}\Big(f(x)d\big((\exp t{\cdot}X) \bullet \mu - \mu\big)\Big),$

(ii) *If there is an $\varepsilon > 0$ such that $\int fd\big((\exp t{\cdot}X)\bullet\mu-\mu\big) \geq tk$ for some positive number k and all $0 \leq t \leq \varepsilon$, then*

$$\langle d\psi(1), X \rangle \geq k.$$

Proof.   (i) We compute

$$\langle d\psi(1), X \rangle = X(\psi) = \frac{d}{dt}|_{t=0}\psi(\exp t{\cdot}X)$$

$$= \frac{d}{dt}|_{t=0} \int f(x \exp -t{\cdot}X)d\mu(x)$$

$$= \int \lim_{t \to 0} \frac{1}{t}\Big(f(x)d\big((\exp t{\cdot}X) \bullet \mu - \mu\big)\Big).$$

(ii) If $\int fd\big((\exp t{\cdot}X) \bullet \mu - \mu\big) \geq tk$ for $t \in [0, \varepsilon]$, then $\langle d\psi(1), X \rangle = \lim_{t \to 0} \int f(x)d\frac{(\exp t{\cdot}X)\bullet\mu-\mu}{t} \geq k$, since $f \geq 0$. ∎

At this point we return to the following set-up of this section: *We consider a Lie group G and an infinitesimally generated subsemigroup S such that $G(S) = G$. We let H be the group of invertible elements in S which we know to be closed and connected by the Unit Group Theorem V.2.8. We fix a vector space complement E for $\mathfrak{h}$ in $\mathfrak{g}$ and a wedge W' surrounding the Lie wedge $W = \mathbf{L}(S)$ and having $\mathfrak{h}$ as edge. We choose a C-H-neighborhood B so that $B \cap \exp^{-1}\overline{S} \subseteq W'$.*

We choose a linear form α on $\mathfrak{g}$ with $\alpha(\mathfrak{h}) = \{0\}$ and $\alpha(X) > 0$ for $0 \neq X \in W' \cap E$. We may assume that α is normalized in such a fashion that the compact base $K_1 = \alpha^{-1}(1) \cap E \cap W$ of the pointed cone $E \cap W$ is contained in B. Let us write E_1 for the hyperplane $\alpha^{-1}(0)$. Then E_1 is a vector space complement for any of the 1-dimensional vector spaces $\mathbb{R}{\cdot}X$ with $X \in K_1$.

We need the following lemma in which K_1 is the compact base of $W \cap E$ just introduced.

VI.4.9.   **Lemma.**   *Let K denote any relatively compact identity neighborhood in G. Then there is a positive number T such that for any number t_0 with $0 < t_0 \leq T$ there is a symmetric open identity neighborhood U such that for all $u \in U$, all $X \in K_1$, and all $t \in]t_0, T[$ one has*

(4) $u \exp -t{\cdot}X \in K \setminus \overline{S}.$

Proof.   We first select an identity neighborhood $D = D^{-1}$ in G with $D^2 \subseteq K$ and then choose $T > 0$ so that

$$\exp([0, T]{\cdot}K_1) \subseteq D.$$

Now for a given positive $t_0 \leq T$, the set $\Gamma = \exp[t_0, T]\cdot K_1$ is compact and contained in D. In any topological group, the product of a closed and of a compact set is closed, whence $\overline{S}\Gamma$ is closed. We claim that $\mathbf{1} \notin \overline{S}\Gamma$; for if $\mathbf{1} = sg$ with $s \in \overline{S}$ and $g = \exp t\cdot X$, then $g \in H$; since $\overline{S} \setminus H$ is an ideal of $\overline{S}$ by Proposition V.0.1, we conclude $\exp[0, 1]\cdot X \subseteq H$ and thus $X \in H \cap K_1 = \emptyset$, an impossibility. Now we select an open symmetric identity neighborhood $U = U^{-1} \subseteq D$ with $U \cap \overline{S}\Gamma = \emptyset$. Then $U\Gamma^{-1} \subseteq DD^{-1} \subseteq K$ and $U\Gamma^{-1} \cap \overline{S} = \emptyset$. Thus, in particular, $u \exp -t_0\cdot X \notin \overline{S}$ for all $u \in U$. Now let $t > t_0$. Then $\exp(t - t_0)\cdot X \in S$; if we had $\exp -t\cdot X \in \overline{S}$, then $\exp -t_0\cdot X = (\exp -t\cdot X)(\exp(t - t_0)\cdot X) \in \overline{S}\cdot S \subseteq \overline{S}$ which is not the case. Hence $\exp -t\cdot X \notin \overline{S}$ and the lemma is proved. ∎

We now select and fix a relatively compact identity neighborhood K in G and a smooth function

(5) $$f \in C^{\infty} \cap C_{00}^{+} \text{ such that } f(K) = \{1\}.$$

Our objective is now to prove the following core lemma which links with condition (ii) of Lemma VI.4.8 above:

VI.4.10. Lemma. *For each $X \in K_1$ there is an $\varepsilon > 0$ and a number $k > 0$ such that*

$$\int fd\big((\exp t\cdot X) \bullet \mu - \mu\big) \geq tk \text{ for all } t \in [0, \varepsilon].$$

Proof. We compute $\int fd((\exp t\cdot X)\bullet\mu - \mu) = \int \big(f(x \exp -t\cdot X) - f(x)\big)\chi_{\overline{S}}(x)dx = \int f(x)\big(\chi_{\overline{S}}(x \exp t\cdot X) - \chi_{\overline{S}}(x)\big)dx$. Since $X \in K_1 \subseteq W$ we observe $\overline{S} \subseteq \overline{S} \exp -t\cdot X$ and thus $\chi_{\overline{S}}(x \exp t\cdot X) - \chi_{\overline{S}}(x) = \chi_{\overline{S}\exp -t\cdot X}(x) - \chi_{\overline{S}}(x) = \chi_{\overline{S}\exp -t\cdot X\setminus\overline{S}}(x)$. Therefore we have

(6) $$\int fd\big((\exp t\cdot X) \bullet \mu - \mu\big) = \int f\chi_{\overline{S}\exp -t\cdot X\setminus\overline{S}}.$$

We now must consider the 1-parameter transformation group

$$(t, g) \mapsto g(\exp -t\cdot X) : \mathbb{R} \times G \to G$$

on G and the subset $\overline{S}$ with its boundary $D = \overline{S} \setminus \operatorname{int} \overline{S}$. From the Infinitesimal Generation Theorem V.1.16(iii) we know that $\operatorname{int} S$ is a dense ideal of S, hence of $\overline{S}$. Since $\operatorname{int} S$ is dense in S, we find an element

(∗) $$u \in U \cap (\operatorname{int} S \setminus \overline{S}\exp X),$$

where U is as in Lemma VI.4.9.

Our next claim is that there is a compact convex cell neighborhood in $E_1 = \alpha^{-1}(0)$ such that

(7) $$u \exp C \subseteq U \cap (\operatorname{int} S \setminus \overline{S}\exp X)$$

and

(8) $$(c, t) \overset{\varphi}{\mapsto} \exp c \exp t\cdot X : C \times \mathbb{R} \to N \overset{\mathrm{def}}{=} \exp C \exp \mathbb{R}\cdot X$$

is a homeomorphism onto a tubular neighborhood of $\mathbf{1}$. Indeed condition (7) is satisfied for all sufficiently small sets C in $\mathfrak{g}$ by continuity in view of condition $(*)$ above. As to condition (8), we first observe that $\overline{\exp \mathbb{R} \cdot X}$ cannot be compact in G, since this relation would imply that this set would be contained in the group H of units of $\overline{S}$ which is not possible because of $X \notin \mathfrak{h}$. Hence by Weil's Lemma (see for instance [MZ55], page 122), $t \mapsto \exp t \cdot X$ is an algebraic and topological isomorphism from $\mathbb{R}$ onto its image. Next, since $\mathfrak{g} = E_1 \oplus \mathbb{R} \cdot X$, we find a compact cell neighborhood C of 0 in E_1 and some $\gamma > 0$ so small that condition (7) is satisfied and

$$\varphi \colon C \times \,] - \gamma, \gamma [\longrightarrow \exp C \exp] - \gamma, \gamma [\cdot X$$

is a diffeomorphism and that

$$(9) \qquad (\exp C)^{-1} \exp C \cap \exp \mathbb{R} \cdot X \subseteq \exp] - \gamma, \gamma [\cdot X.$$

Now suppose that $\exp c_1 \exp t_1 \cdot X = \exp c_2 \exp t_2 \cdot X$ for $t_1, t_2 \in \mathbb{R}$, and $c_1, c_2 \in C$. Then $(\exp c_2)^{-1} \exp c_1 = \exp(t_2 - t_1) \cdot X$, and this implies $|t_2 - t_1| < \gamma$ by (9). But then $\exp c_1 = \exp c_2 \exp(t_2 - t_1) \cdot X$ entails $t_2 = t_1$ and $c_1 = c_2$ since φ is injective on $C \times \,] - \gamma, \gamma [$. The fact that $\varphi \colon C \times \mathbb{R} \to N$ is bijective and a diffeomorphism on $C \times \,] - \gamma, \gamma [$ suffices to prove our claim that φ is a diffeomorphism. From (7), by Lemma VI.4.9, for each $c \in C$, we conclude that the set

$$\{r \in \mathbb{R} \colon u \exp c \exp r \cdot X \in \overline{S}\}$$

is bounded below. Let $\kappa(c)$ be its minimum. Then $s = u \exp c \exp(\kappa(c)) \cdot X \in D \subseteq \overline{S}$, while $s \exp(\kappa(c) - \gamma) \cdot X \notin \overline{S}$ for any $\gamma > 0$. We claim that the set $\{r \in \mathbb{R} \colon s \exp r \cdot X \in \overline{S}\}$ is a half- line; for if $s \exp r \cdot X \in \overline{S}$ and $t' > 0$, then $s \exp(r + t') \cdot X = (s \exp r \cdot X) \exp t' \cdot X \in \overline{S}$, since $\exp t' \cdot X \in S$.

Thus we have defined a function

$$(10) \qquad \kappa \colon C \to \mathbb{R}^+, \qquad \kappa(c) = \min\{r \in \mathbb{R} \colon u \exp c \exp r \cdot X \in \overline{S}\}.$$

We note that

$$(11) \qquad u \exp c \exp r \cdot X \in \overline{S} \Longleftrightarrow \kappa(c) \le r.$$

Finally we define

$$(12) \qquad \Phi \colon C \times \mathbb{R} \to G, \Phi(c, r) = u\varphi(c, t) = u \exp c \exp t \cdot X.$$

Since $\varphi \colon C \times \mathbb{R} \to N$ is a homeomorphism, Φ maps $C \times \mathbb{R}$ homeomorphically onto uN. Then $\Phi\{(c, t) \colon c \in C, \kappa(c) \le t\} = uN \cap \overline{S}$ whence $\{(c, t) \colon \kappa(c) \le t\}$ is closed in $C \times \mathbb{R}$. In other words, κ is lower semicontinuous. Now the set $\{(c, t) \colon \Phi(c, t) \in K\}$ is an open neighborhood A of the graph $\{(c, \kappa(c)) \colon c \in C\}$ of κ in $C \times \mathbb{R}$. Since κ is lower semicontinuous and C is compact, there is an $\varepsilon = \varepsilon(X)$ such that $(c, \kappa(c) - t) \in A$ for all $t \in [0, \varepsilon]$. Then

$$(13) \qquad \Phi(c, r) \in K \quad \text{for any } r \in [\kappa(c) - \varepsilon, \kappa(c)].$$

For any $t \in [0, \varepsilon]$ we shall consider the set

$$M_t = \{(c, r) : \kappa(c) - t \le r < \kappa(c)\}.$$

Then (13) says

(14) $$\Phi(M_t) \subseteq K \text{ for all } 0 \le t \le \varepsilon.$$

We claim that

(15) $$\Phi(M_t) = uN \cap (\overline{S} \exp -tX \setminus \overline{S}).$$

Indeed let $c \in C$, $\kappa(c) - t \le r < t$. Then $\Phi(c, r) = u\varphi(c, r) \notin \overline{S}$ as $r < \kappa(c)$. Further, $u \exp c \exp(r + t) \cdot X \exp -t \cdot X \in \overline{S} \exp -t \cdot X$ since $r + t \ge \kappa(c)$. This shows that the left side of (15) is contained in the right side. Conversely, let $g \in uN \cap (\overline{S} \exp -tX \setminus \overline{S})$. Then there are unique elements $c \in C$ and $r \in \mathbb{R}$ such that $g = u \exp c \exp r \cdot X$. Since $g \notin \overline{S}$ we have $r < \kappa(c)$ by the definition of κ. Since $g \in \overline{S} \exp -t \cdot X$, we note $g \exp t \cdot X \in \overline{S}$, that is $u \exp c \exp(t + r) \cdot X \in \overline{S}$; hence $t + r \ge \kappa(c)$ by the definition of κ. Thus $(c, r) \in M_t$ and therefore $g \in \Phi(M_t)$. The claim is proved.

Now we fix any Lebesgue measure p on E_1. (For instance, select a basis $e_1, \ldots, e_{n-1}$, $n = \dim \mathfrak{g}$, and let p be the product Lebesgue measure on $\mathbb{R} \cdot e_1 \oplus \cdots \oplus \mathbb{R} \cdot e_{n-1} \cong \mathbb{R}^{n-1}$.) Let q denote Lebesgue measure on $\mathbb{R}$. As a lower semicontinuous function, κ is a countable sup of continuous functions; hence it is measurable with respect to p, and

(16) $$(p \times q)(M_t) = \int_C \kappa(c) dp(c) - \int_C \big(\kappa(c) - t\big) dp(c) = tp(C).$$

The diffeomorphism Φ induces a diffeomorphism of M_ε onto $M'_\varepsilon = \Phi(M_\varepsilon)$. This allows us to transport the restriction $m|M'_\varepsilon$ of Haar measure on G to M_ε. In this fashion we obtain a measure m_ε on M_ε. The measures m_ε and Lebesgue measure $p \times q$ are equivalent in such a fashion that there is a positive continuous function $\rho : M_\varepsilon \to \mathbb{R}^+$ such that $m_\varepsilon = \rho(p \times q)$. We set

$$k = \frac{\min \rho(M_\varepsilon)}{p(C)} > 0.$$

Now from (6) and (1) we calculate $\int f d((\exp t \cdot X) \bullet \mu - \mu) \ge \int f \chi_{\Phi(M_t)} = \int (f \circ \Phi) \chi_{M_t} dm_\varepsilon = \int_{M_t} f(\Phi(c, t')) \rho(c, t') dc dt' \ge k \int_{M_t} f(\Phi(c, t')) dc dt'$. But by (14), $\Phi(M_t) \subseteq K$, and by the choice of f in (5), we have $\int_{M_t} f \circ \Phi d(p \times q) = \int \chi_{M_t} d(p \times q) = tp(C)$ in view of (16). It follows that

$$\int f d((\exp t \cdot X) \bullet \mu - \mu) \ge tk$$

with a positive number $k = k(X)$, and this completes the proof of the lemma. ∎

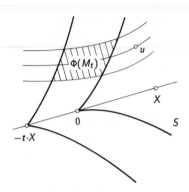

Figure 1

VI.4.11.   Lemma. *Let ψ be a smooth monotone function with respect to an infinitesimally generated semigroup S in a Lie group. Then $d\psi$ is an $\mathbf{L}(S)$-positive (exact!) form on G.*

Proof.   Let $X \in \mathbf{L}(S)$; we must show that $\langle d\psi(x), d\lambda_x(1)(X)\rangle \geq 0$ for all $x \in G$. Now $\langle d\psi(x), d\lambda_x(1)(X)\rangle = \frac{d}{dt}|_{t=0}\psi(x \exp t{\cdot}X))$. But $x \preceq x \exp t.X$ for $t \geq 0$ since $\exp \mathbb{R}^+{\cdot}X \subseteq S$. Thus

$$\lim_{t\searrow 0} \frac{\psi(x \exp t{\cdot}X) - \psi(x)}{t} \geq 0$$

as ψ is monotone. This proves what we had to show.   ∎

   With these somewhat arduous preparations we are now fully prepared for a converse of the Globality Theorem VI.3.8.

VI.4.17.   Theorem.   (Existence Theorem for Strictly Positive Forms) *Let G be a Lie group and S an infinitesimally generated subsemigroup such that $G = G(S)$, that is, G is the subgroup algebraically generated by S. Let H be the group of invertible elements of S and $W = \mathbf{L}(S)$ its Lie wedge. Then H is closed and on G/H there is an exact W-positive form which is strictly positive at $\varepsilon = H$ on G/H. Moreover, there is an exact W-positive form on G which is strictly W-positive at $\mathbf{1}$.*

Proof.   By the Unit Group Theorem V.2.8, the subgroup H is closed in G and connected which allows us to work with the quotient space G/H. Since W is a Lie wedge we may use the machinery of Section 1. For the existence of the required forms, we work in the setting of the preceding lemmas of this section. We continue to use the concepts and the notation introduced earlier. At this point we define a function

(17) $\psi \colon G \to \mathbb{R}$ by $\psi(g) = \displaystyle\int f(xg^{-1})d\mu(x)$

with f as in (5) and with $\mu = \chi_{\overline{S}}{\cdot}m$ for a right Haar measure m on G. Then μ is a monotone measure by Lemma VI.4.5(i). Then ψ is a monotone function by Lemma VI.4.5(ii). By Lemma VI.4.5(iii), the function ψ is smooth. The derivative $d\psi$ is a W-positive exact form on G by Lemma VI.4.11. Now we apply the key

Lemma VI.4.10 and note that for each vector X in the base K_1 of the pointed cone $E \cap W$ we have an $\varepsilon = \varepsilon(X) > 0$ and a $k = k(X) > 0$ such that

$$\int f d(\exp t \cdot X) \bullet \mu - \mu) \geq tk \text{ for all } t \in [0, \varepsilon].$$

By Lemma VI.4.8(ii), we now deduce

(†) $$\langle d\psi(1), X \rangle > 0.$$

Since ψ is monotone, it is constant on cosets gH, $g \in G$ by Remark VI.4.4 and induces a function $\underline{\psi} : G/H \to \mathbb{R}$ with $\psi = \underline{\psi} \circ \pi$. Clearly, $\underline{\psi}$ is smooth. The form $\omega = d\underline{\psi}$ is an exact form on G/H. Then $\overline{d\psi} = \pi^*(\omega)$ and ω is W-positive by Proposition VI.2.2.

Now let Y be any non-zero element of $V = d\pi(1)(W)$. Then there are unique elements $X \in K_1$ and $r > 0$ such that $Y = d\pi(1)(r \cdot X)$. Then $\langle \omega(\varepsilon), Y \rangle = \langle \underline{\psi}(\varepsilon, d\pi(1)(r \cdot X)) = r \langle \pi^*(d\underline{\psi})(1), X \rangle = r \langle d\psi(1), X \rangle > 0$ by (†) above. This shows that ω is strictly positive at ε.

Finally, $d\psi = \pi^*(\omega)$ is an exact W-positive form which is strictly W-positive at 1 by Proposition VI.2.2 again. ∎

We note that, even though this is not expressly emphasized in the formulation of the Existence Theorem VI.4.12, the function $\psi : G/H \to \mathbb{R}$ which yields the desired form $\omega = d\psi$ is in fact a monotone function on G/H.

The following example illustrates the significance of the fact that, in Theorem VI.4.17, the group of units of the semigroup S turned out to be closed.

VI.4.18. **Example.** We let K be a simple, simply connected compact Lie group of rank at least 2 and let T denote a maximal torus in K. Suppose that A is a non-closed dense analytic subgroup of T. Set $G = K \times \mathbb{R}$ and choose a pointed generating invariant cone W in the Lie algebra $\mathfrak{g}$ of G by the First Theorem on Compact Automorphism Groups III.2.1. Then $W_1 = (\mathbf{L}(A) \times \{0\}) + W$ is a generating Lie wedge in $\mathfrak{g}$ with edge $E = \mathbf{L}(A) \times \{0\}$, and W_1 is contained in the global Lie wedge $W_2 = \mathfrak{g} \times \mathbb{R}^+ = \mathbf{L}(G \times R^+)$ in such a way that $-W_2 \cap W_1 = E$. But the analytic subgroup generated by E is $A \times \{0\}$. It is *not* closed, and thus there cannot exist an infinitesimally generated semigroup S with $\mathbf{L}(S) = W_1$ by the Unit Group Theorem V.2.8. In fact, the simply connected Lie group G does not contain *any* preanalytic subsemigroup S with $\mathbf{L}(S) = W_1$ in view of Proposition V.1.7.

5. Globality and its Applications

In this section we begin with a summary of the main characterization theo-
rems on the globality of a Lie wedge W in the Lie algebra $\mathfrak{g}$ of a Lie group G, that
is, the property of being the Lie wedge $L(S)$ of a semigroup S in G. Subsequently,
we shall give some applications and prove some variants of the characterization
theorem which uses an explicit trivialization of the tangent bundle of G. We recall
from Definition VI.3.7 that a Lie wedge W in the Lie algebra $\mathfrak{g}$ of a Lie group G
is called *global in G* if and only if there is a semigroup S in G with $W = \mathbf{L}(S)$.
This is tantamount to saying that $\mathbf{L}(\langle \exp W \rangle) = W$.

The Principal Theorem on Globality

VI.5.1.   Theorem.   (Principal Characterization of Globality) *Let G denote a
finite-dimensional real Lie group and W a Lie wedge in $\mathfrak{g}$. Let A be the analytic
subgroup whose Lie algebra $\mathbf{L}(A)$ is $\langle\langle W \rangle\rangle$ and let A_L denote the group A with its
intrinsic Lie group structure. Then the following three statements are equivalent:*

(1) *W is global in G.*

(2) *W is global in A_L.*

(3) *There is a a closed connected subgroup H of A_L with $\mathfrak{h} = W \cap -W$ and a
1-form ω on A_L which satisfies the following conditions:*

    (i) *ω is exact.*

    (ii) *ω is W-positive.*

    (iii) *ω is strictly W-positive in $\mathbf{1}$.*

Proof. $(3) \Rightarrow (2)$: In view of Proposition VI.2.2, this is exactly the content of
Theorem VI.3.8.

$(1) \Rightarrow (3)$: Suppose that $S = \langle \exp W \rangle$ is the strictly infinitesimally generated semigroup in G for which $\mathbf{L}(S) = W$. We consider $G(S)$ and recall from
Proposition V.2.1 that $\mathbf{L}\big(G(S)\big) = \langle\langle W \rangle\rangle$. Hence $A_L = G(S)$. By Theorem VI.4.17,
there is an exact form ω on $G(S)$ which satisfies the conditions (i), (ii), and (iii).

$(2) \Rightarrow (1)$: If W is global in A_L then there is a subsemigroup S of A_L generated algebraically by $\exp_{A_L}(W)$ and $\mathbf{L}_{A_L}(S) = W$, $G(S) = A_L$. By Proposition

V.1.7, the relation $\mathbf{L}_{A_L}(S) = W$ means that $x \in W$ and $\exp_{A_L} \mathbb{R}^+ \cdot x \subseteq \mathrm{cl}_{A_L} S$ are equivalent. Now $\exp_{A_L} = \exp_G | \mathbf{L}(A)$. Thus S is algebraically generated by $\exp_G(W)$ and $x \in W$ is equivalent with $\exp_G \mathbb{R}^+ \cdot x \subseteq \mathrm{cl}_{G(S)} S$, that is, with $\mathbf{L}_G(S) = W$ by Proposition V.1.7 again. But then W is global in G by definition.∎

Notice that the equivalent conditions of the Principal Theorem do not in general imply that there is an exact W-positive and in $\mathbf{1}$ strictly W-positive form *on the group* G. This is illustrated by the Example V.1.4(ii) of the torus $G = \mathbb{R}^2/\mathbb{Z}^2$ and the global Lie wedge $W = \mathbb{R}^+ \cdot (1, \sqrt{2}) \subseteq \mathbb{R}^2 = \mathfrak{g}$. By Corollary VI.3.10, the existence of an exact, W-positive and in $\mathbf{1}$ strictly W-positive form on G would guarantee the existence of a *closed* subsemigroup T such that $\mathbf{L}(T) = W$ and this is impossible since $\exp W = S = \big(\mathbb{R}^+ \cdot (1, \sqrt{2}) + \mathbb{Z}^2\big)/\mathbb{Z}^2$ is already dense in G.

We shall discuss some applications. Firstly we note that a Lie wedge contained in a global Lie wedge frequently is global as well:

VI.5.2. **Corollary.** *Let G be a Lie group and $W_1 \subseteq W_2$ two Lie wedges in* $\mathfrak{g}$. *Suppose that the following condition is satisfied:*

$$(E) \qquad\qquad W_1 \setminus -W_1 \subseteq W_2 \setminus -W_2.$$

Then W_1 is global if W_2 is global in G and the analytic group with Lie algebra $W_1 \cap -W_1$ is closed in $G(\langle \exp W_1 \rangle)$.

Proof. Let A denote the analytic subgroup with Lie algebra $\mathbf{L}(A) = \langle\!\langle W_1 \rangle\!\rangle$ and A_L the same group with its intrinsic Lie group structure. By the Principal Theorem VI.5.1, there is an exact, W_2-positive and at $\mathbf{1}$ strictly W_2-positive form ω_2 on A_L. Since $W_1 \subseteq W_2$, the form ω_2 is also W_1-positive. If X is in W_1 but not in the edge of W_1, then it is in W_2 but not in the edge of W_2 by (E). Hence $\langle \omega_2(\mathbf{1}), X \rangle > 0$. Thus ω_2 is strictly W_1-positive in $\mathbf{1}$. Hence by the Principal Theorem VI.5.1 again, W_1 is global. ∎

According to (E), the edge of W_1 is the intersection of the edge of W_2 with W_1. Without (E), nothing can be concluded. For instance, if G is the Heisenberg group (Example V.4.4!), W_1 the circular cone of Example VI.1.6, and W_2 the whole Lie algebra $\mathfrak{g}$, then all the hypotheses of Corollary VI.5.2 are satisfied except (E). Indeed W_2 is global while W_1 is not. Example VI.4.18 shows that Corollary VI.5.2 fails if the edge of W_1 does not generate a closed group.

The following result is useful for globality arguments using the factoring of normal subgroups.

VI.5.3. **Corollary.** *Let G be a Lie group and N a closed normal subgroup of G. Let $\pi: G \to G/N$ denote the canonical projection. Moreover, let W be a Lie wedge in $\mathfrak{g}$ with edge $\mathfrak{h} = W \cap -W$. Suppose the following hypotheses:*

(i) *The analytic subgroup H with $\mathbf{L}(H) = \mathfrak{h}$ is closed.*

(ii) $W \cap \mathbf{n} = \mathfrak{h} \cap \mathbf{n}$.

(iii) *There exists a semigroup S_π in G/N with $\mathbf{L}(S_\pi) = d\pi(\mathbf{1})(W)$.*

Then W is global in G.

Proof.   Note that (ii) and Proposition I.2.32 imply that $W' \overset{\text{def}}{=} W + \mathbf{n}$ and $d\pi(1)(W)$ are closed. Then Proposition V.3.1(ii) shows that W' is global in G. In view of the preceding Corollary VI.5.2 it now suffices to show that

$$(*) \qquad\qquad W \setminus -W \subseteq W' \setminus -W'.$$

For a proof of this claim we consider $y \in W \cap -W'$ and show that $-y \in W$. Now we find $x \in W$ and $n \in \mathbf{n}$ with $-y = x + n$. Hence $-n = x + y \in W \cap \mathbf{n} = \mathfrak{h} \cap \mathbf{n}$ by (ii), and so $n \in W$. Thus $-y = x + n \in W$ which we had to show.   ∎

Closed versus exact forms

For our next applications, we would like to replace, under suitable circumstances, Condition (3)(i) in the Principal Theorem by the generally weaker condition that the form be closed. This purpose is served by the following lemma which requires some insight into the workings of cohomology and homology on a group manifold.

VI.5.4.   Lemma. *Let G be a connected finite-dimensional Lie group. Then the following three statements are equivalent:*

(1) *The fundamental group $\pi_1(G)$ is finite.*

(2) *Every closed form on G is exact.*

(3) *Each maximal compact subgroup of G is semisimple.*

Proof. $(1) \Leftrightarrow (3)$: If K is a maximal compact subgroup of G, then all maximal compact subgroups are conjugate to K and G is diffeomorphic to the cartesian product space $K \times \mathbb{R}^n$ for a suitable n. (See [Ho65], Chap. XV, Sec. 3. Theorem 3.1 on p. 180.) Hence $\pi_1(G) \cong \pi_1(K)$. But the fundamental group of a compact Lie group is finite if and only if it is semisimple (See for instance [Bou63], Chap. VII, §7, n° 3, Proposition 5).

$(1) \Leftrightarrow (2)$: Let Z^1 denote the real vector space of all closed 1-forms on G and B^1 the vector subspace of all exact 1-forms. Then Z^1/B^1 is the first de Rham cohomology group $\mathcal{H}^1(G)$. Condition (2) above then is equivalent to $\mathcal{H}^1(G) = \{0\}$. Now de Rham's Theorem says that $\mathcal{H}^n(G)$ is naturally isomorphic to the singular real cohomology group $H^n(G, \mathbb{R})$ (see for instance [GHV72], pp.217 ff.). The first Universal Coefficient Theorem gives us an exact sequence of abelian groups for singular cohomology H^* and singular homology H_*:

$$0 \to \text{Ext}(H_{n-1}(G, \mathbb{Z}), \mathbb{R}) \to H^n(G, \mathbb{R}) \to \text{Hom}(H_n(G, \mathbb{Z}), \mathbb{R}) \to 0.$$

(See for instance [MacL63], p.77, Theorem 4.1.) Since $\mathbb{R}$ is divisible, $\text{Ext}(A, \mathbb{R}) = \{0\}$ for any abelian group A. Thus we note

$$(*) \qquad\qquad H^n(G, \mathbb{R}) \cong \text{Hom}(H_n(G, \mathbb{Z}), \mathbb{R}).$$

For a connected manifold M, the Theorem of Hurewicz gives an exact sequence of groups

$$(**) \qquad 0 \to \pi_1(M)' \to \pi_1(M) \to H_1(M, \mathbb{Z}) \to 0,$$

where the prime indicates the formation of the commutator group. (See for instance [Sp66], Chap.7, Sec.5, Prop.2 for $n = 1$.) Now for a topological group, the fundamental group $\pi_1(G)$ is always commutative by Hilton's Lemma (see for instance [Sp66], Chap.1, Sec.6 Theorem 8 and Corollary 10). Thus $(**)$ shows

$$(\dagger) \qquad\qquad H_1(G, \mathbb{Z}) \cong \pi_1(G).$$

If $\widetilde{G}$ denotes the universal covering group of G, then there is an exact sequence

$$0 \to \pi_1(G) \to \widetilde{G} \to G \to 0,$$

more specifically, the fundamental group of G is isomorphic to a discrete central subgroup of $\widetilde{G}$. Such groups are finitely generated abelian groups (see for instance [Bou63] and [Hoch65]), hence are of the form $F \oplus \mathbb{Z}^n$ with a finite abelian group F. Thus, using $(*)$ and $(\dagger)$, we find $H^1(G, \mathbb{R}) \cong \mathrm{Hom}(\pi_1(G), \mathbb{R}) \cong \mathrm{Hom}(F \oplus \mathbb{Z}^n, \mathbb{R}) \cong \mathrm{Hom}(\mathbb{Z}^n, \mathbb{R}) \cong \mathbb{R}^n$ since $\mathbb{R}$ is torsion free. We recall from abelian group theory that n is called the *rank of* $\pi_1(G)$. Thus we have found

$$(\ddagger) \qquad\qquad \dim_{\mathbb{R}} \mathcal{H}^1(G) = \mathrm{rank}\, \pi_1(G).$$

Since condition (2) is equivalent to $\dim_{\mathbb{R}} \mathcal{H}^1(G) = 0$, the equivalence of (1) and (2) is now obvious from $(\ddagger)$.

■

As an immediate consequence, we have the following corollary to the Principal Theorem:

VI.5.5.   Corollary.   *In a connected Lie group G with a finite fundamental group $\pi_1(G)$, in Theorem VI.5.1, we may replace condition (3)(i) by the following condition:*

(i*) ω *is closed.*   ■

The example of half-space semialgebras is instructive in this context.

VI.5.6.   Corollary.   *Let G be a Lie group and W a half-space Lie-semialgebra in $\mathfrak{g}$, that is, a wedge whose edge $\mathfrak{h}$ is a Lie algebra of codimension 1 (see Corollary II.2.24 and Proposition IV.1.34). Let ω_1 be any linear form on $\mathfrak{g}$ vanishing on $\mathfrak{h}$ and non-negative on W. Let ω be the unique left invariant form on G with $\omega(1) = \omega_1$. Then W is global in G if and only if ω there is a positive smooth function $\varphi: G \to]0, \infty[$ such that $\varphi \cdot \omega$ is exact.*

If the fundamental group of G is finite, then, as a consequence, W is global in G if and only if there is a smooth function φ on G such that $\varphi \cdot \omega$ is closed.

In particular, if $\pi_1(G)$ is finite and $d\omega = 0$, then W is global.

Proof.   Since W is a half space, $W - W = \mathfrak{g}$ and G is itself the analytic group A_L whose Lie algebra is $\langle\!\langle W \rangle\!\rangle = \mathfrak{g}$. Moreover, since W is a half-space, a 1-form μ is W-positive if and only if $\mu(x)$ vanishes on the edge $d\lambda_x(\mathbf{1})(\mathfrak{h})$ of $d\lambda_x(\mathbf{1})(W)$ and takes a positive value on some element $d\lambda_x(\mathbf{1})(X)$ with $X \in W$. Every such form is automatically strictly W-positive. Among all these μ there is exactly one left invariant one up to positive scalar multiples, namely, ω. Moreover, $\mu(x)$ is positive on $d\lambda_x(\mathbf{1})(W \setminus \mathfrak{h})$ if and only if there is a positive number $\varphi(x)$ such that $\mu(x) = \varphi(x)\omega(x)$. Since both μ and ω are smooth, also φ is smooth. By the Principal Globality Theorem VI.5.1, W is global in G if there is such a form μ which is, in addition, exact. This proves the first assertion. The remainder follows from Corollary VI.5.5.   ∎

We shall show that the required conditions for globality are certainly satisfied if W contains the commutator algebra. In fact we shall show more. For this purpose, the following remark becomes relevant:

VI.5.7.   **Lemma.**   *Let ω be a left invariant 1-form such that $\omega(\mathbf{1})$ vanishes on $\mathfrak{g}' = [\mathfrak{g}, \mathfrak{g}]$. Then ω is closed.*

Proof.   If ω is left invariant, then for $X \in \mathfrak{g} = T(G)_{\mathbf{1}}$ we have $\big(d\lambda_g(\mathbf{1})\big)\widehat{\ }(\omega(g)) = \omega(\mathbf{1})$ and thus

$$\langle \omega(\mathbf{1}), X \rangle = \langle \big(d\lambda_g(\mathbf{1})\big)\widehat{\ }\omega(g), X \rangle = \langle \omega(g), d\lambda_g(\mathbf{1})(X) \rangle$$

for all $X \in \mathfrak{g}$. The exterior derivative $d\omega$ is likewise left invariant (see for instance [Bou71], 8.3.5(6))and we have, accordingly,

$$(1) \qquad\qquad \langle d\omega(\mathbf{1}), X \wedge Y \rangle = \langle d\omega(g), d\lambda_g(\mathbf{1})(X) \wedge d\lambda_g(\mathbf{1})(Y) \rangle$$

for all $X, Y \in \mathfrak{g}$. Here we write $\langle d\omega(g), X' \wedge Y' \rangle$ for $d\omega(g)(X', Y')$ with $X', Y' \in T(G)_g$ which is justified by the fact that $d\omega(g)$ is a skew-symmetric bilinear form on $T(G)_g$. By the formula of Maurer and Cartan (see e.g. [Bou75], Chap.III ,§3, n° 14, Proposition 51)

$$(\mathrm{MC}) \qquad\qquad d\omega(\mathbf{1})(X, Y) = \omega([X, Y]) \text{ for all } X, Y \in \mathfrak{g}.$$

Hence, setting $X' = d\lambda_g(\mathbf{1})(X)$ and $Y' = d\lambda_g(\mathbf{1})(Y)$, from (1) and (MC) we derive

$$(2) \qquad\qquad d\omega(g)(X', Y') = \omega(\mathbf{1})([X, Y]) = 0$$

in view of the fact that ω vanishes on $\mathfrak{g}'$. Since (2) holds for arbitrary $X', Y' \in T(G)_g$ for all $g \in G$, we conclude $d\omega = 0$ as asserted.   ∎

This has the following consequence

VI.5.8.   **Proposition.**   *Suppose that G is a connected Lie group with finite fundamental group and that the following condition is satisfied:*

$$(\mathrm{C}) \qquad\qquad\qquad \mathfrak{g}' \cap W \subseteq -W,$$

where $\mathfrak{g}'$ is the commutator algebra of $\mathfrak{g}$. Then W is global. In fact, there is a closed infinitesimally generated subsemigroup S of G with $\mathbf{L}(S) = W$.

Proof.   Let $\omega \in \widehat{\mathfrak{g}}$ be selected such that $\langle \omega, \mathfrak{g}' \rangle = \{0\}$ and $\langle \omega, X \rangle > 0$ for all $X \in W \setminus -W$. Such a functional exists by Condition (C). Then $\omega \in \operatorname{algint} W^*$. By Lemma VI.5.7 above, ω is closed. Thus W is global by Corollary VI.5.5. The existence of a *closed* infinitesimally generated subsemigroup S of G with $\mathbf{L}(S) = W$ follows from Corollary VI.3.9.   ∎

We note that condition (C) says that the commutator algebra $\mathfrak{g}'$ meets the wedge W only in its edge. We have encountered this condition in the Fourth Triviality Theorem II.7.11 and its Corollary II.7.12.

VI.5.9. **Corollary.** *In a Lie group G with finite fundamental group every trivial semialgebra is global.*

Proof. We recall that a Lie semialgebra W is called trivial if $\mathfrak{g}' \subseteq W$ (see for instance Theorem II.7.5 ff.). This obviously implies condition (C) and thus the corollary follows from the preceding proposition. ∎

The tangent bundle of a group

It is very instructive to note that the tangent bundle $T(G)$ of a Lie group G is itself a Lie group. (See for instance [Bou75], Chapter III, §2, n°2.) What is of relevance for us is the standard trivialization of the tangent bundle $T(G)$. Indeed, let $\mathfrak{g}_a$ denote the underlying vector space of the Lie algebra $\mathbf{L}(G) = T(G)_1$ and form the semidirect product $\mathfrak{g}_a \times_{\mathrm{Ad}} G$ with multiplication $(X, g)(X', g') = (X + \mathrm{Ad}(g)(X'), gg')$ (see Proposition V.3.4).

VI.5.10. **Lemma.** *The function $\psi \colon \mathfrak{g}_a \times_{\mathrm{Ad}} G \to T(G)$ given by $\psi(X, g) = d\rho_g(\mathbf{1})$ is an isomorphism of vector bundles which maps $\mathfrak{g} \times \{g\}$ isomorphically onto the tangent space $T(G)_g$ in such a fashion that*

$$(3) \qquad \psi(\mathrm{Ad}(g)(X), g) = d\lambda_g(\mathbf{1})\big(\psi(X, \mathbf{1})\big).$$

Moreover, ψ is an isomorphism of Lie groups.

Proof. After the definition of ψ, formula (1) is just a restatement of $\mathrm{Ad}(g) = d\lambda_g(\mathbf{1}) \circ d\rho_g(\mathbf{1})^{-1} = d\rho_g(\mathbf{1})^{-1} d\lambda_g(\mathbf{1})$. For the last statement we refer to Bourbaki at the source given above. ∎

It suffices for our purposes to have the bundle isomorphism ψ and to visualize, if necessary, the Lie group structure of $T(G)$ as the one obtained by transporting the Lie group structure of $\mathfrak{g}_a \times_{\mathrm{Ad}} G$ via ψ. Lemma VI.5.10 essentially says that if we identify $T(G)$ with $\mathfrak{g}_a \times_{\mathrm{Ad}} G$, then the linear map $d\lambda_g(\mathbf{1}) \colon T(G)_1 = \mathfrak{g} \to T(G)_g$ becomes identified with $(X, 0) \mapsto (\mathrm{Ad}(g)(X), g) \colon \mathfrak{g} \times \{0\} \to \mathfrak{g} \times \{g\}$. One may view this content in the form of the diagram

$$
\begin{array}{ccc}
\mathfrak{g} & \xrightarrow{\ \mathrm{Ad}\,g\ } & \mathfrak{g} \\
\cong & & \cong \\
\mathfrak{g} \times \{0\} & \longrightarrow & \mathfrak{g} \times \{0\} \\
\psi \downarrow & & \downarrow \psi \\
T(G)_1 & \xrightarrow{\ d\lambda_g(\mathbf{1})\ } & T(G)_g.
\end{array}
$$

Accordingly, the cotangent bundle $\widehat{T}(G)$ of G may be identified with $\widehat{\mathfrak{g}} \times G$ via

$$(4) \qquad \widehat{\psi} \colon \widehat{T}(G) \to \widehat{\mathfrak{g}} \times G, \qquad \widehat{\psi}(\omega) = (\omega \circ d\rho_g(\mathbf{1}), g) \ \text{ for } \ \omega \in \widehat{T}(G)_g.$$

If we define the *coadjoint representation* $\widehat{\mathrm{Ad}}: G \to \mathrm{Aut}\,\mathfrak{g}_a$ by

$$\widehat{\mathrm{Ad}}(g) = \big(\mathrm{Ad}(g^{-1})\big)\widehat{},$$

then once again the cotangent bundle $\widehat{T}(G)$ becomes identified with the Lie group $\widehat{\mathfrak{g}}_a \times_{\widehat{\mathrm{Ad}}} G$.

If B is a C-H-neighborhood such that $\exp|B: B \to U$ is a diffeomorphism onto an open identity neighborhood U of $\mathbf{1}$ in G, then $(U, (\exp|B)^{-1}, \mathfrak{g})$ is a coordinate chart of G with domain U. The diffeomorphism $\exp|B: B \to U$ induces isomorphisms $T(B) \to T(U)$ of the tangent bundles and $\widehat{T}(B) \to \widehat{T}(U)$ of the cotangent bundles, and $T(U)$ and $\widehat{T}(U)$ are open subbundles of $T(G)$ and $\widehat{T}(G)$, respectively. The bundles $T(B)$ and $\widehat{T}(B)$ have the standard trivializations $\mathfrak{g} \times B$ and $\widehat{\mathfrak{g}} \times B$ which we have always used in all of Chapter IV for the local theory. These trivializations allow us, for any smooth functions $\varphi: B \to F$ into a Banach space F to identify the derivative $d\varphi(X): T(B)_X \to T(F)_{\varphi(X)}$ with a linear map $\varphi'(X): \mathfrak{g} \to F$. Similar statements apply to $\widehat{T}(B)$. It is very important for us to realize that this standard trivializations of $T(B)$ and $\widehat{T}(B)$ are *not* the ones coming from the trivializations of $T(G)$ and $\widehat{T}(G)$. In fact, if for $X, Y \in B$ as usual we set $\rho_Y(X) = X * Y$, we have bundle isomorphisms

(5)
$$\begin{aligned} R: T(B) &\to \mathfrak{g} \times B, \quad R(X, b) = (d\rho_b^*(0)(X), b), \\ \widehat{R}: \widehat{T}(B) &\to \widehat{\mathfrak{g}} \times B, \quad \widehat{R}(\omega, b) = (\omega \circ d\rho_b^*(0), b). \end{aligned}$$

Then we have a sequence of bundle maps

(6)
$$T(B) \xrightarrow{\;R\;} \mathfrak{g} \times B \xrightarrow{\;\mathrm{id}_{\mathfrak{g}} \times \exp|B\;} \mathfrak{g} \times U \;=\; T(U) \;\subseteq\; T(G),$$

and, accordingly

(6̂)
$$\widehat{T}(B) \xrightarrow{\;\widehat{R}\;} \widehat{\mathfrak{g}} \times B \xrightarrow{\;\mathrm{id}_{\widehat{\mathfrak{g}}} \times \exp|B\;} \widehat{\mathfrak{g}} \times U \;=\; \widehat{T}(U) \;\subseteq\; \widehat{T}(G),$$

in which the first two are isomorphisms in both cases, and in which we understand $T(B)$ to have the standard trivialization $\mathfrak{g} \times G$ and $\widehat{T}(B)$ to have the standard trivialization $\widehat{\mathfrak{g}} \times G$.

If we henceforth identify $T(G)$ with $\mathfrak{g}_a \times_{\mathrm{Ad}} G$ and $\widehat{T}(G)$ with $\widehat{\mathfrak{g}}_a \times_{\widehat{\mathrm{Ad}}} G$ then there is a bijection between smooth functions $X: G \to \mathfrak{g}$ and vector fields on G under which a smooth function $X: G \to \mathfrak{g}$ becomes identified with the vector field $g \mapsto \big(X(g), g\big)$. Likewise, smooth function $\omega: G \to \widehat{\mathfrak{g}}$ become identified with smooth 1-forms $g \mapsto \big(\omega(g), g\big)$ in such a fashion that $\langle\big(\omega(g), g\big), (X, g)\rangle = \langle\omega(g), X\rangle$ for all $X \in \mathfrak{g}$. We note that the left invariant vector fields are in this way represented by the functions $g \mapsto \mathrm{Ad}(g)(X)$, $X \in \mathfrak{g}$, and the left invariant forms by the functions $g \mapsto \widehat{\mathrm{Ad}}(g)(\omega)$, $\omega \in \widehat{\mathfrak{g}}$. On the other hand, the constant functions $G \to \mathfrak{g}$ are identified with the right invariant vector fields while the constant functions $G \to \widehat{\mathfrak{g}}$ are identified with the right invariant 1-forms.

Forms as functions

The tangent bundle $T\big(T(G)\big)$ of the manifold $T(G) = \mathfrak{g}_a \times G$ is $(\mathfrak{g}_a \times \mathfrak{g}_a) \times (\mathfrak{g}_a \times G)$. Let $X: G \to \mathfrak{g}$ be a smooth function and write $\widetilde{X}: G \to T(G)$ for the vector field given by $\widetilde{X}(g) = (X(g), g)$. Then we shall define $X'(g) \in \text{Hom}(\mathfrak{g}_a, \mathfrak{g}_a)$ by $d\widetilde{X}(g)(v, g) = \big((X'(g)(v), v), (X(g), g)\big)$ for $v \in \mathfrak{g}$.

Quite analogously, the tangent bundle $T\big(\widehat{T}(G)\big)$ of the cotangent bundle $\widehat{T}(G) = \widehat{\mathfrak{g}_a} \times G$ is $(\widehat{\mathfrak{g}_a} \times \mathfrak{g}_a) \times (\widehat{\mathfrak{g}_a} \times G)$. If $\omega: G \to \widehat{\mathfrak{g}}$ is a smooth function and $\widetilde{\omega}: G \to \widehat{T}(G)$ the form given by $\widetilde{\omega}(g)(v, g) = \langle \omega(g), v \rangle$ for $v \in \mathfrak{g}$, then we define $\delta\omega(g) \in \text{Hom}(\mathfrak{g}_a, \widehat{\mathfrak{g}_a})$ so that the derivative of $\widetilde{\omega}$ in the point $g \in G$ evaluated at $(v, g) \in T(G)_g$ with $v \in \mathfrak{g}$ is $\big((\delta\omega(g)(v), v), (\omega(g), g)\big)$.

If $f: G \to \mathbb{R}$ is a smooth function, then $df(g)$ is a linear form $\mathfrak{g} \times \{g\} \to \mathbb{R}$; we define the linear form $f'(g) \in \widehat{\mathfrak{g}}$ by $df(g)(v, g) = \langle f'(g), v \rangle$ for $v \in \mathfrak{g}$. We then note

$$(\widetilde{X}f)(g) = df(g)\big(\widetilde{X}(g)\big) = \langle f'(g), X(g) \rangle$$

for smooth functions $X: G \to \mathfrak{g}$ and $f: G \to \mathbb{R}$. We shall briefly write $(Xf)(g) = \langle f'(g), X(g) \rangle$. In particular, for smooth functions $X, Y: G \to \mathfrak{g}$ and $\omega: G \to \widehat{\mathfrak{g}}$ we firstly set $(\omega X)(g) = \langle \omega(g), X(g) \rangle = \langle \widetilde{\omega}(g), \widetilde{X}(g) \rangle$ and then have $X(\omega Y)(g) = \langle (\omega Y)'(g), X(g) \rangle$. But $\langle (\omega X)'(g), v \rangle = \langle \delta\omega(g)(v), X(g) \rangle + \langle \omega(g), X'(g)(v) \rangle$ by the product rule. Thus

$$(7) \qquad X(\omega Y)(g) = \langle \delta\omega(g)\big(X(g)\big), Y(g) \rangle + \langle \omega(g), Y'(g)\big(X(g)\big) \rangle.$$

It is convenient to identify $\text{Hom}(A, \widehat{B})$ with $(A \otimes B)\widehat{}$ so that

$$\langle F, a \otimes b \rangle = F(a \otimes b) = \langle F(a), b \rangle.$$

In particular we shall say that F is *symmetric* if $F(a \otimes b) = F(b \otimes a)$

With this notation, (7) yields

$$
\begin{aligned}
(8) \qquad X(\omega Y)(g) - Y(\omega X)(g) &= \langle \delta\omega(g)\big(X(g)\big), Y(g) \rangle - \langle \delta\omega(g)\big(Y(g)\big), X(g) \rangle \\
&\quad + \langle \omega(g), Y'(g)\big(X(g)\big) - X'(g)\big(Y(g)\big) \rangle \\
&= \langle \delta\omega(g), X(g) \otimes Y(g) - Y(g) \otimes X(g) \rangle \\
&\quad + \langle \omega(g), Y'(g)\big(X(g)\big) - X'(g)\big(Y(g)\big) \rangle.
\end{aligned}
$$

We recall that the *exterior* derivative of a form $\widetilde{\omega}$ is defined by

$$2d\widetilde{\omega}(\widetilde{X}, \widetilde{Y}) = \widetilde{X}(\widetilde{\omega}\widetilde{Y}) - \widetilde{Y}(\widetilde{\omega}\widetilde{X}) - (\widetilde{\omega}[\widetilde{X}, \widetilde{Y}]).$$

Then (8) allows us to write the exterior derivative in terms of the smooth functions which represent vector fields and forms as soon as we know how to treat the bracket $[\widetilde{X}, \widetilde{Y}]$ in terms of functions.

VI.5.11.   **Lemma.**   *With the notation just introduced we have*

$$[\widetilde{X}, \widetilde{Y}](g) = \Big(Y'(g)(X(g)) - X'(g)(Y(g)) - [X(g), Y(g)], g \Big),$$

where $[X(g), Y(g)]$ denotes the bracket in the Lie algebra $\mathfrak{g}$.

Proof.   Let us first assume that $\widetilde{X}$ and $\widetilde{Y}$ are right invariant vector fields, whence X and Y are constant functions. Let $x \colon G \to \mathbb{R}$ and $y \colon G \to \mathbb{R}$ two smooth functions. Then

$$[x \cdot \widetilde{X}, y \cdot \widetilde{Y}] = xy \cdot [\widetilde{X}, \widetilde{Y}] + x\widetilde{X}(y) \cdot \widetilde{Y} + y\widetilde{Y}(y) \cdot \widetilde{X}.$$

(See for instance [Bou71], §8, n° 5, (2).) If we now select a basis $\widetilde{X}_1, \ldots, \widetilde{X}_n$ for the Lie algebra of all right invariant vector fields, then we can represent arbitrary vector fields $\widetilde{X}$ and $\widetilde{Y}$ in the form

$$\widetilde{X} = \sum_{k=1}^{n} x_k \cdot \widetilde{X}_k, \qquad \widetilde{Y} = \sum_{k=1}^{n} y_k \cdot \widetilde{Y}_k.$$

We note that

$$\sum_{j,k=1}^{n} x_j(g) \widetilde{X}_k(y_k)(g) \cdot \widetilde{Y}_k(g) = \sum_{j,k=1}^{n} x_j(g) \langle d(y_k)(g), X_k(g) \rangle \cdot \widetilde{Y}_k(g)$$

$$= \Big(Y(g)'(X(g)), g \Big).$$

If we also observe that $\sum_{j,k=1}^{n} x_j(g) y_k(g) \cdot [\widetilde{X}_k(g), \widetilde{Y}_k(g)] = -\big([X(g), Y(g)], g\big)$ in view of the fact that the bracket of right invariant vector fields is exactly the negative of the bracket in $\mathfrak{g}$ and take all of this information together we obtain the lemma.   ∎

From (8) and Lemma VI.5.11 we now obtain

(9)
$$2d\widetilde{\omega}(\widetilde{X}, \widetilde{Y})(g) = \langle \delta\omega(g), X(g) \otimes Y(g) - Y(g) \otimes X(g) \rangle$$
$$+ \langle \omega(g), [X(g), Y(g)] \rangle.$$

It is useful to realize an alternative description of the derivative $\delta\omega(g)$. Let B be a C-H-neighborhood of $\mathfrak{g}$ and $\exp|B \colon B \to U$ a diffeomorphism onto an identity neighborhood U in G. Define $\omega_g \colon B \to \widehat{\mathfrak{g}}$ by $\omega_g(X) = \omega\big((\exp X)g\big)$ where X is a vector in $\mathfrak{g}$, that is, $\omega_g = \omega \circ \rho_g \circ \exp$. Thus $\delta\omega_g(0) = \delta\omega(g) \circ d\rho_g(\mathbf{1})$, since $d\exp(0) = \mathrm{id}_{\mathfrak{g}}$. With our identification of $T(G)_g$ with $\mathfrak{g} \times \{g\}$, the vector space isomorphism $d\rho_g(0) \colon \mathfrak{g} \to \mathfrak{g}$ is now none other than the identity. Then

(10)
$$\delta\omega(g) = \delta\omega_g(0).$$

Next we invoke $(\widehat{6})$ in order to express $\delta\omega_g(0)$ in terms of the standard trivialization of the tangent bundle of some C-H-neighborhood B of 0. We recall $\widehat{R} \colon \widehat{T}(B) \to$

$\widehat{\mathfrak{g}} \times B$, $\widehat{R}(\alpha, X) = (\alpha \circ d\rho_X^*(0), X)$. This morphism induces a linear isomorphism $d\widehat{R}(\beta, 0): \widehat{T}(B)_0 \to (\widehat{\mathfrak{g}} \times \mathfrak{g}) \times \{(\beta, 0)\}$, which we should like to compute. We write $d\widehat{R}(\beta, 0)\big((\alpha, X), (\beta, 0)\big) = \big((\varphi(\beta)(\alpha, X), X), (\beta, 0)\big)$ with $\varphi(\beta)(\alpha, X) \in \widehat{\mathfrak{g}}$. We must compute this element. Now with (Δ, h) small in $\widehat{\mathfrak{g}} \times B \subseteq \widehat{\mathfrak{g}} \times \mathfrak{g}$ we compute $\widehat{R}(\beta + \Delta, h) - \widehat{R}(\beta, 0) = \big((\beta + \Delta) \circ d\rho_h^*(0) - \beta, h\big)$, and $d\rho_h^*(0) = \mathrm{id}_{\mathfrak{g}} - \frac{1}{2} \cdot \mathrm{ad}\, h + o(h)$ so that $o(h)/\|h\| \to 0$ for $h \to 0$ with a suitable norm on $\mathfrak{g}$. Then $(\beta + \Delta) \circ d\rho_h^*(0) = \beta - \frac{1}{2} \cdot \beta \circ \mathrm{ad}\, h + \Delta + o(\Delta, h)$. Thus we have $\widehat{R}(\beta + \Delta, h) - \widehat{R}(\beta, 0) = (\Delta - \frac{1}{2} \cdot \beta \circ \mathrm{ad}\, h, h) + o(\Delta, h)$, whence

$$(11) \qquad \varphi(\beta)(\alpha, X) = \alpha - \frac{1}{2} \cdot \beta \circ \mathrm{ad}\, X.$$

Thus

$$(12) \qquad d\widehat{R}(\beta, 0)\big((\alpha, X), (\beta, 0)\big) = \big((\alpha - \frac{1}{2} \cdot \beta \circ \mathrm{ad}\, X, X), (\beta, 0)\big).$$

Now we consider the section $b \mapsto (\omega_g(b), b): B \to \mathfrak{g} \times B$ and define a smooth function $\underline{\omega}_g: B \to \widehat{\mathfrak{g}}$ in such a fashion that the bundle isomorphism $\widehat{R}$ transforms the section $b \mapsto (\underline{\omega}_g(b), b): B \to T(B)$ into the section defined by ω_g. Thus $\widehat{R}(\underline{\omega}_g(b), b) = (\omega_g(b), b)$. It follows that $\underline{\omega}_g(b) \circ d\rho_b^*(0) = \omega_g(b)$, or, equivalently,

$$(13) \qquad \underline{\omega}_g(b) = \omega_g(b) \circ d\rho_b^*(0)^{-1}.$$

If we now denote with $\underline{\omega}_g'(b): \mathfrak{g} \to \widehat{\mathfrak{g}}$ the standard derivative at $b \in B$, then the section $b \mapsto (\underline{\omega}_g(b), b): B \to \widehat{T}(B)$ induces the linear map

$$(X, 0) \mapsto \big((\underline{\omega}_g'(0)(X), X), (\underline{\omega}_g(0), 0)\big) : \widehat{T}(B)_0 \longrightarrow (\widehat{\mathfrak{g}} \times \mathfrak{g}) \times \{(\underline{\omega}_g(0), 0)\},$$

while the section $b \mapsto (\omega_g(b), b): B \to \widehat{\mathfrak{g}} \times B$ induces the linear map

$$(X, 0) \mapsto \big((\delta\omega_g(0)(X), X), (\omega_g(0), 0)\big) : \widehat{\mathfrak{g}} \times \{0\} \longrightarrow (\widehat{\mathfrak{g}} \times \mathfrak{g}) \times \{(\omega_g(0), 0)\}.$$

The derivative $d\widehat{R}(\underline{\omega}_g(0), 0)$ transforms the former into the latter; more precisely

$$(14) \qquad d\widehat{R}(\underline{\omega}_g(0), 0)\big((\underline{\omega}_g'(0)(X), X), (\underline{\omega}_g(0), 0)\big) = \big((\delta\omega_g(0)(X), X), (\omega_g(0), 0)\big),$$

where for the next to the last component we recall (13). In view of $\underline{\omega}_g(0) = \omega_g(0)$, equations (12) and (14) yield

$$(15) \qquad \delta\omega_g(0)(X) = \underline{\omega}_g'(0)(X) - \frac{1}{2} \cdot \omega_g(0) \circ \mathrm{ad}\, X.$$

Therefore, in view of $\omega_g(0) = \omega(g)$, we deduce

$$(16) \qquad \delta\omega(g)(X \otimes Y) = \underline{\omega}_g'(0)(X \otimes Y) - \frac{1}{2} \cdot \omega(g)([X, Y]),$$

for vectors $X, Y \in \mathfrak{g}$. It follows at once that for arbitrary smooth functions X and Y from G to $\mathfrak{g}$ we obtain

$$
\begin{aligned}
(17) \qquad & \delta\omega(g)\big(X(g) \otimes Y(g) - Y(g) \otimes X(g)\big) \\
& = \underline{\omega}_g'(0)\big(X(g) \otimes Y(g) - Y(g) \otimes X(g)\big) \\
& \quad - \omega(g)([X(g), Y(g)]).
\end{aligned}
$$

If we compare this with (9) we obtain at once the following result:

VI.5.12.   **Proposition.**   *Let $\widetilde{\omega}$ be a smooth form on a Lie group G (expressed in terms of a smooth function $\omega\colon G \to \mathfrak{g}$ explained above) and define the function $\underline{\omega}_g\colon B \to \widehat{\mathfrak{g}}$ on a suitable Campbell–Hausdorff neighborhood B in $\mathfrak{g}$ by*

$$\underline{\omega}_g(b) = \omega\big((\exp X)g\big) \circ d\rho_X^*(0)^{-1}.$$

Then

(18) $2 \cdot d\widetilde{\omega}(\widetilde{X}, \widetilde{Y})(g) = \{\underline{\omega}_g'(0)\{(X(g) \otimes Y(g) - Y(g) \otimes X(g))\}$

for smooth functions $X, Y\colon G \to \mathfrak{g}$. In particular, $\widetilde{\omega}$ is closed if and only if $\underline{\omega}_g'$ is symmetric for all $g \in G$.   ∎

The significance of this proposition is that the formula (18) reduces the exterior derivative of $\widetilde{\omega}$ exclusively to data which are directly accessible from the smooth function $\omega\colon G \to \widehat{\mathfrak{g}}$ and to the elementary calculus derivative. After this proposition, we have several equivalent ways of thinking about the exterior derivative: Firstly, the original definition of the exterior derivative of a 1-form as a special case of the general definition of exterior derivatives of arbitrary n-forms; secondly after choosing our trivialization of $T(G)$ and $\widehat{T}(G)$ via right translations and after identifying 1-forms with smooth functions $\omega\colon G \to \widehat{\mathfrak{g}}$, we can compute the exterior derivative via (9) above in terms of the derivative $\delta\omega$, and thirdly, we have the formula (18) above which expresses $d\widetilde{\omega}$ in terms of the most immediate data and operations.

It is instructive to contemplate the special case that $\mathfrak{g}$ happens to be exponential and $G = (\mathfrak{g}, *)$ with $\exp_G = \mathrm{id}_\mathfrak{g}$. In this case it is natural to choose a *different* trivialization of $T(G)$, namely, the standard one associated with the topological vector space $\mathfrak{g}$ given by $T(G) = \mathfrak{g} \times \mathfrak{g}$. Accordingly, $\widehat{T}(G) = \widehat{\mathfrak{g}} \times \mathfrak{g}$. We note in passing, that, if we choose the C-H-neighborhood $B = \mathfrak{g}$, the bundle maps R and $\widehat{R}$ of (5) become bundle isomorphisms $\mathfrak{g}_a \times \mathfrak{g} \to \mathfrak{g}_a \rtimes_{\mathrm{Ad}} G = T(G)$ and $\widehat{\mathfrak{g}}_a \times \mathfrak{g} \to \widehat{\mathfrak{g}}_a \rtimes_{\widehat{\mathrm{Ad}}} G = \widehat{T}(G)$. However, in relation to this trivialization, we can once again identify vector fields and 1-forms with smooth functions. We should be aware, however, that one and the same 1-form, say, is represented by different smooth functions depending on which trivialization we consider. For instance, if, in the exponential case we choose the vector space trivialization and proceed with the identification of 1-forms and smooth functions, the derivative $\delta\omega(g)$ is none other than the standard derivative $\omega'(g)$, and formula (9) in this case becomes

(9′) $d\widetilde{\omega}(\widetilde{X}, \widetilde{Y})(g) = \omega'(g)\{\frac{1}{2} \cdot (X(g) \otimes Y(g) - Y(g) \otimes X(g))\}.$

In particular, every constant function has zero exterior derivative, hence is closed. Of course, these constant function represent neither left- nor right-invariant 1-forms in general.

Tangent bundles and wedge fields

For a wedge W in $\mathfrak{g}$ we constructed in Lemma VI.1.5 the invariant wedge field $\Theta\colon G \to \mathcal{W}(G)$ by $\Theta(g) = d\lambda_g(\mathbf{1})(W)$. If we identify $T(G)$ and $\mathfrak{g} \times G$, then this wedge field is given by $\Theta(g) = \operatorname{Ad}(g)(W) \times \{g\}$. Wedge fields in general are in bijective correspondence with functions $\Xi\colon G \to \mathcal{W}(\mathfrak{g})$, any one of which gives a wedge field $g \mapsto \Xi(g) \times \{g\}$.

Now we can rephrase Corollary VI.5.5 as follows:

VI.5.13.   Corollary.   *In a connected Lie group G with a finite fundamental group $\pi_1(G)$, with the notation of Theorem VI.5.1, the wedge W is global in G if and only if there is a smooth function $\omega\colon A_L \to \widehat{\mathfrak{g}}$ satisfying the following conditions*

(i) $\widehat{\operatorname{Ad}}(g^{-1})\omega(g) \in W^*$ *for all $g \in A_L$.*

(ii) $\omega(\mathbf{1}) \in \operatorname{algint} W^*$

(iii) $\underline{\omega}'_g(0)$ *is symmetric for all $g \in A_L$.*

Proof.   It suffices to observe that, in view of the trivialization of the tangent and the cotangent bundle of G and in view of Corollary VI.5.5 and Remark VI.5.11, conditions (i), (ii) and (iii) of Theorem VI.5.1 are equivalent to conditions (i), (ii), and (iii) above, respectively.   ∎

In the spirit of this corollary and earlier remarks on exponential Lie algebras it is not hard to establish

EVI.5.1.   Exercise.   Let $\mathfrak{g}$ be an exponential Lie algebra and $G = (\mathfrak{g}, *)$. Then a Lie wedge W is global in G if there is an element $\omega \in \operatorname{algint} W^*$ such that $(d\lambda_g^*(0)^{-1})\widehat{\ }\omega \in W^*$ for all $g \in G$.   ∎

A *right invariant wedge field* is one defined by a constant function Ξ. As before we speak of invariant wedge fields when we mean wedge fields which are invariant under left translations; they are represented by by functions $g \mapsto \operatorname{Ad}(g)(W)$ with a fixed wedge $W \subseteq \mathfrak{g}$. We shall call such a field the *invariant wedge field generated by W*.

VI.5.14.   Lemma.   *Let W be a wedge in the Lie algebra $\mathfrak{g}$ of a connected Lie group G. Then the following statements are equivalent:*

(I) *The invariant wedge field generated by W is right invariant.*

(II) *W is an invariant wedge.*

(III) $\bigcup_{g \in G} \mathcal{W}(g) = W \times G$.

Proof.   We note first that in view of the preceding remarks conditions (I) and (III) express the same fact. Now since $\operatorname{Ad}(\exp X) = e^{\operatorname{ad} X}$ for all $X \in \mathfrak{g}$, the wedge W is invariant if and only if $\operatorname{Ad}(\exp X)(W) = W$ for all $X \in \mathfrak{g}$ by Definition II.1.9. Since G is connected and therefore generated by $\exp \mathfrak{g}$, this is the case if and only if $\operatorname{Ad}(g)(W) = W$ for all $g \in G$. In view of the preceding remarks, this is exactly condition (I).   ∎

Corollary VI.5.13 above has the following variant for *invariant* wedges:

VI.5.15.   Corollary. *In a connected Lie group G with a finite fundamental group $\pi_1(G)$, with the notation of* Theorem VI.5.1, *an invariant wedge W is global in G if and only if there is a smooth function $\omega\colon A_L \to \widehat{\mathfrak{g}}$ satisfying the following conditions*

   (i) $\omega(g) \in W^*$ *for all* $g \in A_L$.

   (ii) $\omega(1) \in \operatorname{algint} W^*$

   (iii) $\underline{\omega}'_g(0)$ *is symmetric for all* $g \in A_L$.   ∎

Problems for Chapter VI

PVI.1.   Problem. *Find all global invariant cones.* (*For a solution of the case of simple Lie algebras see* [Ol82b]).   ∎

Notes for Chapter VI

Section 1.   The concept of cone fields was used by Vinberg [Vi80] and Ol'shanskiĭ [Ol81]. We use it in order to solve "differential inequalities" via the concepts introduced in Definition VI.1.7. The piecewise smooth curves we obtain serve the purpose of producing a semigroup which is close to the semigroup generated by a given Lie wedge in a Lie algebra. The details are expressed in the Approximation Theorem for Chains VI.1.16. Similar Theorems have been proved by Levichev [Lev86] and comparable results are well-known in control theory.

Section 2.   The positive forms introduced in this section are new in themselves, but, if exact, are the exterior derivatives of the positive functions introduced by Vinberg [Vi80] and Ol'shanskiĭ [Ol81].

Section 3.   The first essential result here is the Escape Theorem VI.3.4 which is new (cf. [Hi87]). The second crucial result is the Globality Theorem VI.3.8 giving sufficient conditions for a Lie wedge to be global (see [Hi87]); it is motivated by results of Ol'shanskiĭ who used the important concept of a positive function in [Ol82b].

Section 4.   The smoothing of the characteristic function of a semigroup to obtain positive functions is due to Ol'shanskiĭ [Ol82b]. The construction of the positive form in the Existence Theorem for Strictly Positive Forms VI.4.17 is from unpublished notes by Hofmann.

Section 5.   The Principal Characterization Theorem of Globality VI.5.12 is a summary of preceding results. Its applications through Corollary VI.5.9 are also new. For a version of Corollary VI.5.2 see [Hi87].

Chapter VII

Embedding semigroups into Lie groups

There are several viewpoints in studying Lie groups. The guiding principle in the earlier chapters of this book is that Lie groups are functorially associated with Lie algebras and exponential mappings. A subsemigroup of a Lie group is then studied via its tangent wedge in the Lie algebra. In this context, the appropriate class of semigroups is that of infinitesimally generated ones.

Alternative viewpoints are those of differential geometry and of topology. Under the first one defines Lie groups as differentiable or analytic groups, under the second as locally Euclidean topological groups. The purpose of this chapter is to consider to what extent a theory of differentiable semigroups or semigroups on manifolds would overlap with that of the semigroups that we have studied. In particular, how limiting has been our restiction to only those semigroups which are subsemigroups of Lie groups? Our goal is to show that under a various rather general hypotheses, we can obtain at least a local embedding of a semigroup into a Lie group, and hence via the log function a local embedding into a Campbell–Hausdorff neighborhood in a completely normable real Lie algebra. Hence our previous restriction to subsemigroups of Lie groups and local subsemigroups in Campbell–Hausdorff neighborhoods is not as restrictive as it might first appear. Also these results suggest that our earlier theory concerning tangent objects of (local) subsemigroups of Lie groups might find applications in the settings considered in this chapter via these (local) embeddings.

Standard references are [CP61] and [CP67] for the algebraic theory of semigroups and [HM66] and [CHK83] for topological semigroups.

Section 1 deals with the local constructions of quotients and local embeddings of semigroups into groups in a topological setting. The techniques are an elaboration of those set forth in [BH87]. In Section 2 we survey and extend work of Graham [Gr83] concerning the rudiments of a theory of differentiable (local) semigroups and their (local) embeddings into Lie groups. In Section 3 we consider cancellable semigroups on manifolds and what can be said about embedding them into Lie groups. Here techniques of Houston and Brown [BH87] and Hofmann and Weiss [HW87] are employed.

Throughout this chapter the field in question for such notions as analytic structures, normed spaces, Lie groups, and so on will always be assumed to be the field of real numbers $\mathbb{R}$.

1. General embedding machinery

The construction of quotients is a technique of wide applicability in mathematics. In this first section of the chapter we develop the basic algebraic and topological constructions for the local embedding of semigroups into a local group. We show that there is functorially associated with a wide class of cancellative local semigroups a local group of quotients. The techniques in this section involve the local construction of quotients along the lines laid down by Houston and Brown in [BH 87].

Algebraic preliminaries

In this subsection we lay the necessary algebraic foundations for the later embedding theorems. We define an appropriate (somewhat technical) setting which is general enough to encompass the various applications we have in mind.

VII.1.1.   Definition. A *cancellative partial semigroup* is a triple (S, T, m) satisfying

(i) $\emptyset \neq S \subseteq T$,

(ii) $m: \operatorname{dom}(m) \to T, \qquad S \times S \subseteq \operatorname{dom}(m) \subseteq T \times T$,

(iii) $a(bc)$ and $(ab)c$ are defined and equal for all possible associations for $a, b, c \in S$ (where $m(a, b) = ab$),

(iv) if $a \in S$ and $c, d \in S \cup S^2$ and $ac = ad$ or $ca = da$, then $c = d$.

We abuse the previous notation and refer to S as a cancellative partial semigroup.

VII.1.2.   Definition. If A, B are subsets of S, a cancellative partial semigroup, then we say A *right-reverses in* B if given $s, t \in A$, then $sB \cap tB \neq \emptyset$, i.e., there exist $b, c \in B$ such that $sb = tc$.

Note. If A reverses in any subset of B, then A reverses in B.

Let U be a fixed non-empty subset of a cancellative partial semigroup S. Our goal is to build a local group of quotients whose members are of the form

$a^{-1}b$ resp. ab^{-1} for $a, b \in U$. The approach taken is a standard one for such constructions: we obtain the local group of quotients as a set of equivalence classes of an appropriate equivalence relation $\equiv$ on $U \times U$ ($a^{-1}b$ resp. ab^{-1} then denotes the equivalence class of (a, b)). Since $(a^{-1}b)(b^{-1}c) = a^{-1}c$ and $(ab^{-1})(bc^{-1}) = ac^{-1}$, we define a partial operation $\cdot$ on $S \times S$ by $(a, b) \cdot (b, c) = (a, c)$. It will then be necessary that $\equiv$ restricted to $U \times U$ is a congruence with respect to this operation.

Let $a, b, c, d \in S$ and let $x, y \in V$, where V is some non-empty subset of S. If S were in a group and if $ax = by$ and $cx = dy$, then $a^{-1}b = xy^{-1} = c^{-1}d$; similarly if $ax = cy$ and $bx = dy$, then $ab^{-1} = ax(bx)^{-1} = cy(dy)^{-1} = cd^{-1}$. These observations motivate the following definition:

VII.1.3. **Definition.** For $a, b, c, d \in S$, we define the *left quotient relation* by

$$(a, b) \underset{V}{\overset{\text{def}}{\equiv}} (c, d) \quad \text{if} \quad ax = by, \ cx = dy \quad \text{where} \quad x, y \in V$$

and the *right quotient relation* by

$$(a, b) \overset{\text{def}}{\equiv}_V (c, d) \quad \text{if} \quad ax = cy, \ bx = dy \quad \text{where} \quad x, y \in V.$$

VII.1.4. **Remark.** Observe that $(a, b) \underset{V}{\equiv} (c, d)$ iff $(a, c) \equiv_V (b, d)$. Thus properties of one yield "dual" properties of the other.

If V is understood and we are considering properties common to both relations, we sometimes omit the subscript and write $(a, b) \equiv (c, d)$.

VII.1.5. **Lemma.** *The relation $\equiv_V$ is reflexive and symmetric. The relation $\underset{V}{\equiv}$ is symmetric, and if $aV \cap bV \neq \emptyset$, then $(a, b) \underset{V}{\equiv} (a, b)$.*

Proof. That $\underset{V}{\equiv}$ and $\equiv_V$ are symmetric is immediate. For $x \in V$, $ax = ax$ and $bx = bx$ yield that $(a, b) \equiv_V (a, b)$. If $ax = by$ for $x, y \in V$, then it follows that $(a, b) \underset{V}{\equiv} (a, b)$. $\blacksquare$

The next lemma is a basic calculational tool.

VII.1.6. **Lemma.** (Malcev's Condition) *Suppose $ax = cy$, $bx = dy$, and $au = cv$ for $a, b, c, d, u, v, x, y \in S$. If $yS \cap vS \neq \emptyset$, then $bu = dv$.*

Proof. There exist $s, t \in S$ such that $ys = vt$. Then $axs = cys = cvt = aut$, so by cancellation $xs = ut$. We now obtain $but = bxs = dys = dvt$. Again by cancellation we obtain $bu = dv$. $\blacksquare$

VII.1.7. **Lemma.** *Suppose V right-reverses in S. Then*
  i) *$(a, a') \underset{V}{\equiv} (b, b')$ and $(b, b') \underset{V}{\equiv} (c, c')$ imply $(a, a') \underset{V}{\equiv} (c, c')$;*
  ii) *$(a, b) \equiv_V (a', b')$ and $(b, c) \equiv_V (b', c')$ imply $(a, c) \equiv_V (a', c')$.*

Proof. Suppose we have $ax = a'x'$, $bx = b'x'$, $bu = b'v$, $cu = c'v$ for $x, x', u, v \in V$. From the first three equalities and Lemma VII.1.6 (Malcev's condition), we conclude $au = a'v$. Thus $(a, a') \underset{V}{\equiv} (c, c')$ and $(a, c) \equiv_V (a', c')$. $\blacksquare$

VII.1.8.   Lemma.   *Let V and V' be non-empty subsets of S such that VV' is also a subset of S, and suppose that V right-reverses in V' and V' right-reverses in S. Let $a, c \in S$ satisfy $aV \cap cV \neq \emptyset$.*

   i) *If (a, b) $_V\equiv (a', b')$ and (b, c) $_V\equiv (b', c')$, then (a, c) $_V\equiv (a', c')$.*

   ii) *$(a, a') \equiv_V (b, b')$ and $(b, b') \equiv_V (c, c')$ imply $(a, a') \equiv_V (c, c')$.*

Proof.   There exist $x_1, x_2, x_3, x_4 \in V$ such that $ax_1 = bx_2$, $a'x_1 = b'x_2$, $bx_3 = cx_4$, and $b'x_3 = c'x_4$. Since V right-reverses in V', there exist $y, y' \in V'$ such that $x_2y = x_3y'$. Then $ax_1y = bx_2y = bx_3y' = cx_4y'$ and similarly $a'x_1y = c'x_4y'$. Note that by hypothesis $x_1y, x_4y' \in S$.

   Also by hypothesis $ax = cx'$ for some $x, x' \in V$. Since V right-reverses in V', there exist $y_1, y_2 \in V'$ such that $x'y_1 = x_4y_2$. Since V' right-reverses in S, there exist $s, t \in S$ such that $y's = y_2t$. Then $x_4y's = x_4y_2t = x'y_1t$. Thus $x_4y'S \cap x'y_1S \neq \emptyset$. Multiplying $ax = cx'$ by y_1, we obtain $a(x'y_1) = c(x'y_1)$. From the preceding paragraph and Lemma VII.1.6 (Malcev's condition), we conclude $a'xy_1 = c'x'y_1$. By cancellation $a'x = c'x'$, whence (a, c) $_V\equiv (a', c')$ and $(a, a') \equiv_V (c, c')$.   ∎

VII.1.9.   Theorem.   *Suppose $V, V', VV' \subseteq S$, V right-reverses in V' and V' right-reverses in S. Let $S \times_V S$ denote the set $\{(a, b) \in S \times S : aV \cap bV \neq \emptyset\}$. Define a partial operation of $S \times_V S$ by $(a, b) \cdot (b, c) = (a, c)$ if $(a, c) \in S \times_V S$. Then $_V\equiv$ is a congruence relation on $S \times_V S$ with respect to this partial operation. Hence there is induced on $Q(S, V) \overset{\text{def}}{=} S \times_V S / {_V\equiv}$ a partial operation $*$ such that the natural mapping $\beta : S \times_V S \to Q(S, V)$ sending (a, b) to the equivalence class $\beta(a, b)$ (thought of as $a^{-1}b$) is a homomorphism.*

   With respect to the operation $*$, $\beta(a, b) * \beta(c, d)$ is defined if and only if there exist $p, q, r \in S$ such that (a, b) $_V\equiv (p, q)$, (c, d) $_V\equiv (q, r)$, and $pV \cap rV \neq \emptyset$. In this case, $\beta(a, b) * \beta(c, d) = \beta(p, r)$.

   The diagonal $\Delta = \{(a, a) : a \in S\}$ is an equivalence class of $_V\equiv$ and acts as an identity on $Q(S, V)$. With respect to this identity, $\beta(b, a)$ is an inverse for $\beta(a, b)$.

Proof.   The relation $_V\equiv$ is reflexive and symmetric by Lemma VII.1.5, transitive by Lemma VII.1.7, and compatible with the partial multiplication $\cdot$ by Lemma VII.1.8. Hence it is a congruence. It then follows directly from this fact that if the multiplication $*$ is defined on $Q(S, V)$ by the conditions given in the theorem, then this partial multiplication is well-defined and the mapping β is a homomorphism.

   Let $a, b \in S \times_V S$. Since $ax = ax$ and $bx = bx$ for $x \in V$, it follows that (a, a) $_V\equiv (b, b)$. Conversely suppose (a, a) $_V\equiv (b, c)$. Then $ax = ay$ and $bx = cy$ for some $x, y \in V$. By cancellation $x = y$ and again by cancellation $b = c$. Thus Δ is an equivalence class.

   We have $\Delta * \beta(a, b) = \beta(a, a) * \beta(a, b) = \beta(a, b)$ and similarly $\beta(a, b) * \Delta = \beta(a, b)$. Thus Δ is an identity. Finally $\beta(a, b) * \beta(b, a) = \beta(a, a) = \Delta$.   ∎

   In order to get an analogous result for $\equiv_V$, we must restrict our attention to smaller sets than $S \times_V S$.

VII.1.10.   Definition.   For $U, V \subseteq S$, we say that (U, V) is a *suitable pair* if there exists a non-empty subset $V' \subseteq S$ such that

   (i) U right-reverses in V,

   (ii) V right-reverses in V',

   (iii) $VV' \subseteq S$ and V' right-reverses in S.

VII.1.11.   **Theorem.** *Let (U, V) be a suitable pair and let $\equiv$ be either the relation $_V{\equiv}$ or $\equiv_V$ restricted to $U \times U$. Then $\equiv$ is an equivalence relation which is a congruence with respect to the partial operation on $U \times U$ defined by $(a, b) \cdot (b, c) = (a, c)$. Hence there is induced on $Q(U, V) \stackrel{def}{=} U \times U/ \equiv$ a partial operation $*$ such that the natural mapping $\beta: U \times U \to Q(U, V)$ sending (a, b) to the equivalence class $\beta(a, b)$ of (a, b) with respect to $\equiv$ is a homomorphism. With respect to this operation $\beta(a, b) * \beta(c, d)$ is defined if and only if there exist $p, q, r \in U$ such that $(a, b) \equiv (p, q)$ and $(c, d) \equiv (q, r)$; in this case the product is $\beta(p, r)$. The diagonal $\Delta(= \Delta_U) = \{(x, x): x \in U\}$ is an equivalence class of $\equiv$ and acts as an identity in $Q(U, V)$ and $\beta(b, a)$ is an inverse for $\beta(a, b)$.*

Proof.   The case $\equiv$ is $_V{\equiv}$ is a restriction of the setting of Theorem VII.1.9 to $U \times U$ and hence follows from that theorem.

   Let V' be as in Definition VII.1.10. By Lemma VII.1.5, $\equiv_V$ is reflexive and symmetric, and by Lemma VII.1.8, $\equiv_V$ is transitive. Hence $\equiv_V$ is an equivalence relation. By Lemma VII.1.7 it is compatible with the operation $\cdot$ on $U \times U$ and hence is a congruence relation. It then follows directly from this fact that if the multiplication $*$ is defined on $Q(U, V)$ by the conditions given in the theorem, then this partial multiplication is well-defined and the mapping β is a homomorphism.

   Let $a, b \in U$. Since U right-reverses in V, there exist $x, y \in V$ such that $ax = by$. It follows that $(a, a) \equiv_V (b, b)$. Conversely suppose $(a, a) \equiv (b, c)$. Then there exist $x, y \in V$ such that $ax = by$ and $ax = cy$. By cancellation $b = c$. Thus Δ is an equivalence class for $\equiv_V$.

   The proofs that Δ is an identity and that $\beta(b, a)$ is an inverse for $\beta(a, b)$ are the same as in Theorem VII.1.9.   ∎

   We remark that although the operation $\cdot$ is associative, this does not immediately yield that $*$ is associative, since $\cdot$ is only a partial operation. Indeed problems associated with associativity in $Q(S, V)$ or $Q(U, V)$ seem difficult. However, the associativity of $\cdot$ does give rise to many associative triples in $Q(U, V)$.

VII.1.12.   **Proposition.** *Let (U, V) be a suitable pair, $a \in U$. If*

$$\gamma_1, \gamma_2, \gamma_3, \gamma_1 * \gamma_2 \in \beta(\{a\} \times U) \cap \beta(U \times \{a\}),$$

then $\gamma_1 * (\gamma_2 * \gamma_3)$ and $(\gamma_1 * \gamma_2) * \gamma_3$ are defined and equal.

Proof.   Let $\gamma_1 = \beta(x, a)$, $\gamma_2 = \beta(a, y) = \beta(q, a)$, $\gamma_3 = \beta(a, r)$, $\gamma_1 * \gamma_2 = \beta(w, a)$. Then $\gamma_1 * \gamma_2 = \beta(x, a) * \beta(a, y) = \beta(x, y)$ and similarly $\gamma_2 * \gamma_3 = \beta(q, r)$. Note that $\beta(a, y) = \beta(q, a)$ implies $(a, y) \equiv (q, a)$, hence by definition $(y, a) \equiv (a, q)$, i.e., $\beta(y, a) = \beta(a, q)$. Then

$$\gamma_1 = \beta(x, a) = \beta(x, y) * \beta(y, a) = (\gamma_1 * \gamma_2) * \beta(a, q) = \beta(w, a) * \beta(a, q) = \beta(w, q).$$

From this we deduce

$$\gamma_1 * (\gamma_2 * \gamma_3) = \beta(w, q) * \beta(q, r) = \beta(w, r)$$
$$= \beta(w, a) * \beta(a, r) = (\gamma_1 * \gamma_2) * \gamma_3.$$

    ∎

It is well-known that a necessary and sufficient condition for a semigroup S to be embeddable in a group G of right quotients (i.e. every element of G is of the form st^{-1} for some $s, t \in S$) is that it be cancellative and right-reversible in itself (sometimes called Ore's condition; see Chapter 1 of [CP]).

VII.1.13.   Proposition. *Assume that a semigroup S is cancellative and right reversible (in itself). Then (S, S) is an suitable pair. Let G be the group of right quotients in which S embeds (we think of S as a subset of G) and define $\mu: S \times S \to G$ by $\mu(a, b) = ab^{-1}$. Then $\mu(a, b) = \mu(c, d)$ if and only if $(a, b) \equiv_S (c, d)$. Furthermore, the operation $*$ on $Q(S, S)$ in Theorem VII.1.8 is globally defined and the mapping induced from $(Q(S, S), *)$ to G is an isomorphism (hence in particular $*$ is associative). The embedding from S to $\beta(S, S)$ given by $s \mapsto \beta(sa, a)$ is independent of a and corresponds to the embedding of S into G.*

Proof.   To see (S, S) is suitable, take $V' = S$.

Suppose that $ab^{-1} = cd^{-1}$. Pick $x, y \in S$ such that $bx = dy$. Then

$$ax(bx)^{-1} = ab^{-1} = cd^{-1} = cy(dy)^{-1}.$$

By cancellation $ax = cy$, so $(a, b) \equiv_S (c, d)$. The argument reverses to obtain the converse. Since the mapping μ is a homomorphism with respect to the partial multiplication $\cdot$ on $S \times S$, it follows that there will be induced a bijective homomorphism from $Q(S, S)$ onto G. This will then be an isomorphism as soon as the multiplication on $Q(S, S)$ is known to be global. Let $(a, b), (c, d) \in S \times S$. Pick $x, y \in S$ such that $bx = cy$. One verifies directly that $(a, b) \equiv_S (ax, bx)$ and similarly $(c, d) \equiv_S (cy, dy)$. Then $\beta(a, b) * \beta(c, d) = \beta(ax, dy)$.   ∎

Since under μ, the element $\beta(sa, a)$ corresponds to $(sa)a^{-1} = s$ in G, we conclude that the embedding of S into $\beta(S, S)$ given by $s \mapsto \beta(sa, a)$ is independent of $a \in S$ and is a monomorphism.

Local embeddings

We turn now to the topological setting and consider first of all the problem of the existence of suitable pairs. Again we posit a rather technical setting that will encompass our later applications.

VII.1.14.   Definition. A *cancellative partial topological semigroup* denotes a triple (S, T, m) together with a T_3-topology on T satisfying

(i) (S, T, m) is a cancellative partial semigroup (cf. Definition VII.1.1),

(ii) S is open in T, the domain of m is open in $T \times T$, and $m: \operatorname{dom} m \to T$ is continuous, and

(†) there exists a dense open subset S° of S such that given $a \in S$, $b \in S^\circ$, and V open containing b, there exist open sets U containing a and N containing ab (resp. ba) such that $x \in U$ implies $N \subseteq xV$ (resp. $N \subseteq Vx$).

The condition (†) is a translation into semigroup language of the group condition that multiplication is continuous at (a^{-1}, ab) and at (ba, a^{-1}).

VII.1.15. **Lemma.** *Let (S, T, m) be a cancellative partial topological semigroup. Then for $a \in S$, the mapping $x \mapsto ax \colon S^\circ \to T$ is open. A similar result holds for right translations.*

Proof. Let V be an open set in S° and let $b \in V$. By (†) there exist open sets U containing a and N containing ab such that $N \subseteq xV$ for $x \in U$. In particular, $ab \in N \subseteq aV$. Since N is open and ab was an arbitrary element of aV, this shows aV is open. ∎

VII.1.16. **Proposition.** *Let S be a cancellative partial topological semigroup. Then for any non-empty open set V in S and any $s \in S$, there exists an open set U containing s such that U right-reverses in V.*

Proof. Fix $b \in V \cap S^\circ$. By (†) of Definition VII.1.14 there exist open sets U and N with $s \in U$ and $sb \in N$ such that $x \in U$ implies $N \subseteq xV$. Clearly if $c, d \in U$, then $cV \cap dV \neq \emptyset$. ∎

VII.1.17. **Corollary.** *Let $Z \subseteq S$ be an open set, and fix $a, b, c \in S$ such that $bc \in Z$. There exist open sets U, V, V' containing a, b, c respectively such that U right reverses in V, V right-reverses in V', V' right-reverses in Z, and $VV' \subseteq Z$. If $b \in S^\circ$, then V can be chosen so that $V \subseteq S^\circ$.*

Proof. Pick open sets V_1 containing b and W_1 containing c such that $V_1 W_1 \subseteq Z$. By Proposition VII.1.16 there exists an open set V', $c \in V'$ such that V' right-reverses in Z. We may assume $V' \subseteq W_1$ (by replacing it with $V' \cap W_1$ if necessary, since this smaller set will still right-reverse in Z). Again using Proposition VII.1.16, we obtain an open set $V \subseteq V_1$ containing b so that V right reverses in V'. If $b \in S^\circ$, then we can replace V by $V \cap S^\circ$. A final application of Proposition VII.1.16 yields an open set U containing a such that U right-reverses in V. ∎

The existence of suitable pairs in ample numbers is thus established. Using the alternate hypothesis of Definition VII.1.14, one sees readily that both Proposition VII.1.16 and Corollary VII.1.17 possess dual versions dealing with the existence of left reversing open sets.

VII.1.18. **Definition.** Let (U, V) be a suitable pair of open sets in S, and let $\equiv$ denote either the congruence $_V\!\equiv$ or $\equiv_V$ on $U \times U$. Let $Q(U, V)$ denote the quotient space $(U \times U)/\!\equiv$ endowed with the quotient topology, and let β denote the quotient map, $\beta \colon U \times U \to Q(U, V)$. Fix $a \in U$, and define $a^\triangle \colon U \to Q(U, V)$ by $a^\triangle(x) = \beta(x, a)$, $a^\nabla \colon U \to Q(U, V)$ by $a^\nabla(x) = \beta(a, x)$.

VII.1.19. **Lemma.** *The maps $a^\triangle$ and a^∇ are continuous and injective on U. If $V \subseteq S^\circ$, then these mappings are homeomorphisms of $U \cap S^\circ$ onto open subsets of $Q(U, V)$.*

Proof. Continuity of these maps is immediate, since $Q(U, V)$ has the quotient topology, while injectivity is a consequence of cancellation.

It must be shown that each map is open on $U \cap S^\circ$, and we begin with a^∇ and $_V\!\equiv$. Fix H, an open set contained in $U \cap S^\circ$, and let $(c, d) \in \beta^{-1}(a^\nabla(H))$.

Then there exists $b \in H$ such that $(a,b) \;_V\!\equiv (c,d)$; hence for some elements $x, y \in V$, $ax = by$ and $cx = dy$. By (†) there exists an open set U_y containing y such that $ax = by \in Hy'$ for all $y' \in U_y$. Since $y \in V \subseteq S^\circ$, we can assume without loss of generality that $U_y \subseteq S^\circ$. Then again by (†) there exist open sets N containing dy and U_d containing d such that $N \subseteq d'U_y$ for all $d' \in U_d$. We have $cx = dy \in N$. By continuity pick U_c open containing c such that $U_c \cdot x \subseteq N$.

Now let $(c', d') \in U_c \times U_d$. Then $c'x = d'y'$ for some $y' \in U_y$ by the choice of U_c and N. Then $ax = b'y'$ for some $b' \in H$ since $ax = by \in Hy'$. Thus $(c', d') \;_V\!\equiv (a, b')$. This shows that $U_c \times U_d \subseteq \beta^{-1}(a^\nabla(H))$, and thus the latter is open. Since β is a quotient mapping, $a^\nabla(H)$ is open.

The proof that $a^\triangle$ is open may be deduced from the preceding. If $(b, a) \;_V\!\equiv (d, c)$, then $(a, b) \;_V\!\equiv (c, d)$, so C and D exist as in the preceding paragraph (for a given open set $H \subseteq U$). It follows immediately that $(d, c) \in D \times C$, and $\beta(D \times C) \subseteq a^\triangle(H)$.

The case that $\equiv$ is $\equiv_V$ involves analogous arguments which we leave to the reader.   ■

VII.1.20.   Definition. A *local group* is a system (G, e, θ, m) consisting of a topological space G, an element $e \in G$, a function $\theta: \mathrm{dom}\,\theta \to G$, and a function m (partial multiplication) defined on a subset $\mathrm{dom}(m) \subseteq G \times G$ into G. Furthermore, there must exist an open set Γ containing e such that $\Gamma \subseteq \mathrm{dom}\,\theta$ and $\Gamma \times \Gamma \subseteq \mathrm{dom}(m)$ and such that the following conditions are also satisfied:

   i) the restriction of m to $\Gamma \times \Gamma$ is continuous;

   ii) if $a, b, c, m(a, b), m(b, c) \in \Gamma$, then $m(m(a, b), c)$ and $m(a, m(b, c))$ are defined and equal;

   iii) if $a \in \Gamma$, then $m(a, e) = a = m(e, a)$;

   iv) the set $\{e\}$ is closed.

   v) θ restricted to Γ is continuous and $m(a, \theta(a)) = e = m(\theta(a), a)$ for $a, \theta(a) \in G$.

A local group is a *locally Euclidean local group* if e has an open neighborhood homeomorphic in the relative topology to Euclidean n-space for some n. A local group G is a *local Lie group* provided G is an analytic manifold, m restricted to $\Gamma \times \Gamma$ is analytic, and θ restricted to Γ is analytic.

The local group is generally denoted more compactly by G. The element e is the identity element. We usually write ab (or some other multiplicative notation) for $m(a, b)$ and a^{-1} for $\theta(a)$. A *standard neighborhood* in a local group is an open set B containing the identity, such that B is contained in some Γ satisfying the previous conditions, all quadruple products of elements of B are defined and associative, and B is symmetric ($B = B^{-1}$).

EVII.1.1.   Exercise. Show that if U is an open subset of a local group and $e \in U$, then U with the appropriate restrictions of θ and m is again a local group.

EVII.1.2.   Exercise. Show that a local group has a basis of neighborhoods at the identity that are standard neighborhoods.

Many of the properties of topological groups have corresponding local properties in local groups.

EVII.1.3. **Exercise.** Show that the axioms of a local group are sufficiently strong to conclude that a standard neighborhood is Hausdorff (even T_3).

EVII.1.4. **Exercise.** Show that if B is a standard neighborhood, $g \in B$ and $U \subseteq B$ is an open set, then gU and Ug are open in G. Show further that θ restricted to B is a homeomorphism and anti-automorphism of order 2.

VII.1.21. **Definition.** Let G and H be local groups. We say that G and H are *locally isomorphic* if there exist an open set U containing e_G, an open set V containing e_H and a homeomorphism $h: U \to V$ such that for $a, b \in U$, we have $ab \in U$ iff $h(a)h(b) \in V$ and in this case $h(ab) = h(a)h(b)$. The function h is called a *local isomorphism.*

One of the principal results of [Ja 57] is the local version of the solution of Hilbert's fifth problem; to wit, a locally Euclidean local group is locally isomorphic to a local Lie group (Theorem 107). In this case there is a corresponding finite dimensional real Lie algebra and a locally defined exponential mapping. One may take one of the Lie groups corresponding to this Lie algebra and show that the original local Lie group is locally isomorphic to this group (since the Campbell–Hausdorff multiplication on a neighborhood of 0 in the Lie algebra is locally isomorphic both to that of the local Lie group and the Lie group via the corresponding exponential mappings).

VII.1.22. **Theorem.** *If (U, V) is a suitable pair of open sets in a cancellative partial topological semigroup and $V \subseteq S^\circ$, then $Q(U, V)$ is a local group. Furthermore, if some point of $U \cap S^\circ$ has a Euclidean neighborhood, then $Q(U, V)$ is locally isomorphic to a Lie group.*

Proof. Assume that we know already that $Q(U, V)$ is a local group and we are given that some point $a \in U \cap S^\circ$ has a Euclidean neighborhood. Then a has a basis of Euclidean neighborhoods, in particular one (say W) contained in $U \cap S^\circ$. By Lemma VII.1.19 $a^\triangle(W)$ is a Euclidean neighborhood containing $\beta(a, a)$, the identity of $Q(U, V)$. Thus $Q(U, V)$ is locally Euclidean, and in view of the above remarks, locally embeddable as a neighborhood of the identity in a Lie group.

It remains to show that $Q(U, V)$ is a local group. Recall that β denotes the quotient map of $U \times U$ onto $Q(U, V)$. Fix any $a \in W = U \cap S^\circ$. In Definition VII.1.20, let $e = \beta(a, a)$, and $\Gamma = a^\triangle(W) \cap a^\nabla(W)$. By Lemma VII.1.19, Γ is open, hence is an open neighborhood of e in $Q(U, V)$. The function m is the multiplication $*$ induced on $Q(U, V)$ by the congruence $\equiv$ (see Theorem VII.1.11); m will be suppressed in favor of $*$ for ease of notation. That the operation is defined and associative (in the sense of Definition VII.1.20) for all triples in Γ follows directly from Proposition VII.1.12. Define θ on all of $Q(U, V)$ by $\theta(\beta(x, y)) = \beta(y, x)$. That e is an identity and θ is inversion on all of $Q(U, V)$ follows from Theorem VII.1.11.

The following diagram is commutative:

$$
\begin{array}{ccc}
W \times \{a\} \times \{a\} \times W & \xrightarrow{\text{proj}} & W \times W \\
{\scriptstyle a^\triangle \times a^\nabla} \downarrow & & \downarrow {\scriptstyle \beta} \\
a^\triangle(W) \times a^\nabla(W) & \xrightarrow[\quad m \quad]{} & Q(U, V).
\end{array}
$$

Since a^∇ and $a^\triangle$ are open mappings by Lemma VII.1.19, it follows that the left-hand vertical mapping is open, hence a quotient mapping. Since the bottom horizontal arrow is a projection it is continuous. It follows that m is continuous on $a^\triangle(W) \times a^\nabla(W)$, hence on $\Gamma \times \Gamma$.

Let $\Theta: U \times U \to U \times U$ be defined by $\Theta(b, c) = (c, b)$. Then $\theta \circ \beta = \beta \circ \Theta$ is continuous. Since β is a quotient mapping, it follows that θ is continuous.

Finally, $\{e\}$ is a closed set, since $\beta^{-1}(e)$ is precisely the diagonal of $U \times U$, and U is a Hausdorff space, so that this set is closed.   ∎

Admissible sets and local semigroups

In this section we apply the preceding results to derive embedding theorems for certain special classes of local semigroups and of semigroups. In the remainder of this first section we restrict our attention to the relation $\equiv_V$ since in a neighborhood of an identity either construction gives rise to local quotients which can be represented as both left and right quotients. Hence it does not really matter which construction we use.

VII.1.23.   Definition. A subset A of a topological space X is called an *admissible subset* of X if the interior of A in X is dense in A. If X is understood we sometimes refer to A simply as an *admissible set*. A T_3-space X is called a *Euclidean manifold with generalized boundary* if there exists an n such that each point of X possesses a neighborhood which is homeomorphic to some admissible set in $\mathbb{R}^n$.

We consider an important special setting where condition (†) of Definition VII.1.14 arises in a natural way.

VII.1.24.   Proposition. *Let (S, T, m) satisfy conditions* (i) *and* (ii) *of Definition VII.1.14. Let T be a Euclidean manifold with generalized boundary and let T° denote the set of points in T where T is locally homeomorphic to Euclidean n-space $\mathbb{R}^n$. Then $S(S \cap T^\circ) \subseteq T^\circ$ and condition* (†) *is satisfied.*

Proof.   It follows directly from Definition VII.1.23 that T° is open and dense in T and hence $S^\circ = S \cap T^\circ$ is open and dense in S. Let $a \in S$, $b \in S^\circ$. Let A be an open set containing a and V an open set containing b such that $AV \subseteq B$, where B is an open set around ab which is homeomorphic to an admissible subset of $\mathbb{R}^n$. It is no loss of generality to assume that V is chosen homeomorphic to an open n-cell, that the closure is contained in S° and is homeomorphic to a closed n-cell, and that $a\overline{V}$ is also contained in B. Since left translation by a is one-to-one and $\overline{V}$ is compact, it follows by invariance of domain that aV is homeomorphic to an open subset of $\mathbb{R}^n$, and hence $ab \in T^\circ$.

Denote the compact (sphere) boundary of V by F. Since left translation is one-to-one, $ab \notin aF$. Choose disjoint open sets N and W with $ab \in N$ and $aF \subseteq W$. Since $ab \in T^\circ$, we can choose N connected. By compactness arguments, there exists an open set $U \subseteq A$ containing a such that $Ub \subseteq N$ and $UF \subseteq W$. If

$x \in U$ and if N meets the complement of xV, then by the connectedness of N, N must contain a boundary point of xV. By invariance of domain, the boundary of xV is xF. Since $xF \subseteq W$, this is impossible. Thus $N \subseteq xV$. The dual version follows similarly. ∎

VII.1.25. Definition. A *local semigroup (with identity)* is a system (S, e, m) such that S is a T_3-space, $e \in S$, and $m: \mathrm{dom}(m) \to S$ is a function (partial multiplication) such that $\mathrm{dom}(m) \subseteq S \times S$. Furthermore, there must exist an open set Σ containing e such that $\Sigma \times \Sigma \subseteq S$ and such that the following conditions are satisfied:

    i) the restriction of m to $\Sigma \times \Sigma$ is continuous;

    ii) if $a, b, c, m(a, b), m(b, c) \in \Sigma$, then $m(m(a, b), c)$ and $m(a, m(b, c))$ are defined and equal;

    iii) if $a \in \Sigma$, then $m(a, e) = a = m(e, a)$.

The local semigroup is a *cancellative local semigroup (with identity)* if the multiplication is cancellative on Σ. We say that the local semigroup satisfies (†) if

    (†) there exists a dense open subset Σ° of Σ such that given $a \in \Sigma$, $b \in \Sigma^\circ$, and V open containing b, there exist open sets U containing a and N containing ab (resp. ba) such that $x \in U$ implies $N \subseteq xV$ (resp. $N \subseteq Vx$).

An open neighborhood N of the identity in a local semigroup is called a *standard neighborhood* if all quadruple products of elements of N are defined and associative and N is contained in one of the neighborhoods Σ guaranteed in the definition of a local semigroup.

We note by continuity of multiplication at the identity that the standard neighborhoods form a basis at the identity.

Again note that if T is any open set with $e \in T \subseteq S$, then T can replace S in the definition of a local semigroup. By a mild abuse of notation we generally refer to the local semigroup simply as S.

VII.1.26. Theorem. *Let S be a local semigroup satisfying the cancellation property and (†) on Σ. Then there exist an open neighborhood $U \subseteq \Sigma$ of e and an open set $V \subseteq \Sigma^\circ$ such that (U, V) is a suitable pair, $Q(U, V)$ is local group and the mapping $e^\triangle: U \to Q(U, V)$ given by $s \mapsto \beta(s, e)$ is a continuous monomorphism which restricted to $\Sigma^\circ \cap U$ is a topological embedding onto an open subset of $Q(U, V)$. If U is identified with its image $\beta(U \times \{e\})$ in $Q(U, V)$, then every member of $Q(U, V)$ has a representation of the form st^{-1} for $s, t \in U$.*

If, additionally, each neighborhood of e contains a point p such that $\Sigma p \cap \Sigma \subseteq \Sigma^\circ$ and the mapping $x \mapsto xp: \Sigma \to \Sigma p$ is a homeomorphism, then the mapping $s \mapsto \beta(s, e)$ is a topological embedding on all of U.

Proof. By continuity at e, pick an open set S_1 such that $e \in S_1 \subseteq (S_1)^3 \subseteq \Sigma$. Let $S_1^\circ = S \cap \Sigma^\circ$ and let the multiplication be m restricted to $\{(x, y): x, y \in \Sigma, \ m(x, y) \in \Sigma\}$. Then S_1, Σ, and this restriction satisfy the conditions of Definition VII.1.14, and hence form a cancellative partial topological semigroup.

Pick $b \in S_1^\circ$. Letting $Z = S_1$ and $a = c = e$ in Corollary VII.1.17, it follows directly from that corollary that there exist an open set $U_1 \subseteq S_1$ containing

e and an open set $V \subseteq S_1^\circ$ containing b such that (U_1, V) is a suitable pair. Pick an open set U with $e \in U \subseteq U^2 \subseteq U_1$ and $Ub \subseteq V$. Then (U, V) is still a suitable pair.

That $Q(U, V)$ is a local group follows from Theorem VII.1.22. By Lemma VII.1.19 $e^\triangle$ is a continuous injection. Now let $s, t, st \in U$. Then $(st, t) \equiv_V (s, e)$ since $st(b) = s(tb)$ and $t(b) = e(tb)$. Thus $\beta(s, e) * \beta(t, e) = \beta(st, t) * \beta(t, e) = \beta(st, e)$; this shows that the injection of U into $Q(U, V)$ is a monomorphism. By Lemma VII.1.19 the restriction of $e^\triangle$ to $S_1^\circ \cap U$ is a homeomorphism onto an open subset of $Q(U, V)$. Any member of $Q(U, V)$ is of the form $\beta(a, b)$ for $a, b \in U$. Then $\beta(a, b) = \beta(a, e) * \beta(e, b)$. It follows from Theorem VII.1.22 and its proof that $\beta(e, b)$ is the inverse of $\beta(b, e) = e^\triangle(b)$. Thus every element of $Q(U, V)$ can be represented in the form ab^{-1}, where a, b are in the embedded image of U.

Finally suppose that the last condition is satisfied. Let $s \in U$ and let B be an open set containing s. Pick an open set U' containing e such that $sU' \subseteq U \subseteq \Sigma$. By hypothesis there exists $p \in U'$ such that $\Sigma p \cap \Sigma \subseteq \Sigma^\circ$ and $x \mapsto xp: \Sigma \to \Sigma p$ is a homeomorphism. Thus there exists an open set $W \subseteq U \cap \Sigma^\circ$ containing sp such that $x \in \Sigma$ and $xp \in W$ implies $x \in B$. By Lemma VII.1.19, $p^\triangle(W) = \beta(W \times \{p\})$ is open in $Q(U, V)$. Suppose $\beta(t, e) \in p^\triangle(W)$. Then $(t, e) \equiv (q, p)$ for some $q \in W$. So there exist $v, w \in V$ such that $tv = qw$, $ev = v = pw$. Then $qw = tv = tpw$. By cancellation, $q = tp$. By choice of W, we conclude that $t \in B$. This shows $x \mapsto \beta(x, e)$ is an embedding at s, an arbitrarily chosen member of U.   ∎

VII.1.27.   Corollary.   *Let S be a locally compact locally cancellative local semigroup with identity and suppose that S is homeomorphic to an admissible subset of $\mathbb{R}^n$. Then S is locally embeddable into a finite-dimensional Lie group G such that the image of the embedding is an admissible subset of G. Alternately, S is locally topologically isomorphic to some local subsemigroup on an admissible subset of a finite-dimensional real Lie algebra with the restricted Campbell–Hausdorff multiplication.*

Proof.   Let Σ be a standard neighborhood such that the multiplication is cancellative on Σ; we further assume that $\overline{\Sigma}$ is compact and contained in S. One verifies directly that (Σ, S, m) is a cancellative partial semigroup satisfying (ii) of Definition VII.1.14. Let Σ° be the points at which Σ is locally Euclidean; then since Σ is open in S, it follow that Σ° is open and dense in Σ. By Proposition VII.1.24, the condition (†) is also satisfied. Hence (Σ, S, m) is a cancellative partial topological semigroup and a cancellative local semigroup. Pick open sets U and V as in Theorem VII.1.26. Then $Q(U, V)$ is a local group, the mapping $s \mapsto \beta(s, e)$ is a continuous monomorphism on U, and the image of $\Sigma^\circ \cap U$ is open (by continuity the closure of its image contains the identity $\beta(e, e)$). By Theorem VII.1.22, $Q(U, V)$ is locally embeddable as a neighborhood of the identity in a Lie group G. We then pick the open set W containing e, having its closure contained in U, and such that the composition of the continuous embedding of U into $Q(U, V)$ and $Q(U, V)$ into G is defined on $\overline{W}$. Since the closure is compact, the composition is a topological embedding on $\overline{W}$ and hence on W. Also the image of $\Sigma^\circ \cap W$ will be open in G and dense in the image of W (since this is true in $Q(U, V)$). The final statement of the corollary follows directly from the fact that the exponential mapping is locally a topological isomorphism from some neighborhood of 0 in $\mathbf{L}(G)$ equipped with the Campbell–Hausdorff multiplication to some neighborhood of the identity in G.

One simply composes with the inverse of this mapping. ∎

The next proposition gives a global version of the preceding results.

VII.1.28. **Proposition.** *Let S be a cancellative right-reversible semigroup with a dense open subset S° satisfying $SS^\circ \subset S^\circ$ and*

(†) *given $a, b \in S$, and V open containing b, there exist open sets U containing a and N containing ab (resp. ba) such that $x \in U$ implies $N \subseteq xV$ (resp. $N \subseteq Vx$).*

(Note by Proposition VII.1.24 that these are satisfied if S is a Euclidean manifold with generalized boundary.) Suppose there exists $p \in S^\circ$ such that the mapping $x \mapsto xp \colon S \to Sp$ is a homeomorphism. Then $G = Q(S, S)$ is a topological group and for $a \in S$ the mapping $s \mapsto \beta(sa, a)$ is a topological and isomorphic embedding of S onto an admissible subset of G which is independent of a.

Proof. The algebraic part of the proposition follows from Proposition VII.1.13.

Suppose $(a, b) \equiv_S (c, d)$. Then there exist $x, y \in S$ such that $ax = cy$ and $bx = dy$. Let $z \in S^\circ$. Then $axz = cyz$ and $bxz = cyz$ and $xz, yz \in S^\circ$ by hypothesis. Hence $(a, b) \equiv_{S^\circ} (c, d)$. Thus the equivalence relations on S given by $\equiv_S$ and $\equiv_{S^\circ}$ are the same, and so $G = Q(S, S) = Q(S, S^\circ)$. It is convenient to use the latter equality to apply our earlier results.

We consider left translations in G by $\beta(a, b)$. Replacing (a, b) by the equivalent element (ax, bx) for $x \in S^\circ$ if necessary, we see that we can assume without loss of generality that $b \in S^\circ$. By Lemma VII.1.19, $\beta(\{b\} \times S^\circ) = b^\nabla(S^\circ)$ is an open set containing the identity of G. On this neighborhood we have for $g = \beta(b, s)$,

$$\beta(a, b) * \beta(b, s) = \beta(a, s) = a^\nabla(s) = a^\nabla \circ (b^\nabla)^{-1}(g).$$

Thus translation is a homeomorphism on this neighborhood. Similarly a right translation is a homeomorphism on some neighborhood of the identity. By Theorem VII.1.22 the multiplication in G is continuous at the identity. From these facts we conclude that

$$(x, y) \mapsto x * y = g * \big((g^{-1} * x) * (y * h^{-1})\big) * h$$

is continuous at an arbitrary (g, h) in $G \times G$, and hence multiplication is continuous. Inversion is continuous by Theorem VII.1.22.

By Proposition VII.1.13 the mapping $s \mapsto \beta(sp, p)$ is a monomorphism from S into G. It is clearly continuous. Since $\beta(sp, p) = p^\triangle \rho_p(s)$, and since $\rho_p(S) \subseteq S^\circ$, it now follows from hypothesis and Lemma VII.1.19 that the monomorphism is a topological embedding. By Lemma VII.1.15 the right translation function $\rho_p \colon S^\circ \to S^\circ$ is an open mapping and $p^\triangle$ is open on S° by VII.1.19. Hence the image of S° is open in G, so the image of S is an admissible subset of G. ∎

The reader is referred to Theorem 4.1 of [McK70] for the special case of the preceding theorem that $S^\circ = S$. The condition (†) is stated in an alternate (but equivalent) form in the hypotheses.

EVII.1.5. **Exercise.** Show that if S is a topological semigroup which topologically and algebraically embeds as a subsemigroup of a group of right quotients on an admissible subset, then S is cancellative and right reversible and S°, the elements of S embedding in the interior of the image of S, satisfies the hypotheses of Proposition VII.1.28. Hence Proposition VII.1.28 gives necessary and sufficent conditions for a topological semigroup to embed as an admissible subset of a topological group of right quotients.

For the formulation of the next exercise, in which we present a more intrinsic approach to the existence of the set S^0, we first introduce suitable concepts. Let S be a topological semigroup. A point $b \in S$ is a *pseudo-interior point* if given $a, c \in S$ and an open set V containing b, there exist open sets U, W, and N with $a \in U$, $c \in W$, and $abc \in N$ such that $x \in U$, $z \in W$ implies $N \subseteq xVz$. The *pseudo-interior* S^0 of S consists of those points which possess neighborhoods made up entirely of pseudo-interior points.

EVII.1.6.   **Exercise.**   Under the present circumstances, prove the following assertions:

(i) If $s \in S$ and if $U \subseteq S^0$ is open, then sU and Us are open.

(ii) S^0 is an open ideal.

(iii) The hypotheses of Proposition VIII.1.28 can be replaced by the following hypothesis:

(‡) S is a cancellative right-reversible topological semigroup in which the pseudo-interior S^0 is dense.

Local homomorphisms

In this section we consider the morphisms of local semigroups and how these relate to the embedding procedures that we have developed.

VII.1.29.   **Definition.**   Let S and T be local semigroups with identities e and f resp. A *local homomorphism* is a continous function $\alpha: N \to T$, where N is some neighborhood of e, such that $\alpha(e) = f$ and for $x, y, xy \in N$, we have $\alpha(xy) = \alpha(x)\alpha(y)$.

Note that the restriction of a local homomorphism to any smaller neighborhood of e is again a local homomorphism. Thus if we wish to restrict down to smaller neighborhoods in the domain or codomain where certain properties hold, we may do so and simply restrict the local homomorphism.

VII.1.30.   **Definition.**   Let S be a local semigroup. A *local group of quotients* for S is a mapping $i: U \to G$, where U is an open set containing e in S, G is a local group, and i is both a local homomorphism and a homeomorphism onto an admissible subset of G.

VII.1.31.   **Remark.**   Let $i: U \to G$ be a local group of quotients for S. Pick a set B open in G contained in $i(U) \cap \Gamma$, where Γ is a standard neighborhood in G.

(This is possible since $i(U)$ is admissible and contains the identity.) For $i(s) \in B$ we have that $i(s)^{-1}i(U)$ and $i(U)i(s)^{-1}$ are neighborhoods of the identity. Hence the elements in some neighborhood of the identity in G can be written as left quotients and right quotients of the embedded image of U. This justifies the terminology "local group of quotients."

VII.1.32. **Proposition.** *Let S be a local semigroup. Then S admits a local group of quotients iff e has a standard neighborhood Σ with a dense open subset Σ° such that*

> *i) the multiplication is cancellative on Σ,*
>
> *ii) the multiplication satisfies (†) on Σ,*
>
> *iii) $\Sigma\Sigma^\circ \cap \Sigma \subseteq \Sigma^\circ$,*
>
> *iv) the mapping $x \mapsto xp: \Sigma \to \Sigma p$ is a homeomorphism for $p \in \Sigma$.*

Proof. The proof in one direction follows directly from Theorem VII.1.26. Conversely suppose S admits a local group of quotients, $i: U \to G$. We choose a standard neighborhood $\Sigma \subseteq U$ such that $i(\Sigma)$ is contained in a standard neighborhood Γ of G. Now the multiplication is cancellative on Γ, (by the same argument as in groups), thus on $i(\Sigma)$, and hence on Σ since i is an isomorphism onto $i(\Sigma)$.

Let Σ° be all elements in Σ which map into the interior of $i(\Sigma)$. (Note that $i(\Sigma)$ is admissible in G since it is open in the admissible set $i(U)$.) Let W be an open set contained in $i(\Sigma)$. Then sW is open for $s \in i(\Sigma)$ (see Exercise VII.1.4). Thus $sW \subseteq i(U)$ and $i(\Sigma)$ is open in $i(U)$ yield that $sW \cap i(\Sigma)$ is open in sW, hence open. It follows that $i(\Sigma)i(\Sigma)^\circ \cap i(\Sigma) \subseteq i(\Sigma)^\circ$, and hence the corresponding containment holds in S.

Since translations by elements of Γ are homeomorphisms on Γ, the restriction of right translation by $i(p)$ to $i(\Sigma)$ is a homeomorphism to $i(\Sigma p)$, and hence the corresponding property holds in S.

Finally we check condition (†). Suppose that $a \in i(\Sigma)$, $b \in i(\Sigma)^\circ$, and V is an open set containing b. Pick an open set W' containing the identity of G such that $W'W' \subseteq \Gamma$ and $W'b \subseteq V \cap \Gamma$. Pick another open set W containing the identity such that $W^{-1}W \subseteq W'$. Then $W \subseteq wW'$ for $w \in W$ and aWb is an open set containing ab. If $s \in i(\Sigma) \cap aW$, then $s = aw$ for some $w \in W$. We then have $aWb \subseteq awW'b \subseteq sV$. Thus (†) is satisfied when one pulls back to S. ∎

VII.1.33. **Proposition.** *Let S be a local semigroup, $i: U \to G$ a local group of quotients for S, and let α be a local homomorphism to a local group H. Then there exists a local homomorphism γ from G to H such that $\gamma \circ i = \alpha$ on some neighborhood of e. If δ is another local homomorphism from G to H such that $\delta \circ i = \alpha$ on some neighborhood of e, then δ and γ agree on some neighborhood of the identity $i(e)$.*

Proof. Pick an open set V containing e such that $V^2 \subseteq U$, α is defined on V^2 and $\alpha(V)$ and $i(V)$ are contained in standard neighborhoods of G and H respectively. Since $i(U)$ is an admissible subset of G and $i(V)$ is open in U, $i(V)$ is also an admissible subset of G. Pick $s \in V$ such that $i(s)$ is in the interior of $i(V)$. Then there exists an open set W containing $i(e)$ such that $W \subseteq i(V)(i(s))^{-1} \cap i(V)^{-1}i(s) \cap i(s)^{-1}i(V)$. Pick an open neighborhood $N \subseteq V$ of e such that $i(N) \subseteq W$.

Define $\gamma: W \to H$ by $\gamma(i(x)i(s)^{-1}) = \alpha(x)\alpha(s)^{-1}$ for $x \in V$. Every member of W can be so represented since $W \subseteq i(V)i(s)^{-1}$. Cancellation and the fact that i is injective guarantee that γ is well-defined. Note that γ can be written alternately as $\gamma(g) = \alpha(i^{-1}(g \cdot i(s)))\alpha(s)^{-1}$; thus γ is the composition of continuous functions and hence continuous. If $x \in N$ then $\gamma(i(x)) = \alpha(i^{-1}(i(x)i(s)))\alpha(s)^{-1} = \alpha(x)\alpha(s)\alpha(s)^{-1} = \alpha(x)$.

We now show that γ is a homomorphism. Suppose that $g \in W$ has a representation $g = i(y)^{-1}i(z)$ for $y, z \in V$. Then we can also write $g = i(x)i(s)^{-1}$ for some $x \in V$. Then by cross multiplying, $i(z)i(s) = i(y)i(x)$, so $zs = yx$ by the fact i is an injective homomorphism. It follows that $\alpha(z)\alpha(s) = \alpha(y)\alpha(x)$ and hence $\gamma(g) = \alpha(x)\alpha(s)^{-1} = \alpha(y)^{-1}\alpha(z)$. Now let $g, h, gh \in W$. Then by choice of W, there exist $u, v \in V$ such that $g = i(u)^{-1}s$ and $h = i(s)^{-1}i(v)$. We have just seen that

$$\gamma(g)\gamma(h) = \alpha(u)^{-1}\alpha(s)\alpha(s)^{-1}\alpha(v) = \alpha(u)^{-1}\alpha(v).$$

On the other hand

$$\gamma(gh) = \gamma(i(u)^{-1}i(s)i(s)^{-1}i(v)) = \gamma(i(u)^{-1}i(v)) = \alpha(u)^{-1}\alpha(v).$$

Now suppose δ is a local homomorphism from G to H such that $\delta \circ i = \alpha$ on some neighborhood Q of e. Pick a open set $M \subseteq Q \cap N$ such that $i(M)i(M)^{-1}$ is a subset of the domains of δ and γ. We have seen previously that $i(M)i(M)^{-1}$ is a neighborhood of $i(e)$. Then for $x, y \in M$,

$$\delta(i(x)i(y)^{-1}) = \delta(i(x))(\delta(i(y)))^{-1} = \alpha(x)(\alpha(y))^{-1}$$

and similarly $\gamma(i(x)i(y)^{-1}) = \alpha(x)(\alpha(y))^{-1}$. ∎

The next corollary states that a local group of quotients is unique up to local isomorphism.

VII.1.34.   Corollary.   *Let $i: U \to G$ and $j: V \to H$ be local groups of quotients for S. Then there exists a local topological isomorphism γ from G to H such that $\gamma \circ i = j$ on some neighborhood of e.*

Proof.   By the preceding proposition there exist local homomorphisms γ from G to H and λ from H to G with the appropriate commuting properties. By the uniqueness property their compositions in both directions must agree locally with the appropriate identity mapping. ∎

VII.1.35.   Remark.   If S is a local semigroup that admits a local group of quotients, then one can choose the local group of quotients to be $Q(U, V)$ for some suitable pair (U, V) in S.

Proof.   By Proposition VII.1.32 S satisfies the hypotheses of Theorem VII.1.26. It follows from the conclusion of that theorem that $e^{\triangle}: U \to Q(U, V)$ is a local group of quotients. By the preceding corollary the local group of quotients is unique up to local isomorphism. ∎

The next proposition gives the sense in which embedding in local quotient groups is functorial.

VII.1.36. **Proposition.** *Let S and T be local semigroups with local groups of quotients $i_S: U_S \to Q(S)$ and $i_T: U_T \to Q(T)$ and let $\alpha: S \to T$ be a local homomorphism. Then there exists a local homomorphism $Q(\alpha): Q(S) \to Q(T)$ such that $Q(\alpha) \circ i_S = i_T \circ \alpha$ on some neighborhood of e_S. Any two such local homomorphisms agree on some neighborhood of $i_S(e_S)$. If $\alpha(V)$ has non-empty interior in T for each open set in S containing e, then $Q(\alpha)$ is a locally open mapping (i.e., its restriction to some neighborhood of the identity if open).*

Proof. We apply Proposition VII.1.33 to the local homomorphism $i_T \circ \alpha$ to obtain the first part of the proposition.

We now assume the additional hypothesis and first show that $Q(\alpha)$ is open at the identity. Let W be a standard neighborhood of $Q(S)$ such that W is contained in the domain of $Q(\alpha)$ and $Q(\alpha)(W)$ is contained in a standard neighborhood in $Q(T)$. Pick V open in S containing e such that $N = i_S(V)(i_S(V))^{-1} \subseteq W$ and V is contained in the open set where $Q(\alpha) \circ i_S$ and $i_T \circ \alpha$ agree. It follows from commutativity that $Q(\alpha)(N) = i_T(\alpha(V)(i_T(\alpha(V)))^{-1}$. The latter is a neighborhood of the identity in $Q(T)$ since $\alpha(V)$ has interior in T by hypothesis and i_T is an embedding onto an admissible subset. By standard translation arguments for (local) groups, $Q(\alpha)$ is open at all points of W. ∎

We turn now to the global situation.

VII.1.37. **Lemma.** *Let S be a cancellative partial semigroup and let (U,V) be a suitable pair in S. If $\alpha: S \to G$ is a homomorphism (wherever it makes sense) into a group G, then there exists a unique homomorphism $\gamma: Q(U,V) \to G$ such that $\gamma(\beta(a,b)) = \alpha(a)(\alpha(b))^{-1}$.*

Proof. Define $\mu: U \times U \to G$ by $\mu(a,b) = \alpha(a)(\alpha(b))^{-1}$. One verifies easily that μ is a homomorphism from $U \times U$ with the partial product $(a,b) \cdot (b,c) = (a,c)$ in G. Suppose that $(a,b) \equiv (c,d)$. Then there exist $x, y \in V$ such that $ax = cy$ and $bx = dy$. Then

$$\mu(a,b) = \alpha(a)(\alpha(b))^{-1} = \alpha(a)\alpha(x)(\alpha(b)\alpha(x))^{-1} = \alpha(ax)(\alpha(bx))^{-1}$$

and similarly $\mu(c,d) = \alpha(cy)\alpha(dy)^{-1}$. Thus $\mu(a,b) = \mu(c,d)$. Since $Q(U,V)$ consists of the equivalence classes of $\equiv$ endowed with the quotient operation, it follows that there exists a unique homomorphism $\gamma: Q(U,V) \to G$ such that $\gamma(\beta(a,b)) = \alpha(a)(\alpha(b))^{-1}$. ∎

VII.1.38. **Proposition.** *Let S be as in Proposition VII.1.28 and let $\alpha: S \to H$ be a continuous homomorphism into a topological group H. Then there exists a unique continuous homomorphism $\gamma: Q(S,S) \to H$ such that α is equal to γ composed with the embedding $s \mapsto \beta(sa, a): S \to Q(S,S)$.*

Proof. The existence of the homomorphism γ follows from the preceding lemma since (S,S) is a suitable pair. Since $Q(S,S)$ is given the quotient topology, γ is continuous. The uniqueness follows from the fact that the embedded image of S group generates $Q(S,S)$. ∎

Canonical embeddings

We recall some basic facts about local Lie groups. One may consult N. Bourbaki, Groupes et algebrès de Lie, Chapter III for a statement of these results in the most general setting. According to Lie's Fundamental Theorems each local Lie group G has associated to it in a functorial way a completely normable Lie algebra $\mathbf{L}(G)$. Conversely, a completely normable Lie algebra gives rise to a local Lie group $(\mathbf{L}(G), *)$, namely the locally defined Campbell–Hausdorff multiplication on some neighborhood of 0. These operations are inverse operations in the sense that G is locally analytically isomorphic to $(\mathbf{L}(G), *)$ (via the local inverse of the exponential mapping), and conversely if L is a completely normable Lie algebra, then the Lie algebra of $(L, *)$ is naturally isomorphic (as a completely normable Lie algebra) to L. The local isomorphism from G to $(\mathbf{L}(G), *)$ is sometimes called a "canonical chart". In this section we adapt the idea of a canonical chart to the local semigroup setting.

VII.1.39.   **Definition.** A *canonical embedding* for a local semigroup S is a mapping $i : U \to L$ which is a local group of quotients (see Definition VII.1.30) for which L is a completely normable Lie algebra equipped with the local group structure arising from the Campbell–Hausdorff multiplication.

The existence of a canonical embedding is closely tied to the existence of a local embedding into a local Lie group.

VII.1.40.   **Proposition.** *Let S be a local semigroup. Then a canonical embedding exists for S iff S has a local group of quotients which is a local Lie group.*

Proof.   Since a completely normable Lie algebra equipped locally with the Campbell–Hausdorff multiplication is a local Lie group, one implication is immediate. Conversely suppose that S has a local group of quotients G where G is a local Lie group. Then the composition of the local embedding of S into G with the local embedding of G into $(\mathbf{L}(G), *)$ by means of the local inverse of the exponential mapping gives the desired canonical embedding.   ∎

As the proof of Proposition VII.1.40 shows, a canonical embedding may be thought of as arising by restricting the "log" function on the local group of quotients to the embedded image of the local semigroup.

The next proposition shows that associating a Lie algebra with a local semigroup by means of a canonical embedding is a functorial construction.

VII.1.41.   **Proposition.** *Let $i: U \to \mathcal{L}(S)$ and $j: V \to \mathcal{L}(T)$ be canonical embeddings for S and T respectively. If α is a local homomorphism from S to T, then there exists a unique continuous Lie algebra homomorphism $\mathcal{L}(\alpha): \mathcal{L}(S) \to \mathcal{L}(T)$ such that $\mathcal{L}(\alpha) \circ i = j \circ \alpha$ on a neighborhood of e in S.*

$$
\begin{array}{ccc}
S & \xrightarrow{\ \ \alpha\ \ } & T \\
{\scriptstyle i}\downarrow & & \downarrow{\scriptstyle j} \\
\mathcal{L}(S) & \xrightarrow[\ \mathcal{L}(\alpha)\ ]{} & \mathcal{L}(T)
\end{array}
$$

Proof. By Proposition VII.1.36 there exists a local homomorphism $Q(\alpha)$ from the Campbell–Hausdorff local group $\mathcal{L}(S)$ to $\mathcal{L}(T)$. One extends $Q(\alpha)$ by

$$\mathcal{L}(\alpha)(x) = nQ(\alpha)((1/n)x)$$

where $(1/n)x$ is in the domain of $Q(\alpha)$. $\mathcal{L}(\alpha)$ agrees with $Q(\alpha)$ on a neighborhood of 0 (since Campbell–Hausdorff multiplication is just addition on lines), hence is continuous at 0, and it follows from the Trotter Product Formula and Commutator Formula that $\mathcal{L}(\alpha)$ is a continuous Lie algebra homomorphism (see Corollary A.1.8 of Appendix A).

If F were another Lie algebra homomorphism from $\mathcal{L}(S)$ to $\mathcal{L}(T)$ such that $F \circ i = j \circ \alpha$ on some neighborhood of e in S, then F would define a local homomorphism of the Campbell–Hausdorff local groups (since the multiplication is defined as a series involving Lie brackets). By Proposition VII.1.36 again, F would have to agree with $Q(\alpha)$ and hence $\mathcal{L}(\alpha)$ on some neighborhood of 0. Since they are both linear, they must agree. ∎

The essential uniqueness of a canonical embedding (if it exists) now follows.

VII.1.42. **Corollary.** *Let* $i: U \to L_1$ *and* $j: V \to L_2$ *be canonical embeddings for a local semigroup* S. *Then there exists a unique isomorphism of completely normable Lie algebras* $\Gamma: L_1 \to L_2$ *such that* $\Gamma \circ i = j$.

Proof. Apply the preceding proposition in both directions to the identity mapping restricted to $U \cap V$. Note that the uniqueness property implies the compositions in both directions must be the appropriate identity mappings. ∎

2. Differentiable semigroups

In this section we introduce the concept of a differentiable semigroup. Our main purpose is not to develop a coherent theory of such objects, but only to indicate how they fit into the framework of a Lie theory of semigroups that we have developed in earlier chapters. The interested reader should consult the work of G. Graham ([Gr79], [Gr83], [Gr84]), where the foundations of a theory of differentiable semigroups are laid. The basic approach of this section and many of the results are drawn from his work.

Admissible sets and strong derivatives

In contrast with the group case, standard and basic examples of topological semigroups do not have manifolds for their underlying spaces. Thus one is confronted at the beginning with the problem of finding an appropriate setting in which to develop a theory of differentiable semigroups that will encompass a sufficiently broad scope of examples. We follow the lead of Graham and consider semigroups with underlying space a differentiable manifold with generalized boundary. If one strengthens the definition of the derivative of a function at a point, one is able to extend the differential calculus and geometry of manifolds without boundary to include these spaces. The development then follows along more or less standard lines as found in, for example, [La72] or [Dieu71].

We first introduce some fixed notation for this section:

E, F real Banach spaces
$B_\delta(a)$ the open ball of radius δ centered at a
$L(E, F)$ the Banach space of continuous linear maps from E to F
$L_k(E, F)$ the Banach space of continuous k-multilinear maps
         from E^k to F

We now extend Definition VII.1.23.

VII.2.1.   Definition.   For the Banach space E, an *E-manifold with generalized boundary*, or simply *E-manifold*, is a T_3-space M such that if $p \in M$, then there is an open set $U \subseteq M$ about p and a homeomorphism φ from U onto an admissible subset of E.

For most local questions, an E-manifold may be assumed to be an admissible subset of E. For example, differentiability of maps between manifolds is defined in terms of the differentiability of maps $f: A \subseteq E \to F$, where A is an admissible set. An appropriate version of differentiability for admissible sets is the notion of a strong derivative, a strengthening of the notion of a Fréchet derivative.

VII.2.2. **Definition.** Let $f: A \subseteq E \to F$ and let $a \in A$. A linear map $T \in L(E, F)$ is a *strong derivative* of f at a if for each $\varepsilon > 0$, there exists $\delta > 0$ such that

$$|f(y) - f(x) - T(y - x)| < \varepsilon |y - x| \qquad \text{whenever} \qquad x, y \in B_\delta(a) \cap A.$$

If A is admissible, then f has at most one strong deriviative at a, denoted by $df(a)$ if it exists.

We illustrate the notion of the strong derivative with the following proposition:

VII.2.3. **Proposition.** *Let S be an admissible subset of E containing 0 and let $m: S \times S \to E$ sending $(x, y) \to x * y$ be a multiplication function satisfying $0 * x = x * 0 = x$ for all $x \in S$. Then m has a strong derivative at $(0, 0)$ iff given $\varepsilon > 0$, there exists $\delta > 0$ such that for any $a, b, x, y \in S \cap N_\delta(0)$, we have*

$$\|(x * a - x * b) - (a - b)\| \leq \varepsilon \|a - b\| \quad and \quad \|(a * y - b * y) - (a - b)\| \leq \varepsilon \|a - b\|.$$

In this case the strong derivative is $dm(0, 0)(x, y) = x + y$.

Proof. Suppose first that m is strongly differentiable at $(0, 0)$. Then for $\varepsilon > 0$, there exists $\delta > 0$ such that for $x, y \in N_\delta(0) \cap S$,

$$\|x * 0 - y * 0 - dm(0, 0)\big((x, 0) - (y, 0)\big)\| = \|x - y - dm(0, 0)(x - y, 0)\| \leq \varepsilon \|(x - y, 0)\|.$$

Multiplying through by any $r > 0$, we obtain $\|r(x - y) - dm(0, 0)(r(x - y), 0)\| \leq \varepsilon \|r(x - y)\|$. Since x and y range over a set with interior, $r(x - y)$ ranges over the whole Banach space. Thus the linear mapping $v \mapsto dm(0, 0)(v, 0)$ differs in norm from the identity mapping by any preassigned ε. If follows that $dm(0, 0)(v, 0) = v$. Similarly $dm(0, 0)(0, w) = w$ and by linearity, $dm(0, 0)(v, w) = v + w$.

For $\varepsilon > 0$ and for $x, a, b \in N_\delta(0) \cap S$, we now obtain

$$\|x * a - x * b - dm(0, 0)\big((x, a) - (x, b)\big)\| = \|x * a - x * b - (a - b)\| \leq \varepsilon \|a - b\|,$$

and the dual inequality follows analogously.

Assume now the converse and let $\varepsilon > 0$. Choose $\delta > 0$ as hypothesized. Then for $x, y, a, b \in N_\delta(0)$, we have $\|x * y - a * b - \big((x - a) + (y - b)\big)\| \leq \|x * y - x * b - (y - b)\| + \|x * b - a * b - (x - a)\| \leq \varepsilon (\|x - a\| + \|y - b\|)$. It follows that m has strong derivative $dm(0, 0)(u, v) = u + v$. $\blacksquare$

The notion of a strong derivative can frequently substitute for the Mean Value Theorem and allow one to obtain results under the hypothesis that the derivative exists only at a point instead of in an entire neighborhood. The next proposition illustrates this principle. It follows almost directly from the definition of the strong derivative. A more significant illustration of the principle is the Inverse Function Theorem, which can be derived from strong differentiability at a single point (see e.g. [Nj84]).

VII.2.4.   Proposition. *Let A be an admissible subset of E and let $f: A \to F$ be strongly differentiable at $a \in A$. Then for $\varepsilon > 0$, there exists $\delta > 0$ such that*

$$|f(x) - f(y)| < (|df(a)| + \varepsilon)|x - y| \qquad \text{whenever} \qquad x, y \in B_\delta(a) \cap A.$$

In particular, f is Lipschitz continuous on a neighborhood of a.   ∎

VII.2.5.   Definition. Let A be an admissible subset of E and let $f: A \to F$. Then f is *strongly differentiable*, or C_s^1, if $df(x)$ exists for each $x \in A$. Inductively, f is *k-times strongly differentiable*, or C_s^k if f is C_s^1 and df is C_s^{k-1}. If f is C_s^k, then the jth *strong derivative* of f $(j \leq k)$ is the map $d^j f = d(d^{j-1}): A \to L_j(E, F)$. We say f is C_s^∞ if f is C_s^k for all positive integers k. If f is C_s^∞ on A, then the derivatives $d^j f(x)$ can be used to define the Taylor's series for f at x. If this series converges to $f(y)$ for every y in some neighborhood of x in A, then we say f is *analytic* at x. Finally, f is *analytic* or C_s^ω if it is analytic at every point of A.

On open subsets of E the usual notion of C^k-differentiability coincides with that of C_s^k-differentiability. This fact has tended to obscure the role of the strong derivative in the development of differential calculus. However, the recognition that it is the notion of the strong derivative that is crucial to the proof of the inverse and implicit function theorems has led to a more systematic development of this approach (see [Gr84], [Lea61], [Na78], [Nj74]). Since the statements and proofs of the main results that we need consist of appropriate modifications of standard results to this setting, we content ourselves with stating those that will be applicable to our situation without proof. The proofs of VII.2.4 through VII.2.11 may be found in [Gr84] and sometimes in other of the references just cited.

VII.2.6.   Proposition. *Let A be an admissible subset of E and let $f: A \to F$ be a map. If $df(x)$ exists for each $x \in A$, then $df: A \to L(E, F)$ is continuous.*   ∎

A finite Cartesian product of Banach spaces is again a Banach space with any of the various equivalent product norms. Since strong differentiability is invariant under change to an equivalent norm, one may choose a convenient one.

VII.2.7.   Proposition. *Let $A \subseteq E$ be an admissible set and let $f: A \to F_1 \times \ldots \times F_n$. Then f is C_s^k iff each component function $f_i: A \to F_i$, $i = 1, \ldots n$, is C_s^k, and in this case*

$$\big(df(x)\big)_i = df_i(x) \qquad \text{for each} \qquad x \in A, \, i = 1, \ldots, n.$$

∎

VII.2.8.   Proposition. *(Chain rule) Let $A \subseteq E$ and $B \subseteq F$ be admissible sets and let $f: A \to B$ and $g: B \to G$ be C_s^k maps, where G is a Banach space. Then $g \circ f$ is a C_s^k map and*

$$d(g \circ f)(x) = dg\big(f(x)\big) \circ df(x) \qquad \text{for each} \qquad x \in A.$$

∎

The following is a version of the inverse function theorem for admissible sets and C_s^k maps. Several of the salient points in the proof appear also in our proof of VII.2.11.

VII.2.9.   **Proposition.** (Inverse Function Theorem) *Let $A \subseteq E$ be an admissible set and let $f: A \to F$ be a C_s^k map. If $df(a)$ is an isomorphism onto F for some $a \in A$, then there is an open set U about a such that $f|U \cap A$ is a homeomorphism onto the admissible set $f(U \cap A)$, $f|U \cap A^\circ$ is an open map, and if $g = (f|U \cap A)^{-1}$, then g is C_s^k and $dg\big(f(x)\big) = df(x)^{-1}$ for each $x \in U \cap A$.* ∎

Notation. Let A and B be topological spaces and let $f : A \times B \to F$ be a map. For each $a \in A$, let $f^a: B \to F$ be defined by $f^a(b) = f(a, b)$. Similarly $f_b: A \to F$ is defined by $f_b(a) = f(a, b)$.

VII.2.10.   **Proposition.** (Product Rule) *Let $A \subseteq E$ and $B \subseteq F$ be admissible sets and let $f: A \times B \to G$ be a C_s^k map. Then the maps f^a and f_b are C_s^k maps and for each $a \in A$ and $b \in B$,*

$$df(a, b)(v, w) = df_b(a)(v) + df^a(b)(w)$$

for each $v \in E$ and $w \in F$. ∎

In the next theorem we consider a general setting that encompasses as a special case the notion we introduce shortly of a differentiable local semigroup. We include a proof to show the flavor of the calculus of admissible sets and strong derivatives, and because this result most closely pertains to the goals of this section.

VII.2.11.   **Theorem.** (Parameterized Mapping Theorem) *Let $A \subseteq E$ and $B \subseteq F$ be admissible sets, and let $f: A \times B \to F$ be a map. Let (a, b) be a fixed point in $A \times B$ and suppose $df^a(b): F \to F$ exists and is an isomorphism. Let U denote an A–open set containing a and let V denote a B–open set containing b.*

(i) *U and V can be chosen so that for each $x \in U$, the map $f^x|V$ is one–to–one and a topological embedding.*

Assume additionally that f is C_s^1 differentiable. Then U, V can be chosen to satisfy, in addition,

(ii) *if $p \in U$ and $v \in V$, then $df^p(v)$ is invertible;*

(iii) *if $a_1 \in U$, $b_1 \in B_1 \subseteq V$, B_1 is open in F, and $c_1 = f(a_1, b_1)$, then there exists an A–open set U_1 containing a_1 and W open in F containing c_1 such that $W \subseteq f(x, B_1)$ for all $x \in U_1$;*

(iv) *for $p \in U$, $f^p|V$ is an embedding carrying open sets to open sets and admissible sets to admissible sets, and if $f^p|V$ is C_s^k, then so is its inverse.*

Proof.   (i) Since $df^a(b)$ is an isomorphism, there exists $m > 0$ such that $df^a(b)(v) > m$ if $v \in F$ and $|v| = 1$. For $\varepsilon = m/2$, there exist an A-open set U containing a and a B-open set V containing b such that if $x = (p, v)$ and $y = (q, w)$ are in $U \times V$ with $x \neq y$, then

$$\frac{|f(q, w) - f(p, v) - df(a, b)\big((q, w) - (p, v)\big)|}{|(q, w) - (p, v)|} < \frac{m}{2}.$$

Let $p \in U$ and let $v, w \in V$ with $v \neq w$. Let $x = (p, v)$ and $y = (p, w)$. Then

$$\left| \frac{f^p(w) - f^p(v)}{|w - v|} - df^a(b)\Big(\frac{w - v}{|w - v|}\Big) \right| = \frac{|f(p, w) - f(p, v) - df(a, b)(0, w - v)|}{|(p, w) - (p, v)|} < \frac{m}{2}.$$

It follows that

$$\frac{|f^p(w) - f^p(v)|}{|w - v|} > |df^a(b)\left(\frac{w - v}{|w - v|}\right)| - \frac{m}{2} > \frac{m}{2} > 0.$$

Hence $f^p(w) \neq f^p(v)$ and $f^p|V$ is one-to-one.

Note that if $f^p(w) \neq f^p(v)$ in $f^p(V)$, then from the preceding inequality we have

$$|w - v| < (m/2)^{-1}|f^p(w) - f^p(v)|.$$

Hence $(f^p|V)^{-1}$ is continuous on $f^p(V)$ and $f^p|V$ is a homeomorphism.

(ii) If f is C_s^1 differentiable, then by Proposition VII.2.4, df is continuous. Let $i_0 \colon F \to E \times F$ be given by $i_0(y) = (0, y)$. Define $\Gamma \colon L(E \times F, F) \to L(F, F)$ by $\Gamma(\varphi) = \varphi \circ i_0$. Note that Γ is a linear map, and hence C_s^∞. By the Product Rule VII.2.9, we have for each $(x, y) \in A \times B$,

$$df^x(y) = df(x, y) \circ i_0 = \Gamma\bigl(df(x, y)\bigr) = (\Gamma \circ df(x, y).$$

Since Γ is continuous and the set of invertible operators is open in the space $L(F, F)$ (a standard result for Banach spaces), we conclude that U and V can be chosen so that if $p \in U$ and $v \in V$, then $df^p(v)$ is invertible.

(iii) Let $a_1 \in U$ and let $T = df^{a_1}(b_1) = df(a_1, b_1) \circ i_0$; by part (ii) T is invertible. Define $g(x, h) = T^{-1}(f(x, b_1 + h) - c_1)$ (where $c_1 = f(a_1, b_1)$). Then g maps the open neighborhood $U^* = \{(x, y - b_1) \colon (x, y) \in U\}$ of $(a_1, 0) \in E \times F$ into F and $g(a, 0) = 0$. Furthermore, g is strongly differentiable at $(a_1, 0)$ and by the Product and Chain Rules (VII.2.10 and VII.2.8)

$$(2.1) \qquad dg(a_1, 0)(0, w) = \bigl(T^{-1} \circ df(a_1, b_1)\bigr)(0, w) = T^{-1} \circ df^{a_1}(b_1)(w) = w.$$

Since g is strongly differentiable at $(a_1, 0)$, there exists an A–open set U_1 containing a_1 and $r > 0$ such that for $x \in U_1$ and $h, k \in \overline{B}_r$, the closed r–ball about 0 in F, we have by (2.1)

$$(2.2) \quad |g(x, h) - g(x, k) - (h - k)| = |g(x, h) - g(x, k) - dg(a, 0)\bigl((x, h) - (x, k)\bigr)|$$
$$\leq \frac{1}{3}|(x, h) - (x, k)| = \frac{1}{3}|h - k|$$

and also $b_1 + \overline{B}_r \subseteq B_1$ and

$$(2.3) \qquad\qquad\qquad\qquad |g(x, 0)| < \frac{r}{3}.$$

Let $y \in \overline{B}_r$, $\zeta \in B_{r/3}$ and define

$$(2.4) \qquad\qquad\qquad \Lambda_{x,\zeta}(y) = y - g(x, y) + \zeta.$$

Then $\Lambda_{x,\zeta}$ maps $\overline{B}_r$ into itself since setting $h = 0$ and $k = y$ in (2.2), we find

$$|\Lambda_{x,\zeta}(y)| \leq |\zeta| + |g(x, 0)| + \frac{1}{3}|y| < r,$$

where we have used (2.3) in the second estimate. Also $\Lambda_{x,\zeta}$ is a contraction, since by (2.2) and (2.3),

$$|\Lambda_{x,\zeta}(y_1) - \Lambda_{x,\zeta}(y_2)| \leq \frac{1}{3}|y_1 - y_2|.$$

By the contraction property for complete metric spaces, such as $\overline{B}_r$, $\Lambda_{x,\zeta}$ has a unique fixed point $y \in \overline{B}_r$. Then

$$y = \Lambda_{x,\zeta}(y) = y - g(x,y) + \zeta = y - T^{-1}(f(x, b_1 + y) - c_1) + \zeta.$$

We conclude that $c_1 + T(\zeta) = f(x, b_1 + y)$. The set $W = c_1 + T(B_{r/3})$ is an open neighborhood of c_1 since T is invertible, and we have just shown that $W \subseteq f(x, B_1)$ for $x \in U_1$ (since $b_1 + \overline{B}_r \subseteq B_1$).

(iv) For $p \in U$ and $v \in N \subseteq V \subseteq B$, N open in F, it follows directly from (iii) that $f^p(N)$ is open in F. It then is immediate that f^p carries admissible sets to admissible sets. It follows from the Inverse Function Theorem (VII.2.9) that if $f^p|V$ is C_s^k, then so is its inverse. ∎

Differentiable local semigroups

VII.2.12. Definition. A C_s^k *local semigroup* (*over* E) ($1 \leq k \leq \omega$) consists of a local semigroup (S, e, m) together with a chart γ which is a homeomorphism from S such that $\gamma(S)$ is an admissible subset of a Banach space E and such that the mapping $(\gamma(x), \gamma(y)) \mapsto \gamma(m(x,y))$ from $\gamma(\Sigma) \times \gamma(\Sigma)$ to E is C_s^k for some standard neighborhood Σ. If the local semigroup is a local group and $\gamma(S)$ is open in E, then we call it a C_s^k *local group*. A function h from some open subset U of S to a local semigroup T is a C_s^k *map* if the composition $\delta \circ h \circ \gamma^{-1}$ from $\gamma(U) \to F$ is of class C_s^k, where δ is the chart for T.

Note that if U is any open set containing e, then U with the appropriate restrictions of multipication and γ is again a C_s^k local semigroup. We shall feel free to pass to such restrictions whenever it is convenient to do so.

VII.2.13. Remark. If S is a C_s^k local semigroup, then left translation by $a \in \Sigma$, $\lambda_a : \Sigma \to S$, defined by $\lambda_a(b) = ab$, is a C_s^k map by the Product Rule (VII.2.10). Similarly, right translation by a, denoted ρ_a, is C_s^k map. ∎

VII.2.14. Proposition. *Let G be a C_s^k local group. Then the inversion mapping $g \mapsto g^{-1}$ is C_s^k on some neighborhood of e.*

Proof. The result is a direct consequence of the Implicit Function Theorem (see [Bou75], Chapter III, §1.1, Lemma 2 or Proposition II.2.2 of [Gr83]). ∎

We frequently identify a local C_s^k semigroup with its image under the chart γ and the induced multiplication. By translating if necessary, we can assume that γ carries the identity of the semigroup to 0. Then we obtain a C_s^k local semigroup with multiplication defined on an admissible subset of a Banach space and with identity 0.

We come now to the central result of this section.

VII.2.15.   Theorem. *Let S be a C_s^k local semigroup. Then there is a local group of quotients h from some neighborhood T of e in S onto some admissible subset of a C_s^k local group G such that h is a diffeomorphism of class C_s^k onto $h(T)$.*

Proof.   As in the remarks preceding this theorem, we can transfer the structure of S by means of the chart γ to an admissible subset of E. If we prove the theorem for $\gamma(S)$, we will have proved it for S by composing the h in the conclusion of the theorem with γ. Thus we assume without loss of generality that S is an admissible subset of the Banach space E.

Let Σ be a standard neighborhood of e such that $m: \Sigma \times \Sigma \to E$ is a C_s^k map. Note that λ_e is the identity map on Σ, and hence $d\lambda_e = 1_E$. In particular, the derivative is an isomorphism. We apply the Parameterized Mapping Theorem (VII.2.11) to $(a, b) = (e, e)$, $A = B = \Sigma$, $E = F$, and f the multiplication function. Then there exist S-open sets U', V' containing e and contained in Σ satisfying the conclusions of Theorem VII.2.11. Similarly since ρ_e is the identity map on Σ, we can consider the left-right dual situation and find T-open sets U'', V'' containing e in this setting. Pick an S-open set O containing e such that $O \cup O^2 \subseteq U' \cap V' \cap U'' \cap V''$. It follows directly from part (i) of the Parameterized Mapping Theorem that the triple (O, S, m) is a cancellative partial semigroup in the sense of Definition VII.1.1 and from part (iii) of the Parameterized Mapping Theorem that condition (†) of Definition VII.1.11 is satisfied (where O° is the interior of O in E). Part (iv) of the Parameterized Mapping Theorem implies that for each open set $V \subseteq O^\circ$ and for each $y \in O$, $\lambda_y(V) = yV$ is open in E. Hence $OO^\circ \cap O \subseteq O^\circ$. By part (i) of VII.2.11 the mapping $x \mapsto xp: O \to Op$ is a homeomorphism. By Theorem VII.1.23 there exist an S-open neighborhood U of e and an open set $V \subseteq O^\circ$ such that (U, V) is a suitable pair, $Q(U, V)$ is a local group, and the mapping $e^\triangle: U \to Q(U, V)$ given by $s \mapsto \bar{s} = \beta(s, e)$ is a continuous monomorphism and a topological embedding.

Let W be a standard neighborhood of $\bar{e}$ such that all products of length 5 of elements of W are defined and associative and let $Q \subseteq U \cap O$ be an open neighborhood of e such that $e^\triangle(Q) \subseteq W$. Let $a \in Q \cap O^\circ$ be chosen so that $a^2 \in Q$; then $\bar{a}, \bar{a}^2 \in W$. Set $b = a^2$. Note that $a \in O^\circ$ implies $b \in O^\circ O^\circ \cap O \subseteq O^\circ$. Pick an open ball $B_1 \subseteq Q$ around b such that $e^\triangle(B_1) * (\bar{b})^{-1} = \beta(B_1 \times \{b\}) \subseteq W$. Pick an open ball B_a around a such that $B_a^2 \subseteq B_1$. By the Inverse Function Theorem (VII.2.9), there exists an open ball B_2 around b and contained in B_1 such that $\lambda_a^{-1}|B_2$ and $\rho_a^{-1}|B_2$ are C_s^k maps from B_2 to open sets, each of which is contained in B_a.

By Theorem VII.1.26 the mapping $x \mapsto \bar{x} * (\bar{b})^{-1} = \beta(x, b) = b^\triangle(x)$ is a homeomorphism from Q to the neighborhood $G = e^\triangle(V) * (\bar{b})^{-1}$ of $\bar{e}$ which carries B_2 homeomorphically to the open set $\Gamma = b^\triangle(B_2)$. Let δ denote the inverse of $b^\triangle$. We claim that δ is the chart that makes G into a C_s^k differentiable group. Let $x, y \in B_2$. Note that

$$\bar{x} * (\bar{a})^{-1} * \bar{a} = \bar{x} = \overline{\rho_a^{-1}(x)a} = \overline{\rho_a^{-1}(x)} * \bar{a}.$$

Thus $\bar{x} * (\bar{a})^{-1} = \overline{\rho_a^{-1}(x)}$. Similarly $(\bar{a})^{-1} * \bar{x} = \overline{\lambda_a^{-1}(x)}$. We then have

$$\beta(x, b) * \beta(y, b) = \bar{x} * (\bar{b})^{-1} * \bar{y} * (\bar{b})^{-1} = \bar{x} * (\bar{a})^{-1} * (\bar{a})^{-1} * \bar{y} * (\bar{b})^{-1}$$

$$= \overline{\rho_a^{-1}(x)} * \overline{\lambda_a^{-1}(y)} * (\bar{b})^{-1} = \overline{\rho_a^{-1}(x)\lambda_a^{-1}(x)} * (\bar{b})^{-1}.$$

Thus $\delta\big(\delta^{-1}(x) * \delta^{-1}(y)\big) = \delta\big(\beta(x,b) * \beta(y,b)\big) = \rho_a^{-1}(x)\lambda_a^{-1}(y)$; the latter is a C_s^k map since it is a composition of such mappings.

It remains to show that the embedding of Q into G and its inverse are C_s^k maps in some neighborhood of e. By definition this is true if and only if $\delta \circ e^{\triangle}$ and its inverse are C_s^k maps. Let T be an open set about e such that $e^{\triangle}(T) \subseteq b^{\triangle}(B_2)$. Then the composition $\delta \circ e^{\triangle}$ on T sends s to $\delta\big(\beta(s,e)\big) = \delta(\overline{s}) = \delta(\overline{s}*\overline{b}*(\overline{b})^{-1}) = sb$, and the inverse is then $s \mapsto (\rho_b)^{-1}(s)$. Since right translation by b and its inverse are C_s^k maps (see Theorem VII.2.11.iv), this completes the proof. $\blacksquare$

VII.2.16. **Theorem.** *Let S be a local semigroup on a locally compact admissible subset of $\mathbb{R}^n$ such that the multiplication is strongly differentiable at the identity. Then S embeds locally (topologically and algebraically) onto an admissible subset of a finite-dimensional Lie group.*

Proof. The proof follows immediately from Proposition VII.1.24 once we show that S is locally cancellable. We may assume without loss of generality that 0 is the identity for S. Cancellation now follows immediately from the alternate characterization of strong differentiability in Proposition VII.2.3 for any $\varepsilon < 1$. $\blacksquare$

Differentiable local groups

It is not our objective here to present the details of a theory of differentiable (local) groups; this the reader can find elsewhere. However, we do sketch the theory as it pertains to our current situation. A particularly appropriate treatment for our purposes is that of G. Birkhoff. [Bi38].

Suppose that one is given a multiplication m on some open set containing 0 in a Banach space E such that $0 * x = x * 0 = x$ for all x in the open set. Then Birkhoff refers to this setup as an "analytical group" if the function m has a strong derivative at $(0,0)$ (he uses the alternate characterization of having a strong derivative given in Proposition VII.2.3). He proceeds to show that there is a neighborhood G of 0 in which every element has an inverse and that the mapping $x \mapsto x^{-1}$ has derivative at 0 given by $v \mapsto -v$. Hence, in particular, we are in the setting of a local group.

A major goal of the paper is to reparametrize the local group G in such a way that the one parameter local subgroups are "straightened", i.e. correspond to the lines in E. The transformation that does this is the "canonical transformation" T and it is shown to be a "distortion", i.e., a function defined on a neighborhood of 0, carrying 0 to 0, and having a strong derivative $dT(0)$ equal to the identity. By the Inverse Function Theorem (the single point version of VII.2.9), it follows that T^{-1} is defined on some neighborhood of 0 and is also a distortion. We assume (by restriction if necessary) that G is chosen so that $T(G)$ within this neighborhood. The image $T(G)$ of the local group G is given the induced structure so that T is an isomorphism. It is shown that $T(G)$ is also an "analytic group", which is characterized by the property that the multiplication restricted to any 1-dimensional subspace is just addition. (Such a local group is said to be under "canonical parameters.") Let us denote the multiplication on $T(G)$ by $\circ$.

Next the construction of the Lie algebra for G is undertaken. It is shown that the formulas

$$x + y = \lim_{\lambda \to 0} \frac{1}{\lambda} (\lambda \cdot x * \lambda \cdot y),$$

$$[x, y] = \lim_{\lambda, \mu \to 0} \frac{1}{\lambda \mu} \left((\lambda \cdot x)^{-1} * (\mu \cdot y)^{-1} * \lambda \cdot x * \mu \cdot y \right)$$

characterize addition in terms of the group multiplication and give rise to a continuous Lie product making E a continuous Lie algebra. We denote E equipped with this Lie product by $\mathbf{L}(G)$.

If one computes the Campbell–Hausdorff multiplication in $\mathbf{L}(G)$, then it is shown to be (on some neighborhood of 0) precisely the multiplication ∘ that was induced on $T(G)$ by the canonical transformation T. Hence T can be viewed as a "log" function and its inverse T^{-1} is the restriction of an exponential mapping. It follows that the multiplication in the canonical coordinates is actually analytic and that $T(G)$ is a local Lie group.

Consider now the case that the multiplication on G is C_s^1. For $a \in G$, we have on some neighborhood of a, $T = \lambda_{T(a)} T \lambda_{a^{-1}}$. By hypothesis $\lambda_{a^{-1}}$ is a C_s^1 map, we have seen that T is strongly differentiable at 0, and $\lambda_{T(a)}$ is C_s^1 since the multiplication in $T(G)$ is analytic. Thus by the Chain Rule, T is strongly differentiable at a. Hence T is a C_s^1.

For the case that $k \geq 2$ (including $k = \infty$ and $k = \omega$) it is a (non-trivial) result that any continuous local homomorphism between two C_s^k local groups is a C_s^k homomorphism in some neighborhood of the identity. This is shown in [Bou75] for the case of local Lie (i.e., C_s^ω) groups (see §4, n° 1 and Theorem 1 of §8, n° 1) and it is remarked in Exercise §8.6 that the proof extends to the C_s^k case for $k \geq 2$. Thus if G is C_s^k, then the canonical transformation T is C_s^k.

If S is a C_s^k local semigroup, then by Theorem VII.2.15 there exists a C_s^k local embedding into a C_s^k local group of quotients. We can compose this embedding with the canonical transformation T of the earlier paragraphs to obtain

VII.2.17.   **Theorem.**   *Let S is a C_s^k ($k \geq 1$) local semigroup (over E). There exists a continuous Lie algebra structure on E and a C_s^k diffeomorphism χ from some neighborhood U of e which is a canonical embedding. In particular, S admits a C_s^k embedding into a local Lie group.*   ∎

A diffeomorphism χ satisfying the properties of Theorem VII.2.17 is called a *canonical chart* for S. The theorem asserts that every C_s^k local semigroup admits a canonical chart.

Theorem VII.2.17 shows that the types of local semigroups which we considered in earlier chapters (namely those that were subsemigroups of the local Campbell–Hausdorff group in a completely normable Lie algebra) were not so restrictive after all, in that any C_s^k local semigroup with identity has a neighborhood of the identity diffeomorphic and isomorphic to such a semigroup.

Differentiable manifolds with generalized boundary

In our discussion of the global Lie theory of semigroups in Chapters V and VI we considered subsemigroups of Lie groups. As an alternate approach to an analog of Lie theory for semigroups, one could develop a theory of differentiable semigroups. In the next two sections we lay the appropriate groundwork for such a theory, but refer the reader to the previously cited works of Graham for proofs and further basic results concerning this approach.

VII.2.18.   Definition. At *atlas* for an E-manifold M is a collection $\mathcal{A}$ of functions (called *charts*) satisfying:

(i) each $\varphi \in \mathcal{A}$ is a homeomorphism from an open subset $\operatorname{dom}(\varphi)$ of M onto an admissible subset $\operatorname{im}(\varphi)$ of E, and

(ii) $M = \bigcup \operatorname{dom}(\varphi)$ $(\varphi \in \mathcal{A})$.

If φ and ψ are members of the atlas $\mathcal{A}$, then the domain $\varphi(\operatorname{dom}(\varphi \cap \operatorname{dom}(\psi))$ of $\psi \circ \varphi^{-1}$ is an open subset of $\operatorname{im}(\varphi)$ and hence is an admissible subset of E. An atlas $\mathcal{A}$ is a C_s^k *atlas* if $\psi \circ \varphi^{-1}$ is a C_s^k map for each $\varphi, \psi \in \mathcal{A}$. Any C_s^k atlas extends to a unique maximal C_s^k atlas. A C_s^k *differentiable structure* for M is a maximal C_s^k atlas. A C_s^k *manifold* (*with generalized boundary*) is a pair $(M, \mathcal{D})$ (frequently denoted simply M) where M is an E-manifold and $\mathcal{D}$ is a C_s^k differentiable structure on M.

Let $p \in M$, a C_s^k manifold. If for some chart φ at p, $\varphi(p)$ is an interior point of $\operatorname{im}(\varphi)$, then $\psi(p)$ is an interior point of $\operatorname{im}(\psi)$ for each chart ψ at p by the Inverse Function Theorem. The *boundary* of M, denoted ∂M, is the set of all $q \in M$ such that $\varphi(q)$ is not in $\big(\operatorname{im}(\varphi)\big)^\circ$ for each chart φ at q. If $\partial M = \emptyset$, then M is a *manifold without boundary*. In any case, $M \setminus \partial M$ is an open dense subset of M and is a manifold without boundary. A C_s^k E-manifold M is a *manifold with smooth boundary* if there exists an atlas $\mathcal{A}$ for the differentiable structure such that $\operatorname{im}(\varphi)$ is an open subset of a half-space in E for each $\varphi \in \mathcal{A}$ (a half-space in E is a set of the form $\alpha^{-1}\big([0, \infty)\big)$ for some $\alpha \in L(E, \mathbb{R})$).

Let M and N be C_s^k manifolds and let $f \colon M \to N$. Then f is C_s^r *differentiable* $(r \le k)$ is f is continuous and $\psi \circ f \circ \varphi^{-1}$ is C_s^r differentiable for each chart φ on M and each chart ψ on N.

Note. If $A \subseteq E$ is an admissible subset, then the inclusion $i \colon A \to E$ is a C_s^∞ atlas for A; thus each admissible subset is a C_s^∞ E-manifold.

We turn now to the construction of the tangent bundle for a differentiable manifold with generalized boundary.

VII.2.19.   Definition. Let M be a C_s^k E-manifold and let $p \in M$. If φ and ψ are charts at p and if $v, w \in E$, then (φ, v) is *p-equivalent* to (ψ, w) if $d(\psi \circ \varphi^{-1})\big(\varphi(p)\big)v = w$. The Chain Rule and Inverse Function Theorem yield that p-equivalence is an equivalence relation. Let $T_p M$ denote the set of equivalence classes $[(\varphi, v)]_p$, where φ is a chart at p and $v \in E$. The map $\widehat{\varphi}_p \colon E \to T_p M$

defined by $\widehat{\varphi}_p(v) = [(\varphi, v)]_p$ is a bijection of sets. Moreover, if T_pM is given the unique vector space structure such that $\widehat{\varphi}$ is an isomorphism of vector spaces and if ψ is a chart at p, then $\widehat{\psi}_p$ is an isomorphism. The *tangent space* of M at p is the set T_pM with the unique vector space structure such that $\widehat{\varphi}_p$ is an isomorphism for each chart φ at p.

Let M and N be C_s^k manifolds and let $f: M \to N$ be a C_s^k map. If $p \in M$, then the *(strong) derivative* of f at p is the map $df(p): T_pM \to T_{f(p)}N$ defined by

$$df(p) = \widehat{\psi}_{f(p)} \circ d(\psi \circ f \circ \varphi^{-1})(\varphi(p)) \circ (\widehat{\varphi}_p)^{-1},$$

where φ is a chart at p and ψ is a chart at $f(p)$. The definition of $df(p)$ is independent of the choice of charts by the Chain Rule.

The tangent spaces at the various points give rise to the tangent bundle. Let $TM = \{(p, v): p \in M \text{ and } v \in T_pM\}$. For each chart $\varphi: U \to A \subseteq E$, define $T\varphi$ from $TU = \{(p, v) \in TM: p \in U\}$ onto $A \times E$ by

$$T\varphi(p, v) = (\varphi(p), d\varphi(p)v) = (\varphi(p), (\widehat{\varphi}_p)^{-1}v).$$

The map $T\varphi$ is a bijection of sets. Moreover, the collection of sets of the form $(T\varphi)^{-1}(W)$, where φ is a chart on M and W is an open subset of $E \times E$ is a base for a topology on TM. With this topology the collection of all maps $T\varphi$ is a C_s^{k-1} atlas for TM as an $E \times E$ manifold. The *tangent bundle* of M is the C_s^{k-1} map $\pi = \pi_M: TM \to M$ defined by $\pi(p, v) = p$. It is customary to call TM the tangent bundle of M and suppress mention of the map π.

For each $p \in M$, $\pi^{-1}(p) = T_pM$, which is isomorphic to E as a vector space. Moreover, T_pM is a topological vector space with its relative topology as a subset of TM. If φ is a chart at p, then the topology of T_pM is the norm topology for the unique norm on T_pM such that $\widehat{\varphi}_p$ is an isometry. Use of a different chart may produce a different norm, but the two norms will be equivalent and hence produce the same topology. Thus T_pM is equivalent to E as a topological vector space.

Note. If $A \subseteq E$ is an admissible set, then A is a C_s^∞ manifold and the tangent bundle of A may be identified with $\pi_1: A \times E \to A$, and hence is a trivial bundle.

We turn now to the topic of product manifolds and submanifolds. Let M be a C_s^k E-manifold and N a C_s^k F-manifold. For each chart φ on M and ψ on N, define $\varphi \times \psi$ on $M \times N$ by $\varphi \times \psi(p, q) = (\varphi(p), \psi(q)) \in E \times F$. The collection of all such maps form a C_s^k atlas for $M \times N$, and the differentiable structure generated by this atlas is called the *product structure*. For each $(p, q) \in M \times N$, $T_{(p,q)}M \times N$ is isomorphic to $T_pM \times T_qN$ by the map $(\widehat{\varphi}_p \times \widehat{\psi}_q) \circ ((\widehat{\varphi \times \psi})_{(p,q)})^{-1}$.

Let M and N be C_s^k manifolds and let $f: M \to N$ be a C_s^k map. Then f is an *immersion* if for each $p \in M$, $df(p)$ is one-to-one and the image of $df(p)$ is closed and complemented in $T_{f(p)}N$; f is an *embedding* if f is an immersion and a homeomorphism onto its image (with the relative topology). If f^{-1} exists and is also a C_s^k map, then f is called a *diffeomorphism*.

A subset P of M is an *immersed submanifold* of M if P can be given the structure of a C_s^k manifold in such a way that the inclusion map $i: P \to M$ is a C_s^k

immersion. A subset P is an *embedded submanifold* or simply a *submanifold* of M if P can be given the structure of a C_s^k manifold in such a way that the inclusion map is a C_s^k embedding.

Note. An admissible subset A of M, i.e., a subset A such that the interior A° of A in M is dense in A, is a submanifold of M; a C_s^k atlas for A arises by taking the charts of M and restricting them to A. In this case $TA = \bigcup\{T_pM : p \in A\}$.

VII.2.20.   Definition. Let M be a C_s^k manifold and let $\pi : TM \to M$ be its tangent bundle. A *vector field on M* is a section of π, i.e., a map $X : M \to TM$ such that $\pi \circ X$ is the identity on M. If X is a vector field on M and $p \in X$, then $X(p)$ is denoted X_p. If $A \subseteq E$ is an admissible set, then A is a C_s^∞ E-manifold and the tangent bundle of A may be identified with $\pi_1 : A \times E \to A$. If $Y = Y_1 \times Y_2 : A \to A \times E$ is a vector field on A, then $Y_1(a) = a$ for all $a \in A$, and Y is completely determined by $Y_2 : A \to E$. Y_2 is called the *principal part* of the vector field. A vector field X on M is a C_s^r vector field if $X : M \to TM$ is a C_s^r map. Since TM is a C_s^{k-1} manifold, $r \le k - 1$. If Y is a vector field on an admissible $A \subseteq E$, then Y is C_s^r if and only if Y_2 is a C_s^r map.

Let M and N be C_s^k manifolds, let $f : M \to N$ be a C_s^k diffeomorphism, and let X be a C_s^{k-1} vector field on M. Then $Tf(X)$ defined by $Tf(X)(p) = df\big(f^{-1}(p)\big)\big(X\big(f^{-1}(p)\big)\big)$ for each $p \in N$ is a C_s^{k-1} vector field on N, called the *push-forward* of X.

Let M be a C_s^k manifold and for each $r \le k-1$, let $V^r(M)$ be the set of all C_s^r vector fields on M. If addition and scalar multiplication are defined pointwise, then $V^r(M)$ becomes a vector space when endowed with these operations. Let $\mathcal{F}^0(M)$ be the set of all continuous real-valued maps on M and let $\mathcal{F}^r(M)$ be the set of all real-valued C_s^r maps. Note that $\mathcal{F}^r$ is a commutative ring with respect to the pointwise operations. For each $X \in V^{k-1}$ and each $f \in \mathcal{F}^k(M)$, define $Xf : M \to \mathbb{R}$ by $Xf(p) = df(p)(X_p)$. Thus X may be thought of as a map from $\mathcal{F}^k(M)$ into $\mathcal{F}^{k-1}(M)$. Moreover, X is a derivation, i.e., a linear operator satisfying $X(fg) = (Xf)g + f(Xg)$ for all $f, g \in \mathcal{F}^k(M)$. If $A \subseteq M$ is an admissible set, then $X|A : \mathcal{F}^k(A) \to \mathcal{F}^{k-1}(A)$ is a derivation which we also denote by X.

We turn now to the construction of the Lie bracket of two vector fields.

VII.2.21.   Proposition. *Let M be a C_s^k manifold ($k \ge 2$) and let $X, Y \in V^{k-1}(M)$. Then there is a unique C_s^{k-2} vector field $[X, Y]$ on M such that for each open set U and each $f \in \mathcal{F}^k(U)$, we have $[X, Y]f = X(Yf) - Y(Xf)$.*   ∎

For an admissible subset A of E with tangent bundle $\pi_1 : A \times E \to A$, let $X_2 : A \to E$ and $Y_2 : A \to E$ denote the principal parts of X and Y. For each $a \in A$, set

$$[X, Y]_2(a) = dY_2(a)\big(X_2(a)\big) - dX_2(a)\big(Y_2(a)\big).$$

It follows directly from the definition that $[X, Y]$ is C_s^{k-2} differentiable. It is a standard computation to show that $[X, Y]$ behaves in the desired manner on functions (see e.g. Proposition I.5.2 of [Gr83]). In the case that M is an arbitrary manifold one carries out the preceding construction for charts and argues that the definition given for $[X, Y]$ is independent of the particular chart that one chooses and has the desired effect on functions.

Let M and N be C_s^k manifolds and let $f: M \to N$ be a C_s^k map. If f is a diffeomorphism, then each vector field X on M induces a vector field $Tf(X)$ on N, but if f is not a diffeomorphism this may be impossible. We say that a vector field X on M is *f-related* to a vector field Y on N if $df(p)(X_p) = Y_{f(p)}$ for each $p \in M$. The following fact is standard (see e.g. [Gr83], Proposition I.5.5).

VII.2.22.   Proposition. *Let M and N be C_s^k manifolds ($k \geq 2$) and let $f: M \to N$ be a C_k^s map. Suppose that $X_1, X_2 \in V^{k-1}(M)$ and $Y_1, Y_2 \in V^{k-1}(N)$ and that X_i is f-related to Y_i for $i = 1, 2$. Then $[X_1, X_2]$ is f-related to $[Y_1, Y_2]$.* ∎

Differentiable semigroups

VII.2.23.   Definition. A C_s^k *semigroup* ($1 \leq k \leq \infty$) is a semigroup S on a C_s^k manifold (with generalized boundary) such that the multiplication $m: S \times S \to S$ is a C_s^k map, where $S \times S$ carries the product structure. A C_s^k *monoid* is a C_s^k semigroup with a two-sided identity e.

VII.2.24.   Remark. Let S be a C_s^k local semigroup ($1 \leq k \leq \omega$) with a chart $\chi: S \to E$. We make S into a C_s^k manifold by taking the chart χ for an atlas. We henceforth (equivalently) view S as a C_s^k manifold (instead of a local semigroup with a single chart). This allows us to treat S in the framework of differentiable manifolds. If S is a C_s^k monoid, then any chart at e gives rise to a C_s^k local semigroup on the domain of the chart, and this local semigroup is an open submanifold of S.

Note that if S is a C_s^k (local) semigroup, then for any $0 < r < k$, there is generated a C_s^r (local) semigroup by taking all the C_s^k charts on S as an atlas. These considerations hold in particular in the case that S is a (local) group.

If S is a C_s^k semigroup, then it is a topological semigroup. If $a \in S$, then the left translation map $\lambda_a: S \to S$ defined by $\lambda_a(x) = ax$ and the right translation map $\rho_a: S \to S$ are both C_s^k maps. If G is a Lie group and S is a subsemigroup of G with dense interior, then S is a C_s^ω semigroup with respect to the relative C_s^ω structure.

VII.2.25.   Example. Let S be the semigroup of all $n \times n$ real matrices under multiplication. The identity is a chart which makes S into an C_s^ω semigroup since multiplication is polynomial. However, S cannot be a subsemigroup of a group since S in not cancellative.

VII.2.26.   Proposition. *Let S be a local semigroup which admits a canonical embedding into a completely normable Lie algebra E. Then there is a chart defined on some neighborhood of e which makes that neighborhood into a C_s^ω local semigroup. Furthermore, any two such analytic structures agree on some neighborhood of e.*

Proof.   Let $i: U \to E$ be the canonical embedding. Then the local Campbell–Hausdorff multiplication on E is analytic since it is given by a power series. If one uses i as an atlas for an analytic structure on U, then it follows immediately that the multiplication on U is analytic since the canonical embedding is an isomorphism on U.

Suppose $j: V \to E$ is another canonical embedding. Then by Corollary VII.1.41, there exists a unique continuous Lie algebra isomorphism $\Gamma: E \to L$ such that $\Gamma \circ i = j$ on some neighborhood N of e. Since the isomorphism is linear, it is an analytic map. Thus the two induced analytic structures agree on N.   ■

There are various ways of associating with a differentiable group its corresponding Lie algebra. One can take the set of tangent vectors at the identity and equip them with an "appropriate" Lie product. One can take the set of left invariant (or right invariant) vector fields with Lie product the usual bracket product of vector fields. One can take the set of one parameter subgroups and use the Trotter addition formula and commutator formula (see Appendix A) to define vector addition and Lie product. We wish to adapt some of these constructions to the semigroup setting. Here, however, the Lie algebra associated with a semigroup should not be viewed as the set of vectors tangent to the identity, since we have seen that the appropriate tangent objects for semigroups are wedges. Rather we think of the Lie algebra as corresponding to the tangent space of the identity of some appropriate (local) group that the semigroup in some sense locally generates. Hence in the semigroup case we will denote Lie algebra by $\mathcal{L}(S)$ and in the group case by $\mathbf{L}(G)$.

VII.2.27.   Lemma.   *Let S be a C_s^k monoid or local semigroup. Then every canonical embedding which is a C_s^r diffeomorphism for $1 \le r \le k$ induces on $T_1(S)$ a continuous Lie algebra structure that is independent of the canonical embedding.*

Proof.   Let $\varphi: U \to L$ be a C_s^r canonical embedding, where U is an open neighborhood of e in S. By Theorem VII.2.17 there exists at least once such. We identify $T(L)$ with $L \times L$ and $T_0(L)$ with L. Then $T(\varphi)(e): T_e(S) \to L$ is an isomorphism of completely normed spaces since φ is a local C_s^r embedding onto an admissible subset. We use the inverse of $T(\varphi)(e)$ to pull the Lie algebra structure on L back to $T_e(S)$. It follows in a straightforward manner from Proposition VII.1.41 and the chain rule that this Lie algebra structure is independent of the canonical embedding that one uses.   ■

VII.2.28.   Definition.   Let S be a C_s^k monoid or local semigroup. Then $T_e(S)$ equipped with the Lie algebra structure arising as in Lemma VII.2.27 is denoted $\mathcal{L}(S)$ and called the Lie algebra *infinitesimally generated by S*.

We identify as is customary the tangent bundle of $\mathcal{L}(S) = T_e(S)$ with $\mathcal{L}(S) \times \mathcal{L}(S)$ and the set of tangent vectors at 0 with $\mathcal{L}(S)$. If $j: U \to \mathcal{L}(S)$ is a differentiable canonical embedding, then $dj(e)$ is an isomorphism from $T_e(S) = \mathcal{L}(S) \to T_0\big(\mathcal{L}(S)\big) = \mathcal{L}(S)$.

VII.2.29.   Definition.   A *log* function for S is C_s^k canonical embedding into $\mathcal{L}(S)$ such that $d(\log)(e) = 1_{\mathcal{L}(S)}$.

We come now to a major theorem.

VII.2.30.   Theorem. *Let S and T be C_s^k monoids or local semigroups (over E) for $1 \le k \le \omega$.*

(i) *There is a C_s^k canonical chart into E for S.*

(ii) *There is a log function from S into $\mathcal{L}(S)$, and any two agree on some neighborhood of e.*

(iii) *There is an analytic structure on some neighborhood of e in S such that the multiplication is analytic on this neighborhood and such that the C_s^k differentiable structure generated by this analytic structure agrees with the restriction of the given C_s^k structure. Furthermore, this analytic structure is unique in the sense that any other analytic structure on a neighborhood of e that makes multiplication analytic agrees with this one on some neighborhood of e.*

(iv) *Let $f\colon S \to T$ to a continuous local homomorphism. Then f is a C_s^k map on some neighborhood of e. Furthermore, $\mathcal{L}(f) = df(e)\colon \mathcal{L}(S) \to \mathcal{L}(T)$ is the unique continuous Lie algebra homomorphism such that $\log_T \circ f = \mathcal{L}(f) \circ \log_S$ on some neighborhood of e.*

$$
\begin{array}{ccc}
S & \xrightarrow{\ \ f\ \ } & T \\
\log \downarrow & & \downarrow \log \\
\mathcal{L}(S) & \xrightarrow[\mathcal{L}(f)=df(e)]{} & \mathcal{L}(T)
\end{array}
$$

Proof.   We will show the implications (i) $\Leftrightarrow$ (ii), (ii) $\Rightarrow$ (iv), and (i) $\Rightarrow$ (iii). For C_s^k local groups, the results of Birkhoff mentioned in the subsection on local groups establish (i) for the case $k = 1$ and, as was mentioned there, (i) continues to hold for $k \ge 2$ by results that can be found in Bourbaki. Alternately, (at least for $k > 2$) one can use the local flows associated with the local left (or right) invariant vector fields to establish the existence of a local exponential function (this is essentially the approach given in [Pon57], Chapters VI and IX). The local inverse of this exponential function is then a log function; hence (ii) and then the other implications result.

Once the theorem is established for local groups, it is established for local semigroups by first C_s^k embedding in a local group (Theorem VII.2.15) and then composing with a C_s^k canonical chart for the local group. This establishes (i) for local C_s^k semigroups and then the other statements can be deduced.

(i) implies (ii): Let $j\colon U \to E$ be a C_s^k canonical chart on some neighborhood of e, i.e., a canonical embedding into the Campbell–Hausdorff multiplication for some Lie algebra structure on E. Then $dj(e)\colon \mathcal{L}(S) \to E$ is an isomorphism of Lie algebras from Lemma VII.2.27 and the definition of the Lie algebra structure on $\mathcal{L}(S)$, and hence is also a local isomorphism of the corresponding Campbell–Hausdorff multiplications (see Proposition VII.1.41). The composition $\bigl(dj(e)\bigr)^{-1} \circ j$ is then a canonical embedding into $\mathcal{L}(S)$; we call it "log" and show that it is indeed a log function.

We have $d\log(e)\colon T_e(S) = \mathcal{L}(S) \to T_0\bigl(\mathcal{L}(S)\bigr) = \mathcal{L}(S)$. By definition $dj(e) \circ \log = j$, so that $T\bigl(dj(e)\bigr) \circ T(\log) = T(j)$ when we pass to the tangent bundles. Restricting to the tangent vectors at the identities, we obtain $dj(e) \circ d(\log)(e) = dj(e)$ which implies $d(\log)(e) = 1_{\mathcal{L}(S)}$ since $dj(e)$ is an isomorphism.

Suppose that Log is another log function. By Corollary VII.1.42, there exists $\Gamma \colon \mathcal{L}(S) \to \mathcal{L}(S)$ such that $\mathrm{Log} = \Gamma \circ \log$ on some neighborhood of e. Passing to the derivatives an 0, we have

$$1_{\mathcal{L}(S)} = d(\mathrm{Log})(e) = \Gamma \circ d(\log)(e) = \Gamma \circ 1_{\mathcal{L}(S)} = \Gamma.$$

Thus $\mathrm{Log} = 1_{\mathcal{L}(S)} \circ \log = \log$ on some neighborhood of e.

(ii) implies (i): Since $T_e(S) = \mathcal{L}(S)$ and S is a manifold over E, there exists an isomorphism Γ from $\mathcal{L}(S)$ to E. Use this isomorphism to define a Lie algebra structure on E. Then Γ will induce an isomorphism of the local groups defined by Campbell–Hausdorff multiplication on $\mathcal{L}(S)$ and E and hence the composition $\Gamma \circ \log$ is a canonical chart.

(ii) implies (iv): Let $\log_S$ and $\log_T$ be log functions for S and T respectively. By Proposition VII.1.41 there exists a unique continuous Lie algebra homomorphism $\mathcal{L}(f) \colon \mathcal{L}(S) \to \mathcal{L}(T)$ such that $\log_T \circ f = \mathcal{L}(f) \circ \log_S$ on some neighborhood of $e \in S$. Since the right-hand side of the equation is a C_s^k map, so is the left-hand side. Thus $f = (\log_T)^{-1} \circ \log_T \circ f$ is a C_s^k map.

We take the derivative at e of the map $\log_T \circ f = \mathcal{L}(f) \circ \log_S$ and obtain

$$
\begin{aligned}
df(e) &= 1_{\mathcal{L}(T)} \circ df(e) = d(\log_T)\big(f(e)\big) \circ df(e) \\
&= d\big(\mathcal{L}(f)\big)(0) \circ d(\log_S)(e) = \mathcal{L}(f) \circ 1_{\mathcal{L}(S)} = \mathcal{L}(f).
\end{aligned}
$$

(i) implies (iii): Let $\chi \colon U \to E$ be some canonical chart from some open set U with $e \in U \subset S$ into $(E, *)$, where $*$ is the local Campbell–Hausdorff multiplication arising from some continuous Lie algebra structure on E. Let χ be an atlas for a C_s^ω structure on U. Then as in Proposition VII.2.26, the multiplication on U is analytic with respect to this C_s^ω structure. Since χ was a chart, it follows that the relative C_s^k structure on U agrees with the C_s^k structure generated by this analytic structure.

Suppose now that S admits another analytic structure on some neighborhood of e that makes multiplication analytic. Then again by (i) there is a canonical chart (which in this case is C_s^ω) $\varphi \colon V \to E$. So the analytic structure restricted to V must be the one induced by using this canonical chart as an atlas. By Proposition VII.2.26, this analytic structure must agree with that of the preceding paragraph on some neighborhood of e. ∎

We close this section by considering briefly the approach of defining the infinitesimally generated Lie algebra of a differentiable semigroup in terms of invariant vector fields. This was the earlier approach of Graham.

VII.2.31. **Definition.** Let S be a C_s^k semigroup. A vector field X on S is *left-invariant* if $d\lambda_a(b)(X_b) = X_{ab}$ for each $a, b \in S$. Equivalently, a vector field X is left-invariant if X is λ_a related to itself for all $a \in S$.

As in the group case all left invariant vector fields on a C_s^k monoid arise by fixing some tangent vector v at the identity and considering the vector field $s \mapsto d\lambda_s(v)$ for $s \in S$. This is a C_s^{k-1} vector field since multiplication is C_s^k. For the case $k \geq 2$ we can form the Lie product of two left invariant vector fields, and by Proposition VII.2.22 this Lie product is again λ_a related to itself for all $a \in S$ and is hence also left invariant. We thus obtain a bracket multiplication on $T_e(S)$ by taking two vectors, passing to the corresponding left invariant vector fields, taking the Lie bracket of these vector fields, and then picking the vector in $T_e(S)$ of this vector field. The next proposition shows that this approach of defining a bracket multiplication agrees with our earlier one.

VII.2.32. **Proposition.** *Let S be a C_s^k monoid for $k \geq 2$. Then the set of left invariant vector fields is linearly isomorphic to the tangent space T_eS and induces on $T_e(S)$ a bracket multiplication which agrees with that of $\mathcal{L}(S)$.*

Proof. Only the last assertion remains to be discussed. We know from Theorem VII.2.30 that we have a log function $f: U \to \mathcal{L}(S)$ defined on some neighborhood of e into $T_e(S) = \mathcal{L}(S)$ which is a homomorphism into $(\mathcal{L}(S), *)$, where $*$ is the local Campbell–Hausdorff multiplication corresponding to the Lie algebra structure. For v in $T_e(S)$ one verifies directly from the Chain Rule and the fact that f is a local embedding that the restriction to U of the left invariant vector field corresponding to v is f-related to the local left invariant vector field on some Campbell–Hausdorff neighborhood B of 0 corresponding to $df(e)(v)$. Since the restriction to U is also i-related to its global extension, where $i: U \to S$ is the inclusion, it now follows from VII.2.22 that $df(e) = d(\log)(e) = 1_{\mathcal{L}(S)}$ preserves the Lie products. It is a now a standard fact that if one forms the local left invariant vector fields in a Campbell–Hausdorff neighborhood B of 0 in a Lie algebra and uses them as we discussed earlier to define a Lie product on the tangent vectors at the identity, which we identify with the Lie algebra, then the induced bracket product agrees with the original Lie product (see the discussion at the beginning of the subsection "Invariant vector fields" in Section IV.5). Thus the induced bracket multiplication arising from the left invariant vector fields agrees with that we defined earlier on $\mathcal{L}(S)$. ∎

EVII.2.1. **Exercise.** Let S be a C_s^k semigroup (or monoid). Show that the multiplication on $T(S)$ given by

$$u_s \cdot v_t \overset{\text{def}}{=} d\rho_t(s)(u_s) + d\lambda_s(t)(v_t),$$

where $u_s \in T_s(S)$ and $v_t \in T_t(S)$ makes $T(S)$ into a C_s^{k-1} semigroup, and that the projection from $T(S)$ to S is a C_s^{k-l} homomorphism. Show that if $f: S \to Q$ is a C_s^k homomorphism, then $T(f): T(S) \to T(Q)$ is a C_s^{k-1} homomorphism.

Applications

One of the goals of this chapter has been to find settings where the Lie theory of semigroups developed in earlier chapters can be applied. The embedding theorems that we have obtained in this section allow such applications. In this subsection we demonstrate how this machinery can be applied to problems that have been previously studied in the area of differentiable semigroups. The two types of problems that we consider are concerned with the existence of one parameter semigroups and the existence of units when the boundary of the semigroup is smooth in a neighborhood of the identity **1**. In most cases we are able to derive results that strengthen those previously obtained.

We begin with some results on the existence of one parameter semigroups. The central thread in what follows is that the results of Chapter IV allow us to identify local one parameter semigroups in local subsemigroups of Campbell–Hausdorff neighborhoods; then one can use the local isomorphism to lift these results to any local semigroup admitting a local embedding into a Campbell–Hausdorff neighborhood.

VII.2.33. **Definition.** Let S be a local semigroup, let U be a neighborhood of **1**, and let $\varphi : U \to L$ be a *Campbell–Hausdorff embedding*, that is, a homeomorphism into a completely normable Lie algebra L equipped with the Campbell–Hausdorff multiplication such that $\varphi(xy) = \varphi(x) * \varphi(y)$ for $x, y, xy \in U$. Then S is said to *admit a tangent wedge (with respect to φ)* if $\mathbf{L}\big(\varphi(U)\big) \cap B \subseteq \varphi(U)$ for some neighborhood B of 0, and S is said to be *infinitesimally closed (with respect to φ)* if $\varphi(U) \cap B$ is closed in B for some neighborhood B of 0.

The set $\mathbf{L}\big(\varphi(U)\big)$ is called the *tangent wedge of S with respect to φ* and is denoted $\mathbf{L}_\varphi(S)$.

VII.2.34. **Remark.** The neighborhood U is a local semigroup, and thus $S_U \overset{\text{def}}{=} \varphi(U)$ is a local semigroup in L with respect to the Campbell–Hausdorff multiplication. Thus by Theorem IV.1.27, $\mathbf{L}_\varphi(S)$ is a Lie wedge. By Remark IV.1.17, $\mathbf{L}_\varphi(S)$ is independent of restrictions or extensions of φ.

VII.2.35. **Proposition.** *Let S be a local semigroup, and let $\varphi : U \to L$ be a Campbell–Hausdorff embedding.*

   (i) *If S is infinitesimally closed, then it admits a tangent wedge.*

   (ii) *If U is locally compact, then S is infinitesimally closed.*

   (iii) *Suppose there exists a homeomorphism $\psi : U \to E$ into a Banach space E such that $\psi(U) \cap W$ is closed in W for some open set W containing $\psi(\mathbf{1})$ and such that $\psi \circ \varphi^{-1}$ is C_s^1 differentiable at 0. Then S is infinitesimally closed.*

Proof. (i) The set $S_U = \varphi(U)$ is a local semigroup in L. Pick an open set B containing 0 small enough so that $B \cap S_U$ is closed in B. We also pick B small enough so that it is a Campbell–Hausdorff neighborhood and $S_U \cap B$ is a local semigroup with respect to B (see Proposition IV.1.12.). By Proposition IV.1.21, $\mathbb{R}^+ \cdot X \cap B \subseteq S_U$ if $X \in \mathbf{L}(S_U) = \mathbf{L}_\varphi(S)$. Thus S admits a tangent wedge.

(ii) Pick a compact neighborhood Q of **1** contained in U, and pick an open set B containing 0 such that $B \cap S_U \subseteq \varphi(Q)$. Then since $\varphi(Q)$ is compact and hence closed in L, it follows that $B \cap S_U = B \cap \varphi(Q)$ is closed in B.

(iii) Pick an open set B containing 0 in L small enough so that $\psi \circ \varphi^{-1}$ restricted to $B \cap S_U$ is Lipschitz (see Proposition VII.2.4) and so that the closure of the image of $B \cap S_U$ under $\psi \circ \varphi^{-1}$ is contained in W. Let $\{X_n\}$ be a sequence in $B \cap S_U$ converging to some $X \in B$. The image of this sequence under $\psi \circ \varphi^{-1}$ is Cauchy by the Lipschitz condition, hence converges in the Banach space E to some Y. By hypothesis $Y \in W \cap \psi(U)$. By continuity of $\varphi^{-1} \circ \psi$,

$X_n \to \varphi^{-1}\big(\psi(Y)\big) \in S_U$. Hence $X = \varphi^{-1}\big(\psi(Y)\big) \in S_U$. Thus $B \cap S_U$ is closed in B. ∎

We note the relationship of canonical embeddings (Definition VII.1.39) and Campbell–Hausdorff embeddings.

VII.2.36.   Remark. Let S be a local semigroup which admits a canonical embedding. Then any canonical embedding is a Campbell–Hausdorff embedding. It follows from Corollary VII.1.42 that given any two canonical embeddings, there is an isomorphism of normable Lie algebras that carries the tangent wedge of S in one to the tangent wedge of S in the other. Hence the choice of the tangent wedge is independent of the canonical embedding up to natural isomorphism. Similary by Corollary VII.1.42, S is infinitesimally closed with respect to some canonical embedding if and only if it is infinitesimally closed with respect to every one of them.

We recall some basic facts about one parameter semigroups. A continuous mapping $\sigma\colon \mathbb{R}^+ \to S$ is a *one parameter semigroup* if $\sigma(0) = \mathbf{1}$ and $\sigma(s+t) = \sigma(s)\sigma(t)$ for all $s, t \in \mathbb{R}^+$. A continuous mapping $\sigma\colon [0,T] \to S$ is a *local one parameter semigroup* if $\sigma(0) = \mathbf{1}$ and $\sigma(s+t) = \sigma(s)\sigma(t)$ for all $s, t, s+t \in [0,T]$. If two local one parameter semigroups agree on any interval containing 0, then they agree on every interval containing 0 which is contained in the domain of both. If S is a topological semigroup with $\mathbf{1}$, then any local one parameter semigroup σ into S may be extended uniquely to a one parameter semigroup; this extension is given by $\sigma(t) = \big(\sigma(\tfrac{1}{n}t)\big)^n$ where $(1/n)t$ is in the domain of σ.

VII.2.37.   Lemma. *Let S be a (local) semigroup. Let $\varphi\colon U \to L$ be a Campbell–Hausdorff embedding which admits a tangent wedge. Assigning to each $X \in \mathbf{L}_\varphi(S)$ the mapping $t \mapsto \varphi^{-1}(tX)$ is a one-to-one correspondence between $\mathbf{L}_\varphi(S)$ and the (local) one parameter subsemigroups of S. (We identify two local semigroups if they agree on an interval $[0,T]$ for some $T > 0$.)*

Proof.   Note that U and hence $S_U = \varphi(U)$ is a local semigroup. Since U is a neighborhood of $\mathbf{1}$ in S, its local one parameter semigroups agree with those of S (up to extension and restriction). The local one parameter semigroups of U and S_U are carried on to one another by φ and φ^{-1} respectively. Thus it suffices to identify the local one parameter semigroups of S_U. Since the Campbell–Hausdorff multiplication is just addition on 1-dimensional subspaces, the local one parameter semigroups of L are given by $t \mapsto t{\cdot}X$ for $X \in L$. To complete the proof, we must show that $t \mapsto t{\cdot}X$ is a local one parameter semigroup in S_U precisely when $X \in \mathbf{L}_\varphi(S)$.

Pick an open set B containing 0 small enough so that $S_U \cap B$ is a local semigroup with respect to B (see Proposition IV.1.12) and so that $\mathbf{L}(S_U) \cap B \subseteq S_U$ (since φ admits a tangent wedge). Then $\mathbb{R}^+{\cdot}X \cap B \subseteq S_U$ if $X \in \mathbf{L}(S_U) = \mathbf{L}_\varphi(S)$. Conversely if $[0,T]{\cdot}X \subseteq S_U$, then $X = \lim n\big((1/n)X\big) \in \mathbf{L}(S_U)$. This completes what we needed to show.

In the case that S is a semigroup, then it is standard that each local one parameter semigroup extends uniquely to a one parameter semigroup into S (see the remarks before this proposition). The correspondence in this case thus follows from the local case. ∎

In the case that S is a semigroup, the preceding results allow us to organize the one parameter semigroups into a wedge in L and define an exponential mapping.

VII.2.38.   Proposition.   *Suppose that S is a topological semigroup with $\mathbf{1}$ and that $\varphi: U \to L$ is a Campbell–Hausdorff embedding which admits a tangent wedge. Then there exists an unique function $\exp_\varphi: \mathbf{L}_\varphi(S) \to S$ such that $\exp = \exp_\varphi$ extends $\varphi^{-1}|(\varphi(U) \cap \mathbf{L}_\varphi(S))$ and such that the one parameter semigroups of S are all of the form $t \mapsto \exp(t \cdot X): \mathbb{R}^+ \to S$ for $X \in \mathbf{L}_\varphi(S)$. The function $\exp$ is continuous, and if S and φ are C_s^k for $1 \le k \le \omega$, then so is $\exp$. Finally addition in $\mathbf{L}_\varphi(S)$ is related to multiplication in S by*

$$\exp\big(t \cdot (X + Y)\big) = \lim_n \big(\exp(\tfrac{t}{n} \cdot X) \exp(\tfrac{t}{n} \cdot Y)\big)^n.$$

Proof.   Define $\exp_\varphi: \mathbf{L}_\varphi(S) \to S$ by $\exp(X) = \big(\varphi^{-1}((1/n) \cdot X)\big)^n$, where $(1/n) \cdot X$ is in $\varphi(S)$. (It follows from Definition VII.2.33 that $(1/n) \cdot X \in \varphi(S)$ for large enough n.) From Lemma VII.2.37, $\exp(X)$ is just $\sigma(1)$, where σ is the one parameter semigroup corresponding to X. It then follows that $\exp(tX) = \sigma(t)$; hence the one parameter semigroups of S may be written alternately in the form $t \mapsto \exp(tX)$. Given any $X \in \mathbf{L}_\varphi(S)$, there exists a neighborhood V of X such that the function $\exp|V$ is given by $Y \mapsto \big(\varphi^{-1}(\tfrac{1}{n} \cdot Y)\big)^n$ for n large enough. This is a compostion of continuous functions on V, hence is continuous. Similarly, if S and φ are C_s^k, then so is each of the three functions in the composition and hence their composition.

We establish the last assertion. Let $X, Y, X + Y \in \mathbf{L}_\varphi(S) \cap \varphi(U)$. By the Trotter Product Formula (see (A3) of Appendix 1) and the fact that φ^{-1} is a topological isomorphism on $\varphi(U)$ which restricted to $\varphi(U) \cap \mathbf{L}_\varphi(S)$ is equal to $\exp$, we obtain

$$\begin{aligned}
\exp(X + Y) &= \varphi^{-1}(X + Y) = \varphi^{-1}\big(\lim_n n \cdot (\tfrac{1}{n} \cdot X * \tfrac{1}{n} \cdot Y)\big) \\
&= \lim_n \big(\varphi^{-1}(\tfrac{1}{n} \cdot X) \varphi^{-1}(\tfrac{1}{n} \cdot Y)\big)^n \\
&= \lim_n \big(\exp(\tfrac{1}{n} \cdot X) \exp(\tfrac{1}{n} \cdot Y)\big)^n.
\end{aligned}$$

Now for arbitrary $X, Y \in \mathbf{L}_\varphi(S)$, there exists a positive integer m such that $(1/m) \cdot X, (1/m) \cdot Y, (1/m) \cdot (X + Y) \in \varphi(U)$. Then since $\exp$ is a homomorphism on each ray

$$\begin{aligned}
\exp(X + Y) &= \big(\exp(\tfrac{1}{m} \cdot X + \tfrac{1}{m} \cdot Y)\big)^m \\
&= \big(\lim_n \big(\exp(\tfrac{1}{nm} \cdot X) \exp(\tfrac{1}{nm} \cdot Y)\big)^n\big)^m \\
&= \lim_k \big(\exp(\tfrac{1}{k} \cdot X) \exp(\tfrac{1}{k} \cdot Y)\big)^k.
\end{aligned}$$

Finally one applies the preceding with tX replacing X and tY replacing Y.   ∎

VII.2.39.   Theorem. *Let S be a local semigroup for which $\mathbf{1}$ has a neighborhood homeomorphic to an admissible subset of $\mathbb{R}^n$. The following are equivalent:*

(1) *S is locally cancellative.*

(2) *The multiplication is C_s^1 at $(\mathbf{1}, \mathbf{1})$ for some local C_s^1 structure.*

(3) *The multiplication is analytic on some neighborhood of $\mathbf{1}$.*

(4) *S admits a canonical embedding $\varphi\colon U \to L$ into a finite-dimensional real Lie algebra.*

Proof.   (1)$\Rightarrow$(4): Immediate from Corollary VII.1.27.

(4)$\Rightarrow$(3): The canonical embedding gives a chart for which multiplication is analytic, since the multiplication in the Lie algebra is given by the Campbell–Hausdorff series.

(3)$\Rightarrow$(2): Immediate.

(2)$\Rightarrow$(1): Local cancellation is obtained as in Theorem VII.2.16.   ∎

The next result is a generalization of the main result of M. Anderson [An 88a].

VII.2.40.   Theorem. *Let S be a local semigroup in which $\{\mathbf{1}\}$ is not open, which has a Campbell–Hausdorff embedding $\varphi : U \to L$ into a finite-dimensional Lie algebra L, and which admits a tangent wedge with respect to φ. Then S has a non-trivial local one parameter semigroup. Furthermore, if $\mathbf{1}$ is not in the interior of the set of units, then there exists a local one parameter semigroup meeting the local group of units only at $\mathbf{1}$.*

In particular, if S has a compact neighborhood homeomorphic to an admissible subset of $\mathbb{R}^n$ and satisfies any of the equivalent conditions of Theorem VII.2.39, then S has a non-trivial local one parameter semigroup, and if $\mathbf{1}$ is not in the interior of the set of units, then there exists a local one parameter semigroup meeting the local group of units only at $\mathbf{1}$.

Proof.   By Lemma VII.2.37 the local one parameter semigroups correspond to the members of $\mathbf{L}_\varphi(S)$. If $\mathbf{L}_\varphi(S) = \{0\}$, then by the Surrounding Wedge Theorem IV.2.11, there exists a neighborhood B of 0 such that $B \cap \varphi(U) = \{0\}$ (see also the remarks after IV.2.11). But this is impossible since φ is a homeomorphism and $\mathbf{1}$ is not an isolated point. So there exists $0 \neq X \in \mathbf{L}_\varphi(S)$. Then the local one parameter semigroup of S which corresponds to X (by Lemma VII.2.37) is the desired one.

Suppose that $\mathbf{1}$ is not in the interior of the units. If $\mathbf{L}_\varphi(S)$ is a subspace of L, then again by the Surrounding Wedge Theorem IV.2.11 there exists a symmetric neighborhood B of 0 such that $\varphi(U) \cap B \subseteq \mathbf{L}_\varphi(S)$. Conversely, if B is chosen small enough, $\mathbf{L}_\varphi(S) \cap B \subseteq \varphi(U)$, since S admits a tangent wedge. Thus $\varphi(U) \cap B = \mathbf{L}_\varphi(S) \cap B$. Thus since $\mathbf{L}_\varphi(S)$ is a subspace, every member $X \in \varphi(U) \cap B$ has an inverse (namely $-X$) in $\varphi(U) \cap B$ with respect to the Campbell–Hausdorff multiplication. It follows that the open set $\varphi^{-1}\big(\varphi(U) \cap B\big)$ in S consists entirely of units, a contradiction. Thus $\mathbf{L}_\varphi(S)$ is not a subspace. Let $X \in \mathbf{L}_\varphi(S) \setminus -\mathbf{L}_\varphi(S)$. Then since $-X \notin \mathbf{L}_\varphi(S)$, it follows that there exists an interval $[0, T]$ such that the local one parameter semigroup corresponding to X is defined on $[0, T]$ and meets the units only in $\mathbf{1}$.

If S has a compact neighborhood homeomorphic to an admissible subset of $\mathbb{R}^n$ and satisfies any of the conditions of the preceding theorem, then by (4) S admits a canonical embedding. This is a Campbell–Hausdorff embedding, and is infinitesimally closed and hence admits a tangent wedge by Proposition VII.2.35. Since no point of an admissible set is isolated, it follows that $\mathbf{1}$ is not isolated in S. Hence the previous part of this theorem applies. ∎

We turn now to differentiable semigroups. In the following we assume that $k \geq 1$.

VII.2.41. **Definition.** Let S be a C_s^k local semigroup or monoid for $1 \leq k \leq \omega$, and $\log : U \to \mathcal{L}(S)$ an associated C_s^k log function from some neighborhood U of the identity $\mathbf{1}$ in S into the Lie algebra $\mathcal{L}(S)$ infinitesimally generated by S. We note that the existence of such a log function is guaranteed by Theorem VII.2.30. We define the *tangent wedge* $\mathbf{L}(S)$ to be the tangent wedge of the local semigroup $S_U = \log U$.

The implication (3) implies (2) in the following theorem is a slight variant of the main result of [An 88b]. Earlier slightly less general versions were obtained by J. P. Holmes [Ho 88]. We remark that their approach is direct, not via the machinery of local group embeddings and the local Lie theory.

VII.2.42. **Theorem.** *Suppose that S is a C_s^k local semigroup which possesses a chart $\varphi : U \to \mathcal{L}(S)$ such that $\varphi(U) \cap V$ is closed in V for some neighborhood V of $\varphi(\mathbf{1})$. Then S is infinitesimally closed. Furthermore, in this case the following are equivalent:*

(1) *$X \in \mathbf{L}(S)$;*

(2) *there exists a (unique) local one parameter semigroup in S with tangent vector X at $\mathbf{1}$;*

(3) *there exists a curve $\alpha : [0, T] \to S$ with $\alpha(0) = \mathbf{1}$ and $\alpha'(0) = X$.*

If S is a C_s^k semigroup with $\mathbf{1}$, then the mapping $\exp : \mathbf{L}(S) \to S$ is C_s^k, and the one parameter semigroups of S are all C_s^k and can be represented uniquely in the form $t \mapsto t \cdot X$ for $X \in \mathbf{L}(S)$.

Proof. By definition of charts the compositions $\log \circ \varphi^{-1}$ and $\varphi \circ \log^{-1}$ are diffeomorphisms of appropriate admissible sets containing $\varphi(\mathbf{1})$ and 0 respectively. It then follows from Proposition VII.2.35 that S is infinitesimally closed and hence admits a tangent wedge.

$(1) \Rightarrow (2)$: Let $X \in \mathbf{L}(S)$. By Lemma VII.2.37, the mapping $\sigma(t) = \log^{-1}(t \cdot X)$ is a local one parameter semigroup in S. Note that $\log(\sigma(t)) = t \cdot X$ has tangent vector X at 0. Since the derivative of the log function is the identity on $\mathcal{L}(S)$ (see Definition VII.2.29), it follows from the chain rule that $\sigma'(0) = X$.

$(2) \Rightarrow (3)$: Immediate. Take the curve to be the local one parameter semigroup.

$(3) \Rightarrow (1)$: The composition $\log \circ \alpha$ is a curve in $\mathcal{L}(S)$. By the chain rule and the fact that the derivative of $\log$ at $\mathbf{1}$ is the identity, the composition has tangent vector X at 0. By Definition I.5.1, $X \in \mathbf{L}(\log(U)) = \mathbf{L}(S)$.

The last part of the theorem follows from Proposition VII.2.38 (noting that $t \mapsto \exp(t \cdot X)$ is the composition of a C_s^ω mapping and a C_s^k mapping). ∎

We consider now a final alternate application of the machinery of local semigroups in Chapter IV and the results of this chapter on local embeddings. We say that the boundary of a C_s^k manifold M with generalized boundary is *smooth near* $p \in \partial M$ if there exists a chart $\psi: U \to E$ such that $\psi(p) = 0$ and $\psi(U) = B \cap W$ for some half-space W and some open ball B around 0. We derive a strengthened version of earlier results of Graham [Gr83] and Holmes [Ho87] concerning the case that the boundary is smooth near $\mathbf{1}$.

VII.2.43.   **Theorem.**   *Let S be a C_s^k local semigroup with $\mathbf{1} \in \partial S$. The following are equivalent:*

    (1) S is smooth near $\mathbf{1}$.

    (2) S is infinitesimally closed and $\mathbf{L}(S)$ is a half-space.

    (3) S is infinitesimally closed and $\mathbf{L}(S)$ is a half-space Lie semialgebra.

    (4) There exists a neighborhood U of $\mathbf{1}$ such that $U \cap \partial S$ consists of units (i.e., $\mathbf{1}$ is in the interior of the set of units in ∂S) and a chart $\varphi: U \to \mathcal{L}(S)$ such that $\varphi(U) \cap V$ is closed in V for some neighborhood V of $\varphi(\mathbf{1})$.

    (5) If $\log : U \to \mathcal{L}(S)$, then $\log(U) \cap B = \mathbf{L}(S) \cap B$ for some Campbell–Hausdorff neighborhood B, $\mathbf{L}(S)$ is a half-space Lie semialgebra, and the local group of units $H(\log(U) \cap B)$ is $B \cap H(\mathbf{L}(S))$.

Proof. $(1) \Rightarrow (2)$: By Proposition VII.2.35(iii), S is infinitesimally closed. Let $\psi: U \to E$ be the chart guaranteed by the definition of smooth near $\mathbf{1}$. Since $\log \circ \psi^{-1}$ and its inverse are differentiable, it follows that the derivative at 0 is an isomorphism of Banach spaces and from Proposition IV.1.19 that the tangent set of $\psi(U)$ is carried to the tangent set of $\log(U)$, which is just $\mathbf{L}(S)$. Since the tangent set of $\psi(U)$ is clearly a half-space, it follows that $\mathbf{L}(S)$ is also a half space.

    $(2) \Leftrightarrow (3)$: Follows from Proposition IV.1.35.

    $(3) \Rightarrow (5)$: Choose a Campbell–Hausdorff neighborhood B such that $S_U = \log(U) \cap B$ is a local semigroup with respect to B (see Proposition IV.1.12) and is closed in B (see Remark VII.2.36). By Proposition IV.1.35 there is a closed half-space W such that $B \cap W = S_U = \log(U) \cap B$, and the proof of that proposition shows $W = \mathbf{L}(S_U) = \mathbf{L}(S)$. The final assertion of (5) follows from the final assertation of Proposition IV.1.35.

    $(5) \Rightarrow (1)$: Let ψ be the restriction of $\log$ to $\log^{-1}(\log(U) \cap B)$.

    $(5) \Rightarrow (4)$: Since $H(\log(U) \cap B) = B \cap H(\mathbf{L}(S)) = B \cap \partial \mathbf{L}(S)$ and $\log(U) \cap B = \mathbf{L}(S) \cap B$, it follows that 0 is in the interior of the set of units in $\partial \log(U)$. Since $\log$ is a local topological isomorphism, it follows that the same is true for $\mathbf{1}$ in S. From the preceding cycle we have that (5) implies (3), so that S is infinitesimally closed. Hence $\log(U) \cap B$ is closed in B if B is chosen small enough.

    $(4) \Rightarrow (2)$: By Proposition VII.2.35(iii), S is infinitesimally closed. Let $\log: U \to \mathcal{L}(S)$ be a log function (see Definition VII.2.29 and Theorem VII.2.30). We assume that U is chosen open and small enough so that $\partial U = U \cap \partial S$ consists of units (such a neighborhood exists by hypothesis). It follows from Corollary VII.1.42 that S is infinitesimally closed with respect to the canonical embedding "$\log$"; hence $\log(U) \cap B$ is closed in B for all Campbell–Hausdorff neighborhoods B chosen small enough. We have that $\log(U)$ is a local semigroup in $\mathcal{L}(S)$, so by Proposition IV.1.12, $\log(U) \cap B$ is a local semigroup with respect to B for all B small enough. So we fix some convex symmetric Campbell–Hausdorff neighborhood

B such that $S_B = B \cap \log(U)$ is a local semigroup with respect to B and is closed in B.

It follows from the fact that log is a canonical embedding that $\log(U)$ and hence S_B are admissible subsets of $\mathcal{L}(S)$, that $0 = \log(\mathbf{1})$ is in ∂S_B, and that $\partial S_B = S_B \cap \partial(\log(U)) = S_B \cap \log(\partial U)$ consists of units of S_B. Pick a symmetric Campbell–Hausdorff neighborhood B' such that $B' * B' \subseteq B$. Again $\log(U) \cap B' = S_B \cap B'$ is an admissible subset of $\mathcal{L}(S)$. Pick $Q \neq \emptyset$ open in $\mathcal{L}(S)$ such that $Q \subseteq S_B \cap B'$, and pick $p \in Q$.

We claim that $-Q \cap S_B = \emptyset$. For if there existed $x \in S_B \cap (-Q)$, then $y = -x \in Q \subseteq S_B$. Thus

$$0 = x + y = x * y \in x * Q \subseteq B' * B' \cap S_B * S_B \subseteq B \cap S_B * S_B \subseteq S_B.$$

Since $x * Q$ is open, we have $0 \in \operatorname{int}(S_B)$, a contradiction to $0 \in \partial S_B$.

Suppose that $\mathbf{L}(S) = \mathbf{L}(S_B)$ is not a half-space. Then $H \overset{\text{def}}{=} H(\mathbf{L}(S)) = \mathbf{L}(S) \cap -\mathbf{L}(S)$ is not a hyperplane. Hence $\mathbb{R} \cdot p + H \neq \mathcal{L}(S)$. Since Q is open, it follows that there exists $q \in Q$ such that $q \notin \mathbb{R} \cdot p + H$. Let $F = \mathbb{R} \cdot p + \mathbb{R} \cdot q$. It follows that $F \cap H = \{0\}$.

For each y in the F-open set $F \cap -Q$, consider the line segment from p to y. The segment lies in $B \cap F$, is connected, and meets S_B and the complement of S_B (since $-Q \cap S_B = \emptyset$). Hence there exists z_y on the line segment such that $z_y \in \partial S_B$. Since ∂S_B consists of units, z_y is a unit of S_B. Since $\{z_y : y \in F \cap -Q\}$ is an uncountable set in the second countable space F, there exists some $z \in \{z_y\}$ which is a limit point of the set. We can then choose a sequence $\{z_n\}$ contained in $\{z_y\}$ such that $z_n \to z$. It follows that $z_n * (-z) \to 0$, and is a sequence of units in S_B. Since F is locally compact, there exists a sequence of positive real numbers $\{r_n\}$ such that $r_n \cdot (z_n * (-z)) \to w$ for some subsequence of $\{z_n * (-z)\}$ and some $w \in F \setminus \{0\}$. Then $w \in \mathbf{L}(S_B)$ (by definition) and similarly $-w \in \mathbf{L}(S_B)$. Thus $w \in H$, since $\mathbf{L}(S_B) = \mathbf{L}(S)$. But $w \in F \cap H$ implies $w = 0$, a contradiction. This completes the proof. ∎

VII.2.44. **Corollary.** *Let S be a C_x^k semigroup with smooth compact connected boundary ∂S and with $\mathbf{1} \in \partial S$. Then ∂S is precisely the group of units of S.*

Proof. Since translations by units are diffeomorphisms and since $\mathbf{1} \in \partial S$, it follows that the group of units H is contained in ∂S. If a net in H converges $s \in S$, then by compactness of ∂S the inverses cluster to some t, which by continuity must equal s^{-1}. Thus H is closed. By Theorem VII.2.43, $\mathbf{1}$ is in the interior of H in ∂S. By translation H is open in ∂S. By connectivity $H = \partial S$. ∎

3. Cancellative semigroups on manifolds

In the preceding section a C_s^ω or analytic semigroup S was defined to be a semigroup on a analytic manifold with boundary such that the multiplication is an analytic function from $S \times S$ to S. In this section we mainly restrict our attention to manifolds with empty boundary. Hilbert's fifth problem, to show that a locally Euclidean topological group is a Lie group, evidently has a semigroup version in view of the above definition: is a locally Euclidean semigroup an analytic semigroup, i.e., does it admit the structure of an analytic manifold so that multiplication is analytic and the underlying topology agrees with the original topology?

The conjecture in this generality is too broad, however, since the existence of too many commuting idempotents is incompatible with differentiability. Indeed, the real line with the operation $xy = \min\{x, y\}$ is a canonical counterexample to this conjecture. One sure way to remove this annoyance is to require cancellation, an axiom sufficiently strong to deny the existence of any non-identity idempotents. The amended conjecture now becomes: is a cancellative semigroup on a Euclidean manifold an analytic semigroup? The answer here is yes, and is provided in Corollary VII.3.14 below. Of course, there are analytic semigroups (even linear ones) which are not cancellative — the matrix semigroup of $n \times n$ matrices under multiplication, for example — so that this result is not a characterization.

As always in the study of semigroups, when cancellation occurs, the possibility of group embedding looms, and this situation is no exception. However, global embedding is not obtained herein; rather, the local embedding theorems of Section VII.1 permit the use of the analytic structure within a related Lie group to establish such a structure within S.

The problem of which cancellative semigroups are algebraically embeddable in groups is a complex one. We refer the reader to Chapter 12 of [CP67] for the main results in this area. Necessary and sufficient conditions do exist, but they are complicated and difficult to apply, except in special cases where the group can be constructed as the group of right or left quotients (the case considered in Proposition VII.1.28). The question of algebraic embeddability is an unseen presence that haunts our steps at every turn in the topological setting. If a semigroup is algebraically embeddable, then it is also topologically embeddable (Theorem VII.3.27). Without the assumption of algebraic embeddability, the best results we are able to obtain are homomorphic embeddings that are local homeomorphisms.

The foundational work on which the material in this subsection is based was done in the dissertation of R. Houston [Hou73]. An improved version of these results appeared in [BH87]. Hofmann and Weiss [HW88] found a more direct method of obtaining the analytic structure with the introduction of what they called the "double Lie sheaf." The approach here is a refinement of their approach.

Left quotients and partial right translations

Throughout this subsection S denotes a cancellative topological semigroup that satisfies

($\ddagger$) *given $a, b \in S$ and V open containing b, there exist open sets U containing a and N containing ab (resp. ba) such that $x \in U$ implies $N \subseteq xV$ (resp. $N \subseteq Vx$).*

(This is condition ($\dagger$) of Definition VII.1.14 for $S = S^\circ$.) Note by Lemma VII.1.15 that left and right translations are then open mappings. We use this fact and other facts about ($\ddagger$) freely in the following.

In Section VII.1 we constructed $Q(U, V)$ as a set of right or left quotients. A major motivation for considering right quotients is that a cancellative semigroup S right reverses in itself if and only if S embeds in a group of right quotients. For the local theory (where one does not demand closure) one can use either left or right quotients. In Section VII.1 we concentrated more on right quotients, but in this section left quotients will be more convenient, and we restrict our attention to them.

Recall from Section VII.1 that in a semigroup S we defined for $a, b, c, d \in S$ and $V \subseteq S$

$$(a, b) \,_V\!\equiv (c, d) \qquad \text{if} \qquad au = bv, \ cu = dv \quad \text{where} \quad u, v \in V.$$

We collect some of the results of that section.

VII.3.1. **Theorem.** *Suppose $V \subseteq S$ is an open set such that V right reverses in V' and V' right reverses in S for some V'. Then $Q(S, V) = \left(S \times_V S/ \,_V\!\equiv\right)$ is a local group with respect to the operation $\beta(a, b) * \beta(c, d) = \beta(p, r)$ if there exist $p, q, r \in S$ such that $(a, b) \,_V\!\equiv (p, q)$ and $(b, c) \,_V\!\equiv (q, r)$.*

Furthermore, if U is any open set that right reverses in V, then the inclusion $j_U : U \times U \hookrightarrow S \times_V S$ induces an embedding of $\beta(U, V)$ onto an open subset of $Q(S, V)$. If additionally U left reverses in S, then $Q(U, V_1)$ is identical to $Q(U, V)$ for any suitable pair (U, V_1).

Proof. That $\,_V\!\equiv$ is a congruence on $S \times_V S$ and gives rise to $Q(S, V)$ is the content of Theorem VII.1.9. Let U be an open subset of S which right reverses in V. Since $\,_V\!\equiv$ on $U \times U$ is just the restriction of $\,_V\!\equiv$ on $S \times_V S$, there is a natural injection j_U sending the equivalence class of an element (x, y) in $U \times U$ to its equivalence class in $S \times_V S$. The injection j_U is continuous since $Q(U, V)$ has the quotient topology.

Let W be an open set in $Q(U, V)$ and let $\beta(a, b) \in W$. There exists an open set $A \subseteq U$ containing a such that $b^\triangle(A) \subseteq W$. There is also a mapping

$b^\triangle \colon S \times \{b\} \to Q(S, V)$, which is an open mapping by Lemma VII.1.19 (the result was stated there for $Q(U, V)$, but the proof applies equally well to $Q(S, V)$). Thus the image of A under $b^\triangle$ is also open in $Q(S, V)$; but this image is easily seen to be precisely what one obtains by applying j_U to the image of A in $Q(U, V)$. It now follows that $j_U(W)$ is open in $Q(S, V)$, that is, j_U is an open mapping, and hence a homeomorphism onto an open subset of $Q(S, V)$.

That the mapping j_U is a homomorphism is easily deduced from the definition of multiplication on $Q(U, V)$ and $Q(S, V)$. The fact that $Q(U, V)$ is a local group now ensures that $Q(S, V)$ is also (or, alternately, one can observe that the proof of Theorem VII.1.22 remains valid for $Q(S, V)$).

Suppose now that U left reverses in S. Let $(a, b)\;_V\!\equiv (c, d)$ for $a, b, c, d \in U$. Then $au = bv$ and $cu = dv$ for some $u, v \in V$. Suppose that U also right reverses in V_1. Then $au' = bv'$ for some $u', v' \in V_1$. Pick $s, t \in S$ such that $sa = tc$. Then $sbv = sau = tcu = tdv$, so by cancellation $sb = td$. Thus $tcu' = sau' = sbv' = tdv'$. Again by cancellation $cu' = dv'$, i.e., (a, b) and (c, d) are related for the left quotient relation defined by V_1. The argument is reversible, so $_{V_1}\!\equiv$ and $_V\!\equiv$ agree on $U \times U$, and hence $Q(U, V) = Q(U, V_1)$. ∎

VII.3.2.   Corollary.   *Let S be a cancellative semigroup satisfying $(\ddagger)$. Then $Q(U_1, V_1)$ and $Q(U_2, V_2)$ are locally isomorphic for suitable pairs (U_1, V_1) and (U_2, V_2).*

Proof.   By Proposition VII.1.16 and its dual we can pick an open set U that simultaneously right reverses in V_1 and V_2 and left reverses in S. By Theorem VII.3.1 $Q(U_i, V_i)$ and $Q(U, V_i)$ are both locally isomorphic to $Q(S, V_i)$ for $i = 1, 2$ and hence to each other. Since also $Q(U, V_1)$ and $Q(U, V_2)$ are locally isomorphic by VII.3.1, the conclusion follows. ∎

Corollary VII.3.2 shows that associated with S is a local group which is unique up to local isomorphism. We wish now to define a local right action of this group on S. Note that since translation by an element of S is an open mapping, then ρ_s, right translation by s, has range the open left ideal Ss. We consider a more general notion of a right translation.

VII.3.3.   Definition.   A *partial right translation* is an injective open mapping ρ with domain and range non-empty open left ideals of S satisfying $(xy)\rho = x(y\rho)$ for all $x \in \operatorname{dom}\rho$ and $y \in S$. Note that we compose partial right translations on the right. It is an elementary exercise (see, for example, [CP61]) that, if ρ is a partial right translation, then so is ρ^{-1}, and that, if ρ and μ are partial right translations which compose ($\operatorname{ran}(\mu) \cap \operatorname{dom}(\rho) \neq \varnothing$), then $\mu\rho$ is a partial right translation with appropriately curtailed domain.

We want to define a local right action of the local group $Q(S, V)$ of Theorem VII.3.1 on S in such a way that each member of $Q(S, V)$ acts as a partial right translation on S. To this end we define $\rho(a, b)$ for $(a, b) \in S \times_V S$ by $x\rho(a, b) = y$ if there exist $u, v \in V$ such that $au = bv$ and $xu = yv$, i.e. if $(a, b)\;_V\!\equiv (x, y)$.

VII.3.4.   Lemma.   *The function $\rho(a, b)$ is well-defined, is a partial right translation, and is equal to $\rho_u(\rho_v)^{-1}$ if $au = bv$. If $\rho(a, b)$ and $\rho(c, d)$ agree at a point, they are equal.*

Proof. Since $(a, b) \in S \times_V S$, there exist $u, v \in S$ such that $au = bv$. Suppose also that $au' = bv'$ for $u', v' \in V$. Then if $xu = yv$ and $xu' = zv'$ we conclude from Malcev's Condition (VII.1.6) that $xu' = yv'$, so by cancellation $y = z$. Thus $\rho(a, b)$ is well-defined.

By definition $\big(x\rho(a, b)\big)v = yv = xu$, so $x\rho(a, b) = x\rho_u \circ \rho_v{}^{-1}$. Conversely if $x\rho_u \circ \rho_v{}^{-1} = y$, then $xu = yv$, so $x\rho(a, b) = y$. Thus $\rho(a, b) = \rho_u \circ \rho_v{}^{-1}$, and hence $\rho(a, b)$ is a partial right translation.

Suppose $s\rho(a, b) = t = s\rho(c, d)$. Then $au = bv$, $su = tv$ for $u, v \in V$, and $cu' = dv'$, $su' = tv'$ for $u', v' \in V$. By Malcev's Condition $cu = dv$, so from the preceding, $\rho(a, b) = \rho_u \circ \rho_v{}^{-1} = \rho(c, d)$. ∎

We now make precise the notion of a local right action.

VII.3.5. **Definition.** Let X be a space and G a local group. A *local right action* of G on X is a continuous function $(x, g) \mapsto xg$ defined on an open subset of $X \times G$ containing $X \times \{e\}$ with values in X such that

(i) for each $x \in X$, $xe = x$;

(ii) there exists a neighborhood Ω of $X \times \{e\} \times \{e\}$ in $X \times G \times G$ such that for all $(x, g, g') \in \Omega$, we have that gg', $(xg)g'$, and $x(gg')$ are defined and $(xg)g' = x(gg')$.

The local right action is *locally simply transitive* if given $x \in X$, there exists an open set N_x in G containing e such that the function $g \mapsto xg \colon N_x \to X$ is defined on all of N_x and is a homeomorphism onto an open set containing x.

VII.3.6. **Theorem.** *The function $\beta(a, b) \mapsto \rho(a, b)$ from $Q(S, V)$ to the partial semigroup of partial right translations under the operation of composition (where defined) is a monomorphism onto its image. Furthermore, the partial function $\big(s, \beta(a, b)\big) \to s\rho(a, b)$ is a locally simply transitive local right action of $Q(S, V)$ on S.*

Proof. Let us consider first of all the mapping $(a, b) \mapsto \rho(a, b)$ with domain $S \times_V S$. If $(a, b) \;_V\!\equiv (c, d)$, then $au = bv$ and $cu = dv$ for some $u, v \in V$. By definition $p\rho(a, b) = q$ iff $pu = qv$ iff $p\rho(c, d) = q$. Thus $\rho(a, b) = \rho(c, d)$. Hence the function $\beta(a, b) \mapsto \rho(a, b)$ is well-defined.

Conversely, suppose that $\rho(a, b) = \rho(c, d)$. Let $t = s\rho(a, b) = s\rho(c, d)$. Then there exist $u, v, u', v' \in V$ such that $au = bv$, $su = tv$, $cu' = dv'$, and $su' = tv'$. By Malcev's Condition (VII.1.6), $cu = dv$, and hence $\beta(a, b) = \beta(c, d)$. Thus the mapping is a bijection with its image.

Suppose now that $\beta(a, b) * \beta(c, d)$ is defined. Then by Theorem VII.1.9 there exist $(p, q), (q, r), (p, r) \in S \times_V S$ such that $\beta(a, b) = \beta(p, q)$, $\beta(c, d) = \beta(q, r)$, and the product is $\beta(p, r)$. Then $p\big(\rho(p, q)\rho(q, r)\big) = q\rho(q, r) = r$ and $p\rho(p, r) = r$. By the last part of Lemma VII.3.4 we have $\rho(p, q)\rho(q, r) = \rho(p, r)$. Thus the bijection is a homomorphism.

We now verify that a local right action is defined. Indeed we have

$$s\big(\beta(a, b) * \beta(c, d)\big) = s\big(\rho(a, b)\rho(c, d)\big) = \big(s\rho(a, b)\big)\rho(c, d) = \big(s(\beta(a, b))\beta(c, d).$$

(Note that the monomorphism gives us associativity for all triples in $Q(S, V)$ whenever both sides are defined, not just in a neighborhood of the identity). Also $se = s\beta(s, s) = s\rho(s, s) = s$ directly from the definition of $\rho(s, s)$.

We conclude by proving the openess and continuity assertations. Let $q = p\beta(a, b)$ and let W be an open set containing q. Then there exist $u, v \in V$ such that $au = bv$ and $pu = qv \in Wv$. Since Wv is open, there exist open sets P containing p and U_u containing u such that $PU_u \subseteq Wv$. By ($\ddagger$) there exists an open set N containing au and an open set A containing a such that $N \subseteq a'U_u$ for all $a' \in A$. Since $bv = au \in N$, pick an open set B containing b such that $Bv \subseteq N$.

Now let $p' \in P$, $a' \in A$ and $b' \in B$. Then $b'v \in N \subseteq a'U_u$. Hence there exists $u' \in U$ such that $a'u' = b'v$. Now $p'u' \in PU_u \subseteq Wv$, so $p'u' = q'v$ for some $q' \in W$. Thus $p'\beta(a', b') = q' \in W$. This establishes both the openess of the domain and the continuity.

Finally, let O be a neighborhood of e, and let $s \in S$. There exists W open containing s such that $\beta(\{s\} \times W) \subseteq O$ (since $\beta(s, s) = e$). Then $\beta(s, t) \to s\beta(s, t) = t : \beta(\{s\} \times W \to W$ is one-to-one and is continuous by the preceding paragraphs. The inverse, $t \mapsto \beta(s, t) : W \to \beta\{s\} \times W$ is also continuous. Hence the mapping is a homeomorphism. This shows that the right action is locally simply transitive.   ∎

Suppose now that G is a topological group that is locally isomorphic to $Q(S, V)$. Then we can use the local isomorphism to define a locally transitive right local action of G on S in the obvious way.

VII.3.7.   Corollary.   *Let G be a topological group which is locally isomorphic to $Q(S, V)$. Then the local isomorphism defines a locally simply transitive local right action of G on S. There exists $\mathcal{D} \subseteq S \times \mathcal{N}(e)$ (where $\mathcal{N}(e)$ denotes the set of open subsets containing e_G) such that the following conditions are satisfied:*

(i) *$g \mapsto ag : N \to aN$ is a homeomorphism onto the open set aN for all $(a, N) \in \mathcal{D}$;*

(ii) *$ae = a$;*

(iii) *Let $a \in S$. If we set $\mathcal{D}_a = \{W \in \mathcal{N}(e) : (a, W) \in \mathcal{D}\}$, then $\mathcal{D}_a$ is a basis of open neighborhoods of e such that $V \subseteq W$, $V \in \mathcal{O}(S)$, and $W \in \mathcal{D}_a$ imply $V \in \mathcal{D}_a$.*

Proof.   Let α be a local isomorphism from $Q(S, V)$ to G. Pick an open set Γ containing the identity in $Q(S, V)$ such that α is defined on $(\Gamma)^2$. We define sg for $s \in S$ and $g \in \alpha(\Gamma)$ to be $s(\alpha)^{-1}(g)$. This clearly defines a locally simply transitive local right action since the original one was.

Let $\mathcal{D}$ be all pairs (a, N) such that $e \in N \subseteq \alpha(\Gamma)$ and $g \mapsto ag : N \to aN$ is a homeomorphism onto an open subset of S. Then (i), (ii), and (iii) are automatically satisfied.   ∎

The double cover and analytic structures

Every locally simply transitive local right action of a topological group G on a topological space X gives rise to a topology on $X \times G$ finer than the product topology such that the projection into X is a covering projection, the projection into G is a local homeomorphism, and the mapping $((x, h), g) \mapsto (xg, hg)$ is again a locally simply transitive local right action. In the case that X is a semigroup on which G acts by partial right translations, then one may induce on $X \times G$ a continuous semigroup multiplication such that the projections are homeomorphisms.

Let S be a cancellative semigroup which satisfies condition ($\ddagger$), let G be a Haudorff topological group which is locally isomorphic to $Q(S, V)$. Then there is a locally simply transitive local right action of G on S. We define $\mathcal{D}$ as in Corollary VII.3.7. For every $(a, N) \in \mathcal{D}$ and $h \in G$, let

$$B(a, N, h) = \{(ag, hg): g \in N\} \subseteq S \times G.$$

VII.3.8. **Lemma.** *The set $\mathcal{B} = \{B(a, N, h): (a, N) \in \mathcal{D}, h \in G\}$ is a base for a Hausdorff topology τ on $S \times G$. The coordinatewise multiplication is continuous with respect to this topology.*

Proof. Let $(a, U), (b, V) \in \mathcal{D}$, $h_1, h_2 \in G$ and $(s, g) \in B(a, U, h_1) \cap B(b, V, h_2)$. There exist $g_1 \in U$ and $g_2 \in V$ such that $(s, g) = (ag_1, h_1 g_1) = (bg_2, h_2 g_2)$. Pick an open set W containing e such that $g_1 W \subseteq U$ and $g_2 W \subseteq V$. We show $(s, g) \in B = B(s, W, g) \subseteq B(a, U, h_1) \cap B(b, V, h_2)$. Clearly $(s, g) = (se, ge) \in B$. Let $(t, h) \in B$. Then there exists $k \in W$ such that

$$(t, h) = (sk, gk) = (ag_1 k, h_1 g_1 k) \in B(a, U, h_1)$$

since $g_1 k \in g_1 W \subseteq U$. Similarly $(t, h) \in B(b, V, h_2)$.

We next establish the Hausdorff axiom. Suppose $(s, g) \neq (t, h)$. If $s \neq t$, pick an open set N containing e such that $(s, N), (t, N) \in \mathcal{D}$ and $sN \cap tN = \emptyset$ (which we can do since S is assumed Hausdorff). Then $B(s, N, g) \cap B(t, N, h) = \emptyset$. Similarly if $g \neq h$, pick N such that $(s, N), (t, N) \in \mathcal{D}$ and $gN \cap hN = \emptyset$. Then again $B(s, N, g) \cap B(t, N, h) = \emptyset$.

Suppose $(st, gh) \in B(st, N, gh)$. Since stN is an open set containing st, there exist open sets U_s containing s and U_t containing t such that $U_s U_t \subseteq stN$. Similarly there exist open sets V_g containing g and V_h containing h such that $V_g V_h \subseteq ghN$. Pick W open containing e such that $(s, W), (t, W) \in \mathcal{D}$, $sW \subseteq U_s$, $tW \subseteq U_t$, $gW \subseteq V_g$, and $hW \subseteq V_h$. Then it follows that $B(s, W, g) \cdot B(t, W, h) \subseteq B(st, N, gh)$, and hence multiplication is continuous. ∎

VII.3.9. **Definition.** The space $\Sigma = (S \times G, \tau)$ is called *double cover* of S and G. The projections onto the factors are denoted $\sigma_S: \Sigma \to S$ and $\psi_S: \Sigma \to G$.

VII.3.10. **Remark.** The topology τ on Σ is finer than the product topology, and hence the projections σ and ψ are continuous homomorphisms.

VII.3.11. **Theorem.** *The following items obtain:*

(i) *the mapping $\psi_S: \Sigma \to G$ is a local homeomorphism;*

(ii) *the mapping $\sigma_S: \Sigma \to S$ is a covering projection with evenly covered neighborhoods sN for $(s, N) \in \mathcal{D}$ and local sections $sg \mapsto (sg, hg): sN \to B(s, N, h)$.*

Proof.   (i) Let $(s,g) \in \Sigma$ and pick $(s,N) \in \mathcal{D}$. Then $gh \mapsto (sh, gh) \colon gN \to B(s,N,g)$ is a local section for ψ_S with domain and range open sets. It follows that the local section is a homeomorphism, and hence ψ_S is a local homeomorphism.

(ii) Let $s \in S$ and pick $(s,N) \in \mathcal{D}$. Then $sg \mapsto (sg, hg) \colon sN \to B(s,N,g)$ is a local section for σ_S with domain and range open sets. It follows that the local section is a homeomorphism and hence the restriction of σ_S to $B(s,N,g)$ is a homeomorphism for each $g \in G$. Since these sets are open, pairwise disjoint, and the union is the inverse image of sN, it follows that sN is evenly covered. Hence σ_S is a covering projection.   ∎

We turn now to considerations of analyticity.

VII.3.12.   Lemma. (i) *Let $f \colon X \to Y$ be a local homeomorphism. Given an analytic structure on Y, there exists a unique analytic structure on X such that f is a local diffeomorphism. Given an analytic structure on X, there exists at most one analytic structure on Y such that f is a local diffeomorphism.*

(ii) *Let $f_i \colon X_i \to Y_i$ be surjective analytic mappings which are local diffeomorphisms for $i = 1,2$. Let $g \colon X_1 \to X_2$ and $h \colon Y_1 \to Y_2$ be continuous mappings such that the following diagram commutes:*

$$\begin{array}{ccc} X_1 & \xrightarrow{\ g\ } & X_2 \\ {\scriptstyle f_1}\downarrow & & \downarrow{\scriptstyle f_2} \\ Y_1 & \xrightarrow[\ h\ ]{} & Y_2 \end{array}$$

Then h is an analytic mapping iff g is.

Proof.   (i) Take for charts on X the compostion of a chart on Y with the restriction of f to some open set where it is a homeomorphism. One verifies that these relate analytically and hence form an atlas for an analytic structure on X. Conversely, given any analytic structure on X for which f is a local diffeomorphism, such compositions will form an atlas of charts for the analytic structure. Hence this structure is unique in the sense specified. If X and Y are initially endowed with analytic structures such that f is a local diffeomorphism, then the compositions of the charts on X with the inverses of the restrictions of f to open sets where it is a local diffeomorphism form an atlas of charts on Y that determine its analytic structure. Hence it is also unique in the sense specified.

(ii) If one appropriately restricts domains and codomains to small enough open sets, then the vertical mappings in the given diagram become diffeomorphisms. The lemma is easily established in this special case. But since g and h are analytic iff they are locally analytic, the general case follows from this special one.   ∎

VII.3.13.   Theorem. *Suppose that G is a Lie group. Then there exist unique analytic structures on Σ and S such that ψ_S and σ_S are local diffeomorphisms. With respect to these analytic structures, multiplication is analytic on Σ and S.*

Proof.   By Lemma VII.3.12 (in light of VII.3.11) there exists a unique analytic structure on Σ such that ψ_S is a local diffeomorphism. Since the diagram

$$\begin{array}{ccc} \Sigma \times \Sigma & \xrightarrow{\ m\ } & \Sigma \\ {\scriptstyle \psi\times\psi}\downarrow & & \downarrow{\scriptstyle \psi} \\ G \times G & \xrightarrow[\ m\ ]{} & G \end{array}$$

commutes, again from Lemma VII.3.12, it follows that multiplication is analytic.

Let φ be a chart on G with $\operatorname{dom}\varphi$ a neighborhood of e. For $s \in S$, let $(s, N) \in \mathcal{D}$ such that $N \subseteq \operatorname{dom}\varphi$. Define $\varphi_{s,N}(sg) = \varphi(g)$. This defines a chart for each $s \in S$. We show that they analyticallly relate. Let $p \in \operatorname{dom}_{s,N} \cap \operatorname{dom}_{t,M}$. Then $p = sg_1 = tg_2$ for some $g_1 \in N$ and $g_2 \in M$. We then have $s = tg_2 g_1^{-1}$. For $sg \in \operatorname{dom}\varphi_{s,N} \cap \operatorname{dom}\varphi_{t,M}$, we have

$$(\varphi_{t,M} \circ \varphi_{s,N}^{-1})(\varphi(g)) = \varphi_{t,M}\big((\varphi_{s,N})^{-1}(\varphi_{s,N}(sg))\big)$$
$$= \varphi_{t,M}(sg) = \varphi_{t,M}(tg_2 g_1^{-1} g) = \varphi\big((g_2 g_1^{-1})g\big).$$

Since left translation (by $g_2 g_1^{-1}$) in G is analytic, this is an analytic mapping. Hence the charts are analytically related and define an analytic structure.

For $(sh, gh) \in B(s, N, g)$, where $n \subseteq \operatorname{dom}\varphi$, we have

$$\varphi_{s,N}\big(\sigma_S(sh, gh)\big) = \varphi_{s,N}(sh) = \varphi(h) = \varphi\big(g^{-1}(gh)\big) = (\varphi \circ \lambda_{g^{-1}} \circ \psi_S)(sh, gh).$$

Again since ψ_S is a local diffeomorphism and $\lambda_{g^{-1}}$ is a diffeomorphism, it follows that $\varphi \circ \lambda_{g^{-1}} \circ \psi_S$ is a chart on $B(s, N, g)$. Thus $\varphi_{s,N} \circ \sigma_S$ is a chart and it follows that σ_S is a local diffeomorphism. Again by (i) of Lemma VII.3.12 the analytic structure on S is the unique one so that σ_S is a local diffeomorphism and by (ii) the multiplication on S is analytic since it is on Σ. ∎

VII.3.14. **Corollary.** *Let S be a cancellative semigroup on a Euclidean manifold. Then S admits an analytic structure for which the multiplication is analytic.*

Proof. By Proposition VII.1.24 S satisfies condition (‡). By Proposition VII.1.22 $Q(U, V)$ is locally isomorphic to a Lie group, and by Proposition VII.3.1, $Q(S, V)$ is locally isomorphic to a Lie group. The corollary now follows from the preceding theorem. ∎

We consider now the extent to which the construction of Σ is functorial. Of course in this case one needs first of all to select the group G in a functorial way. If S is locally connected, then $Q(U, V)$ will be locally connected (since the mapping from $U \times U$ to $Q(U, V)$ is open by Lemma VII.1.19), and hence if G is a group locally isomorphic to $Q(U, V)$, it will be locally connected. Thus the identity component will be open in G and hence also locally isomorphic to $Q(U, V)$. Then the universal covering group of G, which we denote G^S, will be uniquely determined in terms of S. We let $\Sigma(S)$ denote the Σ corresponding to the choice $G = G^S$.

VII.3.15. **Proposition.** *Let $f: S \to T$ be a continuous homomorphism of locally connected cancellative semigroups satisfying (‡), and suppose that there exist simply connected groups G^S and G^T such that $Q(S, V)$ (resp. $Q(T, W)$) is locally isomorphic to G^S (resp. G^T). Then there exist unique continuous homomorphisms $G(f): G^S \to G^T$ and $\Sigma(f): \Sigma(S) \to \Sigma(T)$ such that the following diagram commutes:*

$$
\begin{array}{ccc}
G^S & \xrightarrow{\;\;G(f)\;\;} & G^T \\[4pt]
\psi \uparrow & & \uparrow \psi \\[4pt]
\Sigma(S) & \xrightarrow{\;\;\Sigma(f)\;\;} & \Sigma(T) \\[4pt]
\sigma \downarrow & & \downarrow \sigma \\[4pt]
S & \xrightarrow{\;\;f\;\;} & T.
\end{array}
$$

      If G^S and G^T are Lie groups, then f is analytic with respect to the induced analytic structures on S and T.

Proof.   Let $a, b \in S$. By Corollary VII.1.17 there exists a suitable pair (U_T, V_T) in T such that $f(a) \in U_T$ and $f(b) \in V_T$. Again by VII.1.17 pick a suitable pair (U, V) in S such that $a \in U$ and $b \in V$. Let $V_S = V \cap f^{-1}(V_T)$ and pick U_1 containing a such that U_1 right reverses in V_S (Proposition VII.1.16). Let $U_S = U_1 \cap f^{-1}U_T$. Then (U_S, V_S) is a suitable pair such that $f(U_S) \subseteq U_T$ and $f(V_S) \subseteq V_T$. It is then immediate that for $a, b, c, d \in U$, $(a, b) \equiv (c, d)$ implies $(f(a), f(b)) \equiv (f(c), f(d))$. Hence there is induced a continuous local homomorphism from $Q(U_S, V_S)$ to $Q(U_T, V_T)$. This gives rise to a continuous local homomorphism from G^S to G^T, since these are locally isomorphic to $Q(U_S, V_S)$ and $Q(U_T, V_T)$ respectively. Since G^S is a universal covering group, the local homomorphism extends to a continuous homomorphism $G(f) \colon G^S \to G^T$. We define $\Sigma(f)(s, g) = (f(s), G(f)(g))$. Clearly $f \circ \sigma_S = \sigma_T \circ \Sigma(f)$ and $G(f) \circ \psi_S = \psi_T \circ \Sigma(f)$. We establish continuity. Let $B(f(s), N, G(f)(g))$ contain $\Sigma(f)(s, g)$. Since $f(s\beta(s, s')) = f(s') = f(s)\beta(f(s), f(s'))$ and $\beta(\{s\} \times U_S) = s^{\nabla}(U_S)$ is open in $Q(U, V)$, there exists a neighborhood M of the identity in G^S such that $f(sh) = f(s)G(f)(h)$ for all $h \in M$ by the definition of $G(f)$. We assume that M is also chosen small enough so that $G(f)(M) \subseteq N$. Then for $(sh, gh) \in B(s, M, g)$ we have

$$\Sigma(f)(sh, gh) = (f(sh), G(f)(gh))$$
$$= (f(s)G(f)(h), G(f)(g)G(f)(h)) \in B(f(s), N, G(f)(g)).$$

Hence $\Sigma(f)$ is continuous.

      If G^S and G^T are Lie groups, then any continuous homomorphism is analytic, and hence $G(f)$ is analytic. It then follows from Lemma VII.3.12.ii that $\Sigma_S(f)$ and then f are analytic.   ■

VII.3.16.   **Remark.**   Note that in the proof of the preceding result we actually verified that a continuous homomorphism preserves the local right action in the sense that given $s \in S$, there exists N open in G^S such that $f(sg) = f(s)G(f)(g)$ for $g \in N$. This is actually a local feature and depends only on the local action of the local group, not on the existence of G^S. Note also that this preservation extends to a neighborhood of s by means of the observation that

$$f((sg)h) = f(s(gh)) = f(s)G(f)(gh) = f(s)(G(f)(g)G(f)(h))$$
$$= (f(s)G(f)(g))G(f)(h) = f(sg)G(f)(h).$$

Connected semigroup coverings

      In this subsection let S be a cancellative topological semigroup which satisfies (‡) and, in addition, is connected and locally connected. We assume also that there exists a universal covering group G^S which is locally isomorphic to $Q(S, V)$. Then $\Sigma = S \times G^S$ again denotes the double cover from the previous subsection.

If $A \subseteq \Sigma$ is connected, $\kappa(A)$ is the connected component of A ($\kappa(x)$ denotes the component of the single point x). Since the product of connected sets is connected, the components give rise to a congruence defined by $x\kappa y$ if and only if $\kappa(x) = \kappa(y)$. The factor semigroup Σ/κ is denoted $\Xi = \Xi(\Sigma)$ and $\kappa: \Sigma \to \Xi$ is the natural homomorphism. We come now to a crucial lemma.

VII.3.17. **Lemma.** *The semigroup Ξ consisting of the connected components of Σ is a group. Therefore there exists a uniquely determined component C of Σ which is a subsemigroup.*

Proof. Let $\lambda, \mu \in \Xi$. It must be shown that there exist components $\nu, \xi \in \Xi$ such that $\lambda\nu = \mu = \xi\lambda$. Choose $(s, g) \in \Sigma$ with $\kappa(s, g) = \lambda$. Since S is connected and locally connected, it is a standard theorem of covering spaces that each component of Σ maps onto S. In particular, $s^2 \in \sigma_S(\mu)$, i.e., there exists an element $h \in G^S$ such that $\kappa(s^2, h) = \mu$. We then define $\nu = \kappa(s, g^{-1}h)$ and $\xi = (s, hg^{-1})$. Then

$$\lambda\nu = \kappa(s, g)\kappa(s, g^{-1}h) = \kappa(s^2, gg^{-1}h) =$$
$$= \kappa(s^2, hgg^{-1}) = \kappa(s, hg^{-1})\kappa(s, g) = \xi\lambda.$$

Since equations can be solved on both sides, Ξ is a group. Then the identity element C of Ξ satisfies $C^2 \subseteq C$ and it is the only component with this property. ∎

As we saw in the previous section, the universal covering group G^S exists since S is locally connected and we have assumed the existence of an appropriate G. Recall that $\Sigma(S)$ denotes the double cover constructed from G^S.

VII.3.18. **Definition.** The unique connected component of $\Sigma(S)$ which is a subsemigroup is denoted by $C(S)$. The restrictions of σ_S and ψ_S to $C(S)$ are again denoted by σ_S and ψ_S and we depend on context to indicate which one is being considered.

There now results a connected version of Proposition VII.3.15.

VII.3.19. **Theorem.** (i) *Let S be a cancellative topological semigroup satisfying ($\ddagger$) on a connected, locally connected space, and let G^S be a simply connected topological group which is locally isomorphic to $Q(S, V)$. Then there exist a connected semigroup $C(S)$ and local isomorphisms $\sigma_S: C(S) \to S$ and $\psi_S: C(S) \to G^S$, where σ_S is also a covering projection. If G^S is a Lie group, then there exist unique analytic structures on S and $C(S)$ such that ψ_S and σ_S are local analytic isomorphisms.*

(ii) *Let $f: S \to T$ be a continuous homomorphism, where S and T both satisfy the hypotheses of part (i). Then there exists a continuous semigroup homomorphism $C(f): C(S) \to C(T)$ such that the following diagram commutes:*

$$
\begin{array}{ccc}
G^S & \xrightarrow{\;\;G(f)\;\;} & G^T \\
\psi \big\uparrow & & \big\uparrow \psi \\
C(S) & \xrightarrow{\;\;C(f)\;\;} & C(T) \\
\sigma \big\downarrow & & \big\downarrow \sigma \\
S & \xrightarrow{\;\;\;f\;\;\;} & T.
\end{array}
$$

If G^S is a Lie group, then $C(f)$ is an analytic mapping.

Proof.   Since S is connected and locally connected and since $\sigma_S: \Sigma(S) \to S$ is a covering projection, it follows that the restriction of σ_S to any component is still a covering projection and maps onto S. Let $C(S)$ be the component guaranteed by Lemma VII.3.17 which is a subsemigroup. Since S is locally connected, $C(S)$ is open in $\Sigma(S)$. Hence the restriction of ψ_S to $C(S)$ (which we again denote ψ_S) is still a local homeomorphism. The rest of part (i) follows from Theorems VII.3.11 and VII.3.13.

Part (ii) follows from Proposition VII.3.15 and the fact that $\Sigma(f)$ must carry $C(S)$ to a connected subsemigroup of $\Sigma(T)$, and hence must carry it into $C(T)$.   ∎

Note that while $\psi_S: \Sigma(S) \to G^S$ is surjective, this will not be the case in general for $\psi_S: C(S) \to G^S$.

It is not clear a priori whether $C(S)$ is a proper subset of $\Sigma(S)$. This is shown to be the case by a calculation involving fibers and weights.

VII.3.20.   **Lemma.**   *Let X be a connected, locally connected space with a countable base and let $p: \widetilde{X} \to X$ be a covering projection, where $\widetilde{X}$ is connected. Then $\widetilde{X}$ has a countable base.*

Proof.   The proof involves picking a countable base $\mathcal{U}$ of open sets in X which are evenly covered, writing the inverse image of each such set as a disjoint union of open sets each of which map homeomorphically onto the original open set, and using connectedness arguments to show the collection of open sets in $\widetilde{X}$ obtained in this way forms a countable basis. We refer the reader to Proposition 1.3 of [HW88] or Lemma 3 of Chapter III, §IX of [Che46] for details.   ∎

We have previously defined a local right action of G^S on Σ. We now define a left action by $g \cdot (s, h) = (s, gh)$. This is clearly a left action, and it is immediate from the definition of the topology on Σ that $(s, h) \mapsto (s, gh)$ is a homeomorphism that carries each fiber onto itself. It follows that each $g \in G^S$ acts as a deck transformation on Σ. We consider the stabilizer group G_C of C consisting of all $h \in G^S$ such that $h \cdot C(S) \subseteq C(S)$. Since $\big(h, (s, g)\big) \mapsto (s, hg)$ defines a left action of G^S on Σ, it follows that G_C is a subgroup. Since G_C acts transitively on the fibers of the connected covering space $\sigma_S: C(S) \to S$, it follows by standard results of covering spaces that it is is group of deck transfomations for this covering space. Furthermore, since for $(s, g) \in C(S)$, the collection $\{B(s, N, hg): h \in G_C\}$ is a collection of pairwise disjoint open sets in $C(S)$ and $C(S)$ has a countable base by the preceding lemma, it follows that G_C is countable.

VII.3.21.   **Proposition.**   *Define $\varphi: G^S \to \Xi$ by $\varphi(g) = g \cdot C$, where $C = C(S)$. Then φ is a surjective homomorphism with kernel the countable central subgroup G_C.*

Proof.   Let $g, h \in G$ and define $\lambda = h \cdot C$, $\mu = C = e_\Xi$. Then

$$\varphi(g)\varphi(h) = (g \cdot \mu)\lambda = g \cdot (\mu\lambda) = g \cdot \lambda = g \cdot (h \cdot C) = (gh) \cdot C = \varphi(gh),$$

i.e., φ is a homomorphism.

Let $(s, h) \in C$. If C' is any other component of Σ, then $(s, k) \in C'$ for some $k \in G^S$ (since $\sigma_S: C' \to S$ is surjective). Then $(s, k) = (kh^{-1}) \cdot (s, k)$, so $C' = (kh^{-1}) \cdot C$, i.e., $C' = \varphi(kh^{-1})$. This proves that φ is surjective.

It follows from the definition of G_C that it is the kernel of φ. Hence it is a normal subgroup. We remarked earlier that it is countable as a consequence of Lemma VII.3.20. Since for $h \in G_C$, $\{ghg^{-1}: g \in G^S\}$ is a connected subset of the countable completely regular space G_C, it follows that this set is a singleton, i.e., G_C is central. ∎

At this point one might conjecture that the central subgroup G_C of G is closed. This conjecture appears plausible, but remains unproved in general. However, it is known to be valid for several important cases.

VII.3.22.   **Proposition.**   *We have a commuting diagram*

$$
\begin{array}{ccc}
C(S) & \xrightarrow{\;\psi\;} & G^S \\
\sigma \downarrow & & \downarrow \nu \\
S & \xrightarrow{\;\gamma\;} & G^S/G_C,
\end{array}
$$

and the homomorphism γ_S is a local homeomorphism into the topological group G^S/G_C in the case that G_C is discrete in G^S. This happens if S is simply connected or if G^S is a Lie group such that $\mathbf{L}(G^S)$ has trivial center.

Proof.   If σ_S identifies two points of $C(S)$, then they are in the same fiber. Hence their G^S-coordinates differ by a left translate of some member of G_C, i.e., they are in the same coset of G_C. It follows that there is induced a homomorphism $\gamma_S: S \to G^S/G_C$ making the diagram commute. If G_C is discrete, then it is closed, and hence G^S/G_C, is a topological group. Since the mappings $G^S \to G^S/G_C$, $\psi_S: C(S) \to G^S$ and $\sigma_S: C(S) \to S$ are all local homeomorphisms, it follows that γ_S is also.

If S is simply connected, then σ_S is a homeomorphism. Hence $G_C = \{e\}$, and $\gamma_S = \psi_S$. If G^S is a Lie group such that $\mathbf{L}(G^S)$ has no center, then any central subgroup of G^S must be discrete (since the adjoint representation will be a local homeomorphism whose kernel is the center). Hence in this case G_C will be discrete. ∎

Problem. Under which conditions is G_C closed?

The free group on S

There is an inclusion functor from the category of groups and homomorphisms to the category of semigroups and homomorphisms. The adjoint to this inclusion functor assigns to a semigroup S a group $G(S)$ and a homomorphism $\gamma_S: S \to G(S)$ with the property that given any group H and any homomorphism $\alpha: S \to H$, there exists a unique homomorphism $\alpha': G(S) \to H$ such that $\alpha' \circ \gamma_S = \alpha$:

$$
\begin{array}{ccc}
S & \xrightarrow{\ \ \gamma\ \ } & G(S) \\
{\scriptstyle \mathrm{id}_S}\downarrow & & \downarrow{\scriptstyle \alpha'} \\
S & \xrightarrow{\ \ \alpha\ \ } & H.
\end{array}
$$

The homomorphism $\gamma_S \colon S \to G(S)$ is called the *free group on* S. It is unique up to an isomorphism which commutes with the mappings from S.

Alternately $G(S)$ can be constructed by first forming the free group $F(S)$ with alphabet S and dividing out the smallest congruence $\cong$ so that the natural inclusion $S \to G(S) = F(S)/\cong$ is a homomorphism. We refer to Chapter 12 of [CP67] for a detailed treatment of $G(S)$.

We need at this point some additional facts about local right actions.

VII.3.23.   Definition.   Let X be a space and suppose we are given a local right action of a local group G on X. For $x \in X$, we define the G-*orbit* of x, denoted $\mathrm{Orb}(x)$ to be the set of all points $y \in X$ such that there exist $g_1, \dots, g_n \in G$ and $x_0, \dots, x_n \in X$ with $x_0 = x$, $x_i = x_{i-1} g_i$ for $1 \le i \le n$ (in particular, $x_{i-1} g_i$ is defined), and $x_n = y$. We abbreviate the last conditions by writing $y = x g_1 \cdots g_n$.

VII.3.24.   Lemma.   *Suppose that X is a space and that there is given a locally simply transitive local right action of a local group G on S. Then for $x \in X$, the orbit of x is both open and closed and hence contains the component of x.*

Proof.   Let Q denote the orbit of x in X. Let $y \in Q$. Then $y = x g_1 \cdots g_n$ for some $g_1, \dots, g_n \in G$. Since the action is locally simply transitive, there exists a neighborhood N of e such that yN is defined and a neighborhood of y. Then for any $z \in yN$, we have $z = x g_1 \cdots g_n h$ for some $h \in N$. Thus $yN \subseteq Q$. This shows that Q is open.

If $y \in \overline{Q}$, choose a symmetric neighborhood N of e and a neighborhood U of y so that UN is defined and yN is a neighborhood of y. Then $yN \cap U \cap Q \ne \varnothing$, so there exists $z = yh^{-1} \in Q \cap U$ for some $h \in N$. Then $y = zh = x g_1 \cdots g_n h$ for some $g_1, \dots, g_n \in G$ since $z \in Q$. Thus $y \in Q$, and we conclude that Q is closed. Since Q is open and closed, we conclude that Q contains the component of x.   ∎

VII.3.25.   Proposition.   *Let S be a connected, locally connected cancellative topological semigroup satisfying ($\ddagger$) and suppose there exists a universal covering group G^S locally isomorphic to $Q(S, V)$. Then $\gamma_S \colon S \to G^S/G_C$ is the free group on S. If there exists a homomorphism from S into some group which is one-to-one on some open set in S, then G_C is discrete in G^S and γ_S is a local homeomorphism.*

Proof.   Let $\theta \colon S \to H$ be a homomorphism into a group H. We wish to construct a homomorphism $\theta' \colon G^S/G_C \to H$ so that $\theta' \circ \gamma_S = \theta$. This will establish that $\gamma_S \colon S \to G^S/G_C$ is the free semigroup on S since it shows that the desired universal property is satisfied. Note that the uniqueness of θ' follows from the fact that $\gamma_S(S)$ generates $G(S)$ by Proposition VII.3.22.

Find a suitable pair (U, V) according to Theorem VII.3.1 and the proof of Corollary VII.3.2 so that U left reverses in S. Then the proof of Lemma VII.1.37, appropriately modified for $_V \equiv$, applies to $\theta \colon S \to H$ to yield a local homomorphism $\delta \colon Q(U, V) \to H$ such that $\delta\big(\beta(a, b)\big) = \theta(a)^{-1}\theta(b)$. When δ is composed with the local isomorphism from G^S, one obtains a local homomorphism from G^S to H. Since G^S is a universal covering group, this local homomorphism extends to a homomorphism $\Delta\colon G^S \to H$.

By Corollary VII.3.7 the local isomorphism between G^S and $Q(S,V)$ defines a locally simply transitive local right action of G^S on S. Let $s \in S$ and $g \in G$ such that sg is defined. Then there exists a $\beta(a,b) \in Q(S,V)$ which corresponds to g under the local isomorphism and $u, v \in V$ such that $au = bv$ and $su = (sg)v$ (see Theorem VII.3.6). Then

$$\theta(s)\Delta(g) = \theta(s)\delta\big(\beta(a,b)\big) = \theta(s)\theta(a)^{-1}\theta(b)$$
$$= \theta(s)\theta(u)\theta(v)^{-1} = \theta(s)\theta(s)^{-1}\theta(sg) = \theta(sg).$$

Thus the local action is preserved if we use the homomorphism Δ to define a right action of G^S on H (note the analogous situation in Remark VII.3.16).

If we consider additionally the right action of multiplication of G^S on itself, then there is induced a local right action of G^S on $S \times G$ which sends $\big((s,h),g\big)$ to (sg, hg). The topology τ on $\Sigma = S \times G^S$ is defined precisely in such a way so that this local right action is locally simply transitive (see the remarks preceding Lemma VII.3.8).

Suppose $g \in G_C$. Then by definition of G_C (see the remarks preceding Proposition VII.3.21) for $(s,h) \in C(S)$, we have $(s,gh) \in C(S)$. Since $C(S)$ is a component in Σ, it follows from Lemma VII.3.24 that there exist $g_1, \dots, g_n \in G$ such that $(s, gh) = (sg_1 \cdots g_n, hg_1 \cdots g_n)$. By Proposition VII.3.21 the group G_C is central, so $hg = gh = hg_1 \cdots g_n$. By cancellation $g = g_1 \cdots g_n$. Thus $\Delta(g) = \Delta(g_1) \cdots \Delta(g_n)$. We have also that $s = sg_1 \cdots g_n$ and hence $\theta(s) = \theta(sg_1 \cdots g_n)$. We have seen in an earlier paragraph that $\theta(sg') = \theta(s)\Delta(g')$ for $g' \in G$ such that sg' is defined, and by induction $\theta(sg_1 \cdots g_n) = \theta(s)\Delta(g_1) \cdots \Delta(g_n)$. Again by cancellation of $\theta(s)$,

$$e_H = \Delta(g_1) \cdots \Delta(g_n) = \Delta(g).$$

Since g was an arbitrary element of G_C, it follows that the kernel of Δ contains G_C. Hence there exists a mapping $\theta'\colon G^S/G_C \to H$ such that $\theta' \circ \nu = \Delta$, where ν is the natural homomorphism from G^S to G^S/G_C.

To complete the first part of the proof, we must show that $\theta' \circ \gamma_S = \theta$. We have the following diagram:

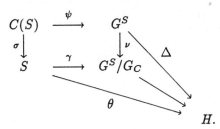

Let $s \in S$ and pick $(s,h) \in C(S)$ (which we can do since σ is surjective). Since $C(S)$ is a semigroup, $(s^2, h^2) \in C(S)$. Again by Lemma VII.3.24 there exist $g_1, \dots, g_n \in G$ such that $(s^2, h^2) = (sg_1 \cdots g_n, hg_1 \cdots g_n)$. Then, as we have argued previously,

$$\theta(s)\theta(s) = \theta(s^2) = \theta(s)\Delta(g_1) \cdots \Delta(g_n).$$

By cancellation, $\theta(s) = \Delta(g_1)\cdots\Delta(g_n)$. Similarly by working in the second coordinate we obtain $\Delta(h) = \Delta(g_1)\cdots\Delta(g_n)$. Thus

$$\theta(s) = \Delta(h) = \theta' \circ \nu(h) = \theta' \circ \nu \circ \psi(s, h)$$
$$= \theta' \circ \gamma_S \circ \sigma(s, h) = \theta' \circ \gamma_S(s).$$

Suppose now that the homomorphism $\theta: S \to H$ is injective on some open set U containing some $s \in S$. Pick an open set N containing the identity in G^S such that sN is defined and contained in U and $g \mapsto sg: N \to sN$ is a homeomorphism. For $e \neq g \in N$, we have seen in a previous paragraph that $\theta(sg) = \theta(s)\Delta(g)$ and by hypothesis $\theta(sg) \neq \theta(s)$. Hence it must be the case that $\Delta(g)$ is not the identity of H. Since we have already established that the kernel of Δ contains G_C, it follows that $g \notin G_C$. Thus $G_C \cap N = \{e\}$ and it follows that G_C is discrete. That γ_S is a local homeomorphism now follows from Proposition VII.3.22. ∎

VII.3.26.   Lemma.   *Suppose that S is a cancellative topological semigroup which contains an ideal I satisfying*

    (i) *There is a topological and algebraic embedding $\delta: I \to G$ onto an open subset of G;*

    (ii) *There exists $a \in I$ such that the mapping $s \mapsto as: S \to I$ is a homeomorphism onto an admissible subset of I.*

    Then δ extends to a topological and algebraic embedding $\gamma: S \to G$ such that $\gamma(S)$ is an admissible subset of G.

Proof.   Define $\gamma: S \to G$ by $\gamma(s) = \delta(a)^{-1}\delta(as)$. It is immediate that γ is an extension of δ since δ is a homomorphism on I. We can write γ as the composition of left translation by a followed by δ followed by left translation by $\delta(a)^{-1}$. Since each of these mappings is a homeomorphic embedding, it follows that the composition is also. Also aS is a subset of I with interior, so $\delta(aS)$ is has interior in the open set $\delta(I)$ and hence has interior in G. Thus $\gamma(aS)$, the left translate of $\delta(aS)$ by $\delta(a)^{-1}$, also has interior in G.

It remains to verify that γ is a homomorphism. Let $s, t \in S$. We have

$$\gamma(t)\delta(a) = \delta(a)^{-1}\delta(at)\delta(a) = \delta(a)^{-1}\delta(ata)$$
$$= \delta(a)^{-1}\delta(a)\delta(ta) = \delta(ta).$$

Thus $\gamma(s)\gamma(t)\delta(a) = \delta(a)^{-1}\delta(as)\delta(ta)$. On the other hand

$$\gamma(st)\delta(a) = \delta(a)^{-1}\delta(ast)\delta(a) = \delta(a)^{-1}\delta(asta) = \delta(a)^{-1}\delta(as)\delta(ta).$$

By cancellation $\gamma(s)\gamma(t) = \gamma(st)$. ∎

VII.3.27.   Theorem.   *Let S be a topological semigroup on a connected Euclidean manifold with generalized boundary and suppose that S can be algebraically embedded in a group. Then $G(S)$, the free group on S, admits the structure of a Lie group, and*

    (i) *if the boundary of S is empty, then $\gamma_S: S \to G(S)$ is an embedding onto an open subsemigroup of $G(S)$;*

 (ii) *if there exists a in the interior of S such that $s \mapsto as \colon S \to S$ is an embedding onto an admissible subset of S, then the homomorphism $\gamma \colon S \to G(S)$ is an embedding onto a subsemigroup which is an admissible subset of $G(S)$.*

Proof. Since S can be embedded in a group, it is cancellative. Let us suppose first that the boundary of S is empty. By Proposition VII.1.24 S satisfies ($\ddagger$). By Proposition VII.1.22 for a suitable pair (U, V) in S, we have that $Q(U, V)$ is locally isomorphic to a Lie group, and by Propositon VII.3.1 $Q(S, V)$ is locally isomorphic to a Lie group G^S, which we might as well take to be simply connected.

 Since S embeds in a group, by Proposition VII.3.25 the subgroup G_C is discrete. Hence $G(S) = G^S/G_C$ is a Lie group, since it is the image of G^S modulo a discrete subgroup. Also by Proposition VII.3.25 the mapping $\gamma \colon S \to G(S)$ is a local homeomorphism. By the universal property of $G(S)$, the mapping γ is injective if and only if S embeds injectively in some group. Thus γ is injective and hence a homeomorphism onto an open subset.

 In case (ii) let us consider the set I of interior points of S. By Proposition VII.1.24 and its dual I is an ideal of S. We can apply case (i) to I and conclude that $\gamma_I \colon I \to G(I)$ is a topological isomorphism onto an open subsemigroup of the Lie group $G(I)$. By Lemma VII.3.26 there exists $\gamma_S \colon S \to G(I)$ such that γ_S is a topological embedding and an algebraic isomorphism onto a subsemigroup of $G(I)$ which is an admissible subset and such that γ_S is an extension of γ_I.

 We show that $\gamma_S \colon S \to G(I)$ is the free group on S. Let $\theta \colon S \to H$ be a homomorphism into a group H. Then θ restricted to I is also a homomorphism, so there exists $\theta' \colon G(I) \to H$ such that $\theta' \circ \gamma_S(s) = \theta(s)$ for all $s \in I$. If $t \in S$, then for $s \in I$

$$\theta(s)\theta'\big(\gamma_S(t)\big) = \theta'\big(\gamma_S(s)\big)\theta'\big(\gamma_S(t)\big)$$
$$= \theta'\big(\gamma_S(st)\big) = \theta(st) = \theta(s)\theta(t).$$

By cancellation $\theta'\big(\gamma_S(t)\big) = \theta(t)$. This completes the proof.   ■

VII.3.28.   **Proposition.** *Let S be a connected subsemigroup of a connected Lie group G such that the interior of S in G is dense in S. Then $G(S)$ is the largest connected covering group of G in which S lifts.*

Proof. The interior of S is an ideal by Proposition V.0.15. Hence the hypotheses of part (ii) of Theorem VII.3.27 are satisfied since translations are homeomorphisms. Let $\pi \colon H \to G$ be any covering group of G in which S lifts (including the identity mapping on G), i.e., there exists a continuous isomorphism j from S into H such that $\pi\big(j(s)\big) = s$ for all $s \in S$. Then by the universal property of the Lie group $G(S)$ there exists $J \colon G(S) \to H$ such that $J \circ \gamma_S = j$. One verifies directly that j is continuous, one-to-one, and open on the interior of S since it has continuous inverse π. Thus J must be continuous, one-to-one, and open on $\gamma_S(\operatorname{int} S)$, which embeds as an open subset of $G(S)$ as in Theorem VII.3.27. It follows that J is continuous and a local homeomorphism since this is a true on an open set (the properties translate around). Hence the kernel subgroup is discrete, and thus J is a covering projection. We note that S does lift to $G(S)$ by the embedding γ_S. Finally observe that $\gamma_S(S)$ is contained in the identity component of $G(S)$ which will again satisfy the same universal properties that $G(S)$ satisfies. Since the free group on S is unique, it follows that $G(S)$ is the identity component, i.e., $G(S)$ is connected. This completes the proof.   ∎

VII.3.29.   Example. If $S = \mathrm{Sl}(2, \mathbb{R})^+$ (cf. Proposition V.4.19(i)), then the inclusion $i \colon S \to \mathrm{Sl}(2, \mathbb{R})$ is not the inclusion into the free group on S, the lifting $j \colon S \to \widetilde{\mathrm{Sl}}(2, \mathbb{R})$ however is. Typically, j does not factor through i.

Problems for Chapter VII

PVII.1.   Problem. *Consider the countable central subgroup G_C of* Proposition VII.3.21. *Is it always closed?*   ∎

PVII.2.   Problem. *Let G denote a simply connected Lie group and let S denote an infinitesimally generated subsemigroup. (Assume S to be strictly infinitesimally generated, or closed, or satisfying any other additional condition preventing undue pathology.) Is S also simply connected?*   ∎

PVII.3.   Problem. *Suppose that S is a finite-dimensional compact topological semigroup. Let T be the closure of the subsemigroup generated by all one-parameter subsemigroups of S starting at the identity. Is T locally cancellative in a neighborhood of the identity? If T is locally connected, is some neighborhood of the identity homeomorphic to an admissible subset of $\mathbb{R}^n$?*   ∎

PVII.4.   Problem. *Suppose that S is a cancellative topological semigroup on a connected locally euclidean space. Is there an open subsemigroup T of a Lie group and a continuous homomorphism $\varphi \colon S \to T$ which is a covering map (or a local homeomorphism)?*   ∎

PVII.5.   Problem. *Can the hypothesis of* Theorem VII.2.17 *be weakened to the assumption that multiplication is C_s^1 at the identity?*   ∎

Notes on Chapter VII

Section 1.   The theory of local quotients and local embeddings developed here elaborates the ideas of Houston [Hou73] and Brown and Houston [BH87] along the lines of [Law 89]. For global embeddings see also [McK70] and [BF71].

Section 2.   The first part of this section summarizes the calculus needed for treating manifolds with rather general boundaries. This calculus is needed for the study of differentiable semigroups on manifolds with generalized boundaries. As background reference we refer, for instance, to Bourbaki [Bou71], Lang [La72] or Dieudonné [Dieu71]. The idea to use this form of calculus for the foundation of a theory of differentiable semigroups is due to Graham [Gr79], [Gr83] and [Gr84]. Differentiable semigroups were also considered by Holmes (see [Hol74], [Hol88]) and Anderson (see [An88a], [An88b]). The approach taken here employs the local embedding machinery in Section 1 above as opposed to Graham's approach using invariant vector fields. This allows for considerable strengthening of previous results in several places. The presentation is from unpublished notes by Lawson. Theorem VII.2.43 is a definitive discussion of a differentiable semigroup with smooth boundary at the identity. Some aspects of this were treated by Holmes [Hol87].

Section 3. The study of cancellative semigroups on locally euclidean spaces was initiated in the dissertation of Houston [Hou73]. The basic results with a solution of Hilbert's Fifth Problem under these circumstances are published by Brown and Houston [BH87]. They rest on the Solution of Hilbert's Fifth Problem for local groups by Jacoby [Jac57]. A further analysis of the situation was given by Hofmann and Weiss [HW87]. The presentation here is a generalization and refinement of the material in these sources. All results in this chapter which are not in the indicated sources appear here for the first time and originate in unpublished notes by Lawson. This applies, in particular, to the material pertaining to the free group on a semigroup.

Appendix

1. The Campbell–Hausdorff formalism

In our study of subsemigroups of Lie groups we work extensively in the associated Lie algebras. Since the exponential mapping is locally one-to-one, the group multiplication can be pulled back locally to a neighborhood of 0 in the Lie algebra, and it is frequently useful to work with this induced multiplication. The Baker–Campbell–Hausdorff formula gives an explicit internal characterization of this multiplication in terms of the Lie algebra structure. In this section we review certain basic properties of this multiplication. These properties are rather standard, and proofs may be found in most texts on Lie groups (see e.g. [Bou75], [Hoch65]).

The Baker–Campbell–Hausdorff formula is derived by formally computing the terms of the power series for $\log(\exp x \cdot \exp y)$ in two non-commuting variables x and y and rearranging the answer in terms of Lie brackets (where $[x, y] = xy - yx$). More specifically, consider the algebra of all formal power series in two non-commuting variables x, y over the field of rational numbers. Define the exponential valuation by $|0| = 0$ and $|\alpha| = 2^{-i}$ if $\alpha \neq 0$ and i is the degree of the first non-zero term of α. This valuation may be viewed as a norm on the algebra of power series, and hence gives rise to a topology. Let J denote the ideal of all power series with constant term 0; elements of J can be substituted into other power series. The set $1 + J$ forms a group with respect to multiplication of power series. Let L denote the closure with respect to the topology arising from the exponential valuation of the Lie algebra generated by x and y, where the Lie bracket operation on the algebra of power series is given by $[\alpha, \beta] = \alpha\beta - \beta\alpha$. For two elements $\alpha, \beta \in J$, the element $\log(\exp \alpha \exp \beta) = \log(1 + (\exp \alpha \exp \beta - 1))$ is a well-defined power series over the rationals (where $\exp$ and $\log(1 + (\bullet))$ have their usual power series definitions).

We designate this result by $H(\alpha, \beta)$ or $\alpha * \beta$. It follows from this definition that J with respect to the operation $*$ is a group with identity element 0 and inverse $-\alpha$ of α. Further, $\exp: J \to 1 + J$ is an isomorphism to the multiplicative group $1 + J$ with inverse $\log$. Verifications of the preceding assertions are relatively straightforward. The computation of the series $H(x, y)$ in terms of Lie brackets alone is considerably more difficult. In the algebra of formal power series, we have

$$\exp x \cdot \exp y = 1 + W \quad \text{where} \quad W = \sum_{r+s \geq 1} \frac{x^r}{r!} \frac{y^s}{s!}.$$

The Baker–Campbell–Hausdorff series H is defined by

$$H = H(x, y) = \log(\exp x \cdot \exp y) = \sum_{n > 0} H_n = \sum_{\substack{r,s \geq 0 \\ r+s > 0}} H_{r,s}$$

where

$$H_n = \sum_{\substack{r+s=n \\ r,s \geq 0}} H_{r,s} \; ; \quad H_{r,s} = \sum_{m > 0} \frac{(-1)^{m-1}}{m} \sum_{\substack{r_1 + \cdots + r_m = r \\ s_1 + \cdots + s_m = s \\ r_i + s_i \geq 1}} \left(\prod_{i=1}^{m} \frac{x^{r_i}}{r_i!} \frac{y^{s_i}}{s_i!} \right)$$

Upon rearrangement of the terms with respect to Lie brackets, one obtains

A.1.1.   **Theorem.**   (Dynkin) $H_{r,s} = H'_{r,s} + H''_{r,s}$ *where*

$$H'_{r,s} = \frac{1}{r+s} \sum_{m>0} \frac{(-1)^{m-1}}{m} \cdot \sum_{\substack{r_1 + \cdots + r_m = r \\ s_1 + \cdots + s_{m-1} = s-1 \\ r_i + s_i \geq 1}} \left(\prod_{i=1}^{m-1} \frac{(\operatorname{ad} x)^{r_i}}{r_i!} \frac{(\operatorname{ad} y)^{s_i}}{s_i!} \right) \frac{(\operatorname{ad} x)^{r_m}}{r_m!} (y)$$

$$H''_{r,s} = \frac{1}{r+s} \sum_{m>0} \frac{(-1)^{m-1}}{m} \cdot \sum_{\substack{r_1 + \cdots + r_{m-1} = r-1 \\ s_1 + \cdots + s_{m-1} = s \\ r_i + s_i \geq 1}} \left(\prod_{i=1}^{m-1} \frac{(\operatorname{ad} x)^{r_i}}{r_i!} \frac{(\operatorname{ad} y)^{s_i}}{s_i!} \right) (x)$$

where $(\operatorname{ad} a)(b) = [a, b] = ab - ba$. *In particular,* $H(x, y) \in L$.   ∎

Define the power series η in two commuting variables over $\mathbb{R}$ by

$$\eta(x, y) = -\log(2 - \exp(x + y)) = -\log(1 - (\exp(x + y) - 1))$$

$$= \sum_{m \geq 1} \frac{1}{m} (\exp(x + y) - 1)^m = \sum_{m \geq 1} \frac{1}{m} \sum_{\substack{r_1, \cdots, r_m \\ s_1, \cdots, s_m \\ r_i + s_i \geq 1}} \frac{x^{r_1}}{r_1!} \frac{y^{s_1}}{s_1!} \cdots \frac{x^{r_m}}{r_m!} \frac{y^{s_m}}{s_m!}$$

$$= \sum_{m \geq 1} \frac{1}{m} \sum_{\substack{r_1 + \cdots + r_m = r \\ s_1 + \cdots + s_m = s \\ r_i + s_i \geq 1}} \frac{1}{r_1! s_1! \ldots r_m! s_m!} x^r y^s = \sum_{r,s \geq 0} \eta_{r,s} x^r y^s$$

where

$$\eta_{r,s} = \sum_{m \geq 1} \frac{1}{m} \sum_{\substack{r_1 + \cdots + r_m = r \\ s_1 + \cdots + s_m = s \\ r_i + s_i \geq 1}} \frac{1}{r_1! \cdots r_m! s_1! \cdots s_m!}.$$

Note that all sums in $\eta_{r,s}$ are finite. If x and y are real numbers with $0 \leq x + y <$ log 2, then $0 \leq \exp(x + y) - 1 < 1$, and thus

$$\sum_{r,s \geq 0} \eta_{r,s} x^r y^s = -\log(2 - \exp(x + y)) < \infty.$$

A.1.2.   Definition.   A (*complex*) *Dynkin algebra* is a completely normable real (complex) Lie algebra $\mathfrak{g}$ for which the Lie product is continuous. (Note that all finite dimensional Lie algebras are Dynkin algebras.) For any norm $\| \cdot \|$ on $\mathfrak{g}$ compatible with the topology there exists an $M > 0$ such that $\| [x, y] \| \leq M \|x\| \|y\|$, where $[\bullet, \bullet]: L \times L \to L$ is the Lie product (namely, pick $M > (1/\varepsilon^2)$, where ε is chosen by continuity of the Lie bracket to satisfy $\| [x, y] \| \leq 1$ for $\|x\|, \|y\| \leq \varepsilon$). If one renorms by scalar multiplying the old norm by M to obtain a new norm, then one obtains another norm compatible with the topology which satisfies $\| [x, y] \| \leq \|x\| \|y\|$; such norms will be called *standard norms*. The vector space endomorphism $x \mapsto [y, x]: L \to L$ is denoted $\operatorname{ad} y$, and by the Jacobi identity $\operatorname{ad} y$ is a derivation; such derivations are called *inner derivations*).

We consider the convergence of the Baker–Campbell–Hausdorff series on the Dynkin algebra $\mathfrak{g}$. Applying repeatedly the inequality $\| [x, y] \| \leq M \|x\| \|y\|$ to the formulas of Theorem A.1.1, we obtain

$$\|H'_{r,s}\| \leq \frac{1}{r + s} \sum_{m > 0} \frac{1}{m} \sum_{\substack{r_1 + \cdots + r_m = r \\ s_1 + \cdots + s_{m-1} = s-1 \\ r_i + s_i \geq 1}} \frac{1}{r_1! \cdots r_m! s_1! \cdots s_{m-1}!} M^{r+s} \|x\|^r \|y\|^s$$

$$\|H''_{r,s}\| \leq \frac{1}{r + s} \sum_{m > 0} \frac{1}{m} \sum_{\substack{r_1 + \cdots + r_{m-1} = r-1 \\ s_1 + \cdots + s_{m-1} = s \\ r_i + s_i \geq 1}} \frac{1}{r_1! \cdots r_{m-1}! s_1! \cdots s_{m-1}!} M^{r+s} \|x\|^r \|y\|^s.$$

Adding (and noting the two summands do not overlap), we obtain

$$\|H_{r,s}\| \leq \frac{M^{r+s}}{r + s} \eta_{r,s} \|x\|^r \|y\|^s \leq \eta_{r,s} \|Mx\|^r \|My\|^s.$$

From the preceding paragraph, we obtain by summing over r and s

A.1.3.   Proposition.   *The Baker–Campbell–Hausdorff power series* $H(x, y)$ *converges for* $\|x\| + \|y\| < (\log 2)/M$.   ∎

A careful inspection of the Baker–Campbell–Hausdorff series and the preceding convergence argument yields additional useful information. Note first of all that $H_{0,s} = H_{r,0} = 0$ for $s \neq 1 \neq r$ since $(ad\,x)^{r-1}(x) = 0 = (ad\,y)^{s-1}(y)$ in this case. Thus the series is of the form

$$H(x,y) = x + y + \frac{1}{2}[x,y] + R(x,y)$$

where the remainder $R(x,y)$ consists of higher order terms involving mixed Lie brackets of degree at least 3. In the process for obtaining the inequality for $\|H_{r,s}\|$ for $r, s > 0$ and $r + s \geq 3$, if we include the next to last reduction, we have

$$\|H_{r,s}\| \leq \frac{M^{r+s-1}}{r+s} \eta_{r,s} \|x\|^{r-1} \|y\|^{s-1} \| [x,y] \| \leq \eta_{r,s} \|Mx\|^r \|My\|^s.$$

These observations lead to the following

A.1.4.   Corollary.   *Let B be a ball of radius $r < (\log 2)/2M$. Then there exists $K > 0$ such that for $x, y \in B$*

(A1) $x * y = x + y + \dfrac{1}{2}[x,y] + R(x,y)$ *where* $\|R(x,y)\| \leq K \| [x,y] \| (\|x\| + \|y\|)$.

There exists an open ball B' such that for $x, y \in B'$

(A2) $\|x * y - x - y\| \leq \| [x,y] \| \leq M \|x\| \|y\|,$

and $M = 1$ if the norm is a standard norm.

Proof.   Since the series $\sum \eta_{r,s} \|Mx\|^r \|My\|^s$ converges on the open ball of radius $(\log 2)/2M$ and dominates the series

$$\sum (M^{r+s-1}/r + s)\eta_{r,s} \|x\|^{r-1} \|y\|^{s-1} \| [x,y] \|,$$

the latter series also converges. For each term in this latter series with $r + s \geq 3$, we can factor out $\| [x,y] \|$ and either $\|x\|$ or $\|y\|$ (choose one arbitrarily if both are possible). Then the series remaining after this factorization converges for $\|x\| = r = \|y\|$ to some value K. Thus the sum of the terms with $r + s \geq 3$ is bounded by $K\| [x,y] \|(\|x\| + \|y\|)$. Since by the inequalities preceding the corollary the terms of this series dominate in norm the corresponding terms of $R(x,y)$ in the Baker–Campbell–Hausdorff series, we have the desired inequality, and formula (A1) obtains.

It follows from formula (A1) that

$$\|x * y - x - y\| \leq \frac{1}{2}\| [x,y] \| + K\| [x,y] \|(\|x\| + \|y\|).$$

If $\|x\| + \|y\|$ is chosen less than $1/(2K)$, then (A2) follows. If the norm is standard, then $M = 1$.   ∎

The basic properties of the Baker–Campbell–Hausdorff multiplication on the formal power series algebra can be directly carried over locally to those neighborhoods of 0 in a Dynkin algebra on which one has convergence. Thus the Baker–Campbell–Hausdorff multiplication is defined by $x * y = H(x,y)$ whenever $H(x,y)$ converges.

A.1.5.   **Proposition.**   *For all elements* $x, y, z \in \mathfrak{g}$ *satifying*

$$\|x\|, \|y\|, \|z\| < \frac{1}{3M} \log 2,$$

the following conditions hold:

   (i) $x * (y * z) = (x * y) * z$,

   (ii) $x * 0 = x = 0 * x$,

   (iii) $x * (-x) = 0 = (-x) * x$,

   (iv) $x * y = x + y$ *if* $[x, y] = 0$.

Thus the operation $*$ *gives rise to a local group on small neighborhoods of* 0.

Proof.   Note that (iv) follows immediately from the Baker–Campbell–Hausdorff formula, and that (ii) and (iii) follow from (iv). The Baker–Campbell–Hausdorff multiplication is assocative on J, the ideal of formal power series with 0 constant term, since J with this multiplication is isomorphic to $1 + J$ with the usual multiplication of power series (the isomorphism being the exponential mapping). This associativity property carries over to neighborhoods on which one has absolute convergence. We omit the details.   ∎

A.1.6.   **Definition.**   A neighborhood B of 0 in $\mathfrak{g}$ is called a *Baker–Campbell–Hausdorff neighborhood* (or *C-H-neighborhood*) if $x * y$ is defined and continuous on $B \times B$, all triple products are defined and associative, and B is symmetric (i.e. $B = -B$) and full (i.e. $rB \subseteq B$ for $0 \le r \le 1$).

The Baker–Campbell–Hausdorff series tells us how to obtain the $*$-multiplication from the Lie algebra operations. The inequalities of Corollary A.1.4 lead to formulas giving the Lie algebra operation in terms of the $*$-multiplication.

A.1.7.   **Proposition.**   *Suppose that* $x, y \in \mathfrak{g}$, $x = \lim_n n x_n$, $y = \lim_n n y_n$. *Then the following conclusions hold:*

(A3) $x + y = \lim\limits_{n} n(x_n * y_n)$ (Trotter Product Formula);

(A4) $[x, y] = \lim\limits_{n} n^2 (x_n * y_n * (-x_n) * (-y_n))$ (Commutator Formula);

(A5) $[x, y] = \lim\limits_{n} n^2 (x_n * y_n - y_n * x_n)$.

Proof.   By (A2) of Proposition A.1.4 we have

$$\|n(x_n * y_n) - n(x_n + y_n)\| \le Mn\|x_n\| \, \|y_n\| \to M\|x\| \cdot 0 = 0.$$

It follows that $\lim n(x_n * y_n) = \lim n(x_n + y_n) = x + y$.

We next observe that $\lim n^2 R(x_n, y_n) = 0$ since by Proposition A.1.4,

$$\|n^2 R(x_n, y_n)\| \le K\|n^2 [x_n, y_n]\|(\|x_n\| + \|y_n\|) = K\| [nx_n, ny_n] \|(\|x_n\| + \|y_n\|),$$

which converges to $K\| [x, y] \| \cdot 0 = 0$.

To establish (A5), we have that $\lim n^2 (x_n * y_n - y_n * x_n) = \lim(n^2 x_n + n^2 y_n + n^2(1/2)[x_n, y_n] + n^2 R(x_n, y_n) - n^2 y_n - n^2 x_n - n^2(1/2)[y_n, x_n] - n^2 R(y_n, x_n))$
$= \lim(1/2)[nx_n, ny_n] - (1/2)[ny_n, nx_n] = \lim[nx_n, ny_n] = [x, y]$.

Finally we observe from formula (A2) that

$$\|n^2 (x_n * y_n * (-x_n) * (-y_n)) - n^2(x_n * y_n - y_n * x_n)\| \le n^2 \| [x_n * y_n, -(y_n * x_n)] \|$$
$= \| [n(x_n * y_n), -n(y_n * x_n)] \| \to \| [x + y, -(y + x)] \|$, where (A3) is used at the end. But $[x + y, -(y + x)] = -[x + y, x + y] = 0$. Thus (A4) follows from (A5).   ∎

A.1.8.   Corollary.   *Let $B \subseteq \mathfrak{g}$ and $B' \subseteq \mathfrak{h}$ be C-H-neighborhoods, and let T be a continuous $*$-homomorphism from B into B'. Then T extends uniquely to a continuous Lie algebra homomorphism from $\mathfrak{g}$ into $\mathfrak{h}$. Conversely, if T is a continuous Lie algebra homomorphism from $\mathfrak{g}$ into $\mathfrak{h}$ with $T(B) \subseteq B'$, then T restricted to B is a continuous $*$-homomorphism into B'.*

Proof.   Extend T by the rule $T(x) = nT((1/n)x)$, where $(1/n)x \in B$. Since T is a $*$-homomorphism on B, it is locally linear there on lines through the origin (since on them the $*$-multiplication is just addition). It follows that the extension is well-defined and agrees with the original T on B. That T preserves addition and the Lie products follows from Proposition A.1.7. Since it is continuous at 0, it is continuous everywhere. Since a continuous additive homomorphism is a linear transformation, one direction follows. The converse follows directly from the definition of the C-H-multiplication $*$ and its convergence on B and B'.   ∎

The mapping $x \mapsto \operatorname{ad} x$ is a Lie algebra homomorphism from $\mathfrak{g}$ onto the Lie algebra of inner derivations of $\mathfrak{g}$ with kernel the center of $\mathfrak{g}$. Various power series mappings may be applied to the algebra of continuous linear transformations on $\mathfrak{g}$, in particular to the inner derivations. The first that we consider is the exponential mapping which sends T to $I + T + (1/2!)T^2 + (1/3!)T^3 + \dots$ ($\operatorname{ad} x$ to $I + \operatorname{ad} x + (1/2!)(\operatorname{ad} x)^2 + \dots$). For the case of $\operatorname{ad} x$, this is denoted by $e^{\operatorname{ad} x}$ or $\exp(\operatorname{ad} x)$. The following standard formulas relate $\exp(\operatorname{ad} x)$ to the C-H-multiplication. See Chapter II.2 for additional discussion.

A.1.9.   Proposition.   *Let $x, y \in B$, a C-H-neighborhood. Then*

$$(\text{A6}) \qquad x * y * (-x) = y + \operatorname{ad} x \; y + \frac{1}{2!}(\operatorname{ad} x)^2 y + \dots = \sum_{n=0}^{\infty} \frac{(\operatorname{ad} x)^n y}{n!} = e^{\operatorname{ad} x} y.$$

In the algebra of formal power series in one variable T over $\mathbb{Q}$, we define

$$f(T) = \frac{1 - e^{-T}}{T} = \sum_{n=0}^{\infty} \frac{(-1)^n}{(n+1)!} T^n$$

$$g(T) = \frac{1}{f(T)} = 1 + \frac{1}{2}T + \sum_{n=1}^{\infty} \frac{b_{2n}}{(2n)!} T^{2n},$$

where b_{2n} are the Bernoulli numbers (which may be computed by inverting the power series f).

In the power series of non-commuting variables u, v

$$-u * (u + v) = \big(f(\operatorname{ad} u)\big)(v) + R_1$$

$$u * v = u + \big(g(\operatorname{ad} u)\big)(v) + R_2,$$

where R_1 and R_2 denote the sums of the bihomogeneous terms whose degree in v is greater than 1.

We thus have for the C-H-neighborhoods B for $x \in B, y \in L$,

$$(\text{A7}) \qquad\qquad -x * (x + hy) = hf(\operatorname{ad} x)(y) + R_1(h),$$

$$(\text{A8}) \qquad\qquad x * (hy) = x + hg(\operatorname{ad} x)(y) + R_2(h),$$

where

$$\lim_{h \to 0} \frac{\|R_i(h)\|}{h} = 0 \qquad \text{for } i = 1, 2.$$

2. Compactly embedded subalgebras

Dense analytic subgroups

One of the main complications in the study of invariant cones is the fact that the group of inner automorphisms of a Lie algebra need not be closed. Even though we don't know if there occur Lie algebras with non-closed groups of inner automorphisms *and* invariant cones, we still have to provide the means to deal with non-closed groups of inner automorphisms. We start with some very general results concerning the closures of analytic subgroups in a connected Lie group.

A.2.1.   **Lemma.** *Let G be a connected Lie group and G' its commutator subgroup. If Z denotes the center of G and Z_0 the identity component of Z, then $G'Z_0$ is closed.*

Proof.   The subgroup $G'Z_0$ of $G'Z$ is pathwise connected, hence analytic, and thus carries a topology which makes it into a Lie group. We denote $G'Z_0$ with its Lie group topology by $(G'Z_0)_L$. Note that this topology may be finer than the topology induced by G. The identity homomorphism $\varphi \colon (G'Z_0)_L \to G'Z_0$ is continuous and bijective. All we have to show is that φ is open. To do that, using the open mapping theorem for locally compact groups, it suffices to show that $G'Z_0$ is locally compact. Note that the group Z_0 is closed in G, since Z is closed in G. Hence Z_0 is locally compact and by [HeRo63, Thm.5.25] we only have to show that the factor group $G'Z_0/Z_0$ is locally compact. Consider the morphism $\overline{\varphi} \colon (G'Z_0)_L/Z_0 \to G'Z/Z$ defined by $\overline{\varphi}(gZ_0) = gZ_0$ and the surjective homomorphism $\psi \colon G'Z_0/Z_0 \to G'Z/Z$ defined by $\psi(gZ_0) = gZ$. We claim that $G'Z/Z$ is a closed subgroup of G/Z hence a Lie group. In fact, the center Z of G is the kernel of the adjoint representation Ad, so the factor group G/Z has a faithful representation. Therefore [Hoch65, p.224] implies that the commutator group $(G/Z)' = G'Z/Z$ of G/Z is closed. But now the open mapping theorem implies that the composed map $\pi = \psi \circ \overline{\varphi} \colon (G'Z_0/Z_0)_L \to G'Z/Z$ is open. Moreover $\ker \psi = (G'Z_0 \cap Z)/Z_0 \subseteq Z/Z_0$ is discrete, so that ψ is a covering map. Therefore also $\overline{\varphi}$ is open, hence a homeomorphism. This shows that $G'Z_0/Z_0$ is locally compact and the proof is complete.

∎

A.2.2.   **Lemma.**   *Let G be a connected Lie group and G' its commutator subgroup. If C denotes the maximal compact connected central subgroup of G, then $G'C$ is closed.*

Proof.   Note first that the closure $\overline{G'C}$ of $G'C$ is contained in $G'Z_0$ by Lemma A.2.1. Therefore on the Lie algebra level we have $\mathbf{L}(G') + \mathbf{L}(C) = \mathbf{L}(G'C) \subseteq \mathbf{L}(G'Z_0) = \mathbf{L}(G') + \mathbf{L}(Z)$. Thus we have a vector space complement $V \subseteq \mathbf{L}(Z)$ of $\mathbf{L}(G') + \mathbf{L}(C)$ in $\mathbf{L}(G') + \mathbf{L}(Z)$ such that $\mathbf{L}(\overline{G'C}) = \mathbf{L}(G'C) \oplus V$. We have to show that $V = \{0\}$. Recall that Z_0 is closed and C is the maximal torus in Z_0. Therefore $V \cap \mathbf{L}(C) = \{0\}$ implies that $R \overset{\text{def}}{=} \exp V$ is closed, central and isomorphic to $\mathbb{R}^n$ with $n = \dim V$. We say that R is a central vector subgroup. If $n > 1$ we can factor an $n - 1$-dimensional vector subgroup of $\exp V$ and derive a contradiction in the factor group. Thus we may assume from now on that $\dim V = 1$. Note that $\overline{G'C} = G'CR$ because of $\mathbf{L}(\overline{G'C}) = \mathbf{L}(G'C) \oplus V$. Let $(G'C)_L$ denote the group $G'C$ together with its intrinsic Lie group topology, then the morphism $\pi : (G'C)_L \to \overline{G'C}/R$ of Lie groups, defined by $\pi(g) = gR$ induces an isomorphism on the Lie algebra level. Since $(G'C)_L$ is connected this shows that π is surjective and has discrete kernel. A discrete normal subgroup in a Lie group is a finitely generated abelian group. Therefore $\ker \pi = G'C \cap R \subseteq R$ is then finitely generated and since R is isomorphic to $\mathbb{R}$, it is even cyclic. In particular $\ker \pi$ is closed in R and hence also in $\overline{G'C}$. Now the open mapping theorem shows that π is open. What we really want to show is that $G'C$ is locally compact, hence closed which implies that $V = \{0\}$ contradicting our assumptions. To show that $G'C$ is locally compact it already suffices to show that $G'C/(G'C \cap R)$ is locally compact, since $G'C \cap R$ is locally compact. We use the map π to show that the continuous homomorphism $\psi : G'C/(G'C \cap R) \to \overline{G'C}/R$ defined by $\psi\big(g(G'C \cap R)\big) = gR$ is a homeomorphism. For this purpose we note that $\pi = \psi \circ \overline{\varphi}$, where $\overline{\varphi}$ is the obvious homomorphism from $(G'C)_L to G'C/(G'C \cap R)$. Moreover ψ is open and surjective since π is open and surjective and $\overline{\varphi}$ is continuous. But obviously the kernel of ψ is $(G'C \cap R)/(G'C \cap R)$ which means that ψ is a homeomorphism. This shows that $G'C/(G'C \cap R)$ is locally compact, which is what we wanted to show.   ■

It is well known that a dense analytic subgroup A of a connected Lie group G is normal. In fact one even knows that the commutator subgroups A' and G' are equal (see [Hoch65, p.190]). Also one knows (loc.cit.) that there is a closed abelian subgroup T of G such that $G = AT$. However, we need a sharper result:

A.2.3.   **Theorem.**   *Let G be a connected Lie group and T a maximal torus in G. If A is a dense analytic subgroup of G, then we have $G = AT$. Moreover the maximal tori in G are all conjugate under the elements of A.*

Proof.   We show first that it suffices to show that $G = AT$ for one specific maximal torus in G. In fact the maximal tori in G are conjugate under G (see [Hoch65]) and if S is a second maximal torus in G, then we find an element $g = at$ with $a \in A, t \in T$ and $gTg^{-1} = S$. But then $aTa^{-1} = atTt^{-1}a^{-1} = gTg^{-1} = S$. Therefore S is conjugate to T under A and $AS = AaTa^{-1} = ATa^{-1} = Ga^{-1} = G$.

As in Lemma A.2.2 let C be the maximal compact connected subgroup of Z then C is contained in each maximal torus of G. Therefore it suffices to prove the claim for AC instead of A, that is, to assume that C is contained in A. If that is the case we may consider G/C; for a proof of the theorem it then suffices to show that $G/C = (A/C)(T/C)$. It is therefore no loss of generality if we henceforth assume $C = \{\mathbf{1}\}$.

As a first step we prove the claim for the special case that G' is semisimple. In that case the Lie algebra $\mathbf{L}(G)$ is reductive and the radical R of G is the identity component Z_0 of the center Z of G. We claim that we already have $A = G$.

To prove this claim, we recall that $G' = A' \subseteq A$. We note first that $\mathbf{L}(G') = \mathbf{L}(A') \subseteq \mathbf{L}(A) \subseteq \mathbf{L}(G) = \mathbf{L}(G') \oplus \mathbf{L}(R)$. Hence we can find a vector space complement $V \subseteq \mathbf{L}(R)$ for $\mathbf{L}(G')$ in $\mathbf{L}(A)$ such that

$$(\dagger) \qquad \mathbf{L}(A) = \mathbf{L}(G') \oplus V.$$

Since $C = \{\mathbf{1}\}$, the radical R is a vector group. Then $E \stackrel{\mathrm{def}}{=} \exp V \subseteq A$ is a central vector subgroup of G and is, in particular, closed and normal. The factor group A/E is a dense analytic subgroup of G/E. Also $(G/E)' = (G'E)/E$ is semisimple as the homomorphic image of a semisimple Lie group. If we can show $A/E = G/E$ then $A = G$ follows. It is, therefore, no loss of generality to assume that $E = \{\mathbf{1}\}$ and $V = \{0\}$. Then $\mathbf{L}(A) = \mathbf{L}(G')$ by $(\dagger)$, and thus $A = G'$. Since G' is closed by Lemma A.2.2 and A is dense in G we conclude $A = G$. This finishes the proof of the claim in the special case.

Now we prove the claim in the general case. As before we know from Lemma A.2.2 that G' is closed in G. Let R be the radical of $G' = A' \subseteq A$. Since the radical is characteristic subgroup we find that R is a solvable connected normal subgroup of G and hence contained in the radical $\mathrm{rad}(G)$ of G. Moreover R is closed in G so that we may consider the factor group G/R. The commutator of this factor group is $(G/R)' = G'R/R = G'/R$. But this means that $(G/R)'$ is semisimple so that we may apply the special case to the groups G/R and AR/R. Hence if K is a Lie subgroup of G such that $R \subseteq K$ and K/R is a maximal torus in G/R then we have $G/R = (AR/R)(K/R)$ and therefore also $G = AK$. Now K is a Lie group with a solvable closed normal subgroup R such that K/R is compact. But then the results of [HM63, p.31] show that there is a maximal torus T_1 in K such that $K = RT_1$. The torus T_1 is contained in *some* maximal torus T of G and then we calculate $G = AK = ART_1 = AT_1 \subseteq AT$ hence $AT = G$ which concludes the proof. ∎

Since every maximal torus in a connected Lie group is contained in a compact subgroup and since all maximal compact subgroups are conjugate, the following is then an immediate consequence of Theorem A.2.3.

A.2.4.   Corollary.   *Let G be a connected Lie group and K a maximal compact subgroup of G. If A is a dense analytic subgroup of G then we have $G = AK$. Moreover all maximal compact subgroups of G are conjugate under elements of A.* ∎

We will give some further consequences of Theorem A.2.3, but first we need to establish two lemmas.

A.2.5.   Lemma.   *Let G be a connected Lie group and T a maximal torus in G. Let N be a closed normal subgroup of G containing the commutator subgroup G'. Denote by R_N the radical of N and assume that the center $Z(G/R_N)$ of N/R_N is finite. Then the group TN/N is the maximal torus of the abelian Lie group G/N.*

Proof. Let C be that closed subgroup of G containing N for which C/N is *the* maximal torus of G/N. If R is the radical of G and S a Levi complement, then we have $S \subseteq G' \subseteq N$ whence $G = NR$ and $G/N = R/(N \cap R)$ as well as $C/N = (C \cap R)/(N \cap R)$. Since the radical R_N of N is a characteristic subgroup of N, we know that R_N is a normal solvable subgroup of G and hence contained in R. Thus $R_N \subseteq N \cap R$. Since S is also a Levi complement for R_N in N we have $N = SR_N$. If $x \in N \cap R$, then $x = sr$ with $s \in S$ and $r \in R$. But then $s = xr^{-1} \in R \cap S \subseteq Z(S)$ where $Z(S)$ is the center of S. This shows that $N \cap R \subseteq R_N Z(S)$. But then $(N \cap R)/R_N \subseteq (R_N Z(S))/R_N \subseteq Z(N/R_N)$, whence $(N \cap R)/R_N$ is finite. This implies that $(C \cap R)/R_N$ is compact since $((C \cap R)/R_N)/((N \cap R)/R_N)$ is isomorphic as a Lie group to $((C \cap R)N)/N$ which in turn we have identified as a torus. Thus there is a compact subgroup K of $C \cap R$ such that $C \cap R = KR_N$ (see [Hoch65, p.186]). Now let T_R be a maximal compact subgroup of R containing K, then T_R is a torus since R is connected and solvable. Moreover we find $(T_R(N \cap R))/(N \cap R) = (T_R N)/N$ is compact. But C/N is the maximal torus in G/N, whence $(T_R N)/N \subseteq (C/N)$. This shows $T_R N \subseteq C$. But we also have $C = (N \cap R)N = KR_N N \subseteq T_R N$ so that $C = T_R N$. Now consider the maximal torus T of G. The conjugacy of maximal tori shows that we can find an element $g \in G$ such that $T_R \subseteq g^{-1}Tg$. Hence we have $C = gCg^{-1} = gT_R g^{-1}gNg^{-1} \subseteq TN$. But $(TN)/N$ is compact and hence contained in C/N. Thus $C = TN$ which is what we had to show. ∎

A.2.6. **Lemma.** *Consider a connected Lie group G for which the center $Z(\overline{G'}/\mathrm{rad}(\overline{G'}))$ of $\overline{G'}/\mathrm{rad}(\overline{G'})$ is finite. Let K be any compact connected subgroup of G containing a maximal torus of G. Moreover suppose that A is a dense analytic subgroup of G containig the closure $\overline{G'}$ of the commutator subgroup G' of G. Then the analytic subgroup E generated by $\mathbf{L}(A) \cap \mathbf{L}(K)$ is dense in K and has finite index in $A \cap K$.*

Proof. We set $N = \overline{G'}$ and apply Lemma A.2.5 in order to see that the maximal torus of the connected abelian group G/N is $(KN)/N$. Moreover we apply Theorem A.2.3 to see that $G = AK$. Thus we also see that $(A/N)((KN)/N) = G/N$. On the Lie algebra level this implies $\mathbf{L}(G/N) = \mathbf{L}(A/N) + \mathbf{L}(KN/N)$ and we can write $\mathbf{L}(A/N) = V \oplus (\mathbf{L}(A/N) \cap \mathbf{L}(KN/N))$ with a vector space complement V of $\mathbf{L}(KN/N)$ in $\mathbf{L}(G/N)$. The image of V in G/N under the exponential function of G/N is a closed vector subgroup H/N which is contained in A/N. But then $(A/N) \cap (KN/N)$ is the projection of A/N into KN/N along H/N and is, therefore, an analytic dense subgroup of KN/N. By the modular law we have $A \cap KN = (A \cap K)N$, and since N is connected, the group $(A \cap K)N$ is a dense analytic subgroup of KN. The isomorphism $\varphi: K/(N \cap K) \to KN/N$ defined by $\varphi(k(N \cap K)) = kN$ maps the group $(A \cap K)/(N \cap K)$ isomorphically onto $((A \cap K)N)/N$. Hence $(A \cap K)/(N \cap K)$ is a dense analytic subgroup of the torus KN/N. Now $N \cap K$ is a compact Lie group and thus the factor group $(N \cap K)/(N \cap K)_0$, where $(N \cap K)_0$ is the identity component of $N \cap K$, is finite. Note that then $K/(N \cap K)_0$ is a finite covering of $K/(N \cap K)$ and hence itself a torus. This shows that $(A \cap K)/(N \cap K)_0$ is dense in $K/(N \cap K)_0$ and only has finitely many components. But then the arc component of the identity in $(A \cap K)/(N \cap K)_0$ is itself already dense in $K/(N \cap K)_0$. Since $(N \cap K)_0$ is connected this implies that the arc component of the identity in $A \cap K$ is dense in K. But this arc component is the analytic subgroup E generated by $\mathbf{L}(A) \cap \mathbf{L}(K)$. Note that it follows also from the above that there are only finitely many arc components in $A \cap K$ which means that E has finite index in $A \cap K$. $\blacksquare$

A.2.7. **Theorem.** *Let* $\pi: G \to \mathrm{Gl}(n)$ *be a continuous real or complex representation of a connected Lie group* G. *Let* $\mathbf{L}(\pi): \mathbf{L}(G) \to \mathrm{gl}(n)$ *be the induced representation on the Lie algebra level. Let* K *be a compact subgroup of* $\overline{\pi(G)}$ *containing a maximal torus of* $\overline{\pi(G)}$. *Let* H *be the arc component of the identity in* $\pi^{-1}(K)$. *Then* H *is the analytic subgroup generated by* $\mathbf{L}(\pi)^{-1}\big(\mathbf{L}(K)\big)$, *and* $\overline{\pi(H)} = K$.

Proof. The subgroup $G^* = \overline{\pi(G)}$ of $\mathrm{Gl}(n)$ is a connected linear Lie group. Hence its commutator subgroup $G^{*'}$ is closed (see [Hoch65, p.224]). The subgroup $A = \pi(G)$ is a dense analytic subgroup of G^* containing $G^{*'} = A'$. Since G^* is linear, all of its semisimple factors have finite center (see [Hoch65, p.221]). Hence $(G^{*'})/\mathrm{rad}(G^{*'})$ has finite center. Thus we can apply Lemma A.2.6 to show that the analytic subgroup E generated in G^* by $\mathbf{L}(A) \cap \mathbf{L}(K)$, that is, the arc component of the identity in $A \cap K$, is dense in K. Note that $\pi^{-1}(K) = \pi^{-1}(A \cap K)$ and recall that H is the identity component of this group. Then $\mathbf{L}(H)$ consists of all elements $x \in \mathbf{L}(G)$ with $\exp tx \in H$ for all $t \in \mathbb{R}$, that is, $\pi(\exp tx) \in K$ for all $t \in \mathbb{R}$. This in turn is equivalent to $\pi(\exp tx) \in A \cap K$ for all $t \in \mathbb{R}$, so that we find $\mathbf{L}(H) = \mathbf{L}(\pi)^{-1}\big(\mathbf{L}(K)\big) = \mathbf{L}(\pi)^{-1}\big(\mathbf{L}(A \cap K)\big)$. Therefore we have $\pi(H) = E$ and thus, in particular, $\overline{\pi(H)} = K$ since E is dense in K. $\blacksquare$

ρ-compactness

The results of the previous subsection allow us to associate with any representation $\rho: L \to \mathrm{gl}(n)$ of a Lie algebra L a concept of "compactness for a subalgebra" of L.

A.2.8. **Definition.** Let $\rho : L \to \mathrm{gl}(n)$ be a representation of a (finite dimensional) Lie algebra. For a subalgebra $M \subseteq L$ and a subgroup K of $\mathrm{Gl}(n)$ we set

(i) $M^\rho = \overline{\langle e^{\rho(M)} \rangle} =$ closure in $\mathrm{Gl}(n)$ of the analytic subgroup generated by all $e^{\rho(x)}$ with $x \in M$.

(ii) $K_\rho = \rho^{-1}\big(\mathbf{L}(K_a)\big) = \{x \in L : e^{t\rho(x)} \in K \text{ for all } t \in \mathbb{R}\}$ where K_a is the arc component of the identity in K, that is, the largest analytic subgroup of $\mathrm{Gl}(n)$ contained in K.

We shall say that a subalgebra M of L is $\rho - compact$ if M^ρ is compact.

The following remarks are almost immediate.

A.2.9. **Remark.** Let $\rho : L \to \mathrm{gl}(n)$ denote a representation. Then

(i) The assignments $M \to M^\rho$ and $K \to K_\rho$ are monotone.

(ii) $\{0\}^\rho = \{1\}$ and $\{1\}_\rho = ker(\rho)$.

(iii) M^ρ is a closed connected Lie subgroup of $\mathrm{Gl}(n)$.

(iv) If $M \subseteq L$ is a subalgebra of L and $K \subseteq \mathrm{Gl}(n)$ is a closed subgroup of $\mathrm{Gl}(n)$, then $M^\rho \subseteq K$ if and only if $M \subseteq K_\rho$.

(v) $M \subseteq M^\rho{}_\rho$ and $K_\rho{}^\rho \subseteq K$ for all subalgebras and all closed subgroups of $\mathrm{Gl}(n)$, respectively.   ∎

A.2.10.   Remark.   Under the same hypotheses as in Remark A.2.9 the following propositions hold:

(vi) $M + ker(\rho) = M^\rho{}_\rho$ if the analytic subgroup $\langle e^{\rho(M)} \rangle$ generated by $\rho(M)$ in $\mathrm{Gl}(n)$ is closed.

(vii) $K = K_\rho{}^\rho$ if and only if the analytic subgroup generated by $\mathbf{L}(K) \cap \rho(L)$ in K is dense in K.

Proof.   (vi) We have $x \in M^\rho{}_\rho$ if and only if $e^{t\rho(x)} \in \langle e^{\rho(M)} \rangle^-$ for all $t \in \mathbb{R}$. This means $\rho(x) \in \mathbf{L}(\langle e^{\rho(M)} \rangle^-)$. If $\langle e^{\rho(M)} \rangle$ is closed, then $\rho(M) = \mathbf{L}(\langle e^{\rho(M)} \rangle) = \mathbf{L}(\langle e^{\rho(M)} \rangle^-)$, and the assertion follows.

(vii) The analytic subgroup $\langle e^{\rho(K_\rho)} \rangle$ is the one generated by $\rho(K_\rho) = \rho(\rho^{-1} \mathbf{L}(K)) = \mathbf{L}(K) \cap \rho(L)$. Its closure is $K_\rho{}^\rho$, whence the assertion.   ∎

It is clear that maximal ρ-compact subalgebras exist in any finite dimensional Lie algebra, but the following theorem, relating maximal ρ-compact subalgebras of L to maximal compact subgroups of L^ρ is no longer on the surface, since it needs the previous results.

A.2.11.   Theorem.   *Let $\rho : L \to \mathrm{gl}(n)$ be a representation of a finite dimensional Lie algebra.*

(i) *If K is a compact subgroup of L^ρ containing a maximal torus, then $K = K_\rho{}^\rho$. If, in addition, K is a maximal compact subgroup of L^ρ, then K_ρ is a maximal ρ-compact subalgebra of L.*

(ii) *If M is a maximal ρ-compact subalgebra of L, then $M = M^\rho{}_\rho$ and M^ρ is a maximal compact subgroup of L^ρ. The analogous statement with "maximal" replaced by "maximal abelian" remains true.*

Proof.   (i) Let K be a compact subgroup of L^ρ containing a maximal torus. We denote by G the simply connected Lie group with $L = \mathbf{L}(G)$. Then there is a unique representation $\pi: G \to \mathrm{Gl}(n)$ with $\mathbf{L}(\pi) = \rho$. Let H be the analytic subgroup of G generated by $K_\rho = \mathbf{L}(\pi)^{-1}(\mathbf{L}(K))$. We note that $\pi(H)$ is the analytic subgroup of $\mathrm{Gl}(n)$ generated by $e^{\mathbf{L}(\pi)(K_\rho)} = e^{\rho(K_\rho)}$. Hence we have $K_\rho{}^\rho = \overline{\pi(H)}$. But by Theorem A.2.7, also $\overline{\pi(H)} = K$, so that $K = K_\rho{}^\rho$. Now suppose that K is a maximal compact subgroup of L^ρ and that M is a ρ-compact subalgebra of L containing K_ρ. Then $K = K_\rho{}^\rho \subseteq M^\rho$ and M^ρ is compact by definition. The maximality of K then implies that $K = M^\rho$. In particular, we find $M^\rho \subseteq K$ which is, by Remark A.2.9, equivalent to $M \subseteq K_\rho$. But since $K_\rho{}^\rho = K$ is compact, the algebra K_ρ is ρ-compact so that $M \subseteq K_\rho$ shows that K_ρ is a maximal ρ-compact algebra.

(ii) Let M be a maximal ρ-compact subalgebra of L and K a maximal compact subgroup of L^ρ containing M^ρ. Then $M \subseteq K_\rho$ by Remark A.2.9. Moreover we know that K_ρ is ρ-compact since $K_\rho{}^\rho \subseteq K$ by Remark A.2.9 so that $K_\rho{}^\rho$ is compact. The maximality of M shows that $M = K_\rho$. Hence $M^\rho =$

$K_\rho{}^\rho = K$ in view of (i) above. Thus M^ρ is a maximal compact subgroup of L^ρ. Note that we also know $M \subseteq M^\rho{}_\rho$ by Remark A.2.9. Further, $M^\rho{}_\rho{}^\rho = M^\rho$, which follows from Remark A.2.9 as in the case of any Galois connection. Therefore $M^\rho{}_\rho$ is a ρ-compact subalgebra of L. Now again by the maximality of M, we conclude that $M = M^\rho{}_\rho$. The same proof works verbatim for the abelian case. ∎

Recall that $\operatorname{Aut} L$ is a closed subgroup of $\operatorname{Gl}(L)$, hence is a Lie group. Its Lie algebra is the Lie algebra $\operatorname{Der} L$ of all derivations of L. The group $\operatorname{Inn} L$ is the analytic subgroup generated in $\operatorname{Aut} L$ by the Lie algebra $\operatorname{ad} L$ of all inner derivations.

A.2.12. **Lemma.** *For a representation $\rho\colon L \to \operatorname{gl}(n)$ of a Lie algebra L the following propositions hold:*

(i) $\rho(e^{\operatorname{ad} x}y) = e^{\operatorname{ad}\rho(x)}\rho(y)$ *for all $x, y \in L$.*

(ii) $(e^{\operatorname{ad} x}M)^\rho = e^{\operatorname{ad}\rho(x)}(M^\rho)$ *for all $x \in L$ and each subalgebra M of L.*

(iii) *If $\pi\colon G \to \operatorname{Gl}(n)$ is any representation of a Lie group G with $\mathbf{L}(G) = L$ and $\mathbf{L}(\pi) = \rho$, then*
$$(e^{\operatorname{ad} x}M)^\rho = \pi(\exp_G x)M^\rho\pi(\exp_G x)^{-1} \text{ for } x \in L.$$

Proof. (i) We compute
$$\rho(e^{\operatorname{ad} x}y) = \rho\Big(\sum_{n=0}^{\infty} \frac{1}{n!}(\operatorname{ad} x)^n y\Big)$$
$$= \sum_{n=0}^{\infty} \frac{1}{n!}\big(\operatorname{ad}\rho(x)\big)^n\rho(y)$$
$$= e^{\operatorname{ad}\rho(x)}\rho(y).$$

(ii) $(e^{\operatorname{ad} x}M)^\rho = \overline{\langle\exp\rho(e^{\operatorname{ad} x}M)\rangle} = \overline{\langle\exp e^{\operatorname{ad}\rho(x)}\rho(M)\rangle}$ by (i) above. Now for two matrices S and T we always have $e^{\operatorname{ad} S}T = e^S T e^{-S}$. Hence $e^{\operatorname{ad}\rho(x)}\rho(M) = e^{\rho(x)}\rho(M)e^{-\rho(x)}$. For each automorphism I of a Banach algebra we have $I(e^X) = e^{I(X)}$ for all X in the Banach algebra. This shows $(e^{\operatorname{ad} x}M)^\rho = \overline{\langle e^{\rho(x)}e^{\rho(M)}e^{-\rho(x)}\rangle} = e^{\rho(x)}\overline{\langle e^{\rho(M)}\rangle}e^{-\rho(x)} = e^{\operatorname{ad}\rho(x)}M^\rho$, which in turn implies (ii).

(iii) The claim follows from (ii) and the fact that $\pi(\exp_G x) = e^{\mathbf{L}(\pi)(x)} = e^{\rho(x)}$ holds for all $x \in L$. ∎

A.2.13. **Theorem.** *Let $\rho\colon L \to \operatorname{gl}(n)$ denote a representation of the Lie algebra L. Then the maximal ρ-compact subalgebras of L are conjugate under inner automorphisms of L. In particular, if M is a maximal ρ-compact and N is any ρ-compact subalgebra of L, then there is an inner automorphism $\gamma \in \operatorname{Inn}(L)$ such that $\gamma(N) \subseteq M$.*

Proof. Let G be the simply connected Lie group with Lie algebra $L = \mathbf{L}(G)$ and $\pi\colon G \to \operatorname{Gl}(n)$ be the representation with $\mathbf{L}(\pi) = \rho$. Moreover let M_1 and M_2 be two maximal ρ-compact subalgebras of L. Then $M_1{}^\rho$ and $M_2{}^\rho$ are two maximal compact subgroups of L^ρ by Theorem A.2.11(ii). From Corollary A.2.4 it follows that $L^\rho = \pi(G)M_1{}^\rho = \pi(G)M_2{}^\rho$ and that $M_1{}^\rho$ and $M_2{}^\rho$ are conjugate under elements of $\pi(G)$. This means that there is a $g \in G$ such that $M_2{}^\rho = \pi(g)M_1{}^\rho\pi(g)^{-1}$.

Now we find elements $x_1, \ldots, x_n$ in L such that $g = \exp_G x_1 \cdots \exp_G x_n$. We use Lemma A.2.12(iii) to show that $M_2{}^\rho = (e^{\operatorname{ad} x_1} \cdots e^{\operatorname{ad} x_n} M_1)^\rho$. Now Theorem A.2.11(ii) applies and shows $M_2 = \gamma M_1$ with $\gamma = e^{\operatorname{ad} x_1} \cdots e^{\operatorname{ad} x_n} \in L$. The last assertion is now clear since any ρ-compact algebra is contained in a maximal one. ∎

Compact and ρ-compact elements

A.2.14.   Definition. An element x in a Hausdorff topological group G is called *compact* if the closed subgroup $\overline{\langle x \rangle}$ generated by x in G is compact. The set of all compact elements in G will be denoted by $\operatorname{comp} G$.

A.2.15.   Theorem. *Let G be a Lie group with a dense analytic subgroup A satisfying $\exp \mathbf{L}(A) \subseteq \operatorname{comp} G$. Then G is compact.*

Proof.   We prove this claim by induction with respect to the dimension of G. So let G be a counterexample to this theorem with minimal dimension; we shall derive a contradiction. A 1-dimensional Lie group cannot be a counterexample to the theorem, and thus the dimension of G is at least two. If $A = G$ then $\exp \mathbf{L}(G) \subseteq \operatorname{comp} G$ and G must be compact because of the existence of a manifold factor in non-compact Lie groups (see [Hoch65, p.180]). Hence, in a counterexample, $A \neq G$. But $G' = A' \subseteq A$ so that a counterexample cannot be semisimple.

We claim that G cannot be solvable. In fact, if G is solvable then A is solvable as well, so that we can find a non-trivial abelian characteristic analytic subgroup I in A. Then the closure N of I in G is abelian and normal in G. But $I \subseteq \exp \mathbf{L}(A) \subseteq \operatorname{comp} G$, whence $I \subseteq N \cap \operatorname{comp} G = \operatorname{comp} N$. However, in a connected abelian Lie group the set of compact elements is just the maximal torus and is, in particular, closed. Hence N is compact. Now AN is a dense analytic subgroup of G contained in $N \operatorname{comp} G$ and $N \operatorname{comp} G$ is contained in $\operatorname{comp} G$ since the subgroup generated by nx with $n \in N$ and $x \in \operatorname{comp} G$ is contained in $N\overline{\langle x \rangle}$, which is compact. Therefore AN/N is a dense analytic subgroup of G/N with $\exp_{G/N}(AN/N) \subseteq \operatorname{comp} G/N$. But the dimension of G/N is less than the dimension of G since we assumed I to be non-trivial. Thus G/N has to be compact since G/N cannot be a counterexample. This contradicts the assumption that G is not compact. We have now shown that G cannot be solvable.

Now let R denote the radical of A. Then R is normal in G and since G is not solvable, the closure $\overline{R}$ of R in G is not all of G. Note that $\exp \mathbf{L}(R) \subseteq \exp \mathbf{L}(A) \subseteq \operatorname{comp} G$ so that $\exp \mathbf{L}(R) \subseteq \operatorname{comp} \overline{R}$. Since G was a counterexample of minimal dimension, the group $\overline{R}$ cannot be a counterexample, hence $\overline{R}$ is compact. If we replace N in the above argument by $\overline{R}$ then we find that $\overline{R}$ is trivial, that is, that A is semisimple. But then $G' = A' = A$ by an earlier observation. This shows that the radical of G has to be abelian and that A is a normal Levi subgroup. From this we can conclude that $\mathbf{L}(G)$ is reductive. Since A is dense in G but not all of G, it cannot be compact. Therefore $\mathbf{L}(A)$ contains a subalgebra isomorphic to $\operatorname{sl}(2, \mathbb{R})$, hence it also contains a two dimensional non-abelian solvable subalgebra which generates an analytic subgroup H with $\exp \mathbf{L}(H) \subseteq \exp \mathbf{L}(A) \subseteq \operatorname{comp} G$.

Again we conclude that $\exp \mathbf{L}(H) \subseteq \operatorname{comp} \overline{H}$ and that $\overline{H}$ is compact. But every solvable compact Lie group is abelian and this contradiction proves the claim. ∎

A.2.16. **Definition.** Let $\rho: L \to \operatorname{gl}(n)$ be a representation of a finite-dimensional Lie algebra L. An element $x \in L$ will be called $\rho - compact$ if $\overline{e^{\mathbb{R}\rho(x)}}$ is compact in $\operatorname{Gl}(n)$, i.e., if $\mathbb{R} \cdot x$ is a ρ-compact subalgebra of L in the sense of Definition A.2.8. The set of all ρ-compact elements of L will be denoted by $\operatorname{comp}_\rho L$.

We consider the following lemma as familiar:

A.2.17. **Lemma.** *Let V be a finite-dimensional real vector space and $\varphi: V \to V$ an endomorphism of V. Let G be the closure of $e^{\mathbb{R} \cdot \varphi}$ in $\operatorname{Gl}(n)$. Then the following conditions are equivalent:*

(1) *G is compact.*

(2) *φ is semisimple and has purely imaginary spectrum.* ∎

We now have the following characterization theorem for ρ-compact subalgebras of a Lie algebra L:

A.2.18. **Theorem.** *Let M be a subalgebra of a finite dimensional Lie algebra L and $\rho: L \to \operatorname{gl}(V)$ a representation on a finite-dimensional vectorspace V. Let $M_{\max}$ denote an arbitrary maximal ρ-compact subalgebra of L. Then the following statements are equivalent:*

(1) *M is ρ-compact.*

(2) *$M \subseteq \operatorname{comp}_\rho L$.*

(3) *For each $x \in M$, the endomorphism $\rho(x)$ of V is semisimple and has purely imaginary spectrum.*

(4) *There is an inner automorphism $\gamma \in \operatorname{Inn} L$ of L such that $\gamma(M) \subseteq M_{\max}$.*

(5) *There is a positive definite quadratic form $(\bullet \mid \bullet)$ on V such that*

$$\bigl(\rho(x)(u) \mid v\bigr) = -\bigl(u \mid \rho(x)(v)\bigr) \text{ for all } x \in M, \quad v, u \in V.$$

Proof. The equivalence of (2) and (3) is an immediate consequence of Lemma A.2.17. Condition (2) translates into the equivalent condition

(2′) $e^{\rho(M)} \subseteq \operatorname{comp} L^\rho$.

By Theorem A.2.15 this condition is equivalent to

(1′) M^ρ is compact.

But (1) and (1′) are obviously equivalent, so that (1), (2) and (3) are all equivalent. The equivalence of (1) and (4) is a consequence of Theorem 1.12. It remains to establish the equivalence of (5) with the other conditions. First, we claim that (5) is equivalent to

(5′) $\bigl(g(u) \mid g(v)\bigr) = \bigl(u \mid v\bigr)$ for all $g \in M^\rho$ and $u, v \in V$.

Since $e^{\rho(M)}$ generates a dense analytic subgroup of M^ρ, condition (5′) is equivalent to

(5″) $\bigl(e^{\rho(x)}u \mid e^{\rho(x)}v\bigr) = \bigl(u \mid v\bigr)$ for all $x \in M$ and $u, v \in V$.

We define a function $f\colon \mathbb{R} \to \mathbb{R}$ by $f(t) = \left(e^{t\rho(x)}u \mid e^{t\rho(x)}v\right)$. Then f is differentiable and $f'(t) = \left(\rho(x)e^{t\rho(x)}u \mid e^{t\rho(x)}v\right) + \left(e^{t\rho(x)}u \mid \rho(x)e^{t\rho(x)}v\right)$. Now (5″) is equivalent to the constancy of f for all $x \in M$ and $u, v \in V$, that is, the vanishing of $f'(t)$. But the vanishing of $f'(t)$ for all t, x, u and v is equivalent to condition (5), which now shows the equivalence of (5) and (5′). Finally we note that (5′) says that M^ρ is a closed subgroup of some orthogonal group, and this statement is equivalent to (1′). ∎

A.2.19. **Corollary.** *Let $\rho\colon L \to \mathrm{gl}(V)$ be a representation of L on a finite-dimensional vector space V. If M is ρ-compact subalgebra, then V is a semisimple M-module, that is, every $\rho(M)$-invariant vector subspace V_1 has an invariant vector space complement V_2 such that $V = V_1 \oplus V_2$.*

Proof.   It suffices to select a scalar product on V according to Theorem A.2.18 and to take for V_2 the orthogonal complement of V_1. ∎

For the applications of the preceding concepts in the classification of invariant cones the most important representation is the adjoint representation. We introduce a special notation for this case.

A.2.20. **Definition.** Let L be a finite-dimensional Lie algebra. A subalgebra M of L is said to be *compactly embedded* if it is ad-compact, where $\mathrm{ad}\colon L \to \mathrm{gl}(L)$ is the adjoint representation of L. An element $x \in L$ is called *compact* if it is ad-compact. The set of compact elements in L is denoted by $\mathrm{comp}\,L$. Finally we say that the Lie algebra L is *compact* if it is compactly embedded in itself.

With this notation the following corollary is an immediate consequence of Theorem A.2.18.

A.2.21. **Corollary.** *Let M be a subalgebra of a finite dimensional Lie algebra L. Let M_{max} denote an arbitrary maximal compactly embedded subalgebra of L. Then the following statements are equivalent:*

(1) *M is compactly embedded.*

(2) *$M \subseteq \mathrm{comp}\,L$.*

(3) *For each $x \in M$, the endomorphism $\mathrm{ad}\,x$ of L is semisimple and has purely imaginary spectrum.*

(4) *There is an inner automorphism $\gamma \in \mathrm{Inn}\,L$ of L such that $\gamma(M) \subseteq M_{\mathrm{max}}$.*

(5) *There is a positive definite quadratic form $(\bullet \mid \bullet)$ on L such that*

$$\left((\mathrm{ad}\,x)(u) \mid v\right) = -\left(u \mid (\mathrm{ad}\,x)(v)\right) \text{ for all } x \in M, \quad u, v \in L.$$

∎

Corollary A.2.21 in turn implies the following corollary since any compactly embedded subalgebra of a Lie algebra is a compact Lie algebra.

A.2.22. **Corollary.** *If K is a compactly embedded subalgebra of L, then there exists a vector subspace P of L such that $L = K \oplus P$ and $[K, P] \subseteq P$, that is, P is a K-module complement for K.* ∎

The interior of comp L

Before we can characterize the interior of comp L in an algebraic way, we have to establish some invariance properties of comp L.

A.2.23. **Proposition.** *Let L be a finite-dimensional Lie algebra. Then the set* comp L *of all compact elements of L' is invariant under scalar multiplication and under all inner automorphisms. If $M_{\max}$ denotes a maximal compactly embedded subalgebra, then* comp $L = (\operatorname{Inn} L) \cdot M_{\max}$.

Proof. We have $x \in$ comp L if and only if $\mathbb{R} \cdot x$ is compactly embedded, which in turn is equivalent to $\mathbb{R} \cdot x \subseteq$ comp L. Moreover, Corollary A.2.21 shows that comp $L \subseteq (\operatorname{Inn} L) \cdot M_{\max}$. The reverse inclusion is clear from the same result. Hence comp $L = (\operatorname{Inn} L) \cdot M_{\max}$ and the invariance of comp L under inner automorphisms is a trivial consequence. ∎

Note that the above argument can be modified to show that comp L is invariant under arbitrary automorphisms of L.

A.2.24. **Lemma.** *Let K be a compactly embedded subalgebra of L and $x \in K$. Then we have*

(i) $(\operatorname{ad} x)^{-1}(K) = K + \ker(\operatorname{ad} x)$.

(ii) *The linear map $T: L \times K \to L$, defined by $T(u, h) = h + [u, x]$ is surjective if and only if $\ker(\operatorname{ad} x) \subseteq K$.*

(iii) *The function $F: L \times K \to L$, defined by $F(u, v) = e^{\operatorname{ad} u} v$ is open at the point $(0, x)$ if $\ker(\operatorname{ad} x) \subseteq K$.*

Proof. (i) We write $L = K \oplus P$ according to Corollary A.2.22. For $y \in L$ we write $y = k + p$ with $k \in K$ and $p \in P$. Now $y \in (\operatorname{ad} x)^{-1}(K)$ if and only if $[x, y] \in K$. This means $[x, p] = [x, y] - [x, k] \in K$ since $[x, k] \in [K, K] \subseteq K$. But $[x, p] \in [K, P] \subseteq P$ by Corollary A.2.22, whence $[x, p] \in K$ if and only if $[x, p] = 0$ because of $K \cap P = \{0\}$. Hence $y \in (\operatorname{ad} x)^{-1}(K)$ if and only if $p \in \ker(\operatorname{ad} x)$. Thus $(\operatorname{ad} x)^{-1}(K) = K \oplus (P \cap \ker(\operatorname{ad} x)) \subseteq K + \ker(\operatorname{ad} x)$. But $(\operatorname{ad} x)(K + \ker(\operatorname{ad} x)) \subseteq [x, K] \subseteq K$ which shows that $K + \ker(\operatorname{ad} x) \subseteq (\operatorname{ad} x)^{-1}(K)$.

(ii) It is clear that $K \subseteq \operatorname{im} T$. We let $p: L \to L/K$ denote the quotient map. Then T is surjective if and only if $p \circ T: L \times K \to L/K$ is surjective. Now $(u, h) \in \ker(p \circ T)$ if and only if $[u, x] \in K$, which by (i) is equivalent to $u \in (\operatorname{ad} x)^{-1}(K) = K + \ker(\operatorname{ad} x)$. The surjectivity of $p \circ T$ is equivalent to the relation $\dim(\operatorname{im} p \circ T) = \dim L/K = \dim L - \dim K$. So $\dim L - \dim K = \dim(\operatorname{im} p \circ T) = \dim(L \times K) - \dim(\ker p \circ T) = \dim L + \dim K - \dim(K + \ker(\operatorname{ad} x)) - \dim K = \dim L - \dim(K + \ker(\operatorname{ad} x))$. Hence $p \circ T$ is surjective if and only if $\dim(K + \ker(\operatorname{ad} x)) = \dim K$, that is, if and only if $\ker(\operatorname{ad} x) \in K$.

(iii) The differential $dF_{(0,x)}: L \times K \to L$ of F at the point $(0, x)$ is given by $dF_{(0,x)}(u, h) = h + [u, x]$. Thus $dF_{(0,x)} = T$ with T as in (ii). Hence $dF_{(0,x)}$ is surjective if and only if $\ker(\operatorname{ad} x) \subseteq K$. Using the Implicit Function Theorem we see this condition to imply that F is open.

∎

A.2.25.   **Theorem.**   *Let L be a finite-dimensional Lie algebra. Then an element $x \in L$ is in the interior $\operatorname{int}(\operatorname{comp} L)$ if and only if the centralizer $Z(x, L) = \ker(\operatorname{ad} x)$ of x in L is contained in $\operatorname{comp} L$.*

Proof.   First we show that $x \in \operatorname{int}(\operatorname{comp} L)$ whenever $\ker(\operatorname{ad} x)$ is compactly embedded. Here we apply Lemma A.2.24(iii) with $K = \ker(\operatorname{ad} x)$. Whenever $y \in \operatorname{comp} L$ then $e^{\operatorname{ad} u} y \in \operatorname{comp} L$ for all $u \in L$ by Proposition A.2.23. Now Lemma A.2.24(iii) implies that $\operatorname{im} F$ is a neighborhood of x, and by what we just saw, this neighborhood is contained in $\operatorname{comp} L$. Conversely, we show that $\ker(\operatorname{ad} x) \subseteq \operatorname{comp} L$ whenever $x \in \operatorname{int}(\operatorname{comp} L)$. To this end we note that the set $V = (\operatorname{comp} L) - x$ is a neighborhood of 0. If $y \in V \cap \ker(\operatorname{ad} x)$, then $y = c - x$ with $c \in \operatorname{comp} L$ and $[x, y] = 0$. Now $[x, c] = [x, x + y] = 0$. Hence x and c are two commuting compact elements, whence $y = c - x$ is compact (indeed $(\mathbb{R} y)^{\operatorname{ad}} \subseteq (\mathbb{R} c)^{\operatorname{ad}} (\mathbb{R} x)^{\operatorname{ad}}$, and $(\mathbb{R} c)^{\operatorname{ad}}$ as well as $(\mathbb{R} x)^{\operatorname{ad}}$ is compact). Thus $V \cap \ker(\operatorname{ad} x) \subseteq \operatorname{comp} L$. Hence $\mathbb{R} \cdot (V \cap \ker(\operatorname{ad} x)) \subseteq \operatorname{comp} L$ by Proposition A.2.23. However, since V is a neighborhood of 0 and $\ker(\operatorname{ad} x)$ is a vector space we have $\mathbb{R} \cdot (V \cap \ker(\operatorname{ad} x)) = \ker(\operatorname{ad} x)$. This proves the claim. Since $\ker(\operatorname{ad} x)$ is clearly the centralizer $Z(x, L)$ of x in L, the proof of the theorem is complete. ∎

Compactly embedded Cartan algebras

A.2.26.   **Lemma.**   *Any compactly embedded solvable subalgebra of a finite dimensional Lie algebra is abelian.*

Proof.   If S is a solvable compactly embedded subalgebra of L, then S^{ad} is a compact connected solvable subgroup of the Lie group L^{ad} and is, therefore, abelian. Hence $(S + Z(L))/Z(L)$ is abelian, where $Z(L)$ denotes the center of L. Thus $S + Z(L)$ is a nilpotent compactly embedded subalgebra of L in which $Z(L)$ has an S-invariant complement V by Corollary A.2.19. Now $S + Z(L) = V \oplus Z(L)$ with $[S, V] \subseteq V$. Hence $[S + Z(L), V] \subseteq V$ and V is an ideal which is isomorphic to $(S + Z(L))/Z(L)$. Thus V is abelian which implies that $S + Z(L)$ is abelian as well. In particular, S is abelian. ∎

The next result illuminates the significance of compactly embedded Cartan algebras in the study of $\operatorname{comp} L$.

A.2.27.   **Theorem.**   *Let H be a Cartan subalgebra of a finite-dimensional real Lie algebra L. Let $x \in H$ be any regular element of L with $H = L^0(x) = \{y \in L \mid (\operatorname{ad} x)^n y = 0 \text{ for some } n \in \mathbb{N}\}$. Then the following two statements are equivalent:*

(1) $x \in \operatorname{int}(\operatorname{comp} L)$.

(2) H *is a compactly embedded subalgebra.*

Moreover, every compactly embedded Cartan algebra of L is abelian.

Proof. $(1) \Rightarrow (2)$: Let $x \in \mathrm{int}(\mathrm{comp}\, L)$. Since H is nilpotent the group H^{ad} is a nilpotent connected Lie group in which $(\mathbb{R} \cdot x)^{\mathrm{ad}}$ is a compact connected subgroup. But compact connected subgroups in nilpotent Lie groups are central. Hence x is central in H modulo $Z(L)$, that is, $[x, H] \subseteq Z(L) \subseteq Z(H)$. By Corollary A.2.21 the linear map $\mathrm{ad}\, x$ is semisimple, so that we have $H = V \oplus Z(H)$ with an $\mathrm{ad}\, x$-invariant vector subspace V. Now $[x, V] \subseteq [x, H] \cap V \subseteq Z(H) \cap V = \{0\}$, whence $x \in Z(H)$. In other words, $H \subseteq \ker(\mathrm{ad}\, x)$. But we know from Theorem A.2.25 that $\ker(\mathrm{ad}\, x) \subseteq \mathrm{comp}\, L$ which shows that $H \subseteq \mathrm{comp}\, L$, so that (2) holds by Corollary A.2.21.

$(2) \Rightarrow (1)$: We know from the first part of the proof that any compactly embedded Cartan algebra is abelian, whence $\ker(\mathrm{ad}\, x) = L^0(x) = H \subseteq \mathrm{comp}\, L$. But this implies (1) because of Theorem A.2.25. ∎

Note that not all Cartan algebras of L will generally be compactly embedded, but if one of them is, then all will be *abelian*, as a quick complexification argument shows.

We now know that compactly embedded Cartan algebras exist precisely when the set $\mathrm{comp}\, L$ of compact elements of L has interior points. We shall make this more precise.

A.2.28. **Proposition.** *For an element x in a finite-dimensional real Lie algebra L, the following conditions are equivalent:*

(1) $x \in \mathrm{int}(\mathrm{comp}\, L)$.

(2) *Let $\mathcal{C}$ denote the family of all compactly embedded Cartan subalgebras of L. Then*

$$\ker(\mathrm{ad}\, x) = \bigcup \mathcal{C}.$$

Proof. $(1) \Rightarrow (2)$: Recall first that all compactly embedded Cartan algebras of L are abelian. Thus any compactly embedded Cartan algebra containing x must be contained in $\ker(\mathrm{ad}\, x)$, which means that the right hand side of (2) is always contained in the left hand side. Next we note that $\ker(\mathrm{ad}\, x)$ is a compact algebra by Corollary A.2.21 and Theorem A.2.25, which shows that it is the the union of all its Cartan algebras. But every element of $\mathbb{R} \cdot \mathrm{ad}\, x$ is semisimple by Corollary A.2.21 so that we may apply [Bou75, VII, p.16, Proposition 10] and conclude that the Cartan algebras of $\ker(\mathrm{ad}\, x)$ are exactly those Cartan subalgebras of L which are contained in $\ker(\mathrm{ad}\, x)$. Therefore it only remains to show that all Cartan algebras of $\ker(\mathrm{ad}\, x)$ contain x. But x is central in $\ker(\mathrm{ad}\, x)$, so that all Cartan algebras of $\ker(\mathrm{ad}\, x)$ do indeed contain x.

$(2) \Rightarrow (1)$: Each compactly embedded Cartan subalgebra of L is contained in $\mathrm{comp}\, L$. Hence the right hand side of (2) is contained in $\mathrm{comp}\, L$ and thus $\ker(\mathrm{ad}\, x) \subseteq \mathrm{comp}\, L$. By Theorem A.2.25 this is equivalent to (1). ∎

A.2.29. **Corollary.** *In any finite-dimensional real Lie algebra L, with the notation of Proposition A.2.28, we have*

$$\mathrm{int}(\mathrm{comp}\, L) \subseteq \bigcup \mathcal{C} \subseteq \mathrm{comp}\, L,$$

and the regular elements contained in the middle set are all in int(comp L). *In particular,* int(comp L) *is dense in the middle set.*

Proof.   The containments are clear from the preceding. If x is regular and $x \in H$ where H is compactly embedded, then $H = L^0(x) = \ker(\operatorname{ad} x)$ and $H \subseteq \operatorname{comp} L$ whence $x \in \operatorname{int}(\operatorname{comp} L)$ by Theorem A.2.27.   ∎

The Weyl group

A.2.30.   **Proposition.**   *Any two compactly embedded Cartan subalgebras are conjugate under inner automorphisms.*

Proof.   If H_1 and H_2 are two compactly embedded Cartan algebras, then we find two maximal compactly embedded subalgebras K_1 and K_2 with $H_1 \subseteq K_1$ and $H_2 \subseteq K_2$. By Corollary A.2.21 there is an inner automorphism γ such that $\gamma(K_1) = (K_2)$. Hence $\gamma(H_1)$ and H_2 are two Cartan algebras in the compact Lie algebra K_2. Hence there is an element $x \in K_2$ such that $e^{\operatorname{ad} x}\gamma(H_1) = H_2$ (see [Bou82, §2]).   ∎

A.2.31.   **Definition.**   Let M be a subalgebra of a Lie algebra L. We set

(i) $$Z(M,L)^* = \{\alpha \in L^{ad} : \alpha\beta = \beta\alpha \text{ for all } \beta \in M^{ad}\}$$
$$= \text{centralizer of } M^{ad} \text{ in } L^{ad}.$$

(ii) $$N(M,L)^* = \{\alpha \in L^{ad} : \alpha M^{ad}\alpha^{-1} = M^{ad}\}$$
$$= \text{normalizer of } M^{ad} \text{ in } L^{ad}.$$

A.2.32.   **Lemma.**   *If M is a compactly embedded abelian subalgebra of L, then $N(M,L)^*/Z(M,L)^*$ is finite.*

Proof.   The group M^{ad} is a compact abelian group acting linearly on L, hence on the complexification $L_{\mathbb{C}} = \mathbb{C} \otimes L$. There is a finite set R of characters χ of M^{ad} and a decomposition of $L_{\mathbb{C}}$ into a direct sum of isotypic M^{ad}-submodules $V_{\chi} = \{v \in L : \alpha(v) = \chi(\alpha) \cdot v \text{ for all } \alpha \in M^{ad}\}$. If $\nu \in N(M,L)^*$ and χ is an arbitrary character of M^{ad}, then $(\chi \cdot \nu)(\alpha) = \chi(\nu\alpha\nu^{-1})$ defines a new character, and $(\chi, \nu) \mapsto \chi \cdot \nu$ defines an action of $N(M,L)^*$ on the right on the character group of M^{ad}. This action leaves the set R invariant. In fact $\nu\beta\nu^{-1}(\nu(v)) = \nu(\chi(\beta)v) = \chi(\beta)(\nu(v))$ for all $\beta \in M^{ad}$. Choosing $\beta = \nu^{-1}\alpha\nu$ we find that $\alpha(\nu(v)) = (\chi(\nu^{-1}\alpha\nu))\nu(v) = ((\chi \cdot \nu^{-1})(\alpha))\nu(v)$. Thus we have a homomorphism from $N(M,L)^*$ to the finite set of permutations of R. An element $\nu \in N(M,L)^*$ is in the kernel of that homomorphism if and only if $\chi \cdot \nu = \chi$ for $\chi \in R$. This means that for all $v \in V$ and all $\chi \in R$ we have $\nu\alpha\nu^{-1}(v) = \chi(\nu\alpha\nu^{-1}) \cdot v = \chi(\alpha) \cdot v = \alpha(v)$. It follows that $\nu\alpha\nu^{-1} = \alpha$ for all $\alpha \in M^{ad}$ which means that $\nu \in Z(M,L)^*$. But this proves the claim.   ∎

A.2.33. **Lemma.** *If H is a compactly embedded Cartan algebra of L, then $\mathbf{L}(H^{\mathrm{ad}})$ is a Cartan algebra of $\mathbf{L}(L^{\mathrm{ad}})$.*

Proof. Since $\mathbf{L}(H^{\mathrm{ad}})$ is abelian, we have to show that $\mathbf{L}(H^{\mathrm{ad}})$ is its own normalizer. Thus let X be an element of $\mathbf{L}(L^{\mathrm{ad}})$ such that $[X, \mathbf{L}(H^{\mathrm{ad}})] \subseteq \mathbf{L}(H^{\mathrm{ad}})$. This is tantamount to saying that $(\exp tX)H^{\mathrm{ad}}(\exp tX)^{-1} \subseteq H^{\mathrm{ad}}$, and this, due to the density of $e^{\mathrm{ad}\,H}$ in H^{ad}, is equivalent to $(\exp tX)e^{\mathrm{ad}\,H}(\exp tX)^{-1} \subseteq H^{\mathrm{ad}}$ for all $t \in \mathbb{R}$. This means

$$[X, \mathrm{ad}\,H] \subseteq \mathbf{L}(H^{\mathrm{ad}}).$$

By Theorem A.2.11 the group H^{ad} is a maximal torus in L^{ad}. Therefore Theorem A.2.3 applies and shows that $L^{\mathrm{ad}} = \langle e^{\mathrm{ad}\,L} \rangle H^{\mathrm{ad}}$. Translated to the Lie algebra level this means $\mathbf{L}(L^{\mathrm{ad}}) = \mathrm{ad}\,L + \mathbf{L}(H^{\mathrm{ad}})$. Thus we may write $X = \mathrm{ad}\,x + Y$ with suitable elements $x \in H$ and $Y \in \mathbf{L}(H^{\mathrm{ad}})$. Since $\mathrm{ad}\,H \subseteq \mathbf{L}(H^{\mathrm{ad}})$ and H^{ad} is abelian, we conclude that $[Y, \mathrm{ad}\,H] = \{0\}$. Thus we obtain $\mathrm{ad}[x, H] = [\mathrm{ad}\,x, \mathrm{ad}\,H] \subseteq \mathbf{L}(H^{\mathrm{ad}})$. In particular $\mathrm{ad}[x, H]$ and $\mathrm{ad}\,H$ are both subsets of $\mathbf{L}(H^{\mathrm{ad}})$, hence commute, which shows that $\mathrm{ad}[[x, H], H] = [\mathrm{ad}[x, H], \mathrm{ad}\,H] = \{0\}$. Now we see that $[[x, H], H] \subseteq \ker \mathrm{ad} = Z(L) \subseteq H$ since H is a Cartan algebra, which implies that $[x, H]$ is in the normalizer of H, hence contained in H. But then, again since H is its own normalizer, we find $x \in H$ so that $X \in \mathrm{ad}\,H + \mathbf{L}(H^{\mathrm{ad}}) = \mathbf{L}(H^{\mathrm{ad}})$ which is what we had to show. ∎

A.2.34. **Lemma.** *Let G be a group with a group A acting as a group of automorphisms on G. Suppose that H and N are subgroups of G such that N is normal in G with $A \cdot N \subseteq N$. Then we have $H = \mathrm{Fix}(A, G) = \{g \in G \mid a \cdot g = g$ for all $a \in A\}$ if the following three conditions hold:*

(i) $\mathrm{Fix}(A, N) = \{1\}$.

(ii) $\mathrm{Fix}(A, G/N) \subseteq HN/N$.

(iii) $H \subseteq \mathrm{Fix}(A, G)$.

Proof. Let $f \in \mathrm{Fix}(A, G)$. Then $fN \in \mathrm{Fix}(A, G/N)$, so by (ii) we have $f = hn$ with $h \in H$ and $n \in N$. Then $hn = f = a \cdot f = (a \cdot h)(a \cdot n) = h(a \cdot n)$ for all $a \in A$ in view of (iii). Thus $n = a \cdot n$ for all $a \in A$, whence $n = 1$ by (i), and so $f = h \in H$. ∎

A.2.35. **Proposition.** *Let T be a torus subgroup of a connected Lie group G such that $\mathbf{L}(T)$ is a Cartan algebra of $\mathbf{L}(G)$. Then $Z(T, G) = T$, where $Z(T, G)$ is the centralizer of T in G.*

Proof. We prove this in several steps.

Step 1: The first step is to show the proposition for semisimple G. In that case the claim is a consequence of [Wa72, Proposition 1.4.1.4] since, by Corollary A.2.21 and [Wa72, Proposition 1.3.3.4] the algebra $\mathbf{L}(T)$ is fundamental in the sense of [Wa72].

Step 2: Now we show that the assertion is true for solvable G. We proceed by induction with respect to the dimension of G. Let N be a minimal closed connected non-trivial normal subgroup of G. Then N is abelian and the dimension of $\mathbf{L}(N)$ is either one or two since $\mathbf{L}(N)$ is an irreducible G-module. If T acts trivially on $\mathbf{L}(N)$, then $\mathbf{L}(N)$ is in $Z(\mathbf{L}(T), \mathbf{L}(G))$; but $\mathbf{L}(T)$ is a Cartan algebra so that this centralizer is $\mathbf{L}(T)$ itself. Hence $\mathbf{L}(N) \subseteq \mathbf{L}(T)$ and thus $N \subseteq T$. Now we apply the induction hypothesis to G/N and find that $Z(T/N, G/N) = T/N$, which implies $Z(T, G) \subseteq T$ and thus $Z(T, G) = T$. Therefore we now assume that T does not act trivially on $\mathbf{L}(N)$. Then the dimension of $\mathbf{L}(N)$ cannot be one since T is compact and connected. Thus the dimension of $\mathbf{L}(N)$ is two and T acts irreducibly on $\mathbf{L}(N)$. The fixed point set of T is a vector space, hence must be singleton. We want to apply Lemma A.2.34 with T acting on G by inner automorphism as A and H, and N as normal subgroup of G. Condition (i) has just been shown; condition (ii) holds by induction hypotheses, and condition (iii) is the trivial inclusion $T \subseteq Z(T, G)$. Now Lemma A.2.34 shows that $T = Z(T, G)$ which is what we wanted to show.

Step 3: Now let R be the radical of G. Then RT is a closed solvable subgroup of G and $\mathbf{L}(T)$ is still a Cartan subalgebra of $\mathbf{L}(RT)$. We know from Step 2 that $Z(T, RT) = T$. By Step 1 we have $Z(RT/R, G/R) = T/R$. Now we observe that we can apply Lemma A.2.34 with $A = T$, acting on G by inner automorphisms, $N = R$ and $H = T$. The conclusion is $Z(T, G) = T$ which completes the proof. ∎

A.2.36.   Corollary.   *If H is a compactly embedded Cartan algebra in L, the $Z(H, L)^* = H^{ad}$.*

Proof.   Since $\mathbf{L}(H^{\mathrm{ad}})$ is a Cartan algebra of $\mathbf{L}(L^{\mathrm{ad}})$ by Lemma A.2.33, Proposition A.2.35 applies and establishes the claim. ∎

A.2.37.   Theorem.   *Let H be a compactly embedded Cartan algebra in a finite dimensional Lie algebra L. The $N(H, L)^*/H^{ad}$ is a finite group $\mathcal{W}(H, L)$ acting on H as follows: If $w = \nu H^{ad}$ with $\nu \in N(H, L)^*$, then $w \cdot x = \nu(x)$, and $e^{ad(w \cdot x)} = \nu e^{ad\, x} \nu^{-1}$.*

Proof.   Corollary A.2.36 and Lemma A.2.32 show that the group $\mathcal{W}(H, L)$ is finite. The dense analytic subgroup $\langle e^{\mathrm{ad}\, L} \rangle$ of L^{ad} intersects the maximal torus H^{ad} in a dense subgroup whose identity path component is the analytic subgroup generated by $\mathrm{ad}\, H$ (see Theorem A.2.11). This subgroup is invariant under inner automorphisms by elements of $N(H, L)^*$ since $\langle e^{\mathrm{ad}\, L} \rangle$ is normal in L^{ad}. Furthermore,
$$\nu e^{\mathrm{ad}\, x} \nu^{-1}(y) = \sum_{n=0}^{\infty} \frac{1}{n!} \nu(\mathrm{ad}\, x)^n \nu^{-1}(y) = \sum_{n=0}^{\infty} \frac{1}{n!} \big(\mathrm{ad}\, \nu(x)\big)^n(y) = e^{\mathrm{ad}\, \nu(x)}(y).$$
This shows that $e^{\mathrm{ad}\, \nu(H)} \subseteq \nu e^{\mathrm{ad}\, H} \nu^{-1} \subseteq H^{\mathrm{ad}}$, so that $e^{\mathrm{ad}\, \nu(H)} \subseteq H^{\mathrm{ad}}{}_{\mathrm{ad}}$. But according to Theorem A.2.11(ii) we know that $H = H^{\mathrm{ad}}{}_{\mathrm{ad}}$, whence $\nu(H) = H$. Thus the well defined action $(w, x) \mapsto w \cdot x = \nu(x): \mathcal{W}(H, L) \times H \to H$ satisfies $e^{\mathrm{ad}(w \cdot x)} = \nu e^{\mathrm{ad}\, x} \nu^{-1}$. ∎

A.2.38.   Definition.   The group $\mathcal{W}(H, L)$ is called the *Weyl group* of the compactly embedded Cartan algebra H in L.

A.2.39.   Lemma.   *If ν is an inner automorphism of a Lie algebra L and if C is a closed subgroup of L^{ad}, then $(\nu^{-1} C \nu)_{ad} = \nu(C_{ad})$.*

Proof.   We have $\nu(x) \in C_{\mathrm{ad}}$ if and only if $e^{t\, \mathrm{ad}\, \nu(x)} \in C$ for all $t \in \mathbb{R}$. But $\nu(e^{t\, \mathrm{ad}\, x})\nu^{-1} = e^{t\, \mathrm{ad}\, \nu(x)}$ as we have seen in the proof of Theorem A.2.37. Finally we note that $x \in (\nu^{-1} C \nu)_{\mathrm{ad}}$ if and only if $e^{t\, \mathrm{ad}\, x} \in \nu^{-1} C \nu$ for all $t \in \mathbb{R}$, which is equivalent to $\nu e^{t\, \mathrm{ad}\, x} \nu^{-1} \in C$ for all $t \in \mathbb{R}$. ∎

A.2.40. **Theorem.** *Let H be a compactly embedded Cartan algebra in a finite dimensional Lie algebra L. Then there exists a unique maximal compactly embedded subalgebra $K(H)$ of L such that $H \subseteq K(H)$. Moreover, $N(H,L)^* \subseteq K(H)^{ad}$, and if γ is any inner automorphism of L, then we have $K\big(\gamma(H)\big) = \gamma\big(K(H)\big)$.*

Proof. Let K be any maximal compactly embedded subalgebra of L containing H. By Theorem A.2.37, $N(H,L)^*$ is a compact subgroup of L^{ad}, hence is contained in a maximal compact subgroup C of L^{ad}. Then C_{ad} is a maximal compactly embedded subalgebra of L containing H by Theorem A.2.11. Hence by Theorem A.2.13 there is an inner automorphism γ of L such that $C_{ad} = \gamma(K)$. Thus $\gamma(H)$ and H are Cartan algebras of C_{ad}. Therefore Proposition A.2.30 and the surjectivity of the exponential function for compact Lie groups imply that there is an $x \in C_{ad}$ such that $\kappa\gamma(H) = H$ where $\kappa = e^{ad\,x}$. Then obviously $\kappa\gamma \in N(H,L)^* \subseteq C$ and hence $(\kappa\gamma)(C_{ad}) = \big((\kappa\gamma)^{-1}C(\kappa\gamma)\big)_{ad} = C_{ad}$ by Lemma A.2.39. But $\kappa \in C_{ad}{}^{ad} = C$ by Theorem A.2.11 since C is maximal compact, so that Lemma A.2.39 again shows $\kappa(C_{ad}) = C_{ad}$. As a consequence, we have $\gamma(C_{ad}) = C_{ad}$ and thus $K = C_{ad}$. This shows that C_{ad} is the only maximal compactly embedded subalgebra of L containing H. Finally we note that, if γ is an inner automorphism of L, then $\gamma\big(K(H)\big)$ is a maximal compactly embedded subalgebra of L containing $\gamma(H)$, hence must be equal to $K\big(\gamma(H)\big)$. This completes the proof of the theorem. ∎

A.2.41. **Corollary.** *The Weyl group $\mathcal{W}(H,L)$ of a compactly embedded Lie algebra H in a Lie algebra L is the classical Weyl group of the torus H^{ad} in the compact group $K(H)^{ad}$.*

Proof. By Theorem A.2.11 the group $K(H)^{ad}$ is a maximal compact subgroup of L^{ad} containing H^{ad}. The proof of Theorem A.2.40 shows that $N(H,L)^* \subseteq K(H)^{ad}$ and so the normalizer of H^{ad} in $K(H)^{ad}$ contains $N(H,L)^*$ which implies the corollary. ∎

Notes on the Appendix

Appendix 1. In this section we collect the essential information on the formalism of the Baker-Campbell-Hausdorff-Dynkin Series to the extend we need it in the book. The standard references are [Hoch65] or [Bou75].

Appendix 2. The material in this section of the Appendix is taken from [HH86d].

Reference material

Bibliography

[An87a]   Anderson, M., One parameter submonoids in locally compact differentiable monoids, Preprint (1987).

[An87b]   Anderson, M., One parameter submonoids in locally complete differentiable monoids, Preprint (1987).

[At82]   Atiyah, M. F., Convexity and commuting hamiltonians, Bull. London Math. Soc. **14** (1982), 1–15.

[Ba81]   Barker, G., Theory of cones, Lin. Alg. and Appl. **39** (1981), 263–291.

[Bi38]   Birkhoff, G., Analytic groups, Trans. Amer. Math. Soc. **43** (1938), 61–101.

[Bi73]   Birkhoff, G., *Lattice Theory*, Providence, 1973.

[BJKS82]   Bonnard, B., V. Jurdjevic, I. Kupka, and G. Sallet, Transitivity of families of invariant vector fields on semidirect products of Lie groups, Trans. Amer. Math. Soc. **271** (1982), 525–535.

[Bo84]   Bonnard, B., Controllabilité de systèmes méchaniques sur les groupes de Lie, SIAM J. Contr. and Opt. **22** (1984), 711–722.

[BoFe34]   Bonnesen, T., and W. Fenchel, *Theorie der konvexen Körper*, Springer, Berlin, 1934.

[Bon69]   Bony, J., Principe du maximum, inégalité de Harnack et unicité du problème de Cauchy pour les opérateurs elliptiques dégénérés, Ann. Inst. Fourier **19** (1969), 277–304.

[Bor69]   Borel, A., *Linear Algebraic Groups*, Benjamin, New York, Amsterdam, 1969.

[Bou55]   Bourbaki, N., *Topologie générale*, Chap. V–VIII, Hermann, Paris, 1955.

[Bou58]   Bourbaki, N., *Topologie générale*, Chap. IX, Hermann, Paris, 1958.

[Bou61]   Bourbaki, N., *Fonctions d'une variable réelle*, Chap. I–VII, Hermann, Paris, 1961.

[Bou63]   Bourbaki, N., *Intégration*, Chap. VII, Hermann, Paris, 1963.

[Bou67]   Bourbaki, N., *Théories spectrales*, Hermann, Paris, 1967.

[Bou70]   Bourbaki, N., *Algèbre*, Chap. VIII, Hermann, Paris, 1970.

[Bou71]   Bourbaki, N., *Variétés différentielles et analytiques*, Hermann, Paris, 1971.

[Bou75]   Bourbaki, N., *Groupes et algèbres de Lie*, Chap. I–VIII, Hermann, Paris, 1975.

[Bou82]   Bourbaki, N., *Groupes et algèbres de Lie*, Chap. IX, Masson, Paris, 1982.

[Bre70]   Brezis, H., On a characterization of flow-invariant sets, Comm. Pure Appl. Math. **23** (1970), 261–263.

[Bro72]   Brockett, R., System theory on group manifolds and coset spaces, SIAM J. Contr. and Opt. **10** (1972), 265–284.

[Bro73]   Brocket, R., Lie algebras and Lie groups in control theory, in: Geom. Methods in Systems Theory, Reidel (1973), 43–82.

[BF71]   Brown, D., and M. Friedberg, Linear representations of certain compact semigroups, Trans. Amer. Math. Soc. **160** (1971), 453–465.

[BH87]   Brown, D., and R. Houston, Cancellative semigroups on manifolds, Semigroup Forum **35** (1987), 279–302.

[Bru85]   Brunet, M., The metaplectic semigroup and related topics, Reports on Math. Phys. **22** (1985), 149–170.

[BK79]   Brunet, M., and P. Kramer, Semigroups of length increasing transformations, Reports on Math. Phys. **15** (1979), 287–304.

[BK80]   Brunet, M., and P. Kramer, Complex extensions of the representations of the symplectic group associated with the canonical commutation relations, Reports on Math. Phys. **17** (1980), 205–215.

[CHK83]   Carruth, J. H., J. A. Hildebrant, and R. J. Koch, *The Theory of Topological Semigroups*, Vol. I, Marcel Dekker, New York, 1983.

[CHK86]   Carruth, J. H.,J. A. Hildebrant, and R. J. Koch, *The Theory of Topological Semigroups*, Vol. II, Marcel Dekker, New York, 1986.

[Che46]   Chevalley, C., *Theory of Lie Groups*, Princeton Univ. Press, Princeton, 1946.

[CP61]   Clifford, A.H., and G. B. Preston, *The Algebraic Theory of Semigroups*, Vol. I, Amer. Math. Soc., 1961.

[CP67]   Clifford, A.H., and G. B. Preston, *The Algebraic Theory of Semigroups*, Vol. II, Amer. Math. Soc., 1967.

[Dieu71]   Dieudonné, J., *Eléments d'Analyse*, Tome I, Gauthier Villars, Paris, 1971.

[Dix57]   Dixmier, J., L'application exponentielle dans le groupes de Lie résolubles, Bull. Soc. Math. France **85** (1957), 113–121.

[Do76]   Dobbins, J. G., Well-bounded semigroups in locally compact groups, Math. Zeit. **148** (1976), 155–167.

[EG87]   El Assoudi, R., and J. P. Gauthier, Controllability of right invariant systems on real simple Lie groups of type F_4, G_2, C_n and B_n, Preprint (1987).

[Fe53]   Fenchel, W., Convex Cones, Sets and Functions, Lecture Notes, Princeton, 1953.

[Fu63]   Fuchs, L., *Partially Ordered Algebraic Systems*, Pergamon, Oxford 1963.

[Ga59]   Gantmacher, F., *Matrizenrechnung*, VEB Deutscher Verlag der Wissenschaften, Berlin, 1959.

[GKS84]   Gauthier, J. P., I. Kupka, and G. Sallet, Controllability of right invariant systems on real simple Lie groups, System and Control Letters **5** (1984), 187–190.

[GHKLM80]   Gierz, G., K. H. Hofmann, K. Keimel, J. D. Lawson, M. Mislove and D. S. Scott, *A Compendium of Continuous Lattices*, Springer, Berlin, Heidelberg, New York, Tokyo, 1980.

[Gl52]   Gleason, A. M., Groups without small subgroups, Ann. Math. **56** (1952), 193–212.

[Gor86]   Goryainov, V., Semigroups of conformal mappings, Math. USSR Sbornik **57** (1986), 463–483.

[Go69]   Goto, M., On an arcwise connected subgroup of a Lie group, Proc. Amer. Math. Soc. **20** (1969), 157–162.

[Gr79]   Graham, G., Differentiability and Semigroups, Dissertation, Univ. of Houston, 1979.

[Gr83]   Graham, G., Differentiable semigroups, in: Lecture Notes in Math. **998** (1983), 57–127.

[Gr84]   Graham, G., Differentiable manifolds with generalized boundary, Czech. Math. J. **34** (1984), 46–63.

[GdeV88]   Graham, G., and E. de Vun, Semigroups with commuting threads, Semigroup Forum, to appear.

[GHV72]   Greub, W., S. Halperin, and R. Vanstone, *Connections, Curvature and Cohomology*, Vol. I, Acad. Press, New York and London, 1972.

[GS82]   Guillemin, V., and S. Sternberg, Convexity properties of the moment map, Inv. Math. **67** (1982), 491–513.

[GL84]   Guts, A., and A. Levichev, On the foundations of relativity theory, Sov. Math. Dokl. **30** (1984), 253–257.

[Heck80]   Heckman, G., Projection of orbits and asymptotic behaviour of multiplicities for compact Lie groups, Thesis Rijksuniversiteit Leiden, 1980.

[He78]   Helgason, S., *Differential Geometry, Lie Groups, and Symmetric Spaces*, Acad. Press, Orlando, 1978.

[HeHe61]   Henney, D., and A. Henney, One parameter semigroups, Math. Japon.**6** (1961), 39–43.

[HeRo63]   Hewitt, E., and K. A. Ross, *Abstract Harmonic Analysis*, Vol.I, Springer, Berlin, Göttingen, Heidelberg, 1963.

[Hey77]   Heyer, H., *Probability Measures on Locally Compact Groups*, Springer, Berlin, Heidelberg, New York, 1977.

[Hi86a]   Hilgert, J., Infinitesimally generated subsemigroups of motion groups, Rocky Mountain J. of Math., to appear.

[Hi86b]   Hilgert, J., Invariant Lorentzian orders on simply connected Lie groups, Arkiv för Mat., to appear.

[Hi86c]   Hilgert, J., Maximal semigroups and the support of Gauss-semigroups, in: Probability and Bayesian Statistics, R. Viertl ed., Plenum Press, New York, 1987, 257–262.

[Hi87a]   Hilgert, J., Subsemigroups of Lie groups, Habilitationsschrift, TH Darmstadt, 1987.

[Hi87b]   Hilgert, J., Maximal semigroups and controllability in products of Lie groups, Arch. Math. **49** (1987), 189–195.

[Hi87c]   Hilgert, J., Spectrally ordered Lie algebras, Preprint **1107**, TH Darmstadt (1987).

[Hi88]   Hilgert, J., Some Lie theory of semigroups based on examples, Semesterberichte Tübingen, WS 1987/88 (1988).

[HH84]   Hilgert, J., and K. H. Hofmann, Lie theory for semigroups, Semigroup Forum **30** (1984), 243–251.

[HH85a]   Hilgert, J., and K. H. Hofmann, Lie semialgebras are real phenomena, Math. Ann. **270** (1985), 97–103.

[HH85b]   Hilgert, J., and K. H. Hofmann, Old and new on Sl(2), Manus. Math. **54** (1985), 17–52.

[HH85c]   Hilgert, J., and K. H. Hofmann, Semigroups in Lie groups, semialgebras in Lie algebras, Trans. Amer. Math. Soc. **288** (1985), 481–504.

[HH85d]   Hilgert, J., and K. H. Hofmann, Lorentzian cones in real Lie algebras, Monatsh. Math. **100** (1985), 183–210.

[HH86a]   Hilgert, J., and K. H. Hofmann, Invariant cones in real Lie algebras, in: Aspects of Positivity in Functional Analysis, R. Nagel ed., North Holland (1986), 209–216.

[HH86b]   Hilgert, J., and K. H. Hofmann, On Sophus Lie's Fundamental Theorem, J. Funct. Anal. **67** (1986), 1–27.

[HH86c]   Hilgert, J., and K. H. Hofmann, On the automorphism group of cones and wedges, Geom. Dedicata **21** (1986), 205–217.

[HH86d]   Hilgert, J., and K. H. Hofmann, Compactly embedded Cartan algebras and invariant cones in Lie algebras, Adv. in Math., to appear.

[HH88]   Hilgert, J., and K. H. Hofmann, Invariant cones in Lie algebras, Semigroup Forum, **37** (1988),241–252.

[HHL85]   Hilgert, J., K. H. Hofmann, and J.D. Lawson, Controllability of systems on nilpotent Lie groups, Beitr. Alg. Geom. **20** (1985), 185–190.

[Hil50]   Hille, E., Lie theory of semigroups of linear transformations, Bull. Amer. Math. Soc. **56** (1950), 89–114.

[HP57]   Hille, E., and R. S. Phillips, *Functional Analysis and Semigroups*, Amer. Math. Soc. Coll. Publ. **31**, Providence, 1957.

[Hir73]   Hirschorn, R., Topological semigroups, sets of generators and controllabillity, Duke J. Math. **40** (1973), 937–947.

[Hoch65]   Hochschild, G., *The Structure of Lie Groups*, Holden Day, San Francisco, 1965.

[Hoch71]   Hochschild, G., *Introduction to Affine Algebraic Groups*, Holden Day, San Francisco, 1971.

[Hof60]   Hofmann, K. H., Topologische Halbgruppen mit dichter submonogener Unterhalbgruppe, Math. Zeit. **74** (1960), 232–276.

[Hof63]   Hofmann, K. H., Einführung in die Theorie der Lie Gruppen, Teil II, Vorlesungsausarbeitung von Falko Lorenz, Universität Tübingen, 1963.

[Hof65]   Hofmann, K.H., Lie algebras with subalgebras of codimension one, Illinois J. Math. **9** (1965), 639–643.

[Hof76]   Hofmann, K. H., Topological semigroups, History, Theory, Applications, Jber. Deutsch. Math. Verein. **78** (1976), 9–59.

[HK86]   Hofmann, K.H., and V. Keith, Invariant quadratic forms on finite dimensional Lie algebras, Bull. Austral. Math. **33** (1986), 21–36.

[HL81]   Hofmann, K. H., and J. D. Lawson, The local theory of semigroups in nilpotent Lie groups, Semigroup Forum **23** (1981), 343–357.

[HL83a]   Hofmann, K. H., and J. D. Lawson, Foundations of Lie semigroups, in: Lecture Notes in Math. **998** (1983), 128–201.

[HL83b]   Hofmann, K. H., and J. D. Lawson, Divisible subsemigroups of Lie groups, J. London Math. Soc. **27** (1983), 427–437.

[HL83c]   Hofmann, K.H., and J. D. Lawson, On Sophus Lie's Fundamental Theorems, I and II, Indag. Math. **45** (1983), 453–466, and **46** (1984),255–266.

[HL88]   Hofmann, K.H., and J. D. Lawson, Generating local semigroups in Lie groups, Rocky Mountain J. of Math., to appear.

[HM63]   Hofmann, K. H., and P. S. Mostert, Splitting in topological groups, Mem. Amer. Math. Soc. **43** (1963).

[HM66]   Hofmann, K. H., and P. S. Mostert, *Elements of Compact Semigroups*, Charles E. Merrill, Columbus, 1966.

[HM68]   Hofmann, K.H., and P. S. Mostert, One dimensional coset spaces, Math. Ann. **178** (1968), 44–52.

[HMu78]   Hofmann, K. H., and A. Mukherjea, On the density of the image of the exponential function, Math. Ann. **234** (1978), 263–273.

[HR88]   Hofmann, K. H., and W. Ruppert, Congruences and foliations of semigroups in Lie groups, Monatsh. Math. **106** (1988), 179–204.

[HW88]   Hofmann, K. H., and W. Weiss, More on cancellative semigroups on manifolds, Semigroup Forum **37** (1988), 93–111.

[Hol74]   Holmes, J. P., Differentiable semigroups, Coll. Math. **32** (1974), 99–104.

[Hol87]   Holmes, J. P., Differentiable manifolds with smooth boundary, Semigroup Forum **36** (1987), 211–222.

[Hol88]   Holmes, J.P., One parameter subsemigroups in locally compact differentiable semigroups, Houston J. Math., to appear.

[Hör]   Hörmander, L., Pseudodifferentiable operators of principal type, in: Singularities in Boundary Value Problems, H. Garnir ed., Reidel, Dordrecht, 1981.

[Hou73]   Houston, R., Cancellative semigroups on manifolds, Dissertation, Univ. of Houston, 1973.

[Hu75]   Humphreys, J., *Linear Algebraic Groups*, Springer, New York, Heidelberg, Berlin, 1975.

[Ih86]   Ihringer, S., Keile und Halbgruppen, Dissertation, TH Darmstadt, 1986.

[Ja62]   Jacobson, N., *Lie Algebras*, Interscience Publishers, New York, London, 1962.

[Jac57]   Jacoby, R., Some theorems on the structure of locally compact local groups, Ann. Math. **66** (1957), 36–69.

[Joh73]   Johansen, S., The imbedding problem for finite Markov chains, in: Geom. Methods in Systems Theory, Reidel (1973), 227–236.

[JK81a]   Jurdjevic, V., and I. Kupka, Control systems on semi-simple Lie groups and their homogeneous spaces, Ann. Inst. Fourier **31** (1981), 151–179.

[JK81b]   Jurdjevic, V., and I. Kupka, Control systems subordinated to a group action: Accessibility, J. Diff. Eq. **39** (1981), 180–211.

[JS72]   Jurdjevic, V., and H. Sussmann, Control systems on Lie groups, J. Diff. Eq. **12** (1972), 313–329.

[KP84]   Kac, V., and D. H. Petereson, Unitary structure in representations of infinite dimensional groups and a convexity theorem, Inv. Math. **76** (1984), 1–15.

[Ke67]   Keimel, K., Eine Exponentialfunktion für kompakte abelsche Halbgruppen, Math. Zeit. **96** (1967), 7–25.

[Ko73]   Kostant, B., On convexity, the Weyl group, and the Iwasawa decomposition. Ann. Sci. Ec. Norm. Sup. **6** (1973), 413–455.

[Kre74]   Krener, A. J., A generalisation of Chow's Theorem and the Bang-Bang Theorem to nonlinear systems, SIAM J. Control **17** (1974), 670–676.

[KR82]   Kumaresan, S., and A. Ranjan, On invariant convex cones in simple Lie algebras, Proc. Ind. Acad. Sci. Math. **91** (1982), 167–182.

[La72]   Lang, S., *Differentiable Manifolds*, Addison Wesley, London 1972.

[Lan60]   Langlands, R., On Lie semigroups, Canad. J. Math. **12** (1960), 686–693.

[Law86]   Lawson, J. D., Computing tangent wedges of semigroups, in: Kochfest, J. A. Hildebrant ed., Baton Rouge, 1986.

[Law87a]   Lawson, J. D., Maximal subsemigroups of Lie groups that are total, Proc. Edinburgh Math.Soc.**30** (1987), 479–501.

[Law87b]   Lawson, J. D., Fields of tangent cones and Hofmann cones, Semigroup Forum **35** (1987), 1–27.

[Law87a]   Lawson, J. D., Embedding local semigroups into groups, Proc. of the New York Acad. Sci., to appear.

[LM74]   Lawson, J. D., and B. Madison, Quotients of k-semigroups, Semigroup Forum **9** (1974), 1–19.

[Lea61]   Leach, E., A note on inverse function theorems, Proc. Amer. Math. Soc. **12** (1961), 694–697.

[Lev85]   Levichev, A., Sufficient conditions for the nonexistence of closed causal curves in homogeneous space-times, Izvestia Phys. **10** (1985), 118–119.

[Lev86]   Levichev, A., Lie algebras admitting elliptic semialgebras, Funct. Anal. and Appl. **20** (1986), 146–148.

[Lev87]   Levichev, A., Left invariant orders on special affine groups, Sib. Math. J. **28** (1987), 152–156.

[Loe56]   Loewner, Ch., On some transformation semigroups, J. Rat. Mech. and Anal. **5** (1956), 791–804.

[Loe59a]   Loewner, Ch., A theorem on the partial order derived from a certain transformation semigroup, Math. Zeit. **72** (1959), 53–60.

[Loe59b]   Loewner, Ch., On some transformation semigroups invariant under Euclidean and non Euclidean isometries, J. Math. and Mech. **8** (1959), 393–409.

[Loe64]   Loewner, Ch., On semigroups and geometry, Bull. Amer. Math. Soc. **70** (1964), 1–15.

[MacL63]   Mac Lane, S., *Homology*, Springer, Berlin, 1963.

[Mar81]   Markus, L., Controllabillity of multitrajectories in Lie groups, in Lecture Notes in Math. **898** (1981), 250–265.

[Maz33]   Mazur, S., Über konvexe Mengen in linearen normierten Räumen, Stud. Math. **4** (1933), 70–84.

[McC81]   McCrudden, M., On n-th roots and infinitely divisible elements in a connected Lie group, Math. Proc. Cambridge Phil. Soc. **89** (1981), 293–299.

[McC84]   McCrudden, M., On the supports of absolutely continuous Gauss measures on connected Lie groups, Monatsh. Math. **98** (1984), 295–310.

[McCW83] McCrudden, M., and R. M. Wood, On the supports of absolutely continuous Gauss measures on SL(2,$\mathbb{R}$), in: Lecture Notes in Math. **1064** (1983), 379–397.

[McK70]   McKilligan, S., Embedding topological semigroups in topological groups, Proc. Edinburgh Math. Soc. **17** (1970), 127–138. Ergaenze

[Mi87]   Mizony, M., Semi-groupes de Lie et fonctions de Jacobi de deuxieme espèce, Thèse d'état, Université de Lyon I, 1987.

[MR83]   Medina, A. and Ph. Revoy, Sur une géometrie Lorentzienne du groupe oscillateur, Sém. Géom. Diff. Montpellier (1983).

[MR84]   Medina, A. and Ph. Revoy, Algèbre de Lie et produit scalaire invariant, Sém. Géom. Diff. Montpellier (1983).

[MZ55]   Montgomery, D. and L. Zippin, *Topological Transformation Groups*, Interscience Publishers, New York, 1955.

[Mo57]   Mostow, G. D., On the fundamental group of a homogeneous space, Ann. Math. **66** (1957), 249–255.

[Nach65] Nachbin, *Topology and Order*, Van Nostrand, Princeton, 1965.

[Na78]   Nashed, M., Generalized inverse mapping theorems and related applications of generalized inverses in nonlinear analysis, in: Nonlinear equations in abstract spaces, V. Lakshmikantham ed., Acad. Press, New York (1978), 217–252.

[Nj74]   Nijenhuis, A., Strong derivatives and inverse mapping theorems, Amer. Math. Monthly **81** (1974), 969–981.

[Ol78]   Ol'shanskiĭ, G. I., Unitary representations of the infinite dimensional classical groups U(p,∞), SO$_0$(p,∞), Sp(p,∞) and the corresponding motion groups, Funct. Anal. and Appl. **12** (1978), 32–44.

[Ol80]   Ol'shanskiĭ, G. I., Construction of unitary representations of the infinite dimensional classical groups, Sov. Math. Dokl. **21** (1980), 66–70.

[Ol81]   Ol'shanskiĭ, G. I., Invariant cones in Lie algebras, Lie semigroups and the holomorphic discrete series, Funct. Anal. and Appl. **15** (1981), 275–285.

[Ol82a]   Ol'shanskiĭ, G. I., Convex cones in symmetric Lie algebras, Lie
            semigroups, and invariant causal (order) structures on pseudo-
            Riemannian symmetric spaces, Sov. Math. Dokl. **26** (1982), 97–
            101.

[Ol82b]   Ol'shanskiĭ, G. I., Invariant orderings on simple Lie groups, the
            solution to E. B. Vinberg's problem, Funct. Anal. and Appl. **16**
            (1982), 311–313.

[Ol84]   Ol'shanskiĭ, G. I., Infinite dimensional classical groups of finite
            R-rank: Description of representations and asymptotic theory,
            Funct. Anal. and Appl. **18** (1982), 28–42.

[Ol85]   Ol'shanskiĭ, G. I., Unitary repesentations of the infinite symmetric
            group: a semigroup approach, in: Representations of Lie groups
            and Lie algebras, Akad. Kiado, Budapest (1985).

[Os73]   Ostrowski, A. M., *Solution of Equations in Euclidean and Banach
            spaces*, Acad. Press, New York, 1973.

[Pa81]   Paneitz, S., Invariant convex cones and causality in semisimple Lie
            algebras and groups, J. Funct. Anal. **43** (1981), 313–359.

[Pa84]   Paneitz, S., Determination of invariant convex cones in simple Lie
            algebras, Arkiv för Mat. **21** (1984), 217–228.

[Pog77]   Poguntke, D., Well-bounded semigroups in connected groups, Semi-
            group Forum **15** (1977), 159–167.

[Pon57]   Pontrjagin, L. S., *Topologische Gruppen*, Teil I und II, Teubner,
            Leibzig, 1957.

[Ra52]   Rådström, H., Convexity and norm in topological groups, Arkiv för
            Mat. **2** (1952), 99–137.

[Ra59]   Rådström, H., One-parameter semigroups of subsets of a real linear
            space, Arkiv för Mat. **4** (1959), 87–97.

[Re72]   Redheffer, R.M., The theorem of Bony and Brezis on flow invariant
            sets, Amer. Math. Monthly **79** (1972), 740–747.

[Rock70]   Rockafellar, R., *Convex Analysis*, Princeton Univ. Press, Princeton,
            1970.

[Rot80]   Rothkrantz, L. J. M., Transformatiehalfgroepen van nietcompacte
            hermitesche symmetrische Ruimten, Dissertation, Univ. of Ams-
            terdam, 1980.

[Ru73]   Rudin, W., *Functional Analysis*, Mc Graw-Hill, New York, 1973.

[Se76]   Segal, I.E., *Mathematical cosmology and Extragalactic Astronomy*, Acad. Press New York, 1976.

[Si82]   Siebert, E., Absolute continuity, singularity, and supports of Gauss semigroups on a Lie group, Monatsh. Math. **83** (1982), 239–253.

[Sp84]   Spindler, K., Über lineare Lie Gruppen, die einen Keil invariant lassen, Diplomarbeit, TH Darmstadt, 1984.

[Sp88]   Spindler, K., Invariant cones in Lie algebras, Dissertation, TH Darmstadt, 1988.

[St35]   Straszewicz, S., Über exponierte Punkte abgeschlossener Punktmengen, Fund. Math. **24** (1935), 139–143.

[SJ72]   Sussmann, H., and V. Jurdjevic, Contollabillity of nonlinear systems, J. Diff. Eq. **12** (1972), 95–116.

[Te84]   Terp, Ch., Über simultane Eigenvektoren von Halbgruppen und topologischen Gruppen, die einen Kegel invariant lassen, Diplomarbeit, TH Darmstadt, 1984.

[Ti59]   Tits.J., Sur une classe de groupes de Lie résoluble, Bull. Soc. Belg. **11** (1959), 100–115.

[Ti67]   Tits, J., Tabellen zu den einfachen Lie Gruppen und ihren Darstellungen, Lecture Notes in Math. **40** (1967).

[Vi80]   Vinberg, E. B., Invariant cones and orderings in Lie groups, Funct. Anal. and Appl. **14** (1980), 1–13.

[Wa72]   Warner, G., *Harmonic Analysis on Semi-Simple Lie Groups*, Vol. I, Springer, Berlin, Heidelberg, New York, 1972.

[Wr57]   Wright, F., Topological abelian semigroups, Amer. J. Math. **79** (1957), 744–796.

[Ya50]   Yamabe, H., On arcwise connected subgroups of a Lie group, Osaka Math. J. **2** (1950), 13–14.

Special symbols

Index